BIOSTATISTICAL ANALYSIS

FOURTH EDITION

JERROLD H. ZAR

Department of Biological Sciences
Northern Illinois University

PRENTICE HALL
Upper Saddle River, New Jersey 07458

Library of Congress Cataloging in Publication Data

Zar, Jerrold H.
  Biostatistical analysis / Jerrold H. Zar. — 4th ed.
    p. cm.
  Includes bibliographical references (p.   ) and index.
  ISBN 0-13-081542-X (alk. paper)
  1. Biometry. I. Title
QH323.5.Z37 1999
570′.1′5195—dc21   98-34062
                                                  CIP

Editorial/production supervision: *Interactive Composition Corporation*
Cover director: *Jayne Conte*
Cover designer: *Bruce Kenselaar*
Manufacturing manager: *Trudy Pisciotti*
Editor: *Teresa Ryu*
Senior editor: *Sheri L. Snavely*
Editorial assistants: *Nancy Bauer* and *Lisa Tarabokjia*

© 1999, 1996, 1984, 1974 by Prentice-Hall, Inc.
Simon & Schuster/A Viacom Company
Upper Saddle River, New Jersey 07458

Printed in the United States of America

10 9 8 7 6

ISBN 0-13-081542-X

Prentice-Hall International (UK) Limited, *London*
Prentice-Hall of Australia Pty. Limited, *Sydney*
Prentice-Hall Canada Inc., *Toronto*
Prentice-Hall Hispanoamericana, S. A., *Mexico*
Prentice-Hall of India Private Limited, *New Delhi*
Prentice-Hall of Japan, Inc., *Tokyo*
Simon & Schuster Asia Pte. Ltd., *Singapore*
Editora Prentice-Hall do Brasil, Ltda., *Rio de Janeiro*

CONTENTS

27 CIRCULAR DISTRIBUTIONS: HYPOTHESIS TESTING (continued)

APPENDIX A   ANALYSIS OF VARIANCE HYPOTHESIS TESTING   App1

APPENDIX B   STATISTICAL TABLES AND GRAPHS   App11

APPENDIX B *STATISTICAL TABLES AND GRAPHS (continued)*

ANSWERS TO EXERCISES   *Ans1*

LITERATURE CITED   *L1*

INDEX   *I1*

PREFACE

A great portion of contemporary biological inquiry requires a basic appreciation and knowledge of statistical techniques, as has become apparent to biological researchers, journal editors, and college curriculum planners. Reflecting the magnificent diversity of scientific endeavors that can be found within the biological sciences, this book presents a broad collection of data-analysis techniques, which will address the statistical needs of the majority of biological investigators.

Now in its fourth edition, this book has been called upon to fulfill two purposes. First, it has served as an introductory textbook, assuming no prior knowledge of statistics. Secondly, it has functioned as a reference work, covering a sufficient variety of concepts and procedures to satisfy a large portion of the biological disciplines that require statistical analysis, and being consulted long after formal instruction has ended.

Colleges and universities have long offered a diverse array of introductory statistics courses, some without emphasis on particular fields in which data might be collected and some—like those for which this book will be explicitly useful—focusing on statistical methods of utility to a specific field (in this case, biology). Walker (1929: 148, 151–163) reported that, although the teaching of probability has a much longer history, the first statistics course at a U.S. university or college probably was at Columbia, in the economics department, in 1880; followed in 1887 by the first in psychology, at Pennsylvania; in 1889 by the first in anthropology, at Clark; in 1897 by the first in biology, at Harvard; in 1898 by the first in mathematics, at Illinois; and in 1900 by the first in education, at Columbia. By the end of the nineteenth century only about a dozen institutions offered courses dealing with statistical methods, but by 1929 all universities and most large colleges in the country offered such instruction. Specifically in biology, the first courses with statistical content were probably those taught by Charles B. Davenport at Harvard (1887–1899) and at Chicago (1889–1904), and his *Statistical Methods in Biological Variation* may have been the first American book focused on statistics (ibid.: 159)

In order to be useful as a reference, as well as to allow for differences in content among courses for which it might be used, this book contains much more material than would be included in a one-academic-term course. Therefore, I have been asked to recommend what I consider the basic topics for an introductory treatment. With no authoritarian intent, I suggest these book sections as such a core treatment of biostatistical methods, to be augmented by (or substituted by) others of the instructor's preference: 1.1–1.4, 2.1–2.4, 3.1, 3.2, 3.4, 4.1, 4.4–4.6, 6.1–6.4, 7.1–7.6, 8.1–8.5, 8.9, 8.10, 9.1–9.3, 9.5, 10.1–10.4, 10.6, 11.1–11.7, 12.1–12.7, 14.1, 15.1, 17.1–17.7, 18.1–18.3, 19.1–19.3, 19.5, 19.9, 20.1–20.4, 22.1, 22.2, 22.4, 22.5, 23.1, 23.3, 23.6.

The material in this book requires no previous mathematical competence beyond very elementary algebra, although the discussions include some topics that appear seldom if at all, in other general texts. Also, cognizance is taken of the increased use of computer capability in academic and nonacademic institutions of all sizes. There are statistical procedures that are of importance but which involve computations so demanding that they practically, if not actually, preclude noncomputer execution. The principles of some of these are presented with the assumption that computer programs (software) will perform the laborious computations, but with the realization that the biologist must enter into the interpretation of the results of the computer's calculations. The data in the examples and exercises are largely fictional and are intended to demonstrate statistical procedures, not biological principles.

A final contribution toward achieving a book with self-sufficiency for most biostatistical needs is the inclusion of a thorough set of statistical tables, the majority of which are more extensive than those found in other introductory or advanced texts, and including many not found in any other texts.

A book of this nature requires and benefits from the assistance of many people. For the preparation of three editions I have been indebted to the library services of the University of Illinois (Urbana), Northern Illinois University, and the latter's collections networks. I also gratefully acknowledge the cooperation of the computer services at Northern Illinois University, which assisted in running many of the computer programs I prepared to generate some of the statistical appendix tables. For the tables taken from previously published sources, thanks are here given for the permission to reprint them; full acknowledgement of each source is found immediately following the appearance of the reprinted material. Additionally, I am pleased to recognize the editorial and production staff at Prentice Hall and Interactive Composition Corporation for their valued professional assistance in transforming my manuscript into the published product.

Over many years, my teachers, students, and colleagues have aided in guiding me to the material that is presented in this volume. Available space precludes mention of all those providing input and influence to this writing endeavor. However, special recognition must be made of the late S. Charles Kendeigh, University of Illinois (Urbana), who, through considerate mentorship, first alerted me to the need for quantitative analysis of biological data that led me to produce the first edition; the late Edward Batschelet, University of Zurich, who, with enthusiasm, patience, and kindness, provided me with encouragement and inspiration on statistical matters throughout the preparation of much of the first two editions; and the ever supportive and stimulating Arthur W. Ghent, University of Illinois (Urbana), who—from pre-publishing days through the current book edition—has offered statistical and biological commentary both enlightening and challenging. Major new material in this edition benefited from the substantive contributions of Carol J. Feltz, Northern Illinois University. And, prior to publication, the book drew upon the expertise and acumen of reviewers William H. Baltosser (University of Arkansas, Little Rock), Michael C. Grant (University of Colorado, Boulder), Karl Kornacker (The Ohio State University), John C. Hunter (State University of New York, Brockport), David Moriarty (California State Polytechnic University), Edward C. Murphy (University of Alaska, Fairbanks), Karen Olmstead (University of South Dakota), Nancy Ostiguy (California State University, Sacramento), Donald Price (University of Hawaii), Russell I. Shoemaker (Cochise College), William R. Sise (Humboldt State University), Ken Yasukawa

(Beloit College), and George Zimmerman (Richard Stockton College). Finally, I acknowledge my wife, Carol, for her prolonged patience during the preparation of the four editions of this book over a period of more than twenty-five years.

J. H. Zar
DeKalb, Illinois

1

INTRODUCTION

Many investigations in the biological sciences are quantitative, with observations consisting of numerical facts called *data*. (One numerical fact is a *datum*.*) As biological entities are counted or measured, it becomes apparent that some objective methods are necessary to aid the investigator in presenting and analyzing research data.

The word "statistics" is derived from the Latin for "state," indicating the historical importance of governmental data gathering, which related principally to demographic information (including census data and "vital statistics"), and often to their use in military recruitment and tax collecting.[†]

The term "statistics" is often encountered as a synonym for "data": One hears of college enrollment statistics (how many senior students, how many students from each geographic location, etc.), statistics of a baseball game (how many runs scored, how many strike-outs, etc.), labor statistics (numbers of workers unemployed, numbers employed in various occupations), and so on. Hereafter, this use of the word "statistics" will not appear in this book. Instead, "statistics" will be used in its other common manner: to refer to the *analysis and interpretation of data with a view toward objective evaluation of the reliability of the conclusions based on the data*. Statistics applied to

---

*Contemporary usage (e.g., among computer users) employs "data" as a singular noun, as a synonym for numerical information.

[†]Peters (1987: 79) and Walker (1929: 32) attribute the first use of the term "statistics" to a German professor, Gottfried Achenwall (1719–1772), who used the German word *Statistik* in 1749, and the first published use of the English word to John Sinclair (1754–1835) in 1791.

biological problems is simply called *biostatistics* or, sometimes, *biometry*[‡] (the latter term literally meaning "biological measurement"). Although the field of statistics has roots extending back hundreds of years, its development began in earnest in the late nineteenth century, and a major impetus from early in this development has been the need to examine biological data.

Before data can be analyzed, they must be collected, and statistical considerations can aid in the design of experiments and in the setting up of hypotheses to be tested. Many biologists attempt the analysis of their research data only to find that too few data were collected to enable reliable conclusions to be drawn, or that much extra effort was expended in collecting data that cannot be of ready aid in the analysis of the experiment. Thus, a knowledge of basic statistical principles and procedures is important even before an experiment is begun.

Once the data have been obtained, we may organize and summarize them in such a way as to arrive at their orderly and informative presentation. Such procedures are often termed *descriptive statistics*. For example, a tabulation might be made of the heights of all members of a senior English class, indicating an average height for each sex, or for each age. However, it might be desired to make some generalizations from these data. We might, for example, wish to make a reasonable estimate of the heights of all seniors in the university. Or we might wish to conclude whether the males in the university are on the average taller than the females. The ability to make such generalized conclusions, inferring characteristics of the whole from characteristics of its parts, lies within the realm of *inferential statistics*.

1.1 TYPES OF BIOLOGICAL DATA

A characteristic that may differ from one biological entity to another is termed a *variable* (or *variate**). Different kinds of variables may be encountered by biologists, and it is desirable to be able to distinguish among them. The classification used here is that which is standardly employed (Senders, 1958; Siegel, 1956; Stevens, 1946, 1968).

However, slavish adherence to this taxonomy can be misleading (e.g., see Velleman and Wilkinson, 1993, and the references cited therein), for not all data fit neatly into these categories and some may be treated differently depending upon the questions asked of them.

Data on a Ratio Scale.   Consider that we are studying a group of plants, that the heights of a group of plants constitute a variable of interest, and that the number of leaves per plant is another variable under study. Thanks to measuring devices at the

---

[‡]The term "biometry" apparently was conceived between 1892 and 1901 by Karl Pearson, along with the name, *Biometrika*, for the still-important English journal he helped found, and this term was first published in the inaugural issue of that journal in 1901 (Snedecor, 1954). The word had been used decades earlier with other meanings, and even well into the twentieth century it was mostly associated with quantitative genetics (Armitage, 1985).

*"Variate" was first used by R. A. Fisher (1925: 5; David, 1995).

biologist's command, it is possible to assign a numerical value to the height of each plant, and counting the leaves allows a numerical value to be assigned to the number of leaves on each plant. Regardless of whether the height measurements are recorded in centimeters, inches, or other units, and regardless of whether the leaves are counted in a number system using base 10 or any other base, there are two fundamentally important characteristics of these data.

First, there is a constant size interval between any adjacent units on the measurement scale. That is, the difference in height between a 36 cm and a 37 cm plant is the same as the difference between a 39 cm and a 40 cm plant, and the difference between eight and ten leaves is equal to the difference between nine and eleven leaves. (This may seem simpleminded, but it is very important, as we shall see on examining the other scales of measurement.)

Second, it is important that there exists a zero point on the measurement scale and that there is a physical significance to this zero. This enables us to say something meaningful about the ratio of measurements. We can say that a 30 cm (11.8 in.) tall plant is half as tall as a 60 cm (23.6 in.) plant, and that a plant with forty-five leaves has three times as many leaves as a plant with fifteen.

Measurement scales having a constant interval size and a true zero point are said to be *ratio scales* of measurement. Besides lengths and numbers of items, ratio scales include weights (mg, lb, etc.), volumes (cc, cu ft, etc.), capacities (ml, qt, etc.), rates (cm/sec, mph, mg/min, etc.), and lengths of time (hr, yr, etc.).

Data on an Interval Scale.   Some measurement scales possess a constant interval size but not a true zero; they are called *interval scales*. An outstanding example is that of the two common temperature scales: Celsius (C) and Fahrenheit (F). We can see that the same difference exists between 20°C (68°F) and 25°C (77°F) as between 5°C (41°F) and 10°C (50°F); i.e., the measurement scale is composed of equal-sized intervals. But it cannot be said that a temperature of 40°C (104°F) is twice as hot as a temperature of 20°C (68°F); i.e., the zero point is arbitrary.*(Temperature measurements on the absolute, or Kelvin [K], scale can be referred to a physically meaningful zero and thus constitute a ratio scale.)

Some interval scales encountered in biological data collection are *circular scales*. Time of day and time of the year are examples of such scales. The interval between 2:00 PM (i.e., 1400 hr) and 3:30 PM (1530 hr) is the same as the interval between 8:00 AM (0800 hr) and 9:30 AM (0930 hr). But one cannot speak of ratios of times of day because the zero point (midnight) on the scale is arbitrary, in that one could just as well set up a scale for time of day which would have noon, or 3:00 PM, or any other time as the zero point. Circular biological data are occasionally compass points, as if one records

---

*The German-Dutch physicist, Gabriel Daniel Fahrenheit (1686–1736), invented the thermometer in 1714, employing a scale on which salt water froze at zero degrees, pure water froze at 32 degrees, and pure water boiled at 212 degrees. In 1742 the Swedish astronomer, Anders Celsius (1701–1744), devised a temperature scale with 100 degrees between the freezing and boiling points of water (the so-called "centigrade" scale) first by referring to zero degrees as boiling and and 100 degrees as freezing, and a year later reversing these two reference points (Asimov, 1982: 177).

the compass direction in which an animal or plant is oriented. As the designation of north as $0°$ is arbitrary, this circular scale is a form of interval scale of measurement. Some special statistical procedures are available for circular data; these are discussed in Chapters 26 and 27.

Data on an Ordinal Scale. The preceding paragraphs on ratio and interval scales of measurement discussed data between which we know numerical differences. For example, if man A weighs 90 kg and man B weighs 80 kg, then man A is known to weigh 10 kg more than B. But our data may, instead, be a record only of the fact that man A weighs more than man B (with no indication of how much more). Thus, we may be dealing with relative differences rather than with quantitative differences. Such data consist of an ordering or ranking of measurements and are said to be on an *ordinal* scale of measurement ("ordinal" being from the Latin word for "order"). One may speak of one biological entity being shorter, darker, faster, or more active than another; the sizes of five cell types might be labeled 1, 2, 3, 4, and 5, to denote their magnitudes relative to each other; or success in learning to run a maze may be recorded as A, B, or C.

It is often true that biological data expressed on the ordinal scale could have been expressed on the interval or ratio scale had exact measurements been obtained (or obtainable). Sometimes data that were originally on interval or ratio scales will be changed to ranks; for example, examination grades of 99, 85, 73, and 66% (ratio scale) might be recorded as A, B, C, and D (ordinal scale), respectively.

Ordinal scale data contain and convey less information than ratio or interval data, for only relative magnitudes are known. Consequently, quantitative comparisons are impossible (e.g., we cannot speak of a grade of C being half as good as a grade of A, or of the difference between cell sizes 1 and 2 being the same as the difference between sizes 3 and 4). However, we will see that many useful statistical procedures are, in fact, applicable to ordinal data.

Data on a Nominal Scale. Sometimes the variable under study is classified by some quality it possesses rather than by a numerical measurement. In such cases the variable may be called an *attribute*, and we are said to be using a *nominal scale* of measurement. Genetic phenotypes are commonly encountered biological attributes; the possible manifestation of an animal's eye color may be blue or brown, and if human hair color were the attribute of interest, we might record black, brown, blonde, or red. On a nominal scale ("nominal" is from the Latin word for "name"), animals might be classified as male or female, or as left- or right-handed. Or plants might be classified as dead or alive, or as with or without thorns. Taxonomic categories also form a nominal classification scheme (e.g., a plant might be classified as pine, spruce, or fir). Sometimes data from an ordinal, interval, or ratio scale of measurement may be recorded in nominal-scale categories. For example, heights may be recorded as tall or short, or performance on an examination as pass or fail.

As will be seen, statistical methods useful with ratio, interval, or ordinal data generally are not applicable to nominal data, and we must, therefore, be able to identify such situations when they occur.

Continuous and Discrete Data.   When we spoke above of plant heights, we were dealing with a variable that could be any conceivable value within any observed range; this is referred to as a *continuous variable*. That is, if we measure a height of 35 cm and a height of 36 cm, an infinite number of heights is possible in the range from 35 to 36 cm: a plant might be 35.07 cm tall or 35.988 cm tall, or 35.3263 cm tall, etc., although, of course, we do not have devices sensitive enough to detect this infinity of heights. A continuous variable is one for which there is a possible value between any other two possible values.

However, when speaking of the number of leaves on a plant, we are dealing with a variable that can take on only certain values. It might be possible to observe 27 leaves, or 28 leaves, but 27.43 leaves and 27.9 leaves are values of the variable that are impossible to obtain. Such a variable is termed a *discrete* or *discontinuous variable* (also known as a *meristic variable*). The number of white blood cells in 1 mm^3 of blood, the number of giraffes visiting a water hole, and the number of eggs laid by a grasshopper are all discrete variables. The possible values of a discrete variable generally are consecutive integers, but this is not necessarily so. If the leaves on our plants are always formed in pairs, then only even integers are possible values of the variable. And the ratio of number of wings to number of legs of insects is a discrete variable that may only have the value of 0, 0.3333..., or 0.6666... (i.e., $\frac{0}{6}$, $\frac{2}{6}$, or $\frac{4}{6}$, respectively).*

Ratio-, interval-, and ordinal-scale data may be either continuous or discrete. Nominal-scale data by their nature are discrete.

1.2   ACCURACY AND SIGNIFICANT FIGURES

Accuracy is the nearness of a measurement to the actual value of the variable being measured. *Precision* is not a synonymous term, but refers to the closeness to each other of repeated measurements of the same quantity.

If we report that the hind leg of a frog is 8 cm long, we are stating the number 8 (a value of a continuous variable) as an estimate of the frog's true leg length. This estimate was made using some sort of a measuring device. Had the device been capable of more accuracy, we might have concluded that the leg was 8.3 cm long, or perhaps 8.32 cm long. When recording values of continuous variables, it is important to designate the accuracy with which the measurements have been made. By convention, the value 8 denotes a measurement in the range of 7.50000... to 8.49999..., the value 8.3 designates a range of 8.25000... to 8.34999..., and the value 8.32 implies that the true value lies within the range of 8.31500... to 8.32499.... That is, the reported value is the midpoint of the implied range, and the size of this range is designated by the last decimal place in the measurement. The value of 8 cm implies a range of accuracy of 1 cm, 8.3 cm implies a range of 0.1 cm, and 8.32 cm implies a range of 0.01 cm. Thus, to record a value of 8.0

---

*The ellipsis (...) may be read as "and so on." Here, they indicate that $\frac{2}{6}$ and $\frac{4}{6}$ are repeating decimal fractions, which could just as well have been written as 0.3333333333333... and 0.6666666666666..., respectively.

implies greater accuracy of measurement than does the recording of a value of 8, for in the first instance the true value is said to lie between 7.95000 ... and 8.049999 ... (i.e., within a range of 0.1 cm), whereas 8 implies a value between 7.50000 ... and 8.49999 ... (i.e., within a range of 1 cm). To state 8.00 cm implies an accuracy in measurement which ascertains the frog's limb length to be between 7.99500 ... and 8.00499 ... cm (i.e., within a range of 0.01 cm). Those digits in a number that denote the accuracy of the measurement are referred to as *significant figures*. Thus, 8 has one significant figure, 8.0 and 8.3 each have two significant figures, and 8.00 and 8.32 each have three.

In working with exact values of discrete variables, the preceding considerations do not apply. That is, it is sufficient to state that our frog has four limbs or that its left lung contains thirteen flukes. The use of 4.0 or 13.00 would be inappropriate, for as the numbers involved are exactly 4 and 13, there is no question of accuracy or significant figures.

But there are instances where significant figures and implied accuracy come into play with discrete data. An entomologist may report that there are 72,000 moths in a particular forest area. In doing so, it is probably not being claimed that this is the exact number but an estimate of the exact number, perhaps accurate to two significant figures. In such a case, 72,000 would imply a range of accuracy of 1000, so that the true value might lie anywhere from 71,500 to 72,500. If the entomologist wished to convey the fact that this estimate is believed to be accurate to the nearest 100 (i.e., to three significant figures), rather than to the nearest 1000, it would be better to present the data in the form of *scientific notation*, as follows: If the number 7.2×10^4 ($= 72,000$) is written, a range of accuracy of 0.1×10^4 ($= 1000$) is implied, and the true value is assumed to lie between 71,500 and 72,500. But if 7.20×10^4 were written, a range of accuracy of 0.01×10^4 ($= 100$) would be implied, and the true value would be assumed to be in the range of 71,950 to 72,050. Thus, the accuracy of large values (and this applies to continuous as well as discrete variables) can be expressed succinctly using scientific notation.

Calculators and computers typically yield results with more significant figures than are justified by the data. However, it is good practice—to avoid rounding error—to retain many significant figures for all steps until the last in a sequence of calculations, and on attaining the result of the final step to round off to the appropriate number of figures.

1.3 FREQUENCY DISTRIBUTIONS

When collecting and summarizing large amounts of data, it is often helpful to record the data in the form of a *frequency table*. Such a table simply involves a listing of all the observed values of the variable being studied and how many times each value is observed. Consider the tabulation of the frequency of occurrence of sparrow nests in each of several different locations. This is illustrated in Example 1.1, where the observed kinds of nest sites are listed, and for each kind the number of nests observed is recorded. The distribution of the total number of observations among the various categories is termed a *frequency distribution*. Example 1.1 is a frequency table for nominal data, and these data may also be presented graphically by means of a *bar graph* (Fig. 1.1),

EXAMPLE 1.1   The location of sparrow nests. A frequency table of nominal data.

Nest site	*Number of nests observed*
A. Vines	56
B. Building eaves	60
C. Low tree branches	46
D. Tree and building cavities	49

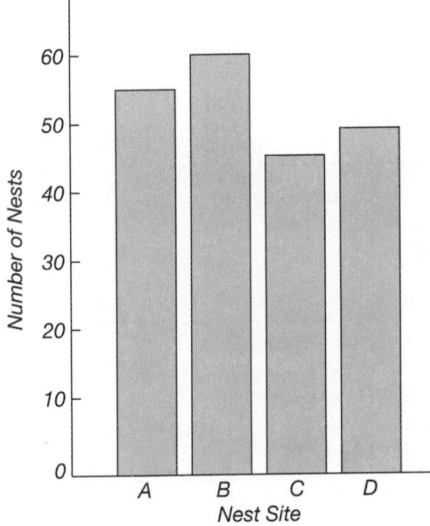

Figure 1.1 A bar graph of the sparrow nest data of Example 1.1. An example of a bar graph for nominal data.

where the height of each bar is proportional to the frequency in the class represented. The widths of all bars in a bar graph should be equal so that the eye of the reader is not distracted from the differences in bar heights; this also makes the area of each bar proportional to the frequency it represents. Also, the frequency scale on the vertical axis should begin at zero to avoid the apparent differences among bars. If, for example, a bar graph of the data of Example 1.1 were constructed with the vertical axis representing frequencies of 45 to 60 rather than 0 to 60, the results would appear as in Fig. 1.2. Huff (1954) illustrates other techniques that can mislead the readers of graphs.

A frequency tabulation of ordinal data might appear as in Example 1.2, which presents the observed numbers of sunfish collected in each of five categories, each category being a degree of skin pigmentation. A bar graph (Fig. 1.3) can be prepared for this frequency distribution just as for nominal data.

In preparing frequency tables of interval- and ratio-scale data, we make a procedural distinction between discrete and continuous data. Example 1.3 shows discrete data which

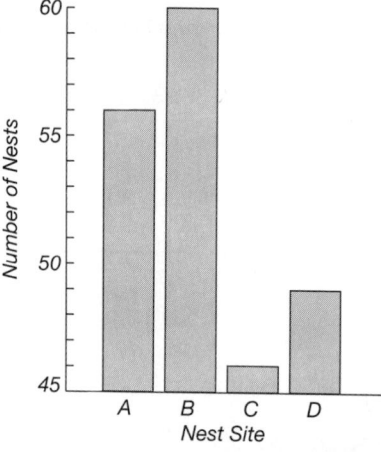

Figure 1.2 A bar graph of the sparrow nest data of Example 1.1, drawn with the vertical axis starting at 45. Compare this with Fig. 1.1, where the axis starts at 0.

EXAMPLE 1.2   Numbers of sunfish, tabulated according to amount of black pigmentation. A frequency table of ordinal data.

Pigmentation class	Amount of pigmentation	Number of fish
0	No black pigmentation	13
1	Faintly speckled	68
2	Moderately speckled	44
3	Heavily speckled	21
4	Solid black pigmentation	8

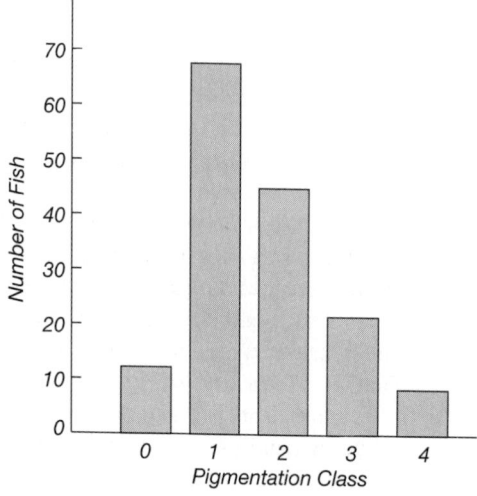

Figure 1.3 A bar graph of the sunfish pigmentation data of Example 1.2. An example of a bar graph for ordinal data.

EXAMPLE 1.3   Frequency of occurrence of various litter sizes in foxes. A frequency table of discrete, ratio-scale data.

Litter size	Frequency
3	10
4	27
5	22
6	4
7	1

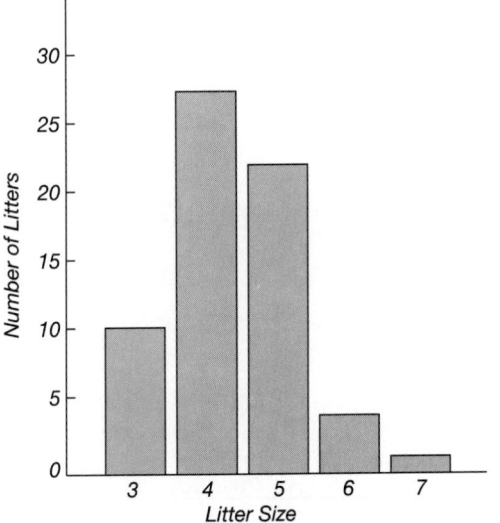

Figure 1.4   A bar graph of the fox litter data of Example 1.3. An example of a bar graph for discrete, ratio-scale data.

are frequencies of litter sizes in foxes, and Fig. 1.4 presents this frequency distribution graphically.

Example 1.4a shows discrete data that are the numbers of aphids found per clover plant. These data create quite a lengthy frequency table, and it is not difficult to imagine sets of data whose tabulation would result in an even longer list of frequencies. Thus, for purposes of preparing bar graphs, we often cast data into a frequency table by grouping them.

Example 1.4b is a table of the data from Example 1.4a arranged by grouping the data into size classes. The bar graph for this distribution appears as Fig. 1.5. Such grouping results in the loss of some information and is generally utilized only to make frequency tables and bar graphs easier to read, and not for calculations performed on the data. There have been several "rules of thumb" proposed to aid in deciding how many classes data might reasonably be grouped, for the use of too few groups will obscure the general shape of the distribution. But such "rules" or recommendations are only rough guides, and the choice is generally left to good judgment, bearing in mind that from ten to twenty groups are useful for most biological work. (See also Doane, 1976.) In general, groups

EXAMPLE 1.4a Number of aphids observed per clover plant. A frequency table of discrete, ratio-scale data.

Number of aphids on a plant	Number of plants observed	Number of aphids on a plant	Number of plants observed
0	3	20	17
1	1	21	18
2	1	22	23
3	1	23	17
4	2	24	19
5	3	25	18
6	5	26	19
7	7	27	21
8	8	28	18
9	11	29	13
10	10	30	10
11	11	31	14
12	13	32	9
13	12	33	10
14	16	34	8
15	13	35	5
16	14	36	4
17	16	37	1
18	15	38	2
19	14	39	1
		40	0
		41	1

Total number of observations = 424

should be established that are equal in the size interval of the variable being measured. (For example, the group size interval in Example 1.4b is four aphids per plant.)

Because continuous data, contrary to discrete data, can take on an infinity of values, one is essentially always dealing with a frequency distribution tabulated by groups. If the variable of interest were a weight, measured to the nearest 0.1 mg, a frequency table entry of the number of weights measured to be 48.6 mg would be interpreted to mean the number of weights grouped between 48.5500... and 48.6499... mg (although in a frequency table this class interval is usually written as 48.55–48.65). Example 1.5 presents a tabulation of 130 determinations of the amount of phosphorus, in milligrams per gram, in dried leaves. (Ignore the last two columns of this table until Section 1.4).

In presenting this frequency distribution graphically, one can prepare a *histogram*, which is the name given to a bar graph based on continuous data. This is done in Fig. 1.6a; note that rather than indicating the range on the horizontal axis, we indicate only the midpoint of the range, a procedure that results in less crowded printing on the graph. Note also that adjacent bars in a histogram are often drawn touching each other, whereas in the other bar graphs discussed they generally are not.

EXAMPLE 1.4b   Number of aphids observed per clover plant.   A frequency table grouping the discrete, ratio-scale data of Example 1.4a.

Number of aphids on a plant	Number of plants observed
0–3	6
4–7	17
8–11	40
12–15	54
16–19	59
20–23	75
24–27	77
28–31	55
32–35	32
36–39	8
40–43	1

Total number of observations = 424

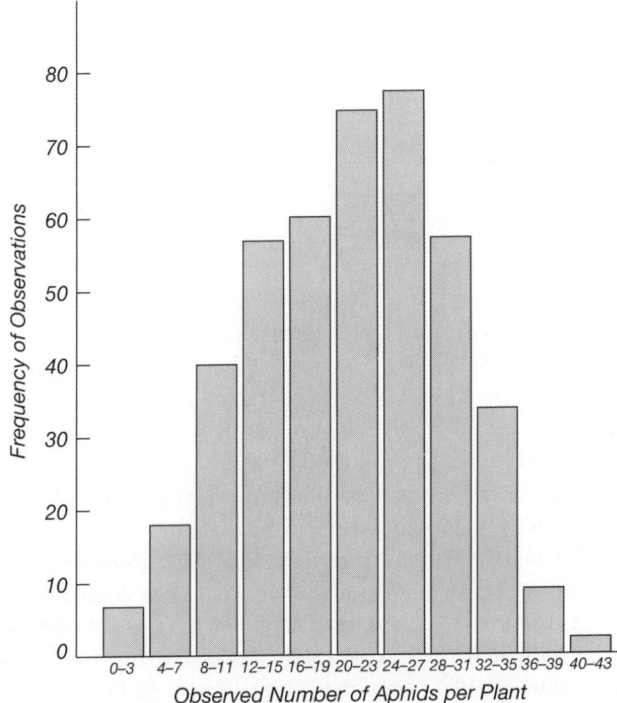

Figure 1.5   A bar graph of the aphid data of Example 1.4b. An example of a bar graph for grouped discrete, ratio-scale data.

EXAMPLE 1.5 **Determinations of the amount of phosphorus in leaves. A frequency table of continuous data.**

Phosphorus mg/g of leaf)	Frequency (i.e., number of determinations)	Cumulative frequency Starting with low values	Starting with high values
8.15–8.25	2	2	130
8.25–8.35	6	8	128
8.35–8.45	8	16	122
8.45–8.55	11	27	114
8.55–8.65	17	44	103
8.65–8.75	17	61	86
8.75–8.85	24	85	69
8.85–8.95	18	103	45
8.95–9.05	13	116	27
9.05–9.15	10	126	14
9.15–9.25	4	130	4

Total frequency = 130

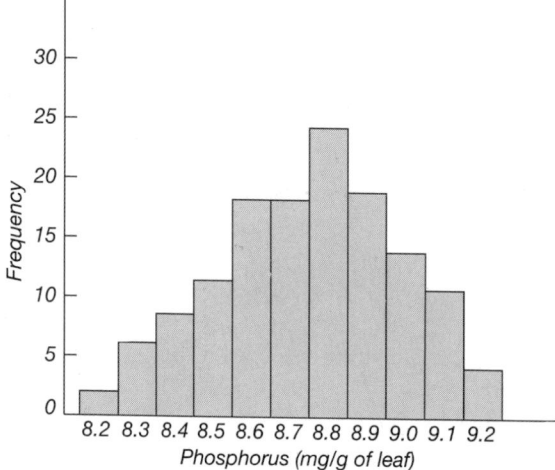

Figure 1.6a A histogram of the leaf phosphorus data of Example 1.5. An example of a histogram for continuous data.

Often a *frequency polygon* is drawn instead of a histogram. This is done by plotting the frequency of each class as a dot (or other symbol) at the class midpoint and then connecting each adjacent pair of dots by a straight line (Fig. 1.6b). It is, of course, the same as if the midpoints of the tops of the histogram bars were connected by straight lines. Instead of plotting frequencies on the vertical axis, one can plot *relative frequencies*, or proportions of the total frequency. This enables different distributions to be readily compared and even plotted on the same axes. Sometimes, as in Fig. 1.6b, frequency is indicated on one vertical axis and the corresponding relative frequency on the other.

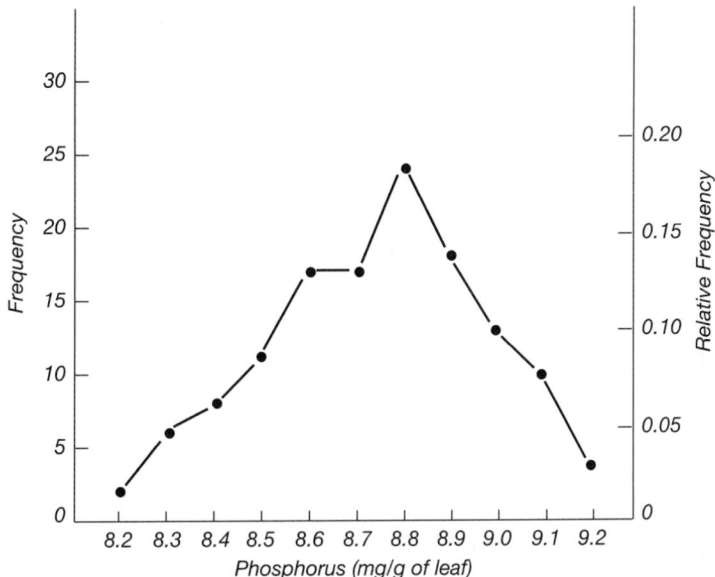

Figure 1.6b   A frequency polygon for the leaf phosphorus data of Example 1.5.

Frequency polygons are also commonly used for discrete distributions, but one can argue against their use when dealing with ordinal data, as the polygon implies to the reader a constant size interval horizontally between points on the polygon. Frequency polygons should not be employed for nominal scale data.

If we have a frequency distribution of values of a continuous variable that falls into a large number of class intervals, the data may be grouped as was demonstrated with discrete variables. This results in fewer intervals, but each interval is, of course, larger. The midpoints of these intervals may then be used in the preparation of a histogram or frequency polygon. The user of frequency polygons is cautioned that such a graph is simply an aid to the eye in following trends in frequency distributions, and one should not attempt to read frequencies between points on the polygon. Also note that the method presented for the construction of histograms and frequency polygons requires that the class intervals be equal. Lastly, the vertical axis (e.g., the frequency scale) on frequency polygons and bar graphs generally should begin with zero, especially if graphs are to be compared with one another. If this is not done, the eye may be misled by the appearance of the graph (as shown for nominal-scale data in Figs. 1.1 and 1.2).

1.4 CUMULATIVE FREQUENCY DISTRIBUTIONS

A frequency distribution informs us of how many observations occurred for each value (or group of values) of a variable. That is, examination of the frequency table of Example 1.3 (or its corresponding bar graph or frequency polygon) would yield information such as

"How many fox litters of four were observed?", the answer being twenty-seven. But if it is desired to ask questions such as "How many litters of four or more were observed?", or "How many fox litters of five or fewer were observed?", we are speaking of *cumulative frequencies*. To answer the first question, we sum all frequencies for litter sizes four and up, and for the second question, we sum all frequencies from the smallest litter size up through a size of five. We arrive at answers of fifty-four and fifty-nine, respectively.

In Example 1.5, the phosphorus concentration data are cast into two cumulative frequency distributions, one with cumulation commencing at the low end of the measurement scale and one with cumulation being performed from the high values toward the low values. The choice of the direction of cumulation is immaterial, as can be demonstrated. If one desired to calculate the number of phosphorus determinations less than 8.55 mg/g, namely 27, a cumulation starting at the low end might be used, whereas the knowledge of the frequency of determinations greater than 8.55 mg/g, namely 103, can be readily obtained from the cumulation commencing from the high end of the scale. But one can easily calculate any frequency from a low-to-high cumulation (e.g., 27) from its complementary frequency from a high-to-low cumulation (e.g., 103), simply by knowing that the sum of these two frequencies is the total frequency (i.e., 130); therefore, in practice it is not necessary to calculate both sets of cumulations.

Cumulative frequency distributions are useful in determining medians, percentiles, and other quantiles, as discussed in Sections 3.2 and 3.3. They are not often presented in bar graphs, but *cumulative frequency polygons* (sometimes called *ogives*) are not uncommon. (See Figs. 1.7 and 1.8.) Relative frequencies (proportions of the total frequency)

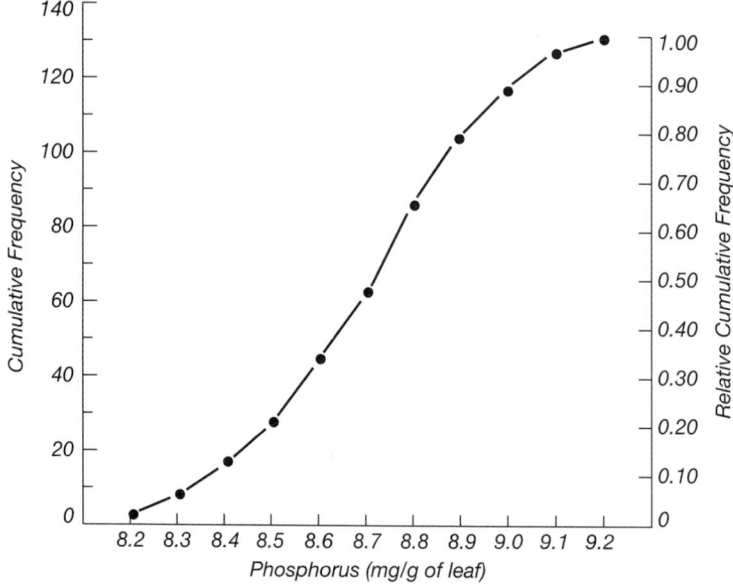

Figure 1.7 Cumulative frequency polygon of the leaf phosphorus data of Example 1.5, with cumulation commencing from the lowest to the highest values of the variable.

can be plotted instead of (or, as in Figs. 1.7 and 1.8, in addition to) frequencies on the vertical axis of a cumulative frequency polygon. This enables different distributions to be readily compared and even plotted on the same axes.

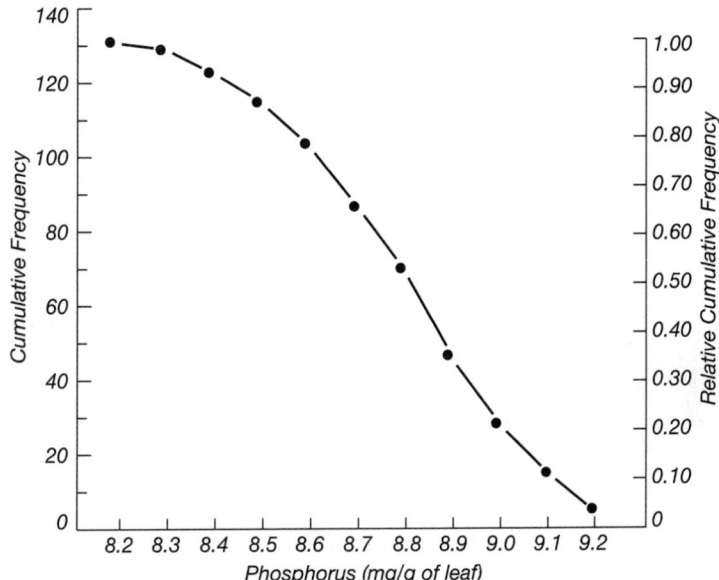

Figure 1.8 Cumulative frequency polygon of the leaf phosphorus data of Example 1.5, with cumulation commencing from the highest to the lowest values of the variable.

2

POPULATIONS AND SAMPLES

The primary objective of statistical analysis is to infer characteristics of a group of data by analyzing the characteristics of a small sampling of the group. This generalization from the part to the whole requires the consideration of such important concepts as population, sample, parameter, statistic, and random sampling. These topics are discussed in this chapter.

2.1 POPULATIONS

Basic to statistical analysis is the desire to draw conclusions about a group of measurements of a variable being studied. This entire collection of measurements about which one wishes to draw conclusions is the *population* or *universe*. For example, an investigator may desire to draw conclusions about the tail lengths of bobcats in Montana. All Montana bobcat tail lengths are, therefore, the population under consideration. If a study is concerned with the blood glucose concentration in children of a certain age, then the blood glucose levels in all children of that age comprise the population of interest.

Populations are often very large, such as the body weights of all grasshoppers in Kansas or the eye colors of all female New Zealanders, but occasionally populations of interest may be relatively small, such as the ages of men who have traveled to the moon or the heights of women who have swum the English Channel.

2.2 SAMPLES FROM POPULATIONS

If the population under study is very small, it might be practical to obtain all the measurements in the population. If one wishes to draw conclusions about the ages of all men who have traveled to the moon, it would not be unreasonable to attempt to collect the ages of the small number of individuals under consideration. Generally, however, populations of interest are so large as to render the obtaining of all the measurements unfeasible. For example, we could not reasonably expect to determine the body weight of each grasshopper in Kansas. What can be done in such cases is to obtain a subset of all the measurements in the population. This subset of measurements comprises a *sample*, and from the characteristics of samples conclusions can be drawn about the characteristics of the populations from which the samples came.

Often one samples a population that does not physically exist. Suppose an experiment is performed in which a food supplement is administered to forty guinea pigs, and the sample data consist of the growth rates of these forty animals. Then the population about which conclusions might be drawn is the growth rates of all the guinea pigs that conceivably might have been administered the same food supplement under identical conditions. Such a population is said to be "imaginary" and is also referred to as "hypothetical" or "potential."

2.3 RANDOM SAMPLING

Samples from populations can be obtained in a number of ways; however, to reach valid conclusions about populations by induction from samples, statistical procedures typically assume that the samples are obtained in a *random* fashion. To sample a population randomly requires that each member of the population has an equal and independent chance of being selected. That is, not only must each measurement in the population have an equal chance of being chosen as a member of the sample, but the selection of any member of the population must in no way influence the selection of any other member. Throughout this book, "sample" will always imply "random sample."

It is sometimes possible to assign each member of a population a unique number and to draw a sample by choosing a set of such numbers at random. This is equivalent to having all members of a population in a hat and drawing a sample from them while blindfolded. Appendix Table B.41 provides 10,000 random digits for this purpose. In this table, each digit from 0 to 9 has an equal and independent chance of appearing anywhere in the table. Similarly, each combination of two digits, from 00 to 99, is found at random in the table, as is each three-digit combination, from 000 to 999, etc.

Assume that a random sample of 200 names is desired from a telephone directory having 274 pages, three columns of names per page, and ninety-eight names per column. Entering Table B.41 at random (i.e., do not always enter the table at the same place), one might decide first to arrive at a random combination of three digits. If this three-digit number is 001 to 274, it can be taken as a randomly chosen page number (if it is 000 or larger than 274, simply skip it and choose another three-digit number, e.g., the next one on the table). Then one might examine the next digit in the table; if it is a 1, 2, or 3, let it denote a page column (if a digit other than 1, 2, or 3 is encountered, it is ignored, passing to the next digit that is 1, 2, or 3). Then one could look at the next two-digit number in the table; if it is from 01 to 98, let it represent a randomly selected name within that column. This three-step procedure would be performed a total of 200 times to obtain the desired random sample. One can proceed in any direction in the random number table: left to right, right to left, upward, downward, or diagonally. But the direction should be decided on before looking at the table. Computers are capable of rather quickly generating random numbers (sometimes called "pseudo-random" numbers because the number generation is not perfectly random), and this is how Table B.41 was derived.

Very often it is not possible to assign a number to each member of a population, and random sampling then involves biological, rather than simply mathematical, considerations. That is, the techniques for sampling Montana bobcats or Kansas grasshoppers require quite a bit of knowledge about the particular organism to insure that the sampling is random.

2.4 Parameters and Statistics

Several measures help to describe or characterize a population. For example, generally a preponderance of measurements occurs somewhere around the middle of the range of a population of measurements. Thus, some indication of a population "average" would express a useful bit of descriptive information. Such information is called a *measure of central tendency*, and several such measures (e.g., the mean and the median) will be discussed in Chapter 3.

It is also important to describe how dispersed the measurements are around the "average." That is, we can ask whether there is a wide spread of values in the population or whether the values are rather concentrated around the middle. Such a descriptive property is called a *measure of dispersion*, and several such measures (e.g., the range and the standard deviation) will be discussed in Chapter 4.

A quantity such as a measure of central tendency or a measure of dispersion is called a *parameter* when it describes or characterizes a population, and we shall be very interested in discussing parameters and drawing conclusions about them. Section 2.2 pointed out, however, that one seldom has data for entire populations, but nearly always has to rely on samples to arrive at conclusions about populations. Thus, one rarely is able to calculate parameters. However, by random sampling of populations, parameters can be estimated very well, as we shall see throughout this book. An estimate of a population parameter is called a *statistic*.* It is statistical convention to represent population parameters by Greek letters and sample statistics by Latin letters; the following chapters will demonstrate this custom for specific examples.

---

*This use of the terms "parameter" and "statistic" was defined by R. A. Fisher as early as 1925 (David, 1995).

The statistics one calculates will vary from sample to sample for samples taken from the same population. Because one uses sample statistics as estimates of population parameters, it behooves the researcher to arrive at the "best" estimates possible. As for what properties to desire in a "good" estimate, consider the following.

First, it is desirable that if we take an indefinitely large number of samples from a population, the long run average of the statistics obtained will equal the parameter being estimated. That is, for some samples a statistic may underestimate the parameter of interest, and for others it may overestimate that parameter; but in the long run the estimates that are too low and those that are too high will "average out." If such a property is exhibited by a statistic, we say that we have an *unbiased* statistic or an unbiased estimator.

Second, it is desirable that a statistic obtained from any single sample from a population be very close to the value of the parameter being estimated. This property of a statistic is referred to as *precision,** *efficiency*, or *reliability*. As one commonly secures only one sample from a population, it is important to arrive at a close estimate of a parameter from a single sample.

Third, consider that one can take larger and larger samples from a population (the largest sample being the entire population itself). As the sample size increases, a *consistent* statistic will become a better estimate of the parameter it is estimating. Indeed, if the sample were the size of the population, then the best estimate would be obtained: the parameter itself.

In the chapters that follow, the statistics recommended as estimates of parameters are "good" estimates in the sense that they possess a desirable combination of unbiasedness, efficiency, and consistency.

---

*The precision of a sample statistic, as defined here, should not be confused with the precision of a measurement, defined in Section 1.2.

3

MEASURES OF CENTRAL TENDENCY

In samples, as well as in populations, one generally finds a preponderance of values somewhere around the middle of the range of observed values. The description of this concentration near the middle is an *average*, or a *measure of central tendency* to the statistician. It is also termed a *measure of location*, for it indicates where, along the measurement scale, the sample or population is located.

Various measures of central tendency are useful parameters, in that they describe a property of populations. This chapter discusses the characteristics of these parameters and the sample statistics that are good estimates of them.

3.1 THE ARITHMETIC MEAN

The most widely used measure of central tendency is the *arithmetic mean*,* usually referred to simply as the *mean*,† which is the measure most commonly called an "average."

Each measurement in a population may be referred to as an X_i, (read "X sub i") value. Thus, one measurement might be denoted as X_1, another as X_2, another as X_3, and so on. The subscript i might be any integer value up through N, the total number of

---

*As an adjective, "arithmetic" is pronounced with the accent on the third syllable. In early literature on the subject, the adjective "arithmetical" was employed.

†The term "mean" (the arithmetic mean, as well as the geometric and harmonic means of Section 3.5) dates from ancient Greece (Walker, 1929: 183).

X values in the population.* The mean of the population is denoted by the Greek letter μ (lowercase mu), and is calculated as the sum of all the X_i values divided by the size of the population.

The calculation of the population mean can be abbreviated concisely by the formula

$$\mu = \frac{\sum_{i=1}^{N} X_i}{N}. \tag{3.1}$$

The Greek letter Σ (capital sigma) means "summation"[†] and $\sum_{i=1}^{N} X$ means "summation of all X_i values from X_1 through X_N." Thus, for example, $\sum_{i=1}^{4} X_i = X_1 + X_2 + X_3 + X_4$ and $\sum_{i=3}^{5} X_i = X_3 + X_4 + X_5$. Since, in statistical computations, summations are nearly always performed over the entire set of X_i values, this book will assume $\sum X_i$ to mean "sum X_i's over all values of i," simply as a matter of printing convenience, and $\mu = \sum X_i / N$ would therefore designate the same calculation as would $\mu = \sum_{i=1}^{N} X_i / N$.

The most efficient, unbiased, and consistent estimate of the population mean, μ, is the sample mean, denoted as $\bar{X}$ (read as "X bar"). Whereas the size of the population (which we generally do not know) is denoted as N, the size of a sample is indicated by n, and $\bar{X}$ is calculated as

$$\bar{X} = \frac{\sum_{i=1}^{n} X_i}{n} \quad \text{or} \quad \bar{X} = \frac{\sum X_i}{n}, \tag{3.2}$$

which is read "the sample mean equals the sum of all measurements in the sample divided by the number of measurements in the sample."[‡] Example 3.1 demonstrates the calculation of the sample mean. Note that the mean has the same units of measurement as do the individual observations. The question of how many decimal places should be reported for the mean will be answered at the end of Section 6.3; until then we shall simply record the mean with one more decimal place than the data.

If, as in Example 3.1, a sample contains multiple identical data for several values of the variable, then it may be convenient to record the data in the form of a frequency

---

*Charles Babbage (1792–1871) was an English mathematician and inventor, who conceived principles used by modern computers—well before the advent of electronics—and who, in 1832, proposed the modern convention of italicizing letters to denote quantities (Cajori, 1929: 2, 6).

†Swiss mathematician Leonhard Euler, in 1755, was the first to use Σ to denote summation (Cajori, 1928: 2, 61).

‡The modern symbols for plus and minus arose in Germany during the 1480s, with Johann Widman the first to use them in print (Cajori, 1928: 222–223). The modern equal sign was invented by English mathematician Robert Recorde, who published it in 1577 along with the first appearance of "+" and "−" in an English work; and it became standard about the start of the eighteenth century (Cajori, 1928: 164, 232–233). Many other symbols were used for mathematical operations, before and after these introductions (e.g., ibid.: 229–245). Using a horizontal line to express division derives from its use, in denoting fractions, by Arabic author Al-Ḥaṣṣâr in the twelfth century, though it was not consitently employed for several more centuries (ibid.: 269, 310). The solidus ("/") was recommended for division by the English writer Augustus De Morgan in 1845 (ibid.: 312–313), and the Swiss author Johann Heinrich Rahn proposed, in 1659, the symbol "÷" which was previously often used by authors as a minus sign (ibid.: 211, 270).

EXAMPLE 3.1   A sample of 24 from a population of butterfly wing lengths.

X_i (in centimeters): 3.3, 3.5, 3.6, 3.6, 3.7, 3.8, 3.8, 3.8, 3.9, 3.9, 3.9, 4.0, 4.0, 4.0, 4.0, 4.1, 4.1, 4.1, 4.2, 4.2, 4.3, 4.3, 4.4, 4.5.

$$\sum X_i = 95.0 \text{ cm}$$

$$n = 24$$

$$\bar{X} = \frac{\sum X_i}{n} = \frac{95.0 \text{ cm}}{24} = 3.96 \text{ cm}$$

---

table, as in Example 3.2. Then X_i can be said to denote each of k different measurements and f_i can denote the frequency with which that X_i occurs in the sample. The sample mean may then be calculated, using the sums of the products of f_i and X_i, as[*]

$$\bar{X} = \frac{\sum_{i=1}^{k} f_i X_i}{n}. \tag{3.3}$$

Example 3.2 demonstrates this calculation for the same data as in Example 3.1.

EXAMPLE 3.2   The data from Example 3.1 recorded as a frequency table.

X_i (cm)	f_i	$f_i X_i$ (cm)
3.3	1	3.3
3.4	0	0
3.5	1	3.5
3.6	2	7.2
3.7	1	3.7
3.8	3	11.4
3.9	3	11.7
4.0	4	16.0
4.1	3	12.3
4.2	2	8.4
4.3	2	8.6
4.4	1	4.4
4.5	1	4.5
$\sum f_i = 24$		$\sum f_i X_i = 95.0$ cm

$$k = 13$$

$$\sum_{i=1}^{k} f_i = n = 24$$

$$\bar{X} = \frac{\sum_{i=1}^{k} f_i X_i}{n} = \frac{95.0 \text{ cm}}{24} = 3.96 \text{ cm}$$

$$\text{median} = 3.95 \text{ cm} + \left(\tfrac{1}{4}\right)(0.1 \text{ cm})$$

$$= 3.95 \text{ cm} + 0.025 \text{ cm}$$

$$= 3.975 \text{ cm}$$

---

[*]Denoting the multiplication of two quantities (e.g., a amd b) by their adjacent placement (i.e., ab) derives from very old practices as far back as Hindu manuscripts from the seventh century (Cajori, 1928: 77, 250). Modern usage also includes Gottfried Wilhelm Leibniz's 1698 recommendation of a dot: $a \cdot b$ (ibid.: 267) and William Oughtred's 1631 suggestion of St. Andrew's cross: $a \times b$ (ibid.: 251). The 1659 use of an asterisk-like symbol "$*$" (ibid.: 212–213) did not persist but resurfaced in electronic computer languages of the latter half of the twentieth century.

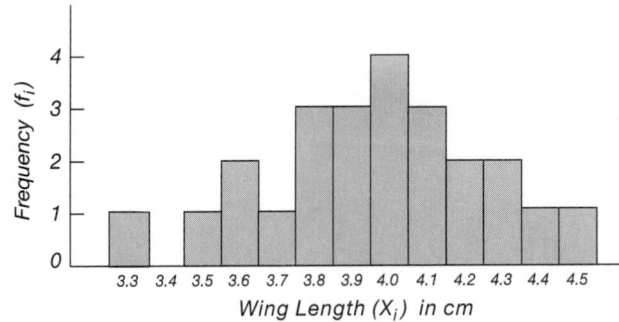

Figure 3.1   A histogram of the data in Example 3.2. The mean (3.96 cm) is the center of gravity of the histogram, and the median (3.975 cm) divides the histogram into two equal areas.

If data are plotted as a histogram (Fig. 3.1), the mean is the *center of gravity* of the histogram.* That is, if the histogram were made of a solid material, it would balance horizontally with the fulcrum at $\bar{X}$. The mean is applicable to ratio or interval scale data.

3.2 THE MEDIAN

The median is typically defined as the middle measurement in an ordered set of data.[†] That is, there are just as many observations larger than the median as there are smaller. The sample median is the best estimate of the population median. In a symmetrical distribution (Figs. 3.2a and 3.2b) the sample median is also an unbiased and consistent estimate of μ, but it is not as efficient a statistic as $\bar{X}$, and should not be used as a substitute for $\bar{X}$. (If the frequency distribution is asymmetrical, the median is a poor estimate of the mean.)

The median ($\mathcal{M}$) of a sample of data may be found by first arranging the measurements in order of magnitude. The order may be either ascending or descending, but ascending order is most commonly used as is done with the samples in Examples 3.1, 3.2, and 3.3. Then, we define the sample median as

$$\mathcal{M} = X_{(n+1)/2}. \tag{3.4}$$

If the sample size (n) is odd, then the subscript in Equation 3.4 will be an integer and will indicate which datum is the middle measurement in the ordered sample. For the data of species A in Example 3.3, $n = 9$ and the sample median is $\mathcal{M} = X_{(n+1)/2} = X_{(9+1)/2} = X_5 = 40$ mo. If n is even, then the subscript in Equation 3.4 will be a half-integer, a number midway between two integers. This indicates that there is not a middle value in the ordered list of data; instead, there are two middle values, and the median is defined as the midpoint between them. For the species B data in Example 3.3, $n = 10$ and

---

*The concept of the mean as the center of gravity was used by L. A. J. Quetelet in 1846 (Walker, 1929: 73).

[†]The concept of the median was conceived as early as 1816, by K. F. Gauss; enunciated and reinforced by others, including F. Galton in 1869 and 1874; and independently discovered and promoted by G. T. Fechner beginning in 1874 (Walker, 1929: 83–88, 184).

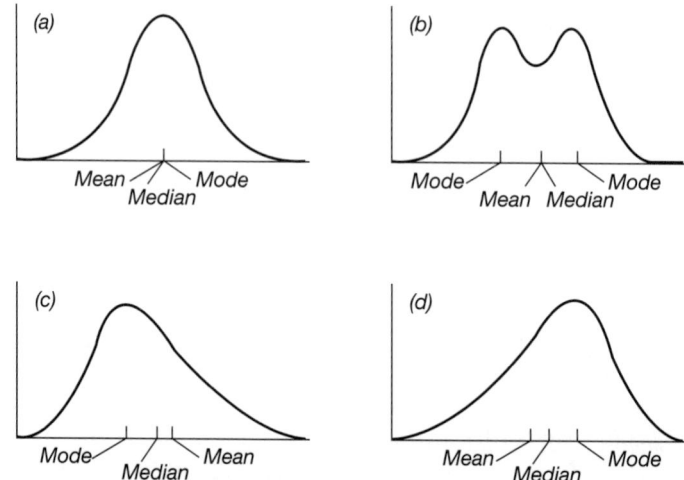

Figure 3.2 Frequency distributions showing measures of central tendency. Values of the variable are along the abscissa (horizontal axis), and the frequencies are along the ordinate (vertical axis). Distributions (a) and (b) are symmetrical, (c) is positively skewed, and (d) is negatively skewed. Distributions (a), (c), and (d) are unimodal, and distribution (b) is bimodal. In a unimodal asymmetric distribution, the median lies about one-third the distance between the mean and the mode.*

EXAMPLE 3.3   Life expectancy for two hypothetical species of birds in captivity.

Species A X_i (mo)	*Species B* X_i (mo)
34	34
36	36
37	37
39	39
40	40
41	41
42	42
43	43
79	44
	45

$n = 9$

$M = X_{(n+1)/2} = X_{(9+1)/2}$

$\quad = X_5 = 40$ mo

$\bar{X} = 43.4$ mo

$n = 10$

$M = X_{(n+1)/2} = X_{(10+1)/2}$

$\quad = X_{5.5} = 40.5$ mo

$\bar{X} = 40.1$ mo

---

*An interesting relationship between the mean, median, and standard deviation is shown in Equation 4.14.

$X_{(n+1)/2} = X_{(10+1)/2} = X_{5.5}$, which signifies that the median is midway between X_5 and X_6, namely $\mathcal{M} = (40 \text{ mo} + 41 \text{ mo})/2 = 40.5$ mo.

Note that the median has the same units as each individual measurement. If data are plotted as a frequency histogram (e.g., Fig. 3.1), the median is the value of X that divides the area of the histogram into two equal parts. The sample median is, in general, a more efficient estimate of the population median for larger sample sizes.

If we find the middle value(s) in an ordered set of data to be among identical observations (referred to as *tied* values), as in Example 3.1 or 3.2, a difficulty arises. If we apply Equation 3.4 to these twenty-four data, then we conclude the median to be $X_{12.5} = 4.0$ cm. But four data are tied at 4.0 cm, and eleven measurements are less than 4.0 cm and nine are greater. Thus, 4.0 cm does not fit the definition above of the median as that value for which there is the same number of data larger and smaller. Therefore, a better definition of the median of a set of data is that value for which no more than half the data are smaller and no more than half are larger.

When the sample median falls among tied observations, we may interpolate to better estimate the population median. Using the data of Example 3.2, we desire to estimate a value below which 50% of the observations in the population lie. Fifty percent of the observations in the sample would be twelve observations. As the first 7 classes in the frequency table include 11 observations and 4 observations are in class 4.0 cm, we know that the desired sample median lies within the range of 3.95 to 4.05 cm. Assuming that the four observations in class 4.0 cm are distributed evenly within the 0.1 cm range of 3.95 to 4.05 cm, then the median will be $\left(\frac{1}{4}\right)(0.1 \text{ cm}) = 0.025$ cm into this class. Thus, the median = 3.95 cm + 0.025 cm = 3.975 cm. In general, for the sample median within a class interval containing tied observations,

$$\mathcal{M} = \left(\begin{array}{c}\text{lower limit} \\ \text{of interval}\end{array}\right) + \left(\frac{0.5n - \text{cum. freq.}}{\text{no. of observations in interval}}\right)\left(\begin{array}{c}\text{interval} \\ \text{size}\end{array}\right), \qquad (3.5)$$

where "cum. freq." refers to the cumulative frequency of the previous classes.* By using this procedure, the calculated median will be the value of X that divides the area of the histogram of the sample into two equal parts. As another example, refer back to Example 1.5, where, by Equation 3.5: median = $\mathcal{M} = 8.75$ mg/g + {[(0.5)(130) − 61]/24}{0.10 mg/g} = 8.75 mg/g + 0.02 mg/g= 8.77 mg/g.

The median expresses less information than does the mean, for it does not take into account the actual value of each measurement, but only considers the rank of each measurement. Still, it may offer certain advantages in some situations. First, it is plain from the two samples in Example 3.3 that extremely high (or extremely low) measurements will not affect the median as much as they affect the mean (causing the sample median to be called a "resistant" statistic). Thus, when we deal with skewed populations, we may prefer the median to the mean to express central tendency.

Note that in Example 3.3 the researcher would have to wait 79 mo to compute a mean life expectancy for species A (45 mo for species B), whereas the median could be

---

*This procedure was enunciated in 1878 by the German scholar, Gustav Theodor Fechner (1801–1887) (Walker, 1929: 86).

determined in only 40 mo (41 mo for species B). Also, to calculate a median one does not need to have accurate data for all members of the sample. If we did not have the first three data for species A accurately recorded, but could state them as "less than 39 mo," then the median could have been determined just as readily, although calculations of the mean would not have been possible. Lastly, the median can be determined not only for interval and ratio scale data, but also for data on the ordinal scale, data for which the use of the mean usually would not be considered appropriate.

3.3 OTHER QUANTILES

Just as the median is the value above and below which lies half the set of data, one can define measures above or below which lie other fractional parts of the data. For example, if the data are divided into four equal parts, we speak of *quartiles*.

One-fourth of all the ranked observations are smaller than the first quartile, one-fourth lie between the first and second quartiles, one-fourth lie between the second and third quartiles, and one-fourth are larger than the third quartile. The second quartile is identical to the median. As with the median, the first and third quartiles might be one of the data or the midpoint between two of the data. The first quartile, Q_1, is

$$Q_1 = X_{(n+1)/4}; \tag{3.6}$$

if the subscript, $(n + 1)/4$, is not an integer or half-integer, then it is rounded up to the nearest integer or half-integer. The second quartile is the median $\mathcal{M}$, and the subscript on X for the third quartile, Q_3, is

$$n + 1 - \text{subscript on } X \text{ for } Q_1. \tag{3.7}$$

Examining the data in Example 3.3: For species $A, n = 9, (n + 1)/4 = 2.5$, and $Q_1 = X_{2.5} = 36.5$ mo; and $Q_3 = X_{10-2.5} = X_{7.5} = 42.5$ mo. For species $B, n = 10, (n + 1)/4 = 2.75$ (which we round up to 3), and $Q_1 = X_3 = 37$ mo, and $Q_3 = X_{11-3} = X_8 = 43$ mo.

Similarly, values that partition the ordered data set into eight equal parts (or as equal as n will allow) are called *octiles*. The first octile, $\mathcal{O}_1$, is

$$\mathcal{O}_1 = X_{(n+1)/8}; \tag{3.8}$$

and if the subscript, $(n + 1)/8$, is not an integer or half-integer, then it is rounded up to the nearest integer or half-integer. The second, fourth, and sixth octiles are the same as quartiles; i.e., $\mathcal{O}_2 = Q_1$, $\mathcal{O}_4 = Q_2 = \mathcal{M}$, and $\mathcal{O}_6 = Q_3$. The subscript on X for the third octile, $\mathcal{O}_3$, is

$$2(\text{subscript on } X \text{ for } Q_1) - \text{subscript on } X \text{ for } \mathcal{O}_1; \tag{3.9}$$

the subscript on X for the fifth octile, $\mathcal{O}_5$, is

$$n + 1 - \text{subscript on } X \text{ for } \mathcal{O}_3; \tag{3.10}$$

and the subscript on X for the seventh octile, $\mathcal{O}_7$, is

$$n + 1 - \text{subscript on } X \text{ for } \mathcal{O}_1. \tag{3.11}$$

Thus, for the data of Example 3.3: For species A, $n = 9$, $(n + 1)/8 = 1.5$ and $\mathcal{O}_1 = X_{1.5} = 35$ mo; $2(2.5) - 1.5 = 3.5$, so $\mathcal{O}_3 = X_{3.5} = 38$ mo; $n + 1 - 3.5 = 6.5$, so $\mathcal{O}_5 = X_{6.5} = 41.5$ mo; and $n + 1 - 1.5 = 8.5$, so $\mathcal{O}_7 = 61$. For species B, $n = 10$, $(n + 1)/8 = 1.25$ (which we round up to 1.5) and $\mathcal{O}_1 = X_{1.5} = 35$ mo; $2(3) - 1.5 = 4.5$, so $\mathcal{O}_3 = X_{4.5} = 39.5$ mo; $n + 1 - 4.5 = 6.5$, so $\mathcal{O}_5 = X_{6.5} = 41.5$ mo; and $n + 1 - 1.5 = 9.5$, so $\mathcal{O}_7 = 44.5$ mo.

Besides the median, quartiles, and octiles, ordered data may be divided into fifths, tenths, or hundredths by quantities that are respectively called *quintiles*, *deciles*, and *centiles* (the latter also called *percentiles*). Measures that divide a group of ordered data into equal parts are collectively termed *quantiles*.[*] The expression "LD$_{50}$," used in some areas of biological research, is simply the 50th percentile of the lethal doses, or the median lethal dose. That is, 50% of the experimental subjects survived this dose, whereas 50% did not. Likewise, "LC$_{50}$" is the median lethal concentration, or the 50th percentile of the lethal concentrations.

3.4 THE MODE

The *mode* is commonly defined as the most frequently occurring measurement in a set of data.[†] In Example 3.2, the mode is 4.0 cm. But it is perhaps better to define a mode as a measurement of relatively great concentration, for some frequency distributions may have more than one such point of concentration, even though these concentrations might not contain precisely the same frequencies. Thus, a sample consisting of the data: 6, 7, 7, 8, 8, 8, 8, 8, 8, 9, 9, 10, 11, 12, 12, 12, 12, 12, 13, 13, and 14 mm would be said to have two modes: at 8 mm and 12 mm. (Some authors would refer to 8 mm as the "major mode" and call 12 mm the "minor mode.") A distribution in which each different measurement occurs with equal frequency is said to have no mode. If two consecutive values of X have frequencies great enough to declare the X values modes, the mode of the distribution is said to be the midpoint of these two X's; e.g., the mode of 3, 5, 7, 7, 7, 8, 8, 8, and 10 liters is 7.5 liters. A distribution with two modes is said to be *bimodal* (e.g., Fig. 3.2b) and may indicate a combination of two distributions with different modes (e.g., heights of men and women). Modes are readily discerned from histograms or frequency polygons.

The sample mode is the best estimate of the population mode. When we sample a symmetrical unimodal population, the mode is an unbiased and consistent estimate of the mean and median (Fig. 3.2a), but it is relatively inefficient and should not be so used.

---

[*]Sir Francis Galton developed the concept of percentiles, quartiles, deciles, and other quantiles in writings from 1869 to 1885 (Walker, 1929: 86–87, 177, 179). The term "quantile" was introduced in 1940 by M. G. Kendall (David, 1995).

[†]The term "mode" was introduced by Karl Pearson in 1894 (Walker, 1929: 184).

As a measure of central tendency, the mode is affected by skewness less than is the mean or the median, but it is more affected by sampling and grouping than these other two measures. The mode, but neither the median nor the mean, may be used for data on the nominal, as well as the ordinal, interval, and ratio scales of measurement. In a unimodal asymmetric distribution (Figs. 3.2c and 3.2d), the median lies about one-third the distance between the mean and the mode.

The mode is not often used in biological research, although it is often interesting to report the number of modes detected in a population, if there are more than one.

3.5 OTHER MEASURES OF CENTRAL TENDENCY

The *range midpoint*, or *midrange*, is also a measure of central tendency, being half-way between the highest and lowest values in the set of data. It is not to be considered a good estimate of the mean and is a seldom-used measure, for it utilizes relatively little information from the data (although the so-called "mean" daily temperature is often reported as the mean of the minimum and maximum, and is thus a range midpoint). The mean of any two symmetrically located percentiles, such as the mean of the first and third quartiles (i.e., the 25th and 75th percentiles), may be used in the same fashion as the range midpoint as a measure of central tendency (see Dixon and Massey, 1969: 133–134), and it is not as adversely affected by aberrantly extreme values. But such a procedure is seldom encountered. As such measures are based on quantiles, they may be applied to either ratio, interval, or ordinal data.

The *geometric mean* is the nth root* of the product of the n data:

$$\bar{X}_G = \sqrt[n]{X_1 X_2 X_3 \dots X_n} = \sqrt[n]{\prod_{i=1}^{n} X_i}, \tag{3.12}$$

Capital Greek pi, Π, means "take the product" in an analogous fashion as Σ indicates "take the sum." The geometric mean may also be calculated as the antilogarithm of the arithmetic mean of the logarithms of the data (where the logarithms may be in any base); this is much more feasible computationally:

$$\bar{X}_G = \text{antilog} \left(\frac{\log X_1 + \log X_2 + \dots + \log X_n}{n} \right) = \text{antilog} \frac{\sum_{i=1}^{n} \log X_i}{n}. \tag{3.13}$$

The geometric mean is appropriate only when all the data are positive. If the data are all equal, then the geometric and arithmetic means are identical; otherwise,[†] $\bar{X}_G < \bar{X}$. This measure finds use in averaging ratios where it is desired to give each ratio equal weight,

---

*Denoting the nth root as $\sqrt[n]{}$ was suggested by Albert Girard as early as 1629, but this symbol was not generally used until well into the eighteenth century (Cajori, 1928: 371–372).

[†]The symbols "<" (meaning "less than") and ">" (meaning "greater than") were invented in 1631 by the English writer Thomas Harriot (Cajori, 1928: 199).

and in averaging percent changes, discussions of which are found in Croxton, Cowden, and Klein (1967: 178–182).

The *harmonic mean* is the reciprocal of the arithmetic mean of the reciprocals of the data:

$$\bar{X}_H = \cfrac{1}{\cfrac{1}{n}\sum\cfrac{1}{X_i}} = \cfrac{n}{\sum\cfrac{1}{X_i}}. \tag{3.14}$$

It is occasionally used when dealing with averaging rates, as described by Croxton, Cowden, and Klein (1967: 182–188). If all the data are identical, then the harmonic mean is the same as the arithmetic mean (and also the same as the geometric mean). If they are positive but not identical, then $\bar{X}_H < \bar{X}_G < \bar{X}$.

The geometric and harmonic means are appropriate only for ratio scale data. They are rarely encountered and the term "mean" typically implies "arithmetic mean."

3.6 THE EFFECT OF CODING DATA

Often in the manipulation of data, considerable time and effort can be saved if *coding* is employed. Coding is the conversion of the original measurements into easier-to-work-with values by simple arithmetic operations. Generally coding employs a *linear transformation* of the data, such as multiplying (or dividing) or adding (or subtracting) a constant. The addition or subtraction of a constant is sometimes termed a translation of the data (i.e., changing the origin), whereas the multiplication or division by a constant causes an expansion or contraction of the scale of measurement. The first set of data in Example 3.4 are coded by subtracting a constant value of 840 g. Not only is each coded value equal to $X_i - 840$ g, but the mean of the coded values is equal to $\bar{X} - 840$ g. Thus, the easier-to-work-with coded values may be used to calculate a mean that then is readily converted to the mean of the original data, simply by adding back the coding constant. In Sample 2 of Example 3.4, the observed data are coded by dividing each observation by 1000 (i.e., by multiplying by 0.001).* The resultant mean only needs to be multiplied by the coding factor of 1000 (i.e., divided by 0.001) to arrive at the mean of the original data. As the other measures of central tendency have the same units as the mean, they are affected by coding in exactly the same fashion. For calculations more involved than computing means, the advantages of coding will become more apparent. In general, linear transformations of ratio or interval scale data will not affect the hypothesis tests to be described later.

In general, if we code X by addition of a constant, A, the coded X is

$$[X_i] = X_i + A. \tag{3.15}$$

---

*In 1593, the German astronomer Christoph Clavius (1537–1612) became the first to use a decimal point to separate units from tenths; in 1617, the Scottish mathematician John Napier (1550–1617) used both points and commas for this purpose (Cajori, 1928: 322-323), and the comma is still used in some parts of the world. In some countries a raised dot has been used—a symbol Americans occasinaly employ to denote multiplication.

EXAMPLE 3.4   Coding data to facilitate calculations.

| Sample 1 (Coding by Subtraction: $A = -840$ g) | | Sample 2 (Coding by Division: $M = 0.001$ liters/ml) | |
X_i (g)	$[X_i] = X_i - 840$ g	X_i (ml)	$[X_i] = (X_i)(0.001 \text{ liters/ml}) = [X_i]$ liters
842	2	8,000	8.000
844	4	9,000	9.000
846	6	9,500	9.500
846	6	11,000	11.000
847	7	12,500	12.500
848	8	13,000	13.000
849	9		

$$\sum X_i = 5922 \text{ g} \qquad \sum [X_i] = 42 \text{ g} \qquad\qquad \sum X_i = 63,000 \text{ ml} \qquad \sum [X_i] = 63.000 \text{ liters}$$

$$\bar{X} = \frac{5922 \text{ g}}{7} \qquad [\bar{X}] = \frac{42 \text{ g}}{7} \qquad\qquad \bar{X} = 10,500 \text{ ml} \qquad [\bar{X}] = 10.500 \text{ liters}$$

$$= 846 \text{ g} \qquad\qquad = 6 \text{ g}$$

$$\bar{X} = [\bar{X}] - A \qquad\qquad\qquad \bar{X} = \frac{[\bar{X}]}{M}$$

$$= 6 \text{ g} - (-840 \text{ g}) \qquad\qquad = \frac{10.500 \text{ liters}}{0.001 \text{ liters/ml}}$$

$$= 846 \text{ g} \qquad\qquad\qquad\qquad = 10,500 \text{ ml}$$

In Sample 1 of Example 3.4, $A = -840$ g. The mean of a set of data thus coded is

$$[\bar{X}] = \bar{X} + A; \tag{3.16}$$

so if one has calculated $[\bar{X}]$ using coded data, it is a simple matter to determine what the sample mean would have been if the data had not been coded, namely

$$\bar{X} = [\bar{X}] - A. \tag{3.17}$$

If one codes X by multiplying by a constant, M, then each coded datum is

$$[X_i] = M X_i. \tag{3.18}$$

In Sample 2 of Example 3.4, $M = 1/1000 = 0.001$ liters/ml. The mean of the coded data is

$$[\bar{X}] = M \bar{X}. \tag{3.19}$$

Knowing $[\bar{X}]$, one can determine that the mean of the uncoded data is

$$\bar{X} = \frac{[\bar{X}]}{M}. \tag{3.20}$$

Coding affects the median and mode in the same way as the mean is affected.

EXERCISES

3.1 If $X_1 = 3.1$ kg, $X_2 = 3.4$ kg, $X_3 = 3.6$ kg, $X_4 = 3.7$ kg, and $X_5 = 4.0$ kg, what is the value of

(a) $\sum_{i=1}^{4} X_i$,

(b) $\sum_{i=2}^{4} X_i$,

(c) $\sum_{i=1}^{5} X_i$,

(d) $\sum X_i$?

3.2 (a) Calculate the mean of the five weights in Exercise 3.1.

(b) Calculate the median of these weights.

3.3 The ages in years, of the faculty members of a university biology department are 32.2, 37.5, 41.7, 53.8, 50.2, 48.2, 46.3, 65.0, and 44.8.

(a) Calculate the mean age of these nine faculty members.

(b) Calculate the median of the ages.

(c) If the person 65.0 years of age retires and is replaced on the faculty with a person 46.5 years old, what is the new mean age?

(d) What is the new median age?

3.4 Consider the following frequency tabulation of leaf weights (in grams):

X_i	f_i
1.85–1.95	2
1.95–2.05	1
2.05–2.15	2
2.15–2.25	3
2.25–2.35	5
2.35–2.45	6
2.45–2.55	4
2.55–2.65	3
2.65–2.75	1

Using the midpoints of the indicated ranges of X_i,

(a) calculate the mean leaf weight using Equation 3.2, and

(b) calculate the mean leaf weight using Equation 3.3.

(c) Calculate the median leaf weight using Equation 3.4, and

(d) calculate the median using Equation 3.5.

(e) Determine the mode of the frequency distribution.

4

MEASURES OF DISPERSION AND VARIABILITY

In addition to a measure of central tendency, it is generally desirable to have a *measure of dispersion* of data. A measure of dispersion, or a *measure of variability*, as it is sometimes called, is an indication of the spread of measurements around the center of the distribution, or the opposite of how clustered the measurements are around the center. Measures of dispersion of a population are parameters of the population, and the sample measures of dispersion that estimate them are statistics.

4.1 THE RANGE

The difference between the highest and lowest measurements in a group of data is termed the *range*.* If sample measurements are arranged in increasing order of magnitude, as if the median were about to be determined, then

$$\text{sample range} = X_n - X_1. \tag{4.1}$$

Sample 1 in Example 4.1 is a hypothetical set of ordered data in which $X_1 = 1.2$ g and $X_n = 2.4$ g. Thus, the range may be expressed as 1.2 to 2.4 g, or as $2.4 \text{ g} - 1.2 \text{ g} = 1.2$ g. Note that the range has the same units as the individual measurements. Sample 2 in Example 4.1 has the same range as Sample 1.

---

*This statistical term dates from an 1848 paper by H. Lloyd (David, 1995).

EXAMPLE 4.1   Calculation of measures of dispersion for two hypothetical samples.

Sample 1

X_i (g)	$X_i - \bar{X}$ (g)	$|X_i - \bar{X}|$ (g)	$(X_i - \bar{X})^2$ (g^2)
1.2	−0.6	0.6	0.36
1.4	−0.4	0.4	0.16
1.6	−0.2	0.2	0.04
1.8	0.0	0.0	0.00
2.0	0.2	0.2	0.04
2.2	0.4	0.4	0.16
2.4	0.6	0.6	0.36

$\sum X_i = 12.6$ g $\sum(X_i - \bar{X}) = 0.0$ g $\sum |X_i - \bar{X}| = 2.4$ g $\sum(X_i - \bar{X})^2$

$$= 1.12 \text{ g}^2$$

$$= \text{"sum of squares"}$$

$$\bar{X} = \frac{12.6 \text{ g}}{7} = 1.8g$$

$$\text{range} = X_7 - X_1 = 2.4 \text{ g} - 1.2 \text{ g} = 1.2 \text{ g}$$

$$\text{quartile deviation} = \frac{Q_3 - Q_1}{2} = \frac{2.2 \text{ g} - 1.4 \text{ g}}{2} = 0.4 \text{ g}$$

$$\text{mean deviation} = \frac{\sum |X_i - \bar{X}|}{n} = \frac{2.4 \text{ g}}{7} = 0.34 \text{ g}$$

$$s^2 = \frac{\sum(X_i - \bar{X})^2}{n - 1} = \frac{1.12 \text{ g}^2}{6} = 0.1867 \text{ g}^2$$

$$s = \sqrt{0.1867 \text{ g}^2} = 0.43 \text{ g}$$

Sample 2

X_i (g)	$X_i - \bar{X}$ (g)	$|X_i - \bar{X}|$ (g)	$(X_i - \bar{X})^2$ (g^2)
1.2	−0.6	0.6	0.36
1.6	−0.2	0.2	0.04
1.7	−0.1	0.1	0.01
1.8	0.0	0.0	0.00
1.9	0.1	0.1	0.01
2.0	0.2	0.2	0.04
2.4	0.6	0.6	0.36

$\sum X_i = 12.6$ g $\sum(X_i - \bar{X}) = 0.0$ g $\sum |X_i - \bar{X}| = 1.8$ g $\sum(X_i - \bar{X})^2$

$$= 0.82 \text{ g}^2$$

$$= \text{"sum of squares"}$$

$$\bar{X} = \frac{12.6 \text{ g}}{7} = 1.8 \text{ g}$$

$$\text{range} = X_7 - X_1 = 2.4 \text{ g} - 1.2 \text{ g} = 1.2 \text{ g}$$

$$\text{quartile deviation} = \frac{Q_3 - Q_1}{2} = \frac{2.0 \text{ g} - 1.6 \text{ g}}{2} = 0.2 \text{ g}$$

$$\text{mean deviation} = \frac{\sum |X_i - \bar{X}|}{n} = \frac{1.8 \text{ g}}{7} = 0.26 \text{ g}$$

$$s^2 = \frac{\sum(X_i - \bar{X})^2}{n - 1} = \frac{0.82 \text{ g}^2}{6} = 0.1367 \text{ g}^2$$

$$s = \sqrt{0.1367 \text{ g}^2} = 0.37 \text{ g}$$

The range is a relatively crude measure of dispersion, inasmuch as it does not take into account any measurements except the highest and the lowest. Furthermore, as it is unlikely that a sample will contain both the highest and lowest values in the population, the sample range usually underestimates the population range; therefore, it is a biased and inefficient estimator. Nonetheless, it is considered useful by some to present the sample range as an estimate (although a poor one) of the population range. For example, taxonomists are often concerned with having an estimate of what the highest and lowest values in a population are expected to be. Whenever the range is specified in reporting data, however, it is usually a good practice to report another measure of dispersion as well. The range is applicable to ordinal, interval, and ratio-scale data.

4.2 Dispersion Measured with Quantiles

Because the sample range is a biased and inefficient estimate of the population range, being sensitive to extremely large and small measurements, alternative measures of dispersion may be desired. For example, we may use the distance between quantiles (Section 3.3). The distance between Q_1 and Q_3, the first and third quartiles (i.e., the 25th and 75th percentiles), is known as the *semi-quartile range* (or *interquartile range*). More commonly one encounters the use of the semi-interquartile range:

$$\text{semi-interquartile range} = \frac{Q_3 - Q_1}{2},\tag{4.2}$$

also known as the *quartile deviation*.*

If the distribution of data is symmetrical, then 50% of the measurements lie within one quantile deviation above and below the median. For Sample 1 in Example 4.1, $Q_1 = 1.4$ g, $Q_3 = 2.2$ g, and the quantile deviation is $(2.2 \text{ g} - 1.4 \text{ g})/2 = 0.4$ g. And for Sample 2, $Q_1 = 1.6$ g, $Q_3 = 2.0$ g, and the quantile deviation is $(2.0 \text{ g} - 1.6 \text{ g})/2 = 0.2$ g.

Instead of distance between the 25th and 75th percentiles, distances between other quantiles (e.g., 10th and 90th percentiles) may be used (see Dixon and Massey, 1969: 134–139). Quantile-based measures of dispersion are valid for either ordinal, interval, or ratio-scale data.

4.3 The Mean Deviation

As is evident from the two samples in Example 4.1, the range conveys no information about how clustered about the middle of the distribution the measurements are. As the mean is so useful a measure of central tendency, one might express dispersion in terms of deviations from the mean. The sum of all deviations from the mean, i.e., $\sum (X_i - \bar{X})$,

---

*This measure was proposed in 1846 by L. A. J. Quetelet (1796–1874); Sir Francis Galton (1822–1911) later called it the "quartile deviation" (Walker, 1929: 84) and, in 1882, the "interquartile range" (David, 1995).

will always equal zero, however, so that such a summation would be useless as a measure of dispersion (as seen in Example 4.1).

Summing the absolute values of the deviations from the mean results in a quantity that is an expression of dispersion about the mean. Dividing this quantity by n yields a measure known as the *mean deviation*, or *mean absolute deviation*,[*] of the sample; this measure has the same units as do the data. In Example 4.1, Sample 1 is more variable (or more dispersed, or less concentrated) than Sample 2. Although the two samples have the same range, the mean deviations, calculated as

$$\text{sample mean deviation} = \frac{\sum |X_i - \bar{X}|}{n}, \tag{4.3}$$

express the differences in dispersion.[†] A different kind of mean deviation can also be defined by using the sum of the absolute deviations from the median instead of from the mean.

4.4 THE VARIANCE

Another method of eliminating the signs of the deviations from the mean is to square the deviations. The sum of the squares of the deviations from the mean is called the *sum of squares*, abbreviated SS, and is defined as follows:[‡]

$$\text{population SS} = \sum (X_i - \mu)^2 \tag{4.4}$$

$$\text{sample SS} = \sum (X_i - \bar{X})^2. \tag{4.5}$$

It can be seen from the above two equations that as a measure of variability, or dispersion, the sum of squares considers how far the X_i's deviate from the mean. In Sample 1 of Example 4.1, the sample mean is 1.8 g and it is seen (in the last column) that

$$SS = (1.2 - 1.8)^2 + (1.4 - 1.8)^2 + (1.6 - 1.8)^2 + (1.8 - 1.8)^2 + (2.0 - 1.8)^2$$

$$+ (2.2 - 1.8)^2 + (2.4 - 1.8)^2$$

$$= 0.36 + 0.16 + 0.04 + 0.00 + 0.04 + 0.16 + 0.36$$

$$= 1.12$$

---

[*]The term "mean deviation" is apparently due to Karl Pearson (1857–1936) (Walker, 1929: 55) and "mean absolute deviation," in 1972, to D. F. Andrews, P. J. Bickel, F. R. Hampel, P. J. Huber, W. H. Rogers, and J. W. Tukey (David, 1995).

[†]K. Weierstrass, in 1841, was the first to denote the absolute value of a quantity by enclosing it within two vertical lines (Cajori, 1929: 123); i.e., $|a| = a$ and $|-a| = a$.

[‡]The use of numerals as exponents was introduced by René Descartes in 1637, though many other kinds of notation were employed before and after that (Cajori, 1928: 1, 205); one alternative was Augustus De Morgan's 1845 notation: $a \wedge b$ to indicate a^b (ibid.: 358), a convention sometimes re-emerging in modern computer programming. The use of parentheses or brackets to group quantities dates back to the mid-sixteenth century, though it was not common mathematical notation until more than two centuries later (ibid.: 1, 392).

(where the units are grams^2).* The sum of squares may also be visualized as a measure of the average extent to which the data deviate from each other, for (using the same seven data as above)

$$
\begin{aligned}
\text{SS} = &[(1.2 - 1.4)^2 + (1.2 - 1.6)^2 + (1.2 - 1.8)^2 + (1.2 - 2.0)^2 + (1.2 - 2.2)^2 \\
&+ (1.2 - 2.4)^2 + (1.4 - 1.6)^2 + (1.4 - 1.8)^2 + (1.4 - 2.0)^2 + (1.4 - 2.2)^2 \\
&+ (1.4 - 2.4)^2 + (1.6 - 1.8)^2 + (1.6 - 2.0)^2 + (1.6 - 2.2)^2 + (1.6 - 2.4)^2 \\
&+ (1.8 - 2.0)^2 + (1.8 - 2.2)^2 + (1.8 - 2.4)^2 + (2.0 - 2.2)^2 + (2.0 - 2.4)^2 \\
&+ (2.2 - 2.4)^2]/7 \\
= &[0.04 + 0.16 + 0.36 + 0.64 + 1.00 + 1.44 + 0.04 + \cdots + 0.04 + 0.16 + 0.04]/7 \\
= &7.84/7 = 1.12
\end{aligned}
$$

(again in grams^2).

The mean sum of squares is called the *variance* (or *mean square*,† the latter being short for *mean squared deviation*), and for a population is denoted by σ^2 ("sigma squared," using the lowercase Greek letter):

$$
\sigma^2 = \frac{\sum(X_i - \mu)^2}{N}. \tag{4.7}
$$

The best estimate of the population variance, σ^2, is the sample variance, s^2:

$$
s^2 = \frac{\sum(X_i - \bar{X})^2}{n - 1}. \tag{4.8}
$$

If, in Equation 4.7, we replace μ by $\bar{X}$ and N by n, the result is a quantity that is a biased estimate of σ^2. The dividing of the sample sum of squares by $n - 1$ (called the *degrees of freedom*, abbreviated DF) rather than by n, yields an unbiased estimate, and it is Equation 4.8 that should be used to calculate the sample variance.

---

*Owing to an important concept in statistics, that known as *least squares*, the sum of squared deviations from the mean is smaller than the sum of squared deviations from any other quantity (e.g., the median). Indeed, if Equation 4.5 is applied using some quantity in place of the mean, the resultant "sum of squares" would be

$$
SS + nd^2, \tag{4.6}
$$

where d is the difference between the mean and the quantity used. For the population sum of squares (defined in Equation 4.4), the relationship would be $SS + Nd^2$.

†The term "mean square" dates back at least to an 1875 publication of Sir George Biddel Airy (1801–1892), Astronomer Royal of England (Walker, 1929: 54). The term "variance" was introduced in 1918 by English statistician Sir Ronald Aylmer Fisher (1890–1962) (ibid.: 189).

If all observations are equal, then there is no variability and $s^2 = 0$. And, s^2 becomes increasingly large as the amount of variability, or dispersion, increases. Because s^2 is a mean sum of squares, it can never be a negative quantity.

The variance expresses the same type of information as does the mean deviation, but it has certain very important properties relative to probability and hypothesis testing that make it distinctly superior. Thus, the mean deviation is very seldom encountered in biostatistical analysis.

The calculation of s^2 can be tedious for large samples, but it can be facilitated by the use of the equality

$$\text{sample SS} = \sum X_i^2 - \frac{\left(\sum X_i\right)^2}{n}. \tag{4.9}$$

This formula is much simpler to work with than is Equation 4.5. Example 4.2 demonstrates its use to obtain a sample sum of squares.

Because sample variance equals sample SS divided by DF,

$$s^2 = \frac{\sum X_i^2 - \dfrac{\left(\sum X_i\right)^2}{n}}{n-1}. \tag{4.10}$$

This last formula is often referred to as a "working formula," or "machine formula," because of its computational advantages. There are, in fact, two major advantages in calculating SS by Equation 4.9 rather than by Equation 4.5. First, fewer computational steps are involved, a fact that decreases chance of error. On many calculators the summed quantities, $\sum X_i$ and $\sum X_i^2$, can both be obtained with only one pass through the data, whereas Equation 4.5 requires one pass through the data to calculate $\bar{X}$ and at least one more pass to calculate and sum the squares of the deviations, $X_i - \bar{X}$. Second, there may be a good deal of rounding error in calculating each $X_i - \bar{X}$, a situation that leads to decreased accuracy in computation, but which is avoided by the use of Equation 4.9.*

For data recorded in frequency tables,

$$\text{sample SS} = \sum f_i X_i^2 - \frac{\left(\sum f_i X_i\right)^2}{n}, \tag{4.11}$$

where f_i is the frequency of observations with magnitude X_i. But with a calculator it is often faster to use Equation 4.9 for each individual observation, disregarding the class groupings.

---

*Computational formulas advantageous on calculators may not prove accurate on computers (Wilkinson and Dallal, 1977), largely because computers may use fewer significant figures. (Also see Ling, 1974.)

EXAMPLE 4.2 "Machine formula" calculation of variance, standard deviation, and coefficient of variation. (These are the data of Example 4.1)

	Sample 1		Sample 2
X_i (g)	X_i^2 (g^2)	X_i (g)	X_i^2 (g^2)
1.2	1.44	1.2	1.44
1.4	1.96	1.6	2.56
1.6	2.56	1.7	2.89
1.8	3.24	1.8	3.24
2.0	4.00	1.9	3.61
2.2	4.84	2.0	4.00
2.4	5.76	2.4	5.76

Sample 1:

$$\sum X_i = 12.6 \text{ g} \qquad \sum X_i^2 = 23.80 \text{ g}^2$$

$$n = 7$$

$$\bar{X} = \frac{12.6 \text{ g}}{7} = 1.8 \text{ g}$$

$$SS = \sum X_i^2 - \frac{\left(\sum X_i\right)^2}{n}$$

$$= 23.80 \text{ g}^2 - \frac{(12.6 \text{ g})^2}{7}$$

$$= 23.80 \text{ g}^2 - 22.68 \text{ g}^2$$

$$= 1.12 \text{ g}^2$$

$$s^2 = \frac{SS}{n-1}$$

$$= \frac{1.12 \text{ g}^2}{6} = 0.1867 \text{ g}^2$$

$$s = \sqrt{0.1867 \text{ g}^2} = 0.43 \text{ g}$$

$$V = \frac{s}{\bar{X}} = \frac{0.43 \text{ g}}{1.8 \text{ g}}$$

$$= 0.24 = 24\%$$

Sample 2:

$$\sum X_i = 12.6 \text{ g} \qquad \sum X_i^2 = 23.50 \text{ g}^2$$

$$n = 7$$

$$\bar{X} = \frac{12.6 \text{ g}}{7} = 1.8 \text{ g}$$

$$SS = 23.50 \text{ g}^2 - \frac{(12.6 \text{ g})^2}{7}$$

$$= 0.82 \text{ g}^2$$

$$s^2 = \frac{0.82 \text{ g}^2}{6} = 0.1367 \text{ g}^2$$

$$s = \sqrt{0.1367 \text{ g}^2} = 0.37 \text{ g}$$

$$V = \frac{0.37 \text{ g}}{1.8 \text{ g}}$$

$$= 0.21 = 21\%$$

The variance has square units. If measurements are in grams, their variance will be in grams squared, or if the measurements are in cubic centimeters, their variance will be in terms of cubic centimeters squared, even though such squared units have no physical interpretation. The question of how many decimal places to report for the variance will be considered at the end of Section 6.3.

4.5 THE STANDARD DEVIATION

The *standard deviation*[*] is the positive square root[†] of the variance; therefore, it has the same units as the original measurements. Thus, for a population,

$$\sigma = \sqrt{\frac{\sum X_i^2 - \frac{\left(\sum X_i\right)^2}{N}}{N}}, \tag{4.12}$$

And for a sample,[‡]

$$s = \sqrt{\frac{\sum X_i^2 - \frac{\left(\sum X_i\right)^2}{n}}{n-1}}. \tag{4.13}$$

Examples 4.1 and 4.2 demonstrate the calculation of s. This quantity frequently is abbreviated SD, and on rare occasions is called the *root mean square deviation* or *root mean square*. Remember that the standard deviation is, by definition, always a nonnegative quantity.[§] The end of Section 6.3 will explain how to determine the number of decimal places that may appropriately be recorded for the standard deviation.

---

[*]It was the great English statistician Karl Pearson (1857–1936) who coined the term *standard deviation* and its symbol, σ, in 1893, prior to which this quantity was called the "mean error" (Eells, 1926; Walker, 1929: 54–55, 183, 188). In early literature (e.g., by G. U. Yule in 1919), it was termed "root mean square deviation" and acquired the symbol s, and (particularly in the fields of education and psychology) it was occasionally computed using deviations from the median (or even the mode) instead of from the mean (Eells, 1926).

[†]The square-root sign ("$\sqrt{}$") originated in Germany, apparently in the early sixteenth century (Cajori, 1928: 366–369); sometime prior to 1637, René Descartes combined this with a vinculum (a horizontal bar placed above quantities to group them as we do with parentheses or brackets) (ibid.: 1, 208, 375) to obtain the symbol "$\sqrt{}$"; however (ibid.: 1, 372), Gottfried Wilhelm Leibniz preferred "$\sqrt{()}$" which is still occasionally seen.

[‡]The sample s is actually a slightly biased estimate of the population σ, in that on the average it is a slightly low estimate, especially in small samples. But this fact is generally considered to be offset by the statistic's usefulness. Correction for this bias is sometimes possible (e.g., Bliss, 1967: 131; Dixon and Massey, 1969: 136; Gurland and Tripathi, 1971; Tolman, 1971), but it is rarely employed.

[§]It can be shown that the median of a distribution ($\mathcal{M}$) is never more than one standard deviation away from the mean (μ); that is,

$$|\mathcal{M} - \mu| \leq \sigma \tag{4.14}$$

(O'Cinneide, 1990). This is a special case, where $p = 50$, of the relationship,

$$\mu - \sigma\sqrt{\frac{1 - p/100}{p/100}} \leq X_p \leq \mu + \sigma\sqrt{\frac{p/100}{1 - p/100}}, \tag{4.15}$$

where X_p is the pth percentile of the distribution (Dharmadhikari, 1991).

4.6 The Coefficient of Variation

The *coefficient of variation** or *coefficient of variability*, is defined as

$$V = \frac{s}{\bar{X}} \text{ or } V = \frac{s}{\bar{X}} \cdot 100\%. \tag{4.16}$$

As $s/\bar{X}$ is generally a small quantity, it is frequently multiplied by 100% in order to express V as a percentage. (The coefficient of variation is often abbreviated as CV.)

As a measure of variability, the variance and standard deviation have magnitudes that are dependent on the magnitude of the data. Elephants have ears that are perhaps 100 times larger than those of mice. If elephant ears were no more variable, relative to their size, than mouse ears, relative to their size, the standard deviation of elephant ear lengths would be 100 times as great as the standard deviation of mouse ear lengths (and the variance of the former would be $100^2 = 10,000$ times the variance of the latter). The coefficient of variation expresses sample variability relative to the mean of the sample (and is on rare occasion referred to as the "relative standard deviation"). It is called a measure of *relative variability* or *relative dispersion*.

Because s and $\bar{X}$ have identical units, V has no units at all, a fact emphasizing that it is a relative measure, divorced from the actual magnitude or units of measurement of the data. Thus, had the data in Example 4.2 been measured in pounds, kilograms, or tons, instead of grams, the calculated V would have been the same. The coefficient of variation of a sample, namely V, is an estimate of the coefficient of variation of the population from which the sample came (i.e., an estimate of σ/μ). The coefficient of variation may be calculated only for ratio scale data; it is, for example, not valid to calculate coefficients of variation of temperature data measured on the Celsius or Fahrenheit temperature scales. Simpson, Roe, and Lewontin (1960: 89–95) present a good discussion of V and its biological application, especially with regard to zoomorphological measurements.

4.7 Indices of Diversity

For nominal scale data there is no mean or median to serve as a reference for discussion of dispersion. Instead, we can invoke the concept of *diversity*, the distribution of observations among categories. Consider that sparrows are found to nest in four different types of location (vines, eaves, branches, and cavities). If, out of twenty nests observed, five are found at each of the four locations, then we would say that there was great diversity in nesting sites. If, however, seventeen nests were found in cavities and only one in each of the other three locations, then we would consider the situation to be one of very low nest-site diversity. In other words, observations distributed evenly among categories display high diversity, whereas a set of observations where the bulk of the data occurs in very few of the categories is one exhibiting low diversity.

---

*The term *coefficient of variation* was introduced by the statistical giant, Karl Pearson (1857–1936), in 1895 (Walker, 1929: 178). In early literature the term was variously applied to the ratios of different measures of dispersion and different measures of central tendency (Eells, 1926).

Among the quantitative descriptions of diversity available are those based on *information theory*. The underlying considerations of these measures can be visualized by considering *uncertainty* to be synonymous with diversity. If seventeen out of twenty nest sites were to be found in cavities, then one would be relatively certain of being able to predict the location of a randomly encountered nest site. However, if nests were found to be distributed evenly among the various locations (a situation of high nest-site diversity), then there would be a good deal of uncertainty involved in predicting the location of a nest site selected at random. If a set of nominal scale data may be considered to be a random sample, then a quantitative expression appropriate as a measure of diversity is that of Shannon (1948):

$$H' = - \sum_{i=1}^{k} p_i \log p_i \qquad (4.17)$$

(often referred to as the Shannon-Wiener diversity index or the Shannon-Weaver index). Here, k is the number of categories and p_i is the proportion of the observations found in category i. Denoting n to be sample size, and f_i to be the number of observations in category i, then $p_i = f_i/n$. Some mathematical manipulation arrives at the equivalent function:

$$H' = \frac{n \log n - \sum_{i=1}^{k} f_i \log f_i}{n}, \qquad (4.18)$$

a formula that is easier to use than Equation 4.17 because it eliminates the necessity of calculating the proportions (p_i). Published tables of $n \log n$ and $f_i \log f_i$ are available (e.g., Brower, Zar, and von Ende, 1998: 181; Lloyd, Zar, and Karr, 1968). Any logarithmic base may be used to compute H'; bases 10, e, and 2 (in that order of commonness) are the most frequently encountered. A value of H' (or of any other measure of this section except evenness measures) calculated using one logarithmic base may be converted to that of another base; Table 4.1 gives factors for doing this for bases 10, e, and 2. Unfortunately, H' is known to be an underestimate of the diversity in the sampled population (Bowman et al., 1971). However, this bias decreases with increasing sample size.

Ghent (1991) demonstrates a relationship between H' and testing hypotheses for equal abundance among the k categories.

The magnitude of H' is affected not only by the distribution of the data but also by the number of categories, for, theoretically, the maximum possible diversity for a set of data consisting of k categories is

$$H'_{\text{max}} = \log k. \qquad (4.19)$$

Therefore, some users of Shannon's index prefer to calculate

$$J' = \frac{H'}{H'_{\text{max}}} \qquad (4.20)$$

TABLE 4.1 Multiplication Factors for Converting among Diversity Measures (H, H', H_{max}, or H'_{max}) Calculated Using Different Logarithmic Bases*

To convert to:	To convert from:		
	Base 2	Base e	Base 10
Base 2	1.0000	1.4427	3.3219
Base e	0.6931	1.0000	2.3026
Base 10	0.3010	0.4343	1.0000

For example, if $H' = 0.255$ using base 10; H' would be $(0.255)(3.3219) = 0.847$ using base 2.

*The measures J and J' are unaffected by change in logarithmic base.

instead of (or in addition to) H', thus expressing the observed diversity as a proportion of the maximum possible diversity. The quantity J' has been termed *evenness* (Pielou, 1966) and may also be referred to as *homogeneity* or *relative diversity*. The measure $1 - J'$ may then be viewed as a measure of *heterogeneity* or *dominance*. As k is typically an underestimate of the number of categories in the population, the sample eveness, J', is typically an overestimate of the population evenness. (That is, J' is a biased statistic.) Example 4.3 demonstrates the calculation of H' and J'.

If a set of data may not be considered a random sample, then Equation 4.18 (or 4.19) is not an appropriate diversity measure (Pielou, 1966). Examples of such situations may be when we have, in fact, data comprising an entire population, or data that are a sample obtained nonrandomly from a population. In such a case, one may use the information-theoretic diversity measure of Brillouin (1962: 7–8):*

$$H = \frac{\log \left(\dfrac{n!}{\prod\limits_{i=1}^{k} f_i!} \right)}{n}, \tag{4.21}$$

where Π (capital Greek pi) means to take the product, just as Σ means to take the sum. Equation 4.21 may be written, equivalently, as

$$H = \frac{\log \dfrac{n!}{f_1! f_2! \dots f_k!}}{n} \tag{4.22}$$

or as

$$H = \frac{(\log n! - \sum \log f_i!)}{n}. \tag{4.23}$$

---

*$n!$ is read as "n factorial" and implies the product, $(n)(n-1)(n-2)\dots(2)(1)$. The factorial symbol ("!") was proposed by Christian Kramp of Strasbourg in 1808 (Cajori, 1928: 2, 72).

EXAMPLE 4.3   Indices of diversity for nominal scale data. The nesting sites of sparrows.

Category (i)	Observed Frequencies (f_i)

	Sample 1
Vines	5
Eaves	5
Branches	5
Cavities	5

$$H' = \frac{n \log n - \sum f_i \log f_i}{n} = [20 \log 20 - (5 \log 5 + 5 \log 5 + 5 \log 5 + 5 \log 5)]/20$$

$$= [26.0206 - (3.4949 + 3.4949 + 3.4949 + 3.4949)]/20$$

$$= 12.0410/20 = 0.602$$

$$H'_{max} = \log 4 = 0.602$$

$$J' = \frac{0.602}{0.602} = 1.00$$

	Sample 2
Vines	1
Eaves	1
Branches	1
Cavities	17

$$H' = \frac{n \log n - \sum f_i \log f_i}{n} = [20 \log 20 - (1 \log 1 + 1 \log 1 + 1 \log 1 + 17 \log 17)]/20$$

$$= [26.0206 - (0 + 0 + 0 + 20.9176)]/20$$

$$= 5.1030/20 = 0.255$$

$$H'_{max} = \log 4 = 0.602$$

$$J' = \frac{0.255}{0.602} = 0.42$$

	Sample 3
Vines	2
Eaves	2
Branches	2
Cavities	34

$$H' = \frac{n \log n - \sum f_i \log f_i}{n} = [40 \log 40 - (2 \log 2 + 2 \log 2 + 2 \log 2 + 34 \log 34)]/40$$

$$= [64.0824 - (0.6021 + 0.6021 + 0.6021 + 52.0703)]/40$$

$$= 10.2058/40 = 0.255$$

$$H'_{max} = \log 4 = 0.602$$

$$J' = \frac{0.255}{0.602} = 0.42$$

Table B.40 gives logarithms of factorials to ease this calculation. Other such tables are available, as well (e.g., Brower, Zar, and von Ende 1998: 183; Lloyd, Zar, and Karr, 1968; Pearson and Hartly, 1966: Table 51).* Ghent (1991) discusses the relationship between H and the test of hypotheses about equal abundance among k categories.

The maximum possible Brillouin diversity for a set of n observations distributed among K categories is

$$H_{\max} = \frac{\log n! - (k - d)\log c! - d\log(c + 1)!}{n},$$ (4.28)

where c is the integer portion of n/k, and d is the remainder. The Brillouin-based evenness measure is, therefore,

$$J = \frac{H}{H_{\max}},$$ (4.29)

with $1 - J$ being a dominance measure. In as much as we consider that we have data from an entire population, k is a population measurement, rather than an estimate of one, and J is not a biased estimate as is J'.

For further considerations of these and other diversity measures, see Brower, Zar, and von Ende (1998: Chapter 5B).

4.8 THE EFFECT OF CODING DATA

In Section 3.6 it was shown how coding data may facilitate statistical computations. Such benefits are even more apparent when calculating SS, s^2, and s, because of the labor, and concomitant chances of error, involved in the squaring of unwieldy numbers.

When data are coded by adding or subtracting a constant, all the above measures of dispersion except the coefficient of variation are not changed from what they were for uncoded data. This is because these measures involve deviations, and deviations are

---

*For moderate to large n (or f_i), "Stirling's approximation" is excellent (see note after Table B.40):

$$n! = \sqrt{2\pi n}(n/e)^n = \sqrt{2\pi}\sqrt{n}e^{-n}n^n,$$ (4.24)

of which this is an easily usable derivation:

$$\log n! = (n + 0.5)\log n - 0.434294n + 0.399090.$$ (4.25)

An approximation with only half the error of the above is

$$n! = \sqrt{2\pi}\left(\frac{n + 0.5}{e}\right)^{n+0.5}$$ (4.26)

and

$$\log n! = (n + 0.5)\log(n + 0.5) - 0.434294(n + 0.5) + 0.399090.$$ (4.27)

This is named for James Stirling, who published something similar to the latter approximation formula in 1730, making an arithmetic improvement in the approximation earlier known by Abraham de Moivre (Kemp, 1989; Pearson, 1924; Walker, 1929: 16).

not changed by translation. Sample 1 in Example 4.4 demonstrates these relationships. To arrive at the desired coefficient of variation, simply decode $\bar{X}$ and s before calculating V.

When coding by multiplication or division, however, the measures of dispersion are affected, for the magnitudes of the deviations will be changed. The standard deviation, range, and mean deviation are changed in the same manner as the measures of central tendency (Section 3.6). However, the variance changes as the square of the coding constant, whereas the coefficient of variation is unchanged, as is shown in Sample 2 of Example 4.4.

When calculating information-theoretic diversity indices, coding by multiplication or division does not affect the results (see Samples 2 and 3 in Example 4.3). Coding by addition or subtraction should not be employed.

Table 4.2 summarizes the effect of coding on sample statistics considered thus far. A coded datum may be defined as

$$[X_i] = MX_i + A, \tag{4.30}$$

where M is a multiplication coding constant and A is an addition coding constant. (In Example 4.4, we see that $M = 1$ and $A = -840$ g in Sample 1; $M = 0.01$ and $A = 0$ in Sample 2.)

TABLE 4.2 The Effect of Coding Data on Sample Statistics, Where $[X_i] = MX_i + A$

Statistic	Value without coding	Value using coding
Mean, $\bar{X}$	$\bar{X} = \dfrac{[\bar{X}] - A}{M}$	$[\bar{X}] = M\bar{X} + A$
Median and mode	same as mean	
Sum of squares, SS	$SS = \dfrac{[SS]}{M^2}$	$[SS] = M^2 SS$
Variance, s^2	$s^2 = \dfrac{[s^2]}{M^2}$	$[s^2] = M^2 s^2$
Standard deviation, s	$s = \dfrac{[s]}{M}$	$[s] = Ms$
Range and mean deviation	same as standard deviation	
Coefficient of variation, V (if $A = 0$)*	$V = [V]$	$[V] = V$
Shannon diversity index, H' (if $A = 0$)*	$H' = [H']$	$[H'] = H'$
Shannon evenness index, J' (if $A = 0$)*	$J' = [J']$	$[J'] = J'$

*If $A \neq 0$, one cannot convert between coded and uncoded statistics.

EXAMPLE 4.4   Coding data to facilitate the calculation of measures of dispersion.

Sample 1 (Coding by Subtraction: $A = -840$ g)

	Without coding X_i	Using coding $[X_i]$	
X_i (g)	X_i^2 (g^2)	$[X_i]$ (g)	$[X_i]^2$ (g^2)
842	708,964	2	4
843	710,649	3	9
844	712,336	4	16
846	715,716	6	36
846	715,716	6	36
847	717,409	7	49
848	719,104	8	64
849	720,801	9	81

$\sum X_i = 6765$ g $\sum X_i^2 = 5,720,695$ g^2 $\sum [X_i] = 45$ g $\sum [X_i]^2 = 295$ g^2

$$s^2 = \frac{5720695 \text{ g}^2 - \dfrac{(6765 \text{ g})^2}{8}}{7}$$

$$[s^2] = \frac{295 \text{ g}^2 - \dfrac{(45 \text{ g})^2}{8}}{7}$$

$$= 5.98 \text{ g}^2$$

$$= 5.98 \text{ g}^2$$

$$s = 2.45 \text{ g}$$

$$[s] = 2.44 \text{ g}$$

$$\bar{X} = 845.6 \text{ g}$$

$$[\bar{X}] = 5.6 \text{ g}$$

$$V = \frac{s}{\bar{X}} = \frac{2.45 \text{ g}}{845.6 \text{ g}}$$

$$= 0.0029 = 0.29\%$$

Sample 2 (Coding by Division: $M = 0.01$)

	Without coding X_i	Using coding $[X_i]$	
X_i (sec)	X_i^2 (sec^2)	$[X_i]$ (sec)	$[X_i]^2$ (sec^2)
800	640,000	8.00	64.00
900	810,000	9.00	81.00
950	902,500	9.50	90.25
1100	1,210,000	11.00	121.00
1250	1,562,500	12.50	156.25
1300	1,690,000	13.00	169.00

$\sum X_i = 6300$ sec $\sum X_i^2 = 6,815,000$ sec^2 $\sum [X_i] = 63.00$ sec $\sum [X_i]^2 = 681.50$ sec^2

$$s^2 = \frac{6815000 \text{ sec}^2 - \dfrac{(6300 \text{ sec})^2}{6}}{5}$$

$$[s^2] = \frac{681.50 \text{ sec}^2 - \dfrac{(63.00 \text{ sec})^2}{6}}{5}$$

$$= 40,000 \text{ sec}^2$$

$$= 4 \text{ sec}^2$$

$$s = 200 \text{ sec}$$

$$[s] = 2.00 \text{ sec}$$

$$\bar{X} = 1050 \text{ sec}$$

$$[\bar{X}] = 10.50 \text{ sec}$$

$$V = 0.19 = 19\%$$

$$[V] = 0.19 = 19\%$$

EXERCISES

4.1 Five body weights, in grams, collected from a population of rodent body weights are:

66.1, 77.1, 74.6, 61.8, 71.5.

(a) Compute the "sum of squares" and the variance of these data using Equations 4.5 and 4.8, respectively.
(b) Compute the "sum of squares" and the variance of these data by using Equations 4.9 and 4.10, respectively.

4.2 Consider the following data, which are a sample of amino acid concentrations (mg/100 ml) in arthropod hemolymph:

240.6, 238.2, 236.4, 244.8, 240.7, 241.3, 237.9.

(a) Determine the range of the data.
(b) Calculate the "sum of squares" of the data.
(c) Calculate the variance of the data.
(d) Calculate the standard deviation of the data.
(e) Calculate the coefficient of variation of the data.

4.3 The following frequency distribution of tree species was observed in a random sample from a forest:

Species	Frequency
White oak	44
Red oak	3
Shagbark hickory	28
Black walnut	12
Basswood	2
Slippery elm	8

(a) Use the Shannon index to express the tree species diversity.
(b) Compute the maximum Shannon diversity possible for the given number of species and individuals.
(c) Calculate the Shannon evenness for these data.

4.4 Assume the data in Exercise 4.3 were an entire population (e.g., all the trees planted around a group of buildings).
(a) Use the Brillouin index to express the tree species diversity.
(b) Compute the maximum Brillouin diversity possible for the given number of species and individuals.
(c) Calculate the Brillouin evenness measure for these data.

5

PROBABILITIES

Everyday concepts of "likelihood," "predictability," and "certainty" are formalized by that branch of mathematics called *probability*. Although earlier work on the subject was done by writers such as Giralamo Cardano (1501–1576) and Galileo Galilei (1564–1642), the investigation of probability as a branch of mathematics sprang in earnest from 1654 correspondence between two great French mathematicians, Blaise Pascal (1623–1662) and Pierre de Fermat (1601–1665). These two men were stimulated by the desire to predict outcomes in the games of chance popular among the French nobility of the mid-seventeenth century; we still use the devices of such games (e.g., dice and cards) to demonstrate the basic concepts of probability.*

A thorough discourse on probability is well beyond the scope and intent of this book, but aspects of probability are of biological interest and considerations of probability theory underlie the many procedures for statistical hypothesis testing discussed in the following chapters. Therefore, this chapter will introduce probability concepts that bear the most pertinence to biology and biostatistical analysis. Although mastery of this chapter is not essential to apply the statistical procedures in the remainder the book, occasionally later reference will be made to it.

Worthwhile presentations of probability specifically for the biologist are found in Batschelet (1976: 441–474); Eason, Coles, and Gettinby (1980: 395–414); and Mosimann (1968).

---

*The first published work on the subject of probability and gaming was by the Dutch astronomer, physicist, and mathematician, Christiaan (also Christianus) Huygens (1629–1695), in 1657 (Asimov, 1982: 138; David, 1962: 113, 133). This, in turn, aroused the interest of other major minds: the Swiss mathematician, Jacques (also known as Jakob and James) Bernoulli (1654–1705) and his brother Nicolas' son Nicolas (1687–1759) who edited Jacques' 1713 book which was the first devoted entirely to probability, and others such as Abraham de Moivre (1667–1754), Pierre Rémond de Montmort (1678–1719), and Pierre Simon, Marquis de Laplace (1749–1827). For more detailed history of the subject, see David (1962) and Walker (1929: 5–13).

5.1 COUNTING POSSIBLE OUTCOMES

Suppose a phenomenon can occur in any one of k different ways, but in only one of those ways at a time. For example, a coin has two sides and when tossed will land with either the "head" side (H) up or the "tail" side (T) up, but not both. Or, a die has six sides and when thrown will land with either the 1, 2, 3, 4, 5, or 6 side up.*

We shall now refer to each possible outcome (i.e., H or T with the coin; or 1, 2, 3, 4, 5, or 6 with the die) as an *event*.

If something can occur in any one of k_1 different ways, and something else can occur in any one of k_2 different ways, then the number of possible ways for both things to occur is $k_1 \times k_2$. For example, suppose that two coins are tossed, say a silver one and a copper one. There are two possible outcomes of the toss of the silver coin (H; T) and two possible outcomes of the toss of the copper coin (H; T). Therefore, $k_1 = 2$ and $k_2 = 2$ and there are $(k_1)(k_2) = (2)(2) = 4$ possible outcomes of the toss of both coins: both heads, silver head and copper tail, silver tail and copper head, and both tails (i.e., H,H; H,T; T,H; T,T).

Or, consider tossing of a coin together with throwing a die. There are two possible coin outcomes ($k_1 = 2$) and six possible die outcomes ($k_2 = 6$), so there are $(k_1)(k_2) = (2)(6) = 12$ possible outcomes of the two phenomena together:

$$H,1; \ H,2; \ H,3; \ H,4; \ H,5; \ H,6; \ T,1; \ T,2; \ T,3; \ T,4; \ T,5; \ T,6.$$

If two dice are thrown, we can count six possible outcomes for the first die and six for the second, so there are $(k_1)(k_2) = (6)(6) = 36$ possible outcomes when two dice are thrown:

$$1,1; \ 1,2; \ 1,3; \ 1,4; \ 1,5; \ 1,6; \qquad 2,1; \ 2,2; \ 2,3; \ 2,4; \ 2,5; \ 2,6;$$

$$3,1; \ 3,2; \ 3,3; \ 3,4; \ 3,5; \ 3,6; \qquad 4,1; \ 4,2; \ 4,3; \ 4,4; \ 4,5; \ 4,6;$$

$$5,1; \ 5,2; \ 5,3; \ 5,4; \ 5,5; \ 5,6; \qquad 6,1; \ 6,2; \ 6,3; \ 6,4; \ 6,5; \ 6,6$$

The above counting rule is extended readily to determine the number of ways more than two things can occur together. If one thing can occur in any one of k_1 ways, a second thing in any one of k_2 ways, a third thing in any of k_3 ways, and so on, through in nth thing in any one of k_n ways, then the number of ways for all n things to occur together is

$$(k_1)(k_2)(k_3) \cdots (k_n).$$

---

*What we recognize as metallic coins originated shortly after 650 B.C.—perhaps in ancient Lydia (located on the Aegean Sea in what is now western Turkey). From the beginning, the obverse and reverse sides of coins have had different designs, in earliest times with the obverse commonly depicting animals and, later, deities and rulers (Sutherland, 1992). Dice have long been used for both games and religion. They date from nearly 3000 years B.C., with the modern conventional arrangement of dots on the six faces of a cubic die (1 opposite 6, 2 opposite 5, and 3 opposite 4) becoming dominant around the middle of the fourteenth century B.C. (David, 1962: 10). Of course, the arrangement of the numbers 1 through 6 on the six faces has no effect on the outcome of throwing a die.

Thus, if three coins are tossed, each toss resulting in one of two possible outcomes, then there is a total of

$$(k_1)(k_2)(k_3) = (2)(2)(2) = 2^3 = 8$$

possible outcomes for the three tosses together:

H,H,H; H,H,T; H,T,H; H,T,T; T,H,H, T,H,T; T,T,H; T,T,T.

Similarly, if three dice are thrown, there are $(k_1)(k_2)(k_3) = (6)(6)(6) = 6^3 = 216$ possible outcomes; if two dice and three coins are thrown, there are $(k_1)(k_2)(k_3)(k_4)(k_5) = (6)(6)(2)(2)(2) = (6^2)(2^3) = 288$ outcomes; and so on. Example 5.1 gives two biological examples of counting possible outcomes.

EXAMPLE 5.1   Counting possible outcomes.

(a) A linear arrangement of three deoxyribonucleic acid (DNA) nucleotides is called a triplet. A nucleotide may contain any one of four possible bases: adenine (A), cytosine (C), guanine (G), and thymine (T). How many different triplets are possible?

As the first nucleotide in the triplet may be any one of the four bases (A; C; G; T), the second may be any one of the four, and the third may be any one of the four, there is a total of

$$(k_1)(k_2)(k_3) = (4)(4)(4) = 4^3 = 64 \text{ possible outcomes;}$$

that is, there are 64 possible triplets:

A, A, A;   A, A, C;   A, A, G;   A, A, T;
A, C, A;   A, C, C;   A, C, G;   A, C, T;
A, G, A;   A, G, C;   A, G, G;   A, G, T;
and so on.

(b) If a diploid cell contains three pairs of chromosomes, and one member of each pair is found in each gamete, how many different gametes are possible?

As the first chromosome may occur in a gamete in one of two forms, as may the second and the third chromosomes,

$$(k_1)(k_2)(k_3) = (2)(2)(2) = 2^3 = 8.$$

Let us designate one of the pairs of chromosomes as "long," with the members of the pair being L_1 and L_2; one pair as "short," indicated as S_1 and S_2; and one pair as "midsized," labeled M_1 and M_2. Then, the eight possible outcomes are

L_1, M_1, S_1; L_1, M_1, S_2; L_1, M_2, S_1; L_1, M_2, S_2;
L_2, M_1, S_1; L_2, M_1, S_2; L_2, M_2, S_1; L_2, M_2, S_2.

5.2 PERMUTATIONS

Linear Arrangements.   A *permutation** is an arrangement of objects in a specific sequence. A horse (H), cow (C), and sheep (S) could be arranged linearly in six different ways: H,C,S; H,S,C; C,H,S; C,S,H; S,H,C; S,C,H. This set of outcomes may be examined

---

*The term "permutation" was invented by Jacques Bernoulli in his landmark 1713 book on probability (Walker, 1929: 9).

by noting that there are three possible ways to fill the first position in the linear order; but once an animal is placed in this position there are only two ways to fill the second position; and after animals are placed in the first two positions there is only one possible way to fill the third position. Therefore, $k_1 = 3, k_2 = 2$, and $k_3 = 1$, so that by the method of counting of Section 5.1 there are $(k_1)(k_2)(k_3) = (3)(2)(1) = 6$ ways to align these three animals. We may say that there are six permutations of three distinguishable objects.

In general, if there are n linear positions to fill with n objects, the first position may be filled in any one of n ways, the second may be filled in any one of $n - 1$ ways, the third in any one of $n - 2$ ways, and so on until the last position, which may be filled in only one way. That is, the filling of n positions with n objects results in $_nP_n$ permutations, where

$$_nP_n = n(n - 1)(n - 2) \cdots (3)(2)(1). \tag{5.1}$$

This equation may be written more simply in *factorial* notation, as:

$$_nP_n = n!, \tag{5.2}$$

where "n factorial" is the product of n and each smaller positive integer; that is,

$$n! = n(n - 1)(n - 2) \cdots (3)(2)(1). \tag{5.3}$$

Example 5.2 demonstrates such computation of the numbers of permutations.

EXAMPLE 5.2   The number of permutations of distinct objects.

In how many sequences can six different slides be shown on a slide projector?
$$_nP_n = 6! = (6)(5)(4)(3)(2)(1) = 720$$

Circular Arrangements.   The numbers of permutations considered above are for objects arranged on a line. If objects are arranged on a circle, there is no "starting position" as there is on a line, and the number of permutations is

$$_nP'_n = \frac{n!}{n} = (n - 1)!. \tag{5.4}$$

(Observe that the notation $_nP'_n$ is used here for circular permutations to distinguish it from the symbol $_nP_n$ used for linear permutations.)

Referring again to a horse, a cow, and a sheep, there are $_nP'_n = \frac{n!}{n} = (n - 1)! = (3 - 1)! = 2! = 2$ distinct ways in which the three animals could be seated around a circular (or triangular) table, or arranged around the shore of a pond:

<pre>
        H   H
                 or
   S   C   C   S
</pre>

In this example, there is an assumed orientation of the observer, so clockwise and counter-clockwise patterns are treated as different. That is, the animals are observed arranged around the top of the table, or observed from above the surface of the pond. But, either one of these arrangements would look like the other one if observed from under the table or under the water; and if we did not wish to count the results of these two mirror-image observations as different, we would speak of there being one possible permutation, not two. For example, consider each of the above two diagrams to represent three beads on a circular string, one bead in the shape of a horse, one in the shape of a cow, and the other in the shape of a sheep. The two arrangements of H, C, and S shown above are not really different, for there is no specific way of viewing the circle; one of the two arrangements turns into the other if the circle is turned over. If $n > 2$ and the orientation of the circle is not specified, then the number of permutations of n objects on a circle is

$$_nP_n'' = \frac{n!}{2n} = \frac{(n-1)!}{2}. \tag{5.5}$$

Fewer than n Positions. If one has n objects, but fewer than n positions in which to place them, then there would be considerably fewer numbers of ways to arrange the objects than in the case where there are positions for all n. For example, there are $_4P_4 = 4! = (4)(3)(2)(1) = 24$ ways of placing a horse (H), cow (C), sheep (S), and pig (P) in four positions on a line. However, there are only twelve ways of linearly arranging two of these four animals:

H,C; H,S; H,P; C,H; C,S; C,P; S,H; S,C; S,P; P,H; P,C; P,S.

The number of linear permutations of n objects taken X at a time is

$$_nP_X = \frac{n!}{(n-X)!}. \tag{5.6}$$

For the above example,

$$_4P_2 = \frac{4!}{(4-2)!} = \frac{4!}{2!} = \frac{(4)(3)(2)(1)}{(2)(1)} = 12.$$

Equation 5.2 is a special case of Equation 5.6, where $X = n$; it is important to know that 0! is defined to be 1.*

If the arrangements are circular, instead of linear, then the number of them possible is

$$_nP_X' = \frac{n!}{(n-X)!X}. \tag{5.7}$$

So, for example, there are only $4!/[(4-2)!2] = 6$ different ways of arranging two out of our four animals around a circular table:

H	H	H	C	C	S
C	S	P	S	P	P

_____

*Why is 0! defined to be 1? In general, $n! = n[(n-1)!]$; e.g., $5! = 5(4!)$, $4! = 4(3!)$, $3! = 3(2!)$, and $2! = 2(1!)$. Thus, $1! = 1(0!)$, which is so only if $0! = 1$.

for C seated at the table opposite H is the same arrangement as H seated across from C, S seated with H is the same as H with S, and so on. Example 5.3 demonstrates this further. Equation 5.4 is a special case of Equation 5.7, where $X = n$; and recall that 0! is defined as 1.

EXAMPLE 5.3   **The number of permutations of n objects taken X at a time. In how many different ways can a sequence of four slides be made from a collection of six slides?**

$$_nP_X = {_6}P_4 = \frac{6!}{(6-4)!} = \frac{6!}{2!} = \frac{(6)(5)(4)(3)(2)(1)}{(2)(1)} = (6)(5)(4)(3) = 360$$

If $n > 2$, then for every circular permutation viewed from above there is a mirror-image of that permutation, which would be observed from below. If these two mirror-images are not to be counted as different (e.g., if we are dealing with beads of different shapes or colors on a string), then the number of circular permutations is

$$_nP''_X = \frac{n!}{2(n-X)!X}. \tag{5.8}$$

If Some of the Objects Are Indistinguishable.   If our group of four animals consisted of two horses (H), a cow (C), and a sheep (S), the number of permutations of the four animals would be twelve:

H,H,C,S; H,H,S,C; H,C,H,S; H,C,S,H; H,S,H,C; H,S,C,H;

C,H,H,S; C,H,S,H; C,S,H,H; S,H,H,C; S,H,C,H; S,C,H,H.

If n_i represents the number of like individuals in category i (in this case the number of animals in species i), then in this example $n_1 = 2, n_2 = 1$, and $n_3 = 1$, and we can write the number of permutations as

$$_nP_{n_1,n_2,n_3} = \frac{n!}{n_1!n_2!n_3!} = \frac{4!}{2!1!1!} = 12.$$

If the four animals were two horses (H) and two cows (C), then there would be only six permutations:

H,H,C,C; C,C,H,H; H,C,H,C; C,H,C,H; H,C,C,H; C,H,H,C.

In this case, $n = 4, n_1 = 2$, and $n_2 = 2$, and the number of permutations is calculated to be $_nP_{n_1,n_2} = n!/(n_1!n_2!) = 4!/(2!2!) = (4)(3)(2)/[(2)(2)] = 6$.

In general, if n_1 members of the first category of objects are indistinguishable, as are n_2 of the second category, n_3 of the third category, and so on through n_k members of the kth category, then the number of different permutations is

$$_nP_{n_1,n_2,\cdots,n_k} = \frac{n!}{n_1!n_2!\cdots n_k!} \text{ or } \frac{n!}{\displaystyle\prod_{i=1}^{k} n_i!}, \tag{5.9}$$

where the capital Greek letter pi (Π) denotes taking the product just as the capital Greek sigma (Σ, introduced in Section 3.1) indicates taking the sum. This is shown further in Example 5.4.

EXAMPLE 5.4   Permutations with categories containing indistinguishable members.

There are twelve potted plants, six of one species, four of a second species, and two of a third species. How many different sequences of species are possible?

$$_nP_{n_1,n_2,n_3} = \frac{n}{\prod n_i!}$$

$$= \,_{12}P_{6,4,2} = \frac{12!}{6!4!2!}$$

$$= \frac{(12)(11)(10)(9)(8)(7)(6)(5)(4)(3)(2)(1)}{(6)(5)(4)(3)(2)(1)(4)(3)(2)(1)(2)(1)} = 13,860.$$

Note that the above calculation could have been simplified by writing

$$\frac{12!}{6!4!2!} = \frac{(12)(11)(10)(9)(8)(7)6!}{6!(4)(3)(2)(2)} = \frac{(12)(11)(10)(9)(8)(7)}{(4)(3)(2)(2)} = 13,860.$$

Here, "(1)" is dropped; also, "6!" appears in both the numerator and denominator, thus canceling each other out.

5.3 COMBINATIONS

In Section 5.2 we considered groupings of objects where the sequence within the groups was important. In many instances, however, only the components of a group, not their arrangement within the group, are important. We saw that if we select two animals from among a horse (H), cow (C), sheep (S), and pig (P), there are twelve ways of arranging the two on a line:

$$\text{H,C; H,S; H,P; C,H; C,S; C,P; S,H; S,C; S,P; P,H; P,C; P,S.}$$

However, some of these arrangements contain exactly the same kinds of animals, only in different order: e.g., H,C and C,H; H,S and S,H. If the groups of two are important to us, but not the sequence of objects within the groups, then we are speaking of *combinations*, rather than permutations. Designating the number of combinations of n objects taken X at a time as $_nC_X$, we have

$$_nC_X = \frac{_nP_X}{X!} = \frac{n!}{X!(n-X)!}. \tag{5.10}$$

So for the present example, $n_1 = 4$, $n_2 = 2$, and

$$_4C_2 = \frac{4!}{2!(4-2)!} = \frac{4!}{2!2!} = \frac{(4)(3)(2)(1)}{(2)(1)(2)(1)} = \frac{(4)(3)}{2} = 6,$$

the six combinations of the four animals taken two at a time being

$$H,C; \ H,S; \ H,P; \ C,S; \ C,P; \ S,P.$$

Example 5.5 demonstrates the determination of numbers of combinations for another set of data.

EXAMPLE 5.5   Combinations of n objects taken X at a time.

Of a total of ten dogs, eight are to be used in a laboratory experiment. How many different combinations of eight animals may be formed from the ten?

$$_nC_X = {_{10}C_8} = \frac{10!}{8!(10-8)!} = \frac{10!}{8!2!} = \frac{(10)(9)(8)(7)(6)(5)(4)(3)(2)(1)}{(8)(7)(6)(5)(4)(3)(2)(1)(2)(1)}$$

$$= 45.$$

It should be noted that the above calculations with factorials could have been simplified by writing

$$_{10}C_8 = \frac{10!}{8!2!} = \frac{(10)(9)8!}{8!2!} = \frac{(10)(9)}{2} = 45,$$

so that "8!" appears in both the numerator and denominator, thus canceling each other out.

It may be noted that

$$_nC_n = 1, \tag{5.11}$$

meaning that there is only one way of selecting all n items; and

$$_nC_1 = n, \tag{5.12}$$

indicating that there are n ways of selecting n items one at a time. Also,

$$_nC_X = {_nC_{n-X}}, \tag{5.13}$$

which means that if we select X items from a group of n, we have at the same time selected the remaining $n-X$ items; that is, an exclusion is itself a selection. For example, if we selected two out of five persons to write a report, we have simultaneously selected three of the five to refrain from writing. Thus:

$$_5C_2 = \frac{5!}{2!(5-2)!} = \frac{5!}{2!3!} = 10 \quad \text{and} \quad {_5C_{5-2}} = {_5C_3} = \frac{5!}{3!(5-3)!} = \frac{5!}{3!2!} = 10,$$

meaning that there are ten ways to select two out of five persons to perform a task and ten ways to select three out of five persons to be excluded from that task. This question may be addressed by applying Equation 5.9, reasoning that we are asking how many distinguishable arrangements there are of two writers and three nonwriters: $_5P_{2,3} = 5!/(2!3!) = 10$.

The product of combinatorial outcomes may also be employed to address questions such as in Example 5.4. This is demonstrated in Example 5.6.

From Equation 5.10 it may be noted that, as $_nC_X = {_nP_X}/X!$:

$$_nP_X = X!\,_nC_X. \tag{5.14}$$

EXAMPLE 5.6   Products of combinations.

An alternate method of answering the question of Example 5.4.

There are twelve potted plants, six of one species, four of a second species, and two of a third. How many different sequences of species are possible?

There are twelve positions in the sequence, which may filled by the six members of the first species in this many ways:

$$_{12}C_6 = \frac{12!}{(12-6)!6!} = 924.$$

The remaining six positions in the sequence may be filled by the four members of the second species in this many ways:

$$_6C_4 = \frac{6!}{(6-4)!4!} = 15.$$

And, the remaining two positions may be filled by the two members of the third species in only one way:

$$_2C_2 = \frac{2!}{(2-2)!2!} = 1.$$

As each of the ways of filling positions with members of one species exists in association with each of the ways of filling positions with members of each other species, the total different sequences of species is:

$$(924)(15)(1) = 13,860.$$

It is common mathematical convention to indicate the number of combinations of n objects taken X at a time as $\binom{n}{X}$ instead of $_nC_X$, so for the above problem we could have written[*]

$$\binom{n}{X} = \binom{4}{2} = 6.$$

Binomial coefficients, which are discussed in Section 23.1, take this form.

5.4 SETS

A *set* is a defined collection of items. For example, a set may be a group of four animals, a collection of eighteen amino acids, an assemblage of twenty-five students, or a group of three genetic traits. Each item in a set is termed an *element*. If a set of animals includes these four elements: horse (H), cow (C), sheep (S), and pig (P), and a second set consists of the elements P, S, H, and C, then we say that the two sets are *equal*, as they contain exactly the same elements. The sequence of elements within sets is immaterial in defining equality or inequality of sets.

---

[*]This parenthetical notation was introduced in 1778 by Swiss mathematician Leonhard Euler (1707–1783) (Cajori, 1929:62). Some authors use the symbol C_X^n instead of $_nC_X$ for combinations and P_X^n instead of $_nP_X$ for permutations, but we shall avoid that in this book so as not to confuse n with an exponent; some use ^nC_X and ^nP_X.

If a set consisted of animals H and P, it would be declared a *subset* of the above set (H, C, S, P). A subset is a set, all of whose elements are elements of a larger set. Therefore, the determination of combinations of X items taken from a set of n items (Section 5.3) is really the counting of possible subsets of items from the set of n items.

In an experiment (or other phenomenon that yields results to observe), there is a set (usually very large) of possible outcomes. Let us refer to this set as the *outcome set*.[*]

Each element of the set is one of the possible outcomes of the experiment. For example, if an experiment consists of tossing two coins, the outcome set consists of four elements: H,H; H,T; T,H; T,T, as these are all of the possible outcomes.

A subset of the outcome set is called an *event*. If the outcome set were the possible rolls of a die: 1, 2, 3, 4, 5, 6, an event might be declared to be "even-numbered rolls" (i.e., 2, 4, 6), and another event might be defined as "rolls greater than 4" (i.e., 5, 6). In tossing two coins, one event could be "the two coins land differently" (i.e., T,H; H,T), and another event could be "heads do not appear" (i.e., T,T). If the two events in the same outcome set have some elements in common, the two events are said to intersect; and the *intersection* of the two events is that subset composed of those common elements. For example, the event "even-numbered rolls" of a die (2, 4, 6) and the event "rolls greater than 4" (5, 6) have an element in common (namely, the roll 6); therefore 6 is the intersection of the two events. For the events "even-numbered rolls" (2, 4, 6) and "rolls less than 5" (1, 2, 3, 4), the intersection subset consists of those elements of the events that are both even-numbered and less than 5 (namely, 2, 4).[†]

If two events have no elements in common, they are said to be *mutually exclusive*, and the two sets are said to be *disjoint*. The set that is the intersection of disjoint sets contains no elements and is often called the *empty set* or the *null set*. For example, the events "odd-numbered rolls" and "even-numbered rolls" are mutually exclusive and there are no elements common to both of them.

If we ask what elements are found in either one event or another, or in both of them, we are speaking of the *union* of the two events. The union of the events "even-numbered rolls" and "rolls less than 5" is that subset of the outcome set that contains elements found in either set (or both sets), namely 1, 2, 3, 4, 6.[‡]

Once a subset has been defined, all other elements in the outcome set are said to be the *complement* of that subset. So, if an event is defined as "even-numbered rolls" of a die (2, 4, 6), the complementary subset consists of "odd-numbered rolls" (1, 3, 5). If subset is "rolls less than 5" (1, 2, 3, 4), the complement is the subset consisting of rolls 5 or greater (5, 6).

The above considerations may be presented by what are known as *Venn diagrams*,[§] shown in Fig. 5.1.

---

[*]Also called the *sample space*.

[†]The mathematical symbol for "intersection" is "∩," so the intersection of set A (consisting of 2, 4, 6) and set B (consisting of 5, 6) is $A \cap B$ (consisting of 6).

[‡]"Union" is denoted by the mathematical symbol "∪." Thus, if set A consists of even-numbered rolls of a die, and set B is odd-numbered rolls, then the union of the two sets, namely $A \cup B$, is 2, 4, 6, 1, 3, 5.

[§]Named for the English logician, John Venn (1834–1923).

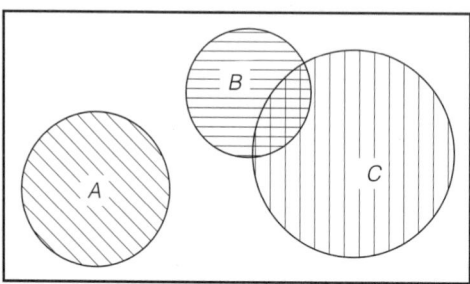

Figure 5.1 A Venn diagram showing the relationships among the outcome set represented by the rectangle and the subsets represented by circles A, B, and C. Subsets B and C intersect, with no intersection with A.

The rectangle in this diagram denotes the outcome set, the set of all possible outcomes from an experiment or other producer of observations. The circle on the left represents a subset of the outcome set that we shall refer to as event A, the circle in the center signifies a second subset of the outcome set that we shall refer to as event B, and the circle on the right depicts a third subset of the outcome set that we shall call event C. If, for example, an outcome set (the rectangle) is the number of vertebrate animals in a forest, subset A might be animals without legs (namely, snakes), subset B might be mammals, and subset C might be flying animals. Figure 5.1 demonstrates graphically what is meant by union, intersection, mutually exclusive, and complementary, sets: The union of B and C (the areas with any horizontal or vertical shading) represents all birds and mammals; the intersection of B and C (the area with both horizontal and vertical shading) represents flying mammals (i.e., bats); the portion of C with only vertical shading represents birds; A is mutually exclusive relative to the union of B and C, and the unshaded area (representing all other vertebrates—namely, amphibians and turtles) is complementary to A, B, and C (and is also mutually exclusive of A, B, and C).

5.5 PROBABILITY OF AN EVENT

As in Section 1.3, we shall define the *relative frequency* of an event as the proportion of the total observations of outcomes that event represents. Consider an outcome set with two elements, such as the possible results from tossing a coin (H; T) or the sex of a person (male; female). If n is the total number of coin tosses and f is the total number of heads observed, then the relative frequency of heads is f/n. Thus, if heads are observed 52 times in 100 coin tosses, the relative frequency is $52/100 = 0.52$ (or 52%). If 275 males occur in 500 human births, the relative frequency of males is $f/n = 275/500 = 0.55$ (or 55%). In general, we may write

$$\text{relative frequency of an event} \; = \; \frac{\text{frequency of that event}}{\text{total number of all events}} = \frac{f}{n}. \tag{5.15}$$

The value of f may, of course, range from 0 to n, and the relative frequency may, therefore, range from 0 to 1 (or 0% to 100%). A biological example is given as Example 5.7.

The *probability* of an event is the likelihood of that event expressed either by the relative frequency observed from a large number of data or by knowledge of the

---

EXAMPLE 5.7   Relative frequencies.

A sample of 852 vertebrate animals is taken randomly from a forest. The sampling was done *with replacement*, meaning that the animals were taken one at a time, returning each one to the forest before the next one was selected. This is done to prevent the sampling procedure from altering the relative frequency in the sampled population. If the sample size is very small compared to the population size, replacement is not necessary. (Recall that random sampling assumes that each individual animal is equally likely to become a part of the sample.)

Vertebrate subset	Number	Relative frequency
amphibians	53	0.06
turtles	41	0.05
snakes	204	0.24
birds	418	0.49
mammals	136	0.16
total	852	1.00

---

system under study. In Example 5.7 the relative frequencies of vertebrate groups have been observed from randomly sampling forest animals. If, for the sake of the present example, we assume that each animal has the same chance of being caught as part of our sample (an unrealistic assumption in nature), we may estimate the probability, P, that the next animal captured will be a snake ($P = 0.24$). Or, using the data of the preceding paragraph we can estimate that the probability that a human birth will be a male is 0.55, or that the probability of tossing a coin that lands head side up is 0.52. A probability may sometimes be predicted on the basis of knowledge about the system (e.g., the structure of a coin or of a die, or the Mendelian principles of heredity). If we assume that there is no reason why a tossed coin should land "heads" more or less often than "tails," we say there is an equal probability of each outcome: $P(H) = \frac{1}{2}$ and $P(T) = \frac{1}{2}$ states that "the probability of heads is 0.5 and the probability of tails is 0.5."

Probabilities, like relative frequencies, can range from 0 to 1. A probability of 0 means that the event is impossible. For example, in tossing a coin, $P(\text{neither H nor T}) = 0$, or in rolling a die, $P(\text{number} > 6) = 0$. A probability of 1 means that an event is certain. For example, in tossing a coin, $P(\text{H or T}) = 1$; or in rolling a die, $P(1 \leq \text{number} \leq 6) = 1$.*

5.6 ADDING PROBABILITIES

If Events Are Mutually Exclusive.   If two events (call them A and B) are mutually exclusive (e.g., legless vertebrates and mammals are disjoint sets in Fig. 5.1), then the

---

*A concept related to probability is the *odds* for an event, namely the ratio of the probability of the event occurring and the probability of that event not occurring. For example, if the probability of a male birth is 0.55 (and, therefore, the probability of a female birth is 0.45), then the odds in favor of male births are 0.55/0.45, expressed as "11 to 9."

probability of either event A or event B is the sum of the probabilities of the two events:

$$P(A \text{ or } B) = P(A) + P(B). \tag{5.16}$$

For example, if the probability of a tossed coin landing head up is $\frac{1}{2}$ and the probability of its landing tail up is $\frac{1}{2}$, then the probability of either head or tail up is

$$P(H \text{ or } T) = P(H) + P(T) = \frac{1}{2} + \frac{1}{2} = 1. \tag{5.17}$$

And, for the data in Example 5.7, the probability of selecting, at random, a reptile would be $P(\text{turtle or snake}) = P(\text{turtle}) + P(\text{snake}) = 0.05 + 0.24 = 0.29$.

This rule for adding probabilities may be extended for more than two mutually exclusive events. For example, the probability of rolling a 2 on a die is $\frac{1}{6}$, the probability of rolling a 4 is $\frac{1}{6}$, and the probability of rolling a 6 is $\frac{1}{6}$; so the probability of rolling an even number is

$$P(\text{even number}) = P(2 \text{ or } 4 \text{ or } 6) = P(2) + P(4) + P(6)$$

$$= \frac{1}{6} + \frac{1}{6} + \frac{1}{6} = \frac{3}{6} = \frac{1}{2}.$$

If Events Are Not Mutually Exclusive.   If two events are not mutually exclusive— i.e., they intersect (e.g., mammals and flying vertebrates are not disjoint sets in Fig. 5.1)— then the addition of the probabilities of the two events must be modified. For example, if we roll a die, the probability of rolling an odd number is

$$P(\text{odd number}) = P(1 \text{ or } 3 \text{ or } 5) = P(1) + P(3) + P(5)$$

$$= \frac{1}{6} + \frac{1}{6} + \frac{1}{6} = \frac{3}{6} = \frac{1}{2};$$

and the probability of rolling a number less than 4 is

$$P(\text{number} < 4) = P(1 \text{ or } 2 \text{ or } 3) = P(1) + P(2) + P(3)$$

$$= \frac{1}{6} + \frac{1}{6} + \frac{1}{6} = \frac{3}{6} = \frac{1}{2}.$$

The probability of rolling either an odd number or a number less than 4 obviously is *not* calculated by Equation 5.16, for that equation would yield

$$P(\text{odd number or number} < 4) \overset{?}{=} P(\text{odd}) + P(\text{number} < 4)$$

$$= P[(1 \text{ or } 3 \text{ or } 5) \text{ or } (1 \text{ or } 2 \text{ or } 3)]$$

$$= [P(1) + P(3) + P(5)] + [P(1) + P(2) + P(3)]$$

$$= \left(\frac{1}{6} + \frac{1}{6} + \frac{1}{6}\right) + \left(\frac{1}{6} + \frac{1}{6} + \frac{1}{6}\right) = 1,$$

and that would mean that we are certain ($P = 1$) to roll either an odd number or a number less than 4, which would mean that a roll of 4 or 6 is impossible!

The invalidity of the last calculation is due to the fact that the two elements (namely 1 and 3) that lie in both events are counted twice. The subset of elements consisting of rolls 1 and 3 is the intersection of the two events and its probability needs to be subtracted from the above computation so that $P(1$ or $3)$ is counted once, not twice. Therefore, for two intersecting events, A and B, the probability of either A or B is

$$P(A \text{ or } B) = P(A) + P(B) - P(A \text{ and } B). \tag{5.18}$$

In the above example,

$$P(\text{odd number or number} < 4) = P(\text{odd number}) + P(\text{number} < 4)$$

$$- P(\text{odd number and number} < 4)$$

$$= P[(1 \text{ or } 3 \text{ or } 5) \text{ or } (1 \text{ or } 2 \text{ or } 3)]$$

$$- P(1 \text{ or } 3)$$

$$= [P(1) + P(3) + P(5)] + [P(1) + P(2) + P(3)]$$

$$- [P(1) + P(3)]$$

$$= \left(\frac{1}{6} + \frac{1}{6} + \frac{1}{6}\right) + \left(\frac{1}{6} + \frac{1}{6} + \frac{1}{6}\right) - \left(\frac{1}{6} + \frac{1}{6}\right) = \frac{4}{6} = \frac{2}{3}.$$

It may be noted that Equation 5.16 is a special case of Equation 5.18, where $P(A$ and $B) = 0$. Example 5.8 demonstrates these probability calculations with a different set of data.

If three events are not mutually exclusive, the situation is more complex, yet straightforward. As seen in Fig. 5.2, there may be three two-way intersections, shown with vertical shading (A and B; A and C; and B and C), and a three-way intersection, shown with horizontal shading (A and B and C). If we add the probabilities of the three events, A, B, and C, as $P(A) + P(B) + P(C)$, we are adding the two-way intersections twice. So, we can subtract $P(A$ and $B)$, $P(A$ and $C)$, and $P(B$ and $C)$. Also, the three-way intersection is added three times in $P(A) + P(B) + P(C)$, and subtracted three times by subtracting the three two-way intersections; thus, $P(A$ and B and $C)$ must be added back. Therefore, for three events, not mutually exclusive,

$$P(A \text{ or } B \text{ or } C) = P(A) + P(B) + P(C)$$

$$- P(A \text{ and } B) - P(A \text{ and } C) - P(B \text{ and } C) \tag{5.19}$$

$$+ P(A \text{ and } B \text{ and } C).$$

5.7 MULTIPLYING PROBABILITIES

If two or more events intersect (as A and B in Fig. 5.1 and A, B, and C in Fig. 5.2), the probability associated with the intersection is the product of the probabilities of the

EXAMPLE 5.8   Adding probabilities of intersecting events.

A deck of playing cards is composed of fifty-two cards, with thirteen cards in each of four suits called clubs, diamonds, hearts, and spades. In each suit there is one card each of the following thirteen denominations: ace (A), 2, 3, 4, 5, 6, 7, 8, 9, 10, jack (J), queen (Q), king (K). What is the probability of selecting at random a diamond from the deck of fifty two cards?

The event in question (diamonds) is a subset with thirteen elements; therefore,

$$P(\text{diamond}) = \frac{13}{52} = \frac{1}{4} = 0.250.$$

What is the probability of selecting at random a king from the deck?

The event in question (king) has four elements; therefore,

$$P(\text{king}) = \frac{4}{52} = \frac{1}{13} = 0.077.$$

What is the probability of selecting at random a diamond or a king?

The two events (diamonds and kings) intersect, with the intersection having one element (the king of diamonds); therefore,

$$P(\text{diamond or king}) = P(\text{diamond}) + P(\text{king}) - P(\text{diamond and king})$$

$$= \frac{13}{52} + \frac{4}{52} - \frac{1}{52}$$

$$= \frac{16}{52} = \frac{4}{13} = 0.308.$$

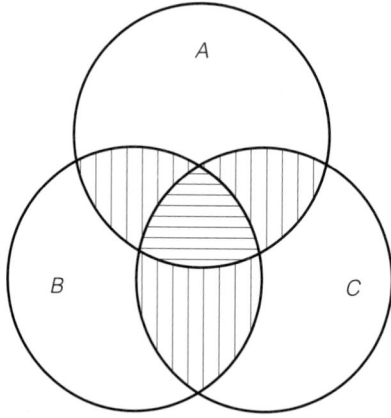

Figure 5.2 A Venn diagram showing three intersecting sets: *A*, *B*, and *C*.* Here there are three two-way intersections (vertical shading) and one three-way intersection (horizontal shading).

individual events. That is,

$$P(A \text{ and } B) = [P(A)][P(B)], \tag{5.20}$$

$$P(A \text{ and } B \text{ and } C) = [P(A)][P(B)][P(C)], \tag{5.21}$$

and so on.

---

*This configuration of three overlapping circles is sometimes termed a *ballantine* (Cohen and Cohen, 1983:88).

For example, the probability of a tossed coin landing heads is $\frac{1}{2}$. If two coins are tossed, the probability of *both* coins landing heads is

$$P(H, H) = [P(H)][P(H)] = \left(\frac{1}{2}\right)\left(\frac{1}{2}\right) = \left(\frac{1}{4}\right) = 0.25.$$

This can be verified by examining the outcome set:

$$H,H; \quad H,T; \quad T,H; \quad T,T,$$

where $P(H, H)$ is one outcome out of four equally likely outcomes. The probability that 3 tossed coins will land heads is

$$P(H, H, H) = [P(H)][P(H)][P(H)] = \left(\frac{1}{2}\right)\left(\frac{1}{2}\right)\left(\frac{1}{2}\right) = \left(\frac{1}{8}\right) = 0.125.$$

Note, however, that if one or more coins have already been tossed, the probability that the next coin toss (of the same or a different coin) will be heads is simply $\frac{1}{2}$.

EXERCISES

5.1 A person may receive a grade of either high (H), medium (M), or low (L) on a hearing test, and a grade of either good (G) or poor (P) on a sight test.
(a) How many different outcomes are there if both tests are taken?
(b) What are these outcomes?

5.2 A menu lists three meats, four salads, and two desserts. In how many ways can a meal of one meat, one salad, and one dessert be selected?

5.3 If an organism (e.g., human) has twenty three pairs of chromosomes in each diploid cell, how many different gametes are possible for the individual to produce by assortment of chromosomes?

5.4 In how many ways can five animal cages be arranged on a shelf?

5.5 In how many ways can twelve different amino acids be arranged into a polypeptide chain of five amino acids?

5.6 An octapeptide is known to contain four of one amino acid, two of another, and two of a third. How many different amino-acid sequences are possible?

5.7 Students are given a list of nine books and told that they will be examined on the contents of five of them. How many combinations of five books are possible?

5.8 The four human blood types below are genetic phenotypes that are mutually exclusive events. Of 5400 individuals examined, the following frequency of each blood type is observed. What is the relative frequency of each blood type?

Blood type	Frequency
O	2672
A	2041
B	486
AB	201

5.9 An aquarium contains the following numbers of tropical freshwater fishes. What is the relative frequency of each species?

Species	Number
Paracheirodon innesi, neon tetra	11
Cheirodon axelrodi, cardinal tetra	6
Pterophyllum scalare, angelfish	4
Pterophyllum altum, angelfish	2
Pterophyllum dumerilii, angelfish	2
Nannostomus marginatus, one-lined pencilfish	2
Nannostomus anomalus golden pencilfish	2

5.10 Use the data of Exercise 5.8, assuming that each of the 5400 has an equal opportunity of being encountered.
(a) Estimate the probability of encountering a person with type A blood.
(b) Estimate the probability of encountering a person who has either type A or type AB blood.

5.11 Use the data of Exercise 5.9, assuming that each individual fish has the same probability of being encountered.
(a) Estimate the probability of encountering an angelfish of the species *Pterophyllum scalare*.
(b) Estimate the probability of encountering a fish belonging to the angelfish genus *Pterophyllum*.

5.12 Either allele *A* or *a* may occur at a particular genetic locus. An offspring receives one of its alleles from each of its parents. If one parent possesses alleles *A* and *a* and the other parent possesses *a* and *a*:
(a) What is the probability of an offspring receiving an *A* and an *a*?
(b) What is the probability of an offspring receiving two *a* alleles?
(c) What is the probability of an offspring receiving two *A* alleles?

5.13 In a deck of playing cards (see Example 5.8 for a description):
(a) What is the probability of selecting a queen of clubs?
(b) What is the probability of selecting a black (i.e., club or spade) queen?
(c) What is the probability of selecting a black face card (i.e., a black jack, queen, or king)?

5.14 A cage contains six rats, two of them white (W) and four of them black (B); a second cage contains four rats, two white and two black; and a third cage contains five rats, three white and two black. If one rat is selected randomly from each cage:
(a) What is the probability that all three rats selected will be white?
(b) What is the probability that exactly two of the three will be white?
(c) What is the probability of selecting at least two white rats?

6

THE NORMAL DISTRIBUTION

Commonly, a distribution of interval or ratio scale data is observed to have a preponderance of values around the mean with progressively fewer observations toward the extremes of the range of values (see, e.g., Fig. 1.5). If n is large, the frequency polygons of many biological data distributions are "bell-shaped" and look something like Fig. 6.1.

Fig. 6.1 is a frequency curve for a *normal distribution*.* Not all bell-shaped curves are normal, however; a *normal distribution* is defined as one in which height of the curve at X_i is as expressed by the relation:

$$Y_i = \frac{1}{\sigma\sqrt{2\pi}} e^{-(X_i-\mu)^2/2\sigma^2}.$$

(6.1)

The height of the curve, Y_i, is referred to as the *normal density*. It is not a frequency, for in a normally distributed population the frequency of occurrence of a measurement

---

*The normal distribution is sometimes called the *Gaussian distribution*, after [Johann] Karl Friedrich Gauss (1777–1855), a phenomenal German mathematician contributing to many fields of mathematics and for whom the unit of magnetic induction ("gauss") is named. Gauss discussed this distribution in 1809, but the influential French mathematician and astronomer Pierre Simon, Marquis de Laplace (1749–1827) mentioned it in 1774, and it was first announced in 1733 by mathematician Abraham de Moivre (1667–1754; also spelled De Moivre and Demoivre), who was born in France but emigrated to England at age twenty-one (after three years in prison) to escape religious persecution as a Protestant (David, 1962: 161–178; Pearson, 1924; Stigler, 1980; Walker, 1934). This situation has been cited as an example of "Stigler's Law of Eponymy," which states that "no scientific discovery is named after its original discoverer" (Stigler, 1980). The adjective "normal" had first been used for the distribution by Sir Francis Galton in 1877 (Kruskal, 1978), and Karl Pearson recommended the routine use of that term to avoid "an international question of priority" although it "has the disadvantage of leading people to believe that all other distributions of frequency are in one sense or another 'abnormal'" (Pearson, 1920).

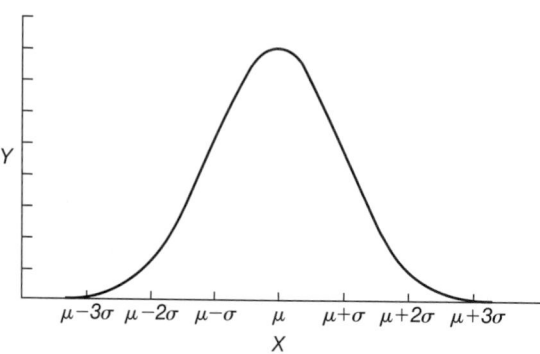

Y

$\mu-3\sigma$ $\mu-2\sigma$ $\mu-\sigma$ μ $\mu+\sigma$ $\mu+2\sigma$ $\mu+3\sigma$

X

Figure 6.1   A normal distribution.

exactly equal to X_i (e.g., exactly equal to 12.5000 cm, or exactly equal to 12.50001 cm) is zero. Equation 6.1 contains two mathematical constants: π (lowercase Greek pi),[*] which equals 3.14159...; and e (the base of Naperian, or natural, logarithms),[†] which equals 2.71828.... There are also two parameters (μ and σ^2) in the equation. Thus, for any given standard deviation, σ, there are an infinite number of normal curves possible, depending on μ. Fig. 6.2a shows normal curves for $\sigma = 1$ and $\mu = 0, 1$, and 2. Likewise, for any given mean, μ, an infinity of normal curves is possible, each with a different value of σ. Fig. 6.2b shows normal curves for $\mu = 0$ and $\sigma = 1, 1.5$, and 2.

A normal curve with $\mu = 0$ and $\sigma = 1$ is said to be a *standardized normal curve*. Thus, for a standardized normal distribution,

$$Y_i = \frac{1}{\sqrt{2\pi}} e^{-X_i^2/2}. \tag{6.2}$$

---

[*]π, the ratio between the circumference and the diameter of a circle, is a symbol introduced in 1706 by William Jones (1675–1749) (Beckmann, 1977: 141; Walker, 1934). Pi is an irrational number, meaning it cannot be expressed as the ratio between two integers; and the digits appear so randomly that they have been recommended in place of random digits otherwise generated (see Dodge, 1996). The lowercase Greek pi was first used for the ratio of a circle's circumference to its diameter by William Jones in 1706—after it had been used for over 50 years to denote the circumference (Cajori, 1929: 9)—but it did not gain popularity for this purpose until after Leonhard Euler began using it 30 years later (Blatner, 1997: 78). (See also Section 26.1.) To twenty decimal places its value is 3.14159 26535 89792 33846 (although it may be noted that ten decimal places are sufficient to obtain, from the diameter, the circumference of a circle as large as the earth's equator to within about a centimeter). Beckmann (1977), Dodge (1996), and Blatner (1997) present a fascinating history of π and its calculation. By 2000 B.C., the Babylonians knew its value to within 0.02. Archimedes of Syracuse (287–212 B.C.) was the first to present a procedure to calculate π to any desired accuracy (and computed it accurate to the third decimal place). Many methods have since been developed, and pi was determined to six decimal places of accuracy (but not widely known) by around 500 A.D., to 20 decimal places by around 1600, to 100 in 1706, to 607 in 1853; and 100 decimal places were reached, using a mechanical calculating machine, in 1948 before electronic computers joined the challenge in 1949. In the computer era, one million digits were reached in 1973, and the 1980s witnessed calculations accurate to millions then hundreds of millions of digits; Blatner (1997) reported that π had recently been computed to 51 billion digits.

[†]e (as is π) is an irrational number, which to twenty decimal places is 2.71828 18284 59045 23536. The symbol, e, for this quantity was introduced by the great Swiss mathematician, Leonhard Euler (1707–1783) Asimov, 1982: 181), in 1727 or 1728 (Cajori, 1929: 13). Johnson and Leeming (1990) discuss the randomness of the digits of e, and Maor (1993) presents a history of this number and its mathematical ramifications.

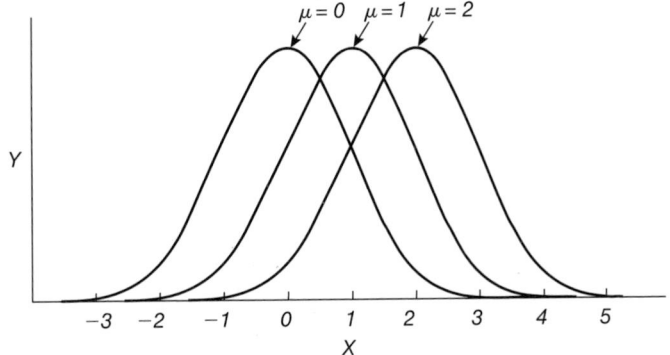

Figure 6.2a Normal distribution with $\sigma = 1$, varying in location with different means.

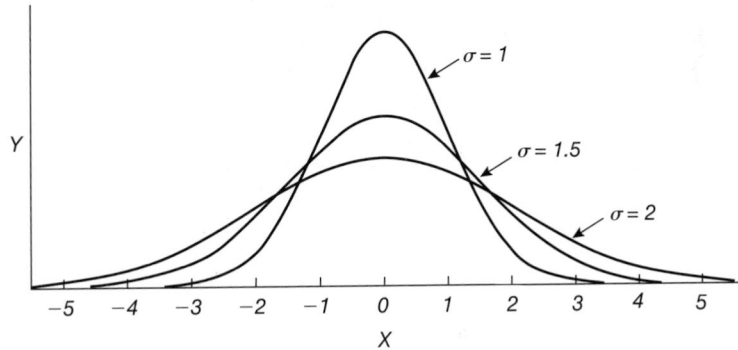

Figure 6.2b Normal distributions with $\mu = 0$, varying in spread with different standard deviations.

6.1 SYMMETRY AND KURTOSIS

Statisticians refer to $\sum(X_i - \mu)^p/N$ as the "pth moment about the mean."* For any population distribution, the first moment about the mean, $\sum(X_i - \mu)/N$, is zero, for $\sum(X_i - \mu)$ is zero. The second moment about the mean, $\sum(X_i - \mu)^2/N$, has already been defined as the population variance, σ^2 (Section 4.4). The third moment about the mean, $\sum(X_i - \mu)^3/N$, tells us about the symmetry of the distribution. A sample statistic based on this parameter is

$$k_3 = \frac{n\sum(X_i - \bar{X})^3}{(n-1)(n-2)}, \tag{6.3}$$

---

*This statistical concept of moments was developed by K. Pearson around 1893 (although moments were described by others long before). He called the quantities "moment coefficients" (Walker, 1929: 71, 74, 184–185).

which may be calculated by this "machine formula":

$$k_3 = \frac{n \sum X_i^3 - 3 \sum X_i \sum X_i^2 + 2(\sum X_i)^3/n}{(n-1)(n-2)}. \qquad (6.4)$$

As k_3 has cubed units, the following statistic is more commonly used,[*] as it has no units:

$$g_1 = \frac{k_3}{s^3} = \frac{k_3}{\sqrt{(s^2)^3}}. \qquad (6.5)$$

A g_1 not significantly different from 0 indicates that the sample comes from a population distributed symmetrically around the mean, one in which the mean and median are identical and the portion of the frequency polygon to the left of the mean is a mirror image of the portion to the right of the mean. A g_1 significantly less than 0 indicates that the sample comes from a population that is skewed to the left, exhibiting a mean less than the median (as in Fig. 3.2d); and a g_1 significantly greater than 0 indicates sampling from a population whose distribution is skewed[†] to the right, in which the mean is larger than the median (as in Fig. 3.2c). The sample statistic, g_1, is an estimate of the population parameter, γ_1 (called "gamma one"). Testing for statistical significance of asymmetry will be described in Section 7.14.

As not all symmetrical distributions are normal, statisticians have desired ways of assessing whether the shape of a distribution reflects normality. Employing the fourth power of the deviations from the mean provides a measure called *kurtosis*. This may be done by first calculating

$$k_4 = \frac{\sum(X_i - \bar{X})^4 n(n+1)/(n-1) - 3\left[\sum(X_i - \bar{X})^2\right]^2}{(n-2)(n-3)}, \qquad (6.6)$$

which may be computed by the "machine formula,"

$$k_4 = \frac{(n^3+n^2)\sum X^4 - 4(n^2+n)\sum X^3 \sum X - 3(n^2-n)(\sum X^2)^2 + 12n \sum X^2(\sum X)^2 - 6(\sum X)^4}{n(n-1)(n-2)(n-3)} \qquad (6.7)$$

(Bennett and Franklin, 1954: 81). As k_4 has units to the fourth power, the following statistic—which has no units—is more commonly used as a measure of kurtosis:

$$g_2 = \frac{k_4}{s^4}, \qquad (6.8)$$

where, of course, $s^4 = (s^2)^2$. The sample statistic, g_2, is an estimate of the population parameter, γ_2 ("gamma two").

Kurtosis of a population is often characterized in terms of "peakedness" or "tailedness," but it may best be described as dispersion around these two values: $\mu - \sigma$ and $\mu + \sigma$ (Moors, 1986). A distribution having many more values than does a normal distribution around $\mu - \sigma$ and $\mu + \sigma$ will have $\gamma_2 < 0$ (Moors, 1986) and is said to be *platykurtic* (see Fig. 6.3b). Such a distribution might be the composite of two normal

---

[*]This is related to an 1893 suggestion by K. Pearson (ibid.: 74).

[†]Karl Pearson introduced the term "skewness" in 1895 (David, 1995).

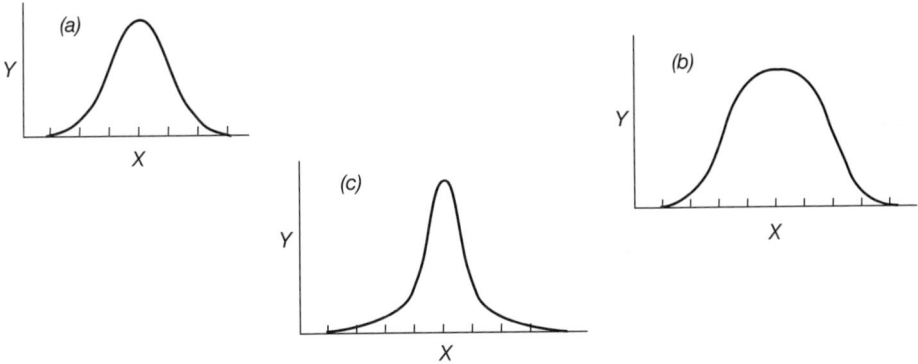

Figure 6.3 Symmetric frequency distributions. Distribution a is mesokurtic ("normal"), b is platykurtic, and c is leptokurtic.

populations with the same variance but different means. In contrast, a distribution that has fewer values than does a normal distribution around $\mu - \sigma$ and $\mu + \sigma$ (i.e., that has more concentration around the mean, and/or in the "tails" far from the mean) is called *leptokurtic* (see Fig. 6.3c) and will have a $\gamma_2 > 0$. Such a distribution might be the composite of two normal populations with the same μ but with different σ's. A normal distribution is said to be *mesokurtic.**

When calculating sample statistics, a g_2 not significantly different from 0 indicates that the sample was obtained from a mesokurtic population, a g_2 significantly less than 0 indicates the sampled population is platykurtic, and a g_2 significantly greater than 0 indicates an underlying leptokurtic distribution. Testing for statistical significance of departure from mesokurtosis will be described in Section 7.14.

If there are multiple observations, f_i, of a given X_i, then substitute $f_i X_i$ for X_i in the above formulas for k_3 and k_4, as shown in Example 6.1.

Because taking large or small numbers to their third or fourth powers can lead to serious rounding errors, computer algorithms may use formulas for k_3 and k_4 other than those shown above, and for large or small X_i's Equations 6.3 and 6.6 may be preferable to the machine formulas shown. Also, coding data may be used to reduce rounding error. (Coding will alter k statistics but not g statistics.)

Beta Measures of Symmetry and Kurtosis. Some authors speak of a population parameter called $\sqrt{\beta_1}$ as a measure of symmetry (where $\sqrt{\beta_1} = 0$ indicates symmetry as does $\gamma_1 = 0$) and a parameter designated β_2 as a measure of kurtosis (with $\beta_2 = 3$ indicating mesokurtosis).[†]

---

*The terms "platykurtic," "leptokurtic," and "mesokurtic" were introduced by Karl Pearson in 1906; in 1905 he had introduced the term "kurtosis" and the terms "isokurtic" and "allokurtic," respectively, to refer to distributions that were symmetric and skewed (David, 1995; Walker, 1929: 182).

[†]The beta (β) functions were introduced by K. Pearson in 1895 (David, 1995).

EXAMPLE 6.1   The heights of the first seventy graduate students to enroll in my biostatistics course*, and measures of symmetry and kurtosis.

Height (X_i) (in.)	Observed frequency (f_i)	$f_i X_i$ (in.)	$f_i X_i^2$ (in.^{2})	$f_i X_i^3$ (in.^{3})	$f_i X_i^4$ (in.^{4})
63	2	126	7,938	500,094	31,505,922
64	2	128	8,192	524,288	33,554,432
65	3	195	12,675	823,875	53,551,875
66	5	330	21,780	1,437,480	94,873,680
67	4	268	17,956	1,203,052	80,604,484
68	6	408	27,744	1,886,592	128,288,256
69	5	345	23,805	1,642,545	113,335,605
70	8	560	39,200	2,744,000	192,080,000
71	7	497	35,287	2,505,377	177,881,767
72	7	504	36,288	2,612,736	188,116,992
73	10	730	53,290	3,890,170	283,982,410
74	6	444	32,856	2,431,344	179,919,456
75	3	225	16,875	1,265,625	94,921,875
76	2	152	11,552	877,952	66,724,352
	$\sum f_i =$ $n = 70$	$\sum f_i X_i =$ 4,912 in.	$\sum f_i X_i^2 =$ 345,438 in.^{2}	$\sum f_i X_i^3 =$ 24,345,130 in.^{3}	$\sum f_i X_i^4 =$ 1,719,341,106 in.^{4}

$$\text{SS} = \sum f_i X_i^2 - \frac{\sum (f_i X_i)^2}{n} = 345,438 \text{ in.}^2 - \frac{(4,912 \text{ in.})^2}{(70)} = 755.9429 \text{ in.}^2$$

$$s^2 = \frac{\text{SS}}{n-1} = \frac{755.9429 \text{ in.}^2}{69} = 10.9557 \text{ in.}^2$$

$$k_3 = \frac{n \sum f_i X_i^3 - 3 \sum f_i X_i \sum f_i X_i^2 + 2(\sum f_i X_i)^3/n}{(n-1)(n-2)}$$

$$= \frac{70(24,345,130 \text{ in.}^3) - 3(4,912 \text{ in.})(345,438 \text{ in.}^2) + 2(4,912 \text{ in.}^3)/70}{(69)(68)}$$

$$= \frac{1,704,159,100 \text{ in.}^3 - 5,090,374,368 \text{ in.}^3 + 3,386,156,529 \text{ in.}^3}{4,692}$$

$$= \frac{-58,739}{4,692} = -12.5190 \text{ in.}^3$$

$$g_1 = \frac{k_3}{\sqrt{(s^2)^3}} = \frac{k_3}{\sqrt{(10.9557)^3}} = \frac{-12.5190}{36.2627} = -0.3452$$

$$k_4 = \frac{(n+1)[n \sum f_i X_i^4 - 4 \sum f_i X_i \sum f_i X_i^3 + 6(\sum f_i X_i)^2 \sum f_i X_i^2/n - 3(\sum f_i X_i)^4/n^2]}{(n-1)(n-2)(n-3)}$$

$$- \frac{3(\text{SS})^2}{(n-2)(n-3)}$$

EXAMPLE 6.1   (continued)

$$= 71[70(1,719,341,106 \text{ in.}^4) - 4(4,912 \text{ in.})(24,345,130 \text{ in.}^3)$$

$$\frac{+6(4,912 \text{ in.})^2(345,438 \text{ in.}^2)/70 - 3(4,912) \text{ in.})^4/70^2]}{(69)(68)(67)}$$

$$-\frac{3(755.9429)^2}{(68)(67)}$$

$$= \frac{91,185,300 \text{ in.}^4}{314,364} - \frac{1,714,349 \text{ in.}^4}{4,556} = 290.0628 \text{ in.}^4 - 376.2838 \text{ in.}^4 = -86.2210 \text{ in.}^4$$

$$g_2 = \frac{k_4}{(s^2)^2} = \frac{-86.2210 \text{ in.}^4}{(10.9557 \text{ in.}^2)^2} = -0.7183$$

If $\sqrt{b_1}$ or b_2 is desired:

$$\sqrt{b_1} = \frac{(n-2)g_1}{\sqrt{n(n-1)}} = \frac{(68)(-0.3452)}{\sqrt{70(69)}} = \frac{-23.4736}{69.4982} = -0.3378$$

$$b_2 = \frac{(n-2)(n-3)g_2}{(n+1)(n-1)} + \frac{3(n-1)}{n+1} = \frac{(68)(67)(-0.7183)}{(71)(69)} + \frac{3(69)}{71}$$

$$= \frac{-3272.5748}{4899} + \frac{207}{71} = -0.6680 + 2.9155 = 2.2475$$

*Metric units of measurements are used throughout this book, with the sole exception of the data in Example 6.1 which were self-reported in English units.

The sample estimates of these measures are related to the g statistics as

$$\sqrt{b_1} = \frac{(n-2)g_1}{\sqrt{n(n-1)}} \tag{6.9}$$

$$b_2 = \frac{(n-2)(n-3)g_2}{(n+1)(n-1)} + \frac{3(n-1)}{n+1}, \tag{6.10}$$

and $\sqrt{b_1} = 0$ and $b_2 = 3(n-1)/(n+1)$ indicate symmetry and mesokurtosis, respectively (D'Agostino, Belanger, and D'Agostino, 1990).[†]

These statistics are demonstrated at the end of Example 6.1 and will be used in hypothesis testing in Section 7.14.

Quantile Measures of Symmetry and Kurtosis.   Denoting the ith quartile as Q_i (as in Section 4.2), Q_1 is the first quartile (i.e., the 25th percentile), Q_3 is the third quartile (i.e., the 75th percentile), and Q_2 is the second quartile (i.e., the 50th percentile,

---

[†]Statistics referred to as m_3 and m_4 are analogous to k_3 and k_4, respectively, and may be used for direct computation of $\sqrt{b_1}$ and b_2 (as k_3 and k_4 are used to compute g_1 and g_2). (See Bliss, 1967: 143; D'Agostino, Belanger, and D'Agostino, 1990; Fisher, 1958: 72).

namely the median). The so-called Bowley coefficient of skewness (Bowley, 1920: 116; Groeneveld and Meeden, 1984) is

$$\text{skewness} = \frac{Q_3 + Q_1 - 2Q_2}{Q_3 - Q_1},$$

$$(6.11)$$

a measure without units that may range from -1, for a distribution with extreme left skewness; to 0, for a symmetrical distribution; to 1, for a distribution with extreme right skewness.

A kurtosis measure based on quantiles was proposed by Moors (1988), using octiles: O_1 (i.e., the first octile) is the 12.5th percentile; O_3 (the third octile) is the 37.5th percentile, O_5 is the 62.5th percentile, and O_7 is the 87.5th percentile. Also, $O_2 = Q_1, O_4 = Q_2$, and $O_6 = Q_3$. The measure is

$$\text{kurtosis} = \frac{(O_7 - O_5) + (O_3 - O_1)}{O_6 - O_2} = \frac{(O_7 - O_5) + (O_3 - O_1)}{Q_3 - Q_1}$$

$$(6.12)$$

which has no units and may range from zero, for extreme platykurtosis; to 1.233, for mesokurtosis; to infinity, for extreme leptokurtosis.

These measures of symmetry and kurtosis are shown in Example 6.2.

6.2 PROPORTIONS OF A NORMAL DISTRIBUTION

If a normal population of 1000 body weights has a mean, μ, of 70 kg, one-half of the population (500 weights) is larger than 70 kg and one-half is smaller. This is true simply because the normal distribution is symmetrical. But if we desire to ask what portion of the population is larger than 80 kg, we need to know σ, the standard deviation of the population. If $\sigma = 10$ kg, then 80 kg is one standard deviation larger than the mean, and the portion of the population in question is the shaded area in Fig. 6.4a. If, however, $\sigma = 5$ kg, then 80 kg is two standard deviations above μ, and we are referring to a relatively small portion of the population, as shown in Fig. 6.4b.

Appendix Table B.2 enables us to determine proportions of normal distributions. For any X_i value from a normal population with mean μ, and standard deviation σ, the value

$$Z = \frac{X_i - \mu}{\sigma}$$

$$(6.13)$$

tells us how many standard deviations from the mean the X_i value is located. Carrying out the calculation of Equation 6.13 is known as *normalizing*, or *standardizing*, X_i; and Z is known as a *normal deviate*, or a standard score.* The mean of a set of standard scores is 0, and the variance is 1.

---

*This standard normal curve was introduced in 1899 by W. F. Sheppard (Walker, 1929: 188), and the term "normal deviate" was first used, in 1907, by F. Galton (David, 1995).

EXAMPLE 6.2   Quantile measures of symmetry and kurtosis for the data of Example 6.1.

$n = 70$

To obtain the quartiles of the data:

Q_1:   subscript on X for $Q_1 = (n+1)/4 = (70+1)/4 = 17.75$; round up to 18; $Q_1 = X_{18} = 68$ in.

Q_2: $Q_2 = M = X_{(n+1)/2} = X_{(70+1)/2} = X_{35.5} = 70$ in.

Q_3:   subscript on X for $Q_3 = n + 1 -$ subscript on X for $Q_1 = 70 + 1 - 18 = 53$; $Q_3 = X_{53} = 73$ in.

$$\text{skewness} = \frac{Q_3 + Q_1 - 2Q_2}{Q_3 - Q_1} = \frac{73 \text{ in.} + 68 \text{ in.} - 2(70 \text{ in.})}{73 \text{ in.} - 68 \text{ in.}} = 0.20$$

To obtain the octiles of the data:

$\mathcal{O}_1$:   subscript on X for $\mathcal{O}_1 = (n + 1)/8 = (70 + 1)/8 = 8.875$; round up to 9; $\mathcal{O}_1 = X_9 = 66$ in.

$\mathcal{O}_2$: $\mathcal{O}_2 = Q_1 = X_{18} = 68$ in.

$\mathcal{O}_3$:   subscript on X for $\mathcal{O}_3 = 2$(subscript on X for Q_1)$-$ subscript on X for $\mathcal{O}_1 = 2(18) - 9 = 27$; $\mathcal{O}_3 = X_{27} = 69$ in.

$\mathcal{O}_4$: $\mathcal{O}_4 = Q_2 = M = 71$ in.

$\mathcal{O}_5$:   subscript on X for $\mathcal{O}_5 = n + 1 -$ subscript on X for $\mathcal{O}_3 = 70 + 1 - 27 = 44$; $\mathcal{O}_5 = X_{44} = 72$ in.

$\mathcal{O}_6$: $\mathcal{O}_6 = Q_3 = X_{53} = 73$ in.

$\mathcal{O}_7$:   subscript on X for $\mathcal{O}_7 = n + 1 -$ subscript on X for $\mathcal{O}_1 = 70 + 1 - 9 = 62$; $\mathcal{O}_7 = X_{62} = 74$ in.

$$\text{kurtosis} = \frac{(\mathcal{O}_7 - \mathcal{O}_5) + (\mathcal{O}_3 - \mathcal{O}_1)}{\mathcal{O}_6 - \mathcal{O}_2} = \frac{(74 \text{ in.} - 72 \text{ in.}) + (69 \text{ in.} - 66 \text{ in.})}{73 \text{ in.} - 68 \text{ in.}} = 1.00$$

Table B.2 tells us what proportion of a normal distribution lies beyond a given value of Z.* If $\mu = 70$ kg, $\sigma = 10$ kg, and $X_i = 70$ kg, then $Z = (70 \text{ kg} - 70 \text{ kg})/10 \text{ kg} = 0$, and by consulting Table B.2 we see that $P(X_i > 70 \text{ kg}) = P(Z > 0) = 0.5000$.[†] That is, 0.5000 (or 50.00%) of the distribution is larger than 70 kg. To determine the proportion of the distribution that is greater than 80 kg in weight, $Z = (80 \text{ kg} - 70 \text{ kg})/10 \text{ kg} = 1$, and $P(X_i > 80 \text{ kg}) = P(Z > 1) = 0.1587$ (or 15.87%). This could be stated as being the probability of drawing at random a measurement, X_i, greater than 80 kg from a population with $\mu = 70$ kg and $\sigma = 10$ kg. What, then, is the probability of obtaining, at random, a measurement, X_i, which is less than 80 kg? $P(X_i > 80 \text{ kg}) = 0.1587$, so

---

*The first tables of areas under the normal curve were published in 1799 by the French physicist, C. Kramp (Walker, 1929: 58) Today, some calculators and computer programs determine normal probabilities (e.g., see Boomsma and Molenaar, 1994; Guenther, 1977).

[†]Read $P(X_i > 70 \text{ kg})$ as "the probability of an X_i greater than 70 kg"; $P(Z > 0)$ is read as "the probability of a Z greater than 0."

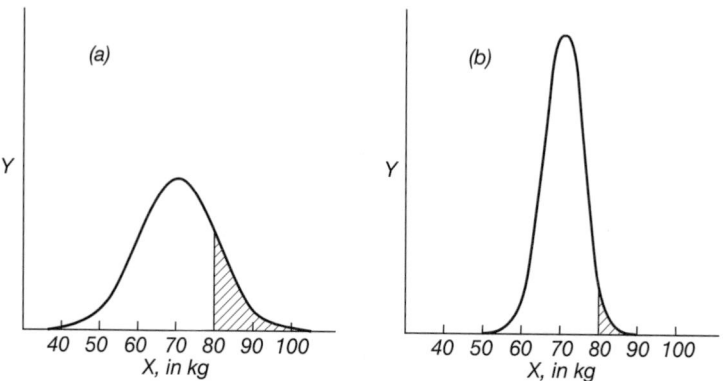

Figure 6.4 Two normal distributions with $\mu = 70$ kg. The shaded areas are the portions of the curves that lie above $X = 80$ kg. For distribution a, $\mu = 70$ kg and $\sigma = 10$ kg; for distribution b, $\mu = 70$ kg and $\sigma = 5$ kg.

$P(X_i < 80$ kg$) = 1.0000 - 0.1587 = 0.8413$; that is, if 15.87% of the population is greater than X_i, then $100\% - 15.87\%$ of the population is less than X_i.* Example 6.3a presents calculations for determining proportions of a normal distribution lying between a variety of limits.

     Note that Table B.2 contains no negative values of Z. However, if we are concerned with proportions in the left half of the distribution, we are simply dealing with areas of the curve that are mirror images of those present in the table. This is demonstrated in Example 6.3b.†

     Using the preceding considerations of the table of normal deviates (Table B.2), we can obtain the following information:

68.27% of the measurements in a normal population lie within the range of $\mu \pm \sigma$,
95.44% lie within $\mu \pm 2\sigma$,
99.73% lie within $\mu \pm 3\sigma$,

50% lie within $\mu \pm 0.67\sigma$,
95% lie within $\mu \pm 1.96\sigma$,
97.5% lie within $\mu \pm 2.24\sigma$,
99% lie within $\mu \pm 2.58\sigma$,
99.5% lie within $\mu \pm 2.81\sigma$,
99.9% lie within $\mu \pm 3.29\sigma$.

---

    *The statement that "$P(X_i > 80$ kg$) = 0.1587$, therefore $P(X_i < 80) = 1.0000 - 0.1587$" does not take into account the case of $X_i = 80$ kg. But, as we are considering the distribution at hand to be a continuous one, the probability of X_i being *exactly* 80.000 … kg (or being *exactly* any other stated value) is practically nil, so these types of probability statements offer no practical difficulties.

    †Some old literature avoided referring to negative Z's by expressing the quantity, $Z + 5$, called a *probit*. This term was introduced in 1934 by C. I. Bliss (David, 1995).

EXAMPLE 6.3a   **Calculating proportions of a normal distribution of bone lengths, where $\mu = 60$ mm and $\sigma = 10$ mm.**

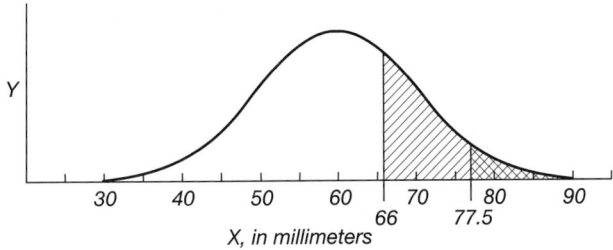

Y

X, in millimeters

1. What proportion of the population of bone lengths is larger than 66 mm?

$$Z = \frac{X_i - \mu}{\sigma} = \frac{66 \text{ mm} - 60 \text{ mm}}{10 \text{ mm}} = 0.60$$

$$P(X_i > 66 \text{ mm}) = P(Z > 0.60) = 0.2743 \text{ or } 27.43\%$$

2. What is the probability of picking, at random from this population, a bone larger than 66 mm? This is simply another way of stating the quantity calculated in part (1). The answer is 0.2743.

3. If there are 2000 bone lengths in this population, how many of them are greater than 66 mm?

$$(0.2743)(2000) = 549$$

4. What proportion of the population is smaller than 66 mm?

$$P(X_i < 66 \text{ mm}) = 1.0000 - P(X_i > 66 \text{ mm}) = 1.0000 - 0.2743 = 0.7257$$

5. What proportion of this population lies between 60 and 66 mm? Of the total population, 0.5000 is larger than 60 mm and 0.2743 is larger than 66 mm. Therefore, $0.5000 - 0.2743 = 0.2257$ of the population lies between 60 and 66 mm. That is, $P(60 \text{ mm} < X_i < 66 \text{ mm}) = 0.5000 - 0.2743 = 0.2257$.

6. What portion of the area under the normal curve lies to the right of 77.5 mm?

$$Z = \frac{77.5 \text{ mm} - 60 \text{ mm}}{10 \text{ mm}} = 1.75$$

$$P(X_i > 77.5 \text{ mm}) = P(Z > 1.75) = 0.0401 \text{ or } 4.01\%$$

7. If there are 2000 bone lengths in the population, how many of them are larger than 77.5 mm?

$$(0.0401)(2000) = 80$$

8. What is the probability of selecting at random from this population a bone measuring between 66 and 77.5 mm in length?

$$P(66 \text{ mm} < X_i < 77.5 \text{ mm}) = P(0.60 < Z < 1.75) = 0.2743 - 0.0401 = 0.2342$$

EXAMPLE 6.3b Calculating proportions of a normal distribution of sucrose concentrations, where $\mu = 65$ mg/100 ml and $\sigma = 25$ mg/100 ml.

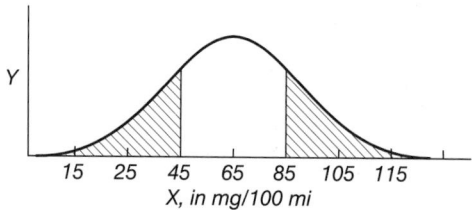

1. What proportion of the population is greater than 85 mg/100 ml?

$$Z = \frac{(X_i - \mu)}{\sigma} = \frac{85 \text{ mg/100 ml} - 65 \text{ mg/100 ml}}{25 \text{ mg/100 ml}} = 0.8$$

$$P(X_i > 85 \text{ mg/100 ml}) = P(Z > 0.8) = 0.2119 \text{ or } 21.19\%$$

2. What proportion of the population is less than 45 mg/100 ml?

$$Z = \frac{45 \text{ mg/100 ml} - 65 \text{ mg/100 ml}}{25 \text{ mg/100 ml}} = -0.80$$

$$P(X_i < 45 \text{ mg/100 ml}) = P(Z < -0.80) = P(Z > 0.80) = 0.2119$$

That is, the probability of selecting from this population an observation less than 0.80 standard deviations below the mean is equal to the probability of obtaining an observation greater than 0.8 standard deviations above the mean.

3. What proportion of the population lies between 45 and 85 mg/100 ml?

$$P(45 \text{ mg/100 ml} < X_i < 85 \text{ mg/100 ml}) = P(-0.80 < Z < 0.80)$$

$$= 1.0000 - P(Z < -0.80 \text{ or } Z > 0.80)$$

$$= 1.0000 - (0.2119 + 0.2119)$$

$$= 1.0000 - 0.4238$$

$$= 0.5762$$

6.3 THE DISTRIBUTION OF MEANS

If random samples of size n are drawn from a normal population, the means of these samples will conform to normal distribution. The distribution of means from a non-normal population will not be normal but will tend toward normality as n increases in size.* Furthermore, the variance of the distribution of means will decrease as n increases; in fact, the variance of the population of all possible means of samples of size n from a

---

*This result is known as the *central limit theorem*.

population with variance σ^2 is

$$\sigma_{\bar{X}}^2 = \frac{\sigma^2}{n}. \tag{6.14}$$

The quantity $\sigma_{\bar{X}}^2$ is called the *variance of the mean*. A distribution of sample statistics is called a *sampling distribution*; therefore, we are discussing the sampling distribution of means.

Since $\sigma_{\bar{X}}^2$ has square units, its square root, $\sigma_{\bar{X}}$, will have the same units as the original measurements (and, therefore, the same units as the mean, μ, and the standard deviation, σ). This value, $\sigma_{\bar{X}}$, is the *standard deviation of the mean*. The standard deviation of a statistic is referred to as a *standard error*; thus, $\sigma_{\bar{X}}$ is frequently called the *standard error of the mean* (sometimes abbreviated SEM), or simply the *standard error* (sometimes abbreviated SE)*:

$$\sigma_{\bar{X}} = \sqrt{\frac{\sigma^2}{n}} \quad \text{or} \quad \sigma_{\bar{X}} = \frac{\sigma}{\sqrt{n}}. \tag{6.15}$$

Just as $Z = (X_i - \mu)/\sigma$ (Equation 6.13) is a normal deviate that refers to the normal distribution of X_i values,

$$Z = \frac{\bar{X} - \mu}{\sigma_{\bar{X}}} \tag{6.16}$$

is a normal deviate referring to the normal distribution of means ($\bar{X}$ values). Thus, we can ask questions such as: What is the probability of obtaining a random sample of nine measurements with a mean larger than 50.0 mm from a population having a mean of 47.0 mm and a standard deviation of 12.0 mm? This and other examples of the use of normal deviates for the sampling distribution of means are presented in Example 6.4.

As seen from Equation 6.15, to determine $\sigma_{\bar{X}}$ one must know σ^2 (or σ), which is a population parameter. Since we very seldom can calculate population parameters, we must rely on estimating them from random samples taken from the population. The best estimate of $\sigma_{\bar{X}}^2$, the population variance of the mean, is

$$s_{\bar{X}}^2 = \frac{s^2}{n}, \tag{6.17}$$

the sample variance of the mean. Thus,

$$s_{\bar{X}} = \sqrt{\frac{s^2}{n}} \quad \text{or} \quad s_{\bar{X}} = \frac{s}{\sqrt{n}} \tag{6.18}$$

is an estimate of $\sigma_{\bar{X}}$ and is the sample standard error of the mean. Example 6.5 demonstrates the calculation of $s_{\bar{X}}$.

The importance of the standard error in hypothesis testing and related procedures will be evident in Chapter 7. At this point, however, it can be noted that the magnitude of $s_{\bar{X}}$ is helpful in determining the precision to which the mean and some measures of

---

*This relationship between the standard deviation of the mean and the standard deviation was published by Karl Friedrich Gauss in 1809 (Walker, 1929: 23). The term "standard error" was introduced in 1897 by G. U. Yule (David, 1995).

EXAMPLE 6.4   Proportions of a sampling distribution of means.

1. If a population has $\mu = 47.0$ mm and $\sigma = 12.0$ mm, what is the probability of drawing from it a random sample of nine measurements that has a mean larger than 50.0 mm?

$$\sigma_{\bar{X}} = \frac{12.0 \text{ mm}}{\sqrt{9}} = 4.0 \text{ mm}$$

$$Z = \frac{\bar{X} - \mu}{\sigma_{\bar{X}}} = \frac{50.0 \text{ mm} - 47.0 \text{ mm}}{4.0 \text{ mm}} = 0.75$$

$$P(\bar{X} > 50.0 \text{ mm}) = P(Z > 0.75) = 0.2266$$

2. What is the probability of drawing a sample of twenty-five measurements from the preceding population and finding that the mean of this sample is less than 40.0 mm?

$$\sigma_{\bar{X}} = \frac{12.0 \text{ mm}}{\sqrt{25}} = 2.4 \text{ mm}$$

$$Z = \frac{40.0 \text{ mm} - 47.0 \text{ mm}}{2.4 \text{ mm}} = -2.92$$

$$P(\bar{X} < 40.0 \text{ mm}) = P(Z < -2.92) = P(Z > 2.92) = 0.0018$$

3. If 500 random samples of size twenty-five are taken from the preceding population, how many of them would have means larger than 50.0 mm?

$$\sigma_{\bar{X}} = \frac{12.0 \text{ mm}}{\sqrt{25}} = 2.4 \text{ mm}$$

$$Z = \frac{50.0 \text{ mm} - 47.0 \text{ mm}}{2.4 \text{ mm}} = 1.25$$

$$P(\bar{X} > 50.0 \text{ mm}) = P(Z > 1.25) = 0.1056$$

Therefore, $(0.1056)(500) = 53$ samples would be expected to have means larger than 50.0 mm.

---

variability may be reported. Although different practices have been followed by many, we shall employ the following (Eisenhart, 1968). We shall state the standard error to two significant figures (e.g., 2.7 mm in Example 6.5). Then the standard deviation and the mean will be reported with the same number of decimal places (e.g., $\bar{X} = 137.6$ mm in Example 6.5*). The variance may be reported with twice the number of decimal places as the standard deviation.

---

*In Example 6.5, s is written with more decimal places than the Eisenhart recommendations indicate because it is an intermediate, rather than a final, result; and rounding off intermediate computations may lead to serious rounding error. Indeed, some authors routinely report extra decimal places, even in final results, with the consideration that readers of the results may use them as intermediates in additional calculations.

EXAMPLE 6.5   The calculation of the standard error of the mean, $s_{\bar{X}}$. The following are data for systolic blood pressure, in mm of mercury.

121	$n = 12$
125	
128	$\bar{X} = \dfrac{1651 \text{ mm}}{12} = 137.6 \text{ mm}$
134	
136	$SS = 228,111 \text{ mm}^2 - \dfrac{(1651 \text{ mm})^2}{12}$
138	
139	$= 960.9167 \text{ mm}^2$
141	
144	$s^2 = \dfrac{960.9167 \text{ mm}^2}{11} = 87.3561 \text{ mm}^2$
145	
149	$s = \sqrt{87.3561 \text{ mm}^2} = 9.35 \text{ mm}$
151	
$\sum X = 1651 \text{ mm}$	$s_{\bar{X}} = \dfrac{s}{\sqrt{n}} = \dfrac{9.35 \text{ mm}}{\sqrt{12}} = 2.7 \text{ mm}$ *or*
$\sum X^2 = 228,111 \text{ mm}^2$	$s_{\bar{X}} = \sqrt{\dfrac{s^2}{n}} = \sqrt{\dfrac{87.3561 \text{ mm}^2}{12}} = \sqrt{7.2797 \text{ mm}^2} = 2.7 \text{ mm}$

6.4 INTRODUCTION TO STATISTICAL HYPOTHESIS TESTING

A major goal of statistical analysis is to draw inferences about a population by examining a sample from that population. A very common example of this is the desire to draw conclusions about one or more population means.

We begin by making a concise statement about the population mean, a statement called a *null hypothesis* (abbreviated H_0)* because it expresses the concept of "no difference." For example, a null hypothesis about a population mean (μ) might assert that μ is not different from zero (i.e., μ is equal to zero); and this would be written as

$$H_0\text{: } \mu = 0.$$

Or, we could hypothesize that the population mean is not different from (i.e., is equal to) 3.5 cm, or not different from 10.5 kg, in which case we would write H_0: $\mu = 3.5$ cm or H_0: $\mu = 10.5$ kg, respectively.

If it is concluded that it is likely that a null hypothesis is false, then an *alternate hypothesis* (abbreviated H_A) is assumed to be true. One states a null hypothesis and an alternate hypothesis for each statistical test performed, and all possible outcomes are accounted for by this pair of hypotheses. So, for the examples above:

$$H_0\text{: } \mu = 0, \quad H_A\text{: } \mu \neq 0;$$

$$H_0\text{: } \mu = 3.5 \text{ cm}, \quad H_A\text{: } \mu \neq 3.5 \text{ cm};$$

$$H_0\text{: } \mu = 10.5 \text{ kg}, \quad H_A\text{: } \mu \neq 10.5 \text{ kg}.$$

---

*E. S. Pearson (1947) credits the introduction of the symbol, "H_0," to J. Neyman and himself and the origin of the term, "null hypothesis," to R. A. Fisher; David (1995) credits a 1935 Fisher paper with introducing the latter term.

It must be emphasized that statistical hypotheses are to be stated *before* data are collected to test them. To propose hypotheses after examination of data can invalidate a statistical test. One may, however, legitimately formulate hypotheses *after* inspecting data if a new set of data is then collected with which to test the hypotheses.

Statistical Testing and Probability. Statistical testing of a null hypothesis about the mean of a population (μ) involves determining the mean of a random sample from that population $\bar{X}$. Then we determine the probability, *if H_0 is true*, of an $\bar{X}$ at least as far from μ as the $\bar{X}$ in the sample. This is accomplished by the considerations of Section 6.3 and is demonstrated in Example 6.6.

Here, a manufacturer has produced a device that is to sound an alarm when the concentration of carbon monoxide (CO) in the air is at 10.00 mg/m^3; and we wish to know whether the device works as intended. Known amounts of carbon monoxide are introduced into a chamber initially containing no CO, and it is recorded at what CO concentration the alarm sounds. This is done eighteen times, with the resultant data (the eighteen values of X_i) shown in Example 6.6. These eighteen data have a mean of $\bar{X} = 10.43$ mg/m^3 and they represent a sample (we presume a random sample) of a very large number of data, namely the very large number of alarm-triggering CO concentrations that would result from repeating this experiment a very large number of times. This large number of X_i's is the statistical population. Although one almost never knows the actual parameters of a sampled population, for this introduction to statistical testing let us suppose that the variance of the population for this example is known to be $\sigma^2 = 1.0434$ (mg/m^3)^2. Thus, for the population of means that could be drawn from this population of measurements, the standard error of the mean is $\sigma_{\bar{X}} = \sqrt{1.0434 \ (\text{mg/m}^3)^2/18} = \sqrt{0.0580 \ (\text{mg/m}^3)^2} = 0.24$ mg/m^3.

As we wish to know whether the mean of a very large number of repetitions of this experiment equals 10.00 mg/m^3, the appropriate null and alternate hypotheses are H_0: $\mu = 10.00$ mg/m^3 and H_A: $\mu \neq 10.00$ mg/m^3, respectively. And, what we ask is the following:

> If we have a normal population with $\mu = 10.00$ mg/m^3 and $\sigma_{\bar{X}} = 0.24$ (mg/m^3), what is the probability of obtaining a random sample with a mean ($\bar{X}$) at least as far from 10.00 mg/m^3 as 10.43 mg/m^3? Another way to state this would be: What is $P(\bar{X} \geq 10.43$ mg/m^3 or $\bar{X} \leq 9.57$ mg/m^3)?

Reflecting upon Section 6.3 it can be seen that such probabilities may be ascertained by computation of Z (by Equation 6.16), so Z is referred to as a *test statistic*. The above null hypothesis is tested in Example 6.6, and the two normal-distribution tail regions of interest are shown in Fig. 6.5.

Statistical Errors in Hypothesis Testing. One needs an objective criterion for rejecting or not rejecting the null hypothesis for a statistical test. Theoretically, a very large (or very small) sample mean might be obtained—and a very large absolute value of Z thereby computed—even when H_0 is true; however, the larger the $|Z|$ the smaller the probability that H_0 is true. So we can ask: "How small a probability (or, how

EXAMPLE 6.6   Testing the hypotheses H_0: $\mu = 10.00$ mg/m^3 and H_A: $\mu \neq 10.00$ mg/m^3.

The variable, X, is the carbon monoxide concentration in air, and eighteen measurements are obtained, as follows: 10.25, 10.37, 10.66, 10.47, 10.56, 10.22, 10.44, 10.38, 10.63, 10.40, 10.39, 10.26, 10.32, 10.35, 10.54, 10.33, 10.48, 10.68 mg/m^3.

For these data the sample mean is $\bar{X} = 10.43$ mg/m^3; and for the sake of this example the population standard error of the mean is said to be known to be $\sigma_{\bar{X}} = 0.24$ mg/m^3.

$$Z = \frac{\bar{X} - \mu}{\sigma_{\bar{X}}} = \frac{10.43 \text{ mg/m}^3 - 10.00 \text{ mg/m}^3}{0.24 \text{ mg/m}^3} = 1.79$$

Using Table B.2:

$$P(\bar{X} \geq 10.43 \text{ mg/m}^3) = P(Z \geq 1.79) = 0.0367$$

and

$$P(Z \leq -1.79) = 0.0367.$$

Therefore,

$$P(\bar{X} \geq 10.43 \text{ mg/m}^3 \text{ or } \bar{X} \leq 9.57 \text{ mg/m}^3) = 0.0367 + 0.0367 = 0.0734.$$

As $0.0734 > 0.05$, do not reject H_0.

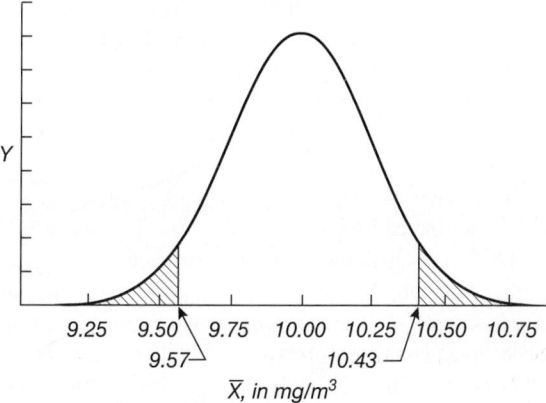

Y

9.25   9.50   9.75   10.00   10.25   10.50   10.75
9.57⌐   10.43⌐
$\bar{X}$, in mg/m^3

Figure 6.5 The normal distribution of means referred to in the hypothesis testing of Example 6.6a, with a hypothesized population mean, μ, of 10.00 mg/m^3 and a presumed population standard error of the mean, $\sigma_{\bar{X}}$, of 0.24 mg/m^3. It is found that there is a probability of 0.0734 of the mean of a sample of eighteen being in the shaded portions of the two tails.

large a $|Z|$) will be required to reject the null hypothesis?" As explained below, a probability of 5% or less is commonly used as the criterion for rejection of H_0. The probability used as the criterion for rejection is called the *significance level*,* denoted by α (the lowercase Greek letter, alpha). The value of the test statistic (in this case, Z) corresponding to α is termed the *critical value* of the test statistic. In Appendix B.2 it is seen that $P(Z \geq 1.96) = 0.025$; and, inasmuch as the normal distribution is symmetrical, $P(Z \leq -1.96) = 0.025$. Therefore, the critical value for testing the above H_0 at a 5%

---

*David (1995) credits R. A. Fisher as the first to refer to "level of significance," in 1925, when Fisher also formally recommended use of the 5% level (Cowles and Davis, 1982).

level of significance is $Z = 1.96$. As the test statistic in Example 6.6 (namely, $Z = 1.79$) is not as large as the critical value, the null hypotheses is not rejected.

It is very important to realize that a true null hypothesis occasionally will be rejected, which of course means that we have committed an error in drawing a conclusion about the sampled population. Moreover, this error will be committed with a frequency of α. That is, if H_0 is in fact a true statement about a statistical population, it will be concluded (erroneously) to be false 5% of the time. The rejection of a null hypothesis when it is in fact true is what is known as a *Type I error* ("Type 1 error," also called an alpha error or an "error of the first kind"). On the other hand, if H_0 is in fact false, a statistical test will sometimes not detect this fact, and we shall thus reach an erroneous conclusion by not rejecting H_0. The probability of committing this error, of not rejecting the null hypothesis when it is in fact false, is represented by β (the lowercase Greek letter, beta). This error is referred to as a *Type II error* ("Type 2 error," also called a beta error, or an "error of the second kind"). The *power* of a statistical test is defined as $1 - \beta$; i.e., power is the probability of rejecting the null hypothesis when it is in fact false and should be rejected.*

Whereas the probability of committing a Type I error is α, the specified significance level, the probability of committing a Type II error is β, a value that generally we neither specify nor know. What we do know is that for a given sample size, n, the value of α is related inversely to the value of β. That is, lower probabilities of committing a Type I error are associated with higher probabilities of committing a Type II error. Both types of error may be reduced simultaneously by increasing n. Thus, for a given α, larger samples will result in statistical testing with greater power $(1 - \beta)$.

Table 6.1 summarizes these two types of statistical errors. Since, for a given n, one cannot minimize both of them, it is appropriate to ask what the acceptable combination of the two might be. By experience, and hence by convention, an α of 0.05 is usually considered to be a "small enough" chance of committing a Type I error, while not being so small as to result in "too large a chance" of a Type II error. But there is nothing sacrosanct about the 0.05 level. Although it is the most widely used significance level, researchers may decide for themselves whether it is more important to minimize one type of error or the other. In some situations, for example, a 5% chance of an incorrect rejection of H_0 may be felt to be unacceptably high, so the 1% level of significance is sometimes employed. It is necessary, of course, to state the significance level used when reporting the results of a statistical test. Indeed, rather than simply stating whether the null hypothesis is rejected, it is good practice to state also the test statistic itself and the best estimate of its exact probability. (In Example 6.6, it is asserted that $Z = 1.79$ and $P = 0.0734$, in addition to expressing the conclusion that H_0 is not rejected.) In this way, readers of the research results may draw their own conclusions, even if their choice of significance level is different from the author's. Bear in mind, however, that

---

*The distinction between these two fundamental kinds of statistical errors, and the concept of power, date back to the pioneering work, in England, of Jerzy Neyman (1894–1981; Russian-born, of Polish roots, emigrating as an adult to Poland and then to England, and spending the last half of his life in the United States) and the English statistician, Egon S. Pearson (1895–1980) (Lehmann and Reid, 1982; Neyman and Pearson, 1928a; Pearson, 1947). Their naming of the two types of errors, and of power, dates from 1933 (David, 1995).

TABLE 6.1 The Two Types of Errors in
Hypothesis Testing

	If H_0 is true	If H_0 is false
If H_0 is rejected:	Type I error	No error
If H_0 is not rejected:	No error	Type II error

the choice of α is to be made before even seeing the data. Otherwise there is a great risk of having the choice influenced by examination of the data, introducing bias instead of objectivity into the proceedings. The best practice generally is to decide on the null hypothesis, alternate hypothesis, and significance level before commencing with data collection. It is conventional to refer to rejection of H_0 at the 5% significance level as denoting a "significant" difference (in the present example, a significant difference between $\bar{X}$ and 10.00 mg/m^3) and rejection at the 1% level as indicating a "highly significant difference."[*]

As the significance level selected is somewhat arbitrary, if test results are very near that level (e.g., between 0.04 and 0.06 if $\alpha = 0.05$ is used), then it may be wiser to repeat the analysis with additional data than to declare emphatically that the null hypothesis is or is not a reasonable statement about the sampled population.

More on Statistical Power.   The power of a statistical testing procedure was defined above as the probability that the test will correctly reject the null hypothesis when it is false. For many hypothesis tests it is possible to compute the power under specified conditions. For the situation in Example 6.6, for instance, it was hypothesized that the population mean, μ, is 10.00 mg/m^3; and we could ask how powerful the test would be in rejecting H_0 if in reality H_0 is 10.50 mg/m^3. The normal curve in Fig. 6.6a has a mean of 10.50 mg/m^3 and the same standard error as the curve in Fig. 6.5. If, in the sampled population, $\mu = 10.50$ mg/m^3, then the power of the testing procedure of Example is the proportion of the area under the curve of Fig. 6.6a that lies below 9.57 mg/m^3 and above 10.43 mg/m^3. This is computed in Example 6.7a to be 0.61, which means that there was only a 61% chance of rejecting H_0 even though it is false. Fig. 6.6b presents the normal curve of Fig. 6.6a (for which the mean of 10.50 mg/m^3 exists in the population that was sampled) together with the normal curve of Fig. 6.5 (for which the mean of 10.00 mg/m^3 is what was hypothesized), so the relationship between the two may be more clearly observed.

If the population mean, μ, is really larger than 10.50 mg/m^3, then the curve would be ever farther to the right than in Fig. 6.6a and the power would be greater than 0.61, as shown in Example 6.7b. This demonstrates the general principle that the farther a population characteristic is from that stated in the null hypothesis, the greater will be the power to reject the null hypothesis.

---

[*]Many authors attach an asterisk (*) to a test statistic if it is associated with a probability ≤ 0.05 and two asterisks (**) if the probability is ≤ 0.01.

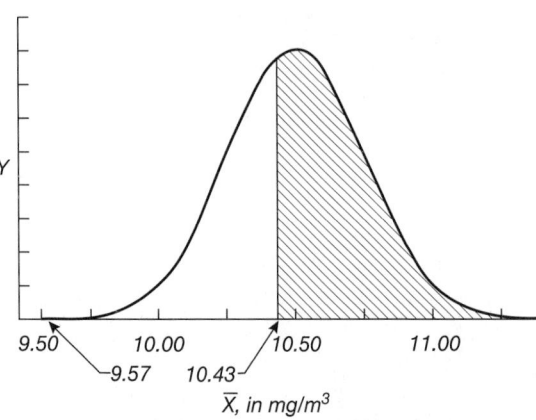

Y

9.50   10.00   10.50   11.00

-9.57   10.43

$\overline{X}$, in mg/m³

Figure 6.6a   A normal distribution of means with a mean of 10.50 mg/m³ and the same standard error as in Fig. 6.5. The shaded area (61% of the area under the curve) is the power of the test performed in Example 6.6b.

One-Tailed Testing.   As in Example 6.6, we have thus far considered the testing of null hypotheses declaring no difference between a population parameter and a hypothesized value, where the alternate hypothesis embodies difference in either of two directions from that value. There are instances, however, where the researcher's interest is in whether there is in the population a difference *in a specified direction* between a population parameter and a hypothesized value. Indeed, the data in Example 6.6 might have been collected with the express purpose of asking whether the mean CO concentration that sets off the alarm is *greater than* (rather than simply different from) 10.00 mg/m³. As the null hypothesis is to contain the concept of "no difference," the appropriate H_0 and H_A may be stated as

$$H_0: \mu \leq 10.00 \text{ mg/m}^3;$$

$$H_A: \mu > 10.00 \text{ mg/m}^3.$$

These hypotheses would be preferred to those in Example 6.6 if, for example, the measurement device is to be employed in environmental-safety monitoring and our concern is whether the carbon monoxide concentration *exceeds* a specified level (namely, 10.00 mg/m³), which is expressed in H_A. The above null hypothesis is rejected if the calculated Z is in the extreme α (e.g., 5%) of the *right-hand tail* (i.e., if Z is positive and $P(Z) \leq \alpha$). For the data in Example 6.6, this probability is 0.0367; and, as this is < 0.05, H_0 is rejected.

If, instead, our interest were specifically whether the mean of the sampled population is *less than* a hypothesized value, then the relevant statistical hypotheses would be

$$H_0: \mu \geq 10.00 \text{ mg/m}^3;$$

$$H_A: \mu < 10.00 \text{ mg/m}^3;$$

and H_0 would be rejected only if the calculated Z is in the extreme α (e.g., 5%) of the *left-hand tail* (i.e., if Z is *negative* and $P(Z) \leq \alpha$).

Statistical testing that examines differences in only one of two possible directions is called *one-tailed testing*. It may be seen in Appendix B.2 that, if one employs the

EXAMPLE 6.7 **Calculating the power of the hypothesis test of Example 6.6.**

H_0: $\mu = 10.00$ mg/m^3; H_A: $\mu \neq 10.00$ mg/m^3; $\sigma_{\bar{X}} = 0.24$ mg/m^3.

(a)   What is the power if the population mean is actually 10.50 mg/m^3?

$$\text{Power} = P(\bar{X} \leq 9.57 \text{ mg/m}^3 \text{ or } \bar{X} \geq 10.43 \text{ mg/m}^3)$$

For $P(\bar{X} \leq 9.57 \text{ mg/m}^3)$:

$$Z = \frac{9.57 \text{ mg/m}^3 - 10.50 \text{ mg/m}^3}{0.24 \text{ mg/m}^3} = -3.88$$

Consulting Table B.2:

$$P(Z \leq -3.88) = P(Z \geq 3.88) = 0.0001$$

For $P(\bar{X} \geq 10.43 \text{ mg/m}^3)$:

$$Z = \frac{10.43 \text{ mg/m}^3 - 10.50 \text{ mg/m}^3}{0.24 \text{ mg/m}^3} = -0.29$$

$$P(Z \geq -0.29) = 1 - P(Z \leq -0.29) = 1 - P(Z \geq 0.29) = 1 - 0.3859 = 0.6141$$

So, power $= 0.0001 + 0.6141 = 0.6142$.

(b)   What is the power if the population mean is actually 10.75 mg/m^3?

$$\text{Power} = P(\bar{X} \leq 9.57 \text{ mg/m}^3 \text{ or } \bar{X} \geq 10.43 \text{ mg/m}^3)$$

For $P(\bar{X} \leq 9.57 \text{ mg/m}^3)$:

$$Z = \frac{9.57 \text{ mg/m}^3 - 10.75 \text{ mg/m}^3}{0.24 \text{ mg/m}^3} = -4.92$$

$$P(Z \leq -4.92) = P(Z \geq 4.92) = 0.0000$$

For $P(\bar{X} \geq 10.43 \text{ mg/m}^3)$:

$$Z = \frac{10.43 \text{ mg/m}^3 - 10.75 \text{ mg/m}^3}{0.24 \text{ mg/m}^3} = -1.33$$

$$P(Z \geq -1.33) = 1 - P(Z \leq -1.33) = 1 - P(Z \geq 1.33) = 1 - 0.0918 = 0.9082$$

So, power $= 0.0000 + 0.9082 = 0.9082$.

5% level of significance, the one-tailed critical value of Z is 1.645. The one-tailed critical value (let's call it $Z_{\alpha(1)}$) is always less than the two-tailed critical value ($Z_{\alpha(2)}$); for example, for the 5% significance level $Z_{\alpha(1)} = 1.645$ and $Z_{\alpha(2)} = 1.960$. Thus, a researcher may be tempted to use one-tailed testing if he or she wishes to declare significance and the calculated Z lies between $Z_{\alpha(1)}$ and $Z_{\alpha(2)}$ (as in Example 6.6). But, this would not be a valid use of this testing procedure, for recall that *statistical hypotheses are to be declared before examining the data*, and they should reflect the question of interest about the population. Another example of one-tailed testing of a mean is found in Exercise 6.6, parts b and c.

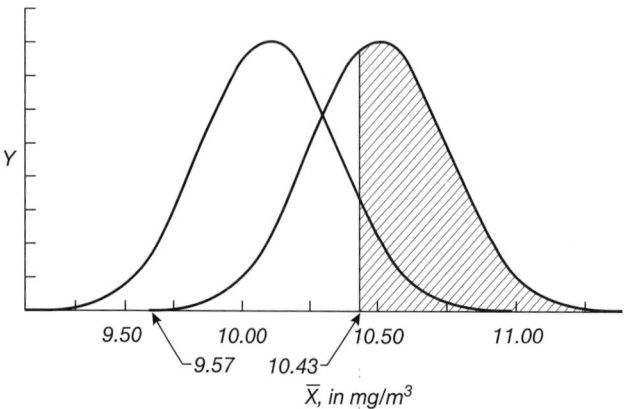

Figure 6.6b The normal distributions of Figures 6.5 and 6.6a.

6.5 ASSESSING DEPARTURES FROM NORMALITY

It is sometimes desired to test the hypothesis that a sample came from a population whose members follow a normal distribution. Example 6.1 and Fig. 6.7 present a frequency distribution of sample data, and we may wish to ask if the data came from a population that is normally distributed.

"Outliers." Occasionally a set of data will have one or more observations that are so extreme in value, relative to the other data in the sample, that we suspect they should not be a part of the sample. Such data are often called "outliers." Sometimes it is clear that an outlier is the result of incorrect recording of data. For example, if Example 6.1 contained a student height of 680 inches, or 7.2 inches, or 160 inches, we would immediately suspect an error. We might surmise that there was a decimal-point error in recording the first two (which might have actually been heights of 68 inches and 72 inches, respectively) and that the third height was measured in centimeters and mistakenly recorded as if it were inches. In other instances it is known that a measurement was faulty (e.g., a measuring instrument malfunctioned or a technician contaminated the item to be measured).

It is not appropriate to discard data simply because they appear (to someone) to be unreasonably extreme. However, there might be a very obvious reason such as the above for correcting or discarding a datum. If there are not such conspicuous grounds, then an outlier might be objectively rejected as erroneous by statistical methods that are referred to in the later discussions on comparing variables among populations (the end of Section 10.1) or analyzing the relationship between variables (the end of Section 16.2).

Goodness of Fit Testing of Normality.   Goodness of fit procedures for testing the null hypothesis that a sample came from a normal population are based on the methods to be discussed in Chapter 21. Methods such as chi-square, or Kolmogorov-Smironov goodness of fit procedures, with appropriate modifications, can be used to test the hypothesis of normality in the population distribution (e.g., Zar, 1984: 88–93). These

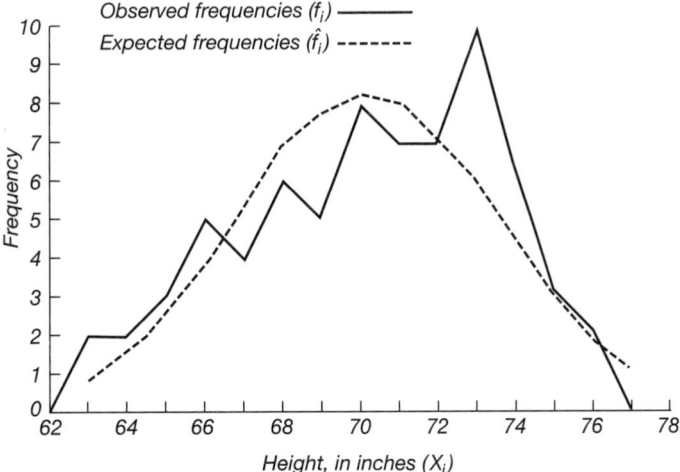

Figure 6.7   The frequency polygon for the student height data in Example 6.1 (solid line) with the frequency curve that would be expected if the data followed a normal distribution (broken line).

methods perform poorly, however, in that they possess very low power; and they are not recommended.

Graphical Assessment of Normality.   The graphical representation of a normal distribution as a frequency curve is presented in Fig. 6.1. Plotting a cumulative frequency distribution for normally distributed data results in an S-shaped, or sigmoid, curve, as shown in Fig. 6.8, using the data of Example 6.1. It is instructive to plot the cumulative frequency distribution sigmoid curve. (The data in Fig. 6.9 lean slightly toward platykurtosis.) A negatively skewed distribution will show an upward curve, as the lower portion of a sigmoid curve, and a positively skewed distribution will result in a shape resembling the upper portion of a sigmoid curve.

Assessing Normality Using Symmetry and Kurtosis Measures.   In a thorough review of many available methods, D'Agostino (1986) concluded that the most desirable procedure for testing hypotheses about normality is that described by D'Agostino and Pearson (1973). The null hypothesis of population normality is tested using the statistic

$$K^2 = Z_{g_1}^2 + Z_{g_2}^2, \tag{6.19}$$

where Z_{g_1} and Z_{g_2} are as given in Section 7.14. The significance of K^2 is determined by employing the statistical distribution recorded in Appendix Table B.1. This is the so-called chi-squared distribution, χ^2, about which much more will be said in later chapters. For this particular test, we consult Table B.1 for degrees of freedom (ν) of 2 (i.e., the second line in the body of the table), as shown in Example 6.8; and the probability that H_0 is a correct statement about the sampled population is indicated by

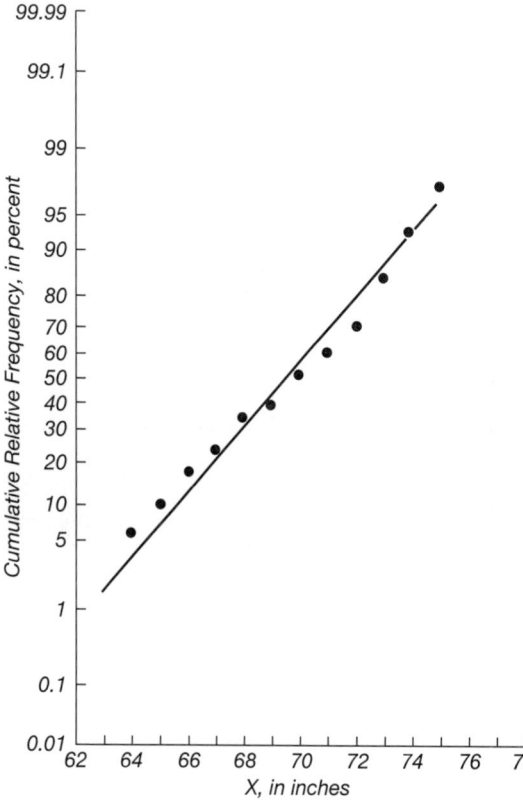

Figure 6.9 The cumulative relative frequency distribution for the data of Example 6.1, plotted with the normal probability scale as the ordinate. The expected frequencies (i.e., the frequencies from a normal distribution) would fall on the straight line shown.

the probabilities along the top margin of the table. In this example, $K^2 = 3.1396$, which for $\nu = 2$ lies between 2.773 and 4.605 which are in the columns headed 0.25 and 0.10. Thus, the probability of a K^2 at least as large as 3.1396 may be expressed as $0.10 < P < 0.25$. According to D'Agostino (personal communication), this test works well for $n \geq 20$.

If the null hypothesis is rejected, then it may be desired to employ the testing described in Section 7.14 to ask whether the non-normality is due to departure from symmetry or from mesokurtosis, or both.

Other Methods of Assessing Normality. Shapiro and Wilk (1965) presented a test for normality involving the calculation of a statistic they called W. The computation requires an extensive table of constants, however, because a different set of $n/2$ constants is required for each sample size, n. The authors provide a table of these constants and also a table of critical values of W, for n as large as 50. The power of W has been shown to be excellent (D'Agostino, 1986; Shapiro, Wilk, and Chen, 1968) when testing for departures from normality. However, the performance of this test is adversely affected in the common situation where there are tied data (i.e., data that are identical, such as occur in Example 6.1 where there are two observa-

---

EXAMPLE 6.8   The D'Agostino-Pearson K^2 test for normality applied to the data of Example 6.1.

H_0:   The sampled population is normally distributed.

H_A: The sampled population is not normally distributed.

From Example 7.12: $Z_{g_1} = -1.2291$
From Example 7.13: $Z_{g_2} = 1.2763$
Therefore,

$$K^2 = Z_{g_1}^2 + Z_{g_2}^2 = (-1.2291)^2 + (1.2763)^2$$

$$= 1.5107 + 1.6289 = 3.1396$$

Considering K^2 to be from a χ^2 distribution with $v = 2$, $0.10 < P < 0.25$ $[P = 0.21]$
Therefore, do not reject H_0.

---

tions of height 63 in., two of 64 in., and so on) (Pearson, D'Agostino, and Bowman, 1977).

The computation of W for normality testing, especially when $n > 50$, can be very cumbersome. An alternative procedure is that of D'Agostino (1971a, 1971b), which involves the computation of a statistic he calls D, applicable as a powerful test for departure from normality. In presenting this test, D'Agostino (1971a, 1971b, 1972) converted D to another quantity, Y, and provided critical values of Y. The performance of this test is not as good as that of the Shapiro and Wilk procedure, but it is not as affected by tied data.

D'Agostino (1986) explains why the D'Agostino-Pearson Test (Equation 6.19) is, on balance, preferable to either D or W.

EXERCISES

6.1 The following body weights were measured in thirty-seven animals:

Weight (X_i) (kg)	Frequency (f_i)
4.0	2
4.3	3
4.5	5
4.6	8
4.7	6
4.8	5
4.9	4
5.0	3
5.1	1

(a) Calculate the symmetry measure, g_1.
(b) Calculate the kurtosis measure, g_2
(c) Calculate the skewness measure based on quantiles.
(d) Calculate the kurtosis measure based on quantiles.

6.2 A normally distributed population of lemming body weights has a mean of 63.5 g and a standard deviation of 12.2 g.
(a) What proportion of this population is 78.0 g or larger?
(b) What proportion of this population is 78.0 g or smaller?
(c) If there are 1000 weights in the population, how many of them are 78.0 g or larger?
(d) What is the probability of choosing at random from this population a weight smaller than 41.0 g?

6.3 (a) Considering the population of Exercise 6.2, what is the probability of selecting at random a body weight between 60.0 and 70.0 g?
(b) What is the probability of a body weight between 50.0 and 60.0 g?

6.4 (a) What is the standard deviation of all possible means of samples of size ten which could be drawn from the population in Exercise 6.2?
(b) What is the probability of selecting at random from this population a sample of ten weights that has a mean greater than 65.0 g?
(c) What is the probability of the mean of a sample of ten being between 60.0 and 62.0 g?

6.5 Using the data of Exercise 6.1, apply the D'Agostino-Pearson K^2 test for normality.

6.6 Twenty women were given a pharmaceutical preparation for several weeks and each woman's change in body weight was recorded. It is found that the mean change in body weight was -1.1 kg (i.e., there was a mean weight loss of 1.1 kg). For purposes of this exercise, let us say that we know that the twenty data came from a population of weight changes that had a variance of 7.79 kg^2. State the appropriate null and alternative hypotheses, and perform the appropriate statistical test at $\alpha = 0.05$ for each of these three situations:
(a) The drug administered is intended to reduce migraine headaches. Is there a significant effect on body weight?
(b) The drug administered is intended to reduce body weight. Is it effective in doing so?
(c) The drug administered is intended to increase body weight. Is it effective in doing so?

7

ONE-SAMPLE HYPOTHESES

This chapter will continue the discussion of Section 6.4 on how to draw inferences about population parameters by testing hypotheses about them using appropriate sample estimates. It will consider hypotheses about each of several population parameters: population mean, median, variance (or standard deviation), and coefficient of variation. The chapter will also introduce procedures for expressing the confidence one can have in estimating parameters from sample statistics.

7.1 TWO-TAILED HYPOTHESES CONCERNING THE MEAN

Section 6.4 introduced the concept of statistical testing using a pair of statistical hypotheses, the null and alternate hypotheses, as statements that a population mean (μ) is equal to some specified value (let's call it μ_0):

H_0: $\mu = \mu_0$;

H_A: $\mu \neq \mu_0$.

For example, let us consider the body temperatures of twenty-five intertidal crabs that we exposed to air at 24.3°C (Example 7.1). We may wish to ask whether the mean body temperature of members of this species of crab is the same as the ambient air temperature of 24.3°C. Therefore,

H_0: $\mu = 24.3$°C, and

H_A: $\mu \neq 24.3$°C,

where the null hypothesis states that the mean of the population of data from which this sample of twenty-five came is 24.3°C (i.e., μ is "no different from 24.3°C"), and the alternate hypothesis is that the population mean is not equal to (i.e., μ is different from) 24.3°C.

EXAMPLE 7.1   The two-tailed t test for difference between a population mean and a hypothesized population mean.

Body temperatures (measured in °C) of twenty-five intertidal crabs placed in air at 24.3°C: 25.8, 24.6, 26.1, 22.9, 25.1, 27.3, 24.0, 24.5, 23.9, 26.2, 24.3, 24.6, 23.3, 25.5, 28.1, 24.8, 23.5, 26.3, 25.4, 25.5, 23.9, 27.0, 24.8, 22.9, 25.4.

H_0: $\mu = 24.3$°C

H_A: $\mu \neq 24.3$°C

$\alpha = 0.05$

$n = 25$

$\bar{X} = 25.03$°C

$s^2 = 1.80(°C)^2$

$$s_{\bar{X}} = \sqrt{\frac{1.80(°C)^2}{25}} = 0.27°C$$

$$t = \frac{\bar{X} - \mu}{s_{\bar{X}}} = \frac{25.03°C - 24.3°C}{0.27°C} = 2.704$$

$\nu = 24$

$t_{0.05(2),24} = 2.064$

As $|t| > t_{0.05(2),24}$, reject H_0 and conclude that the sample of twenty-five body temperatures came from a population whose mean is not 24.3°C.

$$0.01 < P < 0.02 \quad [P = 0.012]$$

In Section 6.2 (Equation 6.16), $Z = (\bar{X} - \mu)/\sigma_{\bar{X}}$ was introduced as a normal deviate, and it was shown how one can determine the probability of obtaining a sample with mean $\bar{X}$ from a population with a specified mean μ. And Section 6.4 discussed how the normal deviate can be used to test hypotheses about a population mean. Note, however, that the calculation of Z requires the knowledge of $\sigma_{\bar{X}}$, which we typically do not have. The best we can do is to calculate $s_{\bar{X}}$ as an estimate of $\sigma_{\bar{X}}$. If n is very, very large, then $s_{\bar{X}}$ is a good estimate of $\sigma_{\bar{X}}$, and we can calculate Z using this estimate. However, for most biological situations n is insufficiently large to consider $s_{\bar{X}}$ as an accurate estimate of $\sigma_{\bar{X}}$; but we can use, in place of the normal distribution (Z), a distribution known as t, the development of which was a major breakthrough in statistical methodology:*

$$t = \frac{\bar{X} - \mu}{s_{\bar{X}}}. \tag{7.1}$$

---

*The t statistic is also referred to as "Student's t." William Sealy Gosset (1876–1937), was an English statistician with the title "brewer" in the Guiness brewery of Dublin, who used the pseudonym "Student" to publish many noteworthy developments in statistical theory and practice, including ("Student," 1908) the presentation of the distribution that often bears his name. (See Irwin, 1978; Pearson, 1939; Pearson, Plackett, and Barnard, 1990.) Gosset referred to his distribution as z; and between 1922 and 1925 R. A. Fisher (e.g., 1925a, 1925b: 106–113, 117–125; 1928) helped develop its potential in statistical testing while modifying it; Gosset called the modification "t" (Eisenhart, 1979). Although Gosset was a modest man, he is referred to as "one of the most original minds in contemporary science" by R. A. Fisher (1939a), himself one of the most insightful and influential statisticians of all time.

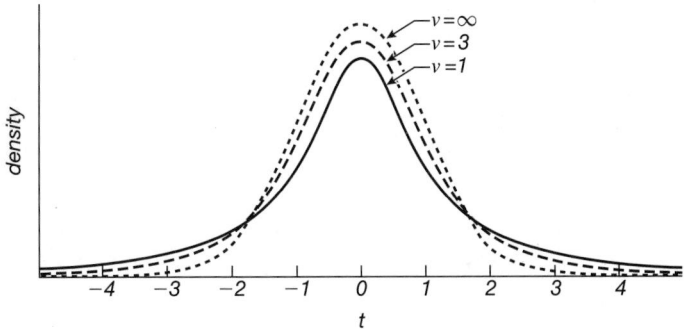

Figure 7.1   The t distribution for various degrees of freedom, ν. For $\nu = \infty$, the t distribution is identical to the normal distribution.

As do other distributions to be encountered among statistical methods, the t distribution has different shapes for different values of what is known as *degrees of freedom* (denoted by ν, the lowercase Greek nu).* For hypotheses concerning a mean,

$$\nu = n - 1. \tag{7.2}$$

Recall that n is the size of the sample (i.e., the number of data from which $\bar{X}$ has been calculated). The influence of ν on the shape of the t distribution is shown in Fig. 7.1. This distribution is leptokurtic (see Section 6.1), having a greater concentration of values around the mean and in the tails than does a normal distribution; but as n (and, therefore, ν) increases, the t distribution tends to resemble a normal distribution more closely, and for $\nu = \infty$ (i.e., for an infinitely large sample[†]), the t and normal distributions are identical; that is, $t_{\alpha, \infty} = Z_\alpha$.

The mean of the sample of twenty-five data shown in Example 7.1 is 25.03°C, and the sample variance is $1.80(°C)^2$. These statistics are estimates of the mean and variance of the population from which this sample came. However, this is only one of a very large number of samples of size twenty-five that could have been taken at random from the population. The distribution of the means of all possible samples with $n = 25$ is the t distribution for $\nu = 24$, which is represented by the curve of Fig. 7.2. In this figure, the mean of the t distribution (i.e., $t = 0$) represents the mean hypothesized in H_0; (i.e.,

---

*In early writings of the t distribution (during the 1920s and 1930s), the symbol n or f was used for degrees of freedom. This was often confusing because these letters had commonly been used to denote other quantities in statistics so Maurice G. Kendall (1943: 292) recommended ν.

[†]The modern symbol for infinity (∞) was introduced in 1655 by English mathematician John Wallis (1616–1703) (Cajori, 1929:44), who also introduced the notation of negative and fractional exponents, which his countryman, Sir Isaac Newton (1642–1727) wrote out in a form similar to modern usage (Cajori, 1928: 354–355):

$$X^{-a} = \frac{1}{X^a}; \quad X^{\frac{1}{a}} = \sqrt[a]{X}.$$

The notion of fractional powers was conceived by French writer Nicole Oresme (ca. 1323–1382), a bishop in Normandy (Cajori, 1928: 91).

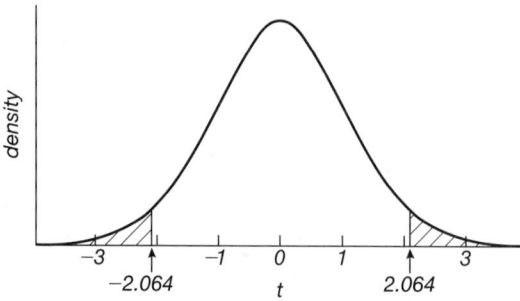

Figure 7.2 The t distribution for $\nu = 24$, showing the critical region (shaded area) for a two-tailed test using $\alpha = 0.05$. (The critical value of t is 2.064.)

$\mu = \mu_0 = 24.3°C$), for, by Equation 7.1, $t = 0$ when $\bar{X} = \mu$. The shaded areas in this figure represent the extreme 5% of the total area under the curve (2.5% in each tail). Thus, an $\bar{X}$ so far from μ that it lies in either of the shaded areas has a probability of less than 5% of occurring by chance alone, and we assume that it occurred because H_0 is, in fact, false. As explained in Section 6.4 regarding the Z distribution, because an extreme t value in either direction from μ will cause us to reject H_0, we are said to be considering a "two-tailed" (or "two-sided") test.

For $\nu = 24$, we can consult Appendix Table B.3 to find the following two-tailed probabilities (denoted as "$\alpha(2)$") of various values of t:

ν	$\alpha(2)$:	0.50	0.20	0.10	0.05	0.02	0.01
24		0.685	1.318	1.711	2.064	2.492	2.797

Thus, for example, for a two-tailed α of 0.05, the shaded areas of the curve begin at 2.064 t units on either side of μ. Therefore, we can state:

$$P(|t| \geq 2.064) = 0.05.$$

That is, 2.064 and -2.064 are the *critical values* of t; and if t (calculated from Equation 7.1) is equal to or greater than 2.064, or is equal to or less than -2.064, that will be considered reasonable cause to reject H_0 and consider H_A to be a true statement. That portion of the t distribution beyond the critical values (i.e., the shaded areas in the figure) is called the *critical region*.* For the sample of twenty-five body temperatures (see Example 7.1), $t = 2.704$. As 2.704 lies within the critical region (i.e., $2.704 > 2.064$), H_0 is rejected, and we conclude that the mean body temperature of crabs under the conditions of our experiment is not 24.3°C.

To summarize, the hypotheses for the two-tailed test are

$$H_0: \mu = \mu_0 \quad \text{and} \quad H_A: \mu \neq \mu_0,$$

where μ_0 denotes the hypothesized value to which we are comparing the population mean. (In the above example, $\mu_0 = 24.3°C$.) The test statistic is calculated by Equation 7.1, and if its absolute value is larger than the two-tailed critical value of t from Appendix Table B.3, we reject H_0 and assume H_A to be true. The critical value of t can be

---

*David (1995) traces the first use of this term to J. Neyman and E. S. Pearson in 1933.

abbreviated as $t_{\alpha(2), \nu}$, where $\alpha(2)$ refers to the two-tailed probability of α. Thus, for the preceding example, we could write $t_{0.02(2), 24} = 2.064$. In general, for a two-tailed t test,

$$\text{if } |t| \geq t_{\alpha(2), \nu}, \text{ then reject } H_0.$$

Example 7.1 presents the computations for the analysis of the crab data. A t of 2.704 is calculated, which for 24 degrees of freedom lies between the tabled critical values of $t_{0.02(2), 24} = 2.492$ and $t_{0.01(2), 24} = 2.797$. Therefore, if the null hypothesis, H_0, is a true statement about the population we sampled, the probability of $\bar{X}$ being at least this far from μ is between 0.01 and 0.02; that is $0.01 < P(|t| \geq 2.704) < 0.02$.* As this probability is less than 0.05, we reject H_0 and declare it is not a true statement. For a consideration of the types of errors involved in rejecting or accepting the null hypothesis, refer to Section 6.4.

Frequently, the hypothesized value in the null and alternate hypotheses is zero. For example, the weights of twelve rats might be measured before and after the animals are placed on a regimen of forced exercise for one week. The change in weight of the animals (i.e., weight after minus weight before) could be recorded, and it might have been found that the mean weight change was -0.65 g (i.e., the mean weight change is a 0.65 g weight loss). If we wished to infer whether such exercise causes any significant change in rat weight, we could state H_0: $\mu = 0$ and H_A: $\mu \neq 0$; Example 7.2 summarizes the t test for this H_0 and H_A. This test is two tailed, for a large $\bar{X} - \mu$ difference in either direction will constitute grounds for rejecting the veracity of H_0.[†]

The theoretical basis of the t testing utilized throughout this chapter assumes that sample data came from a normal population, assuring that the mean at hand came from a normal distribution of means. Fortunately, the t test is *robust*,[‡] meaning that its validity is not seriously affected by moderate deviations from this underlying assumption. The test also assumes—as other statistical tests typically do—that the data are a random sample (see Section 2.3).

A common situation in which one is dealing with a population known to be nonnormal is the case where the data are percentages or proportions. Such data are known to be binomial, rather than normal, and the treatment of such data is discussed in Section 13.3.

The effect of nonnormality is greater for smaller α, and the effect decreases as n increases (Ractliffe, 1968). For symmetric distributions there is little effect of departure from normality (i.e., from mesokurtosis); for asymmetric distributions the test performs best with strong leptokurtosis present and poorly with platykurtosis and mesokurtosis; and the adverse effect of non-normality is much less for two-tailed testing than for one-tailed (Section 7.2) testing (Cicchitelli, 1989).

---

*Some calculators and computer programs have the capability of determining the probability of a given t (e.g., sec. Boomsma and Molenaar, 1994; Guenther, 1977). For the present example, we would thereby find that $P(|t| \geq 2.704) = 0.012$.

[†]Data that result from the differences between pairs of data (such as measurements before and after an experimental treatment) are discussed further in Chapter 9.

[‡]The term "robustness" was introduced by Box in 1953 (David, 1995).

EXAMPLE 7.2   A two-tailed test for significant difference between a population mean and a hypothesized population mean of zero.

Weight change of twelve rats after being subjected to a regimen of forced exercise. Each weight change (in g) is the weight after exercise minus the weight before.

1.7	H_0: $\mu = 0$
0.7	H_A: $\mu \neq 0$
−0.4	
−1.8	
0.2	$\alpha = 0.05$
0.9	
−1.2	$n = 12$
−0.9	$\bar{X} = -0.65$ g
−1.8	$s^2 = 1.5682$ g^2
−1.4	
−1.8	$s_{\bar{X}} = \sqrt{\dfrac{1.5682 \text{ g}^2}{12}} = 0.36$ g
−2.0	

$$t = \frac{\bar{X} - \mu}{s_{\bar{X}}} = \frac{-0.65 \text{ g}}{0.36 \text{ g}} = -1.81$$

$$\nu = n - 1 = 11$$

$$t_{0.05(2), 11} = 2.201$$

Since $|t| < t_{0.05(2), 11}$, do not reject H_0.

$$0.05 < P < 0.10 \quad [P = 0.098]$$

It is important to appreciate that a sample used in statistical testing such as that discussed here must consist of truly replicated data. In Example 7.1, we desired to draw conclusions about a population of measurements representing a large number of animals (i.e., crabs). Therefore, the sample must consist of measurements (i.e., body temperatures) from n (i.e., twenty-five) animals; it would *not* be valid to obtain twenty-five body temperatures from a single animal. And, in Example 7.2, twelve individual rats must be used; it would *not* be valid to employ data obtained from subjecting the same animal to the experiment twelve times. Such invalid attempts at replication are discussed by Hurlbert (1984), who calls them *pseudoreplication*.

7.2 One-Tailed Hypotheses Concerning the Mean

In Section 7.1, we spoke of the hypotheses H_0: $\mu = \mu_0$ and H_A: $\mu \neq \mu_0$, because we were willing to consider a large deviation of $\bar{X}$ in either direction from μ_0 as grounds for rejecting H_0. However, in many instances, our interest lies only in whether $\bar{X}$ is significantly larger (or significantly smaller) than μ_0, and this is termed a "one-tailed" (or "one-sided") test situation. For example, one might be testing a drug hypothesized to cause weight reduction in humans. The investigator is interested only in whether a weight *loss* occurs after the drug is taken. (In Example 7.2, using a two-sided test, we

---

EXAMPLE 7.3   A one-tailed t test for the hypotheses H_0: $\mu \geq 0$ and H_A: $\mu < 0$.

The data are weight changes of humans, tabulated after administration of a drug proposed to result in weight loss. Each weight change (in kg) is the weight after minus the weight before drug administration.

0.2	$n = 12$
−0.5	$\bar{X} = -0.61$ kg
−1.3	$s^2 = 0.4008$ kg^2
−1.6	
−0.7	
0.4	$s_{\bar{X}} = \sqrt{\dfrac{0.4008 \text{ kg}^2}{12}} = 0.18$ kg
−0.1	
0.0	$t = \dfrac{\bar{X} - \mu}{s_{\bar{X}}} = \dfrac{-0.61 \text{ kg}}{0.18 \text{ kg}} = -3.389$
−0.6	
−1.1	$\nu = n - 1 = 11$
−1.2	
−0.8	$t_{0.05(1), 11} = 1.796$.

If $t \leq -t_{0.05(1), 11}$, reject H_0.

Conclusion: reject H_0.

$0.0025 < P(t \leq -3.389) < 0.005$ $[P = 0.0030]$

---

were interested in determining whether either weight loss or weight gain had occurred.) If there is either weight gain or no weight change, the drug will be considered a failure. Therefore, for this one-sided test, we should state H_0: $\mu \geq 0$ and H_A: $\mu < 0$. Here, the null hypothesis states that there is no mean weight loss (i.e., the mean weight change is greater than or equal to zero), and the alternate hypothesis states that there is a mean weight loss (i.e., the mean weight change is less than zero). By examining the alternate hypothesis, H_A, we see that H_0 will be rejected if t is in the left-hand critical region of the t distribution. In general,

$$\text{for } H_A: \mu < \mu_0,$$

$$\text{if } t \leq -t_{\alpha(1), \nu}, \text{ then reject } H_0.^*$$

Example 7.3, summarizes such a set of twelve weight change data tested against this pair of hypotheses. From Appendix Table B.3 we find that $t_{0.05(1), 11} = 1.796$, and the critical region for this test is shown in Fig. 7.3. From this figure, and by examining Appendix Table B.3, we see that $t_{\alpha(1), \nu} = t_{2\alpha(2), \nu}$ or $t_{\alpha(2), \nu} = t_{\alpha/2(1), \nu}$; that is, for example, the critical value of t for one-sided test at $\alpha = 0.05$ is the same as the critical value of t for a two-sided test at $\alpha = 0.10$.

If we are interested in whether $\bar{X}$ is significantly *greater* than some value, μ_0, the hypotheses for the one-tailed test are H_0: $\mu \leq \mu_0$ and H_A: $\mu > \mu_0$. For example, a drug manufacturer might advertise that a product dissolves completely in gastric juice

---

*For one-tailed testing of this H_0, probabilities of t up to 0.25 are indicated in Appendix Table B.3. If $t = 0$, then $P = 0.50$; so if $-t_{0.25(1), \nu} < t < 0$, then $0.25 < P < 0.50$; and if $t > 0$, then $P > 0.50$.

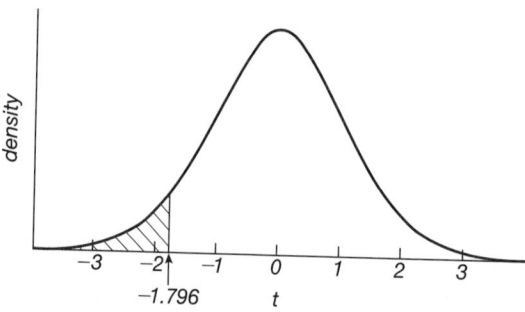

Figure 7.3 The distribution of t for $\nu = 11$, showing the critical region (shaded area) for a one-tailed test using $\alpha = 0.05$. (The critical value of t is -1.796.)

within 45 sec. The hypotheses appropriate for testing this claim are H_0: $\mu \leq 45$ sec and H_A: $\mu > 45$ sec, because we are not particularly interested in the possibility that the product dissolves faster than is claimed, but we wish to determine whether its dissolving time is longer than advertised. Thus, the rejection region would be in the right-hand tail, rather than in the left-hand tail (the latter being the case in Example 7.3). The details of such a test are shown in Example 7.4. In general,

$$\text{for } H_A\text{: } \mu > \mu_0,$$

$$\text{if } t \geq t_{\alpha(1),\nu}, \text{ then reject } H_0.^*$$

EXAMPLE 7.4   The one-tailed t test for the hypotheses H_0: $\mu \leq 45$ sec and H_A: $\mu > 45$ sec.
Dissolving times (in sec) of a drug in gastric juice: 42.7, 43.4, 44.6, 45.1, 45.6, 45.9, 46.8, 47.6.

H_0: $\mu \leq 45$ sec	$s_{\bar{X}} = 0.58$ sec
H_A: $\mu > 45$ sec	$t = \dfrac{45.21 \text{ sec} - 45 \text{ sec}}{0.58 \text{ sec}} = 0.36$
$\alpha = 0.05$	$\nu = 7$
$n = 8$	$t_{0.05(1),7} = 1.895$
$\bar{X} = 45.21$ sec	If $t \geq t_{0.05(1),7}$, reject H_0.
SS $= 18.8288$ sec^2	Conclusion: do not reject H_0.
$s^2 = 2.6898$ sec^2	

$$P(t \geq 0.36) > 0.25 \quad [P = 0.36]$$

7.3 Confidence Limits for the Population Mean

We learned in Section 7.1 that 5% of all possible sample means from a population with mean μ will yield t values—where $t = (\bar{X} - \mu)/s_{\bar{X}}$—that are either larger than $t_{0.05(2),\nu}$, or smaller than $-t_{0.05(2),\nu}$ (i.e., $|t| > t_{0.05(2),\nu}$). This means that 95% of all t values

---
*For this H_0, if $t = 0$, then $P = 0.50$; therefore, if $0 < t < t_{0.25(1),\nu}$, then $0.25 < P < 0.50$, and if $t < 0$, then $P > 0.50$.

obtainable lie between the limits of $-t_{0.05(2), \nu}$ and $t_{0.05(2), \nu}$; that is,

$$P\left[-t_{0.05(2), \nu} \le \frac{\bar{X} - \mu}{s_{\bar{X}}} \le t_{0.05(2), \nu}\right] = 0.95. \tag{7.3}$$

It follows from this that

$$P[\bar{X} - t_{0.05(2), \nu}\, s_{\bar{X}} \le \mu \le \bar{X} + t_{0.05(2), \nu}\, s_{\bar{X}}] = 0.95. \tag{7.4}$$

Equation 7.4 is read as "the probability that the interval between $\bar{X} - t_{0.05(2), \nu}\, s_{\bar{X}}$ and $\bar{X} + t_{0.05(2), \nu}\, s_{\bar{X}}$ includes μ is 0.95." This interval is called the *confidence interval* (abbreviated CI) for μ. What it tells us is that we can, knowing $\bar{X}$, $s_{\bar{X}}$, and ν, say that we are 95% confident that the stated interval encompasses μ. In general, the confidence interval for μ can be stated as

$$P[\bar{X} - t_{\alpha(2), \nu}\, s_{\bar{X}} \le \mu \le \bar{X} + t_{\alpha(2), \nu}\, s_{\bar{X}}] = 1 - \alpha. \tag{7.5}$$

The quantity $\bar{X} - t_{\alpha(2), \nu}\, s_{\bar{X}}$ is called the *lower confidence limit* (abbreviated L_1); and $\bar{X} + t_{\alpha(2), \nu}\, s_{\bar{X}}$ is the *upper confidence limit* (abbreviated L_2); therefore, the two confidence limits can be stated concisely as

$$\bar{X} \pm t_{\alpha(2), \nu}\, s_{\bar{X}} \tag{7.6}$$

(reading "$\pm$" to be "plus or minus"[*]). When referring to a confidence interval, we call the quantity $1 - \alpha$ (namely, $1 - 0.05 = 0.95$ in the present example) the *confidence level* (or the *confidence coefficient*, or the *coverage probability*).[†]

Although $\bar{X}$ is the best estimate of μ, it is still only an estimate, and the calculation of the confidence interval for μ allows us to express the precision of the estimate. Example 7.5a refers to the data of Example 7.1 and demonstrates the determination of the 95% confidence limits for the mean of the population from which the sample came. As the 95% confidence limits are computed to be 24.27°C and 25.59°C, the 95% confidence interval may be expressed as $P(24.27°C \le \mu \le 25.59°C) = 95\%$, meaning that we are 95% confident that the two temperatures, 24.27°C and 25.59°C, encompass the population mean. This also implies that if all possible samples of size n (i.e., twenty-five) were taken from population, and a 95% confidence interval were calculated from each sample, 95% of these intervals would contain μ.

Note that the smaller $s_{\bar{X}}$ is, the smaller will be the confidence interval, meaning that we estimate μ more precisely when $s_{\bar{X}}$ is small. Also, it is obvious from the calculation of $s_{\bar{X}}$ (see Equation 6.18) that a large n will result in a small $s_{\bar{X}}$. In general, a parameter estimate from a large sample is more precise than an estimate of the same parameter from a small sample.

---

[*]The symbol "$\pm$" was first used by British mathematician William Oughtred in 1631 (Cajori, 1928: 245).

[†]We owe the development of confidence intervals to Jerzy Neyman, beginning in 1934 (Peters, 1987: 189–190), although the concept had been enunciated a hundred years before. Neyman introduced the terms "confidence interval" and "confidence coefficient" in 1934 (David, 1995). On rare occasion, the biologist may see reference to "fiducial intervals," a concept developed by R. A. Fisher beginning in 1930 and identical to confidence intervals in many, but not all, situations (Pfanzagl, 1978).

EXAMPLE 7.5   Computation of confidence intervals and confidence limits for the mean, using the data of Example 7.1.

(a) At the 95% confidence level:

$\bar{X} = 25.03°C$

$s_{\bar{X}} = 0.27°C$

$t_{0.05(2),24} = 2.064$

$\nu = 24$

95% confidence interval $= \bar{X} \pm t_{0.05(2),24}\, s_{\bar{X}}$

$$= 25.03°C \pm (2.064)(0.27°C)$$

$$= 25.03°C \pm 0.56°C$$

95% confidence limits: $L_1 = 25.03°C - 0.56°C = 24.47°C$

$$L_2 = 25.03°C + 0.56°C = 25.59°C$$

(b) At the 99% confidence level:

$t_{0.01(2),24} = 2.797$

99% confidence interval $= \bar{X} \pm t_{0.01(2),24}\, s_{\bar{X}}$

$$= 25.03° \pm (2.797)(0.27°C)$$

$$= 25.03°C \pm 0.76°C$$

99% confidence limits: $L_1 = 25.03°C - 0.76°C = 24.27°C$

$$L_2 = 25.03°C + 0.76°C = 25.79°C$$

If, instead of a 95% confidence interval, we wished to state an interval having a 99% probability of containing μ, then $t_{0.01(2),24}$ would have been used. As $t_{0.01(2),24} = 2.797$, the confidence limits would be calculated as shown in Example 7.5b, and we could state $P(24.27°C \leq \mu \leq 25.79°C) = 99\%$. Note that if we increase the level of confidence we concomitantly increase the width of the confidence interval, evincing the trade-off between confidence and utility. Indeed, if we increase the confidence to 100%, then the confidence interval would be $-\infty < \mu < \infty$, and we would have great confidence in a statement that is useless. Note also that it is the two-tailed t value that is utilized for confidence interval computation, as we are setting limits on both sides of μ.

7.4 REPORTING VARIABILITY ABOUT THE MEAN

It is very important to provide the reader of a research paper with some information concerning the variability of the data reported. But authors of such papers are often unsure of appropriate ways of doing so, and not infrequently do so improperly.

If we wish to describe the population that has been sampled, then the sample mean ($\bar{X}$) and the standard deviation (s) may be reported. The range might also be reported,

TABLE 7.1   Tail Lengths (in mm) of Field Mice from Different Localities.

Location	n	$\bar{X} \pm \text{SD}$ (range in parentheses)
Bedford, Indiana	18	56.22 ± 1.33 (44.8 to 68.9)
Rochester, Minnesota	12	59.61 ± 0.82 (43.9 to 69.8)
Fairfield, Iowa	16	60.20 ± 0.92 (52.4 to 69.2)
Pratt, Kansas	16	53.93 ± 1.24 (46.1 to 63.6)
Mount Pleasant, Michigan	13	55.85 ± 0.90 (46.7 to 64.8)

TABLE 7.2   Evaporative Water Loss of a Small Mammal at Various Air Temperatures. Sample Statistics Are Mean ± Standard Deviation, with Range in Parentheses.

	Air Temperature (°C)				
	16.2	24.8	30.7	36.8	40.9
Sample size	10	13	10	8	9
Evaporative water loss (mg/g/hr)	0.611 ± 0.164 (0.49 to 0.88)	0.643 ± 0.194 (0.38 to 1.13)	0.890 ± 0.212 (0.64 to 1.39)	1.981 ± 0.230 (1.50 to 2.36)	3.762 ± 0.641 (3.16 to 5.35)

but in general it should not be stated without being accompanied by another measure of variability, such as s. Such statistics are frequently presented as in Table 7.1 or 7.2.

If it is the author's intention to provide the reader with a statement about the precision of estimation of the population mean, the use of the standard error ($s_{\bar{X}}$) is appropriate. A typical presentation is shown in Table 7.3a. This table might instead be set up to show confidence intervals, rather than standard errors, as shown in Table 7.3b. The standard error is always smaller than the standard deviation. But this is not a reason to report the former in preference to the latter. The determination should be made on the basis of whether the desire is to describe variability within the population or precision of estimating the population mean.

There are three very important points to note about Tables 7.1, 7.2, and 7.3a+b. First, n should be stated somewhere in the table, either in the caption or in the body of the table. (Thus, the reader has the needed information to convert from SD to SE or from SE to SD, if so desired.) One should always state n when presenting sample statistics ($\bar{X}, s, s_{\bar{X}}$, range, etc.), and if a tabular presentation is prepared it is very good practice to include n somewhere in the table, even if it is mentioned elsewhere in the paper.

Second, the measure of variability is clearly indicated. Not infrequently, an author will state something such as "the mean is 54.2 ± 2.7 g," with no explanation of what "± 2.7" denotes. This renders the statement worthless to the reader, because "± 2.7"

TABLE 7.3a   Enzyme Activities in the Muscle of Various Animals. Data Are $\bar{X} \pm$ SE, with n in Parentheses.

Animal	Enzyme Activity (μmole/min/g of tissue)	
	Isomerase	Transketolase
Mouse	0.76 ± 0.09 (4)	0.39 ± 0.04 (4)
Frog	1.53 ± 0.08 (4)	0.18 ± 0.02 (4)
Trout	1.06 ± 0.12 (4)	0.24 ± 0.04 (4)
Crayfish	4.22 ± 0.30 (4)	0.26 ± 0.05 (4)

TABLE 7.3b   Enzyme Activities in the Muscle of Various Animals. Data are $\bar{X} \pm$ 95% Confidence Limits.

Animal	n	Enzyme Activity (μmole/min/g of tissue)	
		Isomerase	Transketolase
Mouse	4	0.76 ± 0.28	0.39 ± 0.13
Frog	4	1.53 ± 0.25	0.18 ± 0.05
Trout	4	1.06 ± 0.38	0.24 ± 0.11
Crayfish	4	4.22 ± 0.98	0.26 ± 0.15

will be assumed by some to indicate $\pm$ SD, by others to indicate $\pm$ SE, by others to indicate the 95% (or 99%, or other) confidence interval, and by others to indicate the range.* There is no widely accepted convention; one *must* state explicitly what quantity is meant by this type of statement. If such statements of "$\pm$" values appear in a table, then the explanation is best included somewhere in the table (either in the caption or in the body of the table), even if it is stated elsewhere in the paper.

Third, the units of measurement of the variable must be clear. There is little information conveyed by stating that the tail lengths of twenty-four birds have a mean of 8.42 and a standard error of 0.86, if the reader does not know whether the tail lengths were measured in centimeters, or inches, or some other unit. Whenever data appear in tables, the units of measurement should be stated somewhere in the table. Keep in mind that a table should be self-explanatory; one should not have to refer back and forth between the table and the text to determine what the tabled values represent.

Frequently, the types of information given in Tables 7.1, 7.2, and 7.3a+b are presented in graphs, rather than in tables. In such cases, the measurement scale is typically indicated on the vertical axis, and the mean is indicated in the body of the

---

*In older literature the $\pm$ symbol referred to yet another measure, known as the "probable error" (which has fallen into disuse). In a normal curve, the probable error (PE) is 0.6745 times the standard error, so that $\bar{X} \pm$ PE includes 50% of the distribution. The term "probable error" was first used, in 1815, by German astronomer Friedrich Wilhelm Bessel (1784–1846) (Walker, 1929: 24, 51, 186).

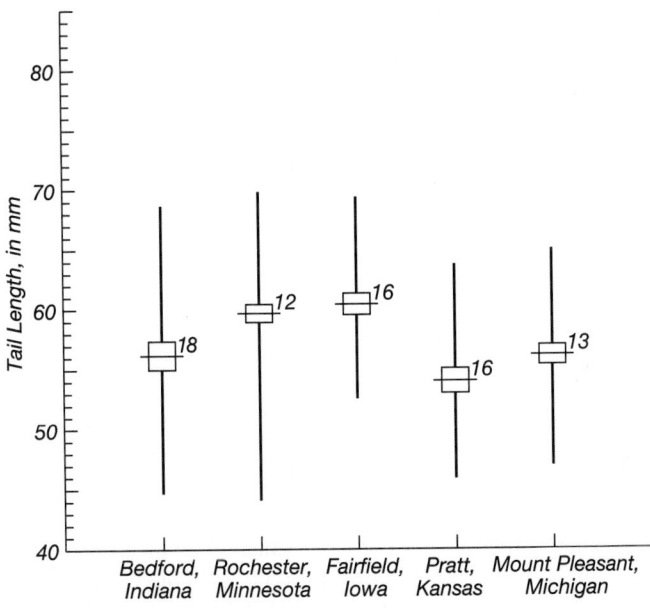

Figure 7.4 Tail lengths of male field mice from different localities, indicating the mean, the mean ± standard deviation (vertical rectangle), and the range (vertical line), with the sample size indicated for each location. The date are from Table 7.1.

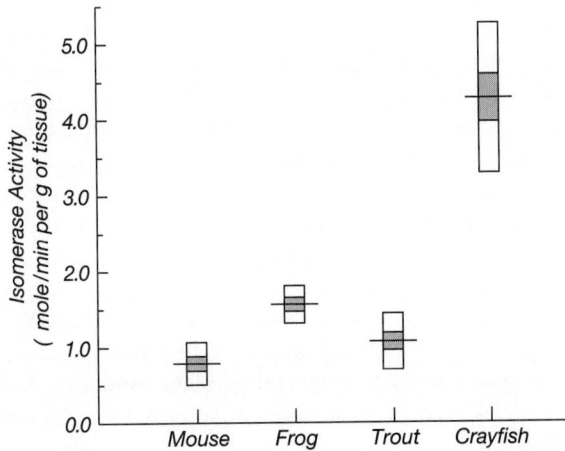

Figure 7.5 Levels of muscle isomerase in various animals. Shown is the mean ± standard error (shaded rectangle), and ± the 95% confidence interval (open rectangle). For each sample, $n = 4$. The data are from Tables 7.3a+b.

graph by a short horizontal line or some other symbol. The standard deviation, standard error, or a confidence interval for the mean is commonly indicated on such graphs via a vertical line or rectangle. Often the range is also included, and in such instances the SD or SE may be indicated by a vertical rectangle and the range by a vertical line. Some authors will indicate a confidence interval (generally 95%) in addition to the range and either SD or SE. Figures 7.4, 7.5, and 7.6 demonstrate how various combinations of

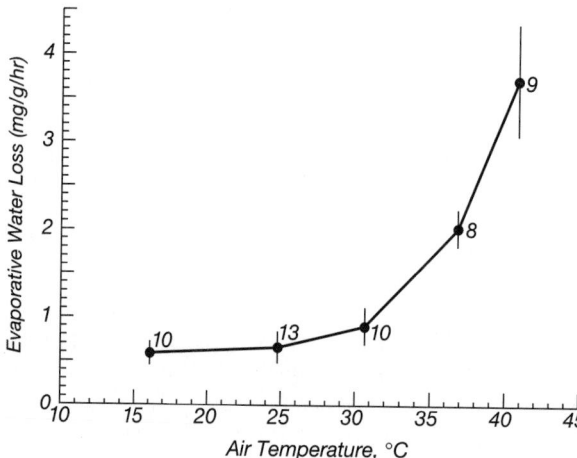

Figure 7.6 Evaporative water loss of a small mammal at various air temperatures. Shown at each temperature is the mean ± the standard deviation, and the sample size. The data are from Table 7.2.

these statistics may be presented graphically.* Note that when the horizontal axis on the graph represents an interval or ratio scale variable (as in Fig. 7.6), adjacent means may be connected by straight lines to aid in the recognition of trends.

In graphical presentation of data, as in tabular presentation, care must be taken to indicate clearly the following either on the graph or in the caption: the sample size (n), the units of measurement, and what measures of variability (if any) are indicated (e.g., SD, SE, range, 95% confidence interval).

Some authors present $\bar{X} \pm 2s_{\bar{X}}$ in their graphs. An examination of the t table (Appendix Table B.3) will show that, except for small samples, this expression will approximate the 95% confidence interval for the mean. But for small samples, the true confidence interval is, in fact, greater than $\bar{X} \pm 2s_{\bar{X}}$. Thus, the general use of this expression is not to be encouraged, and the calculation of the accurate confidence interval is the wiser practice.

A word of caution is in order for those who determine confidence limits, or SDs or SEs, for two or more means and, by observing whether or not the limits overlap, attempt to determine whether there are differences among the population means. Such a procedure is not generally valid (see Andrews, Snee, and Sarner, 1980; Browne, 1979; Simpson, Roe, and Lewontin, 1960: 350–354). The proper methods for testing for differences between means are discussed in the next several chapters. Andrews, Snee and Sarner (1980) explain how graphical display of means can be used validly to demonstrate significant differences among means.

---

*Instead of the mean and a measure of variability based on the variance, one may present tabular or graphical descriptions of samples using the median and quartiles (e.g., McGill, Tukey, and Larsen, 1978), or the median and its confidence interval. Thus, a graphical presentation such as in Fig. 7.4 could have the range indicated by the vertical line, the median by the horizontal line, and the semiquartile range (Section 4.2) by the vertical rectangle.

7.5 SAMPLE SIZE AND ESTIMATION OF THE POPULATION MEAN

A commonly asked question is, "How large a sample must be taken to achieve a desired precision* in estimating the mean of a population?" The answer is obviously related to the concept of a confidence interval; for a confidence interval expresses the precision of a sample statistic, and the precision increases (i.e., the confidence interval becomes narrower) as the sample size increases.

Let us write Equation 7.6 as $\bar{X} \pm d$, which is to say that $d = t_{\alpha(2), \nu} s_{\bar{X}}$. We shall refer to d as the half-width of the confidence interval, which means that μ is estimated to within $\pm d$. Now, the number of data we must collect to calculate a confidence interval of specified width depends upon: (1) the width desired (for a narrower confidence interval— i.e., more precision in estimating μ—requires a larger sample; (2) the variability in the population (which is estimated by s^2, and larger variability requires larger sample size); and (3) the confidence level specified (for greater confidence—e.g., 99% vs. 95%— requires a larger sample size).

If we have a sample estimate (s^2) of the variance of a normal population, then we can estimate the required sample size as

$$n = \frac{s^2 t^2_{\alpha(2), (n-1)}}{d^2}. \tag{7.7}$$

In this equation, s^2 is the sample variance, estimated with $\nu = n - 1$ degrees of freedom, d is the half-width of the desired confidence interval, and $1 - \alpha$ is the confidence level for the confidence interval. Two-tailed critical values of Student's t, with $\nu = n - 1$ degrees of freedom, are found in Appendix Table B.3.

There is a basic difficulty in solving Equation 7.7, however; the value of $t_{\alpha(2), (n-1)}$ depends upon n, the unknown sample size. The solution may be achieved by iteration—a process of trial and error with progressively more accurate approximations—as shown in Example 7.6. We begin the iterative process of estimation with an initial guess; the closer this initial guess is to the finally determined n, the faster we shall arrive at the final estimate. Fortunately, the procedure works well even if this initial guess is far from the final n (although the process is faster if it is a high, rather than a low, guess).

The reliability of this estimate of n depends upon the accuracy of s^2 as an estimate of the population variance, σ^2. As its accuracy improves with larger samples, one should use s^2 obtained from a sample with a size that is not a very small fraction of the n calculated from Equation 7.7.

7.6 POWER AND SAMPLE SIZE IN TESTS CONCERNING THE MEAN

Sample Size Required.   If we are to perform a one-sample test as described in Section 7.1 or 7.2, then it is desirable to know how many data should be collected to

---

*Recall from Section 2.4 that the precision of a sample statistic is the closeness with which it estimates the population parameter; it is not to be confused with the concept of the precision of a measurement (defined in Section 1.2), which is the nearness of repeated measurements to each other.

EXAMPLE 7.6   Determination of sample size needed to achieve a stated precision in estimating a population mean, using the data of Example 7.3.

If we specify that we wish to estimate μ with a 95% confidence interval no wider than 0.5 kg, then $d = 0.25$ kg, $1 - \alpha = 0.95$, and $\alpha = 0.05$. From Example 7.3 we have an estimate of the population variance: $s^2 = 0.4008$ kg^2.

Let us guess that a sample of 40 is necessary; then,

$$t_{0.05(2),39} = 2.023$$

So we estimate (by Equation 7.7):

$$n = \frac{(0.4008)(2.023)^2}{(0.25)^2} = 26.2.$$

Next, we might estimate $n = 27$, for which $t_{0.05(2),26} = 2.056$ and we calculate

$$n = \frac{(0.4008)(2.056)^2}{(0.25)^2} = 27.1.$$

Therefore, we conclude that a sample size greater than 27 is required to achieve the specified confidence interval.

---

detect a specified difference with a specified power. We can estimate the required sample size, n, if we have an estimate, s^2, of the population variance, σ^2.

We may specify that we wish to perform a t test with a probability of α of committing a Type I error, and a probability of β of committing a Type II error; and we can state that we want to be able to detect a difference between μ and μ_0 as small as δ (where μ is the actual population mean and μ_0 is the mean specified in the null hypothesis).* To test at the α significance level with $1 - \beta$ power, the minimum sample size required to detect δ is

$$n = \frac{s^2}{\delta^2}(t_{\alpha,\nu} + t_{\beta(1),\nu})^2, \tag{7.8}$$

where α can be either $\alpha(1)$ or $\alpha(2)$, respectively, depending on whether a one-tailed or two-tailed test is to be used. However, ν depends on n, so n cannot be calculated directly but must be obtained by iteration† (i.e., by a series of estimations, each estimation coming closer to the answer than that preceding). This is demonstrated in Example 7.7.

Equation 7.8 provides better estimates of n when s^2 is a good estimate of the population variance, σ^2, and the latter estimate improves when s^2 is calculated from larger samples. Therefore, it is most desirable that s^2 be obtained from a sample with a size that is not a small fraction of the estimate of n.

Minimum Detectable Difference.   By rearranging Equation 7.8, we can ask how small a δ (the difference between μ and μ_0) can be detected by the t test with $1 - \beta$

---

*δ is lower-case Greek delta.

†If the population variance, σ^2, were actually known, rather than estimated by s^2, then Z_α would be substituted for t_α in this and the other computations in this section, and n would be determined in one step instead of iteratively.

EXAMPLE 7.7   Estimation of required sample size to test H_0: $\mu = \mu_0$.

How large a sample is needed to reject the null hypothesis of Example 7.2, when sampling from the population in that example? We wish to test at the 0.05 level of significance with a 90% chance of detecting a population mean different from $\mu_0 = 0$ by as little as 1.0 g. In Example 7.2, $s^2 = 1.5682$ g^2.

Let us guess that a sample size of twenty would be required. Then, $v = 19$, $t_{0.05(2),19} = 2.093$, $\beta = 1 - 0.90 = 0.10$, $t_{0.10(1),19} = 1.328$, and we use Equation 7.8 to calculate

$$n = \frac{1.5682}{(1.0)^2}(2.093 + 1.328)^2 = 18.4$$

We now use $n = 19$ as an estimate, in which case $v = 18$, $t_{0.05(2),18} = 2.101$, $t_{0.10(1),18} = 1.330$, and

$$n = \frac{1.5682}{(1.0)^2}(2.101 + 1.330)^2 = 18.5$$

Thus, we conclude that a sample size of at least nineteen is required.

---

power, at the α level of significance, using a sample of specified size n:

$$\delta = \sqrt{\frac{s^2}{n}}\left(t_{\alpha,v} + t_{\beta(1),v}\right), \tag{7.9}$$

where $t_{\alpha,v}$, can be either $t_{\alpha(1),v}$ or $t_{\alpha(2),v}$, depending on whether a one-tailed or two-tailed test is to be performed. The estimation of δ is demonstrated in Example 7.8.

EXAMPLE 7.8   Estimation of minimum detectable difference in a one-sample t test.

In the two-tailed test of Example 7.2, what is the smallest difference (i.e., difference between μ and μ_0) that is detectable 90% of the time using a sample of twenty-five data and a significance level of 0.05?

Using Equation 7.9:

$$\delta = \sqrt{\frac{1.5682}{25}}(t_{0.05(2),24} + t_{0.10(1),24})$$

$$= (0.25)(2.064 + 1.318)$$

$$= 0.85 \text{ g}$$

Power of One-Sample Testing.   If our desire is to express the probability of correctly rejecting a false H_0, then we seek to estimate the power of the t test. Equation 7.8 can be rearranged to give

$$t_{\beta(1),v} = \frac{\delta}{\sqrt{\dfrac{s^2}{n}}} - t_{\alpha,v}. \tag{7.10}$$

As shown in Example 7.9, for a stipulated δ, α, σ^2, and sample size, we can express $t_{\beta(1),v}$. Consulting Appendix Table B.3 allows us to convert $t_{\beta(1),v}$, to β, but only roughly

EXAMPLE 7.9   Estimation of power of a one-sample t test.

What is the probability of detecting a true difference (i.e., a difference between μ and μ_0) of at least 1.0 g for the experiment of Example 7.2?

For $n = 12$, $\nu = 11$, $t_{0.05(2),11} = 2.201$, and $s^2 = 1.5682$ g^2, and we use Equation 7.10 to find

$$t_{\beta(1),11} = \frac{1.0}{\sqrt{\dfrac{1.5682}{12}}} - 2.201$$

$$= 0.57.$$

Consulting Appendix Table B.3 tells us that, for $\nu = 11$, $\beta > 0.25$, so we can say the power is $1 - \beta < 0.75$. By considering 0.57 to be a normal deviate and consulting Appendix Table B.2, we conclude $\beta = 0.28$ and that the power of the test $1 - \beta = 0.72$. [The exact probabilities, by computer, are $\beta = 0.29$ and power $= 0.71$.]

(e.g., $\beta > 0.25$ in Example 7.9). However, $t_{\beta(1),\nu}$ may be considered to be approximated by $Z_{\beta(1)}$, so Appendix Table B.2 may be used to determine β.* Then, the power of the test is $1 - \beta$, as shown in Example 7.9.

When the concept of power was introduced in the discussion of "Statistical Errors in Hypothesis Testing" in Section 6.4, it was stated that, for a given sample size (n), α is inversely related to β; that is, the lower the probability of committing a Type I error, the greater the probability of committing a Type II error. It was also noted that α and β can be lowered simultaneously by increasing n. Power is also greater for one-tailed than for two-tailed tests, but recall (from the end of Section 6.4, and from Section 7.2) that power is *not* the criterion for performing a one-tailed instead of a two-tailed test. These relationships are shown in Table 7.4. Power is also greater when s^2 is smaller. Often a smaller s^2 is obtained by narrowing the definition of the population of interest. For example, the data of Example 7.2 may vary as much as they do because the sample contains animals of different ages, or of different strains, or of both sexes. It may be wiser to limit the hypothesis, and the sampling, to animals of the same sex and strain and of a narrow range of ages.

7.7 SAMPLING FINITE POPULATIONS

In general we assume that a sample from a population is a very small portion of the totality of data in that population. Essentially, we consider that the population is infinite in size, so that the removal of a relatively small number of data from the population does not noticeably affect the probability of selecting further data.

However, if the sample size, n, is an appreciable portion of the population size, N (say, at least 5%), then we are said to be sampling a *finite population*. In such a case,

---

*Some calculators and computer programs yield β given $t_{\beta,\nu}$. Approximating $t_{\beta(1),\nu}$, by $Z_{\beta(1)}$ apparently yields a β that is an underestimate (and a power that is an overestimate) of no more than 0.01 for ν at least 11 and no more than 0.02 for ν at least 7.

TABLE 7.4 Relationship Between α, β, Power $(1 - \beta)$, and n, for the Data of Example 7.9.

	Two-Tailed Test				One-Tailed Test		
n	α	β	$1 - \beta$	n	α	β	$1 - \beta$
10	0.10	0.25	0.75	10	0.10	0.14	0.86
10	0.05	0.40	0.60	10	0.05	0.25	0.75
10	0.01	0.76	0.24	10	0.01	0.61	0.39
12	0.10	0.18	0.82	12	0.10	0.09	0.91
12	0.05	0.29	0.71	12	0.05	0.18	0.82
12	0.01	0.73	0.27	12	0.01	0.48	0.52
15	0.10	0.10	0.90	15	0.10	0.05	0.95
15	0.05	0.18	0.82	15	0.05	0.10	0.90
15	0.01	0.45	0.55	15	0.01	0.32	0.68
20	0.10	0.04	0.96	20	0.10	0.02	0.98
20	0.05	0.08	0.92	20	0.05	0.04	0.96
20	0.01	0.24	0.76	20	0.01	0.16	0.84

$\bar{X}$ is a substantially better estimate of μ the closer n is to N; specifically,

$$s_{\bar{X}} = \sqrt{\frac{s^2}{n}}\sqrt{1 - \frac{n}{N}} = \sqrt{\left(\frac{s^2}{n}\right)\left(1 - \frac{n}{N}\right)}, \tag{7.11}$$

where n/N is the *sampling fraction* and $1 - n/N$ is referred to as the *finite population correction*.[*]

Obviously, from Equation 7.11, when n is very small compared to N, then the sampling fraction is almost zero, the finite population correction will be nearly one, and $s_{\bar{X}}$ will be nearly $\sqrt{s^2/n}$, just as we have used (Equation 6.18) when assuming the population size, N, to be infinite. As n becomes closer to N, the correction becomes smaller, and $s_{\bar{X}}$ becomes smaller, which makes sense intuitively. If $n = N$, then $1 - n/N = 0$ and $s_{\bar{X}} = 0$, meaning there is no error at all in estimating μ if the sample consists of the entire population; that is, $\bar{X} = \mu$ if $n = N$. In computing confidence intervals when sampling finite populations (i.e., when n is not a negligibly small fraction of N), Equation 7.11 should be used instead of Equation 6.18.

If we are determining the sample size required to estimate the population mean with a stated precision (Section 7.5), and the sample size is an appreciable fraction of the population size, then the required sample size is calculated as

$$m = \frac{n}{1 + (n - 1)/N} \tag{7.12}$$

(Cochran, 1977: 77–78), where n is from Equation 7.7.

---

[*]One may also calculate $1 - n/N$ as $(N - n)/N$.

7.8 Confidence Limits for the Population Median

The sample median, $\mathcal{M}$ (Section 3.2), is used as the best estimate of M, the population median.[*] Confidence limits for M may be determined by considering the binomial distribution, as discussed in Section 23.9.

7.9 Hypotheses Concerning the Median

In Example 7.2 we examined a sample of weight change data in order to ask whether the mean change in the sampled population was different from zero. Analogously, one may test H_0: $M = M_0$ against H_A: $M \neq M_0$, where M_0 can be zero or any other hypothesized population median.[†]

A simple method for testing this two-tailed hypothesis is to determine the confidence limits for the population median, as discussed in Section 23.9, and reject H_0 (with probability $\leq \alpha$ of a Type I error) if $M_0 \leq L_1$ or $M_0 \geq L_2$. This is essentially a binomial test (Section 23.6), where we consider the number of data $< M_0$ as being in one category and the number of data $> M_0$ being in the second category. If either of these two numbers is less than or equal to the critical value in Appendix Table B.27, then H_0 is rejected. (Data equal to M_0 are ignored in this test.)

For one-tailed hypotheses about the median, the binomial test may also be employed. For H_0: $M \geq M_0$ vs. H_A: $M < M_0$, H_0 is rejected if the number of data less than M_0 is $\leq$ the one-tailed critical value, $C_{\alpha(1),n}$. For H_0: $M \leq M_0$ vs. H_A: $M > M_0$, H_0 is rejected if the number of data greater than M_0 is $\geq n - C_{\alpha(1),n}$.

As an alternative to the binomial test, for either two-tailed or one-tailed hypotheses, one may use the more powerful Wilcoxon signed-rank test. The Wilcoxon procedure is applied as a one-sample median test by ranking the data as described in Section 9.5 and assigning a minus sign to each rank associated with a datum $< M_0$ and a plus sign to each associated with a datum $> M_0$. The sum of the ranks with a plus sign is called T_+ and the sum of the ranks with a minus sign is T_-, with the test then proceeding as described in Section 9.5. The Wilcoxon test assumes that the sampled population is symmetric (in which case the median and mean are identical and this procedure becomes a hypothesis test about the mean as well as about the median, but the one-sample t test is typically a more powerful test about the mean).

7.10 Confidence Limits for the Population Variance

Confidence intervals may be determined for many other parameters, in order to express the precision of estimates of those parameters.

---

[*]Here M represents the Greek capital letter mu.

[†]As in Section 7.8, M, denoting the population median, is the capital Greek mu.

The sampling distribution of means is a symmetrical distribution, approaching the normal distribution as n increases. But the sampling distribution of variances is not symmetrical, and neither the normal nor the t distribution may be employed to set confidence limits around σ^2 or to test hypotheses about σ^2. However, theory states that

$$\chi^2 = \frac{\nu s^2}{\sigma^2},$$ (7.13)

where χ^2 represents a statistical distribution* that, like t, varies with the degrees of freedom, ν. By employing the χ^2 distribution we can define an interval within which there is a $1 - \alpha$ chance of including σ^2. Appendix Table B.1 tells us the probability of a calculated χ^2 being greater than that in the Table. If we desire to know the two χ^2 values that enclose $1 - \alpha$ of the chi-square curve, we want the portion of the curve between $\chi^2_{(1-\alpha/2),\nu}$ and $\chi^2_{\alpha/2,\nu}$ (for a 95% confidence interval, this would mean the area between $\chi^2_{0.975,\nu}$ and $\chi^2_{0.025,\nu}$). It follows from Equation 7.13 that

$$\chi^2_{(1-\alpha/2),\nu} \leq \frac{\nu s^2}{\sigma^2} \leq \chi^2_{\alpha/2,\nu},$$ (7.14)

and

$$\frac{\nu s^2}{\chi^2_{\alpha/2,\nu}} \leq \sigma^2 \leq \frac{\nu s^2}{\chi^2_{(1-\alpha/2),\nu}}.$$ (7.15)

Since $\nu s^2 = SS$, we can also write Expression 7.15 as

$$\frac{SS}{\chi^2_{\alpha/2,\nu}} \leq \sigma^2 \leq \frac{SS}{\chi^2_{(1-\alpha/2),\nu}}.$$ (7.16)

Referring back to the data of Example 7.1, we would calculate the 95% confidence interval for σ^2 as follows. As $\nu = 24$ and $s^2 = 1.80(°C)^2$, $SS = \nu s^2 = 43.20(°C)^2$. From Appendix Table B.1, we find $\chi^2_{0.025,24} = 39.364$ and $\chi^2_{0.975,24} = 12.401$. Therefore, $L_1 = SS/\chi^2_{\alpha/2,\nu} = 43.20(°C)^2/39.364 = 1.10(°C)^2$, and $L_2 = SS/\chi^2_{(1-\alpha),\nu}/12.401 = 3.48(°C)^2$. Note that the confidence limits, $1.10(°C)^2$ and $3.48(°C)^2$, are not symmetrical around s^2; that is, the distance from L_1 to s^2 is not the same as the distance from s^2 to L_2.

To obtain the $1-\alpha$ confidence interval for the population standard deviation, simply use the square roots of the confidence limits for σ^2, so that

$$\sqrt{\frac{SS}{\chi^2_{\alpha/2,\nu}}} \leq \sigma \leq \sqrt{\frac{SS}{\chi^2_{(1-\alpha/2),\nu}}}.$$ (7.17)

For the preceding example, the 95% confidence interval for σ would be $\sqrt{1.10(°C)^2} \leq \sigma \leq \sqrt{3.48(°C)^2}$, or $1.1°C \leq \sigma \leq 1.9°C$.

---

*The Greek letter "chi" (which in lower case is χ) is pronounced as the "ky" in the "sky."

7.11 HYPOTHESES CONCERNING THE VARIANCE

The procedures for testing hypotheses about the population variance come from the consideration of vs^2/σ^2 as a chi-square value (where $v = n - 1$), as introduced in Section 7.10. Consider the pair of two-tailed hypotheses, H_0: $\sigma^2 = \sigma_0^2$ and H_A: $\sigma^2 \neq \sigma_0^2$, where σ_0^2 may be any hypothesized population variance. Then, simply calculate

$$\chi^2 = \frac{vs^2}{\sigma_0^2} \quad \text{or, equivalently,} \quad \chi^2 = \frac{SS}{\sigma_0^2}, \tag{7.18}$$

and if the calculated χ^2 is $\geq \chi^2_{\alpha/2, v}$ or $\leq \chi^2_{(1-\alpha/2), v}$, then H_0 is rejected at the α level of significance. For example, if we wished to test H_0: $\sigma^2 = 1.0(°C)^2$ and H_A: $\sigma^2 \neq 1.0(°C)^2$ for the data of Example 7.1, with $\alpha = 0.05$, we would first calculate $\chi^2 = SS/\sigma_0^2$. In this example, $v = 24$ and $s^2 = 1.80(°C)^2$, so $SS = vs^2 = 43.20(°C)^2$. Also, as σ^2 is hypothesized to be $1.0(°C)^2$, $\chi^2 = SS/\sigma_0^2 = 43.20(°C)^2/1.0(°C)^2 = 43.20$. Two critical values are to be obtained from the chi-square table (Appendix Table B.1): $\chi^2_{(0.05/2), 24} = \chi^2_{0.025, 24} = 39.364$ and $\chi^2_{(1-0.05/2), 24} = \chi^2_{0.975, 24} = 12.401$. As the calculated χ^2 is more extreme than one of these critical values (i.e., the calculated χ^2 is > 39.364), H_0 is rejected, and we conclude that the sample of data was obtained from a population having a variance different from $1.0(°C)^2$.

It is more common to consider one-tailed hypotheses concerning variances. For the hypotheses H_0: $\sigma^2 \leq \sigma_0^2$ and H_A: $\sigma^2 > \sigma_0^2$, H_0 is rejected if the χ^2 calculated from Equation 7.18 is $\geq \chi^2_{\alpha, v}$. For H_0: $\sigma^2 \geq \sigma_0^2$ and H_A: $\sigma^2 < \sigma_0^2$, a calculated χ^2 that is $\leq \chi^2_{(1-\alpha), v}$ is grounds for rejecting H_0. For the data of Example 7.4, a manufacturer might be interested in whether the variability in the dissolving times of the drug is greater than a certain value—say, 1.5 sec. Thus, H_0: $\sigma^2 \leq 1.5$ sec^2 and H_A: $\sigma^2 > 1.5$ sec^2 might be tested, as shown in Example 7.10.

EXAMPLE 7.10   A one-tailed test for the hypotheses H_0: $\sigma^2 \leq 1.5$ sec^2 and H_A: $\sigma^2 > 1.5$ sec^2, using the data of Example 7.4.

$SS = 18.8288$ sec^2

$v = 7$

$s^2 = 2.6898$ sec^2

$\chi^2 = \dfrac{SS}{\sigma_0^2} = \dfrac{18.8288 \text{ sec}^2}{1.5 \text{ sec}^2} = 12.553$

$\chi^2_{0.05, 7} = 14.067$

Since $12.553 < 14.067$, H_0 is not rejected.

$$0.05 < P < 0.10 \quad [P = 0.084]$$

7.12 POWER AND SAMPLE SIZE IN TESTS CONCERNING THE VARIANCE

Sample Size Required. We may ask how large a sample is required to perform the hypothesis tests of Section 7.11 at a specified power. For the hypotheses H_0: $\sigma^2 \leq \sigma_0^2$ vs. H_A: $\sigma^2 > \sigma_0^2$, the minimum sample size is that for which

$$\frac{\chi^2_{1-\beta,\nu}}{\chi^2_{\alpha,\nu}} = \frac{\sigma_0^2}{s^2}, \tag{7.19}$$

and this sample size, n, may be found by iteration (i.e., by a directed trial and error), as shown in Example 7.11. The ratio on the left side of Equation 7.19 increases in magnitude as n increases.

EXAMPLE 7.11   Estimation of required sample size to test H_0: $\sigma^2 \leq \sigma_0^2$ vs. H_A: $\sigma^2 > \sigma_0^2$.

How large a sample is needed to reject $H_0 : \sigma^2 \leq 1.50$ sec^2, using the data of Example 7.10, if we test at the 0.05 level of significance and with a power of 0.90? (Therefore, $\alpha = 0.05$ and $\beta = 0.10$.)

From Example 7.10, $s^2 = 2.6898$ sec^2. As we have specified $\sigma_0^2 = 1.75$ sec^2, $\sigma_0^2/s^2 = 0.558$.

To begin the iterative process of estimating n, let us guess that a sample size of thirty is required. Then,

$$\frac{\chi^2_{0.90,29}}{\chi^2_{0.05,29}} = \frac{19.768}{42.557} = 0.465$$

Because $0.465 < 0.558$, our estimate of n is too low. So we might guess that $n = 50$ is required:

$$\frac{\chi^2_{0.90,49}}{\chi^2_{0.05,49}} = \frac{36.818}{66.339} = 0.555$$

Because 0.555 is a little less than 0.558, $n = 50$ is a little too low and we might guess $n = 55$, for which $\chi^2_{0.90,54}/\chi^2_{0.05,54} = 41.183/70.153 = 0.571$.

Because 0.571 is greater than 0.558, our estimate of n is high, so we try $n = 51$, for which $\chi^2_{0.90,50}/\chi^2_{0.05,50} = 37.689/67.505 = 0.558$.

Therefore, we estimate that a sample size of at least fifty-one is required to perform the hypothesis test with the specified characteristics.

For the hypotheses H_0: $\sigma^2 \geq \sigma_0^2$ vs. H_A: $\sigma^2 < \sigma_0^2$, the minimum sample size is that for which

$$\frac{\chi^2_{\beta,\nu}}{\chi^2_{1-\alpha,\nu}} = \frac{\sigma_0^2}{s^2}, \tag{7.20}$$

Power of the Test. If we test $H_0: \sigma^2 \leq \sigma_0^2$ vs. $H_A: \sigma^2 > \sigma_0^2$, using significance level α and a sample size of n, then the power of the test is

$$1 - \beta = P(\chi^2 \geq \chi_{\alpha,\nu}^2 \sigma_0^2 / s^2). \tag{7.21}$$

Thus, in Example 7.10, $\alpha = 0.05$, $n = 8$, $\nu = 7$, $\chi_{0.05,7}^2 = 14.067$, $s^2 = 2.6898 \sec^2$, $\sigma_0^2 = 1.5 \sec^2$, and the power of the test is

$$P[\chi^2 \geq (14.067)(1.5)/2.6898] = P(\chi^2 \geq 7.845).$$

From Appendix Table B.1 we see that for $\chi^2 \geq 7.845$ with $\nu = 7$, $0.25 < P < 0.50$. By interpolation* between $\chi_{0.25,7}^2$ and $\chi_{0.50,7}^2$, we estimate that the power of the test was 0.36. This is low power. We could ask what power would result if we had a sample of 40 with the above s^2. Since $\chi_{0.05,39}^2 = 54.572$, the power would be estimated as

$$P[\chi^2 \geq (54.572)(1.5)/2.6898] = P(\chi^2 \geq 30.433).$$

Consulting Appendix Table B.1 for $\nu = 39$, $0.75 < P(\chi^2 \geq 30.433) < 0.90$. Or, by interpolation, $P = 0.83.^\dagger$

What has just been shown is the estimation of power after data have been collected. A common desire is to ask what power to expect if data with specified characteristics (i.e., specified n and s^2) were to be obtained.

For the test of $H_0: \sigma^2 \geq \sigma_0^2$ against $H_A: \sigma^2 < \sigma_0^2$, the power is

$$\begin{aligned} 1 - \beta &= P(\chi^2 \leq \chi_{1-\alpha,\nu}^2 \sigma_0^2 / s^2) \\ &= 1 - P(\chi^2 \geq \chi_{1-\alpha,\nu}^2 \sigma_0^2 / s^2) \end{aligned} \tag{7.22}$$

For the two-tailed test, $H_0: \sigma^2 = \sigma_0^2$ vs. $H_A: \sigma^2 \neq \sigma_0^2$, the power is

$$1 - \beta = 1 - P(\chi^2 \geq \chi_{1-\alpha/2,\nu}^2 \sigma_0^2 / s^2) + P(\chi^2 \geq \chi_{\alpha/2,\nu}^2 \sigma_0^2 / s^2). \tag{7.23}$$

7.13 HYPOTHESES CONCERNING THE COEFFICIENT OF VARIATION

Although rarely desired, it is possible to ask whether a sample of data is likely to have come from a population with a specified coefficient of variation, call it $(\sigma/\mu)_0$. This amounts to the testing of the following pair of two-tailed hypotheses: $H_0: \sigma/\mu = (\sigma/\mu)_0$ and $H_A: \sigma/\mu \neq (\sigma/\mu)_0$. Among the testing procedures proposed, that presented by Miller (1991) works well for a sample size of at least 10 if the sampled population is normal and with a coefficient of variation, σ/μ, no greater than 0.67. For one-tailed testing (i.e., $H_0: \sigma/\mu \leq (\sigma/\mu)_0$ vs. $H_A: \sigma/\mu > (\sigma/\mu)_0$, or $H_0: \sigma/\mu \geq (\sigma/\mu)_0$ vs. $H_A: \sigma/\mu < (\sigma/\mu)_0$), Miller's procedure works well for $n > 10$ and normal populations

---

*See beginning of Appendix B for a discussion of interpolation. In this example, the actual probability is 0.35, so the interpolation yielded an excellent result.

†The actual value of $P(\chi^2 \geq 30.433)$ with $\nu = 39$ is 0.84, so the interpolation worked very well.

were $\sigma/\mu \leq 0.33$. The test statistic is

$$Z = \frac{\sqrt{n-1}\,[V - (\mu/\sigma)_0]}{(\mu/\sigma)_0\sqrt{0.5 + (\mu/\sigma)_0^2}}, \tag{7.24}$$

the probability of which may be obtained from Appendix Table B.2; or Z may be compared to the critical values of Z_α, read from the last line of Appendix Table B.3. Miller also showed this procedure to yield results very similar to those from a χ^2 approximation by McKay (1932) which, although applicable for n as small as 5, lacks power at such small sample sizes.

7.14 HYPOTHESES CONCERNING SYMMETRY AND KURTOSIS

Section 6.1 introduced population parameters for symmetry (γ_1 and $\sqrt{\beta_1}$) and for kurtosis (γ_2 and β_2) and their respective sample estimators (g_1, $\sqrt{b_1}$, g_2, and b_2). Procedures will now be considered for testing hypotheses about a population's symmetry and kurtosis. Section 6.5 presents methods for testing overall hypotheses of normality.

Testing Symmetry.   The two-tailed hypotheses, H_0: $\gamma_1 = 0$ and H_A: $\gamma_1 \neq 0$ (or, equivalently, H_0: $\sqrt{b_1} = 0$ and H_A: $\sqrt{b_1} \neq 0$) address the question of whether the population's distribution is symmetrical. The sample's symmetry measure, g_1, is calculated by Equation 6.5. It may then be compared to the critical values in Appendix Table B.22.

Alternatively, the exact probability of H_0 may be very well approximated, if $n \geq 9$, by this method presented by D'Agostino (1970, 1986; D'Agostino, Belanger, and D'Agostino, 1990):

Compute $\sqrt{b_1}$ from g_1 using Equation 6.9. Then compute:

$$A = \sqrt{b_1}\sqrt{\frac{(n+1)(n+3)}{6(n-2)}} \tag{7.25}$$

$$B = \frac{3(n^2 + 27n - 70)(n+1)(n+3)}{(n-2)(n+5)(n+7)(n+9)} \tag{7.26}$$

$$C = \sqrt{2(B-1)} - 1 \tag{7.27}$$

$$D = \sqrt{C} \tag{7.28}$$

$$E = \frac{1}{\sqrt{\ln D}} \tag{7.29}$$

$$F = \frac{A}{\sqrt{\dfrac{2}{C-1}}} \tag{7.30}$$

$$Z_{g_1} = E \ln \left(F + \sqrt{F^2 + 1} \right). \tag{7.31}*$$

Then, consult Appendix Table B.2 to determine the probability of Z being at least as large as the absolute value of the Z_{g_1} that was calculated. Appendix Table B.2 gives one-tailed probabilities, namely probabilities of Z's greater than or equal to given values; so for a two-tailed test such a probability is to be doubled. This is demonstrated in Example 7.12. As a series of computations is involved, it is advisable to carry each step to several decimal places to reduce cumulative rounding error.

For one-tailed testing of $H_0: \gamma_1 \geq 0$ versus $H_A: \gamma_1 < 0$ (or, equivalently, $H_0: \sqrt{\beta_1} \geq 0$ vs. $\sqrt{\beta_1} < 0$), we desire to ask specifically whether the population distribution is skewed to the left, not simply whether the distribution is asymmetrical. The test statistic, Z_{g_1}, is calculated as above, but H_0 is rejected only if the probability of a Z *less than or equal to* the calculated Z_{g_1} is $\leq \alpha$. So, if the latter hypotheses would have been of interest for the data referred to in Example 7.12, we would have found that $P(Z \leq -1.23) = 0.1093$, and—as P is not ≤ 0.05—we would not have rejected H_0 and would not have concluded that the sampled population was skewed to the left. If the one-tailed hypotheses of interest are $H_0: \gamma_1 \leq 0$ and $H_A: \gamma_1 > 0$, then we reject H_0 only if the probability of Z being *greater than or equal to* the calculated Z_{g_1} is $\leq \alpha$.

Testing Kurtosis.   The two-tailed hypotheses, $H_0: \gamma_2 = 0$ and $H_A: \gamma_2 \neq 0$ (or, equivalently, $H_0: \beta_2 = 3$ vs. $H_A: \beta_2 \neq 3$) ask whether the population from which our sample came is mesokurtic. The sample estimate, g_2, of the population kurtosis parameter, γ_2, is calculated by Equation 6.8. It may then compared to the critical values of g_2 in Appendix Table B.23.

If desired, the exact probability of H_0 may be approximated very satisfactorily, for $n \geq 20$, by the method of Anscombe and Glynn (1983), as recommended by D'Agostino (1986; D'Agostino, Belanger, D'Agostino, 1990) and adapted as follows:

Compute:

$$G = \frac{24n(n-2)(n-3)}{(n+1)^2(n+3)(n+5)} \tag{7.34}$$

---

*This book uses the convention of "ln" to denote a natural, or Napierian, logarithm, and "log" to indicate a common, or Briggsian, logarithm. If common logarithms are employed, then Equations 7.29 and 7.31 become

$$E = \frac{1}{\sqrt{2.30259 \log D}} \tag{7.32}$$

and

$$Z_{g_1} = 2.30259 \log \left(F + \sqrt{F^2 + 1} \right), \tag{7.33}$$

respectively.

EXAMPLE 7.12   Two-tailed testing for the hypothesis $H_0: \gamma_1 = 0$, vs. $H_A : \gamma_1 \neq 0$, using the data of Example 6.1.

$n = 70$

$g_1 = -0.3452$

Consulting Appendix Table B.22, $P(g_1 \geq 0.3452) > 0.20$. As $P > 0.05$, do not reject H_0. To determine the exact probability of H_0:

$$\sqrt{b_1} = \frac{(n-2)g_1}{\sqrt{n(n-1)}} = \frac{(68)(-0.3452)}{\sqrt{70(69)}} = -0.337758$$

$$A = \sqrt{b_1}\sqrt{\frac{(n+1)(n+3)}{6(n-2)}} = -0.337758\sqrt{\frac{(71)(73)}{6(68)}} = -1.203833$$

$$B = \frac{3(n^2 + 27n - 70)(n+1)(n+3)}{(n-2)(n+5)(n+7)(n+9)} = \frac{3[70^2 + 27(70) - 70](71)(73)}{(68)(75)(77)(79)} = 3.368090$$

$$C = \sqrt{2(B-1)} - 1 = \sqrt{2(3.368090 - 1)} - 1 = 1.176277$$

$$D = \sqrt{C} = \sqrt{1.176277} = 1.084563$$

$$E = \frac{1}{\sqrt{\ln D}} = \frac{1}{\sqrt{\ln(1.084563)}} = \frac{1}{0.284916} = 3.509806$$

$$F = \frac{A}{\sqrt{\dfrac{2}{C-1}}} = \frac{-1.203833}{\sqrt{\dfrac{2}{1.176277 - 1}}} = \frac{-1.203833}{3.368350} = -0.357395$$

$$Z_{g_1} = E\ln\left(F + \sqrt{F^2 + 1}\right) = 3.509806\left\{\ln\left(-0.357395 + \sqrt{[-0.357395]^2 + 1}\right)\right\}$$

$$= 3.509806\{\ln(0.704552)\} = 3.509806\{-0.350193\} = -1.2291$$

$$P(|Z| \geq 1.23) = P(Z \leq -1.23) + P(Z \geq 1.23)$$

$$= 0.1093 + 0.1093 = 0.2186$$

As $P > 0.05$, do not reject H_0.

$$H = \frac{(n-2)(n-3)|g_2|}{(n+1)(n-1)\sqrt{G}} \tag{7.35}$$

$$J = \frac{6(n^2 - 5n + 2)}{(n+7)(n+9)}\sqrt{\frac{6(n+3)(n+5)}{n(n-2)(n-3)}} \tag{7.36}$$

$$K = 6 + \frac{8}{J}\left[\frac{2}{J} + \sqrt{1 + \frac{4}{J^2}}\right] \tag{7.37}$$

$$L = \frac{1 - \dfrac{2}{K}}{1 + H\sqrt{\dfrac{2}{K-4}}} \tag{7.38}$$

and then compute:

$$Z_{g_2} = \frac{1 - \dfrac{2}{9K} - \sqrt[3]{L}}{\sqrt{\dfrac{2}{9K}}}.$$ (7.39)*

Then, by referring to Appendix Table B.2, determine the probability of Z being at least as large as the calculated Z_{g_2}. As the table states the one-tailed probability of a Z greater than or equal to a given value, that probability is doubled for the two-tailed test. This hypothesis is demonstrated in Example 7.13. The accumulation of rounding errors can be reduced by employing several decimal places at each step of these calculations.

For one-tailed testing of H_0: $\gamma_2 \geq 0$ versus H_A: $\gamma_2 < 0$ (or, equivalently, H_0: $\beta \geq 3$ vs. $\beta_2 < 3$), we desire to ask specifically whether the population distribution is platykurtic (see Section 6.1), not simply whether it is not mesokurtic. The test statistic, Z_{g_2}, is calculated as for the two-tailed test above, but H_0 is rejected only if g_2 is negative *and* the probability of a Z *greater than or equal to* the calculated Z_{g_2} is $\leq \alpha$. So, if the latter hypotheses would have been of interest for the data of Example 7.13, we would have observed that, although g_2 is negative, $P(Z \geq 1.28) = 0.1003$, and we would not have rejected H_0 because P is not ≤ 0.05; and we would not conclude that the sampled population was platykurtic. If the one-tailed hypotheses of interest are H_0: $\gamma_2 \leq 0$ and H_A: $\gamma_2 > 0$, then we would be interested specifically in whether the underlying population was leptokurtic. And we reject H_0 only if g_2 is positive *and* the probability of Z being *greater than or equal to* the calculated Z_{g_2} is $\leq \alpha$.

---

*The cube root of a quantity, Y ($\sqrt[3]{Y}$ is the cube root of Y) may be computed as $Y^{1/3}$ or by using logarithms: $\sqrt[3]{Y}$ = antilogarithm of (log Y)/3.

EXAMPLE 7.13   Two-tailed testing for the hypothesis H_0: $\gamma_2 = 0$ vs. H_A: $\gamma_2 \neq 0$, using the data of Example 6.1.

$n = 70$

$g_2 = -0.7183$

Consulting Appendix Table B.23, $P(g_2 \geq 0.7183) > 0.20$. As $0.20 > 0.05$, do not reject H_0:

To determine the exact probability of H_0.

$$G = \frac{24n(n-2)(n-3)}{(n+1)^2(n+3)(n+5)}$$

$$= \frac{24(70)(68)(67)}{(71)^2(73)(75)} = \frac{7,654,080}{27,599,475} = 0.277327$$

$$H = \frac{(n-2)(n-3)|g_2|}{(n+1)(n-1)\sqrt{G}}$$

$$= \frac{(68)(67)(0.7183)}{(71)(69)\sqrt{0.277327}} = \frac{3272.574800}{2579.903825} = 1.268487$$

EXAMPLE 7.13 (continued)

$$J = \frac{6(n^2 - 5n + 2)}{(n + 7)(n + 9)} \sqrt{\frac{6(n + 3)(n + 5)}{n(n - 2)(n - 3)}}$$

$$= \frac{6[70^2 - 5(70) + 2]}{(77)(79)} \sqrt{\frac{6(73)(75)}{70(68)(67)}}$$

$$= \frac{27,312}{6,083} \sqrt{\frac{32,850}{318,920}} = (4.489890)(0.320942) = 1.440994$$

$$K = 6 + \frac{8}{J} \left[\frac{2}{J} + \sqrt{1 + \frac{4}{J^2}} \right]$$

$$= 6 + \frac{8}{1.440994} \left[\frac{2}{1.440994} + \sqrt{1 + \frac{4}{1.440994^2}} \right]$$

$$= 6 + 5.551723[1.387931 + 1.710658] = 23.202508$$

$$L = \frac{1 - \dfrac{2}{K}}{1 + H\sqrt{\dfrac{2}{K - 4}}} = \frac{1 - 2/23.202508}{1 + (1.268487)\sqrt{2/23.202508 - 4}}$$

$$= \frac{0.913802}{1.409375} = 0.648374$$

$$Z_{g_2} = \frac{1 - \dfrac{2}{9K} - \sqrt[3]{L}}{\sqrt{\dfrac{2}{9K}}}$$

$$= \frac{1 - \dfrac{2}{9(23.202508)} - \sqrt[3]{0.648374}}{\sqrt{\dfrac{2}{9(23.202508)}}} = \frac{1 - 0.009578 - \sqrt[3]{0.0648374}}{0.097867}$$

$$= \frac{1 - 0.009578 - 0.865516}{0.097867} = \frac{0.124906}{0.097867} = 1.2763$$

$$P(|Z| \geq 1.28) = P(Z \leq -1.28) + P(Z \geq 1.28)$$

$$= 0.1003 + 0.1003 = 0.2006$$

As $0.2006 > 0.005$, do not reject H_0.

Testing Symmetry Nonparametrically.   Symmetry of dispersion around the median may be tested nonparametrically by using the Wilcoxon paired-sample test of Section 9.5 (also known as the Wilcoxon signed-rank test). For each datum (X_i) we compute the deviation from the median ($d_i = X_i -$ median) and then analyze the d_i's as in Section 9.5. For the two-tailed test (considering both T_- and T_+ in the Wilcoxon test), the null hypothesis is H_0: The underlying distribution is symmetrical around (i.e., is not skewed from) the median. For a one-tailed test, T_- is the critical value for H_0: The underlying distribution is not skewed to the right of the median; and T_+ is the critical value for H_0: The underlying distribution is not skewed to the left of the median. This test is demonstrated in Example 7.14.

EXAMPLE 7.14   Two-tailed nonparametric testing of symmetry, using the data of Example 6.1.

H_0:   The population of data from which this sample came is distributed symmetrically around its median.

H_A: The population is not distributed symmetrically around its median.

$n = 70$; median $= X_{(70+1)/2} = X_{35.5} = 70.5$ in.

X (in.)	d (in.)	f	$\|d\|$ (in.)	Rank of $\|d\|$	Signed rank of $\|d\|$	(f)(Signed rank)
63	−7.5	2	7.5	69.5	−69.5	−139
64	−6.5	2	6.5	67.5	−67.5	−135
65	−5.5	3	5.5	64	−64	−192
66	−4.5	5	4.5	57.5	−57.5	−287.5
67	−3.5	4	3.5	48.5	−48.5	−194
68	−2.5	6	2.5	35.5	−35.5	−213
69	−1.5	5	1.5	21.5	−21.5	−107.5
70	−0.5	8	0.5	8	−8	−64
71	0.5	7	0.5	8	8	56
72	1.5	7	1.5	21.5	21.5	160.5
73	2.5	10	2.5	35.5	35.5	355
74	3.5	6	3.5	48.5	48.5	291
75	4.5	3	4.5	57.5	57.5	172.5
76	5.5	2	5.5	64	64	128
		70				

$T_- = 1332$

$T_+ = 1163$

$T_{0.05(2), 70} = 907$ (from Appendix Table B.12)

As neither T_- nor $T_+ < T_{0.05(2), 70}$, do not reject H_0. [$P > 0.50$]

EXERCISES

7.1 The following data are the lengths of the menstrual cycle in a random sample of fifteen women. Test the hypothesis that the mean length of human menstrual cycle is equal to a lunar month (a lunar month is 29.5 days).

The data are 26, 24, 29, 33, 25, 26, 23, 30, 31, 30, 28, 27, 29, 26, and 28 days.

7.2 A species of marine arthropod lives in seawater that contains calcium in a concentration of 32 mmole/kg of water. Thirteen of the animals are collected and the calcium concentrations in their coelomic fluid are found to be: 28, 27, 29, 29, 30, 30, 31, 30, 33, 27, 30, 32, and 31 mmole/kg. Test the appropriate hypothesis to conclude whether members of this species maintain a coelomic calcium concentration less than that of their environment.

7.3 Present the following data in a graph that shows the mean, standard error, 95% confidence interval, range, and number of observations for each month.

Table of Caloric Intake (kcal/g of Body Weight) **of Squirrels**

Month	No. of data	Mean	Standard error	Range
January	13	0.458	0.026	0.289–0.612
February	12	0.413	0.027	0.279–0.598
March	17	0.327	0.018	0.194–0.461

7.4 A sample of size eighteen has a mean of 13.55 cm and a variance of 6.4512 cm^2.
  (a) Calculate the 95% confidence interval for the population mean.
  (b) How large a sample would have to be taken from this population to estimate μ to within 1.00 cm, with 95% confidence?
  (c) to within 2.00 cm with 95% confidence?
  (d) to within 2.00 cm with 99% confidence?

7.5 We want to sample a population of lengths and to perform a test of H_0: $\mu = \mu_0$ vs. H_A: $\mu \neq \mu_0$, at the 5% significance level, with a 95% probability of rejecting H_0 when $|\mu - \mu_0|$ is at least 2.0 cm. The estimate of the population variance, σ^2, is $s^2 = 8.44$ cm^2.
  (a) What minimum sample size should be used?
  (b) What minimum sample size would be required if α were 0.01?
  (c) What minimum sample size would be required if $\alpha = 0.05$ and power $= 0.99$?
  (d) If $n = 25$ and $\alpha = 0.05$, what is the smallest difference, $|\mu - \mu_0|$, that can be detected with 95% probability?
  (e) If $n = 25$ and $\alpha = 0.05$, what is the probability of detecting a difference, $|\mu - \mu_0|$, as small as 2.0 cm?

7.6 There are 200 members of a state legislature. The ages of a random sample of fifty of them are obtained, and it is found that $\bar{X} = 53.87$ yr and $s = 9.89$ yr.
  (a) Calculate the 95% confidence interval for the mean age of all members of the legislature.
  (b) If the above $\bar{X}$ and s had been obtained from a random sample of 100 from this population, what would the 95% confidence interval for the population mean have been?

7.7 For the data of Exercise 7.4:
  (a) Calculate the 95% confidence interval for the population variance.
  (b) Calculate the 95% confidence interval for the population standard deviation.
  (c) Using the 5% level of significance, test H_0: $\sigma^2 \leq 4.4000$ cm^2 vs. H_A: $\sigma^2 > 4.4000$ cm^2.
  (d) Using the 5% level of significance, test H_0: $\sigma \geq 3.00$ cm vs. H_A : $\sigma < 3.00$ cm.
  (e) What is the power of the test in part c?
  (f) How large a sample is needed to test H_0: $\sigma^2 \leq 5.0000$ cm^2 if it is desired to test at the 0.05 level of significance with 75% power?

7.8 For the data of Exercise 6.1:
  (a) Test $\gamma_1 \geq 0$ vs. $\gamma_1 < 0$.
  (b) Test $\gamma_2 \geq 0$ vs. $\gamma_2 < 0$.

8

Two-Sample Hypotheses

Among the most commonly employed biostatistical procedures is the comparison of two samples to infer whether differences exist between the two populations sampled. This chapter will consider hypotheses comparing two population means, medians, variances (or standard deviations), coefficients of variation, and indices of diversity. In doing so, we introduce another very important sampling distribution, the F distribution—named for its discoverer, R. A. Fisher (by Snedecor, 1934: 15)—and will demonstrate further use of Student's t distribution.

The objective of many two-sample hypotheses is to make inferences about population parameters by examining sample statistics. Other hypothesis-testing procedures, however, draw inferences about populations without referring to parameters. Such procedures are called *nonparametric* methods, and several will be discussed in this and following chapters.

8.1 Testing for Difference between Two Means

Example 8.1 presents the results of an experiment in which thirteen persons were divided at random into two groups, one group of six and one group of seven.*

The members of the first group were given one kind of drug (called "B"), and the members of the second group were given another kind of drug (called "G"). Blood is to be taken from each person and the time it takes the blood to clot is to be recorded. The

---

*Sir Ronald Aylmer Fisher (1890–1962) introduced the important concept of assigning subjects *at random* to groups for different experimental treatments (Bartlett, 1965).

EXAMPLE 8.1   A two-sample t test for the two-tailed hypotheses, H_0: $\mu_1 = \mu_2$ and H_A: $\mu_1 \neq \mu_2$ (which could also be stated as H_0: $\mu_1 - \mu_2 = 0$ and H_A: $\mu_1 - \mu_2 \neq 0$). The data are human blood-clotting times (in minutes) of individuals given one of two different drugs.

H_0: $\mu_1 = \mu_2$

H_A: $\mu_1 \neq \mu2$

Given drug B	Given drug G
8.8	9.9
8.4	9.0
7.9	11.1
8.7	9.6
9.1	8.7
9.6	10.4
	9.5

$n_1 = 6$ $\qquad$ $n_2 = 7$

$\nu_1 = 5$ $\qquad$ $\nu_2 = 6$

$\bar{X}_1 = 8.75$ min $\qquad$ $\bar{X}_2 = 9.74$ min

$SS_1 = 1.6950$ min^2 $\qquad$ $SS_2 = 4.0171$ min^2

$$s_p^2 = \frac{SS_1 + SS_2}{\nu_1 + \nu_2} = \frac{1.6950 + 4.0171}{5 + 6} = \frac{5.7121}{11} = 0.5193 \text{ min}^2$$

$$s_{\bar{X}_1 - \bar{X}_2} = \sqrt{\frac{s_p^2}{n_1} + \frac{s_p^2}{n_2}} = \sqrt{\frac{0.5193}{6} + \frac{0.5193}{7}} = \sqrt{0.0866 + 0.0742}$$

$$= \sqrt{0.1608} = 0.40 \text{ min}$$

$$t = \frac{\bar{X}_1 - \bar{X}_2}{s_{\bar{X} - \bar{X}_2}} = \frac{8.75 - 9.74}{0.40} = \frac{-0.99}{0.40} = -2.475$$

$$t_{0.05(2), \nu} = t_{0.05(2), 11} = 2.201$$

Therefore, reject H_0.

$$0.02 < P(|t| \geq 2.475) < 0.05 \quad [P = 0.030]$$

---

two-tailed hypotheses, H_0: $\mu_1 - \mu_2 = 0$ and H_A: $\mu_1 - \mu_2 \neq 0$, can be proposed to ask whether, in the population sampled, blood of persons treated with drug B has the same mean clotting time as does blood from persons treated with drug G. These hypotheses are commonly expressed in their equivalent forms: H_0: $\mu_1 = \mu_2$ and H_A: $\mu_1 \neq \mu_2$. The data from this experiment are presented in Example 8.1.

   If the two samples came from normal populations, and if the two populations have equal variances, then a t value may be calculated in a manner analogous to the t test introduced in Section 7.1. The t value for testing the preceding hypotheses concerning the difference between two population means is

$$t = \frac{\bar{X}_1 - \bar{X}_2}{s_{\bar{X}_1 - \bar{X}_2}}. \tag{8.1}$$

The quantity $\bar{X}_1 - \bar{X}_2$ is simply the difference between the two means, and $s_{\bar{X}_1-\bar{X}_2}$ is the standard error of the difference between the sample means.

The quantity $s_{\bar{X}_1-\bar{X}_2}$, along with $s^2_{\bar{X}_1-\bar{X}_2}$, the variance of the difference between the means, is new to us, and we need to consider it further. Both $s^2_{\bar{X}_1-\bar{X}_2}$, and $s_{\bar{X}_1-\bar{X}_2}$ are statistics that can be calculated from the sample data and are estimates of the population parameters, $\sigma^2_{\bar{X}_1-\bar{X}_2}$ and $\sigma_{\bar{X}_1-\bar{X}_2}$, respectively. It can be shown mathematically that the variance of the difference between two independent variables is equal to the sum of the variances of the two variables, so that $\sigma^2_{\bar{X}_1-\bar{X}_2} = \sigma^2_{\bar{x}_1} + \sigma^2_{\bar{x}_2}$. Independence means that there is no correlation between the two variables.* As $\sigma^2_{\bar{X}} = \sigma^2/n$, we can write

$$\sigma^2_{\bar{X}_1-\bar{X}_2} = \frac{\sigma^2_1}{n_1} + \frac{\sigma^2_2}{n_2}. \tag{8.2}$$

As the two-sample t test requires that we assume $\sigma^2_1 = \sigma^2_2$, we can write

$$\sigma^2_{\bar{X}_1-\bar{X}_2} = \frac{\sigma^2}{n_1} + \frac{\sigma^2}{n_2}. \tag{8.3}$$

Thus, to calculate the estimate of $\sigma^2_{\bar{X}_1-\bar{X}_2}$, we must have an estimate of σ^2. Since both s^2_1 and s^2_2 are assumed to estimate σ^2, we compute the *pooled variance*, s^2_p, which is then used as the best estimate of σ^2:

$$s^2_p = \frac{SS_1 + SS_2}{\nu_1 + \nu_2}, \tag{8.4}$$

and

$$s^2_{\bar{X}_1-\bar{X}_2} = \frac{s^2_p}{n_1} + \frac{s^2_p}{n_2}. \tag{8.5}$$

Thus,

$$s_{\bar{X}_1-\bar{X}_2} = \sqrt{\frac{s^2_p}{n_1} + \frac{s^2_p}{n_2}}, \tag{8.6}$$

and Equation 8.1 becomes

$$t = \frac{\bar{X}_1 - \bar{X}_2}{\sqrt{\dfrac{s^2_p}{n_1} + \dfrac{s^2_p}{n_2}}}, \tag{8.7a}$$

which for equal sample sizes (i.e., $n_1 = n_2$, so each sample size may be referred to as n),

$$t = \frac{\bar{X}_1 - \bar{X}_2}{\sqrt{\dfrac{2s^2_p}{n}}}. \tag{8.7b}$$

---

*If there is a correlation between them, then see Chapter 9.

Example 8.1 summarizes the procedure for testing the hypotheses under consideration. The critical value to be obtained from Appendix Table B.3 is $t_{\alpha(2),(\nu_1+\nu_2)}$, the two-tailed t value for the α significance level, with $\nu_1 + \nu_2$ degrees of freedom. We shall also write this as $t_{\alpha(2),\nu}$, defining the pooled degrees of freedom to be

$$\nu = \nu_1 + \nu_2 \quad \text{or, equivalently,} \quad \nu = n_1 + n_2 - 2. \tag{8.8}$$

In the two-tailed test, H_0 will be rejected if either $t \geq t_{\alpha(2),\nu}$ or $t \leq -t_{\alpha(2),\nu}$. Another way of stating this is that H_0 will be rejected if $|t| \geq t_{\alpha(2),\nu}$.

One-tailed hypotheses can be tested in situations where the investigator is interested in detecting a difference in only one direction. For example, a gardener may use a particular fertilizer for a particular kind of plant, and a new fertilizer is advertised as being an improvement. Let us say that plant height at maturity is an important characteristic of this kind of plant, with taller plants being preferable. An experiment was run, raising ten plants on the present fertilizer and eight on the new one, with the resultant eighteen plant heights shown in Example 8.2. If the new fertilizer produces plants that are shorter than, or the same height as, plants grown with the present fertilizer, then we shall decide that the advertising claims are unfounded; therefore, the statements of $\mu_1 > \mu_2$ and $\mu_1 = \mu_2$ belong in the same hypothesis, namely the null hypothesis, H_0. If, however, mean plant height is indeed greater with the newer fertilizer, then it shall be declared to be distinctly better, with the alternate hypothesis (H_A: $\mu_1 < \mu_2$) concluded to be the true statement. The t statistic is calculated by Equation 8.1, just as for the two-tailed test. But this calculated t is then compared with the critical value $t_{\alpha(1),\nu}$, rather than with $t_{\alpha(2),\nu}$.

In other cases, the one-tailed hypotheses, H_0: $\mu_1 \leq \mu_2$ and H_A: $\mu_1 > \mu_2$, may be appropriate. Just as introduced in the one-sample testing of Sections 7.1, and 7.2, the following summary of procedures applies to two-sample t testing:

For H_A: $\mu_1 \neq \mu_2$, if $|t| \geq t_{\alpha(2),\nu}$, then reject H_0.

For H_A: $\mu_1 < \mu_2$, if $t \leq -t_{\alpha(1),\nu}$, then reject H_0.[*]

For H_A: $\mu_1 > \mu_2$, if $t \geq t_{\alpha(1),\nu}$, then reject H_0.[†]

As indicated in Section 6.4, the null and alternate hypotheses are to be decided upon *before* the data are collected.

Note that H_0: $\mu_1 = \mu_2$ may be written H_0: $\mu_1 - \mu_2 = 0$ and H_A: $\mu_1 \neq \mu_2$ as H_A: $\mu_1 - \mu_2 \neq 0$; the generalized two-tailed hypotheses are H_0: $\mu_1 - \mu_2 = \mu_0$ and H_A: $\mu_1 - \mu_2 \neq \mu_0$, tested as

$$t = \frac{|\bar{X}_1 - \bar{X}_2| - \mu_0}{s_{\bar{X}_1 - \bar{X}_2}}, \tag{8.9}$$

where μ_0 may be any hypothesized difference between population means.

---

[*]For this one-tailed hypothesis test, probabilities of t up to 0.25 are indicated in Appendix Table B.3. If $t = 0$, then $P = 0.50$; so if $-t_{0.25(1),\nu} < t < 0$, then $0.25 < P < 0.50$; and if $t > 0$ then $P > 0.50$.

[†]For this one-tailed hypothesis test, $t = 0$ indicates $P = 0.50$; therefore, if $0 < t < t_{0.25(1),\nu}$, then $0.25 < P < 0.50$; and if $t < 0$, then $P > 0.50$.

EXAMPLE 8.2 A two-sample t test for the one-tailed hypotheses, H_0: $\mu_1 \geq \mu_2$ and H_A: $\mu_1 < \mu_2$ (which could also be stated as H_0: $\mu_1 - \mu_2 \geq 0$ and H_A: $\mu_1 - \mu_2 < 0$). The data are heights of plants, each grown with one of two different fertilizers.

H_0: $\mu_1 \geq \mu_2$

H_A: $\mu_1 < \mu_2$

Present fertilizer	Newer fertilizer
48.2 cm	52.3 cm
54.6	57.4
58.3	55.6
47.8	53.2
51.4	61.3
52.0	58.0
55.2	59.8
49.1	54.8
49.9	
52.6	
$n_1 = 10$	$n_2 = 8$
$\nu_1 = 9$	$\nu_2 = 7$
$\bar{X}_1 = 51.91$ cm	$\bar{X}_2 = 56.55$ cm
$SS_1 = 102.23$ cm^2	$SS_2 = 69.20$ cm^2

$$s_p^2 = \frac{102.23 + 69.20}{9 + 7} = \frac{171.43}{16} = 10.71 \text{ cm}^2$$

$$s_{\bar{X}_1 - \bar{X}_2} = \sqrt{\frac{10.71}{10} + \frac{10.71}{8}} = \sqrt{2.41} = 1.55 \text{ cm}$$

$$t = \frac{\bar{X}_1 - \bar{X}_2}{s_{\bar{X}_1 - \bar{X}_2}} = \frac{51.91 - 56.55}{1.55} = \frac{-4.64}{1.55} = -2.99$$

$t_{0.05(1),16} = 1.746$

As t of -2.99 is less than -1.746, H_0 is rejected.

$$0.0025 < P < 0.005 \quad [P = 0.0043]$$

Also, H_0: $\mu_1 \leq \mu_2$ and H_A: $\mu_1 > \mu_2$ may be written as H_0: $\mu_1 - \mu_2 \leq 0$ and H_A: $\mu_1 - \mu_2 > 0$, respectively. The generalized hypotheses for this type of one-tailed test are H_0: $\mu_1 - \mu_2 \leq \mu_0$ and H_A: $\mu_1 - \mu_2 > \mu_0$, for which the t is

$$t = \frac{\bar{X}_1 - \bar{X}_2 - \mu_0}{s_{\bar{X}_1 - \bar{X}_2}}, \tag{8.10}$$

and μ_0 may be any specified value of $\mu_1 - \mu_2$.

Lastly, H_0: $\mu_1 \geq \mu_2$ and H_A: $\mu_1 < \mu_2$ may be written as H_0: $\mu_1 - \mu_2 \geq 0$ and H_A: $\mu_1 - \mu_2 < 0$, and the generalized one-tailed hypotheses of this type are H_0: $\mu_1 - \mu_2 \geq$

μ_0 and H_A: $\mu_1 - \mu_2 < \mu_0$, with the appropriate t statistic being that of Equation 8.10. For example, the gardener collecting the data of Example 8.2 may have decided, because the newer fertilizer is more expensive than the other, that it should be used only if the plants grown with it averaged at least 5.0 cm taller than plants grown with the present fertilizer. Then, $\mu_0 = \mu_1 - \mu_2 = -5.0$ cm and, by Equation 8.10, we would calculate $t = (51.91 - 56.55 + 5.0)/1.55 = 0.36/1.55 = 0.232$, which is not $\geq$ the critical value shown in Example 8.2; so H_0: $\mu_1 - \mu_2 \geq -5.0$ cm is not rejected. The following summary of procedures applies to these general hypotheses:

For H_A: $\mu_1 - \mu_2 \neq \mu_0$, if $|t| \geq t_{\alpha(2),\nu}$, then reject H_0.

For H_A: $\mu_1 - \mu_2 < \mu_0$, if $t \leq -t_{\alpha(1),\nu}$, then reject H_0.

For H_A: $\mu_1 - \mu_2 > \mu_0$, if $t \geq t_{\alpha(1),\nu}$, then reject H_0.

If two sample means are graphed along with the sample standard deviations, sample standard errors, or confidence intervals for the means (as in Figs. 7.4 and 7.5), it is tempting for some to conclude whether the means are significantly different based upon whether the measures of dispersion overlap. The efficacy of such a procedure is discussed by Andrews, Snee, and Sarner (1980), Browne (1979), and Simpson, Roe, and Lewontin (1960: 350–354).

By the procedure of Section 8.9, one can test whether the measurements in one population are a specified amount as large as those in a second population.

Martín Andrés et al. (1995) show how correlation analysis (Section 19.1) may be employed to test the difference between two means.

Violations of the Two-Sample *t* Test Assumptions. The two-sample t test assumes, by dint of its underlying theory, that both samples come at random from normal populations with equal variances. The biological researcher cannot, however, always be assured that these assumptions are correct. Fortunately, numerous studies have shown that the t test is robust enough to stand considerable departures from its theoretical assumptions, especially if the sample sizes are equal or nearly equal, and especially when two-tailed hypotheses are considered (e.g., Boneau, 1960; Box, 1953; Cochran, 1947; Posten, Yeh, and Owen, 1982; Srivastava, 1958).

The larger the samples, the more robust the test. If the underlying populations are markedly skewed, then one should be wary of one-tailed testing, and if there is considerable non-normality in the populations, then very small significance levels (say, $\alpha < 0.01$) should not be depended upon.

The power of the two-tailed t is affected very little by skewness in the sampled populations, but there can be a serious effect on one-tailed tests. The actual power of the test is less than that discussed in Section 8.4 when the sampled populations are platykurtic and greater when the populations are leptokurtic, especially for small sample sizes (Glass, Peckham, and Sanders, 1972).

If the population variances are unequal, then the probability of a Type I error will tend to be greater than the stated α; but if the sample sizes are equal, then the t test is quite robust for moderate to large sample sizes, as shown in Table 8.1. If $n_1 \neq n_2$, then

TABLE 8.1 Maximum Probabilities of Type I Error When Applying the t Test to Two Samples of Various Sizes, $n_1 = n_2 = n$, from Normal Populations Having Various Variance Ratios, σ_1^2/σ_2^2

σ_1^2/σ_2^2 n:	3	5	10	15	16	20	30
For $\alpha = 0.05$:							
3.33 or 0.300	0.059	0.056	0.054	0.052		0.052	0.051
5.00 or 0.200	0.064	0.061	0.056	0.054		0.053	0.052
10.00 or 0.100		0.068	0.059	0.056		0.055	0.053
∞ or 0			0.065	0.060		0.057	0.055
For $\alpha = 0.01$:							
3.33 or 0.300	0.013	0.013	0.012		0.011	0.011	0.011
5.00 or 0.200	0.015	0.015	0.013		0.012	0.011	0.011
10.00 or 0.100	0.020	0.019	0.015		0.013	0.012	0.012
∞ or 0			0.018		0.015	0.014	0.013

These probabilities are gleaned from the extensive analysis of Ramsey (1980).

the probability of a Type I error will be less than α if the larger σ^2 is associated with the larger sample, and this probability will be greater than α if the smaller sample came from the population with the larger variance. The greater the difference between variances the greater will be the departure from α, with there being only a slight effect if the sample sizes are no more than 10% or so from equality. And robustness is compromised more if the smaller variance is associated with the larger sample than if the reverse is true (Kohr and Games, 1974; Posten, 1992; Posten, Yeh, and Owen, 1982; Ramsey, 1980). Ramsey (1980) also referred to an observation by Hsu (1938) of remarkable robustness in the presence of unequal variances if $n_1 = n_2 + 1$ and $\sigma_1^2 > \sigma_2^2$; so—if we have good estimates of the population variances—it is wise to plan experiments that have samples that are unequal in size by 1, where the larger sample comes from the population with the larger variance.

The comparison of two means from normal population without assuming equal variances is known as the "Behrens-Fisher problem," referring to the solution by Behrens (1929) and Fisher (1939b); and numerous other studies of it have ensued (e.g., Cochran, 1964; Cochran and Cox, 1957: 100–102; Dixon and Massey, 1969: 119; Fisher and Yates, 1963: 3–4, 60–61;[*] Gill, 1971; Lee and Fineberg, 1991; Lee and Gurland, 1975; Satterthwaite, 1946). One of the easiest, yet reliable, of such procedures is that attributed to Smith (1936) and also known as "Welch's approximate t";[†] (Davenport and Webster, 1975; Mehta and Srinivasan, 1970; Scheffé, 1970; Wang, 1971; Welch, 1936, 1938). The test statistic is

$$t' = \frac{\bar{X}_1 - \bar{X}_2}{\sqrt{\dfrac{s_1^2}{n_1} + \dfrac{s_2^2}{n_2}}},$$ (8.11)

---

[*]In Fisher and Yates (1963), s refers to the standard error, not the standard deviation.

[†]B[ernard] L[ewis] Welch (1911–1989), English statistician. (See Mardia, 1990.)

and the critical value is Student's t with degrees of freedom of

$$v' = \frac{\left(\dfrac{s_1^2}{n_1} + \dfrac{s_2^2}{n_2} \right)^2}{\dfrac{\left(\dfrac{s_1^2}{n_1} \right)^2}{n_1 - 1} + \dfrac{\left(\dfrac{s_2^2}{n_2} \right)^2}{n_2 - 1}}. \tag{8.12}$$

The degrees of freedom thus computed are usually not integer, in which case the next smaller integer should be used. If $n_1 \neq n_2$, and the population variances are very different, then t' will provide a better test than t. If the population variances are very similar, then t is the better (i.e., more powerful) test. If $n_1 = n_2$, or $s_1^2 = s_2^2$, then t' (Equation 8.11) is identical to t (Equation 8.17). If $n_1 = n_2$ and $s_1^2 = s_2^2$, then $t' = t$ and $v' = v$. Welch (1938) suggested that an improved test is obtained by employing $n_1 - 3$ and $n_2 - 3$ in place of $n_1 - 1$ and $n_2 - 1$, respectively, and this has been confirmed by Fenstad (1983); the two procedures have identical results if $n_1 = n_2$.

Some authors have recommended that the two variances should be compared and concluded to be similar (Section 8.5), prior to employing the t test. However, considering that the t test is so robust, and that the variance-comparison test performs so poorly when the distributions are non-normal (see Section 8.5), the routine test of variances is not recommended. Markowski and Markowski (1990) and Gans (1991) enlarge upon this conclusion, and Moser and Stevens (1992) explain that there is no circumstance when this two-step procedure performs better than using either t or t' (whichever is appropriate).

If there are severe deviations from the normality and/or equality-of-variance assumptions, the nonparametric test of Section 8.10 could be employed, as it is not adversely affected by violations of these assumptions, and some researchers would prefer that procedure to the modified t test above.

Replication of Data.   It is important to employ data that are true replicates of the variable to be tested. In Example 8.1 the purpose of the experiment was to ask whether there is a difference in blood-clotting times between persons administered two different drugs. This necessitates having a blood measurement on each of n_1 individuals in the first sample and n_2 individuals in the second sample. It would *not* be valid to use n_1 measurements from a single person and n_2 measurements from another person, and to do so would be engaging in what Hurlbert (1984) discusses as *pseudoreplication*.

8.2 CONFIDENCE LIMITS FOR POPULATION MEANS

In Section 7.3, we defined the confidence interval for a population mean as $\bar{X} \pm t_{\alpha(2),\nu} s_{\bar{X}}$, where $s_{\bar{X}}$ is the best estimate of $\sigma_{\bar{X}}$ and is calculated as $\sqrt{s^2/n}$. For the two-sample situation, where we assume that $\sigma_1^2 = \sigma_2^2$, the confidence interval for either μ_1 or μ_2 is calculated using s_p^2 (rather than either s_1^2 or s_2^2) as the best estimate of σ^2, and we use the two-tailed tabled t value with $\nu = \nu_1 + \nu_2$ degrees of freedom. Thus, for μ_i (where i is either 1 or 2, referring to either of the two samples), the $1 - \alpha$ confidence interval is

$$\bar{X}_i \pm t_{\alpha(2),\nu}\sqrt{\frac{s_p^2}{n_i}}. \tag{8.13}$$

For the data of Example 8.1, $\sqrt{s_p^2/n_2} = \sqrt{0.5193 \text{ min}^2/7} = 0.27$ min. Thus, the 95% confidence interval for μ_2 would be 9.74 min $\pm$ (2.201)(0.27 min) = 9.74 min $\pm$ 0.59 min, so that L_1 (the lower confidence limit) = 9.15 min and L_2 (the upper confidence limit) = 10.33 min. Thus, we can be 95% confident that the interval of 9.15 to 10.33 minutes includes the population mean, μ_2. This may be written as $P(9.15 \text{ min} \leq \mu_2 \leq 10.33 \text{ min}) = 0.95$. The confidence interval for μ_1 would be 8.75 min $\pm$ (2.201)$\sqrt{0.5193 \text{ min}^2/6}$ = 8.75 min $\pm$ (2.201)(0.29 min) = 8.75 min $\pm$ 0.64 min, so that $L_1 = 8.11$ min and $L_2 = 9.39$ min.

Confidence limits for the difference between the two population means can also be computed. The $1 - \alpha$ confidence interval for $\mu_1 - \mu_2$ is

$$\bar{X}_1 - \bar{X}_2 \pm t_{\alpha(2),\nu}s_{\bar{X}_1 - \bar{X}_2}. \tag{8.14}$$

Thus, for Example 8.1, the 95% confidence interval for $\mu_1 - \mu_2$ is (8.75 min$-$9.74 min) $\pm$ (2.201)(0.40 min) = -0.99 min $\pm$ 0.88 min. Thus, $L_1 = -1.87$ min and $L_2 = -0.11$ min, and we can write $P(-1.87 \text{ min} \leq \mu_1 - \mu_2 \leq -0.11 \text{ min}) = 0.95$. This statement also implies that if one took from the two populations all possible two-sample combinations with $n_1 = 6$ and $n_2 = 7$, and computed a confidence interval from each one, 95% of all the confidence intervals would encompass $\mu_1 - \mu_2$.

If $H_0: \mu_1 = \mu_2$ is not rejected, then both samples are concluded to have come from populations having identical means, the common mean being denoted as μ. The best estimate of μ is the "pooled" or "weighted" or "common" mean:

$$\bar{X}_p = \frac{n_1\bar{X}_1 + n_2\bar{X}_2}{n_1 + n_2}. \tag{8.15}$$

Then the $1 - \alpha$ confidence interval for μ is

$$\bar{X}_p \pm t_{\alpha(2),\nu}\sqrt{\frac{s_p^2}{n_1 + n_2}}. \tag{8.16}$$

If H_0 is not rejected it is the confidence interval of Equation 8.16, rather than those of Equations 8.13 and 8.14, that one would calculate.*

If $\sigma_1^2 \neq \sigma_2^2$. If the two population variances are not equal, then Equation 8.11 might be used to test $H_0: \mu_1 = \mu_2$. If this test rejects the null hypothesis, then the

---

*There is a slightly better expression of the confidence interval for μ when we assume the two population variances are equal and do not conclude that the two population means are different. The population variance common to both populations is estimated by computing a sample variance from the two samples combined (this is called the "common variance"), and this estimate is used in Equation 8.16 instead of s_p^2. Then the degrees of freedom, ν, are $n_1 + n_2 - 1$, slightly larger than the $n_1 + n_2 - 2$ of Equation 8.8.

confidence interval for each of the two population means should be computed as

$$\bar{X}_i \pm t_{\alpha(2), \nu} \sqrt{\frac{s_i^2}{n_i}} \tag{8.17}$$

rather than by Equation 8.13. A confidence interval for the difference between the population means is obtained, then, as follows:

$$\bar{X}_1 - \bar{X}_2 \pm t_{\alpha(2), \nu} \sqrt{\frac{s_1^2}{n_1} + \frac{s_2^2}{n_2}}, \tag{8.18}$$

rather than by Equation 8.14; ν is obtained from Equation 8.12. If $H_0: \mu_1 = \mu_2$ is not rejected, then Equation 8.15 and

$$\bar{X}_p \pm t_{\alpha(2), \nu} \sqrt{\frac{1}{2} \left(\frac{s_1^2}{n_1 + n_2} + \frac{s_2^2}{n_1 + n_2} \right)}, \tag{8.19}$$

with ν from Equation 8.12, can be employed to arrive at a confidence interval for the common mean.

8.3 SAMPLE SIZE AND ESTIMATION OF THE DIFFERENCE BETWEEN TWO POPULATION MEANS

In Section 7.5 it was shown how to determine the size of the sample that is needed to estimate a population mean, by obtaining with stated assurance a confidence interval of specified width. The same type of procedure may be employed to determine the sample size, n, required from each of two populations in order to estimate the difference between the two population means with specified precision. Just as with the one-sample case of Section 7.5, the estimation of this sample size is an iterative procedure, employing a series of successively improving estimates of the required n. As shown in Example 8.3, we use

$$n = \frac{2s_p^2 t_{\alpha(2), 2(n-1)}^2}{d^2}, \tag{8.20}$$

where s_p^2 is the pooled variance, $1 - \alpha$ is the confidence level of the desired confidence interval, critical values are from Appendix Table B.3 where the degrees of freedom are $2(n - 1)$, and d is the half-width of the confidence interval. In general, it takes fewer iterations (i.e., it is more efficient) to employ an initial guess that is too high rather than too low.

It is best to have equal sample sizes (i.e., $n_1 = n_2$) when estimating $\mu_1 - \mu_2$, but occasionally this is impractical. If sample 1 is constrained to have a size n_1, we can, after using the above procedure to calculate n, arrive at the required n_2 by

$$n_2 = \frac{nn_1}{2n_1 - n}, \tag{8.21}$$

EXAMPLE 8.3   Determination of sample size needed to achieve a stated precision in estimating the difference between two population means, using the data of Example 8.1.

If we specify that we wish to estimate $\mu_1 - \mu_2$ by having a 95% confidence interval no wider than 1.0 min, then $d = 0.5$ min, $1 - \alpha = 0.95$, and $\alpha = 0.05$. From Example 8.1 we have an estimate of the population variance, $s_p^2 = 0.5193$ min^2 with $\nu = 11$.

Let us guess that a sample size of 50 is necessary; then, $t_{0.05(2), 98} = 1.984$, so we estimate (by Equation 8.20)

$$n = \frac{2(0.5193)(1.984)^2}{(0.5)^2} = 16.4.$$

Next, we might estimate $n = 17$, for which $t_{0.05(2), 32} = 2.037$, and calculate

$$n = \frac{2(0.5193)(2.037)^2}{(0.5)^2} = 17.2.$$

Next, we try $n = 18$, for which $t_{0.05(2), 34} = 2.032$, and calculate

$$n = \frac{2(0.5193)(2.032)^2}{(0.5)^2} = 17.2.$$

Therefore, we conclude that a sample of at least 18 (i.e., more than 17) should be taken from each of the two populations in order to achieve the specified confidence interval.

If, for some reason (say, there is a limited amount of the first drug available), n_1 is constrained to be no larger than 14, then the necessary n_2 would be determined, from Equation 8.21, to be

$$n_2 = \frac{(n)(n_1)}{2n_1 - n} = 25.2,$$

meaning that we should use n_2 at least 26.

---

as shown in Example 8.3. (If $2n_1 - n \leq 0$, then n_1 must be increased and/or α and/or β and/or d must be altered to obtain a positive n_2.) Note the efficiency in having equal sample sizes: If $n_1 = n_2$, a total of thirty-six data need to be collected in Example 8.3, but if n_1 is limited to 14, then a total of $14 + 26 = 40$ data are to be obtained.

The dependability of Equations 8.20 and 8.21 depends on the accuracy of s_p^2 as an estimate of the population variance, and this improves with the sizes of the samples from which s_p^2 is computed. Therefore, the sizes of those samples should not be a very small fraction of sample sizes estimated by Equations 8.20 and 8.21.

8.4 POWER AND SAMPLE SIZE IN TESTS FOR DIFFERENCE BETWEEN TWO MEANS

Sample Size Required.   Prior to performing a two-sample test for difference between means (Section 8.1), an investigator may ask what size samples to collect. Assuming each sample comes from a normal population, we can estimate the minimum sample size to use to achieve desired test characteristics:

$$n \geq \frac{2s_p^2}{\delta^2}(t_{\alpha, \nu} + t_{\beta(1), \nu})^2 \tag{8.22}$$

(Cochran and Cox, 1957: 19–21).* Here, δ is the smallest population difference we wish to detect: $\delta = \mu_1 - \mu_2$ for the hypothesis test for which Equation 8.1 is used; $\delta = |\mu_1 - \mu_2| - \mu_0$ when Equation 8.9 is appropriate; $\delta = \mu_1 - \mu_2 - \mu_0$ when performing a test using Equation 8.10. Also in Equation 8.22 is $t_{\alpha, \nu}$, which may be either $t_{\alpha(1), \nu}$ or $t_{\alpha(2), \nu}$, depending, respectively, on whether a one-tailed or two-tailed test is to be performed.

Note that the required sample size depends on the following four factors:

1. δ, the minimum detectable difference between population means.[†] If we desire to detect a very small difference between means, then we shall need a larger sample than if we wished to detect only large differences.

2. σ^2, the population variance. If the variability within samples is great, then a larger sample size is required to achieve a given ability of the test to detect differences between means. We need to know the variability to expect among the data; assuming the variance is the same in each of the two populations sampled, σ^2 is estimated by the pooled variance, s_p^2, obtained from similar studies.

3. The significance level, α. If we perform the t test at a low α, then the critical value, $t_{\alpha, \nu}$, will be large and a large n is required to achieve a given ability to detect differences between means. That is, if we desire a low probability of committing a Type I error (i.e., falsely rejecting H_0), then we need large sample sizes.

4. The power of the test, $1 - \beta$. If we desire a test with a high probability of detecting a difference between population means (i.e., a low probability of committing a Type II error), then $\beta(1)$ will be small, $t_{\beta(1)}$ will be large, and large sample sizes are required.

Example 8.4 shows how sample size is estimated. As $t_{\alpha(2), \nu}$, and $t_{\beta(1)}$, depend on n, which is not yet known, Equation 8.22 must be solved iteratively, as we did with Equation 7.8. It matters little if the initial guess for n is inaccurate. Each iterative step will bring the estimate of n closer to the final result (which is declared when two successive iterations fail to change the value of n rounded to the next highest integer). In general, however, fewer iterations are required (i.e., the process is quicker) if one guesses high instead of low.

For a given number of data ($n_1 + n_2$), maximum test power and robustness occur when $n_1 = n_2$ (i.e., the sample sizes are equal). There are occasions, however, when equal sample sizes are impossible or impractical. If, for example, n_1 were fixed, then we would first determine n by Equation 8.22 and then find the required size of the second sample by Equation 8.21, as shown in Example 8.4. Note, from this example, that a total of $45 + 45 = 90$ data are required in the two equal-sized samples to achieve the

---

*The method of Section 10.3 may also be used for estimation of sample size, but if offers no substantial advantage over the present procedure.

[†]δ is lowercase Greek delta. If μ_0 in the statistical hypotheses is not zero (see discussion surrounding Equations 8.9 and 8.10), then δ is the amount by which the absolute value of the difference between the population means differs from μ_0.

EXAMPLE 8.4   Estimation of required sample size for a two-sample t test.

We desire to test for significant difference between the mean blood-clotting times of persons using two different drugs. We wish to test at the 0.05 level of significance, with a 90% chance of detecting a true difference between population means as small as 0.5 min. The within-population variability, based on a previous study of this type (Example 8.1), is estimated to be 0.52 min^2.

Let us guess that sample sizes of 100 will be required. Then, $\nu = 2(n-1) = 198$, $t_{0.05(2),198} = 1.972$, $\beta = 1 - 0.90 = 0.10$, $t_{0.10(1),198} = 1.286$, and we calculate (by Equation 8.22):

$$n \geq \frac{2(0.52)}{(0.5)^2}(1.972 + 1.286)^2 = 44.2.$$

Let us now use $n = 45$ to determine $\nu = 2(n-1) = 88$, $t_{0.05(2),88} = 1.987$, $t_{0.10(1),88} = 1.291$, and

$$n \geq \frac{2(0.52)}{(0.5)^2}(1.987 + 1.291)^2 = 44.7.$$

Therefore, we conclude that each of the two samples should contain at least 45 data.

If n_1 were constrained to be 30, then, using Equation 8.21, the required n_2 would be

$$n_2 = \frac{(44.7)(30)}{2(30) - 44.7} = 88.$$

desired power, whereas a total of $30 + 88 = 118$ data are needed if the two samples are as unequal as in this example.

Minimum Detectable Difference.   Equation 8.22 can be rearranged to ask how small a population difference (δ, defined above) is detectable with a given sample size:

$$\delta \geq \sqrt{\frac{2s_p^2}{n}}(t_{\alpha,\nu} + t_{\beta(1),\nu}). \tag{8.23}$$

The estimation of δ is demonstrated in Example 8.5.

Power of the Test.   Further rearrangement of Equation 8.22 results in

$$t_{\beta(1),\nu} \leq \frac{\delta}{\sqrt{\dfrac{2s_p^2}{n}}} - t_{\alpha,\nu}, \tag{8.24}$$

which is analogous to Equation 7.10 in Section 7.6. On computing $t_{\beta(1),\nu}$, one can consult Appendix Table B.3 to determine $\beta(1)$, whereupon $1 - \beta(1)$ is the power. But this generally will only result in declaring a range of power (e.g., $0.75 <$ power < 0.90). We may, with only slight overestimation of power (as noted in the footnote in Section 7.6) consider $t_{\beta(1)}$ to be approximated by a normal deviate and may thus employ Appendix Table B.2.

EXAMPLE 8.5   Estimation of minimum detectable difference in a two-sample t test.

In two-tailed testing for significant difference between mean blood-clotting times of persons using two different drugs, we desire to use the 0.05 level of significance and sample sizes of 20. What size difference between means do we have a 90% chance of detecting?

Using Equation 8.23 and the sample variance of Example 8.1, we calculate:

$$\delta = \sqrt{\frac{2(0.5193)}{20}} \, (t_{0.05(2),38} + t_{0.10(1),38})$$

$$= (0.2279)(2.024 + 1.304) = 0.76 \text{ min.}$$

The above procedure for estimating power is demonstrated in Example 8.6, along with the following method (which is preferable when performing two-tailed testing at $\alpha = 0.01$ or $\alpha = 0.05$ and will be expanded on in the chapters on analysis of variance). We calculate

$$\phi = \sqrt{\frac{n\delta^2}{4s_p^2}} \tag{8.25}$$

(from Kirk, 1982: 142–144), and ϕ (lowercase Greek phi—pronounced to rhyme with "sky") is then located on the *first page* of Appendix Fig. B.1, along the lower axis (taking care to distinguish between ϕ's for $\alpha = 0.01$ and $\alpha = 0.05$). Along the top margin of the graph are indicated pooled degrees of freedom, ν, for α of either 0.01 or 0.05 (although the symbol ν_2 is used on the graph for a reason that will be apparent in later chapters). By noting where ϕ vertically intersects the curve for the appropriate ν, one can read across to either the left or right axis to find the estimate of power.

Unequal Sample Sizes.   For a given total number of data, $n_1 + n_2$, the two-sample t test has maximum power and robustness when $n_1 = n_2$. However, if $n_1 \neq n_2$, the above procedure for determining minimum detectable difference (Equation 8.23) and power (Equation 8.24 and 8.25) can be performed using the harmonic mean of the two sample sizes (Cohen, 1988: 42):

$$n = \frac{2n_1 n_2}{n_1 + n_2}. \tag{8.26}$$

Thus, for example, if $n_1 = 6$ and $n_2 = 7$, then

$$n = \frac{2(6)(7)}{6 + 7} = 6.46.$$

Power of a Performed Test.   What we estimated above was the power we would expect from a test performed with a specified sample size, n, population variance, σ^2, and minimum detectable difference, δ. This sort of estimation is appropriate prior to collecting data for hypothesis testing. If a hypothesis test is performed, and the null hypothesis rejected, we know the probability of committing a Type I error (e.g., $0.02 < P < 0.05$ in Example 8.1). However, if H_0 is not rejected, we should desire to estimate the probability

EXAMPLE 8.6   Estimation of the power of a two-sample t test.

What would have been the probability of detecting a true difference of 1.0 min between mean blood-clotting times of persons using the two drugs of Example 8.1, if $n_1 = n_2 = 15$, and $\alpha(2) = 0.05$?
For $n = 15$, $\nu = 2(n-1) = 28$, and $t_{0.05(2), 28} = 2.048$. Using Equation 8.24:

$$t_{\beta(1), 28} \leq \frac{1.0}{\sqrt{\dfrac{2(0.5193)}{15}}} - 2.048 = 1.752$$

Consulting Appendix Table B.3, we see that, for one-tailed probabilities and $\nu = 28$: $0.025 < P(t \geq 1.752) < 0.05$, so $0.025 < \beta < 0.05$.
Power $= 1 - \beta$, so $0.95 <$ power < 0.975.
Or, by the normal approximation, we can estimate β by $P(Z \geq 1.752) = 0.04$. So power $= 0.96$. [The exact figures are $\beta = 0.045$ and power $= 0.955$.]
To use Appendix Fig. B.1, we calculate

$$\phi = \sqrt{\frac{n\delta^2}{4s_p^2}} = \sqrt{\frac{(15)(1.0)}{4(0.5193)}} = 2.69.$$

In the first page of Appendix Fig. B.1, we find that $\phi = 2.69$ and $\nu(= \nu_2) = 28$ are associated with a power of about 0.96.

---

of having committed a Type II error. This may be done using Appendix Fig. B.1 by estimating ϕ to be

$$\phi = \sqrt{\frac{nd^2 - 2s_p^2}{4s_p^2}} \tag{8.27}$$

(from Kirk, 1982: 143), where d is the difference in sample means; $d = \bar{X}_1 - \bar{X}_2$. This computation is demonstrated in Example 8.7, where, even though H_0 was rejected, we might ask what the power of the test was. This procedure can be used only for two-tailed tests, where α is either 0.05 or 0.01 (simply because that is the limitation of the available graphs—Appendix Fig. B.1—that are needed).

8.5 TESTING FOR DIFFERENCE BETWEEN TWO VARIANCES

If we have two samples of measurements, each sample taken at random from a normal population, we might ask if the variances of the two populations are equal. Consider the data of Example 8.8, where s_1^2, the estimate of σ_1^2, is 21.87 moths^2, and s_2^2, the estimate of σ_2^2, is 15.36 moths^2. The two-tailed hypotheses can be stated as H_0: $\sigma_1^2 = \sigma_2^2$ and H_A: $\sigma_1^2 \neq \sigma_2^2$, and we can ask, what is the probability of taking two samples from two populations having identical variances and having the two sample variances be as different as are s_1^2 and s_2^2? If this probability is rather low (say $\alpha \leq 0.05$, as in previous chapters), then we reject the veracity of H_0 and conclude that the two samples came from populations having unequal variances. If the probability is greater than α, we state

EXAMPLE 8.7   Estimation of the power of a two-sample t test, estimated after it has been performed.

What was the power of the hypothesis test in Example 8.1?
As $n_1 \neq n_2$, we must calculate n by Equation 8.26:

$$n = \frac{2(6)(7)}{6 + 7} = 6.46.$$

Then, as $d = -0.99$ and $s_p^2 = 0.5193$,

$$\phi = \sqrt{\frac{nd^2 - 2s_p^2}{4s_p^2}} = \sqrt{\frac{(6.46)(-0.99)^2 - (2)(0.5193)}{(4)(0.5193)}} = 1.60$$

and, by consulting the first page of Appendix Fig. B.1 for $\nu(= \nu_2) = 11$, the power is estimated to be 0.55. (That is, the probability of committing a Type II error was 0.45.)

that there is insufficient evidence to conclude that the variances of the two populations are not the same.

The hypotheses may be submitted to the *variance ratio test*, for which one calculates

$$F = \frac{s_1^2}{s_2^2} \quad \text{or} \quad F = \frac{s_2^2}{s_1^2}, \text{ whichever is larger.} \tag{8.28}$$

That is, the larger variance is placed in the numerator and the smaller in the denominator. We then ask whether the calculated ratio of sample variances (i.e., F) deviates so far from 1.0 as to enable us to reject H_0 at the α level of significance. For the data in Example 8.8, the calculated F is 1.42. The critical value, $F_{0.05(2), 10, 7}$, is obtained from Appendix Table B.4 and is found to be 4.76. As $1.42 < 4.76$, we do not reject H_0. The calculated F (namely 1.42) is smaller than the smallest tabled critical value (namely $F_{0.50(2), 10.7} = 1.69$), so the probability of an F at least this large if H_0 is a true statement about the populations is greater than 0.50; i.e., $P(F > 1.42) > 0.50$.*

Note that we consider degrees of freedom associated with the variances in both the numerator and denominator of the variance ratio. Furthermore, it is important to realize that F_{α, ν_1, ν_2} and F_{α, ν_2, ν_1} are not the same value, so one must be careful to refer to the numerator and denominator degrees of freedom in the correct order.

If H_0: $\sigma_1^2 = \sigma_2^2$ is not rejected, then s_1^2 and s_2^2 are assumed to be estimates of the same population variance, σ^2. The best estimate of this σ^2 that underlies both samples is what is called the *pooled variance* (introduced as Equation 8.4):

$$s_p^2 = \frac{SS_1 + SS_2}{\nu_1 + \nu_2} = \frac{\nu_1 s_1^2 + \nu_2 s_2^2}{\nu_1 + \nu_2}. \tag{8.29}$$

One-tailed hypotheses may also be submitted to the variance ratio test. For H_0: $\sigma_1^2 \geq \sigma_2^2$ and H_A: $\sigma_1^2 < \sigma_2^2$, s_2^2 is always used as the numerator of the variance ratio; for H_0: $\sigma_1^2 \leq \sigma_2^2$ and H_A: $\sigma_1^2 > \sigma_2^2$, s_1^2 is always used as the numerator. (A look at the alternate hypothesis tells which variance belongs in the numerator of F.)

---

*Some calculators and computer programs have the capability of determining the probability of a given F (Guenther, 1977). For the present example, we would thereby find that $P(F > 1.42) = 0.66$.

EXAMPLE 8.8   The two-tailed variance ratio test for the hypothesis H_0: $\sigma_1^2 = \sigma_2^2$ and H_A: $\sigma_1^2 \neq \sigma_2^2$. The data are the numbers of moths caught during the night by eleven traps of one style and eight traps of a second style.

H_0: $\sigma_1^2 = \sigma_2^2$

H_A: $\sigma_1^2 \neq \sigma_2^2$

$\alpha = 0.05$

Trap type 1	Trap type 2
41	52
34	57
33	62
36	55
40	64
25	57
31	56
37	55
34	
30	
38	

$n_1 = 11$ $n_2 = 8$

$\nu_1 = 10$ $\nu_2 = 7$

$SS_1 = 218.73 \ \text{moths}^2$ $SS_2 = 107.50 \ \text{moths}^2$

$s_1^2 = 21.87 \ \text{moths}^2$ $s_2^2 = 15.36 \ \text{moths}^2$

$$F = \frac{s_1^2}{s_2^2} = \frac{21.87}{15.36} = 1.42$$

$F_{0.05(2),\,10,\,7} = 4.76$

Therefore, do not reject H_0.

$$P(F \geq 1.42) > 0.50$$

$$s_p^2 = \frac{218.73 \ \text{moths}^2 + 107.50 \ \text{moths}^2}{10 + 7} = 19.19 \ \text{moths}^2$$

The critical value for a one-tailed test is $F_{\alpha(1),\,\nu_1,\,\nu_2}$ from Appendix Table B.4, where ν_1 is the numerator degrees of freedom and ν_2 is the denominator degrees of freedom. Example 8.9 presents the data submitted to the hypothesis test for whether ducks raised in captivity have less variability in clutch size than those breeding in the wild.

The variance ratio test is severely and adversely affected by sampling non-normal populations (e.g., Markowski and Markowski, 1990). Thus, it must be employed with caution and reservation. Better procedures (e.g., Bailer, 1989; Balakrishnan and Ma, 1990) are computationally demanding. See Section 10.6 regarding two-sample testing

EXAMPLE 8.9   A one-tailed variance ratio test for the hypothesis that duck clutch size is less variable in captive than in wild birds.

H_0: $\sigma_1^2 \geq \sigma_2^2$

H_A: $\sigma_1^2 < \sigma_2^2$

$\alpha = 0.05$

Clutch Size of Ducks

Captive	Wild
10	9
11	8
12	11
11	12
10	10
11	13
11	11
	10
	12

$n_1 = 7$	$n_2 = 9$
$\nu_1 = 6$	$\nu_2 = 8$
$SS_1 = 2.86$ eggs^2	$SS_2 = 20.00$ eggs^2
$s_1^2 = 0.48$ eggs^2	$s_2^2 = 2.50$ eggs^2

$F = \dfrac{2.50}{0.48} = 5.21$

$F_{0.05(1), 8, 6} = 4.15$

Therefore, reject H_0.

$$0.025 < P(F \geq 5.21) < 0.05 \quad [P = 0.030]$$

using Bartlett's test. There are nonparametric tests for the difference between the dispersions of two populations but, while they are not based on the assumption of normality, they have other serious drawbacks. For example, the procedure of Siegel and Tukey (1960)—and others—is low in power and requires that the medians of the two populations either are equal or are known.

8.6 CONFIDENCE INTERVAL FOR THE POPULATION VARIANCE RATIO

A $1 - \alpha$ confidence interval for the variance ratio, σ_1^2 / σ_2^2, is defined by its lower confidence limit,

$$L_1 = \left(\frac{s_1^2}{s_2^2} \right) \left(\frac{1}{F_{\alpha(2), \nu_1, \nu_2}} \right), \tag{8.30}$$

and its upper confidence limit

$$L_2 = \left(\frac{s_1^2}{s_2^2}\right) F_{\alpha(2), \nu_2, \nu_1}. \tag{8.31}$$

For the data of Example 8.8, $s_1^2/s_2^2 = 1.42$, $F_{0.05(2), 10, 7} = 4.76$, and $F_{0.05(2), 7, 10} = 3.95$. Therefore, we would calculate $L_1 = 0.298$ and $L_2 = 5.61$, and we could state

$$P\left(0.298 \le \frac{\sigma_1^2}{\sigma_2^2} \le 5.61\right) = 0.95.$$

To calculate a confidence interval for σ_2^2/σ_1^2, simply utilize Equations 8.30 and 8.31 with the two subscripts (1 and 2) reversed.

8.7   SAMPLE SIZE AND POWER IN TESTS FOR DIFFERENCE BETWEEN TWO VARIANCES

Sample Sizes Required. In considering the two-sample hypotheses of Section 8.5, we may ask what minimum sample sizes are required to achieve specified test characteristics. Using the normal approximation recommended by Desu and Raghavarao (1990: 35), the following number of data is needed in each sample to test at the α level of significance with power of $1 - \beta$:

$$n = \left[\frac{Z_\alpha + Z_{\beta(1)}}{\ln\left(\frac{s_1^2}{s_2^2}\right)}\right]^2 + 2. \tag{8.32}$$

For analysts who prefer performing calculations with "common logarithms" (those employing base 10) to using "natural logarithms" (those in base e),* Equation 8.32 may be written equivalently as

$$n = \left[\frac{Z_\alpha + Z_{\beta(1)}}{(2.30259) \log\left(\frac{s_1^2}{s_2^2}\right)}\right]^2 + 2. \tag{8.33}$$

This sample-size estimate assumes that the samples are to be equal in size, which is generally preferable. If, however, it is desired to have unequal sample sizes (which will typically require more total data to achieve a particular power), one may specify that ν_1 is to be m times the size of ν_2; then (after Desu and Raghavarao, 1990: 35):

$$m = \frac{n_1 - 1}{n_2 - 1}, \tag{8.34}$$

---

*In this book, *ln* will denote the natural, or Naperian, logarithm, and *log* will denote the common, or Briggsian, logarithm. These are named for the Scottish mathematician, John Napier (1550–1617), who devised and named logarithms, and the English mathematician, Henry Briggs (1561–1630), who adapted this computational method to base 10. The German astronomer, Johann Kepler (1571–1630), was the first to use the abbreviation "Log" in 1624 (Cajori, 1929: 2, 105), and an Italian mathematician, Bonaventura Cavalieri (1598–1647), was the first to use "log" in 1632 (ibid.: 2, 106).

$$n_2 = \frac{(m+1)(n-2)}{2m} + 2, \tag{8.35}$$

and

$$n_1 = m(n_2 - 1) + 1. \tag{8.36}$$

As in Section 8.5, determination of whether s_1^2 or s_2^2 is placed in the numerator of the variance ratio in Equation 8.32 depends upon the hypothesis test, and Z_α is either a one-tailed or two-tailed normal deviate depending upon the hypothesis to be tested; n_1 and n_2 correspond to s_1^2 and s_2^2, respectively. This procedure is applicable if the variance ratio is > 1. It is demonstrated in Example 8.10.

Power of the Test.   We may also estimate what the power of the variance ratio test would be if specified sample sizes were used. If the two sample sizes are the same (i.e., $n = n_1 = n_2$), then Equations 8.32 and 8.33 may be rearranged, respectively, as follows:

$$Z_{\beta(1)} = \sqrt{n-2} \ln\left(\frac{s_1^2}{s_2^2}\right) - Z_\alpha \tag{8.37}$$

$$Z_{\beta(1)} = \sqrt{n-2}(2.30259) \log\left(\frac{s_1^2}{s_2^2}\right) - Z_\alpha. \tag{8.38}$$

As shown in Example 8.11, after $Z_{\beta(1)}$ is calculated, $\beta(1)$ is determined (for a range of probabilities, from the last line of Appendix Table B.3, or from Appendix Table B.2, or from a calculator or computer that gives probability of a normal deviate); and power $= 1 - \beta(1)$. If the two sample sizes are not the same, then the estimation of power may employ

$$Z_{\beta(1)} = \sqrt{\frac{2m(n_2-2)}{m+1}} \ln\left(\frac{s_1^2}{s_2^2}\right) - Z_\alpha \tag{8.39}$$

or

$$Z_{\beta(1)} = \sqrt{\frac{2m(n_2-2)}{m+1}}(2.30259) \log\left(\frac{s_1^2}{s_2^2}\right) - Z_\alpha, \tag{8.40}$$

where m is as in Equation 8.34.

8.8 TESTING FOR DIFFERENCE BETWEEN TWO COEFFICIENTS OF VARIATION

Procedures have been proposed for testing the null hypothesis that two samples came from populations with identical coefficients of variation. Lewontin (1966) has shown that the variance ratio

$$F = \frac{(s_{\log}^2)_1}{(s_{\log}^2)_2} \tag{8.41}$$

EXAMPLE 8.10   Determination of minimum sample size to perform the two-tailed hypothesis test of Example 8.8 at the 5% level of significance, with 90% power.

$\alpha(2) = 0.05$ and $\beta = 0.10$; $z_{0.05(2)} = 1.9600$ and $z_{0.10(1)} = 1.2816$.

From Example 8.8, $n_1 = 11$, $s_1^2 = 21.87$ moths^2, $n_2 = 8$, $s_2^2 = 15.36$ moths^2, and $s_1^2/s_2^2 = 1.42$.

$$n = \left[\frac{z_{\alpha(2)} + z_{\beta(1)}}{(2.30259) \log \left(\frac{s_1^2}{s_2^2} \right)} \right]^2 + 2$$

$$= \left[\frac{1.9600 + 1.2816}{(2.30259) \log 1.42} \right]^2 + 2$$

$$= \left[\frac{3.2416}{(2.30259)(0.1523)} \right]^2 + 2$$

$$= (9.2436)^2 + 2 = 87.4$$

Therefore, at least 88 data are needed in each sample.

If it is desired that Sample 1 is to be larger than Sample 2, specifically with $\nu_1/\nu_2 = 2$, then (i.e., using $m = 2$)

$$n_2 = \frac{(m+1)(n-2)}{2m} + 2$$

$$= \frac{(3)(86)}{4} + 2 = 66.5$$

and

$$n_1 = m(n_2 - 1) + 1$$

$$= 2(66) + 1 = 133$$

Therefore, sample sizes of 67 and 133 are required. (This is a total of 200 data, whereas a total of only $2(88) = 176$ data is needed if $n_1 = n_2$.)

---

can be used analogously to Equation 8.28. In Equation 8.41, $(s_{\log}^2)_i$ refers to the variance of the logarithms of the data in Sample i, where logarithms to any base may be employed.

A very useful property of coefficients of variation is that they have no units. Thus, V's may be compared even if they are calculated from measurements obtained in different units. Example 8.12 demonstrates two-tailed testing where this is the case. One-tailed testing is also possible, in the same fashion as discussed in Section 8.5.

Unfortunately, we are faced with the requirement of the variance ratio test that the two underlying distributions be normal (or nearly normal). Thus, this test must be applied with caution, for if the two sets of sample data are, in fact, from normal populations, the logarithms of the data *will not* be normally distributed; and the requirement here is that the *logarithms* be normally distributed. A procedure advanced by Miller (1991) avoids this problem and allows testing with the assumption that the data, not the logarithms of

---

EXAMPLE 8.11   Determination of power in performing the two-tailed hypothesis test of Example 8.8 at the 5% level of significance and with specified sample sizes.

$\alpha(2) = 0.05$, $z_{0.05(2)} = 1.9600$.

From Example 8.8, $s_1^2 = 21.87$ moths^2, $s_2^2 = 15.36$ moths^2, and $s_1^2/s_2^2 = 1.42$.

If we specify that each sample size is 60, then

$$Z_{\beta(1)} = \sqrt{n-2}(2.30259)\log\left(\frac{s_1^2}{s_2^2}\right) - Z_\alpha$$

$$= \sqrt{60-2}(2.30259)\log(1.42) - 1.9600$$

$$= (7.6158)(2.30259)(0.1523) - 1.9600$$

$$= 0.7107$$

$0.10 < \beta(1) < 0.25 \quad [\beta(1) = 0.24]$

$$0.90 > \text{power} > 0.75 \quad [\text{power} = 0.76]$$

If we specify $n_1 = 20$ and $n_2 = 30$, then $m = 19/29 = 0.6552$ and

$$Z_{\beta(1)} = \sqrt{\frac{2m(n_2-2)}{m+1}}(2.30259)\log\left(\frac{s_1^2}{s_2^2}\right) - Z_\alpha$$

$$= \sqrt{\frac{2(0.6552)(30-2)}{0.6552+1}}(2.30259)\log(1.42) - 1.9600$$

$$= \sqrt{22.1672}(2.30259)(0.1523) - 1.9600$$

$$= 1.6510 - 1.9600 = -0.3090$$

$\beta(1)$ is $P(Z \geq -0.3090) = 1 - P(Z \leq -0.3090) = 1 - P(Z \geq 0.3090)$

Power $= 1 - \beta(1) = 1 - [1 - P(Z \geq 0.3090)] = P(Z \geq 0.3090)$

$$0.25 \leq P(Z \geq 0.3090) \leq 0.50 \qquad [\text{power} = 0.38]$$

---

the data, are from normal distributions. The test statistic is

$$Z = \frac{V_1 - V_2}{\sqrt{\left(\dfrac{V_p^2}{n_1-1} + \dfrac{V_p^2}{n_2-1}\right)(0.5 + V_p^2)}}, \tag{8.42}$$

where

$$V_p = \frac{(n_1-1)V_1 + (n_2-1)V_2}{(n_1-1) + (n_2-1)} \tag{8.43}$$

is referred to as the "pooled coefficient of variation," which is the best estimate of the population coefficient of variation, σ/μ, that is common to both populations if the null hypothesis of no difference is true.

EXAMPLE 8.12   A two-tailed test for difference between two coefficients of variation.

H_0:   The intrinsic variability of male weights is the same as the intrinsic variability of male heights (i.e., the population coefficients of variation of weight and height are the same, namely H_0: $\sigma_1/\mu_1 = \sigma_2/\mu_2$).

H_A: The intrinsic variability of male weight is not the same as the intrinsic variability of male heights (i.e., the population coefficients of variation of weight and height are not the same, namely H_0: $\sigma_1/\mu_1 \neq \sigma_2/\mu_2$).

(a) The variance-ratio test.

Weight (kg)	Log of weight	Height (cm)	Log of height
72.5	1.86034	183.0	2.26245
71.7	1.85552	172.3	2.23629
60.8	1.78390	180.1	2.25551
63.2	1.80072	190.2	2.27921
71.4	1.85370	191.4	2.28194
73.1	1.86392	169.6	2.22943
77.9	1.89154	166.4	2.22115
75.7	1.87910	177.6	2.24944
72.0	1.85733	184.7	2.26647
69.0	1.83885	187.5	2.27300
		179.8	2.25479

$n_1 = 10$	$n_2 = 11$
$\nu_1 = 9$	$\nu_2 = 10$
$\bar{X}_1 = 70.73$ kg	$\bar{X}_2 = 180.24$ kg
$SS_1 = 246.1610$ kg^2	$SS_2 = 678.9455$ kg^2
$s_1^2 = 27.3512$ kg^2	$s_2^2 = 67.8946$ kg^2
$s_1 = 5.23$ kg	$s_2 = 8.24$ kg
$V_1 = 0.0739$	$V_2 = 0.0457$
$(SS_{\log})_1 = 0.00987026$	$(SS_{\log})_2 = 0.00400188$
$(s_{\log}^2)_1 = 0.0010967$	$(s_{\log}^2)_2 = 0.00040019$

$$F = \frac{0.0010967}{0.00040019} = 2.74$$

$F_{0.05(2),9,10} = 3.78$

Therefore, do not reject H_0.

$$0.10 < P < 0.20 \quad [P = 0.13]$$

EXAMPLE 8.12   (continued)

(b) The Z test.

$$V_p = \frac{(n_1 - 1)V_1 + (n_2 - 1)V_2}{(n_1 - 1) + (n_2 - 1)} = \frac{9(0.0739) + 10(0.0457)}{9 + 10} = \frac{1.1221}{19} = 0.0591$$

$$V_p^2 = 0.003493$$

$$Z = \frac{V_1 - V_2}{\sqrt{\left(\dfrac{V_p^2}{n_1 - 1} + \dfrac{V_p^2}{n_2 - 1}\right)(0.5 + V_p^2)}}$$

$$= \frac{0.0739 - 0.0457}{\sqrt{\left(\dfrac{0.003493}{9} + \dfrac{0.003493}{10}\right)(0.5 + 0.003493)}}$$

$$= \frac{0.0282}{0.0193} = 1.46$$

$$Z_{0.05(2)} = t_{0.05(2),\infty} = 1.960$$

Do not reject H_0.

$$0.10 < P < 0.20 \quad [P = 0.14]$$

This procedure is shown, as a two-tailed test, in Example 8.12b. Recall that critical values of Z may be read from the last line of the table of critical values of t (Appendix Table B.3), so $Z_{\alpha(2)} = t_{\alpha(2),\infty}$. One-tailed testing is also possible, in which case the alternate hypothesis would declare a specific direction of difference and one-tailed critical values ($t_{\infty,\alpha(1)}$) would be consulted.

8.9 NONPARAMETRIC STATISTICAL METHODS

The theory upon which the two-sample t test is based requires that the two sampled populations be normal and have equal variances. Many other common statistical procedures (e.g., analysis of variance and regression, topics to be discussed in later chapters) have similar underlying assumptions. Fortunately, as pointed out in Section 8.1, most of the commonly employed tests are sufficiently robust to allow us to disregard all but severe deviations from the theoretical assumptions.

However, a large body of statistical methods is available that comprises procedures not requiring the estimation of the population variance or mean and not stating hypotheses about parameters. These testing procedures are termed *nonparametric tests.** As these methods also typically do not make assumptions about the nature of the distribution (e.g., normality) of the sampled populations (although they might assume that the sampled populations have the same dispersion or shape), they are sometimes referred to as *distribution-free tests.*

---

*The term "nonparametric" was first used by J. Wolfowitz in 1942 (David, 1995).

Nonparametric tests, such as the two-sample testing procedure described in Section 8.10, may be applied in any situation where we would be justified in employing a parametric test, such as the two-sample t test, as well as in instances when the assumptions of the latter are untenable. However, if either the parametric or nonparametric approach is applicable, then the former will always be more powerful than the latter (i.e., the nonparametric method will have a greater probability of committing a Type II error). Often the difference in power is not great, however, and can be compensated by a small increase in sample size for the nonparametric test.

It is sometimes declared that only nonparametric testing may be employed when dealing with ordinal scale data, but this is not so. There is nothing in the theoretical basis of parametric hypothesis testing that requires interval or ratio scale data. (It might be argued, however, that a population of ordinal scale data is more likely to deviate unacceptably far from normality than is a population of interval or ratio data.) This point is discussed thoroughly by Anderson (1961), Gaito (1959, 1960), Savage (1957), and Stevens (1968).

Some general textbooks include good introductory coverage of nonparametric statistical methods, and there are a number of modern mathematical monographs on this important branch of statistical procedure. In this book, nonparametric testing will be considered in this and following chapters.

8.10 Two-Sample Rank Testing

Although nonparametric procedures have been proposed for testing differences between the dispersion, or variability, of two populations, none has achieved widespread acceptance. However, nonparametric analogues to the two-sample t test are commonly employed. We shall refer to the test originally proposed, for equal sample sizes, by Wilcoxon (1945) and later enlarged upon by Mann and Whitney (1947).* This test procedure is thus called the Wilcoxon-Mann-Whitney test, or, more commonly, the Mann-Whitney test [see Kruskal (1957) for additional history]. Watson's test (Section 26.6) may also be employed when the Mann-Whitney test is applicable, but the latter is easier to perform. Martín Andrés et al. (1995) show that Spearman rank correlation (Section 19.9) may be used instead of the Mann-Whitney test.

The Mann-Whitney Test. For this test, as for many other nonparametric procedures, the actual measurements are not employed, but we use instead the ranks of the measurements. The data may be ranked either from the highest to lowest or from the lowest to the highest values. Example 8.13 ranks the measurements from highest to lowest: The greatest height in either of the two groups is given rank 1, the second greatest height is assigned rank 2, and so on, with the shortest height being assigned rank N, where

$$N = n_1 + n_2. \tag{8.44}$$

---

*Kruskal (1957) identified seven independent developments of the procedure Wilcoxon introduced, two of them prior to Wilcoxon, the earliest being by G. Deuchler in 1914.

We then calculate the Mann-Whitney statistic,

$$U = n_1 n_2 + \frac{n_1(n_1 + 1)}{2} - R_1, \tag{8.45}$$

where n_1 and n_2 are the number of observations in Samples one and two, respectively, and R_1 is the sum of the ranks of the observations in Sample one.* For the two-tailed hypotheses, H_0: male and female students are the same height and H_A : male and female students are not the same height, the calculated U is compared with the two-tailed value of $U_{\alpha(2),n_1,n_2}$, found in Appendix Table B.11. This table assumes that $n_1 < n_2$; if $n_1 > n_2$, simply use $U_{\alpha(2),n_2,n_1}$ as the critical value. The Mann-Whitney statistic can also be calculated as

$$U' = n_2 n_1 + \frac{n_2(n_2 + 1)}{2} - R_2, \tag{8.46}$$

(where R_2 is the sum of the ranks of the observations in Sample two), because the labeling of the two samples as 1 and 2 is purely arbitrary.† For a two-tailed test we must compute both U and U', and the larger of the two is compared to the critical value, $U_{\alpha(2)n_1,n_2}$. If Equation 8.45 has been used to calculate U, then U' can be found quickly as

$$U' = n_1 n_2 - U. \tag{8.47}$$

If Equation 8.46 has been used to compute U', then U is obtainable as

$$U = n_1 n_2 - U'. \tag{8.48}$$

Then, if either U or U' is as great as or greater than $U_{\alpha(2),n_1,n_2}$, H_0 is rejected at the α level of significance. Note that neither parameters nor parameter estimates are employed in the statistical hypotheses or in the calculations of U or U'.

We may assign ranks either from large to small data (as in Example 8.13), or from small to large, calling the smallest datum rank 1, the next largest rank 2, and so on. The value of U obtained using one ranking procedure will be the same as the value of U' using the other procedure. Since for a two-tailed test both U and U' are employed, it makes no difference from which direction the ranks are assigned.

In summary, we note that after ranking the data of the two samples, we calculate U and U' using either Equations 8.45 and 8.47 which requires the determination of R_1, or Equations 8.46 and 8.48 which requires R_2. That is, the sum of the ranks for only one of the samples is needed. However, we may wish to compute both R_1 and R_2 in order to perform the following check on the assignment of ranks (which is especially

---

*The Wilcoxon two-sample test (sometimes referred to as the Wilcoxon rank-sum test) uses a test statistic commonly called W, which is R_1 or R_2; the test is equivalent to the Mann-Whitney test, for $U = R_2 - n_2(n_2 + 1)/2$ and $U' = R_1 - n_1(n_1 + 1)/2$.

†U (or U') is also equal to the number of data in one sample that are exceeded by each datum in the other sample. Note in Example 8.13: For females, ranks 7 and 8 each exceed 6 male ranks and ranks 10, 11, and 12 each exceed all 7 males ranks, for a total of $6 + 6 + 7 + 7 + 7 = 33 = U$; for males, rank 9 exceeds 2 female ranks for a total of $2 = U'$.

EXAMPLE 8.13   The Mann-Whitney test for nonparametric testing of the two-tailed null hypothesis that there is no difference between the heights of male and female students.

H_0:   Male and female students are the same height.

H_A: Male and female students are not the same height.

$\alpha = 0.05$

Heights of males	Heights of females	Ranks of male heights	Ranks of female heights
193 cm	175 cm	1	7
188	173	2	8
185	168	3	10
183	165	4	11
180	163	5	12
178		6	
170		9	
$n_1 = 7$	$n_2 = 5$	$R_1 = 30$	$R_2 = 48$

$$U = n_1 n_2 + \frac{n_1(n_1 + 1)}{2} - R_1$$

$$= (7)(5) + \frac{(7)(8)}{2} - 30$$

$$= 35 + 28 - 30$$

$$= 33$$

$$U' = n_1 n_2 - U$$

$$= (7)(5) - 33$$

$$= 2$$

$U_{0.05(2),7,5} = U_{0.05(2),5,7} = 30$

As $33 > 30$, H_0 is rejected.

$$0.01 < P(U \geq 33 \text{ or } U' \leq 2) < 0.02$$

desirable in the somewhat more complex case of assigning ranks to tied data, as will be shown below):

$$R_1 + R_2 = \frac{N(N + 1)}{2}. \tag{8.49}$$

Thus, in Example 8.13,

$$R_1 + R_2 = 30 + 48 = 78$$

should equal

$$\frac{N(N + 1)}{2} = \frac{12(12 + 1)}{2} = 78.$$

TABLE 8.2   The Appropriate Test Statistic for the One-Tailed Mann-Whitney Test

	H_0: Group 1 $\geq$ Group 2 H_A: Group 1 $<$ Group 2	H_0: Group 1 $\leq$ Group 2 H_A: Group 1 $>$ Group 2
Ranking done *from low to high*	U	U'
Ranking done *from high to low*	U'	U

This provides a check on (although it does not guarantee the accuracy of) the assignment of ranks.

   If the underlying assumptions of the parametric counterpart of a nonparametric test are met, then the parametric procedure will be the more powerful. The Mann-Whitney test is one of the most powerful of nonparametric tests, however; when either the Mann-Whitney test or the two-sample t test is applicable, the former is about 95% as powerful as the latter (Mood, 1954); and when the assumptions of the t test are seriously violated, the Mann-Whitney test may be much more powerful (Hodges and Lehmann, 1956).[*]

   An alternative procedure for nonparametric testing is the application of the two-sample t test (Equation 8.7a) to the ranks of the N data (which is referred to as using the "rank transformation" of the data). This has the same power as the Mann-Whitney test (Nath and Duran, 1981).

   The Mann-Whitney Test with Tied Ranks.   Example 8.14 demonstrates an important consideration encountered in tests requiring the ranking of observations. When two or more observations have exactly the same value, they are said to be *tied*. The rank assigned to each of the tied ranks is the mean of the ranks that would have been assigned to these ranks had they not be tied.[†] For example, in the present set of data, which are ranked from low to high, the third and fourth lowest values are tied at 32 words per minute, so they are each assigned the rank of $(3 + 4)/2 = 3.5$. The eighth, ninth, and tenth observations are tied at 44 words per minute, so each of them receives the rank of $(8 + 9 + 10)/3 = 9$. Once the ranks have been assigned by this procedure, U and U' are calculated as previously described.

   The One-Tailed Mann-Whitney Test.   For one-tailed hypotheses we need to declare which tail of the Mann-Whitney distribution is of interest, as this will determine whether U or U' must be calculated. This consideration is presented in Table 8.2. In Example 8.14 we have data that were ranked from lowest to highest and the alternate hypothesis states that the data in group 1 are greater in magnitude than those in

---

[*]Strictly speaking, Mann-Whitney statistics are affected by both differences between the two populations along the measurement scales—about which the hypothesis test inquires—and differences between the dispersions and shapes of the two populations (e.g., Boneau, 1962); but the latter effect is generally small compared to the former.

[†]Although other procedures have been proposed to deal with ties, assigning the rank mean has predominated for a long time (e.g., Kendall, 1945).

EXAMPLE 8.14   The one-tailed Mann-Whitney test used to determine the effectiveness of high school training on the typing speed of college students. This example also demonstrates the assignment of ranks to tied data.

H_0:   Typing speed is not greater in college students having had high school typing training.

H_A: Typing speed is greater in college students having had high school typing training.

$\alpha = 0.05$

Typing Speed (words per minute)

With training (rank in parentheses)	Without training (rank in parentheses)
44 (9)	32 (3.5)
48 (12)	40 (7)
36 (6)	44 (9)
32 (3.5)	44 (9)
51 (13)	34 (5)
45 (11)	30 (2)
54 (14)	26 (1)
56 (15)	
$n_1 = 8$	$n_2 = 7$
$R_1 = 83.5$	$R_2 = 36.5$

Because ranking was done from low to high and the alternate hypothesis states that the data of group one are larger than the data of group two, use U' as the test statistic (as indicated in Table 8.2).

$$U' = n_2 n_1 + \frac{n_2(n_2 + 1)}{2} - R_2$$

$$= (7)(8) + \frac{(7)(8)}{2} - 36.5$$

$$= 56 + 28 - 36.5$$

$$= 47.5$$

$U_{0.05(1),8,7} = U_{0.05(1),7,8} = 43$

As $47.5 > 43$, reject H_0.

$$0.01 < P < 0.025$$

group 2. Therefore, we need to compute U' and compare it to the one-tailed critical value, $U_{\alpha(1),n_1,n_2}$, from Appendix Table B.11.

The Normal Approximation to the Mann-Whitney Test.   Note that Appendix Table B.11 can be used only if the size of the smaller sample does not exceed twenty and the size of the larger sample does not exceed forty. Fortunately, the distribution of U approaches the normal distribution for larger samples. For large n_1 and n_2 we use the fact that the U distribution has a mean of

$$\mu_U = \frac{n_1 n_2}{2} \tag{8.50}$$

and a standard error of

$$\sigma_U = \sqrt{\frac{n_1 n_2 (N+1)}{12}}, \tag{8.51}$$

where $N = n_1 + n_2$, as used earlier. Thus, if a U, or a U', is calculated from data where either n_1 or n_2 is greater than that in Appendix Table B.11, its significance can be determined by computing

$$Z = \frac{U - \mu_U}{\sigma_U} \tag{8.52}$$

or, using a correction for continuity, by

$$Z_c = \frac{|U - \mu_U| - 0.5}{\sigma_U}. \tag{8.53}$$

The continuity correction is included to account for the fact that Z is a continuous distribution, but U is a discrete distribution. However, it appears to be advisable only if the two-tailed P is about 0.05 or greater (as seen from an expansion of the presentation of Lehmann, 1975: 17).

Recalling that the t distribution with $\nu = \infty$ is identical to the normal distribution, the critical value, Z_α, is equal to the critical value, $t_{\alpha, \infty}$. The normal approximation is demonstrated in Example 8.15. When using the normal approximation for two-tailed testing, only U or U' (not both) need be calculated. If U' is computed instead of U, then U' is simply substituted for U in Equation 8.52 or 8.53, the rest of the testing procedure remaining the same.

One-tailed testing may also be performed using the normal approximation. Here one computes either U or U', in accordance with Table 8.2 and uses it in either Equation 8.54 or 8.55, respectively, inserting the correction term (-0.5) if P is about 0.025 or greater:

$$Z_c = \frac{U - \mu_U - 0.5}{\sigma_U}; \tag{8.54}$$

$$Z_c = \frac{U' - \mu_U - 0.5}{\sigma_U}. \tag{8.55}$$

The resultant Z_c is then compared to the one-tailed critical value, $Z_{\alpha(1)}$, or, equivalently, $t_{\alpha(1), \infty}$; and if $Z \geq$ the critical value, then H_0 is rejected.[‡]

If tied ranks exist and the normal approximation is utilized, the computations are slightly modified as follows. One should calculate the quantity

$$\sum t = \sum (t_i^3 - t_i), \tag{8.56}$$

[‡]By this procedure, Z must be positive in order to reject H_0. If it is negative, then the probability of H_0 being true is $P > 0.50$.

EXAMPLE 8.15   The normal approximation to a one-tailed Mann-Whitney test to determine whether animals raised on a dietary supplement reach a greater body weight than those raised on an unsupplemented diet.

In the experiment, twenty-two animals (group 1) were raised on the supplemented diet, and forty-six were raised on the unsupplemented diet (group 2). The body weights were ranked from 1 (for the smallest weight) to 68 (for the largest weight), and U was calculated to be 282.

H_0:   Body weight of animals on the supplemented diet are not greater than those on the unsupplemented diet.

H_A: Body weight of animals on the supplemented diet are greater than those on the unsupplemented diet.

$n_1 = 22, n_2 = 46, N = 68$

$U = 282$

$$U' = n_1 n_2 - U = (22)(46) - 282 = 1012 - 282 = 730$$

$$\mu_U = \frac{n_1 n_2}{2} = \frac{(22)(46)}{2} = 506$$

$$\sigma_U = \sqrt{\frac{n_1 n_2 (N+1)}{12}} = \sqrt{\frac{(22)(46)(68+1)}{12}} = 76.28$$

$$Z = \frac{U' - \mu_U}{\sigma_U} = \frac{224}{76.28} = 2.94$$

For a one-tailed test at $\alpha = 0.05, t_{0.05(1),\infty} = Z_{0.05(1)} = 1.6449$. As $Z = 2.94 > 1.6449$, reject H_0.   [$P = 0.0016$]

---

where t_i is the number of ties in a group of tied values, and the summation is performed over all groups of ties. Then,

$$\sigma_U = \sqrt{\frac{n_1 n_2}{N^2 - N} \cdot \frac{N^3 - N - \sum t}{12}}, \tag{8.57}$$

and this value is used in place of that from Equation 8.51. (The computation of $\sum t$ is demonstrated, in another context, in Example 10.11.)

The normal approximation is excellent for $\alpha(2) = 0.10$ or 0.05 [or for $\alpha(1) = 0.05$ and 0.025] and is also good for $\alpha(2) = 0.20$ and 0.02 [or for $\alpha(1) = 0.10$ or 0.01]. But for more extreme significance levels it is not reliable, especially if n_1 and n_2 are dissimilar. Buckle, Kraft, and van Eeden (1969) propose another distribution, which they refer to as the "uniform approximation." They show it to be more accurate for $n_1 \neq n_2$, especially when the difference between n_1 and n_2 is great, and especially for small α.

Fix and Hodges (1955) describe an approximation to the Mann-Whitney distribution that is much more accurate than the normal approximation but requires very involved computation. Hodges, Ramsey, and Wechsler (1990) presented a simpler method for an improved normal approximation, one that provides remarkably good results for probabilities of about 0.001 or greater. For this, calculate:

$$A = \frac{20n_1n_2(n_1 + n_2 + 1)}{n_1^2 + n_2^2 + n_1n_2 + n_1 + n_2}, \tag{8.58}$$

$$B = \frac{155(Z_c^2)^2 - 416Z_c^2 - 195}{42}, \tag{8.59}$$

$$C = 1 + \frac{Z_c^2 - 3}{A} + \frac{B}{A^2}, \tag{8.60}$$

and

$$Z_c' = CZ_c. \tag{8.61}$$

This procedure is demonstrated in Example 8.16. Often, as in this example, the difference between Z_c' and Z_c is small. The computational effort for Z_c' is most warranted when the resultant probability is near the α used to conclude statistical significance.

The Mann-Whitney Test for Ordinal Data. The Mann-Whitney test may also be used for ordinal data. Example 8.17 demonstrates this procedure. In this example twenty-five undergraduate students were enrolled in an invertebrate zoology course. Each student was guided through the course by one of two teaching assistants. On the basis of the final grades, we wish to test the null hypothesis that students perform equally well in the course, under both teaching assistants. The variable measured (i.e., the final examination grades) results in ordinal data, and the hypothesis is amenable to testing by the Mann-Whitney test.

EXAMPLE 8.16   The improved normal approximation to the Mann-Whitney test.

For the hypotheses and data of Example 8.15:

$n_1 = 22, n_2 = 46, N = 68, U = 282, U' = 730$

$\mu_U = 506, \quad \sigma_U = 76.2824$

$Z_c = \dfrac{U' - \mu_U - 0.5}{\sigma_U} = \dfrac{730 - 506 - 0.5}{76.2824} = 2.9299$

$Z_c^2 = (2.9299)^2 = 8.5843$

$A = \dfrac{20(22)(46)(68 + 1)}{22^2 + 46^2 + (22)(46) + 22 + 46} = 379.5000$

$B = \dfrac{155(8.5843)^2 - 416(8.5843) - 195}{42} = 182.2836$

$C = 1 + \dfrac{8.5843 - 3}{379.5000} + \dfrac{182.2836}{(379.5000)^2} = 1.0160$

$Z_c' = 1.0160(2.9299) = 2.9768$

For $\alpha = 0.05$, $Z_{0.05(1)} = t_{0.05(1),\infty} = 1.6449$

As $2.9768 > 1.6449$, reject H_0; $0.001 < P < 0.0025$ $[P = 0.0015]$

EXAMPLE 8.17   The Mann-Whitney test for ordinal data.

H_0:   The performance of students is the same under the two teaching assistants.

H_A: Students do not perform equally well under the two teaching assistants.

$\alpha = 0.05$.

Teaching Assistant A		Teaching Assistant B	
Grade	Rank of grade	Grade	Rank of grade
A	3	A	3
A	3	A	3
A	3	B+	7.5
A−	6	B+	7.5
B	10	B	10
B	10	B−	12
C+	13.5	C	16.5
C+	13.5	C	16.5
C	16.5	C−	19.5
C	16.5	D	22.5
C−	19.5	D	22.5
		D	22.5
		D	22.5
		D−	25

$$n_1 = 11 \qquad\qquad n_2 = 14$$

$$R_1 = 114.5 \qquad\qquad R_2 = 210.5$$

$$U = n_1 n_2 + \frac{n_1(n_1 + 1)}{2} - R_1$$

$$= (11)(14) + \frac{(11)(12)}{2} - 114.5$$

$$= 154 + 66 - 114.5$$

$$= 105.5$$

$$U' = n_1 n_2 - U$$

$$= (11)(14) - 105.5$$

$$= 48.5$$

$$U_{0.05(2), 11, 14} = 114$$

As $105.5 < 114$, do not reject H_0.

$$0.10 < P(U \geq 105.5 \text{ or } U \leq 48.5) < 0.20$$

Hypotheses Employing a Specified Difference Other Than Zero.   Using the two-sample t test, one can examine hypotheses such as H_0: $\mu_1 - \mu_2 = \mu_0$, where μ_0 is not zero. Similarly, the Mann-Whitney test can be applied to hypotheses such as H_0: males are at least 5 cm taller than females (a one-tailed hypothesis with data such as those in

Example 8.13) or H_0: the letter grades of students in one course are at least one grade higher that those of students in a second course (a one-tailed hypothesis with data such as those in Example 8.17). In the first hypothesis, one would list all the male heights but list all the female heights after increasing each of them by 5 cm. Then these listed heights would be ranked and the Mann-Whitney analysis would proceed as usual. For testing the second hypothesis, the letter grades for the students in the first course would be listed unchanged, with the grades for the second course increased by one letter grade before listing. Then all the listed grades would be ranked and subjected to the Mann-Whitney test.[*]

When dealing with ratio or interval scale data, it is also possible to propose hypotheses employing a multiplication, rather than an addition, constant. Consider the two-tailed hypothesis H_0: the wings of one species of insect are two times the length of the wings of a second species. We could test it by listing the wing lengths of the first species, listing each of the wing lengths of the second species multiplied by two, and then ranking the data and subjecting the ranks to the Mann-Whitney test. (The parametric t testing procedure, which assumes equal population variance ordinarily would be inapplicable for such a hypothesis, because multiplying the data by a constant changes the variance of the data by the square of the constant.)

8.11 TESTING FOR DIFFERENCE BETWEEN TWO MEDIANS

One can test the null hypothesis that two samples came from a population having the same median by a method called the *median test* (Mood, 1950: 394–395). The procedure is to determine the grand median for all the data in both samples and then to set up a 2×2 contingency table, as shown in Example 8.18. This contingency table can then be analyzed by chi-square (as in Section 23.3) or G (Section 23.7), or, if some frequencies are small (see Section 23.6), the Fisher exact test may be applied (Section 24.10). The median test is only about 64% as powerful as the two-sample t test when applied to data where the latter is applicable (Mood, 1954), and only about 67% as powerful as the Mann-Whitney test of the preceding section.[†]

8.12 THE EFFECT OF CODING

Coding the raw data will not alter the test statistics or conclusions of the preceding sections. In the case of calculating F or t values, this is because the numerator and denominator will be unchanged by coding that employs the addition or subtraction of

---

[*]To increase these grades by one letter each, a grade of "B" would be changed to an "A," a "C" changed to a "B," and so on; a grade of "A" would have to be increased to a grade not on the original scale (e.g., call it a "Z") and, when ranking, we simply have to keep in mind that this new grade is higher than an "A."

[†]As the median test refers to a population parameter in hypothesis testing, it is not a nonparametric test; but it is a distribution-free procedure. Although it does not assume a specific underlying distribution (e.g., normal), it does assume that the two populations have the same shape (a characteristic that is addressed by Schlittgen, 1979).

EXAMPLE 8.18   The two-sample median test, using the data of Example 8.17.

H_0:   The two samples came from populations with identical medians (i.e., the median performance is the same under the two teaching assistants).

H_A: The medians of the two sampled populations are not equal.

$\alpha = 0.05$

The median of all N measurements, where $N = 25$, is $X_{(n+1)/2} = X_{12} = $ grade of C+. The following 2×2 contingency table is analyzed by chi-square:

Number	Sample 1	Sample 2	Total
Above median	6	6	12
Not above median	5	8	13
Total	11	14	25

$$X_c^2 = \frac{n\left(|f_{11}f_{22} - f_{12}f_{21}| - \dfrac{n}{2}\right)^2}{(C_1)(C_2)(R_1)(R_2)} \qquad \text{[Equation 22.7]}$$

$$= 0.031$$

$X_{0.05,1}^2 = 3.841$

Therefore, do not reject H_0.

$$0.75 < P < 0.90 \quad [P = 0.86]$$

---

a constant, whereas coding by multiplication or division will change the numerator and denominator proportionately the same. As coding of any sort will not change the ranks of the data, the Mann-Whitney and median tests remain unaffected.

8.13 TWO-SAMPLE TESTING OF NOMINAL-SCALE DATA

We may compare two samples of nominal data simply by arranging the data in a $2 \times C$ contingency table and proceeding as described in Chapter 23 and demonstrated in Example 23.1.

8.14 TESTING FOR DIFFERENCE BETWEEN TWO DIVERSITY INDICES

If the Shannon index of diversity, H', is obtained for each of two samples, it may be desired to test the null hypothesis that the diversities of the two sampled populations are equal. Hutcheson (1970) proposed a t test for this purpose:

$$t = \frac{H_1' - H_2'}{s_{H_1' - H_2'}}, \tag{8.62}$$

where

$$s_{H_1' - H_2'} = \sqrt{s_{H_1'}^2 + s_{H_2'}^2}. \qquad (8.63)$$

The variance of each H' may be approximated by

$$s_{H'}^2 = \frac{\sum f_i \log^2 f_i - (\sum f_i \log f_i)^2 / n}{n^2} \qquad (8.64)$$

(Basharin, 1959; Lloyd, Zar, and Karr, 1968),* where s, f_1, and n are as defined in Section 4.7. Logarithms to any base may be utilized for this calculation, but those to base 10 are most commonly employed. The degrees of freedom associated with the preceding t are approximated by

$$\nu = \frac{\left(s_{H_1'}^2 + s_{H_2'}^2\right)^2}{\dfrac{\left(s_{H_1'}^2\right)^2}{n_1} + \dfrac{\left(s_{H_2'}^2\right)^2}{n_2}} \qquad (8.65)$$

(Hutcheson, 1970).

Example 8.19 demonstrates these computations. If one is faced with many calculations of $s_{H_2'}^2$, the tables of $f_i \log^2 f_i$ provided by Lloyd, Zar, and Karr (1968) will be helpful. One-tailed as well as two-tailed hypotheses may be tested by this procedure. Also, the population diversity indices may be hypothesized to differ by some value, μ_0, other than zero, in which case the numerator of t would be $|H_1' - H_2'| - \mu_0$.

---

*Bowman et al. (1971) give an approximation [their Equation (11b)] that is more accurate for very small n.

EXAMPLE 8.19   Comparing two indices of diversity.

H_0:   The diversity of plant food items in the diet of Michigan blue jays is the same as the diversity of plant food items in the diet of Louisiana blue jays.

H_A: The diversity of plant food items in the diet of Michigan blue jays is not the same as in the diet of Louisiana blue jays.

$\alpha = 0.05$

	Michigan Blue Jays		
Diet item	f_i	$f_i \log f_i$	$f_i \log^2 f_i$
Oak	47	78.5886	131.4078
Corn	35	54.0424	83.4452
Blackberry	7	5.9157	4.9994
Beech	5	3.4949	2.4429
Cherry	3	1.4314	0.6830
Other	2	0.6021	0.1812
$s_1 = 6$	$n_1 = \sum f_i$ $= 99$	$\sum f_i \log f_i$ $= 144.0751$	$\sum f_i \log^2 f_i$ $= 223.1595$

EXAMPLE 8.19 **(continued)**

$$H_1' = \frac{n \log n - \sum f_i \log f_i}{n} = \frac{197.5679 - 144.0751}{99}$$

$$= 0.5403$$

$$s_{H_1'}^2 = \frac{\sum f_i \log^2 f_i - (\sum f_i \log f_i)^2/n}{n^2} = 0.00137602$$

Louisiana Blue Jays

Diet item	f_i	$f_i \log f_i$	$f_i \log^2 f_i$
Oak	48	80.6996	135.6755
Pine	23	31.3197	42.6489
Grape	11	11.4553	11.9294
Corn	13	14.4813	16.1313
Blueberry	8	7.2247	6.5246
Other	2	0.6021	0.1812
$s_2 = 6$	$n_2 = \sum f_i$	$\sum f_i \log f_i$	$\sum f_i \log^2 f_i$
	$= 105$	$= 145.7827$	$= 213.0909$

$$H_2' = \frac{n \log n - \sum f_i \log f_i}{n} = \frac{212.2249 - 145.7827}{105} = 0.6328$$

$$s_{H_2'}^2 = \frac{\sum f_i \log^2 f_i - \left(\sum f_i \log f_i\right)^2/n}{n^2} = 0.00096918$$

$$s_{H_1'-H_2'} = \sqrt{s_{H_1'}^2 + s_{H_2'}^2} = \sqrt{0.00137602 + 0.00096918} = 0.0484$$

$$t = \frac{H_1' - H_2'}{s_{H_1'-H_2'}} = \frac{-0.0925}{0.0484} = -1.911$$

$$\nu = \frac{\left(s_{H_1'}^2 + s_{H_2'}^2\right)^2}{\dfrac{\left(s_{H_1'}^2\right)^2}{n_1} + \dfrac{\left(s_{H_2'}^2\right)^2}{n_2}} = \frac{(0.00137602 + 0.00096918)^2}{\dfrac{(0.00137602)^2}{99} + \dfrac{(0.00096918)^2}{105}}$$

$$= \frac{0.000005499963}{0.000000028071} = 196$$

$t_{0.05(2),196} = 1.972$

Therefore, do not reject H_0.

$$0.05 < P < 0.10 \quad [P = 0.057]$$

EXERCISES

8.1 Using the following data, test the null hypothesis that male and female turtles have the same mean serum cholesterol concentrations.

Serum Cholesterol (mg/100 ml)

Male	Female
220.1	223.4
218.6	221.5
229.6	230.2
228.8	224.3
222.0	223.8
224.1	230.8
226.5	

8.2 It is proposed that animals with a northerly distribution have shorter appendages than animals from a southerly distribution. Test an appropriate hypothesis (by computing t), using the following wing length data for birds (data are in millimeters).

Northern	Southern
120	116
113	117
125	121
118	114
116	116
114	118
119	123
	120

8.3 If $\bar{X}_1 = 4.6$ kg, $s_1^2 = 3.88$ kg^2, $n_1 = 18$, $\bar{X}_2 = 6.0$ kg, $s_2^2 = 45.3$ kg^2, and $n_2 = 26$, test the hypotheses $H_0: \mu_1 \geq \mu_2$ and $H_A: \mu_1 < \mu_2$.

8.4 If $\bar{X}_1 = 334.6$ g, $\bar{X}_2 = 349.8$ g, SS$_1 = 364.34$ g^2, SS$_2 = 286.78$ g^2, $n_1 = 19$, and $n_2 = 24$, test the hypothesis that the mean weight of population 2 is more than 10 g greater than the mean weight of population 1.

8.5 If the null hypothesis in Exercise 8.1 is rejected, compute the 95% confidence limits for μ_1, μ_2, and $\mu_1 - \mu_2$. If H_0 is not rejected, calculate the 95% confidence limits for the common population mean.

8.6 A sample is to be taken from each of two populations from which previous samples of size fourteen have had SS$_1 = 244.66$ (km/hr)^2 and SS$_2 = 289.18$ (km/hr)^2. What size sample should be taken from each population in order to estimate $\mu_1 - \mu_2$ to within 2.0 km/hr, with 95% confidence?

8.7 Consider the populations described in Exercise 8.6.
   (a) How large a sample should we take from each population if we wish to detect a difference between μ_1 and μ_2 of at least 5.0 km/hr, using a 5% significance level and a t test with 90% power?
   (b) If we take a sample of twenty from one population and twenty-two from the other, what is the smallest difference between μ_1 and μ_2 that we have a 90% probability of detecting with a t test using $\alpha = 0.05$?

(c) If $n_1 = n_2 = 50$, and $\alpha = 0.05$, what is the probability of rejecting H_0: $\mu_1 = \mu_2$ when $\mu_1 - \mu_2$ is as small as 2.0 km/hr?

8.8 A sample of twenty-nine plant heights of members of a certain species had $s^2 = 14.62$ cm^2, and the heights of a sample of twenty-five from a second species had $s^2 = 8.45$ cm^2. Test the null hypothesis that the variances of the two sampled populations are the same.

8.9 If $s_1^2 = 324.46$ sec^2, $s_2^2 = 158.95$ sec^2, $n_1 = 41$, and $n_2 = 36$, test the hypotheses H_0: $\sigma_1^2 \leq \sigma_2^2$ and H_A: $\sigma_1^2 > \sigma_2^2$.

8.10 A sample of thirteen data from a population has a variance of 21.35 g^2, and a sample of fifteen from a second population has a variance of 38.71 g^2.

(a) What is the 95% confidence interval for the ratio of σ_2^2 to σ_1^2?

(b) How large a sample must be taken from each population if we wish to have a 90% chance of rejecting H_0: $\sigma_1^2 \geq \sigma_2^2$ when H_A: $\sigma_1^2 < \sigma_2^2$ is true and we test at the 5% level of significance?

(c) What would be the power of a test of this H_0, with $\alpha = 0.05$, if sample sizes of fifty were used?

8.11 For the data in Exercise 8.8, $\bar{X}_1 = 10.74$ cm and $\bar{X}_2 = 14.32$ cm. Test the null hypothesis that the coefficients of variation of the two sampled populations are the same.

8.12 Using the Mann-Whitney test, test the appropriate hypotheses for the data in Exercise 8.1.

8.13 Using the Mann-Whitney procedure, test the appropriate hypotheses for the data in Exercise 8.2.

8.14 The following data are volumes (in cubic microns) of avian erythrocytes taken from normal (diploid) and intersex (triploid) individuals. Test the hypothesis (using the Mann-Whitney test) that the volume of intersex cells is 1.5 times the volume of normal cells.

Normal	Intersex
248	380
236	391
269	377
254	392
249	398
251	374
260	
245	
239	
255	

9

PAIRED-SAMPLE HYPOTHESES

The two-sample testing procedures discussed in Chapter 8 apply when the two samples are independent, independence implying that each datum in one sample is in no way associated with any specific datum in the other sample. However, there are instances when each observation in Sample 1 is in some way correlated with an observation in Sample 2, so that the data may be said to occur in pairs.

For example, we might wish to test the null hypothesis that the left foreleg and left hindleg lengths of deer are equal. We could make these two measurements on a number of deer, but we would have to remember that the variation among the data might be owing to two possible factors. First, the null hypothesis might be false, there being, in fact, a difference between foreleg and hindleg length. Second, deer are of different sizes, and for each deer the hindleg length is correlated with the foreleg length (i.e., a deer with a large front leg is likely to have a large hind leg). Thus, as Example 9.1 shows, the data can be tabulated in pairs, one pair (i.e., one hindleg measurement and one foreleg measurement) per animal.

9.1 TESTING MEAN DIFFERENCE

The two-tailed hypotheses implied by Example 9.1 are H_0: $\mu_1 - \mu_2 = 0$ and H_A: $\mu_1 - \mu_2 \neq 0$ (which, as pointed out in Section 8.1, could also be stated H_0: $\mu_1 = \mu_2$ and H_A: $\mu_1 \neq \mu_2$). However, we can define a mean population difference, μ_d, as $\mu_1 - \mu_2$, and write the hypotheses as H_0: $\mu_d = 0$ and H_A: $\mu_d \neq 0$. Although the use of either μ_d or $\mu_1 - \mu_2$ is correct, the former will be used hereafter when it implies the paired-sample situation.

EXAMPLE 9.1   The two-tailed paired-sample t test.

H_0: $\mu_d = 0$.

H_A: $\mu_d \neq 0$.

$\alpha = 0.05$

Deer (j)	Hindleg length (cm) (X_{1j})	Foreleg length (cm) (X_{2j})	Difference (cm) ($d_j = X_{1j} - X_{2j}$)
1	142	138	4
2	140	136	4
3	144	147	−3
4	144	139	5
5	142	143	−1
6	146	141	5
7	149	143	6
8	150	145	5
9	142	136	6
10	148	146	2

$n = 10$ $\bar{d} = 3.3$ cm

$\nu = n - 1 = 9$ $t = \dfrac{\bar{d}}{s_{\bar{d}}} = \dfrac{3.3}{0.97} = 3.402$

$s_d^2 = 9.3444$ cm^2

$s_{\bar{d}} = 0.97$ cm $t_{0.05(2),9} = 2.262$

Therefore, reject H_0.

$$0.005 < P(|t| \geq 3.402) < 0.01 \quad [P = 0.008]$$

The test statistic for the null hypothesis is

$$t = \frac{\bar{d}}{s_{\bar{d}}}. \tag{9.1}$$

Therefore, we do not use the original measurements for the two samples, but only the difference within each pair of measurements. One deals, then, with a sample of d_j values, whose mean is $\bar{d}$ and whose variance, standard deviation, and standard error are denoted as s_d^2, s_d, and $s_{\bar{d}}$ respectively. Thus, the *paired-sample t test*, as this procedure may be called, is essentially a one-sample t test, analogous to that described in Sections 7.1 and 7.2. In the paired-sample t test, n is the number of differences (i.e., the number of pairs of data), and $\nu = n - 1$. Note that the hypotheses used in Example 9.1 are special cases of the general hypotheses H_0: $\mu_d = \mu_0$ and H_A: $\mu_d \neq \mu_0$, where μ_0 is usually, but not always, zero.

For one-tailed hypotheses with paired samples, one can test either H_0: $\mu_d \geq \mu_0$ and H_A: $\mu_d < \mu_0$, or H_0: $\mu_d \leq \mu_0$ and H_A: $\mu_d > \mu_0$, depending on the question to be asked. Example 9.2 presents data from an experiment designed to test whether a new fertilizer results in an increase of more than 250 kg/ha in crop yield over the old fertilizer. For testing this hypothesis, eighteen test plots of the crop were set up. It is probably unlikely to find eighteen field plots having exactly the same conditions of soil, moisture, wind, etc., but it should be possible to set up two plots with similar environmental conditions. If so, then the experimenter would be wise to set up nine pairs of plots, applying the new fertilizer to one plot of each pair and the old fertilizer to the other plot of that pair. As Example 9.2 shows, the statistical hypotheses to be tested are H_0: $\mu_d \leq 250$ kg/ha and H_A: $\mu_d > 250$ kg/ha.

EXAMPLE 9.2   A one-tailed paired-sample t test.

H_0: $\mu_d \leq 250$ kg/ha

H_A: $\mu_d > 250$ kg/ha

$\alpha = 0.05$

Plot (j)	**Crop Yield** (kg/ha) With new fertilizer (X_{1j})	With old fertilizer (X_{2j})	d_j
1	2250	1920	330
2	2410	2020	390
3	2260	2060	200
4	2200	1960	240
5	2360	1960	400
6	2320	2140	180
7	2240	1980	260
8	2300	1940	360
9	2090	1790	300

$n = 9$ $\bar{d} = 295.6$ kg/ha

$v = n - 1 = 8$ $t = \dfrac{\bar{d} - 250}{s_{\bar{d}}} = 1.695$

$s_d = 80.6$ kg/ha

$s_{\bar{d}} = 26.9$ kg/ha $t_{0.05(1),8} = 1.860$

Therefore, do not reject H_0

$0.05 < P < 0.10$ $[P = 0.064]$

Paired-sample t-testing requires that each datum in one sample is correlated with one, *but only one*, datum in the other sample. So, in the last example, each yield using new fertilizer is paired with only one yield using old fertilizer; and it would have been inappropriate to have some tracts of land large enough to collect two or more crop yields using each of the fertilizers.

The paired-sample t test does not have the normality and equality of variances assumptions of the two-sample t test, but assumes instead that the differences, d_j, come from a normally distributed population of differences. If there is, in fact, pairwise correlation of data from the two samples, then the paired-sample t test will be more powerful than the two-sample t test. If no such correlation exists, then the two-sample t test will be the more powerful procedure. Hines (1996) showed that, unless n is very tiny, only a small correlation is needed to make the paired-sample test advantageous. If the data from Example 9.1 were subjected (inappropriately) to the two-sample t test, rather than to the paired-sample t test, a difference would not have been concluded, and a Type II error would have been committed.

9.2 Confidence Limits for the Population Mean Difference

In paired-sample testing we deal with a sample of differences, d_j, so confidence limits for the mean of a population of differences, μ_d, may be determined as in Section 7.3. In the manner of Equation 7.6, the $1 - \alpha$ confidence interval for μ_d is

$$\bar{d} \pm t_{\alpha(2),\nu} s_{\bar{d}}. \tag{9.2}$$

For example, for the data in Example 9.1, we can compute the 95% confidence interval for μ_d to be 3.3 cm $\pm(2.262)(0.97$ cm$) = 3.3$ cm ± 2.2 cm; the 95% confidence limits are $L_1 = 1.1$ cm and $L_2 = 5.5$ cm.

Furthermore, we may ask, as in Section 7.5, how large a sample is required to be $1 - \alpha$ confident in estimating μ_d to within $\pm d$ (using Equation 7.7).

9.3 Power and Sample Size in Paired-Sample Testing of Means

By considering the paired-sample t test to be a one-sample t test for a sample of differences, d_j, we may employ the procedures of Section 7.6 to address questions of required sample size, minimum detectable difference, and power (Equations 7.8, 7.9, and 7.10, respectively). For this purpose we simply substitute $\bar{d}$ for $\bar{X}$, and s_d^2 for s^2.

9.4 Testing for Difference between Variances from Two Correlated Populations

The variance ratio test of Section 8.5 assumes that the two samples of data are independent. If, for example, we wished to ask whether human arm lengths are as variable as leg lengths, we could measure a sample of arm lengths from several individuals and a sample of leg lengths from several different individuals. These two samples would be independent, and Equation 8.28 could be used. If, however, one arm length and one leg length were measured from each of several individuals, the arm and leg measurements would be correlated instead of independent (i.e., a person with a long arm would likely have a long leg). In such situations it is appropriate to use a modified variance ratio test that takes into account the amount of correlation; this is presented by Howell (1987: 181) and Snedecor and Cochran (1980: 190–191), based on the procedure of Pitman (1939); we compute

$$t = \frac{(F - 1)\sqrt{n - 2}}{2\sqrt{F(1 - r^2)}}.$$ (9.3)

Here, F is as in Equation 8.28, n is the sample size common to both samples, and r is the correlation coefficient described in Section 19.1 (Equation 19.1). The degrees of freedom associated with this t are $n - 2$. This test is not valid if the two sampled distributions are not normal (Sandvik and Olsson, 1982).

9.5 PAIRED-SAMPLE TESTING BY RANKS

The *Wilcoxon paired-sample test* (Wilcoxon, 1945; Wilcoxon and Wilcox, 1964: 9) is a nonparametric analogue to the paired-sample t test, just as the Mann-Whitney test is a nonparametric procedure analogous to the two-sample t test. The literature refers to the test by a variety of names, but usually in conjunction with Wilcoxon's name[*] and some wording such as "paired sample" or "matched pairs," sometimes together with a phrase like "rank sum" or "signed rank."

Whenever the paired-sample t test is applicable, the Wilcoxon paired-sample test is also applicable. If, however, the d_j values are from a normal distribution, then the latter test has only $3/\pi$ (i.e, 95%) of the power in detecting differences as the former (Conover, 1980: 291; Mood, 1954). But, there are instances when the Wilcoxon paired-sample test is applicable and the parametric paired-sample t test is not, as when one can not assume that the d_j's are from a normal distribution. Section 7.9 introduced the Wilcoxon procedure as a nonparametric one-sample test, but its greatest application is for paired-sample testing. The sign test (Section 23.7) could also be used for this purpose, but it is considerably less powerful.

Example 9.3 demonstrates the use of the Wilcoxon paired-sample test with the ratio scale data of Example 9.1, and it is also commonly used with ordinal scale data. The testing procedure involves the calculation of differences, as does the paired-sample t test. Then one ranks the absolute values of the differences, from low to high, and affixes the sign of each difference to the corresponding rank. As introduced in Section 8.10, the rank assigned to tied observations is the mean of the ranks that would have been assigned to the observations had they not been tied.

Then we sum the ranks with a plus sign (we shall call this sum T_+) and the ranks with a minus sign (calling this sum T_-). For a two-tailed test (as in Example 9.3), we reject H_0 if either T_+ or T_- is *less than or equal to* the critical value, $T_{\alpha(2),n}$, from Appendix Table B.12.

---

[*]Frank Wilcoxon (1892–1965), American (though born in Ireland) chemist and statistician, a major developer of statistical methods based on ranks (Bradley and Hollander, 1978).

EXAMPLE 9.3 **The Wilcoxon paired-sample test applied to the data of Example 9.1.**

H_0: Deer hindleg length is the same as foreleg length.

H_A: Deer hindleg length is not the same as foreleg length.

$\alpha = 0.05$

Deer (j)	Hindleg length (cm) (X_{1j})	Foreleg length (cm) (X_{2j})	Difference ($d_j = X_{1j} - X_{2j}$)	Rank of $|d_j|$	Signed rank of $|d_j|$
1	142	138	4	4.5	4.5
2	140	136	4	4.5	4.5
3	144	147	−3	3	−3
4	144	139	5	7	7
5	142	143	−1	1	−1
6	146	141	5	7	7
7	149	143	6	9.5	9.5
8	150	145	5	7	7
9	142	136	6	9.5	9.5
10	148	146	2	2	2

$n = 10$

$T_+ = 4.5 + 4.5 + 7 + 7 + 9.5 + 7 + 9.5 + 2 = 51$

$T_- = 3 + 1 = 4$

$T_{0.05(2),10} = 8$

Since $T_- < T_{0.05(2),10}$, H_0 is rejected.

$$0.01 < P(T_- \text{ or } T_+ \leq 4) < 0.02$$

Having calculated either T_+ or T_-, the other can be determined as

$$T_- = \frac{n(n+1)}{2} - T_+ \tag{9.4}$$

or

$$T_+ = \frac{n(n+1)}{2} - T_-. \tag{9.5}$$

For one-tailed testing we use one-tailed critical values from Appendix Table B.12 and either T_+ or T_- as follows. For the hypotheses

H_0: Measurements in population 1 $\leq$ measurements in population 2

and H_A: Measurements in population 1 $>$ measurements in population 2

H_0 is rejected if $T_- \leq T_{\alpha(1),n}$. For the opposite hypotheses:

H_0: Measurements in population 1 $\geq$ measurements in population 2

and H_A: Measurements in population 1 $<$ measurements in population 2

reject H_0 if $T_+ \leq T_{\alpha(1),n}$.

For the one-sample median test of Section 7.9, the two-tailed H_0: $M = M_0$ is rejected if favor of H_A: $M \neq M_0$ if either T_+ or $T_- \leq T_{\alpha(2),n}$. For H_0: $M \leq M_0$ vs. H_A: $M > M_0$, reject H_0 if $T_- \leq T_{\alpha(1),n}$; and for H_0: $M \geq M_0$ vs. H_A: $M < M_0$, H_0 is rejected if $T_+ \leq T_{\alpha(1),n}$.

One will obtain a different value of T_+ (call it T_+') or T_- (call it T_-') if rank 1 is assigned to the largest, rather than the smallest, d_i (i.e., the absolute values of the d_j's are ranked from high to low). If this is done, the test statistics are obtainable as

$$T_+ = m(n+1) - T_+' \tag{9.6}$$

and

$$T_- = m(n+1) - T_-', \tag{9.7}$$

where m is the number or ranks with the sign being considered.

Differences of zero are ignored in the analysis (i.e., they are discarded). Pratt (1959) recommended maintaining differences of zero until after ranking, and thereafter ignoring the ranks assigned to the zeros. This procedure may yield slightly better results in some circumstances, though worse results in others (Conover, 1973). If used, then the critical values of Rahe (1974) should be consulted or the normal approximation employed (see the following section) instead of using critical values of T from Appendix Table B.12.

If data are paired, the use of the Mann-Whitney test, instead of the Wilcoxon paired-sample test, may lead to the commission of a Type II error, with the concomitant inability to detect actual population differences.

The Wilcoxon paired-sample test has an underlying assumption that the sampled population of d_j's is symmetrical about the median. Another nonparametric test for paired samples is the sign test (described in Section 24.7), which does not have this assumption but is less powerful if the assumption is met.

The Normal Approximation to the Wilcoxon Paired-Sample Test.   For data consisting of more than 100 pairs (the limit of Appendix Table B.12), the significance of T (where either T_+ or T_- may be used for T) may be determined by considering that for such large samples the distribution of T is closely approximated by a normal distribution with a mean of

$$\mu_T = \frac{n(n+1)}{4} \tag{9.8}$$

and a standard error of

$$\sigma_T = \sqrt{\frac{n(n+1)(2n+1)}{24}}. \tag{9.9}$$

Thus, we can calculate

$$Z = \frac{|T - \mu_T|}{\sigma_T} \tag{9.10}$$

where for T we may use, with identical results, either T_+ or T_-. Then, for a two-tailed test, Z is compared to the critical value, $Z_{\alpha(2)}$, or, equivalently, $t_{\alpha(2),\infty}$ (which for $\alpha = 0.05$ is 1.9600); if Z is greater than or equal to $Z_{\alpha(2)}$, then H_0 is rejected.

A normal approximation with a correction for continuity employs

$$Z_c = \frac{|T - \mu_T| - 0.5}{\sigma_T}. \tag{9.11}$$

As shown at the end of Appendix Table B.12, the normal approximation is better using Z for $\alpha(2)$ from 0.001 to 0.05 and is better using Z_c for $\alpha(2)$ from 0.10 to 0.50.

If there are tied ranks, then use

$$\sigma_T = \sqrt{\frac{n(n+1)(2n+1) - \dfrac{\sum t}{2}}{24}}, \tag{9.12}$$

where

$$\sum t = \sum (t_i^3 - t_i) \tag{9.13}$$

is the correction for ties introduced in using the normal approximation to the Mann-Whitney test (Equation 8.56), applied here to ties of nonzero differences.

If we employ the Pratt procedure for handling differences of zero (described above), then the normal approximation is

$$Z = \frac{\left| T - \dfrac{n(n+1) - m'(m'+1)}{4} \right| - 0.5}{\sqrt{\dfrac{n(n+1)(2n+1) - m'(m'+1)(2m+1) - \dfrac{\sum t}{2}}{24}}} \tag{9.14}$$

(Cureton, 1967), where n is the total number of differences (including zero differences), and m' is the number of zero differences; $\sum t$ is as in Equation 9.13, applied to ties other than those of zero differences. We calculate T_+ or T_- by including the zero differences in the ranking and then deleting from considerations both the zero d_j's and the ranks assigned to them. For T in Equation 9.14, either T_+ or T_- may be used. If neither ranks nor zero d_j's are present, then Equation 9.14 becomes Equation 9.11.

One-tailed testing may also be performed using the normal approximation (Equation 10.9 or 10.10) or Cureton's procedure (Equation 9.14). The calculated Z is compared to $Z_{\alpha(1)}$ (or $t_{\alpha(1),\infty}$), and the direction of the arrow in the alternate hypothesis must be examined. If the arrow points to the left ("<"), then H_0 is rejected if $Z \geq Z_{\alpha(1)}$ and $T_+ < T_-$; if it points to the right (">"), then reject H_0 if $Z \geq Z_{\alpha(1)}$ and $T_+ > T_-$.

Iman (1974a) presents an approximation based on Student's t:

$$t = \frac{T - \mu_T}{\sqrt{\dfrac{n^2(n+1)(2n+1)}{2(n-1)} - \dfrac{(T - \mu_T)^2}{n-1}}}, \tag{9.15}$$

with $n - 1$ degrees of freedom. As shown at the end of Appendix Table B.12, this performs slightly better than the normal approximation (Equation 9.10). The test with a correction for continuity is performed by subtracting 0.5 from $|T - \mu_T|$ in both the numerator and denominator of Equation 9.15. This improves the test for $\alpha(2)$ from 0.001

to 0.10, but the uncorrected t is better for $\alpha(2)$ from 0.20 to 0.50. One-tailed t-testing is effected in a fashion similar to that described for Z in the preceding paragraph.*

9.6 CONFIDENCE LIMITS FOR THE POPULATION MEDIAN DIFFERENCE

In Section 9.2 confidence limits were obtained for the mean of a population of differences. Given a population of differences, one can also determine confidence limits for the population median. This is done exactly as shown in Section 7.8; simply consider the observed differences between members of pairs (d_j) as a sample from a population of such differences.

9.7 PAIRED-SAMPLE TESTING OF NOMINAL-SCALE DATA

Data in a 2 × 2 Table. Enumeration data may be collected from paired samples. If the data are dichotomous (i.e., the nominal scale variable has two possible values), then *McNemar's test* (McNemar, 1947) is applicable.

For example, assume that we wish to test whether two skin lotions are equally effective at relieving a poison ivy rash. The two lotions might be tested on fifty-one persons with poison ivy rash on both arms, applying one lotion to one arm and the other lotion to the second arm. The results of the experiment can be summarized in a 2 × 2 table, as shown in Example 9.4. (As in the more extensive discussion of Sections 23.1 and 23.3, f_{ij} denotes frequency observed in a row i and column j of this table which has 2 rows and 2 columns, R_i is the sum of the data in row i, and C_j is the sum in column j.) Eleven of the patients responded to both lotions and twenty-four responded to neither lotion (i.e., $f_{11} = 11$ and $f_{22} = 24$). The null hypothesis states that the same proportion of persons receive relief with lotion 1 as with lotion 2; this is to say that the sample proportion $(f_{11} + f_{12})/R_i$ is an estimate of the same population proportion as is $(f_{11} + f_{21})/C_j$. H_0 is tested by considering a goodness of fit test (Section 22.1) to a 1:1 ratio, using the observed frequencies f_{12} and f_{21}, as shown in Example 9.4. The null hypothesis may also be stated as H_0: $\psi = 1$, where[†] ψ is the population ratio estimated by f_{12}/f_{21}.

As long as $n = f_{12} + f_{21}$ is not less than 10, chi-square goodness of fit testing may be employed (Section 22.1); otherwise the binomial test (Section 24.6) is called for. For the chi-square goodness of fit to a 1:1 ratio, $\hat{f}_{12} = \hat{f}_{21} = (f_{12} + f_{21})/2$, and, because there is only one degree of freedom, Yates' correction for continuity usually is applied.

---

*When Appendix Table B.12 cannot be used, a slightly improved approximation is effected by comparing the mean of t and Z to the mean of the critical values of t and Z (Iman, 1974a).

[†] ψ is the lowercase Greek letter, psi, pronounced "sigh" (or, sometimes, with the "p" sound preceding).

EXAMPLE 9.4   McNemar's test for paired-sample nominal-scale data, using pairs of data (for the experimental situation described in the test above).

H_0:   The proportion of persons experiencing relief is the same with both lotions.

H_A: The proportion of persons experiencing relief is not the same with both lotions.

$\alpha = 0.05$

	Lotion 1	
	Relief	No relief
Lotion 2		
Relief	11	6
No relief	10	24

$$\chi_c^2 = \frac{(|f_{12} - f_{21}| - 1)^2}{f_{12} + f_{21}} = \frac{(|6 - 10| - 1)^2}{6 + 10} = 0.562$$

$\chi_{0.05,1}^2 = 3.841$

Therefore, do not reject H_0.

$$0.25 < P < 0.50 \quad [P = 0.45]$$

The chi-square (shown later as Equation 22.1) may also be computed as

$$\chi^2 = \frac{(f_{12} - f_{21})^2}{f_{12} + f_{21}}, \tag{9.16}$$

which is equivalent to computing the normal deviate

$$Z = \frac{|f_{12} - f_{21}|}{\sqrt{f_{12} + f_{21}}}, \tag{9.17}$$

and the continuity-corrected chi-square (shown later as Equation 21.3) as

$$\chi_c^2 = \frac{(|f_{12} - f_{21}| - 1)^2}{f_{12} + f_{21}}, \tag{9.18}$$

which is the same as employing

$$Z_c = \frac{|f_{12} - f_{21}| - 1}{\sqrt{f_{12} + f_{21}}}; \tag{9.19}$$

and the calculation of χ_c^2 is demonstrated in Example 9.4. As the application of this continuity correction may result in conservative testing (in that H_0 will be incorrectly rejected less often than α indicates), Haber (1982) has proposed different correction procedures (which are based on considerations beyond the scope of this discussion).

Note that this test employs only two of the four data tabulated (i.e., f_{12} and f_{21}), so the results are the same regardless of the magnitude of the other two data (f_{11} and f_{22}), which are considered tied data and ignored in the analysis. If f_{12} and f_{21} are small, this test does not work well. Therefore, if $(f_{12} + f_{21})/2 \leq 5$, then the binomial test (Section 23.6) is recommended, with $n = f_{12} + f_{21}$ and $X = f_{12}$ (or f_{21}).

Another common type of data amenable to McNemar testing results from the situation where experimental responses are recorded before and after some event, in which case one hears reference to this procedure as the McNemar test for significant changes.

Do not confuse this procedure with that of a 2×2 contingency table analysis (Section 22.3). Contingency table data are analyzed using a null hypothesis of independence between rows and columns, whereas in the case of data amenable to the McNemar test, there is intentional association between the row and column data.

Power and Sample Size.   The ability of the McNemar test to reject H_0 may be estimated by computing

$$Z_{\beta(1)} = \frac{\sqrt{n}\sqrt{p}(\psi - 1) - Z_{\alpha(2)}\sqrt{\psi + 1}}{\sqrt{(\psi + 1) - p(\psi - 1)^2}} \tag{9.20}$$

(Connett, Smith, and McHugh, 1987). Here, n is the number of pairs to be used (i.e., $n = f_{11} + f_{12} + f_{21} + f_{22}$); p is an estimate, as from a pilot study, of the proportion f_{12}/n or f_{21}/n, whichever is smaller; and ψ is the magnitude of difference desired to be detected by the hypothesis test, expressed as the ratio in the population of either f_{12} to f_{21}, or f_{21} to f_{12}, whichever is larger. Then, using Appendix Table B.2, or the last line in Appendix Table B.3 (i.e., for t with $\nu = \infty$), determine $\beta(1)$; and the power of the test is $1 - \beta(1)$. This procedure is demonstrated in Example 9.5.

Similarly, we can estimate the sample size necessary to perform a McNemar test with a specified power:

$$n = \frac{\left[Z_{\alpha(2)}\sqrt{\psi + 1} + Z_{\beta(1)}\sqrt{(\psi + 1) - p(\psi - 1)^2}\right]^2}{p(\psi - 1)^2} \tag{9.21}$$

(Connett, Smith, and McHugh, 1987). This is demonstrated in Example 9.6.

Data in Larger Tables.   The McNemar test may be extended to square tables larger than 2×2 (Bowker, 1948; Maxwell, 1970). What we test is whether the upper right corner of the table is symmetrical with the lower left corner. This is done by ignoring the data along the diagonal containing f_{ii} (i.e., row 1, column 1; row 2, column 2; etc). We compute

$$\chi^2 = \sum_{i=1}^{r}\sum_{j>i} \frac{(f_{ij} - f_{ji})^2}{f_{ij} + f_{ji}}, \tag{9.22}$$

with degrees of freedom of

$$\nu = \frac{r(r - 1)}{2}, \tag{9.23}$$

where r is the number of rows (or, equivalently, the number of columns) in the table of data. This is demonstrated in Example 9.7.

EXAMPLE 9.5   Determination of power of the McNemar test.

Considering the data of Example 9.4 to be from a pilot study, what would be the probability of rejecting H_0 if 200 pairs of data were used, if the test were performed at the 0.05 level of significance, and if the population ratio of f_{21} to f_{12} were at least 2?

From the pilot study, using $n = 51$ pairs of data, $f_{12}/n = 6/51 = 0.1176$ and $f_{21}/n = 10/51 = 0.1961$; so $p = 0.1176$. We specify $\alpha(2) = 0.05$, so $Z_{0.05(2)} = 1.9600$ (from the last line of Appendix Table B.3). And we also specify a new sample size of $n = 200$ and $\psi = 2$. Therefore,

$$Z_{\beta(1)} = \frac{\sqrt{n}\sqrt{p}(\psi - 1) - Z_{\alpha(2)}\sqrt{\psi + 1}}{\sqrt{(\psi + 1) - p(\psi - 1)^2}}$$

$$= \frac{\sqrt{200}\sqrt{0.1176}(2 - 1) - 1.9600\sqrt{2 + 1}}{\sqrt{(2 + 1) - 0.1176(2 - 1)^2}}$$

$$= \frac{(14.1421)(0.3429)(1) - 1.9600(1.7321)}{\sqrt{3 - (0.1176)(1)}}$$

$$= \frac{4.8493 - 3.3949}{\sqrt{2.8824}} = \frac{1.4544}{1.6978} = 0.86.$$

From Appendix Table B.2, if $Z_{\beta(1)} = 0.86$, then $\beta(1) = 0.19$; therefore, power [i.e., $1 - \beta(1)$] is $1 - 0.19 = 0.81$.

From Appendix Table B.3, if $Z_{\beta(1)}$ [i.e., $t_{\beta(1),\infty}$] is 0.86, then $\beta(1)$ lies between 0.25 and 0.10, and the power [i.e., $1 - \beta(1)$] lies between 0.75 and 0.90. [$\beta(1) = 0.19$ and power $= 0.81$.]

EXAMPLE 9.6   Determination of sample size for the McNemar test.

Considering the data of Example 9.4 to be from a pilot study, how many pairs of data would be needed to have a 90% probability of rejecting H_0 if the test were performed at the 0.05 level of significance, and the ratio of f_{21} to f_{12} in the population were at least 2?

As shown in Example 9.5, $p = 0.1176$ and $Z_{\alpha(2)} = 1.9600$. In addition, we specify that $\psi = 2$ and that the power of the test is to be 0.90 [so $\beta(1) = 0.10$]. Therefore, the required sample size is

$$n = \frac{\left[Z_{\alpha(2)}\sqrt{\psi + 1} + Z_{\beta(1)}\sqrt{(\psi + 1) - p(\psi - 1)^2}\right]^2}{p(\psi - 1)^2}$$

$$= \frac{\left[1.9600\sqrt{2 + 1} + 1.2816\sqrt{(2 + 1) - (0.1176)(2 - 1)^2}\right]^2}{(0.1186)(2 - 1)^2}$$

$$= \frac{[1.9600(1.7321) + 1.2816(1.6978)]^2}{0.1176} = \frac{(5.5708)^2}{0.1176} = 263.9.$$

Therefore, at least 264 pairs of data should be used.

Note that Equation 9.22 involves the testing of a series of 1:1 ratios by what is essentially an expansion of Equation 9.16. Each of these 1:1 ratios derives from a unique pairing of the r categories taken two at a time. Recall (Equation 5.10) that the number of ways that r items can be combined two at a time is $_rC_2 = r!/[2(r - 2)!]$. So,

---

EXAMPLE 9.7   McNemar's test for a 3×3 table of nominal-scale data.

H_0:   Of men who adopt a religion different from that of their fathers, a change from religion A to B is as likely as a change from B to A.

H_A: Of men who adopt a religion different from that of their fathers, a change from religion A to B has a likelihood different than does a change from B to A.

Man's Religion	Man's Father's Religion		
	Protestant	*Catholic*	*Jewish*
Protestant	173	20	7
Catholic	15	51	2
Jewish	5	3	24

$r = 3$

$$\chi^2 = \sum_{i=1}^{r} \sum_{j>i} \frac{(f_{ij} - f_{ji})^2}{f_{ij} + f_{ji}}$$

$$= \frac{(20 - 15)^2}{20 + 15} + \frac{(7 - 5)^2}{7 + 5} + \frac{(2 - 3)^2}{2 + 3}$$

$$= 0.7143 + 0.3333 + 0.2000$$

$$= 1.248$$

$$\nu = \frac{r(r-1)}{2} = \frac{3(2)}{2} = 3$$

$\chi^2_{0.05, 3} = 7.815$

Do not reject H_0.

$$0.50 < P < 0.75 \quad [P = 0.74]$$

---

in Example 9.7, where there are three categories, there are $_3C_2 = 3!/[2(3 - 2)!] = 3$ pairings, resulting in three terms in the χ^2 summation. If there were four categories of religion, then the summation would involve $_4C_2 = 4!/[2(4 - 2)!] = 6$ pairings, and 6 χ^2 terms; and so on. For data of this type in a 2×2 table, Equation 9.22 becomes Equation 9.16, and Equation 9.23 yields $\nu = 1$.

Testing for Effect of Treatment Order.   If two treatments are applied sequentially to a group of subjects, we might ask whether the response to each treatment depended on the order in which the treatments were administered. For example, suppose we have two medications for the treatment of poison ivy rash, but, instead of the situation in Example 9.4, they are to be administered orally rather than by external application to the skin; thus both arms receive a given medication at the same time, and the medications must be given at different times.

Gart (1969a) provides the following procedure to test for the difference in response between two sequentially applied treatments and to test whether the order of application had an effect on the response. The following 2×2 contingency table (see Section 23.3) is set up to test for a treatment effect:

	Order of Application of Treatments A and B		
	A, then B	B, then A	Total
Response with first treatment	f_{11}	f_{12}	R_1
Response with second treatment	f_{21}	f_{22}	R_2
Total	C_1	C_2	n

By redefining the rows, the following 2×2 table may be used to test the null hypothesis of no difference in response due to order of treatment application:

	Order of Application of Treatments A and B		
	A, then B	B, then A	Total
Response with treatment A	f_{11}	f_{12}	R_1
Response with treatment B	f_{21}	f_{22}	R_2
Total	C_1	C_2	n

These two contingency tables may be tested by the chi-square test (Section 23.3) or the Fisher exact test (Section 24.10). Example 9.8 demonstrates Gart's test.

If $C_1 = C_2$, which is the typical situation, one-tailed testing may be performed by employing chi-square (as described at the end of Section 23.3), the normal approximation to chi-square (Section 24.11), or the Fisher exact test (Section 23.10). If $C_1 \neq C_2$, the latter procedure is called for to perform a one-tailed test.

EXAMPLE 9.8   Gart's test for effect of treatment and treatment order.

H_0:   The two oral medications have the same effect on relieving poison ivy rash.

H_A: The two oral medications do not have the same effect on relieving poison ivy rash.

	Order of Application of Medications A and B		
	A, then B	B, then A	Total
Response with 1st medication	14	6	20
Response with 2nd medication	4	12	16
Total	18	18	36

Using Equation 22.7:

$$\chi_c^2 = \frac{36 \left[|(14)(12) - (6)(4)| - \frac{36}{2} \right]^2}{(18)(18)(20)(16)} = 5.512$$

$\chi_{0.05,1}^2 = 3.841$; reject H_0.

EXAMPLE 9.8   (continued)

That is, there is a difference in response to the two medications, regardless of the order in which they are administered.

$$0.01 < P < 0.025 \quad [P = 0.019]$$

H_0:   The order of administration of the two oral medications does not affect their abilities to relieve poison ivy rash.

H_A: The order of administration of the two oral medications does affect their abilities to relieve poison ivy rash.

	Order of Application of Medications A and B		
	A, then B	*B, then A*	*Total*
Response with medication A	14	12	26
Response with medication B	4	6	10
Total	18	18	36

Using Equation 22.7:

$$\chi_c^2 = \frac{36 \left[|(14)(6) - (12)(4)| - \dfrac{36}{2} \right]^2}{(18)(18)(26)(10)} = 0.138$$

$\chi_{0.05,1}^2 = 3.841$; do not reject H_0.

That is, the effects of the two medications are not affected by the order in which they are administered.

$$0.50 < P < 0.75 \quad [P = 0.71]$$

EXERCISES

9.1 Concentrations of nitrogen oxides and of hydrocarbons were determined in a certain urban area (recorded in $\mu g/m^3$).

  (a) Test the hypothesis that both classes of air pollutants were present in the same concentration.

Day	*Nitrogen oxides*	*Hydrocarbons*
1	104	108
2	116	118
3	84	89
4	77	71
5	61	66
6	84	83
7	81	88
8	72	76
9	61	68
10	97	96
11	84	81

  (b) Calculate the 95% confidence interval for μ_d.

9.2 Using the data of Exercise 9.1, test the appropriate hypotheses with Wilcoxon's paired-sample test.

9.3 One hundred twenty-two pairs of brothers, one member of each pair overweight and the other of normal weight, were examined for presence of varicose veins. Use the McNemar test for the data below to test the hypothesis that there is no relationship between being overweight and developing varicose veins (i.e., that the same proportion of overweight men as normal weight men possess varicose veins). In the following data tabulation, "v.v." stands for "varicose veins."

| | Overweight | |
Normal Weight	*With v.v.*	*Without v.v.*
With v.v.	19	5
Without v.v.	12	86

$n = 122$

10

MULTISAMPLE HYPOTHESES: THE ANALYSIS OF VARIANCE

When measurements of a variable are obtained for each of two samples, hypotheses such as those described in Chapter 8 are appropriate. However, biologists often collect measurements of a variable as three or more samples, from three or more populations, a situation calling for multisample analyses, as introduced in this chapter.

It is tempting to some to attempt the testing of multisample hypotheses by applying two-sample tests to all possible pairs of samples. In this manner, for example, one might proceed to test the null hypothesis H_0: $\mu_1 = \mu_2 = \mu_3$ by testing each of the following hypotheses by the two-sample t test: H_0: $\mu_1 = \mu_2$, H_0: $\mu_1 = \mu_3$, H_0: $\mu_2 = \mu_3$. But such a procedure, of employing a series of two-sample tests to address a multisample hypothesis, is invalid.

The calculated test statistic, t, and the critical values we find in the t table, are designed to test whether the two sample statistics, $\bar{X}_1$ and $\bar{X}_2$, are likely to have come from the same population (or from two populations with identical means). In properly employing the two-sample test, we could randomly draw two sample means from the same population and wrongly conclude that they are estimates of two different populations' means; but we know that the probability of this error (the Type I error) will be no greater than α. However, consider that three random samples were taken from a single population. In performing the three possible two-sample t tests indicated above, with $\alpha = 0.05$, the probability of wrongly concluding that two of the means estimate different parameters is 14%, considerably greater than α. Similarly, if α is set at 5% and four means are tested, two at a time, by the two-sample t test, there are six pairwise H_0's to be tested in this fashion, and there is a 26% chance of wrongly concluding a difference between one or more of the means. Why is this?

For each two-sample t test performed at the 5% level of significance, there is a 95% probability that we shall correctly conclude not to reject H_0 when the two population means are equal. For the set of three hypotheses, the probability of *correctly* declining to reject all of them is only $0.95^3 = 0.86$. This means that the probability of *incorrectly* rejecting at least one of the H_0's is $1 - 0.95^3 = 0.14$. As the number of means increases, it approaches certainty that performing all possible t tests will conclude that some of the sample means estimate different values of μ, even though the samples come from the same population. Table 10.1 shows the probability of committing Type I errors if multiple t tests are employed to test for difference among more than two means. Two-sample tests, it must be emphasized, should *not* be applied to multisample hypotheses. The appropriate procedures are introduced in the following sections.

TABLE 10.1 Probability of Committing at Least One Type I Error by Using Two-Sample t tests for All K Pairwise Comparisons of k Means*

		Level of Significance, α, Used in the t Tests				
k	K	0.10	0.05	0.01	0.005	0.001
2	1	0.10	0.05	0.01	0.005	0.001
3	3	0.27	0.14	0.03	0.015	0.003
4	6	0.47	0.26	0.06	0.030	0.006
5	10	0.65	0.40	0.10	0.049	0.010
6	15	0.79	0.54	0.14	0.072	0.015
10	45	0.99	0.90	0.36	0.202	0.044
	∞	1.00	1.00	1.00	1.000	1.000

*There are $K = k(k-1)/2$ pairwise comparisons of k means. This is the number of combinations of k items taken two at a time; see Equation 5.10.

10.1 SINGLE-FACTOR ANALYSIS OF VARIANCE

To test the null hypothesis $H_0: \mu_1 = \mu_2 = \ldots = \mu_k$, where k is the number of experimental groups, or samples, we need to become familiar with the topic of *analysis of variance*, often abbreviated ANOVA (or less commonly, ANOV or AOV). Analysis of variance is a large area of statistical methods, owing its name and much of its early development to R. A. Fisher;* in fact, the F statistic was named in his honor by G. W. Snedecor[†] (1934: 15). There are many ramifications of analysis of variance considerations, the most common of which will be discussed in this and subsequent

---

*Sir Ronald Aylmer Fisher (1890–1962), British statistician, who introduced the name and basic concept of the technique. David (1995) traces the first appearance of the term to a 1918 paper by Fisher.

[†]George W. Snedecor (1881–1974), American statistician.

chapters. More complex applications and greater theoretical coverage are to be found in Cochran and Cox (1957), Cox (1958), Scheffé (1959), Guenther (1964), and in books on experimental design. At this point, it may appear strange that a procedure used for testing the equality of *means* should be named analysis of *variance*, but the reason for this terminology soon will become apparent.

Let us assume that we wish to test whether four different feeds result in different body weights in pigs. Since we are to test for the effect of only one *factor* (feed type) on the variable in question (body weight), the appropriate analysis is termed a single-factor (or "single-criterion" or "single-classification" or "one-way") analysis of variance.[*] Furthermore, each type of feed is said to be a *level* of the factor. The design of this experiment should have each experimental animal being assigned at random to receive one of the four feeds, with approximately equal numbers of pigs receiving each feed. The importance of randomness was stressed by Fisher.

Although identical sample sizes are not required for the single-factor ANOVA, the power of the test is heightened by having sample sizes as nearly equal as possible. The performance of the test is also enhanced by having all pigs as similar as possible in all respects except for the experimental factor, diet (i.e., the animals should be of the same breed, sex, and age, should be kept at the same temperature, etc.)

Because the pigs are assigned to the feed groups at random (as with the aid of a random-number table, such as Appendix Table B.41, described in Section 2.3), the single factor ANOVA is said to represent a *completely randomized experimental design*, or "completely randomized design" (sometimes abbreviated "CRD").

Example 10.1 shows the weights of nineteen pigs subjected to this feed experiment, and the null hypothesis to be tested would be H_0: $\mu_1 = \mu_2 = \mu_3 = \mu_4$. Each datum in the experiment may be uniquely represented by the double subscript notation, where X_{ij} denotes datum j in experimental group i. For example, X_{23} denotes the third pig weight in feed group 2, that is, $X_{23} = 74.0$ kg. Similarly, $X_{34} = 96.5$ kg, $X_{41} = 87.9$ kg, etc. We shall let the mean of group i be denoted by $\bar{X}_i$, and the grand mean of all observations will be designated by $\bar{X}$. Furthermore, n_i will represent the size of sample i, and $N = \sum_{i=1}^{k} n_i$ will be the total number of data in the experiment.

Sources of Variation. In the two-sample t test (Section 8.1), where one assumes equality of the variances of the two sampled populations, the common population variance, σ^2, is estimated by the pooled variance, $s_p^2 = (SS_1 + SS_2)/(\nu_1 + \nu_2) = \sum_{i=1}^{2} \left[\sum_{j=1}^{n_i} (X_{ij} - \bar{X}_i)^2 \right] / \sum_{i=1}^{2} (n_i - 1) =$ pooled SS/pooled DF (as in Equation 8.29). Similarly, in the ANOVA under discussion, we assume that $\sigma_1^2 = \sigma_2^2 = \sigma_3^2 = \sigma_4^2$, and we estimate the population variance assumed common to all k groups by a variance obtained using the pooled sum of squares and the pooled degrees of freedom for all the groups:

$$\text{within-groups SS} = \sum_{i=1}^{k} \left[\sum_{j=1}^{n_i} (X_{ij} - \bar{X}_i)^2 \right] \tag{10.1}$$

---

[*]Some authors would here refer to the feed as the "independent variable" and to the weight as the "dependent variable."

EXAMPLE 10.1　A single-factor analysis of variance (Model I).

Nineteen pigs are assigned at random among four experimental groups. Each group is fed a different diet. The data are pig body weights, in kilograms, after being raised on these diets. We wish to ask whether pig weights are the same for all four diets.

H_0:　$\mu_1 = \mu_2 = \mu_3 = \mu_4$.

H_A:　The mean weights of pigs on the four diets are not all equal.

$\alpha = 0.05$

	Feed 1	Feed 2	Feed 3	Feed 4	
	60.8	68.7	102.6	87.9	
	57.0	67.7	102.1	84.2	
	65.0	74.0	100.2	83.1	
	58.6	66.3	96.5	85.7	
	61.7	69.8		90.3	
n_i	5	5	4	5	$N = \sum_{i=1}^{k} n_i = 19$
$\sum_{j=1}^{n_i} X_{ij}$	303.1	346.5	401.4	431.2	
$\bar{X}_i$	60.62	69.30	100.35	86.24	
$\dfrac{\left(\sum_{j=1}^{n_i} X_{ij}\right)^2}{n_i}$	18373.922	24012.450	40280.490	37186.688	$\sum_{i=1}^{k} \dfrac{\left(\sum_{j=1}^{n_i} X_{ij}\right)^2}{n_i}$

$$\sum_{i=1}^{k} \frac{\left(\sum_{j=1}^{n_i} X_{ij}\right)^2}{n_i} = 119853.550$$

$\sum_i \sum_j X_{ij} = 1482.2$

$\sum_i \sum_j X_{ij}^2 = 119981.900$

total DF $= N - 1 = 19 - 1 = 18$

groups DF $= k - 1 = 4 - 1 = 3$

error DF $= N - k = 19 - 4 = 15$

$$C = \frac{\left(\sum_i \sum_j X_{ij}\right)^2}{N} = \frac{2196916.84}{19} = 115627.202$$

total sum of squares $= \sum_i \sum_j X_{ij}^2 - C = 119981.900 - 115627.202 = 4354.698$

groups sum of squares $= \sum_i \dfrac{\left(\sum_j X_{ij}\right)^2}{n_i} - C = 119853.550 - 115627.202 = 4226.348$

error sum of squares $=$ total SS $-$ groups SS $= 4354.698 - 4226.348 = 128.350$

Summary of the Analysis Variance

Source of variation	SS	DF	MS
Total	4354.698	18	
Groups	4226.348	3	1408.783
Error	128.350	15	8.557

EXAMPLE 10.1   (continued)

$$F = \frac{\text{groups MS}}{\text{error MS}} = \frac{1408.783}{8.557} = 165$$

$$F_{0.05(1),3,15} = 3.29$$

Reject H_0.

$$P < 0.0005 \quad [P = 1.045 \times 10^{-11}]$$

Note: All these sums of squares and mean squares have $(\text{kg})^2$ as units. For typographic convenience and ease in reading, however, the units for ANOVA computations are routinely omitted in statistical papers.

and

$$\text{within-groups DF} = \sum_{i=1}^{k}(n_i - 1) = N - k. \tag{10.2}$$

These two quantities are often referred to as the *error sum of squares* and the *error degrees of freedom*, respectively. The former divided by the latter is a statistic that is the best estimate of the variance, σ^2, common to all k populations.

The amount of variability among the k groups is important to our hypothesis testing. This can be denoted as

$$\text{among-groups SS} = \sum_{i=1}^{k} n_i(\bar{X}_i - \bar{X})^2, \tag{10.3}$$

or simply, the *groups sum of squares*, and

$$\text{among-groups DF} = k - 1, \tag{10.4}$$

commonly called the *groups degrees of freedom*.

We also consider the variability present among all N data; that is,

$$\text{total SS} = \sum_{i=1}^{k}\sum_{j=1}^{n_i}(X_{ij} - \bar{X})^2 \tag{10.5}$$

and

$$\text{total DF} = N - 1. \tag{10.6}$$

In summary, each deviation of an observed datum from the grand mean of all data is attributable to a deviation of that datum from its group mean plus the deviation of that group mean from the grand mean; i.e.,

$$(X_{ij} - \bar{X}) = (X_{ij} - \bar{X}_i) + (\bar{X}_i - \bar{X}). \tag{10.7}$$

Furthermore, sums of squares and degrees of freedom are additive, so

$$\text{total SS} = \text{groups SS} + \text{error SS} \tag{10.8}$$

and

$$\text{total DF} = \text{groups DF} + \text{error DF}. \tag{10.9}$$

The total sum of squares may be calculated readily by a machine formula analogous to Equation 4.9:

$$\text{total SS} = \sum_{i=1}^{k} \sum_{j=1}^{n_i} X_{ij}^2 - C, \tag{10.10}$$

where

$$C = \frac{(\sum \sum X_{ij})^2}{N}, \tag{10.11}$$

a quantity referred to as the "correction term." Example 10.1 demonstrates these calculations.

A machine formula for the groups sum of squares is

$$\text{groups SS} = \sum_{i=1}^{k} \frac{\left(\sum_{j=1}^{n_i} X_{ij} \right)^2}{n_i} - C, \tag{10.12}$$

where $\sum_{j=1}^{n_i} X_{ij}$ is the sum of the n_i data from group i.

The error SS may be calculated as

$$\text{error SS} = \sum_{i=1}^{k} \sum_{j=1}^{n_i} X_{ij}^2 - \sum_{i=1}^{k} \frac{\left(\sum_{j=1}^{n_i} X_{ij} \right)^2}{n_i}, \tag{10.13}$$

which is the machine formula for Equation 10.1, but this quantity is most commonly computed by difference:

$$\text{error SS} = \text{total SS} - \text{groups SS}. \tag{10.14}$$

Similarly,

$$\text{error DF} = \text{total DF} - \text{groups DF}, \tag{10.15}$$

which yields the same result as Equation 10.2.

Dividing the groups SS or the error SS by the respective degrees of freedom results in a variance, referred to in ANOVA terminology as a *mean square* (abbreviated MS and short for *mean squared deviation from the mean*). Thus,

$$\text{groups MS} = \frac{\text{groups SS}}{\text{groups DF}} \tag{10.16}$$

and

$$\text{error MS} = \frac{\text{error SS}}{\text{error DF}}, \tag{10.17}$$

the latter quantity occasionally being abbreviated as MSE. A total mean square could also be calculated (as total SS/total DF), but it is not needed in the ANOVA.

Table 10.2 summarizes the single factor ANOVA calculations.*

TABLE 10.2 Summary of the Calculations for a Single-Factor Analysis of Variance

Source of variation	Sum of squares (SS)	Degrees of freedom (DF)	Mean square (MS)
Total $[X_{ij} - \bar{X}]$	$\sum\limits_{i=1}^{k}\sum\limits_{j=1}^{n_i} X_{ij}^2 - C$	$N - 1$	
Groups (i.e. among groups) $[\bar{X}_i - \bar{X}]$	$\sum\limits_{i=1}^{k} \dfrac{\left(\sum\limits_{j=1}^{n_i} X_{ij}\right)^2}{n_i} - C$	$k - 1$	$\dfrac{\text{groups SS}}{\text{groups DF}}$
Error (i.e. within-groups) $[X_{ij} - \bar{X}_i]$	total SS − groups SS	total DF − groups DF (or $N - k$)	$\dfrac{\text{error SS}}{\text{error DF}}$

Note: $C = \left(\sum_{i=1}^{k} \sum_{j=1}^{n_i} X_{ij}\right)^2 \big/ N$; $N = \sum_{i=1}^{k} n_i$; k is the number of groups; n_i is the number of data in group i.

Testing the Null Hypothesis. Statistical theory informs us that if the null hypothesis is a true statement about the populations, then the groups MS and the error MS will each be an estimate of σ^2, the variance common to all k populations. But if the k population means are not equal, then the groups MS in the population will be greater than the population's error MS.[†] Therefore, the test for the equality of means is a one-tailed variance ratio test (introduced in Section 8.5), where the groups MS is always placed in the numerator so as to ask whether it is significantly larger than the error MS:[‡]

$$F = \frac{\text{groups MS}}{\text{error MS}}. \tag{10.18}$$

The critical value for this test is $F_{\alpha(1),(k-1),(N-k)}$. If the calculated F is at least as large as the critical value, then we reject H_0. But remember that all we conclude in such a case is that all the k population means are not equal. To conclude between which means the equalities or inequalities lie, we must turn to the procedures of Chapter 11.

The Case Where $k = 2$. If $k = 2$, then H_0: $\mu_1 = \mu_2$, and either the two-sample t test (Section 8.1) or the single-factor ANOVA may be applied; the conclusions obtained from these two procedures will be identical. The error MS will, in fact, be identical to the pooled variance in the t test; the groups DF will be 1; and the resultant

---

*Occasionally, the following quantity (or its square root) is called the *correlation ratio*: (groups SS)/(total SS). This expresses the proportion of the total variability of X that is accounted for by the effect of the groups and is reminiscent of the "coefficient of determination" introduced in Section 17.3.

†Two decades before R. A. Fisher developed analysis of variance techniques, the Danish applied mathematician, Thorvald Nicolai Thiele (1838–1910) presented the concept of comparing the variance among groups to the variance within groups (Thiele, 1897: 41–44).

‡In practice, the sample's groups MS may be smaller than the sample's error MS, so $F < 1.0$, in which case H_0 is not rejected.

F value from the analysis of variance will be the square of the resultant t value, while $F_{\alpha(1),1,(N-2)} = (t_{\alpha(2),(N-2)})^2$. If a one-tailed test between means is required, or if the hypothesis $H_0: \mu_1 - \mu_2 = \mu_0$ is desired for a μ_0 not equal to zero, then the t test is applicable, whereas the ANOVA is not.

ANOVA Using Means and Variances.   The above discussion assumes that all the data from the experiment to be analyzed are in hand. It may occur, however, that all we have are the means for each group and some measure of variability based on the variances of each group. That is, we may have $\bar{X}_i$ and either SS_i, s_i^2, s_i, or $s_{\bar{X}_i}$, for each group, rather than all the individual values of X_{ij}. For example, we might encounter presentations such as Tables 7.1, 7.2, or 7.3a. If the sample sizes, n_i, are also known, then the single-factor analysis of variance may still be performed, in the following manner.

First, determine the sum of squares or sample variance for each group; recall that

$$s_i^2 = (s_i)^2 = n_i (s_{\bar{X}_i})^2. \tag{10.19}$$

Then, calculate

$$\text{error SS} = \sum_{i=1}^{k} SS_i = \sum_{i=1}^{k} (n_i - 1)s_i^2 \tag{10.20}$$

and

$$\text{groups SS} = \sum_{i=1}^{k} n_i \bar{X}_i^2 - \frac{\left(\sum_{i=1}^{k} n_i \bar{X}_i\right)^2}{\sum_{i=1}^{k} n_i}. \tag{10.21}$$

Knowing the groups SS and error SS, the ANOVA can proceed in the usual fashion.

Two Types of ANOVA.   In Example 10.1, the biologist designing the experiment was interested in whether all of these particular four feeds have the same effect on pig weight. That is, these four feeds were not randomly selected from a feed catalog but were specifically chosen. When the levels of a factor are specifically chosen one is said to have designed a *fixed-effects model*, or a *Model I*, ANOVA. In such a case, the null hypothesis $H_0: \mu_1 = \mu_2 = \mu_3 = \ldots = \mu_k$ is appropriate.

However, there are instances where the levels of a factor to be tested are indeed chosen at random. For example, we might have been interested in the effect of geographic location of the pigs, rather than the effect of their feed. It is possible that our concern might be with certain specific locations, in which case we would be employing a fixed-effects mode ANOVA. But we might, instead, be interested in testing the statement that in general there is a difference in pig weights in animals from different locations. That is, instead of being concerned with only the particular locations used in the study, the intent might be to generalize, considering the locations in our study to be a random sample from all possible locations. In this *random-effects model*, or *Model II*, ANOVA,* all the

---

*Also referred to as a "components of variance model." The terms "fixed-effects," "random-effects," "Model I" and "Model II" for analysis of variance were introduced by Eisenhart (1947).

calculations are identical to those for the fixed-effects model, but the null hypothesis is better stated as H_0: There is no difference in pig weight among geographic locations (or H_0: There is no variability in weights among locations). Examination of Equation 10.18 shows that what the analysis asks is whether the variability among locations is greater than the variability within locations. Example 10.2 demonstrates the ANOVA for a random-effects model. Most biologists will encounter Model I analyses more commonly than Model II situations. When dealing with more than one experimental factor (as in Chapters 12 and 14) the distinction between the two models becomes essential.

Underlying Assumptions.   Recall from Section 8.1 that to test $H_0: \mu_1 = \mu_2$ by the two-sample t test, we had to assume that $\sigma_1^2 = \sigma_2^2$ and that each of the two samples came from a normal population. Similarly, $\sigma_1^2 = \sigma_2^2 = \sigma_3^2 = \sigma_4^2$ should be true in order to apply the analysis of variance to $H_0 : \mu_1 = \mu_2 = \mu_3 = \mu_4$, and each of the k samples should have come from a normal population.* Bartlett's test for homogeneity (Section 10.6) might be used to determine whether the assumption of equal variances is met. However, because Bartlett's test is not very efficient and is badly affected by non-normality, it is generally not worthwhile to use in conjunction with analyses of variance.

Fortunately, the analysis of variance is robust, operating well even with considerable heterogeneity of variances, as long as all n_i are equal or nearly equal (Glass, Peckham, and Sanders, 1972). If the n_i are quite different, then the probability of a Type I error will depart markedly from α, to a degree dependent on the magnitude of the heterogeneity (Box, 1954); if larger variances are associated with the larger samples, the probability of a Type I error will be $< \alpha$, and if they are associated with the smaller samples this probability will be $> \alpha$ (Kohr and Games, 1974; Maxwell and Delaney, 1990: 723–724; see also Horsnell, 1953).

The analysis of variance is also robust with respect to the assumption of the underlying populations' normality. The validity of the analysis is affected only slightly by even considerable deviations from normality (in skewness and/or kurtosis), especially as n increases (Box and Anderson, 1955; Srivastava, 1959; Tiku, 1971). If the underlying populations are very much platykurtic, and sample sizes are small, the actual power of the test will be less than that discussed in Section 10.3; if the populations are greatly leptokurtic, and sample sizes are small, the actual power will be greater than that calculated by the method of that section (Glass, Peckham, and Sanders, 1972).

Thus, the analysis of variance typically may be depended upon unless the data deviate severely from the underlying assumptions. In the latter case, the nonparametric procedure of Section 10.4 is appropriate, as it does not depend upon the sampled populations' being distributed in a particular fashion (such as normal) and is only slightly influenced (relatively) by difference in the populations' dispersions.

Or, one might employ an extension of the Behrens-Fisher procedure introduced at the end of Section 8.1. This is effected as described by Welch (1951), which is similar,

---

*In factorial analyses of variance, to be discussed in Chapters 12 and 14, the underlying assumptions are that the data in each cell (i.e., in each combination of factors) come from a normal population and that all of these populations have the same variance.

EXAMPLE 10.2 A single-factor analysis of variance for a random-effects model (i.e., Model II) experimental design.

A laboratory employs a certain technique for determining the phosphorus content of hay. The question arises: "Do phosphorus determinations differ with the technician performing the analysis?" To answer this question, each of four randomly selected technicians was given five samples from the same batch of hay. The results of the twenty phosphorus determinations (in mg phosphorus/g of hay) are shown.

H_0:   Determinations of phosphorus content do not differ among technicians.

H_A: Determinations of phosphorus content do differ among technicians.

$\alpha = 0.05$

	Technician			
	1	2	3	4
	34	37	34	36
	36	36	37	34
	34	35	35	37
	35	37	37	34
	34	37	36	35
Group sums:	173	182	179	176

$$\sum_i \sum_j X_{ij} = 710$$

$$\sum_i \sum_j X_{ij}^2 = 25234$$

$N = 20$

$C = \dfrac{(710)^2}{20} = 25205.00$

total SS $= 25234 - 25205.00 = 29.00$

groups (i.e., technicians) SS $= \dfrac{(173)^2}{5} + \dfrac{(182)^2}{5} + \dfrac{(179)^2}{5} + \dfrac{(176)^2}{5} - 25205.00$

$$= 25214.00 - 25205.00 = 9.00$$

error SS $= 29.00 - 9.00 = 20.00$

Source of variation	SS	DF	MS
Total	29.00	19	
Groups (technicians)	9.00	3	3.00
Error	20.00	16	1.25

$F = \dfrac{3.00}{1.25} = 2.40$

$F_{0.05(1), 3, 16} = 3.24$

Do not reject H_0.

$$0.10 < P < 0.25 \quad [P = 0.11]$$

but preferable, to the procedure of James (1951) (see Brown and Forsythe, 1974a). The Welch method has been shown by Dijkstra and Werter (1981), Kohr and Games (1974) and Levy (1978a) to perform rather well when population variances are unequal (while the ANOVA typically is better if the variances are equal), although Levy (1978a) found this test to be a little less robust than the ANOVA to departures from normality. The multiple Welch test employs

$$F' = \frac{\sum\limits_{i=1}^{k} c_i (\bar{X}_i - \bar{X}_w)^2}{(k-1)\left[1 + \dfrac{2A(k-2)}{k^2 - 1}\right]}, \tag{10.22}$$

where

$$c_i = \frac{n_i}{s_i^2} \tag{10.23}$$

$$C = \sum_{i=1}^{k} c_i \tag{10.24}$$

$$\bar{X}_w = \frac{\sum\limits_{i=1}^{k} c_i \bar{X}_i}{C} \tag{10.25}$$

$$A = \sum_{i=1}^{k} \frac{(1 - c_i/C)^2}{\nu_i} \quad \text{where } \nu_i = n_i - 1 \tag{10.26}$$

and F' is associated with degrees of freedom of $\nu_1 = k - 1$ and

$$\nu_2 = \frac{k^2 - 1}{3A}, \tag{10.27}$$

which should be rounded to the next lower integer. A similar test, which requires special determination of significance levels, has been developed by Krutchkoff (1988), and other approaches have been advanced (e.g., Bishop and Dudewicz, 1978).

A modified ANOVA advanced by Brown and Forsythe (1974a, b) also works well:

$$F'' = \frac{\text{Groups SS}}{B}, \tag{10.28}$$

where

$$B = \sum_{i=1}^{k} \left(1 - \frac{n_i}{N}\right) s_i^2 \tag{10.29}$$

and F'' is associated with degrees of freedom of $\nu_1 = k - 1$ and

$$\nu_2 = \frac{B^2}{\sum\limits_{i=1}^{k} \dfrac{\left[\left(1 - \dfrac{n_i}{N}\right) s_i^2\right]^2}{\nu_i}}. \tag{10.30}$$

For samples from normal distributions, Brown and Forsythe (1974a) found that F' and F'' worked well for $n_i \geq 10$ but that F' was better when $4 < n_i < 10$, and that F' is preferable to F'' when the extreme means are associated either with small variances or with extreme variances; and F'' is better when variances are equal. Clinch and Keselman (1982) found that both F' and F'' performed well for symmetrical distributions, even when there were unequal sample sizes and variances, and that for skewed distributions F'' worked better (i.e., avoided inflation of Type I error), especially when small sample sizes were associated with large variances.

When the underlying assumptions of the ANOVA are seriously violated, most researchers would opt for the nonparametric analysis of Section 10.4 instead of either the Welch or the Brown and Forsythe procedures above. For $k = 2$, Equations 10.22 and 10.28 are equivalent to each other and to Equation 8.11, and Equations 10.27 and 10.30 are equivalent to each other and to Equation 8.12.

Outliers.   A small number of data that are much more extreme than the rest of the measurements in a sample are called *outliers* (introduced at the beginning of Section 6.5), and they may cause a sample to seriously violate the assumptions of normality and variance equality. If, in the experiment of Example 10.1, a pig weight of 893 kg, or 7.12 kg, or 209 kg, was reported, the researcher would likely suspect an error. Perhaps the first two of these measurements were the result of careless recording of weights of 89.3 kg and 71.2 kg, respectively; and perhaps the third was a weight obtained in pounds and incorrectly recorded as kilograms. If there is an obvious error such as this, then the offending datum might be readily corrected. Or, if it is believed that a greatly disparate datum is the result of erroneous data collection (e.g., an errant technician, a contaminated reagent, or an instrumentation failure), then it might be discarded or replaced. In other cases outliers might be valid data, and their presence may indicate one should not employ statistical analyses that require population normality and variance equality. There are statistical methods to detect outliers, some of which are discussed by Dunn and Clark (1987: 390–391), Hicks (1982: 138–139), and Snedecor and Cochran (1980: 279–282).

Outliers will have little or no influence on analyses employing nonparametric two-sample tests (Sections 8.10 and 8.11) or multisample tests (Section 10.4). In those procedures, the largest data will be treated the same, and the smallest data will be treated the same, regardless of whether they represent accurate or exaggerated measurements.

Blocks and Repeated Measures.   The experimental design in Example 10.1 may be viewed as one having $n = 5$ replicate data (body weights) collected at each one of $k = 4$ treatments (diets), where—except for the application of one of the treatments—each of the $nk = 20$ data is independent of the others. (The fact that there are only nineteen data in this specific example, there being one missing in treatment 3, might be due to the death or illness of one of the experimental animals, or there being only nineteen animals available for the study.) The objective of the experiment is to test the hypothesis about equality of body weights on the k treatments.

There are situations where an experiment testing this null hypothesis is set up so that each observation in one of the treatments has something in common with one observation in each of the other treatments; and each of the n groups of k related data is referred to as a *block*. For example, an experiment for testing the hypothesis of Example 10.1 might have been fashioned with four diets (i.e., $k = 4$) and five litters of animals (i.e., $n = 5$), with one animal from each litter placed on each of the four diets. The concept of blocks is identical, for $k > 2$, to the concept of pairs (Section 9.1), for $k = 2$; it will be discussed in Section 12.4.

Another kind of experimental situation is where each of the several treatments is applied to each of the experimental subjects. (It may be thought of as having blocks of data, where each block consists of the data for one of the subjects.) For example, the null hypothesis of Example 10.1 might have been addressed by having $n = 5$ experimental animals, each raised for a time on each of the $k = 4$ diets, and with its body weight recorded after being on each of the diets. This category of experimental scheme, where there obviously is a relationship among the k measurements on each animal, is referred to as a *repeated measures* experimental design and will be discussed in Section 12.5.

10.2 CONFIDENCE LIMITS FOR POPULATION MEANS

When $k > 2$, confidence limits for μ_i may be computed in a fashion analogous to that for the case where $k = 2$ [Section 8.2, Equation 8.13]. The $1 - \alpha$ confidence interval for μ_i is

$$\bar{X}_i \pm t_{\alpha(2), \nu} \sqrt{\frac{s^2}{n_i}}, \tag{10.31}$$

where s^2 is the error mean square and ν is the error degrees of freedom from the analysis of variance. For example, let us consider the 95% confidence interval for μ_4 in Example 10.1. Here $\bar{X}_4 = 86.24$ kg, $s^2 = 8.557$ kg^2, $n_4 = 5$, and $t_{0.05(2), 15} = 2.131$. Therefore, the lower 95% confidence limit, L_1, is 86.24 kg $- 2.131\sqrt{8.557 \text{ kg}^2/5} = 83.45$ kg, and the upper 95% confidence limit, L_2 is 86.24 kg $+ 2.131\sqrt{8.557 \text{ kg}^2/5} = 89.03$ kg.

Computing a confidence interval for μ_i would only be warranted if that population mean were concluded to be different from each other population mean. And calculation of a confidence interval for each of the k μ's may be performed only if it is concluded that $\mu_1 \neq \mu_2 \neq \ldots \neq \mu_k$. However, the analysis of variance does not enable conclusions as to which population means are different from which. Therefore, we must first perform multiple comparison testing (Chapter 11), after which confidence intervals may be determined for each different population mean. Confidence intervals for differences between means should be calculated as shown in Section 11.3.

10.3 POWER AND SAMPLE SIZE IN ANALYSIS OF VARIANCE

In Section 8.3, dealing with the difference between two means, we saw how to estimate the sample size required to predict such a population difference with a specified level of confidence. When dealing with more than two means, we may also wish to determine the sample size necessary to estimate difference between population means, and the appropriate procedure will be found in Section 11.3.

In Section 8.4, methods were presented for estimating the power of the two-sample t test, the minimum sample size required for such a test, and the minimum difference between population means that is detectable by such a test. Let us now examine such procedures for analysis-of-variance situations, namely for dealing with more than two means. (The following discussion begins with consideration of Model I—fixed-effects model—analyses of variance.)

If H_0 is true for an analysis of variance, then the variance ratio of Equation 10.18 follows the F distribution, this distribution being characterized by the numerator and denominator degrees of freedom (v_1 and v_2, respectively). If, however, H_0 is false, then the ratio of groups MS to error MS follows instead what is known as the *noncentral F distribution*, which is defined by v_1, v_2, and a third quantity known as the *noncentrality parameter*. As power refers to probabilities of detecting a false null hypothesis, statistical discussions of the power of ANOVA testing depend upon the noncentral F distribution.

A number of authors have described procedures for estimating the power of an ANOVA, or the required sample size, or the detectable difference among means (e.g., Cohen, 1988: Ch. 8; Fox, 1956; Odeh and Fox, 1975; Patniak, 1949; Tang 1939; Tiku, 1967, 1972), but the charts prepared by Pearson and Hartley (1951) provide one of the best of the methods and will be described below.

Power of the Test. Prior to performing an experiment, and collecting data from it, it is appropriate and desirable to investigate the power of the proposed test. (Indeed, it is possible that on doing so one would conclude that the power will be so low that the experiment needs to be run with many more data or, perhaps, not run at all.)

Let us specify that an ANOVA involving k groups will be performed at the α significance level, with n data (i.e., replications) per group. We can then estimate the power of the test if we have an estimate of σ^2, the variability within the k populations (e.g., this estimate typically is s^2 from similar experiments), and an estimate of the variability among the populations. From this information we may calculate a quantity called ϕ (lowercase Greek phi*), which is related to the noncentrality parameter.

The variability among populations might be expressed in terms of deviations of the k population means, μ_i, from the overall mean of all populations, μ, in which case

$$\phi = \sqrt{\frac{n \sum_{i=1}^{k} (\mu_i - \mu)^2}{k s^2}} \tag{10.32}$$

---

*Phi" rhymes with "sky."

(e.g., Guenther, 1964: 47; Kirk, 1982: 142–144). Recall that s^2 is error MS. The grand population mean is

$$\mu = \frac{\sum\limits_{i=1}^{k} \mu_i}{k} \tag{10.33}$$

if all the samples are the same size. In practice, we employ the best available estimates of these population means.

Once ϕ has been obtained, we consult Appendix Fig. B.1. This figure consists of several pages, each with a different ν_1 (i.e., groups DF) indicated at the upper left of the graph. Values of ϕ are indicated on the lower axis of the graph for both $\alpha = 0.01$ and $\alpha = 0.05$. Each of the curves on the graph is for a different ν_2 (i.e., error DF), for $\alpha = 0.01$ or 0.05, identified on the top margin of the graph. After turning to the graph for the ν_1 at hand, one locates the point at which the calculated ϕ intersects the curve for the given ν_2 and reads horizontally to either the right or left axis to determine the power of the test. This procedure is demonstrated in Example 10.3.

EXAMPLE 10.3   Estimating the power of an analysis of variance when variability among population means is specified.

A proposed analysis of variance of plant root elongations is to comprise ten roots at each of four chemical treatments. From previous experiments, we estimate σ^2 to be 7.5888 mm^2 and estimate that two of the population means are 8.0 mm, one is 9.0 mm, and one is 12.0 mm. What will be the power of the ANOVA if we test at the 0.05 level of significance?

$k = 4$

$n = 10$

$\nu_1 = k - 1 = 3$

$\nu_2 = k(n-1) = 4(9) = 36$

$\mu = \dfrac{8.0 + 8.0 + 9.0 + 12.0}{4} = 9.25$

$\phi = \sqrt{\dfrac{n \sum (\mu_i - \mu)^2}{ks^2}}$

$\quad = \sqrt{\dfrac{10[(8.0 - 9.25)^2 + (8.0 - 9.25)^2 + (9.0 - 9.25)^2 + (12.0 - 9.25)^2]}{4(7.5888)}}$

$\quad = \sqrt{\dfrac{10(10.75)}{4(7.5888)}}$

$\quad = \sqrt{3.5414}$

$\quad = 1.88$

In Appendix Fig. B.1, we enter the graph for $\nu_1 = 3$ with $\phi = 1.88$, $\alpha = 0.05$, and $\nu_2 = 36$ and read a power of about 0.86. Thus, there will be a 14% chance of committing a Type II error in the proposed analysis.

An alternative, and common, way to determine power is to specify the smallest difference we wish to detect between the two most different population means. Calling this minimum detectable difference δ, we compute

$$\phi = \sqrt{\frac{n\delta^2}{2ks^2}}$$ (10.34)

(Guenther, 1964: 49; Kirk, 1982: 144–145) and proceed to consult Appendix Fig. B.1 as above, and as demonstrated in Example 10.4. This procedure leads us to the statement that the power will be at least that determined from Appendix Fig. B.1 (and, indeed, it typically is greater).

EXAMPLE 10.4   Estimating the power of an analysis of variance when minimum detectable difference is specified.

For the ANOVA proposed in Example 10.3, we do not estimate the population means, but rather specify that, using ten data per sample, we wish to detect a difference between population means of at least 4.0 mm.

$k = 4$

$\nu_1 = 3$

$n = 10$

$\nu_2 = 36$

$\delta = 4.0$ mm

$s^2 = 7.5888$ mm^2

$$\phi = \sqrt{\frac{n\delta^2}{2ks^2}}$$

$$= \sqrt{\frac{10(4.0)^2}{2(4)(7.5888)}}$$

$$= \sqrt{2.6355}$$

$$= 1.62$$

In Appendix Fig. B.1, we enter the graph for $\nu_1 = 3$ with $\phi = 1.62$, $\alpha = 0.05$, and $\nu_2 = 36$ and read a power of about 0.72. That is, there will be a 28% chance of committing a Type II error in the proposed analysis.

Estimating the power of a proposed ANOVA may effect considerable savings in time, effort, and expense. For example, such an estimation might conclude that the power is so very low that the experiment, as planned, ought not to be performed. The proposed experiment might be revised, perhaps by increasing n, so as to render the results more likely to be conclusive. One may also strive to increase power by decreasing s^2, which may be possible by using experimental subjects that are more homogeneous. For instance, if the nineteen pigs in Example 10.1 were not all of the same age and breed, and not all maintained at the same temperature, there might well be more weight variability within the four dietary groups than if all nineteen were the same in all respects except diet.

While it is desirable to estimate power prior to engaging in analysis of variance testing (as described above), it also is useful to ask with what power an ANOVA has been performed. This is especially interesting if the null hypothesis is not rejected, for then we should wish to know how likely the test was to detect true difference among the population means.

As shown in Example 10.5, we compute

$$\phi = \sqrt{\frac{(k-1)(\text{groups MS} - s^2)}{ks^2}} \tag{10.35}$$

(Kirk, 1982: 143) and then proceed as above and as in Examples 10.3 and 10.4.

EXAMPLE 10.5   Estimating the power of an analysis of variance after it has been performed.

In an experiment to determine whether the development time for insect embryos (measured as days elapsed from egg laying to hatching) is the same at three different experimental temperatures, four eggs were placed at each of two temperatures, and five eggs were placed at the third temperature. (Therefore, $k = 3$, $n_1 = n_2 = 4$, and $n_3 = 5$.) A one-way analysis of variance yields the following results:

H_0: $\mu_1 = \mu_2 = \mu_3$

H_A: The mean development time is not the same at all three temperatures.

Source of variation	SS	DF	MS
Total	26.9231	12	
Among groups	10.3731	2	5.1866
Error	16.5500	10	1.6550

$F = 3.13$

$F_{0.05(1),2,10} = 4.10$

Do not reject H_0; $0.05 < P < 0.10$

$$\phi = \sqrt{\frac{(k-1)(\text{groups MS} - s^2)}{ks^2}} = \sqrt{\frac{(3-1)(5.1866 - 1.6550)}{3(1.6550)}} = 1.19$$

On consulting Appendix Fig. B.1, with $\nu_1 = 2$, $\nu_2 = 10$, and $\phi = 1.19$, the power of the ANOVA is estimated to be 0.33; that is, there was a 67% chance of having committed a Type II error in this analysis.

It is apparent from Appendix Fig. B.1 that, for a given α, ν_1, and ν_2, greater values of ϕ are associated with greater power; and, from equations 10.32 and 10.34, that ϕ increases with

1. increased sample size, n;
2. increased difference among population means (as measured either by groups MS, by $\sum(\mu_i - \mu)^2$, or by the minimum detectable difference, δ;
3. a fewer number of groups, k;
4. decreased variability within populations, σ^2, estimated by s^2.

Furthermore, recall that power increases with larger significance levels, α.

Sample Size Required. Prior to performing an analysis of variance, we might ask how many data need to be obtained in order to achieve a desired power. We can specify the power with which we wish to detect a particular difference among the population means and then ask how large the sample from each population must be. This is done, with Equation 10.34, by iteration (i.e., by making an initial guess and repeatedly refining that estimate), as shown in Example 10.6.

How well Equation 10.34 performs depends upon how good an estimate s^2 is of the population variance common to all groups. As the excellence of s^2 as an estimate improves with increased sample size, one should strive to calculate this statistic from a sample with a size that is not a very small fraction of the n estimated from Equation 10.34.

EXAMPLE 10.6   Estimation of required sample size for a one-way analysis of variance.

Let us propose an experiment such as that described in Example 10.5. How many replicate data should be collected so as to have an 80% probability of detecting a difference between population means as small as two days, testing at the 0.05 level of significance? We shall assume that $s^2 = 1.6550$ is a good estimate of σ^2.

Let us guess that $n = 15$ is required; then,

$$\nu_2 = 3(15 - 1) = 42,$$

and, by Equation 10.34,

$$\phi = \sqrt{\frac{(15)(2)^2}{2(3)(1.6550)}} = 2.46.$$

Consulting Appendix Fig. B.1, power ≈ 0.96, which is higher than we specified. Therefore, we can lower our guess, saying perhaps that $n = 10$; then,

$$\nu_2 = 3(10 - 1) = 27,$$

and

$$\phi = \sqrt{\frac{(10)(2)^2}{2(3)(1.6550)}} = 2.01,$$

which estimates a power of about 84%, a little higher than what we specified. Trying $n = 9$, we have

$$\nu_2 = 3(9 - 1) = 24,$$

and

$$\phi = \sqrt{\frac{(9)(2)^2}{2(3)(1.6550)}} = 1.90,$$

estimating a power of 0.79, which is lower than that specified.

Thus, we estimate that using sample sizes of at least ten should result in an ANOVA with a power of at least 0.84.

Minimum Detectable Difference. If we specify the significance level and sample size for an ANOVA, and the power that we desire it to have, and if we have an estimate of σ^2, then we can ask what the smallest detectable difference between population means will be. By entering on Appendix Fig. B.1 the specified α, v_1, and power, we can read a value of ϕ on the bottom axis. Then, by rearrangement of Equation 10.34, the minimum detectable difference is

$$\delta = \sqrt{\frac{2ks^2\phi^2}{n}}. \tag{10.36}$$

Example 10.7 demonstrates this estimation procedure.

EXAMPLE 10.7   Estimation of minimum detectable difference in a one-way analysis of variance.

In an experiment similar to that proposed in Example 10.5, assuming that $s^2 = 1.6550$ (days)^2 is a good estimate of σ^2, how small a difference between μ's can we have 90% confidence of detecting if $n = 20$ and $\alpha = 0.05$ are used?

As $k = 3$ and $n = 20$, $v_2 = 3(20-1) = 57$. For $v_1 = 2$, $v_2 = 57$, $1 - \beta = 0.90$, and $\alpha = 0.05$, Appendix Fig. B.1 gives a ϕ of about 2.1, from which we compute an estimate of

$$\delta = \sqrt{\frac{2ks^2\phi^2}{n}} = \sqrt{\frac{2(3)(1.6550)(2.1)^2}{20}} = 1.48 \text{ days.}$$

Maximum Number of Groups Testable.   For a given α, n, δ, and σ^2, power will decrease as k increases. It may occur that the total number of observations, N, will be limited, and for given ANOVA specifications the number of experimental groups, k, may have to be limited. As Example 10.8 illustrates, the maximum k can be determined by trial-and-error estimation of power, using Equation 10.34.

Random-Effects Analysis of Variance.   If the analysis of variance is a random-effects model (described at the end of Section 10.1), the power, $1-\beta$, may be determined from

$$F_{(1-\beta), v_1, v_2} = \frac{v_2 s^2 F_{\alpha(1), v_1, v_2}}{(v_2 - 2)(\text{groups MS})} \tag{10.37}$$

(after Kirk, 1982: 164–165; Scheffé, 1959: 227; Winer, Brown, and Michels, 1979: 246). This is shown in Example 10.9. As with the fixed-effects ANOVA, power is greater with larger n, larger differences among groups, larger α, and smaller s^2.

To determine required sample size in a random-effects analysis, one can specify values of α, groups MS, s^2, and k. Then, $v_1 = k - 1$ and $v_2 = k(n - 1)$; and, by iterative trial and error, one can apply Equation 10.37 until the desired power (namely, $1 - \beta$) is obtained.

10.4   NONPARAMETRIC ANALYSIS OF VARIANCE

If a set of data is collected according to a completely randomized design where $k > 2$, it is possible to test nonparametrically for difference among groups. This may be done by the

EXAMPLE 10.8   Determination of maximum number of groups to be used in a one-way analysis of variance.

Consider an experiment such as that in Example 10.5. Perhaps we have six temperatures that might be tested, but we have only space and equipment to examine a total of fifty eggs. Let us specify that we wish to test, with $\alpha = 0.05$ and $\beta \le 0.20$ (i.e., power of at least 80%), and to detect a difference as small as 2 days between population means.

If $k = 6$ were used, then $n = 50/6 = 8.3$ (call it 8), $\nu_1 = 5$, $\nu_2 = 6(8 - 1) = 42$, and (by Equation 10.34)

$$\phi = \sqrt{\frac{(8)(2)^2}{2(6)(1.6550)}} = 1.27,$$

for which Appendix Fig. B.1 indicates a power of about 0.60.

If $k = 5$ were used, $n = 50/5 = 10$, $\nu_1 = 4$, $\nu_2 = 5(10 - 1) = 45$, and

$$\phi = \sqrt{\frac{(10)(2)^2}{2(5)(1.6550)}} = 1.55,$$

for which Appendix Fig. B.1 indicates a power of about 0.74.

If $k = 4$ were used, $n = 50/4 = 12.5$ (call it 12), $\nu_1 = 3$, $\nu_2 = 4(12 - 1) = 44$, and

$$\phi = \sqrt{\frac{(12)(2)^2}{2(4)(1.6550)}} = 1.90,$$

for which Appendix Fig. B.1 indicates a power of about 0.87.

Therefore, we conclude that no more than four of the temperatures should be tested in an analysis of variance if we are limited to a total of fifty experimental eggs.

EXAMPLE 10.9   Estimating the power of the random-effects analysis of variance of Example 10.2.

groups MS $= 3.00$; $s^2 = 1.25$; $\nu_1 = 3$, $\nu_2 = 16$

$$F_{\alpha(1), \nu_1, \nu_2} = F_{0.05(1), 3, 16} = 3.24$$

$$
\begin{aligned}
F_{(1-\beta), \nu_1, \nu_2} &= \frac{\nu_2 s^2 F_{\alpha(1), \nu_1, \nu_2}}{(\nu_2 - 2)(\text{groups MS})} \\
&= \frac{(16)(1.25)(3.24)}{(14)(3.00)} = 1.54
\end{aligned}
$$

By consulting Appendix Table B.4, it is seen that an F of 1.54, with degrees of freedom of 3 and 16, is associated with a one-tailed probability between 0.10 and 0.25. (The exact probability is 0.24.) This probability is the power.

---

*Kruskal-Wallis test** (Kruskal and Wallis, 1952), often called an "analysis of variance by ranks." This test may be used in any situation where the parametric single-factor ANOVA of Section 10.1 is applicable, and it will be $3/\pi$ (i.e., 95%) as powerful as the latter (Andrews, 1954). It may be employed in instances where the latter is not applicable, in which

---

*William Henry Kruskal (b. 1919), American statistician, and Wilson Allen Wallis (b. 1912), American statistician and econometrician.

cases it may in fact be the more powerful test. The nonparametric analysis is especially desirable (Krutchkoff, 1988) when the k samples do not come from normal populations, and it may also be applied when the k population variances are somewhat heterogeneous.[*] If $k = 2$, then the Kruskal-Wallis test is identical to the Mann-Whitney test (Section 8.10).

Example 10.10 demonstrates the Kruskal-Wallis test procedure. As in other nonparametric tests, we do not use population parameters in statements of hypotheses, and neither parameters nor sample statistics are used in the test calculations. The Kruskal-Wallis test statistic, H, is calculated as

$$H = \frac{12}{N(N+1)} \sum_{i=1}^{k} \frac{R_i^2}{n_i} - 3(N+1), \tag{10.38}$$

where n_i is the number of observations in group i, $N = \sum_{i=1}^{k} n_i$ (the total number of observations in all k groups), and R_i is the sum of the ranks of the n_i observations in group i.[†] The procedure for ranking data is as presented in Section 8.10 for the

---

[*] The test assumes that the sampled populations have the same dispersions and shapes, but it is typically little affected if this is not the case.

[†] Interestingly, H (or H_c of Equation 10.41) could also be computed as

$$H = \frac{\text{groups SS}}{\text{total MS}}, \tag{10.39}$$

applying the procedures of Section 10.1 to the ranks of the data in order to obtain the groups SS and total MS. The Kruskal-Wallis test is identical to the parametric ANOVA (Section 10.1) performed on the ranks of the N data (Iman, Quade, and Alexander, 1975).

EXAMPLE 10.10   The Kruskal-Wallis single-factor analysis of variance by ranks.

An entomologist is studying the vertical distribution of a fly species in a deciduous forest and obtains five collections of the flies from each of three different vegetation layers: herb, shrub, and tree.

H_0:   The abundance of the flies is the same in all three vegetation layers.

H_A: The abundance of the flies is not the same in all three vegetation layers.

$\alpha = 0.05$

The data are as follows (with ranks of the data in parentheses):[*]

Numbers of Flies/m³ of Foliage

Herbs	Shrubs	Trees
14.0 (15)	8.4 (11)	6.9 (8)
12.1 (14)	5.1 (2)	7.3 (9)
9.6 (12)	5.5 (4)	5.8 (5)
8.2 (10)	6.6 (7)	4.1 (1)
10.2 (13)	6.3 (6)	5.4 (3)
$n_1 = 5$	$n_2 = 5$	$n_3 = 5$
$R_1 = 64$	$R_2 = 30$	$R_3 = 26$

EXAMPLE 10.10   (continued)

$$N = 5 + 5 + 5 = 15$$

$$H = \frac{12}{N(N+1)} \sum_{i=1}^{k} \frac{R_i^2}{n_1} - 3(N+1)$$

$$= \frac{12}{15(16)} \left[\frac{64^2}{5} + \frac{30^2}{5} + \frac{26^2}{5} \right] - 3(16)$$

$$= \frac{12}{240}[1134.400] - 48$$

$$= 56.720 - 48$$

$$= 8.720$$

$$H_{0.05,5,5,5} = 5.780$$

Reject H_0.

$$0.005 < P < 0.01$$

*To check whether ranks were assigned correctly, the sum of the ranks (or sum of the rank sums: $64+30+26 = 120$) is compared to $N(N+1)/2 = 15(16)/2 = 120$. This check will not guarantee that the ranks were assigned properly, but it will often catch errors of doing so.

Mann-Whitney test. A good check (but not a guarantee) of whether ranks have been assigned correctly is to see whether the sum of all the ranks equals $N(N + 1)/2$.

Critical values of H for small sample sizes where $k \leq 5$ are given in Appendix Table B.13. For larger samples and/or for $k > 5$, H may be considered to be approximated by χ^2 with $k - 1$ degrees of freedom. Chi-square, χ^2, is a statistical distribution that is shown in Appendix Table B.1, where probabilities are indicated as column headings and degrees of freedom (ν) designate the rows.

If there are tied ranks, as in Example 10.11, H is a little lower than it should be, and a correction factor may be computed as

$$C = 1 - \frac{\sum t}{N^3 - N}, \tag{10.40}$$

and the corrected value of H is

$$H_c = \frac{H}{C}. \tag{10.41}$$

Here,

$$\sum t = \sum_{i=1}^{m} (t_i^3 - t_i), \tag{10.42}$$

where t_i is the number of ties in the ith group of ties, and m is the number of groups of tied ranks. H_c will differ little from H when the t_i's are very small compared to N.

Kruskal and Wallis (1952) give two approximations that are better than chi-square when the n_i's are small or when significance levels less than 1% are desired; but they

EXAMPLE 10.11   The Kruskal-Wallis test with tied ranks.

A limnologist obtained eight containers of water from each of four ponds. The pH of each water sample was measured. The data are arranged in ascending order within each pond. (One of the containers from pond 3 was lost, so $n_3 = 7$, instead of 8; but the test procedure does not require equal numbers of data in each group.) The rank of each datum is shown parenthetically.

H_0:   pH is the same in all four ponds.

H_A: pH is not the same in all four ponds.

$\alpha = 0.05$

Pond 1	Pond 2	Pond 3	Pond 4
7.68 (1)	7.71(6*)	7.74 (13.5*)	7.71 (6*)
7.69 (2)	7.73 (10*)	7.75 (16)	7.71 (6*)
7.70 (3.5*)	7.74 (13.5*)	7.77 (18)	7.74 (13.5*)
7.70 (3.5*)	7.74 (13.5*)	7.78 (20*)	7.79 (22)
7.72 (8)	7.78 (20*)	7.80 (23.5*)	7.81 (26*)
7.73 (10*)	7.78 (20*)	7.81 (26*)	7.85 (29)
7.73 (10*)	7.80 (23.5*)	7.84 (28)	7.87 (30)
7.76 (17)	7.81 (26*)		7.91 (31)
*Tied ranks.			
$n_1 = 8$	$n_2 = 8$	$n_3 = 7$	$n_4 = 8$
$R_1 = 55$	$R_2 = 132.5$	$R_3 = 145$	$R_4 = 163.5$

$N = 8 + 8 + 7 + 8 = 31$

$$H = \frac{12}{N(N+1)} \sum_{i=1}^{k} \frac{R_i^2}{n_i} - 3(N+1)$$

$$= \frac{12}{31(32)} \left[\frac{55^2}{8} + \frac{132.5^2}{8} + \frac{145^2}{7} + \frac{163.5^2}{8} \right] - 3(32)$$

$$= 11.876$$

Number of groups of tied ranks $= m = 7$.

$$\sum t = \sum (t_i^3 - t_i)$$

$$= (2^3 - 2) + (3^3 - 3) + (3^3 - 3) + (4^3 - 4) + (3^3 - 3) + (2^3 - 2) + (3^3 - 3)$$

$$= 168$$

$$C = 1 - \frac{\sum t}{N^3 - N} = 1 - \frac{168}{31^3 - 31} = 1 - \frac{168}{29760} = 0.9944$$

$$H_c = \frac{H}{C} = \frac{11.876}{0.9944} = 11.943$$

$$\nu = k - 1 = 3$$

$$\chi^2_{0.05,3} = 7.815$$

Reject H_0.

$$0.005 < P < 0.01 \quad [P = 0.0076]$$

EXAMPLE 10.11   (continued)

$$F = \frac{(N-k)H_c}{(k-1)(N-1-H_c)} = \frac{(31-4)(11.943)}{(4-1)(31-1-11.943)} = 5.95$$

$$F_{0.05(1),3,26} = 2.98$$

Reject H_0.

$$0.0025 < P < 0.005 \quad [P = 0.0031]$$

are relatively complicated to use. The chi-square approximation is slightly conservative for $\alpha = 0.05$ or 0.10 (i.e., the true Type I probability is a little less than α) and more conservative for $\alpha = 0.01$ (Gabriel and Lachenbruch, 1969). Among the several other approximations that are better than χ^2 (Iman and Davenport, 1976),

$$F = \frac{(N-k)H}{(k-1)(N-1-H)} \tag{10.43}$$

(which is also the test statistic we would obtain by applying the ANOVA of Section 10.1 to ranks of the data) may be compared to critical values of F for degrees of freedom of $\nu_1 = k - 1$ and $\nu_2 = N - k - 1$ (rather than $N - k$).* This is demonstrated at the end of Example 10.11, where the result with F is found to be inconsequentially different from that with χ^2.

As in the parametric ANOVA, if we reject H_0 by the Kruskal-Wallis test, we do not know which groups differ from which other groups; all we know is that at least one difference among the k groups does exist. To locate the difference(s), a method such as that given in Section 11.6 would have to be employed.

10.5 Testing for Difference among Several Medians

Section 8.11 presented the *median test* for the two-sample case. This procedure may be expanded to multisample considerations (Mood, 1950: 398–399). The method requires the determination of the grand median of all observations in all k samples considered together. The number of data in each sample that are above and not above this median are tabulated, and the significance of the resultant $2 \times k$ contingency table is then analyzed, generally by chi-square (Section 23.1), alternatively by the G test (Section 23.7). For example, if there were four populations being compared, the statistical hypotheses would be H_0: All four populations have the same median, and H_A: All four populations do not have the same median. The median test would be the testing of the following

---

*A slightly better approximation in some cases is to compare

$$\frac{H}{2}\left[1 + \frac{N-k}{N-1-H}\right] \quad \text{to} \quad \frac{(k-1)F_{\alpha(1),k-1,N-k} + \chi^2_{\alpha,k-1}}{2}$$

(Iman and Davenport, 1976).

contingency table:

	Sample 1	Sample 2	Sample 3	Sample 4	Total
Above median	f_{11}	f_{12}	f_{13}	f_{14}	R_1
Not above median*	f_{21}	f_{22}	f_{23}	f_{24}	R_2
Total	C_1	C_2	C_3	C_4	n

*"Not above median" includes data below the median as well as data equal to the median.

This multisample median test is demonstrated in Example 10.12.* Hays (1963: 621) recommends the test as applicable when $n \geq 20$ and $C_j \geq 5$. If H_0 is rejected, then the method of Section 11.8 can be used to attempt to discern which population medians are different from which.

*This test assumes all k population distributions have the same shape, but there is typically little effect on the outcome if this is not so.

EXAMPLE 10.12   The multisample median test.

H_0:   Median elm tree height is the same on all four sides of a building.

H_A: Median elm tree height is not the same on all four sides of a building.

A total of forty eight seedlings were planted; twelve on each of a building's four sides. The heights, after several years of growth, were as follows:

North	East	South	West
7.1 m	6.9 m	7.8 m	6.4 m
7.2	7.0	7.9	6.6
7.4	7.1	8.1	6.7
7.6	7.2	8.3	7.1
7.6	7.3	8.3	7.6
7.7	7.3	8.4	7.8
7.7	7.4	8.4	8.2
7.9	7.6	8.4	8.4
8.1	7.8	8.6	8.6
8.4	8.1	8.9	8.7
8.5	8.3	9.2	8.8
8.8	8.5	9.4	8.9

medians: 7.7 m   7.35 m   8.4 m   8.0 m

grand median $= 7.9$ m

The 2×4 contingency table (with expected frequencies—see Section 6.1—in parentheses):

	North	East	South	West	
Above median	4 (5.7500)	3 (5.7500)	10 (5.7500)	6 (5.7500)	20
Not above median	8 (6.2500)	9 (6.2500)	2 (6.2500)	6 (6.2500)	28
Total	12	12	12	12	48

EXAMPLE 10.12   (continued)

$\chi^2 = 9.600$

$\chi^2_{0.05,3} = 7.815$

Reject H_0.

$$0.01 < P < 0.025 \quad [P = 0.022]$$

10.6 Homogeneity of Variances

If we have three or more samples, and we compute a variance for each, then we can test the hypothesis that all samples came from populations with identical variances (i.e., all the sample variances estimate the same population variance). This statement of the equality, or homogeneity, of variances can be written H_0: $\sigma_1^2 = \sigma_2^2 = \ldots = \sigma_k^2$, where k is the number of samples. *Homoscedasticity* is used as a synonym for homogeneity of variances; variance heterogeneity is then called *heteroscedasticity*.*

The most common method employed to test for homogeneity of variances is *Bartlett's test*[†] (Bartlett, 1937a, 1937b; based on a principle of Neyman and Pearson, 1931). In this procedure, the test statistic is

$$B = (\ln s_p^2) \left(\sum_{i=1}^{k} \nu_i \right) - \sum_{i=1}^{k} \nu_i \ln s_i^2, \tag{10.44}$$

where $\nu_i = n_i - 1$ and n_i is the size of sample i. The pooled variance, s_p^2, is calculated as before as $\sum_{i=1}^{k} SS_i / \sum_{i=1}^{k} \nu_i$. Many biologists prefer to operate with common logarithms (base 10), rather than with natural logarithms (base e);[‡] so Equation 10.44 may be written as

$$B = 2.30259[(\log s_p^2) \left(\sum_{i=1}^{k} \nu_i \right) - \sum_{i=1}^{k} \nu_i \log s_i^2]. \tag{10.45}$$

The distribution of B is approximated by the chi-square distribution,[§] with $k-1$ degrees of freedom (Appendix Table B.1), but a more accurate chi-square approximation is obtained by computing a correction factor,

$$C = 1 + \frac{1}{3(k-1)} \left(\sum_{i=1}^{k} \frac{1}{\nu_i} - \frac{1}{\sum_{i=1}^{k} \nu_i} \right), \tag{10.46}$$

---

*The two terms were introduced by K. Pearson in 1905 (Walker, 1929: 181); since then they have occasionally been spelled "homoskedasticity" and "heteroskedasticity," respectively.

[†]Maurice Stevenson Bartlett (b. 1910), English statistician.

[‡]See footnote in Section 8.7.

[§]A summary of approximations is given by Nagasenker (1984).

with the corrected test statistic being

$$B_c = \frac{B}{C}. \qquad (10.47)$$

Example 10.13 demonstrates these calculations. The null hypothesis for testing the homogeneity of the variances of four populations may be written symbolically as $H_0 : \sigma_1^2 = \sigma_2^2 = \sigma_3^2 = \sigma_4^2$, or, in words, as "the four population variances are homogeneous (i.e., are equal)." The alternate hypothesis can be stated as "The four population

---

EXAMPLE 10.13   Bartlett's test for homogeneity of variances.

Nineteen pigs were divided into four groups, and each group was given a different feed. The data are weights, in kilograms, and we wish to test whether the variance of weights is the same for pigs on all four feeds. (These are the same data as in Example 10.1.)

H_0: $\sigma_1^2 = \sigma_2^2 = \sigma_3^2 = \sigma_4^2$

H_A: The four population variances are heterogeneous (i.e., are not all equal).

$\alpha = 0.05$

	Feed 1	Feed 2	Feed 3	Feed 4	
	60.8	68.7	102.6	87.9	
	57.0	67.7	102.1	84.2	
	65.0	74.0	100.2	83.1	
	58.6	66.3	96.5	85.7	
	61.7	69.8		90.3	
SS_i	37.57	34.26	22.97	33.55	$\sum SS_i = 128.35$
v_i	4	4	3	4	$\sum v_i = 15$
s_i^2	9.39	8.56	7.66	8.39	
$\log s_i^2$	0.9727	0.9325	0.8842	0.9238	
$v_i \log s_i^2$	3.8908	3.7300	2.6526	3.6952	$\sum v_i \log s_i^2 = 13.9686$
$\dfrac{1}{v_i}$	0.250	0.250	0.333	0.250	$\sum \dfrac{1}{v_i} = 1.083$

$s_p^2 = \dfrac{\sum SS_i}{\sum v_i} = \dfrac{128.35}{15} = 8.56$

$\log s_p^2 = 0.9325$

$B = 2.30259 \left[(\log s_p^2) \left(\sum v_i \right) - \sum v_i \log s_i^2 \right]$

$\quad = 2.30259[(0.9325)(15) - 13.9686]$

$\quad = 2.30259(0.0189)$

$\quad = 0.0435$

$C = 1 + \dfrac{1}{3(k-1)} \left(\sum \dfrac{1}{v_i} - \dfrac{1}{\sum v_i} \right)$

$\quad = 1 + \dfrac{1}{3(3)} \left(1.083 - \dfrac{1}{15} \right)$

$\quad = 1.113$

$B_c = \dfrac{B}{C} = \dfrac{0.0435}{1.113} = 0.0391$

$\chi_{0.05, 3}^2 = 7.815$

Do not reject H_0.

$0.995 < P < 0.999 \quad [P = 0.998]$

variances are not homogeneous (i.e., they are not all equal)," or "There is difference (or heterogeneity) among the four population variances." If H_0 is rejected, the further testing of Section 11.9 will allow us to ask which population variances are different from which.

Bartlett's test is powerful if the sampled populations are normal, but it is very badly affected by non-normal populations (Box, 1953; Box and Anderson, 1955; Gartside, 1972). If the population distribution is platykurtic, the true α is less than the stated α (i.e., the test is conservative and the probability of a Type II error is increased); if it is leptokurtic, the true α is greater than the stated α (i.e., the probability of a Type I error is increased). Other variance homogeneity tests have been proposed, but they are also adversely affected by non-normality, or are low in power, or have other serious drawbacks (Brown and Forsythe, 1974c; Conover, Johnson, and Johnson, 1981; Games, Winkler, and Probert, 1972; Hall, 1972; Keselman, Games, and Clinch, 1979; Layard, 1973; Levy, 1978b, Seber, 1977: 147–148). Thus, the Bartlett test remains commendable when the sampled populations are normal, and no procedure is especially good when they are not. As a nonparametric test for differences in population dispersions, Stevens (1989) discusses an extension of the Siegel-Tukey procedure (see the end of Section 8.5) for $k > 2$.

When $k = 2$ and $n_1 = n_2$, Bartlett's test is equivalent to the two-tailed variance-ratio test of Section 8.5. With two samples of unequal size, the two procedures may yield different results, with one more powerful in some cases and the other more powerful in others (Maurais and Ouimet, 1986).

Because of the poor performance of tests for variance homogeneity, and the robustness of analysis of variance for multisample testing among means (Section 10.1), it is not recommended that the former be performed as tests of the underlying assumptions of the latter.

10.7 HOMOGENEITY OF COEFFICIENTS OF VARIATION

The two-sample procedure of Equation 8.42 has been extended by Feltz and Miller (1996) to situations where $k \geq 3$:

$$\chi^2 = \frac{\sum\limits_{i=1}^{k} v_i V_i^2 - \dfrac{\left[\sum\limits_{i=1}^{k} v_i V_i\right]^2}{\sum\limits_{i=1}^{k} v_i}}{V_p^2(0.5 + V_p^2)}, \tag{10.48}$$

where the common coefficient of variation is

$$V_p = \frac{\sum\limits_{i=1}^{k} v_i V_i}{\sum\limits_{i=1}^{k} v_i}. \tag{10.49}$$

This test statistic approximates the chi-square distribution with $k - 1$ degrees of freedom (Appendix Table B.1) and its computation is shown in Example 10.14. When $k = 2$, the test is identical to the two-sample test using Equation 8.42. As with other tests, the power is greater with larger sample size; for a given sample size, the power is greater for smaller coefficients of variation and for greater differences among coefficients of variation. If the null hypothesis of equal population coefficients of variation is not rejected, then V_p is the best estimate of the coefficient of variation common to all k populations.

EXAMPLE 10.14   Testing for homogeneity of coefficients of variation, using the data of Example 10.1

H_0:   The coefficients of variation of all four sampled populations are the same (i.e., $\mu_1/\sigma_1 = \mu_2/\sigma_2 = \mu_3/\sigma_3 = \mu_4/\sigma_4$).

H_A: The coefficients of variation of all four sampled populations are not the same.

	Feed 1	Feed 2	Feed 3	Feed 4
n:	5	5	4	5
ν:	4	4	3	4
$\bar{X}$ (kg):	60.62	69.30	100.35	86.24
s (kg):	3.06	2.93	2.77	2.90
V:	0.0505	0.0423	0.0276	0.0336

$$\sum_{i=1}^{k} \nu_i = 4 + 4 + 3 + 4 = 15$$

$$\sum_{i=1}^{k} \nu_i V_i = (4)(0.0505) + (4)(0.0423) + (3)(0.0276) + (4)(0.0336) = 0.5884$$

$$V_p = \frac{\sum_{i=1}^{k} \nu_i V_i}{\sum_{i=1}^{k} \nu_i} = \frac{0.5884}{15} = 0.0392$$

$$V_p^2 = (0.0392)^2 = 0.001537$$

$$\sum_{i=1}^{k} \nu_i V_i^2 = (4)(0.0505)^2 + (4)(0.0423)^2 + (3)(0.0276)^2 + (4)(0.0336)^2$$

$$= 0.0242$$

$$\chi^2 = \frac{\sum_{i=1}^{k} \nu_i V_i^2 - \frac{\left[\sum_{i=1}^{k} \nu_i V_i\right]^2}{\sum_{i=1}^{k} \nu_i}}{V_p^2(0.5 + V_p^2)} \qquad \begin{array}{l} \nu = k - 1 = 3 \\ \chi^2_{0.05,3} = 7.815 \\ \text{Do not reject } H_0. \end{array}$$

$$= \frac{0.0242 - \frac{(0.5884)^2}{15}}{(0.001537)(0.5 + 0.001537)}$$

$$= \frac{0.00112}{0.000771} = 1.453$$

$$0.50 < P < 0.75 \quad [P = 0.69]$$

10.8 THE EFFECT OF CODING

In the parametric ANOVA, coding the data by addition or subtraction of a constant causes no change in any of the sums of squares or mean squares (recall Section 4.8), so the resultant F and the ensuing conclusions are not affected at all. If the coding is performed by multiplying or dividing all the data by a constant, the sums of squares and the mean squares in the ANOVA each will be altered by an amount equal to the square of that constant, but the F value and the associated conclusions will remain unchanged.

A test utilizing ranks (such as the Kruskal-Wallis procedure) will not be affected at all by coding of the raw data. Thus, the coding of data for analysis of variance, either parametric or nonparametric, may be employed with impunity, and coding frequently renders data easier to manipulate. Neither will coding of data alter the conclusions from the hypothesis tests in Chapter 11 (multiple comparisons) or Chapters 12, 14, 15, or 16 (further analysis of variance procedures).

Bartlett's test is also unaffected by coding. The testing of coefficients of variation is unaffected by coding by multiplication or division, but coding by addition or subtraction may not be used.

10.9 MULTISAMPLE TESTING FOR NOMINAL-SCALE DATA

A $2 \times c$ contingency table may be analyzed to compare frequency distributions of nominal data for two samples. In a like fashion, an $r \times c$ contingency table may be set up to compare frequency distributions of nominal-scale data from r samples. Contingency table procedures are discussed in Chapter 23.

Other procedures have been proposed for multisample analysis of nominal-scale data (e.g., Light and Margolin 1971; Windsor, 1948).

EXERCISES

10.1 The following data are weights of food (in kilograms) consumed per day by adult deer collected at different times of the year. Test the null hypothesis that food consumption is the same for all the months tested.

Feb.	May	Aug.	Nov.
4.7	4.6	4.8	4.9
4.9	4.4	4.7	5.2
5.0	4.3	4.6	5.4
4.8	4.4	4.4	5.1
4.7	4.1	4.7	5.6
	4.2	4.8	

10.2 An experiment is to have its results examined by analysis of variance. The variable is temperature (in degrees Celsius), with twelve measurements to be taken in each of five experimental groups. From previous experiments, we estimate the within-groups variability, σ^2, to be $1.54(°C)^2$. If the 5% level of significance is employed, what is the probability of the ANOVA detecting a difference as small as 2.0°C between population means?

10.3 For the experiment of Exercise 10.2, how many replicates are needed in each of the five groups to detect a difference as small as 2.0°C between population means, with 95% power?

10.4 For the experiment of Exercise 10.2, what is the smallest difference between population means that we are 95% likely to detect with an ANOVA using ten replicates per group?

10.5 Using the Kruskal-Wallis test, test nonparametrically the appropriate hypotheses for the data of Exercise 10.1.

10.6 Three different methods were used to determine the dissolved oxygen content of lake water. Each of the three methods was applied to a sample of water six times, with the following results. Test the null hypothesis that the three methods yield equally variable results ($\sigma_1^2 = \sigma_2^2 = \sigma_3^2$).

Method 1 (mg/kg)	Method 2 (mg/kg)	Method 3 (mg/kg)
10.96	10.88	10.73
10.77	10.75	10.79
10.90	10.80	10.78
10.69	10.81	10.82
10.87	10.70	10.88
10.60	10.82	10.81

10.7 The following statistics were obtained from measurements of the circumferences of trees of four species. Test whether the coefficients of variation of circumferences are the same among the four species.

	Species B	Species A	Species Q	Species H
n:	40	54	58	32
$\bar{X}$ (m):	2.126	1.748	1.350	1.392
s^2 (m^2):	0.488219	0.279173	0.142456	0.203208

11

MULTIPLE COMPARISONS

Using a single-factor analysis of variance, we may test the null hypothesis H_0: $\mu_1 = \mu_2 = \ldots = \mu_k$. However, the rejection of H_0 does not imply that all k means are different from one another, and we know neither how many differences there are nor where differences are located among the k population means. For example, if $k = 3$, and H_0: $\mu_1 = \mu_2 = \mu_3$ is rejected, then we do not know whether H_A: $\mu_1 \neq \mu_2 = \mu_3$, or H_A: $\mu_1 = \mu_2 \neq \mu_3$, or H_A: $\mu_1 \neq \mu_2 \neq \mu_3$ is the appropriate alternate hypothesis.

As explained in the introduction to Chapter 10, it is generally invalid to employ multiple t tests to examine the differences between all possible pairs of means. The present chapter presents statistical procedures that may be used for such a purpose; they are called *multiple comparison tests*.* Although multiple comparisons are very often desired after a Model I analysis of variance, they generally are not applied in Model II situations. Except for the "least significant difference" test, mentioned below, these procedures may be performed even without a preliminary analysis of variance, but this would be atypical protocol.

The multiple-comparison problem has received much attention in the statistical literature, yet there is no agreement as to the "best" procedure to routinely employ. Among the most widely accepted and commonly used methods are the Tukey test (Tukey, 1953) and the Newman-Keuls test (Newman, 1939; Keuls, 1952), which will be discussed in Sections 11.1 and 11.2, respectively. The Duncan test (Duncan, 1955), often referred to as the "Duncan new multiple range test" because it succeeds an earlier procedure (Duncan, 1951), is also encountered; but it has a different theoretical basis, one that

---

*The term "multiple comparisons" was introduced by D. B. Duncan in 1951 (David, 1995).

is not as widely accepted as that of the Tukey and Newman-Keuls procedures, and it is declared by some (e.g., Day and Quinn, 1989) to perform poorly. (It is performed as is the Newman-Keuls test, except that different critical-value tables are required.) We may also encounter a procedure called the "least significant difference" (LSD) test. Although it is not generally recommended, some statisticians favor it for what Saville (1970) has called its "consistency," the characteristic of concluding difference among a pair of means independent of the number and magnitude of the means in the experiment. Other multiple-comparison procedures go by the name of Dunn or Bonferroni tests (see, e.g., Howell, 1987: 349–352).

Yet another technique, Scheffé's test (1953; 1959: Sections 3.4, 3.5) will be discussed in Section 11.5; it is especially suited for a kind of comparison called a multiple contrast. The relative performances of these as well as other, multiple comparison procedures are discussed by many authors.[*] Although not actually required by theory, multiple comparison testing is most commonly performed only if an analysis of variance first rejects a multisample hypothesis of equal means.[†] Thus, these tests are referred to as "a posteriori" tests.

In general, the multiple comparison tests for means have the same underlying assumptions as does the analysis of variance: population normality and homogeneity of variance. Although the Tukey test appears to be robust with respect to departures from these assumptions (Keselman, 1976), the robustness of each of the above procedures is not well-known, with adverse effects on both Type I and Type II errors possible if the assumptions are greatly violated. The homogeneity of variance assumption apparently is the more serious, and parametric multiple comparison testing should not be performed if heteroscedasticity is pronounced. If it is not comfortable to assume normality or homoscedasticity, then a nonparametric analysis of variance (Section 10.4) may be employed, followed by nonparametric multiple comparison testing (Section 11.6 or 11.7).[‡] Sections 11.8 and 11.9 deal with multiple comparisons among medians and variances, respectively.

In all multiple comparison testing, equal sample sizes are desirable for maximum power and robustness, but procedures are presented for analysis with unequal n.

This chapter considers multiple comparisons for the one-way analysis of variance experimental design. Other multiple comparison procedures are found in Sections 12.6 (for two-factor ANOVA), 12.9 (for nonparametric randomized block ANOVA), 12.10 (for dichotomous data in randomized blocks), 14.6 (for multiway ANOVA), 18.6 and 18.7 (for regression), and 19.8 (for correlation).

---

[*]For example, Bancroft (1968: Chapter 8); Day and Quinn (1989); Dunnett (1970); Federer (1955: Section II-1); Harter (1957; 1970: Section 2); Steel and Torrie (1980: Chapter 8); Waldo (1976); and those listed in the review of Miller (1981).

[†]This results (except for Scheffé's test) in the probability of Type I error tending to be less than α (Bernhardson, 1975).

[‡]Howell and Games (1974) and Games and Howell (1976) discuss Behrens-Fisher procedures for when variances are unequal.

11.1 THE TUKEY TEST

A much-used multiple comparison procedure is the Tukey test (also known as the "honestly significant difference test" and, sometimes, the "wholly significant difference test"). It considers the null hypothesis $H_0: \mu_B = \mu_A$ versus the alternate hypothesis $H_A: \mu_B \neq \mu_A$, where the subscripts denote any possible pair of groups. For k groups, $k(k-1)/2$ different pairwise comparisons can be made.* Example 11.1

---

*The number of combinations of k groups taken 2 at a time is (by Equation 5.10):

$$_kC_2 = \frac{k!}{2!(k-2)!} = \frac{k(k-1)(k-2)!}{2!(k-2)!} = \frac{k(k-1)}{2}. \qquad (11.1)$$

EXAMPLE 11.1   Tukey multiple comparison test with equal sample sizes. The data are strontium concentrations (mg/ml) in five different bodies of water.

First an analysis of variance performed.

H_0: $\mu_1 = \mu_2 = \mu_3 = \mu_4 = \mu_5$

H_A: Mean strontium concentrations are not the same in all five bodies of water.

$\alpha = 0.05$.

Grayson's Pond	Beaver Lake	Angler's Cove	Appletree Lake	Rock River
28.2	39.6	46.3	41.0	56.3
33.2	40.8	42.1	44.1	54.1
36.4	37.9	43.5	46.4	59.4
34.6	37.1	48.8	40.2	62.7
29.1	43.6	43.7	38.6	60.0
31.0	42.4	40.1	36.3	57.3
$\bar{X}_1 = 32.1$ mg/ml	$\bar{X}_2 = 40.2$ mg/ml	$\bar{X}_3 = 44.1$ mg/ml	$\bar{X}_4 = 41.1$ mg/ml	$\bar{X}_5 = 58.3$ mg/ml
$n_1 = 6$	$n_2 = 6$	$n_3 = 6$	$n_4 = 6$	$n_5 = 6$

Source of variation	SS	DF	MS
Total	2437.5720	29	
Groups	2193.4420	4	548.3605
Error	244.1300	25	9.7652

$$F = \frac{548.3605}{9.7652} = 56.2$$

As $F_{0.05(1),4,25} = 2.76$, H_0 is rejected.

Because a significant F resulted from the analysis of variance, the Tukey test is now applied on the means ranked in order of magnitude:

Samples ranked by mean (i)	1	2	4	3	5
Ranked sample means ($\bar{X}_i$, in mg/ml)	32.1	40.2	41.1	44.1	58.3

EXAMPLE 11.1   (continued)

To test each H_0: $\mu_B = \mu_A$,

$$SE = \sqrt{\frac{9.7652}{6}} = \sqrt{1.6275} = 1.28$$

As $q_{0.05,25,k}$ does not appear in Appendix Table B.5, $q_{0.05,24,k}$ is utilized (as it is the critical value with the next lower DF).

Comparison (B vs. A)	Difference $(\bar{X}_B - \bar{X}_A)$	SE	q	$q_{0.05,24,5}$	Conclusion
5 vs. 1	$58.3 - 32.1 = 26.2$	1.28	20.47	4.166	Reject H_0: $\mu_5 = \mu_1$
5 vs. 2	$58.3 - 40.2 = 18.1$	1.28	14.14	4.166	Reject H_0: $\mu_5 = \mu_2$
5 vs. 4	$58.3 - 41.1 = 17.2$	1.28	13.44	4.166	Reject H_0: $\mu_5 = \mu_4$
5 vs. 3	$58.3 - 44.1 = 14.2$	1.28	11.09	4.166	Reject H_0: $\mu_5 = \mu_3$
3 vs. 1	$44.1 - 32.1 = 12.0$	1.28	9.38	4.166	Reject H_0: $\mu_3 = \mu_1$
3 vs. 2	$44.1 - 40.2 = 3.9$	1.28	3.05	4.166	Accept H_0: $\mu_3 = \mu_2$
3 vs. 4	Do not test				
4 vs. 1	$44.1 - 32.1 = 9.0$	1.28	7.03	4.166	Reject H_0: $\mu_4 = \mu_1$
4 vs. 2	Do not test				
2 vs. 1	$40.2 - 32.1 = 8.1$	1.28	6.33	4.166	Reject H_0: $\mu_2 = \mu_1$

Overall conclusion: $\mu_1 \neq \mu_2 = \mu_4 = \mu_3 \neq \mu_5$.

demonstrates the testing procedure, utilizing data similar to those in Example 10.1, except that all groups have equal numbers of data (i.e., all the n_i's are equal). Since the single-factor analysis of variance rejects H_0: $\mu_1 = \mu_2 = \mu_3 = \mu_4 = \mu_5$, a multiple comparison test is called for to determine between which population means differences exist. The first step in the analysis is to arrange and number all five sample means in order of increasing magnitude. Then pairwise differences, $\bar{X}_B - \bar{X}_A$, are tabulated. Just as a difference between means divided by the appropriate standard error yields a t value (Section 8.1), a q value in the Tukey test is calculated by dividing a difference between means by

$$SE = \sqrt{\frac{s^2}{n}}, \tag{11.2}$$

where s^2 is the error mean square from the analysis of variance and n is the number of data in each of groups A and B. If this calculated q value,

$$q = \frac{\bar{X}_B - \bar{X}_A}{SE} \tag{11.3}$$

is equal to or greater than the critical value, $q_{\alpha,\nu,k}$, from Appendix Table B.5, then H_0: $\mu_B = \mu_A$ is rejected. The critical value in this test is known as a "Studentized range,"* abbreviated q, and is dependent upon α (the significance level), ν (the error DF for the analysis of variance), and k (the total number of means being tested).

The significance level, α, is the probability of encountering at least one Type I error (i.e., the probability of falsely rejecting at least one H_0) during the course of comparing

---

*E. S. Pearson and H. O. Hartley first used this term in 1953 (David, 1995).

all the pairs of means; it is not the probability of committing a Type I error for a single comparison. (That is, α for these multiple comparison tests is what statisticians refer to as "experimentwise error rate,"[*] rather than "comparisonwise error rate.")[†]

The conclusions reached by multiple comparison testing are dependent upon the order in which the pairwise comparisons are considered. The proper procedure is to compare first the largest mean against the smallest, then the largest against the next smallest, and so on, until the largest has been compared with the second largest. Then one compares the second largest with the smallest, the second largest with the next smallest, and so on. For example, if four means are ranked in ascending order, the sequence of comparisons is as follows: means 4 vs. 1, 4 vs. 2, 4 vs. 3, 3 vs. 1, 3 vs. 2, 2 vs. 1. Another important procedural rule is that if no difference is found between two means, then it is concluded that no difference exists between any means enclosed by these two, and such differences are not tested for. Thus, in Example 11.1, because we conclude no difference to exist between population means 3 and 2, no tests are performed to judge the differences between means 3 and 4, or between means 4 and 2. Our conclusions in Example 11.1 are that Sample 1 came from a population having a mean different from any of the other sampled populations; likewise, the population mean from which Sample 5 came is different from any of the other population means; and Samples 2, 4, and 3 came from populations having identical means. Therefore, we can conclude that $\mu_1 \neq \mu_2 = \mu_4 = \mu_3 \neq \mu_5$. In Example 11.1, each time a null hypothesis was accepted, a single line was drawn beneath the appropriate sample means to visualize the similarity of the population means that they estimate.

If the k group sizes are not equal, a slight modification in the Tukey procedure is necessary. For each comparison involving unequal n, the standard error[‡] is calculated by the following approximation (Kramer, 1956; Tukey, 1953):

$$\text{SE} = \sqrt{\frac{s^2}{2}\left(\frac{1}{n_A} + \frac{1}{n_B}\right)}. \tag{11.4}$$

This is shown in Example 11.2, using the analysis of variance data of Example 10.1.

---

[*]The experimentwise error rate is sometimes called the "familywise error rate," in reference to the entire set ("family") of comparisons being tested.

[†]The relationship between the experimentwise error rate, α_e, and the comparisonwise error rate, α_c, is $1 - \alpha_e = (1 - \alpha_c)^k$. As k increases, the number of pairwise comparisons, $k(k-1)/2$, increases greatly; thus Hamdy and El-Bassiouni (1993) present a procedure based upon α being the probability of *no more than one Type I error*.

[‡]This procedure has been shown to be excellent (e.g., Dunnett, 1980a; Keselman, Murray, and Rogan, 1976; Smith, 1971), often resulting in a probability of Type I error less than the stated α (Somerville, 1993; see Hayter, 1984, for proof). Some authors have used the harmonic mean of all k sample sizes, which usually results in a probability of Type I error differing from the stated α more than does the use of Equation 11.4, and this probability is at times greater than α (Dunnett, 1980a). Multiple comparison testing procedures have been proposed (and compared by Dunnett, 1980b; Tamhane, 1979), for cases where population variances are unequal, and Dunnett (1982) discussed the situation of normality, but if heteroscedasticity or non-normality is great enough to invalidate ANOVA, then we should consider using both nonparametric ANOVA and nonparametric multiple comparison testing.

EXAMPLE 11.2   The Tukey test with unequal sample sizes.

The data (in kg) are those from Example 10.1, where $H_0: \mu_1 = \mu_2 = \mu_3 = \mu_4$ was rejected.

$k = 4$

$s^2 = $ error MS $= 8.557$

error DF $= 15$

Samples ranked by mean (i):	1	2	4	3
Ranked sample means ($\bar{X}_i$):	60.62	69.30	86.24	100.35
Sizes of samples (n_i):	5	5	5	4

Comparison (B vs. A)	Difference ($\bar{X}_B - \bar{X}_A$)	SE	q	$q_{0.05, 15, 4}$	Conclusion
3 vs. 1	$100.35 - 60.62 = 39.73$	1.39*	28.58	4.076	Reject H_0: $\mu_3 = \mu_1$
3 vs. 2	$100.35 - 69.30 = 31.05$	1.39	22.34	4.076	Reject H_0: $\mu_3 = \mu_2$
3 vs. 4	$100.35 - 86.24 = 14.11$	1.39	10.15	4.076	Reject H_0: $\mu_3 = \mu_4$
4 vs. 1	$86.24 - 60.62 = 25.62$	1.31†	19.56	4.076	Reject H_0: $\mu_4 = \mu_1$
4 vs. 2	$86.24 - 69.30 = 16.94$	1.31	12.93	4.076	Reject H_0: $\mu_4 = \mu_2$
2 vs. 1	$69.30 - 60.62 = 8.68$	1.31	6.63	4.076	Reject H_0: $\mu_2 = \mu_1$

*As $n_3 \neq n_1$:

$$ SE = \sqrt{\left(\frac{8.557}{2}\right)\left(\frac{1}{4} + \frac{1}{5}\right)} = \sqrt{(4.2785)(0.25 + 0.20)} = \sqrt{1.9253} = 1.39 $$

†As $n_4 = n_1$:

$$ SE = \sqrt{\frac{8.557}{5}} = \sqrt{1.7114} = 1.31 $$

Thus, we conclude that $\mu_1 \neq \mu_2 \neq \mu_4 \neq \mu_3$.

The null hypothesis $H_0: \mu_B = \mu_A$ may, of course, also be written $H_0: \mu_B - \mu_A = 0$. The hypothesis $H_0: \mu_B - \mu_A = \mu_0$, where $\mu_0 \neq 0$, may also be tested; this is accomplished by replacing $\bar{X}_B - \bar{X}_A$ with $|\bar{X}_B - \bar{X}_A| - \mu_0$ in the numerator of Equation 11.3.

Not infrequently, a multiple comparison test will yield ambiguous results, in the form of conclusions of overlapping sets of similarities. For example, one might arrive at the following:

$$ \underline{\bar{X}_1 \quad \underline{\bar{X}_2 \quad \bar{X}_3 \quad \bar{X}_4}} $$

for an experimental design consisting of four groups of data. Here the four samples appear to have come from populations among which there were two different population means: Samples 1 and 2 were taken from one population and Samples 2, 3, and 4 came from the second population. This is clearly impossible, for we have assigned Sample 2 to both populations. The reason behind such a conclusion is that the multiple range test was not able to determine accurately from which population Sample 2 came (i.e., at least one Type II error has been committed). Therefore, we can state that $\mu_1 \neq \mu_3 = \mu_4$, but we cannot conclude how μ_2 is related to the other population means. Repeating the analysis with a larger number of data would tend to yield more decisive conclusions, as the test would then have more power.

It is also possible for H_0: $\mu_1 = \mu_2 = \ldots = \mu_k$ to be rejected by an analysis of variance and for the subsequent multiple comparison test to fail to detect differences between any pair of means. This occurrence will not be encountered commonly, but it reflects the fact that the analysis of variance is a more powerful test than is the multiple comparison test (i.e., Type II errors are more likely to occur in multiple comparison testing than in performing an analysis of variance). Repeating the experiment with larger sample sizes would tend to result in a multiple comparison analysis more capable of locating differences among means.

11.2 THE NEWMAN-KEULS TEST

The Newman-Keuls test, also commonly referred to as the "Student-Newman-Keuls test" or "SNK test," is performed exactly as is Tukey's test, with one exception. The sample means are ranked, pairwise differences between means are determined, and a standard error is computed, just as in Section 11.1. Then the test statistic, q, is calculated just as is q for Tukey's test (i.e., by Equation 11.3). The difference in the Newman-Keuls procedure lies in the determination of the critical value, which is $q_{\alpha,\nu,p}$, where p is the number of means in the range of means being tested. To compare ranked means $\bar{X}_5$ and $\bar{X}_1$, for example, we are considering a range of 5 means, so $p = 5$; to test mean 5 vs. mean 2, $p = 4$, and so on; this is shown in Example 11.3.

A multiple comparison test of the type that employs different critical values for different ranges of means is called a *multiple range test*. Example 11.3 analyzes the data of Example 11.1 using the Newman-Keuls procedure.

If all k sample sizes are not equal, then SE is calculated using Equation 11.4, rather than Equation 11.2, just as for the Tukey test.

In Examples 11.1 and 11.3, the Tukey and Newman-Keuls tests arrived at the same conclusions. This will not always be the case. The Newman-Keuls test tends to conclude more significant differences than does Tukey's procedure. (That is, the former is more powerful.) While some authors (e.g., Miller, (1981: 44) have criticized the Tukey test as being a little conservative (i.e., too few significant differences are concluded), others (e.g., Einot and Gabriel, 1975; Ramsey, 1978) recommend against the Newman-Keuls test because it may falsely declare significant differences with a probability greater than α (see also Day and Quinn, 1989). While the lack of agreement among statisticians continues, I tend to favor the latter argument and shall recommend (although not vigorously) the Tukey test in preference to the Newman-Keuls.

As a compromise between these two procedures, Tukey proposed employing as critical values the mean of the critical values of the two (i.e., the mean of $q_{\alpha,\nu,k}$ and $q_{\alpha,\nu,p}$). This is referred to as the "wholly significant difference test," a name also occasionally used synonymously with the Tukey test of Section 11.1.

EXAMPLE 11.3   The Newman-Keuls multiple range test applied to the data (in mg/ml) of Example 11.1.

First, an analysis of variance is performed and the sample means are ranked. just as in Example 11.1:

Samples ranked by mean (i)	1	2	4	3	5
Ranked sample means ($\bar{X}_i$)	32.1	40.2	41.1	44.1	58.3

SE = 1.28

Comparison (B vs. A)	Difference ($\bar{X}_B - \bar{X}_A$)	SE	q	p	$q_{0.05, 24, p}$	Conclusion
5 vs. 1	26.2	1.28	20.47	5	4.166	Reject H_0: $\mu_5 = \mu_1$
5 vs. 2	18.1	1.28	14.14	4	3.901	Reject H_0: $\mu_5 = \mu_2$
5 vs. 4	17.2	1.28	13.44	3	3.532	Reject H_0: $\mu_5 = \mu_4$
5 vs. 3	14.2	1.28	11.09	2	2.919	Reject H_0: $\mu_5 = \mu_3$
3 vs. 1	12.0	1.28	9.38	4	3.901	Reject H_0: $\mu_3 = \mu_1$
3 vs. 2	3.9	1.28	3.05	3	3.532	Accept H_0: $\mu_3 = \mu_2$
3 vs. 4	Do not test					
4 vs. 1	9.0	1.28	7.03	3	3.532	Reject H_0: $\mu_4 = \mu_1$
4 vs. 2	Do not test					
2 vs. 1	8.1	1.28	6.33	2	2.919	Reject H_0: $\mu_2 = \mu_1$

Overall conclusion: $\mu_1 \neq \mu_2 = \mu_4 = \mu_3 \neq \mu_5$.

11.3 CONFIDENCE INTERVALS FOLLOWING MULTIPLE COMPARISONS

Once it has been concluded which of three or more sample means are significantly different from which others, we can calculate confidence intervals for each different population mean. If one mean is declared different from all others, then we use Equation 10.31, introduced in Section 10.2:

$$\bar{X}_i \pm t_{\alpha(2), v}\sqrt{\frac{s^2}{n_i}}. \tag{10.31}$$

If two or more means are concluded to be the same, then a pooled sample mean is the best estimate of the underlying population mean:

$$\bar{X}_p = \frac{\sum n_i \bar{X}_i}{\sum n_i}, \tag{11.5}$$

where the summation is over all samples concluded to have come from the same population, and the confidence interval is, then,

$$\bar{X}_p \pm t_{\alpha(2), v}\sqrt{\frac{s^2}{\sum n_i}}, \tag{11.6}$$

again summing over all samples whose means are declared indistinguishable. This is analogous to the two-sample situation handled by Equation 8.16, and it is demonstrated in Example 11.4.

EXAMPLE 11.4   Confidence intervals (CI) for the population means from Example 11.1

It was concluded in Example 11.1 that $\mu_1 \neq \mu_2 = \mu_4 = \mu_3 \neq \mu_5$. Therefore, we may calculate confidence intervals for μ_1 for $\mu_{2,4,3}$ and for μ_5 (where $\mu_{2,4,3}$ indicates the mean of the common population from which Samples 2, 4, and 3 came).

Using Equation 10.31:

$$95\% \text{ CI for } \mu_1 = \bar{X}_1 \pm t_{0.05(2),25}\sqrt{\frac{s^2}{n_1}} = 32.1 \pm (2.060)\sqrt{\frac{9.7652}{6}}$$

$$= 32.1 \text{ mg/ml} \pm 2.6 \text{ mg/ml}$$

Again using Equation 10.31:

$$95\% \text{ CI for } \mu_5 = \bar{X}_5 \pm t_{0.05(2),25}\sqrt{\frac{s^2}{n_5}} = 58.3 \text{ mg/ml} \pm 2.6 \text{ mg/ml}$$

Using Equation 11.5:

$$\bar{X}_p = \bar{X}_{2,4,3} = \frac{n_2\bar{X}_2 + n_4\bar{X}_4 + n_3\bar{X}_3}{n_2 + n_4 + n_3} = \frac{(6)(40.2) + (6)(41.1) + (6)(44.1)}{6+6+6} = 41.8 \text{ mg/ml}$$

Using Equation 11.6:

$$95\% \text{ CI for } \mu_{2,4,3} = \bar{X}_{2,4,3} \pm t_{0.05(2),25}\sqrt{\frac{s^2}{6+6+6}} = 41.8 \text{ mg/ml} \pm 1.5 \text{ mg/ml}$$

Using Equation 11.7:

$$95\% \text{ CI for } \mu_5 - \mu_{2,4,3} = \bar{X}_5 - \bar{X}_{2,4,3} \pm q_{0.05,25,5}\sqrt{\frac{s^2}{2}\left(\frac{1}{n_5} + \frac{1}{n_2 + n_4 + n_3}\right)}$$

$$= 58.3 - 41.8 \pm (4.166)(1.04)$$

$$= 16.5 \text{ mg/ml} \pm 4.3 \text{ mg/ml}$$

Using Equation 11.7:

$$95\% \text{ CI for } \mu_{2,4,3} - \mu_1 = \bar{X}_{2,4,3} - \bar{X}_1 \pm q_{0.05,25,5}\sqrt{\frac{s^2}{2}\left(\frac{1}{n_2 + n_4 + n_3} + \frac{1}{n_1}\right)}$$

$$= 41.8 - 32.1 \pm (4.166)(1.04)$$

$$= 9.7 \text{ mg/ml} \pm 4.3 \text{ mg/ml}$$

If μ_A and μ_B are concluded to be different, the $1 - \alpha$ confidence interval for $\mu_B - \mu_A$ may be computed by Tukey's procedure as

$$(\bar{X}_B - \bar{X}_A) \pm (q_{\alpha,\nu,k})(\text{SE}). \tag{11.7}$$

Here, as in Section 11.1, ν is the error degrees of freedom in the analysis of variance, k is the total number of means, and SE is calculated as in either Equation 11.2 or 11.4, depending on whether or not n_A and n_B are equal. Thus, for example, the 95% confidence interval for $\mu_4 - \mu_1$ in Example 11.2 would be calculated as $(86.24 - 60.62) \pm (4.076)(1.31) = 25.62 \text{ kg} \pm 5.34 \text{ kg}$ and $L_1 = 20.28 \text{ kg}$ and $L_2 = 30.96 \text{ kg}$. This procedure is further demonstrated in Example 11.4, for the data of Example 11.1.

The Tukey procedure for computing confidence intervals for differences between means should be used regardless of whether the Tukey or the Newman-Keuls test was used to determine which means are different.

Sample Size and Estimation of the Difference between Two Population Means. In Section 8.3 it was shown how to estimate the sample size required in order to obtain a confidence interval of specified width for a difference between the population means associated with a two-sample test. In the multisample situation, we speak of all differences between pairs of population means, using a similar procedure but one that employs q rather than the t statistic. As in Section 8.3, iteration is necessary, whereby we determine n such that

$$n = \frac{s^2(q_{\alpha, \text{DF}, k})^2}{d^2}. \tag{11.8}$$

Here, d is the half-width of the $1 - \alpha$ confidence interval, s^2 is the estimate of error variance, and k is the total number of means. DF is the error degrees of freedom that the experiment would have with the estimated n (i.e., $\text{DF} = k(n - 1)$ for a single-factor ANOVA).

11.4 COMPARISON OF A CONTROL MEAN TO EACH OTHER GROUP MEAN

Sometimes the objective of multisample experiments with k samples, or groups, is to determine whether the mean of one group, designated as a "control," differs significantly from each of the means of the $k - 1$ other groups. Dunnett (1955) has provided a procedure for such testing, which differs from the multiple comparison approaches in Sections 11.1 and 11.2, in that the investigator is here not interested in all possible comparisons of pairs of group means, but only in those $k - 1$ comparisons involving the "control" group. Knowing k, the total number of groups in the experiment, and ν, the error degrees of freedom from the analysis of variance for H_0: $\mu_1 = \mu_2 = \ldots = \mu_k$, one obtains critical values from either Appendix Table B.6 or Appendix Table B.7, depending on whether the hypotheses are to be one-tailed or two-tailed, respectively. We shall refer to these tabled values as $q'_{\alpha, \nu, k}$, for they are used in a manner similar to that of the $q_{\alpha, \nu, k}$ values employed in the Tukey test. As in the latter procedure, the error rate, α, denotes the probability of committing a Type I error somewhere among all of the pairwise comparisons made. The standard error for Dunnett's test is

$$\text{SE} = \sqrt{\frac{2s^2}{n}} \tag{11.9}$$

where group sizes are equal, or

$$\text{SE} = \sqrt{s^2\left(\frac{1}{n_A} + \frac{1}{n_{\text{control}}}\right)} \tag{11.10}$$

when group sizes are not equal. The test statistic, analogous to that of Equation 11.3, is

$$q = \frac{\bar{X}_{\text{control}} - \bar{X}_A}{\text{SE}}. \tag{11.11}$$

The Dunnett testing procedure is demonstrated in Example 11.5 for one-tailed hypotheses. The critical values, $q'_{\alpha(1),\nu,k}$ are in Appendix Table B.6. For a two-tailed test, if $|q| \geq q'_{\alpha(2),\nu,k}$, then H_0: $\mu_{\text{control}} = \mu_A$ is rejected. In a one-tailed test H_0: $\mu_{\text{control}} \leq \mu_A$ would be rejected if $q \geq q'_{\alpha(1),\nu,k}$; and H_0: $\mu_{\text{control}} \geq \mu_A$ would be rejected if $|q| \geq q'_{\alpha(1),\nu,k}$ and $\bar{X}_{\text{control}} < \bar{X}_A$ (i.e., if $q \leq -q'_{\alpha(1),\nu,k}$).

The null hypothesis H_0: $\mu_{\text{control}} = \mu_A$ is, of course, a special case of H_0: $\mu_{\text{control}} - \mu_A = \mu_0$, where $\mu_0 = 0$. Other values of μ_0 may appear in the hypothesis, however, and such hypotheses may be tested by placing $|\bar{X}_{\text{control}} - \bar{X}_A| - \mu_0$ in the numerator of the q calculation. In a similar manner, H_0: $\mu_{\text{control}} - \mu_A \leq \mu_0$ or H_0: $\mu_{\text{control}} - \mu_A \geq \mu_0$ may be tested.

EXAMPLE 11.5   Dunnett's test for comparing a control mean to each other group mean.

The yield (in metric tons per hectare) of each of several plots of potatoes has been determined after a season's application of a standard fertilizer. Likewise, the potato yields from several plots were determined for each of four new fertilizers. A manufacturer wishes to sell us one of the four new fertilizers, claiming they will increase potato crop yield. A total of eighty plots is available for use in this experiment.

Optimum allocation of plots among the five fertilizer groups will be such that the control group (let us say it is group 2) has a little less than $\sqrt{k-1} = \sqrt{4} = 2$ times as many data as each of the other groups. Therefore, we use $n_2 = 24$ and $n_1 = n_3 = n_4 = n_5 = 14$.

An analysis of variance was performed and H_0: $\mu_1 = \mu_2 = \mu_3 = \mu_4 = \mu_5$ was rejected; for this analysis, error MS = 10.42 (metric tons/ha)^2 and error DF = 75.

$$\text{SE} = \sqrt{10.42 \left(\frac{1}{14} + \frac{1}{24} \right)} = 1.1 \text{ metric tons/ha}$$

Groups ranked by mean (i):	1	2	3	4	5
Ranked group means ($\bar{X}_i$):	17.3	21.7	22.1	23.6	27.8

As the control group (i.e., the group with the standard fertilizer) is group 2, we wish to test each H_0: $\mu_2 \geq \mu_A$ against H_A: $\mu_2 < \mu_A$.

Comparison (2 vs. A)	Difference ($\bar{X}_2 - \bar{X}_A$)	SE	$|q|$	$q'_{0.05(1),75,5}$	Conclusion
2 vs. 1	$21.7 - 17.3 = 4.4$				As $\bar{X}_2 > \bar{X}_1$, accept H_0: $\mu_2 \geq \mu_1$.
2 vs. 5	$27.8 - 21.7 = 6.1$	1.1	5.55	2.21	Reject H_0: $\mu_2 \geq \mu_5$.
2 vs. 4	$23.6 - 21.7 = 1.9$	1.1	1.73	2.21	Accept H_0: $\mu_2 \geq \mu_4$.
2 vs. 3	Do not test				

Conclusion: Only fertilizer 5 produces a yield greater than the yield from the control fertilizer (fertilizer 2).

When comparisons of group means to a control mean is an investigator's principal desire, the control group ought to contain more observations than the other groups. Dunnett (1955) showed that the optimal size of the control group typically should be a little less than $\sqrt{k-1}$ times the size of each other group.

Sample Size and Estimation of the Difference between One Population Mean and the Mean of a Control Population. This situation is similar to that discussed at the end of Section 11.3, but it pertains to one of the two means of the population being designated as the control. The procedure would use Equation 11.8 substituting the Dunnett's test critical value, $q'_{\alpha(2),\text{DF},k}$ for the Tukey critical value, $q_{\alpha(2),\text{DF},k}$, and inserting 2 into the numerator.

Confidence Intervals for Differences between Control and Other Group Means. Using Dunnett's q' statistic, and the SE of Equation 11.9 or 11.10, we may calculate confidence limits for the difference between the control mean and each of the other group means:

$$1 - \alpha \text{ CI for } \mu_{\text{control}} - \mu_A = (\bar{X}_{\text{control}} - \bar{X}_A) \pm (q'_{\alpha(2),v,k})(\text{SE}). \qquad (11.12)$$

One-tailed confidence limits are also possible. One can express 95% confidence that a difference, $\mu_{\text{control}} - \mu_A$, is no less than (i.e., is at least as large as)

$$(\bar{X}_{\text{control}} - \bar{X}_A) - (q'_{\alpha(1),v,k})(\text{SE}), \qquad (11.13)$$

or it might be desired to state that the difference is no greater than

$$(\bar{X}_{\text{control}} - \bar{X}_A) + (q'_{\alpha(1),v,k})(\text{SE}). \qquad (11.14)$$

11.5 SCHEFFÉ'S MULTIPLE CONTRASTS

Scheffé* (1953; 1959: Sections 3.4, 3.5) has provided a multiple comparison procedure (sometimes called the S test) that can be used to test null hypotheses of the form H_0: $\mu_B - \mu_A = 0$, just as can the Tukey or Newman-Keuls tests (but with less power); it can also be used for the special considerations that will be shown presently.

For H_0: $\mu_B - \mu_A = 0$, one computes

$$S = \frac{|\bar{X}_B - \bar{X}_A|}{\text{SE}}, \qquad (11.15)$$

where

$$\text{SE} = \sqrt{s^2 \left(\frac{1}{n_B} + \frac{1}{n_A} \right)}; \qquad (11.16)$$

and the critical value is

$$S_\alpha = \sqrt{(k-1) F_{\alpha(1),k-1,N-k}}. \qquad (11.17)$$

---

*Henry Scheffé (1907–1977), American statistician.

In the preceding equations, s^2 is the error mean square, and $k - 1$ and $N - k$ are the groups DF and error DF, respectively, from the analysis of variance. Note that this test does not require equal sample sizes.

Example 11.6 demonstrates Scheffé's procedure for the same data as analyzed in Example 11.1. In this case, the use of this test yields the same conclusions as does the Tukey test. Such will not always occur; in fact, it is common to have the former procedure detect fewer differences than the latter. That is, Scheffé's test is more apt to have us commit Type II errors. Because of its relative lack of power it is not generally recommended for the types of multiple comparisons thus far described.

EXAMPLE 11.6   Scheffé's test for the hypotheses and data of Example 11.1.

For $\alpha = 0.05$, the critical value S_α, for each comparison, is

$$\sqrt{(k-1)F_{0.05(1),k-1,N-k}} = \sqrt{(4)(2.76)} = \sqrt{11.04} = 3.32.$$

As $n_B = n_A$ for each comparison, the SE for each comparison is

$$\sqrt{s^2\left(\frac{1}{n_B} + \frac{1}{n_A}\right)} = \sqrt{(9.7652)(0.33)} = \sqrt{3.2547} = 1.80.$$

Comparison (B vs. A)	Difference $(\bar{X}_B - \bar{X}_A)$	SE	S	S_α	Conclusion
5 vs. 1	26.2	1.80	14.56	3.32	Reject H_0: $\mu_5 - \mu_1 = 0$
5 vs. 2	18.1	1.80	10.06	3.32	Reject H_0: $\mu_5 - \mu_2 = 0$
5 vs. 4	17.2	1.80	9.56	3.32	Reject H_0: $\mu_5 - \mu_4 = 0$
5 vs. 3	14.2	1.80	7.89	3.32	Reject H_0: $\mu_5 - \mu_3 = 0$
3 vs. 1	12.0	1.80	6.67	3.32	Reject H_0: $\mu_3 - \mu_1 = 0$
3 vs. 2	3.9	1.80	2.17	3.32	Accept H_0: $\mu_3 - \mu_2 = 0$
3 vs. 4	Do not test				
4 vs. 1	9.0	1.80	5.00	3.32	Reject H_0: $\mu_4 - \mu_1 = 0$
4 vs. 2	Do not test				
2 vs. 1	8.1	1.80	4.50	3.32	Reject H_0: $\mu_2 - \mu_1 = 0$

Overall conclusion: $\mu_1 \neq \mu_2 = \mu_4 = \mu_3 \neq \mu_5$.

However, there are hypotheses for which the S test is better qualified than other procedures. These refer to considerations spoken of as "multiple contrasts." Utilizing once again the data of Example 11.1, one might have proposed the hypothesis H_0: the mean strontium concentration in areas 2, 4, and 3 is no different from the mean concentration in area 5. This hypothesis could be stated concisely as H_0: $(\mu_2 + \mu_4 + \mu_3)/3 - \mu_5 = 0$. The S test then considers that $(\mu_2 + \mu_4 + \mu_3)/3 - \mu_5$ can be expressed as $\mu_2/3 + \mu_4/3 + \mu_3/3 - \mu_5$, so that μ_2, μ_4, μ_3, and μ_5 are preceded by coefficients, c_i, of $c_2 = \frac{1}{3}, c_4 = \frac{1}{3}, c_3 = \frac{1}{3}$, and $c_5 = -1$, respectively. The test statistic, S, is calculated as

$$S = \frac{\left|\sum c_i \bar{X}_i\right|}{\text{SE}}, \tag{11.18}$$

where

$$SE = \sqrt{s^2 \left(\sum \frac{c_i^2}{n_i} \right)}. \tag{11.19}$$

Note that the sum of the coefficients is always zero; i.e., $\sum c_i = 0$.

With $k = 5$ we could consider H_0: $\mu_1 - (\mu_2 + \mu_3 + \mu_4)/3 = 0$, or H_0: $(\mu_1 + \mu_5)/2 - (\mu_2 + \mu_3 + \mu_4)/3 = 0$, H_0: $(\mu_1 + \mu_4)/2 - (\mu_2 + \mu_3)/2 = 0$, or other contrasts. Several of these hypotheses are tested in Example 11.7. In grouping means to be used as contrasts, we should have some reasonable basis for defining the groups. For example, hypotheses are tested in Example 11.7. In grouping means to be used as

---

EXAMPLE 11.7   Scheffé's test for multiple contrasts, using the data of Example 11.1.

For $\alpha = 0.05$, the critical value, S_α, for each contrast is $\sqrt{(k-1) F_{0.05(1), k-1, N-k}}$

$$= \sqrt{(5-1) F_{0.05(1), 4, 25}}$$

$$= \sqrt{4(2.76)}$$

$$= 3.32$$

Contrast	*SE*	*S*	*Conclusion*
$\dfrac{\bar{X}_2 + \bar{X}_3 + \bar{X}_4}{3} - \bar{X}_5$ $= 41.8 - 58.3$ $= -16.5$	$\sqrt{9.7652 \left[\dfrac{\left(\frac{1}{3}\right)^2}{6} + \dfrac{\left(\frac{1}{3}\right)^2}{6} + \dfrac{\left(\frac{1}{3}\right)^2}{6} + \dfrac{(1)^2}{6} \right]} = 1.47$	11.22	Reject H_0: $\dfrac{\mu_2 + \mu_3 + \mu_4}{3} - \mu_5 = 0$
$\bar{X}_1 - \dfrac{\bar{X}_2 + \bar{X}_3 + \bar{X}_4}{3}$ $= 32.1 - 41.8$ $= -9.7$	$\sqrt{9.7652 \left[\dfrac{(1)^2}{6} + \dfrac{\left(\frac{1}{3}\right)^2}{6} + \dfrac{\left(\frac{1}{3}\right)^2}{6} + \dfrac{\left(\frac{1}{3}\right)^2}{6} \right]} = 1.47$	6.60	Reject H_0: $\mu_1 - \dfrac{\mu_2 + \mu_3 + \mu_4}{3}$ $= 0$
$\dfrac{\bar{X}_1 + \bar{X}_5}{2}$ $- \dfrac{\bar{X}_2 + \bar{X}_3 + \bar{X}_4}{3}$ $= 45.2 - 41.8$ $= 3.4$	$\sqrt{9.7652 \left[\dfrac{\left(\frac{1}{2}\right)^2}{6} + \dfrac{\left(\frac{1}{2}\right)^2}{6} + \dfrac{\left(\frac{1}{3}\right)^2}{6} + \dfrac{\left(\frac{1}{3}\right)^2}{6} + \dfrac{\left(\frac{1}{3}\right)^2}{6} \right]}$ $= 1.16$	2.93	Accept H_0: $\dfrac{\mu_1 + \mu_5}{2}$ $- \dfrac{\mu_2 + \mu_3 + \mu_4}{3}$ $= 0$
$\dfrac{\bar{X}_1 + \bar{X}_4}{2} - \dfrac{\bar{X}_2 + \bar{X}_3}{2}$ $= 36.6 - 42.15$ $= -5.55$	$\sqrt{9.7652 \left[\dfrac{\left(\frac{1}{2}\right)^2}{6} + \dfrac{\left(\frac{1}{2}\right)^2}{6} + \dfrac{\left(\frac{1}{2}\right)^2}{6} + \dfrac{\left(\frac{1}{2}\right)^2}{6} \right]} = 1.28$	4.34	Reject H_0: $\dfrac{\mu_1 + \mu_4}{2}$ $- \dfrac{\mu_2 + \mu_3}{2} = 0$

the first body of water in Example 11.1 might be rural and the others urban; or the first four might be standing water and the fifth a river.

If one group of data has been designated a "control" to which all other groups are to be compared, we may compare each group to that control by Dunnett's test (Section 11.4). Combinations of groups may be compared to the control using Scheffé's test, although the test proposed by Shaffer (1977) may be more powerful for simple contrasts and large k.

Confidence Intervals for Contrasts. For the difference between two means, we may calculate a $1 - \alpha$ confidence interval by Tukey's method (Equation 11.7) or by the Scheffé procedure:

$$(\bar{X}_B - \bar{X}_A) \pm S_\alpha \sqrt{s^2 \left(\frac{1}{n_A} + \frac{1}{n_B} \right)},$$ (11.20)

but the former will result in a narrower confidence interval and is, therefore, preferable.

However, the Scheffé procedure enables us to establish confidence limits for a contrast. In general, the $1 - \alpha$ confidence interval for a contrast is

$$\sum c_i \bar{X}_i + S_\alpha \text{SE},$$ (11.21)

(of which Equation 11.20 is a special case). SE is as in Equation 11.19. Example 11.8 demonstrates this for two of the significantly different contrasts of Example 11.7.

EXAMPLE 11.8   Confidence intervals for the significantly different contrasts of Example 11.7.

The critical value, S_α, for each confidence interval is $\sqrt{(k-1) F_{\alpha(1), k-1, N-k}}$. This is the same critical value as found in Example 11.7; that is, $S_{0.05} = 3.32$.

95% confidence interval for $\dfrac{\mu_2 + \mu_3 + \mu_4}{3} - \mu_5$

$$= \left(\frac{\bar{X}_2 + \bar{X}_3 + \bar{X}_4}{3} - \bar{X}_5 \right) \pm S_\alpha \text{SE}$$

$$= -16.5 \pm (3.32)(1.47)$$

$$= -16.5 \text{ mg/ml} \pm 4.9 \text{ mg/ml}$$

$L_1 = -21.4$ mg/ml
$L_2 = -11.6$ mg/ml

95% confidence interval for $\mu_1 - \dfrac{\mu_2 + \mu_3 + \mu_4}{3}$

$$= \left(\bar{X}_1 - \frac{\bar{X}_2 + \bar{X}_3 + \bar{X}_4}{3} \right) \pm S_\alpha \text{SE}$$

$$= -9.7 \pm (3.32)(1.47)$$

$$= -9.7 \text{ mg/ml} \pm 4.9 \text{ mg/ml}$$

$L_1 = -14.6$ mg/ml
$L_2 = -4.8$ mg/ml

11.6 NONPARAMETRIC MULTIPLE COMPARISONS

Let us consider the multisample situation where the nonparametric Kruskal-Wallis test is applied and the null hypothesis is rejected. In such a case, the experimenter usually will desire to determine between which of the samples significant differences occur.

Nonparametric multiple comparisons may be effected in a fashion paralleling the Tukey test (Section 11.1), by using rank sums instead of means. This is demonstrated in Example 11.9. The rank sums from the Kruskal-Wallis test are first arranged in increasing

---

EXAMPLE 11.9   Nonparametric Tukey-type multiple comparisons, using the Nemenyi test.

The data are those from Example 10.10, where the Kruskal-Wallis test rejected the null hypothesis that fly abundance is the same at the three different vegetation heights sampled.

$$\text{SE} = \sqrt{\frac{n(nk)(nk+1)}{12}} = \sqrt{\frac{5(15)(16)}{12}} = \sqrt{100} = 10.00$$

Samples ranked by rank sums (i):   3   2   1
Rank sums (R_i):   26   30   64

Comparison (B vs. A)	Difference $(R_B - R_A)$	SE	q	$q_{0.05,\infty,3}$	Conclusion
1 vs. 3	$64 - 26 = 38$	10.00	3.80	3.314	Reject H_0: Fly abundance is the same at vegetation heights 3 and 1.
1 vs. 2	$64 - 30 = 34$	10.00	3.40	3.314	Reject H_0: Fly abundance is the same at vegetation heights 2 and 1.
2 vs. 3	$30 - 26 = 4$	10.00	0.40	3.314	Accept H_0: Fly abundance is the same at vegetation heights 3 and 2.

Overall conclusion: Fly abundance is the same at vegetation heights 3 and 2 but is different at height 1.

---

order of magnitude. Pairwise differences between rank sums are then tabulated, starting with the difference between the largest and smallest rank sums, and proceeding in the same sequence as described in Section 11.1. The standard error is calculated as

$$\text{SE} = \sqrt{\frac{n(nk)(nk+1)}{12}} \tag{11.22}$$

(Nemenyi, 1963; Wilcoxon and Wilcox, 1964: 10),[†] and the Studentized range (Appendix Table B.5) to be used is $q_{\alpha,\infty,k}$.

[†]Some authors (e.g., Miller 1981: 166) perform this test in an equivalent fashion by considering the difference between mean ranks ($\bar{R}_A$ and $\bar{R}_B$) rather than rank sums (R_A and R_B), in which case the appropriate

The above is a Tukey-type multiple comparison test, a nonparametric analog to the procedure of Section 11.1. If an analog is preferred to the Student-Newman-Keuls multiple range test of Section 11.2, the standard error would be

$$SE = \sqrt{\frac{n(np)(np + 1)}{12}} \qquad (11.24)$$

and $q_{\alpha,\infty,p}$ would be the critical value.[*]

The above multiple comparison testing requires that there be equal numbers of data in each of the k groups. If such is not the case, then we may use the procedure of Section 11.7; but a more powerful test is the following.

Dunn (1964; Hollander and Wolfe, 1973: 125) proposed using a standard error of

$$SE = \sqrt{\frac{N(N + 1)}{12} \left(\frac{1}{n_A} + \frac{1}{n_B} \right)} \qquad (11.26)$$

for a test statistic we shall call

$$Q = \frac{\bar{R}_B - \bar{R}_A}{SE}, \qquad (11.27)$$

where $\bar{R}$ indicates a mean rank (i.e., $\bar{R}_A = R_A/n_A$ and $\bar{R}_B = R_B/n_B$). Critical values for this test, $Q_{\alpha,k}$, are given in Appendix Table B.15. Applying this procedure to the situation of Example 11.9 yields the same conclusions.

If tied ranks are present, then the following is an improvement over Equation 11.26 (Dunn, 1964):

$$SE = \sqrt{\left(\frac{N(N + 1)}{12} - \frac{\sum t}{12(N - 1)} \right) \left(\frac{1}{n_A} + \frac{1}{n_B} \right)}. \qquad (11.28)$$

In the latter equation, $\sum t$ is used in the Kruskal-Wallis test when ties are present and is defined in Equation 10.42. The testing procedure is demonstrated in Example 11.10; note that it is the mean ranks ($\bar{R}_i$), rather than the ranks sums (R_i), that are arranged in order of magnitude.

A procedure developed independently by Steel (1960, 1961b) and Dwass (1960) is somewhat more advantageous than the tests of Nemenyi and Dunn (Critchlow and Fligner, 1991; Miller, 1981: 168–169), but it is less convenient to use and it tends to be very conservative (i.e., having a Type I error much less than the stated α) (Gabriel and Lachenbruch, 1969).

---

standard error would be

$$SE = \sqrt{\frac{k(nk + 1)}{12}}. \qquad (11.23)$$

[*]If mean ranks, rather than rank sums, are used, then

$$SE = \sqrt{\frac{p(np + 1)}{12}}. \qquad (11.25)$$

EXAMPLE 11.10   Nonparametric multiple comparisons with unequal sample sizes. The data are those from Example 10.11, where the Kruskal-Wallis test rejected the null hypothesis that water pH was the same in all four ponds examined.

$\sum t = 168$, as in Example 10.11.

For $n_A = 8$ and $n_B = 8$,

$$SE = \sqrt{\left(\frac{N(N+1)}{12} - \frac{\sum t}{12(N-1)}\right)\left(\frac{1}{n_A} + \frac{1}{n_B}\right)}$$

$$= \sqrt{\left(\frac{31(32)}{12} - \frac{168}{12(30)}\right)\left(\frac{1}{8} + \frac{1}{8}\right)}$$

$$= \sqrt{20.5500} = 4.53$$

For $n_A = 7$ and $n_B = 8$,

$$SE = \sqrt{\left(\frac{31(32)}{12} - \frac{168}{12(30)}\right)\left(\frac{1}{7} + \frac{1}{8}\right)} = \sqrt{22.0179} = 4.69$$

Samples ranked by mean ranks (i):	1	2	4	3
Rank sums (R_i):	55	132.5	163.5	145
Sample sizes (n_i):	8	8	8	7
Mean ranks ($\bar{R}_i$):	6.88	16.56	20.44	20.71

Comparison (B vs. A)	Difference ($\bar{R}_B - \bar{R}_A$)	SE	Q	$Q_{0.05,4}$	Conclusion
3 vs. 1	$20.71 - 6.88 = 13.83$	4.69	2.95	2.639	Reject H_0: Water pH is the same in ponds 1 and 3.
3 vs. 2	$20.71 - 16.56 = 4.15$	4.69	0.88	2.639	Accept H_0: Water pH is the same in ponds 2 and 3.
3 vs. 4	Do not test				
4 vs. 1	$20.44 - 6.88 = 13.56$	4.53	2.99	2.639	Reject H_0
4 vs. 2	Do not test				
2 vs. 1	$16.56 - 6.88 = 9.68$	4.53	2.14	2.639	Accept H_0

Overall conclusion: Water pH is the same in ponds 4 and 3 but is different in pond 1, and the relationship of pond 2 to the others is unclear.

Nonparametric Comparisons of a Control to Other Groups. Subsequent to a Kruskal-Wallis test in which H_0 is rejected, a nonparametric analysis may be performed to seek either one-tailed or two-tailed significant differences between one group (designated as the "control") and each of the other groups of data. This is done in a manner paralleling that of the procedure of Section 11.4, but using group rank sums instead of group means. The standard error to be calculated is

$$SE = \sqrt{\frac{n(nk)(nk+1)}{6}} \tag{11.29}$$

(Wilcoxon and Wilcox, 1964: 11), and one uses as critical values either $q'_{\alpha(1),\infty,k}$ or $q'_{\alpha(2),\infty,k}$ (from Appendix Table B.6 or Appendix Table B.7, respectively) for one-tailed or two-tailed hypotheses, respectively.[*]

The preceding nonparametric test requires equal sample sizes. (If the n's are not all equal, then the procedure suggested by Dunn (1964; Hollander and Wolfe, 1973: 131) may be employed. By this method group B is considered to be the control and uses Equation 11.27, where the appropriate standard error is that of Equation 11.26 or 11.28, depending on whether there are ties or no ties, respectively. We shall refer to critical values for this test, which may be two-tailed or one-tailed, as $Q'_{\alpha,k}$; and they are given in Appendix Table B.16. The test presented by Steel (1959) has drawbacks compared to the procedures above (Miller, 1981: 133).

11.7 NONPARAMETRIC MULTIPLE CONTRASTS

Multiple contrasts, introduced in Section 11.5, can be tested nonparametrically using the Kruskal-Wallis H statistic instead of F. As an analog of Equation 11.18, we compute

$$S = \frac{\left| \sum c_i \bar{R}_i \right|}{\text{SE}}, \tag{11.31}$$

where

$$\text{SE} = \sqrt{\left(\frac{N(N+1)}{12} \right) \left(\sum \frac{c_i^2}{n_i} \right)}, \tag{11.32}$$

unless there are tied ranks, in which cases we use

$$\text{SE} = \sqrt{\left(\frac{N(N+1)}{12} - \frac{\sum t}{12(N-1)} \right) \left(\sum \frac{c_i^2}{n_i} \right)}, \tag{11.33}$$

where $\sum t$ is as in Equation 10.42. The critical value for these multiple contrasts is $\sqrt{H_{\alpha,n_1,n_2,\ldots}}$, using Appendix Table B.13 to obtain the critical value of H. If the needed critical value of H is not on that table, then $\chi^2_{\alpha,(k-1)}$ may be used.

11.8 MULTIPLE COMPARISONS AMONG MEDIANS

If the null hypothesis is rejected in a multisample median test (Section 10.5), then it is usually desirable to ascertain between which groups significant differences exist. A Tukey-type multiple comparison test has been provided by Levy (1979), using

$$q = \frac{f_{1B} - f_{1A}}{\text{SE}}. \tag{11.34}$$

---

[*]If mean ranks, instead of rank sums, are used, then

$$\text{SE} = \sqrt{\frac{k(nk+1)}{6}}. \tag{11.30}$$

As shown in Example 11.11, we employ the values of f_{1j} for each group, where f_{1j} is the number of data in group j that are greater than the grand median. (The values of f_{1j} are the observed frequencies in the first row in the contingency table used in the multisample median test of Section 10.5.) The values of f_{1j} are ranked and pairwise differences among the ranks are examined as in other Tukey-type tests. The appropriate standard error is

$$SE = \sqrt{\frac{n(N+1)}{4N}} \tag{11.35}$$

if N, the total number of data in all groups, is odd, or

$$SE = \sqrt{\frac{nN}{4(N-1)}} \tag{11.36}$$

EXAMPLE 11.11   Tukey-type multiple comparison for differences among medians, using the data of Example 10.12.

Samples ranked by f_{1j} (j) :	2	1	4	3
Ranked f_{1j} :	3	4	6	10
Sample sizes (n_j) :	12	12	12	12

$k = 4$

$N = 12 + 12 + 12 + 12 = 48$

$n = 12$

By Equation 11.36,

$$SE = \sqrt{\frac{(12)(48)}{4(48-1)}} = 1.750$$

H_0:   Median of population B = Median of population A

H_A: Median of population $B \neq$ Median of population A

Comparison	$f_{1B} - f_{1A}$	SE	q	$q_{0.05,4,\infty}$	Conclusion
3 vs. 2	$10 - 3 = 7$	1.750	4.000	3.633	Reject H_0.
3 vs. 1	$10 - 4 = 6$	1.750	3.429	3.633	Accept H_0.
3 vs. 4	Do not test				
4 vs. 2	$6 - 3 = 3$	1.750	1.714	3.633	Accept H_0.
4 vs. 1	Do not test				
1 vs. 2	Do not test				

Overall conclusion: The medians of populations 3 and 2 (i.e., south and east—see Example 10.12) are not the same; but the test lacks the power to allow clear conclusions about the medians of populations 4 and 1.

if N is even. The critical values to be used are $q_{\alpha,\infty,k}$. This multiple comparison procedure, demonstrated in Example 11.11 appears to possess low statistical power. The above procedure is for equal sample sizes (n). If the sample sizes are slightly unequal the test can be used by employing the harmonic mean sample size,

$$n = \frac{k}{\displaystyle\sum_{j=1}^{k} \frac{1}{n_j}}$$

(11.37)

for an approximate result.

11.9 MULTIPLE COMPARISONS AMONG VARIANCES

If the null hypothesis that k population variances are all equal (see Section 10.6) is rejected, then we may wish to determine which of the variances differ from which others. Levy (1975a, 1975c) suggests multiple comparison procedures for this purpose based on a logarithmic transformation of sample variances.

A test analogous to the Tukey test of Section 11.1 is performed by calculating

$$q = \frac{\ln s_B^2 - \ln s_A^2}{\text{SE}},$$

(11.38)

where

$$\text{SE} = \sqrt{\frac{2}{\nu}},$$

(11.39)

if both samples being compared are of equal size. If $\nu_A \neq \nu_B$, we can employ

$$\text{SE} = \sqrt{\frac{1}{\nu_A} + \frac{1}{\nu_B}}.$$

(11.40)

Just as in Sections 11.1 and 11.2, the subscripts A and B refer to the pair of groups being compared; and the sequence of pairwise comparisons must follow that given in those sections. This is demonstrated in Example 11.12.* The critical value for this test is $q_{\alpha,\infty,k}$ (from Appendix Table B.5).

A Newman-Keuls-type test can also be performed using the logarithmic transformation. For this test, we calculate q using Equation 11.38; but the critical value, $q_{\alpha,\infty,p}$, depends on p, the range of variances being tested (just as p is the range of means being tested in Section 11.2).

It must be pointed out that the methods of this section, as well as those of Section 10.6, are valid only if the sampled populations are normal or very close to normal,

---

*Recall (as in Section 10.6) that "ln" refers to natural logarithms (i.e., logarithms using base e). If one prefers using common logarithms ("log"; logarithms in base 10), then

$$q = \frac{2.30259(\log s_B^2 - \log s_A^2)}{\text{SE}}.$$

(11.41)

EXAMPLE 11.12   Tukey-type multiple comparison test for differences among four variances (i.e., $k = 4$).

i	s_i^2	n_i	ν_i	$\ln s_i^2$
1	2.74 g^2	50	49	1.0080
2	2.83 g^2	48	47	1.0403
3	2.20 g^2	50	49	0.7885
4	6.42 g^2	50	49	1.8594

Samples ranked by variances (i):		3	1	2	4
Ranked logarithms of sample variances ($\ln s_i^2$):		0.7885	1.0080	1.0403	1.8594
Sample degrees of freedom (ν_i):		49	49	47	49

Comparison (B vs. A)	Difference ($\ln s_B^2 - \ln s_A^2$)	SE	q	$q_{0.05,\infty,4}$	Conclusions
4 vs. 3	$1.8594 - 0.7885 = 1.0709$	0.202*	5.301	3.633	Reject H_0: $\sigma_4^2 = \sigma_3^2$
4 vs. 1	$1.8594 - 1.0080 = 0.8514$	0.202	4.215	3.633	Reject H_0: $\sigma_4^2 = \sigma_1^2$
4 vs. 2	$1.8594 - 1.0403 = 0.8191$	0.204†	4.015	3.633	Reject H_0: $\sigma_4^2 = \sigma_2^2$
2 vs. 3	$1.0403 - 0.7885 = 0.2518$	0.204	1.234	3.633	Accept H_0: $\sigma_2^2 = \sigma_3^2$
2 vs. 1	Do not test				
1 vs. 3	Do not test				

*As $\nu_4 = \nu_3$: SE $= \sqrt{\dfrac{2}{\nu}} = \sqrt{\dfrac{2}{49}} = 0.202.$

†As $\nu_4 \neq \nu_2$: SE $= \sqrt{\dfrac{1}{\nu_4} + \dfrac{1}{\nu_2}} = \sqrt{\dfrac{1}{49} + \dfrac{1}{47}} = 0.204.$

Overall conclusion: $\sigma_3^2 = \sigma_1^2 = \sigma_2^2 \neq \sigma_4^2$

---

and are severely affected if this assumption is not satisfied. Stevens (1989) discussed nonparametric multiple comparison testing for dispersion.

Comparing a Control Variance to Each Other Group Variance.   If the investigator's intent in multiple comparison testing is to compare each possible pair of variances, then the procedures above are applicable. If, however, it is desired to stipulate that one of the variances (call it the "control," or Sample B, variance) is to be compared with each other variance (but the others are not to be compared with each other), then the Dunnett-type test of Levy (1975b) may be employed for more powerful testing.

Here, in a fashion analogous to that of Section 11.4 for means, we calculate

$$q = \frac{\ln s_{\text{control}}^2 - \ln s_A^2}{\text{SE}}, \tag{11.42}$$

where

$$SE = \sqrt{\frac{4}{\nu}} \qquad\qquad (11.43)$$

if the control sample and Sample A are of equal size. If $\nu_A \neq \nu_{\text{control}}$, then use

$$SE = \sqrt{\frac{2}{\nu_A} + \frac{2}{\nu_{\text{control}}}}. \qquad\qquad (11.44)$$

The appropriate critical value for the two-tailed test (i.e., H_0: $\sigma^2_{\text{control}} = \sigma^2_A$) is $q'_{\alpha(2),\infty,p}$ (from Appendix Table B.7) and for the one-tailed test (i.e., either H_0: $\sigma^2_{\text{control}} \leq \sigma^2_A$ or H_0: $\sigma^2_{\text{control}} \geq \sigma^2_A$) (from Appendix Table B.6). This testing should be used only if it can be assumed that the underlying populations are normally distributed (or very nearly so).

EXERCISES

11.1 **(a)** Apply the Tukey test procedure to the following results from an analysis of variance: $k = 3$ and the three sample means are 14.8, 20.2 and 16.2; the error mean square and degrees of freedom are 8.46 and 21, respectively; there are eight data in each of the three groups.

(b) Employ the Student-Newman-Keuls test for the same data.

(c) Calculate the 95% confidence interval for each different population mean and for each difference between means.

11.2 Assume that in the experiment described in Example 10.1, group 1 was set up to be a control. Use a two-tailed Dunnett's test to compare the control mean with each other mean.

11.3 Use Scheffé's S test on the data of Example 10.1.

(a) Test the hypothesis that the means of the populations represented by groups 1 and 4 are the same as the means of groups 2 and 3.

(b) Test the hypothesis that the means of groups 2 and 4 are the same as the mean of group 3.

11.4 The following ranks result in a significant Kruskal-Wallis test. Employ nonparametric multiple range testing to determine between which of the three groups population differences exist.

Group 1	Group 2	Group 3
8	10	14
4	6	13
3	9	7
5	11	12
1	2	15

12

TWO-FACTOR ANALYSIS OF VARIANCE

Section 10.1 introduced methods for the analysis of the effect of a single factor (such as the type of feed) on a variable (such as the weight of pigs). The present Chapter will show how the effects on a variable of two factors can be assessed simultaneously.*

A simultaneous analysis of the effect of more than one factor on population means is termed a *factorial analysis of variance*, and there are important advantages to such an experimental design. Among them is the simple fact that one experiment can suffice for the analysis, and it is not necessary to perform a one-way ANOVA for each factor. Thus, we may economize with respect to time, effort, and often money. Also, factorial analysis of variance procedures can test for interaction among factors.

The two-factor analysis of variance is introduced in this chapter. Analysis of variance for experiments consisting of more than two factors will be discussed in Chapter 14. Underlying assumptions of such analyses are those discussed in Section 10.1.

There have been attempts to devise nonparametric statistical procedures for experimental designs with two or more factors. Except for the case in Section 12.8, such methods have not been generally acceptable. Scheirer, Ray, and Hare (1976) proposed a procedure that is a multifactor extension of the Kruskal-Wallis test of Section 10.4. However, Toothaker and Chang (1980) found that it performs poorly and should not be used.

---

*Some concepts of two-factor analysis of variance were discussed as early as 1899 (Thiele, 1899).

12.1 TWO-FACTOR ANALYSIS OF VARIANCE WITH EQUAL REPLICATION

Example 12.1 presents data from a two-way analysis of variance, in which the variable under consideration is blood calcium concentration in a population of birds, and the two factors being simultaneously tested are hormone treatment and sex. Because there are two levels in the first factor (hormone-treated and nontreated) and two levels in the second factor (female and male), this experimental design is termed a 2×2 (or 2^2) factorial. The two factors are said to be "crossed" because each level of one factor is found in combination with each level of the second factor.* There are $n = 5$ replicate observations (i.e., calcium determinations on each of five birds) for each of the $2 \times 2 = 4$ combinations of the two factors; therefore, there are a total of $N = 2 \times 2 \times 5 = 20$ data in this experiment. In general, it is advantageous to have equal replication (what is sometimes called a "balanced" or "orthogonal" experimental design), but Section 12.2 will consider cases with unequal numbers of data per cell, and Section 12.3 will discuss analyses with only one datum per combination of factors.

For the general case of the two-way factorial analysis of variance, we can refer to one factor as A and to the other as B. Furthermore, let us have a represent the number of levels in factor A, b the number of levels in factor B, and n the number of replicates. A triple subscript on the variable, as X_{ijl}, will enable us to identify uniquely the value that is replicate l of the combination of level i of factor A and level j of factor B. In Example 12.1, $X_{213} = 21.3$ mg/100 ml, $X_{115} = 12.8$ mg/100 ml, etc. Each combination of a level of factor A with a level of factor B is called a *cell*. The cells may be visualized as the "groups" in a one-factor ANOVA (Section 10.1). There are four cells in Example 12.1: females without hormone treatment, males without hormone treatment, females with hormone treatment, and males with hormone treatment. For the cell formed by the combination of level i of factor A and level j of factor B, $\bar{X}_{ij}$ denotes the cell mean; for the data in Example 12.1, the mean of a cell is the cell total divided by 5, so $\bar{X}_{11} = 14.88$, $\bar{X}_{12} = 12.12$, $\bar{X}_{21} = 32.52$ and $\bar{X}_{22} = 27.78$ (with the units for each mean being mg/100 ml). The mean of all bn data in level i of factor A is $\bar{X}_{i.}$, and the mean of all an data in level j of factor B is $\bar{X}_{.j}$. That is, the mean for the ten non-hormone-treated birds is $\bar{X}_{1.}$, which is an estimate of the population mean, $\mu_{1.}$; the mean for the hormone-treated birds is $\bar{X}_{2.}$, which estimates $\mu_{2.}$; the mean of the female birds is $\bar{X}_{.1}$, which estimates $\mu_{.1}$ and the mean of the male birds is $\bar{X}_{.2}$, which estimates $\mu_{.2}$. There are a total of N data in the experiment, and (just as in the single-factor ANOVA of Section 10.1) the mean of all N data is denoted as $\bar{X}$.

Recall that the total sum of squares is a measure of variability among all the data in a sample. For the two-factor analysis of variance this is conceptually the same as for the single-factor ANOVA (see Equation 10.5):

$$\text{total SS} = \sum_{i=1}^{a} \sum_{j=1}^{b} \sum_{l=1}^{n} (X_{ijl} - \bar{X})^2, \tag{12.1}$$

---

*Two or more factors can exist in an ANOVA without being crossed; see Chapter 15.

EXAMPLE 12.1   Hypotheses and data for a Model I, two-factor analysis of variance with equal replication.

The data are plasma calcium concentrations (in mg/100 ml) of birds of both sexes, half of the birds of each sex being treated with a hormone and half not treated with the hormone.

H_0:   There is no effect of hormone treatment on the mean plasma calcium concentration of birds (i.e., $\mu_{\text{no hormone}} = \mu_{\text{hormone}}$ or $\mu_{1\cdot} = \mu_{2\cdot}$).

H_A: There is an effect of hormone treatment on the mean plasma calcium concentration of birds (i.e., $\mu_{\text{no hormone}} \neq \mu_{\text{hormone}}$ or $\mu_{1\cdot} \neq \mu_{2\cdot}$).

H_0:   There is no difference in mean plasma calcium concentration between female and male birds (i.e., $\mu_{\text{female}} = \mu_{\text{male}}$ or $\mu_{\cdot 1} = \mu_{\cdot 2}$).

H_A: There is difference in mean plasma calcium concentration between female and male birds (i.e., $\mu_{\text{female}} \neq \mu_{\text{male}}$ or $\mu_{\cdot 1} \neq \mu_{\cdot 2}$).

H_0:   There is no interaction of sex and hormone treatment on the mean plasma calcium concentration of birds.

H_A: There is interaction of sex and hormone treatment on the mean plasma calcium concentration of birds.

$\alpha = 0.05$

	No Hormone Treatment		Hormone Treatment	
	Female	*Male*	*Female*	*Male*
	16.5	14.5	39.1	32.0
	18.4	11.0	26.2	23.8
	12.7	10.8	21.3	28.8
	14.0	14.3	35.8	25.0
	12.8	10.0	40.2	29.3
Cell totals:	$\sum_{l=1}^{5} X_{11l} = 74.4$	$\sum_{l=1}^{5} X_{12l} = 60.6$	$\sum_{l=1}^{5} X_{21l} = 162.6$	$\sum_{l=1}^{5} X_{22l} = 138.9$
Cell means:	$\bar{X}_{11} = 14.88$	$\bar{X}_{12} = 12.12$	$\bar{X}_{21} = 32.52$	$\bar{X}_{22} = 27.78$

with

$$\text{total DF} = N - 1. \tag{12.2}$$

Next we may consider the variability among cells (each cell being a combination of a level of factor A and a level of factor B), handling cells as we did treated "groups" in the single-factor ANOVA (see Equation 10.3):

$$\text{cells SS} = \sum_{i=1}^{a} \sum_{j=1}^{b} n(\bar{X}_{ij} - \bar{X})^2; \tag{12.3}$$

and, as the number of cells is ab,

$$\text{cells DF} = ab - 1. \tag{12.4}$$

Furthermore, the quantity analogous to the within-groups SS in the single-factor ANOVA (Equation 10.1) is

$$\text{within-cells SS} = \sum_{i=1}^{a} \sum_{j=1}^{b} \left[\sum_{l=1}^{n} (X_{ijl} - \bar{X}_{ij})^2 \right], \tag{12.5}$$

which has degrees of freedom of

$$\text{within-cells DF} = ab(n-1), \tag{12.6}$$

which may also be calculated as

$$\text{within-cells DF} = \text{total DF} - \text{cells DF}. \tag{12.7}$$

The terms *error SS* and *error DF* are very commonly used for within-cells SS and within-cells DF, respectively.

The calculations indicated above are analogous to those for the one-way analysis of variance (Section 10.1).* But a major desire in the two-factor ANOVA is not to consider differences among the aforementioned cells, but to assess the effects of each of the two factors independently of the other. This is done by considering factor A to be the sole factor in a single-factor ANOVA and then by considering factor B to be the single factor. For factor A this is done as follows:

$$\text{factor } A \text{ SS} = bn \sum_{i=1}^{a} (\bar{X}_{i\cdot} - \bar{X})^2, \tag{12.8}$$

which is associated with degrees of freedom of

$$\text{factor } A \text{ DF} = a - 1; \tag{12.9}$$

and for factor B:

$$\text{factor } B \text{ SS} = an \sum_{j=1}^{b} (\bar{X}_{\cdot j} - \bar{X})^2, \tag{12.10}$$

for which the degrees of freedom are

$$\text{factor } B \text{ DF} = b - 1. \tag{12.11}$$

In general, the variability among cells is not equal to the variability among levels of factor A plus the variability among levels of factor B (i.e., the result of Equation 12.3 is not equal to the sum of the results from Equations 12.8 and 12.10). The amount of variability not accounted for is that due to the effect of *interaction*[†] between factors A and B. This is designated as the $A \times B$ interaction and is readily calculated by difference:

$$A \times B \text{ interaction SS} = \text{cells SS} - \text{factor } A \text{ SS} - \text{factor } B \text{ SS}. \tag{12.12}$$

---

*Also analogous is that this two-factor analysis-of-variance model assumes that the data in each of the cells came at random from a normal population, and that all of those populations have the same variance. The best estimate of that variance is the error MS, which is (error SS)/(error DF).

[†]The term "interaction" was introduced for ANOVA by R. A. Fisher (David, 1995).

The interaction degrees of freedom may also be obtained by difference:

$$A \times B \text{ interaction DF} = \text{cells DF} - \text{factor } A \text{ DF} - \text{factor } B \text{ DF}, \quad (12.13)$$

or as

$$A \times B \text{ interaction DF} = (\text{factor } A \text{ DF})(\text{factor } B \text{ DF})$$
$$= (a-1)(b-1). \quad (12.14)$$

An interaction between two factors means that the effect of one factor is not independent of the presence of a particular level of the other factor. In Example 12.1 no interaction implies that the effect of hormone treatment on plasma calcium between males and females is the same under both hormone treatments.* Therefore, interaction among factors is an effect on the variable (e.g., plasma calcium) that is in addition to the sum of the effects of each factor considered separately.

Just as with one-way ANOVA (Section 10.1), there are machine formulas that allow the computation of sums of squares without first calculating cell, factor, and overall means; and these are demonstrated in Example 12.2.

The total variability is expressed by

$$\text{total SS} = \sum_{i=1}^{a} \sum_{j=1}^{b} \sum_{l=1}^{n} X_{ijl}^2 - C, \quad (12.15)$$

or $11354.31 - C$, where the correction term is

$$C = \frac{\left(\sum_{i=1}^{a} \sum_{j=1}^{b} \sum_{l=1}^{n} X_{ijl} \right)^2}{N}. \quad (12.16)$$

As

$$N = abn, \quad (12.17)$$

or $N = (2)(2)(5) = 20$ in our example, $C = (436.5)^2/20 = 9526.6125$ and total SS $= 11354.31 - 9526.6125 = 1827.6975$.

The variability among cells is

$$\text{cells SS} = \frac{\sum_{i=1}^{a} \sum_{j=1}^{b} \left(\sum_{l=1}^{n} X_{ijl} \right)^2}{n} - C. \quad (12.18)$$

And the variability among levels of factor A is

$$\text{factor } A \text{ SS} = \frac{\sum_{i=1}^{a} \left(\sum_{j=1}^{b} \sum_{l=1}^{n} X_{ijl} \right)^2}{bn} - C. \quad (12.19)$$

---

*Symbolically, the null hypothesis for interaction effect could be stated as H_0: $\mu_{11} - \mu_{12} = \mu_{21} - \mu_{22}$ or H_0: $\mu_{11} - \mu_{21} = \mu_{12} - \mu_{22}$, where μ_{ij} is the population mean of the variable in the presence of level i of factor A and level j of factor B.

EXAMPLE 12.2 **Computations for the Model I two-factor ANOVA of Example 12.1.**

	Female	Male	Treatment totals and means
No hormone treatment	16.5 18.4 12.7 14.0 12.8	14.5 11.0 10.8 14.3 10.0	

$$\sum_{l=1}^{n} X_{11l} = 74.4 \qquad \sum_{l=1}^{n} X_{12l} = 60.6 \qquad \sum_{j=1}^{b}\sum_{l=1}^{n} X_{1jl} = 135.0$$

$$\bar{X}_{11} = 14.88 \qquad\qquad \bar{X}_{12} = 12.12 \qquad\qquad \bar{X}_{1.} = 13.50$$

Hormone treatment	39.1 26.2 21.3 35.8 40.2	32.0 23.8 28.8 25.0 29.3	

$$\sum_{l=1}^{n} X_{21l} = 162.6 \qquad \sum_{l=1}^{n} X_{22l} = 138.9 \qquad \sum_{j=1}^{b}\sum_{l=1}^{n} X_{2jl} = 301.5$$

$$\bar{X}_{21} = 32.52 \qquad\qquad \bar{X}_{22} = 27.78 \qquad\qquad \bar{X}_{2.} = 30.15$$

Grand total

Sex totals $\quad \sum_{i=1}^{a}\sum_{l=1}^{n} X_{i1l} = 237.0 \qquad \sum_{i=1}^{a}\sum_{l=1}^{n} = X_{i2l} = 199.5 \qquad \sum_{i=1}^{a}\sum_{j=1}^{b}\sum_{l=1}^{n} X_{ijl} = 436.5$

Grand mean

Sex means $\qquad\qquad \bar{X}_{.1} = 23.70 \qquad\qquad\qquad \bar{X}_{.2} = 19.95 \qquad\qquad \bar{X} = 21.825$

$$\sum_{i=1}^{2}\sum_{j=1}^{2}\sum_{l=1}^{5} X_{ijl} = 436.5$$

$$\sum_{i=1}^{2}\sum_{j=1}^{2}\sum_{l=1}^{5} X_{ijl}^{2} = 11354.31$$

a = number of hormone groups = 2

b = number of sexes = 2

n = number of replicates per cell = 5

$N = abn = (2)(2)(5) = 20$

$$\text{Total for no hormone} = \sum_{j=1}^{2}\sum_{l=1}^{5} X_{1jl} = 74.4 + 60.6 = 135.0$$

$$\text{Total for hormone} = \sum_{j=1}^{2}\sum_{l=1}^{5} X_{2jl} = 162.6 + 138.9 = 301.5$$

EXAMPLE 12.2   (continued)

$$\text{Total for females} = \sum_{i=1}^{2} \sum_{l=1}^{5} X_{i1l} = 74.4 + 162.6 = 237.0$$

$$\text{Total for males} = \sum_{i=1}^{2} \sum_{l=1}^{5} X_{i2l} = 60.6 + 138.9 = 199.5$$

$$C = \frac{\left(\sum_{i=1}^{a} \sum_{j=1}^{b} \sum_{l=1}^{n} X_{ijl} \right)^2}{N} = \frac{(436.5)^2}{20} = 9526.6125$$

$$\text{total SS} = \sum_{i=1}^{a} \sum_{j=1}^{b} \sum_{l=1}^{n} X_{ijl}^2 - C = 11354.31 - 9526.6125 = 1827.6975$$

$$\text{total DF} = N - 1 = 19$$

$$\text{cells SS} = \sum_{i=1}^{a} \sum_{j=1}^{b} \frac{\left(\sum_{l=1}^{n} X_{ijl} \right)^2}{n} - C$$

$$= \frac{(74.4)^2 + (60.6)^2 + (162.6)^2 + (138.9)^2}{5} - 9526.6125$$

$$= 1461.3255$$

$$\text{cells DF} = ab - 1 = (2)(2) - 1 = 3$$

within-cells (i.e. error) SS = total SS − cells SS

$$= 1827.6975 - 1461.3255 = 366.3720$$

within-cells (error) DF = $ab(n - 1) = (2)(2)(4) = 16$

$$\text{factor } A \text{ (hormone group) SS} = \frac{\sum_{i=1}^{a} \left(\sum_{j=1}^{b} \sum_{l=1}^{n} X_{ijl} \right)^2}{bn} - C$$

$$= \frac{(\text{sum without hormone})^2 + (\text{sum with hormone})^2}{\text{number of data per hormone group}} - C$$

$$= \frac{(135.0)^2 + (301.5)^2}{(2)(5)} - 9526.6125$$

$$= 1386.1125$$

factor A DF = $a - 1 = 1$

$$\text{factor } B \text{ (sex) SS} = \frac{\sum_{j=1}^{b} \left(\sum_{i=1}^{a} \sum_{l=1}^{n} X_{ijl} \right)^2}{an} - C$$

$$= \frac{(\text{sum for females})^2 + (\text{sum for males})^2}{\text{number of data per sex}} - C$$

$$= \frac{(237.0)^2 + (199.5)^2}{(2)(5)} - 9526.6125 = 70.3125$$

factor B DF = $b - 1 = 1$

EXAMPLE 12.2   (continued)

$A \times B$ interaction SS = cells SS − Factor A SS − Factor B SS

$$= 1461.3255 - 1386.1125 - 70.3125 = 4.9005$$

$A \times B$ interaction DF = (factor A DF)(factor B DF)= $(1)(1) = 1$

Analysis of Variance Summary Table

Source of variation	SS	DF	MS
Total	1827.6975	19	
Cells	1461.3255	3	
Factor A (hormone)	1386.1125	1	1386.1125
Factor B (sex)	70.3125	1	70.3125
$A \times B$	4.9005	1	4.9005
Within cells (error)	366.3720	16	22.8982

For H_0: There is no effect of hormone treatment on the mean plasma calcium concentration of birds in the population sampled.

$$F = \frac{\text{hormone MS}}{\text{within-cells MS}} = \frac{1386.1125}{22.8982} = 60.5$$

$F_{0.05(1),1,16} = 4.49$

Therefore, reject H_0.

$$P < 0.0005 \quad [P = 0.00000080]$$

For H_0: There is no difference in mean plasma calcium concentration between male and female birds in the population sampled.

$$F = \frac{\text{sex MS}}{\text{within-cells MS}} = \frac{70.3125}{22.8982} = 3.07$$

$F_{0.05(1),1,16} = 4.49$

Therefore, do not reject H_0.

$$0.05 < P < 0.10 \quad [P = 0.099]$$

For H_0: There is no interaction of sex and hormone treatment affecting the mean plasma calcium concentration of birds in the population sampled.

$$F = \frac{\text{hormone} \times \text{sex interaction MS}}{\text{within-cells MS}} = \frac{4.9005}{22.8982} = 0.214$$

$F_{0.05(1)1,16} = 4.49$

Therefore, do not reject H_0.

$$P > 0.25 \quad [P = 0.65]$$

Simply put, the factor A SS is calculated by considering factor A to be the sole factor in a single-factor analysis of variance of the data. That is, we obtain the sum for each level of factor A (ignoring the fact that the data are also categorized into levels of factor B); the sum of a level is what is in parentheses in Equation 12.19. Then we square each of these level sums and divide the sum of these squares by the number of data per level (i.e., bn). On subtracting the "correction term," C, we arrive at the factor A SS. (If the data were in fact analyzed by a single-factor ANOVA, then the groups SS would indeed be the same as the factor A SS just described, and the groups DF would be the same as factor A DF; but the error SS in the one-way ANOVA would be the within-cells SS plus the factor B and the interaction sums of squares described below, and the error DF would be the sum of the within-cells, factor B, and interaction degrees of freedom).

For factor B computations, we simply ignore the division of the data into levels of factor A and proceed as if factor B were the single factor in a one-way ANOVA:

$$\text{factor } B \text{ SS} = \frac{\sum_{j=1}^{b} \left(\sum_{i=1}^{a} \sum_{l=1}^{n} X_{ijl} \right)^2}{an} - C, \tag{12.20}$$

The cell, column, and row means of Example 12.1 are shown in Example 12.2 and summarized in Table 12.1. Using these means, the effects of each of the two factors, and the presence of interaction, may be visualized by a graph such as Fig. 12.1. We shall refer to the two levels of factor A as A_1 and A_2, and the two levels of factor B as B_1 and B_2. The variable, X, is situated on the vertical axis of the figure and on the horizontal axis we indicate A_1 and A_2. The two cell means for B_1 (14.9 and 32.5 mg/100 ml, which are indicated by black circles) are plotted and connected by a line; and the two cell means for B_2 (12.1 and 27.8 mg/100 ml, which are indicated by black squares) are plotted

TABLE 12.1   Cell, Row, and Column Means of the Data of Example 12.2 (in mg/100 ml)

	Female (B_1)	Male (B_2)	
No hormone (A_1)	14.9	12.1	13.5
Hormone (A_2)	32.5	27.8	30.2
	23.7	20.0	

and connected by a second line. The mean of all the data in each level of factor A is indicated with a plus sign, and the mean of all the data in each level of factor B is denoted by an open circle. Then, the effect of factor A is observed as the vertical distance between the plus signs; the effect of factor B is expressed as the vertical distance between the open circles; and non-parallelism of the lines indicates interaction between factors A

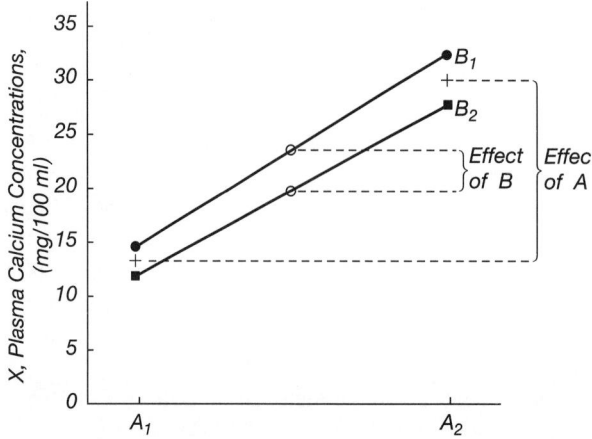

Figure 12.1   The means of the two-factor ANOVA data of Example 12.1, as given in Table 12.1. The A_i are the levels of factor A, the B_i are the levels of factor B. A plus sign indicates the mean of an A_i over all (i. e., both) levels of factor B, and an open circle indicates the mean of a B_i over all (i. e., both) levels of factor A.

and B. Thus, the ANOVA results of Example 12.1 are readily seen in this plot: there is a large effect of factor A (which is found to be significant by the F statistic), a small effect of factor B (which is found to be nonsignificant), and—indicated by only slight departure from parallelism—a very small interaction effect (which is also found to be nonsignificant). Various possible patterns of such plots are shown in Fig. 12.2. Such figures may be drawn for situations with more than two levels within factors. And one may place either factor A or factor B on the horizontal axis; usually the factor with the larger number of levels is placed on this axis, so there are fewer lines to examine.

Table 12.2 summarizes the computing formulas for the two-factor analysis of variance with equal replication.

The Model I ANOVA.   Recall, from Section 10.1, the distinction between fixed and random factors. Example 12.1 is an ANOVA where the levels of both factors are fixed; we did not simply pick these levels at random. A factorial analysis of variance in which all (in this case both) factors are fixed effects is termed a *Model I ANOVA*. In such a model, the null hypothesis of no difference among the levels of a factor is tested using $F =$ factor MS/error MS. As shown in Example 12.2, the appropriate F tests conclude that there is a highly significant effect of the hormone treatment on the plasma calcium content, and that there is not a significantly different mean plasma calcium concentration between males and females.

In addition, we can test for significant interaction in a Model I ANOVA by $F =$ interaction MS/error MS and find, in our present example, that there is no significant interaction between sex and hormone treatment. This is interpreted to mean that the effect of the hormone treatment on calcium is not different in males and females; i.e., the effect of the hormone is not dependent on sex. This concept of interaction (or its converse, independence) is analogous to that employed in the analysis of contingency tables (see Chapter 23).

If, in a two-factor analysis of variance, the effects of one or both factors are significant, the interaction effect may or may not be significant. In fact, it is possible

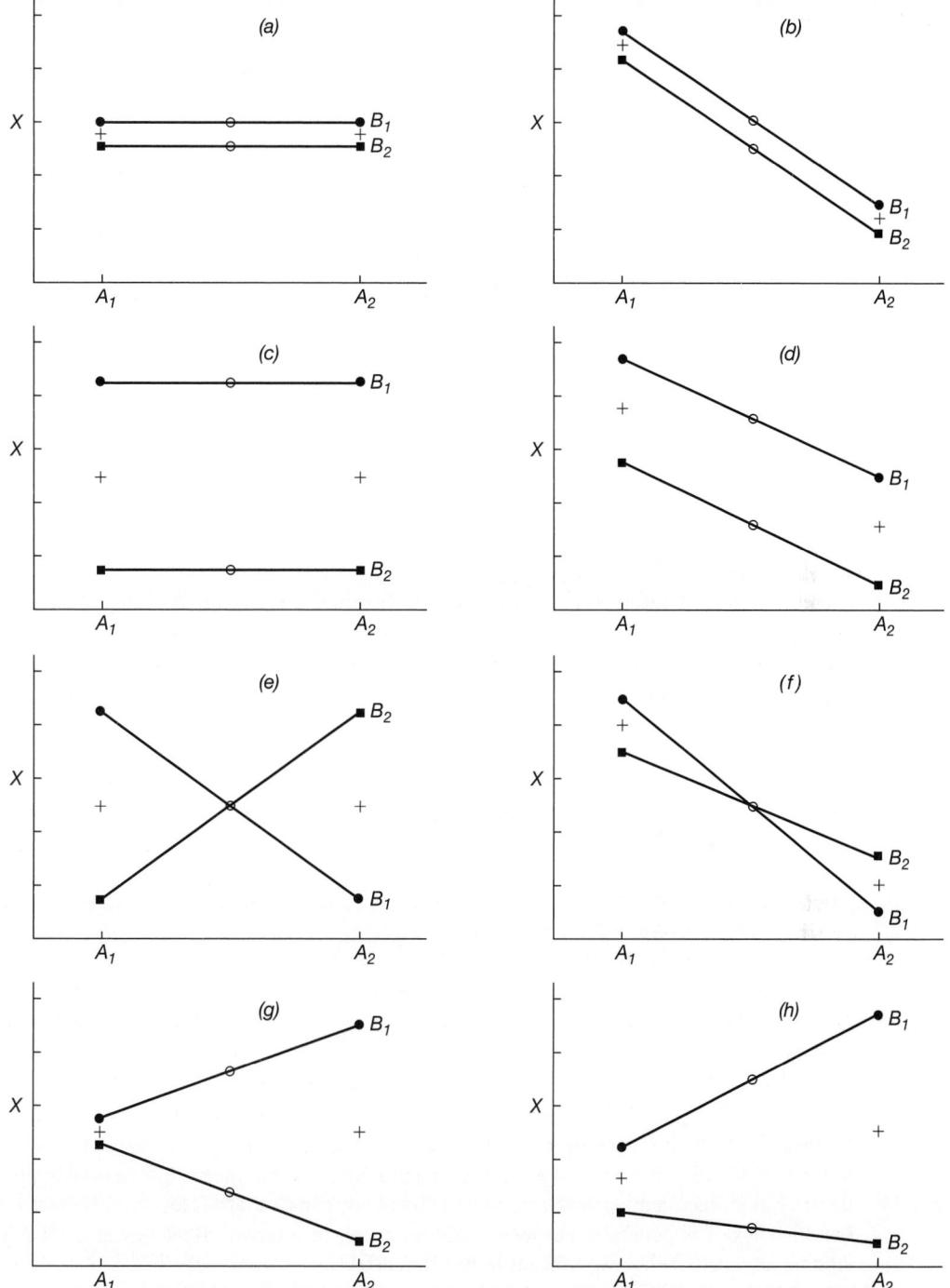

Figure 12.2 Means in a two-factor ANOVA, showing various effects of the two factors and their interaction. (a) No effect of factor A, small effect of factor B (and if there were no effect of B the two lines would coincide), and no interaction of A and B. (b) Large effect of factor A, small effect of factor B, and no interaction (which is the situation in Fig. 12.1). (c) No effect of A, large effect of B, and no interaction. (d) Large effect of A, large effect of B, and no interaction. (e) No effect of A, no effect of B, but interaction between A and B. (f) Large effect of A, no effect of B, with slight interaction. (g) No effect of A, large effect of B, with large interaction. (h) Effect of A, large effect of B, with large interaction.

TABLE 12.2   Two-way Factorial ANOVA with Equal Replication

Source of variation	Sum of squares (SS)	Degrees of freedom (DF)	Mean square (MS)
Total	$\displaystyle\sum_{i=1}^{a}\sum_{j=1}^{b}\sum_{l=1}^{n} X_{ijl}^2 - C$	$N-1$	
Cells	$\displaystyle\frac{\sum_{i=1}^{a}\sum_{j=1}^{b}\left(\sum_{l=1}^{n} X_{ijl}\right)^2}{n} - C$	$ab-1$	
Factor A	$\displaystyle\frac{\sum_{i=1}^{a}\left(\sum_{j=1}^{b}\sum_{l=1}^{n} X_{ijl}\right)^2}{bn} - C$	$a-1$	$\dfrac{\text{factor } A \text{ SS}}{\text{factor } A \text{ DF}}$
Factor B	$\displaystyle\frac{\sum_{j=1}^{b}\left(\sum_{i=1}^{a}\sum_{l=1}^{n} X_{ijl}\right)^2}{an} - C$	$b-1$	$\dfrac{\text{factor } B \text{ SS}}{\text{factor } B \text{ DF}}$
$A \times B$ interaction	Cells SS − factor A SS − factor B SS	$(a-1)(b-1)$	$\dfrac{A \times B \text{ SS}}{A \times B \text{ DF}}$
Within-cells (error)	Total SS − cells SS	$ab(n-1)$ or total DF − cells DF	$\dfrac{\text{error SS}}{\text{error DF}}$

Note: Here, a is the number of levels in factor A, b is the number of levels in factor B, and n is the number of replicates per cell.

$$C = \frac{\left(\sum_{i=1}^{a}\sum_{j=1}^{b}\sum_{l=1}^{n} X_{ijl}\right)^2}{N}$$

and

$$N = abn.$$

to encounter situations where there is a significant interaction even though each of the individual factor effects is judged to be insignificant. A significant interaction implies that the difference among levels of one factor is not constant at all levels of the second factor. Thus, it is generally not meaningful to speak of a factor effect—even if its F is significant—if there is a significant interaction effect.

The Model II ANOVA.   If a factorial design is composed only of factors with random levels, then we are said to be employing a *Model II ANOVA* (a relatively uncommon situation). In such a case, where two factors are involved, the appropriate hypothesis testing for significant factor effects is accomplished by calculating $F = $ factor MS/interaction

TABLE 12.3 Computation of the F Statistic for Tests of Significance in a Two-Factor ANOVA with Replication

Hypothesized effect	Model I (factors A and B both fixed)	Model II (factors A and B both random)	Model III (factor A fixed: factor B random
Factor A	$\dfrac{\text{factor } A \text{ MS}}{\text{error MS}}$	$\dfrac{\text{factor } A \text{ MS}}{A \times B \text{ MS}}$	$\dfrac{\text{factor } A \text{ MS}}{A \times B \text{ MS}}$
Factor B	$\dfrac{\text{factor } B \text{ MS}}{\text{error MS}}$	$\dfrac{\text{factor } B \text{ MS}}{A \times B \text{ MS}}$	$\dfrac{\text{factor } B \text{ MS}}{\text{error MS}}$
$A \times B$ interaction	$\dfrac{A \times B \text{ } MS}{\text{error MS}}$	$\dfrac{A \times B \text{ MS}}{\text{error MS}}$	$\dfrac{A \times B \text{ MS}}{\text{error MS}}$

MS (see Table 12.3). We test for the interaction effect, as before, by $F =$ interaction MS/error MS, and it is generally not useful to declare factor effects significant if there is a significant interaction effect. The Model II ANOVA for designs with more than two factors will be discussed in Chapter 15.

The Model III ANOVA. If a factorial design has both a fixed effect and a random effect factor, then it is said to be a *mixed-model,*[*] or a *Model III ANOVA.* The appropriate F statistics are calculated as shown in Table 12.3.

Pooling Mean Squares. If one does not conclude there to be a significant interaction effect, then the interaction MS and the within-cells (i.e., the error) MS are theoretically estimates of the same population variance. Because of this, some authors suggest the pooling of the interaction and within-cells sums of squares and degrees of freedom in such cases. From these pooled SS and DF values, one can obtain a pooled mean square, which then should be a better estimate of the population random error (i.e., within-cell variability) than either the error MS or the interaction MS alone; and the pooled MS will always be a quantity between the interaction MS and the error MS.

The conservative researcher who does not engage in such pooling can be assured that the probability of a Type I error is at the stated α level. But the probability of a Type II error may be greater than is acceptable to some. The chance of the latter type of error is reduced by the pooling previously described, but confidence in stating the probability of committing a Type I error may be reduced (Brownlee, 1965: 509). Rules of thumb for deciding when to pool have been proposed (e.g., Paull, 1950; Bozivich, Bancroft, and Hartley, 1956), but statistical advice beyond this book should be obtained if such pooling is contemplated. The analyses in this text will proceed according to the conservative, nonpooling approach, which Hines (1996) shows is generally advantageous.

Multiple Comparisons. If significant differences are concluded among the levels of a factor, then the multiple comparison procedures of Section 11.1, 11.2, 11.3, 11.4, or 11.5 may be employed. For such purposes, s^2 is the within-cells MS, ν is the within-cells

---

[*]The term "mixed model" was introduced by A. M. Mood in 1950 (David, 1995).

DF, and the n of Chapter 11 is replaced in the present situation with the total number of data per level of the factor being tested (i.e., what we have noted in this section as bn data per level of factor A and an data per level of factor B). If there is significant interaction between the two factors, then the means of levels should not be compared. Instead, multiple comparison testing may be performed among cell means.

Confidence Limits for Means.   We may compute confidence intervals for population means of levels of a fixed factor by the methods in Section 10.2. That section's error mean square, s^2, is the within-cells MS of the present discussion; the error degrees of freedom, ν, is the within-cells DF; and n in Section 10.2 is replaced in the present context by the total number of data in the level being examined. Confidence intervals for differences between population means are obtained by the procedures of Section 11.3. This is demonstrated in Example 12.3.

EXAMPLE 12.3   Confidence limits for the results of Example 12.2.

We concluded that mean plasma calcium concentration is different between birds with the hormone treatment and those without.

$$\bar{X}_1 = \frac{\text{total for nonhormone group}}{\text{number in nonhormone group}} = \frac{135.0 \text{ mg/100 ml}}{10} = 13.50 \text{ mg/100 ml}$$

$$\bar{X}_2 = \frac{\text{total for hormone group}}{\text{number in hormone group}} = \frac{301.5 \text{ mg/100 ml}}{10} = 30.15 \text{ mg/100 ml}$$

$$95\% \text{ CI for } \mu_1 = \bar{X}_1 \pm t_{0.05(2),\,\nu}\sqrt{\frac{s^2}{bn}}$$

$$= 13.50 \pm t_{0.05(2),\,16}\sqrt{\frac{22.8982}{(2)(5)}}$$

$$= 13.50 \pm (2.120)(1.513)$$

$$= 13.50 \text{ mg/100 ml} \pm 3.21 \text{ mg/100 ml}$$

$$L_1 = 10.29 \text{ mg/100 ml};\ L_2 = 16.71 \text{ mg/100 ml}$$

$$95\% \text{ CI for } \mu_2 = \bar{X}_2 \pm t_{0.05(2),\,\nu}\sqrt{\frac{s^2}{bn}}$$

$$= 30.15 \pm 3.21 \text{ mg/100 ml}$$

$$L_1 = 26.94 \text{ mg/100 ml};\ L_2 = 33.36 \text{ mg/100 ml}$$

$$95\% \text{ CI for } \mu_1 - \mu_2 = \bar{X}_1 - \bar{X}_2 \pm q_{0.05,\,\nu,k}\sqrt{\frac{s^2}{bn}}$$

$$= 13.50 - 30.15 \pm q_{0.05,\,16,2}\sqrt{\frac{22.8982}{(2)(5)}}$$

$$= -16.65 \pm (2.998)(1.513)$$

$$= -16.65 \text{ mg/100 ml} \pm 4.54 \text{ mg/100 ml}$$

$$L_1 = -21.19 \text{ mg/100 ml};\ L_2 = -12.11 \text{ mg/ ml}$$

EXAMPLE 12.3   (continued)

We concluded that mean calcium concentration is not different in males and females. Therefore, $\mu_\delta = \mu_\varphi$, and we would not speak of confidence intervals for each of these two means or for the difference between the means. If we desired, we could pool the means and speak of a confidence interval for the pooled population mean, μ_p:

$$\text{pooled } \bar{X} = \bar{X}_p = \frac{\text{total for females} + \text{total for males}}{\text{number of females} + \text{number of males}}$$

$$= \frac{237.0 + 199.5}{10 + 10} = 21.82 \text{ mg/100 ml}$$

$$95\% \text{ CI for } \mu_p = \bar{X}_p \pm t_{0.05(2),16}\sqrt{\frac{s^2}{20}}$$

$$= 21.82 \pm (2.120)(1.070)$$

$$= 21.82 \text{ mg/100 ml} \pm 2.27 \text{ mg/100 ml}$$

$$L_1 = 19.55 \text{ mg/100 ml}; L_2 = 24.09 \text{ mg/100 ml}$$

12.2 TWO-FACTOR ANALYSIS OF VARIANCE WITH UNEQUAL REPLICATION

The procedures outlined in Section 12.1 for two-factor analysis of variance require that n, the number of replicates per cell, be the same in all cells. In general it is desirable, for optimum power, to design experiments with equal cell sizes, but occasionally this is impossible or impractical. Figure 12.3 shows two-factor experimental designs with various kinds of replication.

Proportional Replication.   A two-factor experimental design exhibits proportional replication if the number of data in the cell in row i and column j is

$$n_{ij} = \frac{(\text{number of data in row } i)(\text{number of data in column } j)}{N}, \tag{12.21}$$

where N is the total number of data in all cells.* (For example, in Fig. 12.3c, there are two data in row 3, column 1; and $(16)(9)/72 = 2$. The proper analysis of variance computes the total SS in a fashion similar to that in Equation 12.1. The appropriate computations of sums of squares, degrees of freedom, and mean squares are shown in Table 12.4.† Once these quantities are obtained, the hypothesis testing proceeds just as in Section 12.1. (See Table 12.3.) Some factorial analysis of variance computer programs accommodate proportionally replicated data.

---

*The number of replicates in each of the ab cells need not be checked against Equation 12.21 to determine whether proportional replication is present. One need check only one cell in each of $a - 1$ levels of factor A and one in each of $b - 1$ levels of factor B (Huck and Layne, 1974).

†If replication is equal, the computations or Table 12.4 are identical to those in Table 12.2.

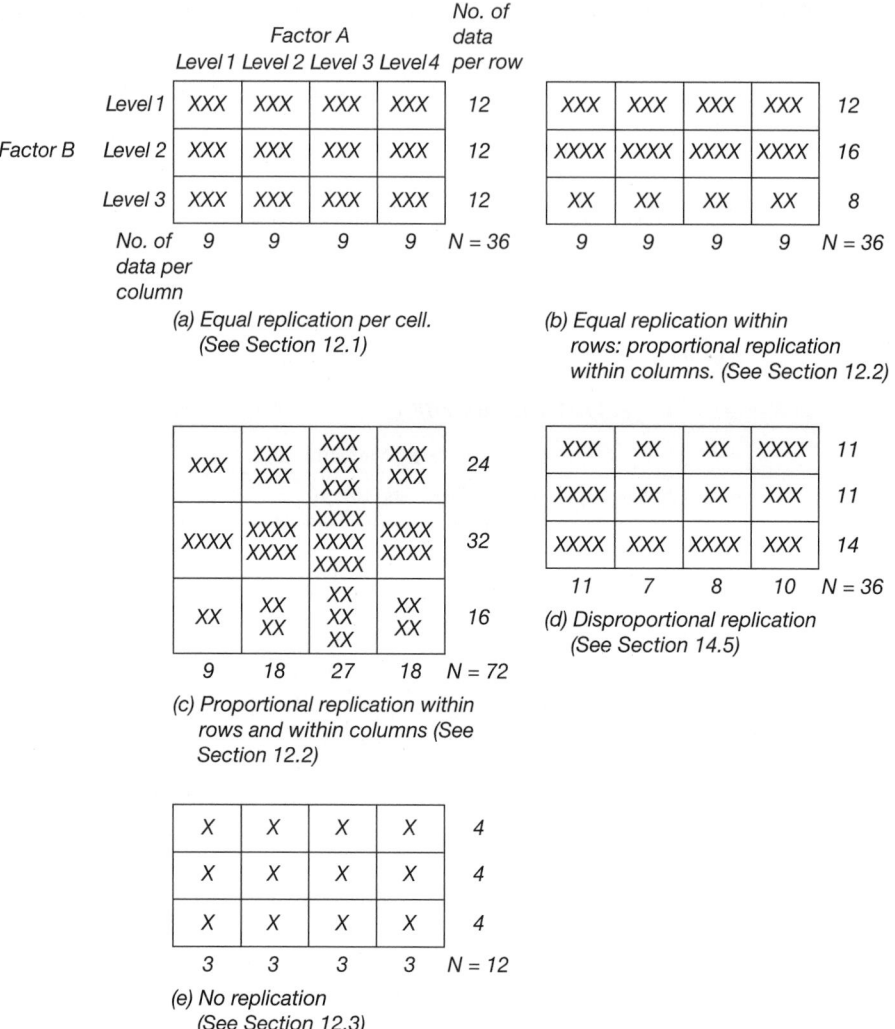

Figure 12.3 Various kinds of replication in a two-factor analysis of variance. In all cases shown, there are four levels of factor *A* and three levels of factor *B*.

Disproportional Replication. We should strive to collect data so that there is equal or proportional replication in the cells. If this is not the case, we may employ a computer program capable of performing factorial analysis of variance with unequal and disproportional replication (Section 14.5). Alternatively, if a few cells have numbers of data in excess of those needed for equal or proportional replication, then data may be deleted, at random, within such cells, so that equality or proportionality is achieved. Then the ANOVA can proceed as usual, as described above or as in Section 12.1.

TABLE 12.4 Two-Way Factorial ANOVA with Unequal but Proportional Replication

Source of variation	Sum of squares (SS)	Degrees of freedom (DF)	Mean square (MS)
Total	$\sum\limits_{i=1}^{a}\sum\limits_{j=1}^{b}\sum\limits_{l=1}^{n_{ij}} X_{ijl}^2 - C$	$N-1$	
Cells	$\sum\limits_{i=1}^{a}\sum\limits_{j=1}^{b} \dfrac{\left(\sum\limits_{l=1}^{n_{ij}} X_{ijl}\right)^2}{n_{ij}} - C$	$ab-1$	$\dfrac{\text{SS}}{\text{DF}}$
Factor A	$\sum\limits_{i=1}^{a} \dfrac{\left(\sum\limits_{j=1}^{b}\sum\limits_{l=1}^{n_{ij}} X_{ijl}\right)^2}{\sum\limits_{j=1}^{b} n_{ij}} - C$	$a-1$	$\dfrac{\text{SS}}{\text{DF}}$
Factor B	$\sum\limits_{j=1}^{b} \dfrac{\left(\sum\limits_{i=1}^{a}\sum\limits_{l=1}^{n_{ij}} X_{ijl}\right)^2}{\sum\limits_{i=1}^{a} n_{ij}} - C$	$b-1$	$\dfrac{\text{SS}}{\text{DF}}$
$A \times B$ Interaction	cells SS − factor A SS − factor B SS	$(a-1)(b-1)$	$\dfrac{\text{SS}}{\text{DF}}$
Within-cells (error)	total SS − cells SS	$\sum\limits_{i=1}^{a}\sum\limits_{j=1}^{b}(n_{ij}-1)$ or total DF − cells DF	$\dfrac{\text{SS}}{\text{DF}}$

Note: Here, a is the number of levels in factor A, b is the number of levels in factor B, and n_{ij} is the number of replicates in the cell formed by level i of factor A and level j of factor B;

$$C = \frac{\left(\sum\limits_{i=1}^{a}\sum\limits_{j=1}^{b}\sum\limits_{l=1}^{n_{ij}} X_{ijl}\right)^2}{N}, \text{ and } N = \sum\limits_{i=1}^{a}\sum\limits_{j=1}^{b} n_{ij}.$$

If one cell is one datum short of the number required for equal proportional replication, a value may be estimated for inclusion in place of the missing datum, as follows (Shearer, 1973):

$$\hat{X}_{ijl} = \frac{aA_i + bB_j - \sum\limits_{i=1}^{a}\sum\limits_{j=1}^{b}\sum\limits_{l=1}^{n_{ij}} X_{ijl}}{N+1-a-b}, \tag{12.22}$$

where $\hat{X}_{ijl}$ is the estimated value for replicate l in level i of factor A and level j of factor B; A_i is the sum of the other data in level i of factor A, B_j is the sum of the other data in level j of factor B, $\sum\sum\sum X_{ijl}$ is the sum of all the known data, and N is the total number of data (including the missing datum) in the experimental design. For example, if datum X_{124} had been missing in Example 12.1, it could have had a quantity inserted in its place, estimated by Equation 12.22, where $a = 2$, $b = 2$, $N = 20$, $A_i = A_1 =$ the sum of all known data from animals receiving no hormone treatment, $B_j = B_2 =$ the sum of all known data from males, and $\sum\sum\sum X_{ijl} =$ the sum of all 19 known data from both hormone treatments and both sexes. After the missing datum has been estimated, it is inserted into the data set and the ANOVA computations may proceed, with the provision that a missing datum is not counted in determining total and within-cells degrees of freedom. (Therefore, if a datum were missing in Example 12.1, the total DF would have been 18 and the within-cells DF would have been 15.)

If more than one datum is missing (but neither more than 10% of the total number of data nor more data than the number of levels of any factor), then Equation 12.22 could be used iteratively to derive estimates of the missing data (e.g., using cell means as initial estimates). The number of such estimates would not enter into the total or within-cells degrees of freedom determinations.

If only a few cells (say, no more than the number of levels in either factor) are each one datum short of the numbers required for equal or proportional replication, then the mean of the data in each such cell may be inserted as an additional datum in that cell. In the latter situation, the analysis proceeds as usual but with the total DF and the within-cells DF each being determined without counting such additional inserted data. Instead of employing these cell means themselves, however, they could be used as starting values for employing Equation 12.22 in iterative fashion. Another procedure for dealing with unequal, and non-proportional, replication is by so-called "unweighted means" analysis, which employs the harmonic mean of the n_{ij}'s. This will not be discussed here.

12.3 TWO-FACTOR ANALYSIS OF VARIANCE
WITHOUT REPLICATION

Occasionally we might encounter a two-factor experimental design in which there is only one datum for each combination of factors (i.e., $n = 1$ for all cells). In such a situation of no replication, each datum may be denoted uniquely by a double subscript, as X_{ij}, where i refers to one of the levels of factor A and j refers to a level of factor B.

For a levels of factor A and b levels of factor B, the appropriate computations of sums of squares, degrees of freedom, and mean squares are shown in Table 12.5. The total SS and DF are obtained just as when there is replication (Section 12.1 and Table 12.2), as are the SS and DF for factor A and for factor B. It should be noted

TABLE 12.5 Two-Way Factorial ANOVA without Replication

Source of variation	Sum of squares (SS)	Degrees of freedom (DF)	Mean square (MS)
Total	$\displaystyle\sum_{i=1}^{a}\sum_{j=1}^{b} X_{ij}^2 - C$	$N-1$	
Factor A	$\dfrac{\displaystyle\sum_{i=1}^{a}\left(\sum_{j=1}^{b} X_{ij}\right)^2}{b} - C$	$a-1$	$\dfrac{\text{factor } A \text{ SS}}{\text{factor } A \text{ DF}}$
Factor B	$\dfrac{\displaystyle\sum_{j=1}^{b}\left(\sum_{i=1}^{a} X_{ij}\right)^2}{a} - C$	$b-1$	$\dfrac{\text{factor } B \text{ SS}}{\text{factor } B \text{ DF}}$
Remainder	total SS − factor A SS − factor B SS	$(a-1)(b-1)$	$\dfrac{\text{remainder SS}}{\text{remainder DF}}$

Note: Here, a is the number of levels in factor A, and b is the number of levels in factor B;

$$C = \frac{\left(\displaystyle\sum_{i=1}^{a}\sum_{j=1}^{b} X_{ij}\right)^2}{N}, \text{ and } N = ab$$

that when $n = 1$, the cells SS of Section 12.1 is identical to the total SS (and the cells DF would be the same as the total DF). Consequently, the within-groups (or error) SS and DF are both zero; with one datum per cell there is no variability within cells.

The part of the total variability not accounted for by the effect of the two factors is:

$$\text{remainder SS} = \text{total SS} - \text{factor } A \text{ SS} - \text{factor } B \text{ SS} \qquad (12.23)$$

with

$$\text{remainder DF} = \text{total DF} - \text{factor } A \text{ DF} - \text{factor } B \text{ DF}. \qquad (12.24)$$

Note that quantities 12.23 and 12.24 are what are referred to as "interaction" quantities when replication is present.

Table 12.5 summarizes the ANOVA calculations just described, and Table 12.6 summarizes the significance tests that may be performed to test hypotheses about each of the factors. Since we do not have measures of both interaction and error variability, the typical test for the significance of interaction is not possible. Testing for the effect of each of the two factors in a Model I analysis (or testing for the effect of the random factor in a Model III design) is not advisable, because there may, in fact, be interaction

TABLE 12.6 Computation of the F Statistic for Tests of Significance in a Two-Factor ANOVA without Replication

	If It Is Assumed That There May Be a Significant Interaction Effect		
Hypothesized effect	Model I (factors A and B both fixed)	Model II (factors A and B both random)	Model III factor A fixed; factor B random)
Factor A	Test with caution*	$\dfrac{\text{factor } A \text{ MS}}{\text{remainder MS}}$	$\dfrac{\text{factor } A \text{ MS}}{\text{remainder MS}}$
Factor B	Test with caution*	$\dfrac{\text{factor } B \text{ MS}}{\text{remainder MS}}$	Test with caution*
$A \times B$ interaction	No test possible	No test possible	No test possible

*Analysis can be performed as in Model II, but with increased chance of Type II error.

	If It Is Correctly Assumed That There Is No Significant Interaction Effect		
Hypothesized effect	Model I	Model II	Model III
Factor A	$\dfrac{\text{factor } A \text{ MS}}{\text{remainder MS}}$	$\dfrac{\text{factor } A \text{ MS}}{\text{remainder MS}}$	$\dfrac{\text{factor } A \text{ MS}}{\text{remainder MS}}$
Factor B	$\dfrac{\text{factor } B \text{ MS}}{\text{remainder MS}}$	$\dfrac{\text{factor } B \text{ MS}}{\text{remainder MS}}$	$\dfrac{\text{factor } B \text{ MS}}{\text{remainder MS}}$
$A \times B$ intersection	No test possible	No test possible	No test possible

between the two factors (and there will be decreased test power); but if a significant difference is concluded, then that conclusion may be accepted. The presence of interaction, also called non-additivity, may be detectable by the testing procedure of Tukey (1949).

12.4 THE RANDOMIZED BLOCK EXPERIMENTAL DESIGN

In Section 10.1, an experiment was performed in which a number of animals were maintained on each of several diets. The variable of interest was body weight, and the null hypothesis was that there is no difference in mean body weight of animals in the various experimental groups (i.e., on the various diets). In that experimental design, the *completely randomized design*, each animal was assumed to be independent of (i.e., unrelated to) each other animal—except for the assignment of groups of animals to the experimental diets.

Another kind of experimental design is that in which one datum in each treatment is related in some way to one datum in each other treatment. For example, each of four animals from the same parents could be assigned to be raised on each of four diets. The body weights of each set of four animals would be said to constitute a *block*. If there were five such blocks in the experiment, there would be a total of twenty animals, and their body weights (X_{ij}) would be tabulated as follows, indicating the group totals (G_i), block totals (B_j), and grand total of all the data $\left(\sum \sum X_{ij} \right)$:

Blocks	1	2	3	4	Totals
			Diets		
1	X_{11}	X_{21}	X_{31}	X_{41}	B_1
2	X_{12}	X_{22}	X_{32}	X_{42}	B_2
3	X_{13}	X_{23}	X_{33}	X_{43}	B_3
4	X_{14}	X_{24}	X_{34}	X_{44}	B_4
5	X_{15}	X_{25}	X_{35}	X_{45}	B_5
Totals	G_1	G_2	G_3	G_4	$\sum \sum X_{ij}$

The expectation that variation among the body weights of the twenty animals might be in part related to genetic factors would lead the experimenter to place siblings in blocks, a procedure that will result in accounting for more of the variability among the data in the experiment, with the desirable outcome of a smaller error mean square and greater statistical power.

The statistical term "block" is conceptually an extension of the term "pair" introduced in Chapter 9. In that chapter, it was shown how many members of two groups might be paired and that the resultant paired-sample testing could be more powerful than the nonpaired two-sample test. We shall now see that the randomized block* ANOVA is an extension of the paired t test just as the one-way ANOVA is an extension of the two-sample t test, and also that the data from such an experimental design can be analyzed by Model III ANOVA considerations.

Consider the experiment represented by the data in Example 12.4. The variable in question is weight gain in guinea pigs; the fixed effect factor whose effect is to be tested is diet. Twenty animals are to be used in this experiment, five on each of four diets. However, the experimenter believes there are some environmental factors that likely would affect weight gain and does not feel that all twenty animal cages can be kept in the laboratory at identical conditions of temperature, light, etc. Therefore, five blocks of experimental units are established, i.e., five groups of guinea pigs. Each block of animals

---

*Randomized blocks were so named by R. A. Fisher, in 1926 (David, 1995).

EXAMPLE 12.4   A randomized block analysis of variance (Model III two-factor analysis of variance without replication).

H_0:   The mean weight gain of guinea pigs is the same on each of four specified diets (i.e., $\mu_1 = \mu_2 = \mu_3 = \mu_4$).

H_A: The mean weight gain of guinea pigs is not the same on each of the four specified diets.

$\alpha = 0.05$

Each guinea pig is housed in a separate cage. A block consists of a group of four animals that we can be reasonably assured will experience identical environmental conditions (light, temperature, draft, noise, etc). Each block has each of its four animals assigned at random to one of the four

EXAMPLE 12.4 (continued)

experimental diets, so that each animal in a given block is to receive a different diet. By consulting a table of random numbers (as Appendix Table B.41), only the digits 1, 2, 3, or 4 need be considered; any others, and any repeats of 1, 2, 3, or 4, are simply ignored. The assignment of diets to animals within blocks is shown, where the assignment in the first block may have resulted from a random digit sequence from Appendix Table B.41 of 34102, the assignment in the second block by a random digit sequence of 153924, the assignment in block 3 by 7352431, etc. Shown in parenthesis is the weight gain (in grams) of each of the twenty animals.

Block 1:	Diet 3	Diet 4	Diet 1	Diet 2
	(4.9)	(8.8)	(7.0)	(5.3)
Block 2:	Diet 1	Diet 3	Diet 2	Diet 4
	(9.9)	(7.6)	(5.7)	(8.9)
Block 3:	Diet 3	Diet 2	Diet 4	Diet 1
	(5.5)	(4.7)	(8.1)	(8.5)
Block 4:	Diet 4	Diet 2	Diet 1	Diet 3
	(3.3)	(3.5)	(5.1)	(2.8)
Block 5:	Diet 1	Diet 4	Diet 3	Diet 2
	(10.3)	(9.1)	(8.4)	(7.7)

The preceding data (weight gains, in grams) are summarized in the following table.

		Diets				
Blocks	1	2	3	4	Totals $\left(B_j = \sum\limits_{i=1}^{a} X_{ij} \right)$	
1	7.0	5.3	4.9	8.8	26.0	
2	9.9	5.7	7.6	8.9	32.1	
3	8.5	4.7	5.5	8.1	26.8	
4	5.1	3.5	2.8	3.3	14.7	
5	10.3	7.7	8.4	9.1	35.5	
Totals: $\left(G_i = \sum\limits_{j=1}^{b} X_{ij} \right)$	40.8	26.9	29.2	38.2	Grand total: $\sum\limits_{i=1}^{a}\sum\limits_{j=1}^{b} X_{ij} = 135.1$	

$$a = 4, \, b = 5, \, N = ab = 20$$

$$\sum_{i=1}^{a}\sum_{j=1}^{b} X_{ij}^2 = 1011.95$$

$$C = \frac{\left(\sum\limits_{i=1}^{a}\sum\limits_{j=1}^{b} X_{ij} \right)^2}{N} = \frac{(135.1)^2}{20} = 912.6005$$

$$\text{total SS} = \sum_{i=1}^{a}\sum_{j=1}^{b} X_{ij}^2 - C = 1011.95 - 912.6005 = 99.3495$$

EXAMPLE 12.4   (continued)

$$\text{diets SS} = \frac{\sum\limits_{i=1}^{a} G_i^2}{b} - C$$

$$= \frac{(40.8)^2 + (26.9)^2 + (29.2)^2 + (38.2)^2}{5} - 912.6005$$

$$= 27.4255$$

$$\text{blocks SS} = \frac{\sum\limits_{j=1}^{b} B_j^2}{a} - C$$

$$= \frac{(26.0)^2 + (32.1)^2 + (26.8)^2 + (14.7)^2 + (35.5)^2}{4} - 912.6005$$

$$= 62.6470$$

$$\text{remainder SS} = \text{total SS} - \text{diets SS} - \text{block SS}$$

$$= 99.3495 - 27.4255 - 62.6470 = 9.2770$$

Analysis of Variance Table

Source of variation	SS	DF	MS
Total	99.3495	19	
Diets	27.4255	3	9.1418
Blocks	62.6470	4	
Remainder	9.2770	12	0.7731

$$\text{To test } H_0, F = \frac{\text{diets MS}}{\text{remainder MS}} = \frac{9.1418}{0.7731} = 11.8$$

$$F_{0.05(1), 3, 12} = 3.49$$

Therefore, reject H_0.

$$0.0005 < P < 0.001 \quad [P = 0.00068]$$

consists of four guinea pigs, one on each of the experimental diets. All members of a block are considered to be at identical conditions (except, of course, for diet).

The assignment of an experimental diet to each of the individuals in a block is done at random; for this purpose, Appendix Table B.41 may be consulted. A random series of digits allows us to assign diets 1 through 4 to the four animals in the first block. Then a second random series of four digits determines the assignment of the four diets to the individuals in the second block, etc. After the data are collected, they are commonly arranged in an array, such as shown in Example 12.4.

The analysis of the randomized block data is performed as a mixed model two-way ANOVA without replication, by consulting Tables 12.5 and 12.6. In the present example, diet is the fixed factor, and block is always the random factor. The null hypothesis of equal weight gain on all diets is readily tested. A null hypothesis of equal weight gain among blocks is usually of no particular interest; furthermore, it is not advisable, because it would require knowledge about an interaction effect. Thus, it is generally not tested.*

Example 12.4 introduces some simplified symbols:

$$G_i = \text{sum (over all blocks) of all data in group } i$$

$$= \sum_{j=1}^{b} X_{ij} \tag{12.25}$$

$$B_j = \text{sum (over all groups) of all data in block } j$$

$$= \sum_{i=1}^{a} X_{ij}. \tag{12.26}$$

Therefore, we can write

$$\text{groups SS} = \frac{\sum_{i=1}^{a} G_i^2}{b} - C \tag{12.27}$$

and

$$\text{blocks SS} = \frac{\sum_{j=1}^{b} B_j^2}{a} - C, \tag{12.28}$$

where the total SS and the quantity C are as in Table 12.5. The sums of squares, degrees of freedom, and mean squares for the randomized block ANOVA are also as in Table 12.5, where factor A is groups and factor B is blocks.

Missing Data in the Randomized Block Design. On occasion, a datum in a randomized block design may be missing due to no fault of the experimental conditions under study. An experimental mouse may be ingested by a pet hawk, a culture tube may accidentally be shattered on a desktop, an experimental plant may be chewed by a hungry guinea pig, etc. In such a situation, one may discard the entire block of data in which the missing datum belonged, but in general it is undesirable to discard data. Alternatively, we may employ the following procedure to estimate what the missing datum would have been (Li, 1964: 228–233).

A datum which is missing from block j in experimental group i may be estimated as

$$\hat{X}_{ij} = \frac{aG_i + bB_j - \sum_{i=1}^{a}\sum_{j=1}^{b} X_{ij}}{(a-1)(b-1)}, \tag{12.29}$$

where G_i is the sum of the known data in group i, B_j is the sum of the known data in block j, and $\sum_{i=1}^{a}\sum_{j=1}^{b} X_{ij}$ is the sum of all known data. The total, groups, and blocks

---

*There is another statistical model that does allow for the latter test (see Samuels, Casella, and McCabe, 1991), but it is seldom encountered.

sums of squares are then calculated using this estimated value in place of the missing datum, except that the groups SS is slightly biased, a situation corrected by subtracting the quantity

$$\text{bias} = \frac{[B_j - (a - 1)\hat{X}_{ij}]^2}{a(a - 1)}. \tag{12.30}$$

Also, 1 must be subtracted from the total DF, which means that the remainder DF is also reduced by 1.

If there are more than one missing data, but not too many (say, no more than 10% of all data are missing, but in no case more than $a - 1$ data missing), then the following procedure may be used. Insert into the data array estimates of all but one of the missing data. Such estimates may be guesses; one might use as a guess, the mean of the known data for the group in which the missing datum would be located, or even the mean of all known data. Then the remaining missing datum is estimated by Equation 12.29, where the group and block sums include all data except that one being estimated. Then that estimate is placed in the data array and one of the other estimates is recalculated using Equation 12.29. This is done for each missing datum in turn. Then, the process is repeated for each of the missing data, thus arriving at somewhat better estimates of all of them, This iterative, or repetitive, process is engaged in until there is no further change in the estimated values. Typically, only two (or sometimes three) cycles of the process are required.

The total SS, blocks SS, and groups SS are then computed using all data, both known and estimated. But this groups SS is biased and should be corrected in the following manner (Glen and Kramer, 1958). Use Equation 12.30 to determine the bias for each block having one missing datum. Then, use Equation 12.31 to determine the bias for each block containing two or more data:

$$\text{bias} = \sum^{m_j} \hat{X}_{ij}^2 + \frac{B_j^2}{a - m_j} - \frac{(B_j + \sum^{m_j} \hat{X}_{ij})^2}{a}, \tag{12.31}$$

where m_j is the number of missing data in block j (m_j must be less than a), and each summation is performed over all m_j missing data in that block.

Finally, the results of all the above bias calculations are added and this sum is subtracted from the groups SS. Using the total SS, the blocks SS, and the corrected groups SS, the remainder SS is calculated. The total DF and the remainder DF each must be reduced by the number of missing data estimated; the groups DF and blocks DF are as usual (i.e., $a - 1$ and $b - 1$, respectively).

12.5 REPEATED-MEASURES EXPERIMENTAL DESIGNS

A *repeated-measures* experimental design, also called a *within-subjects* or *treatment-by-subject* design, is one in which multiple measurements on the same experimental subject comprise the replicate data.*

---

*Kepner and Robinson (1988) discuss nonparametric methods for this experimental design.

In Example 10.1, the desire was to test the null hypothesis of no effect of four different diets on the body weight of animals. If, in the completely randomized experimental design for that hypothesis, there were four independent samples of animals (say, five animals per sample), then all members of a given sample would be fed one of the four diets. In a repeated-measures design there would not be independent samples of animals; instead, each of five animals would have its body weight recorded after being maintained on one of the diets, then each of them would be weighed after being fed another diet for a time, and so on. The array of data would look like the tabulation in Section 12.4, for the randomized complete block design, but where k columns of data represent the dietary treatments and n rows are the repeated measures on the several animals (referred to as "subjects"); the column (group) totals are denoted as G_i, the row (subject) totals as S_j, and the grand total of the data as $\sum \sum X_{ij}$ (which also is $\sum G_i$ and $\sum S_j$):

		Diets			
Subjects	1	2	3	4	*Totals*
1	X_{11}	X_{21}	X_{31}	X_{41}	S_1
2	X_{12}	X_{22}	X_{32}	X_{42}	S_2
3	X_{13}	X_{23}	X_{33}	X_{43}	S_3
4	X_{14}	X_{24}	X_{34}	X_{44}	S_4
5	X_{15}	X_{25}	X_{35}	X_{45}	S_5
Totals	G_1	G_2	G_3	G_4	$\sum \sum X_{ij}$

In this experimental design, the total sum of squares and degrees of freedom are computed as in Table 12.5 where $a = k$ and $b = n$. Then the total SS is apportioned into two parts: the sum of squares among subjects and the sum of squares within subjects. The measure of variability among subjects is

$$\text{subjects SS} = \frac{\sum\limits_{j=1}^{n} S_j^2}{k} - C, \tag{12.32}$$

and the variability within subjects is

$$\text{within-subjects SS} = \text{total SS} - \text{subjects SS}. \tag{12.33}$$

The total degrees of freedom, $N - 1$, are partitioned between

$$\text{subjects DF} = n - 1 \tag{12.34}$$

and

$$\text{within-subjects DF} = \text{total DF} - \text{subjects DF}$$
$$= n(k - 1). \tag{12.35}$$

Then, the within-subjects variability is partitioned into that due to the experimental treatment:

$$\text{groups SS} = \frac{\sum\limits_{i=1}^{k} G_i^2}{n} - C, \tag{12.36}$$

and that unaccounted for by the treatments and the subjects:

$$\text{remainder SS} = \text{within-subjects SS} - \text{groups SS}. \tag{12.37}$$

And the associated degrees of freedom are:

$$\text{groups DF} = k - 1 \tag{12.38}$$

and

$$\text{remainder DF} = \text{within-subjects DF} - \text{groups DF}$$
$$= (k - 1)(n - 1). \tag{12.39}$$

The "remainder" may also be referred to as the "groups × subjects interaction."

The null hypothesis of interest is that of no difference among the means of the treatment groups, and for this $F = $ (groups MS)/(remainder MS) is employed. It is generally inadvisable, on the basis of the underlying statistical theory, to attempt to test for difference among subjects; and that is typically not an object of performing such an experiment.

Example 12.5 shows the results of an experiment in which blood cholesterol was measured in seven subjects after each subject had been treated with one of each of three drugs, one drug at a time (with time allowed between exposure to each drug to allow its effects to disappear from the animal). The hypothesis of interest is that the mean blood cholesterol level is the same regardless of treatment.

EXAMPLE 12.5   A repeated-measures analysis of variance. The data are blood cholesterol concentrations, in milligrams of cholesterol per deciliter of blood (mg/dl).

H_0:   The mean cholesterol level is the same in animals on all three drugs (i.e., $\mu_1 = \mu_2 = \mu_3$).

H_A: The mean cholesterol level is not the same in animals on all three drugs.

Subjects	Drug 1	Drug 2	Drug 3	Totals (S_j)
1	164	152	178	494
2	202	181	222	605
3	143	136	132	411
4	210	194	216	620
5	228	219	245	692
6	173	159	182	514
7	161	157	165	483
Totals (G_i)	1281	1198	1340	3819

EXAMPLE 12.5   (continued)

$$\sum_{i=1}^{k} \sum_{j=1}^{n} X_{ij} = 3819$$

$$\sum_{i=1}^{k} \sum_{j=1}^{n} X_{ij}^2 = 715,393$$

$$C = \frac{(3819)^2}{21} = 694,512.43$$

$$\text{total SS} = \sum_{i=1}^{k} \sum_{j=1}^{n} X_{ij}^2 - C$$

$$= 715,393 - 694,512.43 = 20,880.57$$

$$\text{subjects SS} = \frac{\sum_{j=1}^{n} S_j^2}{k} - C$$

$$= \frac{(494)^2 + (605)^2 + \cdots + (483)^2}{3} - 694,512.43$$

$$= 713,243.67 - 694,512.43 = 18,731.24$$

$$\text{within-subjects SS} = \text{total SS} - \text{subjects SS}$$

$$= 20,880.57 - 18,731.24 = 2,149.33$$

$$\text{drugs SS} = \frac{\sum_{i=1}^{k} G_i^2}{n} - C$$

$$= \frac{(1281)^2 + (1198)^2 + (1340)^2}{7} - C$$

$$= 695,966.43 - 694,512.43 = 1,454.00$$

$$\text{remainder SS} = \text{within-subjects SS} - \text{drugs SS}$$

$$= 2,149.33 - 1,454.00 = 695.33$$

Analysis of Variance Table

Source of variation	SS	DF	MS
Total	20880.57	20	
Subjects	18731.24	6	
Within-Subjects	2149.33	14	
Drugs	1454.00	2	727.00
Remainder	695.33	12	57.94

To test H_0, $F = \dfrac{\text{drugs MS}}{\text{remainder MS}} = \dfrac{727.00}{57.94} = 12.6$

$F_{0.05(1),2,12} = 3.89$

Therefore, reject H_0.

$$0.001 < P < 0.0025 \quad [P = 0.0011]$$

The hypothesis of Example 12.5 could have been tested by using a total of twenty-one animals, randomly allocated seven to each of the drug treatment groups, by employing the one-factor ANOVA of Section 10.1. Or, the rows in the data tabulation of Example 12.5 could have represented siblings, or individuals with some factor other than genetics in common, in which case the randomized complete block design of Section 12.4 would have been appropriate. The repeated-measures design will be more powerful than the one-way ANOVA if there is consequential relationship among the data in each row, and it will be more powerful than the randomized block design if there is a stronger relationship within subjects than within blocks. And the repeated-measures design may be advantageous in its economical requirement for fewer experimental subjects. The computational results of this repeated-measures analysis are identical to those that would result from the same data subjected to a randomized block analysis.

The repeated-measures design is disadvantageous if there are effects of the sequence in which the treatments (diets in the present example) are administered to the subjects. Another disadvantage arises if insufficient time is allowed between the administration of different treatments to avoid *carryover* effects of the previous treatment. Carryover effect may often be counteracted by what is known as *counterbalancing*, whereby, to the extent possible, each subject receives the treatments in a different sequence.

In a repeated-measures experiment, we presume that there are correlations among the repeated measurements, in this case, weight measurements of the same individual on different diets. That is, it is reasonable to suppose that an individual that is heavier than its colleagues on one diet will also be heavier on another diet. But for the probability of a calculated F to be determined from tabled values of F, there should be equal correlations among pairs of groups of data. That is, the correlation between the data of groups 1 and 2 should be the same as the correlation between the data of groups 1 and 3, the same as that between the data of groups 2 and 3, and so on. This characteristic, referred to as *compound symmetry*, coupled with the usual ANOVA assumption of equality of group variances, is related to what statisticians call *sphericity* (Huynh and Feldt, 1970), or *circularity* (Rouanet and Lépine, 1970), an underlying assumption of repeated-measures ANOVA. Violation of the latter assumption is, unfortunately, common but difficult to test for, and results in a Type I error greater than the specified α. Box (1954) and Geisser and Greenhouse (1958) described an approximation procedure, tedious to apply, that does not have this assumption but that usually results in an α that is too low. (See also Crowder and Hand, 1990: 54–58; Girden, 1992: 19–26; Glantz and Slinker, 1990: 400–404; Keppel, 1991: 352–353; Maxwell and Delaney, 1990: 476–477.) An alternative procedure for analyzing repeated-measures experimental designs, one that does not depend upon the sphericity assumption, is *multivariate analysis of variance* (see Chapter 16)—abbreviated MANOVA—which has gained in popularity with the availability of computer programs to handle the relatively complex computations (Girden, 1992: 22–26; Maxwell and Delaney, 1990: Chapters 13 and 14; O'Brien and Kaiser, 1985; Stevens, 1996: Chapter 13).

Two Within-Subjects Factors.   An experimental design might have two factors of interest—say, diet and exercise regimen—with the replication being multiple observations (e.g., body weight) made on the same experimental subject. That is, the body weight of n animals would be measured at each of the ab combinations of a diets and b exercise regimens. This would be analyzed as a three-factor ANOVA, with subjects (i.e., animals) as the third (random-effects) factor. An appropriate computer program will yield the necessary sums of squares; the pertinent degrees of freedom would be $a - 1$ for diets, $b - 1$ for exercise, $(a - 1)(b - 1)$ for the interaction of diet $\times$ exercise (i.e., factor $A \times$ factor B interaction), $(a - 1)(n - 1)$ for the diet $\times$ subject (i.e., $A \times S$) interaction, $(b - 1)(n - 1)$ for the exercise $\times$ subject (i.e., $B \times S$) interaction, and $(a-1)(b-1)(n-1)$ for the diet $\times$ exercise $\times$ subject $(A \times B \times S)$ interaction. The needed mean squares would be obtained by dividing each sum of squares by its associated degrees of freedom; and the hypothesis tests for diet (factor A), exercise treatment (factor B), and the $A \times B$ interaction would be tested as indicated in the second example of Appendix Section A.3.

12.6 MULTIPLE COMPARISONS AND CONFIDENCE INTERVALS IN TWO-FACTOR ANALYSIS OF VARIANCE

If a two-factor analysis of variance reveals a significant effect among levels of a fixed-effects factor having more than two levels, then we can determine between which levels the difference(s) occur(s). This may be done using the Tukey test (Section 11.1) or the Newman-Keuls test (Section 11.2). The appropriate SE is calculated by Equation 11.2, substituting for n the number of data in each level (i.e., there are bn data in each level of factor A and an data in levels of factor B); s^2 is the within-cells MS: and ν is the within-cells degrees of freedom. If there is no replication in the experiment, then we are obliged to use the remainder MS in place of the within-cells MS, and to use the remainder DF as ν.

The calculation of confidence limits for the population mean estimated by each significantly different level mean can be performed by the procedures of Section 11.3, as can the computation of confidence limits for differences between pairs of significantly different level means.

If it is desired to compare a control mean to each of the other level means, Dunnett's test described in Section 11.4, may be used. Section 11.4 also shows how to calculate confidence limits for the difference between such means. Scheffé's procedure for multiple contrasts (Section 11.5) may also be applied to the levels of a factor, where the critical value in Equation 11.17 employs either a or b in place of k (depending, respectively, on whether the levels of factor A or B are being examined), and the within-cells DF is used in place of $N - k$. In all references to Chapter 11, n in the standard error computation is to be replaced by the number of data per level, and s^2 and ν are the within-cells MS and DF, respectively.

Multiple comparison testing and confidence interval determination are appropriate for levels of a fixed-effects factor but are not used with random-effects factors.

If Interaction Is Significant. On concluding that there is a significant interaction between factors A and B, it is generally not meaningful to test for differences among levels of either of the factors. However, it may be desired to perform multiple comparison testing to seek significant differences among cell means. This can be done with any of the above-mentioned procedures, using the equations from Chapter 11, where n (the number of data per cell) is appropriate instead of the number of data per level. For the Scheffé test critical value (Equation 11.17), k is the number of cells (i.e., $k = ab$) and $N - k$ is the within-cells DF.

Randomized Block Analysis of Variance.   For a randomized block experimental design (Section 12.4), we may perform multiple comparison testing among, and calculate confidence intervals for, the means of the levels of the fixed-effects factor, using any of the procedures of Sections 11.1 through 11.5. (This may not be done for block means, unless it is assumed that there is no interaction effect, but even then it is rarely desirable.) For such tests, s^2 refers to the remainder MS and ν refers to the remainder DF.

If there are missing data that had to be estimated, then the standard error for the difference between two group means may be corrected upward (e.g., Li, 1964: 236), but the amount of correction is usually trivial.

Repeated-Measures Analysis of Variance.   Multiple comparison procedures for repeated-measures ANOVA are performed as for the randomized block design. But if the sphericity assumption is violated, the probability of a Type I error may be much greater than α (Maxwell, 1980).

12.7   POWER AND SAMPLE SIZE IN TWO-FACTOR ANALYSIS OF VARIANCE

If the reader is familiar with the concepts and procedures of estimating power, sample size, and minimum detectable difference for a single-factor ANOVA, as discussed in Section 10.3, then it will be seen how the same considerations can be applied to fixed-effects factors in a two-factor analysis of variance. (The handling of the fixed factor in a mixed model ANOVA will be explained at the end of this section.)

We can consider either factor A or factor B (or both, but one at a time). Let us say k' is the number of levels of the factor being examined. (That is $k' = a$ for factor A; $k' = b$ for factor B.) Let us define n' as the number of data in each level. (That is, $n' = bn$ for factor A; $n' = an$ for factor B.) We shall also have s^2 refer to the within-cells MS.

Power of the Test.   We can now generalize Equation 10.32 as

$$\phi = \sqrt{\frac{n' \sum_{m=1}^{k'} (\mu_m - \mu)^2}{k's^2}}, \qquad (12.40)$$

Equation 10.33 as

$$\mu = \frac{\sum_{m=1}^{k'} \mu_m}{k'}, \tag{12.41}$$

and Equation 10.34 as

$$\phi = \sqrt{\frac{n'\delta^2}{2k's^2}}, \tag{12.42}$$

in order to estimate the power of the analysis of variance in detecting differences among the population means of the levels of the factor under consideration.

After any of the computations of ϕ have taken place, either as above, or as below, then we proceed to employ Appendix Fig. B.1 just as we did in Section 10.3, with ν_1 being the factor DF (i.e., $k' - 1$), and ν_2 referring to the within-cells (i.e., error) DF.

Later in this book will be examples of ANOVAs where the appropriate denominator for F is some MS other than the within-cells MS. In such a case, s^2 and ν_2 will refer to the relevant MS and DF.

As demonstrated in Section 10.3, we can estimate the power of an ANOVA that has been performed, an especially desirable capability when H_0 has not been rejected. Using the above notation, with s^2 being the within-cells MS for the analyzed data, Equation 10.35 can be generalized as

$$\phi = \sqrt{\frac{(k' - 1)(\text{factor MS} - s^2)}{k's^2}} \tag{12.43}$$

and for the interaction effect (after Kirk, 1982: 143),

$$\phi = \sqrt{\frac{(A \times B \text{ DF})(A \times B \text{ MS} - s^2)}{[(A \times B \text{ DF}) + 1]s^2}}. \tag{12.44}$$

Sample Size Required.   By using Equation 12.42 with a specified power, significance level, and detectable difference between means, we can determine the necessary minimum number of data per level, n', needed in the experiment. This is done iteratively, as it was in Example 10.6.

Minimum Detectable Difference.   In Example 10.7 we estimated the smallest detectable difference between population means, given the significance level, sample size, and power of a one-way ANOVA. We can pose the same question in the two-factor experiment, generalizing Equation 10.36 as

$$\delta = \sqrt{\frac{2k's^2\phi^2}{n'}}. \tag{12.45}$$

Maximum Number of Levels Testable.   The considerations of Example 10.8 can be applied to the two-factor case by using Equation 12.42 instead of Equation 10.34.

The Mixed-Model ANOVA.   All the preceding considerations of this section can be applied to the fixed factor in a mixed-model (Model III) two-factor analysis of variance with the following modifications.

For factor A fixed, with replication within cells, substitute the interaction MS for the within-cells MS, and use the interaction DF for v_2.

For factor A fixed, with no replication (i.e., a randomized block experimental design), substitute the remainder MS for the within-cells MS, and use the remainder DF for v_2. If there is no replication, then $n = 1$, and $n' = b$. (Recall that if there is no replication, we do not test for interaction effect.)

12.8 Nonparametric Randomized Block or Repeated-Measures Analysis of Variance

Friedman's* test (1937, 1940) is a nonparametric analysis that may be performed on a randomized block experimental design, and it is especially useful with data that do not meet the parametric analysis of variance assumptions of normality and homoscedasticity, namely that the k samples (i.e., the k levels of the fixed-effect factor) come from populations that are each normally distributed and have the same variance. If these assumptions of the parametric ANOVA are met, the Friedman test will be $3k/[\pi(k+1)]$ as powerful as the parametric method (van Elteren and Noether, 1959). (That is, the power of the latter compared to the former ranges from 64%, when $k = 2$; to 72% when $k = 3$; to 95% when $k = \infty$.) If the assumptions are seriously violated, the former should not be used, but the Friedman test is valid. Where $k = 2$, the Friedman test is equivalent to the sign test (Section 24.7).

In Example 12.6, Friedman's test is applied to the data of Example 12.4. The data within each of the b blocks are assigned ranks. The ranks are then summed for each of the a groups, each rank sum being denoted as R_i. The test statistic, χ_r^2, is calculated as

$$\chi_r^2 = \frac{12}{ba(a+1)} \sum_{i=1}^{a} R_i^2 - 3b(a+1). \tag{12.46}$$

Critical values for many values of a and b are given in Appendix Table B.14.

When $a = 2$, the Wilcoxon paired-sample test (Section 9.5) should be used; if $b = 2$, then the Spearman rank correlation (Section 19.9) should be employed. Appendix Table B.14 should always be used when the a and b of an experimental design are contained therein. For a and b beyond this table, the distribution of χ_r^2 may be considered to be approximated by the χ^2 distribution (Appendix Table B.1), with $a - 1$ degrees

---

*Milton Friedman (b. 1912), American economist and 1976 Nobel Prize winner, to whom is attributed the declaration that "There's no such thing as a free lunch."

EXAMPLE 12.6   Friedman's analysis of variance by ranks applied to the randomized block data of Example 12.4.

H_0:   The weight gain of guinea pigs is the same on each of four specified diets.

H_A: The weight gain of guinea pigs is not the same on each of the four specified diets.

$\alpha = 0.05$.

The following data (weight gain, in grams) have been ranked within each block. The ranks are shown in parentheses.

		Diets (i)		
Blocks	1	2	3	4
1	7.0	5.3	4.9	8.8
	(3)	(2)	(1)	(4)
2	9.9	5.7	7.6	8.9
	(4)	(1)	(2)	(3)
3	8.5	4.7	5.5	8.1
	(4)	(1)	(2)	(3)
4	5.1	3.5	2.8	3.3
	(4)	(3)	(1)	(2)
5	10.3	7.7	8.4	9.1
	(4)	(1)	(2)	(3)
Rank sum (R_i)	19	8	8	15

$a = 4$

$b = 5$

$$\chi_r^2 = \frac{12}{ba(a+1)} \sum R_i^2 - 3b(a+1)$$

$$= \frac{12}{(5)(4)(4+1)}[(19)^2 + (8)^2 + (8)^2 + (15)^2] - (3)(5)(4+1)$$

$$= 0.12(714) - 75 = 85.68 - 75 = 10.68$$

$(\chi_r^2)_{0.05,4,5} = 7.800$

Therefore, reject H_0.

$$0.005 < P < 0.01$$

of freedom. However, Iman and Davenport (1980) showed that this commonly used approximation is undesirably conservative (i.e., it has a high likelihood of a Type II error—and, therefore, low power) and that

$$F_F = \frac{(b-1)\chi_r^2}{b(a-1) - \chi_r^2} \tag{12.47}$$

is generally superior. F_F is compared to F (Appendix Table B.4) with degrees of freedom of $a - 1$ and $(a - 1)(b - 1)$.*

Another approach to testing of this experimental design is that of *rank transformation*, by which one ranks all ab data and performs the analysis of variance of Section 12.4 on those ranks (Iman, 1974b; Conover and Iman, 1976; and Iman, Hora, and Conover, 1984); this is demonstrated in Example 12.7. Quade (1979) presented a test that is an

---

EXAMPLE 12.7 The experiment of Example 12.6, tested by analysis of variance with rank transformation.

H_0: The weight gain of guinea pigs is the same on each of four specified diets.

H_A: The weight gain of guinea pigs is not the same on each of the four specified diets.

$\alpha = 0.05$.

The ranks of each of the data are shown in parentheses.

Blocks (j)	Diets (i) 1	2	3	4	Rank totals (B_j)
1	7.0 (10)	5.3 (7)	4.9 (5)	8.8 (16)	38
2	9.9 (19)	5.7 (9)	7.6 (11)	8.9 (17)	56
3	8.5 (15)	4.7 (4)	5.5 (8)	8.1 (13)	40
4	5.1 (6)	3.5 (3)	2.8 (1)	3.3 (2)	12
5	10.3 (20)	7.7 (12)	8.4 (14)	9.1 (18)	64
Rank totals (G_i)	70	35	39	66	210

Analysis of Variance Table

Source of variation	SS	DF	MS
Total	665.0000	19	
Diets	195.4000	3	65.1333
Blocks	400.0000	4	
Remainder	69.6000	12	5.8000

$$F = \frac{65.1333}{5.8000} = 11.23$$

$F_{0.05(1),3,12} = 3.49$

Therefore, reject H_0.

$$0.0005 < P < 0.001 \quad [P = 0.00085]$$

---

*Iman and Davenport (1980) also show that comparing the mean of χ_r^2 and F_F to the mean of the critical values of χ^2 and F provides an improved approximation when Appendix Table B.14 cannot be used.

extension of the Wilcoxon paired-sample test that may be preferable in some circumstances (Iman, Hora, and Conover, 1984). The rank-transformation procedure, however, typically gives results better than those from the Friedman or Quade tests. But, its proponents do not recommend that it be routinely employed as an alternative to the parametric ANOVA when it is suspected that the underlying assumptions of the latter do not apply. Instead, they propose that it be employed along with the usual ANOVA and, if both yield the same conclusion, one can depend upon that conclusion (as is the case with Examples 12.5 and 12.7).

If tied ranks are present, they may be taken into consideration by computing

$$\left(\chi_r^2\right)_c = \frac{\chi_r^2}{C},$$ (12.48)

Marascuilo and McSweeney, 1967),* (Kendall, 1962: Chapter 6), where

$$C = 1 - \frac{\sum t}{b(a^3 - a)}$$ (12.50)

and $\sum t$ is as defined in Equation 10.42.

The *Kendall coefficient of concordance (W)* is another form of Friedman's χ_r^2:

$$W = \frac{\chi_r^2}{b(a - 1)}.$$ (12.51)

It is used as a measure of the agreement of rankings within blocks and is considered further in Section 19.17.

Multiple Observations per Cell.   In the experiment of Example 12.6 there is one datum for each combination of diet and block. Although this is the typical situation, one might also encounter an experimental design in which there are multiple observations recorded for each combination of block and experimental group. Recall that each combination of level of factor A (experimental group) and level of factor B (block) is called a cell; for n replicate data per cell,

$$\chi_r^2 = \frac{12}{ban^2(na + 1)} \sum_{i=1}^{a} R_i^2 - 3b(na + 1)$$ (12.52)

(Marascuilo and McSweeney, 1977: 376–377), with the critical value being $\chi_{\alpha, k-1}^2$. Note that if $n = 1$, Equation 12.52 reduces to Equation 12.46. Benard and van Elteren (1953),

---

*Equivalently,

$$\left(\chi_r^2\right)_c = \frac{\sum_{i=1}^{a} R_i^2 - \frac{\left(\sum_{i=1}^{a} R_i\right)^2}{a}}{\frac{ba(a + 1)}{12} - \frac{\sum t}{a - 1}}$$ (12.49)

and Skillings and Mack (1981) present procedures applicable when there are unequal numbers of data per cell.

12.9 Multiple Comparisons for Nonparametric Randomized Block or Repeated-Measures Analysis of Variance

A multiple comparison analysis applicable to ranked data in a randomized block is similar to the Tukey procedure for ranked data in a one-way ANOVA design (Section 11.6). In this case, Equation 11.3 is used with the difference between rank sums, i.e., $R_B - R_A$, in the numerator and

$$SE = \sqrt{\frac{ba(a+1)}{12}} \tag{12.53}$$

in the denominator* (Nemenyi, 1963); and this is used in conjunction with the appropriate critical value of $q_{\alpha,\infty,k}$.

If the various groups are to be compared one at a time with a control group, then

$$SE = \sqrt{\frac{ba(a+1)}{6}} \tag{12.55}$$

may be used in Dunnett's procedure, in a fashion similar to that explained at the end of Section 11.6.

The preceding multiple comparisons are applicable to the levels of the fixed-effect factor, not to the blocks (levels of the random-effect factor).

Multiple contrasts, as introduced in Sections 11.5 and 11.7, may be performed using rank sums. We employ Equation 11.31, with[†]

$$SE = \sqrt{\frac{ba(a+1)}{12}\left(\sum_i c_i^2\right)}, \tag{12.57}$$

---

*If desired, mean ranks ($\bar{R}_A = R_A/b$ and $\bar{R}_B = R_B/a$) can be used in the numerator of Equation 11.3, in which case the denominator will be

$$SE = \sqrt{\frac{a(a+1)}{12b}}. \tag{12.54}$$

[†]If mean ranks are used,

$$SE = \sqrt{\frac{a(a+1)}{12b}\left(\sum_i c_i^2\right)}. \tag{12.56}$$

unless there are tied ranks, in which case*

$$
\mathrm{SE} = \sqrt{ \left(\dfrac{ \dfrac{a(a+1)}{b} - \dfrac{\sum t}{b^2(a-1)} }{12} \right) \left(\sum_i c_i^2 \right) }
\tag{12.59}
$$

(Marascuilo and McSweeney, 1967). The critical value for the multiple contrasts is $\sqrt{(\chi_r^2)_{\alpha,a,b}}$, using Appendix Table B.14 to obtain $(\chi_r^2)_{\alpha,a,b}$. If the needed critical value is not on that table, then $\sqrt{\chi^2_{\alpha,a-1}}$ may be used as an approximation to it.

Multiple Observations per Cell. If there is replication per cell (as in Equation 12.52), the standard errors of this section are modified by replacing a with an and b with bn wherever they appear. (See Marascuilo and McSweeney, 1977: 378.) Norwood et al. (1989) and Skillings and Mack (1981) present multiple comparison methods applicable when there are unequal numbers of data per cell.

12.10 DICHOTOMOUS NOMINAL-SCALE DATA IN RANDOMIZED BLOCKS OR FROM REPEATED MEASURES

A randomized block or repeated-measures experimental design may contain values of a dichotomous variable (i.e., a variable with two possible values: e.g., "present" or "absent," "dead" or "alive," "true" or "false," "left" or "right," "male" or "female," etc.), in which case Cochran's Q test[†] (Cochran, 1950) may be applied. For such an analysis, one value of the attribute is recorded with a "1," and the other with a "0." In Example 12.8, the data are the occurrence or absence of mosquito attacks on humans wearing one of several types of clothing. The null hypothesis is that the proportion of people attacked is the same for each type of clothing worn.

For a groups and b blocks, where G_i is the sum of the 1's in group i and B_j is the sum of the 1's in block j,

$$
Q = \frac{(a-1)\left[\displaystyle\sum_{i=1}^{a} G_i^2 - \frac{\left(\displaystyle\sum_{i=1}^{a} G_i \right)^2}{a} \right]}{\displaystyle\sum_{j=1}^{b} B_j - \frac{\displaystyle\sum_{j=1}^{b} B_j^2}{a}}.
\tag{12.60}
$$

---

*If mean ranks are used,

$$
\mathrm{SE} = \sqrt{ \left(\dfrac{ \dfrac{a(a+1)}{b^2} - \dfrac{\sum t}{b^3(a-1)} }{12} \right) \left(\sum_i c_i^2 \right) }.
\tag{12.58}
$$

[†]William G[emmell] Cochran, (1909–1980), born in Scotland, and influential in the United States after some early important work in England (Dempster, 1983; Watson 1982).

EXAMPLE 12.8   Cochran's Q test.

H_0:   The proportion of humans attacked by mosquitoes is the same for all five clothing types.

H_A: The proportion of humans attacked by mosquitoes is not the same for all five clothing types.

$\alpha = 0.05$

A person attacked is scored as a "1"; a person not attacked is scored as a "0."

		Clothing Type				
Person (block)	Light, loose	Light, tight	Dark, long	Dark, short	None	Totals (B_j)
1	0	0	0	1	0	1
2*	1	1	1	1	1	*
3	0	0	0	1	1	2
4	1	1	0	1	0	3
5	0	1	1	1	1	4
6	0	1	0	0	1	2
7	0	0	1	1	1	3
8	0	0	1	1	0	2
Totals* (G_i)	1	3	3	6	4	$\sum_{i=1}^{a} G_i = \sum_{j=1}^{b} B_j = 17$

$a = 5;$ $b = 7^*$

$$Q = \frac{(a-1)\left[\sum_{i=1}^{a} G_i^2 - \dfrac{\left(\sum_{i=1}^{a} G_i\right)^2}{a}\right]}{\sum_{j=1}^{b} B_j - \dfrac{\sum_{j=1}^{b} B_j^2}{a}} = \frac{(5-1)\left[1+9+9+36+16 - \dfrac{17^2}{5}\right]}{17 - \dfrac{(1+4+9+16+4+9+4)}{5}}$$

$$= \frac{52.8}{7.6} = 6.947$$

$\nu = a - 1 = 4$

$\chi^2_{0.05,4} = 9.488$

Therefore, do not reject H_0.

$$0.10 < P < 0.25 \quad [P = 0.14]$$

*The data for block 2 are deleted from the analysis, because 1's occur for all clothing. (See test discussion in Section 12.10.)

Note, as shown in Example 12.8, that $\sum B = \sum G$, which is the total number of 1's in the set of data. This test statistic, Q, is distributed approximately as chi-square, with $a-1$ degree of freedom. Tate and Brown (1970) explain that the value of Q is unaffected by having blocks containing either all 0's or all 1's. Thus, any such block may be disregarded in the calculations. They further point out that the approximation of Q to χ^2 is a satisfactory one only if the number of data is large. These authors suggest as a rule of thumb that a should be at least 4 and ba should be at least 24, where b is the number of blocks remaining after all those containing either all 0's or all 1's are disregarded. For sets of data smaller than these suggestions allow, the analysis may proceed but with caution exercised if Q is near a borderline of significance. In these cases it would be better to use the tables of Tate and Brown (1964) or Patil (1975).

If $a = 2$, then Cochran's test is identical to McNemar's test (Section 9.7), except that the latter employs a correction for continuity.

12.11 Multiple Comparisons with Dichotomous Randomized Block or Repeated-Measures Data

Marascuilo and McSweeney (1967) present a multiple comparison procedure that may be used for multiple contrasts as well as for pairwise comparisons for data subjected to the Cochran Q test of Section 12.10. It may be performed using group means, $\bar{R}_i = G_i/b$.

For pairwise comparisons, the test statistic is

$$S = \frac{|\bar{R}_B - \bar{R}_A|}{\text{SE}} \tag{12.61}$$

(which parallels Equation 11.15), where

$$\text{SE} = \sqrt{2 \left(\frac{a \sum\limits_{j} B_j - \sum\limits_{j} B_j^2}{ab^2(a-1)} \right)}. \tag{12.62}$$

For multiple contrasts, the test statistic is that of Equation 11.18, where $\bar{R}_i$ replaces $\bar{X}_i$ and

$$\text{SE} = \sqrt{\left(\frac{a \sum\limits_{j} B_j - \sum\limits_{j} B_j^2}{ab^2(a-1)} \right) \sum\limits_{i} c_i^2} \tag{12.63}$$

The critical value for such multiple comparisons is $S_\alpha = \sqrt{\chi^2_{\alpha, a-1}}$.

12.12 Introduction to Analysis of Covariance

Each of the two factors in a two-way ANOVA generally consists of levels that are nominal-scale categories. In Example 10.1, for instance, the variable of interest was the body weight of pigs, and the factor tested was diet. In a two-factor ANOVA, we might ask about the effect of diet and also introduce the sex of the animal (or the breed) as a second factor, with the levels of sex (or breed) being on a nominal scale.

In the experiment of Example 10.1, we would attempt to employ animals of the same age (and weight), so differences in the measured variable could be attributed to the effect of the diets. However, if the beginning ages (or weights) were markedly not alike, then we might wish to introduce age (or weight) as a second factor. The relationship between ending weight and age (or ending weight and beginning weight) may be thought of as a regression (see Chapter 17), while the relationship between ending weight and diet is a one-way analysis variance (Chapter 10). The concepts of these two kinds of analyses, and their statistical assumptions, are combined in what is known as *analysis of covariance* (abbreviated ANCOVA),* and the factor that acts as an independent variable in regression is called a *concomitant variable.*

This is a large area of statistical methodology beyond the scope of this book but found in many references, including several dealing with experimental design (e.g., Hicks, 1982: 318–328; Huitema, 1980; Keppel, 1991: Chapter 14; Kirk, 1982: Chapter 14; Maxwell and Delaney, 1990: Chapter 9; Ostle and Mensing, 1975: Chapter 13; Pedhazur, 1982: Chapter 13; Scheffé, 1959: Chapter 6; Tabachnik and Fidell, 1996: Chapter 3; Wildt and Ahtola, 1978; Winer, Brown, and Michels, 1991: Chapter 10).

EXERCISES

12.1 A study is made of amino acids in the hemolymph of millipedes. For a sample of four males and four females of each of three species, the following concentrations of the amino acid, alanine (in mg/100 ml), are determined:

	Species 1	Species 2	Species 3
Male	21.5	14.5	16.0
	19.6	17.4	20.3
	20.9	15.0	18.5
	22.8	17.8	19.3
Female	14.8	12.1	14.4
	15.6	11.4	14.7
	13.5	12.7	13.8
	16.4	14.5	12.0

   (a) Test the hypothesis that there is no difference in mean hemolymph alanine concentration among the three species.
   (b) Test the hypothesis that there is no difference between males and females in mean hemolymph alanine concentration.
   (c) Test the hypothesis that there is no interaction between sex and species in the mean concentration of alanine in hemolymph.

---

*This statistical technique was introduced by R. A. Fisher (e.g., 1932: 249–262).

(d) Prepare a graph of the row, column, and cell means, as done in Fig. 12.1 and interpret it in terms of the results of the above hypothesis tests.

(e) If the null hypothesis of part a, above, is rejected, then perform a Tukey test to assess the mean differences among the species.

12.2 Six greenhouse benches were set up as blocks. Within each block one of each of four varieties of house plants was planted. The plant heights (in centimeters) obtained are tabulated as follows. Test the hypothesis that all four varieties of plants reach the same maximum height.

Block	Variety 1	Variety 2	Variety 3	Variety 4
1	19.8	21.9	16.4	14.7
2	16.7	19.8	15.4	13.5
3	17.7	21.0	14.8	12.8
4	18.2	21.4	15.6	13.7
5	20.3	22.1	16.4	14.6
6	15.5	20.8	14.6	12.9

12.3 Consider the data of Exercise 12.2. Nonparametrically test the hypothesis that all four varieties of plants reach the same maximum height.

12.4 A textbook distributor wishes to assess potential acceptance of four general biology textbooks. He asks fifteen biology professors to examine the books and to respond as to which ones they would seriously consider for their courses. In the table, a positive response is recorded as 1 and a negative response as a 0. Test the hypothesis that there is no difference in potential acceptance among the four textbooks.

Professor	Textbook 1	Textbook 2	Textbook 3	Textbook 4
1	1	1	0	0
2	1	1	0	1
3	1	0	0	0
4	1	1	1	1
5	1	1	0	1
6	0	1	0	0
7	0	1	1	0
8	1	1	1	0
9	0	0	1	0
10	1	0	1	0
11	0	0	0	0
12	1	1	0	1
13	1	0	0	1
14	0	1	1	0
15	1	1	0	0

13

DATA TRANSFORMATIONS

For the valid application of parametric analyses of variance and related parametric procedures, such as those using Student's t, certain basic assumptions must be met. First, we must be able to assume that the data for each group were obtained randomly from a normal population. Second, we must assume that the sampled populations all have equal variances (in which case it can be said that the variances are *homoscedastic*, the opposite situation being referred to as *heteroscedastic*). Third, the effects of the factor levels must be assumed to be *additive*.*

The requirements of normality and homoscedasticity have been stated previously, but the assumption of additivity (sometimes referred to as linearity) needs elaboration. Consider the two-way analysis of variance data in Example 13.1A. In this example, the effect of factor A is said to be additive, for each datum in level 2 differs from the corresponding datum in level 1 by the addition of the same amount (10 g). Similarly, each datum in level 3 differs from its corresponding level 2 datum by a constant value (5 g). Examining the effect of the two levels of factor B, we also observe additivity (by a constant value of 10 g). What we are assuming is that in the population, on the average, the effects of the factor levels are additive. Occasionally, however, the effect of a factor may not be additive; multiplicative effects are such an instance. In Example 13.1B, we can see the factor A has an effect such that each datum in level 3 differs from its corresponding member in level 2 by a multiplication factor of 2, and each member of level 2 differs from that in level 1 by a factor of 3. Examining the two levels of factor B, we can see that the data are affected by a multiplication factor of 2. The importance of Example 13.1C will be discussed shortly.

---

*The term "additivity" was introduced in this context by C. Eisenhart in 1947 (David, 1995).

EXAMPLE 13.1   **Additive and multiplicative effects.**

A. A hypothetical two-way analysis of variance design, where the effects of the factors are additive. (Data are in grams.)

		Factor A	
Factor B	*Level 1*	*Level 2*	*Level 3*
Level 1	10	20	25
Level 2	20	30	35

B. A hypothetical two-way analysis of variance design, where the effects of the factors are multiplicative. (Data are in grams.)

		Factor A	
Factor B	*Level 1*	*Level 2*	*Level 3*
Level 1	10	30	60
Level 2	20	60	120

C. The two-way analysis of variance design of Example 13.1B, showing the logarithms (rounded to two decimal places) of the data.

		Factor A	
Factor B	*Level 1*	*Level 2*	*Level 3*
Level 1	1.00	1.48	1.78
Level 2	1.30	1.78	2.08

Experience has shown that analyses of variance and t tests are usually robust enough to perform well even if the data deviate somewhat from the requirements of normality and homoscedasticity. (See Section 10.1.) But severe deviations, especially if the range is great, can lead to spurious conclusions.

In cases of heteroscedastic data, the inequality of variances often may be "corrected," and this correction will usually result simultaneously in data that do not deviate intolerably from the normality and additivity assumptions.

Also, there are types of data for which it is known on theoretical grounds that the population sampled is not normal. Such a problem can often be "corrected," and such correction procedures generally will result in data that have acceptable homoscedasticity and additivity characteristics.

Situations of nonadditivity exist, as Example 13.1B demonstrates, and they may often be "corrected," the resultant data generally also being acceptably homoscedastic and normal.

These "corrections" for heteroscedasticity, nonnormality, and nonadditivity are frequently possible by means of changing, or transforming, the data from their original form (X values) to a different form (let us call them X' values). Many statistical researchers have addressed themselves to general questions of data transformations (e.g., Bartlett, 1947; Box and Cox, 1964; Kendall and Stuart, 1966: 87–96), and an extraordinarily thorough monograph on transformation methodology has been prepared by Thöni (1967). The most commonly employed transformations are described in the following sections. The transformation of data used in regression analysis will be discussed in a later chapter (Section 17.10).

13.1 THE LOGARITHMIC TRANSFORMATION

If the factor effects in an analysis of variance are, in fact, multiplicative rather than additive, then the logarithms of the data will exhibit additivity (as Example 13.1C shows). Instead of the transformation, $X' = \log X$, however,

$$X' = \log(X + 1) \tag{13.1}$$

is preferred on theoretical grounds and is especially preferable when some of the observed values are small numbers (particularly zero) (Bartlett, 1947). Logarithms in base 10 are generally utilized, but any base would be satisfactory.

A second instance when the logarithmic transformation is applicable is when there is heteroscedasticity and the standard deviations are proportional to the means (i.e., there is a constant coefficient of variation). Such a case is shown in Example 13.2. This transformation may also convert a positively skewed distribution into a symmetrical one.

Using the transformed data, we can now proceed with parametric testing. In Example 13.2, we calculate a confidence interval for the mean of population 1. By finding the antilogarithm and subtracting 1, we can express the confidence limits in the units of the original data. But the confidence limits in the original units will not be symmetrical about the mean.[*]

If the distribution of X' is normal, the distribution of X is said to be *lognormal*.[†]

13.2 THE SQUARE ROOT TRANSFORMATION

The square root transformation is applicable when the group variances are proportional to the means; that is, when the variances increase as the means increase. This most often occurs in biological data when samples are taken from a Poisson distribution (i.e., when the data consist of counts of randomly occurring objects or events; see Chapter 25 for discussion of the Poisson distribution). Transforming such data by utilizing their square roots results in a sample whose underlying distribution is normal. However, Bartlett (1936) proposed that

$$X' = \sqrt{X + 0.5} \tag{13.2}$$

---

[*]Thöni (1967: 16) has shown that an unbiased estimate of μ would be obtained by adding $(1 - 1/n)s^2$ to the $\bar{X}$ derived by untransforming $\bar{X}'$, where s^2 is the variance of the transformed data.

[†]The term "lognormal" was introduced by J. H. Gaddam in 1945 (David, 1995).

EXAMPLE 13.2   Data in which there is heterogeneity of variance and the standard deviations are proportional to the means (i.e., the coefficients of variation are the same). Primes identify statistics obtained using the transformed data (e.g., $\bar{X}'$, s', L').

The original data:

Group 1	Group 2
3.1	7.6
2.9	6.4
3.3	7.5
3.6	6.9
3.5	6.3

$$\bar{X}_1 = 3.28 \text{ cm} \qquad \bar{X}_2 = 6.94 \text{ cm}$$
$$s_1^2 = 0.0820 \text{ cm}^2 \qquad s_2^2 = 0.3630 \text{ cm}^2$$
$$s_1 = 0.29 \text{ cm} \qquad s_2 = 0.60 \text{ cm}$$
$$V_1 = 0.09 \qquad V_2 = 0.09$$

The logarithmically transformed data, using Equation (13.1):

Group 1	Group 2
0.61278	0.93450
0.59106	0.86923
0.63347	0.92942
0.66276	0.89763
0.65321	0.86332

$$\bar{X}'_1 = 0.63066 \qquad \bar{X}'_2 = 0.89882$$
$$(s_1^2)' = 0.0008586657 \qquad (s_2^2)' = 0.0010866641$$
$$s'_1 = 0.02930 \qquad s'_2 = 0.03296$$
$$V'_1 = 0.04646 \qquad V'_2 = 0.03667$$
$$s'_{\bar{X}_1} = 0.01310 \qquad s'_{\bar{X}_2} = 0.01474$$

Calculating confidence limits:

95% confidence interval for $\mu'_1 = \bar{X}'_1 \pm (t_{0.05(2),4})(0.01310)$

$$= 0.63066 \pm (2.776)(0.01310)$$

$$= 0.63066 \pm 0.03637$$

$L'_1 = 0.59429$ and $L'_2 = 0.66703$

95% confidence limits for μ_1, in the original units:

$$L_1 = \text{antilog } 0.59429 - 1 = 3.93 - 1 = 2.93 \text{ cm}$$
$$L_2 = \text{antilog } 0.66703 - 1 = 4.65 - 1 = 3.65 \text{ cm}$$

is preferable to $X' = \sqrt{X}$, especially when there are very small data and/or when some of the observations are zero (see Example 13.3). Actually,

$$X' = \sqrt{X + \frac{3}{8}} \tag{13.3}$$

EXAMPLE 13.3   The square root transformation for Poisson data.

Original data:

	Group 1	Group 2	Group 3	Group 4
	2	6	9	2
	0	4	5	4
	2	8	6	1
	3	2	5	0
	0	4	11	2
$\bar{X}_1$	1.4	4.8	7.2	1.8
s_i^2	1.8	5.2	7.2	2.2

Transformed data; by Equation 13.2:

	Group 1	Group 2	Group 3	Group 4
	1.581	2.550	3.082	1.581
	0.707	2.121	2.345	2.121
	1.581	2.915	2.550	1.225
	1.871	1.581	2.345	0.707
	0.707	2.121	3.391	1.581
$\bar{X}_i'$	1.289	2.258	2.743	1.443
$(s_i^2)'$	0.297	0.253	0.222	0.272
$s_{\bar{X}_i}'$	0.244	0.225	0.211	0.233
$(L_1')_i$	0.612	1.633	2.157	0.796
$(L_2')_i$	1.966	2.883	3.329	2.090

On transforming back to original units [e.g., $\bar{X} = (\bar{X}')^2 - 0.5$]:

	Group 1	Group 2	Group 3	Group 4
$\bar{X}_i$	1.2	4.6	7.0	1.6
$(L_1)_i$	−0.1	2.2	4.2	0.1
$(L_2)_i$	3.4	7.8	10.6	3.9

has even better variance-stabilizing qualities than Equation 13.2 (Anscombe, 1948; Kihlberg, Herson, and Schutz, 1972), and Freeman and Tukey (1950) show

$$X' = \sqrt{X} + \sqrt{X + 1} \tag{13.4}$$

to yield similar results but to be preferable for $X \leq 2$.

Equation 13.2 is most commonly employed. Statistical computation may then be performed on the transformed data. The mean can be expressed in terms of the original data by squaring it and then subtracting 0.5, although the resultant statistic is slightly biased.* Budescu and Appelbaum (1981) examined ANOVA for Poisson data

---

*Also, an antilogarithmic transformation to obtain $\bar{X}$ in terms of the original units is known to result in a somewhat biased estimator of μ, the estimator being more unbiased for larger variances of $\bar{X}'$ values.

and concluded that data transformation is not desirable unless the largest variances are found in the largest samples and the largest sample is more than five times the size of the smallest.

13.3 THE ARCSINE TRANSFORMATION

It is known from statistical theory that percentages from 0 to 100% or proportions from 0 to 1, form a binomial, rather than a normal, distribution, the deviation from normality being great for small or large percentages (0 to 30% and 70 to 100%).* If the square root of each proportion, p, in a binomial distribution is transformed to its arcsine (i.e., the angle whose sine is $\sqrt{p}$), then the resultant data will have an underlying distribution that is nearly normal. This transformation,

$$p' = \arcsin \sqrt{p},\qquad(13.5)$$

is performed easily with the aid of Appendix Table B.24. For proportions of 0 to 1.00 (i.e., percentages of 0 to 100%), the transformed values will range between 0 and 90 degrees (although some authors' tables present the transformation in terms of radians[†]).[‡] As a result of empirical studies, it is recommended that data transformation is not warranted for analysis of variance with binomial data unless the largest sample size is more than five times greater than the smallest and the smaller variances are associated with the smaller samples.

The arcsine transformation ("arcsine" is abbreviated "arcsin") frequently is referred to as the "angular transformation," and "inverse sine" or "$\sin^{-1}$" is sometimes written to denote "arcsine."[§]

Example 13.4 demonstrates calculations using data submitted to the arcsine transformation. Transformed values (such as means or confidence limits) may be transformed back to proportions, as

$$p = (\sin p')^2;\qquad(13.6)$$

and Appendix Table B.25 is useful for this purpose. But confidence limits generally will not be symmetrical about the mean when expressed in proportions.[∥]

This transformation is not as good at the extreme ends of the range of possible values (i.e., near 0 and 100%) as it is elsewhere. If, rather than simply having percent-

---

*The symbol for percent, "%," appeared around 1650 (Cajori, 1928:312).

[†]A radian is $180°/\pi = 57.29577951308232\ldots$ degrees.

[‡]The arcsine transformation is applicable only if the data came from a distribution of data that can lie between 0 and 100% (e.g., not if data are percent increases, which can be greater than 100%).

[§]The arcsine of a number is the angle whose sine is that number. (See Section 26.3 for a description of the sine and other trigonometric functions.) The term was initiated in the latter part of the eighteenth century and the abbreviation, "$\sin^{-1}$," was introduced in 1813 by the English astronomer, Sir John Frederick William Herschel (1792–1871) (Cajori, 1929:175–176).

[∥]The mean, $\bar{p}$, that is obtained from $\bar{p}'$ by consulting Appendix Table B.25 is, however, slightly biased. Quenouille (1950) suggests correcting for the bias by adding to $\bar{p}$ the quantity $0.5\,\cos(2\bar{p}')(1 - e^{-2s^2})$, where s^2 is the variance of the p' values.

EXAMPLE 13.4   The arcsine transformation for percentage data.

Original data (p):

Group 1 (%)	Group 2 (%)
84.2	92.3
88.9	95.1
89.2	90.3
83.4	88.6
80.1	92.6
81.3	96.0
85.8	93.7

$$\bar{p} = 84.7\% \qquad \bar{p}_2 = 92.7\%$$
$$s_1^2 = 12.29(\%)^2 \qquad s_2^2 = 6.73(\%)^2$$
$$s_1 = 3.5\% \qquad s_2 = 2.6\%$$

Transformed data (by using Equation 13.5 or Appendix Table B.24) (p'):

Group 1	Group 2
66.58	73.89
70.54	77.21
70.81	71.85
65.96	70.27
63.51	74.21
64.38	78.46
67.86	75.46

$$\bar{p}'_1 = 67.09 \qquad \bar{p}'_2 = 74.48$$
$$(s_1^2)' = 8.0052 \qquad (s_2^2)' = 8.2193$$
$$s'_1 = 2.83 \qquad s'_2 = 2.87$$
$$s'_{\bar{X}_1} = 1.07 \qquad s'_{\bar{X}_2} = 1.08$$

Calculating confidence limits:

95% confidence interval for μ'_1 : $\bar{p}'_1 \pm (t_{0.05(2),6})(1.07) = 67.09 \pm 2.62$

$$L'_1 = 64.47 \text{ and } L'_2 = 69.71$$

By using Appendix Table B.25 to transform backward from L'_1, L'_2, and p'_1:

95% confidence limits for μ_1 : $L_1 = 81.5\%$ and $L_2 = 88.0\%$.

$$\bar{p}_1 = 84.9\%$$

---

ages as one's data, one knows the actual proportions, X/n, then the arcsine transformation is improved by replacing $0/n$ with $1/4n$ and n/n with $1 - 1/4n$ (Bartlett, 1937a). Anscombe (1948) proposed an even better transformation:

$$p' = \arcsin \sqrt{\frac{X + \dfrac{3}{8}}{n + \dfrac{3}{4}}}. \tag{13.7}$$

And a slight modification of the Freeman and Tukey (1950) transformation, namely,

$$p' = \frac{1}{2}\left[\arcsin\sqrt{\frac{X}{n+1}} + \arcsin\sqrt{\frac{X+1}{n+1}} \right], \tag{13.8}$$

yields very similar results, except for small and large proportions where it appears to be preferable. Determination of transformed proportions, p', by either Equation 13.7 or 13.8 facilitated by using Appendix Table B.24.

13.4 OTHER TRANSFORMATIONS

The logarithmic, arcsine, and square root transformation are those most commonly required to handle non-normal, heteroscedastic, or nonadditive data. Other transformations are only rarely called for.

If the standard deviations of groups of data are proportional to the square of the means of the groups, then the *reciprocal transformation*,

$$X' = \frac{1}{X}, \tag{13.9}$$

may be employed. (If counts are being transformed, then

$$X' = \frac{1}{X+1} \tag{13.10}$$

may be used to allow for observations of zero.) See Thöni (1967: 32) for further discussion of the use of this transformation.

If the standard deviations decrease as the group means increase, and/or if the distribution is skewed to the left, then

$$X' = X^2 \tag{13.11}$$

might prove useful.

If the data come from a population with what is termed a "negative binomial distribution," then the use of inverse hyperbolic sines may be called for (see Anscombe, 1948; Bartlett, 1947; Beall, 1940, 1942; Thöni, 1967: 20–24).

Thöni (1967) mentions other, infrequently employed, transformations.

EXERCISES

13.1 Perform the logarithmic transformation on the following data (using Equation 13.1) and calculate the 95% confidence interval for μ. Express the confidence limits in terms of the original units (i.e., ml). The data are 3.67, 4.01, 3.85, 3.92, 3.71, 3.88, 3.74, and 3.82 ml.

13.2 Transform the following proportions by the arcsine transformation (using Appendix Table B.24) and calculate the 95% confidence interval for μ. Express the confidence limits in terms of proportions (using Appendix Table B.25).

$$0.733, 0.804, 0.746, 0.781, 0.772, \text{ and } 0.793$$

13.3 Apply the square root transformation to the following data (using Equation 13.2) and calculate the 95% confidence interval for μ. Transform the confidence limits back to the units of the original data. The data are 4, 6, 3, 8, 10, 3.

14

Multiway Factorial Analysis of Variance

Chapter 12 discussed the analysis of the effects on a variable of two factors acting simultaneously. In such a procedure—a two-way, or two-factor, analysis of variance— one can conclude whether either of the factors has a significant effect on the magnitude of the variable and also whether the interaction of the two factors significantly affects the variable. By expanding the considerations of the two-way analysis of variance, we can assess the effects on a variable of the simultaneous application of three or more factors, this being done by what we may refer to as a multiway analysis of variance.[*]

It is not unreasonable to expect a researcher to perform a one-way or two-way analysis by hand, although computer programs are frequently employed, especially when the experiment consists of a large number of observations. However, it has become uncommon for analyses of variance with more than two factors to be analyzed other than on a computer, owing to considerations of time, ease, and accuracy. Therefore, this chapter will not instruct the reader in the computational mechanics of multiway analyses of variance. Rather, it will presume that established computer programs will be used to perform the necessary mathematical manipulations. We shall consider only the subsequent examination and interpretation of the numerical results of the computer's labor.

This chapter begins with a discussion of analysis of variance with three factors and proceeds to expand this consideration to analyses with more than three factors. We shall also introduce multiway ANOVA with unequal replication, the performance of

---

[*]The concept of the factorial analysis of variance was introduced by the developer of ANOVA, R. A. Fisher (Bartlett, 1965), and Fisher's first use of the term "factorial" was in 1935 (David, 1995).

multiple comparisons, and the computation of confidence intervals for such analyses of variance. Lastly, we shall consider power and sample size determination for multiway analyses of variance.

14.1 THREE-FACTOR ANALYSIS OF VARIANCE

For a particular variable, we may wish to assess the effects of three factors; let us refer to them as factors A, B, and C. For example, we might desire to determine what effect the following three factors have on the rate of oxygen consumption of crabs: species, temperature, and sex. Example 14.1 shows experimental data collected for crabs of both sexes, representing three species, and measured at three temperatures. For each cell (i.e., each combination of species, temperature, and sex) there was an oxygen consumption datum for each of four crabs (i.e., there were four replicates); therefore, seventy-two animals were used in the experiment ($N = 72$).

EXAMPLE 14.1   A three-factor analysis of variance (Model I), where the variable is respiratory rate of crabs (in ml O_2/hr).

1 $\begin{cases} H_0: & \text{Mean respiratory rate is the same in all three crab species (i.e., } \mu_1 = \mu_2 = \mu_3). \\ H_A: & \text{Mean respiratory rate is not the same in all three crab species.} \end{cases}$

2 $\begin{cases} H_0: & \text{Mean respiratory rate is the same at all three experimental temperatures (i.e., } \mu_{low} = \\ & \mu_{med} = \mu_{high}). \\ H_A: & \text{Mean respiratory rate is not the same at all three experimental temperatures.} \end{cases}$

3 $\begin{cases} H_0: & \text{Mean respiratory rate is the same for males and females (i.e., } \mu_\delta = \mu_\female) \\ H_A: & \text{Mean respiratory rate is not the same for males and females (i.e., } \mu_\delta \neq \mu_\female) \end{cases}$

4 $\begin{cases} H_0: & \text{Differences in mean respiratory rate among the three species are independent of (i.e., are} \\ & \text{the same at) the three experimental temperatures; or, differences in mean respiratory rate} \\ & \text{among the three temperatures are independent of (i.e., are the same in) the three species.} \\ & \text{(Testing for } A \times B \text{ interaction.)} \\ H_A: & \text{Differences in mean respiratory rate among the species are not independent of the experi-} \\ & \text{mental temperatures.} \end{cases}$

5 $\begin{cases} H_0: & \text{Differences in mean respiratory rate among the three species are independent of sex (i.e.,} \\ & \text{are the same for both sexes); or, differences in mean respiratory rate between males and} \\ & \text{females are independent of (i.e., are the same in) the three species. (Testing for } A \times C \\ & \text{interaction.)} \\ H_A: & \text{Differences in mean respiratory rate among the species are not independent of sex.} \end{cases}$

6 $\begin{cases} H_0: & \text{Differences in mean respiratory rate among the three experimental temperatures are inde-} \\ & \text{pendent of (i.e., are the same in) the two sexes; or, differences in mean respiration rate} \\ & \text{between the sexes are independent of (i.e., are the same at) the three temperatures. (Testing} \\ & \text{for } B \times C \text{ interaction.)} \\ H_A: & \text{Differences in mean respiratory rate among the three temperatures are not independent of} \\ & \text{sex.} \end{cases}$

7 $\begin{cases} H_0: & \text{Difference in mean respiratory rate among the species (or temperatures, or sexes) are inde-} \\ & \text{pendent of the other two factors. (Testing for } A \times B \times C \text{ interaction.)} \\ H_A: & \text{Differences in mean respiratory rate among the species (or temperature, or sexes) are not} \\ & \text{independent of the other two factors.} \end{cases}$

EXAMPLE 14.1 (continued)

		Species 1			
Low temp.		*Med. temp.*		*High temp.*	
♂	♀	♂	♀	♂	♀
1.9	1.8	2.3	2.4	2.9	3.0
1.8	1.7	2.1	2.7	2.8	3.1
1.6	1.4	2.0	2.4	3.4	3.0
1.4	1.5	2.6	2.6	3.2	2.7

		Species 2			
Low temp.		*Med. temp.*		*High temp.*	
♂	♀	♂	♀	♂	♀
2.1	2.3	2.4	2.0	3.6	3.1
2.0	2.0	2.6	2.3	3.1	3.0
1.8	1.9	2.7	2.1	3.4	2.8
2.2	1.7	2.3	2.4	3.2	3.2

		Species 3			
Low temp.		*Med. temp.*		*High temp.*	
♂	♀	♂	♀	♂	♀
1.1	1.4	2.0	2.4	2.9	3.2
1.2	1.0	2.1	2.6	2.8	2.9
1.0	1.3	1.9	2.3	3.0	2.8
1.4	1.2	2.2	2.2	3.1	2.9

By consulting the computer output of Table 14.1, we can prepare the following summary:

Effects in hypothesis	*Calculated F*	*Critical F**	*Conclusion*	*P*
1. Species (factor A)	24.475	$F_{0.05(1),2,54} \cong 3.18$	Reject H_0	$P \ll 0.0005$
2. Temperature (factor B)	332.024	$F_{0.05(1),2,54} \cong 3.18$	Reject H_0	$P \ll 0.0005$
3. Sex (factor C)	0.239	$F_{0.05(1),1,54} \cong 4.03$	Accept H_0	$P > 0.50$
4. $A \times B$	7.418	$F_{0.05(1),4,54} \cong 2.56$	Reject H_0	$P < 0.0005$
5. $A \times C$	4.986	$F_{0.05(1),2,54} \cong 3.18$	Reject H_0	$0.01 < P < 0.025$
6. $B \times C$	2.360	$F_{0.05(1),2,54} \cong 3.18$	Accept H_0	$0.10 < P < 0.25$
7. $A \times B \times C$	1.485	$F_{0.05(1),4,54} \cong 2.56$	Accept H_0	$0.10 < P < 0.25$

*There are no critical values in Appendix Table B.4 for $\nu_2 = 54$, so the values for the next lower DF, $\nu_2 = 50$, were utilized.

We conclude that the respiratory rates of the three species are not all the same, and that the respiratory rates at the three temperatures are not all the same. Furthermore, it is concluded that the respiratory rate of a species is dependent on temperature and that the respiratory rate of a species is dependent on sex.

TABLE 14.1 Computer Output from a Three-Factor Analysis of Variance
of the Data Presented in Example 14.1

Source of variation	Sum of squares	DF	Mean square
Factor A	1.81750	2	0.90875
Factor B	24.65583	2	12.32791
Factor C	0.00889	1	0.00889
A × B	1.10167	4	0.27542
A × C	0.37028	2	0.18514
B × C	0.17528	2	0.08764
A × B × C	0.22056	4	0.05514
Error	2.00500	54	0.03713

Table 14.1 presents the computer output for the analysis of these experimental
results, such output typically giving the sums of squares, degrees of freedom, and mean
squares pertaining to the hypotheses to be tested. Some computer programs also give the
F values calculated, assuming that the experiment calls for a Model I analysis, which
is the case with most biological data, and some present the probability of each F. The
major work that the computer has performed for us is the calculation of the sums of
squares. We could easily have arrived at the degrees of freedom for each factor as the
number of levels -1 (so, for factor A, DF $= 3 - 1 = 2$; for factor B, DF $= 3 - 1 = 2$;
and for factor C, DF $= 2 - 1 = 1$). The degrees of freedom for each interaction are
$A \times B$ DF $=$ factor A DF $\times$ factor B DF $= 2 \times 2 = 4$; $A \times C$ DF $=$ factor A DF $\times$
factor C DF $= 2 \times 1 = 2$; $B \times C$ DF $=$ factor B DF $\times$ factor C DF $= 2 \times 1 = 2$;
and $A \times B \times C$ DF $=$ factor A DF $\times$ factor B DF $\times$ factor C DF $= 2 \times 2 \times 1 = 4$.
The error DF is, then, the total DF (i.e., $N - 1$) minus all other degrees of freedom.
Each needed mean square is then obtained by dividing the appropriate sum of squares
by its associated DF. As we are dealing with a Model I (fixed-effects model) ANOVA,
the computation of each F value consists of dividing a mean square by the error MS.

We may now consider the testing of the hypotheses stated in Example 14.1. To test
whether oxygen consumption is the same among all three species, we compare the species
F (i.e., 24.475) with the critical value, $F_{0.05(1),2,54} \cong 3.18$; because the former exceeds
the later, the null hypothesis is rejected.* In a similar fashion, we test the hypothesis
concerning each of the other two factors, as well as each of the four hypotheses regarding
the interactions of factors, by comparing the calculated F values with the critical values
from Appendix Table B.4. The procedure and conclusions are given in Example 14.1

Recall from Chapter 12 that the test for a two-way interaction asks whether differences in the variable among levels of one factor are the same at all levels of the second
factor. A test for a three-factor interaction may be thought of as asking if the interaction
between any two of the factors is the same at all levels of the third factor.

It is only for a factorial ANOVA with all factors fixed that we compute all F
values utilizing the error MS. If any of the factors are random effects, then the analysis

---

*As Appendix Table B.4 does not contain critical values of F for $v_2 = 54$, the critical values for the
next lower v_2 (namely, 50) are employed. (Or, the critical value for $v_2 = 54$ could be estimated by harmonic
interpolation, as explained in the introduction to Appendix B, or it could be obtained from an appropriate
computer program.)

becomes more complicated. The proper F calculations for such situations appear in Appendix A.

If there are not equal numbers of replicates in each cell of a factorial analysis of variance design, then the usual ANOVA computations are invalid (see Section 14.5).

A factorial ANOVA experimental design may also include nesting (see Chapter 15 for a discussion of nesting). For example, in Example 14.1 we might have performed two or more respiratory rate determinations on each of the four animals per cell. Some of the available computer programs for factorial analysis of variance also provide for nested (also called hierarchical) experimental designs.

If one or two of the three factors are measured on an interval or ratio scale, then we have an analysis of covariance situation, as described in Section 12.12, and computer programs are available for such analyses. In Example 12.1, for instance, the variable is plasma calcium concentration and two factors are hormone treatment and sex. A third factor might be age, or weight, or hemoglobin concentration, or temperature.

14.2 THE LATIN SQUARE EXPERIMENTAL DESIGN

The randomized block design, described in Section 12.4, is a two-factor analysis of variance in which there is one fixed-effects factor, one random-effects factor, and no replication. The levels of the random-effects factor are termed "blocks," and it is assumed that no variation exists within blocks except that due to differences among levels of the fixed-effects factor.

There are times when two sources of variability may be accounted for by blocking, and such a situation may be described by a Latin square experimental design, shown diagrammatically:

$$\begin{array}{ccc} \text{II} & \text{I} & \text{III} \\ \text{III} & \text{II} & \text{I} \\ \text{I} & \text{III} & \text{II} \end{array}$$

Here, there are not only three horizontal blocks (rows), but also three vertical blocks (columns) of data. There are three levels of the fixed-effects factor (designed I, II, and III), and they are assigned to each row and column at random, except that each level is represented exactly once in each row and each column. In the Latin square design, the number of vertical blocks, the number of horizontal blocks, and the number of levels in the fixed-effects factor must be equal. The design shown is referred to as a 3×3 Latin square and is the smallest possible Latin square experimental design. Twelve different configurations are possible for a 3×3 Latin square analysis, there being 576 possible 4×4 squares, 161,280 5×5 squares, 812,851,200 6×6 squares, and 61,428,210,278,400 7×7 squares; several sets of configurations are listed by Fisher and Yates (1963: 86–89) and Norton (1939), to facilitate setting up Latin square experimental designs.[*]

---

[*]Long before they were employed in analysis of variance, Latin squares were first studied by Swiss mathematician, Leonhard Euler (1707–1783) late in his very productive life (Norton, 1939). He used the french term "quarré latin," and A. Cayley may have been the first to use the English version, in 1890 (David, 1995).

The only hypothesis generally of interest in a Latin square analysis is the one concerning equality among the levels of the fixed-effects factor. We test this hypothesis by considering the experimental design to be a three-factor ANOVA with one fixed and two random factors, with no replication. Appendix A.3 summarizes the procedure used for such a test.

The Latin square analysis of variance is rarely applied with biological data, with the occasional exception of those from certain agricultural field plot studies and other, uncommonly encountered, situations. This experimental design is more fully discussed elsewhere (e.g., Snedecor and Cochran, 1980: Section 14.10; Steel and Torrie, 1980: 221–231; Woolf, 1968: Chapter 10). A good discussion of the handling of missing data in this design is that by Li (1964: 235–237).

14.3 HIGHER-ORDER FACTORIAL ANALYSIS OF VARIANCE

In theory, there is no limit to the number of factors that might be analyzed simultaneously, but the number of possible interactions to be dealt with soon become unwieldy as larger analyses are considered (see Table 14.2), and interpretation becomes very difficult for interactions of more than three or four variables.

TABLE 14.2   Number of Hypotheses Potentially Testable in Factorial Analyses of Variance

	Number of factors			
	2	3	4	5
Main factor	2	3	4	5
2-way interactions	1	3	6	10
3-way interactions		1	4	10
4-way interactions			1	5
5-way interactions				1

Note: The number of mth-order interactions in a k-factor ANOVA is the number of ways k factors can be combined m at a time (see Section 5.3):

$$_kC_m = \frac{k!}{m!(k-m)!}.$$

Once the sums of squares for all factors and interactions have been computed, the degrees of freedom and mean squares are readily determined, as Section 14.1 describes. However, most available computer programs also provide the degrees of freedom and mean squares. If all factors are fixed, then the F required to test each null hypothesis is obtained by dividing the appropriate mean square by the error MS. If, however, any of the factors represent random effects, then the analysis becomes considerably more complex, and in some cases impossible. Appendix A presents the procedures applicable to hypothesis testing in such cases. See Section 14.5 for consideration of analyses with unequal replication.

If any (but not all) of the factors in a multiway ANOVA are measured on an interval or ratio scale, then we have an analysis of covariance situation (see Section 12.12). If all of the factors are on an interval or ratio scale, then a multiple regression (Chapter 20) may be called for.

14.4 BLOCKED AND REPEATED-MEASURES EXPERIMENTAL DESIGNS

We can devise experimental designs with three or more factors where one of the factors is a block (as described at the end of Section 10.1 and in Section 12.4) or is a subject upon which repeated measurements are taken (see the end of Section 10.1 and Section 12.5). In such a situation, the analysis may proceed as a factorial ANOVA with the block or repeated-measures factor considered a random-effects factor, consulting Appendix A for the appropriate F's with which to test the hypotheses of interest.

There are also cases where the same block is applied to some—*but not all*—of the combinations of factors, in which case we have an experimental design that is one of a family known as *split-plot* experimental designs. And, if the same subject is exposed to some—*but not all*—of the combinations of factors, then we have an experimental design that is one of many kinds of *repeated-measures* designs, specifically those repeated-measures experimental designs sometimes known as *mixed within-subjects designs* because there are effects on the variable (e.g., body weight) both among subjects and within subjects. Some relatively common designs of these kinds are described below, and some others are referred to in Appendix A.5. More extensive discussions of these topics are found in texts on experimental design, such as Kirk (1982: Chapter 11) with respect to split-plot designs and Crowder and Hand (1990: Chapter 3), Dunn and Clark (1987: Chapter 11), Girden (1992: Sections 5 and 6), Keppel (1991: Chapter 17), Maxwell and Delaney (1990: Chapter 12), and Winer, Brown, and Michels (1971: Chapter 7) with respect to repeated-measures designs.

One Among-Blocks (or Among-Subjects) Factor and One Within-Blocks (or Within-Subjects) Factor.   Consider an experiment in which we wish to measure animal body weight at three diets ($a = 3$) and four exercise regimens ($b = 4$), where diet and exercise are each fixed-effect factors. Furthermore, consider that there are five different blocks of animals for each diet (for a total of fifteen different blocks), where there are four animals per block, one animal being assigned at random to each of the four exercise regimens.

The same experiment could be performed using $n = 5$ different animals (i.e., "subjects") for each diet (for a total of fifteen experimental animals), where each animal is subjected in random order to each of the four exercise regimens. In the present example, diet would be said to be the "among-subjects factor" and exercise the "within-subjects factor."

The data for the split-plot or repeated-measures experiment just described can be tabulated as is Example 14.2, where X_{ijl} is the body weight in diet i, in exercise j, and in block (or subject) l within diet.

EXAMPLE 14.2   A repeated-measures analysis of variance with one among-subjects factor and one within-subjects factor; there are different subjects at each level of one of the factors. The variable is animal body weight, factor A is diet (with $a = 3$ levels), factor B is exercise regimen (with $b = 4$ levels), and there are $n = 5$ subjects (or blocks) within each dietary treatment.

	Subjects		*Exercises*		
Diets	*(or blocks)*	1	2	3	4
1	1	X_{111}	X_{121}	X_{131}	X_{141}
	2	X_{112}	X_{122}	X_{132}	X_{142}
	3	X_{113}	X_{123}	X_{133}	X_{143}
	4	X_{114}	X_{124}	X_{134}	X_{144}
	5	X_{115}	X_{125}	X_{135}	X_{145}
2	6	X_{211}	X_{221}	X_{231}	X_{241}
	7	X_{212}	X_{222}	X_{232}	X_{242}
	8	X_{213}	X_{223}	X_{233}	X_{243}
	9	X_{214}	X_{224}	X_{234}	X_{244}
	10	X_{215}	X_{225}	X_{235}	X_{245}
3	11	X_{311}	X_{321}	X_{331}	X_{341}
	12	X_{312}	X_{322}	X_{332}	X_{342}
	13	X_{313}	X_{323}	X_{333}	X_{343}
	14	X_{314}	X_{324}	X_{334}	X_{344}
	15	X_{315}	X_{325}	X_{335}	X_{345}

The analysis of variance table would look like this:

Analysis of Variance Table

Source of variation	SS	DF	MS
Total		59	
Subjects (or blocks)		14	
Factor A (Diet)		2	
S/A		12	
Within-subjects		45	
Factor B (Exercise)		3	
$B \times A$		6	
$B \times S/A$		36	

The following hypotheses are tested as indicated:
For factor A:

H_0:   Mean body weight is the same on all three diets.

H_A: Mean body weight is not the same on all three diets.

EXAMPLE 14.2 (continued)

$$F = \frac{\text{diets MS}}{\text{subjects within-diets MS}}, \text{ with DF of 2 and 12.}$$

For factor B:

H_0: Mean body weight is the same on all four exercise regimens.

H_A: Mean body weight is not the same on all four exercise regimens.

$$F = \frac{\text{exercises MS}}{\text{exercises } \times \text{ (subjects within-diets MS)}}, \text{ with DF of 3 and 36.}$$

For $A \times B$ interaction:

H_0: Differences in mean body weight among diets are independent of exercise regimens. (Or, equivalently, differences in mean body weight among exercises are the same at all diets.)

H_A: Differences in mean body weight among diets are not independent of exercise regimens.

$$F = \frac{\text{exercises } \times \text{ diets MS}}{\text{exercises } \times \text{ (subjects within-diets MS)}}, \text{ with DF of 6 and 36.}$$

This analysis of variance is an example where within each of a levels of factor A there are n different subjects (or blocks), with an observation for each subject (or block) at each of b levels of factor B. In performing the ANOVA, the total variability would be divided into two parts: the variability among subjects (or blocks) and the variability within subjects. Then, the variability among subjects (or blocks) would be separated into the variability among levels of factor A and the variability due to subjects (or blocks) within levels of factor A. Also, the variability within subjects would be apportioned to the variability due to factor B, that due to the $A \times B$ interaction, and that due to the interaction of factor B and subjects (or blocks) within factor A. This partitioning of variability is shown in the analysis of variance table of Example 14.2, where factor A is diet, factor B is exercise, and S/A indicates subjects within factor A.

An appropriate computer program can generate the needed sums of squares for this repeated-measures or blocked ANOVA, and the degrees of freedom are:

$$\text{total DF } = anb - 1 = N - 1 \tag{14.1}$$

$$\text{subjects DF } = an - 1 \tag{14.2}$$

$$\text{factor } A \text{ DF } = a - 1 \tag{14.3}$$

$$\text{subjects within factor } A \text{ DF} = a(n - 1)$$

$$\text{or} \tag{14.4}$$

$$\text{subjects within factor } A \text{ DF} = \text{subjects DF } - \text{ factor } A \text{ DF}$$

$$\text{within-subjects DF} = an(b - 1)$$

or

$$\text{within-subjects DF} = \text{total DF} - \text{subjects DF} \tag{14.5}$$

$$\text{factor } B \text{ DF} = b - 1 \tag{14.6}$$

$$B \times A \text{ DF} = (a - 1)(b - 1) \tag{14.7}$$

$$B \times (\text{subjects within } A) \text{ DF} = a(n - 1)(b - 1)$$

or

$$B \times (\text{subjects within } A) \text{ DF} = \text{within-subjects DF}$$
$$- \text{factor } B \text{ DF} - B \times A \text{ DF}. \tag{14.8}$$

The null hypothesis of no difference among levels of factor A would be tested by

$$F = \frac{\text{factor } A \text{ MS}}{\text{subjects within } A \text{ MS}}; \tag{14.9}$$

the null hypothesis of no difference among levels of factor B would be tested by

$$F = \frac{\text{factor } B \text{ MS}}{B \times (\text{subjects within } A \text{ MS})}; \tag{14.10}$$

and the null hypothesis of no interaction between factors A and B would be tested by

$$F = \frac{A \times B \text{ MS}}{B \times (\text{subjects within } A \text{ MS})}. \tag{14.11}$$

It is not typically desired to test hypotheses about differences among subjects, and it is not possible without assumptions that may be unacceptable. In the above computations of F, the mean square (MS) for each factor or interaction is obtained by dividing the sum of squares (SS) for that factor or interaction by its associated degrees of freedom (DF). As always, the degrees of freedom for each calculated F are those accompanying the numerator and the denominator of the computation.

Note that the same hypotheses of Example 14.2 could have been tested with an experiment constructed with multiple subjects or blocks within exercises instead of within diets, as shown in Example 14.3.

In general, the test for factor A (what has been indicated as the factor in which the blocks or subjects are placed) is less powerful than the test for factor B (the factor across whose levels the blocks or subjects run) or than the test for the interaction of factors A and B.

Two Among-Blocks (or Among-Subjects) Factors and One Within-Blocks (or Within-Subjects) Factor. Another type of experimental design is shown in Example 14.4, where, in repeated-measures parlance, there are two among-subjects factors and one within-subjects factor. Here, there are $a = 3$ diets; $b = 4$ exercise regimens; $c = 2$ drug treatments; $n = 3$ animals at each combination of diet and exercise; and, as before, body weight is the variable measured.

EXAMPLE 14.3 A repeated-measures ANOVA with one among-subject factor and one within-subjects factor: factor A is exercise regimen (with $a = 4$ levels), factor B is diet (with $b = 3$ levels), and body weight is the variable measured; there are $n = 5$ subjects (or blocks) within each exercise treatment. This experimental design differs from that in Example 14.2 in having the subjects (or blocks) within levels of exercise instead of within levels of diet.

Exercises	Subjects (or blocks)	Diets 1	2	3
1	1	X_{111}	X_{121}	X_{131}
	2	X_{112}	X_{122}	X_{132}
	3	X_{113}	X_{123}	X_{133}
	4	X_{114}	X_{124}	X_{134}
	5	X_{115}	X_{125}	X_{135}
2	6	X_{211}	X_{221}	X_{231}
	7	X_{212}	X_{222}	X_{232}
	8	X_{213}	X_{223}	X_{233}
	9	X_{214}	X_{224}	X_{234}
	10	X_{215}	X_{225}	X_{235}
3	11	X_{311}	X_{321}	X_{331}
	12	X_{312}	X_{322}	X_{332}
	13	X_{313}	X_{323}	X_{333}
	14	X_{314}	X_{324}	X_{334}
	15	X_{315}	X_{325}	X_{335}
4	16	X_{411}	X_{421}	X_{431}
	17	X_{412}	X_{422}	X_{432}
	18	X_{413}	X_{423}	X_{433}
	19	X_{414}	X_{424}	X_{434}
	20	X_{415}	X_{425}	X_{435}

Analysis of Variance Table

Source of variation	SS	DF	MS
Total		59	
Subjects		19	
Exercises		3	
Subjects within exercises		16	
Within-subjects		40	
Diets		2	
Diets × exercises		6	
Diets × (subjects within exercises)		32	

The total variability is partitioned into variability among subjects and within subjects; the subjects variability is partitioned into that due to factor A, that due to factor B, that due to the $A \times B$ interaction, and that due to subjects within $A \times B$ (abbreviated S/AB); and the within-subjects variability is partitioned into the variability due to these

EXAMPLE 14.4 **A repeated-measures ANOVA with two among-subjects factors and one within-subjects factor. The variable is animal body weight, factor A is diet (with $a = 3$ levels), factor B is exercise regimen (with $b = 4$ levels), factor C is drug treatment (with $c = 2$ levels), and there are $n = 3$ subjects (or blocks) within each of the $ab = 12$ combinations of diet and exercise.**

Diets	Exercises	Subjects (or blocks)	Drugs 1	2
1	1	1	X_{1111}	X_{1121}
		2	X_{1112}	X_{1122}
		3	X_{1113}	X_{1123}
1	2	4	X_{1211}	X_{1221}
		5	X_{1212}	X_{1222}
		6	X_{1213}	X_{1223}
1	3	7	X_{1311}	X_{1321}
		8	X_{1312}	X_{1322}
		9	X_{1313}	X_{1323}
1	4	10	X_{1411}	X_{1421}
		11	X_{1412}	X_{1422}
		12	X_{1413}	X_{1423}
2	1	13	X_{2111}	X_{2121}
		14	X_{2112}	X_{2122}
		15	X_{2113}	X_{2123}
2	2	16	X_{2211}	X_{2221}
		17	X_{2212}	X_{2222}
		18	X_{2213}	X_{2223}
2	3	19	X_{2311}	X_{2321}
		20	X_{2312}	X_{2322}
		21	X_{2313}	X_{2323}
2	4	22	X_{2411}	X_{2421}
		23	X_{2412}	X_{2422}
		24	X_{2413}	X_{2423}
3	1	25	X_{3111}	X_{3121}
		26	X_{3112}	X_{3122}
		27	X_{3113}	X_{3123}
3	2	28	X_{3211}	X_{3221}
		29	X_{3212}	X_{3222}
		30	X_{3213}	X_{3223}
3	3	31	X_{3311}	X_{3321}
		32	X_{3312}	X_{3322}
		33	X_{3313}	X_{3323}
3	4	34	X_{3411}	X_{3421}
		35	X_{3412}	X_{3422}
		36	X_{3413}	X_{3423}

EXAMPLE 14.4   (continued)

The analysis of variance table would look like this:

Analysis of Variance Table

Source of variation	SS	DF	MS
Total		71	
Subjects (or blocks)		35	
Factor A (Diets)		2	
Factor B (Exercises)		3	
$A \times B$		6	
S/AB		24	
Within-subjects		36	
Factor C (Drugs)		1	
$C \times A$		2	
$C \times B$		3	
$C \times A \times B$		6	
$C \times S/AB$		24	

five effects: factor C, $C \times A$, $C \times B$, $C \times A \times B$, and $C \times S/AB$. The degrees of freedom for the various sources of variability are as follows:

$$\text{total DF} = abnc - 1 = N - 1 \tag{14.12}$$

$$\text{subjects DF} = abn - 1 \tag{14.13}$$

$$\text{factor } A \text{ DF} = a - 1 \tag{14.14}$$

$$\text{factor } B \text{ DF} = b - 1 \tag{14.15}$$

$$A \times B \text{ DF} = (a - 1)(b - 1) \tag{14.16}$$

$$\text{subjects within } A \times B \text{ DF} = ab(n - 1)^* \tag{14.17}$$

$$\text{within-subjects DF} = abn(c - 1)^\dagger \tag{14.18}$$

$$\text{factor } C \text{ DF} = c - 1 \tag{14.19}$$

$$A \times C \text{ DF} = (a - 1)(c - 1) \tag{14.20}$$

$$B \times C \text{ DF} = (b - 1)(c - 1) \tag{14.21}$$

$$A \times B \times C \text{ DF} = (a - 1)(b - 1)(c - 1) \tag{14.22}$$

$$C \times (\text{subjects within } A \times B) \text{ DF} = ab(n - 1)(c - 1)^\ddagger \tag{14.23}$$

---

*Also obtainable as subjects DF $- A$ DF $- B$ DF $- A \times B$ DF.

†Also obtainable as total DF$-$ subjects DF.

‡Also obtainable as within-subjects DF $- C$ DF $- A \times C$ DF $- B \times C$ DF $- A \times B \times C$ DF.

The null hypothesis of no difference among levels of factor A would be tested by

$$F = \frac{\text{factor } A \text{ MS}}{\text{subjects within } A \times B \text{ MS}}; \qquad (14.24)$$

the null hypothesis of no difference among levels of factor B would be tested by

$$F = \frac{\text{factor } B \text{ MS}}{\text{subjects within } A \times B \text{ MS}}; \qquad (14.25)$$

the null hypothesis of no interaction between factors A and B would be tested by

$$F = \frac{A \times B \text{ MS}}{\text{subjects within } A \times B \text{ MS}}; \qquad (14.26)$$

the null hypothesis of no difference among levels of factor C would be tested by

$$F = \frac{\text{factor } C \text{ MS}}{C \times (\text{subjects within } A \times B) \text{ MS}}; \qquad (14.27)$$

the null hypothesis of no $C \times A$ interaction would be tested by

$$F = \frac{C \times A \text{ MS}}{C \times (\text{subjects within } A \times B) \text{ MS}}; \qquad (14.28)$$

the null hypothesis of no $C \times B$ interaction would be tested by

$$F = \frac{C \times B \text{ MS}}{C \times (\text{subjects within } A \times B) \text{ MS}}; \qquad (14.29)$$

and the null hypothesis of no $C \times A \times B$ interaction would be tested by

$$F = \frac{C \times A \times B \text{ MS}}{C \times (\text{subjects within } A \times B) \text{ MS}}; \qquad (14.30)$$

where the degrees of freedom for each F are those associated with the numerator and the denominator of the F computation.

One Among-Blocks (or Among-Subjects) Factor and Two Within-Blocks (or Within-Subjects) Factors.   In Example 14.5 we see an experiment in which body weight is measured for each of six animals in each of two diets (factor A) where each animal is weighed at each of the eight combinations of four exercise regimens (factor B) and two drug treatments (factor C).

The total variability is partitioned into that among subjects and that within subjects. Then, the subjects variability is partitioned into the variability due to factor A and the variability due to subjects within levels of factor A. The within-subjects variability is partitioned into the variability due to factor B, the variability due to the $A \times B$ interaction, the variability due to $B \times$ (subjects within levels of factor A), the factor C variability, the $A \times C$ variability, the $C \times$ (subjects within levels of factor A) variability, the $B \times C$ variability, and the $A \times B \times C$ variability, and the variability due to $B \times C \times$ (subjects within levels of factor A). For a levels of factor A, b levels of factor B, c levels of factor C, and n subjects within each level of factor A, the degrees of freedom for these

EXAMPLE 14.5   A repeated-measures ANOVA with one among-subjects factor and two within-subjects factors. The variable is animal body weight, factor A is diet (with $a = 3$ levels), factor B is exercise regimen (with $b = 4$ levels), factor C is drug treatment (with $c = 2$ levels), and there are $n = 6$ subjects within each of two dietary treatments.

		Exercises							
		1		2		3		4	
Diets	Subjects (or blocks)	Drug 1	Drug 2	Drug 1	Drug 2	Drug 1	Drug 2	Drug 1	Drug 2
1	1	X_{1111}	X_{1121}	X_{1211}	X_{1221}	X_{1311}	X_{1321}	X_{1411}	X_{1421}
	2	X_{1112}	X_{1122}	X_{1212}	X_{1222}	X_{1312}	X_{1322}	X_{1412}	X_{1422}
	3	X_{1113}	X_{1123}	X_{1213}	X_{1223}	X_{1313}	X_{1323}	X_{1413}	X_{1423}
	4	X_{1114}	X_{1124}	X_{1214}	X_{1224}	X_{1314}	X_{1324}	X_{1414}	X_{1424}
	5	X_{1115}	X_{1125}	X_{1215}	X_{1225}	X_{1315}	X_{1325}	X_{1415}	X_{1425}
	6	X_{1116}	X_{1126}	X_{1216}	X_{1226}	X_{1316}	X_{1326}	X_{1416}	X_{1426}
2	7	X_{2111}	X_{2121}	X_{2211}	X_{2221}	X_{2311}	X_{2321}	X_{2411}	X_{2421}
	8	X_{2112}	X_{2122}	X_{2212}	X_{2222}	X_{2312}	X_{2322}	X_{2412}	X_{2422}
	9	X_{2113}	X_{2123}	X_{2213}	X_{2223}	X_{2313}	X_{2323}	X_{2413}	X_{2423}
	10	X_{2114}	X_{2124}	X_{2214}	X_{2224}	X_{2314}	X_{2324}	X_{2414}	X_{2424}
	11	X_{2115}	X_{2125}	X_{2215}	X_{2225}	X_{2315}	X_{2325}	X_{2415}	X_{2425}
	12	X_{2116}	X_{2126}	X_{2216}	X_{2226}	X_{2316}	X_{2326}	X_{2416}	X_{2426}

The analysis of variance table would look like this:

Analysis of Variance Table

Source of variation	SS	DF	MS
Total		95	
Subjects		11	
Factor A (Diets)		2	
Subjects within diets (S/A)		10	
Within-subjects		84	
Factor B (Exercises)		3	
$B \times A$		6	
$B \times S/A$		30	
Factor C (Drugs)		1	
$C \times A$		2	
$C \times S/A$		10	
$B \times C$		3	
$B \times C \times A$		3	
$B \times C \times S/A$		24	

sources of variability are as follows:

$$\text{total DF} = abcn - 1 = N - 1 \tag{14.31}$$

$$\text{subjects DF} = an - 1 \tag{14.32}$$

$$\text{factor } A \text{ DF} = a - 1 \tag{14.33}$$

$$\text{subjects within } A \text{ DF} = a(n-1)^* \tag{14.34}$$

$$\text{within-subjects DF} = an(bc-1)^\dagger \tag{14.35}$$

$$\text{factor } B \text{ DF} = b-1 \tag{14.36}$$

$$A \times B \text{ DF} = (a-1)(b-1) \tag{14.37}$$

$$B \times \text{(subjects within } A) \text{ DF} = a(b-1)(n-1) \tag{14.38}$$

$$\text{factor } C \text{ DF} = c-1 \tag{14.39}$$

$$A \times C \text{ DF} = (a-1)(c-1) \tag{14.40}$$

$$C \times \text{(subjects within } A) \text{ DF} = a(c-1)(n-1) \tag{14.41}$$

$$B \times C \text{ DF} = (b-1)(c-1) \tag{14.42}$$

$$A \times B \times C \text{ DF} = (a-1)(b-1)(c-1) \tag{14.43}$$

$$B \times C \times \text{(subjects within } A \text{ DF} = a(b-1)(c-1)(n-1)^\ddagger. \tag{14.44}$$

The null hypothesis of no difference among levels of factor A would be tested by

$$F = \frac{\text{factor } A \text{ MS}}{\text{subjects within } A \text{ MS}}; \tag{14.45}$$

the null hypothesis of no difference among levels of factor B would be tested by

$$F = \frac{\text{factor } B \text{ MS}}{B \times \text{(subjects within } A \text{ MS)}}; \tag{14.46}$$

the null hypothesis of no $B \times A$ interaction would be tested by

$$F = \frac{B \times A \text{ MS}}{B \times \text{(subjects within } A \text{ MS)}}; \tag{14.47}$$

the null hypothesis of no difference among the levels of factor C would be tested by

$$F = \frac{\text{factor } C \text{ MS}}{C \times \text{(subjects within } A \text{ MS)}}; \tag{14.48}$$

the null hypothesis of no $C \times A$ interaction would be tested by

$$F = \frac{C \times A \text{ MS}}{C \times \text{(subjects within } A \text{ MS)}}; \tag{14.49}$$

the null hypothesis of no $B \times C$ interaction would be tested by

$$F = \frac{B \times C \text{ MS}}{B \times C \times \text{(subjects within } A \text{ MS)}}; \tag{14.50}$$

---

*Also obtainable as subjects DF $-A$ DF $-$ subjects with A DF.

†Also obtainable as total DF $-$ subjects DF.

‡Also obtainable as within-subjects DF minus the eight other DF's contributing to the within-subjects DF (i.e., the eight immediately above).

and the null hypothesis of no $A \times B \times C$ interaction would be tested by

$$F = \frac{A \times B \times C \text{ MS}}{B \times C \times (\text{subjects within } A \text{ MS})};\tag{14.51}$$

where the degrees of freedom for each F are those associated with the numerator and the denominator and the F computation.

Assumptions. As indicated in Section 12.5, a statistical assumption, known as *sphericity* (or *circularity*) is required to test hypotheses about within-subjects factors in a split-plot or repeated-measures ANOVA; and this assumption is often violated in practice, resulting in an inflated rate of Type I error. That is the case if the analysis proceeds as described to this point, by what is known as *univariate analysis of variance.* Another analytical technique, *multivariate analysis of variance* (MANOVA), is popular with the increased availability of appropriate computer software, and it does not depend upon the sphericity assumption. (See, e.g., Girden, 1992: 22–26; Maxwell and Delaney, 1990: Chapters 13 and 14; O'Brien and Kaiser, 1985; and Stevens, 1996: Chapter 13.) An alternative to MANOVA would be one of the modifications that have been proposed to the univariate ANOVA, some of which are discussed by the authors cited above.

14.5 FACTORIAL ANALYSIS OF VARIANCE WITH UNEQUAL REPLICATION

Although equal replication is always desirable for optimum power and ease of computation in analysis of variance, it is not essential for the performance of the computations in a single-factor ANOVA (Section 10.1). However, all the techniques thus far discussed for ANOVA designs consisting of two or more factors require equal numbers of data per cell (with the exception of the case of proportional replication described in Section 12.2). For example, the data in Example 14.1 are composed of four replicates in each combination of species, temperature, and sex. If there were five or more replicates in a very small number of cells, then it is not highly criticizable to discard (at random within a cell) those few data necessary to arrive at equal numbers of replicate data. However, a more general approach is available, a procedure by which data suffering from replication inequality can be analyzed and interpreted by analysis of variance considerations. The mathematical manipulations involved are sufficiently complex as to be attempted reasonably only by a computer, but it is worthwhile for the biologist to be aware of the fact that programs for such an analysis are available. (These procedures may employ a type of multiple linear regression—see Section 20.11—and may be referred to as "general linear models.") A nice introduction to regression methods for ANOVA experimental designs is given by Glantz and Slinker (1990.)*

If inequality is due to one or a few cells containing one fewer datum than the others, then a factorial analysis of variance may be performed after inserting an estimate

---

*R. A. Fisher described the relationship between regression and analysis of variance in 1923 (Peters, 1987: 136).

of each missing datum. If one datum is missing, an estimate of its value may be found as follows (Shearer, 1973):

$$\hat{X} = \frac{a A_i + b B_j + c C_l + \cdots - (k-1) \sum X}{N + k - 1 - a - b - c - \cdots} \tag{14.52}$$

where $\hat{X}$ is the estimated value for a missing datum in level i of factor A, level j of factor B, level l of factor C, etc.; a, b, c, etc. are the numbers of levels in factors A, B, C, etc., respectively; A, is the sum of all the other data in level i of factor A, B, is the sum of the other data in level j of factor B, etc.; the summation of $a A_i + b B_j + c C_l + \ldots$ is over all factors; k is the number of factors; $\sum X$ is the sum of all the other data in all levels of all factors; and N is the total number of data (including the missing one) in the experimental design. This estimated value may then be inserted with the other data in the analysis of variance computations.

An alternative method of handling experimental designs with one or a few cells containing one fewer datum than the other is much simpler than, but not as desirable as, the procedure above. For each small cell the mean of the cell's observed data can be inserted as an additional datum. The analysis of variance is then performed as usual, but with the total DF and within-cells DF calculated without including the number of such additional data. (That is, the total and within-cells DF are those appropriate to the set of original observations.) The best estimation procedure for missing data is to use the cell means as starting values for employing Equation 14.52 iteratively, just as with Equation 12.22.

14.6 MULTIPLE COMPARISONS AND CONFIDENCE INTERVALS IN MULTIWAY ANALYSIS OF VARIANCE

As we have seen, for each factor a hypothesis may be tested concerning the equality of the level means in that factor. If the null hypothesis of equality is rejected for a fixed-effects factor, then it may be desirable to ascertain between which levels the difference(s) lie(s). This can be done by the multiple comparison procedures prescribed for two-way analyses of variance in Section 12.6. Also mentioned in that section is the calculation of confidence intervals with respect to level means in a two-factor analysis of variance; those considerations also apply to an ANOVA with more than two factors. It should be remembered that the sample size, n, referred to in Chapter 11 is replaced in the present context by the total number of data per level (i.e., the number of data used to calculate level mean); k is replaced by the number of levels of the factor being tested; s^2 will be replaced by the MS appropriate in the denominator of the F ratio used to test for significance of the factor being examined; the degrees of freedom, ν (in q, q', and t) is the DF associated with this MS; and F in Scheffé's test is the same as in the ANOVA.

14.7 POWER AND SAMPLE SIZE IN MULTIWAY ANALYSIS OF VARIANCE

The principles and procedures of Section 12.7 (for two-way ANOVA) may be readily expanded to multifactor analysis of variance. In Section 12.7, k' is the number of levels of the factor under consideration, n' is the total number of data in each level of that factor, s^2 is the appropriate MS in the denominator of the F used for the desired hypothesis test, ν_2 is the DF associated with that MS, and $\nu_1 = k' - 1$. Then, the power of the ANOVA in detecting differences among level means may be estimated using Equations 12.40–12.42.

Equation 12.42, in place of Equation 10.34, may be used in the fashion shown in Example 10.6, to estimate the minimum number of data per level that would be needed to achieve a specified power, given the significance level and detectable difference desired among means.

Equation 12.45 enables us to estimate the smallest difference among level means detectable with the ANOVA. As shown in Section 12.7, we can also use Equation 12.42 to estimate the maximum number of levels testable.

The Mixed Model ANOVA. The above procedures are applicable when all the factors are fixed effects (i.e., we have a Model I ANOVA). They may also be applied to any fixed-effects factor in a mixed-model ANOVA, but in such cases we must modify our method as follows.

Consider the appropriate denominator for the F calculated to test for the significance of the factor in question. (See Appendix A.) Then, substitute this denominator for the within-cells MS (s^2); and substitute this denominator DF for ν_2.

EXERCISES

14.1 Use an appropriate computer program to test for all factor and interaction effects in the following $4 \times 3 \times 2$ Model I analysis of variance, where a_i is a level of factor A, b_i is a level of factor B, and c_i is a level of factor C.

		a_1			a_2			a_3			a_4		
		b_1	b_2	b_3	b_1	b_2	b_3	b_1	b_2	b_3	b_1	b_2	b_3
		4.1	4.6	3.7	4.9	5.2	4.7	5.0	6.1	5.5	3.9	4.4	3.7
c_1		4.3	4.9	3.9	4.6	5.6	4.7	5.4	6.2	5.9	3.3	4.3	3.9
		4.5	4.2	4.1	5.3	5.8	5.0	5.7	6.5	5.6	3.4	4.7	4.0
		3.8	4.5	4.5	5.0	5.4	4.5	5.3	5.7	5.0	3.7	4.1	4.4
		4.8	5.6	5.0	4.9	5.9	5.0	6.0	6.0	6.1	4.1	4.9	4.3
c_2		4.5	5.8	5.2	5.5	5.3	5.4	5.7	6.3	5.3	3.9	4.7	4.1
		5.0	5.4	4.6	5.5	5.5	4.7	5.5	5.7	5.5	4.3	4.9	3.8
		4.6	6.1	4.9	5.3	5.7	5.1	5.7	5.9	5.8	4.0	5.3	4.7

14.2 Use an appropriate computer program to test for the effects of all factors and interactions in the following $2 \times 2 \times 2 \times 3$ Model I analysis of variance design, where a_i is a level of factor A, b_i is a level of factor B, c_i is a level of factor C, and d_i is a factor of level D.

		a_1					a_2		
		b_1		b_2		b_1		b_2	
	c_1	c_2	c_1	c_2	c_1	c_2	c_1	c_2
	12.2	13.4	12.2	13.1	10.9	12.1	10.1	11.2
d_1	12.6	13.1	12.4	13.0	11.3	12.0	10.2	10.8
	12.5	13.5	12.3	13.4	11.2	11.7	9.8	10.7
	11.9	12.8	11.8	12.7	10.6	11.3	10.0	10.9
d_2	11.8	12.6	11.9	12.5	10.4	11.1	9.8	10.6
	12.1	12.4	11.6	12.3	10.3	11.2	9.8	10.7
	12.6	13.0	12.5	13.0	11.1	11.9	10.0	10.9
d_3	12.8	12.9	12.7	12.7	11.1	11.8	10.4	10.5
	12.9	13.1	12.4	13.2	11.4	11.7	10.1	10.8

14.3 Using an appropriate computer program, test for all factor and interaction effects in the following Model I 3×2 analysis of variance with unequal replication.

	a_1		a_2		a_3	
	b_1	b_2	b_1	b_2	b_1	b_2
	34.1	35.6	38.6	40.3	41.0	42.1
	36.9	36.3	39.1	41.3	41.4	42.7
	33.2	34.7	41.3	42.7	43.0	43.1
	35.1	35.8	41.4	41.9	43.4	44.8
				40.8		44.5

14.4 Using an appropriate computer program, test for all factor and interaction effects in the following data from an experiment in which each of four male and four female turtles had their plasma protein measured (in mg/ml) while they were well fed and after ten and twenty days of fasting.

	Subjects	Fed	Fasted 10 days	Fasted 20 days
Male	1	42.8	42.4	38.9
	2	43.1	42.2	40.3
	3	40.4	40.8	37.5
	4	46.6	45.9	42.9
Female	5	42.2	42.4	39.7
	6	38.7	38.1	35.8
	7	35.3	34.3	32.3
	8	40.5	40.1	37.3

14.5 Using an appropriate computer program, test for all factor and interaction effects in the following data from an experiment in which frogs of two species had their oxygen consumption measured (in ml O_2/g/hr) at two temperatures and two exercise levels. There were two frogs of each species at each temperature, and each of the two were measured both at rest and during forced exercise.

Species	Temperature	Subject	Rest	Exercise
Species 1	Low	1	0.107	0.152
		2	0.114	0.163
	High	3	0.133	0.194
		4	0.140	0.198
Species 2	Low	5	0.098	0.144
		6	0.093	0.136
	High	7	0.118	0.171
		8	0.110	0.165
Species 3	Low	9	0.126	0.182
		10	0.138	0.196
	High	11	0.159	0.196
		12	0.166	0.204
Species 4	Low	13	0.154	0.207
		14	0.141	0.191
	High	15	0.184	0.244
		16	0.192	0.232

14.6 Using an appropriate computer program, test for all factor and interaction effects in the following data from a study where the sodium content (μmol/g dry weight) was measured in freshly laid fish eggs that were collected from two ponds for two species and at three seasons and two water depths.

	Block (pond)	Early Shallow	Early Deep	Middle Shallow	Middle Deep	Late Shallow	Late Deep
Species 1	1	254	257	249	246	236	241
	2	261	268	257	263	249	252
	3	248	253	239	242	226	221
Species 2	4	227	226	221	218	219	226
	5	233	239	229	235	222	224
	6	212	220	214	220	203	205

15

NESTED (HIERARCHICAL) ANALYSIS OF VARIANCE

In Chapters 12 and 14 we dealt with analysis of variance experimental designs that the statistician refers to as *crossed*. A crossed experiment is one where all possible combinations of levels of the factors exist; the cells of data are formed by each level of one factor being in combination with each level of every other factor. Thus, Example 12.1 is a two-factor crossed experimental design, for each sex is found in combination with each hormone treatment. In Example 14.1, each sex is found in combination with each experimental temperature and each species.

In some experimental designs, however, we may have some levels of one factor occurring in combination with the levels of one or more other factors, and other distinctly different levels occurring in combination with others. In Example 15.1, where blood cholesterol concentration is the variable, there are two factors: drug type and drug source. Each drug was obtained from two sources, but the two sources are not the same for all the drugs. Thus, the experimental design is not crossed; rather, we say it is *nested* (or *hierarchical*), with one factor, drug source, being nested within another, drug type. The nested factor, as in the present example, is typically random, so the experiment may be considered to be a kind of one-way ANOVA, where the levels of this variable (drug sources) are the samples, and the individual cholesterol determinations are called subsamples.

Sometimes experiments are designed with nesting in order to test hypotheses about the samples. More typical, however, is the inclusion of a random-effects nested factor in order to account for some within-group variability and thus make the hypothesis testing for one or more factors (usually fixed effects) of primary interest.

EXAMPLE 15.1   A nested (hierarchical) analysis of variance.

The variable is blood cholesterol concentration in women (in mg/100 ml of plasma). This variable was measured after the administration of one of three different drugs, each drug having been obtained from two sources.

	Drug 1		Drug 2		Drug 3	
	Source A	*Source Q*	*Source D*	*Source B*	*Source L*	*Source S*
	102	103	108	109	104	105
	104	104	110	108	106	107

$$n_{ij} \qquad 2 \qquad 2 \qquad 2 \qquad 2 \qquad 2 \qquad 2$$

$$\sum_{l=1}^{n_{ij}} X_{ijl} \qquad 206 \qquad 207 \qquad 218 \qquad 217 \qquad 210 \qquad 212$$

$$\frac{\left(\sum_{l=1}^{n_{ij}} X_{ijl}\right)^2}{n_{ij}} \quad 21218.0 \quad 21424.5 \quad 23762.0 \quad 23544.5 \quad 22050.0 \quad 22472.0 \qquad \sum_{i=1}^{a}\sum_{j=1}^{b}\frac{\left(\sum_{l=1}^{n_{ij}} X_{ijl}\right)^2}{n_{ij}}$$
$$= 134471.0$$

$$n_i \qquad\qquad 4 \qquad\qquad 4 \qquad\qquad 4 \qquad N = 12$$

$$\sum_{j=1}^{b}\sum_{l=1}^{n_{ij}} X_{ijl} \qquad 413 \qquad\qquad 435 \qquad\qquad 422 \qquad \sum_{i=1}^{a}\sum_{j=1}^{b}\sum_{l=1}^{n_{ij}} X_{ijl}$$
$$= 1270$$

$$\bar{X}_i \qquad\qquad 103.25 \qquad\qquad 108.75 \qquad\qquad 105.5$$

$$\frac{\left(\sum_{j=1}^{b}\sum_{l=1}^{n_{ij}} X_{ijl}\right)^2}{n_i} \qquad 42642.25 \qquad 47306.25 \qquad 44521.00 \qquad \sum_{i=1}^{a}\frac{\left(\sum_{j=1}^{b}\sum_{l=1}^{n_{ij}} X_{ijl}\right)^2}{n_i}$$
$$= 134469.50$$

$$C = \frac{\left(\sum_{i=1}^{a}\sum_{j=1}^{b}\sum_{l=1}^{n_{ij}} X_{ijl}\right)^2}{N} = \frac{(1270)^2}{12} = 134408.33$$

$$\text{total SS} = \sum_{i=1}^{a}\sum_{j=1}^{b}\sum_{l=1}^{n_{ij}} X_{ijl}^2 - C = 134480.00 - 134408.33 = 71.67$$

$$\text{among all subgroups SS} = \sum_{i=1}^{a}\sum_{j=1}^{b}\frac{\left(\sum_{l=1}^{n_{ij}} X_{ijl}\right)^2}{n_{ij}} - C = 134471.00 - 134408.33 = 62.67$$

error SS = total SS − among all subgroups SS = 71.67 − 62.67 = 9.00

$$\text{groups SS} = \sum_{i=1}^{a}\frac{\left(\sum_{j=1}^{b}\sum_{l=1}^{n_{ij}} X_{ijl}\right)^2}{n_i} - C = 134469.50 - 134408.33 = 61.17$$

EXAMPLE 15.1   (continued)

subgroups SS = among all subgroups SS − groups SS = 62.67 − 61.17 = 1.50

Source of variation	SS	DF	MS
Total	71.67	11	
Among all subgroups	62.67	5	
Groups	61.17	2	30.58
Subgroups	1.50	3	0.50
Error	9.00	6	1.50

H_0:   There is no difference among the drug sources in affecting blood cholesterol concentration.

H_A: There is difference among the drug sources in affecting blood cholesterol concentration.

$$F = \frac{0.50}{1.50} = 0.33. \quad F_{0.05(1),3,6} = 4.76. \quad \text{Do not reject } H_0. \quad P > 0.50 \quad [P = 0.80]$$

H_0:   There is no difference in cholesterol concentrations owing to the three drugs (i.e., $\mu_1 = \mu_2 = \mu_3$).

H_A: There is difference in cholesterol concentrations owing to the three drugs.

$$F = \frac{30.58}{0.50} = 61.16. \quad F_{0.05(1),2,3} = 9.55. \quad \text{Reject } H_0. \quad 0.0025 < P < 0.005 \quad [P = 0.0037]$$

15.1 NESTING WITHIN ONE MAIN FACTOR

In the experimental design such as in Example 15.1, the primary concern is to detect true differences among levels of the fixed-effects factor (drug type). And we can often employ a more powerful test by nesting a random-effects factor that can account for some of the variability within the groups of interest. The partitioning of the variability in a nested ANOVA may be observed in this example.

Calculations in the Nested ANOVA.   In the hierarchical design described, we can uniquely designate each datum by using a triple subscript notation, where X_{ijl} indicates the lth datum in subgroup j of group i. Thus, in Example 15.1, $X_{222} = 108$ mg/100 ml, $X_{311} = 104$ mg/100 ml, etc. For the general case, there are a groups, numbered 1 through a, and b is the number of subgroups in each group. For Example 15.1, there are three levels of factor A (drug type), and b (the number of levels of factor B, i.e., sources for each drug) is 2. The number of data in subgroup j of group i may be denoted by n_{ij}, and the total number of data in group i is n_t. The total number of observations in the entire experiment is $N = \sum_{l=1}^{k} n_i$ (which could also be computed as $N = \sum_{i=1}^{a} \sum_{j=1}^{b} n_{ij}$). The sum of the data in subgroup j of group i is calculated as $\sum_{l=1}^{n_{ij}} X_{ijl}$; the sum of the data in group i is $\sum_{j=1}^{b} \sum_{l=1}^{n_{ij}} X_{ijl}$; and the mean of group i is

$$\bar{X}_i = \frac{\sum\limits_{j=1}^{b} \sum\limits_{l=1}^{n_{ij}} X_{ijl}}{n_i}. \tag{15.1}$$

The grand mean of all the data is

$$\bar{X} = \frac{\sum\limits_{i=1}^{a} \sum\limits_{j=1}^{b} \sum\limits_{l=1}^{n_{ij}} X_{ijl}}{N}. \tag{15.2}$$

The total sum of squares for this ANOVA design considers the deviations of all the X_{ijl} from $\bar{X}$ and is calculated

$$\text{total SS} = \sum\limits_{i=1}^{a} \sum\limits_{j=1}^{b} \sum\limits_{l=1}^{n_{ij}} X_{ijl}^2 - C, \tag{15.3}$$

where the "correction term" is

$$C = \frac{\left(\sum\limits_{i=1}^{a} \sum\limits_{j=1}^{b} \sum\limits_{l=1}^{n_{ij}} \right)^2}{N}, \tag{15.4}$$

and

$$\text{total DF} = N - 1. \tag{15.5}$$

The variability among groups, i.e., the indication of the deviations $X_i - \bar{X}$, is expressed as

$$\text{groups SS} = \frac{\sum\limits_{i=1}^{a} \left(\sum\limits_{j=1}^{b} \sum\limits_{l=1}^{n_{ij}} X_{ijl} \right)^2}{n_i} - C, \tag{15.6}$$

and

$$\text{groups DF} = a - 1 \tag{15.7}$$

There is a total of ab subgroups in the design, and, considering them as if they were groups in a one-way ANOVA, we can calculate a measure of the deviations $\bar{X}_{ij} - \bar{X}$ as

$$\text{among all subgroups SS} = \sum\limits_{i=1}^{a} \sum\limits_{j=1}^{b} \frac{\left(\sum\limits_{l=1}^{n_{ij}} X_{ijl} \right)^2}{n_{ij}} - C, \tag{15.8}$$

and

$$\text{among all subgroups DF} = ab - 1. \tag{15.9}$$

The variability due to the subgrouping within groups causes the deviation of a subgroup mean from its group mean, $X_{ij} - \bar{X}_i$, and the appropriate sum of squares is the "among subgroups within groups" SS, which will be referred to as

$$\text{subgroups SS} = \text{among all subgroups SS} - \text{groups SS};\qquad(15.10)$$

and

$$\text{subgroups DF} = \text{among all subgroups DF} - \text{groups DF} = a(b-1).\qquad(15.11)$$

The within-subgroups, or error, variability measures the deviations $X_{ijl} - \bar{X}_{ij}$; it is essentially the within-cells variability encountered in Chapters 12 and 14. The appropriate sum of squares is obtained by difference:

$$\text{error SS} = \text{total SS} - \text{among all subgroups SS},\qquad(15.12)$$

with

$$\text{error DF} = \text{total DF} - \text{among all subgroups DF} = N - ab.\qquad(15.13)$$

The summary of this hierarchical analysis of variance is presented in Table 15.1. Recall that MS = SS/DF. Some similarities may be noted between Tables 12.2 and 15.1, but in the latter we cannot speak of interaction between the two factors.

TABLE 15.1 Summary of Hierarchical (Nested) Single-Factor Analysis of Variance Calculations

Source of variation	SS	DF
Total $[X_{ijl} - \bar{X}]$	$\sum\limits_{i=1}^{a}\sum\limits_{j=1}^{b}\sum\limits_{l=1}^{n_{ij}} X_{ijl}^2 - C$	$N - 1$
Among all subgroups $[\bar{X}_{ij} - \bar{X}]$	$\sum\limits_{i=1}^{a}\sum\limits_{j=1}^{b} \dfrac{\left(\sum\limits_{l=1}^{n_{ij}} X_{ijl}\right)^2}{n_{ij}} - C$	$ab - 1$
Groups (i.e., among groups) $[\bar{X}_i - \bar{X}]$	$\sum\limits_{i=1}^{a} \dfrac{\left(\sum\limits_{j=1}^{b}\sum\limits_{l=1}^{n_{ij}} X_{ijl}\right)^2}{n_i} - C$	$a - 1$
Subgroups (i.e., among subgroups within groups) $[\bar{X}_{ij} - \bar{X}_i]$	among all subgroups SS $-$ groups SS	$a(b-1)$
Error (i.e., within subgroups) $[X_{ijl} - \bar{X}_{ij}]$	total SS $-$ among all subgroups SS	$N - ab$

Note: $C = \dfrac{\left(\sum\limits_{i=1}^{a}\sum\limits_{j=1}^{b}\sum\limits_{l=1}^{n_{ij}} X_{ijl}\right)^2}{N}$

a = number of groups; b = number of subgroups within each group; n_i = number of data in group i; n_{ij} = number of data in subgroup j for group i; N = total number of data in entire experiment.

Hypothesis Testing in the Hierarchical ANOVA.   For the data in Example 15.1, we can test the null hypothesis that no difference in cholesterol occurs among subgroups (i.e., the source of the drugs has no effect on the concentration of blood cholesterol). We do this by examining

$$F = \frac{\text{subgroups MS}}{\text{error MS}}. \tag{15.14}$$

For our example, this is $F = 0.50/1.50 = 0.33$; since $F_{0.05(1),3,6} = 4.76$, H_0 is not rejected. (The exact probability of an F at least this large if H_0 is true is 0.80.)

Next, we can test the null hypothesis that there is no difference in cholesterol with the administration of the three different drugs. We calculate

$$F = \frac{\text{groups MS}}{\text{subgroups MS}}, \tag{15.15}$$

which in the present example is $F = 30.58/0.50 = 61.16$. As $F_{0.05(1),2,3} = 9.55$, H_0 is rejected. (The exact probability is 0.0037.)

If one does not reject the null hypothesis of no difference among subgroups within groups, then the subgroups MS might be considered to estimate the same population variance as does the error MS. Thus, some statisticians suggest that in such cases we calculate a pooled mean square by pooling the sums of squares and pooling the degrees of freedom for the subgroups and the error variability, for this will theoretically provide the ability to perform a more powerful test for differences among groups. However, there is no widespread agreement on this matter, so it is usually suggested that the biologist be conservative and not engage in such pooling, at least not without consulting a statistician.

If there are unequal numbers of subgroups in each group, then the analysis becomes more complex, and the preceding calculations are not applicable. This situation is generally submitted to computer analysis, often by the procedure referred to in Section 14.5.

A hierarchical experimental design might have each subgroup composed of sub-subgroups, thus involving an additional step in the hierarchy. For instance, for the data of Example 15.1, the different drugs define the groups, the different sources define the subgroups, and if different technicians or different instruments were used to perform the cholesterol analyses within each subgroup, then these technicians or instruments would define the sub-subgroups. Sokal and Rohlf (1995: 288–292) describe the necessary calculations for a design with sub-subgroups, although one generally resorts to computer calculation for hierarchical designs with more than the two steps in the hierarchy discussed in the preceding paragraphs. See Appendix A for assistance in hypothesis testing for such nested designs, and designs where the groups are not fixed effects and/or the subsamples are not random effects.

15.2 Nesting in Factorial Experiments

Experimental designs are encountered where there are two or more crossed factors as well as one or more nested factors. For example, in Example 12.1, the two crossed factors are sex and hormone treatment, and five animals of each sex were given each hormone treatment. In addition, the experimenter might have obtained three blood samples (statistically, subsamples) from each animal so that individual birds would be samples and the triplicate blood collections would be subsamples (The animals represent a nested, rather than a crossed factor, because the same animal is not found at each combination of the other factors.) The analysis of variance table would then look like that in Example 15.2. The computation of sums of squares could be obtained by computer, and the appropriate hypothesis testing will be that indicated in Appendix A.4. Some available computer programs can operate with data where there is not equal replication, while others cannot.

The concept of nesting could be extended by considering that each subsample of blood in Example 15.2 was subjected to two or more (i.e., replicate) chemical analyses. Then we would have chemical analysis nested within animal, and animal nested within the two crossed factors. Brits and Lemmer (1990) discuss nonparametric ANOVA with nesting within a single factor.

EXAMPLE 15.2   An analysis of variance with a random-effects factor (blood collection) nested within the two-factor crossed experimental design of Example 12.1.

For each of the four combinations of sex and hormone treatment ($a = 2$ and $b = 2$), there are five animals ($c = 5$), from each of which three blood collections are taken ($n = 3$). Therefore, the total number of data collected is $N = abcn = 60$.

Source of variation	SS	DF	MS
Total		$N - 1 = 59$	
Cells		$ab - 1 = 3$	
Hormone treatment (factor A)	*	$a - 1 = 1$	†
Sex (factor B)	*	$b - 1 = 1$	†
$A \times B$	*	$(a - 1)(b - 1) = 1$	†
Among all animals		$abc - 1 = 19$	
Cells		$ab - 1 = 3$	
Animals (within cells) (factor C)	*	$ab(c - 1) = 16$	†
Error (within animals)	*	$abc(n - 1) = 40$	†

*These sums of squares can be obtained from an appropriate computer program; the other sums of squares in the table may not be given by such a program, or MS might be given but not SS.

†The mean squares can be obtained from an appropriate computer program. Or, they may be obtained from the sums of squares and degrees of freedom (as MS = SS/DF). The degrees of freedom might appear in the computer output, or they may have to be determined by hand. The appropriate F statistics are those indicated in Appendix A.4.

H_0:   There is no difference in mean blood calcium concentration between males and females.

H_A: There is a difference in mean blood calcium concentration between males and females.

$$F = \frac{\text{factor } A \text{ MS}}{\text{factor } C \text{ MS}} \qquad F_{0.05(1), 1, 16} = 4.49$$

EXAMPLE 15.2   (continued)

H_0:   The mean blood calcium concentration is the same in birds receiving and not receiving the hormone treatment.

H_A: The mean blood calcium concentration is not the same in birds receiving and not receiving the hormone treatment.

$$F = \frac{\text{factor } B \text{ MS}}{\text{factor } C \text{ MS}} \qquad F_{0.05(1),1,16} = 4.49$$

H_0:   There is no interactive effect of sex and hormone treatment on mean blood calcium concentration.

H_A: There is interaction between sex and hormone treatment in affecting mean blood calcium concentration.

$$F = \frac{A \times B \text{ MS}}{\text{factor } C \text{ MS}} \qquad F_{0.05(1),1,16} = 4.49$$

H_0:   There is no difference in blood calcium concentration among animals within combinations of sex and hormone treatment.

H_A: There is difference in blood calcium concentration among animals within combinations of sex and hormone treatment.

$$F = \frac{\text{factor } C \text{ MS}}{\text{error MS}} \qquad F_{0.05(1),16,40} = 1.90$$

15.3 Multiple Comparisons and Confidence Intervals

Whenever a fixed-effects factor is concluded by an ANOVA to have a significant effect on the variable, we may turn to the question of which of the factor's levels are different from which others. If there are only two levels of the factor, then of course we have concluded that their population means are different by the ANOVA. But if there are more than two levels, then a multiple comparison test must be employed.

The multiple comparison procedures usable in nested experimental designs are found in Chapter 11, with slight modifications as those we saw in Sections 12.6 and 14.6. Simply keep the following in mind when employing the tests of Sections 11.1, 11.2, 11.4, and 11.5:

1. k refers to the number of levels being compared. (In Example 15.1, $k = a$, the number of levels in factor A. In Example 15.2, $k = a$ when comparing levels of factor A, and $k = b$ when testing levels of factor B.)

2. The sample size, n, refers to the total number of data from which a level mean is calculated. (In Example 15.1, the sample size $bn = 4$ would be used in place of n. In Example 15.2 we would use $bcn = 30$ to compare level means for factor A and $acn = 30$ for factor B.)

3. The mean square, s^2, refers to the MS in the denominator of the F ratio appropriate to testing the effect in question in the ANOVA. (In Example 15.1, the subgroups [sources] MS would be used. In Example 15.2, the factor C MS would be used.)

4. The degrees of freedom, v, for the critical value of q or q' are the degrees of freedom associated with the mean square above. (In Examples 15.1 and 15.2, these would be 3 and 16, respectively.)

5. The critical value of F in the Scheffé test has the same degrees of freedom as it does in the ANOVA for the factor under consideration. (In Example 15.1, these are 2 and 3. In Example 15.2 they are 1 and 16.)

Once a multiple comparison test has determined where differences lie among level means, we can express a confidence interval for each different mean, as was done in Sections 11.3, 12.6, and 14.6, keeping in mind the sample sizes, mean squares, and degrees of freedom defined above.

15.4 POWER AND SAMPLE SIZE IN NESTED ANALYSIS OF VARIANCE

In Sections 12.7 and 14.7, considerations of power and sample size for factorial analyses of variance were discussed. The same types of procedures may be employed for a fixed-effects factor within which nesting occurs. As previously used, k' is the number of levels of the factor, n' is the total number of data in each level, and $v = k' - 1$. The appropriate mean square, s^2, is that appearing in the denominator of the F ratio used to test that factor in the ANOVA, and v_2 is the degrees of freedom associated with s^2.

Referring back to Section 12.7, the power of the nested ANOVA to detect differences among level means may be estimated using Equations 12.40–12.43. Equation 12.42 may be used to estimate the minimum number of data per level that would be needed to achieve a specified power, and Equation 12.45 allows the estimation of the smallest detectable difference among level means. Section 12.7 also shows how to estimate the maximum number of levels testable.

EXERCISE

15.1 Three water samples were taken from each of three locations. Two determinations of fluoride content were performed on each of the nine samples. Test the hypothesis that there was no difference in mean fluoride content among samples at a location. Then test the hypothesis that there is no difference in mean fluoride concentration among the locations. Data are in milligrams fluoride per liter of water.

Locations	1			2			3		
Samples	1	2	3	1	2	3	1	2	3
	1.1	1.3	1.2	1.3	1.3	1.4	1.8	2.1	2.2
	1.2	1.1	1.0	1.4	1.5	1.2	2.0	2.0	1.9

16

MULTIVARIATE ANALYSIS OF VARIANCE

Chapters 10, 12, 14, and 15 discussed various experimental designs categorized as analysis of variance (ANOVA), wherein a variable was measured at each of several levels of one or more factors and hypothesis testing asked whether the mean of the variable differed among the levels of the factors. These are examples of *univariate analysis of variance*, which examines the effect(s) of one or more factors on only one variable.

There are also experimental designs where more than one variable is measured on each experimental subject. Consider again Example 10.1 in Chapter 10; there, 19 pigs were allocated at random to four experimental groups, and each group was fed a different diet. Thus, diet was the experimental factor, and there were four levels of the factor. In that example, only one variable was measured on each pig: the total body weight. But other measurements might have been taken from each animal: e.g., the blood cholesterol, the blood pressure, the total body fat, or the fat-free body weight. If two or more measurements are taken on each individual in an ANOVA design, we say we have a *multivariate analysis of variance* (abbreviated MANOVA).*

This chapter presents an introduction to MANOVA. More extensive discussions are found in Bray and Maxwell (1985), Hair, et al. (1995: Chapter 5), Hand and Taylor (1987: Chapter 4), Johnson and Wichern (1998: Chapter 6), Sharma (1996: Chapters 11–12), Stevens (1996: Chapters 4–6), and Tabachnick and Fidell (1996: Chapter 9). Other multivariate considerations are found in Chapter 22.

---

* Sometimes the variables are referred to as *dependent variables* and the factors as *independent variables* or *criterion variables*.

16.1 THE MULTIVARIATE NORMAL DISTRIBUTION

Recall that univariate analysis of variance assumes that the sample of data for each group of data came from a population of data that were normally distributed and univariate normal distributions may be shown graphically as in Figs. 6.1 and 6.2. In such two-dimensional figures, the height of the curve (Y) is plotted against the variable (X), and the highest point of the symmetrical curve is at the mean of X.

The simplest multivariate case is where there are two variables (call them X_1 and X_2), and they can be plotted on a graph depicting three dimensions (Y, X_1, and X_2). The two-variable extension of the concept of normality is a curve called the *bivariate normal distribution*, which is shown in Fig. 16.1. The three-dimensional bivariate normal curve rises like a hill above a plain (where the plain is the plane formed by the two variables, X_1 and X_2), and the highest point of the curve is at the means of X_1 and X_2. Authors do not attempt to plot more than three dimensions, which would have to be done to depict multivariate normal distributions with more than two measured variables. Multivariate normality requires, among other characteristics (e.g., Stevens, 1996: 243), that each of the variables is normally distributed. That is, for any given X_1 there is a normal distribution of Y values and, for each value of X_2, there is a normal distribution of Y. As shown in Fig. 6.2a, univariate normal distributions with smaller standard deviations form narrower curves than those with larger standard deviations. Similarly, the hill-shaped bivariate graph of Fig. 16.1 will be narrow when the standard deviations of X_1 and X_2 are small and broad when they are large.

Rather than drawing bivariate normal graphs such as Fig. 16.1, we may prefer to depict these three-dimensional plots using two dimensions, just as mapmakers represent

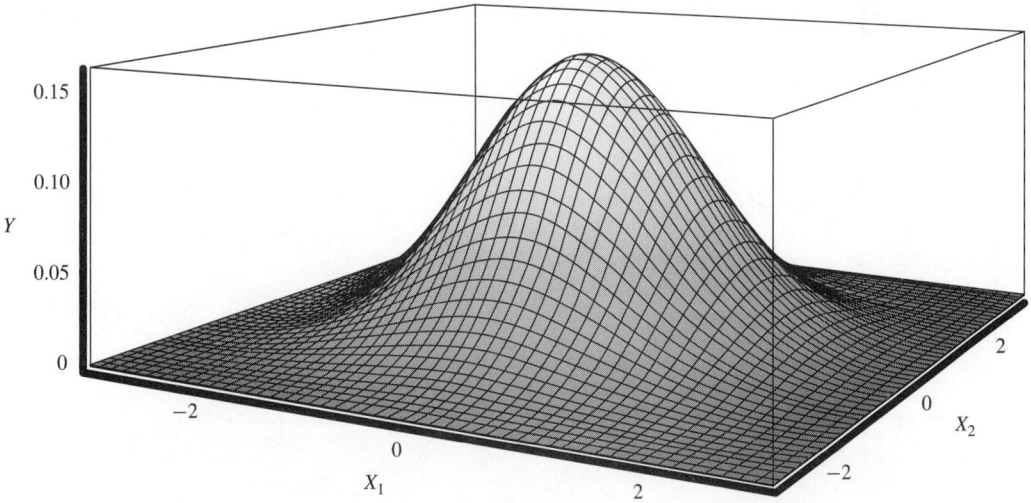

Figure 16.1 A bivariate normal distribution, where X_1 and X_2 have means of 0 and identical standard deviations.

a hill or mountain using contour lines. Fig. 16.2a shows the distribution of Fig. 16.1 with a small plane passing through it parallel to the X_1-X_2 plane; the intersection of the small plane and the distribution delineates a circle (an ellipse if the standard deviations of X_1 and X_2 are not equal). Fig. 16.2b has two small parallel planes passing through the figure, and their intersections create two concentric circles (or ellipses). If Figs. 16.2a and 16.2b are viewed from directly above—looking down, perpendicular to the X_1-X_2 plane—those circles would appear as in Figs. 16.3a and 16.3b, respectively.

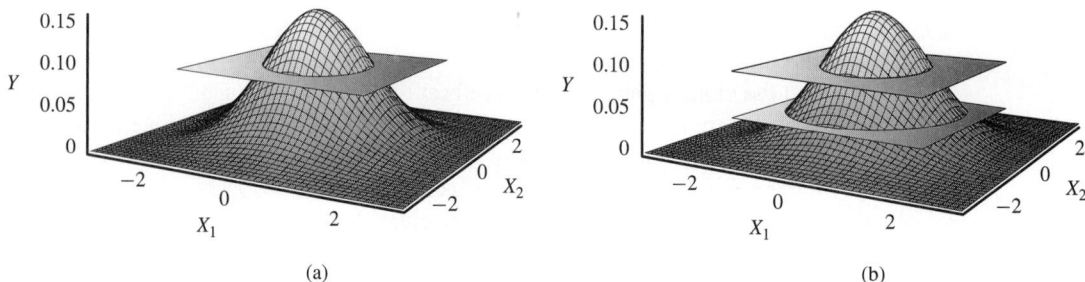

(a)   (b)

Figure 16.2   The bivariate normal distribution of Fig. 16.1, with **(a)** an intersecting plane parallel to the X_1-X_2 plane, and **(b)** two intersection planes parallel to the X_1-X_2 plane.

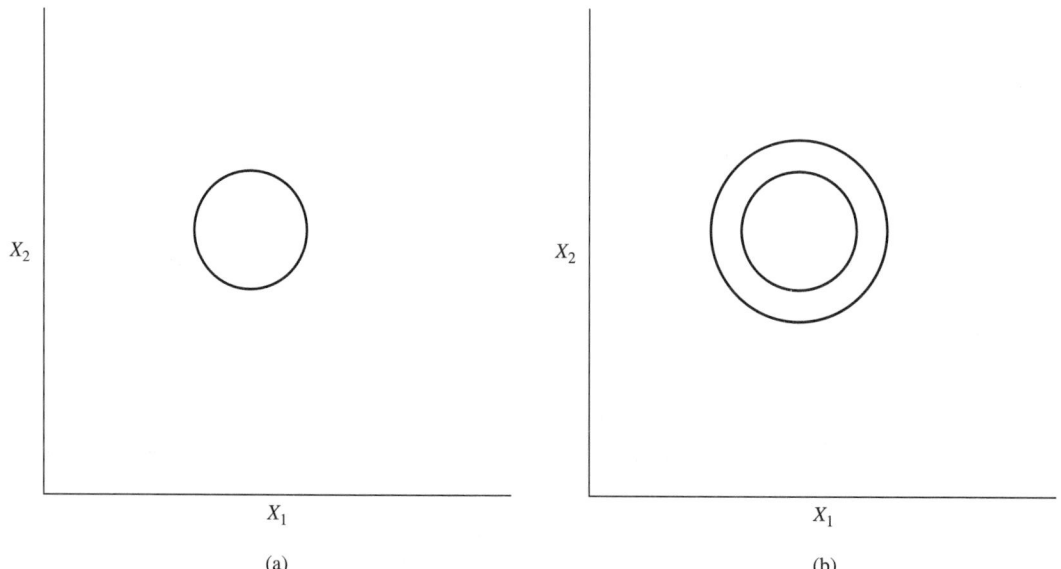

(a)   (b)

Figure 16.3   A representation of Figs. 16.2a and 16.2b, showing the circles defined by the intersecting planes.

Three intersecting planes would result in three circles, and so on. If the standard deviations of X_1 and X_2 are not equal, then parallel planes will intersect the bivariate normal distribution to form ellipses instead of circles. This is shown, for three planes, in Fig. 16.4. In such plots, the largest ellipses (or circles) are formed nearest the tails of the distribution. Hereafter, only ellipses will be discussed, for it is unlikely that the two variables will have identical standard deviations.

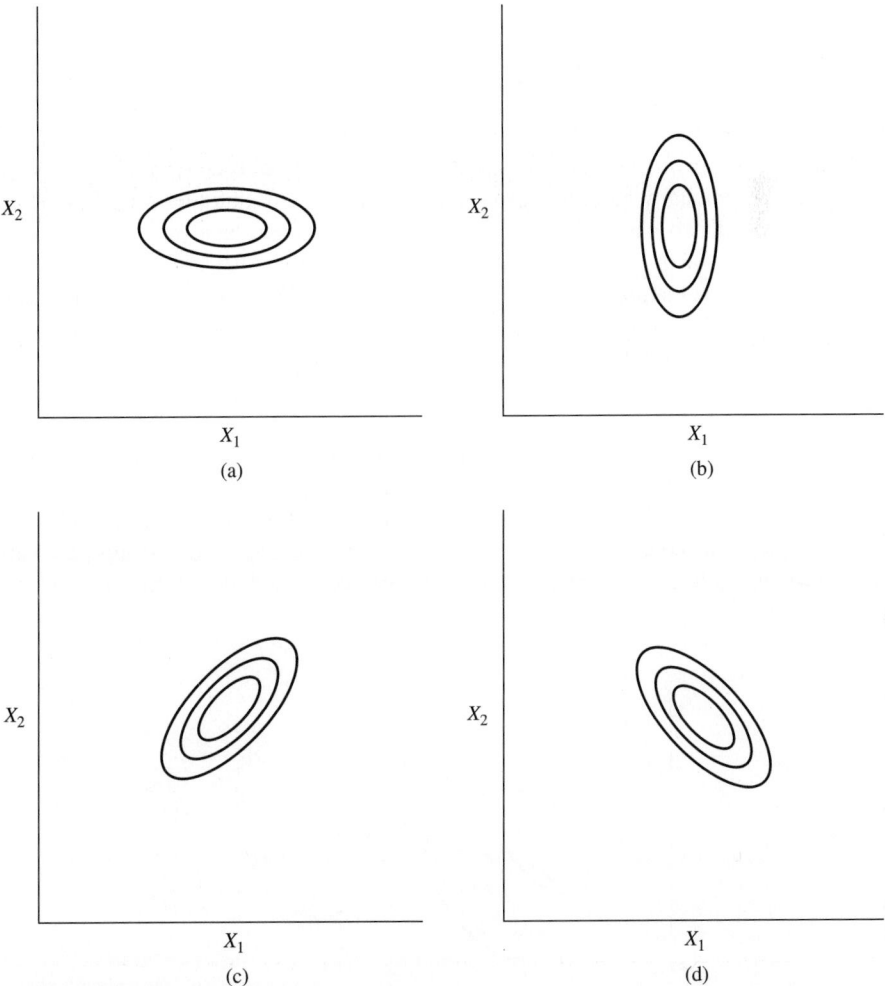

Figure 16.4 Representations of bivariate normal distributions where the standard deviations of X_1 and X_2 are not the same. **(a)** X_1 and X_2 are not correlated. **(b)** X_1 and X_2 are not correlated. **(c)** X_1 and X_2 are positively correlated. **(d)** X_1 and X_2 are negatively correlated.

If the two variables (X_1 and X_2) are not correlated (see Chapter 19 for a thorough discussion of correlation), the latter graphs will show ellipses parallel to one of the axes (Figs. 16.4a and 16.4b). If X_1 and X_2 are positively correlated, the ellipses run from the lower left to the upper right (as in Fig. 16.4c), whereas if they are negatively correlated the ellipses will be oriented from the lower right to the upper left (as in Fig. 16.4d).

16.2 MULTIVARIATE ANALYSIS OF VARIANCE HYPOTHESIS TESTING

At the beginning of the discussion of univariate analysis of variance (Chapter 10) it was explained that, when comparing a variable's mean among more than two groups, to employ multiple t-tests would cause an substantial inflation of α, the probability of a Type I error. In multivariate situations, we desire to compare two or more variables' means among two or more groups, and to do so with multiple ANOVAs would also result in an inflated chance of a Type I error. Multivariate analysis of variance (MANOVA) is a method of comparing the population means of all variables of interest at the same time, while maintaining the chosen magnitude of Type I error. A second desirable trait of MANOVA is that it considers the correlation among multiple variables, which separate ANOVAs can not do. Indeed, if the variables are correlated, MANOVA may provide more powerful testing than performing a series of separate ANOVAs. (If the variables are not correlated, separate ANOVAs may be more powerful than MANOVA.) For example, there are two bivariate distributions depicted in Fig. 16.5, with the distribution of variable X_1 being very similar in the two groups and the distribution of X_2 also being very similar in the two groups. Therefore, a univariate ANOVA (or a two-sample t-test) will be unlikely to conclude a difference between the means of the groups for variable X_1 or for variable X_2, but MANOVA may very well conclude the mean of the two bivariate

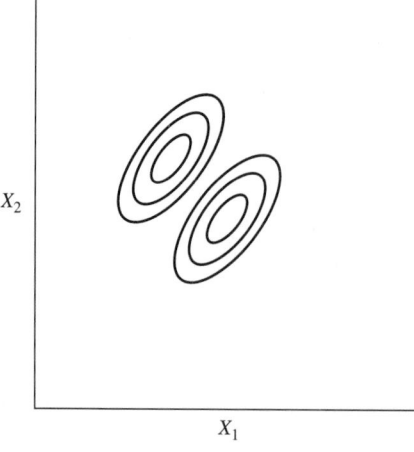

Figure 16.5   Two bivariate normal groups of positively correlated data differing in both dimensions.

distributions to be different. Thirdly, sometimes groups differences for each of several variables are too small to be detected with a series of ANOVAs, but a MANOVA will conclude the groups different by considering the variables jointly.

In univariate ANOVA with k groups, a typical null hypothesis is

$$H_0: \quad \mu_1 = \mu_2 = \cdots = \mu_k,$$

which says that all k population means are the same. And the corresponding alternate hypothesis is

$$H_A: \quad \text{The } k \text{ population means are not all equal.}$$

Recall that H_A does not say that all means are different, only that at least one is different from the others.

Thus, for example, Example 10.1 presented an experiment to ask whether the mean body weight of pigs is the same when the animals are raised on four different feeds. And

$$H_0: \quad \mu_1 = \mu_2 = \mu_3 = \mu_4$$

$$H_A: \quad \text{All four population means are not equal.}$$

In a MANOVA with two variables (X_1 and X_2) and k groups, the null hypothesis may be stated as

$$H_0: \quad \mu_{11} = \mu_{12} = \cdots = \mu_{1k} \text{ and } \mu_{21} = \mu_{22} = \cdots = \mu_{2k},$$

where μ_{ij} denotes the population mean of variable i in group j. This H_0 says that the means of variable 1 are the same for all k groups *and* that the means of variable 2 are the same for all k groups. The corresponding MANOVA alternate hypotheses is

$$H_A: \quad \text{The } k \text{ populations do not have the same means for variable 1 and the same means for variable 2.}$$

Thus, H_0 is rejected if any of the μ_{1j}'s are concluded to differ from each other *or* if any of the μ_{2j}'s are concluded to differ from each other.

Example 16.1 expands the experiment of Example 10.1 by considering two variables: the weight of the animals' body fat in addition to their fat-free body weight. The null hypothesis is that mean weight of total body fat is the same on all four diets *and* the mean fat-free body weight is the same on all of these diets. (It is *not* being hypothesized that mean body-fat weight is the same as mean fat-free body weight!) In this example,

$$H_0: \quad \mu_{11} = \mu_{12} = \mu_{13} = \mu_{14}; \quad \mu_{21} = \mu_{22} = \mu_{23} = \mu_{24}$$

and

$$H_A: \quad \text{The four feeds do not result in the same mean weight of body fat and the same mean fat-free body weight.}$$

If, in the sampled populations, one or more of the six equals signs in H_0 is untrue, then H_0 should be rejected.

There are several methods for comparing means to test MANOVA hypotheses. This chapter will refer to four test statistics employed for this purpose and encountered

EXAMPLE 16.1   A bivariate analysis of variance.

The single-factor experimental design of Example 10.1 with two, instead of one, variables. Nineteen pigs are assigned at random among four groups, each group to receive a different feed. One variable is the weight of the total body fat in kilograms; the second variable is fat-free body weight in kilograms.

H_0: $\mu_{11} = \mu_{12} = \mu_{13} = \mu_{14}$; $\mu_{21} = \mu_{22} = \mu_{23} = \mu_{24}$.

H_A: Pigs raised on these four feeds do not have the same mean weight of body fat and the same mean fat-free body weight.

$\alpha = 0.05$.

	Feed 1		Feed 2		Feed 3		Feed 4	
	Fat-free weight (kg)	Fat weight (kg)	Fat-free weight (kg)	Fat weight (kg)	Fat-free weight (kg)	Fat weight (kg)	Fat-free weight (kg)	Fat weight (kg)
	40.6	20.2	43.6	25.1	57.5	45.1	53.8	34.1
	37.0	20.0	43.5	24.2	58.3	43.8	49.0	35.2
	43.3	21.7	47.9	26.1	56.1	44.1	49.5	33.6
	38.1	20.5	41.1	25.2	53.0	43.5	50.9	34.8
	41.9	19.8	45.7	24.1			50.5	39.8
ΣX	200.9	102.2	221.8	124.7	224.9	176.5	253.7	177.5
$\bar{X}$	40.18	20.44	44.36	24.84	56.22	44.12	50.74	35.50

Computer computation yields the following output:

   Wilks' $\Lambda = 0.0178$, $F = 30.3$,   DF $= 6, 28$, $P \ll 0.0001$.
   Reject H_0.

   Pillai's trace $= 1.0223$, $F = 5.23$,   DF $= 6, 30$, $P = 0.0009$.
   Reject H_0.

   Lawley-Hotelling trace $= 52.9847$, $F = 115$,   DF $= 6, 26$, $P \ll 0.0001$.
   Reject H_0.

   Roy's maximum root $= 52.9421$, $F = 265$,   DF $= 3, 15$, $P \ll 0.0001$.
   Reject H_0.

in MANOVA computer programs. None of these four can be said to be "best" in all situations. Each captures different characteristics of the differences among means; thus, the four have somewhat different abilities to detect differences in various circumstances. The computations of these test statistics are far from simple and—especially for more than two variables—cannot readily be expressed in algebraic equations. (They are represented with much less effort as matrix calculations, which are beyond the scope of this book.) Therefore, we shall depend upon computer programs to calculate these statistics and shall not attempt to demonstrate the numerical manipulations. It will be noted, however, that the necessary calculations involve total, groups, and error sums of squares (SS, introduced in Section 10.1) and sums of crossproducts (introduced in Section 17.2). And, just as mean squares are derived from sums of squares, quantities known as "covariances" are derived from sums of crossproducts.

The four common MANOVA statistics are the following.* They will be certain to agree with each other only when one variable is to be analyzed (a univariate ANOVA situation) or when $k = 2$.

Wilks' lambda. Wilks' Λ (capital Greek lambda), or Wilks' likelihood ratio, is the most commonly encountered MANOVA statistic, dating from the original formulation of the MANOVA procedure (Wilks, 1932). Unlike the typical test statistic (e.g., the F in ANOVA), H_0 is rejected for *small* values of Wilks' Λ. Computer programs may present Λ transformed into a value of F or chi-square, with associated probability (P), and, as elsewhere, large F's or large χ^2's yield small P's. Λ is a quantity, ranging from 0 to 1, which is a measure of the amount of variability among the data that *is not* explained by the effect of the levels of the factor. In Example 16.1, Λ is an expression of the amount of variability among lean body weights, and among fat weights, that is not accounted for by the effect of the four diets. Thus, a measure of the proportion of the variability among the data that *is* explained by the experimental factor is

$$\eta^2 = 1 - \Lambda \qquad (16.1)$$

where η is the lower-case Greek eta).[†]

Pillai's trace, or the Pillai-Bartlett trace.[‡] Many authors recommend this statistic as the best for general use. Large values (Bartlett, 1939; Pillai, 1955) result in the rejection of H_0. As with Wilks' Λ, there is an F transformation for Pillai's trace.

Hotelling-Lawley Trace. This statistic was developed by Lawley (1938) and modified by Hotelling (1951). An F transformation may be encountered for the Lawley-Hotelling trace.

Roy's Maximum Root (also known by similar names, such as Roy's largest, or greatest, root). This statistic (Roy 1945, 1953) may also be approximated by an F distribution.

Wilks' Λ is the most widely used test statistic for MANOVA, but Olson (1974, 1976, 1979) found that Pillai's trace seems to be the most robust and the most desirable for general use. If the four MANOVA test statistics do not indicate the same conclusion about H_0, further scrutiny of the data is called for, such as an examination of scatter plots with the axes as in Figs. 16.5 and 16.6. Correlation among variables will suggest reliance on the conclusion of Pillai's trace; non-correlation will suggest choosing Roy's statistic. In Example 16.1, all four MANOVA test statistics conclude that H_0 is to be rejected.

---

*Each of the four MANOVA statistics is a function of what are called *eigenvalues*, or *roots*, of matrices. Matrix algebra is explained in some texts on multivariate statistics. Various symbols have been used for these four statistics in the literature.

[†]Thus, η^2 has a meaning like that of R^2 in multiple regression or multiple correlation (Section 20.3).

[‡]A *trace* is the result of a specific mathematical operation on a matrix. See Section 20.1 for a little information on matrices.

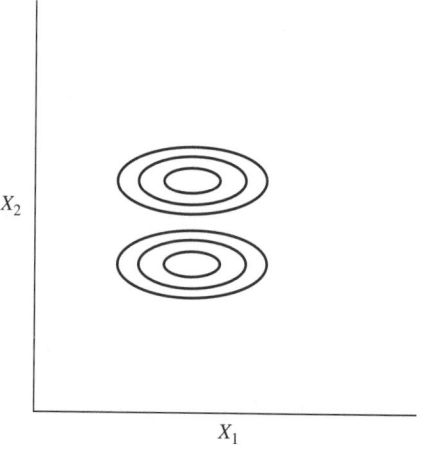

X_2

X_1

Figure 16.6 Two bivariate normal groups of positively correlated data differing in one dimension.

Assumptions. As in ANOVA (Section 10.1), the underlying mathematical foundations of MANOVA depend upon certain assumptions. Although it is unlikely that all of the assumptions will be exactly met for a given set of data, it is important to be cognizant of them and of whether the statistical procedures employed are robust to departures from them.

A very important underlying assumption in MANOVA is that the data represent random samples from the populations of interest and that the observations on each subject are independent. In Example 16.1, the body-fat weights of pigs on each diet are assumed to have come at random from the population of body-fat weights of all pigs on that diet, and the lean body weights on each diet are assumed to have come from randomly from a population of such weights. Also, the body-fat weight of each subject (i.e., each pig) must be independent of the body-fat weight of each other subject, and the lean body weights of all the subjects must be independent of each other. MANOVA is invalidated by departure from the assumption of random and independent data.

There is also an assumption of multivariate normality (see Section 16.1) for each group of data. However, departures from normality have only a slight affect on the Type I error rate.*

MANOVA is especially resistant to non-normality due to skewness, but platykurtosis can severely reduce power. There are nonparametric tests (e.g., analogs of the Kruskal-Wallis and Friedman tests) that can be used if the underlying distributions are far from multivariate normal. And there are procedures for assessing multivariate normality (e.g., Johnson and Wichern, 1998: 194–200; Sharma, 1996: 380–382; Stevens, 1996: 244), although they are seldom needed and applied.

Just as ANOVA assumes that there is equality of variances over all k groups, MANOVA assumes equality of variances across all k groups. Furthermore, MANOVA requires that the correlation between any two variables must be the same in all k groups; this is known as the assumption of equal covariance. If the group sample sizes are

---

*The *central limit theorem* (see first footnote in Section 6.3) also applies to multivariate distributions.

equal, or nearly equal (defined by Stevens, 1996: 238, as the largest $n \leq 1.5$ times the smallest n), then departures from this assumption will have little effect on the Type I error using the Pillai trace, somewhat more effect using the Wilks or Hotelling-Lawley statistics, and considerable adverse effect using Roy's criterion. However, as sample sizes become large, the first three statistics are nearly equivalent in performance (Olson, 1974, 1976, 1979).* With very disparate sample sizes, the adverse effect may be serious using any of the four statistics, causing the hypothesis testing to reject H_0 either more often or less often than it should (depending upon the magnitudes of the variances and correlations and their relationships to the sample sizes). There is an available statistical test for this assumption (Box, 1949), analogous to the Bartlett test of the assumption of homogeneity of variance in ANOVA (Section 10.6), but it is seriously affected by non-normality (Olson, 1974) and thus is not generally recommended. Data transformations (Chapter 13) may be used to reduce non-normality or heterogeneity of variances and covariances.

Although MANOVA is typically robust to departures from the variability and variance-correlations assumptions, the Pillai trace (according to Olson, 1974, 1976) is generally the most robust of the four methods.

Power.   The power of a MANOVA depends upon a complex set of characteristics, including the extent to which the underlying assumptions (see preceding paragraphs) are met. In general, increased sample size is associated with increased power, but power decreases with increase in the number of variables. Thus, if it is desired to employ an experimental design with several variables, larger samples will be needed than would be the case if there were only two variables. Also, as with ANOVA, the power of MANOVA is greater when the population differences between means are larger. The magnitude of correlations between variables can cause MANOVA to be either more or less powerful than separate ANOVAs. Some MANOVA computer programs calculate power.

Differences in power among the four test statistics are typically not great. If the group means differ only one dimension (as shown in Fig. 16.6), a relatively uncommon situation, Roy's statistic is the most powerful procedure, followed—in order of power—by the Hotelling-Lawley trace, Wilks' Λ, and Pillai's trace. If, however, the group means differ along several dimensions, meaning that the variables are correlated (as shown in Fig. 16.5), then the relative power of these statistics is reversed: Pillai's trace being the most powerful, followed by Wilks' Λ, the Hotelling-Lawley trace, and then Roy's statistic. (In intermediate situations, the four statistics probably tend more toward the latter ordering than the former.)

Two-Sample Hypotheses.   In the case of two groups (i.e., $k = 2$), another test statistic encountered is *Hotelling's T^2* (Hotelling, 1931), for which

$$F = \frac{n_1 + n_2 - m - 1}{(n_1 + n_2 - 2)m} T^2,\qquad(16.2)$$

---

*Olson declares that the three may be considered equivalent if the denominator degrees of freedom are at least $10m$ times the numerator DF, where m is the number of variables.

with m and $n_1 + n_2 - m - 1$ degrees of freedom, where m is the number of variables. Any of the four MANOVA test statistics may be used instead of T^2, just as a single-factor ANOVA may be used in place of a two-sample t test (Section 8.1), and all four will yield the same results.

16.3 FURTHER ANALYSIS

When the MANOVA rejects H_0, there are several procedures that might be employed to continue the analysis of difference among groups (e.g., Bray and Maxwell, 1985: 40–45; Hand and Taylor, 1987: Chapter 5; Hummel and Sligo, 1971; Stevens, 1996: 196–199; and Weinfurt, 1995). One approach is to perform a univariate ANOVA on each of the variables (followed, perhaps, by multiple comparisons—see Chapter 11), to test for difference among means for each variable separately. However, this procedure will ignore relationships among the variables, and other criticisms have been raised (Weinfurt, 1995).

In univariate ANOVA, one can reject H_0 and have none of the μ's declared different by further analysis (Chapter 11). Similarly, a MANOVA may reject H_0 with subsequent ANOVAs detecting no differences (either because of lack of power or because the interrelations among variables are important in rejection of the multivariate H_0). Some computer programs perform ANOVAs along with a MANOVA.

If $k = 2$ and H_0 is rejected by MANOVA or Hotelling's T^2 test, then 2-sample t tests and univariate ANOVAs will yield identical results.

16.4 OTHER EXPERIMENTAL DESIGNS

The experimental design described above is a single-factor bivariate analysis of variance; it is the same as the single-factor ANOVA design of Example 10.1 but with measurements taken of two variables, instead of one variable, for each subject.

If more than two variables were measured for each subject, the single-factor multivariate analysis of variance would be analyzed in a similar fashion by an appropriate computer program. For example, the blood cholesterol and tail length might be measured along with weight of body fat and lean body weight. This would result in an analysis with four variables; there would still be only one factor (feed). The null hypothesis would be

H_0: $\mu_{11} = \mu_{12} = \mu_{13} = \mu_{14}$;
       $\mu_{21} = \mu_{22} = \mu_{23} = \mu_{24}$;
       $\mu_{31} = \mu_{32} = \mu_{33} = \mu_{34}$;
       $\mu_{41} = \mu_{42} = \mu_{43} = \mu_{44}$

and the detection of one or more "$\neq$'s" (in place of "$=$'s") between population means causes H_0 to be rejected.

It is also possible to perform MANOVA on data from experimental designs having more than one factor (including factorial designs, hierarchical designs, or repeated-measure designs). These situations are extensions of those in Chapters 12, 14, and 15 and are described in some of the references immediately preceding Section 16.1. While the necessary calculations are difficult to display, they are readily executed by computer.

Analysis of covariance (ANCOVA), introduced in Section 12.12, also may be extended to experimental designs with multiple dependent variables. This is done via techniques known as *multivariate analyses of covariance* (MANCOVA), for which computer routines are available.

EXERCISES

16.1 Using multivariate analysis of variance, analyze the following data for the concentration of three amino acids in centipede hemolymph (mg/100 ml), asking whether the mean concentration of each is the same in males and females:

Male			Female		
Alanine	Aspartic Acid	Tyrosine	Alanine	Aspartic Acid	Tyrosine
7.0	17.0	19.7	7.3	17.4	22.5
7.3	17.2	20.3	7.7	19.8	24.9
8.0	19.3	22.6	8.2	20.2	26.1
8.1	19.8	23.7	8.3	22.6	27.5
7.9	18.4	22.0	6.4	23.4	28.1
6.4	15.1	18.1	7.1	21.3	25.8
6.6	15.9	18.7	6.4	22.1	26.9
8.0	18.2	21.5	8.6	18.8	25.5

16.2 For the experimental design of Example 12.1, perform a multivariate analysis of variance on the following data for plasma concentrations of calcium (in mg/100 ml) and for rate of evaporative water loss (in mg/min):

No Hormone Treatment				Hormone Treatment			
Female		Male		Female		Male	
Plasma Calcium	Water Loss	Plasma Calcium	Water Loss	Plasma Calcium	Water Loss	Plasma Calcium	Water Loss
16.5	76	14.5	80	39.1	71	32.0	65
18.4	71	11.0	72	26.2	70	23.8	69
12.7	64	10.8	77	21.3	63	28.8	67
14.0	66	14.3	69	35.8	59	25.0	56
12.8	69	10.0	74	40.2	60	29.3	52

17

SIMPLE LINEAR REGRESSION

Techniques that consider relationships between two variables are described in this and the following two chapters. Chapter 20 presents the expansion of such techniques to analyze situations where more than two variables may be related to each other.

17.1 REGRESSION VS. CORRELATION*

The relationship between two variables may be one of functional dependence of one on the other. That is, the magnitude of one of the variables (the *dependent variable*) is assumed to be determined by—i.e., is a function of—the magnitude of the second variable (the *independent variable*), whereas the reverse is not true. For example, in the relationship between blood pressure and age in humans, blood pressure may be considered the dependent variable and age the independent variable; we may reasonably assume that although the magnitude of a person's blood pressure might be a function of age, age is not determined by blood pressure. This is not to say that age is the only biological determinant of blood pressure, but we do consider it to be one determining factor. Sometimes, the independent variable is called the "predictor," or "regressor," variable and the dependent variable is called the "response," or "criterion," variable. The term "dependent" does not imply a cause-and-effect relationship between the two variables.

---

*The historical developments of regression and correlation are strongly related, owing their discovery—the latter following the former—to Sir Francis Galton, who first developed these procedures during 1875–1885 (Walker, 1929: 103–104, 187); see also the first footnote in Section 19.1.

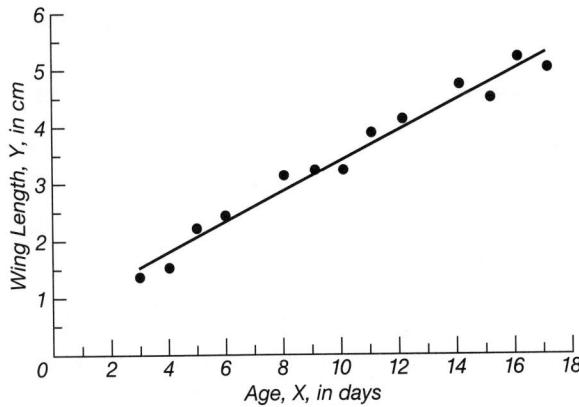

Figure 17.1 Sparrow wing length as a function of age. The data are from Example 17.1.

Such a dependent relationship is termed a *regression*; the term *simple regression* refers to the fact that only two variables are being considered.*

Data amenable to simple regression analysis will consist of a dependent variable that is a random-effect factor and an independent variable that is either a fixed-effect or a random-effect factor. (See the end of Section 10.1 to review these concepts.)

It is very convenient to graph simple regression data, using the ordinate (Y axis) for the dependent variable (conventionally termed Y) and the abscissa (X axis) for the independent variable (X). Thus, as shown in Fig. 17.1, the data of Example 17.1 appear as a scatter of points, each point representing a pair of X and Y values. One pair of X and Y data may be denoted as (X_1, Y_1), another as (X_2, Y_2), another as (X_3, Y_3), etc., resulting in what is called a "scatter plot." (The line in this figure will be explained shortly.)

In many kinds of biological data, however, the relationship between two variables is not one of dependence. In such cases, the magnitude of one of the variables changes as the magnitude of the second variable changes, but it is not reasonable to consider there to be an independent and a dependent variable. In such situations, *correlation*, rather than regression, analyses are called for, and both variables are theoretically to be random-effects factors. An example of data suitable for correlation analysis would be measurements of human arm and leg lengths. It might be found that an individual with long arms will in general possess long legs, so a relationship may be describable; but there is no justification in stating that the length of one limb is dependent upon the length of the other. Correlation techniques involving two variables will be discussed in Chapter 19; if more than two variables are being considered, then the appropriate procedures are found in Chapter 20.

---

*In the case of simple regression, the adjective *linear* may be used to refer to the relationship between the two variables being a straight line, but to a statistician it describes the relationship of the parameters in Section 17.2.

EXAMPLE 17.1   Wing lengths of thirteen sparrows of various ages. The data are plotted in Fig. 17.1.

Age (days) (X)	*Wing length* (cm) (Y)
3.0	1.4
4.0	1.5
5.0	2.2
6.0	2.4
8.0	3.1
9.0	3.2
10.0	3.2
11.0	3.9
12.0	4.1
14.0	4.7
15.0	4.5
16.0	5.2
17.0	5.0

$$n = 13$$

17.2 THE SIMPLE LINEAR REGRESSION EQUATION

The simplest functional relationship of one variable to another in a population is the *simple linear regression*

$$Y_i = \alpha + \beta X_i. \tag{17.1}$$

Here, α and β are population parameters (and, therefore, constants), and this expression will be recognized as the general equation for a straight line.* However, in a population the data are unlikely to be exactly on a straight line, so Y may be said to be related to X by

$$Y_i = \alpha + \beta X_i + \epsilon_i, \tag{17.1a}$$

where ϵ_i is referred to as an "error," or "residual," which is a departure of an actual Y_i from what Equation 17.1 predicts Y_i to be; and the sum of the ϵ_i's is zero.

Consider the data in Example 17.1, where wing length is the dependent variable and age is the independent variable. From a scatter plot of these data (Fig. 17.1), it appears that our sample of measurements from thirteen birds represents a population of data in which wing length is linearly related to age. Thus, we would like to know the values of α and β that would uniquely describe the functional relationship existing in the population.

---

*α and β are often used for these population parameters, and as such should not be confused with the standard use of the same Greek letters to denote the probabilities of a Type 1 and a Type II error, respectively (see Section 6.4). The additive (linear) relationship of the two parameters in Equation 17.1 is what causes statisticians to refer to this as a "linear" relationship. Some examples of nonlinear regression are given in Section 20.15.

If all the data in a scatter diagram such as Fig. 17.1, occurred in a straight line, it would be an unusual situation. Generally, as is shown in this figure, there is considerable variability of data around any straight line we might draw through them. What we seek to define is what is commonly termed the "best fit" line through the data. The criterion for "best fit" that is generally employed utilizes the concept of *least squares*.* Figure 17.2 is an enlarged portion of Fig. 17.1. Each value of X will have a corresponding value of Y lying on the line that we might draw through the scatter of data points. This value of Y is represented as $\hat{Y}$ to distinguish it from the Y value actually observed in our sample.[†] Thus, as Fig. 17.2 illustrates, an observed data point is denoted as (X_i, Y_i), and a point on the regression line is $(X_i, \hat{Y}_i)$.

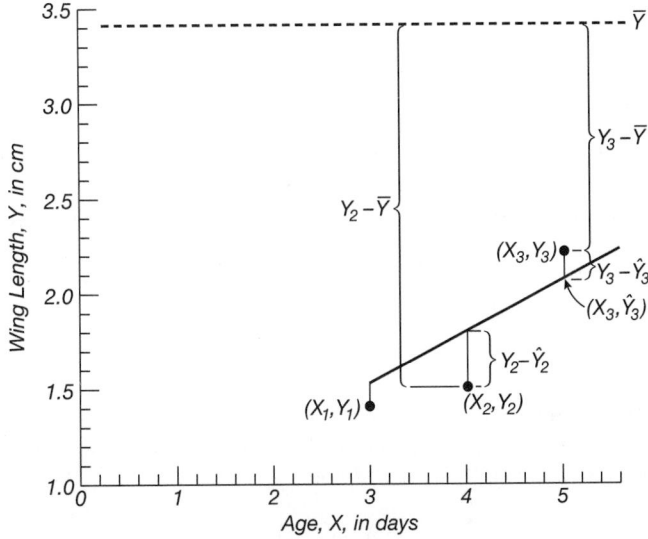

Figure 17.2   An enlarged portion of Fig. 17.1, showing the partitioning of Y deviations.

The criterion of least squares considers the vertical deviation of each point from the line (i.e., the deviation describable as $Y_i - \hat{Y}_i$), and defines the best fit line as that which results in the smallest value for the sum of the squares of these deviations for all values of Y_i and $\hat{Y}_i$. That is $\sum_{i=1}^{n}(Y_i - \hat{Y}_i)^2$ is to be a minimum,[‡] where n is the number of data points comprising the sample. The sum of squares of these deviations is

---

*The discovery of the statistical method known as "least squares" involves what Stigler (1989) calls "one of the most famous priority disputes in the history of science": The French mathematician, Adrien Marie Legendre (1752–1833), clearly published it in 1805, but the brilliant German mathematician and physicist, Karl Friedrich Gauss (1777–1855) claimed—probably truthfully—that he had used it ten years prior to that. (See, also, Eisenhart, 1978). The term "least squares" is property credited to Legendre (David, 1995).

[†]Statisticians refer to $\hat{Y}$ as "Y hat."

[‡]Another way to express this is to say that the correlation between Y_i's and $\hat{Y}_i$'s is to be maximum.

called the *residual sum of squares* (or, sometimes, the *error sum of squares*) and will be discussed later in this chapter.*

The only way to determine the population parameters α and β with complete confidence and accuracy would be to possess all the data for the entire population. Since this is nearly always impossible, we have to estimate these parameters from a sample of n data, where n is the number of pairs of X and Y values. The calculations required to arrive at such estimates, as well as to execute the testing of a variety of important hypotheses, involve the computation of sums of squared deviations from the mean, just as has been encountered before. Recall that the sum of squares of X_i values is defined as $\sum(X_i - \bar{X})^2$, which is more easily obtained on a calculator as $\sum X_i^2 - (\sum X_i)^2/n$. It will be convenient to define $x_i = X_i - \bar{X}$, so that this sum of squares can be abbreviated as $\sum x_i^2$, or, more simply, as $\sum x^2$.

We shall also be required to calculate a quantity referred to as the *sum of the crossproducts* of deviations from the mean:

$$\sum xy = \sum(X_i - \bar{X})(Y_i - \bar{Y}), \tag{17.2}$$

where y denotes a deviation of a Y value from the mean of all Y's just as x denotes a deviation of an X value from the mean of all X's. The sum of the crossproducts, analogously to the sum of squares, has a simple-to-use machine formula:

$$\sum xy = \sum X_i Y_i - \frac{(\sum X_i)(\sum Y_i)}{n}, \tag{17.3}$$

and it is recommended that the latter formula be employed.

The Regression Coefficient.   The parameter β is termed the *regression coefficient*, or the *slope* of the best fit regression line. The best estimate of β is

$$b = \frac{\sum xy}{\sum x^2} = \frac{\sum(X_i - \bar{X})(Y_i - \bar{Y})}{\sum(X_i - \bar{X})^2} = \frac{\sum X_i Y_i - \dfrac{(\sum X_i)(\sum Y_i)}{n}}{\sum X_i^2 - \dfrac{(\sum X_i)^2}{n}}. \tag{17.4}$$

Although the denominator in this calculation is always positive, the numerator may be either positive, negative, or zero, and the value of b theoretically can range from $-\infty$ to $+\infty$, including zero (see Fig. 17.3).

---

*A different definition of a "best fit" line is the line that results in the smallest possible sum of the absolute values of these deviations; that is, $\sum_{i=1}^{n} |Y_i - \hat{Y}_i|$ is to be a minimum. Regression using least absolute deviations (now referred to as "LAD") was introduced in 1757, long before least-squares regression, by Roger Joseph Boscovich (1711–1787), who was born in Yugoslavia and spent most of his life as a Jesuit priest in Rome) and was employed by the great French mathematician, Pierre Simon Laplace (1749–1827), thirty years later (Birkes and Dodge, 1993: Chapter 2; see also Bloomfield and Steiger, 1983). It employs different (and computationally more difficult) statistical procedures than least-squares regression but may be preferable if there are substantial departures from some least-squares assumptions. (See the end of Section 17.2.)

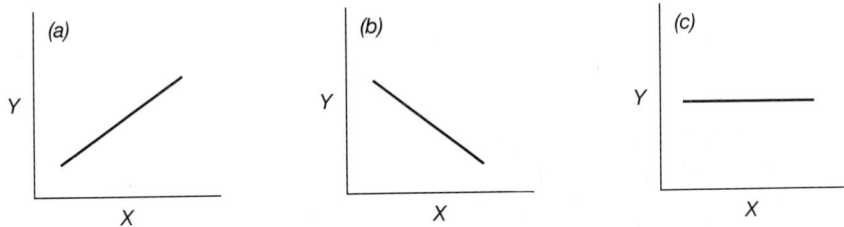

Figure 17.3   The slope of a linear regression line may be (a) positive, (b) negative, or (c) zero.

Example 17.2 demonstrates the calculation of b for the data of Example 17.1. Note that the units of b are the units of Y divided by the units of X. The regression coefficient expresses what change in Y is associated, on the average, with a unit change in X. In the present example, $b = 0.270$ cm/day indicates that there is a mean wing growth of 0.270 cm each day. Determination of the precision of b will be considered in Section 17.4.

EXAMPLE 17.2   The simple linear regression equation calculated by the method of least squares for the data from the thirteen birds of Example 17.1.

$$n = 13 \qquad\qquad \sum Y = 44.4$$

$$\sum X = 130.0 \qquad\qquad \bar{Y} = 3.415$$

$$\bar{X} = 10.0 \qquad\qquad \sum XY = 514.80$$

$$\sum X^2 = 1562.00 \qquad\qquad \sum xy = 514.80 - \frac{(130.0)(44.4)}{13}$$

$$\sum x^2 = 1562.00 - \frac{(130.0)^2}{13} \qquad\qquad = 514.80 - 444.00$$

$$= 1562.00 - 1300.00 \qquad\qquad = 70.80$$

$$= 262.00$$

$$b = \frac{\sum xy}{\sum x^2} = \frac{70.80}{262.00} = 0.270 \text{ cm/day}$$

$$a = \bar{Y} - b\bar{X} = 3.415 \text{ cm} - (0.270 \text{ cm/day})(10.0 \text{ days})$$

$$= 3.415 \text{ cm} - 2.700 \text{ cm}$$

$$= 0.715 \text{ cm}$$

So, the simple linear regression equation is $\hat{Y} = 0.715 + 0.270X$.

The Y Intercept.   An infinite number of lines possess any stated slope, all of them parallel (see Fig. 17.4). However, a line can be defined uniquely, by stating, in addition to β, any one point on the line—i.e., any pair of coordinates, $(X_i, \hat{Y}_i)$. The point conventionally chosen is the point on the line where $X = 0$. The value of Y in the population at this point is the parameter α, which is called the Y *intercept*.

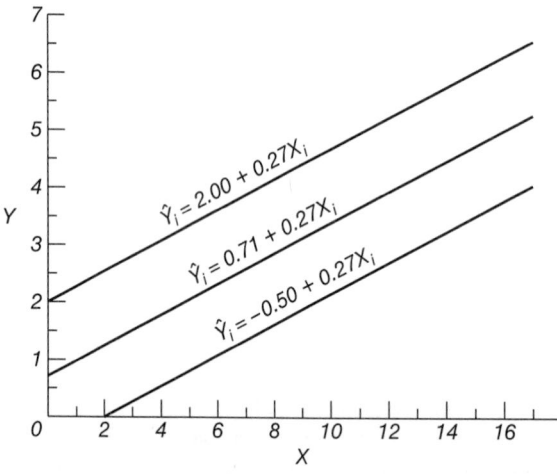

Figure 17.4 For any given slope, there exists an infinite number of possible regression lines, each with a different Y intercept. Three of this infinite number are shown here.

It can be shown mathematically that the point $(\bar{X}, \bar{Y})$ always lies on the best fit regression line. Thus, substituting $\bar{X}$ and $\bar{Y}$ in Equation (17.1), we find that:

$$\bar{Y} = \alpha + \beta \bar{X} \tag{17.5}$$

and

$$\alpha = \bar{Y} - \beta \bar{X}. \tag{17.6}$$

The best estimate of α is

$$a = \bar{Y} - b \bar{X}. \tag{17.7}$$

The calculation of a is shown in Example 17.2. Note that the Y intercept has the same units as any other Y value. (The precision of a is considered in Section 17.4.) The sample regression equation (which estimates the population relationship between Y and X stated in Equation 17.1) is written as

$$\hat{Y}_i = a + b X_i, \tag{17.8}$$

although some authors write

$$\hat{Y}_i = \bar{Y} + b(X_i - \bar{X}), \tag{17.9}$$

which is equivalent.

Figures 17.4 and 17.5 demonstrate that the knowledge of either a or b allows only an incomplete description of a regression function. But by specifying both a and b, a line is uniquely defined. Also, because a and b were calculated using the criterion of least squares, the residual sum of squares from this line is smaller than the residual sum of squares that would result from any other line (i.e, a line with any other a or b) that could be drawn through the data points.

Predicting Values of Y. Knowing the parameter estimates a and b for the linear regression equation, one can predict the value of the dependent variable expected at a

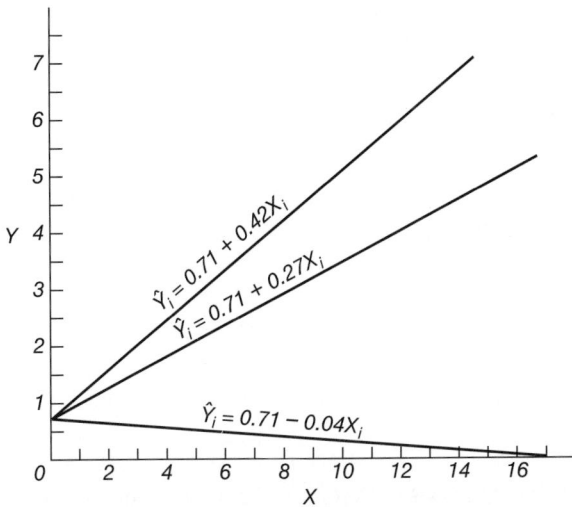

Figure 17.5 For any given Y intercept, there exists an infinite number of possible regression lines, each with a different slope. Three of this infinite number are shown here.

stated value of X_i. For the regression in Example 17.2, the wing length of a sparrow at 13.0 days of age would be predicted to be

$$\hat{Y} = a + bX_i$$

$$= 0.715 \text{ cm} + (0.270 \text{ cm/day})(13.0 \text{ day}) = 4.225 \text{ cm}.$$

The wing length in the population at 7.0 days of age would be estimated to be $\hat{Y} = 0.715 \text{ cm} + (0.270 \text{ cm/day})(7.0 \text{ day}) = 2.605 \text{ cm}$, and so on.

To plot a linear regression line graphically, we need to know only two points that lie on the line. We already know two points, namely $(\bar{X}, \bar{Y})$ and $(0, a)$; however, for ease and accuracy in drawing the line, two points that lie near extreme ends of the observed range of X are most useful. For drawing the line in Fig. 17.1, the values of $\hat{Y}_i$ for $X_i = 3.0$ days and $X_i = 17.0$ days were used. These were found to be $\hat{Y} = 1.525$ and 5.305 cm, respectively. A regression line should always be drawn using predicted points, and never drawn "by eye."

A word of caution is in order concerning predicting $\hat{Y}_i$ values from a regression equation. Generally, it is an unsafe procedure to extrapolate from regression equations— that is, to predict $\hat{Y}_i$ values for X_i values outside the observed range of X_i. It would, for example, be unjustifiable to attempt to predict the wing length of a twenty-day-old sparrow, or a one-day-old sparrow, using the regression calculated for birds ranging from 3.0 to 17.0 days in age. Indeed, applying the equation of Example 17.2 to a one-year-old sparrow would predict a wing nearly one meter long! What the linear regression describes is Y as a function of X *within the range of observed values of X*. For values of X above or below this range, the function may not be the same (i.e., α and/or β may be different); indeed, the relationship may not even be linear in such ranges, even though it is linear within the observed range. If there is good reason to believe that the described function holds for X values outside the range of those observed, then we may cautiously extrapolate. Otherwise, beware. A classic example of nonsensical extrapolation was provided in 1874 by Mark Twain (1950: 156):

> In the space of one hundred and seventy-six years the Lower Mississippi has shortened itself two hundred and forty-two miles. That is an average of a trifle over one mile and a third per year [i.e., a slope of -1.375 mi/yr]. Therefore, any calm person, who is not blind or idiotic, can see that in the Old Oölitic Silurian period, just a million years ago next November, the Lower Mississippi River was upward of one million three hundred thousand miles long, and stuck out over the Gulf of Mexico like a fishing rod. And by the same token any person can see that seven hundred and forty-two years from now, the lower Mississippi will be only a mile and three-quarters long, and Cairo [Illinois] and New Orleans [Louisiana] will have joined their streets together, and be plodding comfortably along under a single mayor and a mutual board of aldermen.*

Section 17.4 discusses the estimation of the error and confidence intervals associated with predicting $\hat{Y}_i$ values.

Assumptions of Regression Analysis. Certain basic assumptions must be met in order to validly test hypotheses about regressions or to set confidence intervals for regression parameters, although these assumptions are not necessary in order to compute validly the regression coefficient, b, and the Y-intercept, a. First, we must assume that for any value of X there exists in the population a normal distribution of Y values. This also means that, for each value of X there exists in the population a normal distribution of ϵ's. Second, we must assume homogeneity of variances; that is, the variances of these population distributions of Y values (and of ϵ's) must all be equal to one another. (Indeed, the residual mean square, to be described shortly, estimates this common variance, just as the error variance estimates the common variance assumed in the analysis of variance in previous chapters.) Third, in the population the mean of the Y's at a given X lies on a straight line with all other mean Y's at the other X's. That is, the actual relationship in the population is linear. Fourth, the values of Y are to have come at random from the sampled population and are to be independent of one another. That is, selecting a particular Y from the population is in no way dependent upon the selection of any other Y. Fifth, the measurements of X are obtained without error. This last requirement, of course, is typically impossible; so what we are doing in practice is assuming that the errors in the X data are negligible, or at least are small compared with the measurement errors in Y. Violations of the first three of these assumptions may sometimes be countered by transformation of data, as will be discussed in Section 17.10. Data in violation of the second assumption may sometimes be analyzed properly by what is known as *weighted regression*, which will not be discussed here.

Regression statistics are known to be robust with respect to at least some of these underlying assumptions (e.g., Jacques and Norusis, 1973), so violations of them are not usually of concern unless they are severe. One kind of datum that causes violation of the assumption of normality and homogeneity of variance is the *outlier*, introduced near

---

*Author Twain concludes by noting that "There is something fascinating about science. One gets such a wholesale return of conjecture out of a trifling investment of fact."

the beginning of Section 6.5, which in regression is a recorded measurement that lies very much apart from the trend of the bulk of the data. (For example, in Fig. 17.1 a data point at $X = 4$ days and $Y = 5$ cm would have been an outlier.) Detecting and dealing with outliers are discussed in some regression treatises (e.g., Dunn and Clark, 1987: 402; Seber, 1977: 164–165).*

Nonparametric regression analysis makes no assumptions about underlying distributions; several varieties exist (including regression using ranks) and are discussed by Birkes and Dodge (1993: Chapter 6), Cleveland, Mallows, and McRae (1993), Daniel (1978: Chapter 10), Härdle (1989), Wang and Scott (1994), and others.

17.3 TESTING THE SIGNIFICANCE OF A REGRESSION

The slope, b, of the regression line computed from the sample data expresses quantitatively the straight-line dependence of Y on X in the sample. But what is really desired is information about the functional relationship (if any) in the population from which the same came. Indeed, the finding of a dependence of Y on X in the sample (i.e., $b \neq 0$) does not necessarily mean that there is a dependence in the population (i.e., $\beta \neq 0$). Consider Fig. 17.6, a scatter plot representing a population of data points with no dependence of Y on X; the best fit regression line would be parallel to the X axis (i.e., the slope, β, would be zero). However, it is possible, by random sampling, to obtain a sample of five data points having the values circled in the figure. By calculating b for this sample of five we would estimate that β was positive, even though it is, in fact, zero.

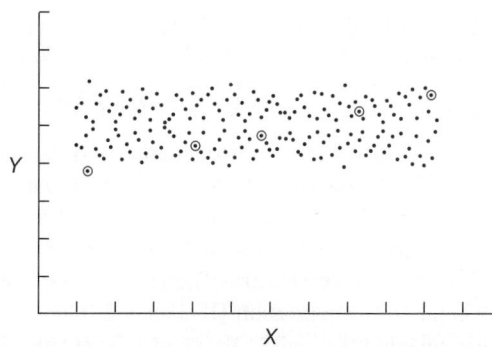

Figure 17.6 A hypothetical population of data points, having a regression coefficient, β, of zero. The circled points are a possible sample of five.

Now, it is not likely to obtain five such points out of this population, but we should desire to assess just how likely it is; therefore, we can set up the null hypothesis, H_0: $\beta = 0$, and the alternate hypothesis, H_A: $\beta \neq 0$. If we could conclude that there is a

---

*The regression method employing least absolute deviations (see fifth footnote in Section 17.2) is robust to outliers and some other departures from normality, and a regression method that performs in a fashion between that of least-squares regression and least-absolute-deviation regression is the *M-regression* described by Huber (1981: Chapter 7; Birkes and Dodge, 1993; based on the concept of Huber, 1964).

reasonable probability (i.e., a probability greater than the chosen level of significance—say, 5%) that the calculated b could have come from sampling a population with a $\beta = 0$, the H_0 is not rejected. If the probability of obtaining the calculated b is small (say, 5% or less), then H_0 is rejected, and H_A is assumed to be true.

Analysis of Variance Testing.   The preceding H_0 may be tested by an analysis of variance (ANOVA) procedure. First, the overall variability of the dependent variable is calculated by computing the sum of squares of deviations of Y_i values from $\bar{Y}$, a quantity termed the *total sum of squares*:

$$\text{total SS} = \sum(Y_i - \bar{Y})^2 = \sum y^2 = \sum Y_i^2 - \frac{\left(\sum Y_i\right)^2}{n}. \qquad (17.10)$$

Then, one determines the amount of variability among the Y_i, values that results from there being a linear regression; this is termed the *linear regression sum of squares*:

$$\text{regression SS} = \sum(\hat{Y}_i - \bar{Y})^2 = \frac{\left(\sum xy\right)^2}{\sum x^2} = \frac{\left(\sum X_i Y_i - \frac{\sum X_i \sum Y_i}{n}\right)^2}{\sum X_i^2 - \frac{\left(\sum X_i\right)^2}{n}}; \qquad (17.11)$$

because $b = \sum xy/\sum x^2$, this can also be calculated as

$$\text{regression SS} = b\sum xy. \qquad (17.12)$$

The value of the regression SS will be equal to that of the total SS only if each data point falls exactly on the regression line, a very unlikely situation. The scatter of data points around the regression line has already been alluded to, and the residual, or error, sum of squares has been defined. Knowing the total and the linear regression sums of squares, we may, by difference, obtain

$$\text{residual SS} = \sum(Y_i - \hat{Y}_i)^2 = \text{total SS} - \text{regression SS}. \qquad (17.13)$$

Table 17.1 presents the analysis-of-variance summary for testing the hypothesis H_0: $\beta = 0$ against H_A: $\beta \neq 0$. Example 17.3 performs such an analysis for the data from Examples 17.1 and 17.2. The degrees of freedom associated with the total variability of Y_i values are $n - 1$. The degrees of freedom associated with the variability among Y_i's due to regression are always 1 in a simple linear regression. The residual degrees of freedom are calculable as residual DF $=$ total DF $-$ regression DF $= n - 2$. Once the regression and residual mean squares are calculated (MS $=$ SS/DF, as usual), H_0 may be tested by determining

$$F = \frac{\text{regression MS}}{\text{residual MS}}, \qquad (17.14)$$

which is then compared to the critical value, $F_{\alpha(1), \nu_1, \nu_2}$, where $\nu_1 = $ regression DF $= 1$, and $\nu_2 = $ residual DF $= n - 2$.

The residual mean square is often written as $s_{Y \cdot X}^2$, a representation denoting that it is the variance of Y after taking into account the dependence of Y on X. The square root of this quantity, i.e. $s_{Y \cdot X}$, is called the *standard error of estimate* (occasionally termed the "standard error of the regression"). In Example 17.3, $s_{Y \cdot X} = \sqrt{0.047701 \text{ cm}^2} = 0.218$ cm. The standard error of estimate is an overall indication of the accuracy with which the fitted regression function predicts the dependence of Y on X. The magnitude of $s_{Y \cdot X}$ is proportional to the magnitude of the dependent variable, Y, making examination of $s_{Y \cdot X}$ a poor method for comparing regressions. Thus, Dapson (1980) recommends using $s_{Y \cdot X} / \bar{Y}$ (a unitless measure) to judge regression fits.

TABLE 17.1 Summary of the Calculations for Testing H_0: $\beta = 0$ against H_A: $\beta \neq 0$ by an Analysis of Variance

Source of variation	Sum of squares (SS)	DF	Mean square (MS)
Total $[Y_i - \bar{Y}]$	$\sum y^2$	$n - 1$	
Linear regression $[\hat{Y}_i - \bar{Y}]$	$\dfrac{\left(\sum xy\right)^2}{\sum x^2}$	1	$\dfrac{\text{regression SS}}{\text{regression DF}}$
Residual $[Y_i - \hat{Y}_i]$	total SS $-$ regression SS	$n - 2$	$\dfrac{\text{residual SS}}{\text{residual DF}}$

Note: To test the null hypothesis, we compute $F =$ regression MS/residual MS. The critical value for the test is $F_{\alpha(1), 1, (n-2)}$.

The proportion (or percentage) of the total variation in Y that is explained or accounted for by the fitted regression is termed the *coefficient of determination*, r^2, which may be thought of as a measure of the strength of the straight-line relationship:

$$r^2 = \frac{\text{regression SS}}{\text{total SS}}. \tag{17.15}$$

For Example 17.3, $r^2 = 0.97$, or 97%. That portion the total variation not explained by the regression is, of course, $1 - r^2$, or residual SS/total SS, and this is called the *coefficient of nondetermination*, a quantity seldom referred to.* In Example 16.3, $1 - r^2 = 1.00 - 0.97 = 0.03$, or 3%. (The quantity r is the correlation coefficient, to be introduced in Chapter 19.)[†]

Another good way to express the accuracy of a regression, or to compare accuracies of several regressions, is to compute confidence intervals for predicted values of Y, as described in Section 17.4.

---

*The standard error of estimate is directly related to the coefficient of nondetermination and to the variability of Y as

$$s_{Y \cdot X} = s_Y \sqrt{(1 - r^2)(n - 1)/(n - 2)}. \tag{17.16}$$

[†]Sutton (1990) shows that

$$r^2 = \frac{F}{F + \nu_2}. \tag{17.17}$$

EXAMPLE 17.3 Analysis of variance testing of H_0: $\beta = 0$ against H_A: $\beta \neq 0$, using the data of Examples 17.1 and 17.2.

$n = 13$

$\sum Y = 44.4$

$\sum Y^2 = 171.30$

$\text{total SS} = \sum y^2 = 171.30 - \dfrac{(44.4)^2}{13}$

$\qquad\qquad\quad = 171.30 - 151.6431$

$\qquad\qquad\quad = 19.656923$

$\text{total DF} = n - 1 = 12$

$\sum xy = 70.80$ (from Example 17.2)

$\sum x^2 = 262.00$ (from Example 17.2)

$\text{regression SS} = \dfrac{\left(\sum xy\right)^2}{\sum x^2} = \dfrac{(70.80)^2}{262.00}$

$\qquad\qquad\qquad = \dfrac{5012.64}{262.00}$

$\qquad\qquad\qquad = 19.132214$

Source of variation	SS	DF	MS
Total	19.656923	12	
Linear regression	19.132214	1	19.132214
Residual	0.524709	11	0.047701

$F = \dfrac{19.132214}{0.047701} = 401.1$

$F_{0.05(1),1,11} = 4.84$

Therefore, reject H_0.

$$P \ll 0.0005 \quad [P = 0.00000000053]$$

$r^2 = \dfrac{19.132214}{19.656923} = 0.97$

$s_{Y \cdot X} = \sqrt{0.047701} = 0.218$ cm

***t* Testing.**   The preceding null hypothesis concerning β can also be tested by using Student's t statistic. In fact, the more general two-tailed hypotheses, H_0: $\beta = \beta_0$ and H_A: $\beta \neq \beta_0$, can be tested in this fashion.* Most frequently, β_0 is zero in these hypotheses, in which case either the analysis of variance or the t test may be employed. But if any value of β other than zero is hypothesized, then the following procedure is applicable, whereas the analysis of variance is not. Also, the t-testing procedure allows for the testing of one-tailed hypotheses: either H_0: $\beta \leq \beta_0$ and H_A: $\beta > \beta_0$, or H_0: $\beta \geq \beta_0$ and H_A: $\beta < \beta_0$.

Since the t statistic is in general calculated as

$$t = \frac{(\text{parameter estimate}) - (\text{parameter value hypothesized})}{\text{standard error of parameter estimate}}, \tag{17.18}$$

we need to compute s_b, the standard error of the regression coefficient.

---

*The use of t for testing regression coefficients emanates from Fisher (1922a).

The variance of b is calculated as

$$s_b^2 = \frac{s_{Y \cdot X}^2}{\sum x^2}.$$ (17.19)

Therefore,

$$s_b = \sqrt{\frac{s_{Y \cdot X}^2}{\sum x^2}},$$ (17.20)

and

$$t = \frac{b - \beta_0}{s_b}.$$ (17.21)

To test H_0: $\beta = 0$ against H_A: $\beta \neq 0$ in Example 17.4, $s_b = 0.0135$ cm/day, so $t = (b - \beta_0)/s_b = (0.270 - 0)/0.0135 = 20.000$. The degrees of freedom for this testing procedure are $n - 2$; thus, the critical value of t, at the 5% significance level, is $t_{0.05(2), 11} = 2.201$, and H_0 is rejected. In a case such as this (i.e., when $\beta_0 = 0$), where either the analysis of variance or the t test may be employed, $t^2 = F$ and $t_{\alpha(2), (n-2)}^2 = F_{\alpha(1), 1, (n-2)}$. The reader may refer to Section 7.2 to review the concepts and procedures involved in one-tailed t testing.

EXAMPLE 17.4   Use of Student's t to test H_0: $\beta = 0$ against H_A: $\beta \neq 0$, employing the data of Examples 17.1 and 17.2.

$n = 13$

$b = 0.270$ cm/day

$$s_b = \sqrt{\frac{s_{Y \cdot X}^2}{\sum x^2}} = \sqrt{\frac{0.047701}{262.00}} = \sqrt{0.00018206} = 0.0135 \text{ cm/day}$$

$$t = \frac{b - 0}{s_b} = \frac{0.270}{0.0135} = 20.000$$

$t_{0.05(2), 11} = 2.201$

Therefore, reject H_0.

$$P \ll 0.01 \quad [P = 0.00000000027]$$

17.4   CONFIDENCE INTERVALS IN REGRESSION

In many cases, knowing the standard error of a statistic allows us to calculate confidence intervals for the parameter being estimated, as

$$\text{confidence interval} = \text{statistic} \pm (t)(\text{SE of statistic}).$$ (17.22)

This was first demonstrated in Section 7.3 for confidence intervals for means, and it has been used repeatedly in succeeding chapters. In addition, the second significant figure of the standard error of a statistic may be used as an indicator of the precision to which that statistic should be reported (as done with the mean in Section 7.4). The standard error of b has been given in Equation 17.20. For the data in Examples 17.2, the second significant figure of $s_b = 0.0135$ cm/day enables us to express b to the third decimal place (i.e., $b = 0.270$ cm/day).

Confidence Interval for the Regression Coefficient.   For the $(1 - \alpha)$ confidence limits of β, we calculate

$$b \pm t_{\alpha(2),(n-2)}s_b. \tag{17.23}$$

Therefore,

$$L_1 = b - t_{\alpha(2),(n-2)}s_b \tag{17.24}$$

and

$$L_2 = b + t_{\alpha(2),(n-2)}s_b. \tag{17.25}$$

For Example 17.2, the 95% confidence interval for β would be $b \pm t_{0.05(2),11}s_b = 0.270 \pm (2.201)(0.0135) = 0.270 \pm 0.030$ cm/day. Thus, $L_1 = 0.270 - 0.030 = 0.240$ cm/day, $L_2 = 0.270 + 0.030 = 0.300$ cm/day; and we can state, with 95% confidence (i.e., we state that there is no greater than a 5% chance that we are wrong), that 0.240 cm/day and 0.300 cm/day form an interval that includes the population regression coefficient, β. Figure 17.7 shows, by the broken lines, these confidence limits for the slope of the regression line. Within these limits, the various possible b values rotate the line about the point $(\bar{X}, \bar{Y})$.

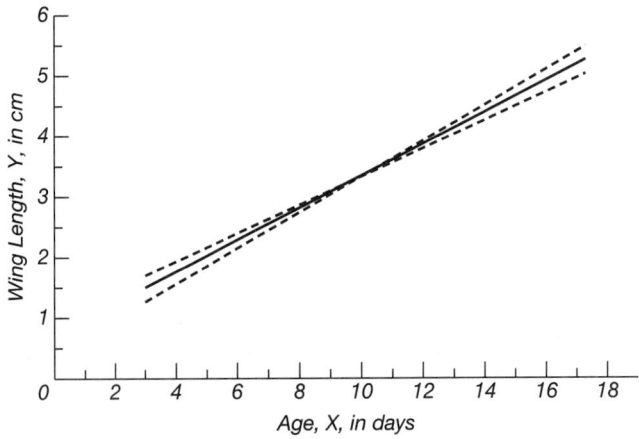

Figure 17.7   The regression line from Fig. 17.1, showing, by broken lines, the lines with slopes equal to the upper and lower 95% confidence limits for β.

Confidence Interval for an Estimated Y. As shown in Section 17.1, a regression equation allows one to estimate the value of $\hat{Y}$, existing in the population at a given value of X_i. The standard error of such a population value is

$$s_{\hat{Y}_i} = \sqrt{s_{Y \cdot X}^2 \left[\frac{1}{n} + \frac{(X_i - \bar{X})^2}{\sum x^2} \right]}. \tag{17.26}$$

Example 17.5A shows how $s_{\hat{Y}_i}$ can be used in Equation 17.22 to calculate confidence intervals. It is apparent from Equation 17.26 that the standard error is a minimum for $X_i = \bar{X}$, and that it increases as estimates are made at values of X_i farther from the mean. If confidence limits were calculated for all points on the regression line, the result would be the *confidence bands* shown in Fig. 17.8.

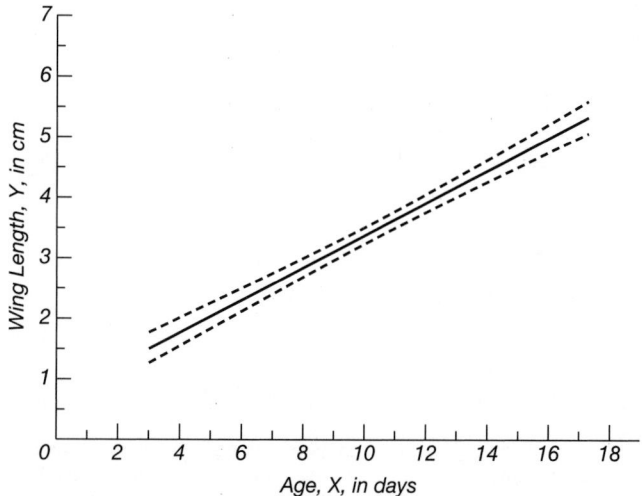

Figure 17.8 The 95% confidence bands (broken lines) for the regression line from Fig. 17.1 (the regression of Example 17.2).

If $X_i = 0$, then $\hat{Y}_i = a$, the Y intercept. Therefore,

$$s_a = \sqrt{s_{Y \cdot X}^2 \left[\frac{1}{n} + \frac{\bar{X}^2}{\sum x^2} \right]}. \tag{17.27}$$

If we predict a value of $\hat{Y}$ that is the mean of m additional measurements at X_i, its standard error would be

$$(s_{\hat{Y}_i})_m = \sqrt{s_{Y \cdot X}^2 \left[\frac{1}{m} + \frac{1}{n} + \frac{(X_i - \bar{X})^2}{\sum x^2} \right]}. \tag{17.28}$$

EXAMPLE 17.5 Standard errors of predicted values of Y.

The regression equation derived in Example 17.2 is used for the following considerations. For this regression, $a = 0.72$ cm, $b = 0.270$ cm/day, $\bar{X} = 10.0$ days, $\sum x^2 = 262.00$ days^2, $n = 13$, $s^2_{Y \cdot X} = 0.0477$ cm^2, and $t_{0.05(2), 11} = 2.201$.

A. Equation 17.26 is used when one wishes to predict the mean value of $\hat{Y}_i$, given X_i, in the entire population. For example, we could ask: "What is the mean wing length of all 13.0-day-old birds in the population under study?"

$$\hat{Y}_i = a + bX_i$$

$$= 0.715 + (0.270)(13.0)$$

$$= 0.715 + 3.510$$

$$= 4.225 \text{ cm}$$

$$s_{\hat{Y}_i} = \sqrt{s^2_{Y \cdot X} \left[\frac{1}{n} + \frac{(X_i - \bar{X})^2}{\sum x^2} \right]} \qquad (17.26)$$

$$= \sqrt{0.047701 \left[\frac{1}{13} + \frac{(13.0 - 10.0)^2}{262.00} \right]}$$

$$= \sqrt{(0.047701)(0.111274)}$$

$$= 0.073 \text{ cm}$$

95% confidence interval $= \hat{Y}_i \pm t_{0.05(2), 11} s_{\hat{Y}_i}$

$$= 4.225 \pm (2.201)(0.073)$$

$$= 4.225 \pm 0.161 \text{ cm}$$

$L_1 = 4.064$ cm

$L_2 = 4.386$ cm

B. Equation 17.28 is used when one proposes taking an additional sample of m individuals from the population and wishes to predict the mean Y value, at a given X, for these m new data. For example, one might ask: "If ten 13.0-day-old birds were taken from the population, what would be their mean wing length?"

$$\hat{Y}_i = 0.715 + (0.270)(13.0) = 4.225 \text{ cm}$$

$$(s_{\hat{Y}_i})_{10} = \sqrt{0.047701 \left[\frac{1}{10} + \frac{1}{13} + \frac{(13.0 - 10.0)^2}{262.00} \right]} \qquad (17.28)$$

$$= \sqrt{(0.047701)(0.211274)}$$

$$= 0.100 \text{ cm}$$

95% confidence interval $= \hat{Y}_i \pm t_{0.05(2), 11} (s_{\hat{Y}_i})_{10}$

$$= 4.225 \pm (2.201)(0.100)$$

$$= 4.225 \pm 0.220 \text{ cm}$$

$L_1 = 4.005$ cm

$L_2 = 4.445$ cm

EXAMPLE 17.5   (continued)

C. Equation 17.29 is used when one wishes to predict the Y value of a single observation taken from the population as a specified X. For example, one could ask: "If one 13.0-day-old bird were taken from the population, what would be its wing length?"

$$\hat{Y}_i = 0.715 + (0.270)(13.0) = 4.225 \text{ cm}$$

$$(s_{\hat{Y}_i})_1 = \sqrt{0.047701 \left[1 + \frac{1}{13} + \frac{(13.0 - 10.0)^2}{262.00} \right]} \qquad (17.29)$$

$$= \sqrt{(0.047701)(1.111274)}$$

$$= 0.230 \text{ cm}$$

$$95\% \text{ confidence interval} = \hat{Y}_i \pm t_{0.05(2),11}(s_{\hat{Y}_i})_1$$

$$= 4.225 \pm (2.201)(0.230)$$

$$= 4.225 \pm 0.506 \text{ cm}$$

$$L_1 = 3.719 \text{ cm}$$

$$L_2 = 4.731 \text{ cm}$$

Note from these three examples that the accuracy of prediction increases as does the number of data upon which the prediction is based. That is, for example, predictions about a mean for the entire population will be more accurate than a prediction about a mean from ten members of the population, which is more accurate than a prediction about a single member of the population.

A special case of Equation 17.28 exists when we ask for the standard error involved in estimating $\hat{Y}_i$ for a single additional measurement at Y_i:

$$(s_{\hat{Y}_i})_1 = \sqrt{s_{Y \cdot X}^2 \left[1 + \frac{1}{n} + \frac{(X_i - \bar{X})^2}{\sum x^2} \right]}. \qquad (17.29)$$

Equation 17.26 is the special case of Equation 17.28 when m approaches infinity. Confidence intervals using Equations 17.28 and 17.29 are sometimes called *prediction intervals*. Examples 17.5B and C demonstrate the uses of the standard errors of predictions.

Testing Hypotheses about Estimated Y Values.   Once we have computed the standard error of a predicted Y, we can test hypotheses about that prediction. For example, we might ask whether the mean population wing length of 13.0-day-old sparrows, call it $\mu_{\hat{Y}_{13.0}}$, is equal to some specified value (two-tailed test) or is greater than (or less than) some specified value (one-tailed test). We simply refer to Equation 17.18, as Example 17.6 demonstrates.

Confidence Interval and Hypothesis Testing for the Residual Mean Square. The sample residual mean square of the population $s_{Y \cdot X}^2$ is an estimate of the residual mean square in the population $\sigma_{Y \cdot X}^2$. Confidence limits may be calculated for $\sigma_{Y \cdot X}^2$ just as they are for the population variance, σ^2, in Section 8.6. Simply use $\nu = n - 2$, rather than $\nu = n - 1$, and replace σ^2 with $\sigma_{Y \cdot X}^2$ and SS with residual SS in Equation 7.15 or

7.16. Also, a confidence interval for the population standard error of estimate, $\sigma_{Y \cdot X}$, may be obtained by analogy to the Equation 7.17. Hypothesis testing for $\sigma_{Y \cdot X}^2$ or $\sigma_{Y \cdot X}$ may be performed by analogy to the procedures of Section 7.11.

EXAMPLE 17.6   Hypothesis testing with an estimated Y value.

H_0:   The mean population wing length of 13.0-day-old birds is not greater than 4 cm (i.e., H_0: $\mu_{\hat{Y}_{13.0}} \leq 4$ cm).

H_A: The mean population wing length of 13.0-day-old birds is greater than 4 cm (i.e., H_A: $\mu_{\hat{Y}_{13.0}} > 4$ cm).

From Example 17.5, $Y_{13.0} = 4.225$ cm and $s_{\hat{Y}_{13.0}} = 0.073$ cm,

$$t = \frac{4.225 - 4}{0.073} = \frac{0.225}{0.073} = 3.082$$

$t_{0.05(1),11} = 1.796$

Therefore, reject H_0.

$$0.005 < P < 0.01 \quad [P = 0.0052]$$

17.5 INVERSE PREDICTION

Situations exist where a biologist wishes to predict the value of the independent variable (X_i) that is to be expected in the population at a specified value of the dependent variable (Y_i), a procedure known as *inverse prediction*. In Example 17.1, for instance, we might ask, "How old is a bird that has a wing 4.5 cm long?" By simple algebraic rearrangement of the linear regression equation (Equation 17.8), we obtain

$$\hat{X}_i = \frac{Y_i - a}{b}. \tag{17.30}$$

From Fig. 17.8, it is clear that, although confidence limits calculated around the predicted $\hat{Y}_i$ are symmetrical above and below $\hat{Y}_i$, confidence limits associated with the predicted $\hat{X}_i$ are not symmetrical to the left and to the right of $\hat{X}_i$. The $1 - \alpha$ confidence limits for the X predicted at a given Y may be calculated as follows, which is demonstrated in Example 17.7:

$$\bar{X} + \frac{b(Y_i - \bar{Y})}{K} \pm \frac{t}{K} \sqrt{s_{Y \cdot X}^2 \left[\frac{(Y_i - \bar{Y})^2}{\sum x^2} + K \left(1 + \frac{1}{n}\right) \right]}. \tag{17.31}$$

where* $K = b^2 - t^2 s_b^2$. This computation is a special case of the prediction of the $\hat{X}$ associated with multiple values of Y at that X. Such a situation would be where, for

---

*It may be recalled that $F_{\alpha(1),1,\nu} = t_{\alpha(2),\nu}^2$. Therefore, we could compute $K = b^2 - F s_b^2$, where $F = t_{\alpha(2),(n-2)}^2 = F_{\alpha(1),1,(n-2)}$.

EXAMPLE 17.7   Inverse prediction.

We wish to estimate, with 95% confidence, the age of a bird with a wing length of 4.5 cm.

Predicted age:

$$\hat{X} = \frac{Y_i - a}{b}$$

$$= \frac{4.5 - 0.715}{0.270}$$

$$= 14.019 \text{ days}$$

To compute 95% confidence interval:

$$t = t_{0.05(2),11} = 2.201$$

$$K = b^2 - t^2 s_b^2$$

$$= 0.270^2 - (2.201)^2 (0.0135)^2$$

$$= 0.0720$$

95% confidence interval:

$$\bar{X} + \frac{b(Y_i - \bar{Y})}{K} \pm \frac{t}{K} \sqrt{s_{Y \cdot X}^2 \left[\frac{(Y_i - \bar{Y})^2}{\sum x^2} + K \left(1 + \frac{1}{n} \right) \right]} \tag{17.31}$$

$$= 10.0 + \frac{0.270(4.5 - 3.415)}{0.0720}$$

$$\pm \frac{2.201}{0.0720} \sqrt{0.047701 \left[\frac{(4.5 - 3.415)^2}{262.00} + 0.0720 \left(1 + \frac{1}{13} \right) \right]}$$

$$= 10.0 + 4.069 \pm 30.569 \sqrt{0.003913}$$

$$= 14.069 \pm 1.912 \text{ days}$$

$$L_1 = 12.157 \text{ days}$$

$$L_2 = 15.981 \text{ days}$$

---

data as in Example 17.1, we had wing length measurements from m birds of the same age and wished to estimate that age. The predicted X would be

$$\hat{X}_i = \frac{\bar{Y}_i - a}{b}, \tag{17.32}$$

where $\bar{Y}_i$ is the mean of the m values of Y_i; and the confidence limits would be calculated as

$$\bar{X} + \frac{b(\bar{Y}_i - \bar{Y})}{K} \pm \frac{t}{K} \sqrt{(s_{Y \cdot X}^2)' \left[\frac{(\bar{Y}_i - \bar{Y})^2}{\sum x^2} + K \left(\frac{1}{m} + \frac{1}{n} \right) \right]}, \tag{17.33}$$

where* $t = t_{\alpha(2),(n+m-3)}$, $K = b^2 - t^2(s_b^2)'$,

$$(s_b^2)' = \frac{(s_{Y \cdot X}^2)'}{\sum x^2},$$ (17.34)

and

$$(s_{Y \cdot X}^2)' = \text{residual SS} + \sum_{j=1}^{m}(Y_{ij} - \bar{Y}_i)^2/(n + m - 3)$$ (17.35)

(Ostle and Mensing, 1975: 180–181; Seber, 1977: 190–191).

17.6 INTERPRETATIONS OF REGRESSION FUNCTIONS

If we calculate the two constants, a and b, that define a linear regression equation, then we have quantitatively described the rate of change of Y with a change in X. However, although we have assumed a mathematical dependence of Y on X, we must not automatically assume that there is a biological cause-and-effect relationship. Causal relationships are concluded only with some insight into the natural phenomenon being investigated and may not be declared by statistical testing alone. Indeed, it is often necessary to determine the interrelationships between the two variables under study and other variables, for an observed dependence may, in fact, be due to the influence of one or more additional variables. (The methods in Chapter 20 are often used in this regard.)

We must also remember that a linear regression function is mathematically nothing more than a straight line forced to fit through a set of data points, and it may not at all describe a natural phenomenon. The biologist may be chagrined when attempting to explain why the observed relationship is well described by a linear function or what biological insights are to be unfolded by the consideration of a particular slope or a particular magnitude of a Y intercept. In other words, although an empirically derived regression function often provides a satisfactory and satisfying description of a natural system, sometimes it does not. Section 20.15 further discusses the fitting of regression models.

Even if a regression function does not help us to explain the functional anatomy of a natural system, it may still be useful in its ability to predict Y, given X. In the sciences, equations may inaccurately represent natural processes yet may be employed advantageously to predict the value of one variable given the value of an associated variable. Thus, predicting $\hat{Y}$ (or $\hat{X}$) values and their standard errors is frequently a useful end in itself.

---

*Alternatively, one may compute $K = b^2 - F(s_b^2)_*$, where $F = t^2_{\alpha(2),(n+m-3)} = F_{\alpha(1),1,(n+m+3)}$.

17.7 REGRESSION WITH REPLICATION AND TESTING FOR LINEARITY

If, in Example 17.1, we had wing measurements for more than one bird for at least some of the recorded ages, then we could, by the procedure of this Section, test the null hypothesis that the population regression is linear.* (Note that true replication requires that there are multiple birds at a given age, not that there are multiple wing measurements on the same bird.) Figure 17.9 presents the data of Example 17.8. A least squares, best fit, linear regression equation can be calculated for any set of at least two data, but neither the equation itself nor the testing for a significant slope (which requires at least three data) indicates whether Y is, in fact, a straight line function of X in the population sampled.

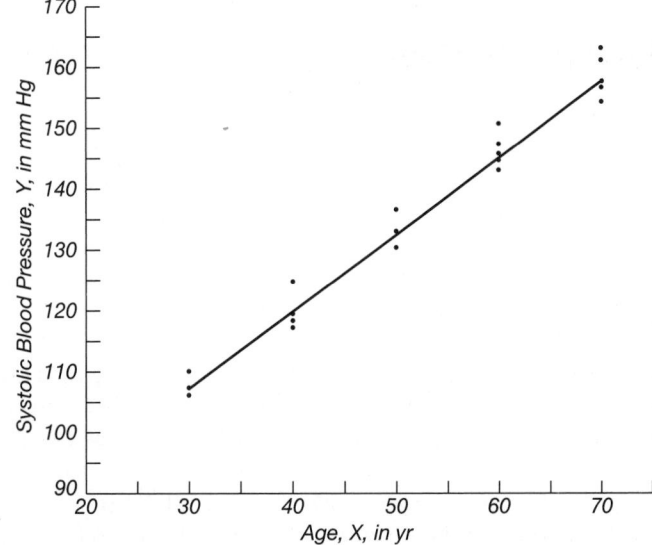

Figure 17.9   A regression where there are multiple values of Y for each value of X.

We occasionally encounter the suggestion that for data such as those in Fig. 17.9 the mean Y at each X be utilized for a regression analysis. However, to do so would be to discard information, and such a procedure is not recommended (Freund, 1971).

Example 17.8 appropriately analyzes data consisting of multiple Y values at each X value, and Fig. 17.9 presents the data graphically. For each of the k unique X_i values we can speak of each of n_i values of Y_{ij}, using the double subscript exactly as in the one-way analysis of variance (Section 10.1). In Example 17.8, $n_1 = 3, n_2 = 4, n_3 = 3,$

---

*Thornby (1972) presents a procedure to test the hypothesis of linearity even when there are not multiple observations of Y. But the computation is rather tedious.

EXAMPLE 17.8 Hypothesis testing for a regression where there are multiple values of Y for each value of X.

Age (yr) X	Systolic blood pressure (mm Hg) Y	n_i
30	108, 110, 106	3
40	125, 120, 118, 119	4
50	132, 137, 134	3
60	148, 151, 146, 147, 144	5
70	162, 156, 164, 158, 159	5

$N = 20$

$\sum\sum X_{ij} = 1050 \qquad \sum\sum Y_{ij} = 2744$

$\sum\sum X_{ij}^2 = 59,100 \qquad \sum\sum Y_{ij}^2 = 383,346 \qquad \sum\sum X_{ij}Y_{ij} = 149,240$

$\sum x^2 = 3975.00 \qquad \sum y^2 = 6869.20 \qquad \sum xy = 5180.00$

$\bar{X} = 52.5 \qquad \bar{Y} = 137.2$

$b = \dfrac{\sum xy}{\sum x^2} = \dfrac{5180.00}{3975.00} = 1.303 \text{ mm Hg/yr}$

$a = \bar{Y} - b\bar{X} = 137.2 - (1.303)(52.5) = 68.79 \text{ mm Hg}$

Therefore, the least squares regression line is $\hat{Y}_{ij} = 68.79 + 1.303 X_{ij}$.

H_0:   The population regression is linear.

H_A: The population regression is not linear.

$$\text{among-groups SS} = \sum_{i=1}^{k} \frac{\left(\sum_{j=1}^{n_i} Y_{ij}\right)^2}{n_i} - \frac{\left(\sum_{i=1}^{k}\sum_{j=1}^{n_i} Y_{ij}\right)^2}{N} = 383,228.73 - 376,476.80$$

$$= 6751.93$$

among-groups DF $= k - 1 = 4$

within-groups SS $=$ total SS $-$ among-groups SS $= 6869.20 - 6751.93 = 117.27$

within-groups DF $=$ total DF $-$ among-groups DF $= 19 - 4 = 15$

deviations-from-linearity SS $=$ among-groups SS $-$ regression SS

$$= 6751.93 - 6750.29 = 1.64$$

deviations-from-linearity DF $=$ among-groups DF $-$ regression DF $= 4 - 1 = 3$

Source of variation	SS	DF	MS
Total	6869.20	19	
Among groups	6751.93	4	
Linear regression	6750.29	1	
Deviations from linearity	1.64	3	0.55
Within groups	117.27	15	7.82

EXAMPLE 17.8   (continued)

$$F = \frac{0.55}{7.82}$$

Since $F < 1.00$, do not reject H_0.

$$P > 0.25$$

H_0: $\beta = 0$.

H_A: $\beta \neq 0$.

total SS $= \sum y^2 = 6869.20$

total DF $= N - 1 = 19$

regression SS $= \dfrac{(\sum xy)^2}{\sum x^2} = \dfrac{(5180.00)^2}{3975.00} = 6750.29$

Source of variation	SS	DF	MS
Total	6869.20	19	
Linear regression	6750.29	1	6750.29
Residual	118.91	18	6.61

$$F = \frac{6750.29}{6.61} = 1021.2$$

$F_{0.05(1),1,18} = 4.41$

Therefore, reject H_0.

$$P \ll 0.0005$$

$r^2 = \dfrac{6750.29}{6869.20} = 0.98$

$s_{Y \cdot X} = \sqrt{6.61} = 2.57$ mm Hg

etc.; and $Y_{11} = 108$ mm, $Y_{12} = 110$ mm, $Y_{32} = 137$ mm, etc. Therefore,

$$\sum xy = \sum_{i=1}^{k} \sum_{j=1}^{n_i} X_{ij} Y_{ij} - \frac{(\sum \sum X_{ij})(\sum \sum Y_{ij})}{N}, \tag{17.36}$$

where $N = \sum_{i=1}^{k} n_i$ the total number of pairs of data. Also,

$$\sum x^2 = \sum_{i=1}^{k} \sum_{j=1}^{n_i} X_{ij}^2 - \frac{(\sum \sum X_{ij})^2}{N}, \tag{17.37}$$

and

$$\text{total SS} = \sum y^2 = \sum_{i=1}^{k} \sum_{j=1}^{n_i} Y_{ij}^2 - C, \tag{17.38}$$

where

$$C = \frac{\left(\sum \sum Y_{ij}\right)^2}{N}, \text{ and } N = \sum n_i. \tag{17.39}$$

It is then a simple matter, as Example 17.8 shows, to calculate the regression coefficient b, the Y intercept a, and the regression and residual sums of squares, using Equations 17.4, 17.7, 17.11, and 17.13, respectively. The total, regression, and residual degrees of freedom are $N - 1$, 1, and $N - 2$, respectively.

As shown in Section 17.3, the analysis of variance for significant slope involves the partitioning of the total variability of Y (i.e., $Y_{ij} - \bar{Y}$) into that variability due to regression ($\hat{Y}_i - \bar{Y}$) and that variability remaining (i.e., residual) after the regression line is fitted ($Y_{ij} - \hat{Y}_i$). However, by considering the k groups of Y values, we can also partition the total variability exactly as we did in the one-way analysis of variance (Section 10.1), by describing variability among groups ($\bar{Y}_i - \bar{Y}$) and within groups ($Y_{ij} - \bar{Y}_i$):

$$\text{among-groups SS} = \sum_{i=1}^{k} \frac{\left(\sum_{j=1}^{n_i} Y_{ij}\right)^2}{n_i} - C, \tag{17.40}$$

$$\text{among-groups DF} = k - 1, \tag{17.41}$$

$$\text{within-groups SS} = \text{total SS} - \text{among-groups SS}, \tag{17.42}$$

$$\text{within-groups DF} = \text{total DF} - \text{among-groups DF} = N - k. \tag{17.43}$$

The variability among groups can, in turn, be partitioned. Part of this variability ($\hat{Y}_i - \bar{Y}$) results from the linear regression being fitted, whereas the remainder is due to the deviation of each group of data from the regression line ($\bar{Y}_i - \hat{Y}_i$), as shown in Fig. 17.10. Therefore,

$$\text{deviations-from-linearity SS} = \text{among-groups SS} - \text{regression SS} \tag{17.44}$$

and

$$\begin{aligned}\text{deviations-from-linearity DF} &= \text{among-groups DF} - \text{regression DF} \\ &= k - 2. \end{aligned} \tag{17.45}$$

Table 17.2 summarizes this partitioning of sums of squares.

Alternatively, and with identical results, we may consider the residual variability ($Y_{ij} - \hat{Y}_i$) to be divisible into two components: within-groups variability ($Y_{ij} - \bar{Y}_i$) and deviations-from-linearity ($\bar{Y}_i - \hat{Y}_i$). This partitioning of sums of squares and degrees of freedom is summarized in Table 17.3.[*]

If the population relationship between Y and X is a straight line (i.e., "H_0: The population regression is linear" is a true statement), then the deviations-from-linearity

---

[*]Some authors refer to deviations-from-linearity as "lack of fit" and to within groups variability as "pure error."

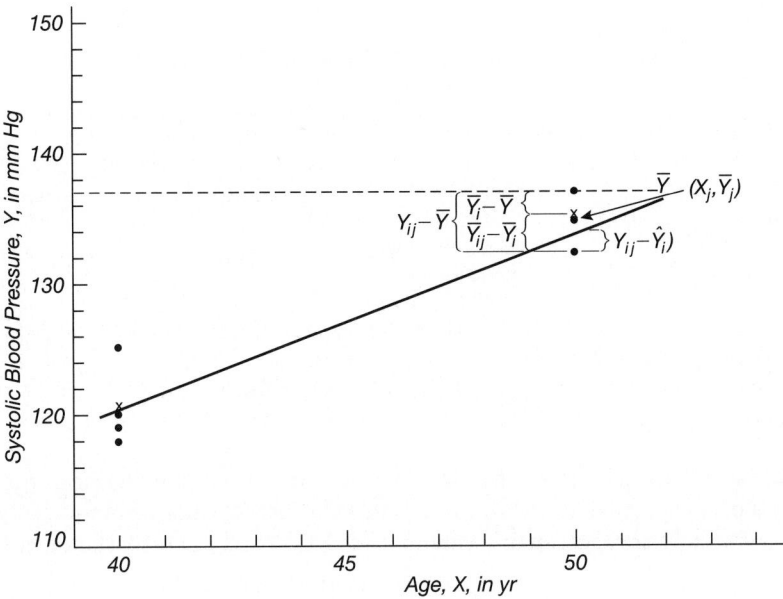

Figure 17.10   An enlarged portion of Fig. 17.9, showing the partitioning of Y deviations.

TABLE 17.2   Summary of the Analyses of Variance Calculations for Testing H_0: the Population Regression is Linear, and for Testing H_0: $\beta = 0$

Source of variation	Sum of squares (SS)	DF	Mean Square (MS)
Total $[Y_{ij} - \bar{Y}]$	$\sum y^2$	$N - 1$	
Linear regression $[\hat{Y}_i - \bar{Y}]$	$\dfrac{\left(\sum xy\right)^2}{\sum x^2}$	1	$\dfrac{\text{regression SS}}{\text{regression DF}}$
Residual $[Y_{ij} - \hat{Y}_i]$	total SS − regression SS	$N - 2$	$\dfrac{\text{residual SS}}{\text{residual DF}}$
Among groups $[\bar{Y}_i - \bar{Y}]$	$\displaystyle\sum_{i=1}^{k} \dfrac{\left(\sum\limits_{j=1}^{n_i} Y_{ij}\right)^2}{n_i} - \dfrac{\left(\sum\limits_{i=1}^{k}\sum\limits_{j=1}^{n_i} Y_{ij}\right)^2}{N}$	$k - 1$	
Linear regression $[\hat{Y}_i - \bar{Y}]$	$\dfrac{\left(\sum xy\right)^2 2}{\sum x^2}$	1	
Deviations from linearity $[\bar{Y}_i - \hat{Y}_i]$	among-groups SS − regression SS	$k - 2$	$\dfrac{\text{deviations SS}}{\text{deviations DF}}$
Within groups $[Y_{ij} - \bar{Y}_i]$	total SS − among-groups SS	$N - k$	$\dfrac{\text{within-groups SS}}{\text{within-groups DF}}$

Note: To test H_0: the population regression is linear, we use $F = $ deviations MS/within-groups MS, with a critical value of $F_{\alpha(1),(k-2),(N-k)}$. If that null hypothesis is not rejected, then H_0: $\beta = 0$ is tested using $F = $ regression MS/residual MS, with a critical value of $F_{\alpha(1),1,(N-2)}$; if the null hypothesis of linearity is rejected, then H_0: $\beta = 0$ is tested using $F = $ regression MS/within-groups MS, with a critical value of $F_{\alpha(1),1,(N-k)}$.

TABLE 17.3 Summary of Analysis of Variance
Partitioning of Sources of Variation for Testing
Linearity, as an Alternative to That in Table 17.2

Source of variation	DF
Total $[Y_{ij} - \bar{Y}]$	$N - 1$
Among groups $[\bar{Y}_i - \bar{Y}]$	$k - 1$
Within groups $[Y_{ij} - \bar{Y}_i]$	$N - k$
Linear regression $[\hat{Y}_i - \bar{Y}]$	1
Residual $[Y_{ij} - \hat{Y}_i]$	$N - 2$
Within groups $[Y_{ij} - \bar{Y}_i]$	$N - k$
Deviations from linearity $[\bar{Y}_i - \hat{Y}_i]$	$k - 2$

Note: Sums of squares and mean squares are as in
Table 17.2.

MS and the within-groups MS will be estimates of the same variance; if the relationship is not a straight line (the H_0 is false), then the deviations-from-linearity MS will be significantly greater than the within-groups MS. Thus, as demonstrated in Example 17.8,

$$F = \frac{\text{deviations-from-linearity MS}}{\text{within-groups MS}} \qquad (17.46)$$

provides a one-tailed test of the null hypothesis of linearity. (If all n_i's are equal, then performing a regression using the k $\bar{Y}$'s will result in the same b and a as will the calculations using all N Y_i's but the significance test for β will be much less powerful and the above test of linearity will be impossible to perform.)

If the null hypothesis of linearity is not rejected, then the deviations-from-linearity MS and the within-groups MS may be considered to be estimates of the same population variance. The latter will be the better estimate, as it is based on more degrees of freedom; but an even better estimate is the residual MS, which is $s_{Y \cdot X}^2$, for it constitutes a pooling of the deviations MS and the within-groups MS. Therefore, if a regression is assumed to be linear, $s_{Y \cdot X}^2$ is the appropriate variance to use in the computation of standard errors (e.g., by Equations 17.20, 17.26–17.29) and confidence intervals resulting from them, and this residual mean square ($s_{Y \cdot X}^2$) is also appropriate in testing the hypothesis H_0: $\beta = 0$ (by either Equation 17.14 or 17.21), as demonstrated in Example 17.8.

If the population regression is concluded not to be linear, then the investigator would do well to consider the procedures of Section 17.10 or 20.14 or of Chapter 21, and to consider the linear regression analysis no further. If, however, we desire to test H_0: $\beta = 0$, then the within-groups MS should be substituted for the residual MS ($s_{Y \cdot X}^2$). It would not be advisable to engage in predictions with the linear regression equation.

17.8 POWER AND SAMPLE SIZE IN REGRESSION

Although there are basic differences between regression and correlation (Section 17.1), a set of data for which there is a statistically significant regression coefficient (i.e., one rejects H_0: $\beta = 0$, as explained in Section 17.3) would also yield a statistically significant correlation coefficient (i.e., one would reject H_0: $p = 0$, to be discussed in Section 19.2). In addition, conclusions about the power of a significance test for a regression coefficient can be obtained by estimating power associated with the significance test for the correlation coefficient that would have been obtained from the same set of data.

After performing a regression analysis for a set of data, we may obtain the sample correlation coefficient, r, either from Equation 18.1, or, more simply, as

$$r = b\sqrt{\frac{\sum x^2}{\sum y^2}}, \tag{17.47}$$

or we may take the square root of the coefficient of determination, r^2, assigning to it the sign of b. Then, with r in hand, the procedures of Section 19.4 may be employed (Cohen, 1988: 76–77) to estimate power and minimum required sample size for the hypothesis test for the regression coefficient, H_0: $\beta = 0$.

17.9 REGRESSION THROUGH THE ORIGIN

Although not of common biological importance, a special type of regression procedure is called for when we are faced with sets of data for which we know, a priori, that in the population Y will be zero when X is zero; i.e., the population Y intercept is known to be zero. Since the point on the graph with coordinates (0, 0) is termed the *origin*, this regression situation is known as regression through the origin. In this type of regression analysis both variables must be measured on a ratio scale, for only such a scale has a true zero (Section 1.1).

For regression through the origin, the linear regression equation would be

$$\hat{Y}_i = bX_i, \tag{17.48}$$

and some of the calculations unique to such a regression are as follows:

$$b = \frac{\sum X_i Y_i}{\sum X_i^2}, \tag{17.49}$$

$$\text{total SS} = \sum Y_i^2, \quad \text{with total DF} = n, \tag{17.50}$$

$$\text{regression SS} = \frac{\left(\sum X_i Y_i\right)^2}{\sum X_i^2}, \quad \text{with regression DF} = 1, \tag{17.51}$$

$$s_{Y \cdot X}^2 = \text{residual MS} = \frac{\sum Y_i^2 - \dfrac{\left(\sum X_i Y_i\right)^2}{\sum X_i^2}}{n - 1} \tag{17.52}$$

$$s_b^2 = \frac{s_{Y \cdot X}^2}{\sum X_i^2}. \tag{17.53}$$

Tests of hypotheses about the slope of the line are performed, as explained earlier in this chapter, with the exception that the above values are used; $n - 1$ is used as degrees of freedom whenever $n - 2$ is used for regressions not assumed to pass through the origin. Some statisticians advise against expressing r^2 for this type of regression. Bissell (1992) discusses potential difficulties with, and alternatives to, this regression model. Interestingly, a regression line forced through the origin does not necessarily pass through point $(\bar{X}, \bar{Y})$.

Confidence Intervals. For regressions passing through the origin, confidence intervals may be obtained in ways analogous to the procedures in Section 17.4. That is, a confidence interval for the population regression coefficient, β, is calculated using Equation 17.53 for s_b^2 and $n - 14$ degrees of freedom in place of $n - 2$. A confidence interval for an estimated $\hat{Y}$ is

$$s_{\hat{Y}} = \sqrt{s_{Y \cdot X}^2 \left(\frac{X_i^2}{\sum X^2} \right)}, \tag{17.54}$$

using the $s_{Y \cdot X}^2$ of Equation 17.52; a confidence interval for $\hat{Y}_i$ predicted as the mean of m additional measurements at X_i is

$$(s_{\hat{Y}})_m = \sqrt{s_{Y \cdot X}^2 \left(\frac{1}{m} + \frac{X_i^2}{\sum X^2} \right)} \tag{17.55}$$

and a confidence interval for the $\hat{Y}_i$ predicted for one additional measurement of X_i is

$$(s_{\hat{Y}})_1 = \sqrt{s_{X \cdot Y}^2 \left(1 + \frac{X_i^2}{\sum X^2} \right)} \tag{17.56}$$

(Seber, 1977: 192).

Inverse Prediction. For inverse prediction (see Section 17.5) with a regression passing through the origin,

$$\hat{X}_i = \frac{Y_i}{b}, \tag{17.57}$$

and the confidence interval for the X_i predicted at a given Y is

$$\bar{X} + \frac{bY_i}{K} \pm \frac{t}{K} \sqrt{s_{Y \cdot X}^2 \left(\frac{Y_i^2}{\sum X_i^2} + K \right)}, \tag{17.58}$$

where $t = t_{\alpha(2), (n-1)}$ and* $K = b^2 - t^2 s_b^2$ (Seber, 1977: 192).

If X is to be predicted for multiple values of Y at that X, then

$$\hat{X}_i = \frac{\bar{Y}_i}{b}, \tag{17.59}$$

---

*Alternatively, $K = b^2 - F s_b^2$ where $F = t_{\alpha(2), (n-1)}^2 = F_{\alpha(1), 1, (n-1)}$.

where $\bar{Y}_i$ is the mean of m values of Y; and the confidence limits would be calculated as

$$\bar{X} + \frac{b\bar{Y}_i}{K} \pm \frac{t}{K}\sqrt{(s_{Y \cdot X}^2)' \left(\frac{\bar{Y}_i^2}{\sum X^2} + \frac{K}{m}\right)}, \tag{17.60}$$

where $t = t_{\alpha(2),(n+m-2)}$ and* $K = b^2 - t^2(s_b^2)'$;

$$(s_b^2)' = \frac{(s_{Y \cdot X}^2)'}{\sum X^2}; \tag{17.61}$$

and

$$(s_{Y \cdot X}^2)' = \frac{\text{residual SS} + \sum_{j=1}^{m}(Y_{ij} - \bar{Y}_i)^2}{n + m - 2} \tag{17.62}$$

(Seber, 1977: 192).

17.10 DATA TRANSFORMATIONS IN REGRESSION

As noted at the end of Section 17.2, the testing of regression hypotheses and the computation of confidence intervals—though not the calculation of a and b—depend upon the assumptions of normality and homoscedasticity, with regard to the values of Y, the dependent variable. Chapter 13 discussed the logarithmic, square root, and arcsine transformations of data to achieve closer approximations to these assumptions. By consciously striving to satisfy one of the assumptions one often (but without guaranty) appeases the others. The same considerations are applicable to regression data.

Transformation of the independent variable will not affect the distribution of Y, so transformations of X generally may be made with impunity, and sometimes they conveniently convert a curved line into a straight line. However, transformations of Y do affect least squares considerations and will therefore be discussed. Acton (1966: Chapter 8) presents a readable discussion of transformations in regression.

If the values of Y are from a Poisson distribution (i.e., the data are counts, especially small counts), then the square root transformation is usually desirable:

$$Y' = \sqrt{Y + 0.5}, \tag{17.63}$$

where the values of the variable after transformation (Y') are then submitted to regression analysis. (Also refer to Section 13.2.)

If the Y values are from a binomial distribution (e.g., they are proportions or percentages), then the arcsine transformation is called for:

$$Y' = \arcsin \sqrt{Y}. \tag{17.64}$$

(See also Section 13.3.) Appendix Table B.24 allows for ready use of this transformation.

---

*Alternatively, $K = b^2 - F(s_b^2)'$, where $F = t_{\alpha(2),(n+m-2)}^2 = F_{\alpha(1),1,(n+m-2)}$.

The most commonly used transformation in regression is the logarithmic transformation (see also Section 13.1), although it is sometimes employed for the wrong reasons. This transformation,

$$Y' = \log Y, \tag{17.65}$$

or

$$Y' = \log(Y + 1), \tag{17.66}$$

is appropriate when there is heteroscedasticity owing to the standard deviation of Y at any X increasing in proportion to the value of X. When this situation exists, it implies that values of Y can be measured more accurately at low than at high values of X. Figure 17.11 shows such data (from Example 17.9) before and after the transformation.

EXAMPLE 17.9   Regression data before and after logarithmic transformation of Y.

Original data (as plotted in Fig. 17.11a), indicating the variance of Y (s_Y^2) at each X:

X	Y	s_Y^2
5	10.72, 11.22, 11.75, 12.31	0.4685
10	14.13, 14.79, 15.49, 16.22	0.8101
15	18.61, 19.50, 20.40, 21.37	1.4051
20	24.55, 25.70, 26.92, 28.18	2.4452
25	32.36, 33.88, 35.48, 37.15	4.2526

Transformed data (as plotted in Fig. 17.11b), indicating the variance of $\log Y$ ($s_{\log Y}^2$) at each X:

X	$\log Y$	$s_{\log Y}^2$
5	1.03019, 1.04999, 1.07004, 1.09026	0.000668
10	1.15014, 1.16997, 1.19005, 1.21005	0.000665
15	1.26975, 1.29003, 1.30963, 1.32980	0.000665
20	1.39005, 1.40993, 1.43008, 1.44994	0.000665
25	1.51001, 1.52994, 1.54998, 1.56996	0.000666

Many scatter plots of data imply a curved, rather than a straight line dependence of Y on X (e.g., Fig. 17.11a). Often, logarithmic or other transformations of the values of Y and/or X will result in a straight line relationship (as Fig. 17.11b) amenable to linear regression techniques. However, if original, nontransformed values of Y agree with our assumptions of normality and homoscedasticity, then the data resulting from any of the preceding transformations will not abide by these assumptions. This is often not considered, and many biologists employing transformations do so simply to straighten out a curved line and neglect to consider whether the transformed data might indeed be analyzed legitimately by least squares regression methods. If a transformation may not be used validly to straighten out a curvilinear regression, then Section 20.15 (or, perhaps Chapter 21) is applicable.

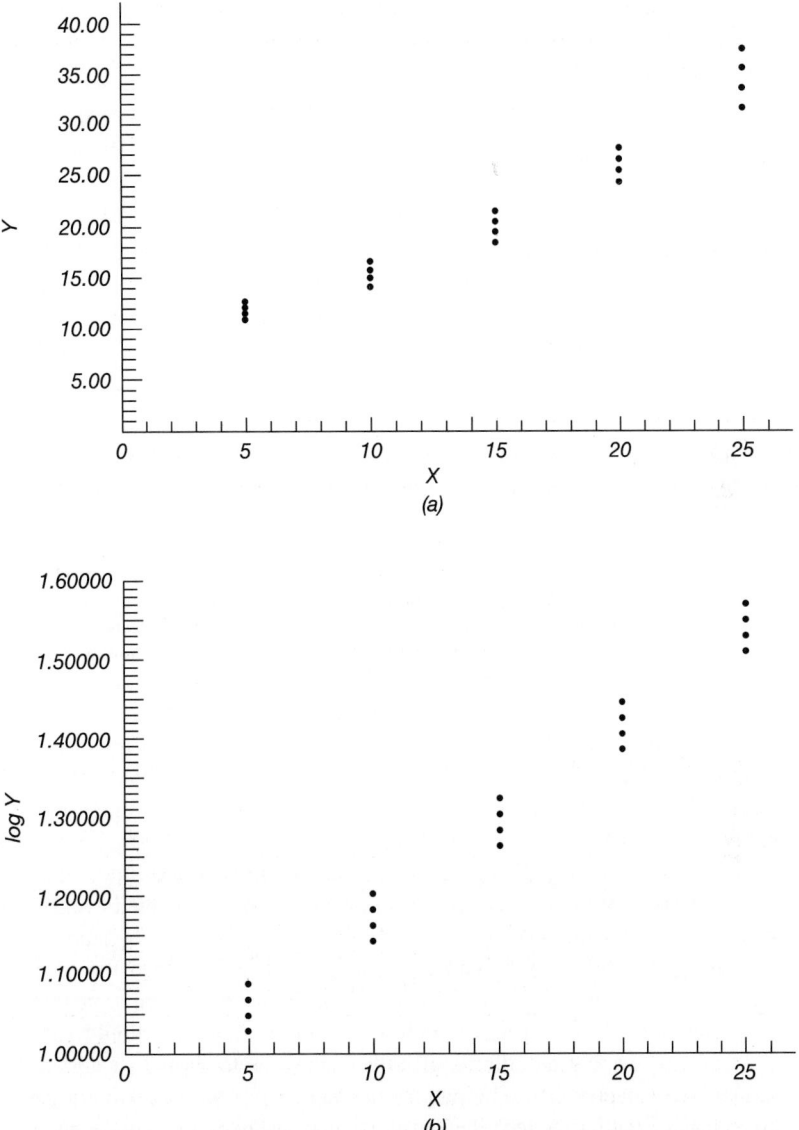

Figure 17.11   Regression data (of Example 17.9), exhibiting an increasing variability of Y with increasing magnitude of X. (a) The original data. (b) The data after logarithmic transformation of Y.

Section 13.4 mentions some other, less commonly employed, data transformations. Iman and Conover (1979) discuss rank transformation (i.e., performing a regression of the ranks of Y on the ranks of X).

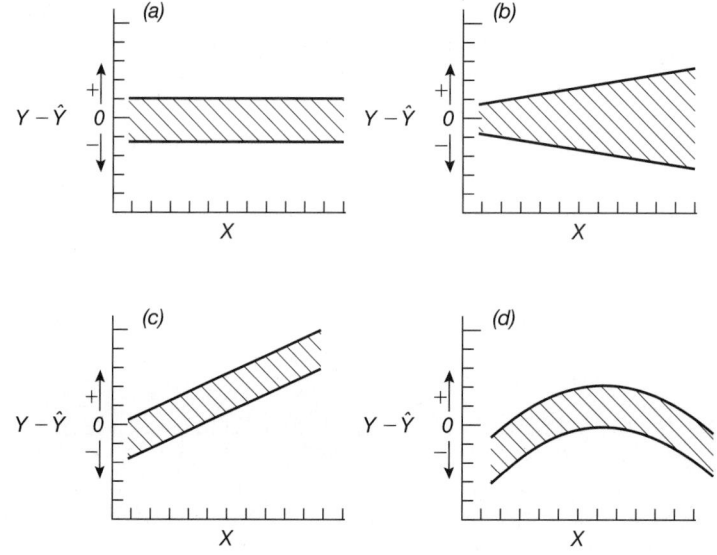

Figure 17.12 The plotting of residuals. (a) Data exhibiting homoscedasticity. (b) Data with heteroscedasticity of the sort in Example 17.9. (c) Data for which there was likely an error in the regression calculations, or an additional variable is needed in the regression model. (d) Data for which a linear regression does not accurately describe the relationship between Y and X, and a curvilinear relationship should be considered.

Examination of Residuals. Since the logarithmic transformation is frequently proposed and employed to try to achieve homoscedasticity, we should consider how we might obtain a justification for such a transformation. If a regression is fitted by least squares, then the sample residuals (i.e., values of $Y_i - \hat{Y}_i$) may be plotted as a function of their corresponding X's, as in Fig. 17.12 (see Draper and Smith, 1981: Chapter 3). If homoscedasticity exists, then the residuals should be distributed evenly above and below zero (i.e., within the shaded area in Fig. 17.12a). If there is heteroscedasticity due to increasing variability in Y with increasing values of X, then the residuals will form a pattern such as in Fig. 17.12b, and a logarithmic transformation might be attempted. If the residuals form a pattern such as in Fig. 17.12c, we should suspect a calculation error or suspect that an additional important variable should be added to the regression model (see Chapter 20). The pattern in Fig. 17.12d indicates that a *linear* regression is an improper model to describe the data; e.g., a quadratic regression (see Section 21.3) might be employed. Glejser (1969) suggests the fitting of the simple linear regression

$$E_i = a + bX_i, \qquad (17.67)$$

where $E_i = |Y_i - \hat{Y}_i|$. A statistically significant b greater than zero indicates Fig. 17.12b to be the case, and the logarithmic transformation may be attempted. Then, after the application of the transformation, a plot of the new residuals (i.e., $\log Y_i - \widehat{\log Y_i}$) should be examined and Equation 17.67 fitted, where $E_i = |\log Y_i - \widehat{\log Y_i}|$. If this regression has a b not significantly different from zero, then we may assume that the transformation was justified. An outlier (discussed at the end of Section 17.2) will appear on plots such as Fig. 17.2 as a point very far outside the pattern indicated by the shaded area.

Tests for normality in the distribution of residuals may be made by using the methods of Section 6.5 (employing $Y_i - \hat{Y}_i$ in place of X_i in that section); graphical examination of normality (as in Fig. 6.9) is often convenient.

17.11 THE EFFECT OF CODING

Either X or Y data, or both, may be coded prior to the application of regression analysis, and coding may facilitate computations, especially when the data are very large or very small in magnitude. As shown in Section 3.6 and 4.8, coding may consist of adding a constant to (or subtracting it from) X, or multiplying (or dividing) X by a constant; or both addition (or subtraction) and multiplication (or division) may be applied simultaneously. Values of Y may be coded in the same fashion, indeed simultaneously with the coding of X values, using either the same or different coding constants. If we let M_X and M_Y represent constants by which X and Y, respectively, are to be multiplied, and let A_X and A_Y be constants then to be added to $M_X X$ and $M_Y Y$ respectively, then the transformed variables, $[X]$ and $[Y]$, are

$$[X] = M_X X + A_X \tag{17.68}$$

and

$$[Y] = M_Y Y + A_Y. \tag{17.69}$$

The slope, b, will not be changed by adding constants to X and/or Y, for such transformations have the effect of simply sliding the scale of one or both axes. But if multiplication factors are used in coding, then the resultant slope, $[b]$, will be equal to $(b)(M_Y/M_X)$, and the slope that would have been calculated had coding not been employed would be $b = [b](M_X/M_Y)$. The effects of coding on other regression statistics are shown in Table 17.4, where bracketed statistics are those resulting from analysis of data coded as in Equations 17.68 and 17.69. Note that coding in no way alters the value of r^2 or the t or F statistics calculated for hypothesis testing (except as noted in the table).

A common situation involving multiplicative coding factors is one where the variables were recorded using certain units of measurement, and it is desire to determine what regression statistics would have resulted if other units of measurement had been used.

For the data in Examples 17.1, 17.2, and 17.3, $a = 0.715$ cm, and $b = 0.270$ cm/day, and $s_{Y \cdot X} = 0.218$ cm. If the wing length data were measured in inches, rather than in centimeters, there would have to be a coding by multiplying by 0.3937 in./cm (for there are 0.3937 inches in one centimeter). By consulting Table 17.4, with $M_Y = 0.3937$ in./cm, $A_Y = 0$, $M_X = 1$, and $A_X = 0$, we can calculate that if a regression analysis were run on these data where X was recorded in inches, the slope would be $[b] = (0.270$ cm/day$)(0.3937$ in./cm$) = 0.106$ in./day; the Y intercept would be

TABLE 17.4   The Effect of Coding on Regression Statistics

Statistic	Value without coding	Value using coding
Regression coefficient, b	$b = [b]M_X/M_Y$	$[b] = bM_Y/M_X$
Y intercept, a	$a = ([a] + [b]A_X - A_Y)/M_Y$	$[a] = aM_Y - [b]A_X + A_Y$
Standard error of estimate, $x_{Y \cdot X}$	$s_{Y \cdot X} = [s_{Y \cdot X}]/M_Y$	$[s_{X \cdot Y}] = M_Y s_{Y \cdot X}$
Coefficient of determination, r^2	$r^2 = [r^2]$	$[r^2] = r^2$
Test statistics,* t and F	$t = [t]; F = [F]$	$[t] = t; [F] = F$
Standard error of b, s_b	$s_b = [s_b]M_X/M_Y$	$[s_b] = s_b M_Y/M_X$
Standard error of a, s_a*	$s_a = [s_a]/M_Y$	$[s_a]s_a M_Y$
Mean of X, $\bar{X}$	$\bar{X} = ([\bar{X}] - A_X)/M_X$	$[\bar{X}] = M_Y \bar{X} + A_X$
Mean of Y, $\bar{Y}$	$\bar{Y} = ([\bar{Y}] - A_Y)M_Y$	$[\bar{Y}] = M_Y \bar{Y} + A_Y$
Sum of squares, SS	$SS = [SS]/M_Y^2$	$[SS] = (SS)(M_Y^2)$
Mean square, MS	$MS = [MS]/M_Y^2$	$[MS] = (MS)(M_Y^2)$

Note: If a regression is fit through the origin (Section 17.9), then coding by addition of a constant may not be used (i.e., $A_X = 0$ and $A_Y = 0$).

*A_X must be zero if s_a is desired or if it is desired to test hypotheses about the Y intercept.

$[a] = (0.715 \text{ cm}) (0.3937 \text{ in./cm}) = 0.281 \text{ in.}$; and the standard error of estimate would be $s_{Y \cdot X} = (0.3937 \text{ in./cm})(0.218 \text{ cm}) = 0.086$.

A relatively common case of biological interest where A_X and/or A_Y are not zero is when we have temperature measurements in degrees Celsius (or Fahrenheit) and wish to determine the regression equation that would have resulted had the data been recorded in degrees Fahrenheit (or Celsius). The appropriate coding constants in Equations 17.68, and 17.69 may be determined by knowing that Celsius and Fahrenheit temperatures are related as follows:

$$\text{degrees Celsius} = \left(\frac{5}{9}\right) (\text{degrees Fahrenheit}) - \left(\frac{5}{9}\right)(32)$$

$$\text{degrees Fahrenheit} = \left(\frac{9}{5}\right) (\text{degrees Celsius}) + 32$$

This is summarized elsewhere (Zar, 1968a), as are the effects of multiplicative coding on logarithmically transformed data (Zar, 1967).

EXERCISES

17.1 The following data are the rates of oxygen consumption of birds, measured at different environmental temperatures:

Temperature (°C)	Oxygen consumption (ml/g/hr)
−18	5.2
−15	4.7
−10	4.5
−5	3.6
0	3.4
5	3.1
10	2.7
19	1.8

(a) Calculate a and b for the regression of oxygen consumption rate on temperature.
(b) Test, by analysis of variance, the hypothesis H_0: $\beta = 0$.
(c) Test, by the t test, the hypothesis H_0: $\beta = 0$.
(d) Calculate the standard error of estimate of the regression.
(e) Calculate the coefficient of determination of the regression.
(f) Calculate the 95% confidence limits for β.

17.2 Utilize the regression equation computed for the data of Exercise 17.1.
(a) What is the mean rate of oxygen consumption in the population for birds at 15°C?
(b) What is the 95% confidence interval for this mean rate?
(c) If we randomly chose one additional bird at 15°C from the population, what would its rate of oxygen consumption be estimated to be?
(d) We can be 95% confident of this value lying between what limits?

17.3 The frequency of electrical impulses emitted from electric fish is measured from three fish at each of several temperatures. The resultant data are as follows:

Temperature (°C)	Impulse frequency (number/sec)
20	225, 230, 239
22	251, 259, 265
23	266, 273, 280
25	287, 295, 302
27	301, 310, 317
28	307, 313, 325
30	324, 330, 338

(a) Compute a and b for the linear regression equation relating impulse frequency to temperature.
(b) Test, by analysis of variance H_0: $\beta = 0$.
(c) Calculate the standard error of estimate of the regression.
(d) Calculate the coefficient of determination of the regression.
(e) Test H_0: The population regression is linear.

18

COMPARING SIMPLE LINEAR REGRESSION EQUATIONS

It is common to possess more than one set of data and to have calculated a regression equation (i.e., a regression line) for each set. We might then ask whether the slopes of these lines are significantly different or whether they might be estimating the same population value of β. Furthermore, if we conclude that the slopes of the several lines are not significantly different, then we might wish to determine whether the several sets of data are likely from the same population (i.e., whether the population Y intercepts, as well as the slopes, are the same). In this chapter, procedures for testing differences among regression lines will be presented as summarized in Fig. 18.1.

18.1 COMPARING TWO SLOPES

A simple method for testing hypotheses about equality of two population regression coefficients involves the use of Student's t in a fashion analogous to that of testing for differences between two population means (Section 8.1). The test statistic is

$$t = \frac{b_1 - b_2}{s_{b_1 - b_2}}, \tag{18.1}$$

where the standard error of the difference between regression coefficients is

$$s_{b_1 - b_2} = \sqrt{\frac{(s_{Y \cdot X}^2)_p}{\left(\sum x^2\right)_1} + \frac{(s_{Y \cdot X}^2)_p}{\left(\sum x^2\right)_2}}, \tag{18.2}$$

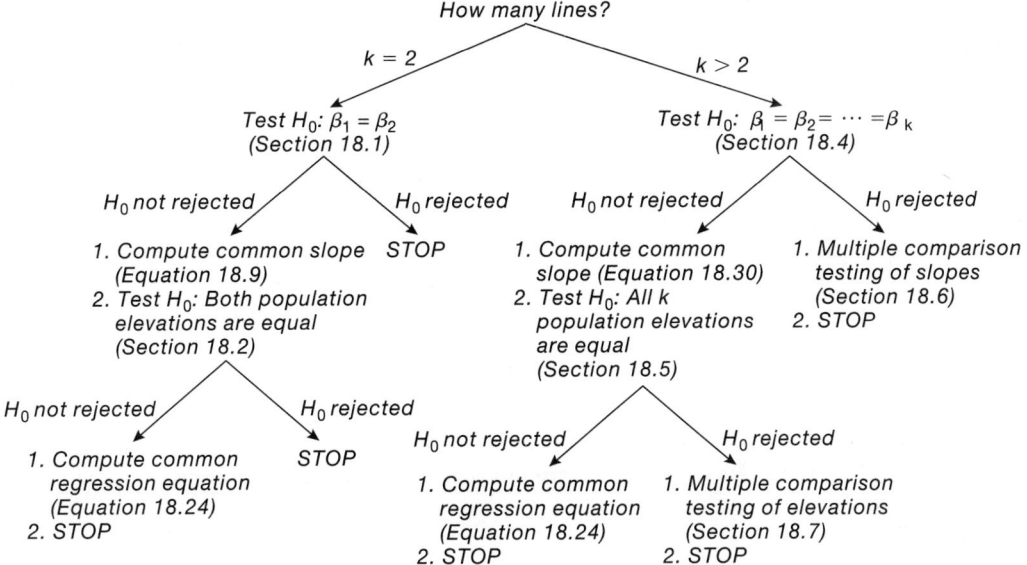

Figure 18.1 Flow chart for the comparison of regression lines.

and the pooled residual mean square is calculated as

$$(s_{Y \cdot X}^2)_p = \frac{(\text{residual SS})_1 + (\text{residual SS})_2}{(\text{residual DF})_1 + (\text{residual DF})_2}, \tag{18.3}$$

the subscripts 1 and 2 referring to the two regression lines being compared. The critical value of t for this test has $(n_1 - 2) + (n_2 - 2)$ degrees of freedom (i.e., the sum of the two residual degrees of freedom), namely

$$\nu = n_1 + n_2 - 4. \tag{18.4}$$

Example 18.1 demonstrates these calculations for testing $H_0: \beta_1 = \beta_2$ against $H_A: \beta_1 \neq \beta_2$. The two regression lines are shown in Fig. 18.2.

Just as the t test for difference between means assumes that $\sigma_1^2 = \sigma_2^2$, the above t test assumes that $(\sigma_{Y \cdot X}^2)_1 = (\sigma_{Y \cdot X}^2)_2$. The presence of the latter condition can be tested by the variance ratio test, $F = (s_{Y \cdot X}^2)_{\text{larger}} / (s_{Y \cdot X}^2)_{\text{smaller}}$; but this is usually not done due to the limitations of this test (see Section 8.5).

The $1 - \alpha$ confidence interval for the difference between two slopes, β_1 and β_2, is

$$(b_1 - b_2) \pm t_{\alpha(2), \nu} s_{b_1 - b_2}, \tag{18.5}$$

where ν is as in Equation 18.4. Thus, for Example 18.1,

$$95\% \text{ confidence interval for } \beta_1 - \beta_2 = (2.97 - 2.17) \pm (t_{0.05(2), 52})(0.1165)$$

$$= 0.80 \pm (2.007)(0.1165)$$

$$= 0.80 \text{ ml/°C} \pm 0.23 \text{ ml/°C};$$

and the lower and upper 95% confidence limits for $\beta_1 - \beta_2$ are $L_1 = 0.57$ ml/°C and $L_2 = 1.03$ ml/°C , respectively.

If H_0: $\beta_1 = \beta_2$ is rejected (as in Example 18.1), we may wish to calculate the point where the two lines intersect. The intersection is at

$$X_I = \frac{a_2 - a_1}{b_1 - b_2},$$ (18.6)

at which the value of $\hat{Y}$ may be computed either as

$$\hat{Y}_I = a_1 + b_1 X_I$$ (18.7)

or

$$\hat{Y}_I = a_2 + b_2 X_I.$$ (18.8)

EXAMPLE 18.1   Testing for difference between two population regression coefficients.

H_0: $\beta_1 = \beta_2$

H_A: $\beta_1 \neq \beta_2$

For Sample 1:

$\sum x^2 = 1470.8712$

$\sum xy = 4363.1627$

$\sum y^2 = 13299.5296$

$n = 26$

$b = \dfrac{4363.1627}{1470.8712} = 2.97$

residual SS $= 13299.5296 - \dfrac{(4363.1627)^2}{1470.8712}$

$= 356.7317$

residual DF $= 26 - 2 = 24$

For Sample 2:

$\sum x^2 = 2272.4750$

$\sum xy = 4928.8100$

$\sum y^2 = 10964.0947$

$n = 30$

$b = \dfrac{4928.8100}{2272.4750} = 2.17$

residual SS $= 10964.0947 - \dfrac{(4928.8100)^2}{2272.4750}$

$= 273.9142$

residual DF $= 30 - 2 = 28$

$(s_{Y \cdot X}^2)_p = \dfrac{356.7317 + 273.9142}{24 + 28} = 12.1278$

$s_{b_1 - b_2} = \sqrt{\dfrac{12.1278}{1470.8712} + \dfrac{12.1278}{2272.4750}} = 0.1165$

$t = \dfrac{2.97 - 2.17}{0.1165} = 6.867$

$\nu = 24 + 28 = 52$

Reject H_0 if $|t| \geq t_{\alpha(2),\nu}$

$t_{0.05(2),52} = 2.007$

Reject H_0.

$P < 0.001$ $[P = 0.0000000081]$

For the two lines in Example 18.1, $a_1 = 10.57$ ml, $a_2 = 24.91$ ml, and the point of intersection is at

$$X_I = \frac{24.91 - 10.57}{2.97 - 2.17} = 17.92°C$$

and

$$\hat{Y}_I = 10.57 + (2.97)(17.92) = 63.79 \text{ ml}.$$

Figure 18.2 illustrates this intersection.

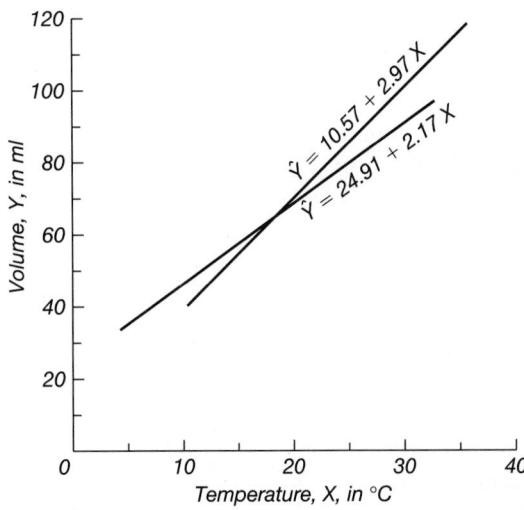

Figure 18.2 The two regression lines of Example 18.1. The two slopes are concluded significantly different and the two lines are found to intersect at $X_I = 17.92°C$ and $\hat{Y}_I = 63.79$ ml.

If H_0: $\beta_1 = \beta_2$ is not rejected (as will be shown in Example 18.2), then an estimate of the population regression coefficient, β, underlying both b_1 and b_2 is called the common (or weighted) regression coefficient:

$$b_c = \frac{\left(\sum xy\right)_1 + \left(\sum xy\right)_2}{\left(\sum x^2\right)_1 + \left(\sum x^2\right)_2} \tag{18.9}$$

or equivalently (but with more chance of rounding error):

$$b_c = \frac{\left(\sum x^2\right)_1 b_1 + \left(\sum x^2\right)_2 b_2}{\left(\sum x^2\right)_1 + \left(\sum x^2\right)_2}. \tag{18.10}$$

Equation 18.1 is a special case of

$$t = \frac{|b_1 - b_2| - \beta_0}{s_{b_1 - b_2}}, \tag{18.11}$$

namely when $\beta_0 = 0$. By using Equation 18.11 we may test the hypothesis that the difference between two population regression coefficients is a specified magnitude; that is, H_0: $\beta_1 - \beta_2 = \beta_0$ may be tested against H_A: $\beta_1 - \beta_2 \neq \beta_0$.

One-tailed testing is also possible, asking whether one population regression coefficient is greater than the other. If we test H_0: $\beta_1 \geq \beta_2$ and H_A: $\beta_1 < \beta_2$, or H_0: $\beta_1 - \beta_2 \geq \beta_0$ vs. H_A: $\beta_1 - \beta_2 < \beta_0$, then H_0 is rejected if $t \leq -t_{\alpha(1),\nu}$; if we test H_0: $\beta_1 \leq \beta_2$ and H_A: $\beta_1 > \beta_2$, or H_0: $\beta_1 - \beta_2 \leq \beta_0$ vs. H_A: $\beta_1 - \beta_2 > \beta_0$, then we reject H_0 if $t \geq t_{\alpha(1),\nu}$. In either case, t is computed by Equation 18.1 (or by Equation 18.11 if $\beta_0 \neq 0$).

An alternative method of testing H_0: $\beta_1 = \beta_2$ is by the analysis of covariance procedure of Section 18.4. However, the preceding t test generally involves less computational effort.

Power and Sample Size in Comparing Regressions. In Section 17.8 it was explained that the procedure for consideration of power in correlation analysis (Section 19.4) could be used to estimate power and sample size in a regression analysis. Section 19.6 presents power and sample size estimation when testing for difference between two correlation coefficients. Unfortunately, utilization of that procedure for the case of comparing two regression coefficients is not valid—unless one has the rare case of $\left(\sum x^2 \right)_1 = \left(\sum x^2 \right)_2$ and $\left(\sum y^2 \right)_1 = \left(\sum y^2 \right)_2$ (Cohen, 1988: 110).

18.2 COMPARING TWO ELEVATIONS

If H_0: $\beta_1 = \beta_2$ is rejected, we may assume that two different populations have been sampled. However, if two population regression lines are not concluded to have different slopes (i.e., H_0: $\beta_1 = \beta_2$ is not rejected), then the two lines are assumed to be parallel. In the latter case, we often wish to determine whether the two population regressions have the same elevation (i.e., the same vertical position on a graph) and thus coincide.

To test the null hypothesis that the elevations of the two population regression lines are the same, we define the following quantities for use in a t test:

Sum of squares of X for "common regression"

$$= A_c = \left(\sum x^2 \right)_1 + \left(\sum x^2 \right)_2 , \tag{18.12}$$

Sum of crossproducts for "common regression"

$$= B_c = \left(\sum xy \right)_1 + \left(\sum xy \right)_2 , \tag{18.13}$$

Sum of squares of Y for "common regression"

$$= C_c = \left(\sum y^2 \right)_1 + \left(\sum y^2 \right)_2 , \tag{18.14}$$

Residual SS for "common regression"

$$= SS_c = C_c - \frac{B_c^2}{A_c} , \tag{18.15}$$

Residual DF for "common regression"

$$= DF_c = n_1 + n_2 - 3 , \tag{18.16}$$

and

Residual MS for "common regression"

$$= (s_{Y \cdot X}^2)_c = \frac{SS_c}{DF_c}. \tag{18.17}$$

Then, the appropriate test statistic is

$$t = \frac{(\bar{Y}_1 - \bar{Y}_2) - b_c(\bar{X}_1 - \bar{X}_2)}{\sqrt{(s_{Y \cdot X}^2)_c \left[\dfrac{1}{n_1} + \dfrac{1}{n_2} + \dfrac{(\bar{X}_1 - \bar{X}_2)^2}{A_c}\right]}} \tag{18.18}$$

and the appropriate critical value of t is that for $\nu = DF_c$. Example 18.2 and Fig. 18.3 consider the regression of human systolic blood pressure on age for men over forty years old. A regression was fitted for data for men in each of two different occupations. The two-tailed null hypothesis is that in the two sampled populations the regression elevations are the same. This also says that blood pressure is the same in both groups, after accounting for the effect of age. In the example, H_0 is rejected, so we conclude that men in these two occupations do not have the same blood pressure. As an alternative to this t-testing procedure, the analysis of covariance of Section 18.4 may be used to test this hypothesis, but it generally involves more computational effort.

EXAMPLE 18.2   Testing for difference between two population regression coefficients and elevations.

For Sample 1:	For Sample 2:
$n = 13$	$n = 15$
$\bar{X} = 54.65$ yr	$\bar{X} = 56.93$ yr
$\bar{Y} = 170.23$ mm Hg	$\bar{Y} = 162.93$ mm Hg
$\sum x^2 = 1012.1923$	$\sum x^2 = 1659.4333$
$\sum xy = 1585.3385$	$\sum xy = 2475.4333$
$\sum y^2 = 2618.3077$	$\sum y^2 = 3848.9333$
$b = 1.57$ mm Hg/yr	$b = 1.49$ mm Hg/yr
$a = 84.6$ mm Hg	$a = 78.0$ mm Hg
residual SS $= 136.2230$	residual SS $= 156.2449$
residual DF $= 11$	residual DF $= 13$

H_0: $\beta_1 = \beta_2$

H_A: $\beta_1 \neq \beta_2$

$$(s_{Y \cdot X}^2)_p = \frac{136.2230 + 156.2449}{11 + 13} = 12.1862$$

EXAMPLE 18.2   (continued)

$$\nu = 11 + 13 = 24$$

$$s_{b_1 - b_2} = 0.1392$$

$$t = \frac{1.57 - 1.49}{0.1392} = 0.575$$

$$t_{0.05(2),24} = 2.064; \text{ do not reject } H_0.$$

$$P > 0.50 \quad [P = 0.57]$$

H_0:   The two population regression lines have the same elevation.

H_A: The two population regression lines do not have the same elevation.

$$A_c = 1012.1923 + 1659.4333 = 2671.6256$$

$$B_c = 1585.3385 + 2475.4333 = 4060.7718$$

$$C_c = 2618.3077 + 3848.9333 = 6467.2410$$

$$b_c = \frac{4060.7718}{2671.6256} = 1.520 \text{ mm Hg/yr}$$

$$SS_c = 6467.2410 - \frac{(4060.7718)^2}{2671.6256} = 295.0185$$

$$DF_c = 13 + 15 - 3 = 25$$

$$(s_{Y \cdot X}^2)_c = \frac{295.0185}{25} = 11.8007$$

$$t = \frac{(170.23 - 162.93) - 1.520(54.65 - 56.93)}{\sqrt{11.8007\left[\dfrac{1}{13} + \dfrac{1}{15} + \dfrac{(54.65 - 56.93)^2}{2671.6256}\right]}} = \frac{10.77}{1.3105} = 8.218$$

$$t_{0.05(2),25} = 2.060; \text{ reject } H_0.$$

$$P < 0.001 \quad [P = 0.0000000072]$$

If it is concluded that two population regressions do not have different slopes but do have different elevations, then the slopes computed from the two samples are both estimates of the common population regression coefficient, and the two regression equations should be written as

$$\hat{Y}_i = a_1 + b_c X_i \tag{18.19}$$

and

$$\hat{Y}_i = a_2 + b_c X_i, \tag{18.20}$$

which for the two lines in Example 18.2 and Fig. 18.3 would be

$$\hat{Y}_i = 84.6 + 1.52 X_i$$

and

$$\hat{Y}_i = 78.0 + 1.52 X_i.$$

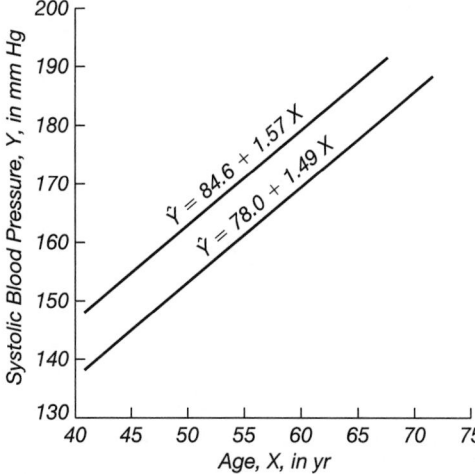

Figure 18.3   The two regression lines of Example 18.2.

(graph labels)

Systolic Blood Pressure, Y, in mm Hg

$\hat{Y} = 84.6 + 1.57\,X$

$\hat{Y} = 78.0 + 1.49\,X$

Age, X, in yr

If it is concluded that two population regressions have neither different slopes nor different elevations, then both sample regressions estimate the same population regression, and we should use a common regression coefficient, b_c, as well as a common Y intercept,

$$a_c = \bar{Y}_p - b_c \bar{X}_p, \tag{18.21}$$

where the pooled sample means of the two variables may be obtained as

$$\bar{X}_p = \frac{n_1 \bar{X}_1 + n_2 \bar{X}_2}{n_1 + n_2} \tag{18.22}$$

and

$$\bar{Y}_p = \frac{n_1 \bar{Y}_1 + n_2 \bar{Y}_2}{n_1 + n_2}. \tag{18.23}$$

Thus, when two samples have been concluded to estimate the same population regression, the regression equation would be

$$\hat{Y}_i = a_c + b_c X_i. \tag{18.24}$$

The above test of difference between elevations may be considered the same as asking whether the two population Y intercepts are different. However, it is not advisable to test H_0: $\alpha_1 = \alpha_2$ using sample estimates a_1 and a_2; the latter test would consider a point on each line that may lie far from the observed range of X's. There are many regressions for which the Y intercept has no importance beyond helping to define the line and in fact may be a sample statistic prone to misleading interpretation. In Fig. 18.3, for example, discussion of the Y intercepts (and testing hypotheses about them) would require a risky extrapolation of the regression lines far below the range of X for which data were obtained. This would assume that the linear relationship that was determined also holds between $X = 40$ yr and $X = 0$, a gravely incorrect assumption in the present case. Additionally, since the Y intercepts are so far from the mean values of X, their standard errors would be very large, and a test of H_0: $\alpha_1 = \alpha_2$ would lack statistical power. For such a test, Equations 18.25 and 18.26 are a special case of Equations 18.27 and 18.28 which follow:

$$t = \frac{a_1 - a_2}{s_{a_1-a_2}},\tag{18.25}$$

where

$$s_{a_1-a_2} = \sqrt{(s_{Y\cdot X}^2)_p \left[\frac{1}{n_1} + \frac{1}{n_2} + \frac{\bar{X}_1^2}{(\sum x^2)_1} + \frac{\bar{X}_2^2}{(\sum x^2)_2}\right]}.\tag{18.26}$$

We may also test one-tailed hypotheses about elevations. For data such as those in Example 18.2 and Fig. 18.3, it might have been the case that one occupation was considered to be more emotionally stressful, and the interest was to determine whether men in that occupation had higher blood pressure than men in the second occupation.

18.3 COMPARING POINTS ON TWO REGRESSION LINES

If the slopes of two regression lines and the elevations of the two lines have not been concluded to be different, then the two lines are estimates of the same population regression line. If the slopes of two lines are not concluded to be different, but their elevations are declared different, then the population lines are assumed to be parallel, and for a given X_i the corresponding $\hat{Y}_i$ on one line is different from that on the other line.

If the slopes of two population regression lines are concluded different, then the lines are intersecting, rather than parallel. In such cases we may wish to test whether a $\hat{Y}$ on one line is the same as the $\hat{Y}$ on the second line at a particular X. For a two-tailed test we can state the null hypothesis as H_0: $\mu_{\hat{Y}_1} = \mu_{\hat{Y}_2}$ and the alternate as H_A: $\mu_{\hat{Y}_1} \neq \mu_{\hat{Y}_2}$. The test statistic is

$$t = \frac{\hat{Y}_1 - \hat{Y}_2}{s_{\hat{Y}_1-\hat{Y}_2}}\tag{18.27}$$

where

$$s_{\hat{Y}_1-\hat{Y}_2} = \sqrt{(s_{Y\cdot X}^2)_p \left[\frac{1}{n_1} + \frac{1}{n_2} + \frac{(X - \bar{X}_1)^2}{(\sum x^2)_1} + \frac{(X - \bar{X}_2)^2}{(\sum x^2)_2}\right]}\tag{18.28}$$

and the degrees of freedom are the pooled degrees of freedom of Equation 18.4. Such a test is demonstrated in Example 18.3. One-tailed testing is also possible. The test should be applied with caution, however, as it assumes that each of the two predicted $\hat{Y}$'s has associated with it the same variance. Therefore, the test works best when the two lines have the same $\bar{X}$, the same $\sum x^2$, and the same n.

EXAMPLE 18.3   Testing for difference between points on the two nonparallel regression lines of Example 18.1 and Fig. 18.2. We are testing whether the volumes (Y) are different in the two groups at $X = 12°$C.

H_0: $\mu_{\hat{Y}_1} = \mu_{\hat{Y}_2}$

H_A: $\mu_{\hat{Y}_1} \neq \mu_{\hat{Y}_2}$

Beyond the statistics given in Example 18.1 we need to know the following:

$a_1 = 10.57$ ml   and $a_2 = 24.91$ ml;

$\bar{X}_1 = 22.93°$C   and $\bar{X}_2 = 18.95°$C.

We then compute:

$\hat{Y}_1 = 10.57 + (2.97)(12) = 46.21$ ml

$\hat{Y}_2 = 24.91 + (2.17)(12) = 50.95$ ml

$$s_{\hat{Y}_1 - \hat{Y}_2} = \sqrt{12.1278 \left[\frac{1}{26} + \frac{1}{30} + \frac{(12 - 22.93)^2}{1470.8712} + \frac{(12 - 18.95)^2}{2272.4750} \right]}$$

$$= \sqrt{2.1135} = 1.45 \text{ ml}$$

$$t = \frac{46.21 - 50.95}{1.45} = -3.269$$

$$\nu = 26 + 30 - 4 = 52$$

$$t_{0.05(2), 52} = 2.007$$

As $|t| > t_{0.05(2), 52}$, reject H_0.

$$0.001 < P < 0.002 \quad [P = 0.0019]$$

18.4 COMPARING MORE THAN TWO SLOPES

One can test H_0: $\beta_1 = \beta_2 = \cdots = \beta_k$, with the alternate hypothesis being that the k regression lines were not derived from samples estimating populations among which the slopes (β) were all equal. This may be done by a procedure known as *analysis of covariance* (which was introduced in Section 12.12).

Analysis of covariance encompasses a large body of statistical methodology too extensive to be covered thoroughly in this book, but the following procedures suffice to test for the homogeneity (i.e., equality) of regression coefficients. Just as an analysis of variance for H_0: $\mu_1 = \mu_2 = \cdots = \mu_k$ assumes that $\sigma_1^2 = \sigma_2^2 = \cdots = \sigma_k^2$, the testing of H_0: $\beta_1 = \beta_2 = \cdots = \beta_k$ proceeds with the assumption that $(\sigma_{Y \cdot X}^2)_1 = (\sigma_{Y \cdot X}^2)_2 = \cdots = (\sigma_{Y \cdot X}^2)_k$. Heterogeneity of the k residual mean squares can be tested by Bartlett's test (Section 10.6), but this generally is not done for the same reasons that the test is not often employed as a prelude to analysis of variance procedures.

The basic calculations necessary to compare k regression lines require quantities already computed: $\sum x^2$, $\sum xy$, $\sum y^2$ (i.e., total SS), and the residual SS and DF for each computed line (Table 18.1). The values of the k residual sums of squares may then

TABLE 18.1 Calculations for Testing for Significant Differences among Slopes and Elevations of k Simple Linear Regression Lines

	$\sum x^2$	$\sum xy$	$\sum y^2$	Residual SS	Residual DF
Regression 1	A_1	B_1	C_1	$SS_1 = C_1 - \dfrac{B_1^2}{A_1}$	$DF_1 = n_1 - 2$
Regression 2	A_2	B_2	C_2	$SS_2 = C_2 - \dfrac{B_2^2}{A_2}$	$DF_2 = n_2 - 2$
$\vdots$	$\vdots$	$\vdots$	$\vdots$	$\vdots$	$\vdots$
Regression k	A_k	B_k	C_k	$SS_k = C_k - \dfrac{B_k^2}{A_k}$	$DF_k = n_k - 2$
"Pooled" regression				$SS_p = \sum\limits_{i=1}^{k} SS_i$	$DF_p = \sum\limits_{i=1}^{k}(n_i - 2)$ $= \sum\limits_{i=1}^{k} n - 2k$
"Common" regression	$A_c = \sum\limits_{i=1}^{k} A_i$	$B_c = \sum\limits_{i=1}^{k} B_i$	$C_c = \sum\limits_{i=1}^{k} C_i$	$SS_c = C_c - \dfrac{B_c^2}{A_c}$	$DF_c = \sum\limits_{i=1}^{k} n_i - k - 1$
"Total" regression*	A_t	B_t	C_t	$SS_t = C_t - \dfrac{B_t^2}{A_t}$	$DF_t = \sum\limits_{i=1}^{k} n_i - 2$

*See Section 18.5 for explanation.

be summed, yielding what we shall call the "pooled" residual sum of squares, SS_p; and the sum of the k residual degrees of freedom is the "pooled" residual degrees of freedom, DF_p. The values of $\sum x^2$, $\sum xy$, and $\sum y^2$ for the regressions may each be summed, and from these sums a residual sum of squares may be calculated. The latter quantity will be termed the "common" residual sum of squares, SS_c.

To test H_0: $\beta_1 = \beta_2 = \cdots = \beta_k$, one may calculate

$$F = \frac{\left(\dfrac{SS_c - SS_p}{k - 1}\right)}{\dfrac{SS_p}{DF_p}},$$ (18.29)

a statistic with numerator and denominator degrees of freedom of $k - 1$ and DF_p, respectively.* Example 18.4 demonstrates this testing procedure for three regression lines calculated from three sets of data (i.e., $k = 3$).

If H_0: $\beta_1 = \beta_2 = \cdots = \beta_k$ is rejected, then one may wish to employ a multiple comparison test to determine which of the k population slopes differ from which others. This is analogous to the multiple comparison testing employed after rejecting H_0: $\mu_1 = \mu_2 = \cdots = \mu_k$ (Chapter 11), and it is presented in Section 18.6.

---

*The quantity $SS_c - SS_p$ is an expression of variability among the k regression coefficients; hence, it is associated with $k - 1$ degrees of freedom.

If H_0: $\beta_1 = \beta_2 = \cdots = \beta_k$ is not rejected, then the common regression coefficient, b_c, may be used as an estimate of the β underlying all k samples:

$$b_c = \frac{\sum\limits_{i=1}^{k} \left(\sum xy\right)_i}{\sum\limits_{i=1}^{k} \left(\sum x^2\right)_i}. \tag{18.30}$$

For Example 18.4, this is $b_c = 2057.66/1381.10 = 1.49$.

EXAMPLE 18.4   Testing for difference among three regression functions.*

	$\sum x^2$	$\sum xy$	$\sum y^2$	n	b	Residual SS	Residual DF
Regression 1	*430.14*	*648.97*	*1065.34*	*24*	1.51	86.21	22
Regression 2	*448.65*	*694.36*	*1184.12*	*29*	1.55	109.48	27
Regression 3	*502.31*	*714.33*	*1186.52*	*30*	1.42	170.68	28
"Pooled" regression						366.37	77
"Common" regression	1381.10	2057.66	3435.98		1.49	370.33	79
"Total" regression	*2144.06*	*3196.78*	*5193.48*	*83*		427.10	81

*The italicized values are calculated from the row data; all other values are derived from them.

To test for differences among slopes: H_0: $\beta_1 = \beta_2 = \beta_3$; H_A: All three β's are not equal.

$$F = \frac{\dfrac{370.33 - 366.37}{3 - 1}}{\dfrac{366.37}{77}} = 0.42$$

As $F_{0.05(1),2,77} \cong 3.13$, do not reject H_0.

$$P > 0.25 \quad [P = 0.66]$$

$$b_c = \frac{2057.66}{1381.10} = 1.49$$

To test for differences among elevations:

H_0:   The three population regression lines have the same elevation.

H_A: The three lines do not have the same elevation.

$$F = \frac{\dfrac{427.10 - 370.33}{3 - 1}}{\dfrac{370.33}{79}} = 6.06$$

As $F_{0.05(1),2,79} \cong 3.13$, reject H_0.

$$0.0025 < P < 0.005 \quad [P = 0.0036]$$

18.5 COMPARING MORE THAN TWO ELEVATIONS

Consider the case where it has been concluded that all k population slopes underlying our k samples of data are equal (i.e., H_0: $\beta_1 = \beta_2 = \cdots = \beta_k$ is not rejected). In this situation, it is reasonable to ask whether all k population regressions are, in fact, identical, i.e., whether they have equal elevations as well as slopes, and thus the lines all coincide.

The null hypothesis of equality of elevations may be tested by a continuation of the analysis of covariance considerations outlined in Section 18.4. We can combine the data from all k samples, and from this compute $\sum x^2$, $\sum xy$, $\sum y^2$, a residual sum of squares, and residual degrees of freedom; the latter will be called the "total" residual sum of squares, SS_t, and "total" residual degrees of freedom, DF_t. (See Table 18.1.) The null hypothesis is tested with the test statistic

$$F = \frac{\dfrac{SS_t - SS_c}{k - 1}}{\dfrac{SS_c}{DF_c}} \tag{18.31}$$

with $k - 1$ and DF_c degrees of freedom. An example of this procedure is offered in Example 18.4.

If the null hypothesis is rejected, we can then employ multiple comparisons to determine the location of significant differences among the elevations, as described in Section 18.6. If it is not rejected, then all k sample regressions are estimates of the same population regression, and the best estimate of that underlying population regression is given by Equation 18.24 using Equations 18.9 and 18.21.

18.6 MULTIPLE COMPARISONS AMONG SLOPES

If an analysis of covariance concludes that k population slopes are not all equal, we may employ a multiple comparison procedure (Chapter 11) to determine which β's are different from which others. For example, the Tukey test (Section 11.1) may be employed to test for differences between each pair of β values, by H_0: $\beta_A = \beta_B$ and H_A: $\beta_A \neq \beta_B$, where A and B can represent any two of the k regression lines.

The test statistic is

$$q = \frac{b_B - b_A}{SE}. \tag{18.32}$$

If $\sum x^2$ is the same for lines A and B, the standard error to be used is

$$SE = \sqrt{\frac{(s_{Y \cdot X}^2)_p}{\sum x^2}}. \tag{18.33}$$

If $\sum x^2$ is different for lines A and B, then use

$$SE = \sqrt{\frac{(s_{Y \cdot X}^2)_p}{2}\left[\frac{1}{(\sum x^2)_A} + \frac{1}{(\sum x^2)_B}\right]}. \tag{18.34}$$

The degrees of freedom for determining the critical value of q are the pooled residual DF (i.e., DF_p in Table 18.1). Although it is not mandatory to have first performed the analysis of covariance before applying the multiple comparison test, such a procedure is commonly followed.

The confidence interval for the difference between the slopes of population regressions A and B is

$$(b_B - b_A) \pm (q_{\alpha, \nu, p})(SE), \tag{18.35}$$

where ν is the pooled residual DF (i.e., DF_p in Table 18.1).

If one of several regression lines is considered to be a control to which each of the other lines is to be compared, then the procedures of Dunnett's test (introduced in Section 11.4) are appropriate. Here,

$$SE = \sqrt{\frac{2(s_{Y \cdot X}^2)_p}{\sum x^2}} \tag{18.36}$$

if $\sum x^2$ is the same for line A (the control line) and line B, and

$$SE = \sqrt{(s_{Y \cdot X}^2)_p\left[\frac{1}{(\sum x^2)_A} + \frac{1}{(\sum x^2)_B}\right]} \tag{18.37}$$

if it is not. Either two-tailed or one-tailed hypotheses may be thus tested.

The $1 - \alpha$ confidence interval for the difference between the slopes of the control line (line A) and another line (line B) is

$$(b_B - b_A) \pm (q'_{\alpha(2), \nu, p})(SE). \tag{18.38}$$

To apply Scheffé's procedure (Section 11.5), we would calculate SE as Equations 18.36 or 18.37.

18.7 MULTIPLE COMPARISONS AMONG ELEVATIONS

If H_0: $\beta_1 = \beta_2 = \cdots = \beta_k$ has been accepted and the null hypothesis of all k elevations being equal has been rejected, then one may apply multiple comparison procedures (analogous to those of Chapter 11) to determine between which elevations differences occur in the populations sampled. The test statistic for the Tukey test (introduced in Section 11.1) is

$$q = \frac{|(\bar{Y}_A - \bar{Y}_B) - b_c(\bar{X}_A - \bar{X}_B)|}{SE}, \tag{18.39}$$

with DF_p degrees of freedom (see Table 18.1), where the subscripts A and B refer to the two lines the elevations of which are being compared, b_c is from Equation 18.30, and

$$SE = \sqrt{\frac{(s^2_{Y \cdot X})_c}{2}\left[\frac{1}{n_A} + \frac{1}{n_B} + \frac{(\bar{X}_A - \bar{X}_B)^2}{\left(\sum x^2\right)_A + \left(\sum x^2\right)_B}\right]}. \qquad (18.40)$$

If we use Dunnett's test to compare the elevation of a control regression line (let's call it line A) with that of another line (line B),

$$SE = \sqrt{(s^2_{Y \cdot X})_c\left[\frac{1}{n_A} + \frac{1}{n_B} + \frac{(\bar{X}_A - \bar{X}_B)^2}{\left(\sum x^2\right)_A + \left(\sum x^2\right)_B}\right]}. \qquad (18.41)$$

Equation (18.41) would also be employed if Scheffé's test were being performed on elevations.

18.8 MULTIPLE COMPARISONS OF POINTS AMONG REGRESSION LINES

If we conclude that the slopes of three or more regression lines are all the same, then we most likely would test for differences among elevations (as shown in Section 18.7). Occasionally, however, when slopes are not concluded to be the same it might be desired to specify a value of X and ask which of the several regression lines differ from the others at that X. This is effected by an extension of Equations 18.27 and 18.28, where for each line we compute the $\hat{Y}$ at the specified X, as

$$\hat{Y}_i = a_i + b_c X \qquad (18.42)$$

and perform a Tukey test of $H_0: \mu_{\hat{Y}_A} = \mu_{\hat{Y}_B}$ as

$$q = \frac{\hat{Y}_B - \hat{Y}_A}{SE}, \qquad (18.43)$$

where

$$SE = \sqrt{\frac{(s^2_{Y \cdot X})_p}{2}\left[\frac{1}{n_A} + \frac{1}{n_B} + \frac{(X - \bar{X}_A)^2}{\left(\sum x^2\right)_A} + \frac{(X - \bar{X}_B)^2}{\left(\sum x^2\right)_B}\right]}, \qquad (18.44)$$

with DF_p degrees of freedom. An analogous Dunnett or Scheffé test would employ

$$SE = \sqrt{(s^2_{Y \cdot X})_p\left[\frac{1}{n_A} + \frac{1}{n_B} + \frac{(X - \bar{X}_A)^2}{\left(\sum x^2\right)_A} + \frac{(X - \bar{X}_B)^2}{\left(\sum x^2\right)_B}\right]}. \qquad (18.45)$$

A special case of this testing is where we wish to test for differences among the Y-intercepts (i.e., the values of $\hat{Y}$ when $X = 0$). Equations 18.43 and 18.44 for the

Tukey test would become

$$q = \frac{a_B - a_A}{\text{SE}}, \tag{18.46}$$

and

$$\text{SE} = \sqrt{\frac{(s_{Y \cdot X}^2)_p}{2} \left[\frac{1}{n_A} + \frac{1}{n_B} + \frac{(\bar{X}_A)^2}{\left(\sum x^2\right)_A} + \frac{(\bar{X}_B)^2}{\left(\sum x^2\right)_B} \right]}, \tag{18.47}$$

respectively. The analogous Dunnett or Scheffé test for Y-intercepts would employ

$$\text{SE} = \sqrt{(s_{Y \cdot X}^2)_p \left[\frac{1}{n_A} + \frac{1}{n_B} + \frac{(\bar{X}_A)^2}{\left(\sum x^2\right)_A} + \frac{(\bar{X}_B)^2}{\left(\sum x^2\right)_B} \right]}. \tag{18.48}$$

18.9 AN OVERALL TEST FOR COINCIDENTAL REGRESSIONS

It is possible to test the null hypothesis that k population regressions are coincident, i.e., that the β's are all identical and the α's are all identical. Here we would calculate

$$F = \frac{\dfrac{\text{SS}_t - \text{SS}_p}{2(k - 1)}}{\dfrac{\text{SS}_p}{\text{DF}_p}} \tag{18.49}$$

with $2(k - 1)$ and DF_p degrees of freedom. If this F is not significant, then all k sample regressions are assumed to estimate the same population regression, and the best estimate of that population regression is that given by Equation 18.24.

Some statistical workers prefer this test to those of the preceding sections in this chapter. However, if the null hypothesis is rejected, it is still necessary to employ the procedures of the previous sections if we wish to determine whether the differences within the regression are due to differences among slopes or elevations.

EXERCISES

18.1 Given:

For Sample 1: $n = 28$, $\sum x^2 = 142.35$, $\sum xy = 69.47$, $\sum y^2 = 108.77$, $\bar{X} = 14.7$, $\bar{Y} = 32.0$.

For Sample 2: $n = 30$, $\sum x^2 = 181.32$, $\sum xy = 97.40$, $\sum y^2 = 153.59$, $\bar{X} = 15.8$, $\bar{Y} = 27.4$.

(a) Test H_0: $\beta_1 = \beta_2$ vs. H_A: $\beta_1 \neq \beta_2$.

(b) If H_0 in part (a) is not rejected, test H_0: The elevations of the two population regressions are the same, vs. H_A: The two elevations are not the same.

18.2 Given:

For Sample 1: $n = 33$, $\sum x^2 = 744.32$, $\sum xy = 2341.37$, $\sum y^2 = 7498.91$.

For Sample 2: $n = 34$, $\sum x^2 = 973.14$, $\sum xy = 3147.68$, $\sum y^2 = 10366.97$.

For Sample 3: $n = 29$, $\sum x^2 = 664.42$, $\sum xy = 2047.73$, $\sum y^2 = 6503.32$.

For the total of all 3 samples: $n = 96$, $\sum x^2 = 3146.72$, $\sum xy = 7938.25$, $\sum y^2 = 20599.33$.

(a) Test H_0: $\beta_1 = \beta_2 = \beta_3$, vs. H_A: All three β's are not equal.

(b) If H_0: in part (a) is not rejected, test H_0: The three population regression lines have the same elevation, vs. H_A: The lines do not have the same elevation.

19

SIMPLE LINEAR CORRELATION

Chapter 17 introduced simple linear regression, the linear dependence of one variable (termed the dependent variable, Y) on a second variable (called the independent variable, X). In simple linear correlation, we also consider the linear relationship between two variables, but neither is assumed to be functionally dependent upon the other. An example of a correlation situation is the relationship between the wing length and tail length of a particular species of birds. Recall that the adjective "simple" refers to the fact that only two variables are considered simultaneously. Chapter 20 discusses correlation involving more than two variables.

19.1 THE CORRELATION COEFFICIENT

Some authors refer to the two variables in a simple correlation analysis as X_1 and X_2. We shall here employ the more common designation of X and Y, which does not, however, imply dependence of Y on X as it does in regression, nor does it imply cause-and-effect relationship between the two variables. The *correlation coefficient* (sometimes called the "simple correlation coefficient"*) is calculated as[†]

$$r = \frac{\sum xy}{\sqrt{\sum x^2 \sum y^2}},$$

(19.1)

which, among other methods (e.g., Symonds, 1926), is readily computed as

$$r = \frac{\sum XY - \dfrac{\sum X \sum Y}{n}}{\sqrt{\left(\sum X^2 - \dfrac{(\sum X)^2}{n}\right)\left(\sum Y^2 - \dfrac{(\sum Y)^2}{n}\right)}}.$$

(19.2)

Although the denominator of Equation 19.1 is always positive, the numerator may be positive, zero, or negative, thus enabling r to be either positive, zero, or negative, respectively. A positive correlation implies that for an increase in the value of one of the variables, the other variable also increases in value; a negative correlation indicates that an increase in value of one of the variables is accompanied by a decrease in value of the other variable. If $\sum xy = 0$, then $r = 0$, and one has a zero correlation, denoting that there is no linear association between the magnitudes of the two variables; that is, a change in magnitude of one does not imply a change in magnitude of the other. Figure 19.1 presents these considerations graphically.

Also important is the fact that the absolute value of the numerator of Equation 19.1 can never be larger than the denominator. Thus, r can never be greater than 1.0 nor

---

*It is also referred to as the "Pearson product-moment correlation coefficient" because of the algebraic expression of the coefficient, and the pioneering work on it, by Karl Pearson (1857–1936) around 1895. This followed the elucidation of correlation by Sir Francis Galton (1822–1911, cousin of Charles Darwin) in 1888 (who published it first with the term "co-relation"). The term "correlation" was first used, although the present-day concept was not developed, by Auguste Bravais (1811–1863), French naval lieutenant, astronomer, and physicist, who is also credited with developing the bivariate normal distribution—described at the end of this section—in 1846 (Rodgers and Nicewander, 1988; Walker, 1929: 96–98.) Galton published the first two-variable scatter plot of data in 1885 (Rodgers and Nicewander, 1988). The symbol, r, can be traced to Galton's 1877–1888 discussions of regression (which he first called "reversion") in heredity studies (and he later used r to indicate the slope of a regression line), and Galton developed correlation from regression. Indeed, in the early history of correlation, correlation coefficients were called "Galton Functions." The term "coefficient of correlation" was used by F. Y. Edgeworth in 1892. The basic concepts of correlation, however, predated Galton's work by several decades. (Pearson, 1920; Rodgers and Nicewander, 1988; Stigler, 1989; Walker, 1929: 92–102, 106, 109–110, 187.)

[†] David (1995) credits K. Pearson with being the first to call this quantity a correlation coefficient, in 1896. The computation depicted in Equation 19.2 was first published by Harris (1910). The correlation coefficient can also be calculated as $r = \dfrac{\sum xy}{(n-1)s_X s_Y}$, where s_X and s_Y are the standard deviations of X and Y, respectively; and another equivalent computation is noted as Equation 26.42. In 1896, K. Pearson gave G. U. Yule credit for pointing out the relationship $r = (s_X/s_Y)b$ (Walker, 1929: 111); it is also the case that $|r| = \sqrt{b_Y b_X}$, where b_Y is the regression coefficient if Y is considered the dependent variable and b_X is the regression coefficient if X is treated as the dependent variable; and, following from Equation 17.15, $|r| = \sqrt{\text{(regression SS)/(total SS)}}$; also see Rodgers and Nicewander (1988). In literature appearing within a couple of decades of Pearson's work, it was—albeit rarely—suggested that the coefficient could also be computed using deviations from the median instead of from the mean (Eells, 1926), which would result in a quantity not only different from r, but without the latter's theoretical and practical advantages.

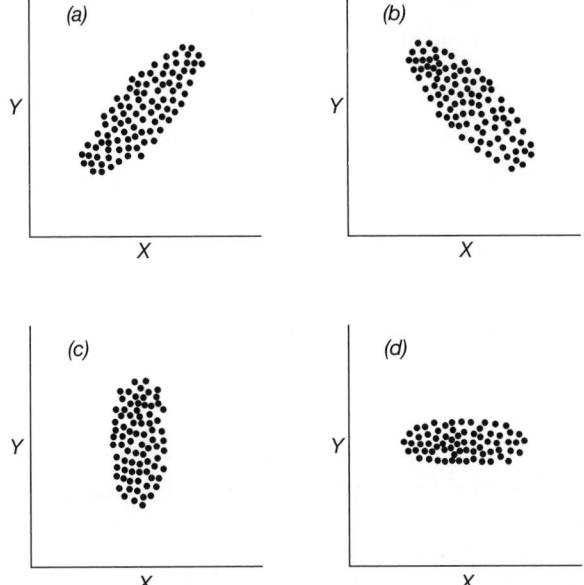

Figure 19.1 Simple linear correlation. (a) Positive correlation. (b) Negative correlation. (c) No correlation. (d) No correlation.

less than -1.0. Inspection of this equation further will reveal also that r has no units of measurement, for the units of both X and Y appear in both the numerator and denominator and thus cancel out arithmetically. A regression coefficient, b, may lie in the range of $-\infty \leq b \leq \infty$, and it expresses the magnitude of a change in Y associated with a unit change in X. But a correlation coefficient is unitless and $-1 \leq r \leq 1$. Thus, the correlation coefficient is not a measure of quantitative change of one variable with respect to the other, but it is a measure of intensity of association between the two variables.

The coefficient of determination, r^2, was introduced in Section 17.3 as a measure of how much of the total variability in Y is accounted for by regressing Y on X. In a correlation analysis, r^2 (occasionally called the "correlation index") may be calculated most simply by squaring the correlation coefficient, r. It may be described as the amount of variability in one of the variables (either Y or X) accounted for by correlating that variable with the second variable.* As in regression analysis, r^2 may be considered to be a measure of the strength of the straight-line relationship.[†] The calculation of r and r^2 is demonstrated in Example 19.1a.

The standard error of the correlation coefficient may be computed as

$$s_r = \sqrt{\frac{1 - r^2}{n - 2}}. \tag{19.3}$$

---

*A term found in older literature is the *coefficient of alienation*: $\sqrt{1 - r^2}$, given by Galton in 1889 and named by T. L. Kelley in 1919 (Walker, 1929: 175).

[†]Ozer (1985) argues that there are circumstances where $|r|$ is a better coefficient of determination than r^2.

EXAMPLE 19.1a   Calculation of the simple correlation coefficient and coefficient of determination. The data are wing and tail lengths among birds of a particular species.

Wing length (cm) (X)	Tail length (cm) (Y)
10.4	7.4
10.8	7.6
11.1	7.9
10.2	7.2
10.3	7.4
10.2	7.1
10.7	7.4
10.5	7.2
10.8	7.8
11.2	7.7
10.6	7.8
11.4	8.3

$n = 12$

$\sum X = 128.2$ cm $\sum Y = 90.8$ cm

$\sum X^2 = 1371.32$ cm^2 $\sum Y^2 = 688.40$ cm^2 $\sum XY = 971.37$ cm^2

$\sum x^2 = 1.7167$ cm^2 $\sum y^2 = 1.3467$ cm^2 $\sum xy = 1.3233$ cm^2

$$\text{correlation coefficient} = r = \frac{1.3233 \text{ cm}^2}{\sqrt{(1.7167 \text{ cm}^2)(1.3467 \text{ cm}^2)}} = 0.870$$

$$\text{coefficient of determination} = r^2 = 0.757$$

The location of the decimal place at which the second significant digit of s_r is located may be noted and r may be expressed rounded off to that decimal place.

Assumptions of Correlation Analysis.   No statistical assumptions need be satisfied in order to compute a correlation coefficient, but there are assumptions underlying the testing of hypotheses about, and the determination of confidence intervals for, correlation coefficients.

In regression we assume that for each X the Y values have come at random from a normal population. However, in correlation, not only are the Y's at each X assumed to be normal, but also the X values at each Y are assumed to have come at random from a normal population. This situation is referred to as sampling from a "bivariate normal distribution." The effect of deviations from the assumption of bivariate normality appears unimportant when there is, in fact, only slight correlation in the population; but if there is substantial population correlation, then there may be a marked adverse effect of such non-normality, this effect not being diminished by increasing sample size (Norris and Hjelm, 1961). If data are to be transformed in correlation analysis, the considerations of Section 17.10 are applicable to X as well as to Y.

19.2 HYPOTHESES ABOUT THE CORRELATION COEFFICIENT

The correlation coefficient, r, that we calculate from a sample is an estimate of a population parameter, namely the correlation coefficient in the population that was sampled. This parameter is denoted by ρ, lowercase Greek rho. If we wish to ask whether there is, in fact, a correlation between Y and X in the population, we can test H_0: $\rho = 0$. We do this, as in Example 19.1b by the familiar Student's t considerations, using

$$t = \frac{r}{s_r},\qquad(19.4)$$

where the standard error of r is calculated by Equation 19.3, and the degrees of freedom are $\nu = n - 2$. The null hypothesis is rejected if

$$|t| \geq t_{\alpha(2),\nu}.$$

Alternatively, this two-tailed hypothesis may be tested using

$$F = \frac{1 + |r|}{1 - |r|}\qquad(19.5)$$

(Cacoullos, 1965), where the critical value is $F_{\alpha(2),\nu,\nu}$. (See Example 19.1b.) Or, critical values of $|r|$ (namely, $r_{\alpha(2),\nu}$) may be read directly from Appendix Table B.17.*

One-tailed hypotheses about the population correlation coefficient may also be tested by the above procedures. For the hypotheses H_0: $\rho \leq 0$ and H_A: $\rho > 0$, compute either t or F (Equations 19.4 or 19.5, respectively) and reject H_0 if r is positive and either $t \geq t_{\alpha(1),\nu}$, or $F \geq F_{\alpha(1),\nu,\nu}$, or $r \geq r_{\alpha(1),\nu}$. To test H_0: $\rho \geq 0$ vs. H_A: $\rho < 0$, reject H_0 if r is negative and either $|t| \geq t_{\alpha(1),\nu}$, or $F \geq F_{\alpha(1),\nu}$, or $|r| \geq r_{\alpha(1),\nu}$.

If we wish to test H_0: $\rho = \rho_0$ for any ρ_0 other than zero, however, Equations 19.4 and 19.5 and Appendix Table B.17 are not applicable. Only for $\rho_0 = 0$ can r be considered to have come from a distribution approximated by the normal, and if the distribution of r is not normal, then the t and F statistics may not be validly employed. Fisher (1915, 1921) dealt with this problem when he proposed a transformation enabling r to be converted to a value, called z, which estimates a population parameter, ζ (lower case Greek

---

*Critical values of r are, by rearranging Equation 19.4,

$$r_{\alpha,\nu} = \sqrt{\frac{t_{\alpha,\nu}^2}{t_{\alpha,\nu}^2 + \nu}},\qquad(19.6)$$

where α may be either one-tailed or two-tailed, and $\nu = n - 2$. If a regression analysis is performed, rather than a correlation analysis, the probability of rejection of H_0: $\beta = 0$ is identical to the probability of rejecting H_0: $\rho = 0$. Also, r is related to b as

$$r = \frac{s_X}{s_Y}b,\qquad(19.7)$$

where s_X and s_Y are the standard deviations of X and Y, respectively.

EXAMPLE 19.1b   Testing H_0: $\rho = 0$ vs. H_A: $\rho \neq 0$. The data are those of Example 19.1a.

From Example 19.1a: $r = 0.870$.
To test H_0: $\rho = 0$ vs. H_A: $\rho \neq 0$:

$$\text{standard error of } r = s_r = \sqrt{\frac{1 - r^2}{n - 2}} = \sqrt{\frac{1 - (0.870)^2}{12 - 2}} = 0.156$$

$r_{0.05(2), 10} = 0.576$ (from Appendix Table B.17)

Therefore, reject H_0.

$$P < 0.001$$

Or:

$$t = \frac{r}{s_r} = \frac{0.870}{0.156} = 5.58$$

$t_{0.05(2), 10} = 2.228$

Therefore, reject H_0.

$$P < 0.001 \quad [P = 0.00012]$$

Or:

$$F = \frac{1 + |r|}{1 - |r|} = \frac{1.870}{0.130} = 14.4$$

$F_{0.05(2), 10, 10} = 3.72$

Therefore, reject H_0.

$$P < 0.001 \quad [P = 0.00014]$$

zeta), that *is* normally distributed. The transformation* is

$$z = 0.5 \ln \left(\frac{1 + r}{1 - r} \right). \tag{19.8}$$

For values of r between 0 and 1, the corresponding values of Fisher's z will lie between 0 and $-\infty$; and for r's from 0 to -1, the corresponding z's will fall between 0 and $-\infty$. For convenience, we may utilize Appendix Table B.18 to avoid having to perform the computation of Equation 19.8 to transform r to z.[†]

For the hypothesis H_0: $\rho = \rho_0$, then, we calculate a normal deviate, as

$$Z = \frac{z - \zeta_0}{\sigma_z}, \tag{19.10}$$

---

*z is also equal to $r + r^3/3 + r^5/5 \ldots$ and is a quantity that mathematicians recognize as the "inverse hyperbolic tangent of r," namely $z = \tanh^{-1} r$. The transformation of z to r, given in Appendix Table B.19, is

$$r = \tanh z \quad \text{or} \quad r = \frac{e^{2z} - 1}{e^{2z} + 1}. \tag{19.9}$$

[†]As noted at the end of Section 19.7, there is a slight, and correctable bias in z. Unless n is very small, however, this correction will be insignificant and may be ignored.

where z is the transform of r, ζ_0 is the transform of the hypothesized coefficient, ρ_0, and the standard error of z is approximated by

$$\sigma_z = \sqrt{\frac{1}{n-3}}. \tag{19.11}$$

In Example 19.1, $r = 0.870$ was calculated. If we had desired to test H_0: $\rho = 0.750$, we would have proceeded as shown in Example 19.2. Recall that the critical value of a normal deviate may be obtained readily from the bottom line of the t table (Appendix Table B.3), for $Z_{\alpha(2)} = t_{\alpha(2),\infty}$. One-tailed hypotheses, H_0: $\rho \leq \rho_0$ and H_A: $\rho > \rho_0$, or H_0: $\rho \geq \rho_0$ and H_A: $\rho < \rho_0$, may also be tested using Equation 19.10 in which case the critical value of the normal deviate would, of course, be $Z_{\alpha(1)} = t_{\alpha(1),\infty}$.

  If the variables in correlation analysis have come from a bivariate normal distribution, as often may be assumed, then we may employ the aforementioned procedures, as well as those that follow. Sometimes, only one of the two variables may be assumed to have been obtained randomly from a normal population. It may be possible to employ a data transformation (see Section 17.10) to remedy this situation. If that cannot be done, then the hypothesis H_0: $\rho = 0$ (or its associated one-tailed hypotheses) may be tested, but none of the other testing procedures of this chapter (except for the methods of Section 19.9) are valid. If neither variable came from a normal population, and data transformations do not improve this condition, then we may turn to the procedures of Section 19.9.

EXAMPLE 19.2   Testing H_0: $\rho = \rho_0$, where $\rho_0 \neq 0$.

  $r = 0.870$

  $n = 12$

  H_0: $\rho = 0.750$; H_A: $\rho \neq 0.750$.

  $z = 1.3331$

  $\zeta_0 = 0.9730$

  $Z = \dfrac{z - \zeta_0}{\sqrt{\dfrac{1}{n-3}}} = \dfrac{1.3331 - 0.9730}{\sqrt{\dfrac{1}{9}}} = \dfrac{0.3601}{0.3333} = 1.0803$

  $Z_{0.05(2)} = t_{0.05(2),\infty} = 1.960$

  Therefore, do not reject H_0.

$$0.20 < P < 0.50 \quad [P = 0.28]$$

19.3 CONFIDENCE INTERVALS FOR THE POPULATION CORRELATION COEFFICIENT

Confidence limits on ρ may be determined by a procedure related to Equation 19.5; the lower and upper confidence limits are

$$L_1 = \frac{(1 + F_\alpha)r + (1 - F_\alpha)}{(1 + F_\alpha) + (1 - F_\alpha)r} \tag{19.12}$$

$$L_2 = \frac{(1 + F_\alpha)r - (1 - F_\alpha)}{(1 + F_\alpha) - (1 - F_\alpha)r},$$ (19.13)

respectively, where $F_\alpha = F_{\alpha(2), \nu, \nu}$ and $\nu = n - 2$ (Muddapur, 1988).* This is shown in Example 19.3.

EXAMPLE 19.3   Setting confidence limits for a correlation coefficient. This example uses the data of Example 19.1a.

$r = 0.870, n = 12, \nu = 10, \alpha = 0.05$.

$F_{0.05(2), 10, 10} = 3.72$; so $1 + F_{0.05(2), 10, 10} = 4.72$, and $1 - F_{0.05(2), 10, 10} = -2.72$; and $(1 + F_{0.05(2), 10, 10})r = (4.72)(0.870) = 4.11$ and $(1 - F_{0.05(2), 10, 10})r = (-2.72)(0.870) = -2.37$.

$$L_1 = \frac{(1 + F_\alpha)r + (1 - F_\alpha)}{(1 + F_\alpha) + (1 - F_\alpha)r} = \frac{4.11 - 2.72}{4.72 - 2.37} = 0.591$$

$$L_2 = \frac{(1 + F_\alpha)r - (1 - F_\alpha)}{(1 + F_\alpha) - (1 - F_\alpha)r} = \frac{4.11 + 2.72}{4.72 + 2.37} = 0.963$$

For the Fisher approximation:

$r = 0.870$, therefore, $z = 1.3331$ (from Appendix Table B.18).

$$\sigma_z = \sqrt{\frac{1}{n - 3}} = 0.3333$$

95% confidence interval for $\zeta = z \pm Z_{0.05(2)}\sigma_z$

$$= z \pm t_{0.05(2), \infty}\sigma_z$$

$$= 1.3331 \pm (1.9600)(0.3333)$$

$$= 1.3331 \pm 0.6533$$

$L_1 = 0.680$

$L_2 = 1.986$

For the 95% confidence limits for ρ, transform L_1 and L_2 computed for ζ (using Appendix Table B.19):
$L_1 = 0.591; L_2 = 0.963$.

Fisher's transformation may be used to approximate these confidence limits, although the confidence interval will generally be larger than that from the above procedure, and the confidence coefficient may occasionally (and undesirably) be less than

---

*Jeyaratnam (1992) asserts that the same confidence limits are obtained by

$$L_1 = \frac{r - w}{1 - rw} \quad \text{and}$$ (19.14)

$$L_2 = \frac{r + w}{1 + rw},$$ (19.15)

where w is $r_{\alpha, \nu}$ from Equation 19.6 using the two-tailed $t_{\alpha(2), \nu}$.

$1 - \alpha$ (Jeyaratnam, 1992). By this procedure, one converts r to z (as with the aid of Appendix Table B.18), and then the $1 - \alpha$ confidence limits may be computed for ζ:

$$z \pm Z_{\alpha(2)}\sigma_z \tag{19.16}$$

or, equivalently,

$$z \pm t_{\alpha(2),\infty}\sigma_z. \tag{19.17}$$

The lower and upper confidence limits, L_1 and L_2, are both z values and may be transformed to r values (Appendix Table B.19 being useful for this purpose). Example 19.3 demonstrates this procedure. Note that, although the confidence limits for ζ are symmetrical, the confidence limits for ρ are not.

19.4 POWER AND SAMPLE SIZE IN CORRELATION

If we test H_0: $\rho = 0$ at the α significance level, with a sample size of n, then we may determine the probability of correctly rejecting H_0 when ρ_0 is in fact a specified value other than zero. This is done (Cohen, 1977: 458) by using the Fisher z transformation for the critical value of r and for the sample r (from Appendix Table B.18 or Equation 19.8); let us call these two transformed values z_α and z, respectively. Then, the power of the test for H_0: $\rho = 0$ is $1 - \beta(1)$, where $\beta(1)$ is the one-tailed probability of the normal deviate

$$Z_{\beta(1)} = (z - z_\alpha)\sqrt{n - 3}, \tag{19.18}$$

(Cohen, 1988: 546), as demonstrated in Example 19.4. This procedure may be used for one-tailed as well as two-tailed hypotheses, so α may be either $\alpha(1)$ or $\alpha(2)$, respectively.

EXAMPLE 19.4   Determination of power of the test of H_0: $\rho = 0$ in Example 19.1b.

$n = 12$; $\nu = 10$

$r = 0.870$, so $z = 1.3331$

$r_{0.05(2),10} = 0.576$, so $z_{0.05} = 0.6565$

$Z_{\beta(1)} = (1.3331 - 0.6565)\sqrt{12 - 3}$

$\qquad = 2.03$

From Appendix Table B.2, $P(Z \geq 2.03) = 0.0212 = \beta$. Therefore, the power of the test is $1 - \beta = 0.98$.

If the desired power is stated, then we can ask how large a sample is required to reject H_0: $\rho = 0$ if it is truly false with a specified $\rho_0 \neq 0$. This can be estimated

(Cohen, 1988: 546) by calculating

$$n = \left(\frac{Z_{\beta(1)} + Z_\alpha}{\zeta_0}\right)^2 + 3, \tag{19.19}$$

where ζ_0 is the Fisher transformation of the ρ_0 specified, and the significance level, α, can be either one-tailed or two-tailed. This procedure is shown in Example 19.5.

EXAMPLE 19.5   Determination of required sample size in testing H_0: $\rho = 0$.

We desire to reject H_0: $\rho = 0$ 99% of the time when $|\rho| \geq 0.5$ and the hypothesis is tested at the 0.05 level of significance. Therefore, $\beta(1) = 0.01$ and (from the last line of Appendix Table B.3) and $Z_{\beta(1)} = 2.3263$; $\alpha(2) = 0.05$ and $Z_{\alpha(2)} = 1.9600$; and, for $r = 0.5$, $z = 0.5493$.
     Then,

$$n = \left(\frac{2.3263 + 1.9600}{0.5493}\right)^2 + 3 = 63.9,$$

so a sample of size at least 64 should be used.

Hypothesizing ρ Other than 0.   For the two-tailed hypothesis H_0: $\rho = \rho_0$, where $\rho_0 \neq 0$, the power of the test is determined from

$$Z_{\beta(1)} = (|z - z_0| - z_\alpha)\sqrt{n - 3}, \tag{19.20}$$

instead of from Equation 19.18; here, z_0 is the Fisher transformation of ρ_0. Either two-tailed or one-tailed hypotheses may be considered, using $\alpha(2)$ or $\alpha(1)$, respectively. Sample size estimation is from Equation 19.19.

19.5 COMPARING TWO CORRELATION COEFFICIENTS

Hypotheses (either one-tailed or two-tailed) about two correlation coefficients may be tested by the use of

$$Z = \frac{z_1 - z_2}{\sigma_{z_1 - z_2}}, \tag{19.21}$$

where

$$\sigma_{z_1 - z_2} = \sqrt{\frac{1}{n_1 - 3} + \frac{1}{n_2 - 3}}. \tag{19.22}$$

If $n_1 = n_2$, then Equation 19.22 reduces to

$$\sigma_{z_1 - z_2} = \sqrt{\frac{2}{n - 3}}, \tag{19.23}$$

where n is the size of each sample. The use of the Fisher z transformation both normalizes the underlying distribution of each of the correlation coefficients, r_1 and r_2, and stabilizes the variances of these distributions (Winterbottom, 1979). The multisample hypothesis test recommended by Paul (1988), and presented in Section 19.7, may be used for the two-tailed two-sample hypothesis above. It tends to result in a probability of Type I error that is closer to the specified α; but, as it also tends to be larger than α, I do not recommend it for the two-sample case. The preferred procedure for testing the two-tailed hypotheses, H_0: $\rho_1 = \rho_2$ vs. H_A: $\rho_1 = \rho_2$ is to employ Equation 19.21, as shown in Example 19.6.* One-tailed hypotheses may be tested using one-tailed critical values, namely $Z_{\alpha(1)}$.

EXAMPLE 19.6   Testing the hypothesis H_0: $\rho_1 = \rho_2$.

For a sample of ninety-eight bird wing and tail lengths, a correlation coefficient of 0.78 was calculated. A sample of ninety-five such measurements from a second bird species yielded a correlation coefficient of 0.84. Let us test for the equality of the two population correlation coefficients.

$$H_0: \rho_1 = \rho_2; \ H_A: \rho_1 \neq \rho_2.$$

$r_1 = 0.78$	$r_2 = 0.84$
$z_1 = 1.0454$	$z_2 = 1.2212$
$n_1 = 98$	$n_2 = 95$

$$Z = \frac{1.0454 - 1.2212}{\sqrt{\dfrac{1}{n_1 - 3} + \dfrac{1}{n_2 - 3}}} = \frac{-0.1758}{0.1463} = -1.202$$

$$Z_{0.05(2)} = t_{0.05(2),\infty} = 1.960$$

Therefore, do not reject H_0.

$$0.20 < P < 0.50 \quad [P = 0.23]$$

The common correlation coefficient may then be computed as

$$z_w = \frac{(n_1 - 3)z_1 + (n_2 - 3)z_2}{(n_1 - 3) + (n_2 - 3)} = \frac{(95)(1.0454) + (92)(1.2212)}{95 + 92} = 1.1319$$

$$r_w = 0.81$$

Occasionally it may be desired to test for equality of two correlation coefficients that are not independent. For example, if Sample 1 in Example 19.6 were data from a group of ninety-eight young birds, and Sample 2 were from ninety-five of these birds when they were older (three of the original birds having died or escaped), the two sets of data should not be considered to be independent. Procedures for computing r_1 and r_2, taking dependence into account, are reviewed by Steiger (1980).

---

*A null hypothesis such as H_0: $\rho_1 - \rho_2 = \rho_0$, where $\rho_0 \neq 0$, might be tested by substituting $|z_1 - z_2| - \zeta_0$ for the numerator in Equation 19.21, but no utility for such a test is apparent.

Common Correlation Coefficient.   As in Example 19.6, a conclusion that $\rho_1 = \rho_2$ would lead us to say that both our samples came from the same population, or from populations with identical correlation coefficients. In such a case, we may combine the information from the two samples to calculate a better estimate of a single underlying ρ. Let us call this estimate the "common," or "weighted," correlation coefficient. We obtain it by converting

$$z_w = \frac{(n_1 - 3)z_1 + (n_2 - 3)z_2}{(n_1 - 3) + (n_2 - 3)} \tag{19.24}$$

to its corresponding r value, r_w, as shown in Example 19.6. If both samples are of equal size (i.e., $n_1 = n_2$), then the previous equation reduces to

$$z_w = \frac{z_1 + z_2}{2}. \tag{19.25}$$

Appendix Table B.19 gives the conversion of z_w to the common correlation coefficient, r_w (which estimates the common population coefficient, ρ). Paul (1988) has shown that if ρ is less than about 0.5, then a better estimate of that parameter utilizes

$$z_w = \frac{(n_1 - 1)z_1' + (n_2 - 1)z_2'}{(n_1 - 1) + (n_2 - 1)}, \tag{19.26}$$

where

$$z_i' = z_i - \frac{3z_i + r_i}{4(n_i - 1)} \tag{19.27}$$

and z_i is as in Equation 19.8 (Hotelling, 1953).

We may test hypotheses about the common correlation coefficient (H_0: $\rho = 0$ vs. H_A: $\rho \neq 0$, or H_0: $\rho = \rho_0$ vs. H_A: $\rho \neq \rho_0$, or similar one-tailed tests) by Equation 19.33 or 19.34.

19.6 POWER AND SAMPLE SIZE IN COMPARING TWO CORRELATION COEFFICIENTS

The power of the above test for difference between two correlation coefficients is estimated as $1 - \beta$, where β is the one-tailed probability of the normal deviate calculated as

$$Z_{\beta(1)} = \frac{|z_1 - z_2|}{\sigma_{z_1 - z_2}} - Z_\alpha \tag{19.28}$$

(Cohen, 1988: 546–547), where α may be either one-tailed or two-tailed and where $Z_{\alpha(1)}$ or $Z_{\alpha(2)}$ is most easily read from the last line of Appendix Table B.3. Example 19.7 demonstrates this calculation for the data of Example 19.6.

If we state a desired power to detect a specified difference between transformed correlation coefficients, then the sample size required to reject H_0 when testing at the α

EXAMPLE 19.7   Determination of the power of the test of H_0: $\rho_1 = \rho_2$ in Example 19.6.

$z_1 = 1.0454 \qquad z_2 = 1.2212$

$\sigma_{z_1 - z_2} = 0.1463$

$Z_\alpha = Z_{0.05(2)} = 1.960$

$Z_{\beta(1)} = \dfrac{|1.0454 - 1.2212|}{0.1463} - 1.960$

$\qquad = 1.202 - 1.960$

$\qquad = -0.76$

From Appendix Table B.2,

$$\beta = P(Z \geq -0.76) = 1 - P(Z \leq -0.76) = 1 - 0.2236 = 0.78$$

Therefore,

$$\text{power} = 1 - \beta = 1 - 0.78 = 0.22.$$

---

level of significance is

$$n = 2 \left(\frac{Z_\alpha + Z_{\beta(1)}}{z_1 - z_2} \right)^2 + 3 \tag{19.29}$$

(Cohen, 1988: 547). This is shown in Example 19.8.

   The test for difference between correlation coefficients is most powerful for $n_1 = n_2$, and the above estimation is for a sample size of n in both samples. Sometimes the size of one sample is fixed and cannot be manipulated, and we then ask how large the second sample must be to achieve the desired power. If n_1 is fixed, and n is determined by Equation 19.29, then (by considering n to be the harmonic mean of n_1 and n_2),

$$n_2 = \frac{n_1(n + 3) - 6n}{2n_1 - n - 3} \tag{19.30}$$

(Cohen, 1988: 137).*

EXAMPLE 19.8   Estimating the sample size necessary for the test of H_0: $\rho_1 = \rho_2$.

Let us say we wish to be 90% confident of detecting a difference, $z_1 - z_2$, as small as 0.5000 when testing H_0: $\rho_1 = \rho_2$ at the 5% significance level. Then $\beta(1) = 0.10$, $\alpha(2) = 0.05$, and

$$n = 2 \left(\frac{1.9600 + 1.2816}{0.5000} \right)^2 + 3$$

$$= 45.0$$

So, sample sizes of at least forty-five should be used.

---

*If the denominator in Equation 19.30 is ≤ 0, then we must either increase n_1 or change the desired power, significance level, or detectable difference in order to solve for n_2.

19.7 Comparing more than Two Correlation Coefficients

If k samples have been obtained and an r has been calculated for each, it is often desirable to conclude whether or not all samples came from populations having identical ρ's. If H_0: $\rho_1 = \rho_2 = \cdots = \rho_k$ is not rejected, then all samples might be combined and one value of r calculated to estimate the single population ρ. As Example 19.9 shows, the testing of this hypothesis involves transforming each r to a z value. One may then calculate

$$\chi^2 = \sum_{i=1}^{k}(n_i - 3)z_i^2 - \frac{\left[\sum_{i=1}^{k}(n_i - 3)z_i\right]^2}{\sum_{i=1}^{k}(n_i - 3)}, \tag{19.31}$$

which may be considered to be a chi-square value with $k - 1$ degrees of freedom.*

Common Correlation Coefficient. If H_0 is not rejected, then all k sample correlation coefficients are assumed to estimate a common population ρ. A "common r" (or "weighted mean of r") may be obtained from transforming the weighted mean z value,

$$z_w = \frac{\sum_{i=1}^{k}(n_i - 3)z_i}{\sum_{i=1}^{k}(n_i - 3)}, \tag{19.32}$$

to its corresponding r value (let's call it r_w), as is shown in Example 19.9. This transformation is that of Equation 19.8 and is given in Appendix Table B.18. And, if H_0 is not rejected, we may test H_0: $\rho = 0$ vs. H_A: $\rho \neq 0$ by the method attributed to Neyman (1959) by Paul (1988):

$$Z = \frac{\sum_{i=1}^{k}n_i r_i}{\sqrt{N}}, \tag{19.33}$$

where $N = \sum_{i=1}^{k}n_i$, and rejecting H_0 if $|Z| \geq Z_{\alpha(2)}$. For one-tailed testing, H_0: $\rho \leq 0$ vs. H_A: $\rho > 0$ is rejected if $Z \geq Z_{\alpha(1)}$; and H_0: $\rho \geq 0$ vs. H_A: $\rho < 0$ is rejected if $Z \leq -Z_{\alpha(1)}$.

For the hypothesis, H_0: $\rho = \rho_0$ vs. H_A: $\rho \neq \rho_0$, the transformation of Equation 19.8 is applied to convert ρ_0 to ζ_0. Then (from Paul, 1988),

$$Z = (z_w - \zeta_0)\sqrt{\sum_{i=1}^{n}(n_i - 3)} \tag{19.34}$$

---

*Equation 19.31 is a computational convenience for

$$\chi^2 = \sum(n_i - 3)(z_i - z_w)^2, \tag{19.31a}$$

where z_w is a weighted mean of z as described below.

EXAMPLE 19.9   Testing a three-sample hypothesis concerning correlation coefficients.

Given the following:

$$n_1 = 24 \qquad n_2 = 29 \qquad n_3 = 32$$
$$r_1 = 0.52 \qquad r_2 = 0.56 \qquad r_3 = 0.87$$

To test:

H_0: $\rho_1 = \rho_2 = \rho_3$.

H_A: All three population correlation coefficients are not equal.

i	r_i	z_i	z_i^2	n_i	$n_i - 3$	$(n_i - 3)z_i$	$(n_i - 3)z_i^2$
1	0.52	0.5763	0.3321	24	21	12.1023	6.9741
2	0.56	0.6328	0.4004	29	26	16.4528	10.4104
3	0.87	1.3331	1.7772	32	29	38.6599	51.5388
Sums:					76	67.2150	68.9233

$$\chi^2 = \sum (n_i - 3)z_i^2 - \frac{\left[\sum (n_i - 3)z_i \right]^2}{\sum (n_i - 3)}$$

$$= 68.9233 - \frac{(67.2150)^2}{76}$$

$$= 9.478$$

$$\nu = k - 1 = 2$$

$$\chi_{0.05,2}^2 = 5.991$$

Therefore, reject H_0.

$$0.005 < P < 0.01 \quad [P = 0.0087]$$

If H_0 had not been rejected, it would have been appropriate to calculate the common correlation coefficient:

$$z_w = \frac{\sum (n_i - 3)z_i}{\sum (n_i - 3)} = \frac{67.2150}{76} = 0.884$$

$$r_w = 0.71$$

---

is computed and H_0 is rejected if $|Z| \geq Z_{\alpha(2)}$. For one-tailed testing, H_0: $\rho \leq \rho_0$ is rejected if $Z \geq Z_{\alpha(1)}$ or H_0: $\rho \geq \rho_0$ is rejected if $Z \leq -Z_{\alpha(1)}$.

If correlations are not independent, then they may be compared as described by Steiger (1980).

Overcoming Bias.   Fisher (1958: 205) and Hotelling (1953) have pointed out that the z transformation is slightly biased, in that each z will be a little inflated. This minor systematic error is likely to have only negligible effects on our previous considerations, but it is inclined to have adverse effects on the testing of multisample hypotheses, for in the latter situations several values of z_i, and therefore several small errors, are being summed. Such a hypothesis test and the estimation of a common correlation coefficient are most improved by correcting for bias when sample sizes are small or there are many samples in the analysis.

Several corrections for bias are available. Fisher recommended subtracting

$$\frac{r}{2(n-1)}$$

from z, whereas Hotelling determined that better corrections to z are available, such as subtracting

$$\frac{3z+r}{4(n-1)}.$$

However, Paul (1988) recommends a test that performs better than one employing such corrections for bias. It uses

$$\chi_P^2 = \sum_{i=1}^{k} \frac{n_i(r_i - r_w)^2}{(1 - r_i r_w)^2} \tag{19.35}$$

with $k-1$ degrees of freedom. The analysis of Example 19.10 demonstrates this test.

If the multisample null hypothesis is not rejected, then ρ, the underlying population correlation coefficient, may be estimated by calculating z_w via Equation 19.32 and converting it to r_w. As an improvement, Paul (1988) determined that $n_i - 1$ should be used in place of $n_i - 3$ in the latter equation if ρ is less than about 0.5. Similarly, to compare ρ to a specified value, ρ_0, $n_i - 1$ would be used instead of $n_i - 3$ in Equation 19.34 if ρ is less than about 0.5.

EXAMPLE 19.10 **The hypothesis testing of Example 19.9, employing correction for bias.**

$$\chi_P^2 = \sum_{i=1}^{k} \frac{n_i(r_i - r_w)^2}{(1 - r_i r_w)^2}$$

$$= \frac{24(0.52 - 0.71)^2}{[1 - (0.52)(0.71)]^2} + \frac{29(0.56 - 0.71)^2}{[1 - (0.56)(0.71)]^2} + \frac{32(0.87 - 0.71)^2}{[1 - (0.87)(0.71)]^2}$$

$$= 2.1774 + 1.7981 + 5.6051$$

$$= 9.5806$$

$$\chi_{0.05,2}^2 = 5.991$$

Therefore, reject H_0.

$$0.005 < P < 0.01 \quad [P = 0.0083]$$

19.8 MULTIPLE COMPARISONS AMONG CORRELATION COEFFICIENTS

If the null hypothesis of the previous section (H_0: $\rho_1 = \rho_2 = \cdots = \rho_k$) is rejected, it is typically of interest to determine which of the k correlation coefficients are different from which others. This can be done, again using Fisher's z transformation (Levy, 1976).

In the fashion of Section 11.1 (where multiple comparisons were made among means), we can test each pair of correlation coefficients, r_B and r_A, by a Tukey-type test, if $n_B = n_A$:

$$q = \frac{z_B - z_A}{\text{SE}}, \tag{19.36}$$

where

$$\text{SE} = \sqrt{\frac{1}{n-3}} \tag{19.37}$$

and n is the size of each sample. If the sizes of the two samples, A and B, are not equal, then we can use

$$\text{SE} = \sqrt{\frac{1}{2}\left(\frac{1}{n_A - 3} + \frac{1}{n_B - 3}\right)}. \tag{19.38}$$

The appropriate critical value for this test is $q_{\alpha,\infty,k}$ (from Appendix Table B.5). This test is demonstrated in Example 19.11.

EXAMPLE 19.11   Tukey-type multiple comparison testing among the three correlation coefficients in Example 19.9.

Samples ranked by correlation coefficient (i):	1	2	3
Ranked correlation coefficients (r_i):	0.52	0.56	0.87
Ranked transformed coefficients (z_i):	0.5763	0.6328	1.3331
Sample size (n_i):	24	29	32

Comparison B vs. A	Difference $z_B - z_A$	SE	q	$q_{0.05,\infty,3}$	Conclusion
3 vs. 1	$1.3331 - 0.5763 = 0.7568$	0.203	3.728	3.314	Reject H_0: $\rho_3 = \rho_1$
3 vs. 2	$1.3331 - 0.6328 = 0.7003$	0.191	3.667	3.314	Reject H_0: $\rho_3 = \rho_2$
2 vs. 1	$0.6328 - 0.5763 = 0.0565$	0.207	0.273	3.314	Accept H_0: $\rho_2 = \rho_1$

Overall conclusion: $\rho_1 = \rho_2 \neq \rho_3$

In a manner analogous to that in Section 11.2, Newman-Keuls-type multiple range testing could be performed instead of the Tukey-type procedure. This would involve computing q as in Equation 19.36 and using as critical values $q_{\alpha,\infty,p}$, where p is the range of correlation coefficients being compared (just as p is the range of means being compared in Section 11.2). It is typically unnecessary in multiple comparison testing to employ the correction for bias described at the end of Section 19.7.

Comparison of a Control Correlation Coefficient to Each Other Correlation Coefficient. The methods above enable us to compare each correlation coefficient with each other coefficient. If, instead, we desire only to compare each coefficient to one particular coefficient (call it the correlation coefficient of the "control" set of data), then a procedure analogous to the Dunnett test of Section 11.4 may be employed (Huitema, 1974).

Let us designate the control set of data as B, and each other group of data, in turn, as A. Then, we compute

$$q = \frac{z_B - z_A}{\text{SE}}, \tag{19.39}$$

for each A, in the same sequence as described in Section 11.4. The appropriate standard error is

$$\text{SE} = \sqrt{\frac{2}{n-3}} \tag{19.40}$$

if samples A and B are of the same size, or

$$\text{SE} = \sqrt{\frac{1}{n_A - 3} + \frac{1}{n_B - 3}} \tag{19.41}$$

if $n_A \neq n_B$. The critical value is $q'_{\alpha(1), \infty, p}$ (from Appendix Table B.6) or $q'_{\alpha(2), \infty, p}$ (from Appendix Table B.7) for the one-tailed or two-tailed test, respectively.

Multiple Contrasts among Correlation Coefficients.   Section 11.5 introduces the concepts and procedures of multiple contrasts among means; these are multiple comparisons involving groups of means. In a similar fashion, multiple contrasts may be examined among correlation coefficients (Marascuilo, 1971: 454–455). We again employ the z transformation and calculate, for each contrast, the test statistic

$$S = \frac{\left| \sum_i c_i z_i \right|}{\text{SE}}, \tag{19.42}$$

where

$$\text{SE} = \sqrt{\sum_i c_i^2 \sigma_{z_i}^2} \tag{19.43}$$

and c_i is a contrast coefficient, as described in Section 11.5. (For example, if we wished to test the hypothesis H_0: $(\rho_1 + \rho_2)/2 - \rho_3 = 0$, then $c_1 = \frac{1}{2}$, $c_2 = \frac{1}{2}$, and $c_3 = -1$.) The critical value for this test is

$$S_\alpha = \sqrt{\chi^2_{\alpha, (k-1)}}. \tag{19.44}*$$

---

*Because $\chi^2_{\alpha, \nu} = \nu F_{\alpha(1), \nu, \infty}$, it is equivalent to write

$$S_\alpha = \sqrt{(k-1) F_{\alpha(1), (k-1), \infty}}, \tag{19.45}$$

but Equation 19.44 is preferable, because it engenders less rounding error in the calculations.

19.9 RANK CORRELATION

If we have data obtained from a bivariate population that is far from normal, then the correlation procedures discussed thus far are generally inapplicable. Instead, we may operate with the ranks of the measurements for each variable. Two different *rank correlation* methods are commonly encountered, that proposed by Spearman (1904), and that of Kendall* (1938).

Example 19.12 demonstrates Spearman's rank correlation procedure. After each measurement of a variable is ranked, as done in previously described nonparametric testing procedures, Equation 19.1 can be applied to the ranks to obtain the *Spearman rank correlation coefficient*, r_s. However, a simpler computation is

$$r_s = 1 - \frac{6 \sum_{i=1}^{n} d_i^2}{n^3 - n},$$
(19.46)[†]

where d_i is a difference between X and Y ranks: $d_i = $ rank of $X_i - $ rank of Y_i.[‡] The value of r_s, as an estimate of the population rank correlation coefficient, ρ_s, may range from -1 to $+1$, and it has no units; however, its value is not to be expected to be the same as the value of r that might have been calculated for the original data instead of their ranks.[§]

Use Appendix Table B.20 to assess the significance of r_s. If n is greater than that provided for in this table, then r_s may be used in place of r in the hypothesis testing procedures of Section 19.2. If either the Spearman or the parametric correlation analysis (Section 19.2) is applicable, the former is $9/\pi^2 = 0.91$ as powerful as the latter (Daniel, 1978: 304; Hotelling and Pabst, 1936).

---

*Charles Edward Spearman (1863–1945), English psychologist and statistician, an important researcher on intelligence and on the statistical field known as factor analysis (Cattell, 1978). Sir Maurice George Kendall (1907–1983), English statistician eminent in statistical theory and social statistics (Bartholomew, 1983). Karl Pearson observed that Sir Francis Galton considered the correlation of ranks even before developing correlation of variables (Walker, 1929: 128).

[†]As the sum of n ranks is $n(n + 1)/2$, Equation 19.1 may be rewritten for rank correlation as

$$r_s = \frac{\sum_{i=1}^{n} (\text{rank of } X_i)(\text{rank of } Y_i) - \frac{n(n + 1)^2}{4}}{\sqrt{\left(\sum_{i=1}^{n} (\text{rank of } X_i)^2 - \frac{n(n + 1)^2}{4}\right)\left(\sum_{i=1}^{n} (\text{rank of } Y_i)^2 - \frac{n(n + 1)^2}{4}\right)}}$$
(19.47)

[‡]Instead of using differences between ranks of pairs of X and Y, one may use the sums of the ranks for each pair, where $S_i = $ rank of $X_i + $ rank of Y_i (Meddis, 1984: 227; Thomas, 1989):

$$r_s = \frac{6 \sum S_i^2}{n^3 - n} - \frac{7n + 5}{n - 1}.$$
(19.48)

[§]This statistic is related to the coefficient of concordance, W, from Section 19.17. Within two groups of ranks,

$$W = (r_s + 1)/2.$$
(19.49)

EXAMPLE 19.12   Spearman rank correlation for the relationship between the scores of ten students on a mathematics aptitude examination and a biology aptitude examination.

Student (i)	Mathematics examination score (X_i)	Rank of X_i	Biology examination score (Y_i)	Rank of Y_i	d_i	d_i^2
1	57	3	83	7	−4	16
2	45	1	37	1	0	0
3	72	7	41	2	5	25
4	78	8	84	8	0	0
5	53	2	56	3	−1	1
6	63	5	85	9	−4	16
7	86	9	77	6	3	9
8	98	10	87	10	0	0
9	59	4	70	5	−1	1
10	71	6	59	4	2	4

$n = 10$

$\sum d_i^2 = 72$

$$r_s = 1 - \frac{6\sum d_i^2}{n^3 - n}$$

$$= 1 - \frac{6(72)}{10^3 - 10}$$

$$= 1 - 0.436$$

$$= 0.564$$

To test H_0: $\rho_s = 0$; H_A: $\rho_s \neq 0$.

$(r_s)_{0.05(2),\,10} = 0.648$ (from Appendix Table B.20)

Therefore, do not reject H_0.

$$P = 0.10$$

Correction for Tied Data.   If there are tied data, then they are assigned average ranks as before (e.g., Section 8.10) and r_s is better calculated either by Equation 19.1 applied to the ranks (Iman and Conover, 1978) or as

$$(r_s)_c = \frac{(n^3 - n)/6 - \sum d_i^2 - \sum t_X - \sum t_Y}{\sqrt{\left[(n^3 - n)/6 - 2\sum t_X\right]\left[(n^3 - n)/6 - 2\sum t_Y\right]}} \tag{19.50}$$

(Kendall, 1962: 38; Thomas, 1989). Here,

$$\sum t_X = \frac{\sum(t_i^3 - t_i)}{12}, \tag{19.51}$$

where t_i is the number of tied values of X in a group of ties, and

$$\sum t_Y = \frac{\sum(t_i^3 - t_i)}{12}, \tag{19.52}$$

where t_i is the number of tied Y's in a group of ties; this is demonstrated in Example 19.13. If $\sum t_X$ and $\sum t_Y$ are zero, then Equation 19.50 is identical to Equation 19.46. Indeed, the two equations differ appreciably only if there are numerous tied data.

EXAMPLE 19.13   The Spearman rank correlation coefficient, computed for the data of Example 19.1.

X	Rank of X	Y	Rank of Y	d_i	d_i^2
10.4	4	7.4	5	−1	1
10.8	8.5	7.6	7	1.5	2.25
11.1	10	7.9	11	−1	1
10.2	1.5	7.2	2.5	−1	1
10.3	3	7.4	5	−2	4
10.2	1.5	7.1	1	0.5	0.25
10.7	7	7.4	5	2	4
10.5	5	7.2	2.5	2.5	6.25
10.8	8.5	7.8	9.5	−1	1
11.2	11	7.7	8	3	9
10.6	6	7.8	9.5	−3.5	12.25
11.4	12	8.3	12	0	0

$n = 12$

$\sum d_i^2 = 42.00$

$$r_s = 1 - \frac{6\sum d_i^2}{n^3 - n}$$

$$= 1 - \frac{6(42.00)}{1716}$$

$$= 1 - 0.147$$

$$= 0.853$$

To test $H_0: \rho_s = 0$; $H_A: \rho_s \neq 0$,

$(r_s)_{0.05(2), 12} = 0.587$ (from Appendix Table B.20)

Therefore, reject H_0.

$$P < 0.001$$

To employ the correction for ties (see Equation 19.50):

among the X's there are two measurements of 10.2 cm and two of 10.8 cm, so

$$\sum t_X = \frac{(2^3 - 2) + (2^3 - 2)}{12} = 1;$$

among the Y's there are two measurements tied at 7.2 cm, 3 at 7.4 cm, and two at 7.8 cm, so

$$\sum t_Y = \frac{(2^3 - 2) + (3^3 - 3) + (2^3 - 2)}{12} = 3;$$

therefore,

$$(r_s)_c = \frac{(12^3 - 12)/6 - 42.00 - 1 - 3}{\sqrt{[(12^3 - 12)/6 - 2(1)][(12^3 - 12)/6 - 2(3)]}} = \frac{240}{282} = 0.851;$$

and the hypothesis test proceeds exactly as above.

Computationally, it is simpler to apply Equation 19.1 to the ranks to obtain $(r_s)_c$ when ties are present.

Other Hypotheses, Confidence Limits, and Power.   If $n \geq 10$ and $\rho_s \leq 0.9$, then the Fisher z transformation may be used for Spearman coefficients, just as it was in Sections 19.2 through 19.6, for testing several additional kinds of hypotheses (including multiple comparisons), estimating power, and setting confidence limits around ρ_s. But in doing so it is recommended that $1.060/(n-3)$ be used instead of $1/(n-3)$ for the variance of z (Fieller, Hartley, and Pearson, 1957, 1961). That is,

$$(\sigma_z)_s = \sqrt{\frac{1.060}{n-3}} \tag{19.53}$$

should be used for the standard error of z (instead of Equation 19.11).

The Kendall Rank Correlation Coefficient.   The Kendall rank correlation method will not be explained here. (See Daniel, 1978: 306–321; Kendall, 1938, 1962; Siegel, 1956: 213–223.) In this procedure, the sample correlation coefficient is usually designated τ (lowercase Greek tau, an exceptional use of a Greek letter to denote a sample statistic). The value of τ for a particular set of data will not necessarily resemble the value of r_s calculated for the same data. The performances of the Spearman and Kendall coefficients are very similar, but the former may be a little better, especially when n is large (Chow, Miller, and Dickinson, 1974), and for a large n the Spearman measure is also easier to calculate than the Kendall. Jolliffe (1981) describes the use of the runs-up-and-down test of Section 25.8 to perform nonparametric correlation testing in situations where r_s and τ are ineffective.

19.10 WEIGHTED RANK CORRELATION

The rank correlation of Section 19.9 gives equal emphasis to each pair of data. There are instances, however, when our interest is predominantly on whether there is correlation among the largest (or smallest) measurements in the two populations. For example, we might wish to declare that the following first set of ranks (A) is more similar to the second set (B) than it is to the third set (C):

$$
\begin{array}{ll}
\text{A:} & 1, 2, 3, 4, 5 \\
\text{B:} & 1, 2, 3, 5, 4 \\
\text{C:} & 2, 1, 3, 4, 5
\end{array}
$$

In such cases we should prefer a procedure that will give stronger weight to inter-sample agreement on which items have the smallest (or largest) ranks, and Quade and Salama (1992) refer to such a method as *weighted rank correlation* (a concept they introduced in Salama and Quade, 1982).

   In Example 19.14 a study has determined the relative importance of eight ecological factors (e.g., aspects of temperature and humidity, diversity of ground cover, abundance

EXAMPLE 19.14   A top-down correlation analysis, where for each of two bird species eight ecological factors are weighted in terms of their importance to the success of the birds in a given habitat.

H_0:   The same ecological factors are most important to both species.

H_A: The same ecological factors are not most important to both species.

Factor (i)	Rank		Savage number (S_i)		
	Species 1	Species 2	Species 1	Species 2	$(S_i)_1 (S_i)_2$
A	1	1	2.718	2.718	7.388
B	2	2	1.718	1.718	2.952
C	3	3	1.218	1.218	1.484
D	4	7	0.885	0.268	0.237
E	5	8	0.635	0.125	0.079
F	6	6	0.435	0.435	0.189
G	7	5	0.268	0.635	0.170
H	8	4	0.125	0.885	0.111
Sum			8.002	8.002	12.610

$$n = 20$$

$$\sum_{i=1}^{n}(S_i)_1 (S_i)_2 = 12.610$$

$$r_T = \frac{\sum_{i=1}^{n}(S_i)_1 (S_i)_2 - n}{(n - S_1)}$$

$$= \frac{12.610 - 8}{8 - 2.718} = 0.873$$

$$0.005 < P < 0.01$$

of each of several food sources) in the success of a particular species of bird in a particular habitat. A similar study ranked the same ecological factors for a second species in that habitat, and the desire is to ask whether the same ecological factors are most important for both species. We desire to ask whether there is a positive correlation between the factors most important to one species and the factors most important to the other species. Therefore, a one-tailed correlation analysis is called for.

A correlation analysis performed on the pairs of ranks would result in a Spearman rank correlation coefficient of $r_s = 0.548$, which is not significantly different from zero. (The one-tailed probability is $0.05 < P < 0.10$.) Iman (1987) and Iman and Conover (1987) propose weighting the ranks by replacing them with the sums of reciprocals known as Savage scores (Savage, 1956). For a given sample size, n, the ith Savage score is

$$S_i = \sum_{j=i}^{n} \frac{1}{j}. \tag{19.54}$$

Thus, for example, if $n = 4$, then $S_1 = 1/1 + 1/2 + 1/3 + 1/4 = 2.083$, $S_2 = 1/2 + 1/3 + 1/4 = 1.083$, $S_3 = 1/3 + 1/4 = 0.583$, and $S_4 = 1/4 = 0.250$. A check on arithmetic is that $\sum_{i=1}^{n} S_i = n$; for this example, $n = 4$ and $2.083 + 1.083 + 0.583 + 0.250 = 3.999$. Table 19.1 gives Savage scores for n of 3 through 20. Scores for larger n are readily

TABLE 19.1 Savage Scores, S_i, for Various Sample Sizes, n.

n $\quad i =$	1	2	3	4	5	6	7	8	9	10
3	1.833	0.833	0.333							
4	2.083	1.083	0.583	0.250						
5	2.283	1.283	0.783	0.450	0.200					
6	2.450	1.450	0.950	0.617	0.367	0.167				
7	2.593	1.593	1.093	0.756	0.510	0.310	0.143			
8	2.718	1.718	1.218	0.885	0.635	0.435	0.268	0.125		
9	2.829	1.829	1.329	0.996	0.746	0.546	0.379	0.236	0.111	
10	2.929	1.929	1.429	1.096	0.846	0.646	0.479	0.336	0.211	0.100
11	3.020	2.020	1.520	1.187	0.937	0.737	0.570	0.427	0.302	0.191
12	3.103	2.103	1.603	1.270	1.020	0.820	0.653	0.510	0.385	0.274
13	3.180	2.180	1.680	1.347	1.097	0.897	0.730	0.587	0.462	0.351
14	3.252	2.251	1.752	1.418	1.168	0.968	0.802	0.659	0.534	0.423
15	3.318	2.318	1.818	1.485	1.235	1.035	0.868	0.725	0.600	0.489
16	3.381	2.381	1.881	1.547	1.297	1.097	0.931	0.788	0.663	0.552
17	3.440	2.440	1.940	1.606	1.356	1.156	0.990	0.847	0.722	0.611
18	3.495	2.495	1.995	1.662	1.412	1.212	1.045	0.902	0.777	0.666
19	3.548	2.548	2.048	1.714	1.464	1.264	1.098	0.955	0.830	0.719
20	3.598	2.598	2.098	1.764	1.514	1.314	1.148	1.005	0.880	0.769

n $\quad i =$	11	12	13	14	15	16	17	18	19	20
11	0.091									
12	0.174	0.083								
13	0.251	0.160	0.077							
14	0.323	0.232	0.148	0.071						
15	0.389	0.298	0.215	0.138	0.067					
16	0.452	0.361	0.278	0.201	0.129	0.062				
17	0.510	0.420	0.336	0.259	0.188	0.121	0.059			
18	0.566	0.475	0.392	0.315	0.244	0.177	0.114	0.056		
19	0.619	0.528	0.445	0.368	0.296	0.230	0.167	0.108	0.053	
20	0.669	0.578	0.495	0.418	0.346	0.280	0.217	0.158	0.103	0.050

computed; but, as rounding errors will be compounded in the summation, it is wise to employ extra decimal places in such calculations. If there are tied ranks, then we may use the mean of the Savage scores for the positions of the tied data. For example, if $n = 4$ and ranks 2 and 3 are tied, then use $(1.083 + 0.583)/2 = 0.833$ for both S_2 and S_3.

The Pearson correlation coefficient of Equation 19.1 may then be calculated using the Savage scores, a procedure that Iman and Conover (1985, 1987) call "top-down correlation"; we shall refer to the top-down correlation coefficient as r_T. Alternatively, if there are no ties among the ranks of either of the two samples, then

$$r_T = \frac{\sum_{i=1}^{n}(S_i)_1(S_i)_2 - n}{(n - S_1)}, \tag{19.55}$$

where $(S_i)_1$ and $(S_i)_2$ are the ith Savage scores in Samples 1 and 2, respectively; this is demonstrated in Example 19.14, where it is concluded that there is significant agreement between the two rankings for the most important ecological factors. (As indicated above, if all factors were to receive equal weight in the analysis of this set of data, a nonsignificant Spearman rank correlation coefficient would have been calculated.)*

Significance-testing of r_T refers to testing H_0: $\rho_T \leq 0$ against H_A: $\rho_T > 0$ and may be effected by consulting Appendix Table B.21, which gives critical values for r_T. For sample sizes greater than those appearing in this table, a one-tailed normal approximation may be employed (Iman and Conover, 1985, 1987):

$$Z = \frac{r_T}{\sqrt{n-1}}. \tag{19.56}$$

The top-down correlation coefficient, r_T, is 1.0 when there is perfect agreement among the ranks of the two sets of data. If the ranks are completely opposite in the two samples, then $r_T = -1.0$ only if $n = 2$; it approaches -0.645 as n increases. If we wished to perform a test that was especially sensitive to agreement at the bottom, instead of the top, of the list of ranks, then the above procedure would be performed by assigning the larger Savage scores to the larger ranks.

If there are more than two groups of ranks, then see the procedure at the end of Section 20.17.

19.11 CORRELATION FOR DICHOTOMOUS NOMINAL-SCALE DATA

Dichotomous nominal-scale data are common in biology (e.g., observations may be recorded as male or female, dead or alive, thorned or thornless), and Chapter 24 is devoted to several aspects of the analysis of such data. If data are recorded for two dichotomous variables, they may be presented in the form of a 2×2 contingency table (see Section 23.3). The data of Example 19.15, for instance, might be cast into a 2×2 table, as shown. We shall set up such tables by having f_{11} and f_{22} be the frequencies of agreement between the two variables (where f_{ij} is the frequency in row i and column j).

Various measures of association of two dichotomous variables have been suggested (Conover, 1980: Section 4.4; Everitt, 1992: Section 3.6; Gibbons, 1976: Section 7.4.2). So-called *contingency coefficients*, such as

$$\sqrt{\frac{\chi^2}{\chi^2 + n}} \tag{19.57}^\dagger$$

---

*Procedures other than the use of Savage scores may be used to assign differential weights to the ranks to be analyzed; some give more emphasis to the lower ranks and some give less (Quade and Salama, 1992). Savage scores are recommended as an intermediate strategy.

†This measure has also been called the "Pearson coefficient of mean square contingency" (Yule, 1917: 64–65).

EXAMPLE 19.15   Correlation for dichotomous nominal-scale data. Data are collected to determine the degree of association, or correlation, between the presence of a plant disease and the presence of a certain species of insect.

Case	Presence of plant disease	Presence of insect
1	+	+
2	+	+
3	−	−
4	−	+
5	+	+
6	−	+
7	−	−
8	+	+
9	−	+
10	−	−
11	+	+
12	−	−
13	+	+
14	−	+

The data may be tabulated in the following 2×2 contingency table:

	Plant Disease		
Insect	Present	Absent	Total
Present	6	4	10
Absent	0	4	4
Total	6	8	14

$$\phi_2 = \frac{f_{11}f_{22} - f_{12}f_{21}}{\sqrt{C_1 C_2 R_1 R_2}}$$

$$= \frac{(6)(4) - (4)(0)}{\sqrt{(6)(8)(10)(4)}}$$

$$= 0.55$$

$$Q = \frac{f_{11}f_{22} - f_{12}f_{21}}{f_{11}f_{22} + f_{12}f_{21}} = \frac{(6)(4) - (4)(0)}{(6)(4) + (4)(0)} = 1.00$$

$$r_n = \frac{(f_{11} + f_{22}) - (f_{12} - f_{21})}{(f_{11} + f_{22}) + (f_{12} + f_{21})} = \frac{(6+4) - (4+0)}{(6+4) + (4+0)} = \frac{10-4}{10+4} = 0.43$$

and

$$\sqrt{\frac{\dfrac{\chi^2}{n}}{1 + \dfrac{\chi^2}{n}}} \tag{19.58}$$

employ the χ^2 statistic of Section 23.3. However, they have drawbacks, among them the lack of the desirable property of ranging between 0 and 1. [They are indeed zero when $\chi^2 = 0$ (i.e., when there is no association between the two variables), but the coefficients can never reach 1, even if there is complete agreement between the two variables.]

The Cramér, or phi, coefficient* (Cramér, 1946: 44),

$$\phi_1 = \sqrt{\frac{\chi^2}{n}}, \tag{19.59}$$

does range from 0 to 1 (as does ϕ_1^2, which may also be used as a measure of association).[†] It is based upon χ^2 (uncorrected for continuity), as obtained from Equation 23.1, or, more readily, from Equation 23.6. Therefore, we can write

$$\phi_2 = \frac{f_{11}f_{22} - f_{12}f_{21}}{\sqrt{C_1 C_2 R_1 R_2}}, \tag{19.61}$$

where R_i is the sum of the frequencies in row i and C_j is the sum of column j. This measure is preferable to Equation 19.59 because it can range from -1 to $+1$, thus expressing not only the strength of an association between variables, but also the direction of the association. If $\phi_2 = 1$, all the data in the contingency table lie in the upper left and lower right cells (i.e., $f_{12} = f_{21} = 0$). In Example 19.15 this would mean there was complete agreement between the presence of both the disease and the insect; either both were always present or both were always absent. If $f_{11} = f_{22} = 0$, all the data lie in the upper right and lower left cells of the contingency table, and $\phi_2 = -1$. The measure ϕ_2 may also be considered as a correlation coefficient, for it is equivalent to the r that would be calculated by assigning a numerical value to members of one category of each variable and another numerical value to members of the second category. For example, if we replace each "+" with 0, and each "−" with 1, in Example 19.15, we would obtain (by Equation 19.1) $r = 0.55$.

The statistic ϕ_2 is also preferred over the previous coefficients of this section because it is amenable to hypothesis testing. The significance of ϕ_2 (i.e., whether it indicates that an association exists in the sampled population) can be assessed by considering the significance of the contingency table. If the frequencies are sufficiently large (see Section 23.6), the significance of χ_c^2 (chi-square with the correction for continuity) may be

---

*Harald Cramér (1893–1985) was a distinguished Swedish mathematician (Leadbetter, 1988). This measure is commonly symbolized by the lowercase Greek phi ϕ—pronounced "fy" as in "simplify" (or, sometimes, "fee")—and is a sample statistic, not a population parameter as a Greek letter typically designates. (It should, of course, not be confused with the quantity used is estimating the power of a statistical test, which is discussed elsewhere in this book.) This measure is what Karl Pearson called "mean square contingency" (Walker, 1929: 133).

[†]ϕ_1 may be used as a measure of association between rows and columns in contingency tables larger than 2×2, as

$$\phi_1 = \sqrt{\frac{\chi^2}{n(k-1)}}, \tag{19.60}$$

where k is the number of rows or the number of columns, whichever is smaller.

determined. The variance of ϕ_2 (which they call V) is given by Kendall and Stuart (1979: 572).

The Yule coefficient of association (Yule 1900, 1912, 1917: 38):*

$$Q = \frac{f_{11}f_{22} - f_{12}f_{21}}{f_{11}f_{12} + f_{12}f_{21}}, \tag{19.62}$$

ranges from -1 (if either f_{11} or f_{22} is zero) to $+1$ (if either f_{12} or f_{21} is zero). The variance of Q is given by Kendall and Stuart (1979: 571).

A better measure is that of Ives and Gibbons (1967). It may be expressed as a correlation coefficient,

$$r_n = \frac{(f_{11} + f_{22}) - (f_{12} + f_{21})}{(f_{11} + f_{22}) + (f_{12} + f_{21})}. \tag{19.63}$$

The interpretation of positive and negative values of r_n (which can range from -1 to $+1$) is just as for ϕ_2.

The expression of significance of r_n involves statistical testing which will be described in Chapter 24. The binomial test (Section 24.6) may be utilized, with a null hypothesis of H_0: $p = 0.5$, using cases of perfect agreement and cases of disagreement as the two categories. Alternatively, the sign test (Section 24.7), the Fisher exact test (Section 24.10), or the chi-square contingency test (Section 23.3) could be applied to the data.

19.12 INTRACLASS CORRELATION

Correlation situations exist where it is impossible to designate one variable as X and one as Y. Consider the data in Example 19.16 where we wish to determine whether there is a relationship between the weights of identical twins. Although we clearly have pairs of data, it is not possible to say that the data in the first column have something in common that is different from what all the data in the second column have in common. Thus, we invoke the use of *intraclass correlation*, a concept generally approached by means of analysis of variance considerations (Model II single-factor ANOVA). Aside from assuming random sampling from a bivariate normal distribution, this procedure also assumes that the population variances are equal.

If we consider each of the pairs in our example as groups in an ANOVA (i.e., $k = 7$), with each group containing two observations (i.e., $n = 2$), then we may calculate mean squares to express variability both between and within the k groups (see Section 10.1). Then the *intraclass correlation coefficient* is defined as

$$r_I = \frac{\text{groups MS} - \text{error MS}}{\text{groups MS} + \text{error MS}}, \tag{19.64}$$

---

*British statistician George Udny Yule (1871–1951) called this coefficient Q in honor of Lambert Adolphe Jacques Quetelet (1796–1874), a pioneering Belgian statistician and astronomer who was a member of more than 100 learned societies, including the American Statistical Association (of which he was the first foreign member elected after its formation in 1839); Quetelet worked on measures of association as early as 1832 (Walker, 1929: 130–131).

this statistic being an estimate of the population intraclass correlation coefficient, ρ_I. To test H_0: $\rho_I = 0$ vs. H_0: $\rho_I \neq 0$, one may utilize

$$F = \frac{\text{groups MS}}{\text{error MS}}, \tag{19.65}$$

a statistic associated with Groups DF and error DF for the numerator and denominator, respectively.[†] If the measurements are equal within each group, then error MS $= 0$, and $r_I = 1$ (a perfect positive correlation). If there is more variability within groups than there is between groups, then r_I will be negative. The smallest it may be, however, is $-1/(n-1)$; therefore, only if $n = 2$ (as in Example 19.16) can r_I be as small as -1.

EXAMPLE 19.16   Intraclass correlation.

Testing for correlation between weights of members of pairs of human twins.

Group (i.e., twin)	Weight of one member of group (kg)	Weight of other member of group (kg)
1	70.4	71.3
2	68.2	67.4
3	77.3	75.2
4	61.2	66.7
5	72.3	74.2
6	74.1	72.9
7	71.1	69.5

Source of variation	SS	DF	MS
Total	220.17	13	
Groups	198.31	6	33.05
Error	21.86	7	3.12

(See Section 10.1 for the MS computation procedures.)

$$r_1 = \frac{\text{groups MS} - \text{error MS}}{\text{groups MS} + \text{error MS}} = 0.827$$

To test H_0: $\rho_1 = 0$, H_A: $\rho_1 \neq 0$,

$$F = \frac{\text{groups MS}}{\text{error MS}} = \frac{33.05}{3.12} = 10.6$$

$$F_{0.05(1),6,7} = 3.87$$

Therefore, reject H_0.

$$0.0025 < P < 0.005 \quad [P = 0.0032]$$

[†]If desired, F may be calculated first, followed by computing

$$r_I = (F - 1)/(F + 1). \tag{19.66}$$

Theoretically, we are not limited to pairs of data (i.e., situations where $n = 2$) to speak of intraclass correlation. Consider, for instance, expanding the considerations of Example 19.16 into a study of weight correspondence among triplets instead of twins. Indeed, n need not even be equal for all groups. We might, for example, ask whether there is any concordance among adult weights of brothers; here, some families might consist of two brothers, some of three brothers, etc. If n is not 2 for all k groups, then

$$r_I = \frac{\text{groups MS} - \text{error MS}}{\text{groups MS} + (n-1)\ \text{error MS}},$$ (19.67)

a calculation for which Equation 19.64 is a special case. If all n's are not equal, then the appropriate n for use in Equation 19.67 may be obtained as

$$n = \frac{\sum\limits_{i=1}^{k} n_i - \dfrac{\sum\limits_{i=1}^{k} n_i^2}{\sum\limits_{i=1}^{k} n_i}}{k-1}.$$ (19.68)

Equation 19.65 is applicable for hypothesis testing in all the preceding cases.

If $n = 2$ in all groups, then we may set confidence limits and test hypotheses as would be done with r, by utilizing the z transformation,* with k in place of n. (See Sections 19.3 through 19.7.) However, the standard error of the resultant z_I will be

$$\sigma_{z_I} = \sqrt{\frac{1}{k - \dfrac{3}{2}}}$$ (19.69)

(Fisher 1958: 215), and the standard error for the difference between two z_i's will be

$$\sigma_{z_1 - z_2} = \sqrt{\frac{1}{k_1 - \dfrac{3}{2}} + \frac{1}{k_2 - \dfrac{3}{2}}},$$ (19.70)

although an exact test is available (Zerbe and Goldgar, 1980).

If $n > 2$ in any groups, then the Fisher transformation is computed as

$$z_I = 0.5 \ln \frac{1 + (n-1)r_I}{1 - r_I}$$ (19.71)

(Fisher 1958: 219), using Equation 19.68 if the n's are unequal. Appendix Table B.18 may be used, instead of Equation 19.71, if r_I is first converted to

$$r' = \frac{n r_I}{2 + (n-2)r_I}$$ (19.72)

---

*The correction for bias in z involves adding $1/(2k-1)$ to z (Fisher, 1958: 216); it is typically negligible unless several z's are being summed, as in the procedure of Section 19.7.

(Rao, Mitra, and Matthai, 1966: 87), and using r' to enter the table. If Appendix Table B.19 is used to convert z_I to r', then r_I is obtainable as

$$r_I = \frac{-2r'}{(n-2)r' - n}.$$ (19.73)

Also, if $n > 2$, then

$$\sigma_{z_I} = \sqrt{\frac{n}{2(n-1)(k-2)}}$$ (19.74)

(Fisher, 1958: 219), and

$$\sigma_{z_1 - z_2} = \sqrt{\frac{\dfrac{n}{k_1 - 2} + \dfrac{n}{k_2 - 2}}{2(n-1)}}$$ (19.75)

(Zerbe and Goldgar, 1980). Nonparametric measures of intraclass correlation have been proposed (e.g., Rothery, 1979).

19.13 CONCORDANCE CORRELATION

If the intent of collecting pairs of data is to assess reproducibility of measurements, an effective technique is that which Lin (1989) refers to as *concordance correlation*. For example, the staff of an analytical laboratory might wish to know whether measurements of a particular substance are the same using two different instruments, or when performed by two different technicians.

In Example 19.17, the concentration of lead was measured in eleven specimens of brain tissue, where each specimen was analyzed by two different atomic-absorption

EXAMPLE 19.17   Reproducibility of analyses of lead concentrations in brain tissue (in micrograms of lead per gram of tissue), using two different atomic-absorption spectrophotometers.

Tissue sample	Tissue Lead (μg/g)	
(i)	*Spectrophotometer A* (X_i)	*Spectrophotometer B* (Y_i)
1	0.22	0.21
2	0.26	0.23
3	0.30	0.27
4	0.33	0.27
5	0.36	0.31
6	0.39	0.33
7	0.41	0.37
8	0.44	0.38
9	0.47	0.40
10	0.51	0.43
11	0.55	0.47

EXAMPLE 19.17   (continued)

$n = 11$

$\sum X = 4.24$ $\qquad\qquad$ $\sum Y = 3.67$ $\qquad\qquad$ $\sum XY = 1.5011$

$\sum X^2 = 1.7418$ $\qquad\qquad$ $\sum Y^2 = 1.2949$ $\qquad\qquad$ $\sum xy = 0.08648$

$\sum x^2 = 0.10747$ $\qquad\qquad$ $\sum y^2 = 0.07045$

$\bar{X} = 0.385$ $\qquad\qquad$ $\bar{Y} = 0.334$

$(n - 1)(\bar{X} - \bar{Y})^2 = (11 - 1)(0.385 - 0.334)^2 = 0.02601$

$$
\begin{aligned}
r_c &= \frac{2\sum xy}{\sum x^2 + \sum y^2 + (n - 1)(\bar{X} - \bar{Y})^2} \\[2mm]
&= \frac{2(0.08648)}{0.10747 + 0.07045 + 0.02601} \\[2mm]
&= \frac{0.17296}{0.20393} \\[2mm]
&= 0.8481
\end{aligned}
$$

To obtain the 95% confidence limits:

$z_c = 1.2493$

$$r = \frac{\sum xy}{\sqrt{\sum x^2 \sum y^2}} = 0.9939$$

$$u = \frac{(n - 1)(\bar{X} - \bar{Y})^2}{\sqrt{\sum x^2 \sum y^2}} = \frac{0.02601}{\sqrt{(0.10747)(0.07045)}} = 0.298921$$

$$
\begin{aligned}
\sigma_{z_c} &= \sqrt{\frac{\dfrac{(1 - r^2)r_c^2}{(1 - r_c^2)r^2} + \dfrac{4r_c^3(1 - r_c)u}{r(1 - r_c^2)^2} - \dfrac{2r_c^4 u^2}{r^2(1 - r_c^2)^2}}{n - 2}} \\[3mm]
&= \sqrt{\frac{\dfrac{(1 - 0.9939^2)(0.8481)^2}{(1 - 0.8481^2)(0.9939)^2} + \dfrac{4(0.8481)^3(1 - 0.8481)(0.298921)}{(0.9939)(1 - 0.8481^2)^2} - \dfrac{2(0.8481)^4(0.298921)^2}{(0.9939)^2(1 - 0.8481^2)^2}}{(11 - 2)}} \\[3mm]
&= \sqrt{\frac{0.031547 + 1.414411 - 1.187625}{9}} \\[3mm]
&= \sqrt{0.028704} \\[2mm]
&= 0.1694
\end{aligned}
$$

$$
\begin{aligned}
\text{95\% confidence interval for } \zeta_c &= z_c \pm Z_{0.05(2)}\sigma_{z_c} \\[2mm]
&= 1.2493 \pm (1.9600)(0.1694) \\[2mm]
&= 1.2493 \pm 0.3320
\end{aligned}
$$

$L_1 = 0.9173; \; L_2 = 1.5813$

For the 95% confidence limits for ρ_c, the above confidence limits for ζ_c (L_1 and L_2) are transformed (using Appendix Table B.19): $L_1 = 0.725; L_2 = 0.919$.

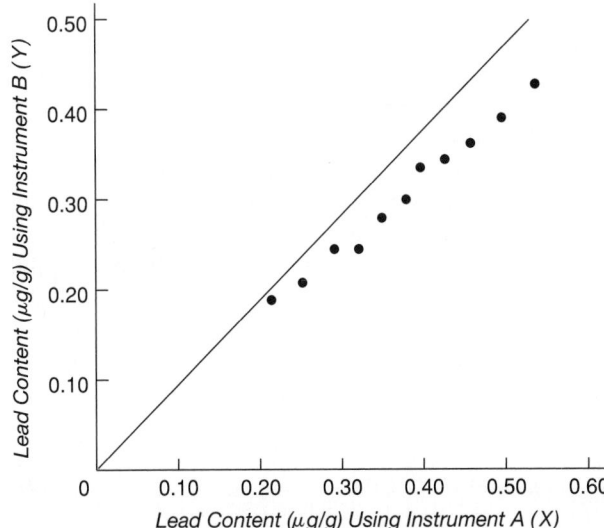

Figure 19.2 Lead concentrations in brain tissue (μg/g), determined by two different analytical instruments. The data are from Example 19.17 and are shown with a 45° line through the origin.

spectrophotometers. These data are plotted in Fig. 19.2. It can be observed that perfect reproducibility of assay would be manifested by the data falling on a 45° line intersecting the origin of the graph (the line as shown in Fig. 19.2).

The concordance correlation coefficient, r_c, is

$$r_c = \frac{2 \sum xy}{\sum x^2 + \sum y^2 + (n-1)(\bar{X} - \bar{Y})^2}.$$
(19.76)

This coefficient can range from -1 to $+1$, and its absolute value can not be greater than the Pearson correlation coefficient, r; so it can be stated that $-1 \leq -|r| \leq r_c \leq |r| \leq 1$; and $r_c = 0$ if, and only if, $r = 0$.

Using the procedure of Section 19.3, a confidence interval may be obtained for the population parameter ρ_c of which r_c is an estimate. To do so, the Fisher transformation (Equation 19.8) is applied to r_c to obtain a transformed value we shall call z_c; and the standard error of z_c is obtained as

$$\sigma_{z_c} = \sqrt{\frac{\dfrac{(1 - r^2)r_c^2}{(1 - r_c^2)r^2} + \dfrac{4r_c^3(1 - r_c)u}{r(1 - r_c^2)^2} - \dfrac{2r_c^4 u^2}{r^2(1 - r_c^2)^2}}{n - 2}},$$
(19.77)

where

$$u = \frac{(n-1)(\bar{X} - \bar{Y})^2}{\sqrt{\sum x^2 \sum y^2}}.$$ (19.78)

This computation is shown in Example 19.17.*

Furthermore, we might ask whether two concordance correlations are significantly different. For example, consider that the between-instrument reproducibility analyzed in Example 19.17 was reported for very experienced technicians, and a set of data was also collected for novice analysts. In order to ask whether the measure of reproducibility (namely, r_c) is different for the highly experienced and the less experienced workers, we can employ the hypothesis testing of Section 19.5. For this, we obtain r_c for the data from the experienced technicians (call it r_1) and another r_c (call it r_2) for the data from the novices. Then, each r_c is transformed to its corresponding z_c (namely z_1 and z_2) and the standard error to be used in Equation 19.21 is

$$\sigma_{z_1-z_2} = \sqrt{\sigma_{z_1}^2 + \sigma_{z_2}^2},$$ (19.79)

where each σ_z^2 is obtained as the square of the σ_{z_c} in Equation 19.77.

Lin (1989) has shown that this method of assessing reproducibility is superior to comparison of coefficients of variation (Section 8.8), to the paired-t test (Section 9.1), to regression (Section 17.2), to Pearson correlation (Section 19.1), and to intraclass correlation (Section 19.12). And he has shown the above hypothesis test to be robust with n as small as 10; however (Lin and Chinchilli, 1996), the two coefficients to be compared should have come from populations with similar ranges of data.

19.14 THE EFFECT OF CODING

Coding of the raw data will have no effect at all on any of the correlation coefficients presented in this chapter. Likewise, there will be no effect of coding on any z transformation or on any hypotheses concerning correlation coefficients or z transformations, nor will the coefficients of determination be altered.

It is interesting to note that if X and Y are transformed to $Z_X = (X - \bar{X})/s_X$ and $Z_Y = (Y - \bar{Y})/s_Y$, respectively, the regression coefficient (with either Z_X or Z_Y considered the dependent variable) will equal the correlation coefficient.

EXERCISES

19.1 Measurements of serum cholesterol (mg/100 ml) and arterial calcium deposition (mg/100 g dry weight of tissue) were made on twelve animals. The data are as follows:

---

*Some statisticians (including Lin, 1989) have employed n in place of $n-1$.

Calcium (X)	Cholesterol (Y)
59	298
52	303
42	233
59	287
24	236
24	245
40	265
32	233
63	286
57	290
36	264
24	239

(a) Calculate the correlation coefficient.

(b) Calculate the coefficient of determination.

(c) Test H_0: $\rho = 0$, vs. H_A: $\rho \neq 0$.

(d) Set 95% confidence limits on the correlation coefficient.

19.2 Using the data from Exercise 19.1:

(a) Test H_0: $\rho \leq 0$; H_A: $\rho > 0$.

(b) Test H_0: $\rho = 0.50$; H_A: $\rho \neq 0.50$.

19.3 Given: $r_1 = -0.44$, $n_1 = 24$, $r_2 = -0.40$, $n_2 = 30$.

(a) Test H_0: $\rho_1 = \rho_2$; H_A: $\rho_1 \neq \rho_2$.

(b) If H_0 in part (a) is not rejected, compute the common correlation coefficient.

19.4 Given: $r_1 = 0.45$, $n_1 = 18$, $r_2 = 0.56$, $n_2 = 16$. Test H_0: $\rho_1 \geq \rho_2$; H_A: $\rho_1 < \rho_2$.

19.5 Given: $r_1 = 0.85$, $n_1 = 24$, $r_2 = 0.78$, $n_2 = 32$, $r_3 = 0.86$, $n_3 = 31$.

(a) Test H_0: $\rho_1 = \rho_2 = \rho_3$, stating the appropriate alternate hypothesis.

(b) If H_0 in part (a) is not rejected, compute the common correlation coefficient.

19.6 (a) Calculate the Spearman rank correlation coefficient for the data of Exercise 19.1.

(b) Test H_0: $\rho_s = 0$ against H_A: $\rho_s \neq 0$.

19.7 Two different laboratories evaluated the efficacy of each of seven pharmaceuticals in treating hypertension in women, ranking them as shown below.

Drug	Lab 1 rank	Lab 2 rank
L	1	1
P	2	3
Pr	3	2
D	4	4
E	5	7
A	6	6
H	7	5

(a) Compute the top-down correlation coefficient.

(b) Test whether there is a significant agreement between the laboratories on which drugs are most effective (H_0: $\rho_T \leq 0$ vs. H_A: $\rho_T > 0$).

19.8 To examine the proposition that the type of school college presidents lead correlates with the type of school from which they received their undergraduate education, the following data were collected and arranged in a contingency table:

College as president	Undergraduate college	
	Public	*Private*
Public	9	5
Private	2	7

(a) Compute the coefficient, r_n, expressing this correlation.

(b) Using the Fisher exact test (Section 24.10), test the significance of the correlation.

19.9 Two samples of blood plasma from the same animal were submitted to each of four testing laboratories. The corticosterone concentrations, in grams per 100 ml, were determined as follows:

Laboratory	Sample 1	Sample 2
1	1.14	1.08
2	1.10	1.07
3	1.04	1.08
4	1.07	1.13

(a) Compute the intraclass correlation coefficient.

(b) Test whether there is a significant correlation between corticosterone determinations from the same laboratory.

19.10 The following avian plasma concentrations (in nanograms per milliliter) were determined by two different assay methods shortly after the blood was collected:

Blood sample	Method A	Method B
1	6.1	5.0
2	8.6	7.7
3	11.0	11.4
4	13.2	13.9
5	16.9	18.5
6	20.5	21.7
7	22.7	25.3
8	25.8	27.9
9	26.7	29.5
10	28.8	32.6
11	31.4	35.9
12	34.3	38.4

(a) Compute the concordance correlation coefficient.

(b) Compute the 95% confidence interval for the concordance correlation coefficient.

20

MULTIPLE REGRESSION AND CORRELATION

The previous three chapters have discussed the analysis of relationships between two variables. We shall now expand these considerations to interrelationship among three or more variables using the procedures of *multiple regression* and *multiple correlation*. If we have simultaneous measurements for more than two variables, and one of the variables is assumed to be dependent upon the others, then we are dealing with a *multiple regression* situation. In order to employ hypothesis testing, or to express confidence intervals, the observed values of the dependent variable (Y) are assumed to have come at random from a normal distribution of Y values existing in the population at the particular observed combination of independent variables. Furthermore, all such normal distributions, at all combinations of values for the dependent variables, are assumed to have the same variance. Other underlying assumptions are that the values of Y are independent of each other and that the error in measuring the X's is small compared to the error in measuring Y.

If none of the variables is assumed to be functionally dependent on any other, then we are dealing with *multiple correlation*, a situation requiring the assumption that each of the variables exhibits a normal distribution at each combination of all the other variables.[*] When dealing with two variables in simple correlation, we spoke of sampling a bivariate normal population; with samples dealing with more than two variables, we speak of a *multivariate normal distribution*. In multiple regression, the data transformation

---

[*]Much development in multiple correlation theory began in the late nineteenth century by several pioneers, including Karl Pearson (1857–1936) and his colleague, George Udny Yule (1871–1951) (Pearson 1967). (Pearson first called partial regression coefficients "double regression coefficients," and Yule later called them "net regression coefficients.") Pearson was the first to use the terms "multiple correlation," in 1908, and "multiple correlation coefficient," in 1914 (David, 1995).

discussions of Section 17.10 apply. In multiple correlation, all variables must be treated as is Y in that discussion.

The computational procedures required for most multiple regression and correlation analyses are difficult enough to preclude hand calculation, but both mainframe and desktop computer capability to perform the necessary operations is widespread. Because of this, we shall not emphasize multiple regression and correlation calculations, but shall concentrate instead on the interpretation of numerical results of the kind that a computer program would provide. Those readers desiring discussion of the details of multiple regression and correlation computations, and an elaboration of other multiple regression and correlation considerations, may consult works such as those by Birkes and Dodge (1993); Chatterjee and Price (1991); Daniel and Wood (1980); Draper and Smith (1981); Dunn and Clark (1987: Chapters 13–14); Efroymson (1960); Glantz and Slinker (1990: Chapters 3–6, 10, 11); Montgomery and Peck (1992); Neter, Wasserman, and Kutner (1990); Pedhazur (1982); Seber (1977); Tabachnick and Fidell (1996: Chapter 5); and Williams (1959).

There are cases where the dependent variable has only two possible values (e.g., male or female, dead or alive, green or red, adult or immature). Such a nominal-scale variable (which is termed "dichotomous" or "binomial") will not follow a normal distribution. For this and other reasons, it should not be analyzed by procedures of this chapter. Instead, *logistic regression* should be used, a technique found in some texts on multivariate statistics and in volumes specifically devoted to this subject. Logistic regression may also be extended to cases where the dependent variable has three or more discrete possible values (i.e., it is a "polytomous" or "multinomial" variable).

20.1 INTERMEDIATE COMPUTATIONAL STEPS

There are certain quantities that a computer program for multiple regression and/or correlation must calculate. Although we shall not concern ourselves with the mechanics of computation, we shall present certain intermediate steps in the calculating procedures here so that the user will not be a complete stranger to them if they appear on the printed computer output. Among the many different programs available for multiple regression and correlation, many do not print all the following intermediate results, or often they will do so only if the user specifically asks for them to appear in the output.

Consider n observations of M variables (the variables being referred to as X_1 through X_M). If we desire to consider one of the M variables as being dependent on the others, then we may eventually designate that variable as Y, but the program will perform most of its computations simply considering all M variables as X's numbered 1 through M (as in Example 20.1a-e).

The sums of the observations of each of the M variables are calculated as

$$\sum_{j=1}^{n} X_{1j} \quad \sum_{j=1}^{n} X_{2j} \quad \cdots \quad \sum_{j=1}^{n} X_{Mj}. \tag{20.1}$$

EXAMPLE 20.1a The $n \times M$ data matrix for a hypothetical multiple regression or correlation ($n = 33$; $M = 5$).

			Variable (i)		
j	1 (°C)	2 (cm)	3 (mm)	4 (min)	5 (ml)
1	6	9.9	5.7	1.6	2.12
2	1	9.3	6.4	3.0	3.39
3	−2	9.4	5.7	3.4	3.61
4	11	9.1	6.1	3.4	1.72
5	−1	6.9	6.0	3.0	1.80
6	2	9.3	5.7	4.4	3.21
7	5	7.9	5.9	2.2	2.59
8	1	7.4	6.2	2.2	3.25
9	1	7.3	5.5	1.9	2.86
10	3	8.8	5.2	0.2	2.32
11	11	9.8	5.7	4.2	1.57
12	9	10.5	6.1	2.4	1.50
13	5	9.1	6.4	3.4	2.69
14	−3	10.1	5.5	3.0	4.06
15	1	7.2	5.5	0.2	1.98
16	8	11.7	6.0	3.9	2.29
17	−2	8.7	5.5	2.2	3.55
18	3	7.6	6.2	4.4	3.31
19	6	8.6	5.9	0.2	1.83
20	10	10.9	5.6	2.4	1.69
21	4	7.6	5.8	2.4	2.42
22	5	7.3	5.8	4.4	2.98
23	5	9.2	5.2	1.6	1.84
24	3	7.0	6.0	1.9	2.48
25	8	7.2	5.5	1.6	2.83
26	8	7.0	6.4	4.1	2.41
27	6	8.8	6.2	1.9	1.78
28	6	10.1	5.4	2.2	2.22
29	3	12.1	5.4	4.1	2.72
30	5	7.7	6.2	1.6	2.36
31	1	7.8	6.8	2.4	2.81
32	8	11.5	6.2	1.9	1.64
33	10	10.4	6.4	2.2	1.82

For simplicity, let us refrain from indexing the $\sum$'s and assume that summations are always performed over all n sets of data. Thus, the sums of the variables could be denoted as:

$$\sum X_1 \ \sum X_2 \ \cdots \ \sum X_M. \tag{20.2}$$

Sums of squares and sums of crossproducts are calculated just as for simple regression, or correlation, for each of the M variables. These sums, generally referred to as *raw sums of squares* and *raw sums of crossproducts*, are often presented in computer output in the form of a matrix, or two-dimensional array (Example 20.1b):

$$
\begin{array}{ccccc}
\sum X_1^2 & \sum X_1 X_2 & \sum X_1 X_3 & \cdots & \sum X_1 X_M \\
\sum X_2 X_1 & \sum X_2^2 & \sum X_2 X_3 & \cdots & \sum X_2 X_M \\
\sum X_3 X_1 & \sum X_3 X_2 & \sum X_3^2 & \cdots & \sum X_3 X_M \\
\vdots & \vdots & \vdots & & \vdots \\
\sum X_M X_1 & \sum X_M X_2 & \sum X_M X_3 & \cdots & \sum X_M^2.
\end{array}
\tag{20.3}
$$

EXAMPLE 20.1b   **A matrix of raw sums of squares and sums of crossproducts, as it might occur as computer output (from the data of Example 20.1a).**

	1	2	3	4	5
1	0.11270E 04	0.13661E 04	0.87270E 03	0.38130E 03	0.30273E 03
2	0.13661E 04	0.26757E 04	0.17218E 04	0.75560E 03	0.71860E 03
3	0.87270E 03	0.17218E 04	0.11466E 04	0.49707E 03	0.47978E 03
4	0.38130E 03	0.75560E 03	0.49707E 03	0.25827E 03	0.21564E 03
5	0.30273E 03	0.71860E 03	0.47978E 03	0.21564E 03	0.21677E 03

Note that in this and in following examples of computer output, numerical results are printed in scientific or exponential, notation. Here, for example, 0.11270E 04 would indicate 0.11270×10^4, or 1127.0. In some instances, exponential notation on computer output will be of the form 0.11270D 04, rather than 0.11270E 04. Both of these notations would be read the same way, but the "D" indicates that the computer program employed "double precision" for its computations, meaning that the calculations were performed with considerably more (although not necessarily double) significant figures than are used when the results are printed with the "E" notation. Also, it is very important to realize that the number of significant figures printed by a computer in no way indicates how many significant figures were used in the calculations.

As $\sum X_i X_k = \sum X_k X_i$, this matrix is said to be symmetrical about the diagonal running from upper left to lower right.* Thus some computer programs print out only a half-matrix, such as

$$
\begin{array}{cccc}
\sum X_1^2 & & & \\
\sum X_2 X_1 & \sum X_2^2 & & \\
\sum X_3 X_1 & \sum X_3 X_2 & \sum X_3^2 & \\
\vdots & \vdots & \vdots & \\
\sum X_M X_1 & \sum X_M X_2 & \sum X_M X_3 & \cdots & \sum X_M^2
\end{array}
\tag{20.4}
$$

for this matrix as well as for those matrices that follow.

If a raw sum of squares, $\sum X_i^2$, is reduced by $(\sum X_i)^2/n$, we have a sum of squares that we have previously symbolized as $\sum x_i^2$ (see Section 17.2). Similarly, a raw sum of crossproducts, $\sum X_i X_k$, if diminished by $\sum X_i \sum X_k/n$, yields $\sum x_i x_k$. These quantities are known as *corrected* sums of squares and sums of crossproducts, respectively, and

---

*Hereafter, we shall refer to the values of a pair of variables as X_i and X_k.

they may be presented as the following matrix (Example 20.1c):

$$
\begin{array}{ccccc}
\sum x_1^2 & \sum x_1 x_2 & \sum x_1 x_3 & \cdots & \sum x_1 x_M \\
\sum x_2 x_1 & \sum x_2^2 & \sum x_2 x_3 & \cdots & \sum x_2 x_M \\
\sum x_3 x_1 & \sum x_3 x_2 & \sum x_3^2 & \cdots & \sum x_3 x_M \\
\vdots & \vdots & \vdots & \vdots & \\
\sum x_M x_1 & \sum x_M x_2 & \sum x_M x_3 & \cdots & \sum x_M^2.
\end{array}
\tag{20.5}
$$

EXAMPLE 20.1c A matrix of corrected sums of squares and sums of crossproducts, as it might appear as computer output (from the data of Example 20.1a).

	1	2	3	4	5
1	0.47218E 03	0.60027E 02	0.80727E 01	0.75636E 01	−0.60984E 02
2	0.60027E 02	0.70622E 02	−0.27091E 01	0.10161E 02	−0.68429E 01
3	0.80727E 01	−0.27091E 01	0.49091E 01	0.35854E 01	−0.47146E 00
4	0.75636E 01	0.10161E 02	0.35854E 01	0.44961E 02	0.80511E 01
5	−0.60984E 02	−0.68429E 01	−0.47146E 00	0.80511E 01	0.14747E 02

From Matrix 20.5, it is simple to calculate a matrix of simple correlation coefficients, for r_{ik} (representing the correlation between variables i and k) $= \sum x_i x_k \big/ \sqrt{\sum x_i^2 \sum x_k^2}$:

$$
\begin{array}{ccccc}
r_{11} & r_{12} & r_{13} & \cdots & r_{1M} \\
r_{21} & r_{22} & r_{23} & \cdots & r_{2M} \\
r_{31} & r_{32} & r_{33} & \cdots & r_{3M} \\
\vdots & \vdots & \vdots & \vdots & \\
r_{M1} & r_{M2} & r_{M3} & \cdots & r_{MM}.
\end{array}
\tag{20.6}
$$

Each element in the diagonal of this matrix (i.e., r_{ii}) is equal to 1.0, for there will always be a perfect positive correlation between a variable and itself (see Example 20.1d).

EXAMPLE 20.1d A matrix of simple correlation coefficients, as it might appear as computer output (from the data of Example 20.1a).

	1	2	3	4	5
1	1.00000	0.32872	0.16767	0.05191	−0.73081
2	0.32872	1.00000	−0.14550	0.18033	−0.21204
3	0.16767	−0.14550	1.00000	0.24134	−0.05541
4	0.05191	0.18033	0.24134	1.00000	0.31267
5	−0.73081	−0.21204	−0.05541	0.31267	1.00000

The final major manipulation necessary before the important regression or correlation statistics of the following sections can be obtained is the computation of the *inverse* of a matrix. Inverting a matrix will not be explained here; it is to two-dimensional algebra what taking the reciprocal is to ordinary, one-dimensional algebra.* While the process of inverting a matrix of moderate size is too cumbersome to be performed easily by hand, it may be readily accomplished by computer. A multiple regression or correlation program may invert the corrected sum of squares and crossproducts matrix, Matrix 20.5, resulting in a symmetrical matrix conventionally symbolized

$$
\begin{matrix}
c_{11} & c_{12} & c_{13} & \cdots & c_{1M} \\
c_{21} & c_{22} & c_{23} & \cdots & c_{2M} \\
\vdots & \vdots & \vdots & & \vdots \\
c_{M1} & c_{M2} & c_{M3} & \cdots & c_{MM}.
\end{matrix}
\tag{20.7}
$$

Or the correlation matrix, Matrix 20.6, may be inverted, yielding a different array of values, which we may designate

$$
\begin{matrix}
d_{11} & d_{12} & d_{13} & \cdots & d_{1M} \\
d_{21} & d_{22} & d_{23} & \cdots & d_{2M} \\
d_{31} & d_{32} & d_{33} & \cdots & d_{3M} \\
\vdots & \vdots & \vdots & & \vdots \\
d_{M1} & d_{M2} & d_{M3} & \cdots & d_{MM}.
\end{matrix}
\tag{20.8}
$$

The program utilized for the computer output presented in this chapter proceeded along the latter route, the inverse correlation matrix being the output shown in Example 20.1e. A program might compute either Matrix 20.7 and 20.8, the choice being irrelevant to our considerations, as the two are interconvertible:

$$
c_{ik} = \frac{d_{ik}}{\sqrt{\sum x_i^2 \sum x_k^2}},
\tag{20.9}
$$

or, equivalently,

$$
c_{ik} = \frac{r_{ik} d_{ik}}{\sum x_i x_k}.
\tag{20.10}
$$

EXAMPLE 20.1e   The inverse matrix of the simple correlation matrix of Example 20.1d.

	1	2	3	4	5
1	0.27610E 01	−0.36091E 00	−0.22915E 00	−0.69375E 00	0.21454E 01
2	−0.36091E 00	0.12501E 01	0.32710E 00	−0.32332E 00	0.12053E 00
3	−0.22915E 00	0.32710E 00	0.11761E 01	−0.35538E 00	0.78178E −01
4	−0.69375E 00	−0.32332E 00	−0.35538E 00	0.15142E 01	−0.10687E 01
5	0.21454E 01	0.12053E 00	0.78178E −01	−0.10687E 01	0.29319E 01

---

*The plural of "matrix" is "matrices." As a shorthand notation, statisticians may refer to an entire matrix by a boldface letter, and the inverse of the matrix by that letter's reciprocal. So, Matrices 20.4, 20.5, and 20.6 might be referred to by the symbols $\mathbf{X}$, $\mathbf{x}$, and $\mathbf{r}$, respectively; and Matrices 20.7 and 20.8 could be written, respectively, as $\mathbf{c} = \mathbf{x}^{-1}$ and $\mathbf{d} = \mathbf{r}^{-1}$.

It is from manipulations of these types of arrays that a computer program can derive the sample statistics and components of analysis of variance described in the following sections. If partial correlation coefficients are desired (Section 20.6), the matrix inversion takes place as shown, If partial regression analysis is desired (Sections 20.2–20.4), then inversion is performed only on the $M - 1$ rows and $M - 1$ columns corresponding to the independent variables in either Matrix 20.5 or 20.6 (see Example 20.1f).

EXAMPLE 20.1f   The inverse of the independent-variable correlation matrix (i.e., if variable 5 is the dependent variable in Example 20.1a, this is the inverse of columns 1 through 4 and rows 1 through 4 of the correlation matrix of Example 20.1d).

	1	2	3	4
1	0.11911E 01	−0.44910E 00	−0.28635E 00	0.88264E −01
2	−0.44910E 00	0.12451E 01	0.32389E 00	−0.27938E 00
3	−0.28635E 00	0.32389E 00	0.11740E 01	−0.32688E 00
4	0.88264E −01	−0.27938E 00	−0.32688E 00	0.11247E 01

20.2 THE MULTIPLE REGRESSION EQUATION

Recall, from Section 17.2, that a simple linear regression for a population of paired variables is the relationship

$$Y_i = \alpha + \beta X_i. \tag{17.1}$$

In this relationship, Y and X represent the dependent and independent variables, respectively; β is the regression coefficient in the population; α (the "Y" intercept") is the value of Y when X is zero.

In many situations, however, Y may be considered dependent upon more than one variable. Thus,

$$Y_j = \alpha + \beta_1 X_{1j} + \beta_2 X_{2j} \tag{20.11}$$

may be proposed, implying that one variable (Y) is linearly dependent upon a second variable (X_1), and that Y is also linearly dependent upon a third variable (X_2). (The double subscript notation, X_{ij}, denotes the jth observation of variable X_i.) In this particular multiple regression model we have one dependent variable and two independent variables.* The two population parameters, β_1 and β_2, are termed *partial regression coefficients*; β_1 expresses how much Y would change for a unit change in X_1, if X_2 were held constant. It is sometimes said that β_1 is a measure of the relationship of Y to X_1 after "controlling for" X_2; that is, it is a measure of the extent to which Y is

---

*"Dependence" refers to mathematical, not necessarily biological, dependence. Sometimes the independent variables are called "predictor" or "regressor" variables; the dependent variable may be referred to as the "response" or "criterion" variable.

related to X_1 after removing the effect of X_2. Similarly, β_2 describes the rate of change of Y as X_2 changes, with X_1 being held constant. They are called partial regression coefficients, then, because each expresses only part of the dependence relationship. The Y intercept, α (sometimes designated β_0) is the value of Y when *both* X_1 and X_2 are zero. Whereas Equation 16.1 mathematically represents a line (which may be presented on a two-dimensional graph), Equation 20.11 defines a plane (which may be plotted on a three-dimensional graph). A regression with m independent variables defines an m-dimensional surface, sometimes referred to as a "response surface" or "hyperplane."

The population data whose relationship is described by Equation 20.11 will probably not all lie exactly on a plane, so this equation may be expressed as

$$Y_j = \alpha + \beta_1 X_{1j} + \beta_2 X_{2j} + \epsilon_j, \tag{20.12}$$

where ϵ_j, the "residual" or "error," is the amount by which Y_j differs from what is predicted by $\alpha + \beta_1 X_{1j} + \beta_2 X_{2j}$; and the sum of all ϵ's is zero.

If we sample the population containing the three variables (Y, X_1, and X_2) in Equation 20.11, we can compute sample statistics to estimate the population parameters in the model. The multiple-regression function derived from a sample of data would be

$$\hat{Y}_j = a + b_1 X_{1j} + b_2 X_{2j}, \tag{20.13}$$

and a, b_1, and b_2 are estimates of the population parameters, α, β_1, and β_2, respectively.

Theoretically, there is no limit to the number of independent variables, X_j, that can be proposed as influencing the dependent variable, Y. The general population model, of which Equation 20.11 is a special case, is

$$Y_j = \alpha + \beta_1 X_{1j} + \beta_2 X_{2j} + \beta_3 X_{3j} + \cdots + \beta_m X_{mj} + \epsilon_j, \tag{20.14}$$

or, more succinctly,

$$Y_j = \alpha + \sum_{i=1}^{m} \beta_j X_{ij} + \epsilon_j, \tag{20.15}$$

where m is the number of independent variables. This model is said to be one of *multiple linear regression* because of the linear (i.e., additive) arrangement of the parameters (α and β_i) in the model. The sample regression equation, containing the statistics used to estimate the population parameters when there are m independent variables would be

$$\hat{Y}_j = a + b_1 X_{1j} + b_2 X_{2j} + b_3 X_{3j} + \cdots + b_m X_{mj}, \tag{20.16}$$

or

$$\hat{Y}_j = a + \sum_{i=1}^{m} b_i X_{ij}. \tag{20.17}$$

At least $m + 2$ data points are required to perform a multiple regression analysis, where m is the number of independent variables determining each data point.

The criterion for defining the "best fit" multiple regression equation is most commonly that of *least squares*, which—as described in Section 17.2 for simple regression—

EXAMPLE 20.1g   Regression statistics computed for the regression model $\hat{Y} = \alpha + \beta_1 X_1 + \beta_2 X_2 + \beta_3 X_3 + \beta_4 X_4$ fit to the data of Example 29.1a. In this computer output, B is a partial regression coefficient estimating β, SE is its standard error, T is the t for testing $H_0: \beta_i = 0$ against $H_A: \beta_i \neq 0$, and DF is the degrees of freedom for this t test. Note that common computer outputting procedure, as below, places subscripts in parentheses, so that $X(1)$ denotes X_1, etc. Also, computer output is commonly all in capital letters.

```
VARIABLE        B            SE            T          DF        B'
  X(1)    -0.12932E 00  0.21287E-01  -0.60750E 01   28   -0.73176E 00
  X(2)    -0.18785E-01  0.56278E-01  -0.33379E 00   28   -0.41108E-00
  X(3)    -0.46215E-01  0.20727E 00  -0.22297E 00   28   -0.26664E-00
  X(4)     0.20876E 00  0.67034E-01  -0.31141E 01   28    0.36451E 00


Y INTERCEPT = 0.29583E 01
COEFFICIENT OF DETERMINATION = 0.65893
STANDARD ERROR OF ESTIMATE = 0.42384E 00

                  ANALYSIS OF VARIANCE TABLE
  SOURCE OF VARIATION      SUM OF SQUARES    DF    MEAN SQUARE
  TOTAL                     0.14747E 02      32
    MULTIPLE REGRESSION     0.97174E 01       4    0.24293E 01
    RESIDUAL                0.50299E 01      28    0.17964E 00


FOR THE ANALYSIS OF VARIANCE, F = 0.13524E 02, WITH DF OF 4 AND 28
```

results in the regression equation with the minimum residual sum of squares[*], i.e. the minimum value of $\sum_{j=1}^{n}(Y_j - \hat{Y}_j)^2$.

From the analysis shown in Example 20.1g, we arrive at a regression function having partial regression coefficients of $b_1 = -0.129$ ml/°C, $b_2 = -0.019$ ml/cm, $b_3 = -0.05$ ml/mm, $b_4 = 0.209$ ml/min, and a Y intercept of $a = 2.96$ ml.[†] Thus we can write the regression function as $\hat{Y} = 2.96 - 0.129X_1 - 0.019X_2 - 0.05X_3 + 0.209X_4$. Therefore b_1 is an estimate of the relationship between Y and X_1 after removing the effects of both X_2 and X_3.

Section 20.4 will explain that, if independent variables are highly correlated with each other, then the interpretation of partial regression coefficients becomes questionable, as does the testing of hypotheses about the coefficients.

---

[*]Another criterion that could be used—with different associated statistical procedures—is that of *least absolute deviations*, which would involve minimizing $\sum_{j=1}^{n}|Y_j - \hat{Y}_j|$ (see Birkes and Dodge, 1993; Chapter 2; Bloomfield and Steiger, 1983). As indicated in a Section 17.2 footnote, this procedure may be beneficial when there are outlier data, and—as indicated in that footnote—an intermediate regression method is what is known as *M-regression*.

[†]By examining the magnitude of the standard errors of the four partial regression coefficients (namely, 0.021287, 0.056278, 0.20727, and 0.067034), we observe that their second significant figures are at the third, third, second, and third decimal places, respectively, making it appropriate to state the four coefficients to those precisions. (See the beginning of Section 17.4.)

20.3 ANALYSIS OF VARIANCE OF MULTIPLE REGRESSION OR CORRELATION

A computer program for multiple regression and correlation analysis will typically include an analysis of variance (ANOVA) of the regression (see Example 20.1g). This analysis of variance is analogous to that in the case of simple regression, consisting of total, multiple regression, and residual sums of squares and degrees of freedom, as well as regression and residual mean squares. The total sum of squares is an expression of the total amount of variability among the Y values $[Y_j - \bar{Y}]$; the regression sum of squares expresses the variability among the Y values attributable to the regression being fit $[\hat{Y}_j - \bar{Y}]$; and the residual sum of squares tells us about the amount of variability of Y still remaining after fitting the regression $[Y_j - \hat{Y}_j]$. The necessary sums of squares, degrees of freedom, and mean squares are summarized in Table 20.1. Remember that the expressions given for sums of squares are the defining equations; the actual computations of these quantities may involve the use of more manageable "machine formulas."

Note how a simple linear regression (i.e., Table 17.1) would conform to Table 20.1 as the special case where $m = 1$.

TABLE 20.1   Definitions of the Appropriate Sums of Squares. Degrees of Freedom, and Mean Squares Used in Multiple Regression or Multiple Correlation Analysis of Variance

Source of variation	Sum of squares (SS)	DF*	Mean square (MS)
Total	$\sum(Y_j - \bar{Y})^2$	$n - 1$	
Regression	$\sum(\hat{Y}_j - \bar{Y}_j)^2$	m	$\dfrac{\text{regression SS}}{\text{regression DF}}$
Residual	$\sum(Y_j - \hat{Y}_j)^2$	$n - m - 1$	$\dfrac{\text{residual SS}}{\text{residual DF}}$

*n = total number of data points (i.e., total number of Y values); m = number of independent variables in the regression model.

If we assume Y to be functionally dependent on each of the X's, then we are dealing with multiple regression. If no such dependence is implied, then any of the $M = m + 1$ variables could be designated as Y for the purposes of utilizing the computer program, this being a case of *multiple correlation*. In either situation, we can test the hypothesis that there is no interrelationship among the variables, as

$$F = \frac{\text{regression MS}}{\text{residual MS}}.$$ (20.18)

The numerator and denominator degrees of freedom for this variance ratio are the regression DF and the residual DF, respectively. For multiple regression this F tests

$$H_0: \beta_1 = \beta_2 = \cdots = \beta_M = 0,$$

which may be written as

$$H_0: \beta_i = 0 \text{ for all } i\text{'s,}$$

against

$$H_A: \beta_i \neq 0 \text{ for one or more } i\text{'s.}$$

The ratio,

$$R^2 = \frac{\text{regression SS}}{\text{total SS}} \tag{20.19}$$

or, equivalently,

$$R^2 = 1 - \frac{\text{residual SS}}{\text{total SS}} \tag{20.20}$$

is the *coefficient of determination* for a multiple regression or correlation, or the *coefficient of multiple determination*.[*] In a regression situation, it is an expression of the proportion of the total variability in Y attributable to the dependence of Y on all the X_i's, as defined by the regression model fit to the data. In the case of correlation, R^2 may be considered to be amount of variability in any one of the M variables that is accounted for by correlating it with the other $M - 1$ variables.

Healy (1984) and others caution against using R^2 as a measure of "goodness of fit" of a given regression model; and one should not attempt to employ R^2 to compare regressions with different m's and different amounts of replication. An acceptable measure of goodness of fit is what is referred to as the *adjusted coefficient of determination*[†]:

$$R_a^2 = 1 - \frac{\text{residual MS}}{\text{total MS}}, \tag{20.22}$$

which is

$$R_a^2 = 1 - \frac{n-1}{n-m-1}(1 - R^2). \tag{20.23}$$

Whereas R^2 always increases when an X_j is added to a regression model, R_a^2 will increase only if an added X_j results in an improved fit of the regression equation to the data. It is also of interest that R_a^2 is a better estimate than R^2 of the population coefficient of determination (ρ^2); and, if ρ^2 is near zero, the calculated R_a^2 may actually be negative.

The square root of the coefficient of determination is referred to as the *multiple correlation coefficient*:[‡]

$$R = \sqrt{R^2}. \tag{20.24}$$

---

[*]R^2 may also be calculated as (Sutton, 1990).

$$R^2 = \frac{F}{F + \nu_2/\nu_1}. \tag{20.21}$$

[†]Apparently first presented by Ezekiel (1930: 225–226), who termed it an "index" of determination to distinguish it from R^2, the "coefficient" of determination.

[‡]G. U. Yule, in 1897, was the first to use R to denote the multiple correlation coefficient (Walker, 1929: 112).

R is also equal to the Pearson correlation coefficient, r, for the correlation of the observed values of Y_j with the respective predicted values, $\hat{Y}_j$. For multiple correlation the F of Equation 20.18 allows us to draw inference about the population multiple correlation coefficient, ρ, by testing H_0: $\rho = 0$ against H_A: $\rho \neq 0$.

In a multiple correlation analysis, Equation 20.18 provides the test for whether the multiple correlation coefficient is zero in the sampled population. In the case of a multiple regression analysis, Equation 20.18 tests the null hypothesis of no dependence of Y on any of the independent variables, X_i; i.e., H_0: $\beta_1 = \beta_2 = \cdots = \beta_m = 0$ (vs. H_A: All m population partial regression coefficients are not equal to zero). Once R^2 has been calculated, the following computation of F may be used as an alternative to Equation 20.18:

$$F = \left(\frac{R^2}{1 - R^2} \right) \left(\frac{\text{residual DF}}{\text{regression DF}} \right), \tag{20.25}$$

and F (from either Equation 20.18 or 20.25) provides a test of H_0: $\rho^2 = 0$ vs. H_A: $\rho^2 \neq 0$.

The square root of the residual mean square is the *standard error of estimate* for the multiple regression:

$$s_{Y \cdot 1, 2, \ldots, m} = \sqrt{\text{residual MS.}} \tag{20.26}$$

As the residual MS is often called the "error MS," $s_{Y \cdot 1, 2, \ldots, m}$ is sometimes termed the "root mean square error." The subscript $(Y \cdot 1, 2, \ldots, m)$ refers to the mathematical dependence of variable Y on the independent variables 1 through m.

The addition of a variable, X_j, to a regression equation increases the regression sum of squares and decreases the residual sum of squares (and, therefore, increases R^2). It is important to ask, however, whether the increase in regression sum of squares is important (i.e., whether the added variable contributes useful information to our analysis). The regression degrees of freedom also increase and the residual degrees of freedom also decrease with the addition of a variable and therefore the regression mean square might decrease and/or the residual mean square might increase, and F might be reduced. This issue is addressed in Sections 20.4 and 20.8.

20.4 HYPOTHESES CONCERNING PARTIAL REGRESSION COEFFICIENTS

In simple regression, we generally desire to test H_0: $\beta = \beta_0$, a two-tailed hypothesis where β_0 is most often zero. If Equation 20.18 yields a significant F (i.e., H_0: $\beta_1 = \beta_2 = \cdots = \beta_m = 0$ is rejected), then we have concluded that at least one β_i is different from zero and its associated X_i contributes to explaining Y. In that case, each of the partial regression coefficients in a multiple regression equation may be submitted to an analogous hypothesis, H_0: $\beta_i = \beta_0$, where, again, the test is usually two-tailed and the constant is most frequently zero. For H_0: $\beta_i = 0$, Student's t may be computed as

$$t = \frac{b_i}{s_{b_i}} \tag{20.27}$$

and it should be apparent, from considerations of Sections 7.1 and 7.2 and the latter part of Section 17.3, how one-tailed tests and cases where $\beta_0 \neq 0$ would be handled.

We may obtain both b_i and s_{b_i} from the computer output shown in Example 20.1g. In the particular computer program employed for this example, the t value is also calculated for each b_i. If it had not been, then Equation 20.27 would have been applied. (Some computer programs present the square of this t value and call it a "partial F value.") The residual degrees of freedom are used for this test.

If the standard errors are not given by the computer program being utilized, then they may be calculated as

$$s_{b_i} = \sqrt{s^2_{Y \cdot 1, 2, \ldots, m} c_{ii}}, \tag{20.28}$$

where $s^2_{Y \cdot 1, 2, \ldots, m}$ is, of course, the square of the standard error of estimate, which is simply the residual mean square, and where c_{ii} is defined in Section 20.1. Knowing s_{b_i} one can obtain a $1 - \alpha$ confidence interval for a partial regression coefficient, β_i as

$$b_i \pm t_{\alpha(2), \nu} s_{b_i}, \tag{20.29}$$

where ν is the residual degrees of freedom.

In general, a significant F value in testing for dependence of Y on all X_i's (by Equation 20.18) will be associated with significance of some of the β_i's being concluded by the t test; but it is possible to have a significant F without any significant t's, or even significant t's without a significant F (Geary and Leser, 1968). The latter situations often indicate a high degree of correlation among the several independent variables. Ordinarily, hypotheses about β_i's will not be appropriate if F is nonsignificant (Cramer, 1972).

If independent variables, say X_1 and X_2, are correlated, then the partial regression coefficients associated with them (b_1 and b_2) may not be assumed to reflect the dependence of Y on X_1 or Y on X_2 that exists in the population. This correlation among independent variables is known as *multicollinearity* (also termed *collinearity* or *intercorrelation* or *non-orthogonality* or *ill-conditioning among variables*) and in practice is of little consequence if it is slight. But if the multicollinearity is substantial, then conclusions regarding the significance of the correlated X_i's are likely to be spurious. One may suspect substantial multicollinearity if the regression coefficients appear unreasonable, such as if they are of unexpected sign or not of sensible magnitude or significance; if there is great change in the remaining b_i's when a variable is added to or deleted from the regression model; or (Yeo, 1984) if there is substantial difference in regression results depending upon the sequence in which the data are entered into the computations. The adverse effect of multicollinearity may be especially pronounced if the range of one or more X_i's is narrow.*

When multicollinearity is present, standard errors of partial regression coefficients (s_{b_i}'s) may be large, meaning that the b_i's are imprecise estimates of the relationships in the population. As a consequence, a b_i may not be declared statistically significant from zero (as by the above t test), even when Y and X_i are related in the population. With

---

*If the intercorrelation is great, we may be unable to calculate the partial regression coefficients at all, for it may not be possible to perform the matrix inversion described in Section 20.1.

highly correlated X_i's the overall F for the regression model can be significant even when the t tests for the individual X_i's are not (Berry and Feldman, 1985: 42–43; Bertrand and Holder, 1988; Hamilton, 1987; Kendall and Stuart, 1979; 367; Routledge, 1990). An additional deleterious effect of multicollinearity is that it may lead to increased roundoff error in the computation of regression statistics. One way to detect some occurrences of multicollinearity is to examine the correlation Matrix 20.6, after which it may be wise to delete one or more intercorrelated independent variables from the regression model and reanalyze the remaining data. Discussion of the analysis and treatment of multicollinearity (including by a procedure known as "ridge regression") may be found in the references cited at the end of the introduction to this chapter.

If any of the partial regression coefficients are found to be nonsignificant (i.e., at least one H_0: $\beta_i = 0$ is not rejected), then the best course to follow is that of Section 20.8. Cohen and Cohen (1983: 118–119) present power analyses for partial regression.

20.5 STANDARDIZED PARTIAL REGRESSION COEFFICIENTS

The quantity

$$b_i' = b_i \left(\frac{s_{X_i}}{s_Y} \right), \quad \text{or, equivalently,} \quad b_i' = b_i \sqrt{\frac{\sum x_i^2}{\sum y^2}}, \tag{20.30}$$

is termed a *standardized* (or *standard*) *partial regression coefficient* (occasionally called a "beta coefficient"). Standard partial regression coefficients are sometimes used as indications of relative importance of the various X_i's in determining the value of Y. The standardized coefficients are unitless; thus, a b_i' with a high absolute value is indicative of its associated X_i having a high degree of influence on Y. Many multiple regression computer programs include these standardized coefficients, and some include their standard errors, $s_{b_i'}$, as well. Tests of hypotheses concerning β_i' are not typically performed, for they would tell the user no more than those tests performed on hypotheses about β_i. Standardized partial regression coefficients suffer from the same problems with intercorrelation as do partial regression coefficients (see Section 20.4).

20.6 PARTIAL CORRELATION

The multiple correlation coefficient, R, reflects the overall interrelationship of all M variables. But we may desire to examine the variables two at a time. We could calculate a simple correlation coefficient, r, for each pair of variables (i.e., what Example 20.1d presents to us). But the problem with considering simple correlations of all variables, two at a time, is that such correlations will fail to take into account the interactions of any of the other variables on the two in question. *Partial correlation* solves this problem because it considers the correlation between each pair of variables while holding constant the value of each of the other variables.* Symbolically, a partial correlation coefficient for a situation considering three variables (sometimes called a first-order partial correlation coefficient) would be $r_{ik \cdot l}$, which refers to the correlation between variables i and k, considering that variable l does not change its value (i.e., we have eliminated any effect of the interaction of variable l on the relationship between variables i and k). For four variables, a partial correlation coefficient, $r_{ik \cdot lp}$ (sometimes called a second-order partial correlation coefficient), expresses the correlation between variables i and k, assuming that variables l and p were held at constant values. In general, a partial correlation coefficient might be referred to as $r_{ik \cdots}$, meaning the correlation between variables i and k, holding all other variables constant (i.e., removing, or "partialling out" the effects of the other variables).

Another way to visualize partial correlation with three variables (i.e., $M = 3$) is as follows. If we perform a regression of variable X_i on variable X_l, a set of residuals ($X_j - \hat{X}_j$) will result; and the regression of X_k on X_l will yield another set of residuals. The correlation between these two sets of residuals will be the partial correlation coefficient, $r_{ik \cdot l}$.

For three variables, partial correlation coefficients may be calculated from simple correlation coefficients as

$$r_{ik \cdot l} = \frac{r_{ik} - r_{il} r_{kl}}{\sqrt{(1 - r_{il}^2)(1 - r_{kl}^2)}}. \tag{20.31}$$

For more than three variables, the calculations become quite burdensome and computer assistance is most welcome. A computer program for multiple regression and correlation providing partial correlation coefficients will generally do so in the form of a matrix, such as in Example 20.1h:

$$\begin{matrix} 1.00 & r_{12 \cdots} & r_{13 \cdots} & \cdots & r_{1M \cdots} \\ r_{21 \cdots} & 1.00 & r_{23 \cdots} & \cdots & r_{2M \cdots} \\ r_{31 \cdots} & r_{32 \cdots} & 1.00 & \cdots & r_{3M \cdots} \\ \vdots & \vdots & \vdots & & \vdots \\ r_{M1 \cdots} & r_{M2 \cdots} & r_{M3 \cdots} & \cdots & 1.00 \end{matrix} \tag{20.32}$$

To test H_0: $\rho_{ik \cdots} = 0$, we may employ

$$t = \frac{r_{ik \cdots}}{s_{r_{ik \cdots}}}, \tag{20.33}$$

---

*The first (in 1892) to extend the concept of correlation to more than two variables was F[rancis] Y[sidro] Edgeworth (1845–1926), a statistician and economist who was born in Ireland and spent most of his career at Oxford University (Stigler, 1978). Karl Pearson was the first to express what we now call multiple and partial correlation coefficients; in 1897 Pearson proposed the term "partial correlation, " in contrast to "total correlation" (i.e., what we now call simple correlation), in preference to what G. U. Yule termed "nett" [sic] and "gross" correlation, respectively (Snedecor, 1954: 7; Walker, 1929: 109, 111, 185).

EXAMPLE 20.1h   A matrix of partial correlation coefficients, as it might appear as computer output (from the data of Example 20.1a).

	1	2	3	4	5
1	1.00000	0.19426	0.12716	0.33929	−0.75406
2	0.19426	1.00000	−0.26977	0.23500	−0.06296
3	0.12716	−0.26977	1.00000	0.26630	−0.04210
4	0.33929	0.23500	0.26630	1.00000	0.50720
5	−0.75406	−0.06296	−0.04210	0.50720	1.00000

---

where

$$s_{r_{ik\cdots}} = \sqrt{\frac{1 - r_{ik\cdots}^2}{n - M}} \tag{20.34}$$

and M is the total number of variables in the multiple correlation. The statistical significance of a partial correlation coefficient (i.e., the test of H_0: $\rho_{ik\cdots} = 0$) may also be determined by employing Appendix Table B.17 for $n - M$ degrees of freedom. One-tailed hypotheses may also be performed as for simple correlation coefficients (Section 19.2). If a multiple regression and a multiple correlation analysis were performed on the same data, we would find that the test for H_0: $\beta_i = 0$ would be identical to the test for H_0: $\rho_{ik\cdots} = 0$ (by either t testing or "partial F" testing), where variable k is the dependent variable. Hypotheses such as H_0: $\rho_{ik\cdots} = \rho_0$, or similar one-tailed hypotheses, where $\rho_0 \neq 0$, may be testing using the z transformation (Section 19.2).

Cohen and Cohen (1983: 118–119) present power analysis for partial correlation analysis. Serlin and Harwell (1993) assess several nonparametric methods for three-variable partial correlation without the assumption of normality.

A related concept, though not as commonly encountered, is that of *semi-partial correlation* (Cohen and Cohen, 1983: 88–90; Pedhazur, 1982: 115–122; also called *part correlation*). When there are more than two variables, semi-partial correlation is correlation between two of them, with the effects of all the other variables removed from only one of the two being correlated; Bush, Rakow, and Gallimore (1980) point out calculation errors in the literature.

20.7 ROUND-OFF ERROR AND CODING DATA

It is important to realize that there may be rounding error associated with regression calculations even when they are performed on a computer (Freund, 1963; Healy, 1963; Longley, 1967; Wampler, 1970). Indeed, the number of significant figures utilized in the calculations by many computers and computer programs is less than the number implied in the output. At each of the many steps summarized in Section 20.1, the computer rounds-off results, and round-off error at an early computational stage will be exacerbated during successive calculations. Such errors may be especially severe if variables have greatly different magnitudes, or if there is considerable intercorrelation (i.e., multicollinearity) among independent variables. Some multiple regression computer programs automatically code input data by subtracting the mean of each variable from each observed value of that variable or by transforming input data to be standardized, or normalized, scores (using the sample mean and standard deviation of each variable instead of μ and σ, respectively, in Equation 6.13). Some programs reduce round-off error by employing extra significant figures in arithmetic computations (often called "double precision"; see footnote in Example 20.1b).

Coding of X_i and/or Y affects multiple regression statistics in a fashion similar to the effects of coding X and/or Y in simple regression (Section 17.11). Each value of X_{ij} might be coded by multiplying by a constant, M_{X_i} and adding a constant, A_{X_i}, so that the coded datum is

$$[X_{ij}] = M_{X_i} X_{ij} + A_{X_i}. \tag{20.35}$$

Similarly, each value of Y_j might be coded by multiplying by M_Y and adding A_Y, so that a coded value of Y_j would be

$$[Y_j] = M_Y Y_j + A_Y. \tag{20.36}$$

The effect of coding on multiple regression and multiple correlation statistics is shown in Table 20.2, where the bracketed statistics are those resulting from analysis of data coded as in Equation 20.35 and 20.36. Note that coding will not affect the values of R, R^2, $r_{ik\cdots}$, or any calculated t or F values thus far presented.

20.8 SELECTION OF INDEPENDENT VARIABLES

In Example 20.1g, we obtained the statistics for the least squares best fit equation for our data. However, the fact that the data we considered consisted of four independent variables does not automatically imply that all four have a significant effect on the magnitude of the dependent variable. The problems facing the user of multiple regression analysis include the determination of which of the independent variables have significant effect on Y in the population sampled. A number of procedures can be used to conclude which is in some objective way the "best" (or at least a very good) regression model. The various methods do not necessarily arrive at the same conclusions (e.g., Chatterjee and Price; 1991; Chapter 9; Draper and Smith, 1981: Chapter 6; Glantz and Slinker, 1990: Chapter 6; Montgomery and Peck, 1992: Chapter 7; Pedhazur 1982: 150–171; Seber, 1977: Chapter 12), but there is no universal agreement among statisticians as to which one is the most advantageous.

One procedure would be to fit a regression equation containing all four independent variables, to fit all four different equations consisting of three of the independent variables, to fit each of the possible six equations consisting of only two of the independent variables, to perform a simple regression analysis utilizing each of the four independent variables separately, and to consider the model having no regression coefficients (i.e., $Y_i = \alpha$). After fitting all sixteen of these regression equations, we might choose the one

TABLE 20.2 The Effect of Coding on Multiple Regression and Correlation Statistics, with Coding Performed as in Equations 20.35 and 20.36

Statistic	Value without coding	Value using coding
Partial regression coefficient, b_i	$b_i = [b_i]M_{X_i}/M_Y$	$[b_i] = b_iM_Y/M_X,$
Standard error of b_i, s_{b_i}	$s_{b_i} = [s_{b_i}]M_{X_i}/M_Y$	$[s_{b_i}] = s_{b_i}M_Y/M_{X_i}$
Y intercept, a	$a = \dfrac{[a] + \sum[b_i]A_{X_i} - A_Y}{M_Y}$	$[a] = aM_Y - \sum[b_i]A_{X_i} + A_Y$
Standard error* of a	$s_a = [s_a]/M_Y$	$[s_a] = s_aM_Y$
Mean of X_i, $\bar{X}_i$	$\bar{X}_i = \dfrac{[\bar{X}_i] - A_{X_i}}{M_{X_i}}$	$[\bar{X}] = M_{X_i}\bar{X}_i + A_{X_i}$
Mean of Y, $\bar{Y}$	$\bar{Y} = \dfrac{[\bar{Y}] - A_Y}{M_Y}$	$[\bar{Y}] = M_Y\bar{Y} + A_Y$
Standard error of estimate, $s_{Y\cdot1,2,...,m}$	$s_{Y\cdot1,2,...,m} = [s_{Y\cdot1,2,...,m}]/M_Y$	$[s_{Y\cdot1,2,...,m}] = M_Ys_{Y\cdot1,2,...,m}$
Coefficient of determination, R^2	$R^2 = [R^2]$	$[R^2] = R^2$
Multiple correlation coefficient, R	$R = [R]$	$[R] = R$
Test statistics,* t and F	$t = [t]; F = [F]$	$[t] = t; [F] = F$
Standardized partial regression coefficient, b_i'	$b_i' = [b_i']$	$[b_i'] = b_i'$
Standard error of b_i', $s_{b_i'}$	$s_{b_i'} = [s_{b_i'}]$	$[s_{b_i'}] = s_{b_i'}$
Correlation coefficient, r_{ik}	$r_{ik} = [r_{ik}]$	$[r_{ik}] = r_{ik}$
Partial correlation coefficient, $r_{ik...}$	$r_{ik...} = [r_{ik...}]$	$[r_{ik...}] = r_{ik...}$
Corrected sum of squares, $\sum x_i^2$	$\sum x_i^2 = \left[\sum x_i^2\right]/M_{X_i}^2$	$\left[\sum x_i^2\right] = \left(\sum x_i^2\right)(M_{X_i}^2)$
Corrected sum of crossproducts, $\sum x_ix_k$	$\sum x_ix_k = \left[\sum x_ix_k\right]/M_{X_i}M_{X_k}$	$\left[\sum x_ix_k\right] = \left(\sum x_ix_k\right)(M_{X_i}M_{X_k})$
Element of inverse corrected sum of squares and sum of crossproducts matrix, c_{ik}	$c_{ik} = [c_{ik}]/M_{X_i}M_{X_k}$	$[c_{ik}] = M_{X_i}M_{X_i}$
Element of inverse correlation matrix, d_{ik}	$d_{ik} = [d_{ik}]$	$[d_{ik}] = d_{ik}$
Sum of squares, SS	$SS=[SS]/M_Y^2$	$[SS]=(SS)(M_Y^2)$
Mean square, MS	$MS=[MS]/M_Y^2$	$[MS]=(MS)(M_Y^2)$

Note: If a regression is fitted through the origin (Section 20.14), then coding by addition of a constant may not be used (i.e., A_{X_i} and A_Y must be zero).

*Each A_{X_i} must be zero if s_a is desired or if it is desired to test hypotheses about the Y intercept.

resulting in the lowest residual mean square, or, equivalently, the largest R_a^2 or smallest standard error of estimate. There are drawbacks to such a procedure, however. First, many regression equations have to be calculated, the number being 2^m. (The reader can study Section 24.1 to see where this number comes from.) Thus, if $m = 5$, there would be a total of 32 regressions to be fit; if $m = 8$, then 256 regressions would be required; if $m = 10$, we would find ourselves trying to choose among 1024 regression equations; etc.; and the consideration of such large numbers of equations becomes formidable, although there is computer software available to ease this effort. A second problem with considering all possible regressions is that of declaring an objective statistical method for determining which one of the many regression equations is to be considered the "best." In addition, this procedure may result in a regression with substantial multi-collinearity.

The following method offers the ability to express whether or not one equation is preferable to another. The aim is to utilize as many variables as are required to provide a good determination of which independent variables in the population effect a significant change in Y, yet to employ as few variables as are necessary for this purpose so as to minimize the time, energy, and finances to be expended in collecting further data or performing further calculations with the regression equation, to optimize statistical estimates (b_i and $\hat{Y}_j$), and, we hope, to simplify the interpretation of the resultant regression equation.

If a multiple regression is fitted using all m independent variables (as we did in Example 20.1g), then we might ask whether the presence of any variable has insignificant influence in the sampled population and thus may be eliminated from the equation. We can examine H_0: $\beta_i = 0$ for each partial regression coefficient; if the t values for such tests are not provided in our computer output (as they are in Example 20.1g), then Equation 20.27 may be applied to calculate them.

If all m of the t values are equal to or greater than the critical value ($t_{\alpha(2),(n-m-1)}$), then we conclude that all the X's have a significant effect on Y and none of them should be deleted from our model. However, if some t values are less than the critical value, then the independent variable associated with the t with the lowest absolute value may be deleted from the model and a new multiple regression equation may be fitted, utilizing the remaining $m - 1$ independent variables. The null hypothesis H_0: $\beta_i = 0$ may then be tested for each partial regression coefficient in this new model, and if some t values are less than the critical value, then one more variable may be deleted and a new multiple regression analysis performed. This procedure is often referred to as "backward elimination." It typically performs as well as the method of comparing all subsets of regressions (Berk, 1978).

As demonstrated in Example 20.2, this procedure is repeated until all b_i's in the equation are concluded to estimate β_i's that are different from zero. (Each time a variable is thus deleted the multiple regression MS decreases slightly and the residual MS increases slightly.) For this model selection procedure, and those of the next two paragraphs, some computer programs equivalently employ F (the square of t) as the test statistic, and some compare this to a value other than the critical value, $F_{\alpha(1),m,(n-m-1)}$.

Another common multistep procedure (often called "forward selection") is to begin with the smallest possible regression model (i.e., one with only one independent variable; in other words, a simple regression) and gradually work up to the multiple regression model incorporating the largest number of significantly important independent variables. The problem here is first to determine which is the "best" simple regression model for the data, then which is the "best" model containing two independent variables, then which is the "best" one containing three X's, etc. Mantel (1970) describes how such a "step-up"

EXAMPLE 20.2   Stepwise regression analysis, utilizing the data from Example 20.1a.

As shown in Example 20.1g, the multiple regression analysis for the model $\hat{Y} = \alpha + \beta_1 X_1 + \beta_2 X_2 + \beta_3 X_3 + \beta_4 X_4$ yields the following statistics:

Variable	b	s_b	t	ν
X_1	-0.12932	0.021287	-6.075	28
X_2	-0.018785	0.056278	-0.334	28
X_3	-0.046215	0.20727	-0.223	28
X_4	0.20876	0.067034	3.114	28

$$a = 2.9583$$

The critical value for testing H_0: $\beta_j = 0$ against H_A: $\beta_j \neq 0$ is $t_{0.05(2),28} = 2.048$. Therefore, H_0 would be rejected for β_1 and β_4, but not for β_2 or β_3. Of the t tests for the latter two, the t for testing the significance of β_3 has the smaller absolute value. Therefore, $\beta_3 X_3$ is deleted from the model, leaving $\hat{Y} = \alpha + \beta_1 X_1 + \beta_2 X_2 + \beta_4 X_4$. For this model with three independent variables, we obtain the following statistics:

Variable	b	s_b	t	ν
X_1	-0.13047	0.020312	-6.423	29
X_2	-0.015424	0.053325	-0.289	29
X_4	0.20450	0.063203	3.236	29

$$a = 2.6725$$

The critical value for testing the significance of these partial regression coefficients is $t_{0.05(2),29} = 2.045$. Therefore H_0: $\beta_j = 0$ would be rejected for β_1 and for β_4, but not for β_2. Therefore, $\beta_2 X_2$ is deleted from the regression model, leaving $\hat{Y} = \alpha + \beta_1 X_1 + \beta_4 X_4$. For this model with two independent variables, we obtain the following statistics:

Variable	b	s_b	t	ν
X_1	-0.13238	0.018913	-6.999	30
X_4	0.20134	0.061291	3.285	30

$$a = 2.5520$$

The critical value for testing H_0: $\beta_j = 0$ against H_0: $\beta_j \neq 0$ is $t_{0.05(2),30} = 2.042$. Therefore, both β_1 and β_4 are concluded to be different from zero, and $\hat{Y} = 2.552 - 0.132 X_1 + 0.201 X_4$ is our final model.

forward-selection procedure can involve more computational effort, and is fraught with more theoretical deficiencies, than is the "step-down" backward elimination procedure described above. Regarding computational effort, the step-up procedure might require as many as $2^m - 1$ regressions to be fit, whereas the step-down method will never involve the fitting of more than m regressions. Also, the forward selection procedure may fail to identify significant independent variables when multicollinearity is present (Mantel, 1970; Routledge, 1990; Chatterjee and Price, 1981: 307; Hamilton, 1987) and may yield erroneous conclusions when dealing with a "dummy variable" (see Section 20.11) with more than two categories (Cohen, 1991).

Finally, there is a popular stepwise regression that employs both the addition and the elimination of independent variables. It begins just as the step-up method; but each time a variable is added all variables in the model are examined to see if any should be eliminated at that step.

Some computer program libraries contain routines for automatically performing addition and/or elimination of variables. However, as long as a program for multiple regression is available, data can be repeatedly submitted to the program with the user selecting a variable for deletion or addition each time as described above. Therefore, no special computer routine for stepwise multiple regression analysis is absolutely necessary.

Some computer programs present other statistics as criteria to select the "best" subset of independent variables. Two such statistics are C_P (introduced by C. L. Mallows* in 1964 and published by Gorman and Toman, 1966), which is closely related to R_a^2 (Kennard, 1971), and "PRESS" ("predicted error sum of squares"). These are described in the references at the end of the introduction to this chapter.

Section 20.4 discussed the problem of *intercorrelation* among independent variables (*multicollinearity*). If substantial intercorrelation is present, then the partial regression coefficient, b_i, associated with an independent variable X_i, depends upon which other independent variables are in the regression model. In the stepwise regression procedure this often means that the magnitude (and even the sign) of a b_i changes as variables are deleted from the model. (Such is not the case in Example 20.2, where b_1, b_2, and b_4 change very little when X_3 is dropped from the model; and then b_1 and b_4 change very little when b_2 is dropped.) The presence of intercorrelation can cause the equation deemed "best" to be very different depending upon whether the "step-up" forward-selection or the "step-down" backward elimination procedure is used, and the step-up method is more likely to yield spurious results.

20.9 Predicting Y Values

Having fitted a multiple regression equation to a set of data, we may desire to calculate the Y value to be expected at a particular combination of X_i values. Consider the a and b_i values determined in Example 20.2 for an equation of the form $\hat{Y} = a + b_1 X_1 + b_2 X_2$. Then the predicted value at $X_1 = 7°C$ and $X_2 = 2.0$ min, for example, would be $\hat{Y} = 2.552 - (0.132)(7) + (0.201)(2.0) = 2.03$ ml. Such predictions may be done routinely, as long as there is, in fact a significant regression (i.e., the F from Equation 20.18 is significant), although, as with simple linear regression (Section 17.2), it is unwise to predict Y for X_i's outside the ranges of the X_i's used to obtain the regression statistics.

In the consideration of the standard error of such a predicted Y, the reader may refer to Section 17.4 for the calculations appropriate when $m = 1$. The following is the

---

*Mallows (1973) credits the conception of this procedure to discussions with Cuthbert Daniel, late in 1993, and he employed the symbol "C" to honor the latter colleague.

standard error of a mean Y predicted from a multiple regression equation:

$$s_{\hat{Y}} = \sqrt{s_{Y \cdot 1,2,\ldots,m}^2 \left[\frac{1}{n} + \sum_{i=1}^{m} \sum_{k=1}^{m} c_{ik} x_i x_k \right]}.$$ (20.37)

In this equation, $x_i = X_i - \bar{X}_i$, where X_i is the value of independent variable i at which Y is to be predicted, and $\bar{X}_i$ is the mean of the observed values of variable i that were used to calculate the regression equation.

Thus, for the value of Y just predicted, we can solve Equation 20.37, as shown in Example 20.3.

EXAMPLE 20.3   The standard error of a predicted Y.

For the equation $\hat{Y} = 2.552 - 0.132X_1 + 0.201X_2$, derived from the data of Example 20.1a, where X_1 is the variable in column 1 of the data matrix, X_2 is the variable in column 4, and Y is the variable in column 5, one obtains the following quantities needed to solve Equation 20.37:

$$s_{Y \cdot 1,2}^2 = 0.16844, \quad n = 33, \quad \bar{X} = 4.4546,$$

$$\bar{X}_2 = 2.5424, \quad \sum x_1^2 = 472.18, \quad \sum x_2^2 = 44.961,$$

$$d_{11} = 1.0027, \quad d_{12} = -0.052051, \quad d_{21} = -0.052051, \quad d_{22} = 1.0027.$$

By employing Equation 20.9 each d_{ik} is converted to a c_{ik}, resulting in:

$$c_{11} = 0.0021236, \quad c_{12} = -0.00035724, \quad c_{21} = -0.00035724, \quad c_{22} = 0.022302.$$

What is the mean population value of Y at $X_1 = 7°C$ and $X_4 = 2.0$ min?

$$\hat{Y} = 2.552 - (0.132)(7) + (0.201)(2.0) = 2.030 \text{ ml}$$

What is the standard error of the mean population value of Y at $X_1 = 7°C$ and $X_4 = 2.0$ min? [Equation 20.37 is used.]

$$s_{\hat{Y}}^2 = 0.16844 \left[\frac{1}{33} + (0.0021236)(7 - 4.4546)^2 + (-0.00035724)(7 - 4.4546)(2.0 - 2.5424) \right.$$

$$\left. + (-0.00035724)(2.0 - 2.5424)(7 - 4.4546) + (0.022302)(2.0 - 2.5424)^2 \right]$$

$$= 0.16844 \left(\frac{1}{33} + 0.0213066 \right)$$

$$= 0.008693 \text{ ml}^2$$

$$s_{\hat{Y}} = \sqrt{0.008693 \text{ ml}^2} = 0.093 \text{ ml}$$

As $t_{0.05(2),30} = 2.042$, the 95% confidence interval for the predicted Y is $2.030 \pm (2.042)(0.093)$ ml $= 2.030 \pm 0.190$ ml.

What is the predicted value of one additional Y value taken from the population at $X_1 = 7°C$ and $X_4 = 2.0$ min?

$$\hat{Y} = 2.552 - (0.132)(7) + (0.201)(2.0) = 2.030 \text{ ml}$$

EXAMPLE 20.3   (continued)

What is the standard error of the predicted value of one additional Y value taken from the population at $X_1 = 7°C$ and $X_4 = 2.0$ min? [Equation 20.39 is used.]

$$s_{\hat{Y}} = \sqrt{0.16844\left[1 + \frac{1}{33} + 0.0213066\right]}$$

$$= 0.421 \text{ ml}$$

As $t_{0.05(2),30} = 2.042$, the 95% confidence interval for the preceding predicted $\hat{Y}$ is $2.03 \pm (2.042)(0.421)$ ml $= 2.03 \pm 0.86$ ml.

What is the predicted value of the mean of 10 additional values of Y taken from the population at $X_1 = 7°C$ and $X_4 = 2.0$ min?

$$\hat{Y} = 2.552 - (0.132)(7) + (0.201)(2.0) = 2.030 \text{ ml}$$

What is the standard error of the predicted value of the mean of 10 additional values of Y taken from the population at $X_1 = 7°C$ and $X_4 = 2.0$ min? [Equation 20.40 is used.]

$$s_{\hat{Y}} = \sqrt{0.16844\left[\frac{1}{10} + \frac{1}{33} + 0.0213066\right]}$$

$$= 0.16 \text{ ml}$$

As $t_{0.05(2),30} = 2.042$, the 95% confidence interval for the predicted Y is $2.03 \pm (2.042)(0.16)$ ml $= 2.03 \pm 0.33$ ml.

A special case of Equation 20.37 is where each $X_i = 0$. The Y in question is then the Y intercept, a, and

$$s_a = \sqrt{s_{Y \cdot 1,2,\ldots,m}^2 \left[\frac{1}{n} + \sum_{i=1}^{m}\sum_{k=1}^{m} c_{ik}\bar{X}_i\bar{X}_k\right]}. \tag{20.38}$$

To predict the value of Y that would be expected if one additional set of X_i were obtained, we may use Equation 20.16, and the standard error of this prediction is

$$(s_{\hat{Y}})_1 = \sqrt{s_{Y \cdot 1,2,\ldots,m}^2 \left[1 + \frac{1}{n} + \sum_{i=1}^{m}\sum_{k=1}^{m} c_{ik}x_i x_k\right]}. \tag{20.39}$$

as Example 20.3 shows. This situation is a special case of predicting the mean Y to be expected from obtaining p additional sets of X_i, where the X_1's in all sets are equal, the X_2's in all sets are equal, etc. Such a calculation is performed in Example 20.3, using

$$(s_{\hat{Y}})_p = \sqrt{s_{Y \cdot 1,2,\ldots,m}^2 \left[\frac{1}{p} + \frac{1}{n} + \sum_{i=1}^{m}\sum_{k=1}^{m} c_{ik}x_i x_k\right]}. \tag{20.40}$$

Adding an independent variable, X_i, to a regression model increases each of the standard errors, $s_{\hat{Y}}$, in this section. Therefore, it is desirable to be assured that all variables included are important in predicting $\hat{Y}$ (see Section 20.8).

20.10 TESTING DIFFERENCE BETWEEN TWO PARTIAL REGRESSION COEFFICIENTS

It is occasionally of interest to test $H_0: \beta_i - \beta_k = \beta_0$. This can be done by using

$$t = \frac{|b_i - b_k| - \beta_0}{s_{b_i - b_k}}, \tag{20.41}$$

or by

$$t = \frac{b_i - b_k}{s_{b_i} - b_k} \tag{20.42}$$

when $\beta_0 = 0$, in which case the null hypothesis is usually written as $H_0: \beta_i = \beta_k$. The standard error of the difference between two partial regression coefficients is

$$s_{b_i - b_k} = \sqrt{s_{Y \cdot 1, 2, \ldots, m}^2 [c_{ii} + c_{kk} - 2c_{ik}]} \tag{20.43}$$

and the degrees of freedom for this test are $n - m - 1$.

20.11 "DUMMY" VARIABLES

It is sometimes useful to introduce into a multiple regression model one or more additional variables, in order to account for the effects of one or more nominal scale variables on the dependent variable, Y. For example, we might be considering fitting the model $\hat{Y}_j = a + b_1 X_{1j} + b_2 X_{2j}$, where Y is human diastolic blood pressure, X_1 is age, and X_2 is body weight. In addition, we might be interested in determining the effect (if any) of sex on blood pressure. Our regression model could then be expanded to $\hat{Y}_j = a + b_1 X_{1j} + b_2 X_{2j} + b_3 X_{3j}$, where X_3 is a "dummy variable," or "indicator variable," with one of two possible values: e.g., set $X_3 = 0$ if the data are for a male and $X_3 = 1$ if the data are for a female. By using this dummy variable, we can determine whether or not sex is a significant determinant of blood pressure (by the considerations of Section 20.4 for testing $H_0: \beta_3 = 0$). If it is, then the use of the model with all three independent variables will yield significantly more accurate Y values than the preceding model with only two independent variables, if the regression equation is used for predicting blood pressure.

If there are three levels of the nominal scale variable, then two dummies would be needed in the regression model. For example, if we were considering the blood pressure of both sexes and of three races of humans, then we might fit the model $\hat{Y}_j = a + b_1 X_{1j} + b_2 X_{2j} + b_3 X_{3j} + b_4 X_{4j} + b_5 X_{5j}$, where X_1, X_2, and X_3 are as before and X_4 and X_5 are used to denote race: e.g., if race 1, then $X_4 = 0$ and $X_5 = 0$; if race 2, then $X_4 = 0$ and $X_5 = 1$; and if race 3, then $X_4 = 1$ and $X_5 = 0$. In general, $L - 1$ dummies are required, where L is the number of levels of the variable to be represented by them. Hardy (1993) discusses many aspects of regression with dummy variables.

When $L > 2$, it is inadvisable to employ stepwise regression by the forward-selection process of Section 20.8 (Cohen, 1991). If the dependent variable, Y, is the dummy variable, appropriate procedures are more complicated (e.g., Glantz and Slinker, 1990; Chapter 11; Montgomery and Peck, 1992: 254–262) and may involve the use of what is known as *logistic regression.*

20.12 INTERACTION OF INDEPENDENT VARIABLES

It may be proposed that two or more independent variables interact in affecting the dependent variable, Y, a concept encountered in Chapters 12 and 14 when discussing factorial analysis of variance. For example, we may propose this regression model:

$$Y_j = \alpha + \beta_1 X_{1j} + \beta_2 X_{2j} + \beta_3 X_{1j} X_{2j} + \epsilon_j. \tag{20.44}$$

The regression analysis would proceed by treating $X_1 X_2$ as a third independent variable (i.e., as if it were X_3); and rejecting H_0: $\beta_3 = 0$ would indicate a significant interaction between X_1 and X_2, meaning that the magnitude of the effect of X_1 on Y is dependent upon X_2 and the magnitude of the effect of X_2 on Y is dependent upon X_1. By using linear regression equations that include interaction terms, a great variety of analysis of variance experimental designs can be analyzed (even those with unequal replication per cell), and this is a technique employed by some computer programs. Many ramifications of interactions in multiple regression are covered by Aiken and West (1991). Interaction, as the joint effect on Y of two or more X's, should not be confused with correlation among X's ("multicollinearity" discussed at the end of Section 20.4).

20.13 COMPARING MULTIPLE REGRESSION EQUATIONS

It is not uncommon to desire to determine whether the multiple regressions from two or more sets of data, all containing the same variables, are estimating the same population regression function. We may test the null hypothesis that all the sample regression equations estimate the same population regression model by an extension of the considerations of Section 18.9. For a total of k regressions, the pooled residual sum of squares, SS_p, is simply the sum of all k residual sums of squares; and the pooled residual degrees of freedom, DF_p, is the sum of all k residual degrees of freedom. We then lump together all data from all k regressions and calculate a regression for this totality of data. The resulting total residual sum of squares and total degrees of freedom will be referred to as SS_t and DF_t, respectively.

The test of the null hypothesis (that there is a single population underlying all k sample regressions) is

$$F = \frac{\dfrac{SS_t - SS_p}{(m+1)(k-1)}}{\dfrac{SS_p}{DF_p}}, \tag{20.45}$$

a statistic with $(m+1)(k-1)$ and DF_p degrees of freedom, Example 20.4 demonstrates this procedure.

We may also employ the concept of parallelism in multiple regression as we did in simple regression. A simple linear regression may be represented as a line on a two-dimensional graph, and two such lines are said to be parallel (i.e., have the same slopes) if the vertical distance between them is constant for all values of the independent variable, meaning that the regression coefficients (i.e., slopes) of the two lines are the same. A multiple regression with two independent variables may be visualized as a plane

---

EXAMPLE 20.4   Comparing multiple regressions.

Let us consider three multiple regressions, each fitted to a different sample of data, and each containing the same dependent variable and the same four independent variables. (Therefore, $m = 4$ and $k = 3$.) The residual sums of squares from each of the regressions are 437.8824, 449.2417, and 411.3548, respectively.

If the residual degrees of freedom for each of the regressions are 41, 32, and 38, respectively (that is, the three sample sizes were 46, 37, and 43, respectively), then the pooled residual sum of squares, SS_p, is 1298.4789, and the pooled residual degrees of freedom, DF_p, is 111.

Then, we combine all 126 data from all three samples and fit to these data a multiple regression having the same variables as the three individual regressions fitted previously. From this multiple regression let us say we have a total residual sum of squares, SS_t, of 1577.3106. The total residual degrees of freedom, DF_t, is 121.

Then we test H_0: All three sample regression functions estimate the same population regression, against H_A: All three sample regression functions do not estimate the same population regression:

$$
\begin{aligned}
F &= \frac{\dfrac{SS_t - SS_p}{(m+1)(k-1)}}{\dfrac{SS_p}{DF_p}} \\[2em]
&= \frac{\dfrac{1577.3106 - 1298.4789}{(5)(2)}}{\dfrac{1298.4789}{111}} \\[2em]
&= 2.38
\end{aligned}
$$

The degrees of freedom associated with F are 10 and 111.

Since $F_{0.05(1),\,10,\,111} \cong 1.93$, reject H_0.

$$0.01 < P < 0.025 \quad [P = 0.013]$$

in three-dimensional space. Two planes are parallel if the vertical distance between them is the same for all combinations of the independent variables, in which case each of the partial regression coefficients for one regression is equal to the corresponding coefficient of the second regression, with only the Y intercepts possibly differing.

In general, two or more multiple regressions are said to be parallel if they all have the same β_1, β_2, β_3, etc. This may be tested by a straightforward extension of the procedure in Section 18.4. The residual sums of squares for all k regressions are summed to give the pooled residual sum of squares, SS_p; the pooled residual degrees of freedom are

$$\mathrm{DF}_p = \sum_{i=1}^{k} n_i - k(m + 1). \tag{20.46}$$

Additionally, we calculate a residual sum of squares for the "combined" regression in the following manner. Each element in a corrected sum of squares and sum of cross-products matrix (Matrix 20.5) is formed by summing all those elements from the k regressions. For example, element $\sum x_1^2$ is formed as $(\sum x_1^2)_1 + (\sum x_1^2)_2 + (\sum x_1^2)_3 + \cdots + (\sum x_1^2)_k$, and element $\sum x_1 x_2$ is formed as $(\sum x_1 x_2)_1 + (\sum x_1 x_2)_2 + \cdots + (\sum x_1 x_2)_k$. The residual sum of squares obtained from the multiple regression analysis using the resulting matrix is the "common" residual sum of squares, SS_c; the degrees of freedom associated with it are

$$\mathrm{DF}_c = \sum_{i=1}^{k} n_i - k - m. \tag{20.47}$$

Then the null hypothesis of all k regressions being parallel is tested by

$$F = \frac{\dfrac{SS_c - SS_p}{k - 1}}{\dfrac{SS_p}{\mathrm{DF}_p}}, \tag{20.48}$$

with $k - 1$ and DF_p degrees of freedom.

If the null hypothesis is not rejected, we then conclude that the independent variables affect the dependent variable in the same manner in all k regressions; we also conclude that all k regressions are parallel. Now we may ask whether the elevations of the k regressions are all the same. Here we proceed by an extension of the method in Section 18.5. All data for all k regressions are pooled together and one overall regression is fitted. The residual sum of square of this regression is the total residual sum of squares, SS_t which is associated with degrees of freedom of

$$\mathrm{DF}_t = \sum_{i=1}^{k} n_t - m - 1. \tag{20.49}$$

Then, the hypothesis of no difference among the k elevations is tested by

$$F = \frac{\dfrac{SS_t - SS_c}{k - 1}}{\dfrac{SS_c}{\mathrm{DF}_c}}, \tag{20.50}$$

with $k - 1$ and DF_c degrees of freedom.

20.14 MULTIPLE REGRESSION THROUGH THE ORIGIN

As an expansion of the simple linear regression model presented in Section 17.9, we might propose a multiple regression model where $\alpha = 0$; that is, when all $X_i = 0$, then $Y = 0$:

$$\hat{Y}_j = \beta_1 X_{1j} + \beta_2 X_{2j} + \cdots + \beta_m X_{mj}. \qquad (20.51)$$

This will be encountered only rarely in biological work, but it is worth noting that some multiple regression computer programs are capable of handling this model.* Striking differences in the computer output will be that total DF $= n$, regression DF$= m$ (the number of parameters in the model), and residual DF $= n - m$. Also, an *inverse pseudocorrelation matrix* may appear in the computer output in place of an inverse correlation or inverse sum of squares and crossproducts matrix. This regression model is legitimate only if each variable (i.e., Y and each X_i) is measured on a ratio scale (as defined in Section 1.1).

20.15 NONLINEAR REGRESSION

Regression models such as

$$Y_i = \alpha + \beta X_i, \qquad (17.1)$$

$$Y_j = \alpha + \beta_1 X_{1j} + \beta_2 X_{2j} + \cdots + \beta_m X_{mj}, \qquad (20.14)$$

or

$$Y_i = \alpha + \beta_1 X_i + \beta_2 X_i^2 + \cdots + \beta_m X_i^m \qquad (21.2)$$

are more completely symbolized as

$$Y_i = \alpha + \beta X_i + \epsilon_i, \qquad (20.52)$$

$$Y_j = \alpha + \beta_1 X_{1j} + \beta_2 X_{2j} + \cdots + \beta_m X_{mj} + \epsilon_j, \qquad (20.53)$$

or

$$Y_i = \alpha + \beta_1 X_i + \beta_2 X_i^2 + \cdots + \beta_m X_i^m + \epsilon_i, \qquad (20.54)$$

respectively, where ϵ is the *residual*, the difference between the value of Y predicted from the equation and the true value of Y in the population. All three of the preceding regression models are termed *linear* models because their parameters (i.e., α, β, and ϵ) appear in an additive fashion. However, cases do arise where the investigator wishes to fit to the data a model that is nonlinear with regard to its parameters. Such models might be those such as "exponential growth,"

$$Y_i = \alpha \beta^{X_i} + \epsilon_i, \qquad (20.55)$$

---

*Hawkins (1980) explains how a regression can be fitted through the origin using the output from a computer program for fitting a regression not assumed to pass through the origin.

or

$$Y = \alpha e^{\gamma X_i} + \epsilon_i; \tag{20.56}$$

"exponential decay,"

$$Y_i = \alpha \beta^{-X_i} + \epsilon_i, \tag{20.57}$$

or

$$Y_i = \alpha e^{-\gamma X_i} + \epsilon_i; \tag{20.58}$$

"asymptotic regression,"

$$Y_i = \alpha - \beta \delta^{X_i} + \epsilon_i \tag{20.59}$$

or

$$Y_i = \alpha - \beta(e^{-\gamma X_i}) + \epsilon_i; \tag{20.60}$$

or "logistic growth,"

$$Y_i = \frac{\alpha}{1 + \beta \delta^{X_i}} + \epsilon_i; \tag{20.61}$$

where the various Greek letters are parameters in the model (see Snedecor and Cochran, 1980: 394, for graphs of such functions). Other nonlinear models would be those in which the residuals were not additive, but, for example, might be multiplicative:

$$Y_i = \beta X_i \epsilon_i. \tag{20.62}$$

Sometimes a nonlinear model may be transformed into a linear one. For example, we may transform

$$Y_i = \alpha X_i^{\beta} \epsilon_i \tag{20.63}$$

by taking the logarithm of each side of the equation, acquiring a model that is linear in its parameters:

$$\log Y_i = \log \alpha + \beta \log X_i + \log \epsilon_i. \tag{20.64}$$

Transformations must be employed with careful consideration, however, so that the assumption of homogeneity of variance is not violated.

Biologists at times wish to fit nonlinear equations, some much more complex than the examples given, and computer programs are available for many of them. Such programs fall into two general groups. First are programs written to fit a particular model or a family of models, and the use of the program is little if any more complicated than the use of a multiple linear regression program (e.g., Zar, 1969). Second are general programs that can handle any of a very wide variety of models. To use the latter type of program, however, requires the user to submit a good deal of information, such as the partial derivatives of the regression function with respect to each parameter in the model (thus, recalling differential calculus and/or consulting with a statistician would be in order).

Nonlinear regression programs typically involve some sort of an iterative procedure, *iteration* being the utilization of a set of parameter estimates to arrive at a set of somewhat better parameter estimates, using the new estimates to derive better estimates, etc. Thus, many of these programs require the user to submit initial estimates of (i.e., to guess the values of) the parameters in the model being fitted.

The program output for a nonlinear regression analysis is basically similar to much of the output from multiple linear regression analyses. Most importantly, the program should provide estimates of the parameters in the model (i.e., the statistics in the regression equation), the standard error of each of these statistics, and an analysis of variance summary including at least the regression and residual SS and DF. If regression and residual MS are not presented in the output, they may be calculated by dividing the appropriate SS by its associated DF. An F test of significance of the entire regression (or correlation) and the coefficient of determination may be obtained by means of Equations 20.18 and 20.19, respectively. Testing whether any of the parameters in the model is equal to any hypothesized value may be effected by t tests similar to those previously used for simple and partial regression coefficients (e.g., Section 20.4). Kvålseth (1985) and others warn that the computation of R^2 may be inappropriate in nonlinear regression.

Further discussions of nonlinear regression are found in Bates and Watts (1988); Berry and Feldman (1985: 51–64); Seber and Wild (1989), Snedecor and Cochran (1980: Chapter 19), and the books cited at the end of the introduction to this chapter. With the increased availability of computer programs for such analyses, biologists will likely consider nonlinear models more often than they have previously.

20.16　Descriptive vs. Predictive Models

Often, it is hoped that a regression model implies a biological dependence (i.e., a cause and effect) in a nature, and that this dependence is confirmed by the mathematical relationship described by the regression equation. However, regression equations are at times useful solely as a means of predicting the value of a variable, if the values of a number of associated variables are known. For example, we may desire to be able to predict the weight (call it variable Y) of a mammal, given the length of the femur (variable X). Perhaps a polynomial regression such as

$$\hat{Y}_i = a + b_1 X_i + b_2 X_i^2 + b_3 X_i^3 + b_4 X_i^4 \tag{20.65}$$

might be found to fit the data rather well. (See Chapter 21 for details of polynomial regression.) Or perhaps we wish to be able to predict a man's blood pressure (call it variable P) as accurately as we can by using measurements of his weight (variable W), his age (variable A), and his height (variable H). By deriving additional regression terms composed of combinations and powers of the three measured independent variables, we might conclude statistical significance of each term in an equation such as

$$\hat{P}_i = a + b_1 W_i + b_2 A_i + b_3 H_i + b_4 W_i^2 + b_5 H_i^2 + b_6 W_i^3$$
$$+ b_7 W_i A_i + b_8 H_i A_i + b_9 W_i^3 A_i. \tag{20.66}$$

Equations such as 20.65 and 20.66 might have statistically significant partial regression coefficients. They might also have associated with them small standard errors of estimate, meaning that the standard error of predicted Y_i's or P_i's would be small. Thus, we would have good regression equations for purposes of prediction; but we are not implying that the fourth power of femur length has any natural significance in determining mammal weights, or that terms such as $H_i A_i$ or $W_i^3 A_i$ have any *biological* significance relative to human blood pressure.

To realize a regression function that describes underlying biological phenomena, the investigator must possess a good deal of knowledge about the interrelationships in nature among the variables in the model. Is it indeed reasonable to assume underlying relationships to be linear, or is there a logical basis for seeking to define a particular nonlinear relationship? (For example, forcing a linear model to fit a set of data in no way "proves" that the underlying biological relationships are, in fact, linear.) Are the variables included in the model meaningful choices? (For example, we might find a significant regression of variable A on variable B, whereas a third variable, C, is actually causing the changes in both A and B.) Statistical analysis is only a tool; it can not substitute for incomplete or fallacious biological information.

20.17 CONCORDANCE: RANK CORRELATION AMONG SEVERAL VARIABLES

The concept of correlation between two variables can be expanded to consider association among more than two, as shown in earlier discussion of multiple correlation. Such association is readily measured nonparametrically by a statistic known as *Kendall's coefficient of concordance** (Kendall, 1962: Chapter 6; Kendall and Babington-Smith, 1939.)[†] To demonstrate, let us expand the considerations of Examples 19.1a and 19.13 to examine whether there is concordance (i.e., association) among the magnitudes of wing, tail, and bill lengths in birds of a particular species. Example 20.5 shows such data, for which we determine the ranks for each of the three variables (just as we did for each of the two variables in Example 19.13).

Several equivalent computational formulas for the coefficient of concordance are found in various texts. One that is easy to use is

$$W = \frac{\sum R_i^2 - \dfrac{\left(\sum R_i\right)^2}{n}}{\dfrac{M^2(n^3 - n)}{12}}, \tag{20.67}$$

---

*Maurice George Kendall (1907–1983), English statistician.

[†]Wallis (1939) introduced this statistic independently, calling it the "correlation ratio," and designating it by η_r^2 (where η is the lowercase Greek eta).

EXAMPLE 20.5   **Kendall's coefficient of concordance.**

Birds (i)	Wing Length (cm) Data	Ranks	Tail Length (cm) Data	Ranks	Bill Length (mm) Data	Ranks	Sums of ranks (R_i)
1	10.4	4	7.4	5	17	5.5	14.5
2	10.8	8.5	7.6	7	17	5.5	21
3	11.1	10	7.9	11	20	9.5	30.5
4	10.2	1.5	7.2	2.5	14.5	2	6
5	10.3	3	7.4	5	15.5	3	11
6	10.2	1.5	7.1	1	13	1	3.5
7	10.7	7	7.4	5	19.5	8	20
8	10.5	5	7.2	2.5	16	4	11.5
9	10.8	8.5	7.8	9.5	21	11	29
10	11.2	11	7.7	8	20	9.5	28.5
11	10.6	6	7.8	9.5	18	7	22.5
12	11.4	12	8.3	12	22	12	36

$M = 3$

$n = 12$

Without correction for ties:

$$W = \frac{\sum R_i^2 - \dfrac{\left(\sum R_i\right)^2}{n}}{\dfrac{M^2(n^3 - n)}{12}}$$

$$= \frac{(14.5^2 + 21^2 + 30.5^2 + \cdots + 36^2) - \dfrac{(14.5 + 21 + 30.5 + \cdots + 36)^2}{12}}{\dfrac{3^2(12^3 - 12)}{12}}$$

$$= \frac{5738.5 - \dfrac{(234)^2}{12}}{\dfrac{15444}{12}}$$

$$= \frac{1175.5}{1287} = 0.913$$

H_0:   There is no association among the three variables.

H_A: There is association among the three variables.

$\chi_r^2 = M(n - 1)W$

    $= (3)(12 - 1)(0.913)$

    $= 30.129$

As $n >$ that in Appendix Table B.14, we use Appendix Table B.1.

$\nu = n - 1 = 11$

$\chi_{0.05, 11}^2 = 19.675$

Reject H_0.

$$0.001 < P < 0.005 \quad [P = 0.0015]$$

EXAMPLE 20.5   (continued)

Incorporating the correction for ties (preferable):

In group 1 (wing length): there are 2 data tied at 10.2 cm

(i.e., $t_1 = 2$); there are 2 data tied at 10.8 cm (i.e., $t_2 = 2$).

In group 2 (tail length): there are 2 data tied at 7.2 cm

(i.e., $t_3 = 2$); there are 3 data tied at 7.4 cm (i.e., $t_4 = 3$); there are 2 data tied at

7.8 cm (i.e., $t_5 = 2$).

In group 3 (bill length): there are 2 data tied at 17 mm (i.e., $t_6 = 2$); there are 2 data tied at

20 mm (i.e., $t_7 = 2$).

Considering all seven groups of ties,

$$\sum t = \sum_{i=1}^{7} (t_i^3 - t_i) \qquad \text{and}$$

$$= (2^3 - 2) + (2^3 - 2) + (2^3 - 2) + (3^3 - 3) + (2^3 - 2) + (2^3 - 2) + (2^3 - 2)$$

$$= 60$$

$$W_c = \frac{1175.5}{\dfrac{15444 - 3(60)}{12}}$$

$$= \frac{1175.5}{1272} = 0.924$$

Then, to test the significance of W_c:

$$(\chi_r^2)_c = M(n-1)W_c$$

$$= (3)(12-1)(0.924)$$

$$= 30.492$$

The same conclusion is reached in this case with W_c as with W, namely: Reject H_0; $0.001 < P < 0.005$. $[P = 0.0013]$

where M is the number of variables being correlated, and n is the number of data per variable.[‡] The numerator is simply the sum of squares of the n rank sums, R_i (analogous to Equation 4.9).

The value of W may range from 0 (when there is no association and, consequently, the R_i's are equal and the sum of squares of R_i is zero) to 1 (when there is complete agreement among the ranking of all groups). In Example 20.5 there is a very high level of concordance ($W = 0.913$), indicating that a bird with a large measurement for one of the variables is likely to have a large measurement for each of the other two variables.

[‡]Another convenient formula is

$$W = \frac{12 \sum R_i^2 - 3M^2 n(n+1)^2}{M^2(n^3 - n)}. \tag{20.68}$$

We can ask whether a calculated sample W is significant; that is, whether it represents an association different from zero in the population that was sampled. A simple way to do this involves the relationship between the Kendall coefficient of concordance, W, and the Friedman chi-square, χ_r^2. Using the notation from this section,

$$\chi_r^2 = M(n-1)W. \tag{20.69}$$

Therefore, we can convert a calculated W to its equivalent χ_r^2 and then employ our table of critical values of χ_r^2 (Appendix Table B.14). If either n or M is larger than that found in this table, then χ_r^2 may be assumed to be approximated by χ^2 with $n-1$ degrees of freedom, and Appendix Table B.1 is used. This is demonstrated in Example 20.5.

The Coefficient of Concordance with Tied Ranks.   If there are tied ranks within any of the M groups, then mean ranks are assigned as in previous discussions (e.g., Section 8.10, Example 8.14). Then W is computed with a correction for ties,

$$W_c = \frac{\sum R_i^2 - \dfrac{\left(\sum R_i\right)^2}{n}}{\dfrac{M^2(n^3 - n) - M\sum t}{12}} \tag{20.70}$$

where

$$\sum t = \sum_{i=1}^{m}(t_i^3 - t_i), \tag{20.71}$$

t_i is the number of ties in the ith group of ties, and m is the number of groups of tied ranks.* This computation of W_c is demonstrated in Example 20.5. W_c will not differ appreciably from W unless the numbers of tied data are great.

The Coefficient of Concordance for Assessing Agreement.   A common use of Kendall's coefficient of concordance is to express the intensity of agreement among several rankings. In Example 20.6, each of three children has been asked to rank the palatability of six flavors of ice cream. We wish to ask whether the three evaluators arrive at the same rankings. The conclusion is that there is concordance (i.e., agreement) among the three.

The Relationship between W and r_s.   It is interesting to note that W is related to the mean value of all possible Spearman rank correlation coefficients that could be obtained from all possible pairs of variables. These correlation coefficients may be listed

---

*Analogous to Equation 20.68 is

$$W_c = \frac{12\sum R_i^2 - 3M^2n(n+1)^2}{M^2(n^3 - n) - M\sum t} \tag{20.72}$$

when ties are present.

EXAMPLE 20.6   Kendall's coefficient of concordance used to assess agreement by three children in ranking palatability of six different ice cream flavors.

H_0:   There is no agreement among the three rankings.

H_A: There is agreement among the three rankings.

	Flavors (i)						
Child	1	2	3	4	5	6	
1	5	1	3	2	4	6	
2	6	2	3	1	5	4	
3	6	3	2	1	4	5	
Rank sum (R_i)	17	6	8	4	13	15	$\sum R_i = 63$

$M = 3$

$n = 6$

$$W = \frac{\sum R_i^2 - \dfrac{\left(\sum R_i\right)^2}{n}}{\dfrac{M^2(n^3 - n)}{12}}$$

$$= \frac{17^2 + 6^2 + 8^2 + 4^2 + 13^2 + 15^2 - \dfrac{63^2}{6}}{\dfrac{3^2(6^3 - 6)}{12}}$$

$$= \frac{137.50}{157.50} = 0.873$$

$$\chi_r^2 = M(n - 1)W$$

$$= (3)(6 - 1)(0.873)$$

$$= 13.095$$

Using Appendix Table B.14, $(\chi_r^2)_{0.05,3.6} = 7.000$. Therefore, reject H_0.

$$P < 0.001$$

---

in a matrix array:

$$
\begin{array}{ccccc}
(r_s)_{11} & (r_s)_{12} & (r_s)_{13} & \cdots & (r_s)_{1M} \\
(r_s)_{21} & (r_s)_{22} & (r_s)_{23} & \cdots & (r_s)_{2M} \\
(r_s)_{31} & (r_s)_{32} & (r_s)_{33} & \cdots & (r_s)_{3M} \\
\vdots & \vdots & \vdots & & \vdots \\
(r_s)_{M1} & (r_s)_{M2} & (r_s)_{M3} & \cdots & (r_s)_{MM}
\end{array}
\tag{20.73}
$$

a form similar to that of Matrix 20.6. As in Matrix 20.6, each element of the diagonal, $(r_s)_{ii}$, is equal to 1.0, and each element below the diagonal is duplicated above the diagonal, as $(r_s)_{ik} = (r_s)_{ki}$. There are $M!/[2(M - 2)!]$ different r_s's possible for M variables.*

In Example 20.5, we are speaking of three r_s's: $(r_s)_{12}$, the r_s for wing length and tail length; $(r_s)_{13}$, the r_s for wing and bill lengths; and $(r_s)_{23}$, the r_s for tail and bill lengths. The Spearman rank correlation coefficient matrix, using correction for ties (Equation 19.50), would be

$$
\begin{array}{lll}
1.000 & & \\
0.852 & 1.000 & \\
0.917 & 0.890 & 1.000.
\end{array}
$$

For Example 20.6, the r_s matrix would be

$$
\begin{array}{lll}
1.000 & & \\
0.771 & 1.000 & \\
0.771 & 0.886 & 1.000
\end{array}
$$

Denoting the mean r_s as $\bar{r}_s$, the relationship with W is

$$
W = \frac{(M - 1)\bar{r}_s + 1}{M}; \tag{20.74}
$$

therefore,

$$
\bar{r}_s = \frac{M W_c - 1}{M - 1}. \tag{20.75}
$$

If there are ties, then the above two equations relate W_c and $(\bar{r}_s)_c$ in the same fashion as W and $\bar{r}_s$ are related. While the possible range of W is 0 to 1, $\bar{r}_s$ may range from $-1/(M - 1)$ to 1. For Example 20.5, $(\bar{r}_s)_c = (0.852 + 0.917 + 0.890)/3 = 0.886$; for Example 20.6, $\bar{r}_s = 0.809$; and the reader can verify that Equations 20.74 and 20.75 hold.

If $M = 2$ (i.e., there are only two variables, or rankings, being correlated, as in Example 19.13), then either r_s or W might be computed; and

$$
W = \frac{\bar{r}_s + 1}{2}, \tag{20.76}
$$

and

$$
r_s = 2W_c - 1. \tag{20.77}
$$

When $M = 2$, the use of r_s is preferable, as it ranges from -1 to 1 and there are more thorough tables of critical values available.

If significant concordance is concluded for each of two groups of data, we may wish to ask if the agreement within each group is the same for both groups. For example, the data in Example 20.6 are for ice cream flavor preference as assessed by three children, and we might have a similar set of data for the preference exhibited by several adults for these same flavors; and if there were significant concordance among the children as well as significant agreement among the adults, we might wish to ask whether the consensus among children is the same as that among adults. A test for this purpose was presented

---

*That is, M things taken two at a time. (See Equation 5.10.)

by Schucany and Frawley (1973), with elaboration by Li and Schucany (1975). However, the hypothesis test is not always conclusive with regard to concordance between the two groups and it has received criticism by Hollander and Sethuraman (1978), who proposed a different procedure. Serlin and Marascuilo (1983) reexamined both approaches as well as multiple comparison testing.

Top-Down Concordance.   Section 19.10 discussed "top-down correlation," a two-sample procedure allowing us to give emphasis to those items ranked high (or low). An analogous situation can occur when there are more than two groups of ranks. For example, for the data of Example 20.6 we might have desired to know whether the children agreed on the most favored ice cream flavors, with our having relatively little interest in whether they agreed on the least appealing flavors. As with the correlation situation, we may employ the Savage scores, S_i, of Equation 19.54 (and Table 19.1) and a concordance test statistic is

$$C_T = \frac{1}{M^2(n - S_1)} \left(\sum_{i=1}^{n} R_i^2 - M^2 n \right),$$ (20.78)

the significance of which may be assessed by

$$\chi_T^2 = M(n - 1)C_T,$$ (20.79)

by comparing it to the chi-square distribution with $n - 1$ degrees of freedom (Iman and Conover, 1987). Here, n and M are as in the concordance computations above: each of M groups has n ranks. Also, R_i is the sum of the Savage scores, across the M groups, at rank position i; and S_1 is Savage score 1 (see Section 19.10). This is demonstrated in Example 20.7. In this example, it is concluded that there is no agreement among the three children regarding the most tasty ice cream flavors. However, we could instead have asked whether there was agreement as to the least tasty flavors. This would have been done by assigning Savage scores in reverse order (i.e., $S_1 = 2.450$ assigned to rank 6, S_2 to rank 5, and so on). If this were done we would have found that $C_T = 0.8222$

EXAMPLE 20.7   Top-down concordance, using the data of Example 20.6 to ask whether there was significant agreement among children regarding the most desirable ice cream flavors. The table of data shows the Savage scores in place of the ranks of Example 20.6.

H_0:   There is no agreement regarding the most preferred flavors.

H_A: There is agreement regarding the most preferred flavors.

Child	Flavors (i)					
	1	2	3	4	5	6
1	0.356	2.450	0.950	1.450	0.617	0.167
2	0.167	1.450	0.950	1.240	0.367	0.617
3	0.167	0.950	1.450	1.240	0.617	0.367
R_i	0.701	4.850	3.350	3.930	1.601	1.151

EXAMPLE 20.7   (continued)

$$C_T = \frac{1}{M^2(n - S_1)} \left(\sum_{i=1}^{n} R_i^2 - M^2 n \right)$$

$$= \frac{1}{3^2(6 - 2.450)} \left[0.701^2 + 4.850^2 + 3.350^2 + 3.930^2 + 1.601^2 + 1.151^2 - (3^2)(6) \right]$$

$$= 0.03130(0.5693)$$

$$= 0.01782$$

$$\chi_T^2 = 3(6 - 1)C_T$$

$$= (15)(0.01782) = 0.267$$

$$\nu = n - 1 = 5$$

$$\chi_{0.05, 5}^2 = 11.070$$

Do not reject. $(0.995 < P < 0.999)$

and $\chi_T^2 = 12.333$, which would have resulted in a rejection of the null hypothesis of no agreement regarding the least-liked flavors $(0.025 < P < 0.05)$.

EXERCISES

20.1 Given the following data:

$Y(g)$	X_1 (m)	X_2 (cm)	X_3 (m^2)	X_4 (cm)
51.4	0.2	17.8	24.6	18.9
72.0	1.9	29.4	20.7	8.0
53.2	0.2	17.0	18.5	22.6
83.2	10.7	30.2	10.6	7.1
57.4	6.8	15.3	8.9	27.3
66.5	10.6	17.6	11.1	20.8
98.3	9.6	35.6	10.6	5.6
74.8	6.3	28.2	8.8	13.1
92.2	10.8	34.7	11.9	5.9
97.9	9.6	35.8	10.8	5.5
88.1	10.5	29.6	11.7	7.8
94.8	20.5	26.3	6.7	10.0
62.8	0.4	22.3	26.5	14.3
81.6	2.3	37.9	20.0	0.5

(a) Fit the multiple regression model $Y = \alpha + \beta_1 X_1 + \beta_2 X_2 + \beta_3 X_3 + \beta_4 X_4$ to the data, computing the sample partial regression coefficients and Y intercept.

(b) By analysis of variance, test the hypothesis that there is no significant multiple regression relationship.

(c) If H_0 is rejected in part (b), compute the standard error of each partial regression coefficient and test each H_0: $\beta_i = 0$.

(d) Calculate the standard error of estimate and the coefficient of determination.

(e) What is the predicted mean population value of Y at $X_1 = 5.2$ m, $X_2 = 21.3$ cm, $X_3 = 19.7$ m^2, and $X_4 = 12.2$ cm?

(f) What are the 95% confidence limits for the $\hat{Y}$ of part (e)?

(g) Test the hypothesis that the mean population value of Y at the X_i's stated in part (e) is greater than 50.0 g.

20.2 Subject the data of Exercise 20.1 to a stepwise regression analysis.

20.3 Analyze the five variables in Exercise 20.1 as a multiple correlation.

(a) Compute the multiple correlation coefficient.

(b) Test the null hypothesis that the population multiple correlation coefficient is zero.

(c) Compute the partial correlation coefficient for each pair of variables.

(d) Determine which of the calculated partial correlation coefficients estimate population partial correlation coefficients that are different from zero.

20.4 The following values were obtained for three multiple regressions of the form $\hat{Y} = a + b_1 X_1 + b_2 X_2 + b_3 X_3$. Test the null hypothesis that each of the three sample regressions estimates the same population regression function.

Regression	Residual sum of squares	Residual degrees of freedom
1	44.1253	24
2	56.7851	27
3	54.4288	21
All data combined	171.1372	

20.5 Each of five research papers was read by each of four faculty reviewers. Each reviewer then ranked the quality of the five papers, as follows:

			Papers		
	1	2	3	4	5
Reviewer 1	5	4	3	1	2
Reviewer 2	4	5	3	2	1
Reviewer 3	5	4	1	2	3
Reviewer 4	5	3	2	4	1

(a) Calculate the Kendall coefficient of concordance.

(b) Test whether the rankings by the four reviewers are in agreement.

21

POLYNOMIAL REGRESSION

A special type of multiple regression is that concerning a *polynomial* expression:

$$Y_i = \alpha + \beta_1 X_i + \beta_2 X_i^2 + \beta_3 X_i^3 + \cdots + \beta_m X_i^m + \epsilon_i, \qquad (21.1)$$

a model with parameters estimated in the expression

$$\hat{Y}_i = a + b_1 X_i + b_2 X_i^2 + b_3 X_i^3 + \cdots + b_m X_i^m, \qquad (21.2)$$

for which a more concise symbolism is:

$$\hat{Y}_i = a + \sum_{j=1}^{m} b_j X_i^j. \qquad (21.3)$$

If $m = 1$, then the polynomial regression reduces to a simple linear regression (with Equations 21.1 and 21.2 becoming Equations 17.1a and 17.8, respectively.)

21.1 POLYNOMIAL CURVE FITTING

As shown in Example 21.1, the data for the fitting of this model really only consist of two variables: the dependent variable, Y, and the independent variable, X. The remaining variables in the model are derived from X; the second independent variable in the model is the square of X, the third is the cube of X, etc.* If Y_i, X_i, X_i^2, $X_i^3, \ldots, X_i^m$, are submitted to a multiple regression computer program, then the statistics,

---

*The transformation of X and/or Y in polynomial regression follows the considerations of Sections 17.10.

b_j, in Equation 21.2 are readily obtained.* Alternatively, there are programs specifically for the analysis of polynomial regression, to which one need only submit Y_i and X_i (e.g., Zar, 1968b).

Regardless of how we perform the fitting of a polynomial regression, we need to determine the maximum power of the polynomial that has statistical significance. To do this, we shall employ a stepwise regression procedure, remembering that the maximum power, m, may be no greater than $n - 1$ if a polynomial is to be fit to the data, and, more practically, no greater than $n - 2$ if statistical analysis is to be performed on the resulting polynomial fit.

We may use stepwise procedures detailed in Section 19.8, beginning with the fitting of a polynomial with a larger m than it is felt will be needed. Then we would examine the significance of b_m, the b_j associated with the largest power in the model. If this term represents a β_j significantly different from zero, we would likely be tempted to try a higher power. If not, the term will be deleted from the model and the new model fitted to the data, with the procedure repeated until the b_j associated with the highest power of X is significantly different from zero.

A second and probably more common method for fitting the polynomial model is to proceed from a small to a large model, using the "forward selection" procedure of Section 20.8. We begin by fitting a linear regression to the data: $Y_i = \alpha + \beta X_i$ (as shown in Example 21.1 and Fig. 21.1a). Then we fit a second-degree polynomial (called a *quadratic* equation) to the data. This model, $Y_i = \alpha + \beta_1 X_i + \beta_2 X_i^2$ (Fig. 21.1b), simply adds one term (called the quadratic term) to the simple regression. To determine whether the addition of this term significantly improves the accuracy of the prediction of Y values, we may use either of two testing procedures. One is to apply the t test to $H_0: \beta_2 = 0$ (see Section 20.4), with failure to reject the null hypothesis indicating that the simple regression model describes the dependence of Y on X sufficiently well and the quadratic term contributes insignificantly to this description. We may reach the same conclusion by calculating

$$F = \frac{\text{(regression SS for higher degree model)} - \text{(regression SS for lower degree model)}}{\text{residual MS for higher degree model}}$$

(21.4)

with a numerator DF of 1 and a denominator DF that is the residual DF for the higher-degree model. This F test is equivalent to the preceding t test, because it also considers $H_0: \beta_j = 0$, where j is the last term in the higher-degree model.

If it is concluded that $\beta_2 \neq 0$, then we may fit a third-degree polynomial (called a *cubic* equation) to the data by adding a cubic term to the model (see Example 21.1 and Fig. 21.1c). Then, $H_0: \beta_3 = 0$ may be tested by either the t or F procedure described above, and we can conclude whether or not the addition of the term $\beta_3 X_i^3$ significantly improves the model. If it does not, then the quadratic equation is assumed to be an appropriate description of the relationship between Y and X. If we conclude

---

*Serious rounding problems can arise when dealing with powers of X_i's, and these can be reduced by coding each X_i by the subtraction of $\bar{X}$, i.e., by using $X_i - \bar{X}$ in place of X_i. (See Section 21.2.)

EXAMPLE 21.1 **Stepwise polynomial regression.**

The following data are submitted to a stepwise polynomial regression analysis:

X (kg)	Y (hr)
1.22	40.9
1.34	41.8
1.51	42.4
1.66	43.0
1.72	43.4
1.93	43.9
2.14	44.3
2.39	44.7
2.51	45.0
2.78	45.1
2.97	45.4
3.17	46.2
3.32	47.0
3.50	48.6
3.53	49.0
3.85	49.7
3.95	50.0
4.11	50.8
4.18	51.1

$n = 19$

First, a linear regression is fit to the data, resulting in:

$a = 37.389$, $b = 3.1269$,   and $s_b = 0.15099$

To test H_0: $\beta = 0$ against H_A: $\beta \neq 0, t = \dfrac{b}{s_b} = 20.709$, with $\nu = 17$.

As $t_{0.05(2),17} = 2.110$, H_0 is rejected.

Then, a quadratic (second-power) regression is fit to the data, resulting in:

$a = 40.302$, $b_1 = 0.66658$, $s_{b_1} = 0.91352$

$b_2 = 0.45397$, $s_{b_2} = 0.16688$

To test H_0: $\beta_2 = 0$ against H_A: $\beta_2 \neq 0, t = 2.720$, with $\nu = 16$.

As $t_{0.05(2),16} = 2.120$, H_0 is rejected.

Then, a cubic (third-power) regression is fit to the data, resulting in:

$a = 32.767$, $b_1 = 10.411$, $s_{b_1} = 3.9030$

$b_2 = -3.3868$, $s_{b_2} = 1.5136$

$b_3 = 0.47011$, $s_{b_3} = 0.18442$

To test H_0: $\beta_3 = 0$ against H_A: $\beta_3 \neq 0, t = 2.549$, with $\nu = 15$.

As $t_{0.05(2),15} = 2.131$, H_0 is rejected.

EXAMPLE 21.1   (continued)

Then, a quartic (fourth-power) regression is fit to the data, resulting in:

$$a = 6.9265, \quad b_1 = 55.835, \quad s_{b_1} = 12.495$$

$$b_2 = -31.487, \quad s_{b_2} = 7.6054$$

$$b_3 = 7.7625, \quad s_{b_3} = 1.9573$$

$$b_4 = -0.67507, \quad s_{b_4} = 0.18076$$

To test H_0: $\beta_4 = 0$ against H_A: $\beta_4 \neq 0, t = 3.735$, with $\nu = 14$.
As $t_{0.05(2), 14} = 2.145$, H_0 is rejected.

Then, a quintic (fifth-power) regression is fit to the data, resulting in:

$$a = 36.239, \quad b_1 = -9.1615, \quad s_{b_1} = 49.564$$

$$b_2 = 23.387, \quad s_{b_2} = 41.238$$

$$b_3 = -14.346, \quad s_{b_3} = 16.456$$

$$b_4 = 3.5936, \quad s_{b_4} = 3.1609$$

$$b_5 = -0.31740, \quad s_{b_5} = 0.23467$$

To test H_0: $\beta_5 = 0$ against H_A: $\beta_5 \neq 0, t = 1.353$, with $\nu = 13$.
As $t_{0.05(2), 13} = 2.160$, do not reject H_0.

Therefore, it appears that a quartic polynomial is the optimum regression function for the data. But to be more certain, we add one more term beyond the quintic to the model (i.e., a sextic, or sixth-power, polynomial regression is fit to the data), resulting in

$$a = 157.88, \quad b_1 = -330.98, \quad s_{b_1} = 192.28$$

$$b_2 = 364.04, \quad s_{b_2} = 201.29$$

$$b_3 = -199.36, \quad s_{b_3} = 108.40$$

$$b_4 = 58.113, \quad s_{b_4} = 31.759$$

$$b_5 = -8.6070, \quad s_{b_5} = 4.8130$$

$$b_6 = 0.50964, \quad s_{b_6} = 0.29560$$

To test H_0: $\beta_6 = 0$ against H_A: $\beta_6 \neq 0, t = 1.724$, with $\nu = 12$.
As $t_{0.05(2), 12} = 2.179$, do not reject H_0.

In concluding that the quartic regression is that of optimum fit to the data, we have: $\hat{Y} = 6.9265 + 55.835X - 31.487X^2 + 7.7625X^3 - 0.67507X^4$. See Fig. 21.1 for graphical presentation of the preceding polynomial equations.

that $\beta_3 \neq 0$ then a fourth-degree polynomial (called a *quartic* equation) is fitted to the data (Fig. 21.1d), we test H_0: $\beta_4 = 0$, and so on.

Once we have failed to reject H_0: $\beta_j = 0$ for the last term in the polynomial, we may cease adding further terms and conclude that the polynomial with $j - 1$ terms is the best model. However, as done in Example 21.1, some would advise carrying the analysis

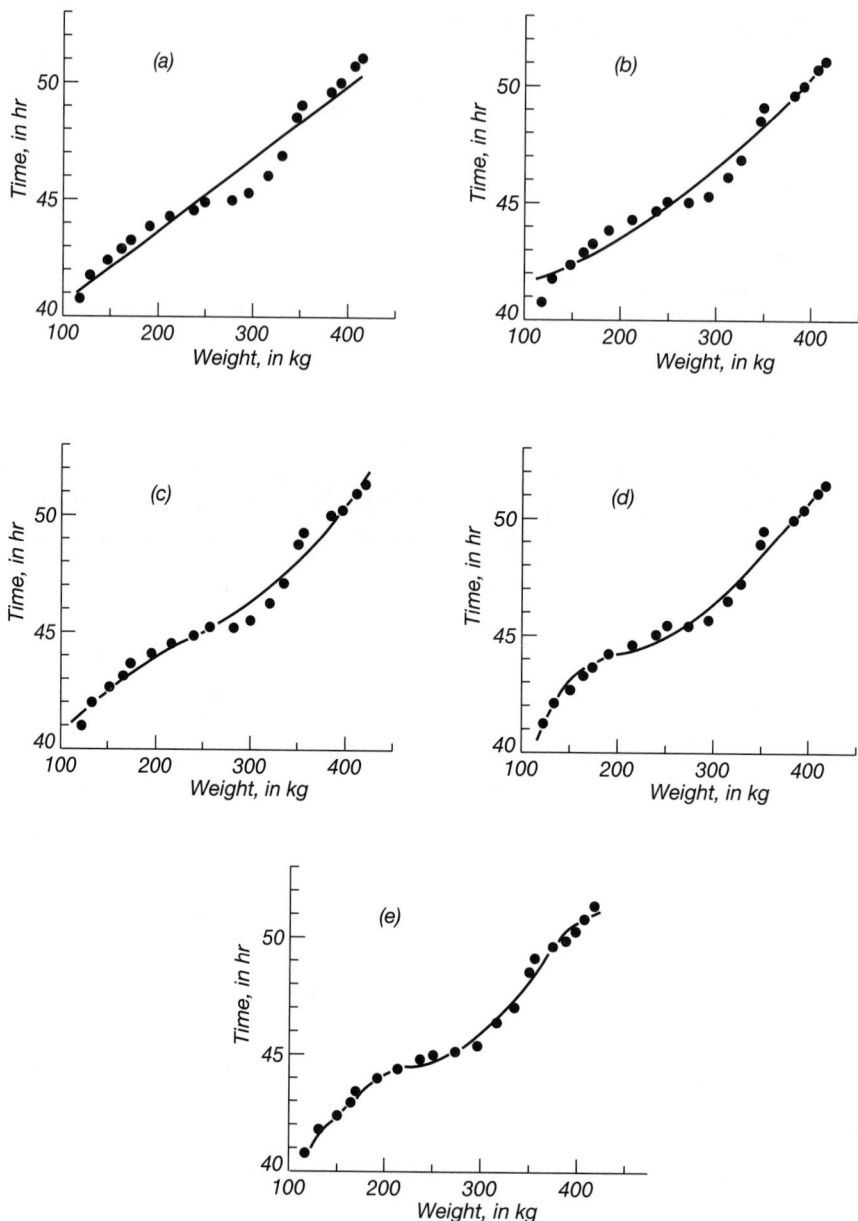

Figure 21.1 Fitting polynomial regression models. Each of the following regressions is fit to the nineteen data points of Example 21.1. (a) Linear: $\hat{Y} = 37.389 + 3.1269X$. (b) Quadratic: $\hat{Y} = 40.302 + 0.66658X + 0.45397X^2$. (c) Cubic: $\hat{Y} = 32.767 + 10.411X - 3.3868X^2 + 0.47011X^3$. (d) Quartic: $\hat{Y} = 6.9265 + 55.835X - 31.487X^2 + 7.7625X^3 - 0.67507X^4$. (e) Quintic $\hat{Y} = 36.239 - 9.1615X + 23.387X^2 - 14.346X^3 + 3.5936X^4 - 0.31740X^5$. The stepwise analysis of Example 21.1 concludes that the quartic equation provides the optimum fit; that is, the quintic expression does not provide a significant improvement in fit over the quartic.

one or two terms beyond the point where the preceding H_0 is not rejected, to be assured that significant terms are not being neglected inadvertently. For example, it is possible to not reject H_0: $\beta_3 = 0$, but by testing further to reject H_0: $\beta_4 = 0$. A polynomial regression may be fit through the origin using the considerations of Section 20.14.

Some authors (e.g., Montgomery and Peck, 1992: 203–204) advise against routine use of polynomial regression, especially for $m > 2$. One reason is that, as X_i will be correlated with powers of X_i (i.e., with X_i^2, X_i^3, and so on), the analysis may suffer from multicollinearity (see the end of Section 20.4).

After arriving at a final equation in a polynomial regression analysis, it may be desired to predict values of Y at a given value of X. This can be done by the procedures of Section 20.9, by which the precision of a predicted $\hat{Y}$ (expressed by a standard error or confidence interval) may also be computed. Indeed, prediction is often the primary goal of a polynomial regression (see Section 20.16) and biological interpretation is generally difficult, especially for m greater than 2 or 3. But it is even less desirable for polynomial regression than for simple or other multiple regression to extrapolate by predicting $\hat{Y}$'s beyond the range of the observed X_i's.

21.2 ROUND-OFF ERROR AND CODING DATA

In a polynomial regression analysis it is very often desirable to code X to avoid extremely large or extremely small ($\ll 1.0$) numbers resulting from raising X to the various powers. Although R^2, t, and F will not be affected by the coding shown below, Table 21.1 shows how other statistics are affected.

Let us code X by multiplying it by a constant, M_X, and code Y by multiplying it by M_Y. If desired, we may code Y by adding a constant, A_Y, whether or not M_Y is employed. Therefore, the coded variables are

$$[X] = M_X X \tag{21.5}$$

and

$$[Y] = M_Y X + A_Y. \tag{21.6}$$

In Table 21.1 the bracketed statistics are those resulting from a polynomial regression analysis using values of X and Y that have been coded. Note that in polynomial regression we should not employ an addition constant, A_X, to code X, as may be done with other kinds of regression (see Sections 17.11 and 20.7).

21.3 QUADRATIC REGRESSION

The most common polynomial regression is the second-order, or *quadratic*, regression:

$$Y_i = \alpha + \beta_1 X_i + \beta_2 X_i^2 + \epsilon_i \tag{21.7}$$

with three population parameters, α, β_1, and β_2, to be estimated by three regression statistics, a, b_1, and b_2, respectively, in the quadratic equation

$$\hat{Y}_i = a + b_1 X_i + b_2 X_i^2. \tag{21.8}$$

TABLE 21.1   The Effect of Coding on Polynomial Regression Statistics, with Coding Performed as in Equations 20.5 and 20.6

Statistic	Value without coding	Value with coding
Partial regression coefficient, b_i	$b_i = [b_i]M_X^i/M_Y$	$[b_1] = b_iM_Y/M_X^i$
Standard error of b_i, s_{b_i}	$s_{b_i} = [s_{b_i}]M_X^i/M_Y$	$[s_{b_i}] = s_{b_i}M_Y/M_X^i$
Y intercept, a	$a = \dfrac{[a] - A_Y}{M_Y}$	$[a] - M_Ya + A_Y$
Standard error of a, s_a	$s_a = [s_a]/M_Y$	$[s_a] = M_Ys_a$
Mean of X^i, $\overline{X^i}$	$\bar{X}^i = [\overline{X^i}]/M_X^i$	$[\overline{X^i}] = M_X^1\overline{X^i}$
Mean of Y, $\bar{Y}$	$\bar{Y} = \dfrac{[Y] - A_Y}{M_Y}$	$[\bar{Y}] = M_Y\bar{Y} + A_Y$
Standard error of estimate, $s_{Y\cdot1,2,\dots,m}$	$s_{Y\cdot1,2,\dots,m}$ $= [s_{Y\cdot1,2,\dots,m}]/M_Y$	$[s_{Y\cdot1,2,\dots,m}]$ $= M_Ys_{Y\cdot1,2,\dots,m}$
Coefficient of determination, R^2	$R^2 = [R^2]$	$[R^2] = R^2$
Test statistics, t and F	$t = [t]; F = [F]$	$[t] = t; [F] = F$
Sum of squares, SS	$SS = [SS]/M_Y^2$	$[SS] = (SS)(M_Y^2)$
Mean square, MS	$MS = [MS]/M_Y^2$	$[MS] = (MS)(M_Y^2)$

See Table 20.2 for the effect of coding on multiple regression matrices and on correlation coefficients.

Note: If a regression is fit through the origin, then A_Y as well as A_X must be zero.

The geometric shape of the curve represented by Equation 21.7 is a *parabola*. An example of a quadratic regression line is shown in Fig. 21.2. If b_2 is negative as shown in Fig. 21.2, the parabola will be concave downward. If b_2 is positive (as shown in Fig. 21.1b), the curve will be concave upward.

Maximum and Minimum Values of Y_i.   A common interest in polynomial regression analysis, especially where $m = 2$ (quadratic), is the determination of a maximum or minimum value of Y_i (Bliss 1970: Section 14.4; Studier, Dapson and Bigelow, 1975). A maximum value of Y_i is defined as one that is greater than those Y_i's that are close to it; and a minimum Y_i is one that is less than those Y_i's close to it. If, in a quadratic regression (Equation 21.8), the coefficient b_2 is negative, then there will be a maximum, as shown in Fig. 21.2. If b_2 is positive, there will be a minimum, (as is implied in Fig. 21.1b). It may be desired to determine what the maximum or minimum value of Y_i is and what the corresponding value of X_i is.

The maximum or minimum of a quadratic equation is at the following value of the independent variable:

$$\hat{X}_0 = \frac{-b_1}{2b_2}. \tag{21.9}$$

Placing $\hat{X}_0$ in the quadratic equation (Equation 21.8), we find that

$$\hat{Y}_0 = a - \frac{b_1^2}{4b_2}. \tag{21.10}$$

Thus, in Fig. 21.2, the maximum is at

$$\hat{X}_0 = \frac{-17.769}{2(-7.74286)} = 1.15 \text{ hr,}$$

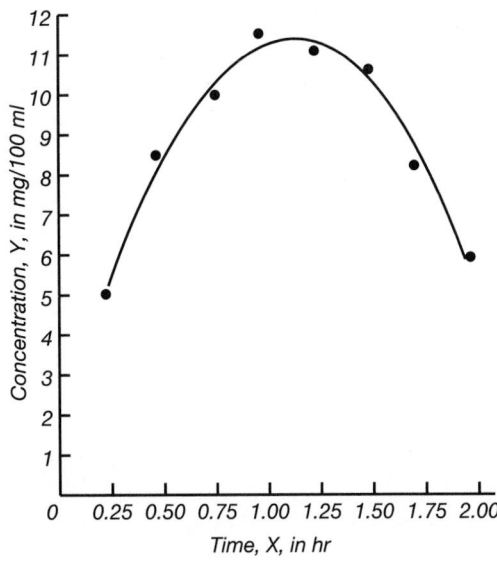

Figure 21.2 Quadratic fit to eight data points resulting in the equation $\hat{Y}_i = 1.39 + 17.769X_i - 7.74286X_i^2$.

at which

$$\hat{Y}_0 = 1.39 - \frac{(17.769)^2}{4(-7.74286)} = 11.58 \text{ mg/100 ml.}$$

A confidence interval for a maximum or minimum $\hat{Y}_0$ may be computed by the procedures of Section 20.9.

EXERCISES

21.1 Subject the following data to a stepwise polynomial regression analysis:

Y (g)	X (cm)
4.5	21.4
4.4	21.7
4.6	22.3
4.7	22.9
4.5	23.2
4.4	23.8
4.5	24.8
4.2	25.4
4.4	25.9
4.2	27.2
3.8	27.4
3.4	28.0
3.1	28.9
3.2	29.2
3.0	29.8

21.2 Consider the following data, where X is temperature (in $°C$) and Y is numbers of eggs per cm^2:

X	Y
3.0	2.8
5.0	4.9
8.0	6.7
14.0	7.6
21.0	7.2
25.0	6.1
28.0	4.7

(a) Fit a quadratic equation to these data.
(b) Test for significance of the quadratic term.
(c) Estimate the mean population value of $\hat{Y}_i$ at $X_i = 10.0°C$ and compute the 95% confidence interval for the estimate.
(d) Determine the values of X and Y at which the quadratic function is maximum.

22

TESTING FOR GOODNESS OF FIT

This chapter will introduce some statistical methods especially suited for the analysis of nominal scale data. As nominal data are counts of items or events in each of several classifications, methods for their analysis are sometimes referred to as *enumeration statistical methods*.

The widely used chi-square[*] statistic was introduced by Karl Pearson[†] in the years surrounding 1900, and its theory and application were subsequently expanded by him and R. A. Fisher (Lancaster, 1969: Chapter 1). Various aspects of chi-square analyses will be discussed in Chapters 22 and 23. The more recently developed log-likelihood ratio approaches to enumeration data analysis are introduced in Sections 22.7 and 23.7. Goodness of fit for ordered data may often be handled best by the Kolmogorov-Smirnov test, considered in Sections 22.8 and 22.9, or by the Watson test of Section 27.1.

---

[*]The symbol for chi-square is χ^2, where the Greek lowercase letter chi, χ, is pronounced as the "ky" in the "sky". Some authors prefer the notation X^2 to χ^2, to avoid using a Greek letter for something other than a population parameter; but this invites confusion with the frequent appearance of X^2 as the square of an observation, X, so the use of χ^2 has much support. David (1995) credits Karl Pearson with the first use of the terms "chi-squared" and "goodness of fit," in 1900. "Chi-squared" is still commonly preferred to "chi-square" by British writers.

[†]Karl Pearson (1857–1936), British mathematician; one of the founders of the field of statistics. He was also a cofounder, in 1901, of the English journal, *Biometrika*, which still influences statistics in many areas. He edited this journal for thirty-five years, succeeded for thirty years by his son, E[gon] S[harpe] Pearson (1895–1980), himself a powerful contributor to statistical theory and practice (see Bartlett, 1981). Walker (1958) has called Karl Pearson's development of statistical thinking and practice "an achievement of fantastic proportions" and has said of his influence on others: "Few men in all the history of science have stimulated so many other people to cultivate and enlarge the fields they had planted."

22.1 CHI-SQUARE GOODNESS OF FIT

It is frequently desired to obtain a sample of nominal scale data and to infer whether the population from which it came conforms to a specified theoretical distribution. For example, a plant geneticist may raise 100 progeny from a cross that is hypothesized to result in a 3:1 phenotypic ratio of yellow-flowered to green-flowered plants. Perhaps a ratio of 84 yellow:16 green is observed, although out of this total of 100 plants, the geneticist's hypothesis would predict a ratio of 75 yellow:25 green. The question to be asked, then, is whether the observed frequencies (84 and 16) deviate significantly from the frequencies expected if the hypothesis were true (75 and 25).

The statistical procedure for attacking the question first involves the concise statement of the hypothesis to be tested. The hypothesis in this case is that the population which was sampled has a 3:1 ratio of yellow-flowered to green-flowered plants. As introduced in Section 6.4, this is referred to as a *null hypothesis* (abbreviated H_0), because it is a statement of "no difference"; in this instance, we are hypothesizing that the population flower color ratio is not different from 3:1. If it is concluded that H_0 is false, then an *alternate hypothesis* (abbreviated H_A) will be assumed to be true. In this case, H_A would be that the population sampled has a flower-color ratio which is *not* 3 yellow:1 green. Recall that one states a null hypothesis and an alternate hypothesis for every statistical test performed, and all possible outcomes are accounted for by the two hypotheses.

The following calculation of a statistic called *chi-square* is used as a measure of how far a sample distribution deviates from a theoretical distribution:

$$\chi^2 = \sum_{i=1}^{k} \frac{(f_i - \hat{f}_i)^2}{\hat{f}_i}. \tag{22.1}*$$

Here, f_i is the frequency, or number of counts, observed in class i, $\hat{f}_i$ is the frequency expected in class i if the null hypothesis is true,[†] and the summation is performed over all k categories of data; χ is lowercase Greek chi. Example 22.1 shows the chi-square calculation for the flower-color data presented. In this sample, there are two categories of data (i.e., $k = 2$): yellow-flowered plants and green-flowered plants. The expected frequency, $\hat{f}_i$, of each class is calculated by multiplying the total number of observations, n, by the proportion of the total that the null hypothesis predicts for the class. Therefore, for the two classes in the example, $\hat{f}_1 = (100)(\frac{3}{4}) = 75$ and $\hat{f}_2 = (100)(\frac{1}{4}) = 25$.

---

*Equation 22.1 can be rewritten as

$$\chi^2 = \sum_{i=1}^{k} \frac{f_i^2}{\hat{f}_i} - n, \tag{22.2}$$

where n is the sum of all the f_i's, namely the total number of observations in the sample. Although this formula renders the calculation of χ^2 a little easier, it has the big disadvantage of not enabling us to examine each contribution to χ^2 [i.e., each $(f_i - \hat{f}_i)^2/\hat{f}_i$], and, as shown in Section 22.3, such an examination is an aid in determining how one might subdivide an overall chi-square analysis into component chi-square analyses for additional data collection. Thus, Equation 22.2 is seldom encountered.

[†]The symbol $\hat{f}$ is pronounced "f hat."

EXAMPLE 22.1   Calculation of chi-square goodness of fit, of data consisting of the colors of 100 flowers, to a hypothesized color ratio of 3:1.

H_0:   The sample data came from a population having a 3:1 ratio of yellow to green flowers.

H_A: The sample data came from a population not having a 3:1 ratio of yellow to green flowers.

The data recorded are the 100 observed frequencies, f_i, in each of the two flower color categories, with the frequencies expected under the null hypothesis, $\hat{f}_i$, in parentheses.

<table>
<tr><td></td><td colspan="3" align="center">Category (flower color)</td></tr>
<tr><td></td><td align="center">Yellow</td><td align="center">Green</td><td align="center">n</td></tr>
<tr><td>f_i</td><td align="center">84</td><td align="center">16</td><td align="center">100</td></tr>
<tr><td>$(\hat{f}_i)$</td><td align="center">(75)</td><td align="center">(25)</td><td></td></tr>
</table>

degrees of freedom $= \nu = k - 1 = 2 - 1 = 1$

$$\chi^2 = \sum \frac{(f_i - \hat{f}_i)^2}{\hat{f}_i} = \frac{(84 - 75)^2}{75} + \frac{(16 - 25)^2}{25}$$

$$= \frac{9^2}{75} + \frac{9^2}{25}$$

$$= 1.080 + 3.240$$

$$= 4.320$$

$\chi^2_{0.05, 1} = 3.841$

Therefore, reject H_0.

$$0.025 < P < 0.05 \quad [P = 0.038]$$

An improved procedure is presented in Section 22.4 (Example 22.4).

It should be apparent, by examining Equation 22.1, that larger disagreement between observed and expected frequencies (i.e., larger $f_i - \hat{f}_i$ values) will result in a larger χ^2 value. Thus, this type of calculation is referred to as a measure of *goodness of fit* (although it might better have been named a measure of "poorness of fit"). A calculated χ^2 value can be as small as zero, in the case of a perfect fit (i.e., each f_i value equals its corresponding $\hat{f}_i$), or very large if the fit is very bad; it can never be a negative value.

It is fundamentally important to appreciate that the chi-square statistic is calculated using the actual frequencies observed. It is not valid to convert the data to percentages and to attempt to submit the percentages to Equation 22.1. An additional consideration in calculating chi-square is described in Section 22.4.

What is meant by statistical significance in goodness of fit testing, as with other statistical tests, is derived from considerations of probability. Consider that the null hypothesis is true, i.e., the geneticist sampled a population of plants in which the yellow-to-green ratio is indeed 3 to 1. What we wish to ask is if it is likely to obtain from such a population a random sample of plants having an 84:16 flower-color ratio. If such a sample ratio can occur reasonably often, then we have no cause to reject H_0. If, however, there is little chance of obtaining a departure at least as great as 84 yellow-flowered and 16 green-flowered plants in a random sample from a population with a 3:1 ratio, then we may infer that the null hypothesis is false and that the alternate hypothesis is true (i.e., the sample came from a population with a color ratio that is not 3:1).

The computation of the probabilities that we require involves such complex mathematics that we are fortunate in having available tables of chi-square probabilities to aid in hypothesis testing. Appendix Table B.1 is a table of χ^2 values having certain probabilities of occurrence if H_0 is true.

By referring to the first line in the body of this table (the line for ν of 1), we can see, for example, that the probability ($P = \alpha$) of a χ^2 equal to or greater than 2.706 is 0.10 (i.e., 10%); this statement can be written concisely as $P(\chi^2 \geq 2.706) = 0.10$. As another example, we see that $P(\chi^2 \geq 3.841) = 0.05$. Now, in Example 22.1 we obtained $\chi^2 = 4.320$. By consulting Appendix Table B.1, we see that this value has associated with it a probability somewhere between 0.025 and 0.05, for $P(\chi^2 \geq 5.024) = 0.025$ and $P(\chi^2 \geq 3.841) = 0.05$. Thus, for this example, one can state that $0.025 < P(\chi^2 \geq 4.320) < 0.05$, or, simply, $0.025 < P < 0.05$. What this table tells us is that if H_0 were true, and if we repeated this same experiment a very large number of times, we could expect to get results that deviate at least this much from the hypothetical frequencies form 2.5% to 5% of the time.*

As explained in Section 6.4, biostatisticians often specify that if the magnitude of a calculated test statistic (such as χ^2) has an associated probability of 5% or less, its occurrence is so unlikely to be due to random sampling alone that we may reasonably conclude that the null hypothesis is false. This is the case for the data in Example 22.1; therefore, we reject H_0 and accept H_A concluding that the population sampled has a flower color ratio other than 3 yellow:1 green.

22.2 CHI-SQUARE GOODNESS OF FIT FOR MORE THAN TWO CATEGORIES

Example 22.1 considered chi-square testing for goodness of fit when there are two categories of data (i.e., $k = 2$). This analysis may be extended readily to any larger number of classes, as Example 22.2 exemplifies. Here, 250 plants were examined ($n = 250$), seeds were classified into four categories ($k = 4$), and the calculated χ^2, using Equation 22.1, is 8.972. (We shall routinely express a calculated chi-square to three decimal places, because that is the accuracy of the table of critical values, Appendix Table B.1. Therefore, to avoid rounding error, we shall perform all intermediate computations, including those of $\hat{f}_i$, to four decimal places.)

---

*Some calculators and computer programs have the capability of determining the exact probability of a given χ^2 (e.g., Guenther, 1977). For the present example, we would thereby find that $P(\chi^2 \geq 4.320) = 0.038$.

EXAMPLE 22.2   Chi-square goodness of fit for $k = 4$.

H_0: The sample comes from a population having a 9:3:3:1 ratio of yellow-smooth to yellow-wrinkled to green-smooth to green-wrinkled seeds.

H_A: The sample comes from a population not having a 9:3:3:1 ratio of the above four seed phenotypes.

The sample data are recorded as observed frequencies, f_i, with the frequencies expected under the null hypothesis, $\hat{f}_i$ in parenthesis.

	Yellow smooth	Yellow wrinkled	Green smooth	Green wrinkled	n
f_i	152	39	53	6	250
$(\hat{f}_i)$	(140.6250)	(46.8750)	(46.8750)	(15.26250)	

$\nu = k - 1 = 3$

$$\chi^2 = \frac{11.3750^2}{140.6250} + \frac{7.8750^2}{46.8750} + \frac{6.1250^2}{46.8750} + \frac{9.6250^2}{15.6250}$$

$$= 0.9201 + 1.3230 + 0.8003 + 5.9290$$

$$= 8.972$$

$\chi^2_{0.05,3} = 7.815$

Therefore, reject H_0.

$$0.025 < P < 0.05 \quad [P = 0.030]$$

It has already been pointed out that larger χ^2 values will result from larger differences between f_i and $\hat{f}_i$, but large calculated χ^2 values may also simply be the result of a large number of classes of data, because the calculation involves the summing over all classes. Thus, in considering the significance of a calculated χ^2, the value of k must in some way be taken into account. What is done is to consider the quantity known as *degrees of freedom** (abbreviated DF, or by lowercase Greek nu, ν). For the goodness of fit discussed in this chapter, DF (i.e., ν) = $k - 1$. Thus, for Example 22.2, DF = $4 - 1 = 3$, while the calculated $\chi^2 = 8.972$. Entering Appendix Table B.1 in the row for 3 DF, we see that $P(\chi^2 \geq 7.815) = 0.05$, and $P(\chi^2 \geq 9.348) = 0.025$. Therefore, $0.025 < P(\chi^2 \geq 8.972) < 0.05$, and we would reject the null hypothesis which states that the sample came from a population having a 9:3:3:1 phenotypic ratio of yellow-smooth:yellow-wrinkled:green-smooth:green-wrinkled seeds. Tabled critical values are frequently denoted as $\chi^2_{\alpha,\nu}$, so we could write $\chi^2_{0.05,3} = 7.815$, $\chi^2_{0.10,3} = 6.251$, etc.

When we say, in a goodness of fit problem such as Example 22.1 or 22.2, that DF = $k - 1$, we are stating that, given the frequencies in any $k - 1$ of the categories, we

---

*This term was introduced by R. A. Fisher, in 1922, while discussing contingency tables (see Chapter 23) (David, 1995).

can readily calculate the frequency in the remaining category. This is true because n is known, and the sum of the frequencies in all k categories equals n. (In other words, one has "freedom" in assigning frequencies to only $k - 1$ of the categories.) Another way of looking at chi-square degrees of freedom is to note that DF equals k minus the number of sample constants used to calculate the expected frequencies. In the present examples, only one constant, n, was so used, so $v = k - 1$.

22.3 SUBDIVIDING CHI-SQUARE ANALYSES

In Example 22.2, the chi-square analysis detected a difference between the observed and expected frequencies too great to be attributed to chance, and the null hypothesis was rejected. This conclusion may be satisfactory in some instances, but in many cases the investigator will wish to perform further analysis.

For the example under consideration, the null hypothesis is that the sample came from a population having a 9:3:3:1 phenotypic ratio. If the chi-square analysis had not led to a rejection of the hypothesis, we would proceed no further. But since H_0 was rejected, we may wish to ask whether the significant disagreement between observed and expected frequencies was concentrated in certain of the classes, or whether the difference was due to the effects of the data in all of the classes. Of the four individual contributions to the chi-square value—0.9201, 1.3230, 0.8003, and 5.9290—that resulting from the last class (the green-wrinkled seeds) contributes a relatively large amount to the size of the calculated χ^2. Thus we see that the nonconformity of the sample frequencies to those expected from a population with a 9:3:3:1 ratio is due largely to the magnitude of the discrepancy between f_4 and $\hat{f}_4$.

This line of thought can be examined as shown in Example 22.3. First, we test H_0: f_1, f_2, and f_3 came from a population having a 9:3:3 ratio. (H_A: The frequencies in the first three categories came from a population having a phenotypic ratio other than 9:3:3.) This null hypothesis is not rejected, indicating that the frequencies in the first

---

EXAMPLE 22.3   Chi-square goodness of fit, subdividing the chi-square analysis of Example 22.2.

H_0: The sample came from a population with a 9:3:3 ratio of the first three phenotypes in Example 22.2.

H_A: The sample came from a population not having a 9:3:3 ratio of the first three phenotypes in Example 22.2.

Seed Characteristics

	Yellow smooth	Yellow wrinkled	Green smooth	n
f_i	152	39	53	244
($\hat{f}_i$)	(146.4000)	(48.80000)	(48.80000)	

EXAMPLE 22.3   (continued)

$\nu = k - 1 = 2$

$$\chi^2 = \frac{5.6000^2}{146.4000} + \frac{-9.8000^2}{48.8000} + \frac{4.2000^2}{48.8000}$$

$$= 0.2142 + 1.9680 + 0.3615$$

$$= 2.544$$

$\chi^2_{0.05,2} = 5.991$

Therefore, do not reject H_0.

$$0.25 < P < 0.50 \quad [P = 0.28]$$

H_0:   The sample came from a population with a 1:15 ratio of green-wrinkled to other seed phenotypes.

H_A: The sample came from a population not having the 1:15 ratio stated in H_0.

Seed Characteristics

	Green wrinkled	Others	n
f_i	6	244	250
$(\hat{f}_i)$	(15.6250)	(234.3750)	

$\nu = k - 1 = 1$

$$\chi^2 = \frac{-9.6250^2}{15.6250} + \frac{9.6250^2}{234.3750}$$

$$= 5.9290 + 0.3953$$

$$= 6.324$$

$\chi^2_{0.05,1} = 3.841$

Therefore, reject H_0.

$$0.01 < P < 0.025 \quad [P = 0.012]$$

(Preferred computation of χ^2, when $k = 2$, is given in Section 22.4.)

---

three categories conform acceptably well to those predicted by H_0. Then we test the frequency of green-wrinkled seeds against the combined frequencies for the other three phenotypes, under the null hypothesis of a 1:15 ratio. The calculated χ^2 value causes us to reject this hypothesis, however, and we draw the conclusion that the nonconformity of the data in Example 22.2 to the hypothesized frequencies is due primarily to the observed frequency of green-wrinkled seeds.

It is not, however, proper to test statistical hypotheses developed after examining the data to be tested. Therefore, the analyses of this section should be considered only a guide to developing additional hypotheses, hypotheses that should then be stated in advance of their being tested with a new set of data.

22.4 CHI-SQUARE CORRECTION FOR CONTINUITY

Chi-square values obtained from actual data, using Equation 22.1, belong to a discrete, or discontinuous, distribution, in that they can take on only certain values. For instance, in Example 22.1 we calculated a chi-square value of 4.320 for $f_1 = 84$, $f_2 = 16$, $\hat{f}_1 = 75$, and $\hat{f}_2 = 25$. If we had observed $f_1 = 83$ and $f_2 = 17$, the calculated chi-square value would have been $(83 - 75)^2/75 + (17 - 25)^2/25 = 0.8533 + 2.5600 = 3.413$; for $f_1 = 82$ and $f_2 = 18$, $\chi^2 = 2.613$; etc. These chi-square values obviously form a discrete distribution, for values between 4.320 and 3.413 or between 3.413 and 2.613 are not possible with the given $\hat{f}_i$ values. However, the theoretical χ^2 distribution, from which Appendix Table B.1 is derived, is a continuous distribution; that is, all values of χ^2 between 2.613 and 4.320 are possible. Thus, our need to determine the probability of a calculated χ^2 can be met only approximately by consulting Appendix Table B.1, and our conclusions are not taking place exactly at the level of α which we set. This situation would be unfortunate were it not for the fact that the approximation is a very good one, except when $\nu = 1$ (and in the instances described in Section 22.5). In the case of $\nu = 1$, it is usually recommended to use the *Yates correction for continuity* (Yates, 1934),* where the absolute value of each deviation of f_i from $\hat{f}_i$ is reduced by 0.5 units. That is,

$$\chi_c^2 = \sum_{i=1}^{2} \frac{(|f_i - \hat{f}_i| - 0.5)^2}{\hat{f}_i}, \tag{22.3}$$

where χ_c^2 denotes the chi-square value calculated with the correction for continuity.

This correction is demonstrated in Example 22.4. Example 22.4a presents the determination of χ_c^2 for the data of 22.1. For this example, the use of χ_c^2 yields the same conclusion as is arrived at without the correction for continuity, but this will not always be the case. Without the continuity correction, the calculated χ^2 may be inflated enough to cause us to reject H_0, whereas the corrected χ_c^2 value might not. In other words, not correcting for continuity may cause us to commit the Type I error with a probability greater than the stated α. The Yates correction should routinely be used when $\nu = 1$; it is not applicable for $\nu > 1$. For very large n, the effect of discontinuity is small, even for $\nu = 1$, and in such cases the Yates correction will change the calculated chi-square very little. Its use remains appropriate with $\nu = 1$, however, regardless of n.

As another example of the use of the correction for continuity, we can reexamine the second null hypothesis of Example 22.3. This is shown as Example 22.4b.

For $k = 2$, if H_0 involves a 1:1 ratio,

$$\chi^2 = \frac{(f_1 - f_2)^2}{n} \tag{22.4}$$

---

*Although English statistician Frank Yates (1902–1994) deserves the credit for suggesting this correction for chi-square testing, it had previously been employed in other statistical contexts (Pearson, 1947). R. A. Fisher associated it with Yates' name in 1936 (David, 1995). This was one of many important contributions Yates made over a distinguished 59-year publishing career, and he was also one of the earliest users of electronic computers to summarize and analyze data (Dyke, 1995).

EXAMPLE 22.4   Chi-square goodness of fit, using the Yates correction for continuity.

(a) For the hypothesis and data of Example 22.1:

Category (flower color)

	Yellow	Green	n
f_i	84	16	100
$(\hat{f}_i)$	(75)	(25)	

$$\nu = k - 1 = 2 - 1 = 1$$

$$\chi_c^2 = \sum_{i=1}^{2} \frac{(|f_i - \hat{f}_i| - 0.5)^2}{\hat{f}_i} = \frac{(|84 - 75| - 0.5)^2}{75} + \frac{(|16 - 25| - 0.5)^2}{25}$$

$$= 0.9633 + 2.8900 = 3.853$$

$$\chi_{0.05,1}^2 = 3.841$$

Therefore, reject H_0.

$$0.025 < P < 0.05 \quad [P = 0.0497]$$

(b) For the second set of hypotheses and data of Example 22.3:

Seed Characteristics

	Green wrinkled	Others	n
f_i	6	244	250
$(\hat{f}_i)$	(15.6250)	(234.3750)	

$$\nu = k - 1 = 1$$

$$\chi_c^2 = \sum_{i=1}^{2} \frac{(|f_i - \hat{f}_i| - 0.5)^2}{\hat{f}_i} = \frac{(9.6250 - 0.5)^2}{15.6250} + \frac{(9.6250 - 0.5)^2}{234.3750}$$

$$= 5.3290 + 0.3553 = 5.684$$

$$\chi_{0.05,1}^2 = 3.841$$

Therefore, reject H_0.

$$0.01 < P < 0.025 \quad [P = 0.017]$$

---

may be used in place of Equation 22.1, and

$$\chi_c^2 \frac{(|f_1 - f_2| - 1)^2}{n} \tag{22.5}$$

may be used instead of Equation 22.3. Note that in these two shortcut equations, $\hat{f}_1$ and $\hat{f}_2$ need not be calculated, thus avoiding the concomitant rounding errors.

22.5 BIAS IN CHI-SQUARE CALCULATIONS

In order for us to assign a probability to the results of a chi-square goodness of fit test, and thereby assess the statistical significance of the test, the calculated χ^2 must be a close approximation to the theoretical distribution that is summarized in Appendix Table B.1. This approximation is quite acceptable as long as the expected frequencies are not too small. If $\hat{f}_i$ values are very small, however, the calculated χ^2 is biased in that it is larger than the theoretical χ^2 it is supposed to estimate, and there is a tendency to reject the null hypothesis with a probability greater than α. This is clearly undesirable, and for decades statisticians have attempted to define in a convenient manner what would constitute $\hat{f}_i$'s that are "too small."

For decades a commonly applied general rule was that no expected frequency should be less than 5.0*, even though it has long been known that is it tolerable to have a few $\hat{f}_i$'s considerably smaller than that (e.g., Cochran, 1952, 1954). By a review of previous recommendations and an extensive empirical analysis, Roscoe and Byars (1971) reached conclusions that provide less restrictive guidelines for chi-square goodness of fit testing. They found that the test is remarkably robust when testing for a uniform distribution—i.e., for H_0: In the population, the frequencies in all k categories are equal—in which case $\hat{f}_i = n/k$. In this situation, it appears that it is acceptable to have expected frequencies as small as 1.0 for testing at α as small as 0.05, or as small as 2.0 for α as small as 0.01. The chi-square test works nearly as well when there is moderate departure from a uniform distribution in H_0, and the average expected frequencies may be as small as those indicated for a uniform distribution. And even with great departure from uniform it appears that the average expected frequency (i.e., n/k) may be as low as 2.0 for testing at α as low as 0.05 and as low as 4.0 for α as small as 0.01. Koehler and Larntz (1980) suggest that the chi-square test is applicable for situations where $k \geq 3, n \geq 10$, and $n^2/k \geq 10$. Users of goodness of fit testing can be comfortable if their data fit both the Roscoe and Byars and the Koehler and Larntz guidelines. These recommendations are for situations where there are more than 2 categories. If $k = 2$, then it is wise to have $\hat{f}_i$'s of at least 5.0, or to use the binomial test as indicated in the next paragraph.

The chi-square calculation can be made if the data for the classes with the offensively low $\hat{f}_i$ values are simply eliminated from H_0 and the subsequent analysis. Or, certain of the classes of data might be meaningfully combined so as to result in all $\hat{f}_i$ values being large enough to proceed with the analysis. Such modified procedures are not to be recommended as routine practice. Rather, the experimenter should strive to obtain a sufficiently large n for the analysis to be performed. When $k = 2$ and each f_i is small, the use of the binomial test (Section 24.6) is preferable to chi-square analysis. [Similarly, use of the multinomial, rather than the binomial, distribution is appropriate when $k > 2$ and the f_i's are small; however, this is a tedious procedure and will not be demonstrated here (Radlow and Alf, 1975).]

---

*Some statisticians were even stricter, recommending a lower limit of 10.0.

22.6 HETEROGENEITY CHI-SQUARE

It is sometimes the case that a number of sets of data are being tested against the same null hypothesis, and we wish to decide whether we may combine all of the sets in order to perform one overall chi-square analysis. As an example, let us examine some of the classic data of Gregor Mendel*(1865). In one series of 10 experiments, Mendel obtained yellow and green pea seeds in the frequencies shown in Example 22.5. The data from each of the ten samples are tested against the null hypothesis that there is a 3:1 ratio of yellow-to-green seeds in the population from which the sample came.[†] Since H_0 is not rejected in any of the ten cases, it is reasonable to test another null hypothesis, that all ten samples could have, in fact, come from the same population. This new hypothesis may be tested by the procedure called *heterogeneity chi-square* analysis (sometimes referred to as "interaction" chi-square analysis or even "homogeneity analysis"). In addition to performing the ten separate chi-squares tests, we total all ten f_i values and total all ten $\hat{f}_i$ values and perform a chi-square test on these totals. But in totaling these values, commonly called *pooling* them, we must assume that all ten samples came from the same population (or from populations having identical seed-color ratios). If this assumption is true, we say that the samples are *homogeneous*. If this assumption is false, the samples are said to be *heterogeneous*, and the chi-square analysis on the pooled data would not be justified. So we are faced with the desirability of testing for heterogeneity, using the null hypothesis that the samples could have come from the same population (i.e., H_0: The samples are homogeneous).

Testing for heterogeneity among replicated goodness of fit tests is based on the fact that the sum of chi-square values is itself a chi-square value. If the samples are indeed homogeneous, then the total of the individual chi-square values should be close to the chi-square for the total frequencies. In Example 22.5, the total chi-square is 7.1899, with a total of 10 DF; and the chi-square of the totals is 0.1367, with 1 DF. The difference between these two chi-squares is itself a chi-square (called the *heterogeneity chi-square*), 7.053, with DF = 10 − 1 = 9.

Consulting Appendix Table B.1, we see that for the heterogeneity chi-square, $\chi^2_{0.05,9} = 16.9$, so H_0 is not rejected. Thus we conclude that the ten samples could have come from the same population and that their frequencies might justifiably be pooled. The Yates correction for continuity may not be applied in a heterogeneity chi-square analysis (Cochran, 1942; Lancaster, 1949). But if we conclude that the sample data may be pooled, we should then analyze these pooled data using the correction for continuity. Thus, for Example 22.5, $\chi^2_c = 0.128$, rather than $\chi^2 = 0.137$, should be used, once it has been determined that the data may be pooled.

---

*Gregor Johann Mendel (1822–1884), Augustinian monk and Austrian school teacher, was a pioneer in biological experimentation and its quantitative analysis—although his data have been called into question by statisticians (Edwards, 1986; Fisher, 1936). His research was unappreciated until sixteen years after his death.

[†]These ten data sets come from what Mendel (1865) collectively called "Experiment 2," in which, he reports, "258 plants yielded 8023 seeds, 6022 yellow and 2001 green; their ratio, therefore, is 3.01:1."

EXAMPLE 22.5   An example of heterogeneity chi-square analysis. The seed color frequencies resulting from pea breeding experiments of Gregor Mendel (1865).

The null hypothesis for each experiment is that the population sampled has a 3:1 ratio of yellow-to-green seeds.

The null hypothesis for heterogeneity chi-square testing is that all ten samples could have come from the same population (i.e., there is homogeneity).

For each experiment, the observed frequencies, f_i, are given, with the frequencies predicted by the null hypothesis, $\hat{f}_i$, in parentheses.

Experiment	Yellow Seeds	Green seeds	Total seeds (n)	Uncorrected chi-square*	DF
1	25 (27.0000)	11 (9.0000)	36	0.5926	1
2	32 (29.2500)	7 (9.7500)	39	1.0342	1
3	14 (14.2500)	5 (4.7500)	19	0.0175	1
4	70 (72.7500)	27 (24.2500)	97	0.4158	1
5	24 (27.7500)	13 (9.2500)	37	2.0270	1
6	20 (19.5000)	6 (6.5000)	26	0.0513	1
7	32 (33.7500)	13 (11.2500)	45	0.3630	1
8	44 (39.7500)	9 (13.2500)	53	1.8176	1
9	50 (48.0000)	14 (16.0000)	64	0.3333	1
10	44 (46.5000)	18 (15.5000)	62	0.5376	1
Total of chi-squares				7.1899	10
Chi-square of totals (i.e., pooled chi-square)	355 (358.5000)	123 (119.5000)	478	0.1367	1
Heterogeneity chi-square				7.0532	9

$\chi^2_{0.05,9} = 16.1919$.

Do not reject the homogeneity null hypothesis. $0.50 < P < 0.75$ [$P = 0.63$]

*In heterogeneity analysis, chi-square is computed *without* correction for continuity.

Example 22.6 demonstrates how one can be misled by pooling heterogeneous samples without testing for acceptable homogeneity. If the six samples shown were pooled blindly, and a chi-square computed ($\chi^2 = 0.2336$), one would not reject the null hypothesis. But such a procedure would have ignored the strong indication obtainable from the heterogeneity analysis ($P < 0.001$) that the samples came from at least two different populations. The appearance of the data in this example suggests that Samples 1, 2, and 3 came from one population, and Samples 4, 5, and 6 came from another, possibilities that can be reexamined with new data.

EXAMPLE 22.6   Hypothetical data for heterogeneity chi-square analysis, demonstrating misleading results from the pooling of heterogeneous samples.

H_0:   The sample population has a 1:1 ratio of right- to left-handed men.

H_A: The sampled population does not have a 1:1 ratio of right- to left-handed men.

Sample frequencies observed, f_i, are listed, with the frequencies predicted by H_0 ($\hat{f_i}$) in parentheses.

Sample	Right-handed	Left-handed	n	Uncorrected chi-square	DF
1	3	11	14	4.5714	1
	(7.0000)	(7.0000)			
2	4	12	16	4.0000	1
	(8.0000)	(8.0000)			
3	5	15	20	5.0000	1
	(10.0000)	(10.0000)			
4	14	4	18	5.5556	1
	(9.0000)	(9.0000)			
5	13	4	17	4.7647	1
	(8.5000)	(8.5000)			
6	17	5	22	6.5455	1
	(11.0000)	(11.0000)			

Total of chi-squares				30.4372	6
Chi-square of totals	56	51	107	0.2336	1
(i.e., pooled chi-square)	(53.5000)	(53.5000)			
Heterogeneity chi-square				30.2036	5

$\chi^2_{0.05,5} = 11.070$.

Reject H_0 for homogeneity. $P < 0.001$ $[P = 0.000013]$

Therefore, one is not justified in performing a goodness of fit analysis on the pooled data.

It is also important to realize that the pooling of homogeneous data may result in a more powerful analysis. Example 22.7 presents hypothetical data for four replicate chi-square analyses. None of the individual chi-square tests detects a significant deviation from the null hypothesis; but on pooling them, the chi-square test performed on the larger number of data is able to reject H_0. The nonsignificant heterogeneity chi-square shows that we are justified in pooling the replicates in order to analyze a single set of data with a large n.

22.7 THE LOG-LIKELIHOOD RATIO

The *log-likelihood ratio* may be used for analyses of the sort for which one would employ chi-square. The log-likelihood ratio,* $\sum f_i \ln(f_i / \hat{f_i})$, may also be written as $\sum f_i \ln f_i - \sum f_i \ln \hat{f}$. Twice this quantity, a value called G, approximates the χ^2

---

*Proposed by Wilks (1935), based upon concepts of Neyman and Pearson (1928a, 1928b).

EXAMPLE 22.7   Hypothetical data for heterogeneity chi-square analysis, demonstrating how nonsignificant sample frequencies can result in significant pooled frequencies.

For each sample, and for the pooled sample:

H_0:   The sampled population has equal frequencies of right- and left-handed men.

H_A: The sampled population does not have equal frequencies of right- and left-handed men.

For heterogeneity testing:

H_0:   All the samples came from the same population.

H_A: The samples came from at least two different populations.

For each sample, the observed frequencies, f_i, are given, together with the expected frequencies, $\hat{f}_i$ in parentheses.

Sample	Right-handed	Left-handed	n	Uncorrected chi-square	DF
1	15 (11.0000)	7 (11.0000)	22	2.9091	1
2	16 (12.0000)	8 (12.0000)	24	2.6667	1
3	12 (8.5000)	5 (8.5000)	17	2.8824	1
4	13 (9.0000)	5 (9.0000)	18	3.5556	1
Total of chi-squares				12.0138	4
Chi-square of totals (pooled chi-square)	56 (40.5000)	25 (40.5000)	81	11.8642	1
Heterogeneity chi-square				0.1496	3

$\chi^2_{0.05,3} = 7.815$.

The homogeneity H_0 is not rejected. $0.975 < P < 0.99$ $[P = 0.985]$

Therefore, one is justified in pooling the four sets of data. On doing so, $\chi^2_c = 11.111$, DF $= 1$.

distribution.* Thus[†]

$$G = 2 \sum f_i \ln \frac{f_i}{\hat{f}_i} = 4.60517 \sum f_i \log \frac{f_i}{\hat{f}_i}, \tag{22.6}$$

or, equivalently,

$$G = 2 \left[\sum f_i \ln f_i - \sum f_i \ln \hat{f}_i \right] = 4.60517 \left[\sum f_i \log f_i - \sum f_i \log \hat{f}_i \right] \tag{22.7}$$

is applicable as a test for goodness of fit, utilizing Appendix Table B.1 with the same number of degrees of freedom as would be used for chi-square testing. Example 22.8

---

*G sometimes appears in the literature as G^2.

[†]As noted in the Section 8.7 footnote, "ln" refers to natural logarithm (in base e) and "log" to common logarithm (in base 10).

EXAMPLE 22.8 Calculation of the G statistic for the log-likelihood ratio goodness of fit test. The data and the hypotheses are the same as in Example 22.2.

	Yellow smooth	Yellow wrinkled	Green smooth	Green wrinkled	n
f_i	152	39	53	6	250
$(\hat{f_i})$	(140.6250)	(46.8750)	(46.8750)	(15.6250)	

$\nu = k - 1 = 3$

$G = 4.60517 \left[\sum f_i \log f_i - \sum f_i \log \hat{f_i} \right]$

$= 4.60517[(152)(2.18184) + (39)(1.59106) + (53)(1.72428)$

$\qquad + (6)(0.77815) - (152)(2.14806) - (39)(1.67094)$

$\qquad - (53)(1.67094) - (6)(1.19382)]$

$= 4.60517[331.63968 + 62.05134 + 91.38684 + 4.66890$

$\qquad - 326.50512 - 65.16666 - 88.55982 - 7.16292]$

$= 4.60517[2.35224]$

$= 10.832$

$\chi^2_{0.05,3} = 7.815$

Therefore, reject H_0.

$$0.01 < P < 0.025 \quad [P = 0.013]$$

---

demonstrates the G test for the data of Example 22.2. In this case, the same conclusion is reached using G and χ^2, but this will not always be so.

Williams (1976) recommends G be used in preference to χ^2 whenever any $|f_i - \hat{f_i}| \geq \hat{f_i}$. The two methods often yield the same conclusions, especially when n is large; when they do not, some statisticians prefer G; others recommend χ^2, for while G may result in a more powerful test in some cases, χ^2 tends to provide a test that operates closer to the stated level of α (e.g., see Chapman, 1976; Hutchinson, 1979; Larntz, 1978; Moore, 1986).

When $\nu = 1$, the Yates correction for continuity is applied in a fashion analogous to that in chi-square analysis in Section 22.4. The procedure is to make each f_i closer to $\hat{f_i}$ by 0.5 and to apply Equation 22.7 (or Equation 22.6) using these modified f_i's. This is demonstrated in Example 22.9.

22.8 KOLMOGOROV-SMIRNOV GOODNESS OF FIT FOR DISCRETE DATA

The preceding sections dealt with goodness of fit tests applicable to nominal scale or ordinal scale data. Sections 22.8 and 22.9 present goodness of fit testing that is better for data in ordered categories.

EXAMPLE 22.9   **The G test for goodness of fit for two categories, for the hypotheses and data of Example 22.1.**

(a) Without the Yates correction for continuity:

Category (flower color)

	Yellow	Green	n
f_i	84	16	100
$(\hat{f}_i)$	(75)	(25)	

$\nu = k - 1 = 2 - 1 = 1$

$G = 4.60517[(84)(1.92428) + (16)(1.20412) - (84)(1.87506) - (16)(1.39794)]$

$\quad = 4.60517[1.03336] = 4.759$

$\chi^2_{0.05, 1} = 3.841$

Therefore, reject H_0.

$$0.025 < P < 0.05 \quad [P = 0.029]$$

(b) With the Yates correction for continuity:

Category (flower color)

	Yellow	Green	n
f_i	84	16	100
$(\hat{f}_i)$	(75)	(25)	
Modified f_i	83.5	16.5	

$\nu = k - 1 = 2 - 1 = 1$

$G_c = 4.60517[(83.5)(1.92169) + (16.5)(1.21748) - (83.5)(1.87506) - (16.5)(1.39794)]$

$\quad = 4.60517[0.916015] = 4.218$

$\chi^2_{0.05, 1} = 3.841$

Therefore, reject H_0.

$$0.025 < P < 0.05 \quad [P = 0.040]$$

In Example 22.10, an experiment was performed in which thirty-five cats were given a choice of five identical containers, each holding a food differing from the others only in moisture content. The experiment was performed in a fashion that ensured that all thirty-five observations were independent: a total of thirty-five cats was used (not, for example, seven cats with each being given five opportunities to choose food); and the cats were tested one at a time, so that no individual's actions would influence another's. The data are observed frequencies, f_i, the numbers of animals choosing each of the five food moistures. Food moisture is recorded on an ordinal scale, for although we can

---

EXAMPLE 22.10   Kolmogorov-Smirnov goodness of fit for discrete ordinal scale data.

H_0:   Cats show no preference among the five food moisture contents (i.e., cats prefer all five equally).

H_A: Cats do show preference among the five food moisture contents (i.e., cats do not prefer all five equally).

Moisture class (X_i)

	(Moist)				*(Dry)*	
	1	2	3	4	5	n
f_i	8	13	6	6	2	35
$\hat{f}_i$	7	7	7	7	7	35
F_i	8	21	27	33	35	
$\hat{F}_i$	7	14	21	28	35	
$|d_i|$	1	7	6	5	0	

$d_{\max} = \text{maximum } |d_i| = 7$

$(d_{\max})_{0.05, 5, 35} = 7$

Therefore, reject H_0.

$$0.02 < P < 0.05$$

---

say that food 1 is moister than food 2 and food 2 is moister than food 3, we cannot say that the difference in moisture between foods 2 and 3 is quantitatively equal to the difference between 1 and 2. That is, we can only speak of relative magnitudes, and not of quantitative measurements, of the foods' moisture contents.

The Kolmogorov-Smirnov goodness of fit test[*] (Kolmogorov, 1933; Smirnov, 1939a,b), also called the Kolmogorov-Smirnov one-sample test, is suitable for assessing goodness of fit of an observed to an expected cumulative frequency distribution. (Section 1.4 introduces the concept of a cumulative frequency distribution.) In Example 22.10, each observed frequency (f_i) is listed with the frequency expected ($\hat{f}_i$) under the null hypothesis that the two distributions are the same, the expected frequencies ($\hat{f}_i$) being calculated just as they would be in a chi-square goodness of fit analysis (Sections 22.1–22.2). The cumulative observed frequencies (F_i) and cumulative expected frequencies ($\hat{F}_i$) are then calculated. The cumulative frequency for category i is the sum of all frequencies from categories 1 through i.

For each category, i, one determines the absolute difference between the two cumulative frequency distributions:

$$|d_i| = |F_i - \hat{F}_i|. \tag{22.8}$$

The largest $|d_i|$, let us call it $d_{\max}$, is the test statistic.

---

[*]The name of the test honors the two Russian mathematicians who worked on its development: Andrei Nikolaevich Kolmogorov (1903–1987) and Nikolai Vasil'evich Smirnov (1900–1966).

Critical values of d_{max} are found in Appendix Table B.8, the use of which requires that in the experiment n, the total number of data, is an even multiple of k, the number of categories.* It should also be noted that the tabled critical values are for the case where all $\hat{f}_i$ are equal; but the table also works well for unequal $\hat{f}_i$, as long as the inequality is not great (Pettitt and Stephens, 1977).

When applicable (i.e., when the categories are ordered), the Kolmogorov-Smirnov test is more powerful than the chi-square test when n is small or when $\hat{f}_i$ values are small, and often in other cases. This difference in power is shown by the data in Example 22.10, for which the Kolmogorov-Smirnov test rejects the null hypothesis at the 5% significance level, whereas χ^2 (which would be calculated to be 9.143 and have 4 DF) would not. Furthermore, results of the Kolmogorov-Smirnov procedure depend upon the order of the categories. If the observed frequencies in Example 22.10 would have been in the sequence 8, 2, 6, 13 and 6, instead of as shown, the computed χ^2 would have been the same but d_{max} would have been 5 instead of 7 and H_0 would not have been rejected at $\alpha = 0.05$.

22.9 KOLMOGOROV-SMIRNOV GOODNESS OF FIT FOR CONTINUOUS DATA

The Kolmogorov-Smirnov goodness of fit procedure was originally developed for use with continuous data (which may be on ratio, interval, or ordinal scales of measurement), instead of with discrete data. Example 22.11 presents the results of data collection where the vertical locations of fifteen moths on a 25-meter tree trunk were recorded as heights above the ground. The vertical distances above the ground are the measurements, X_i, that come from a continuous distribution. For each X_i we record the observed frequency, f_i (i.e., the number of moths at that height). The cumulative observed frequencies, F_i, are then determined from which the cumulative relative frequencies are obtained as

$$\text{rel } F_i = \frac{F_i}{n}, \tag{22.9}$$

where n is the number of measurements taken; rel F_i is simply the proportion of the sample being measurements $\leq X_i$. Then for each X_i we determine the cumulative relative expected frequency, rel $\hat{F}_i$. In the present example, H_0 proposes a uniform distribution over the heights 0 to 25 meters, so rel $\hat{F}_i = X_i/25$ m. [If, for example, H_0 had referred to a uniform distribution over heights 1 to 25 m from the ground, then rel $\hat{F}_i$ would have been $(X_i - 1$ m$)/24$ m.]

To find the test statistic for the Kolmogorov-Smirnov goodness of fit for continuous data, we must calculate both

$$D_i = |\text{rel } F_i - \text{rel } \hat{F}_i| \tag{22.10}$$

---

*If n/k is not a whole number, then we can consult the critical values for the n above and the n below the n in the experiment and, conservatively, use the larger of the two critical values.

EXAMPLE 22.11   Kolmogorov-Smirnov goodness of fit for continuous ratio scale data, vertical distribution of moths on tree trunks.

H_0:   Moths are distributed uniformly from ground level to height of 25 m.

H_A: Moths are not distributed uniformly from ground level to height of 25 m.

Each X_i is a height (in meters) at which a moth was observed on a tree trunk.

i	X_i	f_i	F_i	rel F_i	rel $\hat{F}_i$	D_i	D'_i
1	1.4	1	1	0.0667	0.0560	0.0107	0.0560
2	2.6	1	2	0.1337	0.1040	0.0297	0.0373
3	3.3	1	3	0.2000	0.1320	0.0680	0.0017
4	4.2	1	4	0.2667	0.1680	0.0987	0.0320
5	4.7	1	5	0.3333	0.1880	0.1453	0.0787
6	5.6	2	7	0.4667	0.2240	0.2427	0.1093
7	6.4	1	8	0.5333	0.2560	0.2773	0.2107
8	7.7	1	9	0.6000	0.3080	0.2920	0.2253
9	9.3	1	10	0.6667	0.3720	0.2947	0.2280
10	10.6	1	11	0.7333	0.4240	0.3093	0.2427
11	11.5	1	12	0.8000	0.4600	0.3400	0.2733
12	12.4	1	13	0.8667	0.4960	0.3707	0.3040
13	18.6	1	14	0.9333	0.7440	0.1893	0.1227
14	22.3	1	15	1.0000	0.8920	0.1080	0.0413

$n = 15$

max $D_i = D_{12} = 0.3707$

max $D'_i = D'_{12} = 0.3040$

$D = 0.3707$

$D_{0.05, 15} = 0.33760$

Therefore, reject H_0.

$$0.02 < P < 0.05$$

---

and

$$D'_i = |\text{rel } F_{i-1} - \text{rel } \hat{F}_i| \tag{22.11}$$

for each i. For Equation 22.11, it is important to know that $F_0 = 0$, so $D'_1 = \text{rel } \hat{F}_i$. The test statistic is

$$D = \max[(\max D_i), (\max D'_i)], \tag{22.12}$$

which means "D is the largest D_i or the largest D'_i, whichever is larger." Critical values for this test statistic are referred to as $D_{\alpha,n}$ and are found in Appendix Table B.9. If $D \geq D_{\alpha,n}$, the H_0 is rejected at the α level of significance.

Figure 22.1 demonstrates why both D_i and D_i' are necessarily examined in comparing an observed with a hypothesized cumulative frequency distribution, when the measurement scale is continuous. (See also D'Agostino and Noether, 1973; Fisz, 1963: 12.5A; Gibbons, 1976: Section 3.1). What is required is the maximum deviation between the observed distribution, F (which looks like a staircase when graphed), and the hypothesized distribution, $\hat{F}$. For each $\hat{F}_i$, we must examine the vertical distance, $D_i = |F_i - \hat{F}_i|$, at the left-hand end of each step, as well as the vertical distance, $D_i' = |F_{i-1} - \hat{F}_i|$, at the right-hand end of each step.

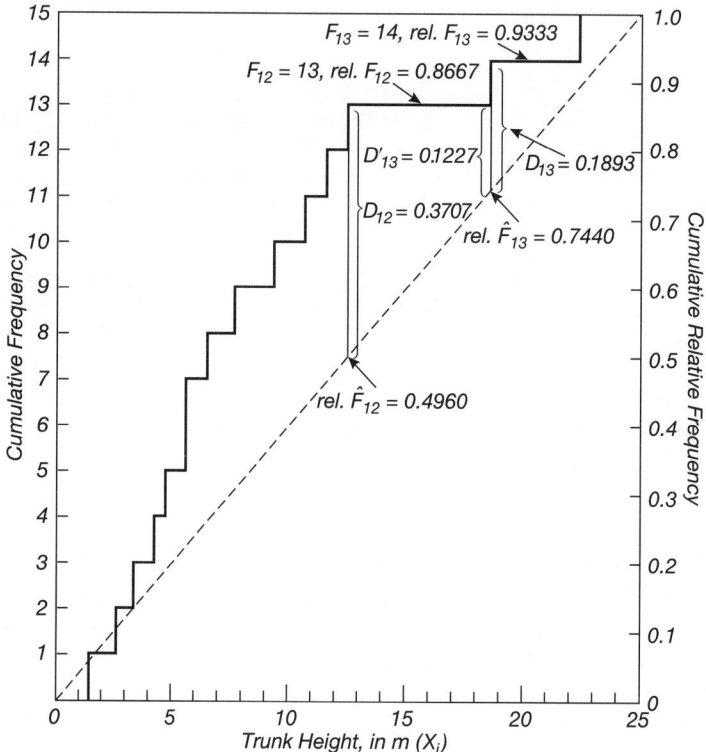

Figure 22.1   Graphical representation of Example 22.11, Kolmogorov-Smirnov goodness of fit testing for continuous data.

A lesser-known, but quite good, test may be used as an alternative to the Kolmogorov-Smirnov test for goodness of fit with continuous data. That is the Watson goodness of fit test; it is discussed in Section 27.6 as being especially suited for data on a circular scale, but it is applicable as well to data such as in the present section.

Correction for Increased Power.   For small sample sizes (say, $n \leq 25$), the power of Kolmogorov-Smirnov testing can be increased impressively by employing the correction expounded by Harter, Khamis, and Lamb (1984) and Khamis (1990, 1993). For each i we determine

$$\text{rel } \mathcal{F}_i = \frac{F_i}{n+1} \tag{22.13}$$

and

$$\text{rel } \mathcal{F}'_i = \frac{F_i - 1}{n-1}. \tag{22.14}$$

Then, we obtain differences similar to D_i and D'_i of Equations 22.10 and 22.11, respectively:

$$D_{0,i} = \left| \text{rel } \mathcal{F}_i - \text{rel } \hat{F}_i \right| \tag{22.15}$$

$$D_{1,i} = \left| \text{rel } \mathcal{F}'_i - \text{rel } \hat{F}_i \right|. \tag{22.16}$$

For these statistics, the subscripts 0 and 1 are referred to as δ (lowercase Greek delta), so the developers of this procedure call it the δ-corrected Kolmogorov-Smirnov goodness of fit test.

The test statistic is either max $D_{0,i}$ or max $D_{1,i}$, whichever leads to the highest level of significance (i.e., the smallest probability). Appendix Table B.10 gives critical values for $D_{\delta,n}$ for various levels of α. This test is demonstrated in Example 22.12. Although in this example the conclusion is the same as with the uncorrected Kolmogorov-Smirnov test (Example 22.11), this is not necessarily so. However, Khamis (1990) reports that if $n > 20$ the results of this corrected Kolmogorov-Smirnov method are practically indistinguishable from those from the uncorrected procedure, and in such cases either the uncorrected or corrected test may be used.

Grouped Data. If continuous data are collected such that $f_i > 1$, then the Kolmogorov-Smirnov test becomes conservative (meaning that the testing is occurring at an α smaller than that which we state; and the probability of a Type II error is inflated). In such cases, the appropriate analysis resembles that for discrete data and the procedures of Section 22.8 should be used. Example 22.13 shows how the data of Example 22.11 would look had the investigator recorded them in five-meter ranges of trunk heights. Note that power is lost (and H_0 is not rejected) by grouping the data, and grouping should be avoided whenever possible.

22.10 SAMPLE SIZE REQUIRED IN KOLMOGOROV-SMIRNOV GOODNESS OF FIT FOR CONTINUOUS DATA

When dealing with continuous data, we may ask how large a sample is needed to be able to detect a difference of a given magnitude between an observed and a hypothesized cumulative frequency distribution. All that need be done is to seek the desired minimum

EXAMPLE 22.12 The δ-corrected Kolmogorov-Smirnov goodness to fit test for the data of Example 22.11.

H_0:   Moths are distributed uniformly from ground level to a height of 25 m.

H_A: Moths are not distributed uniformly from ground level to a height of 25 m.

i	X_i	F_i	rel $\hat{F}_i$	rel $\mathcal{F}_i$	$D_{0,i}$	rel $\mathcal{F}_i'$	$D_{1,i}$
1	1.4	1	0.0560	0.0625	0.0065	0.0000	0.0560
2	2.6	2	0.1040	0.1250	0.0210	0.0714	0.0326
3	3.3	3	0.1320	0.1875	0.0555	0.1429	0.0109
4	4.2	4	0.1680	0.2500	0.0820	0.2143	0.0463
5	4.7	5	0.1880	0.3125	0.1245	0.2857	0.0977
6	5.6	7	0.2240	0.4375	0.2135	0.4286	0.2046
7	6.4	8	0.2560	0.5000	0.2440	0.5000	0.2440
8	7.7	9	0.3080	0.5625	0.2545	0.5714	0.2634
9	9.3	10	0.3720	0.6250	0.2530	0.6429	0.2709
10	10.6	11	0.4240	0.6875	0.2635	0.7143	0.2903
11	11.5	12	0.4600	0.7500	0.2900	0.7857	0.3257
12	12.4	13	0.4960	0.8125	0.3165	0.8571	0.3611
13	18.6	14	0.7440	0.8750	0.1310	0.9286	0.1846
14	22.3	15	0.8920	0.9375	0.0455	1.0000	0.1080

$n = 15$

max $D_{0,i} = D_{0,12} = 0.3165$, which has a probability of $0.05 < P < 0.10$

max $D_{1,i} = D_{1,12} = 0.3611$, which has a probability of $0.02 < P < 0.05$

Therefore, reject H_0; $0.02 < P < 0.05$.

EXAMPLE 22.13 Kolmogorov-Smirnov goodness of fit for continuous, but grouped, data.

The data of Example 22.11, recorded in five-meter segments of tree trunk height. H_0 and H_A as in Example 22.11.

	Trunk Height					
X_i	0–5 m	5–10 m	10–15 m	15–20 m	20–25 m	n
f_i	5	5	3	1	1	15
$\hat{f}_i$	3	3	3	3	3	15
F_i	5	10	13	14	15	
$\hat{F}_i$	3	6	9	12	15	
$|d_i|$	2	4	4	2	0	

d_{max} = maximum $|d_i| = 4$

$(d_{max})_{0.05, 5, 15} = 5$

Do not reject H_0.

$$0.05 < P < 0.10$$

detectable difference in the body of the table of critical values of D (Appendix Table B.9), for the selected significance level, α. For example, to be able to detect a difference as small as 0.30 between an observed and a hypothesized cumulative relative frequency distribution, at a significance level of 0.05, we need a sample of at least 20 (for $D_{0.05,19} = 0.30143$, which is larger than 0.30; and $D_{0.05,20} = 0.29516$, which is smaller than 0.30; and if the desired difference is not on the table we use the nearest smaller one). Thus, in Example 22.11, we estimate that at least twenty moths would have had to be observed to have been able to detect a difference—either D_i or D_i'—as small as 0.30. And, with $n = 20$, this means that $(0.30)(20) = 6$ is the smallest difference between cumulative frequency distributions—$|d_i|$ or $|d_i'|$—that could be detected.

If the desired detectable difference is beyond the $D_{\alpha,n}$ values in Appendix Table B.9 (i.e., $< D_{\alpha,160}$), then we know that the required sample size is greater than 160. This sample size may be estimated by employing the values of d_α at the end of Appendix Table B.9.* If we wish to detect a difference as small as Δ, then the sample size should be at least

$$n = \frac{d_\alpha^2}{\Delta^2}. \tag{22.18}$$

For example, if the collector of data in Example 22.11 had desired to be able to detect a difference, D_i or D_i', as small as 0.10, a sample size of at least 185 moth observations should have been obtained, for

$$n = \frac{(1.35810)^2}{(0.10)^2}$$

$$= 184.4.$$

EXERCISES

22.1 Consult Appendix Table B.1
   (a) What is the probability of computing a χ^2 at least as large as 3.452, if DF $= 2$ and the null hypothesis is true?
   (b) What is $P(\chi^2 \geq 8.668)$ if $\nu = 5$?
   (c) What is $\chi^2_{0.05,4}$?
   (d) What is $\chi^2_{0.01,8}$?
22.2 Each of 126 individuals of a certain mammal species was placed in an enclosure containing equal amounts of each of six different foods. The frequency with which the animals chose

---

*These values at the end of Appendix Table B.9 are

$$d_\alpha = D_{\alpha,n} \sqrt{n}, \tag{22.17}$$

which are asymptotic as n becomes very large.

each of the foods was:

Food item (i)	f_i
1	13
2	26
3	31
4	14
5	28
6	14

(a) Test the hypothesis that there is no preference among the food items.

(b) If the null hypothesis is rejected, ascertain which of the foods were preferred by the animals.

22.3 A sample of hibernating bats consisted of forty-four males and fifty-four females. Test the hypothesis that the hibernating population consists of equal numbers of males and females.

22.4 In attempting to determine whether there is a 1:1 sex ratio among hibernating bats, samples were taken from four different locations in a cave:

Location	Males	Females
1	44	54
2	31	40
3	12	18
4	15	16

By performing a heterogeneity chi-square analysis, determine whether the four samples may justifiably be pooled. If they may, pool them and retest the null hypothesis of equal sex frequencies.

22.5 Test the hypothesis and data of Exercise 22.2 using the log-likelihood G.

22.6 A bird feeder is placed at each of six different heights. It is recorded which feeder was selected by each of eighteen cardinals. Using the Kolmogorov-Smirnov procedure for discrete data, test the null hypothesis that each feeder height is equally desirable to cardinals.

Feeder height	Number observed
1 (lowest)	2
2	3
3	3
4	4
5	4
6 (highest)	2

22.7 A straight line is drawn on the ground perpendicular to the shore of a body of water. Then the locations of ground arthropods of a certain species are measured along a one-meter-wide band on either side of the line. Use the Kolmogorov-Smirnov procedure on the following data to test the null hypothesis of uniform distribution of this species from the water's edge to a distance of ten meters inland.

Distance from water (m)	Numbers observed	Distance from water (m)	Numbers observed
0.3	1	3.4	1
0.6	1	4.8	1
1.0	1	4.9	1
1.1	1	4.1	1
1.2	1	4.6	1
1.4	1	4.7	1
1.6	1	4.9	1
1.9	1	5.3	1
2.1	1	5.8	1
2.2	1	6.4	1
2.4	1	6.8	1
2.6	1	7.5	1
2.8	1	7.7	1
3.0	1	8.8	1
3.1	1	9.4	1

22.8 For a Kolmogorov-Smirnov goodness of fit test with continuous data at the 5% level of significance, how large a sample is necessary to detect a difference as small as 0.25 between cumulative relative frequency distributions?

23

CONTINGENCY TABLES

In many situations, enumeration data are collected simultaneously for two variables, and it is desired to test the hypothesis that the frequencies of occurrence in the various categories of one variable are independent of the frequencies in the second variable. In Example 23.1, such data are tabulated for frequencies of observations of each of four hair colors and each of two sexes. These data are said to be arranged in a *contingency table*.

The number of columns in a contingency table is denoted by c and the number of rows by r, and there are said to be $r \times c$ "cells" in the table. In Example 23.1, $r = 2$ and $c = 4$, so we are said to be considering a 2×4 (read as "two by four") contingency table (i.e., a table with eight cells). Note that the data might have been tabulated just as well as a 4×2 table, having the categories of hair color being the rows and the categories of sex being the columns. But this would in no way change the statistical hypotheses, tests, or conclusions that follow.

The null hypothesis for contingency table testing is that the frequencies of observations found in the rows are independent of the frequencies of observations found in the columns (or, that the column frequencies are independent of the row frequencies). For Example 23.1, H_0 could be stated "hair color relative frequencies are the same for both sexes," which is the same as saying "relative frequencies of males and females (i.e., sex ratios) are the same for each hair color." Then H_A could be stated "the hair color relative frequencies are not the same for males and females," or "the sex ratios are not the same for persons of all hair colors."

It should be realized that there are three ways in which the data of Example 23.1 could have been collected. We might have randomly selected 100 men and tabulated their hair color, and also randomly selected 200 women and tabulated their hair color. In this fashion we would have set, in advance of data collection, the row totals in the

---

EXAMPLE 23.1   A 2×4 contingency table for testing the independence of hair color and sex in humans.

H_0:   Human hair color is independent of sex in the population sampled.

H_A: Human hair color is not independent of sex in the population sampled.

$\alpha = 0.05$

The observed frequency, f_{ij}, in each cell is shown, with the frequency expected if H_0 is true (i.e., $\hat{f}_{ij}$) in parentheses.

Hair color

Sex	*Black*	*Brown*	*Blond*	*Red*	**Total**
Male	32	43	16	9	100 ($= R_1$)
	(29.0000)	*(36.0000)*	*(26.6667)*	*(8.3333)*	
Female	55	65	64	16	200 ($= R_2$)
	(58.0000)	*(72.0000)*	*(53.3333)*	*(16.6667)*	
Total	87	108	80	25	300 ($= n$)
	($= C_1$)	($= C_2$)	($= C_3$)	($= C_4$)	

$$\chi^2 = \sum\sum \frac{(f_{ij} - \hat{f}_{ij})^2}{\hat{f}_{ij}}$$

$$= \frac{(32 - 29.0000)^2}{29.0000} + \frac{(43 - 36.0000)^2}{36.0000} + \frac{(16 - 26.6667)^2}{26.6667} + \frac{(9 - 8.3333)^2}{8.3333}$$

$$+ \frac{(55 - 58.0000)^2}{58.0000} + \frac{(65 - 72.0000)^2}{72.0000} + \frac{(64 - 53.3333)^2}{53.3333} + \frac{(16 - 16.6667)^2}{16.6667}$$

$$= 0.3103 + 1.3611 + 4.2667 + 0.0533 + 0.1552 + 0.6806 + 2.1333 + 0.0267 = 8.987$$

$$\nu = (r - 1)(c - 1) = (2 - 1)(4 - 1) = 3$$

$$\chi^2_{0.05, 3} = 7.815$$

Therefore, reject H_0.

$$0.025 < P < 0.05 \quad [P = 0.029]$$

---

contingency table. Or, we might have decided to observe a random sample of 87 black-haired, 108 brown-haired, 80 blond, and 25 red-headed people, then recording the sex of each person. This would have been an experimental design that set the column totals in advance. Or, we might have selected 300 people at random and then recorded how many of them were black-haired men, black-haired women, brown-haired men, and so on. This would have involved setting, in advance, the total number of data but not the totals in either of the two margins (row totals or column totals).* The analysis proceeds in the same fashion for all of these sampling schemes.

Sections 23.8 and 23.9 will introduce procedures for analyzing contingency tables of more than two dimensions, where frequencies are tabulated simultaneously for more than two variables.

---

*The first two of these three data-collection schemes represent the "Category 2" experimental design to be discussed at the beginning of Section 23.3, and the third is an example of a "Category 1" design.

23.1 CHI-SQUARE ANALYSIS OF CONTINGENCY TABLES

The most common procedure for analyzing contingency table data is by using the chi-square statistic.* Recall that for the computation of chi-square one utilizes observed and expected frequencies (and never proportions or percentages). For the goodness of fit analysis introduced in Section 22.1, f_i denoted the frequency observed in category i of the variable under study. In a contingency table, we have two variables under consideration, and we denote an observed frequency as f_{ij}. By means of the double subscript, f_{ij} refers to the frequency observed in row i and column j of the contingency table. In Example 23.1, the value in row 1 column 1 is denoted as f_{11}, that in row 2 column 3 as f_{23}, and so on. Thus, $f_{11} = 32$, $f_{12} = 43$, $f_{13} = 16, \ldots, f_{23} = 64$, and $f_{24} = 16$.

The total frequency in row i of the table is denoted as R_i and is obtained as $R_i = \sum_{j=1}^{c} f_{ij}$. Thus, $R_1 = f_{11} + f_{12} + f_{13} + f_{14} = 100$, which is the total number of males in the sample, and $R_2 = f_{21} + f_{22} + f_{23} + f_{24} = 200$, which is the total number of females in the sample. The column totals, C_j, are obtained by analogous summations: $C_j = \sum_{i=1}^{r} f_{ij}$. For example, the total number of blonds in the sample data is $C_3 = \sum_{i=1}^{2} f_{i3} = f_{13} + f_{23} = 80$, the total number of redheads is $C_4 = \sum_{i=1}^{2} f_{i4} = 25$, and so on. The total number of observations in all cells of the table is called the grand total, and is $\sum_{i=1}^{r} \sum_{j=1}^{c} f_{ij} = f_{11} + f_{12} + f_{13} + \cdots + f_{23} + f_{24} = 300$, which is simply n, the size of our sample. The computation of the grand total may be written in several other notations: $\sum_i \sum_j f_{ij}$ or $\sum_{i,j} f_{ij}$, or simply $\sum \sum f_{ij}$. When no indices are given on the summation signs, we assume that the summation of all values in the sample is desired.

For chi-square analysis of contingency tables, one uses the formula

$$\chi^2 = \sum \sum \frac{(f_{ij} - \hat{f}_{ij})^2}{\hat{f}_{ij}}. \tag{23.1}$$

In this formula,† similar to Equation 22.1 for chi-square goodness of fit, $\hat{f}_{ij}$ refers to the frequency expected in a row i column j if the null hypothesis is true. If, in Example 23.1, hair color is in fact independent of sex, then $\frac{100}{300} = \frac{1}{3}$ of all black-haired people would

---

*The early development of chi-square analysis of contingency tables is credited to Karl Pearson (1904) and R. A. Fisher (1922). In 1904, Pearson was the first to use the term "contingency table" (David, 1995).

†Just as Equation 22.2 is equivalent to Equation 22.1 for chi-square goodness of fit, the following are mathematically equivalent to Equation 23.1 for contingency tables:

$$\chi^2 = \sum \sum \frac{f_{ij}^2}{\hat{f}_{ij}} - n \tag{23.2}$$

and

$$\chi^2 = n \left(\sum \sum \frac{f_{ij}^2}{R_i C_j} - 1 \right). \tag{23.2a}$$

These formulas are computationally simpler than Equation 23.1, the latter not even requiring the calculation of expected frequencies; however, they do not allow for the examination of the contributions to the computed chi-square, the utility of which will be seen in Section 23.4.

be expected to be males and $\frac{200}{300} = \frac{2}{3}$ would be expected to be females. That is, $\hat{f}_{11} = \frac{100}{300}(87) = 29$ (the expected number of black-haired males), $\hat{f}_{21} = \frac{200}{300}(87) = 58$ (the expected number of black-haired females), $\hat{f}_{12} = \frac{100}{300}(108) = 36$ (the expected number of brown-haired males), etc.

This may also be explained by the probability rule introduced in Section 5.7: The probability of two independent events occurring at once is the product of the probabilities of the two events. Thus, if having black hair is independent of being male, then the probability of a person being both black-haired and male is the probability of a person being black-haired multiplied by the probability of a person being male, namely $\left(\frac{87}{300}\right) \times \left(\frac{100}{300}\right)$, which is 0.0966667. This means that the expected number of black-haired males in a sample of 300 is $(0.0966667)(300) = 29.0000$. In general, the frequency expected in a cell of a contingency table is

$$\hat{f}_{ij} = \left(\frac{R_i}{n}\right)\left(\frac{C_j}{n}\right) n, \tag{23.3}$$

which reduces to

$$\hat{f}_{ij} = \frac{(R_i)(C_j)}{n}, \tag{23.4}$$

and it is in this way that the $\hat{f}_{ij}$ values in Example 23.1 were obtained. Note that one can check for arithmetic errors in one's calculations by observing that $R_i = \sum_{j=1}^{c} f_{ij} = \sum_{j=1}^{c} \hat{f}_{ij}$ and $C_j = \sum_{i=1}^{r} f_{ij} = \sum_{i=1}^{r} \hat{f}_{ij}$. That is, the row totals of the expected frequencies equal the row totals of the observed frequencies, and the column totals of the expected frequencies equal the column totals of the observed frequencies.

Once χ^2 has been calculated, its significance can be ascertained from Appendix Table B.1, but to do so one must determine the degrees of freedom of the contingency table.

In Section 22.2, degrees freedom (DF or ν) was described as the number of categories over which the calculation of χ^2 is summed minus the number of sample constants used to calculate the expected frequencies. In contingency table analyses, the χ^2 calculation is performed over $r \times c$ cells; to calculate all the $\hat{f}_{ij}$ values, we need to know n and at least $r - 1$ of the row totals and $c - 1$ of the column totals. So DF $= rc - 1 - (r - 1) - (c - 1) = rc - r - c + 1$, which is written more simply as

$$DF = (r - 1)(c - 1). \tag{23.5}$$

In Example 23.1, a 2×4 table, $\nu = (2 - 1)(4 - 1) = 3$. As the calculated χ^2 statistic is 8.994, and the critical value, $\chi^2_{0.05, 3} = 7.815$, we have a significant χ^2 test, and H_0 is rejected; we conclude that the proportions of the hair colors in the population we sampled are not the same for both sexes.

It is a good idea to compute expected frequencies and other intermediate results to at least four decimal places and to round to three decimal places only on arriving at the value of χ^2.

23.2 GRAPHING CONTINGENCY TABLE DATA

Contingency table data may often be displayed advantageously by what have become known as *mosaic* displays (Friendly, 1994; Hartigan and Kleiner, 1981). This is demonstrated in Figure 23.1 for the data of Example 23.1.

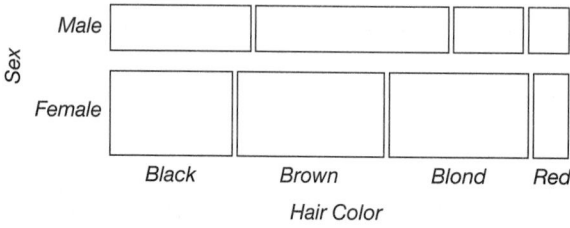

Figure 23.1 Mosaic display for the contingency table of Example 23.1.

   In a mosaic each of the categories of one of the variables is represented by a bar having a thickness expressing the proportion of the total number of data in that category. For the data of Example 23.1, one of the variables is sex (with males and females as the categories) and males comprise one-third (100/300) of the observations. So, as is shown in Figure 23.1, the bar representing males is one-third the thickness of the sum of the two bars.

   Then, for each category of that variable, the bar is divided into sections reflecting the proportion of the observations in that category that lies in each of the categories of the second variable. For the males in Example 23.1, 32% (32/100) have black hair, 43% have brown hair, 16% have blond hair, and 9% have red hair; so the male bar is segmented into four parts, which are 32%, 43%, 16%, and 9% of the bar, respectively. Similarly, the bar for females is apportioned into segments that are 27.5%, 32.5%, 32%, and 8% of the bar, respectively. These eight bar segments (called *tiles*) represent the eight cells in the 2 × 4 contingency table of Example 23.1, and it may be observed graphically that the occurrence of brown and blond hair is strikingly different between men and women. (It will be found in Section 23.5 that the blond category is statistically different from the others.) Either of the two variables may be used to establish the bars, with the other used to define tiles within bars; the choice depends upon which approach will yield the more useful graphical display for the data at hand. The bars may also be drawn vertically instead of horizontally.

   Friendly (1994) suggests that—if there is no natural order to the categories of the second variable—the sequence of the categories be rearranged to make departures from expected frequencies more noticeable. Such a rearrangement strives to have two opposite corners of the graph contain tiles representing the observed frequencies that are most *above* their expected frequencies and the other two opposite corners contain tiles representing the observed frequencies that are the most *below* their expected frequencies. This is shown in Figure 23.2. He also recommends that the tiles denoting observed frequencies prominently above their expected frequencies should be a different color from those depicting observed frequencies substantially below their expected frequencies. If color is not practical, then he proposes—as in Figure 23.2—that the former tiles have a plus sign placed in them and have backslash (⟨⟩) hatching, and that the latter tiles have a minus sign placed in them and have forward-slash (⟨⟩) hatching. Snee (1974) presents different graphical approaches to the examination of $r \times 2$ and $r \times 3$ contingency tables.

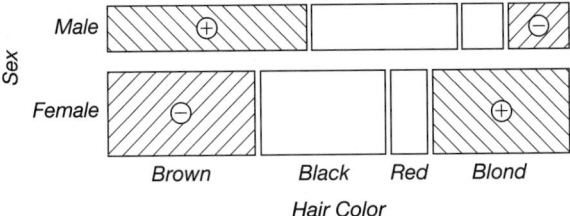

Figure 23.2   The mosaic display of Fig. 23.1, with prominent differences between rows shown with highlighted tiles.

23.3 THE 2 × 2 CONTINGENCY TABLE

The smallest possible contingency table is that consisting of two rows and two columns. It is referred to as a 2×2 ("two by two") table or a fourfold table, and it is often encountered in biological research. By Equation 23.5, the degrees of freedom for a 2×2 tables are $(2 - 1)(2 - 1) = 1$.*

Barnard (1947) described three types of experimental design that result in 2×2 contingency tables, and these have been reiterated by Pearson (1947), Kendall and Stuart (1979: 581–584), and others, including Yates (1984), who made additional helpful distinctions. Let us refer to the information in such a table as

$$
\begin{array}{cc|c}
f_{11} & f_{12} & R_1 \\
f_{21} & f_{22} & R_2 \\
\hline
C_1 & C_2 & n
\end{array}
$$

where f_{ij} denotes the frequency observed in row i and column j, R_i is the total of the frequencies in row i, and C_j is the total for column j.

Category 1.   For what Barnard called a *double dichotomy*, the sample size, n, for a 2×2 table is set at the start of the study; but the totals in neither the row nor column margins are set. Then the n data are collected and the frequency in each of the four cells is tabulated. This sampling situation is based upon what is known as a *multinomial distribution*. The hypothesis to be tested is the familiar contingency table hypothesis of independence between two nominal scale categories.

---

*Pearson (1900) erroneously used $rc - 1$ as the degrees of freedom, but Fisher (1922) published the correct quantity.

Category 2. In the experimental design that Barnard termed a "comparative trial" (and that is also called a "test of homogeneity"), the two totals in one of the margins of a 2×2 table (either the row totals or the column totals) are set before data collection and they represent random samples from two independent populations. The marginal totals that are fixed may be considered to represent sample sizes from the two populations, and the null hypothesis is that there is no difference between proportions in the populations. If, for example, the column totals are fixed, then we are interested in whether the proportions of observations in the two cells of column 1 are significantly different from the proportions in the corresponding cells of column 2. The underlying sampling is of two joint *binomial distributions*, and to distinguish this situation from a kind of experiment under Category 3, below, we shall refer to it as a *binomial comparative trial*.

Category 3. There are situations where both the margins in a 2×2 contingency table are fixed, and the statistical model is the *hypergeometric distribution*. If the n data are not a random sample, then conclusions may be drawn about the sample only, which is sometimes desired, whereas if they are a random sample then conclusions may be drawn about the population sampled. A relatively rare experimental design (sometimes referred to as an *independence trial*) is one in which both of the margins are fixed by the experimenter. A more common situation is where one margin is fixed by the experimenter and the other is fixed by nature, and this may be referred to as a *fixed-margins comparative trial* (sometimes called a "comparative trial," which may confuse it with Category 2 above, or called an "independence trial," not distinguishing it from the other kind of "Category 3" design).

In an ongoing controversy, many authors have asserted that each of these three categories of contingency tables is best subjected to a different method of statistical analysis, with additional disagreement over which method—and which computational formula—is best for each category.*

Several others (e.g., Camilli, 1990; Fisher, 1935; Kendall and Stuart, 1979: 584–585; Little, 1989; Mehta and Hilton, 1993; Upton, 1992; Yates, 1984) argue that the same statistical procedures may and should be employed for all of these experimental designs. The latter convention is followed in this book, and the descriptions of these experimental designs are included to enable the reader to understand reference to them in the literature.

The Double Dichotomy. An example of a 2×2 contingency table with no fixed margins (a "Category 1" situation), is Example 23.2. Here, a random sample of seventy

---

*Contributors to some aspect of these discussions have been numerous, including Barnard (1979); Berkson (1978); Burstein (1981); Camilli and Hopkins (1978; 1979); Conover (1974); D'Agostino, Chase and Belanger (1988); Garside and Mack (1976); Grizzle (1967); Haber (1982); Haviland (1990); Kroll (1989); Liddell (1976, 1980); Martín (1991); Martín, Silva, and Herranz (1992); Overall, Rhoades, and Starbuck (1987); Pearson (1947); Pirie and Hamdan (1972); Plackett (1964); Ramsey (1989); Rhoades and Overall (1982); Richardson (1990); Storer and Kim (1990); Upton (1992); Wilson (1941). The three designs discussed here as Categories 1, 2, and 3 are referred to in some of the literature as III, II, and I, respectively. Contingency tables larger than those of two rows and two columns could be categorized similarly.

EXAMPLE 23.2   A 2×2 contingency table with no fixed margins: a "double dichotomy" experimental design.

H_0:   In the sampled population, handedness is independent of sex.

H_A: In the sampled population, handedness is not independent of sex.

$\alpha = 0.05$

	Boys	Girls	Total
Left-handed	6	12	18
Right-handed	28	24	52
Total	34	36	70

$$\chi_c^2 = \frac{n\left(|f_{11}f_{22} - f_{12}f_{21}| - \frac{n}{2}\right)^2}{R_1 R_2 C_1 C_2}$$

$$= \frac{70\left[|(6)(24) - (12)(28)| - \frac{70}{2}\right]^2}{(18)(52)(34)(36)}$$

$$= 1.5061$$

$\nu = 1; \chi_{0.05,1}^2 = 3.841$

Therefore, do not reject H_0.

$$0.10 < P < 0.25 \quad [P = 0.22]$$

children was obtained in order to study whether handedness is sex-related. This is an illustration of a "Category 1" contingency table because the data were collected by obtaining a random sample of seventy children, without fixing either margin of the contingency table (i.e., without specifying how many children were to be girls and boys or how many were to be left and right-handed), and then tabulating how many children fell into each of the four cells of the table.

To analyze a 2 × 2 table with no fixed margins, the χ^2 of Equation 23.1 could be applied, but

$$\chi^2 = \frac{n(f_{11}f_{22} - f_{12}f_{21})^2}{R_1 R_2 C_1 C_2} \tag{23.6}$$

may be employed as a computational alternative. Although the two equations are mathematically equivalent, Equation 23.6 may yield more accurate results, for as it requires neither the calculation of $\hat{f}_{ij}$ nor of $f_{ij} - \hat{f}_{ij}$, associated rounding errors are avoided.

It is routinely recommended that the chi-square computation be modified by applying the Yates (1934) correction for continuity* (as was done in Section 22.5 for goodness

---

*Pearson (1947) points out that Yates' use of this correction chi-square analysis is predated by about fifteen years by such a correction for other statistical purposes. It should also be noted that the continuity correction should *not* be applied in the rare instance that its inclusion *increases*, instead of decreases, the numerator.

of fit analysis); this correction for Equation 23.6 is

$$\chi_c^2 = \frac{n\left(|f_{11}f_{22} - f_{12}f_{21}| - \frac{n}{2}\right)^2}{R_1 R_2 C_1 C_2}. \tag{23.7}$$

Use of the latter test statistic is actually an approximation to the Fisher exact test of Section 24.10 (e.g., Upton, 1982; Wilson, 1941; Yates, 1984), and is an especially good one if $R_1 = R_2$ or $C_1 = C_2$; and the Fisher exact test is generally preferable to chi-square, especially when frequencies are small. It will be seen that the Fisher test is, in fact, impractical without computer assistance unless the frequencies are small.

Haber (1980) concluded that other correction procedures are better than that of Yates, and he proposed an excellent one based on a principle expounded by Cochran (1942, 1952) and demonstrated in Example 23.3. In Haber's method we determine the smallest of the four expected frequencies, which (using Equation 23.4) would be

$$\hat{f} = \frac{m_1 m_2}{n}, \tag{23.8}$$

EXAMPLE 23.3 The 2×2 contingency table of Example 23.2, using the Haber correction for continuity.

$m_1 = R_1 = 18, \quad m_2 = C_1 = 34$

$\hat{f} = \hat{f}_{11}$, so $\hat{f} = m_1 m_2/n = (18)(34)/70 = 8.74$

$f = f_{11} = 6; d = |f - \hat{f}| = |6 - 8.74| = 2.74$

$2\hat{f} = 2(8.74) = 17.48;$

As $f < 2\hat{f}, D = 2.5$

$$\chi_{c'}^2 = \frac{n^3 D^2}{R_1 R_2 C_1 C_2}$$

$$= \frac{(70)^3 (2.5)^2}{(18)(52)(34)(36)}$$

$$= 1.8712$$

As $\chi_{0.05,1}^2 = 3.841$, do not reject H_0.

$$0.10 < P < 0.25 \quad [P = 0.17]$$

---

where m_1 is the smallest marginal total and m_2 is the smaller total in the opposite margin. (In the present data, the smallest marginal total is that for row 1 and the smaller of the two column totals is C_1. Therefore, $m_1 = R_1 = 18$ and $m_2 = C_1 = 34$). Then, the absolute difference between this expected frequency ($\hat{f}$) and its corresponding observed frequency (f) is $d = |f - \hat{f}|$; and*

    if $f \leq 2\hat{f}$, then define $D =$ the largest multiple of 0.5 that is $< d$;

    if $f > 2\hat{f}$, then define $D = d - 0.5$.

---

*W. Ghent (personal communication) has proposed that a more appropriate criterion is whether $f - 1$, rather than f, is $\leq 2\hat{f}$ or $> 2\hat{f}$.

The chi-square value with the Haber correction is

$$\chi_{c'}^2 = \frac{n^3 D^2}{R_1 R_2 C_1 C_2}.$$ (23.9)

If $f > 2\hat{f}$, then the chi-square with the Haber correction, namely $\chi_{c'}^2$, is the same as the chi-square with the Yates correction, namely χ_c^2. And, if either $C_1 = C_2$ or $R_1 = R_2$, then $\chi_{c'}^2 = \chi_c^2$ (as will be seen to be the case in Example 23.6).

In a 2 × 2 contingency table, $|f - \hat{f}|$ is exactly the same in all four cells. It is interesting to note that Equation 23.9 yields the same results as Equation 23.6 if D is defined as $f - \hat{f}$, and it gives the same results as Equation 23.7 if D is defined as $|f - \hat{f}|$.

The Binomial Comparative Trial.   If the columns (or the rows) of a contingency table represent random samples from independent populations, then the null hypothesis is typically phrased as a comparison of proportions. This is shown in Example 23.4, in which a researcher determined the presence of a specific intestinal parasite in each animal from a random selection of mice of each of two species. The question to be asked is whether the frequency of occurrence of this parasite is the same in the two animal populations. This may be expressed as H_0: $p_1 = p_2$, where p_i is the proportion of parasite-infected animals in population i (with the unusual use of a Roman—not a Greek—letter, p, as a population parameter). Each sample proportion may be expressed as $\hat{p}_i = X_i/n_i$, where X_i is the number of infected animals in sample i and n_i is the total number of animals in that sample. So, in Example 23.4, $X_1 = 18, n_1 = 24, \hat{p}_1 = 18/24 = 0.7500, X_2 = 10, n_2 = 25$, and $\hat{p}_2 = 10/25 = 0.4000$.

This two-tailed null hypothesis may be tested using the above χ_c^2, but the Haber correction (Equation 23.9) is generally even better, and the Fisher exact test (Section 23.10) should be applied whenever possible.

It is instructive to note that the data of Example 23.2 could have represented a "Category 2" situation instead of "Category 1" data collection. The data in this 2 × 2

EXAMPLE 23.4   A 2 × 2 contingency table, with one fixed margin: comparing two proportions (in a "binomial comparative trial").

H_0:   The proportion of the population infected with an intestinal parasite is the same in two species of mouse (i.e., H_0: $p_1 = p_2$).

H_A: The proportion of the population infected with an intestinal parasite is not the same in two species of mouse (i.e., H_A: $p_1 \neq p_2$).

$\alpha = 0.05$

	Species 1	Species 2	Total
With parasite	18	10	28
Without parasite	6	15	21
Total	24	25	49

EXAMPLE 23.4 (continued)

$$\chi_c^2 = \frac{n\left(|f_{11}f_{22} - f_{12}f_{21}| - \frac{n}{2}\right)^2}{R_1 R_2 C_1 C_2}$$

$$= \frac{49\left[|(18)(15) - (10)(6)| - \frac{49}{2}\right]^2}{(24)(25)(28)(21)}$$

$$= 4.779.$$

$\nu = 1; \quad \chi_{0.05,1}^2 = 3.841$

Therefore, reject H_0.

$$0.025 < P < 0.05 \quad [P = 0.029]$$

Or, preferably,

$m_1 = R_2 = 21, m_2 = C_1 = 24$

$\hat{f} = m_1 m_2/n = (21)(24)/49 = 10.29$

$f = f_{21} = 6; d = |f - \hat{f}| = |6 - 10.29| = 4.29$

$2\hat{f} = 2(10.29) = 20.58;$

As $f < 2\hat{f}$, $D = 4.0$,

$$\chi_{c'}^2 = \frac{n^3 D^2}{R_1 R_2 C_1 C_2}$$

$$= \frac{(49)^3 (4.0)^2}{(24)(25)(28)(21)}$$

$$= 5.336.$$

As $\chi_{0.05,1}^2 = 3.841$, reject H_0.

$$0.01 < P < 0.025 \quad [P = 0.021]$$

table might have been collected by obtaining a random sample intentionally consisting of thirty-four boys and a random sample of thirty-six girls, and tabulating how many of each of those two groups were right-and left-handed. Then we would be comparing two proportions—the proportion of left-handed (or right-handed) boys compared to the proportion of left-handed (or right-handed) girls. Or, this could have been a table of data obtained by specifically securing a random sample of eighteen left-handed children and a random sample of fifty-two right-handed children, and recording how many in each of those two groups were boys and how many were girls. Then, the analysis would ask whether the proportion of boys (or girls) was the same among left-handed as among right-handed children.

The Independence Trial. In this kind of data collection, a "Category 3" experimental design, both the row totals and the column totals are set by the experimenter before data are collected. Example 23.5 displays data from an experiment to test for difference between the ability of two species of snail to hold firmly onto a smooth

EXAMPLE 23.5   A 2 × 2 contingency table with two fixed margins: an "independence trial" experimental design.

H_0:   The ability of snails to resist the current is no different between the two species.

H_A: The ability of snails to resist the current is different between the two species.

$\alpha = 0.05$

The four marginal totals are set before performing the experiment, and the four cell frequencies are collected from the experiment.

	Resisted	Yielded	
Species 1	12	7	19
Species 2	2	9	11
	14	16	30

$$\chi_c^2 = \frac{n\left(|f_{11}f_{22} - f_{12}f_{21}| - \frac{n}{2}\right)^2}{R_1 R_2 C_1 C_2}$$

$$= \frac{30\left[|(12)(9) - (7)(2)| - \frac{30}{2}\right]^2}{(14)(16)(19)(11)}$$

$$= 3.999.$$

$\nu = 1$

$\chi_{0.05,1}^2 = 3.841$

Therefore, reject H_0.

$$0.025 < P < 0.05 \quad [P = 0.046]$$

Or, preferably (with the Haber correction),

$m_1 = R_2 = 11, m_2 = C_1 = 14$

$\hat{f} = m_1 m_2/n = (11)(14)/30 = 5.13$

$f = f_{21} = 2; d = |f - \hat{f}| = |2 - 5.13| = 3.13$

$2\hat{f} = 2(5.13) = 10.26;$

As $f < 2\hat{f}, D = 3.0$

$$\chi_{c'}^2 = \frac{n^3 D^2}{R_1 R_2 C_1 C_2}$$

$$= \frac{(30)^3 (3.0)^2}{(14)(16)(19)(11)}$$

$$= 5.191$$

As $\chi_{0.05,1}^2 = 3.841$, reject H_0.

$$0.01 < P < 0.025 \quad [P = 0.023]$$

substrate. A total of twenty-nine aquatic snails, nineteen of one species and ten of a second species, are allowed to attach to a flat rock. The experimental plan is to pass a current of water over the snails in a fashion that each snail is exposed to the same force of water as the others, and to continue the experiment until just over half (i.e., just over 15, namely 16) of all of the snails have yielded their grip on the rock. The data may be tabulated as the 2×2 table of Example 23.5, where both the row totals (the numbers of snails of each species) and the column totals (the number of snails holding onto the substrate and the number yielding their grip) are fixed at the start of the experiment. The null hypothesis shown in this example might also have be stated as "The proportion of snails of species 1 resisting the water current is the same as the proportion of resisting individuals of species 2," or "The ability to resist the current is independent of the species to which the animal belongs."

The Fixed-Margins Comparative Trial. This experimental design is also classified in "Category 3" because both row and column totals are fixed; however, the experimenter sets only one of the margins—and nature sets the other. In Example 23.6 we have a

---

EXAMPLE 23.6   A 2×2 contingency table with two fixed margins: a "fixed-margins comparative trial" experimental design.

H_0:   The incidence of parasitic infection in mice is the same whether or not an experimental substance is ingested. (H_0: $p_1 = p_2$)

H_A: The incidence of parasitic infection in mice is not the same whether or not an experimental substance is ingested. (H_A: $p_1 \neq p_2$)

$\alpha = 0.5$

	Without substance	With substance	Total
With parasite	15	9	24
Without parasite	10	16	26
Total	25	25	50

$$\chi_c^2 = \frac{n\left(|f_{11}f_{22} - f_{12}f_{21}| - \frac{n}{2}\right)^2}{R_1 R_2 C_1 C_2}$$

$$= \frac{50\left[|(15)(16) - (9)(10)| - \frac{50}{2}\right]^2}{(25)(25)(24)(26)}$$

$$= 2.003.$$

$\nu = 1$

$\chi_{0.05, 1}^2 = 3.841$

Therefore, do not reject H_0.

$$0.10 < P < 0.25 \quad [P = 0.16]$$

EXAMPLE 23.6   (continued)

Or, preferably (in this case identically, for $C_1 = C_2$),

$$m_1 = R_2 = 24, m_2 = C_1 = 25$$

$$\hat{f} = m_1 m_2 / n = (24)(25)/50 = 12.00$$

$$f = f_{21} = 15 \text{ or } 9; d = |f - \hat{f}| = |15 - 12.00| = 3.00$$

$$2\hat{f} = 2(12.00) = 24.00;$$

As $f < 2\hat{f}, D = 2.5$

$$\chi_{c'}^2 = \frac{n^3 D^2}{R_1 R_2 C_1 C_2}$$

$$= \frac{(50)^3 (2.5)^2}{(25)(25)(24)(26)}$$

$$= 2.003$$

As $\chi_{0.05,1}^2 = 3.841$, do not reject H_0.

$$0.10 < P < 0.250 \quad [P = 0.16]$$

sample of $n = 50$ mice (all of the same species) and we randomly allocate twenty-five of them to each of two experimental groups (i.e., $C_1 = C_2 = 25$). Then the members of one group are fed a substance that may have an effect on the occurrence of intestinal parasites, and the second group is left untreated. The two-tailed null hypothesis is that there is no effect of ingesting this substance on the occurrence of the parasite. Using p_i to indicate the proportion of the population from which group i came that is infested with the parasite, we could state this as H_0: $p_1 = p_2$. Now, if H_0 is a true statement, then both p_1 and p_2 are the same and are a characteristic of the population of animals that was sampled, a characteristic unaffected by the experiment. And, even though this population proportion is unknown to the experimenter it is fixed before the experiment begins.

Note, as shown in Example 23.6, that when computing $\chi_{c'}^2$ where $R_1 = R_2$ or $C_1 = C_2$, there are two possible values for f (in this case, 15 and 9), and either may be use to calculate d; also, when the row totals or column totals are equal and $f > 2\hat{f}$, $\chi_{c'}^2$ is identical to χ_c^2.

One-Tailed Testing.   The testing shown above for 2×2 tables is applicable to two-tailed hypotheses. If $R_1 = R_2$ or $C_1 = C_2$, one-tailed testing may be performed by employing one-half the chi-square probability (using, for example, $\chi_{0.10,1}^2$ for $\alpha = 0.05$), or by dividing the resultant P by 2 or using one-tailed values for Z in the normal approximation of Section 24.11. In one-tailed testing one must declare, before the data are examined, the direction in which the data must be extreme in order to reject H_0; this is demonstrated in Example 23.7, and additional examples of one-tailed testing are shown in Section 24.3. If neither $R_1 = R_2$ nor $C_1 = C_2$, then this procedure often yields a very crude approximation of the probability of H_0 and is not recommended for general use. As the investigator has control of least one margin in Category 2 and 3 experimental designs, equal row totals, and/or equal column totals should be striven for if

EXAMPLE 23.7   One-tailed testing for the data of Example 23.6.

H_0:   The experimental substance does not reduce the incidence of the parasite.

H_A: The experimental substance reduces the incidence of the parasite.

Expressing the proportion of parasitized animals in each experimental group:

	Without substance	With substance
With parasite	0.60	0.36
Without parasite	0.40	0.64
Total	1.00	1.00

The sample data are in the direction of H_A, but are they significantly so? In Example 23.6, $\chi^2_{c'} = 2.003$ was calculated. As $C_1 = C_2$, we may proceed as follows.

For $\alpha = 0.05$, use $\chi^2_{0.10, 1} = 2.706$ as the critical value; do not reject H_0; $0.10/2 < P < 0.25/2$, or $0.05 < P < 0.125$ $[P = 0.078]$.

(If the hypotheses had been H_0: The experimental substance does not increase the incidence of the parasite and H_A: The experimental substance does increase the incidence of the parasite, then we would have observed that the sample data are *not* in the direction of H_A and would conclude not to reject H_0 without even examining chi-square.)

one-tailed testing is to be performed. Mantel (1974) discusses modifications of chi-square for one-tailed tests, and Haber (1987) describes an alternative, but the best procedure for addressing one-tailed hypotheses is to employ the Fisher exact test of Section 24.10.

23.4 HETEROGENEITY TESTING OF 2×2 TABLES

Just as one may test for heterogeneity of replications of a goodness of fit analysis (Section 22.6), it is sometimes desirable to test for heterogeneity of 2×2 contingency tables. Suppose that in addition to the data set summarized in Example 23.2, the same data collection procedure had been performed three other times. Such data are shown in Example 23.8a. It would then be meaningful to determine whether all four of the data sets are samples that could have come from the same population (i.e., H_0: The four samples are homogeneous; and H_A: The four samples are heterogeneous). In this particular example, each of the four contingency tables was analyzed with chi-square with continuity correction, and in each case there was no rejection of H_0. This failure to reject H_0 could be an indication of low power owing to small sample size, so it would be useful to use the heterogeneity test to ascertain whether it would be reasonable to combine the four sets of data and perform a more powerful test of H_0 using the pooled number of data.

To test for heterogeneity, one first calculates the χ^2 for each of the four samples (recalling from Section 22.6 that a correction for continuity should not be used in a heterogeneity chi-square analysis). These four uncorrected χ^2 values are shown

EXAMPLE 23.8a   A heterogeneity chi-square analysis of four 2 × 2 contingency tables, where Data Set 1 is that of Example 23.2.

H_0:   The four samples are homogeneous.

H_A: The four samples are heterogeneous.

$\alpha = 0.05$

Data Set 1

From the data of Example 23.2: $\chi_c^2 = 1.506,$ DF $= 1,$ $0.10 < P < 0.25$
$\chi^2 = 2.2523$

Data Set 2

	Boys	Girls	Total
Left-handed	4	7	11
Right-handed	25	13	38
Total	29	20	49

$\chi_c^2 = 1.961,$ DF $= 1,$ $0.10 < P < 0.25$
$\chi^2 = 3.0578$

Data Set 3

	Boys	Girls	Total
Left-handed	7	10	17
Right-handed	27	18	45
Total	34	28	62

$\chi_c^2 = 1.087,$ DF $= 1,$ $0.25 < P < 0.50$
$\chi^2 = 1.7653$

Data Set 4

	Boys	Girls	Total
Left-handed	4	7	11
Right-handed	22	14	36
Total	26	21	47

$\chi_c^2 = 1.206,$ DF $= 1,$ $0.25 < P < 0.50$
$\chi^2 = 2.0877$

Data Sets 1–4 pooled

	Boys	Girls	Total
Left-handed	21	36	57
Right-handed	102	69	171
Total	123	105	228

$\chi^2 = 8.9505$
DF $= 1$

χ^2 for Data Set 1:	2.2523	DF $= 1$
χ^2 for Data Set 2:	3.0578	DF $= 1$
χ^2 for Data Set 3:	1.7653	DF $= 1$
χ^2 for Data Set 4:	2.0877	DF $= 1$
Total chi-square:	9.1631	DF $= 4$
Chi-square of totals:	8.9505	DF $= 1$
Heterogeneity chi-square	0.2126	DF $= 3$

EXAMPLE 23.8a   (continued)

For heterogeneity testing (using $\chi^2 = 0.213$):

$$\chi^2_{0.05,3} = 7.815$$

Therefore, do not reject H_0.

$$0.975 < P < 0.99 \quad [P = 0.98]$$

in Example 23.8a; also shown are the contingency table formed from the pooled data and the completion of the heterogeneity analysis. As the heterogeneity χ^2 is 0.213 and $\chi^2_{0.05,3} = 7.815$, the null hypothesis of homogeneity is not rejected (i.e., the four samples are concluded to be homogeneous). This indicates that the four samples may justifiably be pooled. Such pooling, if justified, is generally advantageous as it provides for chi-square analysis with a relatively large n. In Example 23.8b, the pooled data are analyzed, and it is concluded that there is a significant difference in handedness between boys and girls, a difference that was not detected using any of the four smaller samples originally recorded. Note that a correction for continuity (Yates or Haber) should be used on the pooled data, after the heterogeneity analysis has been performed.

EXAMPLE 23.8b   The 2×2 contingency tables analysis, with correction for continuity, for the pooled data of Example 23.8a.

H_0:   Handedness is independent of sex.

H_A: Handedness is not independent of sex.

$\alpha = 0.05$

	Data Sets 1–4 pooled		
	Boys	*Girls*	*Total*
Left-handed	21	36	57
Right-handed	102	69	171
Total	123	105	228

$$\chi^2_{0.05,1} = 3.841$$

$$\chi^2_c = 8.056; \text{ therefore, reject } H_0.$$

$$0.001 < P < 0.005 \quad [P = 0.0045]$$

$$\chi^2_{c'} = 8.497; \text{ therefore, reject } H_0.$$

$$0.001 < P < 0.005 \quad [P = 0.0036]$$

23.5 SUBDIVIDING CONTINGENCY TABLES

In Example 23.1, analysis of a 2×4 contingency table, it was concluded that there was a significant difference in human hair color frequencies between males and females.

EXAMPLE 23.9a   The data of Example 23.1, where for each hair color the percent males and percent females are indicated.

Sex	Black	Brown	Blond	Red	Total
			Hair color		
Male	32	43	16	9	100
	(37%)	(40%)	(20%)	(36%)	
Female	55	65	64	16	200
	(63%)	(60%)	(80%)	(64%)	
Total	87	108	80	25	300

In Example 23.1, the null hypothesis that the four hair colors are independent of sex was rejected.

Examination of the f_{ij} values in this table reveals that the proportion of males in the blond column is decidedly less than in the other columns. (To express the percent males and females in each column, as in Example 23.9a, and in Figures 23.1 and 23.2, helps to elucidate this fact, although percentages may not be used in calculations for a chi-square testing procedure.) Thus, we might suspect that the significant χ^2 calculated in Example 23.1 was due largely to column 3 in the table. We may momentarily ignore the data for column 3 and consider the remaining 2×3 table (Example 23.9b). The nonsignificant χ^2 for this table supports the null hypothesis that these three hair colors

EXAMPLE 23.9b   The 2×3 contingency table formed from columns 1, 2, and 4 of the original 2×4 table. $\hat{f}_{ij}$ values for the cells of the 2×3 table are shown in parentheses.

H_0:   The occurrence of black, brown, and red hair is independent of sex.

H_A: The occurrence of black, brown, and red hair is not independent of sex.

$\alpha = 0.05$

Sex	Black	Brown	Red	Total
		Hair color		
Male	32	43	9	84
	(33.2182)	(41.2364)	(9.5455)	
Female	55	65	16	136
	(53.7818)	(66.7636)	(15.4545)	
Total	87	108	25	220

$\chi^2 = 0.245$ with DF $= 2$

$\chi^2_{0.05,2} = 5.991$

Therefore, do not reject H_0.

$$0.75 < P < 0.90 \quad [P = 0.88]$$

are independent of sex. In Example 23.9c, a 2×2 table is formed by considering blond versus all other hair colors. Here, the null hypothesis of independence is rejected. By the described series of manipulations of the original contingency table, we have confirmed our suspicion that among the four hair colors tested, blond occurs among the sexes with relative frequencies different from those of the other colors. Section 23.2 discusses graphical methods for examining components of contingency tables.

EXAMPLE 23.9c   The 2×2 contingency table formed by combining columns 1, 2, and 4 of the original table.

H_0:   Occurrence of blond and nonblond hair color is independent of sex.

H_A: Occurrence of blond and nonblond hair color is not independent of sex.

$\alpha = 0.05$

	Hair color		
Sex	*Blond*	*Nonblond*	**Total**
Male	16	84	100
Female	64	136	200
Total	80	220	300

$\chi_c^2 = 8.457$

$DF = 1$

$\chi_{0.05,1}^2 = 3.841$

Therefore, reject H_0.

$$0.001 < P < 0.005 \quad [P = 0.0036]$$

Strictly speaking, it is not proper to test statistical hypotheses developed after examining the data to be tested. Therefore, the analysis of a subdivided contingency table should be considered only a guide to developing additional hypotheses; these new hypotheses should then be stated in advance of their being tested with a new set of data.

Other considerations of chi-square contingency table analyses can be found in Bliss (1967: Chapters 3 and 4), Simpson, Roe, and Lewontin (1960: Chapter 13), and Snedecor and Cochran (1980: Chapter 11).

23.6 BIAS IN CHI-SQUARE CONTINGENCY TABLE ANALYSES

Section 22.5 spoke of bias in chi-square goodness of fit testing when expected frequencies are "too small." As with goodness of fit testing, for a long time many statisticians (e.g., Fisher, 1925b) advised that chi-square analysis of contingency tables be employed only if each of the expected frequencies were at least 5.0—even after there was evidence that such analyses worked well even with smaller frequencies (e.g., Cochran, 1952, 1954). Thanks to the review and empirical analysis of Roscoe and Byars (1971), we have more useful guidelines. A secure procedure is to have the average expected frequency, namely $n/(rc)$, be at least 6.0 when testing with α as small as 0.05, and at least 10.0 for $\alpha = 0.01$. Note that requiring an average expected frequency to be at least 6 is typically less restrictive than requiring that each expected frequency be at least 5. These guidelines may also be used when employing a normal approximation in contingency table testing (Section 24.11).

If a 2×2 table has insufficiently large frequencies for a chi-square analysis, then the Fisher exact test (Section 24.10) might profitably be employed. If a contingency table with such frequencies has more than two rows and/or columns, one might simply discard the rows and/or columns with the offensively low $\hat{f}_{ij}$'s, or combine rows and/or columns for the same purpose, but such procedures are not recommended as routine practice; when possible, it would be better to repeat the experiment with a sufficiently large n. Some recommend employing the log-likelihood ratio of Section 23.7 as a test less affected than chi-square by low frequencies. Lawal and Upton (1984) discuss modifications of the chi-square test to allow for the average expected cell frequency to be as small as 0.5.

23.7 THE LOG-LIKELIHOOD RATIO FOR CONTINGENCY TABLES

The log-likelihood ratio was introduced in Section 22.7, where the G statistic was presented as an alternative to chi-square for goodness of fit testing. The G test may also be applied to contingency tables (Neyman and Pearson, 1928a, 1928b; Wilks, 1935), where

$$G = 2 \left[\sum_i \sum_j f_{ij} \ln \left(\frac{f_{ij}}{\hat{f}_{ij}} \right) \right],$$ (23.10)

which, without the necessity of calculating expected frequencies, may readily by computed as

$$G = 2 \left[\sum_i \sum_j f_{ij} \ln f_{ij} - \sum_i R_i \ln R_i - \sum_j C_j \ln C_j + n \ln n \right].$$ (23.11)

If common logarithms (denoted by "log") are used instead of natural logarithms (indicated as "ln"), then use 4.60517 instead of 2 prior to the left bracket. Because G is approximately distributed as χ^2, Appendix Table B.1 may be used with $(r-1)(c-1)$ degrees of freedom. In Example 23.10 the contingency table of Example 23.1 is analyzed using the G statistic, with very similar results.

In the case of a 2×2 table, the Yates correction for continuity is applied by making each f_{ij} 0.5 closer to $\hat{f}_{ij}$. This may be accomplished (without calculating expected frequencies) as follows: If $f_{11}f_{22} - f_{12}f_{21}$ is negative, add 0.5 to f_{11} and f_{22} and subtract 0.5 from f_{12} and f_{21}; if $f_{11}f_{22} - f_{12}f_{21}$ is positive, subtract 0.5 from f_{11} and f_{22} and add 0.5 to f_{12} and f_{21}. This is demonstrated in Example 23.11.

EXAMPLE 23.10   The G test for the contingency table data of Example 23.1.

H_0:   Hair color is independent of sex.

H_A: Hair color is not independent of sex.

$\alpha = 0.05$

Sex	Black	Brown	Blond	Red	Total
Male	32	43	16	9	100
Female	55	65	64	16	200
Total	87	108	80	25	300

Hair color (spanning Black, Brown, Blond, Red)

$$G = 4.60517 \left[\sum \sum f_{ij} \log f_{ij} - \sum R_i \log R_i - \sum C_j \log C_j + n \log n \right]$$

$$= 4.60517[(32)(1.50515) + (43)(1.63347) + (16)(1.20412) + (9)(0.95424)$$

$$+ (55)(1.74036) + (65)(1.81291) + (64)(1.80618) + (16)(1.20412)$$

$$- (100)(2.00000) - (200)(2.30103) - (87)(1.93952) - (108)(2.03342)$$

$$- (80)(1.90309) - (25)(1.39794) + (300)(2.47712)]$$

$$= 4.60517(2.06518)$$

$$= 9.510 \text{ with DF} = 3$$

$\chi^2_{0.05,3} = 7.815$

Therefore, reject H_0.

$$0.01 < P < 0.025 \quad [P = 0.023]$$

Williams (1976) recommends G be used in preference to χ^2 whenever $|f_{ij} - \hat{f}_{ij}| < \hat{f}_{ij}$ for any cell. Both methods commonly result in the same conclusions. When they do not, many statisticians prefer G and recommend its routine use, while others (e.g., see Larntz, 1978; Hutchinson, 1979) conclude χ^2 to be preferable. Cox and Groeneveld (1986) specifically discuss the analysis of 2×2 tables.

23.8 THREE-DIMENSIONAL CONTINGENCY TABLES

In this chapter, thus far, we have considered two-dimensional contingency tables (that is, tables with rows and columns as the two dimensions); each of the two dimensions (row and column) represented a nominal scale variable. However, we may collect and tabulate enumeration data with respect to three or more variables and thus have what are referred to as multidimensional contingency tables (i.e., tables with three or more dimensions). An example would be a study similar to that in Example 23.1, where eye color is a third variable in addition to the variables of hair color and sex. Such tables have been the subject of considerable investigation (e.g., see Bishop, Fienberg, and Holland, 1975; Christensen,

EXAMPLE 23.11   The G test, with the Yates correction for continuity, for the 2×2 contingency table of Example 23.2.

H_0:   In the sampled population, handedness is independent of sex.

H_A: In the sampled population, handedness is not independent of sex.

$\alpha = 0.05$

	Boys	Girls	Total
Left-handed	6 [6.5]	12 [11.5]	18
Right-handed	28 [27.5]	24 [24.5]	52
Total	34	36	70

$f_{11} f_{22} - f_{12} f_{21} = (6)(24) - (12)(28) = -192$; as this is negative, modify f_{11} and f_{22} by adding 0.5 and modify f_{12} and f_{21} by subtracting 0.5 (as shown above in brackets).

$$G_c = 2 \left[\sum f_{ij} \ln f_{ij} - \sum R_i \ln R_i - \sum C_j \ln C_j + n \ln n \right]$$

$$= 2[6.5 \ln 6.5 + 11.5 \ln 11.5 + 27.5 \ln 27.5 + 24.5 \ln 24.5$$

$$- 18 \ln 18 - 52 \ln 52 - 34 \ln 34 - 36 \ln 36 + 70 \ln 70]$$

$$= 2[12.1667 + 28.0870 + 91.1401 + 78.3675$$

$$- 52.0267 - 205.4647 - 119.8963 - 129.0067 + 297.3947]$$

$$= 2[0.7616] = 1.5232$$

$\nu = 1$; $\chi^2_{0.05, 1} = 3.841$

Therefore, do not reject H_0.

$$0.10 < P < 0.25 \quad [P = 0.22]$$

1990; Everitt, 1992: Chapter 4; Fienberg, 1970, 1980; Goodman, 1970; Lancaster, 1969; Upton, 1978), and computer program libraries often include provision for their analysis.

Figure 23.3 shows a three-dimensional contingency table. The three "rows" are species, the four "columns" are geographic locations, and the two "tiers" (or "layers") are presence and absence of a disease. If a sample is obtained containing individuals of these species, from these locations, and with and without the disease in question, then observed frequencies can be recorded in the twenty-four cells of this $3 \times 4 \times 2$ contingency table. We shall refer to the observed frequency in row i, column j, and tier l as f_{ijl}. We shall refer to the number of rows, columns, and tiers as r, c, and t, respectively. The sum of the frequencies in row i will be designated R_i, the sum in column j as C_j, and the sum in tier l as T_l. Friendly (1994) and Hartigan and Kleiner (1981, 1984) discuss mosaic displays for contingency tables with more than two dimensions, and such graphical presentations can make multidimensional contingency table data easier to visualize and interpret than if they are presented only in tabular format.

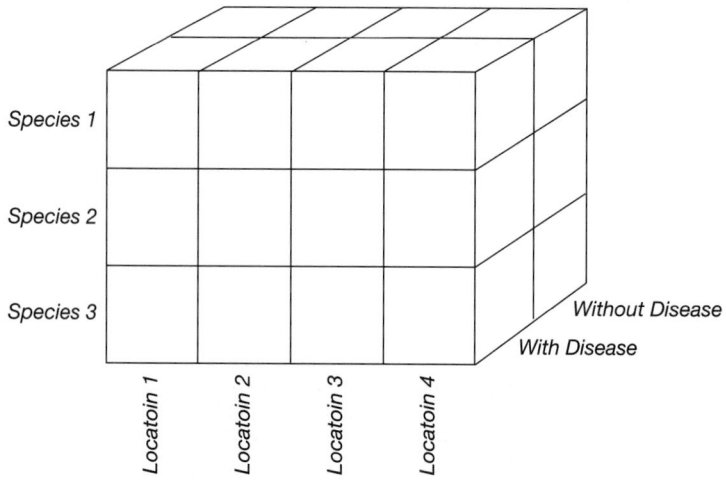

Figure 23.3   A three-dimensional contingency table, where the three rows are species, the four columns are locations, and the two tiers are occurrence of a disease. An observed frequency, f_{ijl}, is recorded in each combination of row, column, and tier.

Example 23.12 presents a $2 \times 2 \times 2$ contingency table where data (f_{ijl}) are collected as described above, but only for two species and two locations. Note that throughout the following discussions the sum of the expected frequencies for a given row, column, or tier equals the sum of the observed frequencies for that row, column, or tier.

Mutual Independence.   We can test more than one hypothesis using multidimensional contingency table data. An overall kind of hypothesis is that which states mutual independence among all the variables. Another way of expressing this H_0 is that there

---

EXAMPLE 23.12   Test for mutual independence in a 2 × 2 × 2 contingency table.

H_0:   Disease occurrence, species, and location are all mutually independent in the population sampled.

H_A: Disease occurrence, species, and location are not all mutually independent in the population sampled.

The observed frequencies (f_{ijl}):

	Disease present		Disease absent		Species totals
	Location 1	Location 2	Location 1	Location 2	($r = 2$)
Species 1	44	12	38	10	$R_1 = 104$
Species 2	28	22	20	18	$R_2 = 88$
Disease totals ($t = 2$):	$T_1 = 106$		$T_2 = 86$		Grand total:
Location totals ($c = 2$):	$C_1 = 130, C_2 = 62$				$n = 192$

EXAMPLE 23.12   (continued)

The expected frequencies ($\hat{f}_{ijl}$):

	Disease present		Disease absent		
	Location 1	*Location 2*	*Location 1*	*Location 2*	**Species totals**
Species 1	38.8759	18.5408	31.5408	15.0425	$R_1 = 104$
Species 2	32.8950	15.6884	26.6884	12.7283	$R_2 = \ 88$
Disease totals:	$T_1 = 106$		$T_2 = 86$		*Grand total:*
Location totals:	$C_1 = 130, C_2 = 62$				$n = 192$

$$\chi^2 = \sum\sum\sum \frac{(f_{ijl} - \hat{f}_{ijl})^2}{\hat{f}_{ijl}}$$

$$\chi^2 = \frac{(44 - 38.8759)^2}{38.8759} + \frac{(12 - 18.5408)^2}{18.5408} + \frac{(38 - 31.5408)^2}{31.5408} + \frac{(10 - 15.0425)^2}{15.0425}$$

$$+ \frac{(28 - 32.8950)^2}{32.8950} + \frac{(22 - 15.6884)^2}{15.6884} + \frac{(20 - 26.6884)^2}{26.6884} + \frac{(18 - 12.7283)^2}{12.7283}$$

$$= 0.6754 + 2.3075 + 1.3228 + 1.6903 + 0.7284 + 2.5392 + 1.6762 + 2.1834$$

$$= 13.123$$

$$\nu = rct - r - c - t + 2 = (2)(2)(2) - 2 - 2 - 2 + 2 = 4$$

$$\chi^2_{0.05,4} = 9.488$$

Reject H_0.

$$0.01 < P < 0.025 \quad [P = 0.011]$$

are no interactions (either three-way or two-way) among any of the variables. For this hypothesis, the expected frequency in row i, column j, and tier l is

$$\hat{f}_{ijl} = \frac{R_i C_j T_l}{n^2}, \tag{23.12}$$

where n is the total of all the frequencies in the entire contingency table.

In Example 23.12 this null hypothesis would imply that presence or absence of the disease occurred independently of species and location. For three dimensions, this null hypothesis is tested by computing

$$\chi^2 = \sum_{i=1}^{r}\sum_{j=1}^{c}\sum_{l=1}^{t} \frac{(f_{ijl} - \hat{f}_{ijl})^2}{\hat{f}_{ijl}}, \tag{23.13}$$

which is a simple extension of the chi-square calculation for a two-dimensional table (by Equation 23.1). The degrees of freedom for this test are the sums of the degrees of freedom for all interactions:

$$\nu = (r - 1)(c - 1)(t - 1) + (r - 1)(c - 1) + (r - 1)(t - 1) + (c - 1)(t - 1), \tag{23.14}$$

which is equivalent to

$$\nu = rct - r - c - t + 2. \tag{23.15}$$

Partial Independence. If the above null hypothesis is not rejected, then we conclude that all three variables are mutually independent and the analysis proceeds no further. If, however, H_0 is rejected, then we may test further to conclude between which variables dependencies and independencies exist. For example, we may test whether one of the three variables is independent of the other two, a situation known as *partial independence*.*

For the hypothesis of rows being independent of columns and tiers, we need total frequencies for rows and total frequencies for combinations of columns and tiers. We designate the total frequency in column j and tier l as $(CT)_{jl}$. Expected frequencies are then computed as

$$\hat{f}_{ijl} = \frac{R_i(CT)_{jl}}{n}, \tag{23.16}$$

and Equation 23.13 is used with degrees of freedom

$$\nu = (r-1)(c-1)(t-1) + (r-1)(c-1) + (r-1)(t-1), \tag{23.17}$$

which is equivalent to

$$\nu = rct - ct - r + 1. \tag{23.18}$$

For the null hypothesis of columns being independent of rows and tiers, we compute expected frequencies using column totals, C_j, and the totals for row and tier combinations, $(RT)_{il}$:

$$\hat{f}_{ijl} = \frac{C_j(RT)_{il}}{n}, \tag{23.19}$$

and

$$\nu = rct - rt - c + 1. \tag{23.20}$$

And, for the null hypothesis of tiers being independent of rows and columns, we use tier totals, T_l, and the totals for row and column combinations, $(RC)_{ij}$:

$$\hat{f}_{ijl} = \frac{T_l(RC)_{ij}}{n}; \tag{23.21}$$

$$\nu = rct - rc - t + 1. \tag{23.22}$$

In Example 23.13, all three hypotheses for partial dependence are tested. In one of the three (the last), H_0 is not rejected; thus we conclude that presence of disease is independent of species and location. However, the hypothesis test of Example 23.12 concluded that all three variables are not independent of each other. Therefore, we suspect that species and location are not independent. The independence of these two variables may be tested using a two-dimensional contingency table, as described earlier in this chapter and demonstrated in Example 23.14. In the present case, the species-location interaction is tested by way of a 2×2 contingency table and we conclude that these two factors are not independent (i.e., species occurrence depends on geographic location).

---

*A different hypothesis is that of *conditional independence*, where two of the variables are said to be independent in each level of the third (but each may have dependence on the third). This is discussed in the references cited at the beginning of this section.

EXAMPLE 23.13   Test for partial independence in a $2 \times 2 \times 2$ contingency table. As the H_0 of overall independence was rejected in Example 23.12, we may test the following three pairs of hypotheses.

H_0:   Species is independent of location and disease.

H_A: Species is not independent of location and disease.

The expected frequencies $(\hat{f}_{ijl})$:

	Disease present		Disease absent		
	Location 1	Location 2	Location 1	Location 2	**Species totals**
Species 1	39.0000	18.4167	31.4167	15.1667	$R_1 = 104$
Species 2	33.0000	15.5833	26.5833	12.8333	$R_2 = 88$
Location and disease totals:	$(CT)_{11}$ $= 72$	$(CT)_{12}$ $= 34$	$(CT)_{21}$ $= 58$	$(CT)_{22}$ $= 28$	Grand total: $n = 192$

$$\chi^2 = \frac{(44 - 39.0000)^2}{39.0000} + \frac{(12 - 18.4167)^2}{18.4167} + \frac{(38 - 31.4167)^2}{31.4167} + \cdots + \frac{(18 - 12.8333)^2}{12.8333}$$

$$= 0.6410 + 2.2357 + 1.3795 + 1.7601 + 0.7576 + 2.6422 + 1.6303 + 2.0801$$

$$= 13.126$$

$$\nu = rct - ct - r + 1 = (2)(2)(2) - (2)(2) - 2 + 1 = 3$$

$$\chi^2_{0.05,3} = 7.815$$

Reject H_0. Species is not independent of location and presence of disease.

$$0.005 < P < 0.001 \quad [P = 0.0044]$$

H_0:   Location is independent of species and disease.

H_A: Location is not independent of species and disease.

The expected frequencies $(\hat{f}_{ijl})$:

	Disease present		Disease absent		
	Species 1	Species 2	Species 1	Species 2	**Location totals**
Location 1	37.9167	33.8542	32.5000	25.7292	$C_1 = 130$
Location 2	18.0833	16.1458	15.5000	12.2708	$C_2 = 62$
Species and disease totals:	$(RT)_{11}$ $= 56$	$(RT)_{12}$ $= 50$	$(RT)_{21}$ $= 48$	$(RT)_{22}$ $= 38$	Grand total: $n = 192$

$$\chi^2 = \frac{(44 - 37.9167)^2}{37.9167} + \frac{(28 - 33.8542)^2}{33.8542} + \cdots + \frac{(18 - 12.2708)^2}{12.2708}$$

$$= 0.9760 + 1.0123 + 0.9308 + 1.2757 + 2.0464 + 2.1226 + 1.9516 + 2.6749$$

$$= 12.990$$

$$\nu = rct - rt - c + 1 = (2)(2)(2) - (2)(2) - 2 + 1 = 3$$

EXAMPLE 23.13   (continued)

$\chi^2_{0.05,3} = 7.815$

Reject H_0. Location is not independent of species and presence of disease.

$$0.001 < P < 0.005 \quad [P = 0.0047]$$

H_0:   Presence of disease is independent of species and location.

H_A: Presence of disease is not independent of species and location.

The expected frequencies ($\hat{f}_{ijl}$):

	Species 1		**Species 2**		
	Location 1	*Location 2*	*Location 1*	*Location 2*	**Disease totals**
Disease present	45.2708	12.1458	26.5000	22.0833	$T_1 = 106$
Disease absent	36.7292	9.8542	21.5000	17.9167	$T_2 = \;\;86$
Species and location totals:	$(RC)_{11}$ $= 82$	$(RC)_{12}$ $= 22$	$(RC)_{21}$ $= 48$	$(RC)_{22}$ $= 40$	*Grand total:* $n = 192$

$$\chi^2 = \frac{(44 - 45.2708)^2}{45.2708} + \frac{(12 - 12.1458)^2}{12.1458} + \cdots + \frac{(18 - 17.9167)^2}{17.9167}$$

$$= 0.0357 + 0.0018 + 0.0849 + 0.0003 + 0.0440 + 0.0022 + 0.1047 + 0.0004$$

$$= 0.274$$

$$\nu = rct - rc - t + 1 = (2)(2)(2) - (2)(2) - 2 + 1 = 3$$

$$\chi^2_{0.05,3} = 7.815$$

Do not reject H_0.

$$0.95 < P < 0.975 \quad [P = 0.96]$$

The Log-Likelihood Ratio.   In the hypothesis testing described above, the log-likelihood ratio (see Section 23.7) may be used in lieu of chi-square. Some authors prefer it and many multidimensional contingency table computer programs employ it.

23.9 LOG-LINEAR MODELS FOR MULTIDIMENSIONAL CONTINGENCY TABLES

The analysis of contingency tables with three or more dimensions (i.e., with three or more variables) is often accomplished (especially by computer) by using what are known as *log-linear models.** A "model" in statistical analysis is an expression of how observed data (in this case observed frequencies, f_{ijl}) are affected by variables and combinations of variables. "Log-linear" refers to a procedure whereby a multiplicative relationship

---

*This term was introduced, in 1969, by Y. M. M. Bishop and S. E. Fienberg (David, 1995).

EXAMPLE 23.14   Test for independence of two variables, following tests for partial dependence.

The hypothesis test of Example 23.12 concluded that all three variables are not mutually independent, while the last test in Example 23.13 concluded that presence of disease is independent of species and location. Therefore, it is desirable (and permissible) to test the following two-dimensional contingency table:

H_0:   Species occurrence is independent of location.

H_A: Species occurrence is not independent of location.

	Location 1	Location 2	Total
Species 1	82	22	104
Species 2	48	40	88
Total	130	62	192

$\chi_c^2 = 11.787$

$\chi_{c'}^2 = 12.690$

$\nu = (r-1)(c-1) = 1$

$\chi_{0.05,1}^2 = 3.841$

Reject H_0.

$$P < 0.001 \quad [P = 0.00037]$$

(e.g., $\hat{f}_{ijl} = R_i C_j T_l / n^2$) is transformed to a linear relationship by the use of logarithms (e.g., $\log \hat{f}_{ijl} = \log R_i + \log C_j + \log T_i - 2 \log n$).

In the terminology of log-linear models, one tests interactions of variables. A null hypothesis that states no interactions of two or more variables is one that implies that all the variables are independent. Hypotheses of partial independence and of conditional independence may be tested using appropriate log-linear models. Of the references cited at the beginning of Section 23.8, those of Fienberg (1970, 1980) and Everitt (1992: Chapter 5) contain accounts of log-linear models especially readable for nonmathematicians, and the first of them is oriented specifically toward biological data.

The expansions of Equations 23.10 and 23.11 to a three-dimensional contingency table are

$$G = 2 \left[\sum_i \sum_j \sum_l f_{ijl} \ln \left(\frac{f_{ijl}}{\hat{f}_{ijl}} \right) \right] \tag{23.23}$$

and

$$G = 2 \left[\sum_i \sum_j \sum_k f_{ijl} \ln f_{ijl} - \sum_i R_i \ln R_i - \sum_j C_j \ln C_j - \sum_l T_l \ln T_l + 2n \ln n \right]. \tag{23.24}$$

As with Equations 23.10 and 23.11, if common logarithms are used, instead of natural logarithms, then use 4.60517 instead of 2 preceding the left bracket.

To test partial independence, we may proceed as follows. To test whether the row variable is independent of the column and tier variables, Equation 23.23 is applied with

$\hat{f}_{ijl}$ obtained by Equation 23.16, or

$$G = 2\left[\sum_i \sum_j \sum_l f_{ijl} \ln f_{ijl} - \sum_i R_i \ln R_i - \sum_j \sum_l (CT)_{jl} \ln(CT)_{jl} + n \ln n\right] \quad (23.25)$$

may be used. To test whether the column variable is independent of the row and tier variables, Equation 23.23 is used with $\hat{f}_{ijl}$ obtained by Equation 23.19, or

$$G = 2\left[\sum_i \sum_j \sum_l f_{ijl} \ln f_{ijl} - \sum_j C_j \ln C_j - \sum_i \sum_l (RT)_{il} \ln(RT)_{il} + n \ln n\right] \quad (23.26)$$

may be employed. And to test whether the tier variable is independent of the row and column variables, use Equation 23.23 with $\hat{f}_{ijl}$ calculated by Equation 23.21, or use

$$G = 2\left[\sum_i \sum_j \sum_l f_{ijl} \ln f_{ijl} - \sum_l T_l \ln T_l - \sum_i \sum_j (RC)_{ij} \ln(RC)_{ij} + n \ln n\right]. \quad (23.27)$$

For the data of Example 23.12, the log-linear procedure would yield $G = 13.238$, and the tests for the three null hypotheses of Example 23.13 would yield $G = 13.233$, $G = 13.068$, and $G = 0.275$, respectively.

EXERCISES

23.1 Consider the following data for the abundance of a certain species of bird.
  (a) Using chi-square, test the null hypothesis that the ratio of numbers of males to females was the same in all four seasons.
  (b) Apply the G test to that hypothesis.

Sex	Spring	Summer	Fall	Winter
Males	163	135	71	43
Females	86	77	40	38

23.2 The following data are frequencies of skunks found with and without rabies in two different geographic areas.
  (a) Using the Yates-corrected chi-square, test the null hypothesis that the incidence of rabies in skunks is the same in both areas.
  (b) Test that hypothesis using the Haber-corrected chi-square.
  (c) Apply the G test to that hypothesis.
  (d) Use the Fisher exact test to test whether the western (W) population is more liable to have rabies than is the eastern (E) population.
  (e) Use the Fisher exact test to test whether the incidence of rabies is the same in skunks from both areas.

Area	With rabies	Without rabies
E	14	29
W	12	38

23.3 Data were collected as in Exercise 23.2, but with the additional tabulation of the sex of each skunk recorded, as follows. Test for mutual independence; and, if H_0 is rejected, test for partial independence.

Area	With rabies		Without rabies	
	Male	*Female*	*Male*	*Female*
E	42	33	55	63
W	84	51	34	48

24

MORE ON DICHOTOMOUS VARIABLES

In this chapter we shall concentrate on nominal scale data that come from a population with only two categories. As examples, members of a mammal litter might be classified as male or female, victims of a disease as dead or alive, trees in an area as "deciduous" or "evergreen," or progeny as color-blind or not color-blind. A nominal scale variable having two categories is said to be *dichotomous*. Such variables have already been discussed in the context of goodness of fit (Chapter 22) and contingency tables (Chapter 23).

The proportion of the population belonging to one of the two categories is denoted as p (here departing from the convention of using Greek letters for population parameters). Therefore, the proportion of the population belonging to the second class is $1 - p$, and the notation $q = 1 - p$ is commonly employed. For example, if 0.5 (i.e., 50%) of a population were male, then we would know that 0.5 (i.e., $1 - 0.5$) of the population were female, and we could write $p = 0.5$ and $q = 0.5$; if 0.4 (i.e., 40%) of a population were male, then 0.6 (i.e., 60%) of the population were female, and we could write $p = 0.4$ and $q = 0.6$.

If we took a random sample of ten from a population were $p = q = 0.5$, then we might expect that the sample would consist of five males and five females. However, we should not be too surprised to find such a sample consisting of six males and four females, or four males and six females, although neither of these combinations would be expected with as great a frequency as samples possessing the population sex ratio of 5:5. It would, in fact, be possible to obtain a sample of ten with nine males and one female, or even one consisting of all males, but the probabilities of such samples being encountered by random chance are relatively low.

If we were to obtain a large number of samples from the population under consideration, the frequency of samples consisting of no males, one male, two males, etc., would be described by the *binomial distribution* (sometimes referred to as the "Bernoulli distribution"*). Let us now examine binomial probabilities.

24.1 BINOMIAL PROBABILITIES

Consider a population consisting of two categories, where p is the proportion of individuals in one of the categories and $q = 1 - p$ is the proportion in the other. Then the probability of selecting at random from this population a member of the first category is p, and the probability of selecting a member of the second category is q.[†]

For example, let us say we have a population of female and male animals, in proportions of $p = 0.4$ and $q = 0.6$, respectively, and we take a random sample of two individuals from the population. The probability of the first being a female is p (i.e., 0.4) and the probability of the second being a female is also p. As the probability of two mutually exclusive events both occurring is the product of the probabilities of the two separate events (Section 5.7), the probability of having two females in a sample of two is $(p)(p) = p^2 = 0.16$; the probability of the sample of two consisting of two males is $(q)(q) = q^2 = 0.36$.

What is the probability of the sample of two consisting of one male and one female? This could occur by the first individual being a female and the second a male (with a probability of pq) or by the first being a male and the second a female (which would occur with a probability of qp). The probability of either of two mutually exclusive outcomes is the sum of the probabilities of each outcome (Section 5.6), so the probability of one female and one male in the sample is $pq + qp = 2pq = 2(0.4)(0.6) = 0.48$. Note that $0.16 + 0.36 + 0.48 = 1.00$.

Now consider another sample from this population, one where $n = 3$. The probability of all three individuals being female is $ppp = p^3 = (0.4)^3 = 0.064$. The probability of two females and one male is ppq (for a sequence of ♀♀♂) + pqp (for ♀♂♀) + qpp (for ♂♀♀), or $3p^2q = 3(0.4)^2(0.6) = 0.288$. The probability of one female and two males is pqq (for ♀♂♂) + qpq (for ♂♀♂) + qqp (for ♂♂♀), or $3pq^2 = 3(0.4)(0.6)^2 = 0.432$. And, finally, the probability of all three being males is $qqq = q^3 = (0.6)^3 = 0.216$. Note that $p^3 + 3p^2q + 3pq^2 + q^3 = 0.064 + 0.288 + 0.432 + 0.216 = 1.000$ (meaning that there is a 100% probability—that is, it is certain—that the three animals will be in one of these three combinations of sexes).

---

*The binomial formula discussed in the following section was first described in 1676 by Sir Isaac Newton (1642–1727), the great English scientist and mathematician; and its first proof, for positive integer exponents, was given by the Swiss mathematician, Jacques (also known as Jakob or James) Bernoulli (1654–1705) in a 1713 publication (Cajori, 1954). Each observed event from a binomial distribution is sometimes called a "Bernoulli trial." David (1995) ascribes the first use of the term "binomial distribution" to G. U. Yule, in 1911.

[†]This assumes "sampling with replacement." That is, each individual in the sample is taken at random from the population and then is returned to the population before the next member of the sample is selected. Sampling without replacement is discussed in Section 24.2. If the population is very large compared to the size of the sample, then sampling with and without replacement are indistinguishable in practice.

If we performed the same exercise with $n = 4$, we would find that the probability of four females is $p^4 = (0.4)^4 = 0.0256$, the probability of three females (and one male) is $4p^3q = 4(0.4)^3(0.6) = 0.1536$, the probability of two females is $6p^2q^2 = 0.3456$, the probability of one female is $4pq^3 = 0.3456$, and the probability of no females (i.e., all four are male) is $q^4 = 0.1296$. (The sum of these five terms is 1.0000, a good arithmetic check.)

If a random sample of size n is taken from a binomial population, then the probability of X individuals being in one category (and, therefore, $n - X$ individuals in the second category) is

$$P(X) = \binom{n}{X} p^X q^{n-X}. \tag{24.1}$$

In this equation, $p^X q^{n-X}$ refers to the probability of sample consisting of X items, each having a probability of p, and $n - X$ items, each with probability q. The *binomial coefficient*,

$$\binom{n}{X} = \frac{n!}{X!(n-X)!}, \tag{24.2}$$

is the number of ways X items of one kind can be arranged with $n - X$ items of a second kind, or, in other words, the number of possible *combinations* of n items divided into one group of X items and a second group of $n - X$ items. (See Section 5.3 for a discussion of combinations; Equation 5.3 explained the factorial notation, "!".) Therefore, Equation 24.1 can be written as

$$P(X) = \frac{n!}{X!(n-X)!} p^X q^{n-X}. \tag{24.3}$$

Thus, $\binom{n}{X} p^X q^{n-X}$ is the Xth term in the expansion of $(p + q)^n$, and Table 24.1 shows this expansion for powers up through 6. Note that for any power, n, the sum of the two exponents in any term is n. Furthermore, the first term will always be p^n, the second will always contain $p^{n-1}q$, the third will always contain $p^{n-2}q^2$, etc., with the last term always being q^n. The sum of all the terms in a binomial expansion will always be 1.0, for $p + q = 1$, and $(p + q)^n = 1^n = 1$.

As for the coefficients of these terms in the binomial expansion, the Xth term of the nth power expansion can be calculated by Equation 24.3. Furthermore, the examination of these coefficients as shown in Table 24.2 has been deemed interesting for centuries.

TABLE 24.1 Expansion of the Binomial, $(p + q)^n$

n	$(p + q)^n$
1	$p + q$
2	$p^2 + 2pq + q^2$
3	$p^3 + 3p^2q + 3pq^2 + q^3$
4	$p^4 + 4p^3q + 6p^2q^2 + 4pq^3 + q^4$
5	$p^5 + 5p^4q + 10p^3q^2 + 10p^2q^3 + 5pq^4 + q^5$
6	$p^6 + 6p^5q + 15p^4q^2 + 20p^3q^3 + 15p^2q^4 + 6pq^5 - q^6$

TABLE 24.2 Binomial Coefficient, $_nC_X$

n	$X = 0$	1	2	3	4	5	6	7	8	9	10	Sum of coefficients
1	1	1										$2 = 2^1$
2	1	2	1									$4 = 2^2$
3	1	3	3	1								$8 = 2^3$
4	1	4	6	4	1							$16 = 2^4$
5	1	5	10	10	5	1						$32 = 2^5$
6	1	6	15	20	15	6	1					$64 = 2^6$
7	1	7	21	35	35	21	7	1				$128 = 2^7$
8	1	8	28	56	70	56	28	8	1			$256 = 2^8$
9	1	9	36	84	126	126	84	36	9	1		$512 = 2^9$
10	1	10	45	120	210	252	210	120	45	10	1	$1024 = 2^{10}$

This arrangement is known as *Pascal's triangle.** We can see from this triangular array that any binomial coefficient is the sum of two coefficients on the line above it, namely,

$$\binom{n}{X} = \binom{n-1}{X-1} + \binom{n-1}{X}.\tag{24.4}$$

This can be more readily observed if we display the triangular array as follows:

```
                1
             1     1
          1     2     1
       1     3     3     1
    1     4     6     4     1
 1     5    10    10     5     1
```

Also note that the sum of all coefficients for the nth power binomial expansion is 2^n. Appendix Table B.26a presents binomial coefficients for much larger n's and X's, and they will be found useful later in this chapter.

Thus, we can calculate probabilities of category frequencies occurring in random samples from binomial population. If, for example, a sample of five (i.e., $n = 5$) is taken from a population composed of 50% males and 50% females (i.e., $p = 0.5$ and $q = 0.5$) then Example 24.1 shows how Equation 24.3 is used to determine the probability of the sample containing 0 males, 1 male, 2 males, 3 males, 4 males, and 5 males. These

---

*Blaise Pascal (1623–1662), French mathematician and physicist and one of the founders of probability theory (in 1654, immediately before abandoning mathematics to become a religious recluse). He had his triangular binomial coefficient derivation published in 1665, although knowledge of the triangular properties appears in Chinese writings as early as 1303 (Cajori, 1954; David, 1962; Struik, 1967: 79). Pascal also invented (at age 19) a mechanical adding and subtracting machine which, though patented in 1649, proved too expensive to be practical to construct (Asimov, 1982: 130–131). His significant contributions to the study of fluid pressures have been honored by naming the international unit of pressure the pascal, which is a pressure of one newton per square meter (where a newton—named for Sir Isaac Newton—is the unit of force representing a one-kilogram mass accelerating at the rate of one meter per second per second). Pascal is also the name given to a modern computer language. The relationship of Pascal's triangle to $_nC_X$ was first published in 1685 by the English mathematician, John Wallis (1616–1703) (David, 1962: 123–124).

probabilities are found to be 0.03125, 0.15625, 0.31250, 0.31250, 0.15625, and 0.03125, respectively. This enables us to state that if we took 100 random samples of five animals each from the population, about three of the sample [i.e., $(0.03125)(100) = 3.125$ of them] would be expected to contain all females, about sixteen [i.e., $(0.15625)(100) = 15.625$] to contain one male and four females, thirty-one [i.e., $(0.31250)(100)$] to consist of two males and three females, etc. If we took 1400 random samples of five, then $(0.03125)(1400) = 43.75$ [i.e., about 44] of them would be expected to contain all females, etc. Figure 24.1a shows graphically the binomial distribution for $p = q = 0.5$, for $n = 5$. Note, from Fig 24.1a and Example 24.1, that when $p = q = 0.5$ the distribution is symmetrical [i.e., $P(0) = P(n)$, $P(1) = P(n-1)$, etc.], and Equation 24.3 becomes

$$P(X) = \frac{n!}{X!(n-X)!} 0.5^n. \tag{24.5}$$

Appendix Table B.26b gives binomial probabilities for $n = 2$ to $n = 20$, for $p = 0.5$.

Example 24.2 presents the calculation of binomial probabilities for the case where $n = 5$, $p = 0.3$, and $q = 1 - 0.3 = 0.7$. Thus, if one were sampling a population consisting of 30% males and 70% females, 0.16807 (i.e., 16.807%) of the samples would be expected to contain no males, 0.36015 to contain one male and four females, etc. Fig. 24.1b presents this binomial distribution graphically, whereas Fig. 24.1c shows the distribution where $p = 0.1$ and $q = 0.9$.

For calculating binomial probabilities for large n, it is often convenient to employ logarithms. For this reason, Appendix Table B.40, a table of logarithms of factorials, is provided. Alternatively, it is useful to note that the denominator of Equation 24.3 cancels out much of the numerator, so that it is possible to simplify the computation of $P(X)$,

---

EXAMPLE 24.1 Computing binomial probabilities, $P(X)$, where $n = 5$, $p = 0.5$, and $q = 0.5$ (following Equation 24.3).

X	$P(X)$
0	$\frac{5!}{0!5!}(0.5^0)(0.5^5) = (1)(1.0)(0.03125) = 0.03125$
1	$\frac{5!}{1!4!}(0.5^1)(0.5^4) = (5)(0.5)(0.0625) = 0.15625$
2	$\frac{5!}{2!3!}(0.5^2)(0.5^3) = (10)(0.25)(0.125) = 0.31250$
3	$\frac{5!}{3!2!}(0.5^3)(0.5^2) = (10)(0.125)(0.25) = 0.31250$
4	$\frac{5!}{4!1!}(0.5^4)(0.5^1) = (5)(0.0625)(0.5) = 0.15625$
5	$\frac{5!}{5!0!}(0.5^5)(0.5^0) = (1)(0.03125)(1.0) = 0.03125$

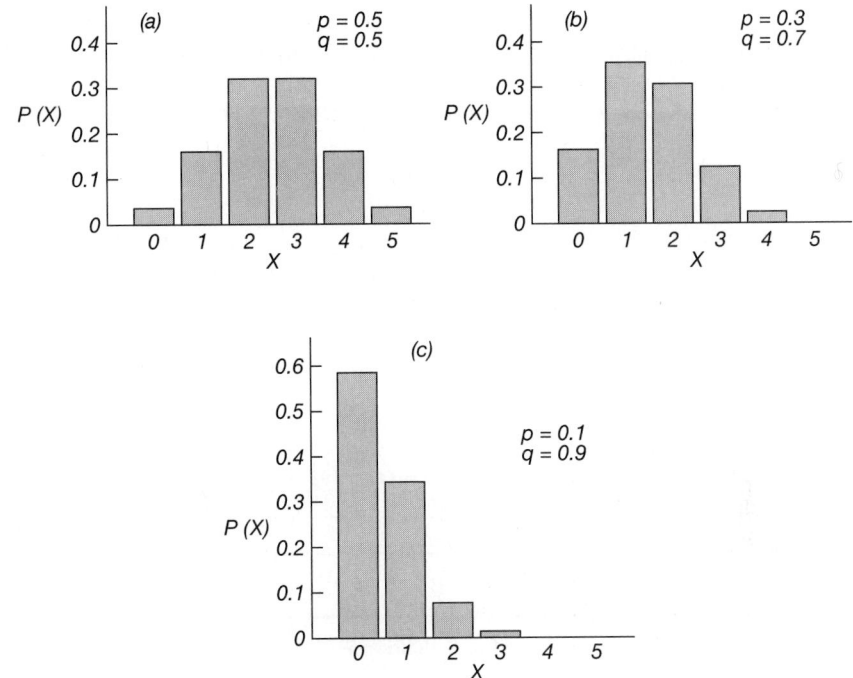

Figure 24.1   The binomial distribution, for $n = 5$. (a) $p = q = 0.5$. (b) $p = 0.3$, $q = 0.7$. (c) $p = 0.1$, $q = 0.9$. These graphs were drawn utilizing the proportions given by Equation 24.1.

especially in the tails of the distribution (i.e., for low X and for high X), as shown in Example 24.3. If p is very small, then the use of the Poisson distribution (Section 25.1), should be considered.*

The mean of a binomial distribution of counts X, is

$$\mu_X = np,\tag{24.6}$$

the variance[†] is

$$\sigma_X^2 = npq,\tag{24.8}$$

and the standard deviation of X is

$$\sigma_X = \sqrt{npq}.\tag{24.9}$$

---

*Raff (1956) and Molenaar (1969a, 1969b) discuss several approximations to the binomial distribution, including the normal and Poisson distributions.

[†]A measure of symmetry (see Section 6.1) for a binomial distribution is

$$\gamma_1 = \frac{q - p}{\sqrt{npq}},\tag{24.7}$$

so it can be seen that $\gamma_1 = 0$ only when $p = q = 0.05$, $\gamma_1 > 0$ implies a distribution skewed to the right (as in Figs. 24.1b and 24.1c) and $\gamma_1 < 0$ indicates a distribution skewed to the left.

EXAMPLE 24.2 Computing binomial probabilities, $P(X)$, where $n = 5, p = 0.4, q = 0.7$ (following Equation 24.3).

X	$P(X)$
0	$\dfrac{5!}{0!5!}(0.3^0)(0.7^5) = (1)(1.0)(0.16807) = 0.16807$
1	$\dfrac{5!}{1!4!}(0.3^1)(0.7^4) = (5)(0.3)(0.2401) = 0.36015$
2	$\dfrac{5!}{2!3!}(0.3^2)(0.7^3) = (10)(0.09)(0.343) = 0.30870$
3	$\dfrac{5!}{3!2!}(0.3^3)(0.7^2) = (10)(0.027)(0.49) = 0.13230$
4	$\dfrac{5!}{4!1!}(0.3^4)(0.7^1) = (5)(0.0081)(0.7) = 0.02835$
5	$\dfrac{5!}{5!0!}(0.3^5)(0.7^0) = (1)(0.00243)(1.0) = 0.00243$

EXAMPLE 24.3 Computing binomial probabilities, $P(X)$, with $n = 400, p = 0.02$, and $q = 0.98$.

(Many calculators can operate with large powers of numbers; otherwise, logarithms may be used.)

X	$P(X)$
0	$\dfrac{n!}{0!(n-0)!}p^0 q^{n-0} = q^n = 0.98^{400} = 0.00031$
1	$\dfrac{n!}{1!(n-1)!}p^1 q^{n-1} = npq^{n-1} = (400)(0.02)(0.98^{399}) = 0.00253$
2	$\dfrac{n!}{2!(n-2)!}p^2 q^{n-2} = \dfrac{n(n-1)}{2!}p^2 q^{n-2} = \dfrac{(400)(399)}{2}(0.02^2)(0.98^{398}) = 0.01028$
3	$\dfrac{n!}{3!(n-3)!}p^3 q^{n-3} = \dfrac{n(n-1)(n-2)}{3!}p^3 q^{n-3} = \dfrac{(400)(399)(398)}{(3)(2)}(0.02^3)(0.98^{397})$
	$\qquad\qquad\qquad\qquad\qquad\qquad\qquad = 0.02784$

and so on.

Thus, if we have a binomially distributed population where p (e.g., the proportion of males) $= 0.5$ and q (e.g., the proportion of females) $= 0.5$ and we take ten samples from that population, the mean of the ten X's (i.e., the mean number of males per sample) would be expected to be $np = (10)(0.05) = 5$ and the standard deviation of the ten X's would be expected to be $\sqrt{npq} = \sqrt{(10)(0.5)(0.5)} = 1.58$. Our concern typically is with the distribution of the expected probabilities rather than the expected X's, as will be explained in Section 24.3.

24.2 THE HYPERGEOMETRIC DISTRIBUTION

Binomial probabilities (Section 24.1) may result from what is known as "sampling with replacement." This means that after an item is randomly removed from the population to be part of the sample it is returned to the population before randomly selecting another item for inclusion in the sample. (This assumes that after the item is returned to the population it has the same chance of being selected again as does any other member of the population; in many biological situations—such as catching a mammal in a trap—this is not so.) Sampling with replacement ensures that the probability of selecting an item belonging to a specific one of the binomial categories remains constant. For, if sampling from an actual population is performed without replacement, then selecting an item from the first category reduces p and increases q (and, if the selected item were from the second category, then q would decrease and p would increase). Binomial probabilities may also arise from sampling "hypothetical" populations (introduced in Section 2.2), such as proportions of heads and tails from all possible coin tosses or of males and females in all possible fraternal twins.

Probabilities associated with sampling without replacement follow the *hypergeometric distribution*, instead of the binomial distribution. The probability of obtaining a sample of n items from a hypergeometric distribution, where the sample consists of X items in one category and $n - X$ items in a second category, is

$$P(X) = \frac{\binom{N_1}{X}\binom{N_2}{n-X}}{\binom{N_T}{n}} \tag{24.10}$$

$$= \frac{N_1!N_2!n!(N_T-n)!}{X!(N_1-X)!(n-X)!(N_2-n+X)!N_T!}. \tag{24.11}$$

Here, N_T is the total number of items in the population, N_1 in category 1 and N_2 in category 2. For example, we could ask what the probability is of forming a sample consisting of three women and two men by taking five people at random from a group of eight women and six men. As $N_1 = 8, N_2 = 6, N_T = 14, n = 5$, and $X = 3$, the probability is

$$P(X) = \frac{\binom{8}{3}\binom{6}{2}}{\binom{14}{5}}$$

$$= \frac{8!\,6!\,5!\,9!}{3!\,5!\,2!\,4!\,14!}$$

$$= \frac{(8\cdot7\cdot6\cdot5\cdot4\cdot3\cdot2)(6\cdot5\cdot4\cdot3\cdot2)(5\cdot4\cdot3\cdot2)(9\cdot8\cdot7\cdot6\cdot5\cdot4\cdot3\cdot2)}{(3\cdot2)(5\cdot4\cdot3\cdot2)(2)(4\cdot3\cdot2)(14\cdot13\cdot12\cdot11\cdot10\cdot9\cdot8\cdot7\cdot6\cdot5\cdot4\cdot3\cdot2)}$$

$$= 0.4196.$$

If the population is very large compared to the size of the sample, then the result of sampling with replacement is indistinguishable from that of sampling without replacement, and the hypergeometric distribution approaches—and is approximated by—the binomial distribution. Table 24.3 compares the binomial distribution with $p = 0.01$ and $n = 100$ to three hypergeometric distributions with the same p and n but with different population sizes. It can be seen that for larger N_T the hypergeometric is closer to the binomial distribution.

TABLE 24.3 The Hypergeometric Distribution Where N_1, the Number of Items of One Category, Is 1% of the Population Size, N_T; and the Binomial Distribution with $p = 0.01$; the Sample Size, n, is 100 in Each Case

X	$P(X)$ for hypergeometric: $N_T = 1000, N_1 = 10$	$P(X)$ for hypergeometric: $N_T = 2000, N_1 = 20$	$P(X)$ for hypergeometric: $N_T = 5000, N_1 = 50$	$P(X)$ for binomial: $p = 0.01$
0	0.34693	0.35669	0.36235	0.36603
1	0.38937	0.37926	0.37347	0.36973
2	0.19447	0.18953	0.18670	0.18486
3	0.05691	0.05918	0.06032	0.06100
4	0.01081	0.01295	0.01416	0.01494
5	0.00139	0.00211	0.00258	0.00290
6	0.00012	0.00027	0.00038	0.00046
> 6	0.00000	0.00001	0.00004	0.00008
Total	1.00000	1.00000	1.00000	1.00000

24.3 SAMPLING A BINOMIAL POPULATION

Let us consider a population of N individuals: Y individuals in one category and $N - Y$ in the second category. Then the proportion of individuals in the first category is

$$p = \frac{Y}{N} \tag{24.12}$$

and the proportion in the second is

$$q = 1 - p \quad \text{or} \quad q = \frac{N - Y}{N}. \tag{24.13}$$

If the sample of n observations is taken from this population, with replacement, and X observations are in category and $n - X$ are in one the other, then the population parameter p is estimated by the sample statistic

$$\hat{p} = \frac{X}{n}, \tag{24.14}$$

which is the proportion of the sample that is in the first category.* The estimate of q is

$$\hat{q} = 1 - p \quad \text{or} \quad \hat{q} = \frac{n - X}{n}, \tag{24.15}$$

which is the proportion of the sample occurring in the second category. In Example 24.4 we have $X = 4$ and $n = 20$, so $\hat{p} = 4/20 = 0.20$ and $\hat{q} = 1 - p = 0.80$.

EXAMPLE 24.4   Sampling a binomial population.

From a population of male and female spiders, a sample of twenty is taken, which contains four males and sixteen females.

$n = 20$

$X = 4$

By Equation 24.14,

$$\hat{p} = \frac{X}{n} = \frac{4}{20} = 0.20.$$

Therefore we estimate that 20% of the population are males and, by Equation 24.15,

$$\hat{q} = 1 - \hat{p} = 1 - 0.20 = 0.80$$

or

$$\hat{q} = \frac{n - X}{n} = \frac{20 - 4}{20} = \frac{16}{20} = 0.80,$$

so we estimate that 80% of the population are females.

The variance of the estimate, $\hat{p}$ (or of $\hat{q}$) is, by Equation 24.17,

$$s_{\hat{p}}^2 = \frac{\hat{p}\hat{q}}{n - 1} = \frac{(0.20)(0.80)}{20 - 1} = 0.008421.$$

If we consider that the sample consists of four 1's and sixteen 0's, then $\sum X = 4$, $\sum X^2 = 4$, and the variance of the twenty 1's and 0's is, by Equation 4.9, $s^2 = (4 - 4^2/20)/(20 - 1) = 0.168421$, and the variance of the mean, by Equation 6.17, is $s_{\bar{X}}^2 = 0.168421/20 = 0.008421$.

The standard error (or standard deviation) of $\hat{p}$ (or of $\hat{q}$) is, by Equation 24.21, $s_{\hat{p}} = \sqrt{0.008421} = 0.092$.

If our sample of twenty were returned to the population (or if the population were extremely large), and we took another sample of twenty, and repeated this multiple sampling procedure many times, we could obtain many calculations of $\hat{p}$, each estimating the population parameter p. If, in the population, $p = 0$, then obviously any sample from that population would have $\hat{p} = 0$; and if $p = 1.0$, then each and every $\hat{p}$ would be 1.0. However, if p is neither 0 nor 1.0, then all the many samples from the population

---

*Placing the symbol "^" above a letter is statistical convention for denoting an estimate of the quantity which that letter denotes. Thus, $\hat{p}$ refers to an estimate of p, and the statistic $\hat{q}$ is a sample estimate of the population parameter q.

would not have the same values of $\hat{p}$. The variance of all possible $\hat{p}$'s is

$$\sigma_{\hat{p}}^2 = \frac{pq}{n}, \tag{24.16}$$

which can be estimated from our sample as

$$s_{\hat{p}}^2 = \frac{\hat{p}\hat{q}}{n-1}. \tag{24.17}$$

This variance is essentially a variance of means, so Equation 24.16 is analogous to Equation 6.14, and Equation 24.17 to Equation 6.17. In Example 24.4 it is shown that the latter is true. The variance of $\hat{q}$ is the same as the variance of $\hat{p}$; i.e.,

$$\sigma_{\hat{q}}^2 = \sigma_{\hat{p}}^2 \tag{24.18}$$

and

$$s_{\hat{q}}^2 = s_{\hat{p}}^2 \tag{24.19}$$

The standard error of $\hat{p}$ (or of $\hat{q}$), also called the standard deviation, is

$$\sigma_{\hat{p}} = \sqrt{\frac{pq}{n}}, \tag{24.20}$$

which is estimated from a sample as

$$s_{\hat{p}} = \sqrt{\frac{\hat{p}\hat{q}}{n-1}}. \tag{24.21}*$$

The possible values of $\sigma_{\hat{p}}^2$, $\sigma_{\hat{q}}^2$, $\sigma_{\hat{p}}$, and $\sigma_{\hat{q}}$ range from a minimum of zero when either $\hat{p}$ or $\hat{q}$ is zero, to a maximum when $p = q = 0.5$; and $s_{\hat{p}}^2$, $s_{\hat{q}}^2$, $s_{\hat{p}}$, and $s_{\hat{q}}$ can range from a minimum of zero when either $\hat{p}$ or $\hat{q}$ is zero, to a maximum when $\hat{p} = \hat{q} = 0.5$.

Sampling Finite Populations†. If n is substantial portion of the entire population of size N, and sampling is without replacement, then a finite population correction is called for (just like that found in Section 7.7) in estimating $\sigma_{\hat{p}}^2$ or $\sigma_{\hat{p}}$:

$$s_{\hat{p}}^2 = \frac{\hat{p}\hat{q}}{n-1}\left(1 - \frac{n}{N}\right) \tag{24.23}$$

and

$$s_{\hat{p}} = \sqrt{\frac{\hat{p}\hat{q}}{n-1}\left(1 - \frac{n}{N}\right)}, \tag{24.24}$$

---

*One often sees

$$s_{\hat{p}} = \sqrt{\frac{\hat{p}\hat{q}}{n}} \tag{24.22}$$

used to estimate $\sigma_{\hat{p}}$. Although it is inaccurate (being an underestimate), when n is large the difference between Equations 24.21 and 24.22 is slight.

†These procedures are from Cochran (1977: 52). When sampling from finite populations, the data follow the hypergeometric (Section 24.2), rather than the binomial, distribution.

when n/N is called the *sampling fraction*, and $1 - n/N$ is the *finite population correction*, the latter also being written as $(N - n)/N$. As N becomes very large compared to n, Equation 24.23 approaches Equation 24.17 and Equation 24.24 approaches 24.21.

We can estimate Y, the total number of occurrences in the population in the first category, as

$$\hat{Y} = \hat{p}N;$$ (24.25)

and the variance and standard error of this estimate are

$$s_{\hat{Y}}^2 = \frac{N(N - n)\hat{p}\hat{q}}{n - 1}$$ (24.26)

and

$$s_{\hat{Y}} = \sqrt{\frac{N(N - n)\hat{p}\hat{q}}{n - 1}},$$ (24.27)

respectively.

24.4 CONFIDENCE LIMITS FOR POPULATION PROPORTIONS

Using a relationship between the F distribution and the binomial distribution (Fisher and Yates, 1963: 3), a confidence interval may be computed for the binomial parameter p (Bliss, 1967: 199–201; Brownlee, 1965: 148–149). As demonstrated in Example 24.5, the lower confidence limit for p is

$$L_1 = \frac{X}{X + (n - X + 1)F_{\alpha(2), \nu_1, \nu_2}},$$ (24.28)

where

$$\nu_1 = 2(n - X + 1)$$

and

$$\nu_2 = 2X.$$

The upper confidence limit for p is

$$L_2 = \frac{(X + 1)F_{\alpha(2), \nu_1' \nu_2'}}{n - X + (X + 1)F_{\alpha(2), \nu_1', \nu_2'}},$$ (24.29)

with

$$\nu_1' = 2(X + 1) = \nu_2 + 2$$

and

$$\nu_2' = 2(n - X) = \nu_1 - 2.$$

EXAMPLE 24.5   Determination of the 95% confidence interval for the binomial population parameter, p.

A random sample of 200 persons contains four with immunity to a certain viral disease. What proportion of the population possesses such immunity?

$$n = 200$$

$$X = 4$$

$$\hat{p} = \frac{X}{n} = \frac{4}{200} = 0.0200$$

For the lower 95% confidence limit:

$$\nu_1 = 2(n - X + 1) = 2(200 - 4 + 1) = 394$$

$$\nu_2 = 2X = 2(4) = 8$$

$$F_{0.05(2), 394, 8} \approx F_{0.05(2), \infty, 8} = 3.67$$

$$L_1 = \frac{X}{X + (n - X + 1)F_{0.05(2), 394, 8}} \approx \frac{4}{4 + (200 - 4 + 1)(3.67)} = 0.00550$$

For the upper 95% confidence limit:

$$\nu_1' = 2(X + 1) = 2(4 + 1) = 10$$

or $$\nu_1' = \nu_2 + 2 = 8 + 2 = 10$$

$$\nu_2' = 2(n - X) = 2(200 - 4) = 392$$

or $$\nu_2' = \nu_1 - 2 = 394 - 2 = 392$$

$$F_{0.05(2), 10, 392} \approx F_{0.05(2), 10, 300} = 2.09$$

$$L_2 = \frac{(X + 1)F_{0.05(2), 10, 392}}{n - X + (X + 1)F_{0.05(2), 10, 392}} \approx \frac{(4 + 1)(2.09)}{200 - 4 + (4 + 1)(2.09)} = 0.0506$$

Therefore, we can state:

$$P(0.00550 \le p \le 0.0506) = 0.95,$$

meaning that we have 95% confidence in stating that the interval formed by 0.00550 and 0.0506 includes the population parameter, p.

Note: By harmonic interpolation (see beginning of Appendix B) we can find that $F_{0.05(2), 394, 8} = 3.69$ and $F_{0.05(2), 10, 392} = 2.10$, which when used in the computation of the above confidence limits yield $L_1 = 0.00547$ and $L_2 = 0.0506$. Appropriate computer programs will give the exact critical values as $F_{0.05(2), 394, 8} = 3.6879$ and $F_{0.05(2), 10, 392} = 2.0811$, which yield $L_1 = 0.00548$ and $L_2 = 0.0504$.

Because Appendix Table B.4 gives critical values of F to only three significant figures, the calculated confidence limits may be in error in the third significant figure. Error is particularly likely for $X = 0$, $X = n - 1$, and $X = n$, for which cases the following exact computations are much preferable (Blyth, 1986):

$$\text{for } X = 0: \quad L_2 = 1 - (\alpha/2)^{1/n}; \tag{24.30}$$

$$\text{for } X = n - 1: \quad L_2 = (1 - \alpha/2)^{1/n}; \tag{24.31}$$

$$\text{for } X = n: \quad L_2 = 1. \tag{24.32}$$

Also:

$$L_1 \text{ for } X = 1 - L_2 \text{ for } n - X; \tag{24.33}$$

$$L_2 \text{ for } X = L_1 \text{ for } n - X. \tag{24.34}$$

For large n the required critical values of F will not be found in Appendix Table B.4 and harmonic interpolation, as explained at the beginning of Appendix B, will be desirable. Some computer routines also provide critical values of F. The useful handbook by Burstein (1971) allows the determination of binomial confidence limits to an accuracy of 0.1% of the exact limits.

The confidence limits for q may be obtained by subtracting each of the confidence limits for p from 1. Thus, for Example 24.5, $q = 1 - p = 0.9800$, and the 95% confidence limits for q are: $L_1 = 1 - 0.0506 = 0.9494$ and $L_2 = 1 - 0.00550 = 0.99450$. Burstein (1971) gives tables and procedures for determining the sample size necessary to estimate p with a desired precision.

Approximations.   Normal and other approximations are for available for calculating confidence limits for proportions (e.g., Cochran, 1977: 57-58; Fleiss, 1981: 14-15; Fujino, 1980; Ghosh, 1979); but such methods often lack accuracy, especially when n is small or p is near 0 or 1, or they are cumbersome. Blyth (1986) compared several normal approximations and found that of Pratt (1968) to be excellent. But, with the availability of the exact method of Equations 24.28 and 24.29, the use of approximate methods may be avoided.

Sample Size Requirements.   If p and q are not very close to 0 or 1, then one can estimate how large a sample is required to estimate p (and q) with an error no greater than δ. This sample size is

$$n = \frac{Z_{\alpha(2)}^2 pq}{\delta^2} \tag{24.35}$$

(Cochran, 1977: 75-76), where $Z_{\alpha(2)}$ is a 2-tailed normal deviate. If the sample sample size, n, is not negligibly small compared to the population size, N, then the required sample size is estimated as

$$m = \frac{n}{1 + \dfrac{n - 1}{N}}, \tag{24.36}$$

where n is as in Equation 24.35 (Cochran, 1977: 75-76). Recall that $t_{\alpha(2),\infty} = Z_{\alpha(2)}$. This sample size is also one which is required to perform a binomial test (Section 24.6) that is capable of detecting a population difference as small as δ from the hypothesized p. Burstein (1971) gives tables allowing for estimation of sample size for p or n not appropriate to the normal approximation.

If one does not have an estimate of p and q, but desires to estimate them with an error of δ, then a sample size n or more should be taken, where, instead of Equation 24.35,

$$n = \frac{Z_{\alpha(2)}^2}{4\delta^2}. \tag{24.37}$$

If n/N is not a negligible fraction, then use the sample size indicated by Equation 24.36, using the n of Equation 24.37.

Confidence Limits with Finite Populations. If n is appreciably large relative to N and sampling is without replacement, then we are said to have sampled a finite population, and the accuracy of estimating p improves greatly as n approaches N. In such cases the lower confidence limit for p is

$$(L_1)_c = \frac{X - 0.5}{n} - \left(\frac{X - 0.5}{n} - L_1 \right) \sqrt{1 - \frac{n}{N}} \tag{24.38}$$

and the upper confidence limit is

$$(L_2)_c = \frac{X'}{n} + \left(L_2 - \frac{X'}{n} \right) \sqrt{1 - \frac{n}{N}}, \tag{24.39}$$

where

$$X' = X + \frac{X}{n} \tag{24.40}$$

(see Burstein, 1975) and L_1 and L_2 are obtained as at the beginning of Section 24.4. The confidence interval is shorter with the finite population correction applied.

24.5 GOODNESS OF FIT FOR THE BINOMIAL DISTRIBUTION

When p Is Hypothesized to Be Known. In some biological situations the population proportions, p and q, might be postulated, as from theory. For example, theory might tell us that 50% of mammalian sperm contain an X chromosome, whereas 50% contain a Y chromosome, and we can expect a 1:1 sex ratio among the offspring. We may wish to test the hypothesis that our sample came from a binomially distributed population with equal sex frequencies. We may do this as follows, by using the goodness of fit test introduced in Chapter 22.

Let us suppose that we have tabulated the sexes of the offspring from fifty-four litters of five animals each (Example 24.6). Setting $p = q = 0.5$, the proportion of each possible litter composition can be computed by the procedures of Example 24.1, using Equation 24.3, or they can be read directly from Appendix Table B.26b. From these proportions, we can tabulate expected frequencies, and then we can subject observed and expected frequencies of each type of litter to a chi-square goodness of fit analysis (see Section 22.1), with $k - 1$ degrees of freedom (k being the number of classes of X). In Example 24.6, we do not reject the null hypothesis, and therefore conclude that the sampled population is binomial with $p = 0.5$.

To avoid bias in this chi-square computation, no expected frequency should be less than 1.0 (Cochran, 1954). If such small frequencies occur, then frequencies in the appropriate extreme classes of X may be pooled to arrive at sufficiently large $\hat{f}_i$ values. Such pooling was not necessary in Example 24.6, as no $\hat{f}_i$ was less than 1.0. But it will be shown in Example 24.7.

EXAMPLE 24.6   Goodness of fit of a binomial distribution, when p is postulated.

The data consist of observed frequencies of females in fifty-four litters of five offspring per litter. $X = 0$ denotes a litter having no females, $X = 1$ a litter having one female, etc.; f is the observed number of litters, and $\hat{f}$ is the number of litters expected if the null hypothesis is true. Computation of the values of $\hat{f}$ requires the values of $P(X)$, as obtained in Example 24.1.

H_0:   The sexes of the offspring reflect a binomial distribution with $p = q = 0.5$.

H_A: The sexes of the offspring do not reflect a binomial distribution with $p = q = 0.5$.

X_i	f_i	$\hat{f}_i$
0	3	$(0.03125)(54) = 1.6875$
1	10	$(0.15625)(54) = 8.4375$
2	14	$(0.31250)(54) = 16.8750$
3	17	$(0.31250)(54) = 16.8750$
4	9	$(0.15625)(54) = 8.4375$
5	1	$(0.03125)(54) = 1.6875$

$\chi^2 = 1.0208 + 0.2894 + 0.4898 + 0.0009 + 0.0375 + 0.2801 = 2.1185$

$\nu = k - 1 = 6 - 1 = 5$

$\chi^2_{0.05,5} = 11.070$

Therefore, do not reject H_0.

$$0.75 < P < 0.90 \quad [P = 0.83]$$

The G statistic (Section 22.7) may be calculated in lieu of chi-square, with the summation being executed over all classes except those where not only $f_i = 0$ but also all more extreme f_i's are zero. The Kolmogorov-Smirnov statistic of Section 22.8 could also be used to determine the goodness of fit. Heterogeneity testing (Section 22.6) may be performed for several sets of data hypothesized to have come from a binomial distribution.

If the preceding null hypothesis had been rejected, we might have looked in several directions for a biological explanation. The rejection of H_0 might have indicated that the population p was in fact, not 0.5. Or, it might have indicated that the underlying distribution was not binomial. The latter possibility may occur when membership of an individual in one of the two possible categories is dependent upon another individual in the sample. In Example 24.6, for instance, identical twins (or other multiple identical births) might have been a common occurrence in the species in question. In that case,

EXAMPLE 24.7   Goodness of fit of a binomial distribution, when p is estimated from the sample data.

The data consist of observed frequencies of left-handed persons in seventy-five samples of eight persons each. $X = 0$ denotes a sample with no left-handed persons, $X = 1$ a sample with one left-handed person, etc.; f is the observed number of samples, and $\hat{f}$ is the number of samples expected if the null hypothesis is true. Each $\hat{f}$ is computed by multiplying 75 by $P(X)$, where $P(X)$ is obtained from Equation 24.3 by substituting $\hat{p}$ and $\hat{q}$ for p and q, respectively.

H_0:   The frequencies of left- and right-handed persons in the population follow a binomial distribution.

H_A: The frequencies of left- and right-handed persons in the population do not follow a binomial distribution.

$$\bar{X} = \frac{\sum f_i X_i}{\sum f_i} = \frac{96}{75} = 1.2800$$

$$\hat{p} = \frac{\bar{X}}{n} = \frac{1.2800}{8} = 0.16 = \text{probability of a person being left-handed}$$

$$\hat{q} = 1 - \hat{p} = 0.84 = \text{probability of a person being right-handed}$$

X_i	f_i	$f_i X_i$	$\hat{f}_i$
0	21	0	$\frac{8!}{0!8!}(0.16^0)(0.84^8)(75) = (0.24788)(75) = 18.59$
1	26	26	$(0.37772)(75) = 28.33$
2	19	38	$(0.25181)(75) = 18.89$
3	6	18	$(0.09593)(75) = 7.19$
4	2 ⎫	8	$(0.02284)(75) = 1.71$ ⎫
5	0 ⎪	0	$(0.00348)(75) = 0.26$ ⎪
6	1 ⎬ 3	6	$(0.00033)(75) = 0.02$ ⎬ 1.99
7	0 ⎪	0	$(0.00002)(75) = 0.00$ ⎪
8	0 ⎭	0	$(0.00000)(75) = 0.00$ ⎭
	75	96	

$$\sum f_i = 75$$

$$\sum f_i X_i = 96$$

$$\chi^2 = \frac{(21 - 18.59)^2}{18.59} + \frac{(26 - 28.33)^2}{28.33} + \frac{(19 - 18.89)^2}{18.89} + \frac{(6 - 7.19)^2}{7.19} + \frac{(3 - 1.99)^2}{1.99}$$

$$= 1.214$$

$$\nu = k - 2 = 5 - 2 = 3$$

$$\chi^2_{0.05,3} = 7.815$$

Therefore, do not reject H_0.

$$0.50 < P < 0.75 \quad [P = 0.7496]$$

if one member of a litter was found to be female, then there would be a greater-than-expected chance of a second member of the litter being female.

When p Is Not Assumed to Be Known.   Commonly, we do not postulate the value of p in the population but estimate it from a sample of data. As shown in Example 24.7, we may do this by calculating

$$\hat{p} = \frac{\sum\limits_{i=1}^{k} f_i X_i / \sum\limits_{i=1}^{k} f_i}{n}. \tag{24.41}$$

It then follows that $\hat{q} = 1 - \hat{p}$.

The values of $\hat{p}$ and $\hat{q}$ may be substituted in Equation 24.3 in place of p and q, respectively. Thus, expected frequencies may be calculated for each X, and a chi-square goodness of fit analysis may be performed just as it was in Example 24.6. In such a procedure, however, DF $= k - 2$, rather than $k - 1$, because two constants (n and $\hat{p}$) must be obtained from the sample, and DF is in general determined as k minus the number of such constants. The G statistic (Section 22.7) may be employed when p is not known, but the Kolmogorov-Smirnov test (Section 22.8) is very conservative in such cases and should be avoided.

The null hypothesis for such a test would be that the sampled population was distributed binomially. This implies that the members of the population are distributed among two categories and that the members occur independently of one another.

24.6 THE BINOMIAL TEST

With the ability to determine binomial probabilities, a simple procedure may be employed for goodness of fit testing of nominal data distributed among two categories. This method is especially welcome as an alternative to chi-square goodness of fit where the expected frequencies are small (see Section 22.5). If p is very small, then the Poisson distribution (Section 25.1) may be used; and it is simpler to employ when n is very large.

One-Tailed Testing.   Animals might be introduced one at a time into a passageway at the end of which each has a choice of turning either to the right or to the left. A substance, perhaps food, is placed out of sight to the left or right; the direction is randomly determined (as by the toss of a coin). We might state a null hypothesis, that there is no tendency for animals to turn in the direction of the food, against the alternative, that the animals prefer to turn toward the food. If we consider p to be the probability of turning toward the food, then the hypothesis (one-tailed) would be stated as H_0: $p \leq 0.5$ and H_A: $p > 0.5$, and such an experiment might be utilized, for example, to determine the ability of the animals to smell the food. We may test H_0 as shown in Example 24.8. In this testing procedure, we determine the probability of obtaining, at random, a distribution of data deviating as much as, or more than, the observed data. In Example 24.8, the most

EXAMPLE 24.8   A one-tailed binomial test.

Twelve animals were introduced, one at a time, into a passageway at the end of which they could turn to the left (where food was placed out of sight) or to the right. We wish to determine if these animals come from a population in which animals would choose the left more often than the right. Then, $n = 12$; X is the number of animals turning left; and, if p is the probability of turning left, H_0: $p \leq 0.5$ and H_A: $p > 0.5$. In the following, $P(X)$ is obtained either from Appendix Table B.26b or by Equation 24.3.

X	$P(X)$
0	0.00024
1	0.00293
2	0.01611
3	0.05371
4	0.12085
5	0.19336
6	0.22559
7	0.19336
8	0.12085
9	0.05371
10	0.01611
11	0.00293
12	0.00024

On performing the experiment, ten of the twelve animals turned to the left and two turned to the right. If H_0 is true, $P(X \geq 10) = 0.01611 + 0.00293 + 0.00024 = 0.01928$. As this probability is less than 0.05, reject H_0.

Alternatively, this test could be performed by using the upper confidence limits as a critical value. By Equation 24.29, with a one-tailed F, we have

$$X = pn = (0.5)(12) = 6,$$

$$v_1' = 2(6 + 1) = 14,$$

$$v_2' = 2(12 - 6) = 12,$$

$$F_{0.05(1), 14, 12} = 2.64, \text{ and}$$

$$L_2 = \frac{(6 + 1)(2.64)}{12 - 6 + (6 + 1)(2.64)} = 0.755.$$

Because the observed $\hat{p}$ (namely $X/n = 10/12 = 0.833$) exceeds the critical value (0.755), we reject H_0.

Or, as a simpler alternative, consult Appendix Table B.27 for $n = 12$ and $\alpha(1) = 0.05$, and find a critical value of $n - C_{0.05(1), n} = 12 - 2 = 10$. As $10 > n - 2$, H_0 is rejected; and $0.01 < P(X \geq 10) < 0.025$.

likely distribution of data in a sample of twelve from a population where p, in fact, was 0.5, would be six left and six right. The samples deviating from a 6:6 ratio even more than our observed sample (having a 10:2 ratio) would be those possessing eleven left, one right, and twelve left, zero right.

The general one-tailed hypotheses are H_0: $p \leq p_0$ and H_A: $p > p_0$, or H_0: $p \geq p_0$ and H_A: $p < p_0$, where p_0 need not be 0.5. The determination of the probability of $\hat{p}$ as extreme as, or more extreme than, that observed is shown in Example 24.8, where the expected frequencies, $P(X)$, are obtained either from Appendix Table B.26b or by Equation 24.3. If the resultant probability is less than or equal to α, then H_0 is rejected. A simple procedure for computing this P when $p_0 = 0.5$ is shown in Section 24.7.

Alternatively, a critical value for the one-tailed binomial test may be found by the confidence limit determinations of Section 24.4 using $X = pn$ and one-tailed values of F. If H_A: $p > 0.5$ (as in Example 24.8), then use Equation 24.29, so that the upper confidence limit is the critical value. If H_A: $p < 0.5$, then use Equation 24.28, so that the lower confidence limit is the critical value. This method is advantageous in that it renders unnecessary the calculation of $P(X)$ for each of several X's, but it lacks the ability to declare the exact probability of H_0 being true.

An even simpler alternative is to consult Appendix Table B.27, which gives the lower confidence limit, $C_{\alpha(1),n}$, for one-tailed probabilities, $\alpha(1)$. If H_A: $p < 0.5$, then H_0 is rejected if $X < C_{\alpha(1),n}$, where $X = np$. If H_A: $p > 0.5$ (as in Example 24.8), then H_0 is rejected if $X > n - C_{\alpha(1),n}$.

Two-Tailed Testing. The above experiment might have been performed without expressing an interest specifically in whether the animals were attracted toward the introduced substance. Thus, there would be no reason for considering a preference for only one of the two possible directions, and we would be dealing with two-tailed hypotheses, H_0: $p = 0.5$ and H_A: $p \neq 0.5$. The testing procedure would be identical to that in Example 24.8, except that we desire to know $P(X \leq 2 \text{ or } X \geq 10)$. This is the probability of a set of data deviating *in either direction* from the expected as much as or more than those data observed. This is shown in Example 24.9. The general two-tailed hypotheses are H_0: $p = p_0$ and H_A: $p \neq p_0$. If $p_0 = 0.5$, a simplified computation of P is shown in Equation 24.5.

Instead of enumerating the several values of $P(X)$ required, one could determine critical values for the two-tailed binomial test using Equations 24.28 and 24.29. If the observed $\hat{p}$ lies outside the interval formed by L_1, and L_2, then H_0 is rejected, as shown in Example 24.9. Note, as in Example 24.9, that if we hypothesize $p = 0.5$, then v_1 is the same for both L_1 and L_2, and v_2 is the same for L_1 and L_2; therefore, $F_{\alpha(2),v_1,v_2}$ is the same for both critical values. Using the critical value approach to the binomial test may be computationally easier than the direct examination of the tails of the binomial distribution, but the latter procedure has the advantage of declaring an exact probability in stating conclusions from the hypothesis test.

A much simpler two-tailed binomial test is possible using Appendix Table B.27, as demonstrated in Example 24.9. If the observed count, $X = pn$, is either $\leq C_{\alpha(2),v}$ or $\geq n - C_{\alpha(2),n}$, then H_0 is rejected.

The Normal Approximation. If the sample size, n, is large, then the procedure just described is very burdensome, unless handled by computer. In such situations a

EXAMPLE 24.9   A two-tailed binomial test.

The experiment is as described in Example 24.8, except that we have no a priori interest in the animals' turning either toward or away from the introduced substance.

H_0: $p = 0.5$.

H_A: $p \neq 0.5$.

$n = 12$
The probabilities of X, for $X = 0$ through $X = 12$, are given in Example 24.8.

$$P(X \geq 10 \text{ or } X \leq 2) = 0.01611 + 0.00293 + 0.00024 + 0.01611 + 0.00293 + 0.00024$$

$$= 0.03856$$

As this probability is less than 0.05, reject H_0.

Alternatively, this test could be performed by using the confidence limits as critical values. By Equations 24.28 and 24.29 we have

$$X = pn = (0.5)(12) = 6,$$

and for L_1 we have

$$\nu_1 = 2(12 - 6 + 1) = 14$$

$$\nu_2 = 2(6) = 12$$

$$F_{0.05(2), 14, 12} = 3.21$$

$$L_1 = \frac{6}{6 + (12 - 6 + 1)(3.21)} = 0.211,$$

and for L_2 we have

$$\nu_1' = 2(6 + 1) = 14$$

$$\nu_2' = 2(12 - 6) = 12$$

$$F_{0.05(2), 14, 12} = 3.21$$

$$L_2 = \frac{(6 + 1)(3.21)}{12 - 6 + (6 + 1)(3.21)} = 0.789.$$

As the observed $\hat{p}$ (namely $X/n = 10/12 = 0.833$) lies outside the range of 0.211 to 0.789, we reject H_0.

A simpler procedure uses Appendix Table B.27 to obtain critical values of $C_{0.05(2), 6} = 2$ and $n - C_{0.05(2), 6} = 12 - 2 = 10$. As $X = 10$, H_0 is rejected; and $0.02 < P(X \leq 2 \text{ or } X > 10) < 0.05$.

normal approximation to the binomial test may be used; however, Ramsey and Ramsey (1988) note that the binomial test is always more powerful than the approximation.

Using the normal approximation for the null hypothesis specifying p_0 as the proportion of a population belonging to one of two categories,

$$Z = \frac{X - np_0}{\sqrt{np_0q_0}} = \frac{n\hat{p} - np_0}{\sqrt{np_0q_0}}, \tag{24.42}$$

where X is the number in the sample observed in that category, and $\hat{p}(= X/n)$ is the proportion of the sample in that category. This procedure is shown in Example 24.10. Recall that the critical value for a normal deviate (Z) is obtained by the equivalence, $Z_\alpha = t_{\alpha,\infty}$.

EXAMPLE 24.10   The normal approximation to the binomial test.

> H_0: $p = 0.5$.
>
> H_A: $p \neq 0.5$.

Here, p is the proportion of females in a population of insects. A sample of sixty insects is taken (i.e., $n = 60$), and the sample contains thirty-five females (i.e., $X = 35$). The hypotheses specify $p_0 = 0.5$; so $q_0 = 1 - p_0 = 0.5$.

We test the null hypothesis by computing

$$Z = \frac{X - np_0}{\sqrt{np_0q_0}} = \frac{36 - (60)(0.5)}{\sqrt{(60)(0.5)(0.5)}} = \frac{6}{3.8730} = 1.549$$

$Z_{0.05(2)} = 1.9600$

Therefore, do not reject H_0.

$$0.05 < P < 0.10 \quad [P = 0.063]$$

Analysis by chi-square goodness of fit to a 1:1 ratio would have yielded $\chi^2 = 2.400$, which is the square of the Z computed above.

One-tailed hypotheses may also be subjected to the normal approximation to the binomial test, in which case $Z_{\alpha(1)}$ would be utilized as the critical value, rather than $Z_{\alpha(2)}$. If H_A refers to the left tail, then H_0 is rejected if $Z \leq -Z_{\alpha(1)}$; if H_A refers to the right tail, then reject H_0 if $Z \geq Z_{\alpha(1)}$.

A correction for continuity can be applied by subtracting 0.5 from the absolute value of the numerator in Equation 24.42. However, the studies of Ramsey and Ramsey (1988) indicate that the noncorrected Z is preferable, and that the normal approximation gives good results, in the following situations: for $\alpha = 0.05$, if neither p_0 nor q_0 is less than 0.01, and $\sqrt{np_0q_0}$ is at least 10; for $\alpha = 0.01$, if neither p_0 nor q_0 is less than 0.10, and $\sqrt{np_0q_0}$ is at least 35. However, if $p_0 = 0.5$, which is the commonest situation, then n may be as small as 27 if $\alpha = 0.05$, or as small as 19 if $\alpha = 0.01$. Compared to the noncorrected Z, the corrected Z (i.e., Z_c) may yield a probability closer to the exact probability, but the P associated with Z_c tends to be greater than the exact P and thus may fail to reject H_0 when it should.

Null hypotheses of this kind could also be expressed and tested in terms of chi-square goodness of fit to a hypothesized ratio. The resultant chi-square would be the square of the Z obtained in the normal approximation procedure above, and associated probabilities and statistical conclusions would be identical by these two procedures. For instance, in Example 24.10 H_0 would refer to a 1:1 ratio of females to males, and

$\chi^2 = Z^2$. If a correction for continuity were employed, then the continuity-corrected χ^2 would be the square of the continuity-corrected Z; i.e., $\chi_c^2 = Z_c$.

24.7 THE SIGN TEST

The concept underlying the binomial test can be developed into a nonparametric paired-sample test. This procedure, the *sign test*,* may also be employed whenever the Wilcoxon paired-sample test (Section 9.5) is appropriate, although it is not as powerful as the latter. The actual differences between members of a pair are not utilized in the sign test; only the direction (or sign) of each difference is tabulated. In Example 24.11, all that is recorded is whether each hindleg length is greater than, equal to, or less than its corresponding foreleg length; we do this by recording $+$, 0, or $-$, respectively. We then ask what the probability is of the observed distribution, or a more extreme distribution, of $+$ and $-$ signs if the null hypothesis is true. (A difference of zero is deleted from the analysis, so n is here defined as the number of differences having a sign.) The analysis proceeds as a binomial test with H_0: $p = 0.5$, and the null hypothesis tested is, essentially, that in the population the median difference is zero (i.e., the population frequencies of positive differences and negative differences are the same).

In performing a binomial test with $p_0 = q_0 = 0.5$, which is always the case with the sign test, the exact probability, P, may be obtained by the following simple considerations. As introduced in the Equation 24.5, for a given n the probability of a specified X is 0.5^n times the binomial coefficient. And binomial coefficients are defined in Equation 24.2 and presented in Table 24.2 and Appendix Table B.26a. In performing the binomial or sign test, we sum binomial terms in one or both tails of the distribution, and if $p_0 = q_0 = 0.5$ then this is the same as multiplying 0.5^n by the sum of the binomial coefficients in the one or two tails. Examining Example 24.11, what is needed is the sum of the probabilities in the two tails defined by $X \leq 2$ and $X \geq 8$. Thus, we may sum the coefficients, $_{10}C_X$, for $X \leq 2$ and $X \geq 8$, and multiply that sum by 0.5^{10} or, equivalently, divide the sum by 2^{10}. For this example, the binomial coefficients are 1, 10, 45, 45, 10, and 1, so the probability of H_0 being a true statment about the sampled population is

$$\frac{1 + 10 + 45 + 45 + 10 + 1}{2^{10}} = \frac{112}{1024} = 0.109375.$$

Actually, this calculation is more accurate than summing the six individual binomial probabilities as shown in Example 24.11, for it avoids rounding error.

Section 24.6 describes a normal approximation that is applicable if n is large enough. The confidence interval approaches to the binomial test (Section 24.6) are also applicable (for any sample size) and are preferable to the normal approximation. An extension of the sign test to nonparametric testing for blocked data from more than two groups is found in the form of the Friedmen test (Section 12.8).

---

*The sign test was reported nearly three centuries ago, making it one of the oldest statistical procedures (Arbuthnott, 1710).

EXAMPLE 24.11   **The sign test for the paired-sample data of Examples 9.1 and 9.3.**

Deer	Hindleg length (cm)	Foreleg length (cm)	Difference
1	142	138	+
2	140	136	+
3	144	147	−
4	144	139	+
5	142	143	−
6	146	141	+
7	149	143	+
8	150	145	+
9	142	136	+
10	148	146	+

H_0:   There is no difference between hindleg and foreleg length in deer.

H_A: There is a difference between hindleg and foreleg length in deer.

$n = 10$, and there are 8 positive differences and 2 negative differences.
Using Appendix Table B.26b for $n = 10$ and $p = 0.50$,

$$P(X \leq 2 \text{ or } X \geq 8)$$

$$= 0.04395 + 0.00977 + 0.00098 + 0.04395 + 0.00977 + 0.00098$$

$$= 0.10940$$

As the probability is greater than 0.05, do not reject H_0.
Using binomial coefficients:

$$\frac{1 + 10 + 45 + 45 + 10 + 1}{2^{10}} = \frac{112}{1024} = 0.109375.$$

Using Appendix Table B.27 for $n = 10$, the critical values are $C_{0.05(2), 10} = 1$ and $n - C_{0.05(2), 10} = 10 - 1 = 9$. As neither $X = 2$ nor $X = 8$ is as small as 1 or as large as 9, H_0 is not rejected; and by consulting Appendix Table B.27 we state $0.10 < P < 0.20$.
Neither the normal approximation, nor chi-square for a 1:1 ratio of $+$'s and $-$'s, should be used for these data, for n is too small.

24.8 Power of the Binomial and Sign Tests

The power of, and required sample size for, the binomial test may be determined by examining the cumulative binomial distribution. As the sign test is essentially a binomial test with p hypothesized to be 0.5 (see Section 24.7), its power and sample size may be assessed in the same manner as for the binomial test.

Power of the Test.   If a binomial test is performed at significance level α and with sample size n, we can determine the power of the test (i.e., the probability of correctly rejecting H_0) as follows. First we determine the critical value(s) of X for the test. For a one-tailed test of H_0: $p \leq p_0$ vs. H_A: $p > p_0$, the critical value is the smallest value of X for which the probability of that X or a larger X is $\leq \alpha$. (In Example 24.8 this is found to be $X = 10$.) For a one-tailed test of H_0: $p \geq p_0$ vs. H_A: $p < p_0$, the critical value is the largest X for which the probability of that X or of a smaller X is $\leq \alpha$. Then we examine the binomial distribution for the observed proportion, $\hat{p}$, from our sample. The power of the test is $\geq$ the probability of an X at least as extreme as the critical value referred to above.* This is demonstrated in Example 24.12.

EXAMPLE 24.12   Determination of the power of the one-tailed binomial test of Example 24.8.

$\hat{p} = X/n = 10/12 = 0.83$, H_0: $p \leq 0.5$, and $X = 10$ is the critical value for the test.

Then $P(X)$ for $X = 10$ through 12 for the binomial distribution having $p = 0.83$ and $n = 12$ is obtained from Equation 24.3 as

X	$P(X)$
10	0.296
11	0.263
12	0.107

Thus, the power of the test is $\geq 0.296 + 0.263 + 0.107 = 0.67$.

For a two-tailed test of H_0: $p = p_0$ vs. H_A: $p \neq p_0$, there are two critical values of X, one that cuts off $\alpha/2$ of the binomial distribution in each tail. Knowing these two X's we examine the binomial distribution for $\hat{p}$, and the power of the test is the probability in the latter distribution that X is at least as an extreme as the critical values. This is demonstrated in Example 24.13.

Cohen (1988: Section 5.4) presents tables to estimate sample size requirements in the sign test.

EXAMPLE 24.13   Determination of the power of the two-tailed sign test of Example 24.11.

$\hat{p} = X/n = 8/10 = 0.80$, and 1 and 9 are the critical values of the test of H_0: $p = 0.50$.

Then $P(X)$ for $X = 0, 1, 9$, and 10 for the binomial distribution having $p = 0.80$ and $n = 10$ is obtained from Equation 24.3, as

X	$P(X)$
0	0.00000
1	0.00000
9	0.26844
10	0.10737

Therefore, the power of the test is $\geq 0.00000 + 0.00000 + 0.26844 + 0.10737 = 0.3758$.

---

*If the critical X delineates a probability of exactly α in the tail of the distribution, then the power = that computed; if the critical value defines a tail $< \alpha$, then the power is $>$ that calculated.

The Normal Approximation.   If p is neither very close to 0 nor very close to 1, then the power of a binomial or sign test may be estimated as

$$
\text{power} = P\left[Z < \frac{p_0 - p}{\sqrt{\dfrac{pq}{n}}} - Z_{\alpha(2)}\sqrt{\dfrac{p_0 q_0}{pq}} \right] + P\left[Z > \frac{p_0 - p}{\sqrt{\dfrac{pq}{n}}} + Z_{\alpha(2)}\sqrt{\dfrac{p_0 q_0}{pq}} \right]
$$

$$(24.43)$$

(Marascuilo and McSweeney, 1977: 62). Here p_0 is the population proportion in the hypothesis to be tested, $q_0 = 1 - p_0$, p is the true population proportion (or our best estimate of it), $q = 1 - p$, $Z_{\alpha(2)} = t_{\alpha(2),\infty}$, and the probabilities of Z are found in Appendix Table B.2, using the considerations of Section 6.2. This is demonstrated in Example 24.14.

EXAMPLE 24.14   Estimation of power in a two-tailed binomial test, using the normal approximation.

We wish to test H_0: $p = 0.5$ vs. H_A: $p \neq 0.5$, using $\alpha = 0.05$, with a sample size of 50, when p is really 0.4.

To employ Equation 24.43 we determine

$$
\frac{p_0 - p}{\sqrt{\dfrac{pq}{n}}} = \frac{0.5 - 0.4}{\sqrt{\dfrac{(0.4)(0.6)}{50}}} = 1.4434;
$$

$$ Z_{0.05(2)} = 1.960; $$

$$
\sqrt{\frac{p_0 q_0}{pq}} = \sqrt{\frac{(0.5)(0.5)}{(0.4)(0.6)}} = 1.0206;
$$

and

$$
\begin{aligned}
\text{power} &= P[Z < 1.4434 - (1.960)(1.0206)] + P[Z > 1.4434 + (1.960)(1.0206)] \\
&= P(Z < -0.56) + P(Z > 3.44) \\
&= P(Z > 0.56) + P(Z > 3.44) \\
&= 0.2877 + 0.0003 \\
&= 0.29
\end{aligned}
$$

For the one tailed test, H_0: $p \leq p_0$ vs. H_A: $p > p_0$, the estimated power is

$$
\text{power} = P\left[Z > \frac{p_0 - p}{\sqrt{\dfrac{pq}{n}}} + Z_{\alpha(1)}\sqrt{\dfrac{p_0 q_0}{pq}} \right]; \qquad (24.44)
$$

and for the one-tailed hypotheses, $H_0: p \geq p_0$ vs. $H_A: p < p_0$,

$$\text{power} = P\left[Z < \frac{p_0 - p}{\sqrt{\dfrac{pq}{n}}} - Z_{\alpha(1)}\sqrt{\dfrac{p_0 q_0}{pq}} \right]. \tag{24.45}$$

Sample Size and Minimum Detectable Difference. The normal approximation can further be used in the following way. We may specify a minimum detectable difference that is of interest by stating p to be a specific distance from the p_0 in the null hypothesis. Then, for a given α, the above procedure may be used to determine the power of the test. If the power thus calculated is below that desired, then the above computations can be effected using a larger value of n. If the power is greater than that desired, then reduce n and recalculate. In this fashion one can, by directed trial and error (which is called "iteration"), estimate the sample size necessary to achieve a given power, given the minimum detectable difference and significance level.

In a similar fashion, we may state α, p_0, and, by varying p, estimate (by trial and error) the value of p that will result in a desired power.

24.9 CONFIDENCE INTERVAL FOR THE POPULATION MEDIAN

The confidence limits for a population median may be obtained by considering a binomial distribution with $p = 0.5$. The procedure thus is related to the binomial and sign tests in earlier sections of this chapter and may conveniently use Appendix Table B.27. That table gives $C_{\alpha,n}$, and from this we can state the confidence interval for a median to be

$$P(X_i \leq \text{population median} \leq X_j) \geq 1 - \alpha, \tag{24.46}$$

where

$$i = C_{\alpha(2),n} + 1 \tag{24.47}$$

and

$$j = n - C_{\alpha(2),n} \tag{24.48}$$

(e.g., MacKinnon, 1964), if the data are arranged in order of magnitude (so that X_i is the ith smallest measurement and X_j is the jth smallest). The confidence limits, therefore, are $L_1 = X_i$ and $L_2 = X_j$. Because of the discreteness of the binomial distribution, the confidence will typically be a little greater than the $1 - \alpha$ specified. This is demonstrated in Example 24.15.

A Large-Sample Approximation. For samples larger than appearing in Appendix Table B.27, an excellent approximation of the lower confidence limit (based on Hollander and Wolfe, 1973: 40), is derived from the normal distribution as

$$L_1 = X_i, \tag{24.49}$$

EXAMPLE 24.15   A confidence interval for a median.

Let us determine a 95% confidence interval for the median of the population from which each of the two sets of data in Example 3.3 came, where the population median was estimated to be 40 mo for species A and 40.5 mo for species B.

For species $A, n = 9$, so (from Appendix Table B.27) $C_{0.05(2),9} = 1$ and $n - C_{0.05(2),9} = 9 - 1 = 8$. The confidence limits are, therefore, X_i and X_j, where $i = 1 + 1 = 2$ and $j = 8$; and we can state

$$P(X_2 \leq \text{population median} \leq X_8) \geq 0.95$$

or

$$P(36 \text{ mo} \leq \text{population median} \leq 43 \text{ mo}) \geq 0.95.$$

For species $B, n = 10$, and Appendix Table B.27 informs us that $C_{0.05(2),10} = 1$; therefore, $n - C_{0.05(2),10} = 10 - 1 = 9$. The confidence limits are X_i and X_j, where $i = 1 + 1 = 2$ and $j = 9$; thus,

$$P(X_2 \leq \text{population median} \leq X_9) \geq 0.95.$$

or

$$P(36 \text{ mo} \leq \text{population median} \leq 44 \text{ mo}) \geq 0.95.$$

---

where

$$i = \frac{n - Z_{\alpha(2)}\sqrt{n}}{2} \tag{24.50}$$

rounded to the nearest integer, and $Z_{\alpha(2)}$ is the two-tailed normal deviate read from Appendix Table B.2. (Recall that $Z_{\alpha(2)} = t_{\alpha(2),\infty}$ and so may be read from the last line of Appendix Table B.3). The upper confidence limit is

$$L_2 = X_{n-i+1}. \tag{24.51}$$

By this method we approximate a confidence interval for the population median with confidence $\geq 1 - \alpha$.

24.10   THE FISHER EXACT TEST

Section 23.3 discussed 2×2 contingency tables, indicating that they are best analyzed by a procedure very commonly called the *Fisher exact test,*[*] which is applicable to

---

[*]Named for Sir Ronald Aylmer Fisher (1890–1962), a monumental English statistician with extremely strong influence in many areas of biostatistics and in statistical theory and method in general (e.g., Rao, 1992). At about the same time he proposed this procedure (Fisher, 1934: Chapter 21; 1935), it was also presented by Yates (1934) and Irwin (1935), so it is sometimes referred to as the Fisher-Yates test or Fisher-Irwin test. Yates (1984) observes that Fisher was probably aware of the exact-test procedure as early as 1926. Although referred to in this book as a statistician, R. A. Fisher also had a strong reputation as a biologist (e.g., Neyman, 1967), publishing—from 1912 to 1962—140 papers on genetics as well as 129 on statistics and 16 on other topics (Barnard, 1990).

either one-tailed or two-tailed hypotheses. This test is based upon the hypergeometric probability (see Section 24.2):

$$P = \frac{\binom{R_1}{f_{11}}\binom{R_2}{f_{21}}}{\binom{n}{C_1}}, \tag{24.52}$$

which is identical to

$$P = \frac{\binom{C_1}{f_{11}}\binom{C_2}{f_{12}}}{\binom{n}{R_1}}. \tag{24.53}$$

From Equation 24.11, both Equations 24.52 and 24.53 reduce to

$$P = \frac{R_1!\,R_2!\,C_1!\,C_2!}{f_{11}!\,f_{21}!\,f_{12}!\,f_{22}!\,n!}, \tag{24.54}$$

and it will be seen that there is advantage in expressing this as

$$P = \frac{\dfrac{R_1!\,R_2!\,C_1!\,C_2!}{n!}}{f_{11}!\,f_{12}!\,f_{21}!\,f_{22}!}. \tag{24.55}$$

The One-Tailed Test.   Consider the data of Example 23.5. If species 1 is naturally found in more rapidly moving waters, it would be reasonable to propose one-tailed hypotheses: H_0: The proportion of snails of species 1 resisting the water current is no greater than (i.e., less than or equal to) the proportion of species 2 withstanding the current, and H_A: The proportion of snails of species 1 resisting the current is greater than the proportion of species 2 resisting the current. The Fisher exact test proceeds as in Example 24.16.

EXAMPLE 24.16   A one-tailed Fisher exact test, using the data of Example 23.5.

H_0:   The proportion of snails of species 1 resisting the water current is no greater than the proportion of species 2 snails resisting the current.

H_A: The proportion of snails of species 1 resisting the water current is greater than the proportion of species 2 snails resisting the current.

	Resisted	Yielded	
Species 1	12	7	19
Species 2	2	9	11
	14	16	30

EXAMPLE 24.16 (continued)

Expressing the proportion of each species resisting the current in the sample:

	Resisted	Yielded	Total
Species 1	0.63	0.37	1.00
Species 2	0.18	0.82	1.00

The sample data are in the direction of H_A, but are they significantly so?

$$P = \frac{\dfrac{R_1!\,R_2!\,C_1!\,C_2!}{n!}}{f_{11}!\,f_{12}!\,f_{21}!\,f_{22}!}$$

$$= \frac{\dfrac{19!\,11!\,14!\,16!}{30!}}{12!\,7!\,2!\,9!}$$

$$= \text{antilog}\,[(\log 19! + \log 11! + \log 14! + \log 16! - \log 30!)$$

$$- (\log 12! + \log 7! + \log 2! + \log 9!)]$$

$$= \text{antilog}\,[16.52362 - 18.24356]$$

$$= \text{antilog}\,[-1.71994]$$

$$= \text{antilog}\,[0.28006 - 2.00000]$$

$$= 0.01906$$

The tables more extreme than that observed are as follows:

13	6	19
1	10	11
14	16	30

$$P = \frac{\dfrac{19!\,11!\,14!\,16!}{30!}}{13!\,6!\,1!\,10!}$$

$$= \text{antilog}\,[16.52362 - (\log 13! + \log 6! + \log 1! + \log 10!)]$$

$$= \text{antilog}\,[-2.68776]$$

$$= 0.00205$$

14	5	19
0	11	11
14	16	30

$$P = \frac{\dfrac{19!\,11!\,14!\,16!}{30!}}{14!\,5!\,0!\,11!}$$

$$= \text{antilog}\,[16.52362 - (\log 14! + \log 5! + \log 0! + \log 11!)]$$

$$= \text{antilog}\,[-4.09713]$$

$$= 0.00008$$

EXAMPLE 24.16   (continued)

Therefore, the probability of H_0 being true is $0.01906 + 0.00205 + 0.00008 = 0.02119$. As this probability is less than 0.05, H_0 is rejected.

Note that if the hypotheses had been H_0: Snail species 2 has no greater ability to resist current than species 1 and H_A: Snail species 2 has greater ability to resist current than species 1, then we would have observed that the sample data are *not* in the direction of H_A and would not reject H_0, without even computing probabilities.

A convenient way to summarize the observed and the more extreme tables and their probabilities is as follows:

f_{11}	P
12	0.01906
13	0.00205
14	0.00008
one-tailed P	0.02119

Instead of computing this exact probability of H_0, one may consult Appendix Table B.28, for $n = 30$, $m_1 = 11$, $m_2 = 14$; and the one-tailed critical values of f, for $\alpha = 0.05$, are 2 and 8. As the observed f in the cell corresponding to $m_1 = 11$ and $m_2 = 14$ is 2, H_0 may be rejected.

The probability of the observed contingency table occurring by random chance, given the row and column totals, may be computed using Equation 24.52, 24.53, 24.54, or 24.55. Then we construct each possible table having an f_{11} value more extreme (i.e., further from f_{22}) than that observed and apply one of these four equations to each such table. As shown in Example 24.16, the null hypothesis is tested by examining the sum of the probabilities of all these tables considered. This procedure yields the exact probability (hence the name of the test) of obtaining this set of tables by chance if the null hypothesis were true; and if this probability is less than or equal to the significance level, α, then H_0 is rejected.

Note that the quantity $R_1! R_2! C_1! C_2!/n!$ appears in each of the probability calculations using Equation 24.55 and therefore need be computed only once. It is only the value of $f_{11}! f_{12}! f_{21}! f_{22}!$ that needs to be computed anew for each table. To undertake these computations, the use of logarithms is advised for all but the smallest tables; and Appendix Table B.40 provides logarithms of factorials. It is also obvious that, unless the four cell frequencies are small, this test calculation is tedious without a computer. Computer programs have been developed to expand exact testing to $r \times c$ tables, where r and/or c are greater than 2.

An alternative to computing the exact probability in the Fisher exact test of 2×2 tables is to consult Appendix Table B.28 to obtain critical values with which to test the null hypotheses for n up to 30. We examine the four marginal frequencies, R_1, R_2, C_1, and C_2; and we designate the smallest of the four as m_1. If m_1 is a row total, then we call the smaller of the two column totals m_2; if m_1 is a column total, then the smaller row total is m_2. In Example 24.16, $m_1 = R_2$ and $m_2 = C_1$; and the one-tailed critical values in Appendix Table B.28, for $\alpha = 0.05$, are 2 and 8. The observed frequency in the cell corresponding to marginal totals m_1 and m_2 is called f; and if f is more extreme than 2 or 8 (i.e., if $f \leq 2$ or $f \geq 8$), then H_0 is rejected. However, to employ tables of critical values results in only a range of probabilities associated with H_0; and a splendid characteristic of the test—namely the exact probability—is lost.

Bennett and Nakamura (1963) published tables for performing an exact test of 2×3 tables where the three column (or row) totals are equal and n is as large as 60.

Additional illustrations of one-tailed Fisher exact testing are in Example 24.17, where the probabilities for each table were obtained by computer (Zar, 1987, using the binomial coefficient method described at the end of this section).

EXAMPLE 24.17   One-tailed Fisher exact testing using data sets in Section 23.2.

(a) For the data of Example 23.2, if

H_0: In the sampled population, left-handedness is not less common among boys than among girls.

H_A: In the sampled population, left-handedness is less common among boys than among girls.

Expressing sample left-handedness as a proportion for each sex:

	Boys	Girls
Left-handed	0.18	0.22
Right-handed	0.82	0.67
Total	1.00	1.00

The sample data are in the direction of H_A, but are they significantly so?

f_{11}	P
6	0.07257
5	0.02772
4	0.00759
3	0.00144
2	0.00018
1	0.00001
0	0.00000
one-tailed P	0.10951

For $\alpha = 0.05$, do not reject H_0.

If the hypotheses had been H_0: In the sampled population, left-handedness is not less common among girls than among boys and H_A: In the sampled population, left-handedness is less common among girls than among boys, we would have observed that the sample data are *not* in the direction of H_A and would not reject H_0, without even computing probabilities.

EXAMPLE 24.17   (continued)

(b) For the data in Example 23.4, if

H_0:   The proportion of the population infected with an intestinal parasite is no greater in species 1 than in species 2 (i.e., H_0: $p_1 \leq p_2$).

H_A: The proportion of the population infected with an intestinal parasite is greater in species 1 than in species 2 (i.e., H_A: $p_1 > p_2$).

Expressing the occurrence of the parasite in the sample as a proportion of animals of each species:

	Species 1	Species 2
With parasite	0.75	0.40
Without parasite	0.25	0.60
Total	1.00	1.00

The sample data are in the direction of H_A, but are they significantly so?

f_{11}	P
18	0.01127
19	0.00222
20	0.00029
21	0.00002
22	0.00000
>22	0.00000
one-tailed P	0.01380

For $\alpha = 0.05$, reject H_0.

If the hypotheses had been H_0: $p_1 \geq p_2$ and H_A: $p_1 < p_2$, then we would have observed that the sample data are *not* in the direction of H_A and would not reject H_0, without even examining calculating probabilities.

(c) For the data in Example 23.6, if

H_0:   The experimental substance does not reduce the incidence of the parasite.

H_A: The experimental substance reduces the incidence of the parasite.

Expressing the proportion of parasitized animals in each experimental group:

	Without substance	With substance
With parasite	0.60	0.36
Without parasite	0.40	0.64
Total	1.00	1.00

EXAMPLE 24.17   (continued)

The sample data are in the direction of H_A, but are they significantly so?

f_{11}	P
15	0.05494
16	0.01818
17	0.00428
18	0.00070
19	0.00008
20	0.00001
21	0.00000
one-tailed P	0.07819

For $\alpha = 0.05$, do not reject H_0.

If the hypotheses had been H_0: The experimental substance does not increase the incidence of the parasite and H_A: The experimental substance does increase the incidence of the parasite, then we would have observed that the sample data are *not* in the direction of H_A and would not reject H_0, without even computing probabilities.

Feldman and Kluger (1963) have demonstrated a simpler computational procedure for obtaining the probabilities of the tables more extreme than the observed table. Let us designate the smallest of the four cell frequencies of a table as f_a and the cell frequency located diagonally from it as f_d. Let f_b be the remaining cell frequency in row 1 and f_c be the remaining cell frequency in row 2. In the next more extreme table f_b will change to $f_b' = f_b + 1$ and f_c will change to $f_c' = f_c + 1$. If P is the probability of one table, the probability of the next more extreme table is

$$P' = \frac{f_a f_d}{f_b' f_c'} P, \tag{24.56}$$

as demonstrated in Example 24.18.

EXAMPLE 24.18   Simplified computations in the one-tailed Fisher exact test of Example 24.17a.

For the original table of data in Example 23.2,

$$f_a = 6, \; f_d = 24, \; f_b = 12, \; f_c = 28, \; \text{and} \; P = 0.07257.$$

Therefore, $f_b' = 12 + 1 = 13$ and $f_c' = 28 + 1 = 29$ and the probability for the next more extreme table is

$$P' = \frac{f_a f_d}{f_b' f_c'} P = \frac{(6)(24)}{(13)(29)}(0.07257) = 0.02772.$$

For the new table with the latter probability,

$$f_a = 5, \; f_d = 23, \; f_b = 13, \; f_c = 29, \; \text{and} \; P = 0.02772.$$

EXAMPLE 24.18   (continued)

Therefore, the next f'_b is $13 + 1 = 14$ and the next f'_c is $29 + 1 = 30$, and the probability for the next more extreme table is

$$P' = \frac{f_a f_d}{f'_b f'_c} P = \frac{(5)(23)}{(14)(30)}(0.02772) = 0.00759.$$

This is repeated until we calculate a probability for each smaller f_a that is possible (namely, through $f_a = 0$ in this case).

The Two-Tailed Test.   Section 23.3 presented testing of two-tailed hypotheses by chi-square. The Fisher exact test may also be employed for two-tailed situations; it is preferable, although more laborious unless computer calculation is employed. Example 24.19 demonstrates the two-tailed Fisher exact test for the data of Example 23.5. The first step in the procedure is to compute the sum of the probabilities of the observed table and all tables more extreme in the same direction as the observed table is extreme. Such calculation is as performed in the one-tailed test of Example 24.16. If either $R_1 = R_2$, or $C_1 = C_2$, then the two-tailed probability is twice the one-tailed probability. Otherwise, it is not, and the probability in the second tail is computed as follows.[†]

We identify the smallest of the four marginal frequencies and call it m_1, and the smaller frequency from the opposite margin we call m_2. In our example, $m_1 = R_2 = 11$ and $m_2 = C_1 = 14$. Then, in the most extreme table considered in the first tail, we identify the frequency associated with m_1 and m_2 and call it f. The latter frequency,

---

[†]Some statisticians recommend that the two-tailed probability should always be determined as two times the one-tailed probability. Others argue against that calculation, and that practice is not employed here; and it can be noted that the second tail may be much smaller than the first and such a doubling procedure could result in a computed two-tailed probability that is greater than 1.

EXAMPLE 24.19   A two-tailed Fisher exact test, using the data and hypotheses of Example 23.5.

The probability of the observed table was found, in Example 24.16, to be 0.01906; and the one-tailed probability was calculated to be 0.02119. For the 2×2 table under consideration, $m_1 = 11$ and $m_2 = 14$; in the one-tailed test (Example 24.16) $f = f_{21} = 0$, so the most extreme table in the second tail is that which contains $m_1 - f_{21} = 11 - 0 = 11$:

3	16	19
11	0	11
14	16	30

the probability of which is

$$P = \frac{\dfrac{19!\,11!\,14!\,16!}{30!}}{3!\,16!\,11!\,0!}$$

$$= 0.000006663$$

EXAMPLE 24.19   (continued)

The less extreme tables that are in the second tail, and have probabilities less than the probability of the observed table, are these two:

4	15	19
10	1	11

| 14 | 16 | 30 |

$P = 0.00029$

5	14	19
9	2	11

| 14 | 16 | 30 |

$P = 0.00440$

The tables in the second tail may be summarized as:

f_{11}	P
3	0.00001
4	0.00029
5	0.00440
P of second tail	0.00470

and the two-tailed P is, therefore, $0.02119 + 0.00470 = 0.02589$. As this is less than 0.05, we may reject H_0. Note that the Haber-corrected chi-square probability in Example 23.5, namely 0.023, is very close to the exact probability shown above.

   If using Appendix Table B.28, $n = 30$, $m_1 = 11$, $m_2 = 14$, and the f corresponding to m_1 and m_2 in the observed table is 2. As the two-tailed critical values of f, for $\alpha = 0.05$, are 2 and 9, H_0 is rejected.

f, is subtracted from m_1 and the remainder is placed in the cell that was occupied by f and a new table is formed retaining the same marginal frequencies. In Example 24.16, the f from the most extreme table is 0, and $m_1 - 0 = 11$. This newly formed table is, then, the most extreme table in the second tail. If the probability of this table is greater than that of the observed table (i.e., the table of the original table), then the probability of the second tail is zero and the two-tailed probability equals the one-tailed probability. If the probability of this table is not greater than the probability of the observed table, the probability of this table contributes to the probability of the second tail and the f being manipulated (in our example, f_{21}) is changed by 1 to make it less extreme (i.e., to make it closer to the f in the same column or in the same row). Then the probability of this new table is computed. This process is repeated as often as necessary to determine the sum of the probabilities of all tables having probabilities at least as small as that of the table originally observed. This procedure is performed in Example 24.19.

   The simplified computational procedure of Feldman and Kluger (1963) may also be used for probabilities in the second tail. The most extreme table is identified, and its probability calculated, just in Example 24.19. Let us say f_a' is the smallest cell frequency in this table, and f_d' is the cell frequency diagonally opposite it. The remaining cell frequency in row 1 is f_b', and the remaining cell frequency in row 2 is f_c'. In the next less extreme table, $f_b = f_b' - 1$ and $f_c = f_c' - 1$. If P' is the probability of a given table, then the probability of the next less extreme table is

$$P = \frac{f_b' f_c'}{f_a f_d} P' \tag{24.57}$$

as demonstrated in Example 24.20.

EXAMPLE 24.20 Simplified computations for the two-tailed Fisher exact test of Example 24.19.

For the most extreme table in the second tail in Example 24.19,

$$f_a' = 0, \ f_d' = 3, \ f_b' = 16, \ f_c' = 11, \ P' = 0.000006663.$$

Therefore, $f_a = 0 + 1 = 1$ and $f_d = 3 + 1 = 4$, and the probability of the next less extreme table is

$$P = \frac{f_b' f_c'}{f_a f_d} P' = \frac{(16)(11)}{(1)(4)} (0.000006663) = 0.0002932.$$

For the new table with the latter probability:

$$f_a' = 1, \ f_d' = 4, \ f_b' = 15, \ f_c' = 10, \ P' = 0.0002932.$$

Therefore, $f_a = 1 + 1 = 2$ and $f_d = 4 + 1 = 5$, and the probability for the next less extreme table is

$$P = \frac{f_b' f_c'}{f_a f_d} P' = \frac{(15)(10)}{(2)(5)} (0.0002932) = 0.004398.$$

This process is repeated until the computed P is greater than 0.01906, the probability of the observed table.

Fisher exact testing of the other two-tailed hypotheses of Section 23.2 is shown in Example 24.21.

Probabilities Using Binomial Coefficients. Ghent (1972), Leslie (1955), Leyton (1968), and Sakoda and Cohen (1957) have shown how the use of binomial coefficients can eliminate much of the laboriousness of Fisher-exact-test computations, and Ghent (1972) and Carr (1980) have expanded these considerations to tables with more than two rows and/or columns. Thanks to Appendix Table B.26a, this computational procedure requires much less effort than the use of logarithms of factorials, and it is at least as accurate. It may be employed for moderately large sample sizes, limited by the number of digits on one's calculator.

Referring back to Equation 24.53, the probability of a given 2×2 table is seen to be the product of two binomial coefficients divided by a third. The numerator of Equation 24.53 consists of one binomial coefficient representing the number of ways C_1 items can be combined f_{11} at a time (or f_{21} at a time, which is equivalent) and a second coefficient expressing the number of ways C_2 items can be combined f_{12} at a time (or, equivalently, f_{22} at a time). And the denominator denotes the number of ways n items can be combined R_1 at a time (or R_2 at a time). Appendix Table B.26a provides

---

EXAMPLE 24.21 Two-tailed Fisher exact testing, using data with which chi-square contingency-table testing was demonstrated in Section 23.2.

(a) For the data and hypotheses of Example 23.2, the probability in one tail was found [in Example 23.17 to be 0.10951. The tables in the second tail have the following probabilities:

f_{11}	P
18	0.00000
17	0.00000
16	0.00006
15	0.00057
14	0.00353
13	0.01508
12	0.04604
P of second tail	0.06538

Therefore, the two-tailed probability is $0.10951 + 0.06538 = 0.17479$ (to which we may compare $P = 0.17$ from the χ_c^2 in Example 23.3).

(b) For the data and hypotheses of Example 23.4, the probability in one tail was found [in Example 24.17b] to be 0.01380. The tables in the second tail have the following probabilities:

f_{11}	P
3, 4, 5	0.00000
6	0.00001
7	0.00011
8	0.00100
9	0.00593
P of second tail	0.00705

Therefore, the two-tailed probability is $0.01380 + 0.00705 = 0.02085$ (to which we may compare $P = 0.021$ from the χ_c^2 in Example 23.4].

(c) For the data and hypotheses of Example 23.6, the probability in one tail was found [in Example 24.17c] to be 0.07819. As $C_1 = C_2$, the two-tailed P is two times the one-tailed P, namely $P = 2(0.07819) = 0.15638$ (to which we may compare $P = 0.16$ from the χ_c^2 in Example 22.6).

---

a large array of binomial coefficients, and the proper selection of those required leads to simple computation of the probability of a 2×2 table. (See Section 5.3 for discussion of combinations).

The procedure is demonstrated in Example 24.22, for the data in Example 24.16. We consider the first row of the contingency table and determine the largest f_{11} and the smallest f_{12} that are possible without exceeding the row totals and column totals. These are $f_{11} = 14$ and $f_{12} = 5$, which sum to the row total of 19. (Other frequencies, such as 15 and 4, also add to 19, but the frequencies in the first column are limited to

EXAMPLE 24.22 The Fisher exact tests of Examples 24.16 and 24.19, employing the binomial-coefficient procedure.

The observed 2×2 contingency table is

12	7	19
2	9	11
14	16	30

The top-row frequencies of all possible contingency tables with the observed row and column totals, and their associated binomial coefficients and coefficient products, are as follows. The observed contingency table is indicated by "*".

f_{11}	f_{12}	Binomial coefficients $C_1 = 14$		$C_2 = 16$		Coefficient product	
14	5	1	×	4,368	=	4,368	
13	6	14	×	8,008	=	112,112	1,157,520
12	7	91	×	11,440	=	1,041,040*	
11	8	364	×	12,870	=	4,684,680	
10	9	1,001	×	11,440	=	11,451,440	
9	10	2,002	×	8,008	=	16,032,016	
8	11	3,003	×	4,368	=	13,117,104	
7	12	3,432	×	1,820	=	6,246,240	
6	13	3,003	×	560	=	1,681,680	
5	14	2,002	×	120	=	240,240	
4	15	1,001	×	16	=	16,016	256,620
3	16	364	×	1	=	364	

$$54,627,300$$

One-tailed probability: $\dfrac{1,157,520}{54,627,300} = 0.02119$

Probability associated with the opposite tail: $\dfrac{256,620}{54,627,300} = 0.00470$

Two-tailed probability: 0.02589

14.) We shall need to refer to the binomial coefficients for what Appendix Table B.26a refers to as "n" = 14 and "n" = 16, for these are the two column totals (C_1 and C_2) in the contingency table in Example 24.16. We record, from Appendix Table B.26a, the binomial coefficient for "n" = C_1 = 14 and "X" = f_{11} = 14 (which is 1), the coefficient for "n" = C_2 = 16 and "X" = f_{12} = 5 (which is 4,368), and the product of the two coefficients (which is 4,368).

     Then, we record the binomial coefficients of the next less extreme table, i.e., the one with f_{11} = 13 and f_{12} = 6 (that is, coefficients of 14 and 8,008) and their product (i.e., 112,112). This process repeated is for each possible table until f_{11} can be no smaller (and f_{12} can be no larger): i.e., $f_{11} = 3$ and $f_{12} = 16$. The sum of all the coefficient products (54,627,300 in this example) is the number of ways n things may be combined R_1 at a time (where n is the total of the frequencies in the 2×2 table); this is the binomial coefficient for n (the total frequency) and "X" $= R_1$ (and in the present example this coefficient is $_{30}C_{11} = 54,627,300$). Determining this coefficient is a good arithmetic check against the sum of the products of the several coefficients of individual contingency tables.

Dividing the coefficient product for a contingency table by the sum of the products yields the probability of that table. For example, the table of observed data in Example 24.16 has $f_{11} = 14$ and $f_{12} = 5$, and we may compute 1,041,040/54,627,300 $= 0.01906$, exactly the probability obtained in Example 24.16 using logarithms of factorials. The probability of the one-tailed test employs the sum of those coefficient products equal to or smaller than the product for the observed table and in the same tail as the observed table. In the present example, this tail would include products 4,368, 112,112, and 1,041,040, the sum of which is 1,157,520, and 1,157,520/54,627,300 $= 0.02119$, which is the probability calculated in Example 24.16. To obtain the probability for the two-tailed test, we add to the one-tail probability the probabilities of all tables in the opposite tail that have coefficient products equal to or less than that of the observed table. In our example these products are 240,240, 16,016, and 364, their sum is 256,620, and 256,620/54,627,300 $= 0.00470$; the probabilities of the two tails are 0.02119 and 0.00470, which sum to the two-tailed probability of 0.02589 (which is what was calculated in Example 24.16).

The binomial coefficient method might be used in conjunction with the other computational procedures described above. For example, if a probability is determined by dividing a coefficient product by $_{R_1}C_n$, then the probability of the next more extreme table could be derived using Equation 24.56, or the probability of the next less extreme table could be derived using Equation 24.57. And if some of the coefficient products, or $_{R_1}C_n$, contained more digits that the available calculating machinery, then that result, or the desired probability, could be computed using logarithms of factorials.

24.11 COMPARING TWO PROPORTIONS

The testing of differences among proportions can be performed as an analysis of a contingency table (Chapter 23). To demonstrate this, let us reexamine the data and null hypothesis of Example 23.4. If two proportions are to be compared, we might use the Fisher exact test (Section 24.10), either one-tailed or two-tailed. Or, a normal approximation of the chi-square test may be employed.

When $\nu = 1$, the square root of χ^2 is a normal deviate, Z, and probabilities associated with a χ^2 from a 2×2 contingency table are the same as those pertaining to the related two-tailed Z. Also, the square root of χ_c^2 is a continuity-corrected normal deviate—let's call it Z_c—and probabilities associated with the resultant Z_c will be identical to those affiliated with χ_c^2. A testing procedure analogous to chi-square employs the normal distribution, where

$$Z = \frac{|\hat{p}_1 - \hat{p}_2|}{\sqrt{\dfrac{\bar{p}\bar{q}}{n_1} + \dfrac{\bar{p}\bar{q}}{n_2}}} \tag{24.58}$$

is equal to the square root of χ^2 (Equation 23.6) and

$$Z_c = \frac{|\hat{p}_1 - \hat{p}_2| - \dfrac{1}{2}\left(\dfrac{1}{n_1} + \dfrac{1}{n_2}\right)}{\sqrt{\dfrac{\bar{p}\bar{q}}{n_1} + \dfrac{\bar{p}\bar{q}}{n_2}}} \tag{24.59}$$

is equal to the square root of χ_c^2 (Equation 23.7). Here,[*]

$$\bar{p} = \frac{X_1 + X_2}{n_1 + n_2} \tag{24.60}$$

or

$$\bar{p} = \frac{n_1 \hat{p}_1 + n_2 \hat{p}_2}{n_1 + n_2}, \tag{24.61}$$

and

$$\bar{q} = 1 - \bar{p}. \tag{24.62}$$

This procedure is generally not as good as employing $\chi_{c'}^2$ (Equation 23.9); however, it is useful when testing H_0: $p_1 - p_2 = p_0$, when p_0 is not zero. If $p_0 = 0$, then H_0: $p_1 - p_2 = p_0$ is identical to H_0: $p_1 = p_2$, which is tested as indicated in Example 24.23. If H_0 and H_A specify p_0 as a proportion other than zero, then subtract p_0 from the numerator in Equation 24.58 or, preferably, in Equation 24.59.

If H_0: $p_1 = p_2$ is rejected, then confidence intervals for $p_1 - p_2$ may be desired. An approximation for them is

$$(\hat{p}_1 - \hat{p}_2) \pm \left[Z_{\alpha(2)} \sqrt{\frac{\bar{p}\bar{q}}{n_1} + \frac{\bar{p}\bar{q}}{n_2}} + \frac{1}{2}\left(\frac{1}{n_1} + \frac{1}{n_2}\right) \right]. \tag{24.63}$$

This is shown in Example 24.24.

If the two proportions to be compared are the means from two sets of proportions, then the two-sample t tests of Section 8.1 could be used, in which case the data should be transformed as indicated in Section 13.3, preferably using Equations 13.7 or 13.8. Or, the nonparametric two-sample testing of Sections 8.10 or 8.11 could be employed.

---

[*]Some authors have used $\sqrt{\hat{p}_1 \hat{q}_1 / n_1 + \hat{p}_2 \hat{q}_2 / n_2}$ as the denominator for the Z statistics in this section (e.g., see Eberhardt and Fligner, 1977); this makes no difference if $n_1 = n_2$ but otherwise may result in Type I error probabilities greater than the stated significance level, α, and is generally not preferred.

EXAMPLE 24.23   The normal approximation for the hypotheses and data of Example 23.4.

H_0: $p_1 = p_2$.

H_A: $p_1 \neq p_2$.

$\alpha = 0.05$

$$\hat{p}_1 = \frac{18}{24} = 0.7500; \quad \hat{p}_2 = \frac{10}{25} = 0.4000$$

$$\bar{p} = \frac{X_1 + X_2}{n_1 + n_2} = \frac{18 + 10}{24 + 25} = 0.5714$$

$$\bar{q} = 1.0000 - \bar{p} = 1.0000 - 0.5714 = 0.4286$$

$$Z_c = \frac{|\hat{p}_1 - \hat{p}_2| - \frac{1}{2}\left(\frac{1}{n_1} + \frac{1}{n_2}\right)}{\sqrt{\frac{\bar{p}\bar{q}}{n_1} + \frac{\bar{p}\bar{q}}{n_2}}}$$

$$= \frac{|0.7500 - 0.4000| - \frac{1}{2}\left(\frac{1}{24} + \frac{1}{24}\right)}{\sqrt{\frac{(0.5714)(0.4286)}{24} + \frac{(0.5714)(0.4286)}{25}}}$$

$$= \frac{0.3092}{0.1414} = 2.187$$

$Z_{0.05(2)} = 1.9600$

Therefore, reject H_0.

$$0.02 < P < 0.05 \quad [P = 0.029]$$

Note that $\sqrt{4.779} = 2.186$; except for rounding error, $\sqrt{\chi_c^2} = Z_c$.

EXAMPLE 24.24   Confidence interval for the difference between the two population proportions of Examples 23.4 and 24.21.

$$95\% \text{ C.I. for } p_1 - p_2 = (\hat{p}_1 - \hat{p}_2) \pm \left[Z_{0.05(2)}\sqrt{\frac{\bar{p}\bar{q}}{n_1} + \frac{\bar{p}\bar{q}}{n_2}} + \frac{1}{2}\left(\frac{1}{n_1} + \frac{1}{n_2}\right) \right]$$

$$= (0.7500 - 0.4000)$$

$$\pm \left[1.9600\sqrt{\frac{(0.5714)(0.4286)}{24} + \frac{(0.5714)(0.4286)}{25}} + \frac{1}{2}\left(\frac{1}{24} + \frac{1}{25}\right) \right]$$

$$= 0.3500 \pm 0.3180$$

$L_1 = 0.0320; \quad L_2 = 0.6680$

24.12 Power and Sample Size in Comparing Two Proportions

Estimating Power.   If the test of H_0: $p_1 = p_2$ vs. H_A: $p_1 \neq p_2$ is to be performed at the α significance level, with n_1 data in sample 1 and n_2 data in sample 2; and if the two samples come from populations actually having proportions of p_1 and p_2, respectively; then an estimate of power is

$$\text{power} = P\left[Z \leq \frac{-Z_{\alpha(2)}\sqrt{\bar{p}\bar{q}/n_1 + \bar{p}\bar{q}/n_2} - (p_1 - p_2)}{\sqrt{p_1 q_1/n_1 + p_2 q_2/n_2}}\right]$$
$$+ P\left[Z \geq \frac{Z_{\alpha(2)}\sqrt{\bar{p}\bar{q}/n_1 + \bar{p}\bar{q}/n_2} - (p_1 - p_2)}{\sqrt{p_1 q_1/n_1 + p_2 q_2/n_2}}\right] \tag{24.64}$$

(Marascuilo and McSweeney, 1977: 111), where

$$\bar{p} = \frac{n_1 p_1 + n_2 p_2}{n_1 + n_2}, \tag{24.65}$$

$$q_1 = 1 - p_1, \tag{24.66}$$

$$q_2 = 1 - p_2, \tag{24.67}$$

and

$$\bar{q} = 1 - \bar{p}. \tag{24.68}$$

The calculation utilizes Appendix Table B.2 and the considerations of Section 6.2 and is demonstrated in Example 24.25.

EXAMPLE 24.25   Estimation of power in a two-tailed test comparing two proportions.

We propose to test H_0: $p_1 = p_2$ vs. H_A: $p_1 \neq p_2$, with $\alpha = 0.05$, $n_1 = 50$, and $n_2 = 45$, where in the sampled populations $p_1 = 0.75$ and $p_2 = 0.50$. The power of the test can be estimated as follows.

We first compute (by Equation 24.65):

$$\bar{p} = \frac{(50)(0.75) + (45)(0.50)}{50 + 45} = 0.6316$$

and

$$\bar{q} = 1 - \bar{p} = 0.3684.$$

Then,

$$\frac{\bar{p}\bar{q}}{n_1} = \frac{(0.6316)(0.3684)}{50} = 0.0047; \quad \frac{\bar{p}\bar{q}}{n_2} = 0.0052;$$

$$\frac{p_1 q_1}{n_1} = \frac{(0.75)(0.25)}{50} = 0.0038; \quad \frac{p_2 q_2}{n_2} = \frac{(0.50)(0.50)}{45} = 0.0056;$$

$$Z_{0.05(2)} = 1.9600;$$

EXAMPLE 24.25   (continued)

and, using Equation 24.64,

$$\text{power} = P\left[Z \leq \frac{-1.9600\sqrt{0.0047 + 0.0052} - (0.75 - 0.50)}{\sqrt{0.0038 + 0.0056}}\right]$$

$$+ P\left[Z \geq \frac{1.9600\sqrt{0.0047 + 0.0052} - (0.75 - 0.50)}{\sqrt{0.0038 + 0.0057}}\right]$$

$$= P(Z \leq -4.59) + P(Z \geq -0.57)$$

$$= P(Z \geq -4.59) + [1 - P(Z \geq 0.57)]$$

$$= 0.0000 + 1.0000 - 0.2843$$

$$= 0.72$$

For the one-tailed test of H_0: $p_1 \geq p_2$ vs H_A: $p_1 < p_2$, the estimated power is

$$\text{power} = P\left[Z \leq \frac{-Z_{\alpha(1)}\sqrt{\bar{p}\bar{q}/n_1 + \bar{p}\bar{q}/n_2} - (p_1 - p_2)}{\sqrt{p_1 q_1/n_1 + p_2 q_2/n_2}}\right]; \tag{24.69}$$

and for the one-tailed hypotheses, H_0: $p_1 \leq p_2$ vs. H_A: $p_1 > p_2$,

$$\text{power} = P\left[Z \geq \frac{Z_{\alpha(1)}\sqrt{\bar{p}\bar{q}/n_1 + \bar{p}\bar{q}/n_2} - (p_1 - p_2)}{\sqrt{p_1 q_1/n_1 + p_2 q_2/n_2}}\right]. \tag{24.70}$$

These power computations are based on approximations to the Fisher exact test (Section 24.10) and tend to produce a conservative result. That is, the power is likely to be greater than that calculated.

Estimating Necessary Sample Size.   Several procedures have been proposed in the statistical literature to estimate the minimum sample sizes that would be required to detect a true population difference between two proportions with specified probabilities of α (Type I error) and β (Type II error; and power $= 1 - \beta$). If the two sample sizes are to be equal, which is generally a preferred experiment design, then the minimum required size of each samples is

$$n = \frac{\left[Z_\alpha\sqrt{2\bar{p}\bar{q}} + Z_{\beta(1)}\sqrt{p_1 q_1 + p_2 q_2}\right]^2}{\delta^2}, \tag{24.71}$$

(Fleiss, 1981: 41), where

$$\bar{p} = \frac{p_1 + p_2}{2}, \tag{24.72}$$

$\bar{q} = 1 - \bar{p}$, δ is the difference between the two proportions (i.e., $p_1 - p_2$), and Z_α is either $Z_{\alpha(2)} = t_{\alpha(2),\infty}$ or $Z_{\alpha(1)} = t_{\alpha(1),\infty}$ (depending upon whether one is testing a two-tailed or a one-tailed alternative hypothesis, respectively).* But, as this procedure is based upon an uncorrected chi-square, it tends to give sample sizes that are too small. Casagrande, Pike, and Smith (1978) and Ury and Fleiss (1980) found the following to be better:

$$n' = \frac{n}{4}\left[1 + \sqrt{1 + \frac{4}{n\delta}}\right]^2,$$ (24.74)

where n is from Equation 24.71. For two-tailed testing, $\delta = |p_1 - p_2|$; for one-tailed testing, $\delta = p_1 - p_2$ if H_A: $p_1 > p_2$, or $\delta = p_2 - p_1$ if H_A: $p_2 > p_1$. The sample-size estimates obtained by this method are somewhat conservative; that is, the numbers of data required are likely to be less than the calculated n_1 and n_2. (Therefore, one may round the results downward.) These calculations are demonstrated in Example 24.26a.

It is desirable to plan data collection so as to obtain equal sample sizes (i.e., $n_1 = n_2$), for this will yield greater power for a given sample size and will require fewer data to achieve a given power. However, this is not always practical, so Fleiss, Tytun, and Ury (1980) and Ury and Fleiss (1980) have shown how we may estimate required sample size when specifying the ratio between the two sample sizes as

$$r = \frac{n_2}{n_1},$$ (24.75)

(where sample 1 is the smaller sample). The required size of sample 1 is

$$n_1 = \frac{n}{4}\left[1 + \sqrt{1 + \frac{2(r + 1)}{rn\delta}}\right]^2,$$ (24.76)

where

$$n = \frac{\left[Z_\alpha\sqrt{(r + 1)\bar{p}'\bar{q}'} + Z_{\beta(1)}\sqrt{rp_1q_1 + p_2q_2}\right]^2}{r\delta^2},$$ (24.77)

$$\bar{p}' = \frac{p_1 + rp_2}{r + 1},$$ (24.78)

$\bar{q}' = 1 - \bar{p}'$, and Z_α and $Z_{\beta(1)}$ are as in Equation 24.71. Then,

$$n_2 = rn_1.$$ (24.79)

This yields a conservative sample-size determination analogous to that of Equation 24.74, and it is demonstrated in Example 24.26b (where results are rounded downward as in Example 24.26a). The computation that is analogous to Equation 24.71, but for unequal sample size, is Equation 24.77; but this tends to err by specifying sample sizes that are too small, so Equation 24.76 is preferred.

---

*Casagrande, Pike, and Smith (1978) report that this procedure gives results very similar to a computation of Cochran and Cox (1957: 27):

$$n = (Z_\alpha + Z_{\beta(1)})^2/[2(\arcsin\sqrt{p_1} - \arcsin\sqrt{p_2})^2].$$ (24.73)

EXAMPLE 24.26 **Estimation of sample size required to detect a specified difference between two population proportions, when testing with specified α and β.**

(a) Equal sample sizes

We wish to test H_0: $p_1 = p_2$ vs. H_A: $p_1 \neq p_2$, at $\alpha = 0.05$, with 90% power (i.e., $\beta = 0.10$). If, in the sampled populations, $p_1 = 0.45$ and $p_2 = 0.25$, how large a sample should be collected from each population? Equation 24.74 is recommended.

$\delta = |0.45 - 0.25| = 0.20$;

$\bar{p} = (0.45 + 0.25)/2 = 0.35$; $\bar{q} = 1 - 0.35 = 0.65$;

$Z_{\alpha(2)} = Z_{0.05(2)} = t_{0.05(2),\infty} = 1.9600$;

$Z_{\beta(1)} = Z_{0.10(1)} = t_{0.10(1),\infty} = 1.2816$;

$$n = \left[1.9600\sqrt{(2)(0.35)(0.65)} + 1.2816\sqrt{(0.45)(0.55) + (0.25)(0.75)} \right]^2 = 117.44;$$

$$n' = (117.44/4)\left\{1 + \sqrt{1 + 4/[(117.44)(0.20)]}\right\}^2 = 127.2.$$

So each of the two sample sizes should be at least 127.

(b) Unequal sample sizes

We wish to perform the test of (a) above, but with three times as many data in sample 2 as in sample 1; i.e., $r = 3$. Equations 24.76 and 24.79 are recommended.

$\bar{p}' = [0.45 + 3(0.25)]/(3 + 1) = 0.30$; $\bar{q}' = 1 - 0.30 = 0.70$;

$$n = \left[1.9600\sqrt{(3 + 1)(0.30)(0.70)} + 1.2816\sqrt{3(0.45)(0.55) + (0.25)(0.75)}\right]^2 / [3(0.20)^2]$$

$$= 76.62;$$

$$n_1 = (76.62/4)\left\{1 + \sqrt{1 + 2(3 + 1)/[3(76.62)(0.20)]}\right\}^2 = 83.15;$$

$n_2 = 3(83.15) = 249.45.$

So the sample sizes should be at least 83 and 249.

Note, in Example 24.26, that the minimum number of data needed for the two sample, combined is less when the sample size are equal ($127 + 127 = 254$) than when the sample sizes are unequal ($83 + 249 = 332$).

Minimum Detectable Difference and Sample Size. If the smallest difference between population proportions (i.e., $p_1 - p_2$) to be detected with a desired power by testing H_0: $p_1 - p_2$ is specified, then the above procedure can estimate the power of the test. If the power thus computed is less than that desired, the calculation can be repeated using larger sample size, a process that can be repeated until the desired power is achieved. (If, during these computations, the estimated power is greater than that desired, then the calculations can be redone with smaller sample sizes.)

Similarly, one may state α and the sample sizes, and by varying p_1 and p_2 estimate (by trial and error) the value of p_1 and p_2 (and, therefore, of $p_1 - p_2$) that will result in a desired power. Ury (1982) describes a different procedure for estimating the p_2 that will result in rejection of H_0 when p_1, n, the desired power, and α are specified.

24.13 COMPARING MORE THAN TWO PROPORTIONS

It is best to test for differences among proportions by contingency table analysis. For example, the null hypothesis of Example 23.1 could be stated as "The proportions of males and females are the same among individuals of each of the four hair colors."

Alternatively, an approximation related to the normal approximation is applicable (if n is large and neither p nor q is very near 1). Using this approximation, one tests H_0: $p_1 = p_2 = \ldots = p_k$ against the alternative hypothesis that all k proportions are not the same, as

$$\chi^2 = \sum_{i=1}^{k} \frac{(X_i - n_i \bar{p})^2}{n_i \bar{p} \bar{q}} \tag{24.80}$$

(Pazer and Swanson, 1972: 187–190). Here,

$$\bar{p} = \frac{\sum_{i=1}^{k} X_i}{\sum_{i=1}^{k} n_i} \tag{24.81}$$

is a pooled proportion, $\bar{q}$ is as in Equation 24.68, and χ^2 has $k - 1$ degrees of freedom. Example 24.27 demonstrates this procedure, which is equivalent to χ^2 testing of a contingency table with two rows (or two columns).

We can instead, test whether k p's are equal not only to each other but to a specified constant, p_0 (i.e., H_0: $p_1 = p_2 = \ldots = p_k = p_0$). This is done by computing

$$\chi^2 = \sum_{i=1}^{k} \frac{(X_i - n_i p_0)^2}{n_i p_0 (1 - p_0)}, \tag{24.82}$$

which is then compared to the critical value of χ^2 for k (rather than $k - 1$) degrees of freedom.

EXAMPLE 24.27   Comparing four proportions, using the data of Example 23.1.

$n_1 = 87, X_1 = 32, \hat{p}_1 = \dfrac{32}{87} = 0.368, \hat{q}_1 = 0.632$

$n_2 = 108, X_2 = 43, \hat{p}_2 = \dfrac{43}{108} = 0.398, \hat{q}_2 = 0.602$

$n_3 = 80, X_3 = 16, \hat{p}_3 = \dfrac{16}{80} = 0.200, \hat{q}_3 = 0.800$

$n_4 = 25, X_4 = 9, \hat{p}_4 = \dfrac{9}{25} = 0.360, \hat{q}_4 = 0.640$

EXAMPLE 24.27 (continued)

$$\bar{p} = \frac{\sum X_i}{\sum n_i} = \frac{32 + 43 + 16 + 9}{87 + 108 + 80 + 25} = \frac{100}{300} = \frac{1}{3}$$

$$\bar{q} = 1 - \bar{p} = \frac{2}{3}$$

$$\chi^2 = \sum \frac{(X_i - n_i \bar{p})^2}{n_i \bar{p}\bar{q}}$$

$$= \frac{[32 - (87)\left(\frac{1}{3}\right)]^2}{(87)\left(\frac{1}{3}\right)\left(\frac{2}{3}\right)} + \frac{[43 - (108)\left(\frac{1}{3}\right)]^2}{(108)\left(\frac{1}{3}\right)\left(\frac{2}{3}\right)} + \frac{[16 - (80)\left(\frac{1}{3}\right)]^2}{(80)\left(\frac{1}{3}\right)\left(\frac{2}{3}\right)} + \frac{[9 - (25)\left(\frac{1}{3}\right)]^2}{(25)\left(\frac{1}{3}\right)\left(\frac{2}{3}\right)}$$

$$= 0.4655 + 2.0417 + 6.4000 + 0.0800$$

$$= 8.987$$

$$\nu = k - 1 = 4 - 1 = 3$$

$$\chi^2_{0.05,3} = 7.815$$

Therefore, reject H_0.

$$0.025 < P < 0.05 \quad [P = 0.029]$$

Note how the calculated χ^2 compares with that in Example 23.1; the two procedures yield the same results for contingency tables with two rows or two columns.

If each of the several proportions to be compared to each other is the mean of a set of proportions, then we can use the multisample testing procedures of Chapters 10, 11, 12, 14, and 15. To do so, the individual data should be transformed as suggested in Section 13.3, preferably by Equation 13.7 or 13.8, if possible.

Finally, it should be noted that comparing several p's yields the same results as if one compared the associated q's.

24.14 MULTIPLE COMPARISONS FOR PROPORTIONS

If the null hypothesis H_0: $p_1 = p_2 = \ldots = p_k$ (see Section 24.13) is rejected, then we may desire to determine specifically which population proportions are different from which others. The following procedure (similar to that of Levy, 1975a) allows for testing analogous to the Tukey or Student-Newman-Keuls tests introduced in Chapter 11. An angular transformation (Section 13.3) of each sample proportion is to be used. If $\hat{p}$, but not X and n, is known, then Equation 13.5 may be used. If, however, X and n are also known, then either Equation 13.7 or 13.8 is preferable. (The latter two equations give similar results, except for small or large $\hat{p}$, where Equation 13.8 is probably better.)

EXAMPLE 24.28 **Tukey-type multiple comparison testing among the four proportions of Example 24.27.**

Samples ranked by proportion (i):		3	4	1	2
Ranked sample proportions ($p_i = X_i/n_i$):		16/80 $= 0.200$	9/25 $= 0.360$	32/87 $= 0.368$	43/108 $= 0.398$
Ranked transformed proportions (p'_i, in degrees):		26.85	37.18	37.42	39.18

Comparison B vs. A	Difference $p'_B - p'_A$	SE	q	$q_{0.05,\infty,4}$	Conclusion
2 vs. 3	$39.18 - 26.85 = 12.33$	2.98	4.14	3.633	Reject H_0: $p_2 = p_3$
2 vs. 4	$39.18 - 37.18 = 2.00$	4.46	0.45	3.633	Accept H_0: $p_2 = p_4$
2 vs. 1	Do not test				
1 vs. 3	$37.42 - 26.85 = 10.57$	3.13	3.38	3.633	Accept H_0: $p_1 = p_3$
1 vs. 4	Do not test				
4 vs. 4	Do not test				

Overall conclusion: $p_4 = p_1 = p_2$ and $p_3 = p_4 = p_1$, which is the kind of ambiguous result described at the end of Section 11.1. By chi-square analysis (Example 23.9) it was concluded that $p_3 \neq p_4 = p_1 = p_2$; it is likely that the present method lacks power for this set of data.

Equation 13.8 is used for the transformations. For sample 3, for example, $X/(n + 1) = 16/81 = 0.198$ and $(X + 1)/(n + 1) = 17/81 = 0.210$, so $p'_3 = \frac{1}{2}[\arcsin\sqrt{0.198} + \arcsin\sqrt{0.210}] = \frac{1}{2}[26.4215 + 27.2747] = 26.848$. If we use Appendix Table B.24 to obtain the two needed arcsines, we have $p'_3 = \frac{1}{2}[26.42 + 27.27] = 26.845$.

As shown in Example 24.28, the multiple comparison procedure is similar to that in Chapter 11 (the Tukey test being in Section 11.1). The standard error for each comparison is, in degrees,[*]

$$\text{SE} = \sqrt{\frac{820.70}{n + 0.5}} \tag{24.83}$$

if the two samples being compared are the same size, or

$$\text{SE} = \sqrt{\frac{410.35}{n_A + 0.5} + \frac{410.35}{n_B + 0.5}} \tag{24.84}$$

if they are not. The critical value is $q_{\alpha,\infty,k}$ (from Appendix Table B.5).

Use of the normal approximation to the binomial is possible in multiple comparison testing (e.g., Marascuilo, 1971: 380–382); but the above procedure is preferable, even though it—and the methods to follow in this section—may lack desirable power.

---

[*]The constant 820.70 square degrees results from $(180°/2\pi)^2$, which follows from the variances reported by Anscombe (1948) and Freeman and Tukey (1950).

Comparison of a Control Proportion to Each Other Proportion. A procedure analogous to the Dunnett test of Section 11.4 may be used as a multiple comparison test where instead of comparing all pairs of proportions we desire to compare one proportion (designated as the "control") to each of the others. Calling the control group B, and each other group, in turn, A, we compute the Dunnett test statistic:

$$q = \frac{p'_B - p'_A}{\text{SE}},\qquad(24.85)$$

Here, the proportions have been transformed as earlier in this section, and the appropriate standard error is

$$\text{SE} = \sqrt{\frac{1641.40}{n + 0.5}}\qquad(24.86)$$

if Samples A and B are the same size, or

$$\text{SE} = \sqrt{\frac{820.70}{n_A + 0.5} + \frac{820.70}{n_B + 0.5}}\qquad(24.87)$$

if $n_A \neq n_B$. The critical value is $q'_{\alpha(1), \infty, p}$ (from Appendix Table B.6) or $q'_{\alpha(2), \infty, p}$ (from Appendix Table B.7) for one-tailed or two-tailed testing, respectively.

Multiple Contrasts among Proportions. The Scheffé procedure for multiple contrasts among means (Section 11.5) may be adapted to proportions by using angular transformations as done earlier in this section. For each contrast, we calculate

$$S = \frac{\left| \sum_i c_i p'_i \right|}{\text{SE}},\qquad(24.88)$$

where

$$\text{SE} = \sqrt{820.70 \sum_i \frac{c_i^2}{n_i + 0.5}},\qquad(24.89)$$

and c_i is a contrast coefficient as described in Section 11.5. (For example, if we wished to test the hypothesis $H_0: (p_1 + p_2 + p_4)/3 - p_3 = 0$, then $c_1 = \frac{1}{3}, c_2 = \frac{1}{3}, c_3 = -1$, and $c_4 = \frac{1}{3}$.)

24.15 TRENDS AMONG PROPORTIONS

In a $2 \times c$ contingency table (2 rows and c columns), the columns may have a natural quantitative sequence; for example, they may represent different age groups, different times, different size groups, different degrees of infection, or different concentrations of a treatment. In Example 24.29 the columns represent three weight classes of elderly women, and the data are the frequencies with which the 104 women in the sample exhibited a particular skeletal condition. The chi-square analysis of Section 23.1 tests

EXAMPLE 24.29   Testing for linear trend in a 2×3 contingency table. The data are the frequencies of occurrence among elderly women of a skeletal condition, tabulated by body weight class.

	Body weight class			
	Light	*Medium*	*Heavy*	*Total*
Condition present	6	16	18	40
Condition absent	22	28	14	64
Total	28	44	32	104
$\hat{p}_j$:	0.2143	0.3636	0.5625	
z_j :	1	2	3	

$$\chi^2 = \sum_{i=1}^{r} \sum_{j=1}^{c} \frac{(f_{ij} - \hat{f}_{ij})^2}{\hat{f}_{ij}}$$

$$= \frac{(6 - 10.7692)^2}{10.7692} + \frac{(16 - 16.9231)^2}{16.9231} + \cdots + \frac{(14 - 19.6923)^2}{19.6923}$$

$$= 2.1121 + 0.0504 + 2.6327 + 1.3200 + 0.0315 + 1.6454$$

$$= 7.7921$$

$$\chi_t^2 = \frac{n \left(n \sum_{j=1}^{c} f_{1,j} z_j - R_1 \sum_{j=1}^{c} C_j z_j \right)^2}{R_1(n - R_1) \left[n \sum_{j=1}^{c} C_j z_j^2 - \left(\sum_{j=1}^{c} C_j z_j \right)^2 \right]},$$

$$= \frac{104\{104[(6)(1) + (16)(2) + (18)(3)] - 40[(28)(1) + (44)(2) + (32)(3)]\}^2}{40(104 - 40)\{104[(28)(1)^2 + (44)(2)^2 + (32)(3)^2] - [(28)(1) + (44)(2) + (32)(3)]^2\}}$$

$$= \frac{104\{9568 - 8480\}^2}{2560\{51168 - 44944\}}$$

$$= 7.726$$

	χ^2	ν		P
Total	7.792	2	$0.01 < P < 0.025$	$[P = 0.020]$
Trend	7.726	1	$0.005 < P < 0.01$	$[P = 0.0054]$
Departure				
from linearity	0.066	1	$0.75 < P < 0.90$	$[P = 0.80]$

the null hypothesis that the occurrence of this condition is independent of weight class (i.e., that the proportion, p, of women with the condition is the same for all three weight classes). It is seen, in Example 24.29, that this hypothesis is rejected, so we conclude that in the sampled population there is a relationship between the two variables (weight class and skeletal condition). In addition, however, we may also ask whether this difference among weight classes follows a linear trend; that is, whether there is a linear increase (or decrease) in frequency of the condition of interest as weight increases. Once arranged in quantitative order, the columns may be designated by sequential scores (z): e.g., $1, 2, 3, \ldots$; or $0, 1, 2, \ldots$; or $-1, 0, 1, \ldots$.

The question of linear trend in a $2 \times c$ contingency table may be addressed by the method promoted by Armitage (1955, 1971: 363–365), by which we subdivide chi-square into component parts, somewhat as we partition sums of squares in analysis of variance. Let us refer to the chi-square of Equation 23.1 as the "total chi-square"; then a portion of this may be identified as being due to a linear trend:

"chi-square for linear trend" =

$$\chi_t^2 = \frac{n \left(n \sum_{j=1}^{c} f_{1,j} z_j - R_1 \sum_{j=1}^{c} C_j z_j \right)^2}{R_1 (n - R_1) \left[n \sum_{j=1}^{c} C_j z_j^2 - \left(\sum_{j=1}^{c} C_j z_j \right)^2 \right]}, \tag{24.90}$$

and the remainder is the portion of the total chi-square that is not due to a linear trend:

$$\text{"chi-square for departure from linear trend"} = \chi_d^2 = \chi^2 - \chi_t^2. \tag{24.91}$$

Associated with these three chi-square values are degrees of freedom of $c - 1$ for χ^2, 1 for χ_t^2 and $c - 2$ for χ_d^2. This procedure may be thought of as a regression of the proportions, $\hat{p}_j$, on the ordinal scores, z_j, where the $\hat{p}_j$'s are weighted by the sample sizes, C_j; or as a regression of n pairs of Y and z, where Y is 1 for each of the observations in row 1 and is 0 for each of the observations in row 2.* Testing for trend among proportions is a more powerful procedure than is the hypothesis test of difference among proportions, so the null hypothesis of no trend might be rejected even when the null hypothesis of equality of proportions is not.

If, in Example 24.29, the data in the second column were for heavy women and the data in the third column were for medium-weight women, the total chi-square would have been the same but χ_t^2 would have been 0.899 [$P = 0.34$], and it would have been concluded that there was no linear trend with body weight.

It is most common to employ consecutive integers for the scores, z, especially when the categories of the column variable are ordinal; however, sometimes it appears that another distribution of scores is warranted. For example, if the columns in Example 24.29, were age classes of women, the three classes being 20–40, 40–60, and 60–80 years, then equally spaced scores of 1, 2, 3 might well be employed (or, with identical results, the equally spaced 2, 4, 6; or equally spaced reflection of the age-group midpoints: 3, 5, 7). But, if the age classes were 20–50, 50–70, and 70–80, then it would

---

*This is equivalent to regression of the two variables in reverse order, or to the correlation of the two variables.

typically be better to use scores not equally spaced: e.g., 2, 5, 7 (or, with identical outcome, a reflection of midpoints: 3.5, 6, 7.5. (See Connor, 1972, for other considerations of grouping categories.)

EXERCISES

24.1 If, in a binomial population, $p = 0.3$ and $n = 6$, what proportion of the population does $X = 2$ represent?

24.2 If, in a binomial population, $p = 0.22$ and $n = 5$, what is the probability of $X = 4$?

24.3 In a random sample of thirty boys, eighteen have curly hair. Determine the 95% confidence limits for the proportion of curly-haired individuals in the population of boys that was sampled.

24.4 Determine whether the following data, where $n = 4$, are likely to have come from a binomial population with $p = 0.25$:

X	f
0	30
1	51
2	33
3	10
4	2

24.5 Determine whether the following data, where $n = 4$, are likely to have come from a binomial population:

X	f
0	20
1	41
2	33
3	11
4	4

24.6 A randomly selected male mouse of a certain species was placed in a cage with a randomly selected male mouse of a second species, and it was recorded which animal exhibited dominance over the other. The experimental procedure was performed, with different pairs of animals, a total of twenty times, with individuals from species 1 being dominant six times and those from species 2 being dominant fourteen times. Test the null hypothesis that there is no difference in the ability of members of either species to dominate.

24.7 A hospital treated 412 skin cancer patients over a period of time. Of these, 197 were female. Using the normal approximation to the binomial test, test the hypothesis that equal numbers of males and females seek treatment for skin cancer.

24.8 Test the null hypothesis of Exercise 22.3, using the binomial test normal approximation.

24.9 Ten students were given a mathematics aptitude test in a quiet room. The same students were given a similar test in a room with background music. Their performances were as follows. Using the sign test, test the hypothesis that the music has no effect on test performance.

Student	Score without music	Score with music
1	114	112
2	121	122
3	136	141
4	102	107
5	99	96
6	114	109
7	127	121
8	150	146
9	129	127
10	130	128

24.10 In a random sample of 1215 animals, 62 exhibited a certain genetic defect. Determine the 95% confidence interval for the proportion of the population displaying this defect.

24.11 Estimate the power of the hypothesis test of Exercise 24.6 if $\alpha = 0.05$.

24.12 Using the normal approximation, estimate the power of the hypothesis test of Exercise 24.7 if $\alpha = 0.05$.

24.13 Seventeen of a sample of forty-five breast-fed babies were found to have an antibody of interest, compared to sixteen of a sample of eighty-five bottle-fed babies. Does this antibody occur in the same proportion in the populations of breast-fed and bottle-fed babies?
   (a) Apply the chi-square procedure with the Yates correction.
   (b) Apply the chi-square procedure with Haber's correction.
   (c) Use the corrected G to test this hypothesis.
   (d) Apply the Fisher exact test to the hypothesis that, in the populations sampled, a greater proportion of breast-fed babies than bottle-fed babies have the antibody.
   (e) Apply the Fisher exact test to the hypothesis that the proportion of the population with the antibody is the same in breast-fed and bottle-fed babies.

24.14 Twelve members of one species of insect and eighteen of another species were placed in a container with insecticide fumes. After half of the thirty insects died there were nine of species 1 alive and five alive of species 2. Are these two species of insect equally susceptible to this insecticide?
   (a) Apply the chi-square procedure with the Yates correction.
   (b) Apply the chi-square procedure with Haber's correction.
   (c) Apply the corrected G.
   (d) Apply the Fisher exact test to the hypothesis that species 1 is less susceptible to the insecticide.
   (e) Apply the Fisher exact test to the hypothesis that the two species are equally susceptible to the insecticide.

24.15 For an analysis such as that in Exercise 24.14, using a significance level of 0.05 and sample sizes of $n_1 = n_2 = 300$, what would the power of the test be if the population proportions were $p_1 = 0.333$ and $p_2 = 0.250$?

24.16 If an analysis such as that in Exercise 24.14 were desired to be performed at the 5% significance level, with 90% power to detect a difference between population proportions of 0.333 and 0.250, what is the minimum size sample that should be taken from each population?

24.17 Using the data of Exercise 23.1, test the null hypothesis that there is the same proportion of males in all four seasons.

24.18 For the same data used in Exercise 24.17, test the hypothesis H_0: $p_1 = p_2 = p_3 = p_4 = 0.05$.

24.19 If the null hypothesis in Exercise 24.18 is rejected, perform a Tukey-type multiple comparison test to conclude which population proportions are different from which.

24.20 A new type of heart valve has been developed and is implanted in sixty-three dogs that have been raised on various levels of exercise. The numbers of valve transplants that succeed are tabulated as follows.

(a) Is the proportion of successful implants the same for dogs on all exercise regimens?

(b) Is there a trend with amount of exercise in the proportion of successful implants?

	Amount of exercise				
Implant	*None*	*Slight*	*Moderate*	*Vigorous*	*Total*
Successful	8	9	17	14	48
Unsuccessful	7	3	3	2	15
Total	15	12	20	16	63

25

TESTING FOR RANDOMNESS

A *random distribution* of objects in space is one in which each portion of the space has the same probability of containing an object and the occurrence of an object in any portion of the space in no way influences the occurrence of any other of the objects in any portion of space. A biological example in one-dimensional space might be the distribution of blackbirds along the top of a fence, an example in two-dimensional space could be the distribution of oak trees in a forest, and an example in three-dimensional space might be the distribution of bacteria in a liquid medium.[*] A random distribution of events in time is one in which each period of time of given length (e.g., an hour) has an equal chance of witnessing an event, and the occurrence of any event is independent of the occurrence of any other event. An example might be the firing of certain nerve fibers.

25.1 POISSON PROBABILITIES

The *Poisson distribution*[†] is important in describing *random* occurrences, when the probability of an occurrence is small. The terms of the Poisson distribution are

$$P(X) = \frac{e^{-\mu}\mu^X}{X!} \tag{25.1a}$$

---

[*]An extensive coverage of the description and analysis of spatial pattern is given by Upton and Fingleton (1985).

[†]Described in 1830 (Dale, 1989) by Simeon Denis Poisson (1781–1840), a French mathematician and physicist (Féron, 1978), although Abraham de Moivre (1667–1754) apparently described it previously in 1718 (David, 1962: 168). It was also described independently by others, including "Student" (W. S. Gosset, 1876–1937) in 1909 (Haight, 1967: 117). Poisson's name may have first been attached to this distribution, in contrast to merely being cited, by Soper (1914) (David, 1995).

or, equivalently,

$$P(X) = \frac{\mu^X}{e^\mu X!}, \tag{25.1b}$$

where $P(X)$ is the probability of X occurrences in a unit of space (or time) and μ is the population mean number of occurrences in a unit of space (or time). Thus,

$$P(0) = e^{-\mu}, \tag{25.2}$$

$$P(1) = e^{-\mu}\mu \tag{25.3}$$

$$P(2) = \frac{e^{-\mu}\mu^2}{2}, \tag{25.4}$$

$$P(3) = \frac{e^{-\mu}\mu^3}{(3)(2)}, \tag{25.5}$$

$$P(4) = \frac{e^{-\mu}\mu^4}{(4)(3)(2)}, \tag{25.6}$$

etc., where $P(0)$ is the probability of no occurrences in the unit space, $P(1)$ is the probability of exactly one occurrence in the unit space, and so on. Figure 25.1 presents some Poisson probabilities graphically.

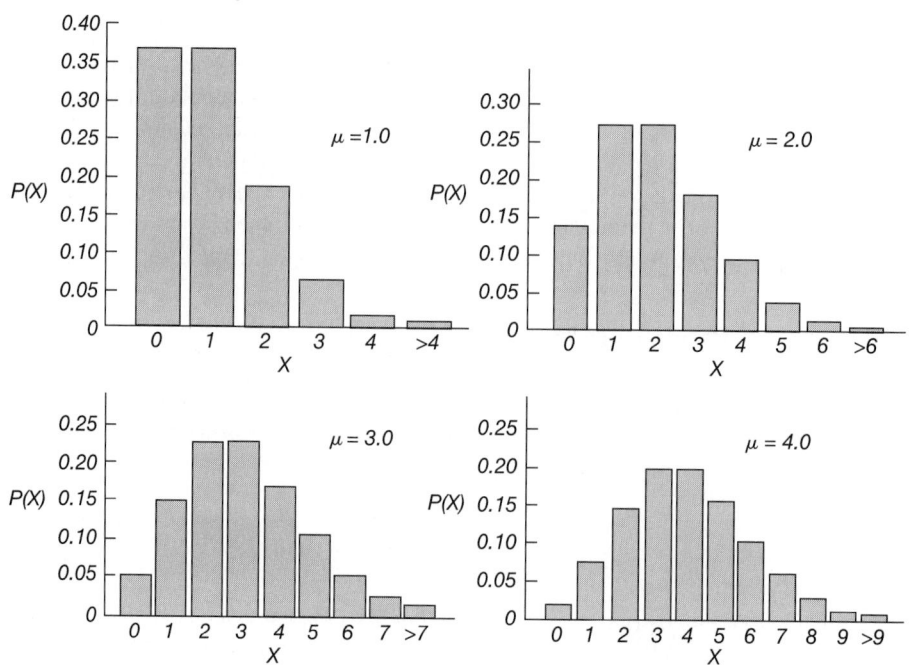

Figure 25.1 The Poisson distribution for various values of μ. These graphs were prepared by using Equation 25.1.

In calculating a series of Poisson probabilities, as represented by the above five equations, a simple computational expedient is available:

$$P(X) = \frac{P(X-1)\mu}{X}. \tag{25.7}$$

Example 25.1 demonstrates these calculations for predicting how many chessboard squares will have no raindrops, how many will receive one raindrop, how many will have three drops, and so on, if a total of 160 raindrops fall randomly on the chessboard.

EXAMPLE 25.1   If a chessboard (with sixty-four equal-area squares) were to receive 160 randomly spaced drops of rain, what would be the expected numbers of drops per square?

$n = 64$

$\mu = \dfrac{160 \text{ drops}}{64 \text{ squares}} = 2.5$ drops/square

Using Equation 25.2: $P(0) = e^{-\mu} = e^{-2.5} = 0.082085$

Thus, 8.2085% of the squares (or about five squares) would be expected to receive no raindrops. The probabilities of receiving 1, 2, 3, ... drops are as follows (using Equation 25.7):

X_i	$P(X_i)$	$\hat{f}_i = [P(X_i)][n]$	number of squares (rounded)
0	0.082085	$(0.082085)(64) = 5.2534$	5
1	$(0.082085)(2.5) = 0.205212$	$(0.205212)(64) = 13.1336$	13
2	$(0.205212)(2.5)/2 = 0.256515$	$(0.256515)(64) = 16.4170$	16
3	$(0.256515)(2.5)/3 = 0.213762$	$(0.213762)(64) = 13.6808$	14
4	$(0.213762)(2.5)/4 = 0.133601$	$(0.133601)(64) = 8.5505$	9
5	$(0.133601)(2.5)/5 = 0.066801$	$(0.066801)(64) = 4.2753$	4
6	$(0.066801)(2.5)/6 = 0.027834$	$(0.027834)(64) = 1.7814$	2
≤ 6	0.985810	63.0920	
≥ 7	$1.000000 - 0.985810 = 0.014190$	$(0.014190)(64) = 0.9082$ or $64 - 63.0920 = 0.9080$	1
		64.0000	64

As the Poisson distribution is appropriate when there is a small probability of a single event, as reflected in a small μ, this distribution is very similar to the binomial distribution where n is large and p is small. For example, Table 25.1 compares the Poisson distribution where $\mu = 1$ with the binomial distribution where $n = 100$ and $p = 0.01$ (and, therefore, $\mu = np = 1$). Thus, the Poisson distribution has importance in describing binomially distributed events having low probability. Another interesting property of the Poisson distribution is that $\sigma^2 = \mu$; that is, the variance and the mean are equal.

TABLE 25.1   The Poisson Distribution Where $\mu = 1$ Compared with the Binomial Distribution Where $n = 100$ and $p = 0.01$ (i.e., with $\mu = 1$) and the Binomial Distribution Where $n = 10$ and $p = 0.1$ (i.e., with $\mu = 1$)

X	$P(X)$ for Poisson: $\mu = 1$	$P(X)$ for binomial: $n = 100, p = 0.01$	$P(X)$ for binomial: $n = 10, p = 0.1$
0	0.36788	0.36603	0.34868
1	0.36788	0.36973	0.38742
2	0.18394	0.18486	0.19371
3	0.06131	0.06100	0.05740
4	0.01533	0.01494	0.01116
5	0.00307	0.00290	0.00149
6	0.00050	0.00046	0.00014
7	0.00007	0.00006	0.00001
>7	0.00001	0.00002	0.00000
Total	1.00000	1.00000	1.00001

25.2 CONFIDENCE LIMITS FOR THE POISSON PARAMETER

Confidence limits for the parameter (which is both the population mean and the population variance) of a population following the Poisson distribution may be obtained as follows. The lower $1 - \alpha$ confidence limit is

$$L_1 = \frac{\chi^2_{(1-\alpha/2),\,\nu}}{2},\tag{25.8}$$

where $\nu = 2X$; and the upper $1 - \alpha$ confidence limit is

$$L_2 = \frac{\chi^2_{\alpha/2,\,\nu}}{2},\tag{25.9}$$

where $\nu = 2(X + 1)$ (Pearson and Hartley, 1966: 81). This is demonstrated in Example 25.2. L_1 and L_2 are the confidence limits for the population mean and for the population variance. Confidence limits for the population standard deviation, σ, are simply the square roots of L_1 and L_2. The confidence limits, L_1 and L_2 (or their square roots), are not symmetrical around the parameter to which they refer. This procedure is a fairly good approximation. If confidence limits are desired to be accurate to more decimal places than given by the available critical values of χ^2, one may engage in the more tedious process of examining the tails of the Poisson distribution (e.g., see Example 25.3) to determine the value of X that cuts off $\alpha/2$ of each tail. Schwertman and

---

EXAMPLE 25.2   Confidence limits for the Poisson parameter.

An oak leaf contains four galls. Assuming that there is a random occurrence of galls on oak leaves in the population, estimate with 95% confidence the mean number of galls per leaf in the population.

The population mean, μ, is estimated as $X = 4$ galls/leaf.

The 95% confidence limits for μ are

$$L_1 = \frac{\chi^2_{(1-\alpha/2),\nu}}{2}, \text{ where } \nu = 2X = 2(4) = 8$$

$$L_1 = \frac{\chi^2_{0.975,8}}{2} = \frac{2.180}{2} = 1.1 \text{ galls/leaf}$$

$$L_2 = \frac{\chi^2_{\alpha/2,\nu}}{2}, \text{ where } \nu = 2(X+1) = 2(4+1) = 10$$

$$L_2 = \frac{\chi^2_{0.025,10}}{2} = \frac{20.483}{2} = 10.2 \text{ galls/leaf}$$

Therefore, we can state

$$P(1.1 \text{ galls/leaf} \leq \mu \leq 10.2 \text{ galls/leaf}) \geq 0.95$$

or

$$P(1.1 \text{ galls/leaf} \leq \sigma^2 \leq 10.2 \text{ galls/leaf}) \geq 0.95;$$

and, using the square roots of L_1 and L_2,

$$P(1.0 \text{ galls/leaf} \leq \sigma \leq 3.2 \text{ galls/leaf}) \geq 0.95.$$

Martinez (1994) discuss various normal approximations to Poisson confidence limits, all of which are inferior to the above procedures.

25.3 GOODNESS OF FIT OF THE POISSON DISTRIBUTION

The goodness of fit of the Poisson distribution to a set of observed data may be tested by chi-square, just as was done with the binomial distribution in Section 24.5. The frequencies in the tails of the distribution should be pooled to arrive at a tabulation having no expected frequencies less than 1.0 (Cochran, 1954), and the degrees of freedom are $k-2$ (k being the number of categories of X remaining after such pooling). Example 25.3 fits the Poisson distribution to a set of biological data. Here, μ is estimated by $\bar{X}$, and Equations 25.2 and 25.7 are employed. The G statistic (Section 22.7) may be used for goodness of fit analysis instead of chi-square. It will give equivalent results when n/k is large; if n/k is very small, G is preferable to χ^2 (Rao and Chakravarti, 1956).

If μ were known for the particular population sampled, or if it were desired to assume a certain value of μ, then the parameter would not have to be estimated by $\bar{X}$, and the degrees of freedom for χ^2 or G goodness of fit testing would be $k-1$. It is only

EXAMPLE 25.3   Fitting the Poisson distributions.

The data are the numbers of sparrow nests in a plot of given size. The plot size used is 8,000 square meters. X_i is the number of nests in a plot; f_i is the frequency with which X_i nests were found in a plot; $P(X_i)$ is the probability of X_i nests in a plot, if nests are distributed randomly; and $\hat{f_i}$ is the expected frequency of plots containing X_i nests, if nests are distributed randomly.

A total of forty 8,000-m^2 plots were surveyed and a total of forty-four nests were observed. Thus, the mean number of nests per plot is $44/40 = 1.1$, which is the best estimate of the population mean, μ. The values of $P(X_i)$ are computed from Equations 25.2 and 25.7.

H_0:   The population of sparrow nests is distributed randomly.

H_A: The population of sparrow nests is not distributed randomly.

$$P(0) = e^{-1.1} = 0.0332871$$

X_i	f_i	$f_i X_i$	$P(X_i)$	$\hat{f_i} = [P(X_i)][n]$
0	9	0	0.332871	13.315
1	22	22	$(0.332871)(1.1) = 0.366158$	14.646
2	6	12	$(0.366158)(1.1)/2 = 0.201387$	8.055
3	2 ⎫	6	$(0.201387)(1.1)/3 = 0.073842$	2.954 ⎫
4	1 ⎬ 3	4	$(0.073842)(1.1)/4 = 0.020307$	0.812 ⎬ 3.983
≥ 5	0 ⎭	0	$(0.020307)(1.1)/5 = 0.005435$	0.217 ⎭
	40	44	1.000000	39.999

$n = \sum f_i = 40$

$\bar{X} = \dfrac{\sum f_i X_i}{\sum f_i} = \dfrac{44}{40} = 1.1$

$\chi^2 = \dfrac{(9 - 13.315)^2}{13.315} + \dfrac{(22 - 14.646)^2}{14.646} + \dfrac{(6 - 8.055)^2}{8.055} + \dfrac{(3 - 3.983)^2}{3.983}$

$\quad = 1.398 + 3.693 + 0.524 + 0.243 = 5.858$

$\nu = k - 2 = 2$

$\chi^2_{0.05, 2} = 5.991$

Therefore, do not reject H_0.

$$0.05 < P < 0.10 \quad [P = 0.053]$$

when this parameter is specified that the Kolmogorov-Smirnov goodness of fit procedure (Section 22.8) may be applied (Massey, 1951).

The null hypothesis in Poisson goodness of fit testing is one of a random distribution of entities in space or time. Rejection of the hypothesis of randomness may result from one of two situations. First, the population distribution may be *uniform*; that is, each unit of space (or time) has the same number of entities. Second, the population may be arranged in what is referred to as a *clustered,* or *aggregated* or *contagious,*[*] distribution. Figures 25.2 and 25.3 present these possibilities diagrammatically. If a population has a random distribution, $\sigma^2 = \mu$, and $\sigma^2/\mu = 1.0$. If the population distribution is more uniform than random (sometimes called "underdispersed"), $\sigma^2 < \mu$, and $\sigma^2/\mu < 1.0$. And if the population is distributed contagiously (sometimes termed "overdispersed"), $\sigma^2 > \mu$, and $\sigma^2/\mu > 1.0$.[†] In these figures, the mean and variance refer to numbers of data per unit length (Fig. 25.2) or per unit area (Fig. 25.3).

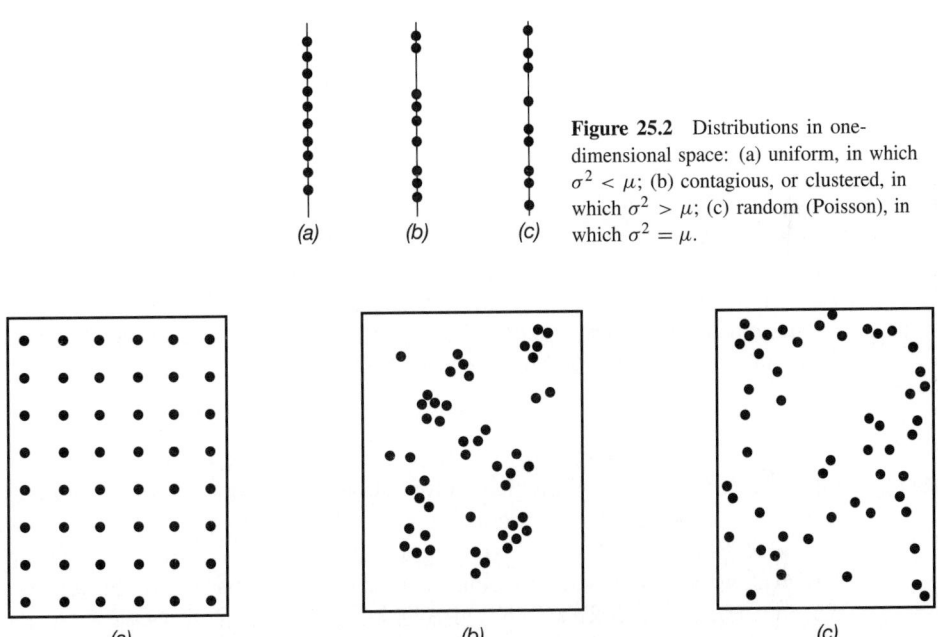

Figure 25.2 Distributions in one-dimensional space: (a) uniform, in which $\sigma^2 < \mu$; (b) contagious, or clustered, in which $\sigma^2 > \mu$; (c) random (Poisson), in which $\sigma^2 = \mu$.

(a)   (b)   (c)

Figure 25.3 Distributions in two-dimensional space: (a) uniform, in which $\sigma^2 < \mu$; (b) contagious, or clustered, in which $\sigma^2 > \mu$; (c) random (Poisson), in which $\sigma^2 = \mu$.

The investigator generally has some control over the size of the space or the length of the time interval that is considered to be the unit of observation. Thus, in Example 25.3, we might have chosen areas twice the size of those used, in which case each f_i would

---

[*]A mathematical distribution that is sometimes used to describe contagious distributions of biological data is the *negative binomial distribution,* which is described, for example, by Ludwig and Reynolds (1988: 24–26, 32–35) and Pielou (1977: 278–281), and by Ross and Preece (1985), who credit a 1930 French paper by G. Polya with the first use of the term "contagious" in this context. David (1995) reported that "negative binomial distribution" is a term first used by M. Greenwood and G. U. Yule, in 1920.

[†]An alternative test for goodness of fit of the Poisson distribution utilizes the test statistic $\chi^2 = SS/\bar{X}$ with $n - 1$ degrees of freedom (or $\chi^2 = SS/\mu$ with $v = n$, if μ is known). This procedure is discussed briefly by Simpson, Roe, and Lewontin (1960: 311–312) and Steel and Torrie (1980: 530–531); it may yield different conclusions than the method of fitting expected to observed frequencies.

have been twice the size as in our example, and $\bar{X}$ would have been 2.20 instead of 1.10. If we wish to consider analyses involving the Poisson distribution, it is desirable to use a sample distribution with a fairly small mean—let us say certainly below 10, preferably below 5, and ideally in the neighborhood of 1. If the mean is too large, then the Poisson too closely resembles the binomial, as well as the normal, distribution. If it is too small, however, then the number of categories, k, with appreciable frequencies will be too small for sensitive analysis.

Graphical testing of goodness of fit is sometimes encountered. The reader may consult Gart (1969b) for such considerations.

25.4 THE BINOMIAL TEST REVISITED

The binomial test was introduced (in Section 24.6) as a goodness of fit test for counts in two categories. If n is large, the binomial test may be unwieldy and its normal approximation can be used. If p is very small, the normal approximation is not applicable, but inasmuch as the Poisson and binomial distributions converge for small p the following procedure may be employed (as in Traut, 1980).

One-Tailed Testing. Let us consider the following example. It is assumed (as from a very large body of previous information) that a certain type of genetic mutation naturally occurs in an insect population with a frequency of 0.0020 (i.e., on average in 20 out of 10,000 insects). On exposing a large number of these insects to a particular chemical, we wish to ask whether that chemical increases the rate of this mutation. Thus, we state H_0: $p \leq 0.0020$ and H_A: $p > 0.0020$. (The general one-tailed hypotheses of this sort would be H_0: $p \leq p_0$ and H_A: $p > p_0$. If we had reason to ask whether some treatment reduced the natural rate of mutations, then the one-tailed test would have used H_0: $p \geq p_0$ and H_A: $p < p_0$.)

If performing this experiment for the hypotheses H_0: $p \leq 0.0020$ and H_A: $p > 0.0020$ yields twenty-eight of the sought mutations in 8000 insects observed, than the sample mutation rate is $\hat{p} = X/n = 28/8000 = 0.0035$. The question is whether the rate of 0.0035 is significantly greater than 0.0020. If we conclude that there is a low probability (i.e., $\leq \alpha$) of a sample rate being at least as large as 0.0035 when the sample is taken at random from a population having a rate of 0.0020, then H_0 is to be rejected. The hypotheses could also be stated H_0: $\mu \leq \mu_0$ and H_A: $\mu > \mu_0$; where $\mu_0 = p_0 n$ (which is $0.0020 \times 8000 = 16$ in this example).

By substituting $p_0 n$ for μ in Equation 25.1 we determine the probability of observing $X = 28$ mutations if our sample came from a population with $p_0 = 0.0020$. To test the hypothesis at hand we determine the probability of observing $X \geq 28$ mutations in a sample. (If the alternate hypothesis being considered were H_A: $p < p_0$, then we would compute the probability of mutations $\leq$ the number observed.) If the one-tailed probability is $\leq \alpha$, then H_0 is rejected at the α level of significance. This process is shown in Examples 25.4 and 25.5a.

EXAMPLE 25.4   Poisson probabilities for performing the binomial test with a very small proportion.

$p_0 = 0.0020$

$n = 8000$

We substitute $p_0 n = (0.0020)(8000) = 16$ for μ in Equation 25.1 to compute the following:[*]

X	P(X)	Cumulative P(X)	X	P(X)	Cumulative P(X)
23	0.02156	0.05825	0	0.00000	0.00000
24	0.01437	0.03669	1	0.00000	0.00000
25	0.00920	0.02232	2	0.00001	0.00001
26	0.00566	0.01312	3	0.00008	0.00009
27	0.00335	0.00746	4	0.00031	0.00040
28	0.00192	0.00411	5	0.00098	0.00138
29	0.00106	0.00219	6	0.00262	0.00400
30	0.00056	0.00113	7	0.00599	0.00999
31	0.00029	0.00057	8	0.01199	0.02198
32	0.00015	0.00028	9	0.02131	0.04329
33	0.00007	0.00013	10	0.03410	0.07739
34	0.00003	0.00006			
35	0.00002	0.00003			
36	0.00001	0.00001			
37[†]	0.00000	0.00000			

*For example, using Equation 25.1a, $P(X = 28) = \dfrac{e^{-16}16^{28}}{28!} = 0.00192$; and $P(X = 29) = \dfrac{e^{-16}16^{29}}{29!} = 0.00106$.

†This series of computations terminates when we reach a $P(X)$ that is zero to the number of decimal places used.

Two-Tailed Testing. If there is no reason, a priori, to hypothesize that a change in mutation rate would be in one specified direction (e.g., an increase) from the natural rate, then a two-tailed test is appropriate. The probability of the observed number of mutations is computed as shown in Example 25.4. Then we calculate and sum all the probabilities (in both tails) that are equal to or smaller than that of the observed. This is demonstrated in Example 25.5b.

Power of the Test. Recall that the power of a statistical test is the probability of that test rejecting a null hypothesis that is in fact a false statement about the population.

EXAMPLE 25.5a   A one-tailed binomial test for a proportion from a Poisson population.

H_0: $p \leq 0.0020$

H_A: $p > 0.0020$

$\alpha = 0.05$

$n = 8000$

$X = 28$

$p_0 n = (0.0020)(8000) = 16$

Therefore, we could state

H_0: $\mu \leq 16$

H_A: $\mu > 16$

From Example 25.4, we see that $P(X = 28) = 0.00192$ and $P(X \geq 28) = 0.00411$. As $0.00411 < 0.05$, reject H_0.

EXAMPLE 25.5b   A two-tailed binomial test for a proportion from a Poisson population, using the information of Example 25.4.

H_0: $p = 0.0020$

H_A: $p \neq 0.0020$

$\alpha = 0.05$

$n = 8000$

$X = 28$

$p_0 n = (0.0020)(8000) = 16$

Therefore, we can state

H_0: $\mu = 16$

H_A: $\mu \neq 16$

From Example 25.4, we see that $P(X = 28) = 0.00192$.
   The sum of the probabilities in one tail that are ≤ 0.00192 is 0.00411; the sum of the probabilities in the other tail that are ≤ 0.00192 is 0.00138.
Therefore, the probability of H_0 being true is $0.00411 + 0.00138 = 0.00549$.
As $0.00549 < 0.05$, reject H_0.

We can determine the power of the above test when it is performed with a sample size of n at a significance level of α. For a one-tailed test, we first determine the critical value of X (i.e., the smallest X that delineates a proportion of the Poisson distribution $\leq \alpha$). Examining the distribution of Example 25.4, for example, for $\alpha = 0.05$, we see that the appropriate X is 24 [for $P(X \geq 24) = 0.037$), while $P(X \geq 23) = 0.058$]. We then examine the Poisson distribution having the sample X, replacing μ with that X in Equation 25.1. The power of the test is $\geq$ the probability of an X at least as extreme as the critical value of X.*

For a two-tailed hypothesis, we identify one critical value of X as the smallest X that cuts off $\leq \alpha/2$ of the distribution in the upper tail and one as the largest X that cuts off $\leq \alpha/2$ of the lower tail. In Example 25.6, these two critical values for $\alpha = 0.05$ (i.e., $\alpha/2 = 0.025$) are $X = 25$ and $X = 8$ [as $P(X \geq 25) = 0.022$ and $P(X \leq 8) = 0.022$]. Then we examine the Poisson distribution having the sample X replace μ in Equation 25.1. As shown in Example 25.6, the power of the two-tailed test

---

*If the critical value delineates exactly α of the tail of the Poisson distribution, then the test's power is exactly what was calculated; if the critical value cuts off $< \alpha$ of the tail, then the power is $>$ that calculated.

EXAMPLE 25.6   Estimation of the power of the small-probability binomial tests of Examples 25.5a and 25.5b, using $\alpha = 0.05$.

Substituting $X = 28$ for μ in Equation 25.1, we compute the following:*

X	$P(X)$	Cumulative $P(X)$	X	$P(X)$	Cumulative $P(X)$
24	0.060	0.798	0	0.000	0.000
25	0.067	0.738	1	0.000	0.000
26	0.072	0.671	2	0.000	0.000
27	0.075	0.599	3	0.000	0.000
28	0.075	0.524	3	0.000	0.000
29	0.073	0.449	5	0.000	0.000
30	0.068	0.376	6	0.000	0.000
31	0.061	0.308	7	0.000	0.000
32	0.054	0.247	8	0.000	0.000
33	0.045	0.193			
34	0.037	0.148			
35	0.030	0.111			
36	0.023	0.081			
37	0.018	0.058			
38	0.013	0.040			
39	0.009	0.027			
40	0.007	0.018			
41	0.004	0.011			
42	0.003	0.007			
43	0.002	0.004			
44	0.001	0.002			
45	0.001	0.001			
46	0.000	0.000			

The critical value for the one-tailed test of Example 25.5a is $X = 24$. The power of this test is $> P(X \geq 24)$ in the above distribution. That is, the power is > 0.798.

EXAMPLE 25.6   (continued)

The critical values for the two-tailed test of Example 25.5b are 25 and 8. The power of this test is $> P(X \geq 25) + P(X \leq 8) = 0.738 + 0.000$. That is, the power is > 0.738.

---

*For example, using Equation 25.1a, $P(X = 24) = \dfrac{e^{-28}28^{24}}{24!} = 0.06010$; and

$P(X = 25) = \dfrac{e^{-28}28^{25}}{25!}$.

---

is at least as large as the probability of X in the latter Poisson distribution being more extreme than either of the critical values. That is, power $\geq P(X \geq$ upper critical value) $+ P(X \leq$ lower critical value).

25.5 COMPARING TWO POISSON COUNTS

If we have two counts, X_1 and X_2, each from a population with a Poisson distribution, we can ask whether they are likely to have come from the same population (or from populations with the same mean). The test of H_0: $\mu_1 = \mu_2$ (against H_A: $\mu_1 \neq \mu_2$) is related to the binomial test with $p = 0.05$; (Przyborowski and Wilenski, 1940; Pearson and Hartley, 1966: 78–79), so that Appendix Table B.27 can be utilized, using $n = X_1 + X_2$. For the two-tailed test, H_0 is rejected if either X_1 or X_2 is $\leq$ the critical value, $C_{\alpha(2),n}$. This is demonstrated in Example 25.7.

EXAMPLE 25.7   A two-sample test with Poisson data.

One fish is found to be infected with thirteen parasites, and a second fish with twenty-two. Assuming parasites are distributed randomly among fish, test whether these two fish are likely to have come from the same population. (If the two were of different species, or sexes, then we could ask whether the two species, or sexes, are equally infected.) The test is two-tailed for hypotheses H_0: $\mu_1 = \mu_2$ and H_A: $\mu_1 \neq \mu_2$.

Using Appendix Table B.27 for $n = X_1 + X_2 = 13 + 22 = 35$, we find a critical value of $C_{0.05(2),35} = 11$. Because neither X_1 nor X_2 is ≤ 11, H_0 is not rejected. Using the smaller of the two X's, we conclude that the probability is between 0.20 and 0.50 that a fish with thirteen parasites and one with twenty-two parasites come from the same Poisson population (or from two Poisson populations having the same means).

Using the normal approximation of Equation 25.10,

$$Z = \frac{|X_1 - X_2|}{\sqrt{X_1 + X_2}} = \frac{|13 - 22|}{\sqrt{13 + 22}} = \frac{9}{5.916} = 1.521$$

$Z_{0.05(2)} = t_{0.05(2),\infty} = 1.960$

Therefore, do not reject H_0.

$$0.10 < P < 0.20 \quad [P = 0.13]$$

For a one-tailed test of H_0: $\mu_1 \leq \mu_2$ against H_A: $\mu_1 > \mu_2$, we reject H_0 if $X_1 > X_2$ and $X_2 \leq C_{\alpha(1),n}$, where $n = X_1 + X_2$. For H_0: $\mu_1 \geq \mu_2$ and H_A: $\mu_1 < \mu_2$, H_0 is rejected if $X_1 < X_2$ and $X_1 \leq C_{\alpha(1),n}$, where $n = X_1 + X_2$.

This procedure results in conservative testing, and if n is at least 5, then a normal approximation should be used (Detre and White, 1970; Przyborowski and Wilenski, 1940; Sichel, 1973). For the two-tailed test:

$$Z = \frac{|X_1 - X_2|}{\sqrt{X_1 + X_2}} \qquad (25.10)$$

is considered a normal deviate, so the critical value is $Z_{\alpha(2)}$ (which can be read as $t_{\alpha(2),\infty}$ at the end of Appendix Table B.3). This is demonstrated in Example 25.7.

For a one-tailed test,

$$Z = \frac{X_1 - X_2}{\sqrt{X_1 - X_2}}. \qquad (25.11)$$

For H_0: $\mu_1 \leq \mu_2$ vs. H_A: $\mu_1 > \mu_2$, H_0: is rejected if $X_1 > X_2$ and $Z \geq Z_{\alpha(1)}$. For H_0: $\mu_1 \geq \mu_2$ vs. H_A: $\mu_1 < \mu_2$, H_0 is rejected if $X_1 < X_2$ and $Z \leq -Z_{\alpha(1)}$.

An alternative normal approximation, based on a square root transformation (Anscombe, 1948) is

$$Z = \left| \sqrt{2X_1 + \frac{3}{4}} - \sqrt{2X_2 + \frac{3}{4}} \right| \qquad (25.12)$$

(Best, 1975). It may be used routinely in place of Equation 25.10, and it has superior power when testing at $\alpha < 0.05$. Equation 25.12 is for a two-tailed test; for one-tailed testing, use

$$Z = \sqrt{2X_1 + \frac{3}{4}} - \sqrt{2X_2 + \frac{3}{4}}, \qquad (25.13)$$

with the same procedure to reject H_0 as with Equation 25.11.

25.6 SERIAL RANDOMNESS OF NOMINAL-SCALE CATEGORIES

Representatives of two different nominal-scale categories may appear serially in space or time, and their randomness of occurrence may be assessed as in the following example. Members of two species of antelopes are observed drinking along a river, and their linear order is as shown in Example 25.8. We may ask whether the order of occurrence of members of the two species is random (as opposed to the animals either forming groups with individuals of the same species or shunning members of the same species). A sequence of like elements, bounded on either side by either unlike elements or no elements, is termed a *run*. Thus, any of the following arrangements of five members of antelope species A and seven members of species B would be considered to consist of five runs:

BAABBBAAABBB,   or   *BBAAAABBBBAB*,   or   *BABAAAABBBBB*,   or
BAAABAABBBBB,   etc.

EXAMPLE 25.8   The two-tailed runs test with elements of two kinds.

Members of two species of antelopes (denoted as species A and B) are drinking along a river in the following order: $AABBAABBBBAAABBBBAABBB$.

H_0:   The distribution of members of the two species along the river is random.

H_A: The sequential distribution of members of the two species is not random.

$n_1 = 9, \quad n_2 = 13, \quad u = 8$

$u_{0.05(2), 9, 13} = 6$ and 17 (from Appendix Table B.29)

As u is neither ≤ 6 nor ≥ 17, do not reject H_0.

$$0.10 \leq P \leq 0.20$$

To test the null hypothesis of randomness, we may use the *runs test*.* If we let n_1 be the total number of elements of the first category (in the present example, the number of antelope of species A), n_2 the number of antelope of species B, and u the number of runs in the entire sequence, then the critical values, $u_{\alpha(2), n_1, n_2}$, can be read from Appendix Table B.29 for cases where both $n_1 \leq 30$ and $n_2 \leq 30$. The critical values in this table are given in pairs; if the u in the sample is $\leq$ the first member of the pair *or* $\geq$ the second, then H_0 is rejected.

If either n_1 or n_2 is larger than 30, then Appendix Table B.29 cannot be employed. But for such large samples, the distribution of u approaches normality, with a mean of

$$\mu_u = \frac{2n_1 n_2}{N} + 1 \tag{25.14}$$

and a standard deviation of

$$\sigma_u = \sqrt{\frac{2n_1 n_2 (2n_1 n_2 - N)}{N^2(N-1)}}, \tag{25.15}$$

where $N = n_1 + n_2$ (Brownlee, 1965: 266–230). And, the statistic

$$Z_c = \frac{|u - \mu_u| - 0.5}{\sigma_u} \tag{25.16}$$

may be considered a normal deviate, with $Z_{\alpha(2)}$ being the critical value for the test. (The 0.5 in the numerator of Z_c is a correction for continuity.)

Using Equation 25.16, the runs test may be extended to data with more than two categories (Wallis and Roberts, 1956: 571), for in general:

$$\mu_u = \frac{N(N+1) - \sum n_i^2}{N} \tag{25.17}$$

---

*From its inception the runs test has also been considered to be a nonparametric test of whether two samples come from the same population (e.g., Wald and Wolfowitz, 1940), but as a two-sample test it has very poor power and the Mann-Whitney test of Section 8.10 is preferable.

and

$$\sigma_u = \sqrt{\frac{\sum n_i^2 \left[\sum n_i^2 + N(N+1)\right] - 2N \sum n_i^3 - N^3}{N^2(N-1)}}, \tag{25.18}$$

where n_i is the number of items in category i, N is the total number of items (i.e., $N = \sum n_i$), and the summations are over all categories. (For two categories, Equations 25.17 and 25.18 are equivalent to Equations 25.14 and 25.15, respectively.) O'Brien (1976) and O'Brien and Dyck (1985) present a runs test, for two or more categories, that utilizes more information from the data, and is more powerful, than the above procedure.

One-tailed Testing.   There are two ways in which a distribution of nominal-scale categories can be nonrandom: (a) the distribution may have fewer runs than would occur at random, in which case the distribution is more clustered, or contagious, than random; (b) the distribution may have more runs than would occur at random, indicating a tendency toward a uniform distribution.

To test for the one-tailed situation of contagion, we state H_0: The elements in the population are not distributed contagiously, vs. H_A: The elements in the population are distributed contagiously; and H_0 would be rejected at the $\alpha(1)$ significance level if $u \leq$ the lower of the pair of critical values in Appendix Table B.29. Thus, had the animals in Example 25.8 been arranged $AAAAA\,BBBBBB\,AAAA\,BBBBBBB$, then $u = 4$ and the one-tailed 5% critical value would be the lower value of $u_{0.05(1),9,13}$, which is 7; as $4 < 7$, H_0 is rejected and the distribution is concluded to be clustered. In using the normal approximation, H_0 is rejected if $Z_c \geq Z_{\alpha(1)}$ and $u \leq \mu_u$.

To test for uniformity, we use H_0: The elements in the population are not uniformly distributed vs. H_A: The elements in the population are uniformly distributed. If $u \geq$ the upper critical value in Appendix Table B.29 for $\alpha(1)$, then H_0 is rejected. If the animals in Example 25.8 had been arranged as $ABABAB\,BABAB\,BABAB\,BAB\,BAB$, then $u = 18$, which is greater than the upper critical value of $u_{0.05(1),9,13}$ (which is 16); therefore, H_0 would have been rejected. If the normal approximation were used, H_0 would be rejected if $Z_c \geq Z_{\alpha(1)}$ and $u \geq \mu_u$.

Centrifugal Patterns.   Occasionally, nonrandomness in the sequential arrangement of two nominal-scale categories is characterized by one of the categories being predominant toward the ends of the series and the other toward the center. In the following sequence, for example:

$$AAAABAABBBBBBABBBBBAABBAAAA$$

the A's are commoner toward the termini of the sequence, and the B's are commoner toward the center of the sequence. Such a situation might be the pattern of two species of plants along a line transect from the edge of a marsh, through the center of the marsh, to the opposite edge. Or one might observe the occurrence of diseased and healthy birds in a row of cages, each cage containing one bird. Ghent (1993) refers to this as a *centrifugal* pattern of A's and a *centripetal* pattern of B's and presents a statistical test to detect such distributions of observations.

25.7 SERIAL RANDOMNESS OF MEASUREMENTS: PARAMETRIC TESTING

The biologist may encounter continuous data that have been collected serially in space or time. For example, rates of conduction might be measured at successive lengths along a nerve. A null hypothesis of no difference in conduction rate as one examines successive portions essentially is stating that all the measurements obtained are a random sample from a population of such measurements.

Example 25.9 presents data consisting of dissolved oxygen measurements of a water solution determined on the same instrument every five minutes. The desire is to conclude whether fluctuations in measurements are random or whether they indicate a nonrandom instability in the measuring device (or in the solution). The null hypothesis that the sequential variability among measurements is random may be subjected to the *mean square successive difference* test, a test that assumes normality in the underlying distribution. In this procedure, we calculate the sample variance, s^2, which is an estimate of the population variance, σ^2, as introduced in Section 4.4:

$$s^2 = \frac{\sum_{i=1}^{n}(X_i - \bar{X})^2}{n - 1}, \tag{4.8}$$

EXAMPLE 25.9   **The mean square successive difference test.**

An instrument for measuring dissolved oxygen is used to record a measurement every five minutes from a container of lake water. It is desired to know whether the differences in measurements are random or whether they are systematic. (If the latter, it could be due to the dissolved oxygen content in the water changing, or the instrument's response changing, or to both.) The data (in ppm) are as follows, recorded in the sequence in which they were obtained: 9.4, 9.3, 9.3, 9.2, 9.3, 9.2, 9.1, 9.3, 9.2, 9.1, 9.1.

H_0:   Consecutive measurements obtained on the lake water with this instrument have random variability.

H_A: Consecutive measurements obtained on the lake water with this instrument have non-random variability and are serially correlated.

$n = 11$

$s^2 = 0.01018$ (ppm)^2

$s_*^2 = \dfrac{(9.3 - 9.4)^2 + (9.3 - 9.3)^2 + (9.2 - 9.3)^2 + \ldots + (9.1 - 9.1)^2}{2(11 - 1)}$

$\quad = 0.00550$

$C = 1 - \dfrac{0.00550}{0.01018} = 1 - 0.540 = 0.460$

$C_{0.05, 11} = 0.452$

Therefore, reject H_0.

$$0.025 < P < 0.05$$

or,

$$s^2 = \frac{\sum\limits_{i=1}^{n} X_i^2 - \dfrac{\left(\sum\limits_{i=1}^{n} X_i\right)^2}{n}}{n-1}. \tag{4.10}$$

If the null hypothesis is true, then another estimate of σ^2 is

$$s_*^2 = \frac{\sum\limits_{i=1}^{n-1} (X_{i+1} - X_i)^2}{2(n-1)} \tag{25.19}$$

(von Neumann et al., 1941). Therefore, the ratio s_*^2/s^2 should equal 1 when H_0 is true. Using Young's (1941) notation, the test statistic is

$$C = 1 - \frac{s_*^2}{s^2}, \tag{25.20}$$

and if this value equals or exceeds the critical value $C_{a,n}$, in Appendix Table B.30, we reject the null hypothesis of serial randomness.* The mean square successive difference test considers the one-tailed alternate hypothesis that measurements are serially correlated.

For n larger than those in Appendix Table B.30, the hypothesis may be tested by a normal approximation:

$$Z = \frac{C}{\sqrt{\dfrac{n-2}{n^2-1}}} \tag{25.22}$$

(von Neumann et al., 1941), with the value of the calculated Z being compared with the critical value of $Z_{\alpha(1)} = t_{\alpha(1),\infty}$. This approximation is very good for $\alpha = 0.05$, for n as small as 10; for $\alpha = 0.10, 0.25$, or 0.025, for n as small as 25; and for $\alpha = 0.01$ and 0.005, for n of at least 100.

25.8 SERIAL RANDOMNESS OF MEASUREMENTS: NONPARAMETRIC TESTING

If we do not wish to assume that a sample of serially obtained measurements came from a normal population, then the procedure of Section 25.7 should not be employed. Instead, there are nonparametric methods that address hypotheses about serial patterns.

---

*Equations 4.8 and 25.19 may be combined so that C can be computed as

$$C = 1 - \frac{\sum\limits_{i=1}^{n-1} (X_i - X_{i+1})^2}{2(\text{SS})}, \tag{25.21}$$

where SS is the numerator of either Equation 4.8 or 4.10.

Runs Up and Down: Two-Tailed Testing. We may wish to test the null hypothesis that successive directions of change in serial data tend to occur randomly, with the alternate hypothesis stating the directions of change occur *either* in clusters (that is, where an increase from one datum to another is likely to be followed by another increase, and a decrease in the magnitude of the variable is likely to be followed by another decrease) *or* with a tendency toward regular alternation of increases and decreases (i.e., an increase is likely to be followed by a decrease, and vice versa). In the series of n data, we note whether datum $i + 1$ is larger than datum i (and denote this as a positive change, indicated as "+") or is smaller than datum i (which is referred to as a negative change, indicated as "−"). By a nonparametric procedure presented by Wallis and Moore (1941), the series of +'s and −'s is examined and we determine the number of runs of +'s and −'s, calling this number u as we did in Section 25.6. Appendix Table B.31 presents pairs of critical values for u, where for the two-tailed test for deviation from randomness one would reject H_0 if u were *either* $\leq$ the first member of the pair *or* $\geq$ the second member of the pair.

For $n > 50$, a normal approximation may be employed (Edgington, 1961; Wallis and Moore, 1941), where

$$\mu_u = \frac{2n - 1}{3} \tag{25.23}$$

and

$$\sigma_u = \sqrt{\frac{16n - 29}{90}} \tag{25.24}$$

in accordance with Equation 25.16. This procedure may be used for ratio, interval, or ordinal data and is demonstrated in Example 25.10. It is most powerful when no adjacent data are the same; if there are identical adjacent data, as in Example 25.10, then indicate the progression from each adjacent datum to the next as "0" and determine the mean of all the u's that would result from all the different conversions of the 0's to either +'s or −'s. Levene (1952) discussed the power of this test.

Runs Up and Down: One-Tailed Testing. In a fashion similar to that in Section 25.6, a one-tailed test would address one of two situations: One is where H_0: The successive positive and negative changes in the series of data are not clustered (i.e., are not contagious), and H_A: The successive positive and negative changes in the series of data are clustered (i.e., are contagious). In that situation, such as with the sequence $+ + + + + − − − + + + + − − − − −$, H_0 would be rejected if $u \leq$ the first member of the tabled pair of critical values; and, if using normal approximation, if $Z_c \geq Z_{\alpha(1)}$ and $u \leq \mu_u$. The other one-tailed circumstance is where H_0: The successive positive and negative changes in the series do not alternate regularly, and H_A: The successive positive and negative changes in the series of data do alternate regularly. In that case, such as with the series $- + - + - + - - + - + + - + - + - + -$, H_o would be rejected if $u \geq$ the second member of the pair in Appendix Table B.31; and, if the normal approximation is used, if $Z_c \geq Z_{\alpha(1)}$ and $u \geq \mu_u$.

EXAMPLE 25.10   Testing of runs up and down.

Data are measurements of temperature in a rodent burrow at noon on successive days.

H_0:   The successive positive and negative changes in temperature measurements are random.

H_A: The successive positive and negative changes in the series of temperature measurements are not random.

Day	Temperature (°C)	Difference
1	20.2	
2	20.4	+
3	20.1	−
4	20.3	+
5	20.5	+
6	20.7	+
7	20.5	−
8	20.4	−
9	20.8	+
10	20.8	0
11	21.0	+
12	21.7	+

$n = 12$

If the difference of 0 is counted as +, then $u = 5$; if the 0 is counted as −, then $u = 7$; mean $u = 6$.

For $\alpha = 0.05$, the critical values are $u_{0.05(2), 12} = 4$ and 11.

H_0 is not rejected; $0.25 < P \leq 0.50$.

Runs Above and Below the Median.   Another method of assessing randomness of ratio, interval, or ordinal scale measurements is by examining the pattern of their distribution with respect to the set of data. We first determine the median of the sample (as explained in Section 3.2). Then we record each datum as being either above (+) or below (−) the median. If a sample datum is equal to the median, it is discarded from the analysis. We then record u, the number of runs, in the resulting sequence of +'s and −'s. The test then proceeds as the run test of Section 25.6. This is demonstrated, for two-tailed hypotheses, in Example 25.11; common one-tailed hypotheses are those inquiring into contagious (i.e., clumped) distributions of data above or below the median.

Although the test for runs up and down and the test for runs above and below the median may both be considered nonparametric alternatives to the parametric means square successive difference test of Section 25.7, the latter runs test often resembles the parametric test more than the former runs test does. The test for runs up and down works well to detect long-term trends in the data, unless there are short-term random fluctuations superimposed upon those trends. The other two tests tend to perform better

---

EXAMPLE 25.11　Runs above and below the median.

The data and hypotheses are those of Example 25.10.
The median of the twelve data is determined to be 20.5°C.
The sequence of data, indicating whether they are above (+) or below (−) the median, is $-\ -\ -\ -$
$0+0-++++$.
For the runs test: $n_1 = 5, n_2 = 5, u = 4$.
The critical values are $u_{0.05(2),5,5} = 2$ and 10; therefore, do not reject H_0; $0.20 < P \le 0.50$.

---

in detecting long-term patterns in the presence of short-term randomness (A. W. Ghent, personal communication).

EXERCISES

25.1 If, in a Poisson distribution, $\mu = 1.5$, what is $P(0)$? What is $P(5)$?

25.2 A solution contains bacterial viruses in a concentration of 5×10^8 bacterial-virus particles per milliliter. In the same solution are 2×10^8 bacteria per milliliter. Assume random distribution of virus among the bacteria.
  (a) What proportion of the bacteria will have no virus particles?
  (b) What proportion of the bacteria will have virus particles?
  (c) What proportion of the bacteria will have at least two virus particles?
  (d) What proportion of the bacteria will have three virus particles?

25.3 A raisin cake is divided into equal-sized slices. The distribution of raisins among these slices is as follows. Test the null hypothesis that the raisins are distributed at random throughout the cake.

X_i	f_i
0	8
1	17
2	18
3	11
4	3
≥ 5	0

25.4 We wish to compile a list of certain types of metabolic human infant diseases that occur in more than 0.01% of the population. A random sample of 25,000 infants reveals five infants with one of these diseases. Should that disease be placed in our list?

25.5 A biologist counts 112 diatoms in a milliliter of lake water, and 134 diatoms are counted in a milliliter of a second collection of lake water. Test the hypothesis that the two water collections came from the same lake (or from lakes with the same mean diatom concentrations).

25.6 An economic entomologist rates the annual incidence of damage by a certain beetle as mild (M) or heavy (H). For a twenty-seven year period he records the following: *H M M M H H M M H M H H H H M M H H H H M M H H M M M M*. Test the null hypothesis that the incidence of heavy damage occurs randomly over the years.

25.7 The following data are the magnitudes of fish kills along a certain river (measured in kilograms of fish killed) over a period of years. Test the null hypothesis that the magnitudes of the fish kills are randomly distributed over time.

Year	Kill (kg)
1955	147.4
1956	159.8
1957	155.2
1958	161.3
1959	173.2
1960	191.5
1961	198.2
1962	166.0
1963	171.7
1964	184.9
1965	177.6
1966	162.8
1967	177.9
1968	189.6
1969	206.9
1970	221.5

25.8 Analyze the data of Exercise 25.7 nonparametrically to test for serial randomness.

25.9 There are four categories (clubs, diamonds, hearts, and spades) in a deck of playing cards, each category with thirteen cards.

(a) If there are thirty runs after shuffling the cards, are the categories randomly distributed?

(b) The largest possible number of runs (namely, fifty-two) will occur when the categories are as uniformly distributed as possible throughout the deck. What is the probability of fifty-two runs if the cards are well shuffled?

(c) The smallest possible number of runs (namely, four) will occur when the categories are as contagiously distributed as possible throughout the deck. What is the probability of four runs if the cards are well shuffled?

26

CIRCULAR DISTRIBUTIONS: DESCRIPTIVE STATISTICS

26.1 DATA ON A CIRCULAR SCALE

In Section 1.1, an interval scale of measurement was defined as a scale with equal intervals but with no true zero point. A special type of interval scale is a circular scale, where not only is there no true zero, but any designation of high or low values is arbitrary. A common example of a circular scale of measurement would be compass direction (Fig. 26.1a), where a circle is said to be divided into 360 equal intervals, called degrees,* and for which the zero point is arbitrary. There is no physical justification for a direction of north to be designated 0 (or 360) degrees, and a direction of 270° cannot be said to be a "larger" direction than 90°.†

Another common circular scale is time of day (Fig. 26.1b), where a day is divided into twenty-four equal intervals, called hours, but where the designation of midnight

---

*A degree is divided into 60 minutes (i.e., $1° = 60'$) and a minute into 60 seconds ($1' = 60''$). A number system based upon 60 is termed "sexagesimal," and we owe the division of the circle into 360 degrees—and the 60-minute hour and 60-second minute—to the ancient Babylonians. The use of the modern symbols (° and ' and ") appears to date from the 1570s (Cajori, 1929: 146).

†Occasionally one will encounter angular measurements expressed in radians instead of in degrees. A radian is the angle that is subtended by an arc of a circle equal in length to the radius of the circle. As a circle's circumference is 2π times the radius, a radian is $360°/2\pi = 180°/\pi = 57.29577951°$ (or 57 deg, 17 min, 44.8062 sec). The term "radian" was first used, in 1873, by James Thomson, brother of Baron William Thomson (Lord Kelvin), the famous Scottish mathematician and physicist (Cajori, 1929: 147). A direction measured clockwise, from 0° at north is called an *azimuth*.

as the zero or starting point is arbitrary. One hour of a day corresponds to 15° (i.e., 360°/24) of a circle, and 1° of a circle corresponds to four minutes of a day. Other time divisions, such as weeks and years (see Fig. 26.1c), also represent circular scales of measurement.

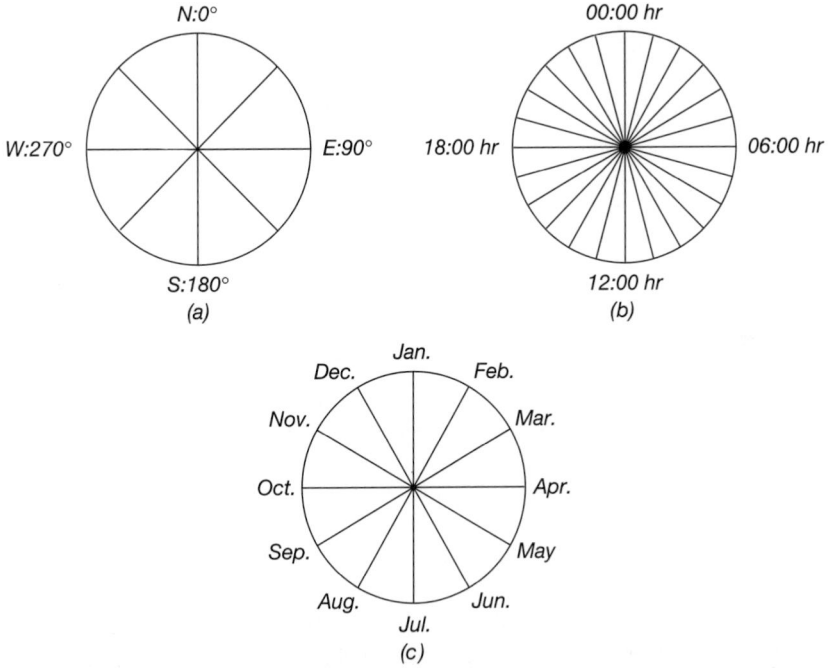

Figure 26.1   Common circular scales of measurement. (a) Compass directions. (b) Times of day. (c) Days of the year (with the first day of each month shown).

In general, we may convert X time units to an angular direction (a, in degrees), where X has been measured on a circular scale having k time units in the full cycle:

$$a = \frac{(360°)(X)}{k}. \tag{26.1}$$

For example, to convert a time of day (X, in hours) to an angular direction, $k = 24$ hr; to convert a day of the week to an angular direction, number the seven days from some arbitrary point (e.g., Sunday = day 1) and use Equation 26.1 with $k = 7$; to convert the Xth day of the year to an angular direction, $k = 365$ (or, $k = 366$ in a leap year); to convert a month of the year, $k = 12$; and so on.* Such conversions are demonstrated in Example 26.1.

---

*Equation 26.1 gives angular directions corresponding to the ends of time periods (e.g., the end of the Xth day of the year). If some other point in a time period is preferred, the equation can be adjusted accordingly. For example, noon can be considered on the Xth day of the year by using $X - 0.5$ in place of X.) If the same point is used in each time period (e.g., always using either noon or midnight), then the statistical procedures of

---

EXAMPLE 26.1   Conversions of times measured on a circular scale to corresponding angular directions.

We use Equation 26.1:

$$a = \frac{(360°)(X)}{k}.$$

1. Given a time of day of 06:00 hr (which is one-fourth of the twenty-four hour clock and should correspond, therefore, to one-fourth of a circle):

$$X = 6 \text{ hr}, \; k = 24 \text{ hr, and}$$

$$a = (360°)(6 \text{ hr})/24 \text{ hr} = 90°.$$

2. Given a time of day of 06:15 hr:

$$X = 6.25 \text{ hr}, \; k = 24 \text{ hr, and}$$

$$a = (360°)(6.25 \text{ hr})/24 \text{ hr} = 93.75°.$$

3. Given the 14th day of February, being the 45th day of the year:

$$X = 45 \text{ days}, \; k = 365 \text{ days, and}$$

$$a = (360°)(45 \text{ days})/365 \text{ days} = 44.38°.$$

Occasionally, angular data may be cyclic on a scale other than 360°. For example, data might be collected on the orientation of plant leaves, where a leaf might be recorded as being aligned along a 0°–180° axis (i.e., a north-south axis), or along a 20°–200° axis, or along a 30°–210° axis, and so on, but without recording whether the stem end or the distal end of the leaf was directed toward 0°, 20°, 30°, and so on. Thus, the measurements, while truly cyclical, would be on a scale of 180° (with the data being known as *axial*). The data could be subjected to Equation 26.1 in order to summarize or analyze them. This would result in $a = (360°)(X)/180° = 2X$, meaning that we would simply double each datum, X, before engaging in such summary or analysis.

Data from circular distributions generally may not be analyzed using the statistical methods presented earlier in this book. This is so for theoretical reasons as well as for empirically obvious reasons stemming from the arbitrariness of the zero point on the circular scale. For example, consider three compass directions: 10°, 30°, and 350°, for which we wish to calculate an arithmetic mean. The arithmetic mean calculation of $(10° + 30° + 350°)/3 = 390°/3 = 130°$ is clearly absurd, for all data are northerly directions and the computed mean is southeasterly.

Statistical methods for describing and analyzing data from circular distributions are relatively new and are still undergoing development. This chapter will introduce some basic considerations useful in calculating descriptive statistics and Chapter 27 will

---

this and the following chapter will be unaffected by the choice of point. (However, graphical procedures, as in Section 26.2, will of course be affected, in the form of a rotation of the graph if Equation 26.1 is adjusted. If, for example, we considered noon on the Xth day of the year, the entire graph would be rotated about half a degree counterclockwise.)

discuss tests of hypotheses. More extensive reviews have been published by Batschelet* (1965, 1972, 1981), Fisher (1993), Mardia (1972a, 1981), Upton and Fingleton (1989: Chapter 9), and the literature cited therein. Statistical methods have also been developed for data that occur on a sphere and are of particular interest to earth scientists (and summarized by Batschelet, 1981: Chapter 11; Fisher, Lewis, and Embleton, 1987; Mardia, 1972: Chapters 8–9, 1981; Upton and Fingleton, 1989: Chapter 10; and Watson, 1983).

26.2 GRAPHICAL PRESENTATION OF CIRCULAR DATA

Circular data are often presented as a scatter diagram, where the scatter is shown on the circumference of a circle. Figure 26.2 shows such a graph for the data of Example 26.2.

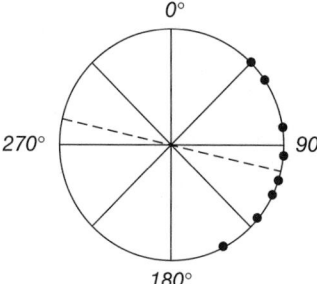

Figure 26.2   A circular scatter diagram for the data of Example 26.2. (The dashed line is explained in Section 26.6.)

EXAMPLE 26.2   A sample of circular data. These data are plotted in Fig. 26.2.

Eight trees are found leaning in the following compass directions: 45°, 55°, 81°, 96°, 110°, 117°, 132°, 154°.

If frequencies of data are too large to be plotted conveniently on a scatter diagram, then a bar graph, or histogram, may be drawn. This is demonstrated in Fig. 26.3, for the data presented in Example 26.3. Recall that in a histogram, the length, as well as the area, of each bar is an indication of the frequency observed at each plotted value of the variable (Section 1.3). Occasionally, as shown in Fig. 26.4, a histogram is seen presented with sectors, instead of bars, comprising the graph; this is sometimes called a *rose diagram*. Here, the radii forming the outer boundaries of the sectors are proportional to the frequencies being represented, but the areas of the sectors are not. Since it is likely that the areas will be judged by the eye to represent the frequencies, the reader of the graph is being

---

*Edward Batschelet (1914–1979), Swiss biomathematician, was one of the most influential writers in developing, explaining, and promulgating circular statistical methods, particularly among biologists.

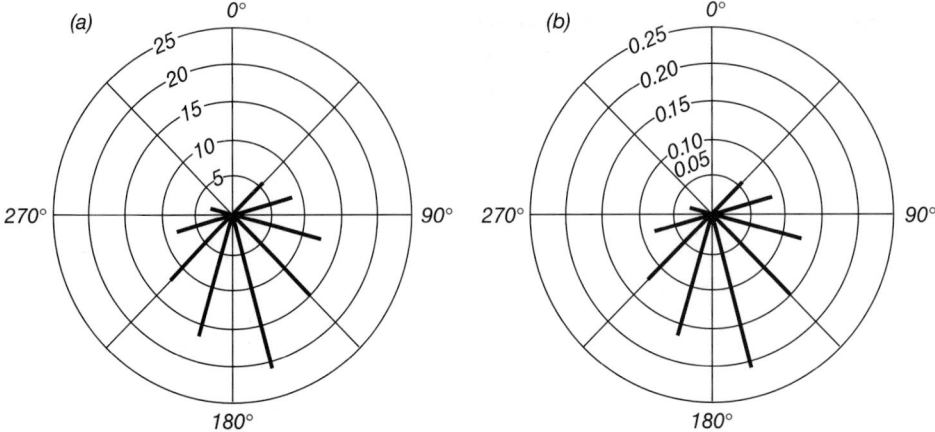

Figure 26.3 (a) Circular histogram for the data of Example 26.3 where the concentric circles represent frequency increments of 5. (b) A relative frequency histogram for the data of Example 26.3 with the concentric circles representing relative frequency increments of 0.05.

EXAMPLE 26.3 **A sample of circular data, presented as a frequency table, where a_i is an angle and f_i is the observed frequency of a_i. These data are plotted in Fig. 26.3.**

a_i (deg)	f_i	Relative f_i
0–30	0	0.00
30–60	6	0.06
60–90	9	0.09
90–120	13	0.12
120–150	15	0.14
150–180	22	0.21
180–210	17	0.16
210–240	12	0.11
240–270	8	0.08
270–300	3	0.03
300–330	0	0.00
330–360	0	0.00
$n = 105$		Total $= 1.00$

---

deceived, and this type of graphical presentation is not recommended. However, a true-area rose diagram can be obtained by plotting the square roots of frequencies as radii.*

And, another manner of expressing circular frequency distributions graphically is shown in Fig. 26.5. Here, the length of each bar of the histogram represents a frequency, as in Fig. 26.3(a), but the bars extend from the circumference of a circle instead of

---

*The earliest user of rose diagrams was the founder of modern nursing and pioneer social and health statistician, Florence Nightingale (1820–1910), in 1858. She employed true-area colored diagrams, which she termed "coxcombs," to indicate deaths from various causes over months of the year (Fisher, 1993: 5–6).

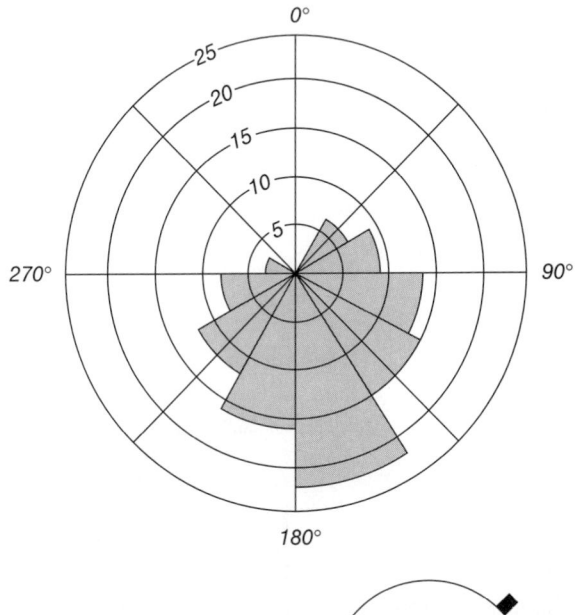

Figure 26.4 A rose diagram of the data of Example 26.3, utilizing sectors instead of bars. This procedure is not recommended unless square roots of the frequencies are employed (see Section 26.2).

Figure 26.5 Circular histogram for the data of Example 26.3, including an arrow depicting the mean angle ($\bar{a}$) and a measure of dispersion (r).

from the center. In addition, an arrow extending from the circle's center toward the circumference indicates both the direction and the length of the mean vector, and this expresses visually both the mean angle and a measure of data concentration (as explained in Sections 26.4 and 26.5).

A histogram of circular data can also be plotted as a linear histogram (see Section 1.3), with degrees on the horizontal axis and frequencies (or relative frequencies) on the vertical axis. But the impression on the eye may vary with the arbitrary location of the origin of the horizontal axis, and (unless the range of data is small—say, no more than 180°) the presentation of Fig. 26.3 or Fig. 26.5 is preferable.

26.3 SINES AND COSINES OF CIRCULAR DATA

A great many of the procedures that follow in this chapter and the next require the determination of two basic trigonometric functions. Let us consider that a circle (perhaps representing a compass face) is drawn on rectangular coordinates (as on common graph

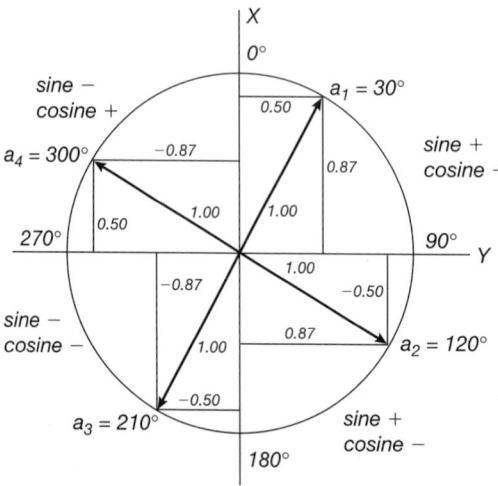

Figure 26.6 A unit circle, showing four points and their polar (a and r) and rectangular (X and Y) coordinates.

paper) with the center as the origin (i.e., zero) of a vertical X axis and a horizontal Y axis; this is what is done in Fig. 26.6.

There are two methods that can be used to locate any point on a plane (such as a sheet of paper). One is to specify X and Y (as done previously in discussing regression and correlation (in Chapters 17 and 19). However, with circular data it is conventional to use a vertical, instead of a horizontal, X axis. This second method specifies both the angle, a, with respect to some starting direction (say, clockwise from the top of the axis X axis, namely "north") and the straightline distance, r, from some reference point (the center of the circle). This pair of numbers, a and r, is known as the "polar coordinates" of a point. Thus, for example, in Fig. 26.6, point 1 is uniquely identified by polar coordinates $a = 30°$ and $r = 1.00$, point 2 by $a = 120°$ and $r = 1.00$, and so on.*

If a is negative, it is expressing a direction *counterclockwise* from zero. It may be added to 360° to yield the equivalent positive angle; thus, e.g., $-60° = 360° - 60° = 300°$. An angle greater than 360° is equivalent to the number of degrees by which it exceeds 360° or a multiple of 360°. So, for example, $450° = 450° - 360° = 90°$ and $780° = 780° - 360° - 360° = 60°$.

The first-mentioned method of locating points on a graph referred to the X and Y axes. By this method, point 1 in Fig. 26.6 is located by the "rectangular coordinates"[†]

---

*If we specify that the radius of the circle is 1 unit, our figure is called a "unit circle."

[†]The familiar system of rectangular coordinates is also known as "Cartesian coordinates," after the French mathematician and philosopher, René Descartes (1596–1650), who wrote under the Latinized version of his name, Renatus Cartesius. His other enduring mathematical introductions included the use of exponents, the square root sign ($\sqrt{\ }$), and, in 1637, the use of letters near the end of the alphabet (e.g., X, Y, and Z) to denote variables and those near the beginning (e.g., a, b, and c) to denote constants (Asimov, 1982: 117). Over time, many different symbols and abbreviations have been used for trigonometric functions. The abbreviations "sin." and "tan." were established in the latter half of the sixteenth century, and the periods were dropped early in the next century (Cajori, 1929: 150, 158). The cosine was first known as the sine of the complement—because

$X = 0.87$ and $Y = 0.50$, point 2 by $X = -0.50$ and $Y = 0.87$, point 3 by $X = -0.87$ and $Y = -0.50$, and point 4 by $X = 0.50$ and $Y = -0.87$. The *cosine* (abbreviated "cos") of an angle is defined as the ratio of the X and the r associated with the circular measurement:

$$\cos a = \frac{X}{r},\qquad(26.2)$$

while the *sine* (abbreviated "sin") of the angle is the ratio of the associated Y and r:

$$\sin a = \frac{Y}{r}.\qquad(26.3)$$

Thus, for example, the sine of a_1 in Fig. 26.6 is $\sin 30° = 0.50/1.00 = 0.50$, and its cosine is $\cos 30° = 0.87/1.00 = 0.87$. Also, $\sin 120° = 0.87/1.00 = 0.87$, $\cos 120° = -0.50/1.00 = -0.50$, and so on. Sines and cosines (two of the most useful "trigonometric* functions") are readily available in published tables, and many electronic calculators give them (and sometimes convert between polar and rectangular coordinates as well). The sines of $0°$ and $180°$ are zero, angles between $0°$ and $180°$ have sines that are positive and the sines are negative for $180° < a < 360°$. The cosine is zero for $90°$ and $270°$, with positive cosines obtained for $0° < a < 90°$ and for $270° < a < 360°$, and negative cosines for angles between $90°$ and $270°$.

At most, only sines and cosines are required for the statistical procedures that follow, although brief mention will be made of a third trigonometric function, the *tangent:*[†]

$$\tan a = \frac{Y}{X} = \frac{\sin a}{\cos a}.\qquad(26.5)$$

We shall see later that rectangular coordinates, X and Y may also be used in conjunction with mean angles just as they are with individual angular measurements.

26.4 THE MEAN ANGLE

If we have a sample consisting of n angles, denoted as a_1 through a_n, then the mean of these angles, $\bar{a}$, is to be an estimate of the mean angle in the sampled population μ_a. To compute the sample mean angle, $\bar{a}$, we first consider the rectangular coordinates of the mean angle:

$$X = \frac{\sum\limits_{i=1}^{n} \cos a_i}{n}\qquad(26.6)$$

---

the cosine of a equals the sine of $(90° - a)$ for angles from $0°$ to $90°$— and the English writer E. Gunter changed "complemental sine" to "cosine" and "complemental tangent" to "cotangent" in 1620 (ibid.: 157).

*"Trigonometry" refers, literally, to the measurement of triangles (such as the triangles that emanate from the center of the circle in Fig. 26.6).

[†]The *cotangent* is the reciprocal of the tangent, namely

$$\cot a = \frac{X}{Y} = \frac{\cos a}{\sin a}.\qquad(26.4)$$

and

$$Y = \frac{\sum\limits_{i=1}^{n} \sin a_i}{n}. \tag{26.7}$$

Then, the quantity

$$r = \sqrt{X^2 + Y^2} \tag{26.8}$$

is computed;* this is the length of the mean vector, which will be further discussed in Section 26.5. The value of $\bar{a}$ is determined as the angle having the following cosine and sine:

$$\cos \bar{a} = \frac{X}{r} \tag{26.9}$$

and

$$\sin \bar{a} = \frac{Y}{r}. \tag{26.10}$$

Example 26.4 demonstrates these calculations. It is also true that

$$\tan \bar{a} = \frac{Y}{X}, \tag{26.11}$$

so we have a check on the calculation of the mean angle, $\bar{a}$. If $r = 0$, the mean angle is undefined and we conclude that there is no mean direction.

If we are dealing with data that are times instead of angles, then the mean time corresponding to the mean angle may be determined by inverting Equation 26.1:

$$\bar{X} = \frac{ka}{360°}. \tag{26.12}$$

For example, to determine a mean time of day, $\bar{X}$, from a mean angle, $\bar{a}$: $\bar{X} = (24 \text{ hr})(\bar{a})/360°$.

Grouped Data. Often circular data are recorded in a frequency table (as in Example 26.3). For such data, the following computations are convenient alternatives to Equations 26.6 and 26.7, respectively:

$$X = \frac{\sum f_i \cos a_i}{n} \tag{26.13}$$

$$Y = \frac{\sum f_i \sin a_i}{n} \tag{26.14}$$

(which are analogous to Equation 3.3 for linear data). In these equations, a_i is the midpoint of the measurement interval recorded (e.g., $a_2 = 45°$ in Example 26.3, which is the midpoint of the second recorded interval, $30-60°$), f_i is the frequency of occurrence

---

*This quantity, r, is not to be confused with a sample correlation coefficient (Section 19.1), with which it bears no relationship but which is denoted by the same symbol.

EXAMPLE 26.4   Calculating the mean angle for the data of Example 26.2.

a_i (deg)	$\sin a_i$	$\cos a_i$
45	0.70711	0.70711
55	0.81915	0.57358
81	0.98769	0.15643
96	0.99452	−0.10453
110	0.93969	−0.34202
117	0.89101	−0.45399
132	0.74315	−0.66913
154	0.43837	−0.89879

$$\sum \sin a_i = 6.52069 \qquad \sum \cos a_i = -1.03134$$

$$Y = \frac{\sum \sin a_i}{n} \qquad X = \frac{\sum \cos a_i}{n}$$

$$= 0.81509 \qquad\qquad = -0.12892$$

$n = 8$

$$r = \sqrt{X^2 + Y^2} = \sqrt{(-0.12892)^2 + (0.81509)^2} = \sqrt{0.68099} = 0.82522$$

$$\cos \bar{a} = \frac{X}{r} = \frac{-0.12892}{0.82522} = -0.15623$$

$$\sin \bar{a} = \frac{Y}{r} = \frac{0.81509}{0.82522} = 0.98772$$

The angle with this sine and cosine is $\bar{a} = 99°$.

of data within that interval (e.g., $f_2 = 6$ in that example), and $n = \sum f_i$. Example 26.5 demonstrates the determination of $\bar{a}$ for the grouped data of Example 26.3.

There is a bias in computing r from grouped data, in that the result is too small. A correction for this is available (Batschelet, 1965: 16–17, 1981: 37–40; Mardia, 1972a: 78–79), which may be applied when the distribution is unimodal and does not deviate greatly from symmetry. For data grouped into intervals of d degrees each,

$$r_c = cr, \tag{26.15}$$

where r_c is the corrected r, and c is a correction factor,

$$c = \frac{\dfrac{d\pi}{360°}}{\sin\left(\dfrac{d}{2}\right)}. \tag{26.16}$$

The correction becomes insignificant for intervals smaller than $30°$. This correction is for the quantity r, and for statistics calculated from it; but the mean angle, $\bar{a}$, requires no correction for grouping.

EXAMPLE 26.5   Calculating the mean angle for the data of Example 26.3.

a_i	f_i	$\sin a_i$	$f_i \sin a_i$	$\cos a_i$	$f_i \cos a_i$
45°	6	0.70711	4.24266	0.70711	4.24266
75°	9	0.96593	8.69337	0.25882	2.32938
105°	13	0.96593	12.55709	-0.25882	-3.36466
135°	15	0.70711	10.60665	-0.70711	-10.60665
165°	22	0.25882	5.69404	-0.96593	-21.25046
195°	17	-0.25882	-4.39994	-0.96593	-16.42081
225°	12	-0.70711	-8.48532	-0.70711	-8.48532
255°	8	-0.96593	-7.72744	-0.25882	-2.07056
285°	3	-0.96593	-2.89779	0.25882	0.77646

$$n = 105 \qquad \sum f_i \sin a_i = 18.28332 \qquad \sum f_i \cos a_i = -54.84996$$

$$Y = \frac{\sum f_i \sin a_i}{n} \qquad\qquad X = \frac{\sum f_i \cos a_i}{n}$$

$$= 0.17413 \qquad\qquad = -0.52238$$

$$r = \sqrt{X^2 + Y^2} = \sqrt{(-0.52238)^2 + (0.17413)^2} = 0.55064$$

$$\cos \bar{a} = \frac{X}{r} = \frac{-0.52238}{0.55064} = -0.94868$$

$$\sin \bar{a} = \frac{Y}{r} = \frac{0.17413}{0.55064} = 0.31623$$

The angle with this cosine and sine is $\bar{a} = 162°$.

26.5 ANGULAR DISPERSION

When dealing with circular data, we wish to have a measure, analogous to those of Chapter 4 for a linear scale, to describe the dispersion of the data.

We can define the *range* in a circular distribution of data as the smallest arc (i.e., the smallest portion of the circle's circumference) that contains all the data in the distribution. For example, in Fig. 26.7a, the range is zero; in Fig. 26.7b the shortest arc is from the data point at 38° to the datum at 60°, making the range 22°; in Fig. 26.7c, the data are found from 10° to 93°, with a range of 83°; in Fig. 26.7d, the data run from 322° to 135°, with a range of 173°; in Fig. 26.7e, the shortest arc containing all the data is that running clockwise from 285° to 171°, namely an arc of 246°; and in Fig. 26.7f the range is 300°. For the data of Example 26.2 the range is 109° (as the data run from 45° to 154°).

Another measure of dispersion is seen by examining Fig. 26.7; the value of r varies inversely with the amount of dispersion in the data. Therefore, r is a measure of concentration. It has no units and it may vary from 0 (when there is so much dispersion that a mean angle cannot be described) to 1.0 (when all the data are concentrated at the same direction). (An r of 0 does not, however, necessarily indicate a uniform distribution. For example, the data of Fig. 26.8 would also yield $r = 0$.)

In Section 3.1 the mean on a linear scale was noted to be the center of gravity of a group of data. Similarly, the tip of the mean vector (i.e., the quantity r), in the direction

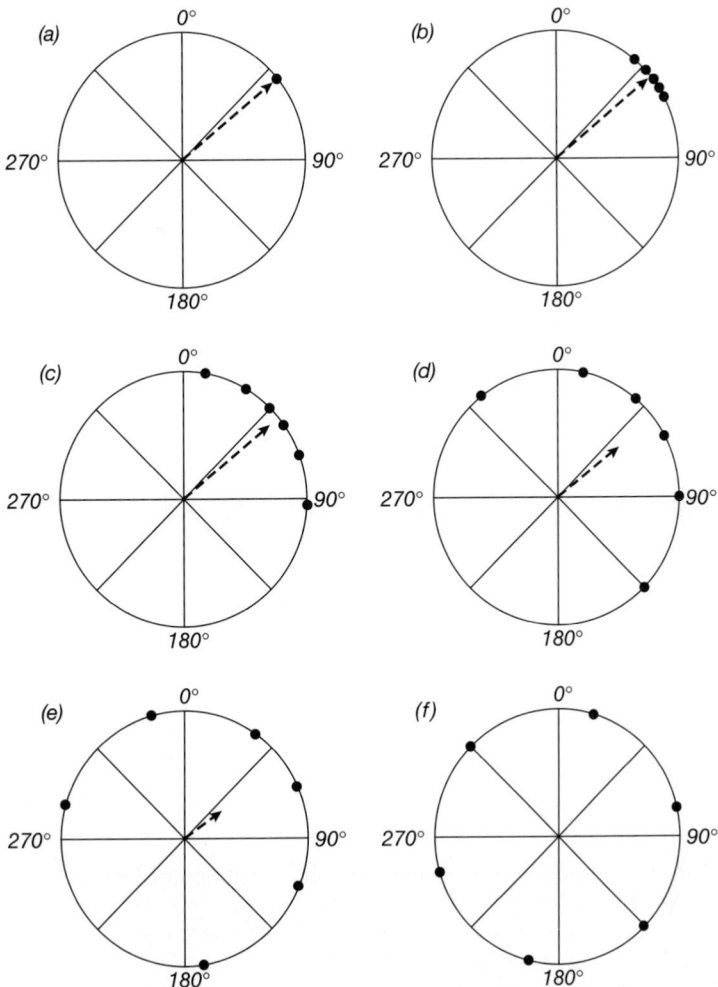

Figure 26.7 Circular distributions with various amounts of dispersion. The broken line indicates the mean angle, which is 50° in each case. Note that the value of r varies inversely with the amount of dispersion, and that the values of s and s_0 vary directly with the amount of dispersion. (a) $r = 1.00$, $s = 0°$, $s_0 = 0°$. (b) $r = 0.99$, $s = 8.10°$, $s_0 = 8.12°$. (c) $r = 0.90$, $s = 25.62°$, $s_0 = 26.30°$. (d) $r = 0.60$, $s = 51.25°$, $s_0 = 57.91°$. (e) $r = 0.30$, $s = 67.79°$, $s_0 = 88.91$. (f) $r = 0.00$, $s = 81.03°$, $s_0 = \infty$. (By the method of Section 26.1, the magnitude of r is statistically significant in Figs. a, b, and c, but not in d, e, and f.)

of the mean angle ($\bar{a}$) lies at the center of gravity. (Consider that each circle in Fig. 26.7 is a disc of material of negligible weight, and each datum is a dot of unit weight. The disc, held parallel to the ground, would balance at the tip of the arrow in the figure. In Fig. 26.7f, $r = 0$ and the center of gravity is the center of the circle.)

Because r is a measure of concentration, $1 - r$ is a measure of dispersion. Lack of dispersion would be indicated by $1 - r = 0$, and maximum dispersion by $1 - r = 1.0$. As a measure of dispersion reminiscent of those for linear data, Mardia (1972a: 45) defined *circular variance*:

$$S^2 = 1 - r. \tag{26.17}$$

Batschelet (1965, 1981: 34) defined *angular variance:*

$$s^2 = 2(1 - r) \tag{26.18}$$

as being a closer analog to linear variance (Equation 4.8). While S^2 may range from 0 to 1, and s^2 from 0 to 2, an S^2 of 1 or an s^2 of 2 does not necessarily indicate a uniform distribution of data around the circle because, as noted above, $r = 0$ does not necessarily indicate a uniform distribution. The variance measure,

$$s_0^2 = -2\ln r, \tag{26.19}$$

is a statistic that ranges from 0 to ∞ (Mardia, 1972: 24). These three dispersion measures are in radians squared. To express them in degrees squared, multiply each by $(180°/\pi)^2$.

Measures analogous to the linear standard deviation include the "mean angular deviation," or simply the *angular deviation*, which is

$$s = \frac{180°}{\pi}\sqrt{2(1 - r)}, \tag{26.20}$$

in degrees.* This ranges from a minimum of zero (e.g., Fig. 26.7a) to a maximum of 81.03° (e.g., Fig. 26.7f).[†] Mardia (1972a: 24, 74) defines *circular standard deviation* as

$$s_0 = \frac{180°}{\pi}\sqrt{-2\ln r} \tag{26.21}$$

degrees; or, employing common, instead of natural, logarithms:

$$s_0 = \frac{180°}{\pi}\sqrt{-4.60517\log r} \tag{26.22}$$

degrees. This is analogous to the standard deviation, s, on a linear scale (Section 4.5) in that it ranges from zero to infinity (see Fig. 26.7). For large r, the values of s and s_0 differ by no more than 2 degrees for r as small as 0.80, by no more than 1 degree for r as small as 0.87, and by no more than 0.1 degree for r as small as 0.97. It is intuitively reasonable that a measure of angular dispersion should have a finite upper limit, so s is the measure preferred in this book. Appendix Tables B.32 and B.33 convert r to s and s_0, respectively. If the data are grouped, then s and s_0 are biased in being too high, so r_c (by Equation 26.15) can be used in place of r. For the data of Example 26.4 (where $r = 0.82522$), $s = 34°$ and $s_0 = 36°$; for the data of Example 26.5 (where $r = 0.55064$), $s = 54°$ and $s_0 = 63°$.

---

*Simply delete the constant, $180°/\pi$, in this and in the following equations if the measurement is desired in radians rather than degrees.

[†]This is a range of 0 to 1.41 radians.

Dispersion measures analogous to the linear mean deviation (Section 4.3) utilize absolute deviations of angles from the mean or median (e.g., Fisher, 1993: 36).

Measures of symmetry and kurtosis on a circular scale, analogous to the g_1 and g_2 that may be calculated on a linear scale (Section 6.1), are discussed by Batschelet (1965: 14–15, 1981: 43–44) and Mardia (1972a: 36–38, 74–76).

26.6 THE MEDIAN AND MODAL ANGLES

In a fashion analogous to considerations for linear scales of measurement (Sections 3.2 and 3.4), one can determine the median and the mode of a set of data on a circular scale.

To find the *median angle*, we first determine which diameter of the circle divides the data into two equal-sized groups. The median angle is the angle indicated by that diameter's radius that is nearer to the majority of the data points. If n is even, the median is nearly always midway between two of the data. In Example 26.2, a diameter extending from $103°$ to $289°$ divides the data into two groups of four each (as indicated by the dashed line in Fig. 26.2). The data are concentrated around $103°$, rather than $289°$, so the sample median is $103°$. If n is odd, the median will almost always be one of the data points or $180°$ opposite from one. If the data in Example 26.2 had been seven in number—with the $45°$ lacking—then the diameter line would have run through $100°$ and $290°$, and the median would have been $110°$. Mardia (1972a: 29–30) shows how the median is estimated, analogously to Equation 3.5, when it lies within a group of tied data. If a sample has the data equally spaced around the circle (as in Fig. 26.7f), then the median, as well as the mean, is undefined.

The *modal angle* is defined as is the mode for linear scale data (Section 3.4). Just as with linear data, there may be more than one mode or there may be no modes.

26.7 CONFIDENCE LIMITS FOR THE POPULATION MEAN AND MEDIAN ANGLES

The confidence limits of the mean of angles may be expressed as

$$\bar{a} \pm d. \tag{26.23}$$

That is, the lower confidence limit is $L_1 = \bar{a} - d$ and the upper confidence limit is $L_2 = \bar{a} + d$. For n as small as 8 the following method may be used (Upton, 1986). For $r \le 0.9$, and $r > \sqrt{\chi^2_{\alpha,1}/2n}$,

$$d = \arccos \left[\frac{\sqrt{\dfrac{2n(2R^2 - n\chi^2_{\alpha,1})}{4n - \chi^2_{\alpha,1}}}}{R} \right] \tag{26.24}$$

and for $r \geq 0.9$,

$$d = \arccos \left[\frac{\sqrt{n^2 - (n^2 - R^2)e^{\chi^2_{\alpha,1}/n}}}{R} \right], \tag{26.25}$$

where

$$R = nr. \tag{26.26}$$

EXAMPLE 26.6   The 95% confidence interval for the data of Example 26.4.

$n = 8$
$\bar{a} = 104°$
$r = 0.87919$
$R = nr = (8)(0.87919) = 7.03352$
Using Equation 26.24:

$$d = \arccos \left[\frac{\sqrt{\dfrac{2n \left(2R^2 - n\chi^2_{\alpha,1} \right)}{4n - \chi^2_{\alpha,1}}}}{R} \right]$$

$$= \arccos \left[\frac{\sqrt{\dfrac{2(8)(2[7.03352]^2 - 8[3.841])}{4(8) - 3.841}}}{7.03352} \right]$$

$$= \arccos(0.88514)$$

$$= 28° \text{ or } 332°$$

The 95% confidence interval is $104° \pm 28°$;
$L_1 = 76°$ and $L_2 = 132°$.

This is demonstrated in Examples 26.6 and 27.3.* As this procedure is only approximate, d—and confidence limits—should not be expressed to fractions of a degree. This procedure is based on the von Mises distribution, a circular analog of the normal distribution.[†] Batschelet (1972: 86; Zar, 1984: 665–666) presents nomograms that yield similar results.

---

*"Arccos" is the abbreviation for "arccosine," also referred to as the "inverse cosine" or "$\cos^{-1}$." The arccosine of X, or arccos (X), is the angle having a cosine of X. As shown in these examples, a given cosine is associated with two different angles; the smaller of the two is to be used.

[†]Richard von Mises (1883–1953), a physicist and mathematician, was born in the Austro-Hungarian Empire and moved to Germany, Turkey, and the United States owing to two world wars (Geiringer, 1978). He introduced this distribution (von Mises, 1918), and it was called "circular normal" by Gumbel, Greenwood, and Durand (1953) because of its similarity to the linear-scale normal distribution. It is described mathematically by Batschelet (1981: 279–282), Fisher (1993: 48–56), and Mardia (1972a: 122–127). Goodness of fit testing for the von Mises distribution is discussed by Fisher (1993: 82–85).

Confidence limits for the median angle may be obtained by the procedure of Section 24.9. The data in the sample are numbered sequentially from 1 through n so that the median fits Equation 3.4.

26.8 DIAMETRICALLY BIMODAL DISTRIBUTIONS

Occasionally populations are encountered having data with two modes lying opposite each other on the diameter of the circle. (Such data are sometimes termed "axial.") For example, Fig. 26.8 shows a distribution having opposite modes at 45° and 225°. If, as in this figure, the distribution is centrally symmetrical (i.e., each observation is matched by an observation 180° away), r computes to be zero and no mean or median angle can be determined. If the diametrically bimodal distribution is not centrally symmetrical, r will not be zero, but it may be so small as to have us conclude that there is no significant direction of orientation of the data (Section 27.1), and the calculated mean may be far from the diameter along which the bulk of the observations lie. However, we can engage in statistical analysis of such a distribution by a procedure involving the doubling of angles.

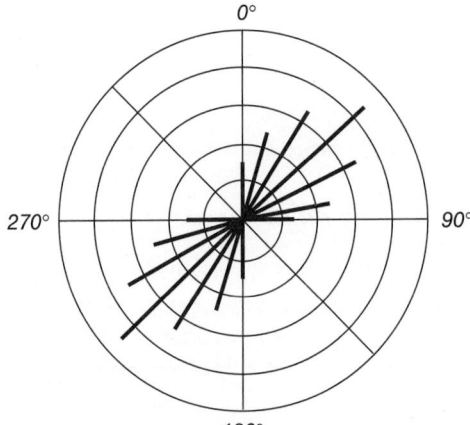

Figure 26.8 A bimodal circular distribution, $H_0 : \rho = 0$ would not be rejected (see Section 26.1).

Example 26.7 shows the data that are graphed in Fig. 26.8. Each angle, a_i, is doubled; if the doubled angle is $< 360°$ it is recorded as $2a_i$, and if it is $\geq 360°$ then 360° is subtracted from it with the result being recorded as $2a_i$. (Also note in this example that Equations 26.13 and 26.14 are used to make the computations easier, because the data are grouped.) Note in Example 26.7 that the angular deviation is one-half the angular deviation of the doubled angles.

EXAMPLE 26.7   **Descriptive statistics for the centrally symmetric distribution shown in Fig. 26.8.**

a_i	f_i	$2a_i$	$\sin 2a_i$	$f_i \sin 2a_i$	$\cos 2a_i$	$f_i \cos 2a_i$
$0°$	15	$0°$	0.00000	0.00000	1.00000	15.00000
$15°$	25	$30°$	0.50000	12.50000	0.86603	21.65075
$30°$	35	$60°$	0.86603	30.31105	0.50000	17.50000
$45°$	45	$90°$	1.00000	45.00000	0.00000	0.00000
$60°$	35	$120°$	0.86603	30.31105	-0.50000	-17.50000
$75°$	25	$150°$	0.50000	12.50000	-0.86603	-21.65075
$90°$	15	$180°$	0.00000	0.00000	-1.00000	-15.00000
$180°$	15	$0°$	0.00000	0.00000	1.00000	15.00000
$195°$	25	$30°$	0.50000	12.50000	0.86603	21.65075
$210°$	35	$60°$	0.86603	30.31105	0.50000	17.50000
$225°$	45	$90°$	1.00000	45.00000	0.00000	0.00000
$240°$	35	$120°$	0.86603	30.31105	-0.50000	-17.50000
$255°$	25	$150°$	0.50000	12.50000	-0.86603	-21.65075
$270°$	15	$180°$	0.00000	0.00000	-1.00000	-15.00000

$$n = 390$$

$$\sum f_i \sin 2a_i = 261.24420 \qquad \sum f_i \cos 2a_i = 0.00000$$

$$Y = \frac{261.24420}{390} \qquad X = \frac{0}{390}$$

$$= 0.66986 \qquad = 0$$

$$r = \sqrt{(0.66986)^2 + (0)^2} = 0.66986$$

$$\cos 2\bar{a} = \frac{0}{0.66986} = 0$$

$$\sin 2\bar{a} = \frac{0.66986}{0.66986} = 1.0000$$

Therefore, $2\bar{a} = 90°$ and $\bar{a} = 45°$, meaning that the bimodal distribution lies along a diameter line oriented at 45° (as can be seen by inspecting Fig. 26.8).

$$s \text{ for doubled angles} = \sqrt{2(1 - 0.66986)} = \sqrt{0.66028} = 0.81258$$

$$s = \frac{0.81258}{2} = 0.41$$

Inasmuch as the data are grouped (in intervals of 15°), Equation 26.15 may be used, in which case we find $s = 0.40$.

26.9 SECOND-ORDER ANALYSIS: THE MEAN OF MEAN ANGLES

If a mean is determined for each of several groups of angles, then we have a set of mean angles. Consider the data in Example 26.8. Here, a mean angle, $\bar{a}$, has been calculated for each of k samples of circular data, using the procedure of Section 26.4. If, now, we desire to determine the grand mean of these several means, it is not appropriate to consider each of the sample means as an angle and employ the method of Section 26.4.

EXAMPLE 26.8   The mean of a set of mean angles.

Under particular light conditions, each of seven butterflies is allowed to fly from the center of an experimental chamber ten times. From the procedures of Section 26.4, the values of $\bar{a}$ and r for each of the seven samples of data are as follows.

$k = 7; \quad n = 10$

Sample (j)	$\bar{a}_j$	r_j	$X_j = r_j \cos \bar{a}_j$	$Y_j = r_j \sin \bar{a}_j$
1	160°	0.8954	−0.84140	0.30624
2	169	0.7747	−0.76047	0.14782
3	117	0.4696	−0.21319	0.41842
4	140	0.8794	−0.67366	0.56527
5	186	0.3922	−0.39005	−0.04100
6	134	0.6952	−0.48293	0.50009
7	171	0.3338	−0.32969	0.05222
			−3.69139	1.94906

$$\bar{X} = \frac{\sum r_j \cos \bar{a}_j}{k} = \frac{-3.69139}{7} = -0.52734$$

$$\bar{Y} = \frac{\sum r_j \sin \bar{a}_j}{k} = \frac{1.94906}{7} = 0.27844$$

$$r = \sqrt{\bar{X}^2 + \bar{Y}^2} = \sqrt{0.35562} = 0.59634$$

$$\cos \bar{a} = \frac{\bar{X}}{r} = \frac{-0.52734}{0.59634} = -0.88429$$

$$\sin \bar{a} = \frac{\bar{Y}}{r} = \frac{0.27844}{0.59634} = 0.46691$$

Therefore, $\bar{a} = 152°$.

To do so would be to assume that each mean had a vector length, r, of 1.0 (i.e., that an angular deviation, s, of zero was the case in each of the k samples), a most unlikely situation. Instead, we shall employ the procedure promulgated by Batschelet* (1978, 1981: Chapter 7), whereby the grand mean has rectangular coordinates

$$\bar{X} = \frac{\sum_{j=1}^{k} X_j}{k} \tag{26.27}$$

and

$$\bar{Y} = \frac{\sum_{j=1}^{k} Y_j}{k}, \tag{26.28}$$

---

*Batschelet refers to the determination of the mean of a set of angles as a first-order analysis and the computation of the mean of a set of means as a second-order analysis.

where X_j and Y_j are the quantities X and Y, respectively, applying Equations 26.6 and 26.7 to sample j; k is the total number of samples. If we do not have X and Y for each sample, but we have $\bar{a}$ and r (polar coordinates) for each sample, then

$$\bar{X} = \frac{\sum_{j=1}^{k} r_j \cos \bar{a}_j}{k} \tag{26.29}$$

and

$$\bar{Y} = \frac{\sum_{j=1}^{k} r_j \sin \bar{a}_j}{k}. \tag{26.30}$$

Having obtained $\bar{X}$ and $\bar{Y}$, we may substitute them for X and Y, respectively, in Equations 26.8, 26.9, and 26.10 (and 26.11, if desired) in order to determine $\bar{a}$, which is the grand mean. For this calculation, all n_j's (sample sizes) should be equal, although a slight departure from this condition will not severely affect the results.

Figure 26.9 shows the individual means and the grand mean for Example 26.8. (By the hypothesis testing of Section 27.1 we would conclude that there is in this example no significant mean direction for Samples 5 and 7. However, the data from these two samples should not be deleted from the present analysis.)

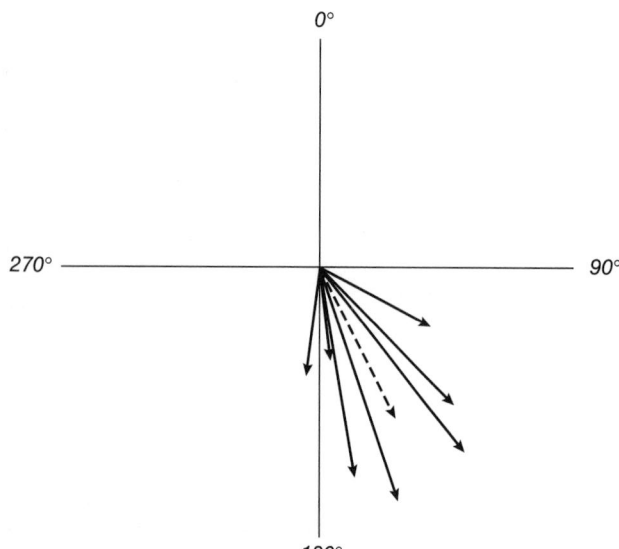

Figure 26.9 The data of Example 26.8. Each of the seven vectors in this sample is itself a mean vector. The mean of these seven means is indicated by the broken line.

26.10 CONFIDENCE LIMITS FOR THE SECOND-ORDER MEAN ANGLE

Section 26.9 explains how to obtain the mean of a set of mean angles. The mean thus computed is a sample estimate of a population mean, μ_a, and it is reasonable to ask how precise an estimate it is. The precision with which we estimate a population mean is typically expressed as a confidence interval for that parameter. For a first-order circular statistical analysis we may find confidence limits for μ_a by the procedure in Section 26.7. For a second-order analysis we may express confidence limits for μ_a if we first conclude (by the method of Section 27.10) that there is a significant directionality in the data.

Batschelet (1981: 144, 262–265) shows geometrically and analytically how the second-order confidence limits are obtained. Here we shall simply present the arithmetic employed:

$$A = \frac{k-1}{\sum x^2} \tag{26.31)\dagger}$$

$$B = -\frac{(k-1)\sum xy}{\sum x^2 \sum y^2} \tag{26.32)\dagger}$$

$$C = \frac{k-1}{\sum y^2} \tag{26.33)\dagger}$$

$$D = \frac{2(k-1)\left[1 - \dfrac{\left(\sum xy\right)^2}{\sum x^2 \sum y^2}\right] F_{\alpha(1),2,k-2}}{k(k-2)} \tag{26.34)\dagger}$$

$$H = AC - B^2 \tag{26.35}$$

$$G = A\bar{X}^2 + 2B\bar{X}\bar{Y} + C\bar{Y}^2 - D \tag{26.36}$$

$$U = H\bar{X}^2 - CD \tag{26.37}$$

$$V = \sqrt{DGH} \tag{26.38}$$

$$W = H\bar{X}\bar{Y} + BD \tag{26.39}$$

$$b_1 = \frac{W+V}{U} \tag{26.40}$$

$$b_2 = \frac{W-V}{U} \tag{26.41}$$

[†]As shown in Example 26.9, $\sum x^2$, $\sum y^2$, and $\sum xy$ are calculated in a fashion analogous to that in Sections 17.2 and 17.3.

The quantities b_1 and b_2 are then examined separately, each yielding one of the confidence limits, as follows:

$$M = \sqrt{1 + b_i^2},\qquad (26.42)$$

after which we determine (from trigonometric tables or by calculator) that angle having

$$\text{sine} = \frac{b_i}{M}\qquad (26.43)$$

and

$$\text{cosine} = \frac{1}{M}.\qquad (26.44)$$

The confidence limit is either the angle thus determined or that angle $+180°$, whichever is nearer the sample mean angle (and, if the angle $+180°$ is greater than $360°$, simply subtract $360°$). This procedure is demonstrated in Example 26.9. The confidence interval thus computed is a little conservative (i.e., the confidence coefficient is a little greater than the stated $1 - \alpha$), and the confidence limits are not necessarily symmetrical about the mean.

EXAMPLE 26.9   **Confidence limits for the mean of a set of mean angles, using the data of Example 26.8, for which $\bar{a} = 152°$.**

j	X_j	X_j^2	Y_j	Y_j^2	$X_j Y_j$
1	-0.84140	0.70795	0.30624	0.09378	-0.25767
2	-0.76047	0.57831	0.14782	0.02185	-0.11241
3	-0.21319	0.04545	0.41842	0.17508	-0.08920
4	-0.67366	0.45382	0.56527	0.31953	-0.38080
5	-0.39005	0.15214	-0.04100	0.00168	0.01599
6	-0.48293	0.23322	0.50009	0.25009	-0.24151
7	-0.32969	0.10870	0.05222	0.00273	-0.01722
	-3.69139	2.27959	1.94906	0.86474	-1.08282

For 95% confidence limits, $\alpha = 0.05$.

$k = 7$

$$\bar{X} = \frac{\sum X_j}{k} = \frac{-3.69139}{7} = -0.52734$$

$$\bar{Y} = \frac{\sum Y_j}{k} = \frac{1.94906}{7} = 0.27844$$

$$\sum x^2 = \sum X_j^2 - \frac{\left(\sum X_j\right)^2}{k} = 2.27959 - \frac{(-3.69139)^2}{7} = 0.33297$$

$$\sum y^2 = \sum Y_j^2 - \frac{\left(\sum Y_j\right)^2}{k} = 0.86474 - \frac{(1.94906)^2}{7} = 0.32205$$

EXAMPLE 26.9   (continued)

$$\sum xy = \sum X_j Y_j - \frac{\sum X_j Y_j}{k} = -1.08282 - \frac{(-3.69139)(1.94906)}{7} = -0.05500$$

$$A = \frac{k-1}{\sum x^2} = \frac{7-1}{0.33297} = 18.01964$$

$$B = -\frac{(k-1)\sum xy}{\sum x^2 \sum y^2} = -\frac{(7-1)(-0.05500)}{(0.33297)(0.32205)} = 3.07741$$

$$C = \frac{k-1}{\sum y^2} = \frac{7-1}{0.32205} = 18.63065$$

$$F_{\alpha(1),2,k-2} = F_{0.05(1),2,5} = 5.79$$

$$D = \frac{2(k-1)\left[1 - \dfrac{\left(\sum xy\right)^2}{\sum x^2 \sum y^2}\right]F_{\alpha(1),2,k-2}}{k(k-2)}$$

$$= \frac{2(7-1)\left[1 - \dfrac{(-0.05500)^2}{(0.33297)(0.32205)}\right](5.79)}{7(7-2)} = 1.92914$$

$$H = AC - B^2 = (18.01964)(18.63065) - (3.07741)^2 = 326.24715$$

$$G = A\bar{X}^2 + 2B\bar{X}\bar{Y} + C\bar{Y}^2 - D$$

$$= (18.01964)(-0.52734)^2 + 2(3.07741)(-0.52734)(0.27844)$$

$$+ (18.63065)(0.27844)^2 - 1.92914 = 3.62258$$

$$U = H\bar{X}^2 - CD$$

$$= (326.24715)(-0.52734)^2 - (18.63065)(1.92914) = 54.78411$$

$$V = \sqrt{DGH} = \sqrt{(1.92914)(3.62258)(326.24715)} = 47.74899$$

$$W = H\bar{X}\bar{Y} + BD$$

$$= (326.24715)(-0.52734)(0.27844) + (3.07741)(1.92914) = -41.96695$$

$$b_1 = \frac{W+V}{U} = \frac{-41.96695 + 47.74899}{54.78411} = 0.10554$$

$$b_2 = \frac{W-V}{U} = \frac{-41.96695 - 47.74899}{54.78411} = -1.63763$$

For b_1: $M = \sqrt{1 + b_1^2} = \sqrt{1 + (0.10554)^2} = 1.00555$

$$\text{sine} = \frac{b_1}{M} = \frac{0.10554}{1.00555} = 0.10496$$

$$\text{cosine} = \frac{1}{M} = \frac{1}{1.00555} = 0.99448$$

The angle with this sine and cosine is $6°$, so one of the confidence limits is either $6°$ or $6° + 180° = 186°$; of the two possibilities, $186°$ is closer to the mean ($152°$).

EXAMPLE 26.9   (continued)

For b_2: $M = \sqrt{1 + b_2^2} = \sqrt{1 + (-1.63763)^2} = 1.91881$

$$\text{sine} = \frac{b_2}{M} = \frac{-1.63763}{1.91881} = -0.85346$$

$$\text{cosine} = \frac{1}{M} = \frac{1}{1.91881} = 0.52116$$

The angle with this sine and cosine is 301°, so the second confidence level is either 301° or 301° + 180° = 481° = 121°; of the two possibilities, 121° is closer to the mean (152°).

EXERCISES

26.1 Twelve nests of a particular bird species were recorded on branches extending in the following directions from the trunks of trees:

Decision		Frequency
N:	0°	2
NE:	45	4
E:	90	3
SE:	135	1
S:	180	1
SW:	225	1
W:	270	0
NW:	315	0

(a) Compute the sample mean direction.
(b) Compute the angular deviation for the data.
(c) Determine 95% confidence limits for the population mean.
(d) Determine the sample median direction.

26.2 A total of fifteen human births occurred as follows:

1:15 AM	4:40 AM	5:30 AM	6:50 AM
2:00 AM	11:00 AM	4:20 AM	5:10 AM
4:30 AM	5:15 AM	10:30 AM	8:55 AM
6:10 AM	2:45 AM	3:10 AM	

(a) Compute the mean time of birth.
(b) Compute the angular deviation for the data.
(c) Determine 95% confidence limits for the population mean time.
(d) Determine the sample median time.

26.3. Five samples of directional data were collected and were as follows.
(a) Determine the mean of the five sample means.

(b) Determine the 95% confidence limits for the second-order mean.

Sample	Sample mean	Sample r
1	230°	0.4542
2	245	0.6083
3	265	0.7862
4	210	0.5107
5	225	0.8639

27

CIRCULAR DISTRIBUTIONS: HYPOTHESIS TESTING

Armed with the procedures in Chapter 26, and the information contained in the basic statistics of circular distributions (e.g., $\bar{a}$ and r), we can now examine a number of statistical methods for testing hypotheses about populations measured on a circular scale.

27.1 TESTING SIGNIFICANCE OF THE MEAN ANGLE: UNIMODAL DISTRIBUTIONS

The Rayleigh Test. One can place more confidence in $\bar{a}$ as an estimate of the population mean angle, μ_a, if s is small, than if it is large. This is identical to stating that $\bar{a}$ is a better estimate of μ_a if r is large than if r is small. What is desired is a method of asking whether there is, in fact, a mean direction among the population of data that were sampled, for even if there is no mean direction (i.e., the circular distribution is uniform) in the population, a random sample might still display a calculable mean. The test we require is that concerning H_0: The sampled population is uniformly* distributed around a circle vs. H_A: The population is not a uniform circular distribution. This may be tested by the *Rayleigh test*[†]. As circular uniformity implies there is no mean direction, the Rayleigh test may be said to test H_0: $\rho = 0$ vs. H_A: $\rho \neq 0$.

---

*In dealing with circular statistics, the terms "uniform distribution" and "random distribution" have been used synonymously in the literature.

[†]Named for Lord Rayleigh (John William Strutt, Third Baron Rayleigh (1842–1919)], a physicist and applied mathematician who gained his greatest fame for discovering and isolating the chemical element argon (winning him the Nobel Prize in physics in 1904), although some of his other contributions to physics were at least as important (Lindsay, 1976). He was a pioneering worker with directional data beginning in 1880 (Fisher, 1993: 10; Rayleigh, 1919; Upton and Fingleton, 1989).

The Rayleigh test asks how large a sample r must be to indicate confidently a nonuniform population distribution. A quantity referred to as "Rayleigh's R" is obtainable as

$$R = nr, \qquad (27.1)$$

and the so-called "Rayleigh's z" may be utilized for testing the null hypothesis of no population mean direction:

$$z = \frac{R^2}{n} \text{ or } z = nr^2. \qquad (27.2)$$

Appendix Table B.34 presents critical values of $z_{\alpha,n}$. An excellent approximation of the probability of Rayleigh's R is*

$$P = \exp\left[\sqrt{1 + 4n + 4(n^2 - R^2)} - (1 + 2n)\right] \qquad (27.4)$$

(derived from Greenwood and Durand, 1955). This calculation is accurate to three decimal places for n as small as 10 and to two decimal places for n as small as 5.[†] The Rayleigh test assumes sampling from a von Mises distribution, a circular analog of the linear normal distribution. (See footnote to Section 26.7.)

If H_0 is rejected by Rayleigh's test, we may conclude that there is a mean population direction (see Example 27.1), and if H_0 is not rejected, we may conclude the population distribution to be uniform around the circle; but only if we may assume that the

---

EXAMPLE 27.1   Rayleigh's test for circular uniformity, applied to the data of Example 26.2.

These data are plotted in Fig. 26.2.

H_0: $\rho = 0$ (i.e., the population is uniformly distributed around the circle).

H_A: $\rho \neq 0$ (i.e., the population is not distributed uniformly around the circle).

Following Example 26.4:

$n = 8$

$r = 0.82522$

$R = nr = (8)(0.82522) = 6.60176$

$z = \dfrac{R^2}{n} = \dfrac{(6.60176)^2}{8} = 5.448$

Using Appendix Table B.34, $z_{0.05,8} = 2.899$. Reject H_0.   [$0.001 < P < 0.002$]

---

*Recall the following notation:

$$\exp[C] = e^C. \qquad (27.3)$$

[†]A simpler, but less accurate, approximation for P is to consider z as a chi-square with 2 degrees of freedom (Mardia, 1972a: 112); this is accurate to two decimal places for n as small as about 15.

population distribution does not have more than one mode. (For example, the data in Example 26.7 and Fig. 26.8 would result in a Rayleigh test failing to reject H_0. While these data have no mean direction, they are not unimodal.)

Section 26.8 explains how axially bimodal data—such as in Fig. 26.8—can be transformed into unimodal data, thereafter to be subjected to Rayleigh testing and other procedures requiring unimodality. What is known as "Rao's spacing test" (Batschelet, 1981: 66–69; Rao, 1976) is particularly appropriate when circular data are neither unimodal nor axially bimodal, and Russell and Levitin (1994) have produced excellent tables for its use.

Modified Rayleigh Test for Uniformity vs. a Specified Mean Angle. The Rayleigh test looks for any departure from uniformity. A modification of that test (Durand and Greenwood, 1958; Greenwood and Durand, 1955) is available for use when the investigator has reason to expect, *in advance*, that if the sampled distribution is not uniform it will have a specified mean direction. In Example 27.2 (and presented graphically in Fig. 27.1), ten birds were released at a site directly west of their home. Therefore, the statistical analysis may include the expectation that such birds tend to fly directly east

---

EXAMPLE 27.2   The V test for circular uniformity under the alternative of nonuniformity and a specified mean direction.

H_0:   The population is uniformly distributed around the circle (i.e., H_0: $\rho = 0$).

H_A: The population is not uniformly distributed around the circle (i.e., H_A: $\rho \neq 0$), but has a mean of $90°$.

a_i (deg)	$\sin a_i$	$\cos a_i$
66	0.91355	0.40674
75	0.96593	0.25882
86	0.99756	0.06976
88	0.99939	0.03490
88	0.99939	0.03490
93	0.99863	0.05234
97	0.99255	0.12187
101	0.98163	0.19081
118	0.88295	0.46947
130	0.76604	0.64279

$n = 10$ $\sum \sin a_i = 9.49762$ $\sum \cos a_i = -0.67216$

$$Y = \frac{9.49762}{10} = 0.94976$$

$$X = -\frac{0.67216}{10} = -0.06722$$

$$r = \sqrt{(-0.06722)^2 + (0.94976)^2} = 0.95213$$

$$\sin \bar{a} = \frac{Y}{r} = 0.99751$$

EXAMPLE 27.2   (continued)

$$\cos \bar{a} = \frac{X}{r} = -0.07060$$

$$\bar{a} = 94°.$$

$$R = (10)(0.95213) = 9.5213$$

$$V = R\cos(94° - 90°)$$

$$= 9.5213\cos(4°)$$

$$= (9.5213)(0.99756)$$

$$= 9.498$$

$$u = V\sqrt{\frac{2}{n}}$$

$$= (9.498)\sqrt{\frac{2}{10}}$$

$$= 4.248$$

Using Appendix Table B.35, $u_{0.05,10} = 1.648$. Reject H_0. $P < 0.0005$.

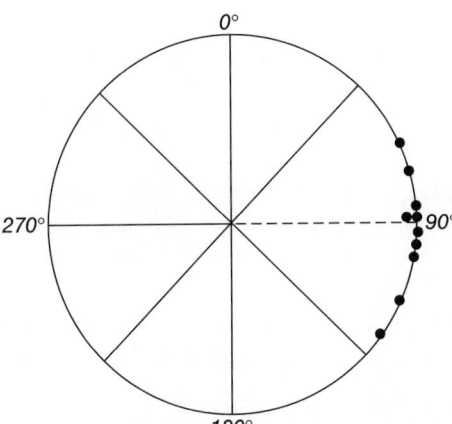

Figure 27.1 The data for the V test of Example 27.2. The broken line indicates the expected mean angle (94°).

(i.e., at an angle of 90°). The testing procedure considers H_0: The population directions are randomly distributed (i.e., H_0: $\rho = 0$) vs. H_A: $\rho \neq 0$, and $\mu_a = 90°$. By using additional information, namely the expected mean angle, this test is more powerful than Rayleigh's test (Batschelet, 1972; 1981: 60).

The preceding hypotheses are tested by a modified Rayleigh test that we shall refer to as the *V test*, in which the test statistic is computed as

$$V = R\cos(\bar{a} - \mu_0), \tag{27.5}$$

where μ_0 is the mean angle predicted. The significance of V may be ascertained from

$$u = V\sqrt{\frac{2}{n}}. \tag{27.6}$$

Appendix Table B.35 gives critical values of $u_{\alpha,n}$, a statistic which, for large sample sizes, approaches a one-tailed normal deviate, $Z_{\alpha(1)}$, especially in the neighborhood of probabilities of 0.05. If the data are grouped, then R may be determined from r_c (Equation 26.15) rather than from r.

Note that if H_0 is not rejected in the V test we do not know whether the population is distributed uniformly or whether it has some mean direction other than μ_0.

One-Sample Test for the Mean Angle.   The Rayleigh test and the V test are methods for testing for random distribution of a population of data around the circle. (See Batschelet, 1981: Chapter 4, for other tests of the null hypothesis of randomness.) If it is desired to test whether the population mean angle is equal to a specified value, say μ_0, then we have a one-sample test situation analogous to that of the one-sample t test for data on a linear scale (Section 7.1). The hypotheses are

$$H_0: \mu_a = \mu_0$$

and

$$H_A: \mu_a \neq \mu_0,$$

and H_0 is tested simply by observing whether μ_0 lies within the $1-\alpha$ confidence interval for μ_a. If μ_a lies outside the confidence interval, then H_0 is rejected. Section 26.7 describes the determination of confidence intervals for the population mean angle, and Example 27.3 demonstrates the hypothesis testing procedure.*

---

*For demonstration purposes (Examples 27.2 and 27.3) we have applied the V test and the one-sample test for the mean angle to the same set of data. In practice this would not be done. Deciding which test to employ would depend, respectively, on whether the intention is to test for circular uniformity or to test whether the population mean angle is a specified value.

EXAMPLE 27.3   **The one-sample test for the mean angle, using the data of Example 27.2.**

H_0:   The population has a mean of 90° (i.e., $\mu_a = 90°$).

H_A: The population mean is not 90° (i.e., $\mu_a \neq 90°$).

The computation of the following is given in Example 27.2:

$r = 0.95$

$\bar{a} = 94°$

Using Equation 25.25, for $\alpha = 0.05$ and $n = 10$:

$R = nr = (10)(0.95) = 9.5$

$\chi^2_{0.05,1} = 3.841$

EXAMPLE 27.3   (continued)

$$d = \arccos \left[\frac{\sqrt{n^2 - (n^2 - R^2)e^{\chi^2_{\alpha,1}/n}}}{R} \right]$$

$$= \arccos \left[\frac{\sqrt{10^2 - (10^2 - 9.5^2)e^{3.841/10}}}{9.5} \right]$$

$$= \arccos[0.9744]$$

$$= 13°$$

Thus, the 95% confidence interval for μ_a is $94° \pm 13°$.
As this confidence interval does contain the hypothesized mean ($\mu_0 = 90°$), do not reject H_0.

27.2 TESTING SIGNIFICANCE OF THE MEDIAN ANGLE: OMNIBUS TEST

The Hodges-Ajne Test.   A simple alternative to the Rayleigh test (Section 27.1) is the so-called *Hodges-Ajne test,** which does not assume sampling from a specific distribution. This is called an "omnibus test" because it works well for unimodal, bimodal, and multimodal distributions. If the underlying distribution is that assumed by the Rayleigh test, then the latter procedure is the more powerful.

Given a sample of circular data, we determine the smallest number of data that occur within a range of 180°. As shown in Example 27.4, this is readily done by drawing a line through the center of the circle (i.e., drawing a diameter) and rotating that line around the center until there is the greatest possible difference between the numbers of data on each side of the line. If, for example, the diameter line were vertical (i.e., through 0° and 180°), there would be ten data on one side of it and fourteen on the other; if the line were horizontal (i.e., through 90° and 270°), then there would be 3.5 points on one side and 20.5 points on the other; and if the diameter were rotated slightly counterclockwise from horizontal (shown as a dashed line in the figure in Example 27.4), then there would be three data on one side and twenty-one on the other, and no line will split the data with fewer data on one side and more on the other. The test statistic, which we shall call m, is the smallest number of data that can be partitioned on one side of a diameter; in Example 27.4, $m = 3$.

The probability of an m at least this small, under the null hypothesis of circular uniformity, is

$$P = 2^{1-n}(n - 2m) \binom{n}{m} \tag{27.7}$$

(Hodges, 1955), using the binomial coefficient notation of Equation 24.2. Instead of

---

*This procedure was presented by Ajne (1968). Shortly thereafter, Bhattacharyya and Johnson (1969) showed that his test is identical to a test given by Hodges (1955) for a different purpose.

EXAMPLE 27.4   The Hodges-Ajne test for circular uniformity.

H_0:   The population is uniformly distributed around the circle.

H_A: The population is not uniformly distributed around the circle.

This sample of twenty-four data is collected: $10°$, $15°$, $25°$, $30°$, $30°$, $30°$, $35°$, $45°$, $50°$, $60°$, $75°$, $80°$, $100°$, $110°$, $255°$, $270°$, $280°$, $280°$, $300°$, $320°$, $330°$, $350°$, $350°$, $355°$.

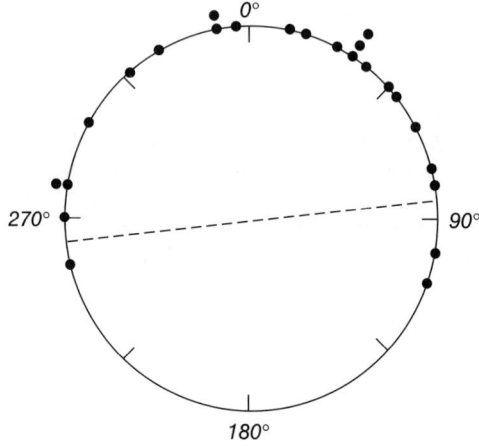

$n = 24; m = 3$

For $\alpha = 0.05$, the critical value (from Appendix Table B.36) is $m_{0.05, 24} = 4$; reject H_0. $0.002 < P \le 0.005$.

$$\text{Exact probability } = P = 2^{1-n}(n - 2m)\binom{n}{m}$$

$$= 2^{1-24}\binom{24}{3} = \frac{2^{-23}24!}{3!(24 - 3)!} = 0.00024.$$

For comparison, the Rayleigh test for these data would yield $\bar{a} = 12°$, $r = 0.563$, $R = 13.513$, $z = 7.609$, $P < 0.001$.

computing this probability, one may refer to Appendix Table B.36, which gives critical values for m as a function of n and α. (It can be seen from this table that in order to test at the 5% significance level one must have a sample of at least nine data.) For $n > 50$, P may be determined by the following approximation:

$$P \approx \frac{\sqrt{2\pi}}{A}e^{-\pi^2/(8A^2)}, \tag{27.8}$$

where

$$A = \frac{\pi\sqrt{n}}{2(n - 2m)} \tag{27.9}$$

(Ajne, 1968); the accuracy of this approximation is indicated at the end of Appendix Table B.36.

Modified Hodges-Ajne Test for Uniformity vs. a Specified Angle. Just as (in Section 27.1) the V test is a modification of the Rayleigh test to test for circular uniformity against an alternative that presumes a specified angle, a test presented by Batschelet (1981: 64–66) is a modification of the Hodges-Ajne test to test nonparametrically for uniformity against an alternative that specifies an angle. For this test, we count the number of data that lie within $90°$ of the specified angle; let us call this number m' and the test statistic is

$$C = n - m'. \tag{27.10}$$

We may proceed by performing a two-tailed binomial test (Section 24.6), with $p = 0.5$ and with C counts in one category and m' counts in the other. As shown in the figure in Example 27.5, this may be visualized as drawing a diameter line perpendicular to the radius extending in the specified angle and counting the data on either side of that line.

EXAMPLE 27.5   The Batschelet test for circular uniformity.

H_0:   The population is uniformly distributed around the circle.

H_A: The population is not uniformly distributed around the circle, but is concentrated around $45°$.

The data are those of Example 27.4.

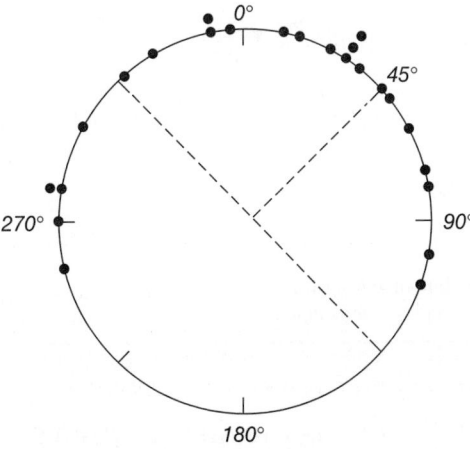

$n = 24;\ p = 0.5;\ m' = 19;\ C = 5$

For the binomial test of Section 24.6, $C_{0.05(2),\,24} = 6$, reject H_0, $0.005 < P \leq 0.01$; by the procedure shown in Example 24.9, the exact probability would be found to be $P = 0.00661$.

27.3 TESTING SIGNIFICANCE OF THE MEDIAN ANGLE: BINOMIAL TEST

We can perform a nonparametric test to conclude whether the population median angle equals a specified value simply by counting the number of observed angles on either side of a diameter through the hypothesized angle and subjecting these data to the binomial test of Section 24.6, with $p = 0.5$. This is demonstrated in Example 27.6.

EXAMPLE 27.6   The significance of the median angle.

The sample data consist of the following directions: 97°, 104°, 121°, 159°, 164°, 172°, 195°, 213°. The median is $(159° + 164°)/2 = 161.5°$.

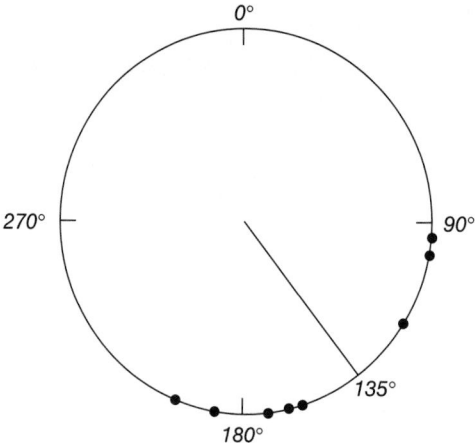

If we wish to test whether the population median is equal to some specific value—say, 135°—we can proceed as follows:

H_0:   The population median angle is 135°.

H_A: The population median angle is not 135°.

Employing the two-tailed binomial test of Section 23.6, we have $n = 9$ and $p = 0.5$. There are three sample directions < 135° and five directions > 135°. It is found that $P = 0.727$. Do not reject H_0.

27.4 TESTING SYMMETRY AROUND THE MEDIAN ANGLE

The symmetry of a distribution around the median may be tested nonparametrically using the Wilcoxon paired-sample test (also known as the Wilcoxon signed-rank test) of Section 9.5. For each angle (X_i) we calculate the deviation of X_i from the median (i.e., $d_i = X_i -$ median), and we then analyze the d_i's as explained in Section 9.5. This is shown in Example 27.7 for a two-tailed test, where H_0: The underlying distribution is not skewed from the median. A one-tailed test could be used to ask whether the distribution was skewed in a specific direction from the median. (T_- would be the test statistic for H_0: The distribution is not skewed clockwise from the median, and T_+ would be the test statistic for H_0: The distribution is not skewed counterclockwise from the median.)

EXAMPLE 27.7   Testing for symmetry around the median angle, for the data of Example 27.6.

H_0:   The underlying distribution is symmetrical around the median.

H_A: The underlying distribution is not symmetrical around the median.

From Example 27.6, the mean = $161.5°$.
Using the Wilcoxon signed-rank test:

X_i	$d_i =$ X_i − median	Rank of $|d_i|$	Signed rank of $|d_i|$
97°	−64.5°	8	−8
104°	−57.5°	7	−7
121°	−40.5°	5	−5
159°	−2.5°	1.5	−1.5
164°	2.5°	1.5	1.5
172°	10.5°	3	3
195°	33.5°	4	4
213°	51.5°	6	6

$T_+ = 1.5 + 3 + 4 + 6 = 14.5$

$T_- = 8 + 7 + 5 + 1.5 = 21.5$

$T_{0.05(2),8} = 3$

Neither T_+ nor T_- is $< T_{0.05(2),8}$, so do not reject H_0.

$P > 0.50$

27.5   TWO-SAMPLE AND MULTISAMPLE TESTING OF MEAN ANGLES

It is common to consider the null hypothesis H_0: $\mu_1 = \mu_2$, where μ_1 and μ_2 are the mean angles for each of two circular distributions (see Example 27.8). Watson and Williams (1956, with an improvement by Stephens, 1972) proposed a test that utilizes the statistic

$$F = K \frac{(N - 2)(R_1 + R_2 - R)}{N - R_1 - R_2}, \tag{27.11}$$

where $N = n_1 + n_2$. In this equation, R is Rayleigh's R calculated by Equation 27.1 with the data from the two samples being combined; R_1 and R_2 are the values of Rayleigh's R

EXAMPLE 27.8   The Watson-Williams test for two samples.

H_0: $\mu_1 = \mu_2$

H_A: $\mu_1 \neq \mu_2$

	Sample 1			**Sample 2**	
a_i (deg)	$\sin a_i$	$\cos a_i$	a_i (deg)	$\sin a_i$	$\cos a_i$
94	0.99756	−0.06976	77	0.97437	0.22495
65	0.90631	0.42262	70	0.93969	0.34202
45	0.70711	0.70711	61	0.87462	0.48481
52	0.78801	0.61566	45	0.70711	0.70711
38	0.61566	0.78801	50	0.76604	0.64279
47	0.73135	0.68200	35	0.57358	0.81915
73	0.95630	0.29237	48	0.74314	0.66913
82	0.99027	0.13917	65	0.90631	0.42262
90	1.00000	0.00000	36	0.58779	0.80902
40	0.64279	0.76604			
87	0.99863	0.05234			

$$n_1 = 11 \qquad \sum \sin a_i = 9.33399 \qquad \sum \cos a_i = 4.39556$$
$$Y = 0.84854, \quad X = 0.39960$$
$$r_1 = 0.93792$$
$$\sin \bar{a}_1 = 0.90470$$
$$\cos \bar{a}_1 = 0.42605$$
$$\bar{a}_1 = 65°$$
$$R_1 = 10.31712$$

$$n_2 = 9 \qquad \sum \sin a_i = 7.07265 \qquad \sum \cos a_i = 5.12160$$
$$Y = 0.78585, \quad X = 0.56907$$
$$r_2 = 0.97026$$
$$\sin \bar{a}_2 = 0.80994$$
$$\cos \bar{a}_2 = 0.58651$$
$$\bar{a}_2 = 54°$$
$$R_2 = 8.73234$$

By combining the twenty data from both samples:

$$\sum \sin a_i = 9.33399 + 7.07265 = 16.40664$$
$$\sum \cos a_i = 4.39556 + 5.12160 = 9.51716$$
$$N = 11 + 9 = 20$$
$$Y = \frac{16.40664}{20} = 0.82033$$
$$X = \frac{9.51716}{20} = 0.47586$$
$$r = 0.94836$$
$$R = 18.96720$$
$$r_w = \frac{10.31712 + 8.73234}{20} = 0.952$$
$$F = K \frac{(N-2)(R_1 + R_2 - R)}{N - R_1 - R_2}$$
$$= (1.0351) \frac{(20-2)(10.31712 + 8.73234 - 18.96720)}{20 - 10.31712 - 8.73234}$$
$$= (1.0351) \frac{1.48068}{0.95054}$$
$$= 1.61$$

EXAMPLE 27.8   (continued)

$F_{0.05(1), 1, 18} = 4.41$

Therefore, do not reject H_0.

$$0.10 < P < 0.25 \quad [P = 0.22]$$

Thus, we conclude that the two sample means estimate the same population mean, and the best estimate of this population mean is obtained by:

$$\sin \bar{a} = \frac{Y}{r} = 0.86500$$

$$\cos \bar{a} = \frac{X}{r} = 0.50177$$

$$\bar{a} = 60°$$

for the two samples considered separately. K is a factor, obtained from Appendix Table B.37, that corrects for bias in the F calculation; in that table one uses the weighted mean of the two vector lengths for the column headed r:

$$r_w = \frac{n_1 r_1 + n_2 r_2}{N} = \frac{R_1 + R_2}{N}. \tag{27.12}$$

The critical value for this test is $F_{\alpha(1), 1, N-2}$. Alternatively,

$$t = \sqrt{K \frac{(N-2)(R_1 + R_2 - R)}{N - R_1 - R_2}} \tag{27.13}$$

may be compared with $t_{\alpha(2), N-2}$. This test may be used for r_w as small as 0.75, if $5 \leq N/2 < 10$; or for r_w as low as 0.70, if $N/2 \geq 10$ (Batschelet, 1981: 97, 321; Mardia, 1972a: 155). The underlying assumptions of the test are discussed at the end of this section.

The data may be grouped as long as the grouping interval is $\leq 10°$. See Batschelet (1972; 1981: Chapter 6) for a review of other two-sample testing procedures. Mardia (1972: 156–158) gives a procedure for an approximate confidence interval for $\mu_1 - \mu_2$.

Multisample Testing.   The Watson-Williams test can be generalized to a multisample test for testing $H_0: \mu_1 = \mu_2 = \ldots = \mu_k$, a hypothesis reminiscent of analysis of variance considerations for linear data (Section 10.1). In multisample tests with circular data (Example 27.9),

$$F = K \frac{(N-k) \left(\sum\limits_{j=1}^{k} R_j - R \right)}{(k-1) \left(N - \sum\limits_{j=1}^{k} R_j \right)}. \tag{27.14}$$

Here, k is the number of samples, R is the Rayleigh's R for all k samples combined, and $N = \sum_{j=1}^{n} n_j$. The correction factor, K, is obtained from Appendix Table B.37,

EXAMPLE 27.9   The Watson-Williams test for three samples.

H_0:   All three samples are from populations with the same mean angle.

H_A: All three samples are not from populations with the same mean angle.

Sample 1			**Sample 2**		
a_1 (deg)	$\sin a_i$	$\cos a_i$	a_i (deg)	$\sin a_i$	$\cos a_i$
135	0.70711	−0.70711	150	0.50000	−0.86603
145	0.57358	−0.81915	130	0.76604	−0.64279
125	0.81915	−0.57358	175	0.08716	−0.99619
140	0.64279	−0.76604	190	−0.17365	−0.98481
165	0.25882	−0.96593	180	0.00000	−1.00000
170	0.17365	−0.98481	220	−0.64279	−0.76604

$n_1 = 6$ $\sum \sin a_i = 3.17510$ $\sum \cos a_i = -4.81662$

$\bar{a}_1 = 147°$

$r_1 = 0.96150$

$R_1 = 5.76894$

$n_2 = 6$ $\sum \sin a_i = 0.53676$ $\sum \cos a_i = -5.25586$

$\bar{a}_2 = 174°$

$r_2 = 0.88053$

$R_2 = 5.28324$

Sample 3		
a_i (deg)	$\sin a_i$	$\cos a_i$
140	0.64279	−0.76604
165	0.25882	−0.96593
185	−0.08715	−0.99619
180	0.00000	−1.00000
125	0.81915	−0.57358
175	0.08716	−0.99619
140	0.64279	−0.76604

$n_3 = 7$ $\sum \sin a_i = 2.36356$ $\sum \cos a_i = -6.06397$

$\bar{a}_3 = 159°$

$r_3 = 0.92976$

$R_3 = 6.50832$

$k = 3$

$N = 6 + 6 + 7 = 19$

For all nineteen data:

$$\sum \sin a_i = 3.17510 + 0.53676 + 2.36356 = 6.07542$$

$$\sum \cos a_i = -4.81662 - 5.25586 - 6.06397 = -16.13645$$

$$Y = 0.31976$$

$$X = -0.84929$$

$$r = 0.90749$$

$$R = 17.24231$$

$$r_w = \frac{5.76894 + 5.28324 + 6.50832}{19} = 0.924$$

EXAMPLE 27.9   (continued)

$$F = K \frac{(N-k)(\sum R_j - R)}{(k-1)(N - \sum R_j)}$$

$$= (1.0546)\frac{(19-3)(5.76894 + 5.28324 + 6.50832 - 17.24231)}{(3-1)(19 - 5.76894 - 5.28324 - 6.50832)}$$

$$= (1.0546)\frac{5.09104}{2.87900}$$

$$= 1.86$$

$$\nu_1 = k - 1 = 2$$

$$\nu_2 = N - k = 16$$

$$F_{0.05(1),2,16} = 3.63$$

Therefore, do not reject H_0.

$$0.10 < P < 0.25 \quad [P = 0.19]$$

Thus, we conclude that the three sample means estimate the same population mean, and the best estimate of that population mean is obtained by:

$$\sin \bar{a} = \frac{Y}{r} = 0.35236$$

$$\cos \bar{a} = \frac{X}{r} = -0.93587$$

$$\bar{a} = 159°$$

using

$$r_w = \frac{\sum_{j=1}^{k} n_j r_j}{N} = \frac{\sum_{j=1}^{k} R_j}{N}. \tag{27.15}$$

The critical value for this test is $F_{\alpha(1),k-1,N-k}$. Equation 27.15 (and, thus this test) may be used for r_w as small as 0.45, if $N/k > 10$; for r_w as small as 0.50, for $N/k > 6$; and for r_w as small as 0.55, for $N/k = 5$ or 6 (Mardia, 1972: 163; Batschelet, 1981: 321). If the data are grouped, the grouping interval should be no larger than 10°. Upton (1976) presents an alternative to the Watson-Williams test that relies on χ^2, instead of F, but the Watson-Williams procedure is a little simpler to use.

The Watson-Williams tests (for two or more samples) are parametric and assume that each of the samples came from a population conforming to what is known as the von Mises distribution, a circular analog to the normal distribution of linear data. (See footnote in Section 26.7.) In addition, the tests assume that the population dispersions are all the same. Fortunately, the tests are robust to departures from these assumptions. But if the underlying assumptions are known to be violated severely (as when the distributions are not unimodal), we should be wary of their use. In the two-sample case, the nonparametric test of Section 27.6 is preferable to the Watson-Williams test when the assumptions of the latter cannot be presumed to prevail.

Stephens (1982) developed a test with characteristics of a hierarchical analysis of variance of circular data, and Harrison and Kanji (1988) and Harrison, Kanji, and Gadsden (1986) present two-factor ANOVA (including the randomized block design).

Batschelet (1981: 122–126), Fisher (1993: 131–133), Mardia (1972a: 158–162, 165–166), and Stephens (1972) discuss testing of equality of population concentrations.

27.6 NONPARAMETRIC TWO-SAMPLE AND MULTISAMPLE TESTING OF ANGLES

If data are grouped—Batschelet (1981: 110) recommends a grouping interval larger than $10°$—then contingency table analysis may be used (as introduced in Section 23.1) as a two-sample test. The runs test of Section 27.19 may be used as a two-sample test, but it is not as powerful for that purpose as are the procedures below, and it is best reserved for testing the hypothesis in Section 27.19.

Watson's Test. Among the nonparametric procedures applicable to two samples of circular data (e.g., see Batschelet, 1972; 1981: Chapter 6; Fisher, 1993: Section 5.3), are the median test of Section 27.7 and the Watson test.

The Watson test, a powerful procedure developed by Watson (1962), is recommended in place of the Watson-Williams two-sample test of Section 27.5 when at least one of the sampled populations is not unimodal or when there are other considerable departures from the assumptions of the latter test. It may be used on grouped data if the grouping interval is no greater than $5°$ (Batschelet, 1981: 115).

The data in each sample are arranged in ascending order, as demonstrated in Example 27.10. For the two sample sizes, n_1 and n_2, let us denote the ith observation in Sample 1 as a_{1i} and the jth datum in Sample 2 as a_{2j}. Then, for the data in Example 27.10, $a_{11} = 35°$, $a_{21} = 75°$, $a_{12} = 45°$, $a_{22} = 80°$, etc. The total number of data is $N = n_1 + n_2$. The cumulative relative frequencies for the observations in Sample 1 are i/n_1, and those for Sample 2 are j/n_2. As shown in the present example, we then define values of d_k (where k runs from 1 through N) as the differences between the two cumulative relative frequency distributions. The test statistic, called the Watson U^2, is computed as

$$U^2 = \frac{n_1 n_2}{N^2} \left[\sum_{k=1}^{N} d_k^2 - \frac{\left(\sum_{k=1}^{N} d_k \right)^2}{N} \right]. \tag{27.16}$$

Critical values of U^2_{α, n_1, n_2} are given in Appendix Table B.38 bearing in mind that $U^2_{\alpha, n_1, n_2} = U^2_{\alpha, n_2, n_1}$.

Watson's U^2 is especially useful for circular data because the starting point for determining the cumulative frequencies is immaterial. It may also be used in any situation with linear data that are amenable to Mann-Whitney testing (Section 8.10), but it is

EXAMPLE 27.10   Watson's U^2 test for nonparametric two-sample testing.

H_0:   The two samples came from the same population, or from two populations having the same direction.

H_A: The two samples did not come from the same population, or from two populations having the same directions.

i	a_{1i} (deg)	$\dfrac{i}{n_1}$	i	a_{2j} (deg)	$\dfrac{j}{n_2}$	$d_k = \dfrac{i}{n_1} - \dfrac{j}{n_2}$	d_k^2
1	35	0.1000			0.0000	0.1000	0.0100
2	45	0.2000			0.0000	0.2000	0.0400
3	50	0.3000			0.0000	0.3000	0.0900
4	55	0.4000			0.0000	0.4000	0.1600
5	60	0.5000			0.0000	0.5000	0.2500
6	70	0.6000			0.0000	0.6000	0.3600
		0.6000	1	75	0.0909	0.5091	0.2592
		0.6000	2	80	0.1818	0.4182	0.1749
7	85	0.7000			0.1818	0.5182	0.2685
		0.7000	3	90	0.2727	0.4273	0.1826
8	95	0.8000			0.2727	0.5273	0.2780
		0.8000	4	100	0.3636	0.4364	0.1904
9	105	0.9000			0.3636	0.5364	0.2877
		0.9000	5	110	0.4546	0.4454	0.1984
10	120	1.0000			0.4546	0.5454	0.2975
		1.0000	6	130	0.5455	0.4545	0.2066
		1.0000	7	135	0.6364	0.3636	0.1322
		1.0000	8	140	0.7273	0.2727	0.0744
		1.0000	9	150	0.8182	0.1818	0.0331
		1.0000	10	155	0.9091	0.0909	0.0083
		1.0000	11	165	1.0000	0.0000	0.0000

$$n_1 = 10 \qquad\qquad n_2 = 11 \qquad\qquad \sum d_k = 7.8272 \quad \sum d_k^2 = 3.5018$$

$$N = n_1 + n_2 = 21$$

$$U^2 = \frac{n_1 n_2}{N^2}\left[\sum d_k^2 - \frac{\left(\sum d_k\right)^2}{N}\right]$$

$$= \frac{(10)(11)}{21^2}\left[3.5018 - \frac{(7.8272)^2}{21}\right]$$

$$= 0.1458$$

$$U^2_{0.05,\,10,\,11} = 0.1856$$

Do not reject H_0.

$$0.10 < P < 0.20$$

generally not recommended as a substitute for the Mann-Whitney test; the latter is easier to perform and has access to more extensive tables of critical values, and the former may declare significance because group dispersions are different.

Watson's Two-Sample Test with Ties. If there are some tied data (i.e., there are two or more observations having the same numerical value), then the Watson two-sample test is modified as demonstrated in Example 27.11. We define t_{1i} as the number of data in Sample 1 with a value of a_{1i} and t_{2j} as the number of data in Sample 2 that have a value of a_{2j}. Additionally, m_{1i} and m_{2j} are the cumulative number of data in Samples 1 and 2,

---

EXAMPLE 27.11   Watson's U^2 test for data containing ties.

H_0:   The two samples came from the same population, or from two populations having the same directions.

H_A: The two samples did not come from the same population, or from two populations having the same directions.

i	a_{1i}	t_{1i}	m_{1i}	$\dfrac{m_{1i}}{n_1}$	j	a_{2j}	t_{2j}	m_{2j}	$\dfrac{m_{2j}}{n_2}$	$d_k = \dfrac{m_{1i}}{n_1} - \dfrac{m_{2j}}{n_2}$	d_k^2	t_k
				0.0000	1	30°	1	1	0.1000	−0.1000	0.0100	1
				0.0000	2	35	1	2	0.2000	−0.2000	0.0400	1
1	40°	1	1	0.0833					0.2000	−0.1167	0.0136	1
2	45	1	2	0.1667					0.2000	−0.0333	0.0011	1
3	50	1	3	0.2500	3	50	1	3	0.3000	−0.0500	0.0025	2
4	55	1	4	0.3333					0.3000	0.0333	0.0011	1
				0.3333	4	60	1	4	0.4000	−0.0667	0.0044	1
				0.3333	5	65	2	6	0.6000	−0.2677	0.0711	2
5	70	1	5	0.4167					0.6000	−0.1833	0.0336	1
				0.4167	6	75	1	7	0.7000	−0.2833	0.0803	1
6	80	2	7	0.5833	7	80	1	8	0.8000	−0.2167	0.0470	3
				0.5833	8	90	1	9	0.9000	−0.3167	0.1003	1
7	95	1	8	0.6667					0.9000	−0.2333	0.0544	1
				0.6667	9	100	1	10	1.0000	−0.3333	0.1111	1
8	105	1	9	0.7500					1.0000	−0.2500	0.0625	1
9	110	2	11	0.9167					1.0000	−0.0833	0.0069	2
10	120	1	12	1.0000					1.0000	0.0000	0.0000	1

$n_1 = 12$ $n_2 = 10$ $\sum t_k d_k = -3.5334$ $\sum t_k d_k^2 = 0.8144$

$N = 12 + 10 - 22$

$$U^2 = \frac{n_1 n_2}{N^2}\left[\sum t_k d_k^2 - \frac{\left(\sum t_k d_k\right)^2}{N}\right]$$

$$= \frac{(12)(10)}{22^2}\left[0.8144 - \frac{(-3.5334)^2}{22}\right]$$

$$= 0.0612$$

$U^2_{0.05, 10, 12} = 0.2246$

Do not reject H_0.

$P > 0.50$

respectively; so the cumulative relative frequencies are m_{1i}/n_1 and m_{2j}/n_2, respectively. As above, d_k, represents a difference between the two cumulative frequency distributions. The test statistic is

$$U^2 = \frac{n_1 n_2}{N^2} \left[\sum_{k=1}^{N} t_k d_k^2 - \frac{\left(\sum_{k=1}^{N} t_k d_k \right)^2}{N} \right]. \tag{27.17}$$

Wheeler and Watson Test.   Another nonparametric test for the null hypothesis of no difference between two circular distributions is one presented by Wheeler and Watson (1964; developed independently by Mardia, 1967). This procedure ranks all N data and for each a calculates what is termed a *uniform score* or *circular rank*:

$$d = \frac{(360°)(\text{rank of } a)}{N}. \tag{27.18}$$

This spaces all of the data equally around the circle. Then,

$$C_i = \sum_{j=1}^{n_i} \cos d_j \tag{27.19}$$

and

$$S_i = \sum_{j=1}^{n_i} \sin d_j, \tag{27.20}$$

where $i = 1$ and 2, representing the two samples. The test statistic is

$$W = 2 \left[\frac{C_1^2 + S_1^2}{n_1} + \frac{C_2^2 + S_2^2}{n_2} \right], \tag{27.21}$$

which may be compared to χ^2 with 2 degrees of freedom (Mardia, 1972).

This test is demonstrated in Example 27.12. It is applicable if n_1 and n_2 are each at least 10 (Fisher, 1993: 123)[*] and if there are no tied data (Batschelet, 1981: 103). If there are ties, or if either sample size is less than 10, then Watson's U^2 test is preferable.

Multisample Testing.   Mardia (1972b) extended the Wheeler and Watson test to more than two samples. The procedure is as before, where all N data from all k samples are ranked and (by Equation 27.18) the circular rank, d, is calculated for each datum. Equations 27.19 and 27.20 are applied to each sample, and

$$W = 2 \sum_{i=1}^{k} \left[\frac{C_i^2 + S_i^2}{n_i} \right], \tag{27.22}$$

which may be compared to χ^2 with $2(k-1)$ degrees of freedom. This procedure is applicable if each n_i is at least 10 (Fisher, 1993: 123)[†]; it should not be used if there are tied data (Batschelet, 1981: 103).

---

[*]A test that is basically the same has available critical values for a few n_i's < 10 (Mardia, 1967, 1969).

[†]Mardia (1970) provides critical values for $k = 3$ and $N \leq 14$.

EXAMPLE 27.12   The Wheeler and Watson two-sample test for the data of Example 27.10.

H_0:   The two samples came from the same population, or from two populations having the same directions.

H_A: The two samples did not come from the same population, or from two populations having the same directions.

$n_1 = 10$, $n_2 = 11$, and $N = 21$

$$\frac{360°}{N} = \frac{360°}{21} = 17.1429°$$

	Sample 1			Sample 2	
Direction (degrees)	Rank of direction	Circular rank (degrees)	Direction (degrees)	Rank of direction	Circular rank (degrees)
35	1	17.14			
45	2	34.29			
50	3	51.43			
55	4	68.57			
60	5	85.71			
70	6	102.86			
			75	7	120.00
			80	8	137.14
85	9	154.29			
			90	10	171.43
95	11	188.57			
			100	12	205.71
105	13	222.86			
			110	14	240.00
120	15	257.14			
			130	16	274.29
			135	17	291.43
			140	18	308.57
			150	19	325.71
			160	20	342.86
			165	21	360.00

$C_1 = -0.2226$
$S_1 = 3.1726$

$C_2 = 0.2226$
$S_2 = -3.1726$

$W = 2\{[(-0.2226)^2 + (3.1726)^2]/10 + [(0.8226)^2 + (-3.1726)^2]/11\}$

$\quad = 2\{1.0115 + 0.9195\} = 2\{1.9310\} = 3.862$

$\nu = 2$

$\chi^2_{0.05,2} = 5.991$

Do not reject H_0.

$$0.10 < P < 0.25.$$

Maag (1966) extended the Watson U^2 test to $k > 2$, but critical values are not available. Comparison of more than two medians may be effected by the procedure in Section 27.7.

If the data are in groups with a grouping interval larger than 10, then an $r \times c$ contingency table analysis may be performed, for r samples in c groups; see Section 23.1 for the computation procedure.

27.7 TWO-SAMPLE AND MULTISAMPLE TESTING OF MEDIAN ANGLES

The following comparison of two or more medians is presented by Fisher (1933: 114), who states it as applicable if each sample size is at least 10 and all data are within $90°$ of the grand median (i.e., the median of all N data from all k samples). If we designate m_i to be the number of data in sample i that lie between the grand median and the grand median $- 90°$, and $M = \sum_{i=1}^{k} m_i$, then

$$\frac{N^2}{M(N-M)} \sum_{i=1}^{k} \frac{m_i^2}{n_i} - \frac{NM}{N-M} \qquad (27.23)$$

is a test statistic that may be compared to χ^2 with $k - 1$ degrees of freedom.[*]

If "H_0: All k population medians are equal" is not rejected, then the grand median is the best estimate of the median of each of the k populations.

27.8 TWO-SAMPLE AND MULTISAMPLE TESTING OF ANGULAR DISTANCES

Angular distance is simply the shortest distance, in angles, between two points on a circle. For example, the angular distance between $95°$ and $120°$ is $25°$, between $340°$ and $30°$ is $50°$, and between $190°$ and $5°$ is $175°$. In general, we shall refer to the angular distance between angles a_1 and a_2 as $d_{a_1-a_2}$. (Thus, $d_{95°-120°} = 25°$, and so on.)

Angular distances are useful in drawing inferences about departures of data from a specified direction. We may observe travel directions of animals trained to travel in a particular compass direction (perhaps "homeward"), or of animals confronted with the odor of food coming from a specified direction. If dealing with times of day we might speak of the time of a physiological or behavioral activity in relation to the time of a particular stimulus.

If the specified angle (e.g., direction or time of day) is μ_0, we may ask whether the mean of a sample of data, $\bar{a}$, significantly departs from μ_0 by testing the one-sample

---

[*]The same results are obtained if m_i is defined as the number of data in sample i that lie between the grand median and the grand median $+ 90°$.

hypothesis, $H_0: \mu_a = \mu_0$, as explained in Section 27.1. However, we may have two samples, Sample 1 and Sample 2, each of which has associated with it a specified angle of interest, μ_1 and μ_2, respectively (where μ_1 and μ_2 need not be the same). We may ask whether the angular distances for Sample 1 ($d_{a_{1i}-\mu_1}$) are significantly different from those for Sample 2 ($d_{a_{2i}-\mu_2}$). As shown in Example 27.13, we rank the angular distances of both samples combined and then perform a Mann-Whitney test (see Section 8.10). This was suggested by Wallraff (1979).

EXAMPLE 27.13   Two-sample testing of angular distances.

Birds of both sexes are transported away from their homes and released, with their directions of travel tabulated. The homeward direction for each sex is 135°.

H_0:   Males and females orient equally well toward their homes.

H_A: Males and females do not orient equally well toward their homes.

Males			Females		
Direction traveled	*Angular distance*	*Rank*	*Direction traveled*	*Angular distance*	*Rank*
145°	10°	6	160°	25°	12.5
155	20	11	135	0	1
130	5	2.5	145	10	6
145	10	6	150	15	9.5
145	10	6	125	10	6
160	25	12.5	120	15	9.5
140	5	2.5			
		46.5			44.5

For the two-tailed Mann-Whitney test:

$n_1 = 7, R_1 = 46.5$

$n_2 = 6, R_2 = 44.5$

$U = n_1 n_2 + \dfrac{n_1(n_1 + 1)}{2} - R_1 = (7)(6) + \dfrac{7(8)}{2} - 46.5 = 23.5$

$U' = n_1 n_2 - U = (7)(6) - 23.5 = 18.5$

$U_{0.05(2),7,6} = U_{0.05(2),6,7} = 36$

Do not reject H_0.

$$P > 0.20$$

The procedure could be performed as a one-tailed, instead of a two-tailed test, if there were reason to be interested in whether the angular distances in one group were greater than those in the other.

Multisample Testing. If more than two samples are involved, then the angular deviation of all of them are pooled and ranked, whereupon the Kruskal-Wallis test (Section 10.4) may be applied, followed if necessary by nonparametric multiple comparison testing (Section 11.6).

27.9 TWO-SAMPLE AND MULTISAMPLE TESTING OF ANGULAR DISPERSION

The Wallraff (1979) procedure of analyzing angular distances (Section 27.8) may be applied to testing for dispersion. The angular distances of concern for Sample 1 are $d_{a_{1i}-\bar{a}_1}$ and those for Sample 2 are $d_{a_{2i}-\bar{a}_2}$. Thus, just as measures of dispersion for linear data may refer to deviations of the data from their mean (Sections 4.3 and 4.4), here we consider the deviations of circular data from their mean.

The angular distances of the two samples are pooled; then they are ranked for application of the Mann-Whitney test, which may be employed for either two-tailed (Example 27.14) or one-tailed testing.

Multisample Testing. If one wishes to compare the dispersions of more than two samples, then the above Mann-Whitney procedure may be expanded by using the Kruskal-Wallis test (Section 10.4), followed if necessary by nonparametric multiple comparisons (Section 11.6).

EXAMPLE 27.14   Two-sample testing for angular dispersion.

The times of day that males and females are born are tabulated. The mean time of day for each sex is determined (to the nearest 5 min) as in Section 26.4. (For males, $\bar{a}_1 = 7{:}55$ AM; for females, $\bar{a}_2 = 8{:}15$ AM.)

H_0: The times of day of male births are as variable as the times of day of female births.

H_A: The times of day of male births do not have the same variability as the times of day of female births.

Male			Female		
Time of day	*Angular distance*	*Rank*	*Time of day*	*Angular distance*	*Rank*
05:10 hr	2:45 hr	11	08:15 hr	0:00 hr	1
06:30	1:25	4	10:20	2:05	8.5
09:40	1:45	6	09:45	1:30	5
10:20	2:25	10	06:10	2:05	8.5
04:20	3:35	13	04:05	4:10	14
11:15	3:20	12	07:50	0:25	2
			09:00	0:45	3
			10:10	1:55	7
		$R_1 = 56$			$R_2 = 49$

EXAMPLE 27.14   (continued)

For the two-tailed Mann-Whitney test:

$$n_1 = 6,\ R_1 = 56$$

$$n_2 = 8,\ R_2 = 49$$

$$U = n_1 n_2 + \frac{n_1(n_1 + 1)}{2} - R_1 = (6)(8) + \frac{6(7)}{2} - 56 = 13$$

$$U' = n_1 n_2 - U = (6)(8) - 13 = 35$$

$$U_{0.05(2),6,8} = 40$$

Do not reject H_0.

$$P = 0.20$$

27.10   PARAMETRIC ONE-SAMPLE SECOND-ORDER ANALYSIS OF ANGLES

A set of n angles, a_i, has a mean angle, $\bar{a}$, and an associated mean vector length, r. This set of data may be referred to as a *first-order sample*. A set of k such means may be referred to as a *second-order sample*. Section 26.9 discussed the computation of the mean of a second-order sample, namely the mean of a set of means. We now wish to test the statistical significance of a mean of means.

For a second-order sample of k mean angles, we can obtain $\bar{X}$ with either Equation 26.27 or 26.29 and $\bar{Y}$ with either Equation 26.28 or 26.30. Assuming that the second-order sample comes from a bivariate normal distribution (i.e., a population in which the X_j's follow a normal distribution, and the Y_j's are normally distributed), a testing procedure due to Hotelling* (1931) may be applied.

The sums of squares and crossproducts of the k means are

$$\sum x^2 = \sum X_j^2 - \frac{(\sum X_j)^2}{k}, \tag{27.24}$$

$$\sum y^2 = \sum Y_j^2 - \frac{(\sum Y_j)^2}{k}, \tag{27.25}$$

and

$$\sum xy = \sum X_j Y_j - \frac{\sum X_j \sum Y_j}{k}, \tag{27.26}$$

where $\sum$ in each instance refers to a summation over all k means (i.e., $\sum = \sum_{j=1}^{k}$).

---

*Harold Hotelling (1895–1973), American mathematical economist and statistician. He owed his life, and thus the achievements of an impressive career, to a zoological mishap. While attending the University of Washington he was called to military service in World War I and appointed to care for mules. One of his charges (named Dynamite) broke Hotelling's leg, thus preventing the young soldier from accompanying his division to France where the unit was annihilated in battle (Darnell, 1988).

Then, we can test the null hypothesis that there is no mean direction (i.e., H_0: $\rho = 0$) in the population from which the second-order sample came by using as a test statistic

$$F = \frac{k(k-2)}{2}\left[\frac{\bar{X}^2\sum y^2 - 2\bar{X}\bar{Y}\sum xy + \bar{Y}^2\sum x^2}{\sum x^2 \sum y^2 - (\sum xy)^2}\right], \quad (27.27)$$

with the critical value being the one-tailed F with degrees of freedom of 2 and $k-2$ (Batschelet, 1978; 1981: 144–150). This test is demonstrated in Example 27.15, using the data from Example 26.8 (and shown in Fig. 26.8).

This test assumes the data are not grouped. The assumption of bivariate normality is a serious one. Although the test appears robust against departures due to kurtosis, the test may be badly affected by departures due to extreme skewness, rejecting a true H_0 far more often than indicated by the significance level, α (Everitt, 1979; Mardia, 1970a).

EXAMPLE 27.15   The second-order analysis for testing the significance of the mean of the sample means in Example 26.8.

H_0:   There is no mean population direction (i.e., $\rho = 0$).

H_A: There is a mean population direction (i.e., $\rho \neq 0$).

$k = 7$, $\bar{X} = -0.52734$, $\bar{Y} = 0.27844$

$\sum X_j = -3.69139$, $\sum X_j^2 = 2.27959$, $\sum x^2 = 0.33297$ (by Equation 27.24)

$\sum Y_j = 1.94906$, $\sum Y_j^2 = 0.86474$, $\sum y^2 = 0.32205$ (by Equation 27.25)

$\sum X_j Y_j = -1.08282$, $\sum xy = -0.05500$ (by Equation 27.26)

$$F = \frac{7(7-2)}{2}\left[\frac{(-0.52734)^2(0.32205) - 2(-0.52734)(0.27844)(-0.05500) + (0.27844)^2(0.33297)}{(0.33297)(0.32205) - (-0.05500)^2}\right]$$

$$= 16.66$$

$F_{0.05(1),2,5} = 5.79$

Reject H_0.

$$0.005 < P < 0.01$$

And, from Example 26.9, we see that the population mean angle is estimated to be $152°$.

27.11 NONPARAMETRIC ONE-SAMPLE SECOND-ORDER ANALYSIS OF ANGLES

The Hotelling testing procedure of Section 27.10 requires that the k $\bar{X}$'s come from a normal distribution, as do the k $\bar{Y}$'s. Although we may assume the test to be robust to some departure from this bivariate normality, there may be considerable non-normality in a sample, in which case a nonparametric method is preferable.

Moore (1980) has provided a nonparametric modification of the Rayleigh test, which can be used to test a sample of mean angles; it is demonstrated in Example 27.16. We first rank the k vector lengths, so that r_1 is the smallest vector length and r_k is the largest. We shall call the ranks i (where i ranges from 1 through k) and compute

$$X = \frac{\sum_{i=1}^{k} i \cos \bar{a}_i}{k} \tag{27.28}$$

$$Y = \frac{\sum_{i=1}^{k} i \sin \bar{a}_i}{k} \tag{27.29}$$

$$R' = \sqrt{\frac{x^2 + Y^2}{k}}. \tag{27.30}$$

EXAMPLE 27.16   **Nonparametric second-order analysis for significant direction in the sample of means of Example 26.8.**

H_0:   The population from which the sample of means came is uniformly distributed around the circle (i.e., $\rho = 0$).

H_A: The population of means is not uniformly distributed around the circle (i.e., $\rho \neq 0$).

Sample rank (i)	r_i	a_i	$i \cos \bar{a}_i$	$i \sin \bar{a}_i$
1	0.3338	171°	−0.98769	0.15643
2	0.3922	186	−1.98904	−0.20906
3	0.4696	117	−1.36197	2.67302
4	0.6962	134	−2.77863	2.87736
5	0.7747	169	−4.90814	0.95404
6	0.8794	140	−4.59627	3.85673
7	0.8954	160	−6.57785	2.39414
			−23.19959	12.70266

$X = \dfrac{\sum i \cos \bar{a}_i}{k} = \dfrac{-23.19959}{7} = -3.31423$

$Y = \dfrac{\sum i \sin \bar{a}_i}{k} = \dfrac{12.70266}{7} = 1.81467$

$R' = \sqrt{\dfrac{X^2 + Y^2}{k}} = \sqrt{\dfrac{(-3.31423)^2 + (1.81467)^2}{7}} = \sqrt{2.03959} = 1.428$

$R'_{0.05, 7} = 1.150$

Therefore, reject H_0.

$$P < 0.002$$

The test statistic, R', is then compared to the appropriate critical value, $R'_{\alpha,n}$, in Appendix Table B.39.

Testing Weighted Angles. The Moore modification of the Rayleigh test can also be used when we have a sample of angles, each of which is weighted. We may then perform the ranking of the angles by the weights, instead of by the vector lengths, r. For example, the data of Example 26.2 could be ranked by the amount of leaning. Or, if we are recording the direction each of several birds flies from a release point, the weights could be the distances flown. (If the birds disappear at the horizon, then there are no weights and the Rayleigh test may be applied.)

27.12 PARAMETRIC TWO-SAMPLE SECOND-ORDER ANALYSIS OF ANGLES

Batschelet (1978, 1981: 150–154) explains how the Hotelling (1931) procedure of Section 27.10 can be extended to consider the hypothesis of equality of the means of two populations of means (assuming each population to be bivariate normal). We proceed as in Section 27.10, obtaining an $\bar{X}$ and $\bar{Y}$ for each of the two samples ($\bar{X}_1$ and $\bar{Y}_1$ for Sample 1, and $\bar{X}_2$ and $\bar{Y}_2$ for Sample 2). Then, we apply Equations 27.24, 27.25, and 27.26 to each of the two samples, obtaining $\left(\sum x^2\right)_1$, $\left(\sum xy\right)_1$, and $\left(\sum y^2\right)_1$ for Sample 1, and $\left(\sum x^2\right)_2$, $\left(\sum xy\right)_2$, and $\left(\sum y^2\right)_2$ for Sample 2.
Then we calculate

$$\left(\sum x^2\right)_c = \left(\sum x^2\right)_1 + \left(\sum x^2\right)_2; \tag{27.31}$$

$$\left(\sum y^2\right)_c = \left(\sum y^2\right)_1 + \left(\sum y^2\right)_2; \tag{27.32}$$

$$\left(\sum xy\right)_c = \left(\sum xy\right)_1 + \left(\sum xy\right)_2; \tag{27.33}$$

and the null hypothesis of the two population mean angles being equal is tested by

$$F = \frac{N-3}{2\left(\frac{1}{k_1}+\frac{1}{k_2}\right)}\left[\frac{\left(\bar{X}_1-\bar{X}_2\right)^2\left(\sum y^2\right)_c - 2\left(\bar{X}_1-\bar{X}_2\right)\left(\bar{Y}_1-\bar{Y}_2\right)\left(\sum xy\right)_c + \left(\bar{Y}_1-\bar{Y}_2\right)^2\left(\sum x^2\right)_c}{\left(\sum x^2\right)_c\left(\sum y^2\right)_c - \left(\sum xy\right)_c^2}\right],$$

$$\tag{27.34}$$

where $N = k_1 + k_2$, and F is one-tailed with 2 and $N-3$ degrees of freedom. This test is shown in Example 27.17, using the data of Fig. 27.2.

The two-sample Hotelling test is robust to departures from the normality assumption (far more so than is the one-sample test of Section 27.10), the effect of non-normality being slight conservatism (i.e., rejecting a false H_0 a little less frequently than indicated by the significance level, α) (Everitt, 1979). The two samples should be of the same size, but departure from this assumption does not appear to have serious consequences (Batschelet, 1981: 202).

EXAMPLE 27.17   Parametric two-sample second-order analysis for testing the difference between mean angles.

We have two samples, each consisting of mean directions and vector lengths, as shown in Fig. 27.2. Sample 1 is the data from Examples 26.8 and 27.15, where

$$k_1 = 7; \quad \bar{X}_1 = -0.52734; \quad \bar{Y}_1 = 0.27844; \quad \bar{a}_1 = 152°;$$

$$\left(\sum x^2\right)_1 = 0.33297; \quad \left(\sum y^2\right)_1 = 0.32205; \quad \left(\sum xy\right)_1 = -0.05500.$$

Sample 2 consists of the following ten data:

j	$\bar{a}_j$	r_j
1	115°	0.9394
2	127	0.6403
3	143	0.3780
4	103	0.6671
5	130	0.8210
6	147	0.5534
7	107	0.8334
8	137	0.8139
9	127	0.2500
10	121	0.8746

Applying the calculations of Examples 26.8 and 27.15, we find:

$$k_2 = 10; \quad \sum r_j \cos a_j = -3.66655; \quad \sum r_j \sin a_j = 5.47197;$$
$$\bar{X}_2 = -0.36660; \quad \bar{Y}_2 = 0.54720; \quad \bar{a}_2 = 124°.$$
$$\left(\sum x^2\right)_2 = 0.20897; \left(\sum y^2\right)_2 = 0.49793; \left(\sum xy\right)_2 = -0.05940$$

Then, we can test

H_0: $\mu_1 = \mu_2$ (The means of the populations from which these two samples came are equal.)

H_A: $\mu_1 \neq \mu_2$ (The two population means are not equal.)

$$N = 7 + 10$$
$$\left(\sum x^2\right)_c = 0.33297 + 0.20897 = 0.54194$$
$$\left(\sum y^2\right)_c = 0.32205 + 0.49793 = 0.81998$$
$$\left(\sum xy\right)_c = -0.05500 + (-0.05940) = -0.11440$$

Using Equation 27.34:

$$F = \frac{(17-3)}{2\left(\frac{1}{7}+\frac{1}{10}\right)} \left[\frac{\begin{array}{c} [-0.52734 - (-0.36660)]^2(0.81998) \\ -2[-0.52734-(-0.36660)](0.27844-0.54720)(-0.11440) \\ +(0.27844-0.54720)^2(0.54194) \end{array}}{(0.54194)(0.81998)-(-0.11440)^2} \right]$$

$$= 4.69$$

$$F_{0.05(1),2,14} = 3.74$$

Reject H_0.

$$0.025 < P < 0.05 \quad [P = 0.028]$$

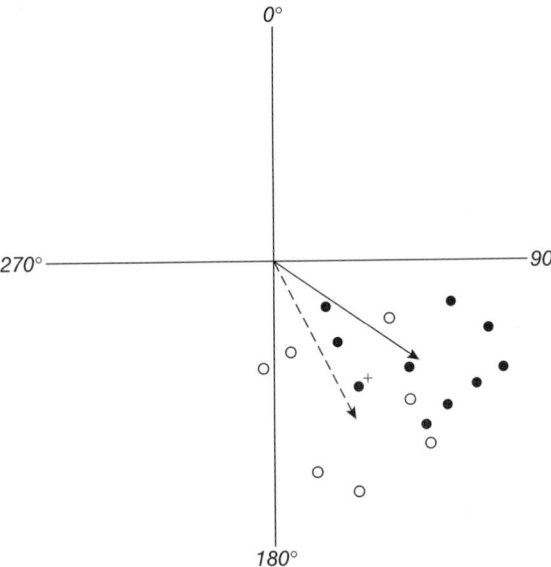

Figure 27.2 The data of Example 27.17. The open circles indicate the ends of the seven mean vectors of Sample 1 (also shown in Fig. 26.9), with the mean of these seven indicated by the broken-line vector. The solid circles indicate the ten data of Sample 2, with their mean shown as a solid-line vector. (The "+" indicates the grand mean of all seventeen data, which is used in Example 27.18.)

27.13 NONPARAMETRIC TWO-SAMPLE SECOND-ORDER ANALYSIS OF ANGLES

The parametric test of Section 27.12 is based on sampled populations being bivariate normal and the two populations having variances and covariances in common, unlikely assumptions to be strictly satisfied in practice. While the test is rather robust to departures from these assumptions, one may be more comfortable in many cases employing a nonparametric test to assess whether two second-order populations have the same directional orientation.

Batschelet (1978; 1981: 154–156) presents the following nonparametric procedure (suggested by Mardia, 1967) as an alternative to the Hotelling test of Section 27.12. First we compute the grand mean vector, pooling all data from both samples. Then, the X coordinate of the grand mean is subtracted from the X coordinate of each of the data in both samples, and the Y of the grand mean is subtracted from the Y of each of the data. (This maneuver determines the direction of each datum from the grand mean.) As shown in Example 27.18, the resulting vectors are then tested by a nonparametric two-sample test (as in Section 27.6). This procedure requires that the data not be grouped.

EXAMPLE 27.18   Nonparametric two-sample second-order analysis, using the data of Example 27.17.

H_0:   The two samples came from the same population, or from two populations with the same directions.

H_A: The two samples did not come from the same population, nor from two populations with the same directions.

EXAMPLE 27.18 (continued)

Total number of vectors $= 7 + 10 = 17$

To determine the grand mean vector (which is shown in Fig. 27.2):

$$\sum r_j \cos a_j = (-3.69139) + (-3.66655) = -7.35794$$

$$\sum r_j \sin a_j = 1.94906 + 5.47197 = 7.42103$$

$$\bar{X} = \frac{-7.35794}{17} = -0.43282$$

$$\bar{Y} = \frac{7.42103}{17} = 0.43653$$

$\bar{X}$ and $\bar{Y}$ are all that we need to define the grand mean; however, if we wish we can also determine the length and direction of the grand mean vector:

$$r = \sqrt{\bar{X}^2 + \bar{Y}^2} = \sqrt{(-0.43282)^2 + (0.43653)^2} = 0.61473$$

$$\cos \bar{a} = \frac{-0.43282}{0.61473} = -0.70408$$

$$\sin \bar{a} = \frac{0.43653}{0.61473} = 0.71012$$

$$\bar{a} = 135°$$

Returning to the hypothesis test, we subtract the above $\bar{X}$ from the X, and the $\bar{Y}$ from the Y, for each of the seventeen data, arriving at seventeen new vectors, as follows:

Sample 1

Datum	X	$X - \bar{X}$	Y	$Y - \bar{Y}$	New a
1	−0.84140	−0.40858	0.30624	−0.13029	184°
2	−0.76047	−0.32765	0.14782	−0.28871	210
3	−0.21319	0.21963	0.41842	−0.01811	20
4	−0.67366	−0.24084	0.56527	0.12874	137
5	−0.39005	0.04277	−0.04100	−0.47753	276
6	−0.48293	−0.05011	0.50009	0.06356	107
7	−0.32969	0.10313	0.05222	−0.38431	290

Sample 2

Datum	X	$X - \bar{X}$	Y	$Y - \bar{Y}$	New a
1	−0.39701	0.03581	0.85139	0.41485	86°
2	−0.38534	0.04748	0.51137	0.07484	75
3	−0.30188	0.13084	0.22749	−0.20904	320
4	−0.15006	0.28276	0.65000	0.21347	48
5	−0.52773	−0.09491	0.62892	0.19239	108
6	−0.46412	−0.03130	0.30140	−0.13513	230
7	−0.24366	0.18916	0.79698	0.36045	68
8	−0.59525	−0.16243	0.55508	0.11855	127
9	−0.15045	0.28237	0.19966	−0.23687	334
10	−0.45045	−0.01763	0.74968	0.31315	92

EXAMPLE 27.18   (continued)

Now, using Watson's two-sample test (Section 27.6) on these new angles:

i	Sample 1 a_{1i}	i/n_1	j	Sample 2 a_{2j}	j/n_2	d_k	d_k^2
		0.0000	1	37°	0.1000	−0.1000	0.0100
		0.0000	2	58	0.2000	−0.2000	0.0400
		0.0000	3	62	0.3000	−0.3000	0.0900
		0.0000	4	85	0.4000	−0.4000	0.1600
		0.0000	5	93	0.5000	−0.5000	0.2500
		0.0000	6	116	0.6000	−0.6000	0.3600
1	128°	0.1429			0.6000	−0.4571	0.2089
		0.1429	7	144	0.7000	−0.5571	0.3104
2	152	0.2857			0.7000	−0.4143	0.1716
3	198	0.4286			0.7000	−0.2714	0.0737
4	221	0.5716			0.7000	−0.1284	0.0165
		0.5716	8	257	0.8000	−0.2284	0.0522
5	275	0.7143			0.8000	−0.0857	0.0073
6	285	0.8571			0.8000	0.0571	0.0033
		0.8571	9	302	0.9000	−0.0429	0.0018
		0.8571	10	320	1.0000	−0.1429	0.0204
7	355	1.0000			1.0000	0.0000	0.0000
	$n_1 = 7$			$n_2 = 10$		$\sum d_k =$ −4.3811	$\sum d_k^2 =$ 1.7761

$$U^2 = \frac{n_1 n_2}{N^2} \left[\sum d_k^2 - \frac{\left(\sum d_k \right)^2}{N} \right]$$

$$= \frac{(7)(10)}{17^2} \left[1.7761 - \frac{(-4.3811)^2}{17} \right]$$

$$= 0.1567$$

$U^2_{0.05,7,10} = 0.1866$

Do not reject H_0.

$$0.05 < P < 0.10$$

27.14 PARAMETRIC PAIRED-SAMPLE TESTING WITH ANGLES

The paired-sample experimental design was introduced in Chapter 9 for linear data, and Section 9.1 showed how the analysis of two samples having paired data could be reduced to a one-sample test employing the differences between members of pairs.

Circular data in two samples might also be paired, in which case the one-sample Hotelling test of Section 27.10 may be used after forming a single sample of data from the differences between the paired angles. If a_{ij} is the jth angle in the ith sample, then a_{1j} and a_{2j} are a pair of data. A single set of rectangular coordinates, X's and Y's, is formed by computing

$$X_j = \cos a_{2j} - \cos a_{1j} \tag{27.35}$$

and

$$Y_j = \sin a_{2j} - \sin a_{1j}. \tag{27.36}$$

Then the procedure of Section 27.10 may be applied, as shown in Example 27.19 (where k is the number of pairs).

Second-Order Data. If each member of a pair of data is a mean angle ($\bar{a}$) from a sample, with an associated vector length (r), then we are dealing with a second-order analysis. The above Hotelling test may be applied if the following computations are used in place of Equations 27.35 and 27.36, respectively:

$$X_j = r_{2j} \cos \bar{a}_{2j} - r_{1j} \cos \bar{a}_{1j} \tag{27.37}$$

$$Y_j = r_{2j} \sin \bar{a}_{2j} - r_{ij} \sin \bar{a}_{1j}. \tag{27.38}$$

EXAMPLE 27.19   The Hotelling test for paired samples of angles.

Ten birds are marked for individual identification, and we record on which side of a tree each bird sits to rest in the morning and in the afternoon. We wish to test the following.

H_0:   The side of a tree on which birds sit is the same in the morning and in the afternoon.

H_A: The side of a tree on which birds sit is not the same in the morning and in the afternoon.

	Morning			**Afternoon**			**Difference**	
Bird (j)	Direction (a_{1j})	$\sin a_{1j}$	$\cos a_{1j}$	Direction (a_{2j})	$\sin a_{2j}$	$\cos a_{2j}$	Y_j	X_j
1	105°	0.9659	−0.2588	205°	−0.4226	−0.9063	−1.3885	−0.6475
2	120	0.8660	−0.5000	210	−0.5000	−0.8660	−1.3660	−0.3660
3	135	0.7071	−0.7072	235	−0.8192	−0.5736	−1.5263	0.1336
4	95	0.9962	−0.0872	245	−0.9063	−0.4226	−1.9025	−0.3354
5	155	0.4226	−0.9063	260	−0.9848	−0.1736	−1.4074	0.7327
6	170	0.1736	−0.9848	255	−0.9659	−0.2588	−1.1395	0.7260
7	160	0.3420	−0.9397	240	−0.8660	−0.5000	−1.2080	0.4397
8	155	0.4226	−0.9063	245	−0.9063	−0.4226	−1.3289	0.4837
9	120	0.8660	−0.5000	210	−0.5000	−0.8660	−1.3660	−0.3660
10	115	0.9063	−0.4226	200	−0.3420	−0.9397	−1.2483	−0.5171

$k = 10$

$\bar{X} = 0.0284$

$\bar{Y} = -1.3981$

EXAMPLE 27.19   (continued)

Using Equations 26.12 through 26.26:

$$\sum x^2 = 2.5761 - \frac{(0.2837)^2}{10} = 2.5681$$

$$\sum y^2 = 19.6717 - \frac{(-13.8814)^2}{10} = 0.4023$$

$$\sum xy = -0.0538 - \frac{(0.2837)(-13.8814)}{10} = 0.3400$$

$$F = \frac{10(10-2)}{2} \left[\frac{(0.0284)^2(0.4023) - 2(0.0284)(-1.3881)(-0.3400) + (-1.3881)^2(2.5681)}{(2.5681)(0.4023) - (-0.3400)^2} \right]$$

$$= 217$$

$$F_{0.05(1),2,8} = 4.46$$

Reject H_0.

$$P \ll 0.0005 \quad [P = 0.00000011]$$

27.15 NONPARAMETRIC PAIRED-SAMPLE TESTING WITH ANGLES

Circular data in a paired-sample experimental design may be tested nonparametrically by forming a single sample of the paired differences, which can then be subjected to the Moore test of Section 27.11. We calculate rectangular coordinates (X_j and Y_j) for each paired difference, as done in Equations 27.35 and 27.36. Then, for each of the j paired differences, we compute

$$r_j = \sqrt{X_j^2 + Y_j^2}, \tag{27.39}$$

$$\cos a_j = \frac{X_j}{r_j}, \tag{27.40}$$

$$\sin a_j = \frac{Y_j}{r_j}. \tag{27.41}$$

Then the values of r_j are ranked, with ranks (i) running from 1 through n, and we complete the analysis using Equations 27.28, 27.29, and 27.30, substituting n for k. The procedure is demonstrated in Example 27.20.

Second-Order Data.   If each member of a pair of circular scale data is a mean angle, a_j, with an associated vector length, r_j, then we modify the above analysis. Calculate X_j and Y_j by Equations 27.37 and 27.38, respectively, instead of by Equations 27.35 and 27.36, respectively. Then apply Equations 27.39 through 27.41 and Equations 27.28 through 27.30 to complete the analysis (using k as the number of paired means).

EXAMPLE 27.20 The Moore test for paired data on a circular scale of measurement.

Ten birds are marked for individual identification, and we record on which side of a tree each bird sits to rest in the morning and in the afternoon. We wish to test the following:

H_0: The side of a tree on which birds sit is the same in the morning and in the afternoon.

H_A: The side of a tree on which birds sit is not the same in the morning and in the afternoon.

| Bird | Direction | Morning | | Afternoon | | | Difference | | | | | Rank of r_j |
(j)	(a_{1j})	$\sin a_{1j}$	$\cos a_{1j}$	Direction (a_{2j})	$\sin a_{2j}$	$\cos a_{2j}$	Y_j	X_j	r_j	$\sin a_j$	$\cos a_j$	(i)
1	105°	0.9659	−0.2588	205°	−0.4226	−0.9063	−1.3885	−0.6475	1.5321	−0.9063	−0.4226	7.5
2	120	0.8660	−0.5000	210	−0.5000	−0.8660	−1.3660	−0.3660	1.4142	−0.9659	−0.2588	4.5
3	135	0.7071	−0.7071	235	−0.8192	−0.5736	−1.5263	0.1335	1.5321	−0.9962	0.0871	7.5
4	95	0.9962	−0.0872	245	−0.9063	−0.4226	−1.9025	−0.3354	1.9318	−0.9848	−0.1736	10
5	155	0.4226	−0.9063	260	−0.9848	−0.1736	−1.4074	0.7327	1.5867	−0.8870	0.4618	9
6	170	0.1736	−0.9848	255	−0.9659	−0.2588	−1.1395	0.7260	1.3511	−0.8434	0.5373	2
7	160	0.3420	−0.9397	240	−0.8660	−0.5000	−1.2080	0.4397	1.2855	−0.9397	0.3420	1
8	155	0.4226	−0.9063	245	−0.9063	−0.4226	−1.3289	0.4837	1.4152	−0.9390	0.3418	6
9	120	0.8660	−0.5000	210	−0.5000	−0.8660	−1.3660	−0.3660	1.4142	−0.9659	−0.2588	4.5
10	150	0.9063	−0.4226	200	−0.3420	−0.9397	−1.2483	−0.5171	1.3512	−0.9238	−0.3827	3

$n = 10$

$$X = \frac{\sum_{i=1}^{n} i\cos a_i}{n} = \frac{1(0.3420) + 2(0.5373) + \cdots + 10(-0.1736)}{10} = -0.0106$$

$$Y = \frac{\sum_{i=1}^{n} i\sin a_i}{n} = \frac{1(-0.9397) + 2(-0.8434) + \cdots + 10(-0.9848)}{10} = -5.1825$$

$$R' = \sqrt{\frac{X^2 + Y^2}{n}} = \sqrt{\frac{(-0.0106)^2 + (-5.1825)^2}{10}} = \sqrt{2.685842} = 1.639$$

$R'_{0.05,10} = 1.144$

Reject H_0.

$$P < 0.002$$

27.16 PARAMETRIC ANGULAR CORRELATION AND REGRESSION

The correlation of two variables, each measured on a linear scale, was discussed in Chapter 19, with linear-scale regression being introduced in Chapter 17. Correlation involving angular data may be of two kinds: Either both variables are measured on a circular scale (a situation sometimes termed "angular-angular," or "spherical," correlation), or one variable is on a circular scale with the other measured on a linear scale (sometimes called an "angular-linear," or "cylindrical," correlation). The study of biological rhythms deals essentially with the rhythmic dependence (i.e., regression) of a linear scale variable (e.g., a measure of biological activity, such as body temperature) on a circular scale variable (namely, time).

Angular-Angular Correlation.   Correlation measures developed for correlation between two angular variables were for years characterized by serious deficiencies, such as not distinguishing between positive and negative relationships (e.g., see the review by Jupp and Mardia, 1980). However, Fisher and Lee (1983) presented a correlation coefficient analogous to the familiar parametric correlation coefficient of Section 19.1[*]; it is[†]

$$r_{aa} = \frac{\sum\limits_{i=1}^{n-1}\sum\limits_{j=i+1}^{n} \sin(a_i - a_j)\sin(b_i - b_j)}{\sqrt{\sum\limits_{i=1}^{n-1}\sum\limits_{j=i+1}^{n} \sin^2(a_i - a_j) \sum\limits_{i=1}^{n-1}\sum\limits_{j=i+1}^{n} \sin^2(b_i - b_j)}}, \tag{27.43}$$

where the ith pair of data is denoted as a_i, b_i[‡].

Upton and Fingleton (1989: 303) give a relatively simple method to test the significance of r_{aa}—that is, to test whether the sample of data came from a population having a correlation coefficient, ρ_{aa}, different from zero. The procedure involves computing r_{aa}

---

[*]Results identical to those from Equation 18.1 may be obtained by

$$r = \frac{\sum\limits_{i=1}^{n-1}\sum\limits_{j=i+1}^{n} (X_i - X_j)(Y_i - Y_j)}{\sqrt{\sum\limits_{i=1}^{n-1}\sum\limits_{j=i+1}^{n} (X_i - X_j)^2 \sum\limits_{i=1}^{n-1}\sum\limits_{j=i+1}^{n} (Y_i - Y_j)^2}}. \tag{27.42}$$

[†]The notation "$\sin^2(a_i - a_j)$" means "$[\sin(a_i - a_j)]^2$."

[‡]Fisher (1993: 151) gives an alternate computation of r_{aa} as

$$\frac{4\left[\left(\sum\limits_{i=1}^{n}\cos a_i \cos b_i\right)\left(\sum\limits_{i=1}^{n}\sin a_i \sin b_i\right) - \left(\sum\limits_{i=1}^{n}\cos a_i \sin b_i\right)\left(\sum\limits_{i=1}^{n}\sin a_i \cos b_i\right)\right]}{\sqrt{\left[n^2 - \left(\sum\limits_{i=1}^{n}\cos(2a_i)\right)^2 - \left(\sum\limits_{i=1}^{n}\sin(2a_i)\right)^2\right]\left[n^2 - \left(\sum\limits_{i=1}^{n}\cos(2b_i)\right)^2 - \left(\sum\limits_{i=1}^{n}\sin(2b_i)\right)^2\right]}}. \tag{27.44}$$

an additional n times for the sample, each time eliminating a different one of the n pairs of a and b data.*

Then a mean and variance of these n additional r_{aa}'s is calculated (let's call the mean $\bar{r}_{aa}$, and the variance $s^2_{r_{aa}}$); and confidence limits for ρ_{aa} are obtained as

$$L_1 = nr_{aa} - (n-1)\bar{r}_{aa} - Z_{\alpha(2)}\sqrt{\frac{s^2_{r_{aa}}}{n}} \qquad (27.45)$$

and

$$L_2 = nr_{aa} - (n-1)\bar{r}_{aa} + Z_{\alpha(2)}\sqrt{\frac{s^2_{r_{aa}}}{n}}. \qquad (27.46)$$

If the confidence interval (i.e., the interval between L_1 and L_2 does *not* include zero, then H_0: $\rho_{aa} = 0$ is rejected in favor of H_A: $\rho_{aa} \neq 0$. The computation of r_{aa}, and testing its significance, is shown in Example 27.21.

---

*This involves what statisticians call the *jackknife* technique (introduced by Quenouille, 1956), named in 1964 by R. G. Miller (David, 1995).

EXAMPLE 27.21   Angular-angular correlation.

We wish to assess the relationship between the orientation of insects and the direction of a light source.

H_0: $\rho_{aa} = 0$; H_A: $\rho_{aa} \neq 0$

i	insect a_i	light b_i
1	145°	120°
2	190°	180°
3	310°	330°
4	210°	225°
5	80°	55°

$n = 5$; the computations proceed as follows:

i	j	$a_i - a_j$	$b_i - b_j$	$\sin(a_i - a_j)$	$\sin(b_i - b_j)$	$\sin(a_i - a_j)$ $\times \sin(b_i - b_j)$	$\sin^2(a_i - a_j)$	$\sin^2(b_i - b_j)$
1	2	−45°	−60°	−0.70711	−0.86603	0.61237	0.50000	0.75001
1	3	−165°	−210°	−0.25882	0.50000	−0.12941	0.06699	0.25000
1	4	−65°	−105°	−0.90631	−0.96593	0.87543	0.82140	0.93302
1	5	65°	65°	0.90631	0.90631	0.82140	0.82140	0.82140
2	3	−120°	−150°	−0.86603	−0.50000	0.43302	0.75001	0.25000
2	4	−20°	−45°	−0.34202	−0.70711	0.24185	0.11698	0.50000
2	5	110°	125°	0.93969	0.81915	0.76975	0.88302	0.67101
3	4	100°	105°	0.98481	0.96593	0.95126	0.96985	0.93302
3	5	230°	275°	−0.76604	−0.99619	0.76312	0.58682	0.99239
4	5	130°	170°	0.76604	0.17365	0.13302	0.58682	0.03015
				Sum:		5.47181	6.10329	6.13100

EXAMPLE 27.21   (continued)

$$r_{aa} = \frac{\displaystyle\sum_{i=1}^{n-1}\sum_{j=i+1}^{n} \sin(a_i - a_j)\sin(b_i - b_j)}{\sqrt{\displaystyle\sum_{i=1}^{n-1}\sum_{j=i+1}^{n} \sin^2(a_i - a_j)\sum_{i=1}^{n-1}\sum_{j=i+1}^{n} \sin^2(b_i - b_j)}}$$

$$= \frac{5.47181}{\sqrt{(6.10329)(6.13100)}} = \frac{5.47181}{\sqrt{37.41927}} = \frac{5.47181}{6.11713} = 0.8945$$

Five r_{aa}'s computed for the above data, each with a different pair of data deleted:

i deleted:	1	2	3	4	5
r_{aa} :	0.90793	0.87419	0.92905	0.89084	0.87393

$\bar{r}_{aa} = 0.89519$; $s^2_{r_{aa}} = 0.0005552$

$nr_{aa} - (n-1)\bar{r}_{aa} = (5)(0.8945) - (5-1)(0.89519) = 0.8917$

$Z_{0.05(2)}\sqrt{\dfrac{s^2_{r_{aa}}}{n}} = 1.9600\sqrt{\dfrac{0.0005552}{5}} = 1.9600(0.0105) = 0.0206$

$L_1 = 0.8917 - 0.0206 = 0.8711$

$L_2 = 0.8917 + 0.0206 = 0.9123$

As this confidence interval does not encompass zero, reject H_0.

Angular-Linear Correlation.   Among the procedures proposed for correlating an angular variable (a) with a linear variable (X) is that deriving from the work of Mardia (1976) and Johnson and Wehrly (1977). Using Equation 19.1, we determine coefficients for the correlation between X and the sine of a (call it r_{XS}), the correlation between X and the cosine of a (call it r_{XC}), and the correlation between the cosine and the sine of a (call it r_{CS}). Then, the angular-linear correlation coefficient is

$$r_{al} = \sqrt{\frac{r^2_{XC} + r^2_{XS} - 2r_{XC}r_{XS}r_{CS}}{1 - r^2_{CS}}}. \tag{27.47}$$

For angular-linear correlation, the correlation coefficient lies between 0 and 1 (i.e., there is no negative correlation). If n is large, then the significance of the correlation coefficient may be assessed by comparing nr^2_{al} to χ^2_2. This procedure is shown in Example 27.22. It is not known how large n must be for the chi-square approximation to give good results, and Fisher (1993: 145) recommends a different (laborious) method for assessing significance of the test statistic.

Regression.   Linear-circular regression, in which the dependent variable (Y) is linear and the independent variable (a) circular, may be analyzed, by the regression methods of Chapter 20, as

$$Y_i = b_0 + b_1 \cos a_i + b_2 \sin a_i \tag{27.48}$$

EXAMPLE 27.22 Angular-linear correlation.

For a sampled population of animals, we wish to examine the relationship between distance traveled and direction traveled.

$H_0: \rho_{al} = 0; \ H_A: \rho_{al} \neq 0$

$\alpha = 0.05$

i	X distance (km)	a_i direction (deg)	$\sin a_i$	$\cos a_i$
1	48	190	−0.17364	−0.98481
2	55	160	0.34202	−0.93969
3	26	210	−0.50000	−0.86603
4	23	225	−0.70711	−0.70711
5	22	220	−0.64279	−0.76604
6	62	140	0.64279	−0.76604
7	64	120	0.86603	−0.50000

$\sum X_i = 300$ kilometers; $\sum X_i^2 = 14{,}958$ km^2

$\sum \sin a_i = -0.17270$ degrees; $\sum \sin^2 a_i = 2.47350$ deg^2

$\sum \cos a_i = -5.52972$ degrees; $\sum \cos^2 a_i = 4.52652$ deg^2

"Sum of squares" of $X = \sum x^2 = 14{,}958 - 300^2/7 = 2100.86$ km^2

"Sum of squares" of $\sin a_i = 2.47350 - (-0.17270)^2/7 = 2.46924$ deg^2

"Sum of squares" of $\cos a_i = 4.52652 - (-5.52972)^2/7 = 0.15826$ deg^2

"Sum of crossproducts" of X and $\cos a_i = (48)(-0.98481) + (55)(-0.93969) + \cdots + (64)(-0.50000) - (300)(-5.52972)/7 = -234.08150 - (-236.98800) = 2.90650$ deg-km

"Sum of crossproducts" of X and $\sin a_i = (48)(-0.17364)+(55)(0.34202)+\cdots+(64)(.86603)- (300)(-0.17270)/7 = 62.35037 - (-7.40143) = 69.75180$ deg-km

"Sum of crossproducts" of $\cos a_i$ and $\sin a_i = (-0.98481)(-0.17364)+(-0.93969)(0.34202)+ \cdots + (-0.50000)(0.86603) - (-5.52972)(-0.17270)/7 = 0.34961 - 0.13643 = -0.21318$ deg-km

Using Equation 19.1:

$r_{XC} = 0.15940; \ r_{XC}^2 = 0.02541$

$r_{XS} = 0.96845; \ r_{XS}^2 = 0.93789$

$r_{CS} = 0.34104; \ r_{CS}^2 = 0.11630$

$$r_{al}^2 = \frac{r_{XC}^2 + r_{XS}^2 - 2r_{XC}r_{XS}r_{CS}}{1 - r_{CS}^2}$$

$$= \frac{0.02541 + 0.93789 - 2(0.15940)(0.96845)(0.34104)}{1 - 0.11630}$$

$$= \frac{0.85801}{0.88370} = 0.97093$$

EXAMPLE 27.22   (continued)

$r_{al} = \sqrt{0.97093} = 0.9854$

$nr_{al}^2 = (7)(0.97093) = 6.797$

$\chi_{0.05,2}^2 = 5.991$; do not reject H_0; $0.025 < P < 0.05$

(Fisher, 1993: 139–140), where b_0 is the Y-intercept and b_1 and b_2 are partial regression coefficients.

In circular-linear regression, where a is the dependent variable and Y the independent variable, the situation is rather more complicated and is discussed by Fisher (1993: 155–168).

Rhytomometry.   The description of biological rhythms may be thought of as a regression (often called *periodic regression*) of the linear variable on time (a circular variable). Excellent discussions of such regression are provided by Batschelet (1972; 1974; 1981: Chapter 8), Bliss (1970: Chapter 17), Bloomfield (1976), and Nelson et al. (1979).

The *period*, or length, of the cycle,* is often stated in advance. Parameters to be estimated in the regression are the *amplitude* of the rhythm (which is the range from the minimum to the maximum value of the linear variable)[†] and the *phase angle*, or *acrophase*, of the cycle (which is the point on the circular time scale at which the linear variable is maximum). If the period is also a parameter to be estimated, then the situation is more complex and one may resort to the broad area of *time series analysis* (e.g., Fisher, 1993: 172–189). Some biological rhythms can be fitted, by least square regression, by a sine (or cosine) curve; and if the rhythm does not conform well to such a symmetrical functional relationship, then a "harmonic analysis" (also called a "Fourier analysis") may be employed.

27.17   NONPARAMETRIC ANGULAR CORRELATION

Angular-Angular Correlation.   A nonparametric correlation procedure proposed by Mardia (1975) employs the ranks of circular measurements as follows. If pair i of circular data is denoted by measurements a_i and b_i, then these two statistics are computed:

$$r' = \frac{\left\{\sum_{i=1}^{n} \cos[C(\text{rank of } a_i - \text{rank of } b_i)]\right\}^2 + \left\{\sum_{i=1}^{n} \sin[C(\text{rank of } a_i - \text{rank of } b_i)]\right\}^2}{n^2}$$

(27.49)

---

*A rhythm with one cycle every twenty-four hours is said to be "circadian" (from the Latin "circa," meaning "about" and "diem," meaning "day"); a rhythm with a seven-day period is said to be "circaseptan"; a rhythm with a fourteen-day period is "circadiseptan"; one with a period of one year is "circannual" (Halberg and Lee, 1974).

[†]The amplitude is often defined as half this range.

$$r'' = \frac{\left\{ \sum\limits_{i=1}^{n} \cos[C(\text{rank of } a_i + \text{rank of } b_i)] \right\}^2 + \left\{ \sum\limits_{i=1}^{n} \sin[C(\text{rank of } a_i + \text{rank of } b_i)] \right\}^2}{n^2}$$

(27.50)

where

$$C = \frac{360°}{n};$$

(27.51)

and Fisher and Lee (1982) show that

$$(r_{aa})_s = r' - r''$$

(27.52)

is, for circular data, analogous to the Spearman rank correlation coefficient of Section 19.9. For n of 8 or more, one may calculate

$$(n - 1)(r_{aa})_s$$

and compare it to the critical value of

$$A + B/n,$$

using A and B from Table 27.1 (which yields excellent approximations to the values given by Fisher and Lee, 1982). This procedure is demonstrated in Example 27.23.

TABLE 27.1 Constants, B, and Critical Values, A, for Nonparametric Angular-Angular Correlation

$\alpha(2)$:	0.20	0.10	0.05	0.02	0.01
$\alpha(1)$:	0.10	0.05	0.025	0.01	0.005
A :	1.61	2.30	2.99	3.91	4.60
B :	1.52	2.00	2.16	1.60	1.60

Fisher and Lee (1982) also describe a nonparametric angular-angular correlation that is analogous to the Kendall rank correlation mentioned in Section 18.9 (see also Upton and Fingleton, 1989).

Angular-Linear Correlation. Mardia (1976) presented a ranking procedure for correlation between a circular and a linear variable, which is analogous to the Spearman rank correlation in Section 18.9. (See also Fisher, 1993: 140–141.) However, its behavior is not entirely satisfactory.

27.18 GOODNESS OF FIT TESTING FOR CIRCULAR DISTRIBUTIONS

Either χ^2 or G may be used to test the goodness of fit of a theoretical to an observed circular frequency distribution. (See Chapter 21 for general aspects of goodness of fit methods.) The procedure is to determine each expected frequency, $\hat{f}_i$, corresponding

EXAMPLE 27.23   Nonparametric angular-angular correlation.

For a population of birds sampled, we wish to correlate the direction toward which they attempt to fly in the morning with that in the evening.

$$H_0: (\rho_{aa})_s = 0; \ H_A: (\rho_{aa})_s \neq 0$$

$\alpha = 0.05$

Bird i	Direction Evening a_i	Direction Morning b_i	Rank of a_i	Rank of b_i	Rank difference	Rank sum
1	30°	60°	4.5	5	−0.5	9.5
2	10°	50°	2	4	−1	6
3	350°	10°	8	2	6	10
4	0°	350°	1	8	−7	9
5	340°	330°	7	7	0	14
6	330°	0°	6	1	5	7
7	20°	40°	3	3	0	6
8	30°	70°	4.5	6	−1.5	10.5

$n = 8; \ C = 360°/n = 45°$

$$r' = \frac{\left\{\sum\limits_{i=1}^{n} \cos[C(\text{rank of } a_i - \text{rank of } b_i)]\right\}^2 + \left\{\sum\limits_{i=1}^{n} \sin[C(\text{rank of } a_i - \text{rank of } b_i)]\right\}^2}{n^2}$$

$$= \left(\{\cos[45°(-0.5)] + \cos[45°(-1)] + \cdots + \cos[45°(-1.5)]\}^2 \right.$$
$$\left. + \{\sin[45°(-0.5)] + \sin[45°(-1)] + \cdots + \sin[45°(-1.5)]\}^2\right)/8^2$$

$$= 0.3417$$

$$r'' = \frac{\left\{\sum\limits_{i=1}^{n} \cos[C(\text{rank of } a_i + \text{rank of } b_i)]\right\}^2 + \left\{\sum\limits_{i=1}^{n} \sin[C(\text{rank of } a_i + \text{rank of } b_i)]\right\}^2}{n^2}$$

$$= \left(\{\cos[45°(9.5)] + \cos[45°(6)] + \cdots + \cos[45°(10.5)]\}^2 \right.$$
$$\left. + \{\sin[45°(9.5)] + \sin[45°(6)] + \cdots + \sin[45°(10.5)]\}^2\right)/8^2$$

$$= 0.0316$$

$$(r_{aa})_s = r' - r'' = 0.3417 - 0.0316 = 0.3101$$
$$(n-1)(r_{aa})_s = (8-1)(0.3101) = 2.17$$

For $\alpha(2) = 0.05$ and $n = 8$, the critical value is estimated to be

$$A + B/n = 2.99 + 2.16/8 = 3.26.$$

As $2.17 < 3.26$, do not reject H_0.

EXAMPLE 27.23   **(continued)**

We may also compute critical values for other significance levels:

     for $\alpha(2) = 0.20 : 1.61 + 1.52/8 = 1.80$;

     for $\alpha(2) = 0.10 : 2.30 + 2.00/8 = 2.55$;

     therefore, $0.10 < P < 0.20$.

to each observed frequency, f_i, in each category, i. For the data of Example 26.3, for instance, we might hypothesize a uniform distribution of data among the twelve divisions of the data. The test of this hypothesis is presented in Example 27.24. Batschelet (1981: 72) recommends grouping the data so that no expected frequency is less than 4 in using chi-square. All of the k categories do not have to be the same size. If they are (as in Example 27.24, where each category is 30° wide), Fisher (1993: 67) recommends that n/k be at least 2.

EXAMPLE 27.24   **Chi-square goodness of fit for the circular data of Example 26.3.**

H_0:   The data in the population are distributed uniformly around the circle.

H_A: The data in the population are not distributed uniformly around the circle.

a_i (deg)	f_i	$\hat{f}_i$
0–30	0	8.7500
30–60	6	8.7500
60–90	9	8.7500
90–120	13	8.7500
120–150	15	8.7500
150–180	22	8.7500
180–210	17	8.7500
210–240	12	8.7500
240–270	8	8.7500
270–300	3	8.7500
300–330	0	8.7500
330–360	0	8.7500

$k = 12$ $n = 105$

$\hat{f}_i = 105/12 = 8.7500$ for all i

$$\chi^2 = \frac{(0 - 8.7500)^2}{8.7500} + \frac{(6 - 8.7500)^2}{8.7500} + \frac{(9 - 8.7500)^2}{8.7500} + \cdots + \frac{(0 - 8.7500)^2}{8.7500}$$

$$= 8.7500 + 0.8643 + 0.0071 + \cdots + 8.7500$$

$$= 66.543$$

$\nu = k - 1 = 11$

$\chi^2_{0.05, 11} = 19.675$

Reject H_0. $P \ll 0.001$. $[P = 0.00000000055]$

Recall that goodness to fit testing by the chi-square or G statistic does not take into account the sequence of categories that occurs in the data distribution. In Sections 22.9 and 22.10, the Kolmogorov-Smirnov test was introduced as an improvement over chi-square when the categories of data are, in fact, ordered. Unfortunately, the Kolmogorov-Smirnov test yields different results for different starting points on a circular scale; however, a modification of this test by Kuiper (1960) provides a goodness of fit test, the results of which are unrelated to the starting point on a circle.

If data are not grouped, the Kuiper test is preferred to the chi-square procedure. It is discussed by Batschelet (1965: 26–27; 1981: 76–79), Fisher (1993: 66–67), and Mardia (1972a: Section 7.2.1).

Another goodness of fit test applicable to circular distributions of ungrouped data is that of Watson (1961; 1962), often referred to as the *Watson one-sample U^2 test*, which is demonstrated in Example 27.25. To test the null hypothesis of uniformity, we first transform each angular measurement, a_i, by dividing it by $360°$:

$$u_i = \frac{a_i}{360°}. \tag{27.53}$$

Then the following quantities are obtained for the set of n values of u_i : $\sum u_i, \sum u_i^2, \bar{u},$ and $\sum iu_i$. The test statistic, called "Watson's U^2," is

$$U^2 = \sum u_i^2 - \frac{\left(\sum u_i\right)^2}{n} - \frac{2}{n}\sum iu_i + (n+1)\bar{u} + \frac{n}{12} \tag{27.54}$$

(Mardia, 1972a: 182). Critical values for this test are $U^2_{a,n,n}$ in Appendix Table B.38. Lockhart and Stephens (1985) discuss the use of Watson's U^2 for goodness of fit to the von Mises distribution and provides tables for that application.

EXAMPLE 27.25   Watson's goodness of fit testing using the data of Example 26.2.

H_0:   The sample data come from a population uniformly distributed around the circle.

H_A: The sample data do not come from a population uniformly distributed around the circle.

i	a_i	u_i	u_i^2	iu_i
1	45°	0.1250	0.0156	0.1250
2	55°	0.1528	0.0233	0.3056
3	81°	0.2250	0.0506	0.6750
4	96°	0.2667	0.0711	1.0668
5	110°	0.3056	0.0934	1.5280
6	117°	0.3250	0.1056	1.9500
7	132°	0.3667	0.1345	2.5669
8	154°	0.4278	0.1830	3.4224

$n = 8$	$\sum u_i =$ 2.1946	$\sum u_i^2 =$ 0.6771	$\sum iu_i =$ 11.6397

EXAMPLE 27.25 (continued)

$$\bar{u} = \frac{\sum u_i}{n} = \frac{2.1946}{8} = 0.2743$$

$$U^2 = \sum u_i^2 - \frac{\left(\sum u_i\right)^2}{n} - \frac{2}{n}\sum i u_i + (n+1)\bar{u} + \frac{n}{12}$$

$$= 0.6771 - \frac{(2.1946)^2}{8} - \frac{2}{8}(11.6397) + (8+1)(0.2743) + \frac{8}{12}$$

$$= 0.6771 - 0.6020 - 2.9099 + 2.4687 + 0.6667$$

$$= 0.2989$$

$$U^2_{0.05,7,7} = 0.1986$$

Therefore, reject H_0. $0.002 < P < 0.005$.

27.19 SERIAL RANDOMNESS OF NOMINAL-SCALE CATEGORIES ON A CIRCLE

When dealing with the occurrence of members of two nominal-scale categories along a linear space or time, the runs test of Section 25.6 is appropriate. A runs test is also available for spatial or temporal measurements that are on a circular scale (Ghent and Zar, 1992). This test may also be employed as a two-sample test, but the tests of Sections 27.5 and 27.6 are more powerful for that purpose; the circular runs test is best reserved for testing the hypothesis of random distribution of members of two categories around a circle.

We define a run on a circle as a sequence of like elements, bounded on each side by unlike elements. Similar to Section 25.6, we let n_1 be the total number of elements in the first category, n_2 the number of elements in the second category, and u the number of runs in the entire sequence of elements. For the runs test on a linear scale (Section 25.6), the number of runs may be even or odd; however, on a circle the number of runs is always even: half of the runs (i.e., $u/2$) consist of elements belonging to one of the categories, and there are also $u/2$ runs of elements of the other category. We shall let $u' = u/2$, namely the number of runs of elements in either one of the categories.

The null hypothesis may be tested by analyzing the following 2×2 contingency table (Stevens, 1939):

u'	$n_1 - u'$	n_1
$n_2 - u'$	$u' - 1$	$n_2 - 1$
n_2	$n_1 - 1$	$n_1 + n_2 - 1$

As demonstrated in Example 27.26, this should be done by the Fisher exact test of Section 24.10, especially if the sample sizes (n_1 and n_2) are small enough to employ the critical values of Appendix Table B.28. For that test, m_1, m_2, n, and f are as defined in Section 24.10 and, for a two-tailed test, the second pair of critical values from that table are used.

EXAMPLE 27.26   The two-tailed runs test on a circle.

Members of the two antelope species of Example 25.8 (referred to as species A and B) are observed drinking on the shore of a pond in the following sequence:

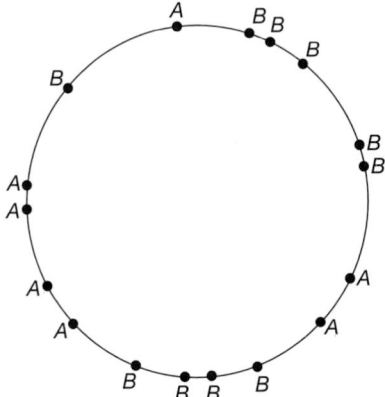

H_0:   The distribution of members of the two species around the pond is random.

H_A: The distribution of members of the two species around the pond is not random.

$n_1 = 7, n_2 = 10, u = 6, u' = 3$

3	4	7
7	2	9
10	6	16

To use Appendix Table B.28, $m_1 = 6, m_2 = 7, f = 4, n = 17$. For a two-tailed test, the critical values of f for $\alpha = 0.05$ are 0 and 5. Therefore, we do not reject H_0; $P \geq 0.20$.

If one or both sample sizes exceed the limits of that table, then a chi-square test with correction for continuity (Section 23.3) may be used; alternatively, in such a case we may employ a normal approximation (Stevens, 1939)[*]:

$$\mu_{u'} = \frac{n_1 n_2}{n_1 + n_2 - 1} \tag{27.55}$$

$$\sigma_{u'} = \sqrt{\frac{n_1 n_2 (n_1 n_2 - n_1 - n_2 + 1)}{(n_1 + n_2 - 1)^2 (n_1 + n_2 - 2)}} \tag{27.56}$$

---

[*]Ghent and Zar (1992) counsel that there have been in the literature several incorrect statements of the correct normal approximation.

$$Z = \frac{|u' - \mu'_u| - 0.5}{\sigma_{u'}} \tag{27.57}$$

and reject H_0 if $Z \geq Z_{\alpha(2)}$ from Appendix Table B.1.

Although this discussion and Example 27.26 depict a distribution around a circle, the testing procedure is appropriate if the physical arrangement of observations is in the shape of an ellipse, a rectangle, or any other closed figure—however irregular—providing that the figure is everywhere wider than the spacing of the elements along its periphery; and it may also be used for data that are conceptually circular, such as clock times or compass directions.

One-Tailed Testing.   For one-tailed testing we use the first pair of critical values in Appendix Table B.28. We can test specifically whether the population is nonrandom due to clustering (also known as contagion) in the following manner. We state H_0: In the population the members of each of the two groups are not clustered (i.e., not distributed contagiously) around the circle and H_A: In the population the members of each of the two groups are clustered (i.e., distributed contagiously) around the circle; and if $f \leq$ the first member of the pair of critical values, then H_0 is rejected. In using the normal approximation, this H_0 is rejected if $Z \geq Z_{\alpha(1)}$ and $u' \leq \mu_{u'}$.

If our interest is specifically whether the population distribution is nonrandom owing to a tendency toward being uniform, then we state H_0: In the population the members of each of the two groups do not tend to be uniformly distributed around the circle vs. H_A: In the population the members of each of the two groups tend to be uniformly distributed around the circle; and if $f \geq$ the second member of the pair of critical values then H_0 is rejected. If the normal approximation is employed, this H_0 is rejected if $Z \geq Z_{\alpha(1)}$ and $u' \geq \mu_{u'}$.

EXERCISES

27.1 Consider the data of Exercise 26.1. Test the null hypothesis that the population is distributed randomly around the circle (i.e., $\rho = 0$).

27.2 Consider the data of Exercise 26.2. Test the null hypothesis that time of birth is distributed randomly around the clock (i.e., $\rho = 0$).

27.3 Trees are planted in a circle to surround a cabin and protect it from prevailing west (i.e., $270°$) winds. The trees suffering the greatest wind damage are the eleven at the following directions.

(a) Using the V test, test the null hypothesis that tree damage is independent of wind direction, versus the alternate hypothesis that tree damage is concentrated around $270°$.

(b) Test H_0: $\mu_a = 270°$ vs. H_A: $\mu_a \neq 270°$.

285°	295°	335°
240	275	260
280	310	300
255	260	

27.4 Test nonparametrically for uniformity, using the data of Exercise 27.1.

27.5 Test nonparametrically for the data and experimental situation of Exercise 27.3.

27.6 The direction of the spring flight of a certain bird species was recorded as follows in eight individuals released in full sunlight and seven individuals released under overcast skies:

Sunny	*Overcast*
350°	340°
340	305
315	255
10	270
20	305
355	320
345	335
360	

Using the Watson-Williams test, test the null hypothesis that the mean flight direction in this species is the same under both cloudy and sunny skies.

27.7 Using the data of Example 27.6, test nonparametrically the hypothesis that birds of the species under consideration fly in the same direction under sunny as well as under cloudy skies.

27.8 Times of arrival at a feeding station of members of three species of hummingbirds were recorded as follows:

Species 1	*Species 2*	*Species 3*
05:40 hr	05:30 hr	05:35 hr
07:15	07:20	08:10
09:00	09:00	08:15
11:20	09:40	10:15
15:10	11:20	14:20
17:25	15:00	15:35
	17:05	16:05
	17:20	
	17:40	

Test the null hypothesis that members of all three species have the same mean time of visiting the feeding station.

27.9 For the data in Exercise 27.6, the birds were released at a site from which their home lies due north (i.e., in a compass direction of $0°$). Test whether birds orient homeward better under sunny skies than under cloudy skies.

27.10 For the data in Exercise 27.6, test whether the variability in flight direction is the same under both sky conditions.

27.11 Consider the data of Exercise 26.3. Test for significance of the mean of the five sample means.

27.12 Nonparametrically test whether the data of Exercise 26.3 came from a population uniformly distributed around the circle.

27.13 Test for significant difference between the mean of the data of Exercise 26.3 and the mean of the data below.

j	a_j	r_j
1	265°	0.5283
2	270	0.4119
3	240	0.6086
4	250	0.8402
5	240	0.7436
6	235	0.7145
7	260	0.6459
8	270	0.7481

27.14 Perform the nonparametric analogue of the test in Exercise 27.13.

27.15 The following data, for each of nine experimental animals, are the time of day when body temperature is greatest and the time of day when heart rate is greatest.
(a) Determine and test the correlation of these two times of day.
(b) Perform nonparametric correlation analysis on these data.

	Time of day	
Animal	Body temperature	Heart rate
i	a_i	b_i
1	09:50	10:40
2	10:20	09:30
3	11:40	11:10
4	08:40	08:30
5	09:10	08:40
6	10:50	09:10
7	13:20	12:50
8	13:10	13:30
9	12:40	13:00

27.16 For a sample of nine human births, the following are the times of day of the births and the ages of the mothers. Test whether there is correlation between these two variables.

Birth	Age (yr)	Time of day (hr:min)
i	X_i	a_i
1	23	06:25
2	22	07:20
3	19	07:05
4	25	08:15
5	28	15:40
6	24	09:25
7	31	18:20
8	17	07:30
9	27	16:10

27.17 Eight men (M) and eight women (W) were asked to sit around a circular conference table; they did so in the following configuration (see figure). Test whether there is evidence that members of the same sex tend to sit next to each other.

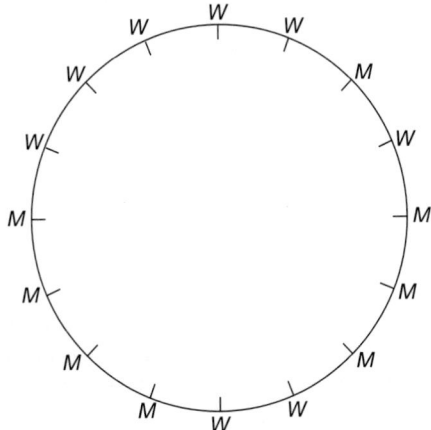

APPENDIX A

ANALYSIS OF VARIANCE HYPOTHESIS TESTING

When an analysis of variance experimental design has more than one factor, then the appropriate F values (and degrees of freedom) depend upon which of the factors are fixed and which are random effects (see Section 12.1), and which factors (if any) are nested within which (see Chapter 15). Given below in Section A.1 is a procedure that enables us to determine the appropriate F's and DF's for a given experimental design. It is a simplification of the procedures outlined by Bennett and Franklin (1954: 413–415), Dunn and Clark (1987: 218–222), Hicks (1982: 214–223), Kirk (1982: 389–394), Scheffé (1959: 284–289), and Winer, Brown, and Michels (1991: 369–377) and is the method used to produce the contents of Sections A.2 through A.5. If we have one of the experimental designs of Sections A.2–A.5, the F's and DF's are given therein; if the design is not found in those pages, then the procedures of Section A.1 may be used to determine the F's and DF's appropriate to the ANOVA.

A.1 DETERMINATION OF APPROPRIATE F'S AND DEGREES OF FREEDOM

We use as an example an analysis of variance design with three factors, A, B, and C, where factor A is a fixed effect, factor B is a random effect, and C is a random effect that is nested within combinations of factors A and B. This is the ANOVA that is the fourth example in Section A.4. The following steps are followed, and a table is prepared as indicated.

1. Assign to each factor a unique letter and subscript. If a factor is nested (or is a block in a split-plot design or is a subject in a mixed within-subjects design)

within one or more other factors, place the subscript(s) of the latter factor(s) in parentheses. For the present example, we would write A_i, B_j, $C_{l(ij)}$.

2. Prepare a table as follows:

   A. Row headings are the factors with their subscripts, the factor interactions (if any) with their subscripts, and a row for error (i.e., within cells) headed "e" with all factor subscripts in parentheses. Factors, interactions, and error will be referred to collectively as "effects."

   B. Column headings are the factor subscripts, with an indication of whether each is associated with a fixed-effects factor or a random-effects factor.

For our example:

Effect	Fixed i	Random j	Random l
A_i			
B_j			
$[AB]_{ij}$			
$C_{l(ij)}$			
$e_{(ijl)}$			

3. The body of the table is filled in as follows:

   A. Examine each column corresponding to a fixed-effects factor. (In our example, only column i is so examined.) For each such column, enter a "0" in each row that the column subscript appears outside parentheses in the row subscript. (In our example, enter "0" in rows A_i and $[AB]_{ij}$ of column i.)

   B. Enter "1" in every other position of the table.

For our example:

Effect	Fixed i	Random j	Random l
A_i	0	1	1
B_j	1	1	1
$[AB]_{ij}$	0	1	1
$C_{l(ij)}$	1	1	1
$e_{(ijl)}$	1	1	1

4. For each row, list all effects that contain all the subscripts of the row heading. In our example:

Effect	Fixed i	Random j	Random l	Effect list
A_i	0	1	1	$A + AB + C + e$
B_j	1	1	1	$B + AB + C + e$
$[AB]_{ij}$	0	1	1	$AB + C + e$
$C_{l(ij)}$	1	1	1	$C + e$
$e_{(ijl)}$	1	1	1	e

5. For each row, examine the list of factors and interactions just prepared, as follows:
   A. Ignore those columns headed by the subscripts in the row heading.
   B. Locate the table row corresponding to each factor or interaction in the list; if there is a zero in that row (ignoring the appropriate columns), then delete that factor or interaction in from the list.

In our example, we examine row A_i, by ignoring column i. Our factor and interaction list consists of B, C, AB, and e. None of these rows contain zeros, so we retain the entire list.

We examine row B_j, by ignoring column j. Our list of factors and interactions consists of B, C, AB, and e. Rows B_j, $C_{l(ij)}$, and e_{ijl} contain no zeros, but row $[AB]_{ij}$ has a zero. Therefore, we delete AB from our list for row B_j.

We similarly examine row $[AB]_{ij}$, by ignoring columns i and j; row $C_{l(ij)}$, by ignoring columns i, j, and l; and row $e_{(ijl)}$, by ignoring columns i, j, and l. No items are deleted from the factor and interaction lists for these rows.

As a result we have the following:

Effect	Effect list
A_i	$A + AB + C + e$
B_j	$B + C + e$
$[AB]_{ij}$	$AB + C + e$
$C_{l(ij)}$	$C + e$
$e_{(ijl)}$	e

6. The appropriate F is determined as follows:
   A. The numerator for F is the mean square (MS) for the effect (factor or interaction) in question. The numerator degrees of freedom (DF) are the DF associated with the numerator MS.
   B. The denominator for F is the mean square for the effect having the same effect list as does the numerator effect, with the exception of not having the numerator effect in the effect list. The denominator degrees of freedom are the DF associated with the denominator MS.

To test for the significance of factor A in our example, MS_A would be the numerator. As factor A has an effect list of A, AB, C, e, we desire in the denominator the mean square for the effect having an effect list of AB, C, e; therefore, the denominator is MS_{AB} and

$$F = \frac{MS_A}{MS_{AB}}.$$

To test for the significance of factor B, place MS_B in the numerator of F. Because factor B has an effect list of B, C, e, the denominator of F needs to be the mean square for the effect having an effect list of C, e, meaning that the denominator is to be MS_C, and

$$F = \frac{MS_B}{MS_C}.$$

To test for the significance of factor interaction AB, we place MS_{AB} in the numerator of F; and, as AB has an effect list of AB, C, e, we need in the denominator the MS for the effect with an effect list of C, e; so we use MS_C in the denominator, and

$$F = \frac{MS_{AB}}{MS_C}.$$

To test for the significance of factor C, place MS_C in the numerator of F. As factor C has an effect list of C, e, the denominator should be the MS for the effect containing only e in its effect list. Therefore,

$$F = \frac{MS_C}{MS_e}.$$

Occasionally, as in the third and fourth examples of Section A.3, there is no single effect that has the effect list required in step 6.B above. If this is the case, then a combination of effect lists may be considered as an approximate procedure (Satterthwaite, 1946).

For the third example in that section, steps 1–5 above yield the following effect list:

Effect	Effect list
A_i	$A + AB + AC + ABC + e$
B_j	$B + BC + e$
C_l	$C + BC + e$
$[AB]_{ij}$	$AB + ABC + e$
$[AC]_{ij}$	$AC + ABC + e$
$[BC]_{jl}$	$BC + e$
$[ABC]_{ijl}$	$ABC + e$
$e_{(ijl)}$	e

To test for the significance of factor A in that example, we place MS_A in the numerator of F and observe that the associated effect list is A, AB, AC, ABC, e, and we require a denominator MS associated with an effect list of AB, AC, ABC, e. We note that if we add the effect lists for effect AB (namely AB, ABC, e) and effect AC (AC, ABC, e) and then subtract that for effect ABC (namely, ABC, e), we have the desired list (AB, AC, ABC, e). Thus, we place in the denominator of F the combination of mean squares associated with the combination of effect lists used, namely $MS_{AB} + MS_{AC} - MS_{ABC}$. Therefore,

$$F = \frac{MS_A}{MS_{AB} + MS_{AC} - MS_{ABC}}.$$

If a combination of mean squares is used as the denominator of F, then the denominator degrees of freedom are

$$\frac{(\text{denominator of } F)^2}{\sum_i [(MS_i)^2/DF_i]},$$

where the summation takes place over all the mean squares in the denominator of F.

For our example, the numerator DF for testing the significance of factor A is DF_A, and the denominator DF is

$$\frac{(MS_{AB} + MS_{AC} - MS_{ABC})^2}{\dfrac{(MS_{AB})^2}{DF_{AB}} + \dfrac{(MS_{AC})^2}{DF_{AC}} + \dfrac{(MS_{ABC})^2}{DF_{ABC}}}.$$

In the following series of tables, mean squares for factors are denoted as MS_A, MS_B, etc.; mean squares for interaction among factors are indicated as MS_{AB}, MS_{AC}, MS_{ABC}, etc.; and the error mean square is denoted by MS_e. Degrees of freedom (DF) are indicated with the same subscripts as the mean squares.

If an analysis of variance has only one replicate per cell, the calculation of MS_e is not possible, the quantity calculated as the "remainder" MS being the MS for the highest-order interaction. If one assumes that the highest-order interaction MS is insignificant, the remainder MS may be used in place of the error MS in the F calculation and other procedures where MS_e is called for.

A.2 TWO-FACTOR ANALYSIS OF VARIANCE

Factors A and B Both Fixed (See Example 12.1.)

Source of variation	F	v_1	v_2
A	MS_A/MS_e	DF_A	DF_e
B	MS_B/MS_e	DF_B	DF_e
AB	MS_{AB}/MS_e	DF_{AB}	DF_e

Factor A Fixed; Factor B Random (See Example 12.4.)

Source of variation	F	v_1	v_2
A	MS_A/MS_{AB}	DF_A	DF_{AB}
B	MS_B/MS_e	DF_B	DF_e
AB	MS_{AB}/MS_e	DF_{AB}	DF_e

Factors A and B Both Random

Source of variation	F	v_1	v_2
A	MS_A/MS_{AB}	DF_A	DF_{AB}
B	MS_B/MS_{AB}	DF_B	DF_{AB}
AB	MS_{AB}/MS_e	DF_{AB}	DF_e

A.3 THREE-FACTOR ANALYSIS OF VARIANCE

Factors A, B and C All Fixed (See Example 14.1.)

Source of variation	F	ν_1	ν_2
A	MS_A/MS_e	DF_A	DF_e
B	MS_B/MS_e	DF_B	DF_e
C	MS_C/MS_e	DF_C	DF_e
AB	MS_{AB}/MS_e	DF_{AB}	DF_e
AC	MS_{AC}/MS_e	DF_{AC}	DF_e
BC	MS_{BC}/MS_e	DF_{BC}	DF_e
ABC	MS_{ABC}/MS_e	DF_{ABC}	DF_e

Factors A and B Fixed; Factor C Random

Source of variation	F	ν_1	ν_2
A	MS_A/MS_{AC}	DF_A	DF_{AC}
B	MS_B/MS_{BC}	DF_B	DF_{BC}
C	MS_C/MS_e	DF_C	DF_e
AB	MS_{AB}/MS_{ABC}	DF_{AB}	DF_{ABC}
AC	MS_{AC}/MS_e	DF_{AC}	DF_e
BC	MS_{BC}/MS_e	DF_{BC}	DF_e
ABC	MS_{ABC}/MS_e	DF_{ABC}	DF_e

Factor A Fixed; Factors B and C Random

Source of variation	F	ν_1	ν_2
A	$MS_A/(MS_{AB} + MS_{AC} - MS_{ABC})$	DF_A	$\dfrac{(MS_{AB} + MS_{AC} - MS_{ABC})^2}{(MS_{AB})^2/DF_{AB} + (MS_{AC})^2/DF_{AC} + (MS_{ABC})^2/DF_{ABC}}$
B	MS_B/MS_{BC}	DF_B	DF_{BC}
C	MS_C/MS_{BC}	DF_C	DF_{BC}
AB	MS_{AB}/MS_{ABC}	DF_{AB}	DF_{ABC}
AC	MS_{AC}/MS_{ABC}	DF_{AC}	DF_{ABC}
BC	MS_{BC}/MS_e	DF_{BC}	DF_e
ABC	MS_{ABC}/MS_e	DF_{ABC}	DF_e

Factors A, B, C All Random

Source of variation	F	v_1	v_2
A	$MS_A/(MS_{AB}+MS_{AC}-MS_{ABC})$	DF_A	$\dfrac{(MS_{AB}+MS_{AC}-MS_{ABC})^2}{(MS_{AB})^2/DF_{AB}+(MS_{AC})^2/DF_{AC}+(MS_{ABC})^2/DF_{ABC}}$
B	$MS_B/(MS_{AB}+MS_{BC}-MS_{ABC})$	DF_B	$\dfrac{(MS_{AB}+MS_{BC}-MS_{ABC})^2}{(MS_{AB})^2/DF_{AB}+(MS_{BC})^2/DF_{BC}+(MS_{ABC})^2/DF_{ABC}}$
C	$MS_C/(MS_{AC}+MS_{BC}-MS_{ABC})$	DF_C	$\dfrac{(MS_{AC}+MS_{BC}-MS_{ABC})^2}{(MS_{AC})^2/DF_{AC}+(MS_{BC})^2/DF_{BC}+(MS_{ABC})^2/DF_{ABC}}$
AB	MS_{AB}/MS_{ABC}	DF_{AB}	DF_{ABC}
AC	MS_{AC}/MS_{ABC}	DF_{AC}	DF_{ABC}
BC	MS_{BC}/MS_{ABC}	DF_{BC}	DF_{ABC}
ABC	MS_{ABC}/MS_e	DF_{ABC}	DF_e

A.4 NESTED ANALYSIS OF VARIANCE

Factor A either Fixed or Random; Factor B Random and Nested within Factor A (See Example 15.1.)

Source of variation	F	v_1	v_2
A	MS_A/MS_B	DF_A	DF_B
B	MS_B/MS_e	DF_B	DF_e

Factor A Either Fixed or Random; Factor B Random and Nested within Factor A; Factor C Random and Nested within Factor B

Source of variation	F	v_1	v_2
A	MS_A/MS_B	DF_A	DF_B
B	MS_B/MS_C	DF_B	DF_C
C	MS_C/MS_e	DF_C	DF_e

Factors A and B Fixed; Factor C Random and Nested within Factors A and B (See Example 15.2.)

Source of variation	F	ν_1	ν_2
A	MS_A/MS_C	DF_A	DF_C
B	MS_B/MS_C	DF_B	DF_C
AB	MS_{AB}/MS_C	DF_{AB}	DF_C
C	MS_C/MS_e	DF_C	DF_e

Factor A Fixed; Factor B Random; Factor C Random and Nested within Factors A and B

Source of variation	F	ν_1	ν_2
A	MS_A/MS_{AB}	DF_A	DF_{AB}
B	MS_B/MS_C	DF_B	DF_C
AB	MS_{AB}/MS_C	DF_{AB}	DF_C
C	MS_C/MS_e	DF_C	DF_e

Factors A and B Random; Factor C Random and Nested within Factors A and B

Source of variation	F	ν_1	ν_2
A	MS_A/MS_{AB}	DF_A	DF_{AB}
B	MS_B/MS_{AB}	DF_B	DF_{AB}
AB	MS_{AB}/MS_C	DF_{AB}	DF_C
C	MS_C/MS_e	DF_C	DF_e

A.5 SPLIT-PLOT, AND MIXED WITHIN-SUBJECTS, ANALYSIS OF VARIANCE

Factor A Fixed and among Subjects; Factor B Fixed and within Subjects; Factor S is Subjects (or Blocks) and Random. (See Examples 14.2 and 14.3.)

Source of variation	F	ν_1	ν_2
A	$MS_A/MS_{S/A}$	DF_A	$DF_{S/A}$
B	$MS_B/MS_{B \times S/A}$	DF_B	$DF_{B \times S/A}$
BA	$MS_{BA}/MS_{B \times S/A}$	DF_{BA}	$DF_{B \times S/A}$

Factor A Fixed and among Subjects; Factor B Fixed and among Subjects; Factor C Fixed and within Subjects; Factor S is Subjects (or Blocks) and Random. (See Example 14.4.)

Source of variation	F	ν_1	ν_2
A	$MS_A/MS_{S/AB}$	DF_A	$DF_{S/AB}$
B	$MS_B/MS_{S/AB}$	DF_B	$DF_{S/AB}$
AB	$MS_{AB}/MS_{S/AB}$	DF_{AB}	$DF_{S/AB}$
C	$MS_C/MS_{C\times S/AB}$	DF_C	$DF_{C\times S/AB}$
CA	$MS_{CA}/MS_{C\times S/AB}$	DF_{CA}	$DF_{C\times S/AB}$
CB	$MS_{CB}/MS_{C\times S/AB}$	DF_{CB}	$DF_{C\times S/AB}$
CAB	$MS_{CAB}/MS_{C\times S/AB}$	DF_{CAB}	$DF_{C\times S/AB}$

Factor A Fixed and among Subjects; Factor B Fixed and within Subjects; Factor C Fixed and within Subjects; Factor S is Subjects (or Blocks) and Random. (See Example 14.5.)

Source of variation	F	ν_1	ν_2
A	$MS_A/MS_{S/A}$	DF_A	$DF_{S/A}$
B	$MS_B/MS_{B\times S/A}$	DF_B	$DF_{B\times S/A}$
BA	$MS_{BA}/MS_{B\times S/A}$	DF_{BA}	$DF_{B\times S/A}$
C	$MS_C/MS_{C\times S/A}$	DF_C	$DF_{C\times S/A}$
CA	$MS_{CA}/MS_{C\times S/A}$	DF_{CA}	$DF_{C\times S/A}$
CB	$MS_{CB}/MS_{C\times S/A}$	DF_{CB}	$DF_{C\times S/A}$
CBA	$MS_{CBA}/MS_{B\times C\times S/A}$	DF_{CBA}	$DF_{B\times C\times S/A}$

APPENDIX B

STATISTICAL TABLES AND GRAPHS

INTERPOLATION

In some of the statistical tables that follow (viz. Appendix Tables B.3, B.4, B.5, B.6, B.7, B.17), a critical value may be required for degrees of freedom not shown on the table, and interpolation may be used to compute an estimate of the required critical value.

Linear Interpolation. Let us say that we desire a critical value for v degrees of freedom, where $a < v < b$; let us call it C_v. We first determine the proportion $p = (v-a)/(b-a)$. Then, the required critical value is determined as $C_v = C_b + p(C_a - C_b)$.

For example, let us consider $F_{0.05(2),1,260}$, which lies between $F_{0.05(2),1,200} = 5.10$ and $F_{0.05(2),1,300} = 5.07$ in Appendix Table B.4. We calculate $p = (260 - 200)/(300 - 200) = 0.600$; then $F_{0.05(2),1,260} = C_v = 5.07 + (0.600)(5.10 - 5.07) = 5.07 + 0.02 = 5.09$.

Linear interpolation cannot be used when $b = \infty$, but harmonic interpolation (below) can be.

Harmonic Interpolation. A more accurate interpolation procedure for critical value determination is one that uses the reciprocals of degrees of freedom. (Because reciprocals of large numbers are small numbers, it is good practice to use 100 times each reciprocal, and this is what will be demonstrated here.) Let us say we desire a critical value for v degrees of freedom, where $a < v < b$; call it C_v. We first determine $p = (100/a - 100/v)/(100/a - 100/b)$. Then, $C_v = C_b + (1 - p)(C_a - C_b)$.

App10

For example, let us consider the above example of desiring $F_{0.05(2),1,260}$. We calculate $p = (100/200 - 100/260)/(100/200 - 100/300) = 0.692$; then, $F_{0.05(2),1,260} = C_v = 5.07 + (1 - 0.692)(5.10 - 5.07) = 5.07 + 0.01 = 5.08$.

Harmonic interpolation is especially useful when $b = \infty$. For example, to determine $t_{0.01(2),2800}$, which lies between $t_{0.01(2),1000} = 2.581$ and $t_{0.01(1),\infty} = 2.5758$ in Appendix Table B.3, we calculate $p = (100/1000 - 100/2800)/(100/1000 - 100/\infty) = 0.0643/0.1000 = 0.6430$. Then, $t_{0.01(2),2800} = C_v = 2.5758 + (1 - 0.6430)(2.581 - 2.5758) = 2.5758 + 0.0019 = 2.578$. (Note that $100/\infty = 0$).

TABLE B.1 Critical Values of the Chi-Square Distribution

ν	α: 0.999	0.995	0.99	0.975	0.95	0.90	0.75	0.50	0.25	0.10	0.05	0.025	0.01	0.005	0.001
1	0.000	0.000	0.000	0.001	0.004	0.016	0.102	0.455	1.323	2.706	3.841	5.024	6.635	7.879	10.828
2	0.002	0.010	0.020	0.051	0.103	0.211	0.575	1.386	2.773	4.605	5.991	7.378	9.210	10.597	13.816
3	0.024	0.072	0.115	0.216	0.352	0.584	1.213	2.366	4.108	6.251	7.815	9.348	11.345	12.838	16.266
4	0.091	0.207	0.297	0.484	0.711	1.064	1.923	3.357	5.385	7.779	9.488	11.143	13.277	14.860	18.467
5	0.210	0.412	0.554	0.831	1.145	1.610	2.675	4.351	6.626	9.236	11.070	12.833	15.086	16.750	20.515
6	0.381	0.676	0.872	1.237	1.635	2.204	3.455	5.348	7.841	10.645	12.592	14.449	16.812	18.548	22.458
7	0.599	0.989	1.239	1.690	2.167	2.833	4.255	6.346	9.037	12.017	14.067	16.013	18.475	20.278	24.322
8	0.857	1.344	1.646	2.180	2.733	3.490	5.071	7.344	10.219	13.362	15.507	17.535	20.090	21.955	26.124
9	1.152	1.735	2.088	2.700	3.325	4.168	5.899	8.343	11.389	14.684	16.919	19.023	21.666	23.589	27.877
10	1.479	2.156	2.558	3.247	3.940	4.865	6.737	9.342	12.549	15.987	18.307	20.483	23.209	25.188	29.588
11	1.834	2.603	3.053	3.816	4.575	5.578	7.584	10.341	13.701	17.275	19.675	21.920	24.725	26.757	31.264
12	2.214	3.074	3.571	4.404	5.226	6.304	8.438	11.340	14.845	18.549	21.026	23.337	26.217	28.300	32.909
13	2.617	3.565	4.107	5.009	5.892	7.042	9.299	12.340	15.984	19.812	22.362	24.736	27.688	29.819	34.528
14	3.041	4.075	4.660	5.629	6.571	7.790	10.165	13.339	17.117	21.064	23.685	26.119	29.141	31.319	36.123
15	3.483	4.601	5.229	6.262	7.261	8.547	11.037	14.339	18.245	22.307	24.996	27.488	30.578	32.801	37.697
16	3.942	5.142	5.812	6.908	7.962	9.312	11.912	15.338	19.369	23.542	26.296	28.845	32.000	34.267	39.252
17	4.416	5.697	6.408	7.564	8.672	10.085	12.792	16.338	20.489	24.769	27.587	30.191	33.409	35.718	40.790
18	4.905	6.265	7.015	8.231	9.390	10.865	13.675	17.338	21.605	25.989	28.869	31.526	34.805	37.156	42.312
19	5.407	6.844	7.633	8.907	10.117	11.651	14.562	18.338	22.718	27.204	30.144	32.852	36.191	38.582	43.820
20	5.921	7.434	8.260	9.591	10.851	12.443	15.452	19.337	23.828	28.412	31.410	34.170	37.566	39.997	45.315
21	6.447	8.034	8.897	10.283	11.591	13.240	16.344	20.337	24.935	29.615	32.671	35.479	38.932	41.401	46.797
22	6.983	8.643	9.542	10.982	12.338	14.041	17.240	21.337	26.039	30.813	33.924	36.781	40.289	42.796	48.268
23	7.529	9.260	10.196	11.689	13.091	14.848	18.137	22.337	27.141	32.007	35.172	38.076	41.638	44.181	49.728
24	8.085	9.886	10.856	12.401	13.848	15.659	19.037	23.337	28.241	33.196	36.415	39.364	42.980	45.559	51.179
25	8.649	10.520	11.524	13.120	14.611	16.473	19.939	24.337	29.339	34.382	37.652	40.646	44.314	46.928	52.620
26	9.222	11.160	12.198	13.844	15.379	17.292	20.843	25.336	30.435	35.563	38.885	41.923	45.642	48.290	54.052
27	9.803	11.808	12.879	14.573	16.151	18.114	21.749	26.336	31.528	36.741	40.113	43.195	46.963	49.645	55.476
28	10.391	12.461	13.565	15.308	16.928	18.939	22.657	27.336	32.620	37.916	41.337	44.461	48.278	50.993	56.892
29	10.986	13.121	14.256	16.047	17.708	19.768	23.567	28.336	33.711	39.087	42.557	45.722	49.588	52.336	58.301
30	11.588	13.787	14.953	16.791	18.493	20.599	24.478	29.336	34.800	40.256	43.773	46.979	50.892	53.672	59.703
31	12.196	14.458	15.655	17.539	19.281	21.434	25.390	30.336	35.887	41.422	44.985	48.232	52.191	55.003	61.098
32	12.811	15.134	16.362	18.291	20.072	22.271	26.304	31.336	36.973	42.585	46.194	49.480	53.486	56.328	62.487
33	13.431	15.815	17.074	19.047	20.867	23.110	27.219	32.336	38.058	43.745	47.400	50.725	54.776	57.648	63.870
34	14.057	16.501	17.789	19.806	21.664	23.952	28.136	33.336	39.141	44.903	48.602	51.966	56.061	58.964	65.247
35	14.688	17.192	18.509	20.569	22.465	24.797	29.054	34.336	40.223	46.059	49.802	53.203	57.342	60.275	66.619

TABLE B.1 (cont.) Critical Values of the Chi-Square Distribution

v	α: 0.999	0.995	0.99	0.975	0.95	0.90	0.75	0.50	0.25	0.10	0.05	0.025	0.01	0.005	0.001
36	15.324	17.887	19.233	21.336	23.269	25.643	29.973	35.336	41.304	47.212	50.998	54.437	58.619	61.581	67.985
37	15.965	18.586	19.960	22.106	24.075	26.492	30.893	36.336	42.383	48.363	52.192	55.668	59.893	62.883	69.346
38	16.611	19.289	20.691	22.878	24.884	27.343	31.815	37.335	43.462	49.513	53.384	56.896	61.162	64.181	70.703
39	17.262	19.996	21.426	23.654	25.695	28.196	32.737	38.335	44.539	50.660	54.572	58.120	62.428	65.476	72.055
40	17.916	20.707	22.164	24.433	26.509	29.051	33.660	39.335	45.616	51.805	55.758	59.342	63.691	66.766	73.402
41	18.576	21.421	22.906	25.215	27.326	29.907	34.585	40.335	46.692	52.949	56.942	60.561	64.950	68.053	74.745
42	19.239	22.138	23.650	25.999	28.144	30.765	35.510	41.335	47.766	54.090	58.124	61.777	66.206	69.336	76.084
43	19.906	22.859	24.398	26.785	28.965	31.625	36.436	42.335	48.840	55.230	59.304	62.990	67.459	70.616	77.419
44	20.576	23.584	25.148	27.575	29.787	32.487	37.363	43.335	49.913	56.369	60.481	64.201	68.710	71.893	78.750
45	21.251	24.311	25.901	28.366	30.612	33.350	38.291	44.335	50.985	57.505	61.656	65.410	69.957	73.166	80.077
46	21.929	25.041	26.657	29.160	31.439	34.215	39.220	45.335	52.056	58.641	62.830	66.617	71.201	74.437	81.400
47	22.610	25.775	27.416	29.956	32.268	35.081	40.149	46.335	53.127	59.774	64.001	67.821	72.443	75.704	82.720
48	23.295	26.511	28.177	30.755	33.098	35.949	41.079	47.335	54.196	60.907	65.171	69.023	73.683	76.969	84.037
49	23.983	27.249	28.941	31.555	33.930	36.818	42.010	48.335	55.265	62.038	66.339	70.222	74.919	78.231	85.351
50	24.674	27.991	29.707	32.357	34.764	37.689	42.942	49.335	56.334	63.167	67.505	71.420	76.154	79.490	86.661
51	25.368	28.735	30.475	33.162	35.600	38.560	43.874	50.335	57.401	64.295	68.669	72.616	77.386	80.747	87.968
52	26.065	29.481	31.246	33.968	36.437	39.433	44.808	51.335	58.468	65.422	69.832	73.810	78.616	82.001	89.272
53	26.765	30.230	32.019	34.776	37.276	40.308	45.741	52.335	59.534	66.548	70.993	75.002	79.843	83.253	90.573
54	27.468	30.981	32.793	35.586	38.116	41.183	46.676	53.335	60.600	67.673	72.153	76.192	81.069	84.502	91.872
55	28.173	31.735	33.570	36.398	38.958	42.060	47.610	54.335	61.665	68.796	73.311	77.380	82.292	85.749	93.168
56	28.881	32.491	34.350	37.212	39.801	42.937	48.546	55.335	62.729	69.919	74.468	78.567	83.513	86.994	94.461
57	29.592	33.248	35.131	38.027	40.646	43.816	49.482	56.335	63.793	71.040	75.624	79.752	84.733	88.236	95.751
58	30.305	34.008	35.913	38.844	41.492	44.696	50.419	57.335	64.857	72.160	76.778	80.936	85.950	89.477	97.039
59	31.021	34.770	36.698	39.662	42.339	45.577	51.356	58.335	65.919	73.279	77.931	82.117	87.166	90.715	98.324
60	31.738	35.535	37.485	40.482	43.188	46.459	52.294	59.335	66.981	74.397	79.082	83.298	88.379	91.952	99.607
61	32.459	36.301	38.273	41.303	44.038	47.342	53.232	60.335	68.043	75.514	80.232	84.476	89.591	93.186	100.888
62	33.181	37.068	39.063	42.126	44.889	48.226	54.171	61.335	69.104	76.630	81.381	85.654	90.802	94.419	102.166
63	33.906	37.838	39.855	42.950	45.741	49.111	55.110	62.335	70.165	77.745	82.529	86.830	92.010	95.649	103.442
64	34.633	38.610	40.649	43.776	46.595	49.996	56.050	63.335	71.225	78.860	83.675	88.004	93.217	96.878	104.716
65	35.362	39.383	41.444	44.603	47.450	50.883	56.990	64.335	72.285	79.973	84.821	89.177	94.422	98.105	105.988
66	36.093	40.158	42.240	45.431	48.305	51.770	57.931	65.335	73.344	81.085	85.965	90.349	95.626	99.330	107.258
67	36.826	40.935	43.038	46.261	49.162	52.659	58.872	66.335	74.403	82.197	87.108	91.519	96.828	100.554	108.526
68	37.561	41.713	43.838	47.092	50.020	53.548	59.814	67.335	75.461	83.308	88.250	92.689	98.028	101.776	109.791
69	38.298	42.494	44.639	47.924	50.879	54.438	60.756	68.334	76.519	84.418	89.391	93.856	99.228	102.996	111.055
70	39.036	43.275	45.442	48.758	51.739	55.329	61.698	69.334	77.577	85.527	90.531	95.023	100.425	104.215	112.317

TABLE B.1 (cont.) Critical Values of the Chi-Square Distribution

α:	0.999	0.995	0.99	0.975	0.95	0.90	0.75	0.50	0.25	0.10	0.05	0.025	0.01	0.00	0.001
ν															
71	39.777	44.058	46.246	49.592	52.600	56.221	62.641	70.334	78.634	86.635	91.670	96.189	101.621	105.432	113.577
72	40.520	44.843	47.051	50.428	53.462	57.113	63.585	71.334	79.690	87.743	92.808	97.353	102.816	106.648	114.835
73	41.264	45.629	47.858	51.265	54.325	58.006	64.528	72.334	80.747	88.850	93.945	98.516	104.010	107.862	116.092
74	42.010	46.417	48.666	52.103	55.189	58.900	65.472	73.334	81.803	89.956	95.081	99.678	105.202	109.074	117.346
75	42.757	47.206	49.475	52.942	56.054	59.795	66.417	74.334	82.858	91.061	96.217	100.839	106.393	110.286	118.599
76	43.507	47.997	50.286	53.782	56.920	60.690	67.362	75.334	83.913	92.166	97.351	101.999	107.583	111.495	119.850
77	44.258	48.788	51.097	54.623	57.786	61.586	68.307	76.334	84.968	93.270	98.484	103.158	108.771	112.704	121.100
78	45.010	49.582	51.910	55.466	58.654	62.483	69.252	77.334	86.022	94.374	99.617	104.316	109.958	113.911	122.348
79	45.764	50.376	52.725	56.309	59.522	63.380	70.198	78.334	87.077	95.476	100.749	105.473	111.144	115.117	123.594
80	46.520	51.172	53.540	57.153	60.391	64.278	71.145	79.334	88.130	96.578	101.879	106.629	112.329	116.321	124.839
81	47.277	51.969	54.357	57.998	61.261	65.176	72.091	80.334	89.184	97.680	103.010	107.783	113.512	117.524	126.083
82	48.036	52.767	55.174	58.845	62.132	66.076	73.038	81.334	90.237	98.780	104.139	108.937	114.695	118.726	127.324
83	48.796	53.567	55.993	59.692	63.004	66.976	73.985	82.334	91.289	99.880	105.267	110.090	115.876	119.927	128.565
84	49.557	54.368	56.813	60.540	63.876	67.876	74.933	83.334	92.342	100.980	106.395	111.242	117.057	121.126	129.804
85	50.320	55.170	57.634	61.389	64.749	68.777	75.881	84.334	93.394	102.079	107.522	112.393	118.236	122.325	131.041
86	51.085	55.973	58.456	62.239	65.623	69.679	76.829	85.334	94.446	103.177	108.648	113.544	119.414	123.522	132.277
87	51.850	56.777	59.279	63.089	66.498	70.581	77.777	86.334	95.497	104.275	109.773	114.693	120.591	124.718	133.512
88	52.617	57.582	60.103	63.941	67.373	71.484	78.726	87.334	96.548	105.372	110.898	115.841	121.767	125.913	134.745
89	53.386	58.389	60.928	64.793	68.249	72.387	79.675	88.334	97.599	106.469	112.022	116.989	122.942	127.106	135.978
90	54.155	59.196	61.754	65.647	69.126	73.291	80.625	89.334	98.650	107.565	113.145	118.136	124.116	128.299	137.208
91	54.926	60.005	62.581	66.501	70.003	74.196	81.574	90.334	99.700	108.661	114.268	119.282	125.289	129.491	138.438
92	55.698	60.815	63.409	67.356	70.882	75.100	82.524	91.334	100.750	109.756	115.390	120.427	126.462	130.681	139.666
93	56.472	61.625	64.238	68.211	71.760	76.006	83.474	92.334	101.800	110.850	116.511	121.571	127.633	131.871	140.893
94	57.246	62.437	65.068	69.068	72.640	76.912	84.425	93.334	102.850	111.944	117.632	122.715	128.803	133.059	142.119
95	58.022	63.250	65.898	69.925	73.520	77.818	85.376	94.334	103.899	113.038	118.752	123.858	129.973	134.247	143.344
96	58.799	64.063	66.730	70.783	74.401	78.725	86.327	95.334	104.948	114.131	119.871	125.000	131.141	135.433	144.567
97	59.577	64.878	67.562	71.642	75.282	79.633	87.278	96.334	105.997	115.223	120.990	126.141	132.309	136.619	145.789
98	60.356	65.694	68.396	72.501	76.164	80.541	88.229	97.334	107.045	116.315	122.108	127.282	133.476	137.803	147.010
99	61.137	66.510	69.230	73.361	77.046	81.449	89.181	98.334	108.093	117.407	123.225	128.422	134.642	138.987	148.230
100	61.918	67.328	70.065	74.222	77.929	82.358	90.133	99.334	109.141	118.498	124.342	129.561	135.807	140.169	149.449
101	62.701	68.146	70.901	75.083	78.813	83.267	91.085	100.334	110.189	119.589	125.458	130.700	136.971	141.351	150.667
102	63.484	68.965	71.737	75.946	79.697	84.177	92.038	101.334	111.236	120.679	126.574	131.838	138.134	142.532	151.884
103	64.269	69.785	72.575	76.809	80.582	85.088	92.991	102.334	112.284	121.769	127.689	132.975	139.297	143.712	153.099
104	65.055	70.606	73.413	77.672	81.468	85.998	93.944	103.334	113.331	122.858	128.804	134.111	140.459	144.891	154.314
105	65.841	71.428	74.252	78.536	82.354	86.909	94.897	104.334	114.378	123.947	129.918	135.247	141.620	146.070	155.528

TABLE B.1 (cont.) Critical Values of the Chi-Square Distribution

ν	α: 0.999	0.995	0.99	0.975	0.95	0.90	0.75	0.50	0.25	0.10	0.05	0.025	0.01	0.00	0.001
106	66.629	72.251	75.092	79.401	83.240	87.821	95.850	105.334	115.424	125.035	131.031	136.382	142.780	147.247	156.740
107	67.418	73.075	75.932	80.267	84.127	88.733	96.804	106.334	116.471	126.123	132.144	137.517	143.940	148.424	157.952
108	68.207	73.899	76.774	81.133	85.015	89.645	97.758	107.334	117.517	127.211	133.257	138.651	145.099	149.599	159.162
109	68.998	74.724	77.616	82.000	85.903	90.558	98.712	108.334	118.563	128.298	134.369	139.784	146.257	150.774	160.372
110	69.790	75.550	78.458	82.867	86.792	91.471	99.666	109.334	119.608	129.385	135.480	140.917	147.414	151.948	161.581
111	70.582	76.377	79.302	83.735	87.681	92.385	100.620	110.334	120.654	130.472	136.591	142.049	148.571	153.122	162.788
112	71.376	77.204	80.146	84.604	88.570	93.299	101.575	111.334	121.699	131.558	137.701	143.180	149.727	154.294	163.995
113	72.170	78.033	80.991	85.473	89.461	94.213	102.530	112.334	122.744	132.643	138.811	144.311	150.882	155.466	165.201
114	72.965	78.862	81.836	86.342	90.351	95.128	103.485	113.334	123.789	133.729	139.921	145.441	152.037	156.637	166.406
115	73.761	79.692	82.682	87.213	91.242	96.043	104.440	114.334	124.834	134.813	141.030	146.571	153.191	157.808	167.610
116	74.558	80.522	83.529	88.084	92.134	96.958	105.396	115.334	125.878	135.898	142.138	147.700	154.344	158.977	168.813
117	75.356	81.353	84.377	88.955	93.026	97.874	106.352	116.334	126.923	136.982	143.246	148.829	155.496	160.146	170.016
118	76.155	82.185	85.225	89.827	93.918	98.790	107.307	117.334	127.967	138.066	144.354	149.957	156.648	161.314	171.217
119	76.955	83.018	86.074	90.700	94.811	99.707	108.263	118.334	129.011	139.149	145.461	151.084	157.800	162.481	172.418
120	77.755	83.852	86.923	91.573	95.705	100.624	109.220	119.334	130.055	140.233	146.567	152.211	158.950	163.648	173.617
121	78.557	84.686	87.773	92.446	96.598	101.541	110.176	120.334	131.098	141.315	147.674	153.338	160.100	164.814	174.816
122	79.359	85.521	88.624	93.320	97.493	102.458	111.133	121.334	132.142	142.398	148.779	154.464	161.250	165.980	176.014
123	80.162	86.356	89.475	94.195	98.387	103.376	112.089	122.334	133.185	143.480	149.885	155.589	162.398	167.144	177.212
124	80.965	87.192	90.327	95.070	99.283	104.295	113.046	123.334	134.228	144.562	150.989	156.714	163.546	168.308	178.408
125	81.770	88.029	91.180	95.946	100.178	105.213	114.004	124.334	135.271	145.643	152.094	157.839	164.694	169.471	179.604
126	82.575	88.866	92.033	96.822	101.074	106.132	114.961	125.334	136.313	146.724	153.198	158.962	165.841	170.634	180.799
127	83.381	89.704	92.887	97.698	101.971	107.051	115.918	126.334	137.356	147.805	154.302	160.086	166.987	171.796	181.993
128	84.188	90.543	93.741	98.576	102.867	107.971	116.876	127.334	138.398	148.885	155.405	161.209	168.133	172.957	183.186
129	84.996	91.383	94.596	99.453	103.765	108.891	117.834	128.334	139.440	149.965	156.508	162.331	169.278	174.118	184.379
130	85.804	92.223	95.451	100.331	104.662	109.811	118.792	129.334	140.482	151.045	157.610	163.453	170.423	175.278	185.571
131	86.613	93.063	96.307	101.210	105.560	110.732	119.750	130.334	141.524	152.125	158.712	164.575	171.567	176.438	186.762
132	87.423	93.904	97.163	102.089	106.459	111.652	120.708	131.334	142.566	153.204	159.814	165.696	172.711	177.597	187.953
133	88.233	94.746	98.021	102.968	107.357	112.573	121.667	132.334	143.608	154.283	160.915	166.816	173.854	178.755	189.142
134	89.044	95.588	98.878	103.848	108.257	113.495	122.625	133.334	144.649	155.361	162.016	167.936	174.996	179.913	190.331
135	89.856	96.431	99.736	104.729	109.156	114.417	123.584	134.334	145.690	156.440	163.116	169.056	176.138	181.070	191.520
136	90.669	97.275	100.595	105.609	110.056	115.338	124.543	135.334	146.731	157.518	164.216	170.175	177.280	182.226	192.707
137	91.482	98.119	101.454	106.491	110.956	116.261	125.502	136.334	147.772	158.595	165.316	171.294	178.421	183.382	193.894
138	92.296	98.964	102.314	107.372	111.857	117.183	126.461	137.334	148.813	159.673	166.415	172.412	179.561	184.538	195.080
139	93.111	99.809	103.174	108.254	112.758	118.106	127.421	138.334	149.854	160.750	167.514	173.530	180.701	185.693	196.266
140	93.926	100.655	104.034	109.137	113.659	119.029	128.380	139.334	150.894	161.827	168.613	174.648	181.840	186.847	197.451

Table B.1 was prepared using Equation 26.4.6 of Zelen and Severo (1964). The chi-square values were calculated to six decimal places and then rounded to three decimal places.

Examples:

$$\chi^2_{0.05,12} = 21.026 \quad \text{and} \quad \chi^2_{0.0,138} = 61.162.$$

For large degrees of freedom (v), critical values of χ^2 can be approximated very well by

$$\chi^2_{\alpha,v} \cong v \left(1 - \frac{2}{9v} + Z_{\alpha(1)} \sqrt{\frac{2}{9v}} \right)^3$$

(Wilson and Hilferty, 1931). It is for this purpose that the values of $Z_{\alpha(1)}$ are given below (from White, 1970).

α: 0.999	0.995	0.99	0.975	0.95	0.90	0.75
$Z_{\alpha(1)}$: -3.09023	-2.57583	-2.32635	-1.95996	-1.64485	-1.28155	-0.67449

α: 0.50	0.25	0.10	0.05	0.025	0.01	0.005	0.001
$Z_{\alpha(1)}$: 0.00000	0.67449	1.28155	1.64485	1.95996	2.32635	2.57583	3.09023

The percent error, i.e., (approximation − true value)/true value × 100%, resulting from the use of this approximation is as follows:

v	α: 0.999	0.995	0.99	0.975	0.95	0.90	0.75	0.50	0.25	0.10	0.05	0.025	0.01	0.005	0.001
30	-0.7	-0.3	-0.2	-0.1	0.0*	0.0*	0.0*	0.0*	0.0*	0.0*	0.0*	0.0*	0.0*	0.1	0.2
100	-0.1	0.0*	0.0*	0.0*	0.0*	0.0*	0.0*	0.0*	0.0*	0.0*	0.0*	0.0*	0.0*	0.0*	0.0*
140	0.0*	0.0*	0.0*	0.0*	0.0*	0.0*	0.0*	0.0*	0.0*	0.0*	0.0*	0.0*	0.0*	0.0*	0.0*

where the asterisk indicates a percent error the absolute value of which is less than 0.05%. Zar (1978) and Lin (1988) discuss this and other approximations for $\chi^2_{\alpha,v}$.

For one degree of freedom, the χ^2 distribution is related to the normal distribution (Appendix Table B.2) and the t distribution (Appendix Table B.3) as

$$\chi^2_{\alpha,1} = \left(Z_{\alpha(2)} \right)^2 = \left(t_{\alpha(2),\infty} \right)^2.$$

For example, $\chi^2_{0.05,1} = 3.841$, and $(Z_{0.05(2)})^2 = (t_{0.05(2),\infty})^2 = (1.9600)^2 = 3.8416$.

The relationship between χ^2 and F (Appendix Table B.4) is

$$\chi^2_{\alpha,v} = v F_{\alpha(1),v,\infty}.$$

For example, $\chi^2_{0.05,9} = 16.919$, and $(9)(F_{0.05(1),9,\infty}) = (9)(1.88) = 16.92$.

TABLE B.2 Proportions of the Normal Curve (One-Tailed)

This table gives the proportion of the normal curve that lies beyond (i.e., is more extreme than) a given normal deviate; e.g., $Z = (X_i - \mu)/\sigma$ or $Z = (\bar{X} - \mu)/\sigma_{\bar{X}}$. For example, the proportion of a normal distribution for which $Z \geq 1.51$ is 0.0655.

Z	0	1	2	3	4	5	6	7	8	9	Z
0.0	0.5000	0.4960	0.4920	0.4880	0.4840	0.4801	0.4761	0.4721	0.4681	0.4641	0.0
0.1	0.4602	0.4562	0.4522	0.4483	0.4443	0.4404	0.4364	0.4325	0.4286	0.4247	0.1
0.2	0.4207	0.4168	0.4129	0.4090	0.4052	0.4013	0.3974	0.3936	0.3897	0.3859	0.2
0.3	0.3821	0.3783	0.3745	0.3707	0.3669	0.3632	0.3594	0.3557	0.3520	0.3483	0.3
0.4	0.3446	0.3409	0.3372	0.3336	0.3300	0.3264	0.3228	0.3192	0.3156	0.3121	0.4
0.5	0.3085	0.3050	0.3015	0.2981	0.2946	0.2912	0.2877	0.2843	0.2810	0.2776	0.5
0.6	0.2743	0.2709	0.2676	0.2643	0.2611	0.2578	0.2546	0.2514	0.2483	0.2451	0.6
0.7	0.2420	0.2389	0.2358	0.2327	0.2297	0.2266	0.2236	0.2207	0.2177	0.2148	0.7
0.8	0.2119	0.2090	0.2061	0.2033	0.2005	0.1977	0.1949	0.1922	0.1894	0.1867	0.8
0.9	0.1841	0.1814	0.1788	0.1762	0.1736	0.1711	0.1685	0.1660	0.1635	0.1611	0.9
1.0	0.1587	0.1562	0.1539	0.1515	0.1492	0.1469	0.1446	0.1423	0.1401	0.1379	1.0
1.1	0.1357	0.1335	0.1314	0.1292	0.1271	0.1251	0.1230	0.1210	0.1190	0.1170	1.1
1.2	0.1151	0.1131	0.1112	0.1093	0.1075	0.1056	0.1038	0.1020	0.1003	0.0985	1.2
1.3	0.0968	0.0951	0.0934	0.0918	0.0901	0.0885	0.0869	0.0853	0.0838	0.0823	1.3
1.4	0.0808	0.0793	0.0778	0.0764	0.0749	0.0735	0.0721	0.0708	0.0694	0.0681	1.4
1.5	0.0668	0.0655	0.0643	0.0630	0.0618	0.0606	0.0594	0.0582	0.0571	0.0559	1.5
1.6	0.0548	0.0537	0.0526	0.0516	0.0505	0.0495	0.0485	0.0475	0.0465	0.0455	1.6
1.7	0.0446	0.0436	0.0427	0.0418	0.0409	0.0401	0.0392	0.0384	0.0375	0.0367	1.7
1.8	0.0359	0.0351	0.0344	0.0336	0.0329	0.0322	0.0314	0.0307	0.0301	0.0294	1.8
1.9	0.0287	0.0281	0.0274	0.0268	0.0262	0.0256	0.0250	0.0244	0.0239	0.0233	1.9
2.0	0.0228	0.0222	0.0217	0.0212	0.0207	0.0202	0.0197	0.0192	0.0188	0.0183	2.0
2.1	0.0179	0.0174	0.0170	0.0166	0.0162	0.0158	0.0154	0.0150	0.0146	0.0143	2.1
2.2	0.0139	0.0136	0.0132	0.0129	0.0125	0.0122	0.0119	0.0116	0.0113	0.0110	2.2
2.3	0.0107	0.0104	0.0102	0.0099	0.0096	0.0094	0.0091	0.0089	0.0087	0.0084	2.3
2.4	0.0082	0.0080	0.0078	0.0075	0.0073	0.0071	0.0069	0.0068	0.0066	0.0064	2.4
2.5	0.0062	0.0060	0.0059	0.0057	0.0055	0.0054	0.0052	0.0051	0.0049	0.0048	2.5
2.6	0.0047	0.0045	0.0044	0.0043	0.0041	0.0040	0.0039	0.0038	0.0037	0.0036	2.6
2.7	0.0035	0.0034	0.0033	0.0032	0.0031	0.0030	0.0029	0.0028	0.0027	0.0026	2.7
2.8	0.0026	0.0025	0.0024	0.0023	0.0023	0.0022	0.0021	0.0021	0.0020	0.0019	2.8
2.9	0.0019	0.0018	0.0018	0.0017	0.0016	0.0016	0.0015	0.0015	0.0014	0.0014	2.9
3.0	0.0013	0.0013	0.0013	0.0012	0.0012	0.0011	0.0011	0.0011	0.0010	0.0010	3.0
3.1	0.0010	0.0009	0.0009	0.0009	0.0008	0.0008	0.0008	0.0008	0.0007	0.0007	3.1
3.2	0.0007	0.0007	0.0006	0.0006	0.0006	0.0006	0.0006	0.0005	0.0005	0.0005	3.2
3.3	0.0005	0.0005	0.0005	0.0004	0.0004	0.0004	0.0004	0.0004	0.0004	0.0003	3.3
3.4	0.0003	0.0003	0.0003	0.0003	0.0003	0.0003	0.0003	0.0003	0.0003	0.0002	3.4
3.5	0.0002	0.0002	0.0002	0.0002	0.0002	0.0002	0.0002	0.0002	0.0002	0.0002	3.5
3.6	0.0002	0.0002	0.0001	0.0001	0.0001	0.0001	0.0001	0.0001	0.0001	0.0001	3.6
3.7	0.0001	0.0001	0.0001	0.0001	0.0001	0.0001	0.0001	0.0001	0.0001	0.0001	3.7
3.8	0.0001	0.0001	0.0001	0.0001	0.0001	0.0001	0.0001	0.0001	0.0001	0.0001	3.8

Table B.2 was prepared using an algorithm of Hastings (1955: 187). Probabilities for values of Z in between those shown in this table may be obtained by either linear or harmonic interpolation.

Critical values of Z may be found in Appendix Table B.3 as $Z_\alpha = t_{\alpha,\infty}$. For example, $Z_{0.05(2)} = t_{0.05(2),\infty} = 1.9600$. These critical values are related to those of χ^2 and F as

$$Z_{\alpha(2)} = t_{\alpha(2),\infty} = \sqrt{F_{\alpha(1),1,\infty}} = \sqrt{\chi^2_{\alpha,1}}.$$

Many computer programs and calculators can generate these proportions. Also, there are many quick and easy approximations. For example, one can compute

$$P = \left[1 - \sqrt{1 - e^{-c^2}} \right] / 2,$$

where setting $c = 0.806z(1 - 0.018z)$ (Hamaker, 1978) yields P dependable to the third decimal place for z as small as about 0.2, and using $c = z/(1.237 + 0.0249z)$ (Lin, 1988) achieves that accuracy for z as small as about 0.1. Hawkes' (1982) formulas are accurate to within 1 in the fifth decimal place, though they require more computation.

Table B.3 was prepared using Equations 26.7.3 and 26.7.4 of Zelen and Severo (1964), except for the values at infinity degrees of freedom, which are adapted from White (1970). Except for the values at infinity degrees of freedom, t was calculated to eight decimal places and then rounded to three decimal places.

Examples:

$$t_{0.05(2), 13} = 2.160 \quad \text{and} \quad t_{0.01(1), 19} = 2.539$$

If a critical value is needed for degrees of freedom not on this table, one may conservatively employ the next smaller ν that is on the table. Or, the needed critical value, for $\nu < 1000$, may be calculated by linear interpolation, with an error of no more than 0.001. If a little more accuracy is desired, or if the needed ν is > 1000, then harmonic interpolation should be used.

Critical values of t for infinity degrees of freedom are related to critical values of Z and χ^2 as

$$t_{\alpha(1),\infty} = Z_{\alpha(1)} \quad \text{and} \quad t_{\alpha(2),\infty} = Z_{\alpha(2)} = \sqrt{\chi^2_{\alpha,1}}.$$

TABLE B.3 Critical Values of the *t* Distribution

	$\alpha(2)$: 0.50	0.20	0.10	0.05	0.02	0.01	0.005	0.002	0.001
ν	$\alpha(1)$: 0.25	0.10	0.05	0.025	0.01	0.005	0.0025	0.001	0.0005
1	1.000	3.078	6.314	12.706	31.821	63.657	127.321	318.309	636.619
2	0.816	1.886	2.920	4.303	6.965	9.925	14.089	22.327	31.599
3	0.765	1.638	2.353	3.182	4.541	5.841	7.453	10.215	12.924
4	0.741	1.533	2.132	2.776	3.747	4.604	5.598	7.173	8.610
5	0.727	1.476	2.015	2.571	3.365	4.032	4.773	5.893	6.869
6	0.718	1.440	1.943	2.447	3.143	3.707	4.317	5.208	5.959
7	0.711	1.415	1.895	2.365	2.998	3.499	4.029	4.785	5.408
8	0.706	1.397	1.860	2.306	2.896	3.355	3.833	4.501	5.041
9	0.703	1.383	1.833	2.262	2.821	3.250	3.690	4.297	4.781
10	0.700	1.372	1.812	2.228	2.764	3.169	3.581	4.144	4.587
11	0.697	1.363	1.796	2.201	2.718	3.106	3.497	4.025	4.437
12	0.695	1.356	1.782	2.179	2.681	3.055	3.428	3.930	4.318
13	0.694	1.350	1.771	2.160	2.650	3.012	3.372	3.852	4.221
14	0.692	1.345	1.761	2.145	2.624	2.977	3.326	3.787	4.140
15	0.691	1.341	1.753	2.131	2.602	2.947	3.286	3.733	4.073
16	0.690	1.337	1.746	2.120	2.583	2.921	3.252	3.686	4.015
17	0.689	1.333	1.740	2.110	2.567	2.898	3.222	3.646	3.965
18	0.688	1.330	1.734	2.101	2.552	2.878	3.197	3.610	3.922
19	0.688	1.328	1.729	2.093	2.539	2.861	3.174	3.579	3.883
20	0.687	1.325	1.725	2.086	2.528	2.845	3.153	3.552	3.850
21	0.686	1.323	1.721	2.080	2.518	2.831	3.135	3.527	3.819
22	0.686	1.321	1.717	2.074	2.508	2.819	3.119	3.505	3.792
23	0.685	1.319	1.714	2.069	2.500	2.807	3.104	3.485	3.768
24	0.685	1.318	1.711	2.064	2.492	2.797	3.091	3.467	3.745
25	0.684	1.316	1.708	2.060	2.485	2.787	3.078	3.450	3.725
26	0.684	1.315	1.706	2.056	2.479	2.779	3.067	3.435	3.707
27	0.684	1.314	1.703	2.052	2.473	2.771	3.057	3.421	3.690
28	0.683	1.313	1.701	2.048	2.467	2.763	3.047	3.408	3.674
29	0.683	1.311	1.699	2.045	2.462	2.756	3.038	3.396	3.659
30	0.683	1.310	1.697	2.042	2.457	2.750	3.030	3.385	3.646
31	0.682	1.309	1.696	2.040	2.453	2.744	3.022	3.375	3.633
32	0.682	1.309	1.694	2.037	2.449	2.738	3.015	3.365	3.622
33	0.682	1.308	1.692	2.035	2.445	2.733	3.008	3.356	3.611
34	0.682	1.307	1.691	2.032	2.441	2.728	3.002	3.348	3.601
35	0.682	1.306	1.690	2.030	2.438	2.724	2.996	3.340	3.591
36	0.681	1.306	1.688	2.028	2.434	2.719	2.990	3.333	3.582
37	0.681	1.305	1.687	2.026	2.431	2.715	2.985	3.326	3.574
38	0.681	1.304	1.686	2.024	2.429	2.712	2.980	3.319	3.566
39	0.681	1.304	1.685	2.023	2.426	2.708	2.976	3.313	3.558
40	0.681	1.303	1.684	2.021	2.423	2.704	2.971	3.307	3.551
41	0.681	1.303	1.683	2.020	2.421	2.701	2.967	3.301	3.544
42	0.680	1.302	1.682	2.018	2.418	2.698	2.963	3.296	3.538
43	0.680	1.302	1.681	2.017	2.416	2.695	2.959	3.291	3.532
44	0.680	1.301	1.680	2.015	2.414	2.692	2.956	3.286	3.526
45	0.680	1.301	1.679	2.014	2.412	2.690	2.952	3.281	3.520
46	0.680	1.300	1.679	2.013	2.410	2.687	2.949	3.277	3.515
47	0.680	1.300	1.678	2.012	2.408	2.685	2.946	3.273	3.510
48	0.680	1.299	1.677	2.011	2.407	2.682	2.943	3.269	3.505
49	0.680	1.299	1.677	2.010	2.405	2.680	2.940	3.265	3.500
50	0.679	1.299	1.676	2.009	2.403	2.678	2.937	3.261	3.496
52	0.679	1.298	1.675	2.007	2.400	2.674	2.932	3.255	3.488
54	0.679	1.297	1.674	2.005	2.397	2.670	2.927	3.248	3.480
56	0.679	1.297	1.673	2.003	2.395	2.667	2.923	3.242	3.473
58	0.679	1.296	1.672	2.002	2.392	2.663	2.918	3.237	3.466
60	0.679	1.296	1.671	2.000	2.390	2.660	2.915	3.232	3.460

TABLE B.3 (cont.) Critical Values of the *t* Distribution

ν	$\alpha(2)$: 0.50 $\alpha(1)$: 0.25	0.20 0.10	0.10 0.05	0.05 0.025	0.02 0.01	0.01 0.005	0.005 0.0025	0.002 0.001	0.001 0.0005
62	0.678	1.295	1.670	1.999	2.388	2.657	2.911	3.227	3.454
64	0.678	1.295	1.669	1.998	2.386	2.655	2.908	3.223	3.449
66	0.678	1.295	1.668	1.997	2.384	2.652	2.904	3.218	3.444
68	0.678	1.294	1.668	1.995	2.382	2.650	2.902	3.214	3.439
70	0.678	1.294	1.667	1.994	2.381	2.648	2.899	3.211	3.435
72	0.678	1.293	1.666	1.993	2.379	2.646	2.896	3.207	3.431
74	0.678	1.293	1.666	1.993	2.378	2.644	2.894	3.204	3.427
76	0.678	1.293	1.665	1.992	2.376	2.642	2.891	3.201	3.423
78	0.678	1.292	1.665	1.991	2.375	2.640	2.889	3.198	3.420
80	0.678	1.292	1.664	1.990	2.374	2.639	2.887	3.195	3.416
82	0.677	1.292	1.664	1.989	2.373	2.637	2.885	3.193	3.413
84	0.677	1.292	1.663	1.989	2.372	2.636	2.883	3.190	3.410
86	0.677	1.291	1.663	1.988	2.370	2.634	2.881	3.188	3.407
88	0.677	1.291	1.662	1.987	2.369	2.633	2.880	3.185	3.405
90	0.677	1.291	1.662	1.987	2.368	2.632	2.878	3.183	3.402
92	0.677	1.291	1.662	1.986	2.368	2.630	2.876	3.181	3.399
94	0.677	1.291	1.661	1.986	2.367	2.629	2.875	3.179	3.397
96	0.677	1.290	1.661	1.985	2.366	2.628	2.873	3.177	3.395
98	0.677	1.290	1.661	1.984	2.365	2.627	2.872	3.175	3.393
100	0.677	1.290	1.660	1.984	2.364	2.626	2.871	3.174	3.390
105	0.677	1.290	1.659	1.983	2.362	2.623	2.868	3.170	3.386
110	0.677	1.289	1.659	1.982	2.361	2.621	2.865	3.166	3.381
115	0.677	1.289	1.658	1.981	2.359	2.619	2.862	3.163	3.377
120	0.677	1.289	1.658	1.980	2.358	2.617	2.860	3.160	3.373
125	0.676	1.288	1.657	1.979	2.357	2.616	2.858	3.157	3.370
130	0.676	1.288	1.657	1.978	2.355	2.614	2.856	3.154	3.367
135	0.676	1.288	1.656	1.978	2.354	2.613	2.854	3.152	3.364
140	0.676	1.288	1.656	1.977	2.353	2.611	2.852	3.149	3.361
145	0.676	1.287	1.655	1.976	2.352	2.610	2.851	3.147	3.359
150	0.676	1.287	1.655	1.976	2.351	2.609	2.849	3.145	3.357
160	0.676	1.287	1.654	1.975	2.350	2.607	2.846	3.142	3.352
170	0.676	1.287	1.654	1.974	2.348	2.605	2.844	3.139	3.349
180	0.676	1.286	1.653	1.973	2.347	2.603	2.842	3.136	3.345
190	0.676	1.286	1.653	1.973	2.346	2.602	2.840	3.134	3.342
200	0.676	1.286	1.653	1.972	2.345	2.601	2.839	3.131	3.340
250	0.675	1.285	1.651	1.969	2.341	2.596	2.832	3.123	3.330
300	0.675	1.284	1.650	1.968	2.339	2.592	2.828	3.118	3.323
350	0.675	1.284	1.649	1.967	2.337	2.590	2.825	3.114	3.319
400	0.675	1.284	1.649	1.966	2.336	2.588	2.823	3.111	3.315
450	0.675	1.283	1.648	1.965	2.335	2.587	2.821	3.108	3.312
500	0.675	1.283	1.648	1.965	2.334	2.586	2.820	3.107	3.310
600	0.675	1.283	1.647	1.964	2.333	2.584	2.817	3.104	3.307
700	0.675	1.283	1.647	1.963	2.332	2.583	2.816	3.102	3.304
800	0.675	1.283	1.647	1.963	2.331	2.582	2.815	3.100	3.303
900	0.675	1.282	1.647	1.963	2.330	2.581	2.814	3.099	3.301
1000	0.675	1.282	1.646	1.962	2.330	2.581	2.813	3.098	3.300
∞	0.6745	1.2816	1.6449	1.9600	2.3263	2.5758	2.8070	3.0902	3.2905

TABLE B.4 Critical Values of the *F* Distribution

Numerator DF = 1

Denom. DF	α(2): 0.50 α(1): 0.25	0.20 0.10	0.10 0.05	0.05 0.025	0.02 0.01	0.01 0.005	0.005 0.0025	0.002 0.001	0.001 0.0005
1	5.83	39.9	161.	648.	4050.	16200.	64800.	405000.	1620000.
2	2.57	8.53	18.5	38.5	98.5	199.	399.	999.	2000.
3	2.02	5.54	10.1	17.4	34.1	55.6	89.6	167.	267.
4	1.81	4.54	7.71	12.2	21.2	31.3	45.7	74.1	106.
5	1.69	4.06	6.61	10.0	16.3	22.8	31.4	47.2	63.6
6	1.62	3.78	5.99	8.81	13.7	18.6	24.8	35.5	46.1
7	1.57	3.59	5.59	8.07	12.2	16.2	21.1	29.2	37.0
8	1.54	3.46	5.32	7.57	11.3	14.7	18.8	25.4	31.6
9	1.51	3.36	5.12	7.21	10.6	13.6	17.2	22.9	28.0
10	1.49	3.29	4.96	6.94	10.0	12.8	16.0	21.0	25.5
11	1.47	3.23	4.84	6.72	9.65	12.2	15.2	19.7	23.7
12	1.46	3.18	4.75	6.55	9.33	11.8	14.5	18.6	22.2
13	1.45	3.14	4.67	6.41	9.07	11.4	13.9	17.8	21.1
14	1.44	3.10	4.60	6.30	8.86	11.1	13.5	17.1	20.2
15	1.43	3.07	4.54	6.20	8.68	10.8	13.1	16.6	19.5
16	1.42	3.05	4.49	6.12	8.53	10.6	12.8	16.1	18.9
17	1.42	3.03	4.45	6.04	8.40	10.4	12.6	15.7	18.4
18	1.41	3.01	4.41	5.98	8.29	10.2	12.3	15.4	17.9
19	1.41	2.99	4.38	5.92	8.18	10.1	12.1	15.1	17.5
20	1.40	2.97	4.35	5.87	8.10	9.94	11.9	14.8	17.2
21	1.40	2.96	4.32	5.83	8.02	9.83	11.8	14.6	16.9
22	1.40	2.95	4.30	5.79	7.95	9.73	11.6	14.4	16.6
23	1.39	2.94	4.28	5.75	7.88	9.63	11.5	14.2	16.4
24	1.39	2.93	4.26	5.72	7.82	9.55	11.4	14.0	16.2
25	1.39	2.92	4.24	5.69	7.77	9.48	11.3	13.9	16.0
26	1.38	2.91	4.23	5.66	7.72	9.41	11.2	13.7	15.8
27	1.38	2.90	4.21	5.63	7.68	9.34	11.1	13.6	15.6
28	1.38	2.89	4.20	5.61	7.64	9.28	11.0	13.5	15.5
29	1.38	2.89	4.18	5.59	7.60	9.23	11.0	13.4	15.3
30	1.38	2.88	4.17	5.57	7.56	9.18	10.9	13.3	15.2
35	1.37	2.85	4.12	5.48	7.42	8.98	10.6	12.9	14.7
40	1.36	2.84	4.08	5.42	7.31	8.83	10.4	12.6	14.4
45	1.36	2.82	4.06	5.38	7.23	8.71	10.3	12.4	14.1
50	1.35	2.81	4.03	5.34	7.17	8.63	10.1	12.2	13.9
60	1.35	2.79	4.00	5.29	7.08	8.49	9.96	12.0	13.5
70	1.35	2.78	3.98	5.25	7.01	8.40	9.84	11.8	13.3
80	1.34	2.77	3.96	5.22	6.96	8.33	9.75	11.7	13.2
90	1.34	2.76	3.95	5.20	6.93	8.28	9.68	11.6	13.0
100	1.34	2.76	3.94	5.18	6.90	8.24	9.62	11.5	12.9
120	1.34	2.75	3.92	5.15	6.85	8.18	9.54	11.4	12.8
140	1.33	2.74	3.91	5.13	6.82	8.14	9.48	11.3	12.7
160	1.33	2.74	3.90	5.12	6.80	8.10	9.44	11.2	12.6
180	1.33	2.73	3.89	5.11	6.78	8.08	9.40	11.2	12.6
200	1.33	2.73	3.89	5.10	6.76	8.06	9.38	11.2	12.5
300	1.33	2.72	3.87	5.07	6.72	8.00	9.30	11.0	12.4
500	1.33	2.72	3.86	5.05	6.69	7.95	9.23	11.0	12.3
∞	1.32	2.71	3.84	5.02	6.64	7.88	9.14	10.8	12.1

TABLE B.4 (cont.) Critical Values of the *F* Distribution

Numerator DF = 2

Denom. DF	α(2): 0.50 α(1): 0.25	0.20 0.10	0.10 0.05	0.05 0.025	0.02 0.01	0.01 0.005	0.005 0.0025	0.002 0.001	0.001 0.0005
1	7.50	49.5	200.	800.	5000.	20000.	80000.	500000.	2000000.
2	3.00	9.00	19.0	39.0	99.0	199.	399.	999.	2000.
3	2.28	5.46	9.55	16.0	30.8	49.8	79.9	149.	237.
4	2.00	4.32	6.94	10.6	18.0	26.3	38.0	61.2	87.4
5	1.85	3.78	5.79	8.43	13.3	18.3	25.0	37.1	49.8
6	1.76	3.46	5.14	7.26	10.9	14.5	19.1	27.0	34.8
7	1.70	3.26	4.74	6.54	9.55	12.4	15.9	21.7	27.2
8	1.66	3.11	4.46	6.06	8.65	11.0	13.9	18.5	22.7
9	1.62	3.01	4.26	5.71	8.02	10.1	12.5	16.4	19.9
10	1.60	2.92	4.10	5.46	7.56	9.43	11.6	14.9	17.9
11	1.58	2.86	3.98	5.26	7.21	8.91	10.8	13.8	16.4
12	1.56	2.81	3.89	5.10	6.93	8.51	10.3	13.0	15.3
13	1.55	2.76	3.81	4.97	6.70	8.19	9.84	12.3	14.4
14	1.53	2.73	3.74	4.86	6.51	7.92	9.47	11.8	13.7
15	1.52	2.70	3.68	4.77	6.36	7.70	9.17	11.3	13.2
16	1.51	2.67	3.63	4.69	6.23	7.51	8.92	11.0	12.7
17	1.51	2.64	3.59	4.62	6.11	7.35	8.70	10.7	12.3
18	1.50	2.62	3.55	4.56	6.01	7.21	8.51	10.4	11.9
19	1.49	2.61	3.52	4.51	5.93	7.09	8.35	10.2	11.6
20	1.49	2.59	3.49	4.46	5.85	6.99	8.21	9.95	11.4
21	1.48	2.57	3.47	4.42	5.78	6.89	8.08	9.77	11.2
22	1.48	2.56	3.44	4.38	5.72	6.81	7.96	9.61	11.0
23	1.47	2.55	3.42	4.35	5.66	6.73	7.86	9.47	10.8
24	1.47	2.54	3.40	4.32	5.61	6.66	7.77	9.34	10.6
25	1.47	2.53	3.39	4.29	5.57	6.60	7.69	9.22	10.5
26	1.46	2.52	3.37	4.27	5.53	6.54	7.61	9.12	10.3
27	1.46	2.51	3.35	4.24	5.49	6.49	7.54	9.02	10.2
28	1.46	2.50	3.34	4.22	5.45	6.44	7.48	8.93	10.1
29	1.45	2.50	3.33	4.20	5.42	6.40	7.42	8.85	9.99
30	1.45	2.49	3.32	4.18	5.39	6.35	7.36	8.77	9.90
35	1.44	2.46	3.27	4.11	5.27	6.19	7.14	8.47	9.52
40	1.44	2.44	3.23	4.05	5.18	6.07	6.99	8.25	9.25
45	1.43	2.42	3.20	4.01	5.11	5.97	6.86	8.09	9.04
50	1.43	2.41	3.18	3.97	5.06	5.90	6.77	7.96	8.88
60	1.42	2.39	3.15	3.93	4.98	5.79	6.63	7.77	8.65
70	1.41	2.38	3.13	3.89	4.92	5.72	6.53	7.64	8.49
80	1.41	2.37	3.11	3.86	4.88	5.67	6.46	7.54	8.37
90	1.41	2.36	3.10	3.84	4.85	5.62	6.41	7.47	8.28
100	1.41	2.36	3.09	3.83	4.82	5.59	6.37	7.41	8.21
120	1.40	2.35	3.07	3.80	4.79	5.54	6.30	7.32	8.10
140	1.40	2.34	3.06	3.79	4.76	5.50	6.26	7.26	8.03
160	1.40	2.34	3.05	3.78	4.74	5.48	6.22	7.21	7.97
180	1.40	2.33	3.05	3.77	4.73	5.46	6.20	7.18	7.93
200	1.40	2.33	3.04	3.76	4.71	5.44	6.17	7.15	7.90
300	1.39	2.32	3.03	3.73	4.68	5.39	6.11	7.07	7.80
500	1.39	2.31	3.01	3.72	4.65	5.35	6.06	7.00	7.72
∞	1.39	2.30	3.00	3.69	4.61	5.30	5.99	6.91	7.60

TABLE B.4 (cont.)   Critical Values of the F Distribution

Numerator DF = 3

Denom. DF	$\alpha(2)$: 0.50 $\alpha(1)$: 0.25	0.20 0.10	0.10 0.05	0.05 0.025	0.02 0.01	0.01 0.005	0.005 0.0025	0.002 0.001	0.001 0.0005
1	8.20	53.6	216.	864.	5400.	21600.	86500.	540000.	2160000.
2	3.15	9.16	19.2	39.2	99.2	199.	399.	999.	2000.
3	2.36	5.39	9.28	15.4	29.5	47.5	76.1	141.	225.
4	2.05	4.19	6.59	9.98	16.7	24.3	35.0	56.2	80.1
5	1.88	3.62	5.41	7.76	12.1	16.5	22.4	33.2	44.4
6	1.78	3.29	4.76	6.60	9.78	12.9	16.9	23.7	30.5
7	1.72	3.07	4.35	5.89	8.45	10.9	13.8	18.8	23.5
8	1.67	2.92	4.07	5.42	7.59	9.60	12.0	15.8	19.4
9	1.63	2.81	3.86	5.08	6.99	8.72	10.7	13.9	16.8
10	1.60	2.73	3.71	4.83	6.55	8.08	9.83	12.6	15.0
11	1.58	2.66	3.59	4.63	6.22	7.60	9.17	11.6	13.7
12	1.56	2.61	3.49	4.47	5.95	7.23	8.65	10.8	12.7
13	1.55	2.56	3.41	4.35	5.74	6.93	8.24	10.2	11.9
14	1.53	2.52	3.34	4.24	5.56	6.68	7.91	9.73	11.3
15	1.52	2.49	3.29	4.15	5.42	6.48	7.63	9.34	10.8
16	1.51	2.46	3.24	4.08	5.29	6.30	7.40	9.01	10.3
17	1.50	2.44	3.20	4.01	5.19	6.16	7.21	8.73	9.99
18	1.49	2.42	3.16	3.95	5.09	6.03	7.04	8.49	9.69
19	1.49	2.40	3.13	3.90	5.01	5.92	6.89	8.28	9.42
20	1.48	2.38	3.10	3.86	4.94	5.82	6.76	8.10	9.20
21	1.48	2.36	3.07	3.82	4.87	5.73	6.64	7.94	8.99
22	1.47	2.35	3.05	3.78	4.82	5.65	6.54	7.80	8.82
23	1.47	2.34	3.03	3.75	4.76	5.58	6.45	7.67	8.66
24	1.46	2.33	3.01	3.72	4.72	5.52	6.36	7.55	8.51
25	1.46	2.32	2.99	3.69	4.68	5.46	6.29	7.45	8.39
26	1.45	2.31	2.98	3.67	4.64	5.41	6.22	7.36	8.27
27	1.45	2.30	2.96	3.65	4.60	5.36	6.16	7.27	8.16
28	1.45	2.29	2.95	3.63	4.57	5.32	6.10	7.19	8.07
29	1.45	2.28	2.93	3.61	4.54	5.28	6.05	7.12	7.98
30	1.44	2.28	2.92	3.59	4.51	5.24	6.00	7.05	7.89
35	1.43	2.25	2.87	3.52	4.40	5.09	5.80	6.79	7.56
40	1.42	2.23	2.84	3.46	4.31	4.98	5.66	6.59	7.33
45	1.42	2.21	2.81	3.42	4.25	4.89	5.55	6.45	7.15
50	1.41	2.20	2.79	3.39	4.20	4.83	5.47	6.34	7.01
60	1.41	2.18	2.76	3.34	4.13	4.73	5.34	6.17	6.81
70	1.40	2.16	2.74	3.31	4.07	4.66	5.26	6.06	6.67
80	1.40	2.15	2.72	3.28	4.04	4.61	5.19	5.97	6.57
90	1.39	2.15	2.71	3.26	4.01	4.57	5.14	5.91	6.49
100	1.39	2.14	2.70	3.25	3.98	4.54	5.11	5.86	6.43
120	1.39	2.13	2.68	3.23	3.95	4.50	5.05	5.78	6.34
140	1.38	2.12	2.67	3.21	3.92	4.47	5.01	5.73	6.28
160	1.38	2.12	2.66	3.20	3.91	4.44	4.98	5.69	6.23
180	1.38	2.11	2.65	3.19	3.89	4.42	4.95	5.66	6.19
200	1.38	2.11	2.65	3.18	3.88	4.41	4.94	5.63	6.16
300	1.38	2.10	2.63	3.16	3.85	4.36	4.88	5.56	6.08
500	1.37	2.09	2.62	3.14	3.82	4.33	4.84	5.51	6.01
∞	1.37	2.08	2.61	3.12	3.78	4.28	4.77	5.42	5.91

TABLE B.4 (cont.) Critical Values of the F Distribution

Numerator DF $= 4$

Denom. DF	$\alpha(2)$: 0.50 $\alpha(1)$: 0.25	0.20 0.10	0.10 0.05	0.05 0.025	0.02 0.01	0.01 0.005	0.005 0.0025	0.002 0.001	0.001 0.0005
1	8.58	55.8	225.	900.	5620.	22500.	90000.	562000.	2250000.
2	3.23	9.24	19.2	39.2	99.2	199.	399.	999.	2000.
3	2.39	5.34	9.12	15.1	28.7	46.2	73.9	137.	218.
4	2.06	4.11	6.39	9.60	16.0	23.2	33.3	53.4	76.1
5	1.89	3.52	5.19	7.39	11.4	15.6	21.0	31.1	41.5
6	1.79	3.18	4.53	6.23	9.15	12.0	15.7	21.9	28.1
7	1.72	2.96	4.12	5.52	7.85	10.1	12.7	17.2	21.4
8	1.66	2.81	3.84	5.05	7.01	8.81	10.9	14.4	17.6
9	1.63	2.69	3.63	4.72	6.42	7.96	9.74	12.6	15.1
10	1.59	2.61	3.48	4.47	5.99	7.34	8.89	11.3	13.4
11	1.57	2.54	3.36	4.28	5.67	6.88	8.25	10.3	12.2
12	1.55	2.48	3.26	4.12	5.41	6.52	7.76	9.63	11.2
13	1.53	2.43	3.18	4.00	5.21	6.23	7.37	9.07	10.5
14	1.52	2.39	3.11	3.89	5.04	6.00	7.06	8.62	9.95
15	1.51	2.36	3.06	3.80	4.89	5.80	6.80	8.25	9.48
16	1.50	2.33	3.01	3.73	4.77	5.64	6.58	7.94	9.08
17	1.49	2.31	2.96	3.66	4.67	5.50	6.39	7.68	8.75
18	1.48	2.29	2.93	3.61	4.58	5.37	6.23	7.46	8.47
19	1.47	2.27	2.90	3.56	4.50	5.27	6.09	7.27	8.23
20	1.47	2.25	2.87	3.51	4.43	5.17	5.97	7.10	8.02
21	1.46	2.23	2.84	3.48	4.37	5.09	5.86	6.95	7.83
22	1.45	2.22	2.82	3.44	4.31	5.02	5.76	6.81	7.67
23	1.45	2.21	2.80	3.41	4.26	4.95	5.67	6.70	7.52
24	1.44	2.19	2.78	3.38	4.22	4.89	5.60	6.59	7.39
25	1.44	2.18	2.76	3.35	4.18	4.84	5.53	6.49	7.27
26	1.44	2.17	2.74	3.33	4.14	4.79	5.46	6.41	7.16
27	1.43	2.17	2.73	3.31	4.11	4.74	5.40	6.33	7.06
28	1.43	2.16	2.71	3.29	4.07	4.70	5.35	6.25	6.97
29	1.43	2.15	2.70	3.27	4.04	4.66	5.30	6.19	6.89
30	1.42	2.14	2.69	3.25	4.02	4.62	5.25	6.12	6.82
35	1.41	2.11	2.64	3.18	3.91	4.48	5.07	5.88	6.51
40	1.40	2.09	2.61	3.13	3.83	4.37	4.93	5.70	6.30
45	1.40	2.07	2.58	3.09	3.77	4.29	4.83	5.56	6.13
50	1.39	2.06	2.56	3.05	3.72	4.23	4.75	5.46	6.01
60	1.38	2.04	2.53	3.01	3.65	4.14	4.64	5.31	5.82
70	1.38	2.03	2.50	2.97	3.60	4.08	4.56	5.20	5.70
80	1.38	2.02	2.49	2.95	3.56	4.03	4.50	5.12	5.60
90	1.37	2.01	2.47	2.93	3.53	3.99	4.45	5.06	5.53
100	1.37	2.00	2.46	2.92	3.51	3.96	4.42	5.02	5.48
120	1.37	1.99	2.45	2.89	3.48	3.92	4.36	4.95	5.39
140	1.36	1.99	2.44	2.88	3.46	3.89	4.32	4.90	5.33
160	1.36	1.98	2.43	2.87	3.44	3.87	4.30	4.86	5.29
180	1.36	1.98	2.42	2.86	3.43	3.85	4.27	4.83	5.26
200	1.36	1.97	2.42	2.85	3.41	3.84	4.26	4.81	5.23
300	1.35	1.96	2.40	2.83	3.38	3.80	4.21	4.75	5.15
500	1.35	1.96	2.39	2.81	3.36	3.76	4.17	4.69	5.09
∞	1.35	1.94	2.37	2.79	3.32	3.72	4.11	4.62	5.00

TABLE B.4 (cont.)   Critical Values of the F Distribution

Numerator DF = 5

Denom. DF	$\alpha(2)$: 0.50 $\alpha(1)$: 0.25	0.20 0.10	0.10 0.05	0.05 0.025	0.02 0.01	0.01 0.005	0.005 0.0025	0.002 0.001	0.001 0.0005
1	8.82	57.2	230.	922.	5760.	23100.	92200.	576000.	2310000.
2	3.28	9.29	19.3	39.3	99.3	199.	399.	999.	2000.
3	2.41	5.31	9.01	14.9	28.2	45.4	72.6	135.	214.
4	2.07	4.05	6.26	9.36	15.5	22.5	32.3	51.7	73.6
5	1.89	3.45	5.05	7.15	11.0	14.9	20.2	29.8	39.7
6	1.79	3.11	4.39	5.99	8.75	11.5	14.9	20.8	26.6
7	1.71	2.88	3.97	5.29	7.46	9.52	12.0	16.2	20.2
8	1.66	2.73	3.69	4.82	6.63	8.30	10.3	13.5	16.4
9	1.62	2.61	3.48	4.48	6.06	7.47	9.12	11.7	14.1
10	1.59	2.52	3.33	4.24	5.64	6.87	8.29	10.5	12.4
11	1.56	2.45	3.20	4.04	5.32	6.42	7.67	9.58	11.2
12	1.54	2.39	3.11	3.89	5.06	6.07	7.20	8.89	10.4
13	1.52	2.35	3.03	3.77	4.86	5.79	6.82	8.35	9.66
14	1.51	2.31	2.96	3.66	4.69	5.56	6.51	7.92	9.11
15	1.49	2.27	2.90	3.58	4.56	5.37	6.26	7.57	8.66
16	1.48	2.24	2.85	3.50	4.44	5.21	6.05	7.27	8.29
17	1.47	2.22	2.81	3.44	4.34	5.07	5.87	7.02	7.98
18	1.46	2.20	2.77	3.38	4.25	4.96	5.72	6.81	7.71
19	1.46	2.18	2.74	3.33	4.17	4.85	5.58	6.62	7.48
20	1.45	2.16	2.71	3.29	4.10	4.76	5.46	6.46	7.27
21	1.44	2.14	2.68	3.25	4.04	4.68	5.36	6.32	7.10
22	1.44	2.13	2.66	3.22	3.99	4.61	5.26	6.19	6.94
23	1.43	2.11	2.64	3.18	3.94	4.54	5.18	6.08	6.80
24	1.43	2.10	2.62	3.15	3.90	4.49	5.11	5.98	6.68
25	1.42	2.09	2.60	3.13	3.85	4.43	5.04	5.89	6.56
26	1.42	2.08	2.59	3.10	3.82	4.38	4.98	5.80	6.46
27	1.42	2.07	2.57	3.08	3.78	4.34	4.92	5.73	6.37
28	1.41	2.06	2.56	3.06	3.75	4.30	4.87	5.66	6.28
29	1.41	2.06	2.55	3.04	3.73	4.26	4.82	5.59	6.21
30	1.41	2.05	2.53	3.03	3.70	4.23	4.78	5.53	6.13
35	1.40	2.02	2.49	2.96	3.59	4.09	4.60	5.30	5.85
40	1.39	2.00	2.45	2.90	3.51	3.99	4.47	5.13	5.64
45	1.38	1.98	2.42	2.86	3.45	3.91	4.37	5.00	5.49
50	1.37	1.97	2.40	2.83	3.41	3.85	4.30	4.90	5.37
60	1.37	1.95	2.37	2.79	3.34	3.76	4.19	4.76	5.20
70	1.36	1.93	2.35	2.75	3.29	3.70	4.11	4.66	5.08
80	1.36	1.92	2.33	2.73	3.26	3.65	4.05	4.58	4.99
90	1.35	1.91	2.32	2.71	3.23	3.62	4.01	4.53	4.92
100	1.35	1.91	2.31	2.70	3.21	3.59	3.97	4.48	4.87
120	1.35	1.90	2.29	2.67	3.17	3.55	3.92	4.42	4.79
140	1.34	1.89	2.28	2.66	3.15	3.52	3.89	4.37	4.74
160	1.34	1.88	2.27	2.65	3.13	3.50	3.86	4.33	4.69
180	1.34	1.88	2.26	2.64	3.12	3.48	3.84	4.31	4.66
200	1.34	1.88	2.26	2.63	3.11	3.47	3.82	4.29	4.64
300	1.33	1.87	2.24	2.61	3.08	3.43	3.77	4.22	4.56
500	1.33	1.86	2.23	2.59	3.05	3.40	3.73	4.18	4.51
∞	1.33	1.85	2.21	2.57	3.02	3.35	3.68	4.10	4.42

TABLE B.4 (cont.)   Critical Values of the *F* Distribution

Numerator DF = 6

Denom. DF	α(2): 0.50 α(1): 0.25	0.20 0.10	0.10 0.05	0.05 0.025	0.02 0.01	0.01 0.005	0.005 0.0025	0.002 0.001	0.001 0.0005
1	8.98	58.2	234.	937.	5860.	23400.	93700.	586000.	2340000.
2	3.31	9.33	19.3	39.3	99.3	199.	399.	999.	2000.
3	2.42	5.28	8.94	14.7	27.9	44.8	71.7	133.	211.
4	2.08	4.01	6.16	9.20	15.2	22.0	31.5	50.5	71.9
5	1.89	3.40	4.95	6.98	10.7	14.5	19.6	28.8	38.5
6	1.78	3.05	4.28	5.82	8.47	11.1	14.4	20.0	25.6
7	1.71	2.83	3.87	5.12	7.19	9.16	11.5	15.5	19.3
8	1.65	2.67	3.58	4.65	6.37	7.95	9.83	12.9	15.7
9	1.61	2.55	3.37	4.32	5.80	7.13	8.68	11.1	13.3
10	1.58	2.46	3.22	4.07	5.39	6.54	7.87	9.93	11.7
11	1.55	2.39	3.09	3.88	5.07	6.10	7.27	9.05	10.6
12	1.53	2.33	3.00	3.73	4.82	5.76	6.80	8.38	9.74
13	1.51	2.28	2.92	3.60	4.62	5.48	6.44	7.86	9.07
14	1.50	2.24	2.85	3.50	4.46	5.26	6.14	7.44	8.53
15	1.48	2.21	2.79	3.41	4.32	5.07	5.89	7.09	8.10
16	1.47	2.18	2.74	3.34	4.20	4.91	5.68	6.80	7.74
17	1.46	2.15	2.70	3.28	4.10	4.78	5.51	6.56	7.43
18	1.45	2.13	2.66	3.22	4.01	4.66	5.36	6.35	7.18
19	1.44	2.11	2.63	3.17	3.94	4.56	5.23	6.18	6.95
20	1.44	2.09	2.60	3.13	3.87	4.47	5.11	6.02	6.76
21	1.43	2.08	2.57	3.09	3.81	4.39	5.01	5.88	6.59
22	1.42	2.06	2.55	3.05	3.76	4.32	4.92	5.76	6.44
23	1.42	2.05	2.53	3.02	3.71	4.26	4.84	5.65	6.30
24	1.41	2.04	2.51	2.99	3.67	4.20	4.76	5.55	6.18
25	1.41	2.02	2.49	2.97	3.63	4.15	4.70	5.46	6.07
26	1.41	2.01	2.47	2.94	3.59	4.10	4.64	5.38	5.98
27	1.40	2.00	2.46	2.92	3.56	4.06	4.58	5.31	5.89
28	1.40	2.00	2.45	2.90	3.53	4.02	4.53	5.24	5.80
29	1.40	1.99	2.43	2.88	3.50	3.98	4.48	5.18	5.73
30	1.39	1.98	2.42	2.87	3.47	3.95	4.44	5.12	5.66
35	1.38	1.95	2.37	2.80	3.37	3.81	4.27	4.89	5.39
40	1.37	1.93	2.34	2.74	3.29	3.71	4.14	4.73	5.19
45	1.36	1.91	2.31	2.70	3.23	3.64	4.05	4.61	5.04
50	1.36	1.90	2.29	2.67	3.19	3.58	3.98	4.51	4.93
60	1.35	1.87	2.25	2.63	3.12	3.49	3.87	4.37	4.76
70	1.34	1.86	2.23	2.59	3.07	3.43	3.79	4.28	4.64
80	1.34	1.85	2.21	2.57	3.04	3.39	3.74	4.20	4.56
90	1.33	1.84	2.20	2.55	3.01	3.35	3.70	4.15	4.50
100	1.33	1.83	2.19	2.54	2.99	3.33	3.66	4.11	4.45
120	1.33	1.82	2.18	2.52	2.96	3.28	3.61	4.04	4.37
140	1.32	1.82	2.16	2.50	2.93	3.26	3.58	4.00	4.32
160	1.32	1.81	2.16	2.49	2.92	3.24	3.55	3.97	4.28
180	1.32	1.81	2.15	2.48	2.90	3.22	3.53	3.94	4.25
200	1.32	1.80	2.14	2.47	2.89	3.21	3.52	3.92	4.22
300	1.32	1.79	2.13	2.45	2.86	3.17	3.47	3.86	4.15
500	1.31	1.79	2.12	2.43	2.84	3.14	3.43	3.81	4.10
∞	1.31	1.77	2.10	2.41	2.80	3.09	3.37	3.74	4.02

TABLE B.4 (cont.)   Critical Values of the F Distribution

Numerator DF $= 7$

Denom. DF	$\alpha(2)$: 0.50 $\alpha(1)$: 0.25	0.20 0.10	0.10 0.05	0.05 0.025	0.02 0.01	0.01 0.005	0.005 0.0025	0.002 0.001	0.001 0.0005
1	9.10	58.9	237.	948.	5930.	23700.	94900.	593000.	2370000.
2	3.34	9.35	19.4	39.4	99.4	199.	399.	999.	2000.
3	2.43	5.27	8.89	14.6	27.7	44.4	71.0	132.	209.
4	2.08	3.98	6.09	9.07	15.0	21.6	31.0	49.7	70.7
5	1.89	3.37	4.88	6.85	10.5	14.2	19.1	28.2	37.6
6	1.78	3.01	4.21	5.70	8.26	10.8	14.0	19.5	24.9
7	1.70	2.78	3.79	4.99	6.99	8.89	11.2	15.0	18.7
8	1.64	2.62	3.50	4.53	6.18	7.69	9.49	12.4	15.1
9	1.60	2.51	3.29	4.20	5.61	6.88	8.36	10.7	12.8
10	1.57	2.41	3.14	3.95	5.20	6.30	7.56	9.52	11.2
11	1.54	2.34	3.01	3.76	4.89	5.86	6.97	8.66	10.1
12	1.52	2.28	2.91	3.61	4.64	5.52	6.51	8.00	9.28
13	1.50	2.23	2.83	3.48	4.44	5.25	6.15	7.49	8.63
14	1.49	2.19	2.76	3.38	4.28	5.03	5.86	7.08	8.11
15	1.47	2.16	2.71	3.29	4.14	4.85	5.62	6.74	7.68
16	1.46	2.13	2.66	3.22	4.03	4.69	5.41	6.46	7.33
17	1.45	2.10	2.61	3.16	3.93	4.56	5.24	6.22	7.04
18	1.44	2.08	2.58	3.10	3.84	4.44	5.09	6.02	6.78
19	1.43	2.06	2.54	3.05	3.77	4.34	4.96	5.85	6.57
20	1.43	2.04	2.51	3.01	3.70	4.26	4.85	5.69	6.38
21	1.42	2.02	2.49	2.97	3.64	4.18	4.75	5.56	6.21
22	1.41	2.01	2.46	2.93	3.59	4.11	4.66	5.44	6.07
23	1.41	1.99	2.44	2.90	3.54	4.05	4.58	5.33	5.94
24	1.40	1.98	2.42	2.87	3.50	3.99	4.51	5.23	5.82
25	1.40	1.97	2.40	2.85	3.46	3.94	4.44	5.15	5.71
26	1.39	1.96	2.39	2.82	3.42	3.89	4.38	5.07	5.62
27	1.39	1.95	2.37	2.80	3.39	3.85	4.33	5.00	5.53
28	1.39	1.94	2.36	2.78	3.36	3.81	4.28	4.93	5.45
29	1.38	1.93	2.35	2.76	3.33	3.77	4.24	4.87	5.38
30	1.38	1.93	2.33	2.75	3.30	3.74	4.19	4.82	5.31
35	1.37	1.90	2.29	2.68	3.20	3.61	4.02	4.59	5.04
40	1.36	1.87	2.25	2.62	3.12	3.51	3.90	4.44	4.85
45	1.35	1.85	2.22	2.58	3.07	3.43	3.81	4.32	4.71
50	1.34	1.84	2.20	2.55	3.02	3.38	3.74	4.22	4.60
60	1.33	1.82	2.17	2.51	2.95	3.29	3.63	4.09	4.44
70	1.33	1.80	2.14	2.47	2.91	3.23	3.56	3.99	4.32
80	1.32	1.79	2.13	2.45	2.87	3.19	3.50	3.92	4.24
90	1.32	1.78	2.11	2.43	2.84	3.15	3.46	3.87	4.18
100	1.32	1.78	2.10	2.42	2.82	3.13	3.43	3.83	4.13
120	1.31	1.77	2.09	2.39	2.79	3.09	3.38	3.77	4.06
140	1.31	1.76	2.08	2.38	2.77	3.06	3.35	3.72	4.01
160	1.31	1.75	2.07	2.37	2.75	3.04	3.32	3.69	3.97
180	1.31	1.75	2.06	2.36	2.74	3.02	3.30	3.67	3.94
200	1.30	1.75	2.06	2.35	2.73	3.01	3.29	3.65	3.92
300	1.30	1.74	2.04	2.33	2.70	2.97	3.24	3.59	3.85
500	1.30	1.73	2.03	2.31	2.68	2.94	3.20	3.54	3.80
∞	1.29	1.72	2.01	2.29	2.64	2.90	3.15	3.47	3.72

TABLE B.4 (cont.)   Critical Values of the F Distribution

Numerator DF $= 8$

Denom. DF	$\alpha(2)$: 0.50 $\alpha(1)$: 0.25	0.20 0.10	0.10 0.05	0.05 0.025	0.02 0.01	0.01 0.005	0.005 0.0025	0.002 0.001	0.001 0.0005
1	9.19	59.4	239.	957.	5980.	23900.	95700.	598000.	2390000.
2	3.35	9.37	19.4	39.4	99.4	199.	399.	999.	2000.
3	2.44	5.25	8.85	14.5	27.5	44.1	70.5	131.	208.
4	2.08	3.95	6.04	8.98	14.8	21.4	30.6	49.0	69.7
5	1.89	3.34	4.82	6.76	10.3	14.0	18.8	27.6	36.9
6	1.78	2.98	4.15	5.60	8.10	10.6	13.7	19.0	24.3
7	1.70	2.75	3.73	4.90	6.84	8.68	10.9	14.6	18.2
8	1.64	2.59	3.44	4.43	6.03	7.50	9.24	12.0	14.6
9	1.60	2.47	3.23	4.10	5.47	6.69	8.12	10.4	12.4
10	1.56	2.38	3.07	3.85	5.06	6.12	7.33	9.20	10.9
11	1.53	2.30	2.95	3.66	4.74	5.68	6.74	8.35	9.76
12	1.51	2.24	2.85	3.51	4.50	5.35	6.29	7.71	8.94
13	1.49	2.20	2.77	3.39	4.30	5.08	5.93	7.21	8.29
14	1.48	2.15	2.70	3.29	4.14	4.86	5.64	6.80	7.78
15	1.46	2.12	2.64	3.20	4.00	4.67	5.40	6.47	7.37
16	1.45	2.09	2.59	3.12	3.89	4.52	5.20	6.19	7.02
17	1.44	2.06	2.55	3.06	3.79	4.39	5.03	5.96	6.73
18	1.43	2.04	2.51	3.01	3.71	4.28	4.89	5.76	6.48
19	1.42	2.02	2.48	2.96	3.63	4.18	4.76	5.59	6.27
20	1.42	2.00	2.45	2.91	3.56	4.09	4.65	5.44	6.09
21	1.41	1.98	2.42	2.87	3.51	4.01	4.55	5.31	5.92
22	1.40	1.97	2.40	2.84	3.45	3.94	4.46	5.19	5.78
23	1.40	1.95	2.37	2.81	3.41	3.88	4.38	5.09	5.65
24	1.39	1.94	2.36	2.78	3.36	3.83	4.31	4.99	5.54
25	1.39	1.93	2.34	2.75	3.32	3.78	4.25	4.91	5.43
26	1.38	1.92	2.32	2.73	3.29	3.73	4.19	4.83	5.34
27	1.38	1.91	2.31	2.71	3.26	3.69	4.14	4.76	5.25
28	1.38	1.90	2.29	2.69	3.23	3.65	4.09	4.69	5.18
29	1.37	1.89	2.28	2.67	3.20	3.61	4.04	4.64	5.11
30	1.37	1.88	2.27	2.65	3.17	3.58	4.00	4.58	5.04
35	1.36	1.85	2.22	2.58	3.07	3.45	3.83	4.36	4.78
40	1.35	1.83	2.18	2.53	2.99	3.35	3.71	4.21	4.59
45	1.34	1.81	2.15	2.49	2.94	3.28	3.62	4.09	4.45
50	1.33	1.80	2.13	2.46	2.89	3.22	3.55	4.00	4.34
60	1.32	1.77	2.10	2.41	2.82	3.13	3.45	3.86	4.19
70	1.32	1.76	2.07	2.38	2.78	3.08	3.37	3.77	4.08
80	1.31	1.75	2.06	2.35	2.74	3.03	3.32	3.70	4.00
90	1.31	1.74	2.04	2.34	2.72	3.00	3.28	3.65	3.94
100	1.30	1.73	2.03	2.32	2.69	2.97	3.25	3.61	3.89
120	1.30	1.72	2.02	2.30	2.66	2.93	3.20	3.55	3.82
140	1.30	1.71	2.01	2.28	2.64	2.91	3.17	3.51	3.77
160	1.29	1.71	2.00	2.27	2.62	2.88	3.14	3.48	3.73
180	1.29	1.70	1.99	2.26	2.61	2.87	3.12	3.45	3.70
200	1.29	1.70	1.98	2.26	2.60	2.86	3.11	3.43	3.68
300	1.29	1.69	1.97	2.23	2.57	2.82	3.06	3.38	3.61
500	1.28	1.68	1.96	2.22	2.55	2.79	3.03	3.33	3.56
∞	1.28	1.67	1.94	2.19	2.51	2.74	2.97	3.27	3.48

TABLE B.4 (cont.)   Critical Values of the *F* Distribution

Numerator DF = 9

Denom. DF	α(2): 0.50 α(1): 0.25	0.20 0.10	0.10 0.05	0.05 0.025	0.02 0.01	0.01 0.005	0.005 0.0025	0.002 0.001	0.001 0.0005
1	9.26	59.9	241.	963.	6020.	24100.	96400.	602000.	2410000.
2	3.37	9.38	19.4	39.4	99.4	199.	399.	999.	2000.
3	2.44	5.24	8.81	14.5	27.3	43.9	70.1	130.	207.
4	2.08	3.94	6.00	8.90	14.7	21.1	30.3	48.5	69.0
5	1.89	3.32	4.77	6.68	10.2	13.8	18.5	27.2	36.3
6	1.77	2.96	4.10	5.52	7.98	10.4	13.4	18.7	23.9
7	1.69	2.72	3.68	4.82	6.72	8.51	10.7	14.3	17.8
8	1.63	2.56	3.39	4.36	5.91	7.34	9.03	11.8	14.3
9	1.59	2.44	3.18	4.03	5.35	6.54	7.92	10.1	12.1
10	1.56	2.35	3.02	3.78	4.94	5.97	7.14	8.96	10.6
11	1.53	2.27	2.90	3.59	4.63	5.54	6.56	8.12	9.48
12	1.51	2.21	2.80	3.44	4.39	5.20	6.11	7.48	8.66
13	1.49	2.16	2.71	3.31	4.19	4.94	5.76	6.98	8.03
14	1.47	2.12	2.65	3.21	4.03	4.72	5.47	6.58	7.52
15	1.46	2.09	2.59	3.12	3.89	4.54	5.23	6.26	7.11
16	1.44	2.06	2.54	3.05	3.78	4.38	5.04	5.98	6.77
17	1.43	2.03	2.49	2.98	3.68	4.25	4.87	5.75	6.49
18	1.42	2.00	2.46	2.93	3.60	4.14	4.72	5.56	6.24
19	1.41	1.98	2.42	2.88	3.52	4.04	4.60	5.39	6.03
20	1.41	1.96	2.39	2.84	3.46	3.96	4.49	5.24	5.85
21	1.40	1.95	2.37	2.80	3.40	3.88	4.39	5.11	5.69
22	1.39	1.93	2.34	2.76	3.35	3.81	4.30	4.99	5.55
23	1.39	1.92	2.32	2.73	3.30	3.75	4.22	4.89	5.43
24	1.38	1.91	2.30	2.70	3.26	3.69	4.15	4.80	5.31
25	1.38	1.89	2.28	2.68	3.22	3.64	4.09	4.71	5.21
26	1.37	1.88	2.27	2.65	3.18	3.60	4.03	4.64	5.12
27	1.37	1.87	2.25	2.63	3.15	3.56	3.98	4.57	5.04
28	1.37	1.87	2.24	2.61	3.12	3.52	3.93	4.50	4.96
29	1.36	1.86	2.22	2.59	3.09	3.48	3.89	4.45	4.89
30	1.36	1.85	2.21	2.57	3.07	3.45	3.85	4.39	4.82
35	1.35	1.82	2.16	2.50	2.96	3.32	3.68	4.18	4.57
40	1.34	1.79	2.12	2.45	2.89	3.22	3.56	4.02	4.38
45	1.33	1.77	2.10	2.41	2.83	3.15	3.47	3.91	4.25
50	1.32	1.76	2.07	2.38	2.78	3.09	3.40	3.82	4.14
60	1.31	1.74	2.04	2.33	2.72	3.01	3.30	3.69	3.98
70	1.31	1.72	2.02	2.30	2.67	2.95	3.23	3.60	3.88
80	1.30	1.71	2.00	2.28	2.64	2.91	3.17	3.53	3.80
90	1.30	1.70	1.99	2.26	2.61	2.87	3.13	3.48	3.74
100	1.29	1.69	1.97	2.24	2.59	2.85	3.10	3.44	3.69
120	1.29	1.68	1.96	2.22	2.56	2.81	3.06	3.38	3.62
140	1.29	1.68	1.95	2.21	2.54	2.78	3.02	3.34	3.57
160	1.28	1.67	1.94	2.19	2.52	2.76	3.00	3.31	3.54
180	1.28	1.67	1.93	2.19	2.51	2.74	2.98	3.28	3.51
200	1.28	1.66	1.93	2.18	2.50	2.73	2.96	3.26	3.49
300	1.27	1.65	1.91	2.16	2.47	2.69	2.92	3.21	3.42
500	1.27	1.64	1.90	2.14	2.44	2.66	2.88	3.16	3.37
∞	1.27	1.63	1.88	2.11	2.41	2.62	2.83	3.10	3.30

TABLE B.4 (cont.)   Critical Values of the *F* Distribution

Numerator DF = 10

Denom. DF	$\alpha(2)$: 0.50 $\alpha(1)$: 0.25	0.20 0.10	0.10 0.05	0.05 0.025	0.02 0.01	0.01 0.005	0.005 0.0025	0.002 0.001	0.001 0.0005
1	9.32	60.2	242.	969.	6060.	24200.	96900.	606000.	2420000.
2	3.38	9.39	19.4	39.4	99.4	199.	399.	999.	2000.
3	2.44	5.23	8.79	14.4	27.2	43.7	69.8	129.	206.
4	2.08	3.92	5.96	8.84	14.5	21.0	30.0	48.1	68.3
5	1.89	3.30	4.74	6.62	10.1	13.6	18.3	26.9	35.9
6	1.77	2.94	4.06	5.46	7.87	10.3	13.2	18.4	23.5
7	1.69	2.70	3.64	4.76	6.62	8.38	10.5	14.1	17.5
8	1.63	2.54	3.35	4.30	5.81	7.21	8.87	11.5	14.0
9	1.59	2.42	3.14	3.96	5.26	6.42	7.77	9.89	11.8
10	1.55	2.32	2.98	3.72	4.85	5.85	6.99	8.75	10.3
11	1.52	2.25	2.85	3.53	4.54	5.42	6.41	7.92	9.24
12	1.50	2.19	2.75	3.37	4.30	5.09	5.97	7.29	8.43
13	1.48	2.14	2.67	3.25	4.10	4.82	5.62	6.80	7.81
14	1.46	2.10	2.60	3.15	3.94	4.60	5.33	6.40	7.31
15	1.45	2.06	2.54	3.06	3.80	4.42	5.10	6.08	6.91
16	1.44	2.03	2.49	2.99	3.69	4.27	4.90	5.81	6.57
17	1.43	2.00	2.45	2.92	3.59	4.14	4.73	5.58	6.29
18	1.42	1.98	2.41	2.87	3.51	4.03	4.59	5.39	6.05
19	1.41	1.96	2.38	2.82	3.43	3.93	4.46	5.22	5.84
20	1.40	1.94	2.35	2.77	3.37	3.85	4.35	5.08	5.66
21	1.39	1.92	2.32	2.73	3.31	3.77	4.26	4.95	5.50
22	1.39	1.90	2.30	2.70	3.26	3.70	4.17	4.83	5.36
23	1.38	1.89	2.27	2.67	3.21	3.64	4.09	4.73	5.24
24	1.38	1.88	2.25	2.64	3.17	3.59	4.03	4.64	5.13
25	1.37	1.87	2.24	2.61	3.13	3.54	3.96	4.56	5.03
26	1.37	1.86	2.22	2.59	3.09	3.49	3.91	4.48	4.94
27	1.36	1.85	2.20	2.57	3.06	3.45	3.85	4.41	4.86
28	1.36	1.84	2.19	2.55	3.03	3.41	3.81	4.35	4.78
29	1.35	1.83	2.18	2.53	3.00	3.38	3.76	4.29	4.71
30	1.35	1.82	2.16	2.51	2.98	3.34	3.72	4.24	4.65
35	1.34	1.79	2.11	2.44	2.88	3.21	3.56	4.03	4.39
40	1.33	1.76	2.08	2.39	2.80	3.12	3.44	3.87	4.21
45	1.32	1.74	2.05	2.35	2.74	3.04	3.35	3.76	4.08
50	1.31	1.73	2.03	2.32	2.70	2.99	3.28	3.67	3.97
60	1.30	1.71	1.99	2.27	2.63	2.90	3.18	3.54	3.82
70	1.30	1.69	1.97	2.24	2.59	2.85	3.11	3.45	3.71
80	1.29	1.68	1.95	2.21	2.55	2.80	3.05	3.39	3.64
90	1.29	1.67	1.94	2.19	2.52	2.77	3.01	3.34	3.58
100	1.28	1.66	1.93	2.18	2.50	2.74	2.98	3.30	3.53
120	1.28	1.65	1.91	2.16	2.47	2.71	2.94	3.24	3.46
140	1.28	1.64	1.90	2.14	2.45	2.68	2.90	3.20	3.42
160	1.27	1.64	1.89	2.13	2.43	2.66	2.88	3.17	3.38
180	1.27	1.63	1.88	2.12	2.42	2.64	2.86	3.14	3.35
200	1.27	1.63	1.88	2.11	2.41	2.63	2.84	3.12	3.33
300	1.26	1.62	1.86	2.09	2.38	2.59	2.80	3.07	3.27
500	1.26	1.61	1.85	2.07	2.36	2.56	2.76	3.02	3.22
∞	1.25	1.60	1.83	2.05	2.32	2.52	2.71	2.96	3.14

TABLE B.4 (cont.)   Critical Values of the *F* Distribution

Numerator DF = 11

Denom. DF	$\alpha(2)$: 0.50 $\alpha(1)$: 0.25	0.20 0.10	0.10 0.05	0.05 0.025	0.02 0.01	0.01 0.005	0.005 0.0025	0.002 0.001	0.001 0.0005
1	9.37	60.5	243.	973.	6080.	24300.	97300.	608000.	2430000.
2	3.39	9.40	19.4	39.4	99.4	199.	399.	999.	2000.
3	2.45	5.22	8.76	14.4	27.1	43.5	69.5	129.	205.
4	2.08	3.91	5.94	8.79	14.5	20.8	29.8	47.7	67.8
5	1.89	3.28	4.70	6.57	9.96	13.5	18.1	26.6	35.5
6	1.77	2.92	4.03	5.41	7.79	10.1	13.1	18.2	23.2
7	1.69	2.68	3.60	4.71	6.54	8.27	10.4	13.9	17.2
8	1.63	2.52	3.31	4.24	5.73	7.10	8.73	11.4	13.8
9	1.58	2.40	3.10	3.91	5.18	6.31	7.63	9.72	11.6
10	1.55	2.30	2.94	3.66	4.77	5.75	6.86	8.59	10.1
11	1.52	2.23	2.82	3.47	4.46	5.32	6.29	7.76	9.05
12	1.49	2.17	2.72	3.32	4.22	4.99	5.85	7.14	8.25
13	1.47	2.12	2.63	3.20	4.02	4.72	5.50	6.65	7.63
14	1.46	2.07	2.57	3.09	3.86	4.51	5.21	6.26	7.13
15	1.44	2.04	2.51	3.01	3.73	4.33	4.98	5.94	6.73
16	1.43	2.01	2.46	2.93	3.62	4.18	4.79	5.67	6.40
17	1.42	1.98	2.41	2.87	3.52	4.05	4.62	5.44	6.12
18	1.41	1.95	2.37	2.81	3.43	3.94	4.48	5.25	5.89
19	1.40	1.93	2.34	2.76	3.36	3.84	4.35	5.08	5.68
20	1.39	1.91	2.31	2.72	3.29	3.76	4.24	4.94	5.50
21	1.39	1.90	2.28	2.68	3.24	3.68	4.15	4.81	5.35
22	1.38	1.88	2.26	2.65	3.18	3.61	4.06	4.70	5.21
23	1.37	1.87	2.24	2.62	3.14	3.55	3.99	4.60	5.09
24	1.37	1.85	2.22	2.59	3.09	3.50	3.92	4.51	4.98
25	1.36	1.84	2.20	2.56	3.06	3.45	3.85	4.42	4.88
26	1.36	1.83	2.18	2.54	3.02	3.40	3.80	4.35	4.79
27	1.35	1.82	2.17	2.51	2.99	3.36	3.75	4.28	4.71
28	1.35	1.81	2.15	2.49	2.96	3.32	3.70	4.22	4.63
29	1.35	1.80	2.14	2.48	2.93	3.29	3.66	4.16	4.56
30	1.34	1.79	2.13	2.46	2.91	3.25	3.61	4.11	4.50
35	1.33	1.76	2.07	2.39	2.80	3.12	3.45	3.90	4.25
40	1.32	1.74	2.04	2.33	2.73	3.03	3.33	3.75	4.07
45	1.31	1.72	2.01	2.29	2.67	2.96	3.25	3.64	3.94
50	1.30	1.70	1.99	2.26	2.63	2.90	3.18	3.55	3.83
60	1.29	1.68	1.95	2.22	2.56	2.82	3.08	3.42	3.68
70	1.29	1.66	1.93	2.18	2.51	2.76	3.00	3.33	3.58
80	1.28	1.65	1.91	2.16	2.48	2.72	2.95	3.27	3.50
90	1.28	1.64	1.90	2.14	2.45	2.68	2.91	3.22	3.44
100	1.27	1.64	1.89	2.12	2.43	2.66	2.88	3.18	3.40
120	1.27	1.63	1.87	2.10	2.40	2.62	2.83	3.12	3.33
140	1.27	1.62	1.86	2.09	2.38	2.59	2.80	3.08	3.28
160	1.26	1.61	1.85	2.07	2.36	2.57	2.78	3.05	3.25
180	1.26	1.61	1.84	2.07	2.35	2.56	2.76	3.02	3.22
200	1.26	1.60	1.84	2.06	2.34	2.54	2.74	3.00	3.20
300	1.26	1.59	1.82	2.04	2.31	2.51	2.70	2.95	3.14
500	1.25	1.58	1.81	2.02	2.28	2.48	2.66	2.91	3.09
∞	1.25	1.57	1.79	1.99	2.25	2.43	2.61	2.84	3.01

TABLE B.4 (cont.) Critical Values of the *F* Distribution

Numerator DF = 12

Denom. DF	$\alpha(2)$: 0.50 $\alpha(1)$: 0.25	0.20 0.10	0.10 0.05	0.05 0.025	0.02 0.01	0.01 0.005	0.005 0.0025	0.002 0.001	0.001 0.0005
1	9.41	60.7	244.	977.	6110.	24400.	97700.	611000.	2440000.
2	3.39	9.41	19.4	39.4	99.4	199.	399.	999.	2000.
3	2.45	5.22	8.74	14.3	27.1	43.4	69.3	128.	204.
4	2.08	3.90	5.91	8.75	14.4	20.7	29.7	47.4	67.4
5	1.89	3.27	4.68	6.52	9.89	13.4	18.0	26.4	35.2
6	1.77	2.90	4.00	5.37	7.72	10.0	12.9	18.0	23.0
7	1.68	2.67	3.57	4.67	6.47	8.18	10.3	13.7	17.0
8	1.62	2.50	3.28	4.20	5.67	7.01	8.61	11.2	13.6
9	1.58	2.38	3.07	3.87	5.11	6.23	7.52	9.57	11.4
10	1.54	2.28	2.91	3.62	4.71	5.66	6.75	8.45	9.94
11	1.51	2.21	2.79	3.43	4.40	5.24	6.18	7.63	8.88
12	1.49	2.15	2.69	3.28	4.16	4.91	5.74	7.00	8.09
13	1.47	2.10	2.60	3.15	3.96	4.64	5.40	6.52	7.48
14	1.45	2.05	2.53	3.05	3.80	4.43	5.12	6.13	6.99
15	1.44	2.02	2.48	2.96	3.67	4.25	4.88	5.81	6.59
16	1.43	1.99	2.42	2.89	3.55	4.10	4.69	5.55	6.26
17	1.41	1.96	2.38	2.82	3.46	3.97	4.52	5.32	5.98
18	1.40	1.93	2.34	2.77	3.37	3.86	4.38	5.13	5.75
19	1.40	1.91	2.31	2.72	3.30	3.76	4.26	4.97	5.55
20	1.39	1.89	2.28	2.68	3.23	3.68	4.15	4.82	5.37
21	1.38	1.87	2.25	2.64	3.17	3.60	4.06	4.70	5.21
22	1.37	1.86	2.23	2.60	3.12	3.54	3.97	4.58	5.08
23	1.37	1.84	2.20	2.57	3.07	3.47	3.89	4.48	4.96
24	1.36	1.83	2.18	2.54	3.03	3.42	3.83	4.39	4.85
25	1.36	1.82	2.16	2.51	2.99	3.37	3.76	4.31	4.75
26	1.35	1.81	2.15	2.49	2.96	3.33	3.71	4.24	4.66
27	1.35	1.80	2.13	2.47	2.93	3.28	3.66	4.17	4.58
28	1.34	1.79	2.12	2.45	2.90	3.25	3.61	4.11	4.51
29	1.34	1.78	2.10	2.43	2.87	3.21	3.56	4.05	4.44
30	1.34	1.77	2.09	2.41	2.84	3.18	3.52	4.00	4.38
35	1.32	1.74	2.04	2.34	2.74	3.05	3.36	3.79	4.13
40	1.31	1.71	2.00	2.29	2.66	2.95	3.25	3.64	3.95
45	1.30	1.70	1.97	2.25	2.61	2.88	3.16	3.53	3.82
50	1.30	1.68	1.95	2.22	2.56	2.82	3.09	3.44	3.71
60	1.29	1.66	1.92	2.17	2.50	2.74	2.99	3.32	3.57
70	1.28	1.64	1.89	2.14	2.45	2.68	2.92	3.23	3.46
80	1.27	1.63	1.88	2.11	2.42	2.64	2.87	3.16	3.39
90	1.27	1.62	1.86	2.09	2.39	2.61	2.83	3.11	3.33
100	1.27	1.61	1.85	2.08	2.37	2.58	2.80	3.07	3.28
120	1.26	1.60	1.83	2.05	2.34	2.54	2.75	3.02	3.22
140	1.26	1.59	1.82	2.04	2.31	2.52	2.72	2.98	3.17
160	1.26	1.59	1.81	2.03	2.30	2.50	2.69	2.95	3.14
180	1.25	1.58	1.81	2.02	2.28	2.48	2.67	2.92	3.11
200	1.25	1.58	1.80	2.01	2.27	2.47	2.66	2.90	3.09
300	1.25	1.57	1.78	1.99	2.24	2.43	2.61	2.85	3.02
500	1.24	1.56	1.77	1.97	2.22	2.40	2.58	2.81	2.97
∞	1.24	1.55	1.75	1.94	2.18	2.36	2.53	2.74	2.90

TABLE B.4 (cont.)   Critical Values of the *F* Distribution

Numerator DF = 13

Denom. DF	α(2): 0.50 α(1): 0.25	0.20 0.10	0.10 0.05	0.05 0.025	0.02 0.01	0.01 0.005	0.005 0.0025	0.002 0.001	0.001 0.0005
1	9.44	60.9	245.	980.	6130.	24500.	98000.	613000.	2450000.
2	3.40	9.41	19.4	39.4	99.4	199.	399.	999.	2000.
3	2.45	5.21	8.73	14.3	27.0	43.3	69.1	128.	204.
4	2.08	3.89	5.89	8.71	14.3	20.6	29.5	47.2	67.1
5	1.89	3.26	4.66	6.49	9.82	13.3	17.9	26.2	34.9
6	1.77	2.89	3.98	5.33	7.66	9.95	12.8	17.8	22.7
7	1.68	2.65	3.55	4.63	6.41	8.10	10.1	13.6	16.8
8	1.62	2.49	3.26	4.16	5.61	6.94	8.51	11.1	13.4
9	1.58	2.36	3.05	3.83	5.05	6.15	7.43	9.44	11.3
10	1.54	2.27	2.89	3.58	4.65	5.59	6.66	8.32	9.80
11	1.51	2.19	2.76	3.39	4.34	5.16	6.09	7.51	8.74
12	1.49	2.13	2.66	3.24	4.10	4.84	5.66	6.89	7.96
13	1.47	2.08	2.58	3.12	3.91	4.57	5.31	6.41	7.35
14	1.45	2.04	2.51	3.01	3.75	4.36	5.03	6.02	6.86
15	1.43	2.00	2.45	2.92	3.61	4.18	4.80	5.71	6.47
16	1.42	1.97	2.40	2.85	3.50	4.03	4.61	5.44	6.14
17	1.41	1.94	2.35	2.79	3.40	3.90	4.44	5.22	5.86
18	1.40	1.92	2.31	2.73	3.32	3.79	4.30	5.03	5.63
19	1.39	1.89	2.28	2.68	3.24	3.70	4.18	4.87	5.43
20	1.38	1.87	2.25	2.64	3.18	3.61	4.07	4.72	5.25
21	1.37	1.86	2.22	2.60	3.12	3.54	3.98	4.60	5.10
22	1.37	1.84	2.20	2.56	3.07	3.47	3.89	4.49	4.97
23	1.36	1.83	2.18	2.53	3.02	3.41	3.82	4.39	4.84
24	1.36	1.81	2.15	2.50	2.98	3.35	3.75	4.30	4.74
25	1.35	1.80	2.14	2.48	2.94	3.30	3.69	4.22	4.64
26	1.35	1.79	2.12	2.45	2.90	3.26	3.63	4.14	4.55
27	1.34	1.78	2.10	2.43	2.87	3.22	3.58	4.08	4.47
28	1.34	1.77	2.09	2.41	2.84	3.18	3.53	4.01	4.40
29	1.33	1.76	2.08	2.39	2.81	3.15	3.49	3.96	4.33
30	1.33	1.75	2.06	2.37	2.79	3.11	3.45	3.91	4.27
35	1.32	1.72	2.01	2.30	2.69	2.98	3.29	3.70	4.02
40	1.31	1.70	1.97	2.25	2.61	2.89	3.17	3.55	3.85
45	1.30	1.68	1.94	2.21	2.55	2.82	3.08	3.44	3.71
50	1.29	1.66	1.92	2.18	2.51	2.76	3.01	3.35	3.61
60	1.28	1.64	1.89	2.13	2.44	2.68	2.91	3.23	3.46
70	1.27	1.62	1.86	2.10	2.40	2.62	2.84	3.14	3.36
80	1.27	1.61	1.84	2.07	2.36	2.58	2.79	3.07	3.29
90	1.26	1.60	1.83	2.05	2.33	2.54	2.75	3.02	3.23
100	1.26	1.59	1.82	2.04	2.31	2.52	2.72	2.99	3.19
120	1.26	1.58	1.80	2.01	2.28	2.48	2.67	2.93	3.12
140	1.25	1.57	1.79	2.00	2.26	2.45	2.64	2.89	3.07
160	1.25	1.57	1.78	1.99	2.24	2.43	2.62	2.86	3.04
180	1.25	1.56	1.77	1.98	2.23	2.42	2.60	2.83	3.01
200	1.24	1.56	1.77	1.97	2.22	2.40	2.58	2.82	2.99
300	1.24	1.55	1.75	1.95	2.19	2.37	2.54	2.76	2.93
500	1.24	1.54	1.74	1.93	2.17	2.34	2.50	2.72	2.88
∞	1.23	1.52	1.72	1.90	2.13	2.29	2.45	2.66	2.81

TABLE B.4 (cont.)   Critical Values of the *F* Distribution

Numerator DF = 14

Denom. DF	α(2): 0.50 α(1): 0.25	0.20 0.10	0.10 0.05	0.05 0.025	0.02 0.01	0.01 0.005	0.005 0.0025	0.002 0.001	0.001 0.0005
1	9.47	61.1	245.	983.	6140.	24600.	98300.	614000.	2460000.
2	3.41	9.42	19.4	39.4	99.4	199.	399.	999.	2000.
3	2.45	5.20	8.71	14.3	26.9	43.2	69.0	128.	203.
4	2.08	3.88	5.87	8.68	14.2	20.5	29.4	46.9	66.8
5	1.89	3.25	4.64	6.46	9.77	13.2	17.8	26.1	34.7
6	1.76	2.88	3.96	5.30	7.60	9.88	12.7	17.7	22.6
7	1.68	2.64	3.53	4.60	6.36	8.03	10.1	13.4	16.6
8	1.62	2.48	3.24	4.13	5.56	6.87	8.43	10.9	13.3
9	1.57	2.35	3.03	3.80	5.01	6.09	7.35	9.33	11.1
10	1.54	2.26	2.86	3.55	4.60	5.53	6.58	8.22	9.67
11	1.51	2.18	2.74	3.36	4.29	5.10	6.02	7.41	8.62
12	1.48	2.12	2.64	3.21	4.05	4.77	5.58	6.79	7.84
13	1.46	2.07	2.55	3.08	3.86	4.51	5.24	6.31	7.23
14	1.44	2.02	2.48	2.98	3.70	4.30	4.96	5.93	6.75
15	1.43	1.99	2.42	2.89	3.56	4.12	4.73	5.62	6.36
16	1.42	1.95	2.37	2.82	3.45	3.97	4.54	5.35	6.03
17	1.41	1.93	2.33	2.75	3.35	3.84	4.37	5.13	5.76
18	1.40	1.90	2.29	2.70	3.27	3.73	4.23	4.94	5.53
19	1.39	1.88	2.26	2.65	3.19	3.64	4.11	4.78	5.33
20	1.38	1.86	2.22	2.60	3.13	3.55	4.00	4.64	5.15
21	1.37	1.84	2.20	2.56	3.07	3.48	3.91	4.51	5.00
22	1.36	1.83	2.17	2.53	3.02	3.41	3.82	4.40	4.87
23	1.36	1.81	2.15	2.50	2.97	3.35	3.75	4.30	4.75
24	1.35	1.80	2.13	2.47	2.93	3.30	3.68	4.21	4.64
25	1.35	1.79	2.11	2.44	2.89	3.25	3.62	4.13	4.54
26	1.34	1.77	2.09	2.42	2.86	3.20	3.56	4.06	4.46
27	1.34	1.76	2.08	2.39	2.82	3.16	3.51	3.99	4.38
28	1.33	1.75	2.06	2.37	2.79	3.12	3.46	3.93	4.30
29	1.33	1.75	2.05	2.36	2.77	3.09	3.42	3.88	4.24
30	1.33	1.74	2.04	2.34	2.74	3.06	3.38	3.82	4.18
35	1.31	1.70	1.99	2.27	2.64	2.93	3.22	3.62	3.93
40	1.30	1.68	1.95	2.21	2.56	2.83	3.10	3.47	3.76
45	1.29	1.66	1.92	2.17	2.51	2.76	3.02	3.36	3.63
50	1.28	1.64	1.89	2.14	2.46	2.70	2.95	3.27	3.52
60	1.27	1.62	1.86	2.09	2.39	2.62	2.85	3.15	3.38
70	1.27	1.60	1.84	2.06	2.35	2.56	2.78	3.06	3.28
80	1.26	1.59	1.82	2.03	2.31	2.52	2.73	3.00	3.20
90	1.26	1.58	1.80	2.02	2.29	2.49	2.69	2.95	3.14
100	1.25	1.57	1.79	2.00	2.27	2.46	2.65	2.91	3.10
120	1.25	1.56	1.78	1.98	2.23	2.42	2.61	2.85	3.03
140	1.24	1.55	1.76	1.96	2.21	2.40	2.58	2.81	2.99
160	1.24	1.55	1.75	1.95	2.20	2.38	2.55	2.78	2.95
180	1.24	1.54	1.75	1.94	2.18	2.36	2.53	2.76	2.93
200	1.24	1.54	1.74	1.93	2.17	2.35	2.52	2.74	2.91
300	1.23	1.53	1.72	1.91	2.14	2.31	2.47	2.69	2.84
500	1.23	1.52	1.71	1.89	2.12	2.28	2.44	2.64	2.79
∞	1.22	1.50	1.69	1.87	2.08	2.24	2.39	2.58	2.72

TABLE B.4 (cont.) Critical Values of the *F* Distribution

Numerator DF = 15

Denom. DF	$\alpha(2)$: 0.50 $\alpha(1)$: 0.25	0.20 0.10	0.10 0.05	0.05 0.025	0.02 0.01	0.01 0.005	0.005 0.0025	0.002 0.001	0.001 0.0005
1	9.49	61.2	246.	985.	6160.	24600.	98500.	616000.	2460000.
2	3.41	9.42	19.4	39.4	99.4	199.	399.	999.	2000.
3	2.46	5.20	8.70	14.3	26.9	43.1	68.8	127.	203.
4	2.08	3.87	5.86	8.66	14.2	20.4	29.3	46.8	66.5
5	1.89	3.24	4.62	6.43	9.72	13.1	17.7	25.9	34.5
6	1.76	2.87	3.94	5.27	7.56	9.81	12.7	17.6	22.4
7	1.68	2.63	3.51	4.57	6.31	7.97	9.98	13.3	16.5
8	1.62	2.46	3.22	4.10	5.52	6.81	8.35	10.8	13.1
9	1.57	2.34	3.01	3.77	4.96	6.03	7.28	9.24	11.0
10	1.53	2.24	2.85	3.52	4.56	5.47	6.51	8.13	9.56
11	1.50	2.17	2.72	3.33	4.25	5.05	5.95	7.32	8.52
12	1.48	2.10	2.62	3.18	4.01	4.72	5.52	6.71	7.74
13	1.46	2.05	2.53	3.05	3.82	4.46	5.17	6.23	7.13
14	1.44	2.01	2.46	2.95	3.66	4.25	4.89	5.85	6.65
15	1.43	1.97	2.40	2.86	3.52	4.07	4.67	5.54	6.26
16	1.41	1.94	2.35	2.79	3.41	3.92	4.47	5.27	5.94
17	1.40	1.91	2.31	2.72	3.31	3.79	4.31	5.05	5.67
18	1.39	1.89	2.27	2.67	3.23	3.68	4.17	4.87	5.44
19	1.38	1.86	2.23	2.62	3.15	3.59	4.05	4.70	5.24
20	1.37	1.84	2.20	2.57	3.09	3.50	3.94	4.56	5.07
21	1.37	1.83	2.18	2.53	3.03	3.43	3.85	4.44	4.92
22	1.36	1.81	2.15	2.50	2.98	3.36	3.76	4.33	4.78
23	1.35	1.80	2.13	2.47	2.93	3.30	3.69	4.23	4.66
24	1.35	1.78	2.11	2.44	2.89	3.25	3.62	4.14	4.56
25	1.34	1.77	2.09	2.41	2.85	3.20	3.56	4.06	4.46
26	1.34	1.76	2.07	2.39	2.81	3.15	3.50	3.99	4.37
27	1.33	1.75	2.06	2.36	2.78	3.11	3.45	3.92	4.29
28	1.33	1.74	2.04	2.34	2.75	3.07	3.40	3.86	4.22
29	1.32	1.73	2.03	2.32	2.73	3.04	3.36	3.80	4.15
30	1.32	1.72	2.01	2.31	2.70	3.01	3.32	3.75	4.09
35	1.31	1.69	1.96	2.23	2.60	2.88	3.16	3.55	3.85
40	1.30	1.66	1.92	2.18	2.52	2.78	3.04	3.40	3.68
45	1.29	1.64	1.89	2.14	2.46	2.71	2.96	3.29	3.55
50	1.28	1.63	1.87	2.11	2.42	2.65	2.89	3.20	3.45
60	1.27	1.60	1.84	2.06	2.35	2.57	2.79	3.08	3.30
70	1.26	1.59	1.81	2.03	2.31	2.51	2.72	2.99	3.20
80	1.26	1.57	1.79	2.00	2.27	2.47	2.67	2.93	3.12
90	1.25	1.56	1.78	1.98	2.24	2.44	2.63	2.88	3.07
100	1.25	1.56	1.77	1.97	2.22	2.41	2.60	2.84	3.02
120	1.24	1.55	1.75	1.94	2.19	2.37	2.55	2.78	2.96
140	1.24	1.54	1.74	1.93	2.17	2.35	2.52	2.74	2.91
160	1.24	1.53	1.73	1.92	2.15	2.33	2.49	2.71	2.88
180	1.23	1.53	1.72	1.91	2.14	2.31	2.48	2.69	2.85
200	1.23	1.52	1.72	1.90	2.13	2.30	2.46	2.67	2.83
300	1.23	1.51	1.70	1.88	2.10	2.26	2.42	2.62	2.77
500	1.22	1.50	1.69	1.86	2.07	2.23	2.38	2.58	2.72
∞	1.22	1.49	1.67	1.83	2.04	2.19	2.33	2.51	2.65

TABLE B.4 (cont.) Critical Values of the *F* Distribution

Numerator DF = 16

Denom. DF	α(2): 0.50 α(1): 0.25	0.20 0.10	0.10 0.05	0.05 0.025	0.02 0.01	0.01 0.005	0.005 0.0025	0.002 0.001	0.001 0.0005
1	9.52	61.3	246.	987.	6170.	24700.	98700.	617000.	2470000.
2	3.41	9.43	19.4	39.4	99.4	199.	399.	999.	2000.
3	2.46	5.20	8.69	14.2	26.8	43.0	68.7	127.	202.
4	2.08	3.86	5.84	8.63	14.2	20.4	29.2	46.6	66.2
5	1.88	3.23	4.60	6.40	9.68	13.1	17.6	25.8	34.3
6	1.76	2.86	3.92	5.24	7.52	9.76	12.6	17.4	22.3
7	1.68	2.62	3.49	4.54	6.28	7.91	9.91	13.2	16.4
8	1.62	2.45	3.20	4.08	5.48	6.76	8.29	10.8	13.0
9	1.57	2.33	2.99	3.74	4.92	5.98	7.21	9.15	10.9
10	1.53	2.23	2.83	3.50	4.52	5.42	6.45	8.05	9.46
11	1.50	2.16	2.70	3.30	4.21	5.00	5.89	7.24	8.43
12	1.48	2.09	2.60	3.15	3.97	4.67	5.46	6.63	7.65
13	1.46	2.04	2.51	3.03	3.78	4.41	5.11	6.16	7.05
14	1.44	2.00	2.44	2.92	3.62	4.20	4.84	5.78	6.57
15	1.42	1.96	2.38	2.84	3.49	4.02	4.61	5.46	6.18
16	1.41	1.93	2.33	2.76	3.37	3.87	4.42	5.20	5.86
17	1.40	1.90	2.29	2.70	3.27	3.75	4.25	4.99	5.59
18	1.39	1.87	2.25	2.64	3.19	3.64	4.11	4.80	5.36
19	1.38	1.85	2.21	2.59	3.12	3.54	3.99	4.64	5.16
20	1.37	1.83	2.18	2.55	3.05	3.46	3.89	4.49	4.99
21	1.36	1.81	2.16	2.51	2.99	3.38	3.79	4.37	4.84
22	1.36	1.80	2.13	2.47	2.94	3.31	3.71	4.26	4.71
23	1.35	1.78	2.11	2.44	2.89	3.25	3.63	4.16	4.59
24	1.34	1.77	2.09	2.41	2.85	3.20	3.56	4.07	4.48
25	1.34	1.76	2.07	2.38	2.81	3.15	3.50	3.99	4.39
26	1.33	1.75	2.05	2.36	2.78	3.11	3.45	3.92	4.30
27	1.33	1.74	2.04	2.34	2.75	3.07	3.40	3.86	4.22
28	1.32	1.73	2.02	2.32	2.72	3.03	3.35	3.80	4.15
29	1.32	1.72	2.01	2.30	2.69	2.99	3.31	3.74	4.08
30	1.32	1.71	1.99	2.28	2.66	2.96	3.27	3.69	4.02
35	1.30	1.67	1.94	2.21	2.56	2.83	3.11	3.48	3.78
40	1.29	1.65	1.90	2.15	2.48	2.74	2.99	3.34	3.61
45	1.28	1.63	1.87	2.11	2.43	2.66	2.90	3.23	3.48
50	1.27	1.61	1.85	2.08	2.38	2.61	2.84	3.14	3.38
60	1.26	1.59	1.82	2.03	2.31	2.53	2.74	3.02	3.23
70	1.26	1.57	1.79	2.00	2.27	2.47	2.67	2.93	3.13
80	1.25	1.56	1.77	1.97	2.23	2.43	2.62	2.87	3.06
90	1.25	1.55	1.76	1.95	2.21	2.39	2.58	2.82	3.00
100	1.24	1.54	1.75	1.94	2.19	2.37	2.55	2.78	2.96
120	1.24	1.53	1.73	1.92	2.15	2.33	2.50	2.72	2.89
140	1.23	1.52	1.72	1.90	2.13	2.30	2.47	2.68	2.84
160	1.23	1.52	1.71	1.89	2.11	2.28	2.44	2.65	2.81
180	1.23	1.51	1.70	1.88	2.10	2.26	2.42	2.63	2.78
200	1.23	1.51	1.69	1.87	2.09	2.25	2.41	2.61	2.76
300	1.22	1.49	1.68	1.85	2.06	2.21	2.36	2.56	2.70
500	1.22	1.49	1.66	1.83	2.04	2.19	2.33	2.52	2.65
∞	1.21	1.47	1.64	1.80	2.00	2.14	2.28	2.45	2.58

TABLE B.4 (cont.)   Critical Values of the *F* Distribution

Numerator DF = 17

Denom. DF	α(2): 0.50 α(1): 0.25	0.20 0.10	0.10 0.05	0.05 0.025	0.02 0.01	0.01 0.005	0.005 0.0025	0.002 0.001	0.001 0.0005
1	9.53	61.5	247.	989.	6180.	24700.	98900.	618000.	2470000.
2	3.42	9.43	19.4	39.4	99.4	199.	399.	999.	2000.
3	2.46	5.19	8.68	14.2	26.8	42.9	68.6	127.	202.
4	2.08	3.86	5.83	8.61	14.1	20.3	29.1	46.5	66.0
5	1.88	3.22	4.59	6.38	9.64	13.0	17.5	25.7	34.2
6	1.76	2.85	3.91	5.22	7.48	9.71	12.5	17.4	22.1
7	1.67	2.61	3.48	4.52	6.24	7.87	9.85	13.1	16.3
8	1.61	2.45	3.19	4.05	5.44	6.72	8.23	10.7	12.9
9	1.57	2.32	2.97	3.72	4.89	5.94	7.16	9.08	10.8
10	1.53	2.22	2.81	3.47	4.49	5.38	6.40	7.98	9.38
11	1.50	2.15	2.69	3.28	4.18	4.96	5.84	7.17	8.34
12	1.47	2.08	2.58	3.13	3.94	4.63	5.40	6.57	7.57
13	1.45	2.03	2.50	3.00	3.75	4.37	5.06	6.09	6.97
14	1.44	1.99	2.43	2.90	3.59	4.16	4.79	5.71	6.49
15	1.42	1.95	2.37	2.81	3.45	3.98	4.56	5.40	6.11
16	1.41	1.92	2.32	2.74	3.34	3.83	4.37	5.14	5.79
17	1.39	1.89	2.27	2.67	3.24	3.71	4.21	4.92	5.52
18	1.38	1.86	2.23	2.62	3.16	3.60	4.07	4.74	5.29
19	1.37	1.84	2.20	2.57	3.08	3.50	3.94	4.58	5.09
20	1.37	1.82	2.17	2.52	3.02	3.42	3.84	4.44	4.92
21	1.36	1.80	2.14	2.48	2.96	3.34	3.74	4.31	4.77
22	1.35	1.79	2.11	2.45	2.91	3.27	3.66	4.20	4.64
23	1.35	1.77	2.09	2.42	2.86	3.21	3.58	4.10	4.52
24	1.34	1.76	2.07	2.39	2.82	3.16	3.52	4.02	4.41
25	1.33	1.75	2.05	2.36	2.78	3.11	3.46	3.94	4.32
26	1.33	1.73	2.03	2.34	2.75	3.07	3.40	3.86	4.23
27	1.33	1.72	2.02	2.31	2.71	3.03	3.35	3.80	4.15
28	1.32	1.71	2.00	2.29	2.68	2.99	3.30	3.74	4.08
29	1.32	1.71	1.99	2.27	2.66	2.95	3.26	3.68	4.02
30	1.31	1.70	1.98	2.26	2.63	2.92	3.22	3.63	3.96
35	1.30	1.66	1.92	2.18	2.53	2.79	3.06	3.43	3.72
40	1.29	1.64	1.89	2.13	2.45	2.70	2.95	3.28	3.54
45	1.28	1.62	1.86	2.09	2.39	2.62	2.86	3.17	3.41
50	1.27	1.60	1.83	2.06	2.35	2.57	2.79	3.09	3.31
60	1.26	1.58	1.80	2.01	2.28	2.49	2.69	2.96	3.17
70	1.25	1.56	1.77	1.97	2.23	2.43	2.62	2.88	3.07
80	1.25	1.55	1.75	1.95	2.20	2.39	2.57	2.81	3.00
90	1.24	1.54	1.74	1.93	2.17	2.35	2.53	2.76	2.94
100	1.24	1.53	1.73	1.91	2.15	2.33	2.50	2.73	2.89
120	1.23	1.52	1.71	1.89	2.12	2.29	2.45	2.67	2.83
140	1.23	1.51	1.70	1.87	2.10	2.26	2.42	2.63	2.78
160	1.23	1.50	1.69	1.86	2.08	2.24	2.40	2.60	2.75
180	1.22	1.50	1.68	1.85	2.07	2.22	2.38	2.58	2.72
200	1.22	1.49	1.67	1.84	2.06	2.21	2.36	2.56	2.70
300	1.22	1.48	1.66	1.82	2.03	2.17	2.32	2.50	2.64
500	1.21	1.47	1.64	1.80	2.00	2.14	2.28	2.46	2.59
∞	1.21	1.46	1.62	1.78	1.97	2.10	2.23	2.40	2.52

TABLE B.4 (cont.) Critical Values of the *F* Distribution

Numerator DF = 18

Denom. DF	$\alpha(2)$: 0.50 $\alpha(1)$: 0.25	0.20 0.10	0.10 0.05	0.05 0.025	0.02 0.01	0.01 0.005	0.005 0.0025	0.002 0.001	0.001 0.0005
1	9.55	61.6	247.	990.	6190.	24800.	99100.	619000.	2480000.
2	3.42	9.44	19.4	39.4	99.4	199.	399.	999.	2000.
3	2.46	5.19	8.67	14.2	26.8	42.9	68.5	127.	202.
4	2.08	3.85	5.82	8.59	14.1	20.3	29.0	46.3	65.8
5	1.88	3.22	4.58	6.36	9.61	13.0	17.4	25.6	34.0
6	1.76	2.85	3.90	5.20	7.45	9.66	12.4	17.3	22.0
7	1.67	2.61	3.47	4.50	6.21	7.83	9.79	13.1	16.2
8	1.61	2.44	3.17	4.03	5.41	6.68	8.18	10.6	12.8
9	1.56	2.31	2.96	3.70	4.86	5.90	7.11	9.01	10.7
10	1.53	2.22	2.80	3.45	4.46	5.34	6.35	7.91	9.30
11	1.50	2.14	2.67	3.26	4.15	4.92	5.79	7.11	8.27
12	1.47	2.08	2.57	3.11	3.91	4.59	5.36	6.51	7.50
13	1.45	2.02	2.48	2.98	3.72	4.33	5.02	6.03	6.90
14	1.43	1.98	2.41	2.88	3.56	4.12	4.74	5.66	6.43
15	1.42	1.94	2.35	2.79	3.42	3.95	4.51	5.35	6.04
16	1.40	1.91	2.30	2.72	3.31	3.80	4.32	5.09	5.72
17	1.39	1.88	2.26	2.65	3.21	3.67	4.16	4.87	5.45
18	1.38	1.85	2.22	2.60	3.13	3.56	4.02	4.68	5.23
19	1.37	1.83	2.18	2.55	3.05	3.46	3.90	4.52	5.03
20	1.36	1.81	2.15	2.50	2.99	3.38	3.79	4.38	4.86
21	1.36	1.79	2.12	2.46	2.93	3.31	3.70	4.26	4.71
22	1.35	1.78	2.10	2.43	2.88	3.24	3.62	4.15	4.58
23	1.34	1.76	2.08	2.39	2.83	3.18	3.54	4.05	4.46
24	1.34	1.75	2.05	2.36	2.79	3.12	3.47	3.96	4.35
25	1.33	1.74	2.04	2.34	2.75	3.08	3.41	3.88	4.26
26	1.33	1.72	2.02	2.31	2.72	3.03	3.36	3.81	4.17
27	1.32	1.71	2.00	2.29	2.68	2.99	3.31	3.75	4.10
28	1.32	1.70	1.99	2.27	2.65	2.95	3.26	3.69	4.02
29	1.31	1.69	1.97	2.25	2.63	2.92	3.22	3.63	3.96
30	1.31	1.69	1.96	2.23	2.60	2.89	3.18	3.58	3.90
35	1.29	1.65	1.91	2.16	2.50	2.76	3.02	3.38	3.66
40	1.28	1.62	1.87	2.11	2.42	2.66	2.90	3.23	3.49
45	1.27	1.60	1.84	2.07	2.36	2.59	2.82	3.12	3.36
50	1.27	1.59	1.81	2.03	2.32	2.53	2.75	3.04	3.26
60	1.26	1.56	1.78	1.98	2.25	2.45	2.65	2.91	3.11
70	1.25	1.55	1.75	1.95	2.20	2.39	2.58	2.83	3.01
80	1.24	1.53	1.73	1.92	2.17	2.35	2.53	2.76	2.94
90	1.24	1.52	1.72	1.91	2.14	2.32	2.49	2.71	2.88
100	1.23	1.52	1.71	1.89	2.12	2.29	2.46	2.68	2.84
120	1.23	1.50	1.69	1.87	2.09	2.25	2.41	2.62	2.78
140	1.22	1.50	1.68	1.85	2.07	2.22	2.38	2.58	2.73
160	1.22	1.49	1.67	1.84	2.05	2.20	2.35	2.55	2.70
180	1.22	1.48	1.66	1.83	2.04	2.19	2.34	2.53	2.67
200	1.22	1.48	1.66	1.82	2.03	2.18	2.32	2.51	2.65
300	1.21	1.47	1.64	1.80	1.99	2.14	2.28	2.46	2.59
500	1.21	1.46	1.62	1.78	1.97	2.11	2.24	2.41	2.54
∞	1.20	1.44	1.60	1.75	1.93	2.06	2.19	2.35	2.47

TABLE B.4 (cont.) Critical Values of the F Distribution

Numerator DF = 19

Denom. DF	$\alpha(2)$: 0.50 $\alpha(1)$: 0.25	0.20 0.10	0.10 0.05	0.05 0.025	0.02 0.01	0.01 0.005	0.005 0.0025	0.002 0.001	0.001 0.0005
1	9.57	61.7	248.	992.	6200.	24800.	99200.	620000.	2480000.
2	3.42	9.44	19.4	39.4	99.4	199.	399.	999.	2000.
3	2.46	5.19	8.67	14.2	26.7	42.8	68.4	127.	201.
4	2.08	3.85	5.81	8.58	14.0	20.2	28.9	46.2	65.7
5	1.88	3.21	4.57	6.34	9.58	12.9	17.4	25.5	33.9
6	1.76	2.84	3.88	5.18	7.42	9.62	12.4	17.2	21.9
7	1.67	2.60	3.46	4.48	6.18	7.79	9.74	13.0	16.1
8	1.61	2.43	3.16	4.02	5.38	6.64	8.13	10.5	12.8
9	1.56	2.30	2.95	3.68	4.83	5.86	7.06	8.95	10.7
10	1.53	2.21	2.79	3.44	4.43	5.31	6.31	7.86	9.23
11	1.49	2.13	2.66	3.24	4.12	4.89	5.75	7.06	8.20
12	1.47	2.07	2.56	3.09	3.88	4.56	5.32	6.45	7.43
13	1.45	2.01	2.47	2.96	3.69	4.30	4.98	5.98	6.84
14	1.43	1.97	2.40	2.86	3.53	4.09	4.70	5.60	6.37
15	1.41	1.93	2.34	2.77	3.40	3.91	4.47	5.29	5.98
16	1.40	1.90	2.29	2.70	3.28	3.76	4.28	5.04	5.66
17	1.39	1.87	2.24	2.63	3.19	3.64	4.12	4.82	5.40
18	1.38	1.84	2.20	2.58	3.10	3.53	3.98	4.63	5.17
19	1.37	1.82	2.17	2.53	3.03	3.43	3.86	4.47	4.97
20	1.36	1.80	2.14	2.48	2.96	3.35	3.76	4.33	4.80
21	1.35	1.78	2.11	2.44	2.90	3.27	3.66	4.21	4.65
22	1.35	1.77	2.08	2.41	2.85	3.21	3.58	4.10	4.52
23	1.34	1.75	2.06	2.37	2.80	3.15	3.50	4.00	4.41
24	1.33	1.74	2.04	2.35	2.76	3.09	3.44	3.92	4.30
25	1.33	1.73	2.02	2.32	2.72	3.04	3.38	3.84	4.21
26	1.32	1.71	2.00	2.29	2.69	3.00	3.32	3.77	4.12
27	1.32	1.70	1.99	2.27	2.66	2.96	3.27	3.70	4.04
28	1.31	1.69	1.97	2.25	2.63	2.92	3.22	3.64	3.97
29	1.31	1.68	1.96	2.23	2.60	2.88	3.18	3.59	3.91
30	1.31	1.68	1.95	2.21	2.57	2.85	3.14	3.53	3.85
35	1.29	1.64	1.89	2.14	2.47	2.72	2.98	3.33	3.61
40	1.28	1.61	1.85	2.09	2.39	2.63	2.87	3.19	3.44
45	1.27	1.59	1.82	2.04	2.34	2.56	2.78	3.08	3.31
50	1.26	1.58	1.80	2.01	2.29	2.50	2.71	2.99	3.21
60	1.25	1.55	1.76	1.96	2.22	2.42	2.61	2.87	3.06
70	1.24	1.54	1.74	1.93	2.18	2.36	2.54	2.78	2.96
80	1.24	1.52	1.72	1.90	2.14	2.32	2.49	2.72	2.89
90	1.23	1.51	1.70	1.88	2.11	2.28	2.45	2.67	2.83
100	1.23	1.50	1.69	1.87	2.09	2.26	2.42	2.63	2.79
120	1.22	1.49	1.67	1.84	2.06	2.22	2.37	2.58	2.73
140	1.22	1.48	1.66	1.83	2.04	2.19	2.34	2.54	2.68
160	1.22	1.48	1.65	1.82	2.02	2.17	2.32	2.51	2.65
180	1.21	1.47	1.64	1.81	2.01	2.15	2.30	2.48	2.62
200	1.21	1.47	1.64	1.80	2.00	2.14	2.28	2.46	2.60
300	1.21	1.46	1.62	1.77	1.97	2.10	2.24	2.41	2.54
500	1.20	1.45	1.61	1.76	1.94	2.07	2.20	2.37	2.49
∞	1.20	1.43	1.59	1.73	1.90	2.03	2.15	2.31	2.42

TABLE B.4 (cont.)　Critical Values of the *F* Distribution

Numerator DF = 20

Denom. DF	α(2): 0.50 α(1): 0.25	0.20 0.10	0.10 0.05	0.05 0.025	0.02 0.01	0.01 0.005	0.005 0.0025	0.002 0.001	0.001 0.0005
1	9.58	61.7	248.	993.	6210.	24800.	99300.	621000.	2480000.
2	3.43	9.44	19.4	39.4	99.4	199.	399.	999.	2000.
3	2.46	5.18	8.66	14.2	26.7	42.8	68.3	126.	201.
4	2.08	3.84	5.80	8.56	14.0	20.2	28.9	46.1	65.5
5	1.88	3.21	4.56	6.33	9.55	12.9	17.3	25.4	33.8
6	1.76	2.84	3.87	5.17	7.40	9.59	12.3	17.1	21.8
7	1.67	2.59	3.44	4.47	6.16	7.75	9.70	12.9	16.0
8	1.61	2.42	3.15	4.00	5.36	6.61	8.09	10.5	12.7
9	1.56	2.30	2.94	3.67	4.81	5.83	7.02	8.90	10.6
10	1.52	2.20	2.77	3.42	4.41	5.27	6.27	7.80	9.17
11	1.49	2.12	2.65	3.23	4.10	4.86	5.71	7.01	8.14
12	1.47	2.06	2.54	3.07	3.86	4.53	5.28	6.40	7.37
13	1.45	2.01	2.46	2.95	3.66	4.27	4.94	5.93	6.78
14	1.43	1.96	2.39	2.84	3.51	4.06	4.66	5.56	6.31
15	1.41	1.92	2.33	2.76	3.37	3.88	4.44	5.25	5.93
16	1.40	1.89	2.28	2.68	3.26	3.73	4.25	4.99	5.61
17	1.39	1.86	2.23	2.62	3.16	3.61	4.09	4.78	5.34
18	1.38	1.84	2.19	2.56	3.08	3.50	3.95	4.59	5.12
19	1.37	1.81	2.16	2.51	3.00	3.40	3.83	4.43	4.92
20	1.36	1.79	2.12	2.46	2.94	3.32	3.72	4.29	4.75
21	1.35	1.78	2.10	2.42	2.88	3.24	3.63	4.17	4.60
22	1.34	1.76	2.07	2.39	2.83	3.18	3.54	4.06	4.47
23	1.34	1.74	2.05	2.36	2.78	3.12	3.47	3.96	4.36
24	1.33	1.73	2.03	2.33	2.74	3.06	3.40	3.87	4.25
25	1.33	1.72	2.01	2.30	2.70	3.01	3.34	3.79	4.16
26	1.32	1.71	1.99	2.28	2.66	2.97	3.28	3.72	4.07
27	1.32	1.70	1.97	2.25	2.63	2.93	3.23	3.66	3.99
28	1.31	1.69	1.96	2.23	2.60	2.89	3.19	3.60	3.92
29	1.31	1.68	1.94	2.21	2.57	2.86	3.14	3.54	3.86
30	1.30	1.67	1.93	2.20	2.55	2.82	3.11	3.49	3.80
35	1.29	1.63	1.88	2.12	2.44	2.69	2.95	3.29	3.56
40	1.28	1.61	1.84	2.07	2.37	2.60	2.83	3.14	3.39
45	1.27	1.58	1.81	2.03	2.31	2.53	2.74	3.04	3.26
50	1.26	1.57	1.78	1.99	2.27	2.47	2.68	2.95	3.16
60	1.25	1.54	1.75	1.94	2.20	2.39	2.58	2.83	3.02
70	1.24	1.53	1.72	1.91	2.15	2.33	2.51	2.74	2.92
80	1.23	1.51	1.70	1.88	2.12	2.29	2.46	2.68	2.85
90	1.23	1.50	1.69	1.86	2.09	2.25	2.42	2.63	2.79
100	1.23	1.49	1.68	1.85	2.07	2.23	2.38	2.59	2.75
120	1.22	1.48	1.66	1.82	2.03	2.19	2.34	2.53	2.68
140	1.22	1.47	1.65	1.81	2.01	2.16	2.31	2.49	2.64
160	1.21	1.47	1.64	1.80	1.99	2.14	2.28	2.47	2.60
180	1.21	1.46	1.63	1.79	1.98	2.12	2.26	2.44	2.58
200	1.21	1.46	1.62	1.78	1.97	2.11	2.25	2.42	2.56
300	1.20	1.45	1.61	1.75	1.94	2.07	2.20	2.37	2.49
500	1.20	1.44	1.59	1.74	1.92	2.04	2.17	2.33	2.45
∞	1.19	1.42	1.57	1.71	1.88	2.00	2.12	2.27	2.37

TABLE B.4 (cont.)   Critical Values of the F Distribution

Numerator DF $= 22$

Denom. DF	$\alpha(2)$: 0.50 $\alpha(1)$: 0.25	0.20 0.10	0.10 0.05	0.05 0.025	0.02 0.01	0.01 0.005	0.005 0.0025	0.002 0.001	0.001 0.0005
1	9.61	61.9	249.	995.	6220.	24900.	99600.	622000.	2490000.
2	3.43	9.45	19.5	39.5	99.5	199.	399.	999.	2000.
3	2.46	5.18	8.65	14.1	26.6	42.7	68.2	126.	201.
4	2.08	3.84	5.79	8.53	14.0	20.1	28.7	45.9	65.3
5	1.88	3.20	4.54	6.30	9.51	12.8	17.2	25.3	33.6
6	1.76	2.83	3.86	5.14	7.35	9.53	12.3	17.0	21.7
7	1.67	2.58	3.43	4.44	6.11	7.69	9.62	12.8	15.9
8	1.61	2.41	3.13	3.97	5.32	6.55	8.02	10.4	12.6
9	1.56	2.29	2.92	3.64	4.77	5.78	6.95	8.80	10.5
10	1.52	2.19	2.75	3.39	4.36	5.22	6.20	7.71	9.06
11	1.49	2.11	2.63	3.20	4.06	4.80	5.64	6.92	8.04
12	1.46	2.05	2.52	3.04	3.82	4.48	5.21	6.32	7.27
13	1.44	1.99	2.44	2.92	3.62	4.22	4.87	5.85	6.68
14	1.42	1.95	2.37	2.81	3.46	4.01	4.60	5.48	6.21
15	1.41	1.91	2.31	2.73	3.33	3.83	4.37	5.17	5.83
16	1.39	1.88	2.25	2.65	3.22	3.68	4.18	4.91	5.52
17	1.38	1.85	2.21	2.59	3.12	3.56	4.02	4.70	5.25
18	1.37	1.82	2.17	2.53	3.03	3.45	3.88	4.51	5.03
19	1.36	1.80	2.13	2.48	2.96	3.35	3.76	4.35	4.83
20	1.35	1.78	2.10	2.43	2.90	3.27	3.66	4.21	4.67
21	1.35	1.76	2.07	2.39	2.84	3.19	3.56	4.09	4.52
22	1.34	1.74	2.05	2.36	2.78	3.12	3.48	3.98	4.39
23	1.33	1.73	2.02	2.33	2.74	3.06	3.41	3.89	4.27
24	1.33	1.71	2.00	2.30	2.70	3.01	3.34	3.80	4.17
25	1.32	1.70	1.98	2.27	2.66	2.96	3.28	3.72	4.07
26	1.32	1.69	1.97	2.24	2.62	2.92	3.22	3.65	3.99
27	1.31	1.68	1.95	2.22	2.59	2.88	3.17	3.58	3.91
28	1.31	1.67	1.93	2.20	2.56	2.84	3.13	3.52	3.84
29	1.30	1.66	1.92	2.18	2.53	2.80	3.08	3.47	3.77
30	1.30	1.65	1.91	2.16	2.51	2.77	3.04	3.42	3.71
35	1.28	1.62	1.85	2.09	2.40	2.64	2.89	3.22	3.48
40	1.27	1.59	1.81	2.03	2.33	2.55	2.77	3.07	3.31
45	1.26	1.57	1.78	1.99	2.27	2.47	2.68	2.96	3.18
50	1.25	1.55	1.76	1.96	2.22	2.42	2.62	2.88	3.08
60	1.24	1.53	1.72	1.91	2.15	2.33	2.52	2.75	2.94
70	1.23	1.51	1.70	1.88	2.11	2.28	2.45	2.67	2.84
80	1.23	1.49	1.68	1.85	2.07	2.23	2.39	2.61	2.77
90	1.22	1.48	1.66	1.83	2.04	2.20	2.35	2.56	2.71
100	1.22	1.48	1.65	1.81	2.02	2.17	2.32	2.52	2.67
120	1.21	1.46	1.63	1.79	1.99	2.13	2.28	2.46	2.60
140	1.21	1.45	1.62	1.77	1.97	2.11	2.24	2.42	2.56
160	1.21	1.45	1.61	1.76	1.95	2.09	2.22	2.39	2.52
180	1.20	1.44	1.60	1.75	1.94	2.07	2.20	2.37	2.50
200	1.20	1.44	1.60	1.74	1.93	2.06	2.19	2.35	2.48
300	1.20	1.43	1.58	1.72	1.89	2.02	2.14	2.30	2.41
500	1.19	1.42	1.56	1.70	1.87	1.99	2.11	2.26	2.37
∞	1.18	1.40	1.54	1.67	1.83	1.95	2.05	2.19	2.30

TABLE B.4 (cont.)   Critical Values of the *F* Distribution

Numerator DF = 24

Denom. DF	α(2): 0.50 α(1): 0.25	0.20 0.10	0.10 0.05	0.05 0.025	0.02 0.01	0.01 0.005	0.005 0.0025	0.002 0.001	0.001 0.0005
1	9.63	62.0	249.	997.	6230.	24900.	99800.	623000.	2490000.
2	3.43	9.45	19.5	39.5	99.5	199.	399.	999.	2000.
3	2.46	5.18	8.64	14.1	26.6	42.6	68.1	126.	200.
4	2.08	3.83	5.77	8.51	13.9	20.0	28.7	45.8	65.0
5	1.88	3.19	4.53	6.28	9.47	12.8	17.1	25.1	33.4
6	1.75	2.82	3.84	5.12	7.31	9.47	12.2	16.9	21.5
7	1.67	2.58	3.41	4.41	6.07	7.64	9.56	12.7	15.8
8	1.60	2.40	3.12	3.95	5.28	6.50	7.95	10.3	12.5
9	1.56	2.28	2.90	3.61	4.73	5.73	6.89	8.72	10.4
10	1.52	2.18	2.74	3.37	4.33	5.17	6.14	7.64	8.96
11	1.49	2.10	2.61	3.17	4.02	4.76	5.58	6.85	7.95
12	1.46	2.04	2.51	3.02	3.78	4.43	5.16	6.25	7.19
13	1.44	1.98	2.42	2.89	3.59	4.17	4.82	5.78	6.60
14	1.42	1.94	2.35	2.79	3.43	3.96	4.55	5.41	6.13
15	1.41	1.90	2.29	2.70	3.29	3.79	4.32	5.10	5.75
16	1.39	1.87	2.24	2.63	3.18	3.64	4.13	4.85	5.44
17	1.38	1.84	2.19	2.56	3.08	3.51	3.97	4.63	5.18
18	1.37	1.81	2.15	2.50	3.00	3.40	3.83	4.45	4.95
19	1.36	1.79	2.11	2.45	2.92	3.31	3.71	4.29	4.76
20	1.35	1.77	2.08	2.41	2.86	3.22	3.61	4.15	4.59
21	1.34	1.75	2.05	2.37	2.80	3.15	3.51	4.03	4.44
22	1.33	1.73	2.03	2.33	2.75	3.08	3.43	3.92	4.31
23	1.33	1.72	2.01	2.30	2.70	3.02	3.35	3.82	4.20
24	1.32	1.70	1.98	2.27	2.66	2.97	3.29	3.74	4.09
25	1.32	1.69	1.96	2.24	2.62	2.92	3.23	3.66	4.00
26	1.31	1.68	1.95	2.22	2.58	2.87	3.17	3.59	3.92
27	1.31	1.67	1.93	2.19	2.55	2.83	3.12	3.52	3.84
28	1.30	1.66	1.91	2.17	2.52	2.79	3.07	3.46	3.77
29	1.30	1.65	1.90	2.15	2.49	2.76	3.03	3.41	3.70
30	1.29	1.64	1.89	2.14	2.47	2.73	2.99	3.36	3.64
35	1.28	1.60	1.83	2.06	2.36	2.60	2.83	3.16	3.41
40	1.26	1.57	1.79	2.01	2.29	2.50	2.72	3.01	3.24
45	1.26	1.55	1.76	1.96	2.23	2.43	2.63	2.90	3.11
50	1.25	1.54	1.74	1.93	2.18	2.37	2.56	2.82	3.01
60	1.24	1.51	1.70	1.88	2.12	2.29	2.46	2.69	2.87
70	1.23	1.49	1.67	1.85	2.07	2.23	2.39	2.61	2.77
80	1.22	1.48	1.65	1.82	2.03	2.19	2.34	2.54	2.70
90	1.22	1.47	1.64	1.80	2.00	2.15	2.30	2.50	2.64
100	1.21	1.46	1.63	1.78	1.98	2.13	2.27	2.46	2.60
120	1.21	1.45	1.61	1.76	1.95	2.09	2.23	2.40	2.53
140	1.20	1.44	1.60	1.74	1.93	2.06	2.19	2.36	2.49
160	1.20	1.43	1.59	1.73	1.91	2.04	2.17	2.33	2.45
180	1.20	1.43	1.58	1.72	1.90	2.02	2.15	2.31	2.43
200	1.19	1.42	1.57	1.71	1.89	2.01	2.13	2.29	2.41
300	1.19	1.41	1.55	1.69	1.85	1.97	2.09	2.24	2.35
500	1.18	1.40	1.54	1.67	1.83	1.94	2.05	2.20	2.30
∞	1.18	1.38	1.52	1.64	1.79	1.90	2.00	2.13	2.23

TABLE B.4 (cont.)   Critical Values of the F Distribution

Numerator DF $= 26$

Denom. DF	$\alpha(2)$: 0.50 $\alpha(1)$: 0.25	0.20 0.10	0.10 0.05	0.05 0.025	0.02 0.01	0.01 0.005	0.005 0.0025	0.002 0.001	0.001 0.0005
1	9.64	62.1	249.	999.	6240.	25000.	99900.	624000.	2500000.
2	3.44	9.45	19.5	39.5	99.5	199.	399.	999.	2000.
3	2.46	5.17	8.63	14.1	26.6	42.6	68.0	126.	200.
4	2.08	3.83	5.76	8.49	13.9	20.0	28.6	45.6	64.9
5	1.88	3.18	4.52	6.26	9.43	12.7	17.1	25.0	33.3
6	1.75	2.81	3.83	5.10	7.28	9.43	12.1	16.8	21.4
7	1.67	2.57	3.40	4.39	6.04	7.60	9.50	12.7	15.7
8	1.60	2.40	3.10	3.93	5.25	6.46	7.90	10.2	12.4
9	1.55	2.27	2.89	3.59	4.70	5.69	6.84	8.66	10.3
10	1.52	2.17	2.72	3.34	4.30	5.13	6.09	7.57	8.89
11	1.48	2.09	2.59	3.15	3.99	4.72	5.54	6.78	7.87
12	1.46	2.03	2.49	3.00	3.75	4.39	5.11	6.19	7.12
13	1.44	1.97	2.41	2.87	3.56	4.13	4.77	5.72	6.53
14	1.42	1.93	2.33	2.77	3.40	3.92	4.50	5.35	6.07
15	1.40	1.89	2.27	2.68	3.26	3.75	4.27	5.04	5.69
16	1.39	1.86	2.22	2.60	3.15	3.60	4.09	4.79	5.37
17	1.38	1.83	2.17	2.54	3.05	3.47	3.92	4.57	5.11
18	1.36	1.80	2.13	2.48	2.97	3.36	3.79	4.39	4.89
19	1.35	1.78	2.10	2.43	2.89	3.27	3.67	4.23	4.70
20	1.35	1.76	2.07	2.39	2.83	3.18	3.56	4.09	4.53
21	1.34	1.74	2.04	2.34	2.77	3.11	3.47	3.97	4.38
22	1.33	1.72	2.01	2.31	2.72	3.04	3.38	3.86	4.25
23	1.32	1.70	1.99	2.28	2.67	2.98	3.31	3.77	4.14
24	1.32	1.69	1.97	2.25	2.63	2.93	3.24	3.68	4.03
25	1.31	1.68	1.95	2.22	2.59	2.88	3.18	3.60	3.94
26	1.31	1.67	1.93	2.19	2.55	2.84	3.13	3.53	3.85
27	1.30	1.65	1.91	2.17	2.52	2.79	3.08	3.47	3.78
28	1.30	1.64	1.90	2.15	2.49	2.76	3.03	3.41	3.71
29	1.29	1.63	1.88	2.13	2.46	2.72	2.99	3.35	3.64
30	1.29	1.63	1.87	2.11	2.44	2.69	2.95	3.30	3.58
35	1.27	1.59	1.82	2.04	2.33	2.56	2.79	3.10	3.35
40	1.26	1.56	1.77	1.98	2.26	2.46	2.67	2.96	3.18
45	1.25	1.54	1.74	1.94	2.20	2.39	2.59	2.85	3.05
50	1.24	1.52	1.72	1.91	2.15	2.33	2.52	2.76	2.95
60	1.23	1.50	1.68	1.86	2.08	2.25	2.42	2.64	2.81
70	1.22	1.48	1.65	1.82	2.03	2.19	2.35	2.56	2.71
80	1.22	1.47	1.63	1.79	2.00	2.15	2.30	2.49	2.64
90	1.21	1.45	1.62	1.77	1.97	2.12	2.26	2.44	2.58
100	1.21	1.45	1.61	1.76	1.95	2.09	2.23	2.41	2.54
120	1.20	1.43	1.59	1.73	1.92	2.05	2.18	2.35	2.48
140	1.20	1.42	1.57	1.72	1.89	2.02	2.15	2.31	2.43
160	1.19	1.42	1.57	1.70	1.88	2.00	2.12	2.28	2.40
180	1.19	1.41	1.56	1.69	1.86	1.98	2.10	2.26	2.37
200	1.19	1.41	1.55	1.68	1.85	1.97	2.09	2.24	2.35
300	1.18	1.39	1.53	1.66	1.82	1.93	2.04	2.18	2.29
500	1.18	1.38	1.52	1.64	1.79	1.90	2.01	2.14	2.24
∞	1.17	1.37	1.50	1.61	1.76	1.86	1.95	2.08	2.17

TABLE B.4 (cont.)   Critical Values of the *F* Distribution

Numerator DF = 28

Denom. DF	α(2): 0.50 α(1): 0.25	0.20 0.10	0.10 0.05	0.05 0.025	0.02 0.01	0.01 0.005	0.005 0.0025	0.002 0.001	0.001 0.0005
1	9.66	62.2	250.	1000.	6250.	25000.	100000.	625000.	2500000.
2	3.44	9.46	19.5	39.5	99.5	199.	399.	999.	2000.
3	2.46	5.17	8.62	14.1	26.5	42.5	67.9	126.	200.
4	2.08	3.82	5.75	8.48	13.9	19.9	28.5	45.5	64.7
5	1.88	3.18	4.50	6.24	9.40	12.7	17.0	24.9	33.2
6	1.75	2.81	3.82	5.08	7.25	9.39	12.1	16.7	21.3
7	1.66	2.56	3.39	4.38	6.02	7.57	9.45	12.6	15.6
8	1.60	2.39	3.09	3.91	5.22	6.43	7.86	10.2	12.3
9	1.55	2.26	2.87	3.58	4.67	5.65	6.80	8.60	10.2
10	1.51	2.16	2.71	3.33	4.27	5.10	6.05	7.52	8.82
11	1.48	2.08	2.58	3.13	3.96	4.68	5.49	6.73	7.81
12	1.46	2.02	2.48	2.98	3.72	4.36	5.07	6.14	7.05
13	1.43	1.96	2.39	2.85	3.53	4.10	4.73	5.67	6.47
14	1.42	1.92	2.32	2.75	3.37	3.89	4.46	5.30	6.01
15	1.40	1.88	2.26	2.66	3.24	3.72	4.23	4.99	5.63
16	1.39	1.85	2.21	2.58	3.12	3.57	4.05	4.74	5.32
17	1.37	1.82	2.16	2.52	3.03	3.44	3.89	4.53	5.05
18	1.36	1.79	2.12	2.46	2.94	3.33	3.75	4.34	4.83
19	1.35	1.77	2.08	2.41	2.87	3.24	3.63	4.18	4.64
20	1.34	1.75	2.05	2.37	2.80	3.15	3.52	4.05	4.47
21	1.33	1.73	2.02	2.33	2.74	3.08	3.43	3.93	4.33
22	1.33	1.71	2.00	2.29	2.69	3.01	3.35	3.82	4.20
23	1.32	1.69	1.97	2.26	2.64	2.95	3.27	3.72	4.08
24	1.31	1.68	1.95	2.23	2.60	2.90	3.20	3.63	3.98
25	1.31	1.67	1.93	2.20	2.56	2.85	3.14	3.56	3.89
26	1.30	1.66	1.91	2.17	2.53	2.80	3.09	3.49	3.80
27	1.30	1.64	1.90	2.15	2.49	2.76	3.04	3.42	3.72
28	1.29	1.63	1.88	2.13	2.46	2.72	2.99	3.36	3.65
29	1.29	1.62	1.87	2.11	2.44	2.69	2.95	3.31	3.59
30	1.29	1.62	1.85	2.09	2.41	2.66	2.91	3.26	3.53
35	1.27	1.58	1.80	2.02	2.30	2.53	2.75	3.06	3.30
40	1.26	1.55	1.76	1.96	2.23	2.43	2.64	2.91	3.13
45	1.25	1.53	1.73	1.92	2.17	2.36	2.55	2.80	3.00
50	1.24	1.51	1.70	1.89	2.12	2.30	2.48	2.72	2.90
60	1.23	1.49	1.66	1.83	2.05	2.22	2.38	2.60	2.76
70	1.22	1.47	1.64	1.80	2.01	2.16	2.31	2.51	2.66
80	1.21	1.45	1.62	1.77	1.97	2.11	2.26	2.45	2.59
90	1.21	1.44	1.60	1.75	1.94	2.08	2.22	2.40	2.53
100	1.20	1.43	1.59	1.74	1.92	2.05	2.19	2.36	2.49
120	1.20	1.42	1.57	1.71	1.89	2.01	2.14	2.30	2.42
140	1.19	1.41	1.56	1.69	1.86	1.99	2.11	2.26	2.38
160	1.19	1.40	1.55	1.68	1.85	1.97	2.08	2.23	2.35
180	1.19	1.40	1.54	1.67	1.83	1.95	2.06	2.21	2.32
200	1.18	1.39	1.53	1.66	1.82	1.94	2.05	2.19	2.30
300	1.18	1.38	1.51	1.64	1.79	1.90	2.00	2.14	2.24
500	1.17	1.37	1.50	1.62	1.76	1.87	1.97	2.10	2.19
∞	1.17	1.35	1.48	1.59	1.72	1.82	1.91	2.03	2.12

TABLE B.4 (cont.) Critical Values of the *F* Distribution

Numerator DF = 30

Denom. DF	α(2): 0.50 α(1): 0.25	0.20 0.10	0.10 0.05	0.05 0.025	0.02 0.01	0.01 0.005	0.005 0.0025	0.002 0.001	0.001 0.0005
1	9.67	62.3	250.	1000.	6260.	25000.	100000.	626000.	2500000.
2	3.44	9.46	19.5	39.5	99.5	199.	399.	999.	2000.
3	2.47	5.17	8.62	14.1	26.5	42.5	67.8	125.	200.
4	2.08	3.82	5.75	8.46	13.8	19.9	28.5	45.4	64.6
5	1.88	3.17	4.50	6.23	9.38	12.7	17.0	24.9	33.1
6	1.75	2.80	3.81	5.07	7.23	9.36	12.0	16.7	21.2
7	1.66	2.56	3.38	4.36	5.99	7.53	9.41	12.5	15.5
8	1.60	2.38	3.08	3.89	5.20	6.40	7.82	10.1	12.2
9	1.55	2.25	2.86	3.56	4.65	5.62	6.76	8.55	10.2
10	1.51	2.16	2.70	3.31	4.25	5.07	6.01	7.47	8.76
11	1.48	2.08	2.57	3.12	3.94	4.65	5.46	6.68	7.75
12	1.45	2.01	2.47	2.96	3.70	4.33	5.03	6.09	7.00
13	1.43	1.96	2.38	2.84	3.51	4.07	4.70	5.63	6.42
14	1.41	1.91	2.31	2.73	3.35	3.86	4.42	5.25	5.95
15	1.40	1.87	2.25	2.64	3.21	3.69	4.20	4.95	5.58
16	1.38	1.84	2.19	2.57	3.10	3.54	4.01	4.70	5.27
17	1.37	1.81	2.15	2.50	3.00	3.41	3.85	4.48	5.01
18	1.36	1.78	2.11	2.44	2.92	3.30	3.71	4.30	4.78
19	1.35	1.76	2.07	2.39	2.84	3.21	3.59	4.14	4.59
20	1.34	1.74	2.04	2.35	2.78	3.12	3.49	4.00	4.42
21	1.33	1.72	2.01	2.31	2.72	3.05	3.40	3.88	4.28
22	1.32	1.70	1.98	2.27	2.67	2.98	3.31	3.78	4.15
23	1.32	1.69	1.96	2.24	2.62	2.92	3.24	3.68	4.03
24	1.31	1.67	1.94	2.21	2.58	2.87	3.17	3.59	3.93
25	1.31	1.66	1.92	2.18	2.54	2.82	3.11	3.52	3.84
26	1.30	1.65	1.90	2.16	2.50	2.77	3.06	3.44	3.75
27	1.30	1.64	1.88	2.13	2.47	2.73	3.00	3.38	3.68
28	1.29	1.63	1.87	2.11	2.44	2.69	2.96	3.32	3.61
29	1.29	1.62	1.85	2.09	2.41	2.66	2.92	3.27	3.54
30	1.28	1.61	1.84	2.07	2.39	2.63	2.88	3.22	3.49
35	1.27	1.57	1.79	2.00	2.28	2.50	2.72	3.02	3.25
40	1.25	1.54	1.74	1.94	2.20	2.40	2.60	2.87	3.08
45	1.24	1.52	1.71	1.90	2.14	2.33	2.51	2.76	2.96
50	1.23	1.50	1.69	1.87	2.10	2.27	2.45	2.68	2.86
60	1.22	1.48	1.65	1.82	2.03	2.19	2.35	2.55	2.71
70	1.21	1.46	1.62	1.78	1.98	2.13	2.28	2.47	2.62
80	1.21	1.44	1.60	1.75	1.94	2.08	2.22	2.41	2.54
90	1.20	1.43	1.59	1.73	1.92	2.05	2.18	2.36	2.49
100	1.20	1.42	1.57	1.71	1.89	2.02	2.15	2.32	2.44
120	1.19	1.41	1.55	1.69	1.86	1.98	2.11	2.26	2.38
140	1.19	1.40	1.54	1.67	1.84	1.96	2.07	2.22	2.33
160	1.18	1.39	1.53	1.66	1.82	1.93	2.05	2.19	2.30
180	1.18	1.39	1.52	1.65	1.81	1.92	2.03	2.17	2.27
200	1.18	1.38	1.52	1.64	1.79	1.91	2.01	2.15	2.25
300	1.17	1.37	1.50	1.62	1.76	1.87	1.97	2.10	2.19
500	1.17	1.36	1.48	1.60	1.74	1.84	1.93	2.05	2.14
∞	1.16	1.34	1.46	1.57	1.70	1.79	1.88	1.99	2.07

TABLE B.4 (cont.)   Critical Values of the F Distribution

Numerator DF = 40

Denom. DF	α(2): 0.50 α(1): 0.25	0.20 0.10	0.10 0.05	0.05 0.025	0.02 0.01	0.01 0.005	0.005 0.0025	0.002 0.001	0.001 0.0005
1	9.71	62.5	251.	1010.	6290.	25100.	101000.	629000.	2510000.
2	3.45	9.47	19.5	39.5	99.5	199.	399.	999.	2000.
3	2.47	5.16	8.59	14.0	26.4	42.3	67.5	125.	199.
4	2.08	3.80	5.72	8.41	13.7	19.8	28.2	45.1	64.1
5	1.88	3.16	4.46	6.18	9.29	12.5	16.8	24.6	32.7
6	1.75	2.78	3.77	5.01	7.14	9.24	11.9	16.4	21.0
7	1.66	2.54	3.34	4.31	5.91	7.42	9.26	12.3	15.2
8	1.59	2.36	3.04	3.84	5.12	6.29	7.68	9.92	12.0
9	1.54	2.23	2.83	3.51	4.57	5.52	6.62	8.37	9.94
10	1.51	2.13	2.66	3.26	4.17	4.97	5.88	7.30	8.55
11	1.47	2.05	2.53	3.06	3.86	4.55	5.33	6.52	7.55
12	1.45	1.99	2.43	2.91	3.62	4.23	4.91	5.93	6.81
13	1.42	1.93	2.34	2.78	3.43	3.97	4.57	5.47	6.23
14	1.41	1.89	2.27	2.67	3.27	3.76	4.30	5.10	5.77
15	1.39	1.85	2.20	2.59	3.13	3.58	4.08	4.80	5.40
16	1.37	1.81	2.15	2.51	3.02	3.44	3.89	4.54	5.09
17	1.36	1.78	2.10	2.44	2.92	3.31	3.73	4.33	4.83
18	1.35	1.75	2.06	2.38	2.84	3.20	3.59	4.15	4.61
19	1.34	1.73	2.03	2.33	2.76	3.11	3.47	3.99	4.42
20	1.33	1.71	1.99	2.29	2.69	3.02	3.37	3.86	4.25
21	1.32	1.69	1.96	2.25	2.64	2.95	3.27	3.74	4.11
22	1.31	1.67	1.94	2.21	2.58	2.88	3.19	3.63	3.98
23	1.31	1.66	1.91	2.18	2.54	2.82	3.12	3.53	3.87
24	1.30	1.64	1.89	2.15	2.49	2.77	3.05	3.45	3.76
25	1.29	1.63	1.87	2.12	2.45	2.72	2.99	3.37	3.67
26	1.29	1.61	1.85	2.09	2.42	2.67	2.93	3.30	3.59
27	1.28	1.60	1.84	2.07	2.38	2.63	2.88	3.23	3.51
28	1.28	1.59	1.82	2.05	2.35	2.59	2.84	3.18	3.44
29	1.27	1.58	1.81	2.03	2.33	2.56	2.79	3.12	3.38
30	1.27	1.57	1.79	2.01	2.30	2.52	2.76	3.07	3.32
35	1.25	1.53	1.74	1.93	2.19	2.39	2.60	2.87	3.09
40	1.24	1.51	1.69	1.88	2.11	2.30	2.48	2.73	2.92
45	1.23	1.48	1.66	1.83	2.05	2.22	2.39	2.62	2.79
50	1.22	1.46	1.63	1.80	2.01	2.16	2.32	2.53	2.69
60	1.21	1.44	1.59	1.74	1.94	2.08	2.22	2.41	2.55
70	1.20	1.42	1.57	1.71	1.89	2.02	2.15	2.32	2.45
80	1.19	1.40	1.54	1.68	1.85	1.97	2.10	2.26	2.38
90	1.19	1.39	1.53	1.66	1.82	1.94	2.06	2.21	2.32
100	1.18	1.38	1.52	1.64	1.80	1.91	2.02	2.17	2.28
120	1.18	1.37	1.50	1.61	1.76	1.87	1.98	2.11	2.21
140	1.17	1.36	1.48	1.60	1.74	1.84	1.94	2.07	2.17
160	1.17	1.35	1.47	1.58	1.72	1.82	1.92	2.04	2.13
180	1.16	1.34	1.46	1.57	1.71	1.80	1.90	2.02	2.11
200	1.16	1.34	1.46	1.56	1.69	1.79	1.88	2.00	2.09
300	1.15	1.32	1.43	1.54	1.66	1.75	1.84	1.94	2.02
500	1.15	1.31	1.42	1.52	1.63	1.72	1.80	1.90	1.98
∞	1.14	1.30	1.39	1.48	1.59	1.67	1.74	1.84	1.90

TABLE B.4 (cont.)   Critical Values of the *F* Distribution

Numerator DF = 50

Denom. DF	α(2): 0.50 α(1): 0.25	0.20 0.10	0.10 0.05	0.05 0.025	0.02 0.01	0.01 0.005	0.005 0.0025	0.002 0.001	0.001 0.0005
1	9.74	62.7	252.	1010.	6300.	25200.	101000.	630000.	2520000.
2	3.46	9.47	19.5	39.5	99.5	199.	399.	999.	2000.
3	2.47	5.15	8.58	14.0	26.4	42.2	67.4	125.	198.
4	2.08	3.80	5.70	8.38	13.7	19.7	28.1	44.9	63.8
5	1.88	3.15	4.44	6.14	9.24	12.5	16.7	24.4	32.5
6	1.75	2.77	3.75	4.98	7.09	9.17	11.8	16.3	20.8
7	1.66	2.52	3.32	4.28	5.86	7.35	9.17	12.2	15.1
8	1.59	2.35	3.02	3.81	5.07	6.22	7.59	9.80	11.8
9	1.54	2.22	2.80	3.47	4.52	5.45	6.54	8.26	9.81
10	1.50	2.12	2.64	3.22	4.12	4.90	5.80	7.19	8.43
11	1.47	2.04	2.51	3.03	3.81	4.49	5.25	6.42	7.43
12	1.44	1.97	2.40	2.87	3.57	4.17	4.83	5.83	6.69
13	1.42	1.92	2.31	2.74	3.38	3.91	4.50	5.37	6.11
14	1.40	1.87	2.24	2.64	3.22	3.70	4.23	5.00	5.66
15	1.38	1.83	2.18	2.55	3.08	3.52	4.00	4.70	5.29
16	1.37	1.79	2.12	2.47	2.97	3.37	3.81	4.45	4.98
17	1.36	1.76	2.08	2.41	2.87	3.25	3.65	4.24	4.72
18	1.34	1.74	2.04	2.35	2.78	3.14	3.52	4.06	4.50
19	1.33	1.71	2.00	2.30	2.71	3.04	3.40	3.90	4.31
20	1.32	1.69	1.97	2.25	2.64	2.96	3.29	3.77	4.15
21	1.32	1.67	1.94	2.21	2.58	2.88	3.20	3.64	4.00
22	1.31	1.65	1.91	2.17	2.53	2.82	3.12	3.54	3.88
23	1.30	1.64	1.88	2.14	2.48	2.76	3.04	3.44	3.76
24	1.29	1.62	1.86	2.11	2.44	2.70	2.98	3.36	3.66
25	1.29	1.61	1.84	2.08	2.40	2.65	2.91	3.28	3.57
26	1.28	1.59	1.82	2.05	2.36	2.61	2.86	3.21	3.49
27	1.28	1.58	1.81	2.03	2.33	2.57	2.81	3.14	3.41
28	1.27	1.57	1.79	2.01	2.30	2.53	2.76	3.09	3.34
29	1.27	1.56	1.77	1.99	2.27	2.49	2.72	3.03	3.28
30	1.26	1.55	1.76	1.97	2.25	2.46	2.68	2.98	3.22
35	1.24	1.51	1.70	1.89	2.14	2.33	2.52	2.78	2.98
40	1.23	1.48	1.66	1.83	2.06	2.23	2.40	2.64	2.82
45	1.22	1.46	1.63	1.79	2.00	2.16	2.31	2.53	2.69
50	1.21	1.44	1.60	1.75	1.95	2.10	2.24	2.44	2.59
60	1.20	1.41	1.56	1.70	1.88	2.01	2.14	2.32	2.45
70	1.19	1.39	1.53	1.66	1.83	1.95	2.07	2.23	2.35
80	1.18	1.38	1.51	1.63	1.79	1.90	2.02	2.16	2.28
90	1.18	1.36	1.49	1.61	1.76	1.87	1.98	2.11	2.22
100	1.17	1.35	1.48	1.59	1.74	1.84	1.94	2.08	2.18
120	1.16	1.34	1.46	1.56	1.70	1.80	1.89	2.02	2.11
140	1.16	1.33	1.44	1.55	1.67	1.77	1.86	1.98	2.06
160	1.15	1.32	1.43	1.53	1.66	1.75	1.83	1.95	2.03
180	1.15	1.32	1.42	1.52	1.64	1.73	1.81	1.92	2.00
200	1.15	1.31	1.41	1.51	1.63	1.71	1.80	1.90	1.98
300	1.14	1.29	1.39	1.48	1.59	1.67	1.75	1.85	1.92
500	1.14	1.28	1.38	1.46	1.57	1.64	1.71	1.80	1.87
∞	1.13	1.26	1.35	1.43	1.52	1.59	1.65	1.73	1.79

TABLE B.4 (cont.) Critical Values of the F Distribution

Numerator DF = 60

Denom. DF	α(2): 0.50 α(1): 0.25	0.20 0.10	0.10 0.05	0.05 0.025	0.02 0.01	0.01 0.005	0.005 0.0025	0.002 0.001	0.001 0.0005
1	9.76	62.8	252.	1010.	6310.	25300.	101000.	631000.	2530000.
2	3.46	9.47	19.5	39.5	99.5	199.	399.	999.	2000.
3	2.47	5.15	8.57	14.0	26.3	42.1	67.3	124.	198.
4	2.08	3.79	5.69	8.36	13.7	19.6	28.0	44.7	63.6
5	1.87	3.14	4.43	6.12	9.20	12.4	16.6	24.3	32.4
6	1.74	2.76	3.74	4.96	7.06	9.12	11.7	16.2	20.7
7	1.65	2.51	3.30	4.25	5.82	7.31	9.12	12.1	15.0
8	1.59	2.34	3.01	3.78	5.03	6.18	7.54	9.73	11.8
9	1.54	2.21	2.79	3.45	4.48	5.41	6.49	8.19	9.72
10	1.50	2.11	2.62	3.20	4.08	4.86	5.75	7.12	8.34
11	1.47	2.03	2.49	3.00	3.78	4.45	5.20	6.35	7.35
12	1.44	1.96	2.38	2.85	3.54	4.12	4.78	5.76	6.61
13	1.42	1.90	2.30	2.72	3.34	3.87	4.44	5.30	6.04
14	1.40	1.86	2.22	2.61	3.18	3.66	4.17	4.94	5.58
15	1.38	1.82	2.16	2.52	3.05	3.48	3.95	4.64	5.21
16	1.36	1.78	2.11	2.45	2.93	3.33	3.76	4.39	4.91
17	1.35	1.75	2.06	2.38	2.83	3.21	3.60	4.18	4.65
18	1.34	1.72	2.02	2.32	2.75	3.10	3.47	4.00	4.43
19	1.33	1.70	1.98	2.27	2.67	3.00	3.35	3.84	4.24
20	1.32	1.68	1.95	2.22	2.61	2.92	3.24	3.70	4.08
21	1.31	1.66	1.92	2.18	2.55	2.84	3.15	3.58	3.93
22	1.30	1.64	1.89	2.14	2.50	2.77	3.07	3.48	3.81
23	1.30	1.62	1.86	2.11	2.45	2.71	2.99	3.38	3.69
24	1.29	1.61	1.84	2.08	2.40	2.66	2.92	3.29	3.59
25	1.28	1.59	1.82	2.05	2.36	2.61	2.86	3.22	3.50
26	1.28	1.58	1.80	2.03	2.33	2.56	2.81	3.15	3.42
27	1.27	1.57	1.79	2.00	2.29	2.52	2.76	3.08	3.34
28	1.27	1.56	1.77	1.98	2.26	2.48	2.71	3.02	3.27
29	1.26	1.55	1.75	1.96	2.23	2.45	2.67	2.97	3.21
30	1.26	1.54	1.74	1.94	2.21	2.42	2.63	2.92	3.15
35	1.24	1.50	1.68	1.86	2.10	2.28	2.47	2.72	2.91
40	1.22	1.47	1.64	1.80	2.02	2.18	2.35	2.57	2.75
45	1.21	1.44	1.60	1.76	1.96	2.11	2.26	2.46	2.62
50	1.20	1.42	1.58	1.72	1.91	2.05	2.19	2.38	2.52
60	1.19	1.40	1.53	1.67	1.84	1.96	2.09	2.25	2.38
70	1.18	1.37	1.50	1.63	1.78	1.90	2.01	2.16	2.28
80	1.17	1.36	1.48	1.60	1.75	1.85	1.96	2.10	2.20
90	1.17	1.35	1.46	1.58	1.72	1.82	1.92	2.05	2.15
100	1.16	1.34	1.45	1.56	1.69	1.79	1.89	2.01	2.10
120	1.16	1.32	1.43	1.53	1.66	1.75	1.84	1.95	2.04
140	1.15	1.31	1.41	1.51	1.63	1.72	1.80	1.91	1.99
160	1.15	1.30	1.40	1.50	1.61	1.69	1.77	1.88	1.95
180	1.14	1.29	1.39	1.48	1.60	1.68	1.75	1.85	1.93
200	1.14	1.29	1.39	1.47	1.58	1.66	1.74	1.83	1.90
300	1.13	1.27	1.36	1.45	1.55	1.62	1.69	1.78	1.84
500	1.13	1.26	1.35	1.42	1.52	1.58	1.65	1.73	1.79
∞	1.12	1.24	1.32	1.39	1.47	1.53	1.59	1.66	1.71

TABLE B.4 (cont.)   Critical Values of the *F* Distribution

Numerator DF = 70

Denom. DF	$\alpha(2)$: 0.50 $\alpha(1)$: 0.25	0.20 0.10	0.10 0.05	0.05 0.025	0.02 0.01	0.01 0.005	0.005 0.0025	0.002 0.001	0.001 0.0005
1	9.77	62.9	252.	1010.	6320.	25300.	101000.	632000.	2530000.
2	3.46	9.48	19.5	39.5	99.5	199.	399.	999.	2000.
3	2.47	5.15	8.57	14.0	26.3	42.1	67.2	124.	198.
4	2.08	3.79	5.68	8.35	13.6	19.6	28.0	44.6	63.4
5	1.87	3.14	4.42	6.11	9.18	12.4	16.6	24.3	32.3
6	1.74	2.76	3.73	4.94	7.03	9.09	11.7	16.1	20.6
7	1.65	2.51	3.29	4.24	5.80	7.28	9.07	12.1	14.9
8	1.59	2.33	2.99	3.77	5.01	6.15	7.49	9.67	11.7
9	1.54	2.20	2.78	3.43	4.46	5.38	6.45	8.13	9.65
10	1.50	2.10	2.61	3.18	4.06	4.83	5.71	7.07	8.28
11	1.46	2.02	2.48	2.99	3.75	4.41	5.16	6.30	7.29
12	1.44	1.95	2.37	2.83	3.51	4.09	4.74	5.71	6.55
13	1.41	1.90	2.28	2.70	3.32	3.84	4.41	5.26	5.98
14	1.39	1.85	2.21	2.60	3.16	3.62	4.14	4.89	5.53
15	1.38	1.81	2.15	2.51	3.02	3.45	3.91	4.59	5.16
16	1.36	1.77	2.09	2.43	2.91	3.30	3.73	4.34	4.85
17	1.35	1.74	2.05	2.36	2.81	3.18	3.57	4.13	4.60
18	1.34	1.71	2.00	2.30	2.72	3.07	3.43	3.95	4.38
19	1.33	1.69	1.97	2.25	2.65	2.97	3.31	3.79	4.19
20	1.32	1.67	1.93	2.20	2.58	2.88	3.20	3.66	4.03
21	1.31	1.65	1.90	2.16	2.52	2.81	3.11	3.54	3.88
22	1.30	1.63	1.88	2.13	2.47	2.74	3.03	3.43	3.76
23	1.29	1.61	1.85	2.09	2.42	2.68	2.95	3.34	3.64
24	1.28	1.60	1.83	2.06	2.38	2.63	2.89	3.25	3.54
25	1.28	1.58	1.81	2.03	2.34	2.58	2.83	3.17	3.45
26	1.27	1.57	1.79	2.01	2.30	2.53	2.77	3.10	3.37
27	1.27	1.56	1.77	1.98	2.27	2.49	2.72	3.04	3.29
28	1.26	1.55	1.75	1.96	2.24	2.45	2.67	2.98	3.22
29	1.26	1.54	1.74	1.94	2.21	2.42	2.63	2.92	3.16
30	1.25	1.53	1.72	1.92	2.18	2.38	2.59	2.87	3.10
35	1.23	1.49	1.66	1.84	2.07	2.25	2.43	2.67	2.86
40	1.22	1.46	1.62	1.78	1.99	2.15	2.31	2.53	2.70
45	1.21	1.43	1.59	1.74	1.93	2.08	2.22	2.42	2.57
50	1.20	1.41	1.56	1.70	1.88	2.02	2.15	2.33	2.47
60	1.19	1.38	1.52	1.64	1.81	1.93	2.05	2.21	2.33
70	1.18	1.36	1.49	1.60	1.75	1.86	1.97	2.12	2.22
80	1.17	1.34	1.46	1.57	1.71	1.82	1.92	2.05	2.15
90	1.16	1.33	1.44	1.55	1.68	1.78	1.88	2.00	2.09
100	1.16	1.32	1.43	1.53	1.66	1.75	1.84	1.96	2.05
120	1.15	1.31	1.41	1.50	1.62	1.71	1.79	1.90	1.98
140	1.14	1.29	1.39	1.48	1.60	1.68	1.76	1.86	1.93
160	1.14	1.29	1.38	1.47	1.58	1.65	1.73	1.83	1.90
180	1.14	1.28	1.37	1.46	1.56	1.64	1.71	1.80	1.87
200	1.13	1.27	1.36	1.45	1.55	1.62	1.69	1.78	1.85
300	1.13	1.26	1.34	1.42	1.51	1.58	1.64	1.72	1.78
500	1.12	1.24	1.32	1.39	1.48	1.54	1.60	1.68	1.73
∞	1.11	1.22	1.29	1.36	1.43	1.49	1.54	1.60	1.65

TABLE B.4 (cont.) Critical Values of the *F* Distribution

Numerator DF = 80

Denom. DF	α(2): 0.50 α(1): 0.25	0.20 0.10	0.10 0.05	0.05 0.025	0.02 0.01	0.01 0.005	0.005 0.0025	0.002 0.001	0.001 0.0005
1	9.78	62.9	253.	1010.	6330.	25300.	101000.	633000.	2530000.
2	3.46	9.48	19.5	39.5	99.5	199.	399.	999.	2000.
3	2.47	5.15	8.56	14.0	26.3	42.1	67.1	124.	198.
4	2.08	3.78	5.67	8.33	13.6	19.5	27.9	44.6	63.3
5	1.87	3.13	4.41	6.10	9.16	12.3	16.5	24.2	32.2
6	1.74	2.75	3.72	4.93	7.01	9.06	11.6	16.1	20.5
7	1.65	2.50	3.29	4.23	5.78	7.25	9.04	12.0	14.8
8	1.59	2.33	2.99	3.76	4.99	6.12	7.46	9.63	11.6
9	1.54	2.20	2.77	3.42	4.44	5.36	6.42	8.09	9.61
10	1.50	2.09	2.60	3.17	4.04	4.80	5.68	7.03	8.23
11	1.46	2.01	2.47	2.97	3.73	4.39	5.13	6.26	7.25
12	1.44	1.95	2.36	2.82	3.49	4.07	4.71	5.68	6.51
13	1.41	1.89	2.27	2.69	3.30	3.81	4.38	5.22	5.94
14	1.39	1.84	2.20	2.58	3.14	3.60	4.11	4.86	5.49
15	1.37	1.80	2.14	2.49	3.00	3.43	3.89	4.56	5.12
16	1.36	1.77	2.08	2.42	2.89	3.28	3.70	4.31	4.81
17	1.35	1.74	2.03	2.35	2.79	3.15	3.54	4.10	4.56
18	1.33	1.71	1.99	2.29	2.70	3.04	3.40	3.92	4.34
19	1.32	1.68	1.96	2.24	2.63	2.95	3.28	3.76	4.15
20	1.31	1.66	1.92	2.19	2.56	2.86	3.18	3.62	3.99
21	1.30	1.64	1.89	2.15	2.50	2.79	3.08	3.50	3.84
22	1.30	1.62	1.86	2.11	2.45	2.72	3.00	3.40	3.72
23	1.29	1.61	1.84	2.08	2.40	2.66	2.93	3.30	3.60
24	1.28	1.59	1.82	2.05	2.36	2.60	2.86	3.22	3.50
25	1.28	1.58	1.80	2.02	2.32	2.55	2.80	3.14	3.41
26	1.27	1.56	1.78	1.99	2.28	2.51	2.74	3.07	3.33
27	1.26	1.55	1.76	1.97	2.25	2.47	2.69	3.00	3.25
28	1.26	1.54	1.74	1.94	2.22	2.43	2.64	2.94	3.18
29	1.25	1.53	1.73	1.92	2.19	2.39	2.60	2.89	3.12
30	1.25	1.52	1.71	1.90	2.16	2.36	2.56	2.84	3.06
35	1.23	1.48	1.65	1.82	2.05	2.22	2.40	2.64	2.83
40	1.22	1.45	1.61	1.76	1.97	2.12	2.28	2.49	2.66
45	1.21	1.42	1.57	1.72	1.91	2.05	2.19	2.38	2.53
50	1.20	1.40	1.54	1.68	1.86	1.99	2.12	2.30	2.43
60	1.18	1.37	1.50	1.63	1.78	1.90	2.02	2.17	2.29
70	1.17	1.35	1.47	1.59	1.73	1.84	1.94	2.08	2.18
80	1.16	1.33	1.45	1.55	1.69	1.79	1.89	2.01	2.11
90	1.16	1.32	1.43	1.53	1.66	1.75	1.84	1.96	2.05
100	1.15	1.31	1.41	1.51	1.63	1.72	1.81	1.92	2.01
120	1.14	1.29	1.39	1.48	1.60	1.68	1.76	1.86	1.94
140	1.14	1.28	1.38	1.46	1.57	1.65	1.72	1.82	1.89
160	1.13	1.27	1.36	1.45	1.55	1.62	1.69	1.79	1.85
180	1.13	1.27	1.35	1.43	1.53	1.61	1.67	1.76	1.83
200	1.13	1.26	1.35	1.42	1.52	1.59	1.66	1.74	1.80
300	1.12	1.24	1.32	1.39	1.48	1.55	1.61	1.68	1.74
500	1.11	1.23	1.30	1.37	1.45	1.51	1.56	1.63	1.68
∞	1.10	1.21	1.27	1.33	1.40	1.45	1.50	1.56	1.60

TABLE B.4 (cont.)   Critical Values of the F Distribution

Numerator DF $= 90$

Denom. DF	$\alpha(2)$: 0.50 $\alpha(1)$: 0.25	0.20 0.10	0.10 0.05	0.05 0.025	0.02 0.01	0.01 0.005	0.005 0.0025	0.002 0.001	0.001 0.0005
1	9.79	63.0	253.	1010.	6330.	25300.	101000.	633000.	2530000.
2	3.46	9.48	19.5	39.5	99.5	199.	399.	999.	2000.
3	2.47	5.15	8.56	14.0	26.3	42.0	67.1	124.	197.
4	2.08	3.78	5.67	8.33	13.6	19.5	27.9	44.5	63.2
5	1.87	3.13	4.41	6.09	9.14	12.3	16.5	24.2	32.1
6	1.74	2.75	3.72	4.92	7.00	9.04	11.6	16.1	20.4
7	1.65	2.50	3.28	4.22	5.77	7.23	9.01	12.0	14.8
8	1.59	2.32	2.98	3.75	4.97	6.10	7.44	9.60	11.6
9	1.53	2.19	2.76	3.41	4.43	5.34	6.40	8.06	9.57
10	1.49	2.09	2.59	3.16	4.03	4.79	5.66	7.00	8.20
11	1.46	2.01	2.46	2.96	3.72	4.37	5.11	6.23	7.21
12	1.43	1.94	2.36	2.81	3.48	4.05	4.69	5.65	6.48
13	1.41	1.89	2.27	2.68	3.28	3.79	4.36	5.19	5.91
14	1.39	1.84	2.19	2.57	3.12	3.58	4.09	4.83	5.45
15	1.37	1.80	2.13	2.48	2.99	3.41	3.86	4.53	5.09
16	1.36	1.76	2.07	2.40	2.87	3.26	3.68	4.28	4.78
17	1.34	1.73	2.03	2.34	2.78	3.13	3.52	4.07	4.53
18	1.33	1.70	1.98	2.28	2.69	3.02	3.38	3.89	4.31
19	1.32	1.68	1.95	2.23	2.61	2.93	3.26	3.73	4.12
20	1.31	1.65	1.91	2.18	2.55	2.84	3.16	3.60	3.96
21	1.30	1.63	1.88	2.14	2.49	2.77	3.06	3.48	3.81
22	1.29	1.62	1.86	2.10	2.43	2.70	2.98	3.37	3.69
23	1.29	1.60	1.83	2.07	2.39	2.64	2.90	3.28	3.57
24	1.28	1.58	1.81	2.03	2.34	2.58	2.84	3.19	3.47
25	1.27	1.57	1.79	2.01	2.30	2.53	2.78	3.11	3.38
26	1.27	1.56	1.77	1.98	2.26	2.49	2.72	3.04	3.30
27	1.26	1.54	1.75	1.95	2.23	2.45	2.67	2.98	3.22
28	1.26	1.53	1.73	1.93	2.20	2.41	2.62	2.92	3.15
29	1.25	1.52	1.72	1.91	2.17	2.37	2.58	2.86	3.09
30	1.25	1.51	1.70	1.89	2.14	2.34	2.54	2.81	3.03
35	1.23	1.47	1.64	1.81	2.03	2.20	2.38	2.61	2.80
40	1.21	1.44	1.60	1.75	1.95	2.10	2.26	2.47	2.63
45	1.20	1.41	1.56	1.70	1.89	2.03	2.17	2.36	2.50
50	1.19	1.39	1.53	1.67	1.84	1.97	2.10	2.27	2.40
60	1.18	1.36	1.49	1.61	1.76	1.88	1.99	2.14	2.25
70	1.17	1.34	1.46	1.57	1.71	1.81	1.92	2.05	2.15
80	1.16	1.33	1.44	1.54	1.67	1.77	1.86	1.98	2.08
90	1.15	1.31	1.42	1.52	1.64	1.73	1.82	1.93	2.02
100	1.15	1.30	1.40	1.50	1.61	1.70	1.78	1.89	1.97
120	1.14	1.28	1.38	1.47	1.58	1.66	1.73	1.83	1.90
140	1.13	1.27	1.36	1.45	1.55	1.62	1.70	1.79	1.86
160	1.13	1.26	1.35	1.43	1.53	1.60	1.67	1.75	1.82
180	1.13	1.26	1.34	1.42	1.51	1.58	1.65	1.73	1.79
200	1.12	1.25	1.33	1.41	1.50	1.56	1.63	1.71	1.77
300	1.11	1.23	1.31	1.38	1.46	1.52	1.58	1.65	1.70
500	1.11	1.22	1.29	1.35	1.43	1.48	1.53	1.60	1.65
∞	1.10	1.20	1.26	1.31	1.38	1.43	1.47	1.52	1.56

TABLE B.4 (cont.) Critical Values of the *F* Distribution

Numerator DF = 100

Denom. DF	α(2): 0.50 α(1): 0.25	0.20 0.10	0.10 0.05	0.05 0.025	0.02 0.01	0.01 0.005	0.005 0.0025	0.002 0.001	0.001 0.0005
1	9.80	63.0	253.	1010.	6330.	25300.	101000.	633000.	2530000.
2	3.47	9.48	19.5	39.5	99.5	199.	399.	999.	2000.
3	2.47	5.14	8.55	14.0	26.2	42.0	67.1	124.	197.
4	2.08	3.78	5.66	8.32	13.6	19.5	27.9	44.5	63.2
5	1.87	3.13	4.41	6.08	9.13	12.3	16.5	24.1	32.1
6	1.74	2.75	3.71	4.92	6.99	9.03	11.6	16.0	20.4
7	1.65	2.50	3.27	4.21	5.75	7.22	8.99	12.0	14.8
8	1.58	2.32	2.97	3.74	4.96	6.09	7.42	9.57	11.6
9	1.53	2.19	2.76	3.40	4.41	5.32	6.38	8.04	9.54
10	1.49	2.09	2.59	3.15	4.01	4.77	5.64	6.98	8.17
11	1.46	2.01	2.46	2.96	3.71	4.36	5.09	6.21	7.19
12	1.43	1.94	2.35	2.80	3.47	4.04	4.67	5.63	6.45
13	1.41	1.88	2.26	2.67	3.27	3.78	4.34	5.17	5.88
14	1.39	1.83	2.19	2.56	3.11	3.57	4.07	4.81	5.43
15	1.37	1.79	2.12	2.47	2.98	3.39	3.85	4.51	5.06
16	1.36	1.76	2.07	2.40	2.86	3.25	3.66	4.26	4.76
17	1.34	1.73	2.02	2.33	2.76	3.12	3.50	4.05	4.50
18	1.33	1.70	1.98	2.27	2.68	3.01	3.36	3.87	4.28
19	1.32	1.67	1.94	2.22	2.60	2.91	3.24	3.71	4.10
20	1.31	1.65	1.91	2.17	2.54	2.83	3.14	3.58	3.93
21	1.30	1.63	1.88	2.13	2.48	2.75	3.04	3.46	3.79
22	1.29	1.61	1.85	2.09	2.42	2.69	2.96	3.35	3.66
23	1.29	1.59	1.82	2.06	2.37	2.62	2.89	3.25	3.55
24	1.28	1.58	1.80	2.02	2.33	2.57	2.82	3.17	3.45
25	1.27	1.56	1.78	2.00	2.29	2.52	2.76	3.09	3.36
26	1.27	1.55	1.76	1.97	2.25	2.47	2.70	3.02	3.27
27	1.26	1.54	1.74	1.94	2.22	2.43	2.65	2.96	3.20
28	1.25	1.53	1.73	1.92	2.19	2.39	2.60	2.90	3.13
29	1.25	1.52	1.71	1.90	2.16	2.36	2.56	2.84	3.06
30	1.25	1.51	1.70	1.88	2.13	2.32	2.52	2.79	3.01
35	1.23	1.47	1.63	1.80	2.02	2.19	2.36	2.59	2.77
40	1.21	1.43	1.59	1.74	1.94	2.09	2.24	2.44	2.60
45	1.20	1.41	1.55	1.69	1.88	2.01	2.15	2.33	2.47
50	1.19	1.39	1.52	1.66	1.82	1.95	2.08	2.25	2.37
60	1.18	1.36	1.48	1.60	1.75	1.86	1.97	2.12	2.23
70	1.16	1.34	1.45	1.56	1.70	1.80	1.90	2.03	2.13
80	1.16	1.32	1.43	1.53	1.65	1.75	1.84	1.96	2.05
90	1.15	1.30	1.41	1.50	1.62	1.71	1.80	1.91	1.99
100	1.14	1.29	1.39	1.48	1.60	1.68	1.76	1.87	1.95
120	1.14	1.28	1.37	1.45	1.56	1.64	1.71	1.81	1.88
140	1.13	1.26	1.35	1.43	1.53	1.60	1.67	1.76	1.83
160	1.13	1.26	1.34	1.42	1.51	1.58	1.64	1.73	1.79
180	1.12	1.25	1.33	1.40	1.49	1.56	1.62	1.70	1.76
200	1.12	1.24	1.32	1.39	1.48	1.54	1.60	1.68	1.74
300	1.11	1.22	1.30	1.36	1.44	1.50	1.55	1.62	1.67
500	1.10	1.21	1.28	1.34	1.41	1.46	1.51	1.57	1.62
∞	1.09	1.18	1.24	1.30	1.36	1.40	1.44	1.49	1.53

TABLE B.4 (cont.)   Critical Values of the F Distribution

Numerator DF = 120

Denom. DF	$\alpha(2)$: 0.50 $\alpha(1)$: 0.25	0.20 0.10	0.10 0.05	0.05 0.025	0.02 0.01	0.01 0.005	0.005 0.0025	0.002 0.001	0.001 0.0005
1	9.80	63.1	253.	1010.	6340.	25400.	101000.	634000.	2540000.
2	3.47	9.48	19.5	39.5	99.5	199.	399.	999.	2000.
3	2.47	5.14	8.55	13.9	26.2	42.0	67.0	124.	197.
4	2.08	3.78	5.66	8.31	13.6	19.5	27.8	44.4	63.1
5	1.87	3.12	4.40	6.07	9.11	12.3	16.4	24.1	32.0
6	1.74	2.74	3.70	4.90	6.97	9.00	11.6	16.0	20.3
7	1.65	2.49	3.27	4.20	5.74	7.19	8.96	11.9	14.7
8	1.58	2.32	2.97	3.73	4.95	6.06	7.39	9.53	11.5
9	1.53	2.18	2.75	3.39	4.40	5.30	6.35	8.00	9.49
10	1.49	2.08	2.58	3.14	4.00	4.75	5.61	6.94	8.12
11	1.46	2.00	2.45	2.94	3.69	4.34	5.07	6.18	7.14
12	1.43	1.93	2.34	2.79	3.45	4.01	4.65	5.59	6.41
13	1.41	1.88	2.25	2.66	3.25	3.76	4.31	5.14	5.84
14	1.39	1.83	2.18	2.55	3.09	3.55	4.04	4.77	5.39
15	1.37	1.79	2.11	2.46	2.96	3.37	3.82	4.47	5.02
16	1.35	1.75	2.06	2.38	2.84	3.22	3.63	4.23	4.72
17	1.34	1.72	2.01	2.32	2.75	3.10	3.47	4.02	4.46
18	1.33	1.69	1.97	2.26	2.66	2.99	3.34	3.84	4.25
19	1.32	1.67	1.93	2.20	2.58	2.89	3.22	3.68	4.06
20	1.31	1.64	1.90	2.16	2.52	2.81	3.11	3.54	3.90
21	1.30	1.62	1.87	2.11	2.46	2.73	3.02	3.42	3.75
22	1.29	1.60	1.84	2.08	2.40	2.66	2.93	3.32	3.62
23	1.28	1.59	1.81	2.04	2.35	2.60	2.86	3.22	3.51
24	1.28	1.57	1.79	2.01	2.31	2.55	2.79	3.14	3.41
25	1.27	1.56	1.77	1.98	2.27	2.50	2.73	3.06	3.32
26	1.26	1.54	1.75	1.95	2.23	2.45	2.68	2.99	3.24
27	1.26	1.53	1.73	1.93	2.20	2.41	2.62	2.92	3.16
28	1.25	1.52	1.71	1.91	2.17	2.37	2.58	2.86	3.09
29	1.25	1.51	1.70	1.89	2.14	2.33	2.53	2.81	3.03
30	1.24	1.50	1.68	1.87	2.11	2.30	2.49	2.76	2.97
35	1.22	1.46	1.62	1.79	2.00	2.16	2.33	2.56	2.73
40	1.21	1.42	1.58	1.72	1.92	2.06	2.21	2.41	2.56
45	1.20	1.40	1.54	1.68	1.85	1.99	2.12	2.30	2.44
50	1.19	1.38	1.51	1.64	1.80	1.93	2.05	2.21	2.34
60	1.17	1.35	1.47	1.58	1.73	1.83	1.94	2.08	2.19
70	1.16	1.32	1.44	1.54	1.67	1.77	1.87	1.99	2.09
80	1.15	1.31	1.41	1.51	1.63	1.72	1.81	1.92	2.01
90	1.15	1.29	1.39	1.48	1.60	1.68	1.76	1.87	1.95
100	1.14	1.28	1.38	1.46	1.57	1.65	1.73	1.83	1.90
120	1.13	1.26	1.35	1.43	1.53	1.61	1.68	1.77	1.83
140	1.13	1.25	1.33	1.41	1.50	1.57	1.64	1.72	1.78
160	1.12	1.24	1.32	1.39	1.48	1.55	1.61	1.69	1.75
180	1.12	1.23	1.31	1.38	1.47	1.53	1.59	1.66	1.72
200	1.11	1.23	1.30	1.37	1.45	1.51	1.57	1.64	1.69
300	1.10	1.21	1.28	1.34	1.41	1.46	1.51	1.58	1.62
500	1.10	1.19	1.26	1.31	1.38	1.42	1.47	1.53	1.57
∞	1.08	1.17	1.22	1.27	1.32	1.36	1.40	1.45	1.48

TABLE B.4 (cont.)   Critical Values of the *F* Distribution

Numerator DF = 140

Denom. DF	α(2): 0.50 α(1): 0.25	0.20 0.10	0.10 0.05	0.05 0.025	0.02 0.01	0.01 0.005	0.005 0.0025	0.002 0.001	0.001 0.0005
1	9.81	63.1	253.	1010.	6340.	25400.	101000.	634000.	2540000.
2	3.47	9.48	19.5	39.5	99.5	199.	399.	999.	2000.
3	2.47	5.14	8.55	13.9	26.2	42.0	67.0	124.	197.
4	2.08	3.77	5.65	8.30	13.5	19.4	27.8	44.4	63.0
5	1.87	3.12	4.39	6.06	9.10	12.3	16.4	24.0	31.9
6	1.74	2.74	3.70	4.90	6.96	8.98	11.5	15.9	20.3
7	1.65	2.49	3.26	4.19	5.72	7.18	8.94	11.9	14.7
8	1.58	2.31	2.96	3.72	4.93	6.05	7.37	9.50	11.5
9	1.53	2.18	2.74	3.38	4.39	5.28	6.33	7.97	9.46
10	1.49	2.08	2.57	3.13	3.98	4.73	5.59	6.92	8.09
11	1.46	2.00	2.44	2.94	3.68	4.32	5.05	6.15	7.11
12	1.43	1.93	2.33	2.78	3.44	4.00	4.63	5.57	6.38
13	1.41	1.87	2.25	2.65	3.24	3.74	4.29	5.11	5.81
14	1.39	1.82	2.17	2.54	3.08	3.53	4.02	4.75	5.36
15	1.37	1.78	2.11	2.45	2.95	3.36	3.80	4.45	5.00
16	1.35	1.75	2.05	2.37	2.83	3.21	3.61	4.20	4.69
17	1.34	1.71	2.00	2.31	2.73	3.08	3.45	3.99	4.44
18	1.33	1.69	1.96	2.25	2.65	2.97	3.32	3.81	4.22
19	1.32	1.66	1.92	2.19	2.57	2.87	3.20	3.66	4.03
20	1.31	1.64	1.89	2.15	2.50	2.79	3.09	3.52	3.87
21	1.30	1.62	1.86	2.10	2.44	2.71	3.00	3.40	3.73
22	1.29	1.60	1.83	2.07	2.39	2.65	2.92	3.29	3.60
23	1.28	1.58	1.81	2.03	2.34	2.59	2.84	3.20	3.49
24	1.27	1.57	1.78	2.00	2.30	2.53	2.77	3.11	3.38
25	1.27	1.55	1.76	1.97	2.26	2.48	2.71	3.03	3.29
26	1.26	1.54	1.74	1.94	2.22	2.43	2.66	2.96	3.21
27	1.26	1.53	1.72	1.92	2.18	2.39	2.60	2.90	3.13
28	1.25	1.51	1.71	1.90	2.15	2.35	2.56	2.84	3.06
29	1.24	1.50	1.69	1.88	2.12	2.32	2.51	2.79	3.00
30	1.24	1.49	1.68	1.86	2.10	2.28	2.47	2.74	2.94
35	1.22	1.45	1.61	1.77	1.98	2.15	2.31	2.53	2.71
40	1.21	1.42	1.57	1.71	1.90	2.05	2.19	2.39	2.54
45	1.19	1.39	1.53	1.66	1.84	1.97	2.10	2.27	2.41
50	1.18	1.37	1.50	1.63	1.79	1.91	2.03	2.19	2.31
60	1.17	1.34	1.46	1.57	1.71	1.81	1.92	2.06	2.16
70	1.16	1.32	1.42	1.53	1.65	1.75	1.84	1.96	2.06
80	1.15	1.30	1.40	1.49	1.61	1.70	1.79	1.90	1.98
90	1.14	1.28	1.38	1.47	1.58	1.66	1.74	1.84	1.92
100	1.14	1.27	1.36	1.45	1.55	1.63	1.70	1.80	1.87
120	1.13	1.26	1.34	1.42	1.51	1.58	1.65	1.74	1.80
140	1.12	1.24	1.32	1.39	1.48	1.55	1.61	1.69	1.75
160	1.12	1.23	1.31	1.38	1.46	1.52	1.58	1.66	1.71
180	1.11	1.22	1.30	1.36	1.45	1.50	1.56	1.63	1.68
200	1.11	1.22	1.29	1.35	1.43	1.49	1.54	1.61	1.66
300	1.10	1.20	1.26	1.32	1.39	1.44	1.49	1.55	1.59
500	1.09	1.18	1.24	1.29	1.35	1.40	1.44	1.49	1.53
∞	1.08	1.16	1.20	1.25	1.30	1.33	1.37	1.41	1.44

TABLE B.4 (cont.)   Critical Values of the F Distribution

Numerator DF = 200

Denom. DF	$\alpha(2)$: 0.50 $\alpha(1)$: 0.25	0.20 0.10	0.10 0.05	0.05 0.025	0.02 0.01	0.01 0.005	0.005 0.0025	0.002 0.001	0.001 0.0005
1	9.82	63.2	254.	1020.	6350.	25400.	102000.	635000.	2540000.
2	3.47	9.49	19.5	39.5	99.5	199.	399.	999.	2000.
3	2.47	5.14	8.54	13.9	26.2	41.9	66.9	124.	197.
4	2.08	3.77	5.65	8.29	13.5	19.4	27.7	44.3	62.9
5	1.87	3.12	4.39	6.05	9.08	12.2	16.4	24.0	31.8
6	1.74	2.73	3.69	4.88	6.93	8.95	11.5	15.9	20.2
7	1.65	2.48	3.25	4.18	5.70	7.15	8.90	11.8	14.6
8	1.58	2.31	2.95	3.70	4.91	6.02	7.33	9.45	11.4
9	1.53	2.17	2.73	3.37	4.36	5.26	6.29	7.93	9.40
10	1.49	2.07	2.56	3.12	3.96	4.71	5.56	6.87	8.04
11	1.46	1.99	2.43	2.92	3.66	4.29	5.01	6.10	7.06
12	1.43	1.92	2.32	2.76	3.41	3.97	4.59	5.52	6.33
13	1.40	1.86	2.23	2.63	3.22	3.71	4.26	5.07	5.76
14	1.38	1.82	2.16	2.53	3.06	3.50	3.99	4.71	5.31
15	1.37	1.77	2.10	2.44	2.92	3.33	3.77	4.41	4.95
16	1.35	1.74	2.04	2.36	2.81	3.18	3.58	4.16	4.64
17	1.34	1.71	1.99	2.29	2.71	3.05	3.42	3.95	4.39
18	1.32	1.68	1.95	2.23	2.62	2.94	3.28	3.77	4.17
19	1.31	1.65	1.91	2.18	2.55	2.85	3.16	3.61	3.98
20	1.30	1.63	1.88	2.13	2.48	2.76	3.06	3.48	3.82
21	1.29	1.61	1.84	2.09	2.42	2.68	2.96	3.36	3.68
22	1.28	1.59	1.82	2.05	2.36	2.62	2.88	3.25	3.55
23	1.28	1.57	1.79	2.01	2.32	2.56	2.81	3.16	3.44
24	1.27	1.56	1.77	1.98	2.27	2.50	2.74	3.07	3.34
25	1.26	1.54	1.75	1.95	2.23	2.45	2.68	2.99	3.24
26	1.26	1.53	1.73	1.92	2.19	2.40	2.62	2.92	3.16
27	1.25	1.52	1.71	1.90	2.16	2.36	2.57	2.86	3.09
28	1.25	1.50	1.69	1.88	2.13	2.32	2.52	2.80	3.02
29	1.24	1.49	1.67	1.86	2.10	2.29	2.48	2.74	2.95
30	1.24	1.48	1.66	1.84	2.07	2.25	2.44	2.69	2.89
35	1.22	1.44	1.60	1.75	1.96	2.11	2.27	2.49	2.66
40	1.20	1.41	1.55	1.69	1.87	2.01	2.15	2.34	2.49
45	1.19	1.38	1.51	1.64	1.81	1.93	2.06	2.23	2.36
50	1.18	1.36	1.48	1.60	1.76	1.87	1.99	2.14	2.26
60	1.16	1.33	1.44	1.54	1.68	1.78	1.88	2.01	2.11
70	1.15	1.30	1.40	1.50	1.62	1.71	1.80	1.92	2.00
80	1.14	1.28	1.38	1.47	1.58	1.66	1.74	1.85	1.93
90	1.13	1.27	1.36	1.44	1.55	1.62	1.70	1.79	1.86
100	1.13	1.26	1.34	1.42	1.52	1.59	1.66	1.75	1.82
120	1.12	1.24	1.32	1.39	1.48	1.54	1.60	1.68	1.74
140	1.11	1.22	1.30	1.36	1.45	1.51	1.56	1.64	1.69
160	1.11	1.21	1.28	1.35	1.42	1.48	1.53	1.60	1.65
180	1.10	1.21	1.27	1.33	1.41	1.46	1.51	1.57	1.62
200	1.10	1.20	1.26	1.32	1.39	1.44	1.49	1.55	1.60
300	1.09	1.18	1.23	1.28	1.35	1.39	1.43	1.48	1.52
500	1.08	1.16	1.21	1.25	1.31	1.35	1.38	1.43	1.46
∞	1.07	1.13	1.17	1.21	1.25	1.28	1.30	1.34	1.36

TABLE B.4 (cont.)   Critical Values of the *F* Distribution

Numerator DF = ∞

Denom. DF	α(2): 0.50 α(1): 0.25	0.20 0.10	0.10 0.05	0.05 0.025	0.02 0.01	0.01 0.005	0.005 0.0025	0.002 0.001	0.001 0.0005
1	9.85	63.3	254.	1020.	6370.	25500.	102000.	637000.	2550000.
2	3.48	9.49	19.5	39.5	99.5	199.	399.	999.	2000.
3	2.47	5.13	8.53	13.9	26.1	41.8	66.8	123.	196.
4	2.08	3.76	5.63	8.26	13.5	19.3	27.6	44.0	62.6
5	1.87	3.11	4.37	6.02	9.02	12.1	16.3	23.8	31.6
6	1.74	2.72	3.67	4.85	6.88	8.88	11.4	15.7	20.0
7	1.65	2.47	3.23	4.14	5.65	7.08	8.81	11.7	14.4
8	1.58	2.29	2.93	3.67	4.86	5.95	7.25	9.33	11.3
9	1.53	2.16	2.71	3.33	4.31	5.19	6.21	7.81	9.26
10	1.48	2.06	2.54	3.08	3.91	4.64	5.47	6.76	7.91
11	1.45	1.97	2.40	2.88	3.60	4.23	4.93	6.00	6.93
12	1.42	1.90	2.30	2.72	3.36	3.90	4.51	5.42	6.20
13	1.40	1.85	2.21	2.60	3.17	3.65	4.18	4.97	5.64
14	1.38	1.80	2.13	2.49	3.00	3.44	3.91	4.60	5.19
15	1.36	1.76	2.07	2.40	2.87	3.26	3.69	4.31	4.83
16	1.34	1.72	2.01	2.32	2.75	3.11	3.50	4.06	4.52
17	1.33	1.69	1.96	2.25	2.65	2.98	3.34	3.85	4.27
18	1.32	1.66	1.92	2.19	2.57	2.87	3.20	3.67	4.05
19	1.30	1.63	1.88	2.13	2.49	2.78	3.08	3.51	3.87
20	1.29	1.61	1.84	2.09	2.42	2.69	2.97	3.38	3.71
21	1.28	1.59	1.81	2.04	2.36	2.61	2.88	3.26	3.56
22	1.28	1.57	1.78	2.00	2.31	2.55	2.80	3.15	3.43
23	1.27	1.55	1.76	1.97	2.26	2.48	2.72	3.05	3.32
24	1.26	1.53	1.73	1.94	2.21	2.43	2.65	2.97	3.22
25	1.25	1.52	1.71	1.91	2.17	2.38	2.59	2.89	3.13
26	1.25	1.50	1.69	1.88	2.13	2.33	2.54	2.82	3.05
27	1.24	1.49	1.67	1.85	2.10	2.29	2.48	2.75	2.97
28	1.24	1.48	1.65	1.83	2.06	2.25	2.44	2.69	2.90
29	1.23	1.47	1.64	1.81	2.03	2.21	2.39	2.64	2.84
30	1.23	1.46	1.62	1.79	2.01	2.18	2.35	2.59	2.78
35	1.20	1.41	1.56	1.70	1.89	2.04	2.18	2.38	2.54
40	1.19	1.38	1.51	1.64	1.80	1.93	2.06	2.23	2.37
45	1.18	1.35	1.47	1.59	1.74	1.85	1.97	2.12	2.23
50	1.16	1.33	1.44	1.55	1.68	1.79	1.89	2.03	2.13
60	1.15	1.29	1.39	1.48	1.60	1.69	1.78	1.89	1.98
70	1.13	1.27	1.35	1.44	1.54	1.62	1.69	1.79	1.87
80	1.12	1.24	1.32	1.40	1.49	1.56	1.63	1.72	1.79
90	1.12	1.23	1.30	1.37	1.46	1.52	1.58	1.66	1.72
100	1.11	1.21	1.28	1.35	1.43	1.49	1.54	1.62	1.67
120	1.10	1.19	1.25	1.31	1.38	1.43	1.48	1.54	1.59
140	1.09	1.18	1.23	1.28	1.35	1.39	1.43	1.49	1.53
160	1.08	1.16	1.21	1.26	1.32	1.36	1.40	1.45	1.49
180	1.08	1.15	1.20	1.24	1.30	1.33	1.37	1.42	1.45
200	1.07	1.14	1.19	1.23	1.28	1.31	1.35	1.39	1.42
300	1.06	1.11	1.15	1.18	1.22	1.25	1.27	1.30	1.33
500	1.05	1.09	1.11	1.14	1.16	1.18	1.20	1.23	1.24
∞	1.00	1.00	1.00	1.00	1.00	1.00	1.00	1.00	1.00

Appendix Table B.4 was prepared using Equations 26.6.4, 26.6.5, 26.6.8, 26.6.11, 26.6.12, 26.4.6, and 26.4.14 of Zelen and Severo (1964). Values of *F* were calculated to a relative error ≤ 10^{-8} and then were rounded to three significant figures.

Examples:

$$F_{0.05(1),2,18} = 3.55, \quad F_{0.01(1),8,10} = 5.06 \quad \text{and} \quad F_{0.05(2),20,40} = 2.07$$

If a critical value is needed for degrees of freedom not on this table, one may conservatively employ the next smaller degrees of freedom that are on the table. Or, the needed critical value may be obtained by linear interpolation, with an error no greater than 0.01 for $\alpha(1) \geq 0.01$ and no greater than 0.02 for $\alpha(1) < 0.01$. If a little more accuracy is desired, or if $\nu_1 > 200$ or $\nu_2 > 500$, harmonic interpolation should be used. Note that $P(F_{\nu_1,\nu_2}) = 1 - P(1/F_{\nu_2,\nu_1})$. *For example:*

$$P\left(F_{2,28} \leq 2.50\right) = 0.10 \quad \text{and} \quad P\left(1/F_{28,2} \leq 2.50\right) = 0.90.$$

F is related to t, Z, and χ^2 as

$$F_{\alpha(1),1,\nu} = \left(t_{\alpha(2),\nu}\right)^2; \ F_{\alpha(1),1,\infty} = \left(Z_{\alpha(2)}\right)^2; \ F_{\alpha(1),\nu,\infty} = \chi^2_{\alpha,\nu}/\nu.$$

Mantel (1966) and George (1987) discuss approximating the F distribution using binomial probabilities.

Table B.5 is reprinted, with permission of the author, from the more extensive Table B.2 of H. L. Harter (1970).

Examples:

$$q_{0.05,24,3} = 5.532 \quad \text{and} \quad q_{0.01,20,5} = 5.294.$$

If a critical value is needed for degrees of freedom not on this table, one may conservatively use the next lower degrees of freedom in the table. The required critical value may be estimated by harmonic interpolation. See Harter (1960) for further considerations of interpolation.

TABLE B.5 Critical Values of the q Distribution

$\alpha = 0.50$

ν	k (or p): 2	3	4	5	6	7	8	9	10
1	1.414	2.338	2.918	3.335	3.658	3.920	4.139	4.327	4.491
2	1.155	1.908	2.377	2.713	2.973	3.184	3.361	3.513	3.645
3	1.082	1.791	2.230	2.545	2.789	2.987	3.152	3.294	3.418
4	1.048	1.737	2.163	2.468	2.704	2.895	3.055	3.193	3.313
5	1.028	1.705	2.124	2.423	2.655	2.843	3.000	3.135	3.253
6	1.015	1.685	2.098	2.394	2.623	2.809	2.964	3.097	3.214
7	1.006	1.670	2.080	2.375	2.601	2.785	2.938	3.070	3.186
8	0.9990	1.659	2.067	2.359	2.584	2.767	2.920	3.051	3.165
9	0.9938	1.651	2.057	2.347	2.572	2.753	2.905	3.035	3.149
10	0.9897	1.645	2.049	2.338	2.561	2.742	2.893	3.023	3.137
11	0.9863	1.639	2.042	2.330	2.553	2.733	2.884	3.014	3.127
12	0.9836	1.635	2.037	2.324	2.546	2.726	2.876	3.005	3.118
13	0.9812	1.631	2.032	2.319	2.540	2.719	2.870	2.998	3.111
14	0.9792	1.628	2.028	2.314	2.535	2.714	2.864	2.993	3.105
15	0.9775	1.625	2.025	2.310	2.531	2.709	2.859	2.987	3.099
16	0.9760	1.623	2.022	2.307	2.527	2.705	2.855	2.983	3.095
17	0.9747	1.621	2.020	2.304	2.524	2.702	2.851	2.979	3.091
18	0.9735	1.619	2.018	2.301	2.521	2.699	2.848	2.976	3.087
19	0.9724	1.618	2.015	2.299	2.518	2.696	2.845	2.973	3.084
20	0.9715	1.616	2.013	2.297	2.516	2.693	2.842	2.970	3.081
24	0.9685	1.611	2.007	2.290	2.509	2.685	2.834	2.961	3.071
30	0.9656	1.606	2.001	2.284	2.501	2.678	2.825	2.952	3.062
40	0.9626	1.602	1.996	2.277	2.494	2.670	2.817	2.943	3.053
60	0.9597	1.597	1.990	2.270	2.486	2.662	2.808	2.934	3.043
120	0.9568	1.592	1.984	2.264	2.479	2.654	2.799	2.925	3.034
∞	0.9539	1.588	1.978	2.257	2.472	2.645	2.791	2.915	3.024

ν	k (or p): 11	12	13	14	15	16	17	18	19
1	4.637	4.767	4.885	4.992	5.091	5.182	5.266	5.345	5.420
2	3.762	3.867	3.963	4.049	4.129	4.203	4.271	4.335	4.395
3	3.528	3.626	3.715	3.797	3.871	3.940	4.004	4.064	4.120
4	3.419	3.515	3.601	3.680	3.752	3.819	3.881	3.939	3.993
5	3.357	3.451	3.535	3.613	3.684	3.749	3.811	3.867	3.920
6	3.317	3.409	3.493	3.569	3.639	3.704	3.764	3.820	3.873
7	3.288	3.380	3.463	3.538	3.608	3.672	3.732	3.788	3.840
8	3.267	3.358	3.440	3.515	3.585	3.648	3.708	3.763	3.815
9	3.250	3.341	3.423	3.498	3.567	3.630	3.689	3.744	3.796
10	3.237	3.328	3.410	3.484	3.552	3.616	3.674	3.729	3.781
11	3.227	3.317	3.398	3.472	3.540	3.604	3.662	3.717	3.769
12	3.219	3.308	3.389	3.463	3.531	3.594	3.652	3.706	3.757
13	3.211	3.300	3.381	3.455	3.523	3.585	3.643	3.698	3.749
14	3.204	3.293	3.375	3.448	3.515	3.578	3.636	3.690	3.741
15	3.199	3.288	3.369	3.442	3.509	3.572	3.630	3.684	3.735
16	3.194	3.283	3.364	3.436	3.504	3.567	3.624	3.678	3.729
17	3.190	3.278	3.359	3.432	3.499	3.562	3.620	3.673	3.724
18	3.186	3.274	3.355	3.428	3.495	3.558	3.615	3.669	3.719
19	3.183	3.271	3.352	3.424	3.491	3.554	3.611	3.665	3.715
20	3.179	3.268	3.348	3.421	3.488	3.550	3.608	3.661	3.712
24	3.170	3.258	3.338	3.410	3.477	3.539	3.596	3.650	3.700
30	3.160	3.248	3.327	3.400	3.466	3.528	3.585	3.639	3.688
40	3.150	3.238	3.317	3.389	3.456	3.517	3.574	3.627	3.677
60	3.141	3.228	3.306	3.378	3.444	3.505	3.562	3.615	3.665
120	3.131	3.217	3.296	3.367	3.433	3.494	3.551	3.603	3.653
∞	3.121	3.207	3.285	3.356	3.422	3.482	3.538	3.591	3.640

TABLE B.5 (cont.)   Critical Values of the q Distribution

$\alpha = 0.50$

ν	k (or p): 20	22	24	26	28	30	32	34	36
1	5.489	5.616	5.731	5.835	5.930	6.017	6.098	6.173	6.244
2	4.451	4.554	4.646	4.730	4.807	4.878	4.943	5.004	5.061
3	4.172	4.269	4.356	4.434	4.507	4.573	4.634	4.691	4.745
4	4.044	4.138	4.222	4.298	4.368	4.432	4.492	4.547	4.599
5	3.970	4.062	4.145	4.220	4.288	4.352	4.410	4.464	4.515
6	3.922	4.014	4.095	4.169	4.237	4.299	4.357	4.411	4.461
7	3.889	3.979	4.060	4.133	4.200	4.262	4.319	4.372	4.422
8	3.864	3.953	4.033	4.106	4.174	4.234	4.291	4.344	4.394
9	3.844	3.933	4.013	4.086	4.152	4.214	4.270	4.323	4.372
10	3.829	3.917	3.997	4.069	4.135	4.196	4.253	4.305	4.354
11	3.816	3.904	3.984	4.056	4.122	4.183	4.239	4.291	4.340
12	3.806	3.894	3.973	4.045	4.110	4.171	4.227	4.279	4.328
13	3.797	3.884	3.964	4.035	4.101	4.162	4.218	4.269	4.318
14	3.789	3.877	3.956	4.027	4.092	4.153	4.209	4.261	4.309
15	3.783	3.870	3.949	4.020	4.085	4.145	4.201	4.253	4.302
16	3.777	3.864	3.942	4.014	4.079	4.139	4.195	4.246	4.295
17	3.772	3.859	3.937	4.009	4.073	4.133	4.189	4.241	4.289
18	3.767	3.854	3.932	4.003	4.069	4.128	4.184	4.236	4.284
19	3.763	3.850	3.928	3.999	4.064	4.124	4.180	4.231	4.279
20	3.759	3.846	3.924	3.995	4.060	4.120	4.175	4.227	4.275
24	3.748	3.834	3.912	3.983	4.048	4.107	4.162	4.214	4.262
30	3.736	3.822	3.899	3.970	4.035	4.094	4.149	4.200	4.248
40	3.724	3.809	3.887	3.957	4.021	4.080	4.135	4.186	4.234
60	3.711	3.797	3.874	3.944	4.008	4.067	4.121	4.172	4.220
120	3.699	3.784	3.861	3.930	3.994	4.053	4.107	4.157	4.204
∞	3.686	3.771	3.847	3.916	3.979	4.037	4.091	4.141	4.188

ν	k (or p): 38	40	50	60	70	80	90	100
1	6.310	6.372	6.637	6.847	7.021	7.169	7.297	7.411
2	5.115	5.165	5.379	5.550	5.690	5.810	5.914	6.006
3	4.795	4.842	5.043	5.202	5.335	5.447	5.544	5.630
4	4.647	4.693	4.888	5.043	5.171	5.280	5.374	5.458
5	4.563	4.608	4.799	4.951	5.077	5.184	5.277	5.359
6	4.500	4.552	4.741	4.891	5.016	5.121	5.213	5.294
7	4.469	4.513	4.700	4.850	4.973	5.078	5.169	5.249
8	4.440	4.484	4.671	4.819	4.941	5.045	5.136	5.215
9	4.418	4.462	4.647	4.794	4.916	5.020	5.110	5.189
10	4.400	4.444	4.629	4.775	4.897	5.000	5.090	5.169
11	4.386	4.429	4.613	4.760	4.881	4.984	5.073	5.152
12	4.374	4.417	4.600	4.747	4.867	4.970	5.059	5.138
13	4.364	4.406	4.590	4.736	4.856	4.959	5.048	5.126
14	4.355	4.397	4.581	4.726	4.846	4.949	5.037	5.116
15	4.347	4.390	4.573	4.718	4.838	4.940	5.029	5.107
16	4.340	4.383	4.566	4.710	4.831	4.932	5.021	5.099
17	4.334	4.377	4.559	4.704	4.824	4.926	5.014	5.092
18	4.329	4.372	4.554	4.698	4.818	4.920	5.008	5.086
19	4.324	4.367	4.549	4.693	4.813	4.914	5.003	5.081
20	4.320	4.363	4.545	4.689	4.808	4.910	4.998	5.076
24	4.307	4.349	4.530	4.674	4.793	4.894	4.982	5.060
30	4.293	4.335	4.515	4.659	4.778	4.878	4.966	5.044
40	4.279	4.321	4.500	4.644	4.762	4.862	4.950	5.027
60	4.264	4.306	4.485	4.627	4.745	4.845	4.932	5.009
120	4.249	4.290	4.469	4.610	4.727	4.827	4.914	4.990
∞	4.232	4.274	4.450	4.591	4.707	4.806	4.892	4.968

TABLE B.5 (cont.) Critical Values of the q Distribution

$\alpha = 0.20$

ν	k (or p): 2	3	4	5	6	7	8	9	10
1	4.353	6.615	8.075	9.138	9.966	10.64	11.21	11.70	12.12
2	2.667	3.820	4.559	5.098	5.521	5.867	6.158	6.409	6.630
3	2.316	3.245	3.833	4.261	4.597	4.872	5.104	5.305	5.481
4	2.168	3.004	3.527	3.907	4.205	4.449	4.655	4.832	4.989
5	2.087	2.872	3.358	3.712	3.988	4.214	4.405	4.570	4.715
6	2.036	2.788	3.252	3.588	3.850	4.065	4.246	4.403	4.540
7	2.001	2.731	3.179	3.503	3.756	3.962	4.136	4.287	4.419
8	1.976	2.689	3.126	3.440	3.686	3.886	4.055	4.201	4.330
9	1.956	2.658	3.085	3.393	3.633	3.828	3.994	4.136	4.261
10	1.941	2.632	3.053	3.355	3.590	3.782	3.944	4.084	4.206
11	1.928	2.612	3.027	3.325	3.557	3.745	3.905	4.042	4.162
12	1.918	2.596	3.006	3.300	3.529	3.715	3.872	4.007	4.126
13	1.910	2.582	2.988	3.279	3.505	3.689	3.844	3.978	4.095
14	1.902	2.570	2.973	3.261	3.485	3.667	3.820	3.953	4.069
15	1.896	2.560	2.960	3.246	3.467	3.648	3.800	3.931	4.046
16	1.891	2.551	2.948	3.232	3.452	3.631	3.782	3.912	4.026
17	1.886	2.543	2.938	3.220	3.439	3.617	3.766	3.895	4.008
18	1.882	2.536	2.930	3.210	3.427	3.604	3.753	3.881	3.993
19	1.878	2.530	2.922	3.200	3.416	3.592	3.740	3.867	3.979
20	1.874	2.524	2.914	3.192	3.407	3.582	3.729	3.855	3.966
24	1.864	2.507	2.892	3.166	3.377	3.549	3.694	3.818	3.927
30	1.853	2.490	2.870	3.140	3.348	3.517	3.659	3.781	3.887
40	1.843	2.473	2.848	3.114	3.318	3.484	3.624	3.743	3.848
60	1.833	2.456	2.826	3.089	3.290	3.452	3.589	3.707	3.809
120	1.822	2.440	2.805	3.063	3.260	3.420	3.554	3.669	3.770
∞	1.812	2.424	2.784	3.037	3.232	3.389	3.520	3.632	3.730

ν	k (or p): 11	12	13	14	15	16	17	18	19
1	12.50	12.84	13.14	13.43	13.68	13.93	14.14	14.35	14.54
2	6.826	7.002	7.162	7.308	7.442	7.566	7.682	7.790	7.891
3	5.637	5.778	5.906	6.023	6.131	6.230	6.323	6.410	6.491
4	5.128	5.253	5.367	5.471	5.566	5.655	5.738	5.815	5.888
5	4.844	4.960	5.066	5.162	5.251	5.334	5.411	5.482	5.550
6	4.663	4.773	4.873	4.965	5.049	5.128	5.201	5.269	5.333
7	4.537	4.643	4.739	4.827	4.908	4.984	5.054	5.120	5.181
8	4.444	4.547	4.640	4.726	4.805	4.877	4.945	5.009	5.069
9	4.372	4.473	4.564	4.647	4.724	4.796	4.862	4.924	4.982
10	4.316	4.414	4.503	4.585	4.660	4.730	4.795	4.856	4.913
11	4.270	4.366	4.454	4.534	4.608	4.677	4.741	4.801	4.857
12	4.231	4.327	4.413	4.492	4.565	4.633	4.696	4.755	4.810
13	4.199	4.293	4.379	4.457	4.529	4.596	4.658	4.716	4.770
14	4.172	4.265	4.349	4.426	4.498	4.564	4.625	4.683	4.737
15	4.148	4.240	4.324	4.400	4.471	4.536	4.597	4.654	4.707
16	4.127	4.218	4.301	4.377	4.447	4.512	4.572	4.628	4.681
17	4.109	4.199	4.282	4.357	4.426	4.490	4.550	4.606	4.659
18	4.093	4.182	4.264	4.339	4.407	4.471	4.531	4.586	4.638
19	4.078	4.167	4.248	4.323	4.391	4.454	4.513	4.569	4.620
20	4.065	4.154	4.234	4.308	4.376	4.439	4.498	4.552	4.604
24	4.024	4.111	4.190	4.262	4.329	4.391	4.448	4.502	4.552
30	3.982	4.068	4.145	4.216	4.281	4.342	4.398	4.451	4.500
40	3.941	4.025	4.101	4.170	4.234	4.293	4.348	4.399	4.447
60	3.900	3.982	4.056	4.124	4.186	4.244	4.297	4.347	4.395
120	3.859	3.938	4.011	4.077	4.138	4.194	4.246	4.295	4.341
∞	3.817	3.895	3.966	4.030	4.089	4.144	4.195	4.242	4.287

TABLE B.5 (cont.)   Critical Values of the q Distribution

$\alpha = 0.20$

ν	k (or p): 20	22	24	26	28	30	32	34	36
1	14.72	15.06	15.36	15.63	15.88	16.11	16.32	16.52	16.71
2	7.986	8.162	8.320	8.463	8.594	8.715	8.827	8.931	9.029
3	6.568	6.709	6.835	6.951	7.057	7.154	7.244	7.328	7.407
4	5.956	6.082	6.195	6.298	6.392	6.479	6.560	6.635	6.706
5	5.613	5.730	5.835	5.931	6.019	6.100	6.175	6.245	6.311
6	5.393	5.504	5.604	5.695	5.779	5.856	5.927	5.994	6.056
7	5.239	5.346	5.442	5.530	5.610	5.684	5.753	5.817	5.877
8	5.125	5.228	5.322	5.407	5.485	5.557	5.624	5.686	5.744
9	5.037	5.138	5.229	5.312	5.388	5.459	5.524	5.585	5.641
10	4.967	5.066	5.155	5.237	5.311	5.380	5.444	5.504	5.559
11	4.910	5.007	5.095	5.175	5.248	5.316	5.379	5.437	5.492
12	4.862	4.958	5.044	5.123	5.196	5.262	5.324	5.382	5.436
13	4.822	4.917	5.002	5.080	5.151	5.217	5.278	5.335	5.388
14	4.787	4.881	4.965	5.042	5.113	5.178	5.238	5.295	5.347
15	4.757	4.850	4.934	5.010	5.080	5.144	5.204	5.260	5.312
16	4.731	4.823	4.906	4.981	5.050	5.114	5.174	5.229	5.281
17	4.708	4.799	4.881	4.956	5.025	5.088	5.147	5.202	5.253
18	4.688	4.778	4.859	4.934	5.002	5.065	5.123	5.177	5.228
19	4.669	4.759	4.840	4.914	4.981	5.044	5.102	5.156	5.206
20	4.652	4.742	4.822	4.895	4.963	5.025	5.082	5.136	5.186
24	4.599	4.687	4.766	4.838	4.904	4.964	5.021	5.073	5.122
30	4.546	4.632	4.710	4.779	4.844	4.903	4.958	5.010	5.058
40	4.493	4.576	4.652	4.720	4.783	4.841	4.895	4.945	4.993
60	4.439	4.520	4.594	4.661	4.722	4.778	4.831	4.880	4.925
120	4.384	4.463	4.535	4.600	4.659	4.714	4.765	4.812	4.857
∞	4.329	4.405	4.475	4.537	4.595	4.648	4.697	4.743	4.786

ν	k (or p): 38	40	50	60	70	80	90	100
1	16.88	17.05	17.74	18.30	18.76	19.15	19.49	19.79
2	9.121	9.207	9.576	9.869	10.11	10.32	10.50	10.66
3	7.481	7.551	7.849	8.086	8.283	8.450	8.596	8.725
4	6.771	6.834	7.100	7.313	7.489	7.639	7.769	7.885
5	6.372	6.430	6.678	6.877	7.041	7.181	7.303	7.411
6	6.115	6.170	6.406	6.595	6.751	6.885	7.001	7.103
7	5.934	5.987	6.214	6.397	6.548	6.676	6.788	6.887
8	5.799	5.851	6.072	6.249	6.395	6.520	6.629	6.725
9	5.695	5.745	5.961	6.134	6.277	6.399	6.506	6.600
10	5.612	5.661	5.873	6.042	6.182	6.302	6.407	6.499
11	5.544	5.592	5.800	5.967	6.105	6.223	6.326	6.416
12	5.487	5.535	5.740	5.904	6.040	6.156	6.257	6.347
13	5.438	5.486	5.689	5.850	5.985	6.100	6.200	6.288
14	5.397	5.444	5.644	5.804	5.937	6.051	6.150	6.237
15	5.361	5.407	5.606	5.764	5.896	6.008	6.106	6.193
16	5.329	5.375	5.572	5.729	5.859	5.971	6.068	6.154
17	5.301	5.347	5.542	5.698	5.827	5.938	6.034	6.119
18	5.276	5.321	5.515	5.670	5.798	5.908	6.004	6.089
19	5.254	5.299	5.491	5.645	5.772	5.881	5.976	6.061
20	5.233	5.278	5.469	5.622	5.749	5.857	5.951	6.035
24	5.169	5.212	5.400	5.549	5.674	5.780	5.872	5.954
30	5.103	5.146	5.329	5.475	5.597	5.701	5.791	5.871
40	5.037	5.078	5.257	5.399	5.518	5.619	5.708	5.786
60	4.969	5.009	5.183	5.321	5.437	5.535	5.621	5.697
120	4.899	4.938	5.106	5.240	5.352	5.447	5.530	5.603
∞	4.826	4.864	5.026	5.155	5.262	5.353	5.433	5.503

TABLE B.5 (cont.) Critical Values of the q Distribution

$\alpha = 0.10$

ν	k (or p): 2	3	4	5	6	7	8	9	10
1	8.929	13.44	16.36	18.49	20.15	21.51	22.64	23.62	24.48
2	4.130	5.733	6.773	7.538	8.139	8.633	9.049	9.409	9.725
3	3.328	4.467	5.199	5.738	6.162	6.511	6.806	7.062	7.287
4	3.015	3.976	4.586	5.035	5.388	5.679	5.926	6.139	6.327
5	2.850	3.717	4.264	4.664	4.979	5.238	5.458	5.648	5.816
6	2.748	3.559	4.065	4.435	4.726	4.966	5.168	5.344	5.499
7	2.680	3.451	3.931	4.280	4.555	4.780	4.972	5.137	5.283
8	2.630	3.374	3.834	4.169	4.431	4.646	4.829	4.987	5.126
9	2.592	3.316	3.761	4.084	4.337	4.545	4.721	4.873	5.007
10	2.563	3.270	3.704	4.018	4.264	4.465	4.636	4.783	4.913
11	2.540	3.234	3.658	3.965	4.205	4.401	4.568	4.711	4.838
12	2.521	3.204	3.621	3.922	4.156	4.349	4.511	4.652	4.776
13	2.505	3.179	3.589	3.885	4.116	4.305	4.464	4.602	4.724
14	2.491	3.158	3.563	3.854	4.081	4.267	4.424	4.560	4.680
15	2.479	3.140	3.540	3.828	4.052	4.235	4.390	4.524	4.641
16	2.469	3.124	3.520	3.804	4.026	4.207	4.360	4.492	4.608
17	2.460	3.110	3.503	3.784	4.004	4.183	4.334	4.464	4.579
18	2.452	3.098	3.488	3.767	3.984	4.161	4.311	4.440	4.554
19	2.445	3.087	3.474	3.751	3.966	4.142	4.290	4.418	4.531
20	2.439	3.078	3.462	3.736	3.950	4.124	4.271	4.398	4.510
24	2.420	3.047	3.423	3.692	3.900	4.070	4.213	4.336	4.445
30	2.400	3.017	3.386	3.648	3.851	4.016	4.155	4.275	4.381
40	2.381	2.988	3.349	3.605	3.803	3.963	4.099	4.215	4.317
60	2.363	2.959	3.312	3.562	3.755	3.911	4.042	4.155	4.254
120	2.344	2.930	3.276	3.520	3.707	3.859	3.987	4.096	4.191
∞	2.326	2.902	3.240	3.478	3.661	3.808	3.931	4.037	4.129

ν	k (or p): 11	12	13	14	15	16	17	18	19
1	25.24	25.92	26.54	27.10	27.62	28.10	28.54	28.96	29.35
2	10.01	10.26	10.49	10.70	10.89	11.07	11.24	11.39	11.54
3	7.487	7.667	7.832	7.982	8.120	8.249	8.368	8.479	8.584
4	6.495	6.645	6.783	6.909	7.025	7.133	7.233	7.327	7.414
5	5.966	6.101	6.223	6.336	6.440	6.536	6.626	6.710	6.789
6	5.637	5.762	5.875	5.979	6.075	6.164	6.247	6.325	6.398
7	5.413	5.530	5.637	5.735	5.826	5.910	5.988	6.061	6.130
8	5.250	5.362	5.464	5.558	5.644	5.724	5.799	5.869	5.935
9	5.127	5.234	5.333	5.423	5.506	5.583	5.655	5.723	5.786
10	5.029	5.134	5.229	5.317	5.397	5.472	5.542	5.607	5.668
11	4.951	5.053	5.146	5.231	5.309	5.382	5.450	5.514	5.573
12	4.886	4.986	5.077	5.160	5.236	5.308	5.374	5.436	5.495
13	4.832	4.930	5.019	5.100	5.176	5.245	5.311	5.372	5.429
14	4.786	4.882	4.970	5.050	5.124	5.192	5.256	5.316	5.373
15	4.746	4.841	4.927	5.006	5.079	5.147	5.209	5.269	5.324
16	4.712	4.805	4.890	4.968	5.040	5.107	5.169	5.227	5.282
17	4.682	4.774	4.858	4.935	5.005	5.071	5.133	5.190	5.244
18	4.655	4.746	4.829	4.905	4.975	5.040	5.101	5.158	5.211
19	4.631	4.721	4.803	4.879	4.948	5.012	5.073	5.129	5.182
20	4.609	4.699	4.780	4.855	4.924	4.987	5.047	5.103	5.155
24	4.541	4.628	4.708	4.780	4.847	4.909	4.966	5.021	5.071
30	4.474	4.559	4.635	4.706	4.770	4.830	4.886	4.939	4.988
40	4.408	4.490	4.564	4.632	4.695	4.752	4.807	4.857	4.905
60	4.342	4.421	4.493	4.558	4.619	4.675	4.727	4.775	4.821
120	4.276	4.353	4.422	4.485	4.543	4.597	4.647	4.694	4.738
∞	4.211	4.285	4.351	4.412	4.468	4.519	4.568	4.612	4.654

TABLE B.5 (cont.) Critical Values of the q Distribution

$\alpha = 0.10$

ν	k (or p): 20	22	24	26	28	30	32	34	36
1	29.71	30.39	30.99	31.54	32.04	32.50	32.93	33.33	33.71
2	11.68	11.93	12.16	12.36	12.55	12.73	12.89	13.04	13.18
3	8.683	8.864	9.029	9.177	9.314	9.440	9.557	9.666	9.768
4	7.497	7.650	7.789	7.914	8.029	8.135	8.234	8.326	8.412
5	6.863	7.000	7.123	7.236	7.340	7.435	7.523	7.606	7.683
6	6.467	6.593	6.708	6.812	6.908	6.996	7.078	7.155	7.227
7	6.195	6.315	6.422	6.521	6.611	6.695	6.773	6.845	6.913
8	5.997	6.111	6.214	6.308	6.395	6.475	6.549	6.618	6.683
9	5.846	5.956	6.055	6.146	6.229	6.306	6.378	6.444	6.507
10	5.726	5.833	5.930	6.017	6.098	6.173	6.242	6.307	6.368
11	5.630	5.734	5.828	5.914	5.992	6.065	6.132	6.196	6.255
12	5.550	5.652	5.744	5.827	5.904	5.976	6.042	6.103	6.161
13	5.483	5.583	5.673	5.755	5.830	5.900	5.965	6.025	6.082
14	5.426	5.524	5.612	5.693	5.767	5.836	5.899	5.959	6.014
15	5.376	5.473	5.560	5.639	5.713	5.780	5.843	5.901	5.956
16	5.333	5.428	5.515	5.593	5.665	5.732	5.793	5.851	5.905
17	5.295	5.389	5.474	5.552	5.623	5.689	5.750	5.806	5.860
18	5.262	5.355	5.439	5.515	5.585	5.650	5.711	5.767	5.820
19	5.232	5.324	5.407	5.483	5.552	5.616	5.676	5.732	5.784
20	5.205	5.296	5.378	5.453	5.522	5.586	5.645	5.700	5.752
24	5.119	5.208	5.287	5.360	5.427	5.489	5.546	5.600	5.650
30	5.034	5.120	5.197	5.267	5.332	5.392	5.447	5.499	5.547
40	4.949	5.032	5.107	5.174	5.236	5.294	5.347	5.397	5.444
60	4.864	4.944	5.015	5.081	5.141	5.196	5.247	5.295	5.340
120	4.779	4.856	4.924	4.987	5.044	5.097	5.146	5.192	5.235
∞	4.694	4.767	4.832	4.892	4.947	4.997	5.044	5.087	5.128

ν	k (or p): 38	40	50	60	70	80	90	100
1	34.06	34.38	35.79	36.91	37.83	38.62	39.30	39.91
2	13.31	13.44	13.97	14.40	14.75	15.05	15.31	15.54
3	9.864	9.954	10.34	10.65	10.91	11.12	11.31	11.48
4	8.493	8.569	8.896	9.156	9.373	9.557	9.718	9.860
5	7.756	7.825	8.118	8.353	8.548	8.715	8.859	8.988
6	7.294	7.358	7.630	7.848	8.029	8.184	8.319	8.438
7	6.976	7.036	7.294	7.500	7.672	7.818	7.946	8.059
8	6.744	6.801	7.048	7.245	7.409	7.550	7.672	7.780
9	6.566	6.621	6.859	7.050	7.208	7.343	7.461	7.566
10	6.425	6.479	6.709	6.895	7.048	7.180	7.295	7.396
11	6.310	6.363	6.588	6.768	6.918	7.047	7.158	7.258
12	6.215	6.267	6.487	6.663	6.810	6.936	7.045	7.142
13	6.135	6.186	6.402	6.575	6.719	6.842	6.949	7.045
14	6.067	6.116	6.329	6.499	6.641	6.762	6.868	6.961
15	6.008	6.057	6.266	6.433	6.573	6.692	6.796	6.888
16	5.956	6.004	6.210	6.376	6.513	6.631	6.734	6.825
17	5.910	5.958	6.162	6.325	6.461	6.577	6.679	6.769
18	5.870	5.917	6.118	6.280	6.414	6.529	6.630	6.719
19	5.833	5.880	6.079	6.239	6.372	6.486	6.585	6.674
20	5.801	5.847	6.044	6.203	6.335	6.447	6.546	6.633
24	5.697	5.741	5.933	6.086	6.214	6.324	6.419	6.503
30	5.593	5.636	5.821	5.969	6.093	6.198	6.291	6.372
40	5.488	5.529	5.708	5.850	5.969	6.071	6.160	6.238
60	5.382	5.422	5.593	5.730	5.844	5.941	6.026	6.102
120	5.275	5.313	5.476	5.606	6.715	5.808	5.888	5.960
∞	5.166	5.202	5.357	5.480	5.582	5.669	5.745	5.812

TABLE B.5 (cont.)   Critical Values of the q Distribution

$\alpha = 0.05$

ν	k (or p): 2	3	4	5	6	7	8	9	10
1	17.97	26.98	32.82	37.08	40.41	43.12	45.40	47.36	49.07
2	6.085	8.331	9.798	10.88	11.74	12.44	13.03	13.54	13.99
3	4.501	5.910	6.825	7.502	8.037	8.478	8.853	9.177	9.462
4	3.927	5.040	5.757	6.287	6.707	7.053	7.347	7.602	7.826
5	3.635	4.602	5.218	5.673	6.033	6.330	6.582	6.802	6.995
6	3.461	4.339	4.896	5.305	5.628	5.895	6.122	6.319	6.493
7	3.344	4.165	4.681	5.060	5.359	5.606	5.815	5.998	6.158
8	3.261	4.041	4.529	4.886	5.167	5.399	5.597	5.767	5.918
9	3.199	3.949	4.415	4.756	5.024	5.244	5.432	5.595	5.739
10	3.151	3.877	4.327	4.654	4.912	5.124	5.305	5.461	5.599
11	3.113	3.820	4.256	4.574	4.823	5.028	5.202	5.353	5.487
12	3.082	3.773	4.199	4.508	4.751	4.950	5.119	5.265	5.395
13	3.055	3.735	4.151	4.453	4.690	4.885	5.049	5.192	5.318
14	3.033	3.702	4.111	4.407	4.639	4.829	4.990	5.131	5.254
15	3.014	3.674	4.076	4.367	4.595	4.782	4.940	5.077	5.198
16	2.998	3.649	4.046	4.333	4.557	4.741	4.897	5.031	5.150
17	2.984	3.628	4.020	4.303	4.524	4.705	4.858	4.991	5.108
18	2.971	3.609	3.997	4.277	4.495	4.673	4.824	4.956	5.071
19	2.960	3.593	3.977	4.253	4.469	4.645	4.794	4.924	5.038
20	2.950	3.578	3.958	4.232	4.445	4.620	4.768	4.896	5.008
24	2.919	3.532	3.901	4.166	4.373	4.541	4.684	4.807	4.915
30	2.888	3.486	3.845	4.102	4.302	4.464	4.602	4.720	4.824
40	2.858	3.442	3.791	4.039	4.232	4.389	4.521	4.635	4.735
60	2.829	3.399	3.737	3.977	4.163	4.314	4.441	4.550	4.646
120	2.800	3.356	3.685	3.917	4.096	4.241	4.363	4.468	4.560
∞	2.772	3.314	3.633	3.858	4.030	4.170	4.286	4.387	4.474

ν	k (or p): 11	12	13	14	15	16	17	18	19
1	50.59	51.96	53.20	54.33	55.36	56.32	57.22	58.04	58.83
2	14.39	14.75	15.08	15.38	15.65	15.91	16.14	16.37	16.57
3	9.717	9.946	10.15	10.35	10.53	10.69	10.84	10.98	11.11
4	8.027	8.208	8.373	8.525	8.664	8.794	8.914	9.028	9.134
5	7.168	7.324	7.466	7.596	7.717	7.828	7.932	8.030	8.122
6	6.649	6.789	6.917	7.034	7.143	7.244	7.338	7.426	7.508
7	6.302	6.431	6.550	6.658	6.759	6.852	6.939	7.020	7.097
8	6.054	6.175	6.287	6.389	6.483	6.571	6.653	6.729	6.802
9	5.867	5.983	6.089	6.186	6.276	6.359	6.437	6.510	6.579
10	5.722	5.833	5.935	6.028	6.114	6.194	6.269	6.339	6.405
11	5.605	5.713	5.811	5.901	5.984	6.062	6.134	6.202	6.265
12	5.511	5.615	5.710	5.798	5.878	5.953	6.023	6.089	6.151
13	5.431	5.533	5.625	5.711	5.789	5.862	5.931	5.995	6.055
14	5.364	5.463	5.554	5.637	5.714	5.786	5.852	5.915	5.974
15	5.306	5.404	5.493	5.574	5.649	5.720	5.785	5.846	5.904
16	5.256	5.352	5.439	5.520	5.593	5.662	5.727	5.786	5.843
17	5.212	5.307	5.392	5.471	5.544	5.612	5.675	5.734	5.790
18	5.174	5.267	5.352	5.429	5.501	5.568	5.630	5.688	5.743
19	5.140	5.231	5.315	5.391	5.462	5.528	5.589	5.647	5.701
20	5.108	5.199	5.282	5.357	5.427	5.493	5.553	5.610	5.663
24	5.012	5.099	5.179	5.251	5.319	5.381	5.439	5.494	5.545
30	4.917	5.001	5.077	5.147	5.211	5.271	5.327	5.379	5.429
40	4.824	4.904	4.977	5.044	5.106	5.163	5.216	5.266	5.313
60	4.732	4.808	4.878	4.942	5.001	5.056	5.107	5.154	5.199
120	4.641	4.714	4.781	4.842	4.898	4.950	4.998	5.044	5.086
∞	4.552	4.622	4.685	4.743	4.796	4.845	4.891	4.934	4.974

TABLE B.5 (cont.)   Critical Values of the q Distribution

$\alpha = 0.05$

ν	k (or p): 20	22	24	26	28	30	32	34	36
1	59.56	60.91	62.12	63.22	64.23	65.15	66.01	66.81	67.56
2	16.77	17.13	17.45	17.75	18.02	18.27	18.50	18.72	18.92
3	11.24	11.47	11.68	11.87	12.05	12.21	12.36	12.50	12.63
4	9.233	9.418	9.584	9.736	9.875	10.00	10.12	10.23	10.34
5	8.208	8.368	8.512	8.643	8.764	8.875	8.979	9.075	9.165
6	7.587	7.730	7.861	7.979	8.088	8.189	8.283	8.370	8.452
7	7.170	7.303	7.423	7.533	7.634	7.728	7.814	7.895	7.972
8	6.870	6.995	7.109	7.212	7.307	7.395	7.477	7.554	7.625
9	6.644	6.763	6.871	6.970	7.061	7.145	7.222	7.295	7.363
10	6.467	6.582	6.686	6.781	6.868	6.948	7.023	7.093	7.159
11	6.326	6.436	6.536	6.628	6.712	6.790	6.863	6.930	6.994
12	6.209	6.317	6.414	6.503	6.585	6.660	6.731	6.796	6.858
13	6.112	6.217	6.312	6.398	6.478	6.551	6.620	6.684	6.744
14	6.029	6.132	6.224	6.309	6.387	6.459	6.526	6.588	6.647
15	5.958	6.059	6.149	6.233	6.309	6.379	6.445	6.506	6.564
16	5.897	5.995	6.084	6.166	6.241	6.310	6.374	6.434	6.491
17	5.842	5.940	6.027	6.107	6.181	6.249	6.313	6.372	6.427
18	5.794	5.890	5.977	6.055	6.128	6.195	6.258	6.316	6.371
19	5.752	5.846	5.932	6.009	6.081	6.147	6.209	6.267	6.321
20	5.714	5.807	5.891	5.968	6.039	6.104	6.165	6.222	6.275
24	5.594	5.683	5.764	5.838	5.906	5.968	6.027	6.081	6.132
30	5.475	5.561	5.638	5.709	5.774	5.833	5.889	5.941	5.990
40	5.358	5.439	5.513	5.581	5.642	5.700	5.753	5.803	5.849
60	5.241	5.319	5.389	5.453	5.512	5.566	5.617	5.664	5.708
120	5.126	5.200	5.266	5.327	5.382	5.434	5.481	5.526	5.568
∞	5.012	5.081	5.144	5.201	5.253	5.301	5.346	5.388	5.427

ν	k (or p): 38	40	50	60	70	80	90	100
1	68.26	68.92	71.73	73.97	75.82	77.40	78.77	79.98
2	19.11	19.28	20.05	20.66	21.16	21.59	21.96	22.29
3	12.75	12.87	13.36	13.76	14.08	14.36	14.61	14.82
4	10.44	10.53	10.93	11.24	11.51	11.73	11.92	12.09
5	9.250	9.330	9.674	9.949	10.18	10.38	10.54	10.69
6	8.529	8.601	8.913	9.163	9.370	9.548	9.702	9.839
7	8.043	8.110	8.400	8.632	8.824	8.989	9.133	9.261
8	7.693	7.756	8.029	8.248	8.430	8.586	8.722	8.843
9	7.428	7.488	7.749	7.958	8.132	8.281	8.410	8.526
10	7.220	7.279	7.529	7.730	7.897	8.041	8.166	8.276
11	7.053	7.110	7.352	7.546	7.708	7.847	7.968	8.075
12	6.916	6.970	7.205	7.394	7.552	7.687	7.804	7.909
13	6.800	6.854	7.083	7.267	7.421	7.552	7.667	7.769
14	6.702	6.754	6.979	7.159	7.309	7.438	7.550	7.650
15	6.618	6.669	6.888	7.065	7.212	7.339	7.449	7.546
16	6.544	6.594	6.810	6.984	7.128	7.252	7.360	7.457
17	6.479	6.529	6.741	6.912	7.054	7.176	7.283	7.377
18	6.422	6.471	6.680	6.848	6.989	7.109	7.213	7.307
19	6.371	6.419	6.626	6.792	6.930	7.048	7.152	7.244
20	6.325	6.373	6.576	6.740	6.877	6.994	7.097	7.187
24	6.181	6.226	6.421	6.579	6.710	6.822	6.920	7.008
30	6.037	6.080	6.267	6.417	6.543	6.650	6.744	6.827
40	5.893	5.934	6.112	6.255	6.375	6.477	6.566	6.645
60	5.750	5.789	5.958	6.093	6.206	6.303	6.387	6.462
120	5.607	5.644	5.802	5.929	6.035	6.126	6.205	6.275
∞	5.463	5.498	5.646	5.764	5.863	5.947	6.020	6.085

TABLE B.5 (cont.)   Critical Values of the q Distribution

$\alpha = 0.025$

v	k (or p): 2	3	4	5	6	7	8	9	10
1	35.99	54.00	65.69	74.22	80.87	86.29	90.85	94.77	98.20
2	8.776	11.94	14.01	15.54	16.75	17.74	18.58	19.31	19.95
3	5.907	7.661	8.808	9.660	10.34	10.89	11.37	11.78	12.14
4	4.943	6.244	7.088	7.716	8.213	8.625	8.976	9.279	9.548
5	4.474	5.558	6.257	6.775	7.186	7.527	7.816	8.068	8.291
6	4.199	5.158	5.772	6.226	6.586	6.884	7.138	7.359	7.554
7	4.018	4.897	5.455	5.868	6.194	6.464	6.695	6.895	7.072
8	3.892	4.714	5.233	5.616	5.919	6.169	6.382	6.568	6.732
9	3.797	4.578	5.069	5.430	5.715	5.950	6.151	6.325	6.479
10	3.725	4.474	4.943	5.287	5.558	5.782	5.972	6.138	6.285
11	3.667	4.391	4.843	5.173	5.433	5.648	5.831	5.989	6.130
12	3.620	4.325	4.762	5.081	5.332	5.540	5.716	5.869	6.004
13	3.582	4.269	4.694	5.004	5.248	5.449	5.620	5.769	5.900
14	3.550	4.222	4.638	4.940	5.178	5.374	5.540	5.684	5.811
15	3.522	4.182	4.589	4.885	5.118	5.309	5.471	5.612	5.737
16	3.498	4.148	4.548	4.838	5.066	5.253	5.412	5.550	5.672
17	3.477	4.118	4.512	4.797	5.020	5.204	5.361	5.496	5.615
18	3.458	4.092	4.480	4.761	4.981	5.162	5.315	5.448	5.565
19	3.442	4.068	4.451	4.728	4.945	5.123	5.275	5.405	5.521
20	3.427	4.047	4.426	4.700	4.914	5.089	5.238	5.368	5.481
24	3.381	3.983	4.347	4.610	4.816	4.984	5.126	5.250	5.358
30	3.337	3.919	4.271	4.523	4.720	4.881	5.017	5.134	5.238
40	3.294	3.858	4.197	4.439	4.627	4.780	4.910	5.022	5.120
60	3.251	3.798	4.124	4.356	4.536	4.682	4.806	4.912	5.006
120	3.210	3.739	4.053	4.276	4.447	4.587	4.704	4.805	4.894
∞	3.170	3.682	3.984	4.197	4.361	4.494	4.605	4.700	4.784

v	k (or p): 11	12	13	14	15	16	17	18	19
1	101.3	104.0	106.5	108.8	110.8	112.7	114.5	116.2	117.7
2	20.52	21.03	21.49	21.91	22.30	22.67	23.01	23.32	23.62
3	12.46	12.75	13.01	13.26	13.48	13.69	13.88	14.06	14.23
4	9.788	10.01	10.20	10.39	10.55	10.71	10.85	10.99	11.11
5	8.490	8.670	8.834	8.984	9.124	9.253	9.374	9.486	9.593
6	7.729	7.887	8.031	8.163	8.286	8.399	8.506	8.605	8.698
7	7.230	7.373	7.504	7.624	7.735	7.839	7.935	8.025	8.111
8	6.879	7.011	7.132	7.244	7.347	7.443	7.532	7.616	7.695
9	6.617	6.742	6.856	6.961	7.058	7.148	7.232	7.311	7.385
10	6.416	6.534	6.643	6.742	6.834	6.920	7.000	7.075	7.146
11	6.256	6.369	6.473	6.568	6.657	6.739	6.815	6.887	6.955
12	6.125	6.235	6.335	6.427	6.512	6.591	6.665	6.734	6.799
13	6.017	6.123	6.220	6.309	6.392	6.468	6.539	6.607	6.670
14	5.926	6.029	6.123	6.210	6.290	6.364	6.434	6.499	6.560
15	5.848	5.949	6.041	6.125	6.203	6.276	6.344	6.407	6.467
16	5.781	5.879	5.969	6.052	6.128	6.199	6.265	6.328	6.386
17	5.722	5.818	5.907	5.987	6.062	6.132	6.197	6.258	6.315
18	5.670	5.765	5.852	5.931	6.004	6.073	6.137	6.197	6.253
19	5.624	5.718	5.803	5.881	5.954	6.020	6.083	6.142	6.198
20	5.583	5.675	5.759	5.836	5.907	5.974	6.036	6.093	6.148
24	5.455	5.543	5.623	5.697	5.764	5.827	5.886	5.941	5.994
30	5.330	5.414	5.490	5.560	5.624	5.684	5.740	5.792	5.841
40	5.208	5.288	5.360	5.426	5.487	5.544	5.597	5.646	5.693
60	5.089	5.164	5.232	5.295	5.352	5.406	5.456	5.503	5.546
120	4.972	5.043	5.107	5.166	5.221	5.271	5.318	5.362	5.403
∞	4.858	4.925	4.985	5.041	5.092	5.139	5.183	5.224	5.262

TABLE B.5 (cont.)   Critical Values of the q Distribution

$\alpha = 0.025$

ν	k (or p): 20	22	24	26	28	30	32	34	36
1	119.2	121.9	124.3	126.5	128.6	130.4	132.1	133.7	135.2
2	23.89	24.41	24.87	25.29	25.67	26.03	26.35	26.66	26.95
3	14.39	14.69	14.95	15.19	15.41	15.62	15.81	15.99	16.15
4	11.23	11.46	11.66	11.84	12.00	12.16	12.30	12.44	12.56
5	9.693	9.878	10.04	10.20	10.34	10.47	10.59	10.70	10.80
6	8.787	8.949	9.097	9.231	9.355	9.469	9.575	9.674	9.767
7	8.191	8.339	8.473	8.595	8.708	8.812	8.909	8.999	9.084
8	7.769	7.907	8.031	8.145	8.250	8.346	8.436	8.520	8.599
9	7.455	7.585	7.702	7.809	7.908	7.999	8.084	8.163	8.237
10	7.212	7.335	7.447	7.549	7.643	7.729	7.810	7.885	7.956
11	7.019	7.137	7.244	7.341	7.431	7.514	7.592	7.664	7.732
12	6.861	6.974	7.078	7.172	7.258	7.338	7.413	7.483	7.548
13	6.730	6.840	6.939	7.031	7.115	7.192	7.265	7.332	7.396
14	6.619	6.726	6.823	6.911	6.993	7.069	7.139	7.204	7.266
15	6.523	6.628	6.723	6.809	6.889	6.962	7.031	7.095	7.155
16	6.441	6.543	6.636	6.721	6.799	6.870	6.938	7.000	7.059
17	6.370	6.469	6.560	6.644	6.720	6.790	6.856	6.917	6.975
18	6.306	6.404	6.493	6.575	6.650	6.720	6.784	6.844	6.900
19	6.250	6.347	6.434	6.514	6.588	6.656	6.719	6.779	6.835
20	6.200	6.295	6.381	6.460	6.532	6.600	6.662	6.720	6.775
24	6.043	6.133	6.215	6.290	6.359	6.423	6.482	6.538	6.589
30	5.888	5.974	6.052	6.123	6.188	6.248	6.305	6.357	6.406
40	5.737	5.818	5.891	5.958	6.020	6.077	6.130	6.179	6.226
60	5.588	5.664	5.733	5.797	5.854	5.908	5.958	6.004	6.048
120	5.442	5.513	5.578	5.637	5.691	5.741	5.788	5.831	5.872
∞	5.299	5.365	5.425	5.480	5.530	5.577	5.620	5.660	5.698

ν	k (or p): 38	40	50	60	70	80	90	100
1	136.6	137.9	143.6	148.1	151.8	154.9	157.7	160.0
2	27.22	27.47	28.55	29.42	30.13	30.74	31.27	31.74
3	16.31	16.46	17.08	17.59	18.00	18.36	18.67	18.95
4	12.68	12.79	13.27	13.65	13.96	14.23	14.47	14.68
5	10.91	11.00	11.40	11.72	11.99	12.21	12.41	12.59
6	9.855	9.938	10.30	10.58	10.81	11.02	11.19	11.35
7	9.164	9.239	9.563	9.822	10.04	10.23	10.38	10.53
8	8.673	8.743	9.044	9.286	9.487	9.660	9.810	9.944
9	8.307	8.373	8.657	8.885	9.076	9.238	9.381	9.507
10	8.023	8.086	8.356	8.574	8.755	8.911	9.046	9.167
11	7.796	7.856	8.116	8.325	8.499	8.648	8.779	8.894
12	7.610	7.668	7.919	8.120	8.289	8.433	8.559	8.671
13	7.455	7.512	7.755	7.950	8.113	8.253	8.375	8.484
14	7.324	7.379	7.615	7.806	7.965	8.101	8.220	8.325
15	7.212	7.265	7.496	7.682	7.837	7.970	8.086	8.189
16	7.115	7.167	7.393	7.574	7.726	7.856	7.969	8.070
17	7.030	7.081	7.302	7.480	7.628	7.756	7.868	7.966
18	6.954	7.005	7.221	7.396	7.543	7.667	7.777	7.874
19	6.887	6.936	7.150	7.322	7.465	7.589	7.696	7.792
20	6.827	6.876	7.086	7.255	7.397	7.518	7.624	7.718
24	6.639	6.685	6.885	7.046	7.180	7.296	7.397	7.486
30	6.453	6.497	6.686	6.839	6.965	7.075	7.171	7.255
40	6.270	6.311	6.489	6.633	6.753	6.855	6.945	7.025
60	6.089	6.127	6.295	6.429	6.540	6.636	6.720	6.795
120	5.910	5.946	6.101	6.225	6.329	6.418	6.495	6.564
∞	5.733	5.766	5.909	6.023	6.118	6.199	6.270	6.333

TABLE B.5 (cont.)   Critical Values of the q Distribution

$\alpha = 0.01$

ν	k (or p): 2	3	4	5	6	7	8	9	10
1	90.03	135.0	164.3	185.6	202.2	215.8	227.2	237.0	245.6
2	14.04	19.02	22.29	24.72	26.63	28.20	29.53	30.68	31.69
3	8.261	10.62	12.17	13.33	14.24	15.00	15.64	16.20	16.69
4	6.512	8.120	9.173	9.958	10.58	11.10	11.55	11.93	12.27
5	5.702	6.976	7.804	8.421	8.913	9.321	9.669	9.972	10.24
6	5.243	6.331	7.033	7.556	7.973	8.318	8.613	8.869	9.097
7	4.949	5.919	6.543	7.005	7.373	7.679	7.939	8.166	8.368
8	4.746	5.635	6.204	6.625	6.960	7.237	7.474	7.681	7.863
9	4.596	5.428	5.957	6.348	6.658	6.915	7.134	7.325	7.495
10	4.482	5.270	5.769	6.136	6.428	6.669	6.875	7.055	7.213
11	4.392	5.146	5.621	5.970	6.247	6.476	6.672	6.842	6.992
12	4.320	5.046	5.502	5.836	6.101	6.321	6.507	6.670	6.814
13	4.260	4.964	5.404	5.727	5.981	6.192	6.372	6.528	6.667
14	4.210	4.895	5.322	5.634	5.881	6.085	6.258	6.409	6.543
15	4.168	4.836	5.252	5.556	5.796	5.994	6.162	6.309	6.439
16	4.131	4.786	5.192	5.489	5.722	5.915	6.079	6.222	6.349
17	4.099	4.742	5.140	5.430	5.659	5.847	6.007	6.147	6.270
18	4.071	4.703	5.094	5.379	5.603	5.788	5.944	6.081	6.201
19	4.046	4.670	5.054	5.334	5.554	5.735	5.889	6.022	6.141
20	4.024	4.639	5.018	5.294	5.510	5.688	5.839	5.970	6.087
24	3.956	4.546	4.907	5.168	5.374	5.542	5.685	5.809	5.919
30	3.889	4.455	4.799	5.048	5.242	5.401	5.536	5.653	5.756
40	3.825	4.367	4.696	4.931	5.114	5.265	5.392	5.502	5.559
60	3.762	4.282	4.595	4.818	4.991	5.133	5.253	5.356	5.447
120	3.702	4.200	4.497	4.709	4.872	5.005	5.118	5.214	5.299
∞	3.643	4.120	4.403	4.603	4.757	4.882	4.987	5.078	5.157

ν	k (or p): 11	12	13	14	15	16	17	18	19
1	253.2	260.0	266.2	271.8	277.0	281.8	286.3	290.4	294.3
2	32.59	33.40	34.13	34.81	35.43	36.00	36.53	37.03	37.50
3	17.13	17.53	17.89	18.22	18.52	18.81	19.07	19.32	19.55
4	12.57	12.84	13.09	13.32	13.53	13.73	13.91	14.08	14.24
5	10.48	10.70	10.89	11.08	11.24	11.40	11.55	11.68	11.81
6	9.301	9.485	9.653	9.808	9.951	10.08	10.21	10.32	10.43
7	8.548	8.711	8.860	8.997	9.124	9.242	9.353	9.456	9.554
8	8.027	8.176	8.312	8.436	8.552	8.659	8.760	8.854	8.943
9	7.647	7.784	7.910	8.025	8.132	8.232	8.325	8.412	8.495
10	7.356	7.485	7.603	7.712	7.812	7.906	7.993	8.076	8.153
11	7.128	7.250	7.362	7.465	7.560	7.649	7.732	7.809	7.883
12	6.943	7.060	7.167	7.265	7.356	7.441	7.520	7.594	7.665
13	6.791	6.903	7.006	7.101	7.188	7.269	7.345	7.417	7.485
14	6.664	6.772	6.871	6.962	7.047	7.126	7.199	7.268	7.333
15	6.555	6.660	6.757	6.845	6.927	7.003	7.074	7.142	7.204
16	6.462	6.564	6.658	6.744	6.823	6.898	6.967	7.032	7.093
17	6.381	6.480	6.572	6.656	6.734	6.806	6.873	6.937	6.997
18	6.310	6.407	6.497	6.579	6.655	6.725	6.792	6.854	6.912
19	6.247	6.342	6.430	6.510	6.585	6.654	6.719	6.780	6.837
20	6.191	6.285	6.371	6.450	6.523	6.591	6.654	6.714	6.771
24	6.017	6.106	6.186	6.261	6.330	6.394	6.453	6.510	6.563
30	5.849	5.932	6.008	6.078	6.143	6.203	6.259	6.311	6.361
40	5.686	5.764	5.835	5.900	5.961	6.017	6.069	6.119	6.165
60	5.528	5.601	5.667	5.728	5.785	5.837	5.886	5.931	5.974
120	5.375	5.443	5.505	5.562	5.614	5.662	5.708	5.750	5.790
∞	5.227	5.290	5.348	5.400	5.448	5.493	5.535	5.574	5.611

TABLE B.5 (cont.) Critical Values of the q Distribution

$\alpha = 0.01$

ν	k (or p): 20	22	24	26	28	30	32	34	36
1	298.0	304.7	310.8	316.3	321.3	326.0	330.3	334.3	338.0
2	37.95	38.76	39.49	40.15	40.76	41.32	41.84	42.33	42.78
3	19.77	20.17	20.53	20.86	21.16	21.44	21.70	21.95	22.17
4	14.40	14.68	14.93	15.16	15.37	15.57	15.75	15.92	16.08
5	11.93	12.16	12.36	12.54	12.71	12.87	13.02	13.15	13.28
6	10.54	10.73	10.91	11.06	11.21	11.34	11.47	11.58	11.69
7	9.646	9.815	9.970	10.11	10.24	10.36	10.47	10.58	10.67
8	9.027	9.182	9.322	9.450	9.569	9.678	9.779	9.874	9.964
9	8.573	8.717	8.847	8.966	9.075	9.177	9.271	9.360	9.443
10	8.226	8.361	8.483	8.595	8.698	8.794	8.883	8.966	9.044
11	7.952	8.080	8.196	8.303	8.400	8.491	8.575	8.654	8.728
12	7.731	7.853	7.964	8.066	8.159	8.246	8.327	8.402	8.473
13	7.548	7.665	7.772	7.870	7.960	8.043	8.121	8.193	8.262
14	7.395	7.508	7.611	7.705	7.792	7.873	7.948	8.018	8.084
15	7.264	7.374	7.474	7.566	7.650	7.728	7.800	7.869	7.932
16	7.152	7.258	7.356	7.445	7.527	7.602	7.673	7.739	7.802
17	7.053	7.158	7.253	7.340	7.420	7.493	7.563	7.627	7.687
18	6.968	7.070	7.163	7.247	7.325	7.398	7.465	7.528	7.587
19	6.891	6.992	7.082	7.166	7.242	7.313	7.379	7.440	7.498
20	6.823	6.922	7.011	7.092	7.168	7.237	7.302	7.362	7.419
24	6.612	6.705	6.789	6.865	6.936	7.001	7.062	7.119	7.173
30	6.407	6.494	6.572	6.644	6.710	6.772	6.828	6.881	6.932
40	6.209	6.289	6.362	6.429	6.490	6.547	6.600	6.650	6.697
60	6.015	6.090	6.158	6.220	6.277	6.330	6.378	6.424	6.467
120	5.827	5.897	5.959	6.016	6.069	6.117	6.162	6.204	6.244
∞	5.645	5.709	5.766	5.818	5.866	5.911	5.952	5.990	6.026

ν	k (or p): 38	40	50	60	70	80	90	100
1	341.5	344.8	358.9	370.1	379.4	387.3	394.1	400.1
2	43.21	43.61	45.33	46.70	47.83	48.80	49.64	50.38
3	22.39	22.59	23.45	24.13	24.71	25.19	25.62	25.99
4	16.23	16.37	16.98	17.46	17.86	18.02	18.50	18.77
5	13.40	13.52	14.00	14.39	14.72	14.99	15.23	15.45
6	11.80	11.90	12.31	12.65	12.92	13.16	13.37	13.55
7	10.77	10.85	11.23	11.52	11.77	11.99	12.17	12.34
8	10.05	10.13	10.47	10.75	10.97	11.17	11.34	11.49
9	9.521	9.594	9.912	10.17	10.38	10.57	10.73	10.87
10	9.117	9.187	9.486	9.726	9.927	10.10	10.25	10.39
11	8.798	8.864	9.148	9.377	9.568	9.732	9.875	10.00
12	8.539	8.603	8.875	9.094	9.277	9.434	9.571	9.693
13	8.326	8.387	8.648	8.859	9.035	9.187	9.318	9.436
14	8.146	8.204	8.457	8.661	8.832	8.978	9.106	9.219
15	7.992	8.049	8.295	8.492	8.658	8.800	8.924	9.035
16	7.860	7.916	8.154	8.347	8.507	8.646	8.767	8.874
17	7.745	7.799	8.031	8.219	8.377	8.511	8.630	8.735
18	7.643	7.696	7.924	8.107	8.261	8.393	8.508	8.611
19	7.553	7.605	7.828	8.008	8.159	8.288	8.401	8.502
20	7.473	7.523	7.742	7.919	8.067	8.194	8.305	8.404
24	7.223	7.270	7.476	7.642	7.780	7.900	8.004	8.097
30	6.978	7.023	7.215	7.370	7.500	7.611	7.709	7.796
40	6.740	6.782	6.960	7.104	7.225	7.328	7.419	7.500
60	6.507	6.546	6.710	6.843	6.954	7.050	7.133	7.207
120	6.281	6.316	6.467	6.588	6.689	6.776	6.852	6.919
∞	6.060	6.092	6.228	6.338	6.429	6.507	6.575	6.636

TABLE B.5 (cont.)   Critical Values of the q Distribution

$\alpha = 0.005$

ν	k (or p): 2	3	4	5	6	7	8	9	10
1	180.1	270.1	328.5	371.2	404.4	431.6	454.4	474.0	491.1
2	19.93	26.97	31.60	35.02	37.73	39.95	41.83	43.46	44.89
3	10.55	13.50	15.45	16.91	18.06	19.01	19.83	20.53	21.15
4	7.916	9.814	11.06	11.99	12.74	13.35	13.88	14.33	14.74
5	6.751	8.196	9.141	9.847	10.41	10.88	11.28	11.63	11.93
6	6.105	7.306	8.088	8.670	9.135	9.522	9.852	10.14	10.40
7	5.699	6.750	7.429	7.935	8.339	8.674	8.961	9.211	9.433
8	5.420	6.370	6.981	7.435	7.797	8.097	8.354	8.578	8.777
9	5.218	6.096	6.657	7.074	7.405	7.680	7.915	8.120	8.303
10	5.065	5.888	6.412	6.800	7.109	7.365	7.584	7.775	7.944
11	4.945	5.727	6.222	6.588	6.878	7.119	7.325	7.505	7.664
12	4.849	5.597	6.068	6.416	6.693	6.922	7.118	7.288	7.439
13	4.770	5.490	5.943	6.277	6.541	6.760	6.947	7.111	7.255
14	4.704	5.401	5.838	6.160	6.414	6.626	6.805	6.962	7.101
15	4.647	5.325	5.750	6.061	6.308	6.511	6.685	6.837	6.971
16	4.599	5.261	5.674	5.977	6.216	6.413	6.582	6.729	6.859
17	4.557	5.205	5.608	5.903	6.136	6.329	6.493	6.636	6.763
18	4.521	3.156	5.550	5.839	6.067	6.255	6.415	6.554	6.678
19	4.488	5.113	5.500	5.783	6.005	6.189	6.346	6.482	6.603
20	4.460	5.074	5.455	5.732	5.951	6.131	6.285	6.418	6.537
24	4.371	4.955	5.315	5.577	5.783	5.952	6.096	6.221	6.332
30	4.285	4.841	5.181	5.428	5.621	5.780	5.914	6.031	6.135
40	4.202	4.731	5.053	5.284	5.465	5.614	5.739	5.848	5.944
60	4.122	4.625	4.928	5.146	5.316	5.454	5.571	5.673	5.762
120	4.045	4.523	4.809	5.013	5.172	5.301	5.410	5.504	5.586
∞	3.970	4.424	4.694	4.886	5.033	5.154	5.255	5.341	5.418

ν	k (or p): 11	12	13	14	15	16	17	18	19
1	506.3	520.0	532.4	543.6	554.0	563.6	572.5	580.9	588.7
2	46.16	47.31	48.35	49.30	50.17	50.99	51.74	52.45	53.12
3	21.70	22.20	22.66	23.08	23.46	23.82	24.15	24.46	24.76
4	15.10	15.42	15.72	15.99	16.24	16.48	16.70	16.90	17.09
5	12.21	12.46	12.69	12.90	13.09	13.27	13.44	13.60	13.75
6	10.63	10.83	11.02	11.20	11.36	11.51	11.65	11.78	11.90
7	9.632	9.812	9.977	10.13	10.27	10.40	10.52	10.64	10.75
8	8.955	9.117	9.265	9.401	9.527	9.644	9.754	9.857	9.953
9	8.466	8.614	8.749	8.874	8.990	9.097	9.198	9.292	9.381
10	8.096	8.234	8.360	8.476	8.583	8.683	8.777	8.865	8.947
11	7.807	7.937	8.055	8.164	8.265	8.359	8.447	8.530	8.608
12	7.575	7.697	7.810	7.914	8.009	8.099	8.183	8.261	8.335
13	7.384	7.502	7.609	7.708	7.800	7.886	7.965	8.040	8.111
14	7.225	7.338	7.442	7.537	7.625	7.707	7.784	7.856	7.924
15	7.091	7.200	7.300	7.392	7.477	7.556	7.630	7.699	7.765
16	6.976	7.081	7.178	7.267	7.349	7.426	7.498	7.566	7.629
17	6.876	6.979	7.072	7.159	7.239	7.314	7.384	7.449	7.511
18	6.788	6.888	6.980	7.064	7.142	7.215	7.283	7.347	7.407
19	6.711	6.809	6.898	6.981	7.057	7.128	7.195	7.257	7.316
20	6.642	6.738	6.826	6.907	6.981	7.051	7.116	7.177	7.235
24	6.431	6.520	6.602	6.677	6.747	6.812	6.872	6.930	6.983
30	6.227	6.310	6.387	6.456	6.521	6.581	6.638	6.691	6.741
40	6.030	6.108	6.179	6.244	6.304	6.360	6.412	6.461	6.507
60	5.841	5.913	5.979	6.039	6.094	6.146	6.194	6.239	6.281
120	5.660	5.726	5.786	5.842	5.893	5.940	5.984	6.025	6.064
∞	5.485	5.546	5.602	5.652	5.699	5.742	5.783	5.820	5.856

TABLE B.5 (cont.) Critical Values of the q Distribution

$\alpha = 0.005$

ν	k (or p): 20	22	24	26	28	30	32	34	36
1	596.0	609.5	621.7	632.6	642.7	652.0	660.6	668.5	676.0
2	53.74	54.89	55.92	56.86	57.73	58.52	59.26	59.95	60.59
3	25.03	25.54	26.00	26.42	26.80	27.15	27.48	27.79	28.07
4	17.28	17.61	17.91	18.19	18.44	18.68	18.89	19.09	19.28
5	13.89	14.14	14.38	14.59	14.79	14.96	15.13	15.29	15.44
6	12.02	12.23	12.43	12.61	12.77	12.92	13.06	13.19	13.32
7	10.85	11.03	11.21	11.36	11.50	11.64	11.76	11.88	11.99
8	10.04	10.22	10.37	10.51	10.64	10.76	10.87	10.97	11.07
9	9.465	9.620	9.761	9.890	10.01	10.12	10.22	10.32	10.41
10	9.026	9.170	9.302	9.422	9.532	9.635	9.730	9.820	9.904
11	8.682	8.818	8.941	9.055	9.159	9.256	9.345	9.430	9.509
12	8.405	8.534	8.652	8.759	8.858	8.950	9.036	9.116	9.191
13	8.178	8.302	8.414	8.516	8.611	8.699	8.781	8.857	8.929
14	7.988	8.107	8.215	8.314	8.404	8.489	8.568	8.641	8.710
15	7.827	7.942	8.046	8.141	8.229	8.311	8.387	8.458	8.524
16	7.689	7.800	7.901	7.994	8.078	8.158	8.231	8.300	8.365
17	7.569	7.677	7.775	7.865	7.948	8.024	8.096	8.163	8.226
18	7.464	7.570	7.665	7.753	7.833	7.908	7.978	8.043	8.104
19	7.372	7.474	7.568	7.653	7.732	7.805	7.873	7.937	7.996
20	7.289	7.390	7.481	7.565	7.642	7.713	7.780	7.842	7.901
24	7.034	7.128	7.213	7.291	7.362	7.429	7.491	7.549	7.603
30	6.788	6.875	6.954	7.026	7.093	7.154	7.212	7.265	7.316
40	6.550	6.631	6.704	6.770	6.832	6.889	6.942	6.991	7.038
60	6.321	6.396	6.462	6.523	6.580	6.632	6.861	6.726	6.769
120	6.101	6.169	6.230	6.286	6.337	6.385	6.428	6.470	6.508
∞	5.889	5.951	6.006	6.057	6.103	6.146	6.186	6.223	6.258

ν	k (or p): 38	40	50	60	70	80	90	100
1	683.0	689.6	717.8	740.2	758.8	774.5	788.2	800.3
2	61.19	61.76	64.19	66.13	67.74	69.10	70.29	71.35
3	28.34	28.60	29.68	30.55	31.27	31.88	32.42	32.90
4	19.46	19.63	20.36	20.93	21.42	21.83	22.18	22.50
5	15.58	15.71	16.27	16.72	17.09	17.41	17.69	17.94
6	13.43	13.54	14.02	14.40	14.71	14.98	15.21	15.43
7	12.09	12.18	12.60	12.93	13.21	13.44	13.65	13.84
8	11.16	11.25	11.63	11.93	12.18	12.39	12.58	12.75
9	10.49	10.58	10.92	11.20	11.43	11.63	11.80	11.96
10	9.983	10.06	10.38	10.64	10.86	11.04	11.20	11.35
11	9.583	9.654	9.957	10.20	10.41	10.59	10.74	10.88
12	9.262	9.328	9.617	9.850	10.04	10.21	10.36	10.49
13	8.997	9.061	9.337	9.560	9.747	9.907	10.05	10.17
14	8.775	8.837	9.103	9.317	9.497	9.652	9.787	9.907
15	8.587	8.647	8.904	9.111	9.285	9.434	9.565	9.680
16	8.425	8.483	8.733	8.933	9.102	9.247	9.373	9.486
17	8.285	8.341	8.583	8.779	8.943	9.084	9.206	9.316
18	8.162	8.217	8.452	8.643	8.803	8.940	9.061	9.167
19	8.053	8.106	8.337	8.523	8.679	8.813	8.931	9.036
20	7.956	8.008	8.234	8.416	8.569	8.700	8.815	8.917
24	7.655	7.704	7.914	8.083	8.226	8.348	8.455	8.551
30	7.364	7.409	7.603	7.760	7.893	8.006	8.105	8.193
40	7.082	7.123	7.302	7.447	7.568	7.672	7.763	7.845
60	6.808	6.846	7.010	7.143	7.252	7.347	7.431	7.504
120	6.545	6.580	6.728	6.846	6.946	7.032	7.107	7.173
∞	6.291	6.322	6.454	6.561	6.649	6.725	6.792	6.850

TABLE B.5 (cont.)　Critical Values of the q Distribution

$\alpha = 0.001$

ν	k (or p): 2	3	4	5	6	7	8	9	10
1	900.3	1351.	1643.	1856.	2022.	2158.	2272.	2370.	2455.
2	44.69	60.42	70.77	78.43	84.49	89.46	93.67	97.30	100.5
3	18.28	23.32	26.65	29.13	31.11	32.74	34.12	35.33	36.39
4	12.18	14.99	16.84	18.23	19.34	20.26	21.04	21.73	22.33
5	9.714	11.67	12.96	13.93	14.71	15.35	15.90	16.38	16.81
6	8.427	9.960	10.97	11.72	12.32	12.83	13.26	13.63	13.97
7	7.648	8.930	9.763	10.40	10.90	11.32	11.68	11.99	12.27
8	7.130	8.250	8.978	9.522	9.958	10.32	10.64	10.91	11.15
9	6.762	7.768	8.419	8.906	9.295	9.619	9.897	10.14	10.36
10	6.487	7.411	8.006	8.450	8.804	9.099	9.352	9.573	9.769
11	6.275	7.136	7.687	8.098	8.426	8.699	8.933	9.138	9.319
12	6.106	6.917	7.436	7.821	8.127	8.383	8.601	8.793	8.962
13	5.970	6.740	7.231	7.595	7.885	8.126	8.333	8.513	8.673
14	5.856	6.594	7.062	7.409	7.685	7.915	8.110	8.282	8.434
15	5.760	6.470	6.920	7.252	7.517	7.736	7.925	8.088	8.234
16	5.678	6.365	6.799	7.119	7.374	7.585	7.766	7.923	8.063
17	5.608	6.275	6.695	7.005	7.250	7.454	7.629	7.781	7.916
18	5.546	6.196	6.604	6.905	7.143	7.341	7.510	7.657	7.788
19	5.492	6.127	6.525	6.817	7.049	7.242	7.405	7.549	7.676
20	5.444	6.065	6.454	6.740	6.966	7.154	7.313	7.453	7.577
24	5.297	5.877	6.238	6.503	6.712	6.884	7.031	7.159	7.272
30	5.156	5.698	6.033	6.278	6.470	6.628	6.763	6.880	6.984
40	5.022	5.528	5.838	6.063	6.240	6.386	6.509	6.616	6.711
60	4.894	5.365	5.653	5.860	6.022	6.155	6.268	6.366	6.451
120	4.771	5.211	5.476	5.667	5.815	5.937	6.039	6.128	6.206
∞	4.654	5.063	5.309	5.484	5.619	5.730	5.823	5.903	5.973

ν	k (or p): 11	12	13	14	15	16	17	18	19
1	2532.	2600.	2662.	2718.	2770.	2818.	2863.	2904.	2943.
2	103.3	105.9	108.2	110.4	112.3	114.2	115.9	117.4	118.9
3	37.34	38.20	38.98	39.69	40.35	40.97	41.54	42.07	42.58
4	22.87	23.36	23.81	24.21	24.59	24.94	25.27	25.58	25.87
5	17.18	17.53	17.85	18.13	18.41	18.66	18.89	19.10	19.31
6	14.27	14.54	14.79	15.01	15.22	15.42	15.60	15.78	15.94
7	12.52	12.74	12.95	13.14	13.32	13.48	13.64	13.78	13.92
8	11.36	11.56	11.74	11.91	12.06	12.21	12.34	12.47	12.59
9	10.55	10.73	10.89	11.03	11.18	11.30	11.42	11.54	11.64
10	9.946	10.11	10.25	10.39	10.52	10.64	10.75	10.85	10.95
11	9.482	9.630	9.766	9.892	10.01	10.12	10.22	10.31	10.41
12	9.115	9.254	9.381	9.498	9.606	9.707	9.802	9.891	9.975
13	8.817	8.948	9.068	9.178	9.281	9.376	9.466	9.550	9.629
14	8.571	8.696	8.809	8.914	9.012	9.103	9.188	9.267	9.343
15	8.365	8.483	8.592	8.693	8.786	8.872	8.954	9.030	9.102
16	8.189	8.303	8.407	8.504	8.593	8.676	8.755	8.828	8.897
17	8.037	8.148	8.248	8.342	8.427	8.508	8.583	8.654	8.720
18	7.906	8.012	8.110	8.199	8.283	8.361	8.434	8.502	8.567
19	7.790	7.893	7.988	8.075	8.156	8.232	8.303	8.369	8.432
20	7.688	7.788	7.880	7.966	8.044	8.118	8.186	8.251	8.312
24	7.374	7.467	7.551	7.629	7.701	7.768	7.831	7.890	7.946
30	7.077	7.162	7.239	7.310	7.375	7.437	7.494	7.548	7.599
40	6.796	6.872	6.942	7.007	7.067	7.122	7.174	7.223	7.269
60	6.528	6.598	6.661	6.720	6.774	6.824	6.871	6.914	6.956
120	6.276	6.339	6.396	6.448	6.496	6.542	6.583	6.623	6.660
∞	6.036	6.092	6.144	6.191	6.234	6.274	6.312	6.347	6.380

TABLE B.5 (cont.) Critical Values of the q Distribution

$\alpha = 0.001$

ν	k (or p): 20	22	24	26	28	30	32	34	36
1	2980.	3047.	3108.	3163.	3213.	3260.	3303.	3343.	3380.
2	120.3	122.9	125.2	127.3	129.3	131.0	132.7	134.2	135.7
3	43.05	43.92	44.70	45.42	46.07	46.68	47.24	47.77	48.26
4	26.14	26.65	27.10	27.51	27.89	28.24	28.57	28.88	29.16
5	19.51	19.86	20.19	20.48	20.75	21.01	21.24	21.46	21.66
6	16.09	16.38	16.64	16.87	17.08	17.28	17.47	17.64	17.81
7	14.04	14.29	14.50	14.70	14.88	15.05	15.20	15.35	15.49
8	12.70	12.91	13.09	13.26	13.42	13.57	13.71	13.84	13.96
9	11.75	11.93	12.10	12.25	12.39	12.53	12.65	12.77	12.87
10	11.03	11.20	11.36	11.50	11.63	11.75	11.87	11.97	12.07
11	10.49	10.65	10.79	10.92	11.04	11.16	11.26	11.35	11.45
12	10.06	10.20	10.34	10.46	10.57	10.68	10.78	10.87	10.96
13	9.704	9.843	9.969	10.09	10.19	10.29	10.39	10.47	10.55
14	9.414	9.546	9.666	9.776	9.878	9.972	10.06	10.14	10.22
15	9.170	9.296	9.411	9.517	9.613	9.703	9.788	9.867	9.940
16	8.963	9.084	9.194	9.295	9.388	9.475	9.556	9.631	9.702
17	8.784	8.900	9.007	9.104	9.194	9.277	9.355	9.429	9.497
18	8.628	8.741	8.844	8.938	9.025	9.106	9.181	9.251	9.318
19	8.491	8.601	8.701	8.792	8.876	8.955	9.028	9.096	9.161
20	8.370	8.477	8.574	8.663	8.745	8.821	8.892	8.959	9.021
24	7.999	8.097	8.185	8.267	8.342	8.411	8.476	8.537	8.594
30	7.647	7.735	7.816	7.890	7.958	8.021	8.080	8.135	8.188
40	7.312	7.393	7.466	7.533	7.594	7.651	7.704	7.754	7.801
60	6.995	7.067	7.133	7.193	7.248	7.299	7.347	7.392	7.433
120	6.695	6.760	6.818	6.872	6.921	6.966	7.008	7.048	7.085
∞	6.411	6.469	6.520	6.568	6.611	6.651	6.689	6.723	6.756

ν	k (or p): 38	40	50	60	70	80	90	100
1	3415.	3448.	3589.	3701.	3794.	3873.	3941.	4002.
2	137.0	138.3	143.7	148.0	151.6	154.7	157.4	159.7
3	48.72	49.16	51.02	52.51	53.75	54.81	55.72	56.53
4	29.43	29.68	30.78	31.65	32.37	32.98	33.52	34.00
5	21.86	22.03	22.82	23.45	23.97	24.41	24.80	25.15
6	17.96	18.10	18.73	19.22	19.64	20.00	20.31	20.58
7	15.62	15.74	16.27	16.69	17.04	17.35	17.61	17.85
8	14.07	14.18	14.64	15.01	15.32	15.59	15.82	16.02
9	12.97	13.07	13.49	13.82	14.10	14.34	14.55	14.74
10	12.16	12.25	12.63	12.94	13.20	13.42	13.61	13.78
11	11.53	11.62	11.97	12.25	12.49	12.70	12.88	13.04
12	11.03	11.11	11.44	11.71	11.94	12.13	12.29	12.45
13	10.63	10.70	11.01	11.27	11.48	11.66	11.82	11.97
14	10.30	10.37	10.66	10.91	11.11	11.28	11.43	11.57
15	10.01	10.08	10.37	10.59	10.79	10.96	11.10	11.23
16	9.769	9.833	10.11	10.34	10.52	10.68	10.82	10.95
17	9.562	9.623	9.888	10.10	10.29	10.44	10.58	10.70
18	9.381	9.440	9.696	9.904	10.08	10.23	10.36	10.48
19	9.221	9.279	9.528	9.730	9.899	10.04	10.17	10.29
20	9.081	9.137	9.379	9.575	9.740	9.881	10.01	10.12
24	8.648	8.700	8.921	9.100	9.250	9.380	9.494	9.596
30	8.237	8.283	8.484	8.647	8.783	8.901	9.004	9.096
40	7.845	7.887	8.067	8.214	8.337	8.442	8.535	8.618
60	7.473	7.510	7.671	7.802	7.911	8.005	8.088	8.161
120	7.121	7.153	7.296	7.411	7.507	7.590	7.662	7.726
∞	6.787	6.816	6.941	7.041	7.124	7.196	7.259	7.314

TABLE B.6 Critical Values of q' for the One-Tailed Dunnett's Test

	$\alpha = 0.05$								
ν	$k: 2$	3	4	5	6	7	8	9	10
5	2.02	2.44	2.68	2.85	2.98	3.08	3.16	3.24	3.30
6	1.94	2.34	2.56	2.71	2.83	2.92	3.00	3.07	3.12
7	1.89	2.27	2.48	2.62	2.73	2.82	2.89	2.95	3.01
8	1.86	2.22	2.42	2.55	2.66	2.74	2.81	2.87	2.92
9	1.83	2.18	2.37	2.50	2.60	2.68	2.75	2.81	2.86
10	1.81	2.15	2.34	2.47	2.56	2.64	2.70	2.76	2.81
11	1.80	2.13	2.31	2.44	2.53	2.60	2.67	2.72	2.77
12	1.78	2.11	2.29	2.41	2.50	2.58	2.64	2.69	2.74
13	1.77	2.09	2.27	2.39	2.48	2.55	2.61	2.66	2.71
14	1.76	2.08	2.25	2.37	2.46	2.53	2.59	2.64	2.69
15	1.75	2.07	2.24	2.36	2.44	2.51	2.57	2.62	2.67
16	1.75	2.06	2.23	2.34	2.43	2.50	2.56	2.61	2.65
17	1.74	2.05	2.22	2.33	2.42	2.49	2.54	2.59	2.64
18	1.73	2.04	2.21	2.32	2.41	2.48	2.53	2.58	2.62
19	1.73	2.03	2.20	2.31	2.40	2.47	2.52	2.57	2.61
20	1.72	2.03	2.19	2.30	2.39	2.46	2.51	2.56	2.60
24	1.71	2.01	2.17	2.28	2.36	2.43	2.48	2.53	2.57
30	1.70	1.99	2.15	2.25	2.33	2.40	2.45	2.50	2.54
40	1.68	1.97	2.13	2.23	2.31	2.37	2.42	2.47	2.51
60	1.67	1.95	2.10	2.21	2.28	2.35	2.39	2.44	2.48
120	1.66	1.93	2.08	2.18	2.26	2.32	2.37	2.41	2.45
∞	1.64	1.92	2.06	2.16	2.23	2.29	2.34	2.38	2.42

	$\alpha = 0.01$								
ν	$k: 2$	3	4	5	6	7	8	9	10
5	3.37	3.90	4.21	4.43	4.60	4.73	4.85	4.94	5.03
6	3.14	3.61	3.88	4.07	4.21	4.33	4.43	4.51	4.59
7	3.00	3.42	3.66	3.83	3.96	4.07	4.15	4.23	4.30
8	2.90	3.29	3.51	3.67	3.79	3.88	3.96	4.03	4.09
9	2.82	3.19	3.40	3.55	3.66	3.75	3.82	3.89	3.94
10	2.76	3.11	3.31	3.45	3.56	3.64	3.71	3.78	3.83
11	2.72	3.06	3.25	3.38	3.48	3.56	3.63	3.69	3.74
12	2.68	3.01	3.19	3.32	3.42	3.50	3.56	3.62	3.67
13	2.65	2.97	3.15	3.27	3.37	3.44	3.51	3.56	3.61
14	2.62	2.94	3.11	3.23	3.32	3.40	3.46	3.51	3.56
15	2.60	2.91	3.08	3.20	3.29	3.36	3.42	3.47	3.52
16	2.58	2.88	3.05	3.17	3.26	3.33	3.39	3.44	3.48
17	2.57	2.86	3.03	3.14	3.23	3.30	3.36	3.41	3.45
18	2.55	2.84	3.01	3.12	3.21	3.27	3.33	3.38	3.42
19	2.54	2.83	2.99	3.10	3.18	3.25	3.31	3.36	3.40
20	2.53	2.81	2.97	3.08	3.17	3.23	3.29	3.34	3.38
24	2.49	2.77	2.92	3.03	3.11	3.17	3.22	3.27	3.31
30	2.46	2.72	2.87	2.97	3.05	3.11	3.16	3.21	3.24
40	2.42	2.68	2.82	2.92	2.99	3.05	3.10	3.14	3.18
60	2.39	2.64	2.78	2.87	2.94	3.00	3.04	3.08	3.12
120	2.36	2.60	2.73	2.82	2.89	2.94	2.99	3.03	3.06
∞	2.33	2.56	2.68	2.77	2.84	2.89	2.93	2.97	3.00

Values in Table B.6 are reprinted, with the permission of the author and publisher, from the tables of C. W. Dunnett (1955, *J. Amer. Statist. Assoc.* 50: 1096–1121).

Examples:

$$q'_{0.05(1),16,4} = 2.23 \quad \text{and} \quad q'_{0.01(1),24,3} = 2.77.$$

If a critical value is required for degrees of freedom not on this table, one may conservatively use the critical value with the next lower degrees of freedom. Or, the critical value may be estimated by harmonic interpolation.

Values in Table B.7 are reprinted, with the permission of the author and editor, from the tables of C. W. Dunnett (1964, *Biometrics* 20: 482–491).

Examples:

$$q'_{0.05(2),30,5} = 2.58 \quad \text{and} \quad q'_{0.01(2),20,4} = 2.54.$$

If a critical value is required for degrees of freedom not on this table, one may conservatively use the critical value with the next lower degrees of freedom. Or, the critical value may be estimated by harmonic interpolation.

TABLE B.7 Critical Values of q' for the Two-Tailed Dunnett's Test

						$\alpha = 0.05$								
ν	k: 2	3	4	5	6	7	8	9	10	11	12	13	16	21
5	2.57	3.03	3.29	3.48	3.62	3.73	3.82	3.90	3.97	4.03	4.09	4.14	4.26	4.42
6	2.45	2.86	3.10	3.26	3.39	3.49	3.57	3.64	3.71	3.76	3.81	3.86	3.97	4.11
7	2.36	2.75	2.97	3.12	3.24	3.33	3.41	3.47	3.53	3.58	3.63	3.67	3.78	3.91
8	2.31	2.67	2.88	3.02	3.13	3.22	3.29	3.35	3.41	3.46	3.50	3.54	3.64	3.76
9	2.26	2.61	2.81	2.95	3.05	3.14	3.20	3.26	3.32	3.36	3.40	3.44	3.53	3.65
10	2.23	2.57	2.76	2.89	2.99	3.07	3.14	3.19	3.24	3.29	3.33	3.36	3.45	3.57
11	2.20	2.53	2.72	2.84	2.94	3.02	3.08	3.14	3.19	3.23	3.27	3.30	3.39	3.50
12	2.18	2.50	2.68	2.81	2.90	2.98	3.04	3.09	3.14	3.18	3.22	3.25	3.34	3.45
13	2.16	2.48	2.65	2.78	2.87	2.94	3.00	3.06	3.10	3.14	3.18	3.21	3.29	3.40
14	2.14	2.46	2.63	2.75	2.84	2.91	2.97	3.02	3.07	3.11	3.14	3.18	3.26	3.36
15	2.13	2.44	2.61	2.73	2.82	2.89	2.95	3.00	3.04	3.08	3.12	3.15	3.23	3.33
16	2.12	2.42	2.59	2.71	2.80	2.87	2.92	2.97	3.02	3.06	3.09	3.12	3.20	3.30
17	2.11	2.41	2.58	2.69	2.78	2.85	2.90	2.95	3.00	3.03	3.07	3.10	3.18	3.27
18	2.10	2.40	2.56	2.68	2.76	2.83	2.89	2.94	2.98	3.01	3.05	3.08	3.16	3.25
19	2.09	2.39	2.55	2.66	2.75	2.81	2.87	2.92	2.96	3.00	3.03	3.06	3.14	3.23
20	2.09	2.38	2.54	2.65	2.73	2.80	2.86	2.90	2.95	2.98	3.02	3.05	3.12	3.22
24	2.06	2.35	2.51	2.61	2.70	2.76	2.81	2.86	2.90	2.94	2.97	3.00	3.07	3.16
30	2.04	2.32	2.47	2.58	2.66	2.72	2.77	2.82	2.86	2.89	2.92	2.95	3.02	3.11
40	2.02	2.29	2.44	2.54	2.62	2.68	2.73	2.77	2.81	2.85	2.87	2.90	2.97	3.06
60	2.00	2.27	2.41	2.51	2.58	2.64	2.69	2.73	2.77	2.80	2.83	2.86	2.92	3.00
120	1.98	2.24	2.38	2.47	2.55	2.60	2.65	2.69	2.73	2.76	2.79	2.81	2.87	2.95
∞	1.96	2.21	2.35	2.44	2.51	2.57	2.61	2.65	2.69	2.72	2.74	2.77	2.83	2.91

						$\alpha = 0.01$								
ν	k: 2	3	4	5	6	7	8	9	10	11	12	13	16	21
5	4.03	4.63	4.98	5.22	5.41	5.56	5.69	5.80	5.89	5.98	6.05	6.12	6.30	6.52
6	3.71	4.21	4.51	4.71	4.87	5.00	5.10	5.20	5.28	5.35	5.41	5.47	5.62	5.81
7	3.50	3.95	4.21	4.39	4.53	4.64	4.74	4.82	4.89	4.95	5.01	5.06	5.19	5.36
8	3.36	3.77	4.00	4.17	4.29	4.40	4.48	4.56	4.62	4.68	4.73	4.78	4.90	5.05
9	3.25	3.63	3.85	4.01	4.12	4.22	4.30	4.37	4.43	4.48	4.53	4.57	4.68	4.82
10	3.17	3.53	3.74	3.88	3.99	4.08	4.16	4.22	4.28	4.33	4.37	4.42	4.52	4.65
11	3.11	3.45	3.65	3.79	3.89	3.98	4.05	4.11	4.16	4.21	4.25	4.29	4.30	4.52
12	3.05	3.39	3.58	3.71	3.81	3.89	3.96	4.02	4.07	4.12	4.16	4.19	4.29	4.41
13	3.01	3.33	3.52	3.65	3.74	3.82	3.89	3.94	3.99	4.04	4.08	4.11	4.20	4.32
14	2.98	3.29	3.47	3.59	3.69	3.76	3.83	3.88	3.93	3.97	4.01	4.05	4.13	4.24
15	2.95	3.25	3.43	3.55	3.64	3.71	3.78	3.83	3.88	3.92	3.95	3.99	4.07	4.18
16	2.92	3.22	3.39	3.51	3.60	3.67	3.73	3.78	3.83	3.87	3.91	3.94	4.02	4.13
17	2.90	3.19	3.36	3.47	3.56	3.63	3.69	3.74	3.79	3.83	3.86	3.90	3.98	4.08
18	2.88	3.17	3.33	3.44	3.53	3.60	3.66	3.71	3.75	3.79	3.83	3.86	3.94	4.04
19	2.86	3.15	3.31	3.42	3.50	3.57	3.63	3.68	3.72	3.76	3.79	3.83	3.90	4.00
20	2.85	3.13	3.29	3.40	3.48	3.55	3.60	3.65	3.69	3.73	3.77	3.80	3.87	3.97
24	2.80	3.07	3.22	3.32	3.40	3.47	3.52	3.57	3.61	3.64	3.68	3.70	3.78	3.87
30	2.75	3.01	3.15	3.25	3.33	3.39	3.44	3.49	3.52	3.56	3.59	3.62	3.69	3.78
40	2.70	2.95	3.09	3.19	3.26	3.32	3.37	3.41	3.44	3.48	3.51	3.53	3.60	3.68
60	2.66	2.90	3.03	3.12	3.19	3.25	3.29	3.33	3.37	3.40	3.42	3.45	3.51	3.59
120	2.62	2.85	2.97	3.06	3.12	3.18	3.22	3.26	3.29	3.32	3.35	3.37	3.43	3.51
∞	2.58	2.79	2.92	3.00	3.06	3.11	3.15	3.19	3.22	3.25	3.27	3.29	3.35	2.42

TABLE B.8 Critical Values of d_{max} for the Kolmogorov-Smirnov Goodness of Fit Test for Discrete or Grouped Data

k	n	$\alpha < 0.50$	0.20	0.10	0.05	0.02	0.01	0.005	0.002	0.001
3	3	2	2	2	3	3	3	3	3	3
3	6	2	3	3	3	4	4	4	5	5
3	9	2	3	4	4	4	5	5	5	6
3	12	3	4	4	4	5	5	6	6	7
3	15	3	4	4	5	6	6	6	7	7
3	18	3	4	5	5	6	6	7	7	8
3	21	3	4	5	6	6	7	7	8	8
3	24	3	5	5	6	7	7	8	8	9
3	27	3	5	6	6	7	8	8	9	9
3	30	4	5	6	7	8	8	9	9	10
3	33	4	5	6	7	8	8	9	10	10
3	35	4	5	6	7	8	9	9	10	11
3	39	4	6	7	7	8	9	10	11	11
3	42	4	6	7	8	9	9	10	11	12
3	45	4	6	7	8	9	10	10	11	12
3	48	4	6	7	8	9	10	11	12	12
3	51	4	6	7	8	10	10	11	12	13
3	54	4	6	8	9	10	11	11	12	13
3	57	5	7	8	9	10	11	12	13	13
3	60	5	7	8	9	10	11	12	13	14
3	63	5	7	8	9	10	11	12	13	14
3	66	5	7	8	9	10	11	12	13	14
3	69	5	7	8	9	11	12	13	14	14
3	72	5	7	8	9	11	12	13	14	15
3	75	4	7	8	10	11	12	13	14	15
3	78	5	7	9	10	11	12	13	14	15
3	81	5	7	9	10	11	13	13	15	16
3	84	5	7	9	10	12	13	14	15	16
3	87	4	7	9	10	12	13	14	15	16
3	90	4	7	9	10	12	13	14	15	16
3	93	4	7	9	11	12	13	14	15	16
3	96	4	7	9	10	12	13	14	15	16
3	99	4	7	9	10	12	13	14	15	16
4	4	2	2	3	3	3	3	4	4	4
4	8	2	3	4	4	4	5	5	5	5
4	12	3	4	4	5	5	6	6	6	7
4	16	3	4	5	5	6	6	7	7	8
4	20	3	4	5	6	6	7	7	8	8
4	24	4	5	6	6	7	8	8	9	9
4	28	4	5	6	7	7	8	9	9	10
4	32	4	5	6	7	8	9	9	10	10
4	36	4	6	7	7	8	9	10	10	11
4	40	4	6	7	8	9	9	10	11	12
4	44	5	6	7	8	9	10	11	11	12
4	48	5	6	7	8	10	10	11	12	13
4	52	5	7	8	9	10	11	11	12	13
4	56	5	7	8	9	10	11	12	13	13
4	60	5	7	8	9	10	11	12	13	14
4	64	5	7	8	9	11	12	13	14	14
4	68	5	7	9	10	11	12	13	14	15
4	72	5	7	9	10	11	12	13	14	15
4	76	5	8	9	10	11	12	13	14	15
4	80	5	8	9	10	11	12	13	15	15
4	84	5	7	9	10	12	13	14	15	16
4	88	5	7	9	10	12	13	14	15	16

TABLE B.8 (cont.) Critical Values of d_{max} for the
Kolmogorov-Smirnov Goodness of Fit Test for Discrete or Grouped
Data

k	n	α < 0.50	0.20	0.10	0.05	0.02	0.01	0.005	0.002	0.001
4	92	5	7	9	10	12	13	14	16	16
4	96	5	7	9	10	12	13	14	16	17
4	100	5	8	9	11	12	13	14	16	17
5	5	2	3	3	3	4	4	4	4	4
5	10	3	3	4	4	5	5	5	6	6
5	15	3	4	5	5	6	6	7	7	7
5	20	4	5	5	6	7	7	7	8	8
5	25	4	5	6	6	7	8	8	9	9
5	30	4	5	6	7	8	8	9	10	10
5	35	4	6	7	7	8	9	10	10	11
5	40	5	6	7	8	9	10	10	11	12
5	45	5	6	7	8	9	10	11	12	12
5	50	5	7	8	9	10	11	11	12	13
5	55	5	7	8	9	10	11	12	13	14
5	60	5	7	8	9	11	12	12	13	14
5	65	5	7	9	10	11	12	13	14	14
5	70	6	8	9	10	11	12	13	14	15
5	75	6	8	9	10	12	13	14	15	15
5	80	5	8	9	11	12	13	14	15	16
5	85	5	8	9	11	12	13	14	15	16
5	90	6	8	10	11	12	13	14	15	16
5	95	6	8	9	11	12	13	14	16	17
5	100	5	8	9	11	12	14	15	16	17
6	6	2	3	3	4	4	4	4	5	5
6	12	3	4	4	5	5	6	6	6	7
6	18	3	4	5	6	6	7	7	8	8
6	24	4	5	6	6	7	8	8	9	9
6	30	4	6	6	7	8	9	9	10	10
6	36	5	6	7	8	9	9	10	11	11
6	42	5	6	7	8	9	10	11	11	12
6	48	5	7	8	9	10	11	11	12	13
6	54	5	7	8	9	10	11	12	13	14
6	60	6	7	9	10	11	12	13	13	14
6	66	6	8	9	10	11	12	13	14	15
6	72	6	8	9	10	12	13	13	14	15
6	78	6	8	9	11	12	13	14	15	16
6	84	6	8	9	11	12	13	14	15	16
6	90	5	8	10	11	13	14	15	16	16
6	96	6	8	10	11	13	14	15	16	17
7	7	3	3	4	4	4	5	5	5	5
7	14	3	4	5	5	6	6	7	7	7
7	21	4	5	6	6	7	7	8	8	9
7	28	4	5	6	7	8	8	9	10	10
7	35	5	6	7	8	9	9	10	11	11
7	42	5	6	7	8	9	10	11	12	12
7	49	5	7	8	9	10	11	12	12	13
7	56	6	7	8	9	11	12	12	13	14
7	63	6	8	9	10	11	12	13	14	15
7	70	6	8	9	10	12	13	13	15	15
7	77	6	8	9	11	12	13	14	15	16
7	84	6	8	10	12	12	13	14	15	16
7	91	6	8	10	11	13	14	15	16	17
7	98	6	8	10	11	13	14	15	16	17
8	8	3	3	4	4	5	5	5	5	6
8	16	3	4	5	6	6	7	7	7	8

TABLE B.8 (cont.) Critical Values of d_{max} for the Kolmogorov-Smirnov Goodness of Fit Test for Discrete or Grouped Data

k	n	α < 0.50	0.20	0.10	0.05	0.02	0.01	0.005	0.002	0.001
8	24	4	5	6	7	7	8	8	9	9
8	32	4	6	7	7	8	9	10	10	11
8	40	5	6	7	8	9	10	11	11	12
8	48	5	7	8	9	10	11	12	12	13
8	56	6	7	9	10	11	12	12	13	14
8	64	6	8	9	10	11	12	13	14	15
8	72	6	8	9	11	12	13	14	15	15
8	80	6	8	10	11	12	13	14	15	16
8	88	6	8	10	11	13	14	15	16	17
8	95	6	9	10	11	13	14	15	16	17
9	9	3	4	4	4	5	5	5	6	6
9	18	4	5	5	6	7	7	7	8	8
9	27	4	6	6	7	8	8	9	10	10
9	36	5	6	7	8	9	10	10	11	11
9	45	5	7	8	9	10	11	11	12	13
9	54	6	7	9	10	11	11	12	13	14
9	63	6	8	9	10	11	12	13	14	15
9	72	6	8	10	11	12	13	14	15	16
9	81	6	8	10	11	13	13	14	15	16
9	90	6	9	10	11	13	14	15	16	17
9	99	6	9	10	12	13	14	15	17	18
10	10	3	4	4	5	5	5	6	6	6
10	20	4	5	6	6	7	7	8	8	9
10	30	4	6	7	7	8	9	9	10	11
10	40	5	7	8	8	9	10	11	11	12
10	50	6	7	8	9	10	11	12	13	13
10	60	6	8	9	10	11	12	13	14	15
10	70	6	8	10	11	12	13	14	15	15
10	80	6	9	10	11	13	14	14	16	16
10	90	6	9	10	12	13	14	15	16	17
10	100	6	9	10	12	13	14	15	17	18
11	11	3	4	4	5	5	6	6	6	7
11	22	4	5	6	6	7	8	8	9	9
11	33	5	6	7	8	9	9	10	11	11
11	44	5	7	8	9	10	11	11	12	13
11	55	6	8	9	10	11	12	12	13	14
11	66	6	8	9	11	12	13	14	15	15
11	77	6	9	10	11	12	13	14	15	16
11	88	6	9	10	12	13	14	15	16	17
11	99	6	9	10	12	13	14	16	17	18
12	12	3	4	5	5	6	6	6	7	7
12	24	4	5	6	7	8	8	9	9	10
12	36	5	6	7	8	9	10	10	11	12
12	48	6	7	8	9	10	11	12	13	13
12	60	6	8	9	10	11	12	13	14	15
12	72	6	9	10	11	12	13	14	15	16
12	84	7	9	10	11	13	14	15	16	17
12	96	7	9	10	12	13	14	15	17	18
13	13	3	4	5	5	6	6	6	7	7
13	26	4	6	6	7	8	8	9	10	10
13	39	5	7	8	8	9	10	11	11	12
13	52	6	8	9	10	11	12	12	13	14
13	65	6	8	9	11	12	13	14	15	15
13	78	7	9	10	11	13	14	14	16	16
13	91	7	9	11	12	13	14	15	16	17

TABLE B.8 (cont.)　Critical Values of d_{max} for the
Kolmogorov-Smirnov Goodness of Fit Test for Discrete or Grouped
Data

k	n	α < 0.50	0.20	0.10	0.05	0.02	0.01	0.005	0.002	0.001
14	14	3	4	5	5	6	6	7	7	7
14	28	5	6	7	7	8	9	9	10	10
14	42	5	7	8	9	10	10	11	12	12
14	56	6	8	9	10	11	12	13	14	14
14	70	7	8	10	11	12	13	14	15	16
14	84	7	9	10	12	13	14	15	16	17
14	98	7	9	11	12	13	15	16	17	18
15	15	4	4	5	6	6	7	7	7	8
15	30	5	6	7	8	8	9	10	10	11
14	45	6	7	8	9	10	11	12	12	13
15	60	6	8	9	10	12	12	13	14	15
15	75	7	9	10	11	13	14	14	15	16
15	90	7	9	11	12	13	14	15	16	17
16	16	4	5	5	6	6	7	7	8	8
16	32	5	6	7	8	9	9	10	11	11
16	48	6	7	8	9	10	11	12	13	13
16	64	6	8	10	11	12	13	14	15	15
16	80	7	9	10	11	13	14	15	16	17
16	96	7	9	11	12	14	15	16	17	18
17	17	4	5	5	6	7	7	7	8	8
17	34	5	6	7	8	9	10	10	11	11
17	51	6	8	9	10	11	12	12	13	14
17	68	7	9	10	11	12	13	14	15	16
17	85	7	9	10	12	13	14	15	16	17
18	18	4	5	6	6	7	7	8	8	8
18	36	5	7	7	8	9	10	10	11	12
18	54	6	8	9	10	11	12	13	14	14
18	72	7	9	10	11	13	13	14	15	16
18	90	7	9	11	12	13	14	15	17	17
19	19	4	5	6	6	7	7	8	8	9
19	38	5	7	8	8	9	10	11	11	12
19	57	6	8	9	10	11	12	13	14	15
19	76	7	9	10	11	13	14	15	16	16
19	95	7	9	11	12	14	15	16	17	18
20	20	4	5	6	6	7	8	8	9	9
20	40	5	7	8	9	10	10	11	12	12
20	60	6	8	9	10	12	13	13	14	15
20	80	7	9	10	12	13	14	15	16	17
20	100	7	9	11	12	14	15	16	17	18
21	21	4	5	6	7	7	8	8	9	9
21	42	5	7	8	9	10	11	11	12	13
21	63	7	8	10	11	12	13	14	15	15
21	84	7	9	11	12	13	14	15	16	17
22	22	4	5	6	7	7	8	8	9	9
22	44	6	7	8	9	10	11	12	12	13
22	66	7	9	10	11	12	13	14	15	16
22	88	7	9	11	12	13	15	15	17	17
23	23	4	5	6	7	8	8	9	9	9
23	46	6	7	8	9	10	11	12	13	13
23	69	7	9	10	11	12	13	14	15	16
23	92	7	9	11	12	14	15	16	17	18
24	24	4	6	6	7	8	8	9	9	10
24	48	6	8	9	10	11	11	12	13	14
24	72	7	9	10	11	13	14	14	16	16
24	96	7	10	11	12	14	15	16	17	18

TABLE B.8 (cont.)   Critical Values of d_{max} for the Kolmogorov-Smirnov Goodness of Fit Test for Discrete or Grouped Data

k	n	α < 0.50	0.20	0.10	0.05	0.02	0.01	0.005	0.002	0.001
25	25	4	6	6	7	8	8	9	10	10
25	50	6	8	9	10	11	12	12	13	14
25	75	7	9	10	11	13	14	15	16	16
25	100	7	10	11	12	14	15	16	17	18
26	26	5	6	7	7	8	9	9	10	10
26	52	6	8	9	10	11	12	13	13	14
26	78	7	9	11	12	13	14	15	16	17
27	27	5	6	7	7	8	9	9	10	10
27	54	6	8	9	10	11	12	13	14	14
27	81	7	9	11	12	13	14	15	16	17
28	28	5	6	7	7	8	9	9	10	11
28	56	6	8	9	10	11	12	13	14	15
28	84	7	9	11	12	13	14	15	16	17
29	29	5	6	7	8	8	9	10	10	11
29	58	6	8	9	10	12	12	13	14	15
29	87	7	10	11	12	14	15	16	17	17
30	30	5	6	7	8	9	9	10	10	11
30	60	7	8	10	11	12	13	13	14	15
30	90	7	9	11	12	14	15	16	17	18
31	31	5	6	7	8	9	9	10	11	11
31	62	7	9	10	11	12	13	14	15	15
31	93	7	9	11	12	14	15	16	17	18
32	32	5	6	7	8	9	9	10	11	11
32	64	7	9	10	11	12	13	14	15	16
32	96	7	10	11	12	14	15	16	17	18
33	33	5	6	7	8	9	10	10	11	11
33	66	7	9	10	11	12	13	14	15	16
33	99	7	10	11	13	14	15	16	17	18
34	34	5	7	7	8	9	10	10	11	12
34	68	7	9	10	11	13	13	14	15	16
35	35	5	7	8	8	9	10	11	11	12
35	70	7	9	10	11	13	14	14	16	16
36	36	5	7	8	8	9	10	11	11	12
36	72	7	9	10	11	13	14	15	16	16
37	37	5	7	8	9	10	10	11	12	12
37	74	7	9	11	12	13	14	15	16	17
38	38	5	7	8	9	10	10	11	12	12
38	76	7	9	11	12	13	14	15	16	17
39	39	6	7	8	9	10	10	11	12	12
39	78	7	9	11	12	13	14	15	16	17
40	40	6	7	8	9	10	11	11	12	13
40	80	7	9	11	12	13	14	15	16	17
41	41	6	7	8	9	10	11	11	12	13
41	82	7	10	11	12	13	14	15	16	17
42	42	6	7	8	9	10	11	11	12	13
42	84	7	10	11	12	14	15	15	17	17
43	43	6	7	8	9	10	11	12	12	13
43	86	8	10	11	12	13	15	16	17	18
44	44	6	7	8	9	10	11	12	13	13
44	88	7	10	11	12	14	15	16	17	18
45	45	6	8	9	9	10	11	12	13	13
45	90	7	10	11	12	14	15	16	17	18
46	46	6	8	9	10	11	11	12	13	13
46	92	7	10	11	13	14	15	16	17	18
47	47	6	8	9	10	11	11	12	13	14

TABLE B.8 (cont.)   Critical Values of d_{max} for the
Kolmogorov-Smirnov Goodness of Fit Test for Discrete or Grouped
Data

k	n	$\alpha < 0.50$	0.20	0.10	0.05	0.02	0.01	0.005	0.002	0.001
47	94	8	10	11	13	14	15	16	17	18
48	48	6	8	9	10	11	12	12	13	14
48	96	7	10	11	13	14	15	16	18	18
49	49	6	8	9	10	11	12	12	13	14
49	98	8	10	11	13	14	15	16	18	19
50	50	6	8	9	10	11	12	13	13	14
50	100	7	10	11	13	14	15	16	18	19

These values were determined by the method used in Pettitt and Stephens (1977) by modifying a computer program provided by Pettitt.

Examples:

$$(d_{max})_{0.05, 30, 90} = 11 \quad \text{and} \quad (d_{max})_{0.01, 25, 100} = 15.$$

The table above is applicable when all expected frequencies are equal; it also works well with expected frequencies that are slightly to moderately unequal (Pettitt and Stephens 1977).

Note: The values of d_{max} shown have probabilities slightly less than their column headings. Therefore, for example, $0.02 < P(d_{max} = 12$, for $k = 30$, and $n = 90) < 0.05$.

TABLE B.9 Critical Values of D for the Kolmogorov-Smirnov Goodness of Fit Test for Continuous Distributions

n	α:	0.50	0.20	0.10	0.05	0.02	0.01	0.005	0.002	0.001
1		0.75000	0.90000	0.95000	0.97500	0.99000	0.99500	0.99750	0.99900	0.99950
2		0.50000	0.68377	0.77639	0.84189	0.90000	0.92929	0.95000	0.96838	0.97764
3		0.43529	0.56481	0.63604	0.70760	0.78456	0.82900	0.86428	0.90000	0.92063
4		0.38209	0.49265	0.56522	0.62394	0.68887	0.73424	0.77639	0.82217	0.85047
5		0.34319	0.44698	0.50945	0.56328	0.62718	0.66853	0.70543	0.75000	0.78137
6		0.31447	0.41037	0.46799	0.51926	0.57741	0.61661	0.65287	0.69571	0.72479
7		0.29312	0.38148	0.43607	0.48342	0.53844	0.57581	0.60975	0.65071	0.67930
8		0.27567	0.35831	0.40962	0.45427	0.50654	0.54179	0.57429	0.61368	0.64098
9		0.26082	0.33910	0.38746	0.43001	0.47960	0.51332	0.54443	0.58210	0.60846
10		0.24809	0.32260	0.36866	0.40925	0.45662	0.48893	0.51872	0.55500	0.58042
11		0.23709	0.30829	0.35242	0.39122	0.43670	0.46770	0.49639	0.53135	0.55588
12		0.22748	0.29577	0.33815	0.37543	0.41918	0.44905	0.47672	0.51047	0.53422
13		0.21901	0.28470	0.32549	0.36143	0.40362	0.43247	0.45921	0.49189	0.51490
14		0.21146	0.27481	0.31417	0.34890	0.38970	0.41762	0.44352	0.47520	0.49753
15		0.20465	0.26589	0.30397	0.33760	0.37713	0.40420	0.42934	0.46011	0.48182
16		0.19844	0.25778	0.29472	0.32733	0.36571	0.39201	0.41644	0.44637	0.46750
17		0.19277	0.25039	0.28627	0.31796	0.35528	0.38086	0.40464	0.43380	0.45440
18		0.18757	0.24360	0.27851	0.30936	0.34569	0.37062	0.39380	0.42224	0.44234
19		0.18277	0.23735	0.27136	0.30143	0.33685	0.36117	0.38379	0.41156	0.43119
20		0.17833	0.23156	0.26473	0.29408	0.32866	0.35241	0.37451	0.40165	0.42085
21		0.17421	0.22617	0.25858	0.28724	0.32104	0.34426	0.36588	0.39243	0.41122
22		0.17036	0.22115	0.25283	0.28087	0.31394	0.33666	0.35782	0.38382	0.40223
23		0.16676	0.21646	0.24746	0.27490	0.30728	0.32954	0.35027	0.37575	0.39380
24		0.16338	0.21205	0.24242	0.26931	0.30104	0.32286	0.34318	0.36817	0.38588
25		0.16021	0.20790	0.23768	0.26404	0.29516	0.31657	0.33651	0.36104	0.37843
26		0.15721	0.20399	0.23320	0.25908	0.28962	0.31063	0.33022	0.35431	0.37139
27		0.15437	0.20030	0.22898	0.25438	0.28438	0.30502	0.32426	0.34794	0.36473
28		0.15169	0.19680	0.22497	0.24993	0.27942	0.29971	0.31862	0.34190	0.35842
29		0.14914	0.19348	0.22117	0.24571	0.27471	0.29466	0.31327	0.33617	0.35242
30		0.14672	0.19032	0.21756	0.24170	0.27023	0.28986	0.30818	0.33072	0.34672
31		0.14442	0.18732	0.21412	0.23788	0.26596	0.28529	0.30333	0.32553	0.34129
32		0.14222	0.18445	0.21085	0.23424	0.26189	0.28094	0.29870	0.32058	0.33611
33		0.14012	0.18171	0.20771	0.23076	0.25801	0.27677	0.29428	0.31584	0.33115
34		0.13811	0.17909	0.20472	0.22743	0.25429	0.27279	0.29005	0.31131	0.32641
35		0.13618	0.17659	0.20185	0.22425	0.25073	0.26897	0.28600	0.30697	0.32187
36		0.13434	0.17418	0.19910	0.22119	0.24732	0.26532	0.28211	0.30281	0.31751
37		0.13257	0.17188	0.19646	0.21826	0.24404	0.26180	0.27838	0.29882	0.31333
38		0.13086	0.16966	0.19392	0.21544	0.24089	0.25843	0.27480	0.29498	0.30931
39		0.12923	0.16753	0.19148	0.21273	0.23786	0.25518	0.27135	0.29128	0.30544
40		0.12765	0.16547	0.18913	0.21012	0.23494	0.25205	0.26803	0.28772	0.30171
41		0.12613	0.16349	0.18687	0.20760	0.23213	0.24904	0.26482	0.28429	0.29811
42		0.12466	0.16158	0.18468	0.20517	0.22941	0.24613	0.26173	0.28097	0.29465
43		0.12325	0.15974	0.18257	0.20283	0.22679	0.24332	0.25875	0.27778	0.29130
44		0.12188	0.15796	0.18053	0.20056	0.22426	0.24060	0.25587	0.27468	0.28806
45		0.12056	0.15623	0.17856	0.19837	0.22181	0.23798	0.25308	0.27169	0.28493
46		0.11927	0.15457	0.17665	0.19625	0.21944	0.23544	0.25038	0.26880	0.28190
47		0.11803	0.15295	0.17481	0.19420	0.21715	0.23298	0.24776	0.26600	0.27896
48		0.11683	0.15139	0.17301	0.19221	0.21493	0.23059	0.24523	0.26328	0.27611
49		0.11567	0.14987	0.17128	0.19028	0.21277	0.22828	0.24277	0.26065	0.27335
50		0.11453	0.14840	0.16959	0.18841	0.21068	0.22604	0.24039	0.25809	0.27067
51		0.11344	0.14697	0.16796	0.18659	0.20864	0.22386	0.23807	0.25561	0.26807
52		0.11237	0.14558	0.16637	0.18482	0.20667	0.22174	0.23582	0.25319	0.26555
53		0.11133	0.14423	0.16483	0.18311	0.20475	0.21968	0.23364	0.25085	0.26309
54		0.11032	0.14292	0.16332	0.18144	0.20289	0.21768	0.23151	0.24857	0.26070
55		0.10934	0.14164	0.16186	0.17981	0.20107	0.21574	0.22944	0.24635	0.25837

TABLE B.9 (cont.) Critical Values of *D* for the Kolmogorov-Smirnov Goodness of Fit Test for Continuous Distributions

n	α: 0.50	0.20	0.10	0.05	0.02	0.01	0.005	0.002	0.001
56	0.10839	0.14040	0.16044	0.17823	0.19930	0.21384	0.22742	0.24419	0.25611
57	0.10746	0.13919	0.15906	0.17669	0.19758	0.21199	0.22546	0.24208	0.25390
58	0.10655	0.13801	0.15771	0.17519	0.19590	0.21020	0.22355	0.24003	0.25175
59	0.10566	0.13686	0.15639	0.17373	0.19427	0.20844	0.22169	0.23803	0.24966
60	0.10480	0.13573	0.15511	0.17231	0.19267	0.20673	0.21987	0.23608	0.24761
61	0.10396	0.13464	0.15385	0.17091	0.19112	0.20506	0.21809	0.23418	0.24562
62	0.10314	0.13357	0.15263	0.16956	0.18960	0.20343	0.21636	0.23232	0.24367
63	0.10234	0.13253	0.15144	0.16823	0.18812	0.20184	0.21467	0.23051	0.24177
64	0.10155	0.13151	0.15027	0.16693	0.18667	0.20029	0.21302	0.22873	0.23991
65	0.10079	0.13052	0.14913	0.16567	0.18525	0.19877	0.21141	0.22700	0.23810
66	0.10004	0.12954	0.14802	0.16443	0.18387	0.19729	0.20983	0.22531	0.23633
67	0.09931	0.12859	0.14693	0.16322	0.18252	0.19584	0.20829	0.22365	0.23459
68	0.09859	0.12766	0.14587	0.16204	0.18119	0.19442	0.20678	0.22204	0.23289
69	0.09789	0.12675	0.14483	0.16088	0.17990	0.19303	0.20530	0.22045	0.23123
70	0.09721	0.12586	0.14381	0.15975	0.17863	0.19167	0.20386	0.21890	0.22961
71	0.09653	0.12499	0.14281	0.15864	0.17739	0.19034	0.20244	0.21738	0.22802
72	0.09588	0.12413	0.14183	0.15755	0.17618	0.18903	0.20105	0.21589	0.22646
73	0.09523	0.12329	0.14087	0.15649	0.17499	0.18776	0.19970	0.21444	0.22493
74	0.09460	0.12247	0.13993	0.15544	0.17382	0.18650	0.19837	0.21301	0.22343
75	0.09398	0.12167	0.13901	0.15442	0.17268	0.18528	0.19706	0.21161	0.22196
76	0.09338	0.12088	0.13811	0.15342	0.17155	0.18408	0.19578	0.21024	0.22053
77	0.09278	0.12011	0.13723	0.15244	0.17045	0.18290	0.19453	0.20889	0.21912
78	0.09220	0.11935	0.13636	0.15147	0.16938	0.18174	0.19330	0.20757	0.21773
79	0.09162	0.11860	0.13551	0.15052	0.16832	0.18060	0.19209	0.20628	0.21637
80	0.09106	0.11787	0.13467	0.14960	0.16728	0.17949	0.19091	0.20501	0.21504
81	0.09051	0.11716	0.13385	0.14868	0.16626	0.17840	0.18974	0.20376	0.21373
82	0.08997	0.11645	0.13305	0.14779	0.16526	0.17732	0.18860	0.20253	0.21245
83	0.08944	0.11576	0.13226	0.14691	0.16428	0.17627	0.18748	0.20133	0.21119
84	0.08891	0.11508	0.13148	0.14605	0.16331	0.17523	0.18638	0.20015	0.20995
85	0.08840	0.11442	0.13072	0.14520	0.16236	0.17421	0.18530	0.19898	0.20873
86	0.08790	0.11376	0.12997	0.14437	0.16143	0.17321	0.18423	0.19784	0.20753
87	0.08740	0.11311	0.12923	0.14355	0.16051	0.17223	0.18319	0.19672	0.20635
88	0.08691	0.11248	0.12850	0.14274	0.15961	0.17126	0.18216	0.19562	0.20520
89	0.08643	0.11186	0.12779	0.14195	0.15873	0.17031	0.18115	0.19453	0.20406
90	0.08596	0.11125	0.12709	0.14117	0.15786	0.16938	0.18016	0.19347	0.20294
91	0.08550	0.11064	0.12640	0.14040	0.15700	0.16846	0.17918	0.19242	0.20184
92	0.08504	0.11005	0.12572	0.13965	0.15616	0.16755	0.17822	0.19138	0.20076
93	0.08459	0.10947	0.12506	0.13891	0.15533	0.16666	0.17727	0.19037	0.19969
94	0.08415	0.10889	0.12440	0.13818	0.15451	0.16579	0.17634	0.18937	0.19865
95	0.08371	0.10833	0.12375	0.13746	0.15371	0.16493	0.17542	0.18838	0.19761
96	0.08328	0.10777	0.12312	0.13675	0.15291	0.16408	0.17452	0.18741	0.19660
97	0.08286	0.10722	0.12249	0.13606	0.15214	0.16324	0.17363	0.18646	0.19560
98	0.08245	0.10668	0.12187	0.13537	0.15137	0.16242	0.17275	0.18552	0.19461
99	0.08204	0.10615	0.12126	0.13469	0.15061	0.16162	0.17189	0.18460	0.19364
100	0.08163	0.10563	0.12067	0.13403	0.14987	0.16081	0.17104	0.18368	0.19268
102	0.08084	0.10460	0.11949	0.13273	0.14841	0.15925	0.16938	0.18190	0.19081
104	0.08008	0.10361	0.11836	0.13146	0.14700	0.15773	0.16777	0.18017	0.18900
106	0.07933	0.10264	0.11725	0.13023	0.14562	0.15625	0.16620	0.17848	0.18723
108	0.07861	0.10170	0.11618	0.12904	0.14429	0.15482	0.16467	0.17685	0.18551
110	0.07790	0.10079	0.11513	0.12787	0.14299	0.15342	0.16319	0.17525	0.18384
112	0.07722	0.09990	0.11411	0.12674	0.14172	0.15207	0.16174	0.17370	0.18222
114	0.07655	0.09903	0.11312	0.12564	0.14049	0.15074	0.16034	0.17219	0.18063
116	0.07590	0.09818	0.11215	0.12457	0.13929	0.14945	0.15897	0.17072	0.17909
118	0.07527	0.09736	0.11121	0.12352	0.13812	0.14820	0.15763	0.16929	0.17759
120	0.07465	0.09656	0.11029	0.12250	0.13697	0.14697	0.15633	0.16789	0.17612

TABLE B.9 (cont.)   Critical Values of D for the Kolmogorov-Smirnov Goodness of Fit Test for Continuous Distributions

n	α: 0.50	0.20	0.10	0.05	0.02	0.01	0.005	0.002	0.001
122	0.07404	0.09577	0.10940	0.12150	0.13586	0.14578	0.15506	0.16652	0.17469
124	0.07345	0.09501	0.10852	0.12053	0.13477	0.14461	0.15382	0.16519	0.17329
126	0.07288	0.09426	0.10767	0.11958	0.13371	0.14347	0.15261	0.16389	0.17193
128	0.07232	0.09353	0.10684	0.11866	0.13268	0.14236	0.15142	0.16262	0.17060
130	0.07177	0.09282	0.10602	0.11775	0.13166	0.14128	0.15027	0.16138	0.16930
132	0.07123	0.09213	0.10523	0.11687	0.13068	0.14021	0.14914	0.16017	0.16802
134	0.07071	0.09144	0.10445	0.11600	0.12971	0.13918	0.14804	0.15898	0.16678
136	0.07019	0.09078	0.10369	0.11516	0.12876	0.13816	0.14696	0.15782	0.16556
138	0.06969	0.09013	0.10294	0.11433	0.12784	0.13717	0.14590	0.15669	0.16437
140	0.06920	0.08949	0.10221	0.11352	0.12693	0.13620	0.14487	0.15558	0.16321
142	0.06872	0.08887	0.10150	0.11273	0.12604	0.13524	0.14385	0.15449	0.16207
144	0.06825	0.08826	0.10080	0.11195	0.12517	0.13431	0.14286	0.15343	0.16095
146	0.06778	0.08766	0.10012	0.11119	0.12432	0.13340	0.14189	0.15238	0.15986
148	0.06733	0.08707	0.09944	0.11044	0.12349	0.13250	0.14094	0.15136	0.15879
150	0.06689	0.8650	0.09879	0.10971	0.12267	0.13163	0.14001	0.15036	0.15774
152	0.06646	0.08593	0.09814	0.10900	0.12187	0.13077	0.13909	0.14938	0.15671
154	0.06603	0.08538	0.09751	0.10830	0.12109	0.12993	0.13820	0.14842	0.15570
156	0.06561	0.08484	0.09689	0.10761	0.12032	0.12910	0.13732	0.14747	0.15471
158	0.06520	0.08430	0.09628	0.10693	0.11956	0.12829	0.13645	0.14655	0.15374
160	0.06480	0.08378	0.09569	0.10627	0.11882	0.12749	0.13561	0.14564	0.15278
d_α	0.83255	1.07298	1.22387	1.35810	1.51743	1.62762	1.73082	1.85846	1.94947
A_α	−0.042554	0.002557	0.052556	0.112820	0.205662	0.284642	0.370673	0.494581	0.595698

Table B.9 was prepared using Equation 3.0 of Birnbaum and Tingey (1951). The values of $D_{\alpha,n}$ were computed to eight decimal places and then rounded to five decimal places.

Examples:

$$D_{0.05,12} = 0.37543 \quad \text{and} \quad D_{0.01,55} = 0.21574.$$

For large n, critical values of $D_{\alpha,n}$ can be approximated by

$$D_{\alpha,n} \cong \sqrt{\frac{-\ln(\alpha/2)}{2n}}$$

(Smirnov, 1939a) or, more accurately, by either

$$D_{\alpha,n} \cong \sqrt{\frac{-\ln(\alpha/2)}{2n}} - \frac{0.16693}{n}$$

or

$$D_{\alpha,n} \cong \sqrt{\frac{-\ln(\alpha/2)}{2n}} - \frac{0.16693}{n} - \frac{A_\alpha}{\sqrt{n^3}}$$

(Miller, 1956), where

$$A_\alpha = 0.09037 \left[-\log\left(\frac{\alpha}{2}\right) \right]^{3/2} + 0.01515 \left[\log\left(\frac{\alpha}{2}\right) \right]^2 + 0.08467 \left(\frac{\alpha}{2}\right) - 0.11143.$$

For the significance levels, α, in Table B.9, the appropriate A_α's are given at the end of the table.

The accuracy of each of these three approximations is given as follows as percent error, where percent error $= |(\text{approximate } D_{\alpha,n} - \text{true } D_{\alpha,n})|/\text{true } D_{\alpha,n} \times 100\%$.

For the first approximating equation:

n	α: 0.50	0.20	0.10	0.05	0.02	0.01	0.005	0.002	0.001
20	4.4%	3.6%	3.4%	3.3%	3.2%	3.3%	3.3%	3.5%	3.6%
50	2.8	2.3	2.1	1.9	1.9	1.8	1.8	1.8	1.9
100	2.0	1.6	1.4	1.3	1.2	1.2	1.2	1.2	1.2
160	1.6	1.3	1.1	1.0	1.0	0.9	0.9	0.9	0.9

For the second approximating equation:

n	α: 0.50	0.20	0.10	0.05	0.02	0.01	0.005	0.002	0.001
10	0.6%	0.0*	0.5%	0.9%	1.4%	1.9%	2.3%	2.9%	3.3%
20	0.3	0.0*	0.2	0.4	0.7	0.9	1.1	1.4	1.6
50	0.1	0.0*	0.1	0.2	0.3	0.4	0.4	0.5	0.6
100	0.1	0.0*	0.0*	0.1	0.1	0.2	0.2	0.3	0.3
160	0.0*	0.0*	0.0*	0.1	0.1	0.1	0.1	0.2	0.2

For the third approximating equation:

n	α: 0.50	0.20	0.10	0.05	0.02	0.01	0.005	0.002	0.001
5	0.1%	0.2%	0.0*	0.1%	0.1%	0.1%	0.3%	0.5%	0.5%
10	0.1	0.0*	0.0*	0.0*	0.0*	0.1	0.0*	0.1	0.1
20	0.0*	0.0*	0.0*	0.0*	0.0*	0.0*	0.0*	0.0*	0.0*

In the above, the asterisk indicates a percent error the absolute value of which is less than 0.05%.

The first approximation may also be written as

$$D_{\alpha,n} = \frac{d_\alpha}{\sqrt{n}},$$

where d_α is given at the end of Table B.9.

TABLE B.10 Critical Values of D_δ for the δ-Corrected Kolmogorov-Smirnov Goodness of Fit Test for Continuous Distributions

n	δ	α: 0.50	0.20	0.10	0.05	0.02	0.01	0.005	0.002	0.001
3	0	0.23875	0.35477	0.41811	0.46702	0.53456	0.57900	0.61428	0.65000	0.67065
	1	0.38345	0.53584	0.63160	0.70760	0.78456	0.82900	0.86428	0.90000	0.92063
4	0	0.23261	0.33435	0.39075	0.44641	0.50495	0.54210	0.57722	0.62216	0.65046
	1	0.33126	0.46154	0.53829	0.60468	0.68377	0.73409	0.77639	0.82217	0.85047
5	0	0.22665	0.31556	0.37359	0.42174	0.47692	0.51576	0.54981	0.58934	0.61682
	1	0.29930	0.41172	0.48153	0.54273	0.61133	0.65692	0.69887	0.74881	0.78133
6	0	0.21803	0.30244	0.35522	0.40045	0.45440	0.48988	0.52240	0.56231	0.58954
	1	0.27516	0.37706	0.44074	0.49569	0.55969	0.60287	0.64167	0.68777	0.71966
7	0	0.20935	0.28991	0.33905	0.38294	0.43337	0.46761	0.49932	0.53714	0.56345
	1	0.25645	0.35066	0.40892	0.46010	0.51968	0.55970	0.59646	0.64081	0.67126
8	0	0.20148	0.27828	0.32538	0.36697	0.41522	0.44819	0.47834	0.51499	0.54065
	1	0.24149	0.32925	0.38365	0.43160	0.48732	0.52519	0.56000	0.60194	0.63114
9	0	0.19475	0.26794	0.31325	0.35277	0.39922	0.43071	0.45983	0.49525	0.51997
	1	0.22919	0.31157	0.36287	0.40794	0.46067	0.49652	0.52953	0.56963	0.59760
10	0	0.18913	0.25884	0.30221	0.34022	0.38481	0.41517	0.44329	0.47747	0.50148
	1	0.21881	0.29668	0.34525	0.38798	0.43809	0.47220	0.50373	0.54207	0.56889
11	0	0.18381	0.25071	0.29227	0.32894	0.37187	0.40122	0.42835	0.46147	0.48475
	1	0.20981	0.28388	0.33008	0.37084	0.41864	0.45127	0.48146	0.51823	0.54408
12	0	0.17878	0.24325	0.28330	0.31869	0.36019	0.38856	0.41484	0.44696	0.46954
	1	0.20190	0.27269	0.31686	0.35588	0.40167	0.43298	0.46197	0.49737	0.52225
13	0	0.17410	0.23639	0.27515	0.30935	0.34954	0.37703	0.40254	0.43372	0.45570
	1	0.19487	0.26279	0.30520	0.34265	0.38668	0.41680	0.44475	0.47889	0.50292
14	0	0.16976	0.23010	0.26767	0.30081	0.33980	0.36649	0.39127	0.42159	0.44298
	1	0.18859	0.25395	0.29478	0.33086	0.37331	0.40238	0.42936	0.46237	0.48563
15	0	0.16575	0.22430	0.26077	0.29296	0.33083	0.35679	0.38090	0.41043	0.43129
	1	0.18293	0.24600	0.28541	0.32026	0.36128	0.38940	0.41551	0.44748	0.47001
16	0	0.16204	0.21895	0.25439	0.28570	0.32256	0.34784	0.37132	0.40012	0.42043
	1	0.17778	0.23879	0.27692	0.31065	0.35039	0.37764	0.40296	0.43398	0.45591
17	0	0.15859	0.21397	0.24847	0.27897	0.31489	0.33953	0.36245	0.39055	0.41041
	1	0.17308	0.23221	0.26918	0.30189	0.34045	0.36691	0.39151	0.42168	0.44298
18	0	0.15537	0.20933	0.24296	0.27270	0.30775	0.33181	0.35419	0.38164	0.40106
	1	0.16875	0.22617	0.26208	0.29386	0.33134	0.35707	0.38101	0.41037	0.43114
19	0	0.15235	0.20498	0.23781	0.26685	0.30108	0.32459	0.34647	0.37332	0.39235
	1	0.16475	0.22060	0.25553	0.28646	0.32295	0.34801	0.37133	0.39995	0.42020
20	0	0.14948	0.20089	0.23298	0.26137	0.29484	0.31784	0.33924	0.36553	0.38414
	1	0.16104	0.21544	0.24947	0.27961	0.31518	0.33962	0.36237	0.39031	0.41008
21	0	0.14678	0.19705	0.22844	0.25622	0.28898	0.31149	0.33246	0.35822	0.37645
	1	0.15758	0.21064	0.24384	0.27325	0.30796	0.33182	0.35404	0.38134	0.40067
22	0	0.14422	0.19343	0.22416	0.25136	0.28346	0.30552	0.32607	0.35133	0.36921
	1	0.15435	0.20616	0.23859	0.26732	0.30123	0.32456	0.34628	0.37298	0.39189
23	0	0.14179	0.19001	0.22012	0.24679	0.27825	0.29989	0.32005	0.34483	0.36239
	1	0.15132	0.20197	0.23367	0.26176	0.29494	0.31776	0.33902	0.36516	0.38365
24	0	0.13949	0.18677	0.21630	0.24245	0.27333	0.29456	0.31435	0.33868	0.35593
	1	0.14848	0.19804	0.22906	0.25656	0.28904	0.31138	0.33221	0.35782	0.37597
25	0	0.13730	0.18370	0.21268	0.23835	0.26866	0.28951	0.30895	0.33285	0.34980
	1	0.14579	0.19433	0.22472	0.25166	0.28349	0.30539	0.32580	0.35091	0.36875
26	0	0.13522	0.18077	0.20924	0.23445	0.26423	0.28472	0.30382	0.32733	0.34398
	1	0.14326	0.19084	0.22063	0.24704	0.27825	0.29973	0.31980	0.34441	0.36187
27	0	0.13323	0.17799	0.20596	0.23074	0.26001	0.28016	0.29895	0.32206	0.33845
	1	0.14086	0.18753	0.21676	0.24267	0.27330	0.29439	0.31405	0.33825	0.35541
28	0	0.13133	0.17533	0.20283	0.22721	0.25600	0.27582	0.29430	0.31704	0.33319
	1	0.13858	0.18440	0.21309	0.23853	0.26861	0.28933	0.30864	0.33242	0.34927
29	0	0.12951	0.17280	0.19985	0.22383	0.25217	0.27168	0.28987	0.31227	0.32818
	1	0.13641	0.18142	0.20961	0.23461	0.26417	0.28452	0.30351	0.32688	0.34349

TABLE B.10 (cont.) Critical Values of D_δ for the δ-Corrected Kolmogorov-Smirnov Goodness of Fit Test for Continuous Distributions

n	δ	α: 0.50	0.20	0.10	0.05	0.02	0.01	0.005	0.002	0.001
30	0	0.12777	0.17037	0.19700	0.22061	0.24851	0.26772	0.28564	0.30770	0.32335
	1	0.13435	0.17859	0.20630	0.23088	0.25994	0.27996	0.29863	0.32162	0.33794
31	0	0.12610	0.16805	0.19427	0.21752	0.24501	0.26393	0.28157	0.30333	0.31875
	1	0.13238	0.17589	0.20314	0.22732	0.25591	0.27561	0.29399	0.31662	0.33268
32	0	0.12450	0.16582	0.19166	0.21457	0.24165	0.26030	0.27771	0.29915	0.31436
	1	0.13051	0.17332	0.20014	0.22393	0.25207	0.27146	0.28955	0.31184	0.32764
33	0	0.12295	0.16368	0.18915	0.21173	0.23843	0.25683	0.27399	0.29513	0.31015
	1	0.12871	0.17086	0.19726	0.22069	0.24840	0.26750	0.28532	0.30728	0.32286
34	0	0.12147	0.16162	0.18674	0.20901	0.23534	0.25348	0.27042	0.29127	0.30609
	1	0.12699	0.16850	0.19451	0.21759	0.24490	0.26371	0.28127	0.30296	0.31827
35	0	0.12004	0.15964	0.18442	0.20639	0.23237	0.25027	0.26698	0.28757	0.30219
	1	0.12534	0.16625	0.19188	0.21462	0.24154	0.26008	0.27740	0.29873	0.31388
36	0	0.11866	0.15774	0.18218	0.20387	0.22951	0.24718	0.26368	0.28401	0.29844
	1	0.12375	0.16408	0.18935	0.21278	0.23831	0.25660	0.27368	0.29472	0.30968
37	0	0.11733	0.15590	0.18003	0.20144	0.22676	0.24421	0.26050	0.28057	0.29481
	1	0.12223	0.16200	0.18692	0.20904	0.23522	0.25326	0.27011	0.29087	0.30558
38	0	0.11604	0.15413	0.17796	0.19910	0.22410	0.24134	0.25743	0.27726	0.29135
	1	0.12076	0.16000	0.18459	0.20642	0.23225	0.25005	0.26668	0.28717	0.30172
39	0	0.11480	0.15242	0.17595	0.19684	0.22154	0.23857	0.25447	0.27406	0.28800
	1	0.11935	0.15808	0.18234	0.20389	0.22938	0.24696	0.26334	0.28361	0.29799
40	0	0.11360	0.15076	0.17402	0.19465	0.21907	0.23589	0.25161	0.27098	0.28474
	1	0.11798	0.15622	0.18018	0.20145	0.22663	0.24399	0.26020	0.28019	0.29439
41	0	0.11243	0.14916	0.17215	0.19254	0.21667	0.23331	0.24884	0.26799	0.28162
	1	0.11667	0.15443	0.17810	0.19910	0.22397	0.24112	0.25713	0.27688	0.29092
42	0	0.11130	0.14761	0.17034	0.19050	0.21436	0.23081	0.24617	0.26511	0.27880
	1	0.11540	0.15270	0.17608	0.19684	0.22141	0.23835	0.25418	0.27369	0.28758
43	0	0.11021	0.14611	0.16858	0.18852	0.21212	0.22839	0.24358	0.26232	0.27564
	1	0.11417	0.15103	0.17414	0.19465	0.21893	0.23568	0.25132	0.27061	0.28433
44	0	0.10915	0.14466	0.16688	0.18661	0.20995	0.22604	0.24108	0.25962	0.27277
	1	0.11298	0.14942	0.17226	0.19253	0.21654	0.23310	0.24857	0.26764	0.28120
45	0	0.10812	0.14325	0.16524	0.18475	0.20785	0.22377	0.23865	0.25700	0.27004
	1	0.11183	0.14786	0.17044	0.19049	0.21423	0.23060	0.24590	0.26476	0.27815
46	0	0.10712	0.14188	0.16364	0.18295	0.20581	0.22157	0.23629	0.25445	0.26737
	1	0.11072	0.14635	0.16868	0.18851	0.21199	0.22818	0.24331	0.26198	0.27524
47	0	0.10615	0.14055	0.16208	0.18120	0.20383	0.21943	0.23401	0.25199	0.26477
	1	0.10964	0.14488	0.16697	0.18659	0.20982	0.22584	0.24081	0.25928	0.27242
48	0	0.10520	0.13926	0.16058	0.17950	0.20190	0.21735	0.23179	0.24959	0.26225
	1	0.10859	0.14346	0.16532	0.18473	0.20772	0.22357	0.23839	0.25666	0.26966
49	0	0.10428	0.13800	0.15911	0.17785	0.20003	0.21534	0.22963	0.24726	0.25991
	1	0.10757	0.14208	0.16371	0.18293	0.20568	0.22137	0.23604	0.25413	0.26701
50	0	0.10339	0.13678	0.15769	0.17624	0.19822	0.21337	0.22753	0.24500	0.25742
	1	0.10659	0.14074	0.16216	0.18117	0.20370	0.21924	0.23376	0.25167	0.26441

These critical values were kindly provided by H. J. Khamis, by the method described in Khamis (1990).

TABLE B.11   Critical Values of the Mann-Whitney U Distribution

n_1	n_2	$\alpha(2)$: 0.20 $\alpha(1)$: 0.10	0.10 0.05	0.05 0.025	0.02 0.01	0.01 0.005	0.005 0.0025	0.002 0.001	0.001 0.0005
1	1	—	—	—	—	—	—	—	—
	2	—	—	—	—	—	—	—	—
	3	—	—	—	—	—	—	—	—
	4	—	—	—	—	—	—	—	—
	5	—	—	—	—	—	—	—	—
	6	—	—	—	—	—	—	—	—
	7	—	—	—	—	—	—	—	—
	8	—	—	—	—	—	—	—	—
	9	9	—	—	—	—	—	—	—
	10	10	—	—	—	—	—	—	—
	11	11	—	—	—	—	—	—	—
	12	12	—	—	—	—	—	—	—
	13	13	—	—	—	—	—	—	—
	14	14	—	—	—	—	—	—	—
	15	15	—	—	—	—	—	—	—
	16	16	—	—	—	—	—	—	—
	17	17	—	—	—	—	—	—	—
	18	18	—	—	—	—	—	—	—
	19	18	19	—	—	—	—	—	—
	20	19	20	—	—	—	—	—	—
	21	20	21	—	—	—	—	—	—
	22	21	22	—	—	—	—	—	—
	23	22	23	—	—	—	—	—	—
	24	23	24	—	—	—	—	—	—
	25	24	25	—	—	—	—	—	—
	26	25	26	—	—	—	—	—	—
	27	26	27	—	—	—	—	—	—
	28	27	28	—	—	—	—	—	—
	29	27	29	—	—	—	—	—	—
	30	28	30	—	—	—	—	—	—
	31	29	31	—	—	—	—	—	—
	32	30	32	—	—	—	—	—	—
	33	31	33	—	—	—	—	—	—
	34	32	34	—	—	—	—	—	—
	35	33	35	—	—	—	—	—	—
	36	34	36	—	—	—	—	—	—
	37	35	37	—	—	—	—	—	—
	38	36	38	—	—	—	—	—	—
	39	36	38	39	—	—	—	—	—
1	40	37	39	40	—	—	—	—	—
2	2	—	—	—	—	—	—	—	—
	3	6	—	—	—	—	—	—	—
	4	8	—	—	—	—	—	—	—
	5	9	10	—	—	—	—	—	—
	6	11	12	—	—	—	—	—	—
	7	10	14	—	—	—	—	—	—
	8	14	15	16	—	—	—	—	—
	9	16	17	18	—	—	—	—	—
	10	17	19	20	—	—	—	—	—
	11	19	21	22	—	—	—	—	—
	12	20	22	23	—	—	—	—	—
	13	22	24	25	26	—	—	—	—
	14	23	25	27	28	—	—	—	—
2	15	25	27	29	30	—	—	—	—

TABLE B.11 (cont.) Critical Values of the Mann-Whitney U Distribution

n_1	n_2	$\alpha(2)$: 0.20 $\alpha(1)$: 0.10	0.10 0.05	0.05 0.025	0.02 0.01	0.01 0.005	0.005 0.0025	0.002 0.001	0.001 0.0005
2	16	27	29	31	32	—	—	—	—
	17	28	31	32	34	—	—	—	—
	18	30	32	34	36	—	—	—	—
	19	31	34	36	37	38	—	—	—
	20	33	36	38	39	40	—	—	—
	21	34	37	39	41	42	—	—	—
	22	36	39	41	43	44	—	—	—
	23	37	41	43	45	46	—	—	—
	24	39	42	45	47	48	—	—	—
	25	41	44	47	49	50	—	—	—
	26	42	46	48	51	52	—	—	—
	27	44	47	50	52	53	54	—	—
	28	45	49	52	54	55	56	—	—
	29	47	51	54	56	57	58	—	—
	30	48	53	55	58	59	60	—	—
	31	50	54	57	60	61	62	—	—
	32	51	56	59	62	63	64	—	—
	33	53	58	61	64	65	66	—	—
	34	55	59	63	65	67	68	—	—
	35	56	61	64	67	69	70	—	—
	36	58	63	66	69	71	72	—	—
	37	59	64	68	71	73	74	—	—
	38	61	66	70	73	75	76	—	—
	39	62	68	71	75	76	77	—	—
2	40	64	69	73	77	78	79	—	—
3	3	8	9	—	—	—	—	—	—
	4	11	12	—	—	—	—	—	—
	5	13	14	15	—	—	—	—	—
	6	15	16	17	—	—	—	—	—
	7	15	19	20	21	—	—	—	—
	8	19	21	22	24	—	—	—	—
	9	22	23	25	26	27	—	—	—
	10	24	26	27	29	30	—	—	—
	11	26	28	30	32	33	—	—	—
	12	28	31	32	34	35	36	—	—
	13	30	33	35	37	38	39	—	—
	14	32	35	37	40	41	42	—	—
	15	35	38	40	42	43	44	—	—
	16	37	40	42	45	46	47	—	—
	17	39	42	45	47	49	50	51	—
	18	41	45	47	50	52	53	54	—
	19	43	47	50	53	54	56	57	—
	20	45	49	52	55	57	58	60	—
	21	48	52	55	58	60	61	62	63
	22	50	54	57	60	62	64	65	66
	23	52	56	60	63	65	67	68	69
	24	54	59	62	66	68	69	71	72
	25	56	61	65	68	70	72	74	75
	26	58	63	67	71	73	75	77	78
	27	60	66	70	74	76	78	79	80
	28	63	68	72	76	79	80	82	83
	29	65	70	74	79	81	83	85	86
3	30	67	73	77	81	84	86	88	89

TABLE B.11 (cont.)   Critical Values of the Mann-Whitney U
Distribution

n_1	n_2	$\alpha(2)$: 0.20 $\alpha(1)$: 0.10	0.10 0.05	0.05 0.025	0.02 0.01	0.01 0.005	0.005 0.0025	0.002 0.001	0.001 0.0005
3	31	69	75	79	84	87	89	91	92
	32	71	77	82	87	89	91	94	95
	33	73	80	84	89	92	94	96	98
	34	76	82	87	92	95	97	99	101
	35	78	84	89	94	97	100	102	103
	36	80	87	92	97	100	103	105	106
	37	82	89	94	100	103	105	108	109
	38	84	91	97	102	105	108	111	112
	39	86	94	99	105	108	111	113	115
3	40	89	96	102	107	111	114	116	118
4	4	13	15	16	—	—	—	—	—
	5	16	18	19	20	—	—	—	—
	6	19	21	22	23	24	—	—	—
	7	20	24	25	27	28	—	—	—
	8	25	27	28	30	31	32	—	—
	9	27	30	32	33	35	36	—	—
	10	30	33	35	37	38	39	40	—
	11	33	36	38	40	42	43	44	—
	12	36	39	41	43	45	46	48	—
	13	39	42	44	47	49	50	51	52
	14	41	45	47	50	52	53	55	56
	15	44	48	50	53	55	57	59	60
	16	47	50	53	57	59	60	62	63
	17	50	53	57	60	62	64	66	67
	18	52	56	60	63	66	67	69	71
	19	55	59	63	67	69	71	73	74
	20	58	62	66	70	72	75	77	78
	21	61	65	69	73	76	78	80	82
	22	63	68	72	77	79	82	84	85
	23	66	71	75	80	83	85	88	89
	24	69	74	79	83	86	89	91	93
	25	72	77	82	87	90	92	95	97
	26	74	80	85	90	93	96	98	100
	27	77	83	88	93	96	99	102	104
	28	80	86	91	96	100	103	106	108
	29	83	89	94	100	103	106	109	111
	30	85	92	97	103	107	110	113	115
	31	88	95	100	106	110	113	117	119
	32	91	98	104	110	114	117	120	122
	33	94	101	107	113	117	120	124	126
	34	96	104	110	116	120	124	127	130
	35	99	107	113	120	124	127	131	133
	36	102	110	116	123	127	131	135	137
	37	105	113	119	126	131	134	138	141
	38	107	116	122	130	134	138	142	144
	39	110	118	125	133	137	141	145	148
4	40	113	121	129	136	141	145	149	152
5	5	20	21	23	24	25	—	—	—
	6	23	25	27	28	29	30	—	—
	7	24	29	30	32	34	35	—	—
	8	30	32	34	36	38	39	40	—
	9	33	36	38	40	42	43	44	45
	10	37	39	42	44	46	47	49	50

TABLE B.11 (cont.) Critical Values of the Mann-Whitney U Distribution

n_1	n_2	$\alpha(2)$: 0.20 $\alpha(1)$: 0.10	0.10 0.05	0.05 0.025	0.02 0.01	0.01 0.005	0.005 0.0025	0.002 0.001	0.001 0.0005
5	11	40	43	46	48	50	52	53	54
	12	43	47	49	52	54	56	58	59
	13	47	50	53	56	58	60	62	63
	14	50	54	57	60	63	64	67	68
	15	53	57	61	64	67	69	71	72
	16	57	61	65	68	71	73	75	77
	17	60	65	68	72	75	77	80	81
	18	63	68	72	76	79	81	84	86
	19	67	72	76	80	83	86	88	90
	20	70	75	80	84	87	90	93	95
	21	73	79	83	88	91	94	97	99
	22	77	82	87	92	96	98	102	104
	23	80	86	91	96	100	103	106	108
	24	84	90	95	100	104	107	110	113
	25	87	93	98	104	108	111	115	117
	26	90	97	102	108	112	115	119	121
	27	94	100	106	112	119	120	123	126
	28	97	104	110	116	120	124	128	130
	29	100	107	113	120	124	128	132	135
	30	104	111	117	124	128	132	136	139
	31	107	115	121	128	133	136	141	144
	32	110	118	125	132	137	141	145	148
	33	114	122	128	136	141	145	150	153
	34	117	125	132	140	145	149	154	157
	35	120	129	136	144	149	153	158	161
	36	124	132	140	148	153	158	163	166
	37	127	136	144	152	157	162	167	170
	38	130	140	147	156	161	166	171	175
	39	134	143	151	160	165	170	176	179
5	40	137	147	155	164	169	174	180	184
6	6	27	29	31	33	34	35	—	—
	7	29	34	36	38	39	40	42	—
	8	35	38	40	42	44	45	47	48
	9	39	42	44	47	49	50	52	53
	10	43	46	49	52	54	55	57	58
	11	47	50	53	57	59	60	62	64
	12	51	55	58	61	63	65	68	69
	13	55	59	62	66	68	70	73	74
	14	59	63	67	71	73	75	78	79
	15	63	67	71	75	78	80	83	85
	16	67	71	75	80	83	85	88	90
	17	71	76	80	84	87	90	93	95
	18	74	80	84	89	92	95	98	100
	19	78	84	89	94	97	100	103	106
	20	82	88	93	98	102	105	108	111
	21	86	92	97	103	107	110	114	116
	22	90	96	102	108	111	115	119	121
	23	94	101	106	112	116	120	124	126
	24	98	105	111	117	121	125	129	132
6	25	102	109	115	121	126	130	134	137

TABLE B.11 (cont.)   Critical Values of the Mann-Whitney U Distribution

n_1	n_2	$\alpha(2)$: 0.20 $\alpha(1)$: 0.10	0.10 0.05	0.05 0.025	0.02 0.01	0.01 0.005	0.005 0.0025	0.002 0.001	0.001 0.0005
6	26	106	113	119	126	131	134	139	142
	27	110	117	124	131	135	139	144	147
	28	114	122	128	135	140	144	149	152
	29	118	126	132	140	145	149	154	157
	30	122	130	137	145	150	154	159	163
	31	125	134	141	149	154	159	164	168
	32	129	138	146	154	159	164	169	173
	33	133	142	150	158	164	169	174	178
	34	137	147	154	163	169	174	179	183
	35	141	151	159	168	173	179	185	188
	36	145	155	163	172	178	184	190	194
	37	149	159	167	177	183	188	195	199
	38	153	163	172	182	188	193	200	204
	39	157	167	176	186	193	198	205	209
6	40	161	172	181	191	197	203	210	214
7	7	36	38	41	43	45	46	48	49
	8	40	43	46	49	50	52	54	55
	9	45	48	51	54	56	58	60	61
	10	49	53	56	59	61	63	65	67
	11	54	58	61	65	67	69	71	73
	12	58	63	66	70	72	75	77	79
	13	63	67	71	75	78	80	83	85
	14	67	72	76	81	83	86	89	91
	15	72	77	81	86	89	92	95	97
	16	76	82	86	91	94	97	101	103
	17	81	86	91	96	100	103	106	109
	18	85	91	96	102	105	108	112	115
	19	90	96	101	107	111	114	118	120
	20	94	101	106	112	116	120	124	126
	21	99	106	111	117	122	125	129	132
	22	103	110	116	123	127	131	135	138
	23	108	115	121	128	132	136	141	144
	24	112	120	126	133	138	142	147	150
	25	117	125	131	139	143	148	153	156
	26	121	129	136	144	149	153	158	162
	27	126	134	141	149	154	159	164	168
	28	130	139	146	154	160	164	170	174
	29	135	144	151	160	165	170	176	179
	30	139	149	156	165	170	176	181	185
	31	144	153	161	170	176	181	187	191
	32	148	158	166	175	181	187	193	197
	33	153	163	171	181	187	192	199	203
	34	157	168	176	186	192	198	204	209
	35	162	172	181	191	198	203	210	215
	36	166	177	186	196	203	209	216	221
	37	171	182	191	202	208	215	222	227
	38	175	187	196	207	214	220	227	232
	39	180	191	201	212	219	226	233	238
7	40	184	196	206	217	225	231	239	244
8	8	45	49	51	55	57	58	60	62
	9	50	54	57	61	63	65	67	68
8	10	56	60	63	67	69	71	74	75

TABLE B.11 (cont.) Critical Values of the Mann-Whitney U Distribution

n_1	n_2	$\alpha(2)$: 0.20 $\alpha(1)$: 0.10	0.10 0.05	0.05 0.025	0.02 0.01	0.01 0.005	0.005 0.0025	0.002 0.001	0.001 0.0005
8	11	61	65	69	73	75	77	80	82
	12	66	70	74	79	81	84	87	89
	13	71	76	80	84	87	90	93	95
	14	76	81	86	90	94	96	100	102
	15	81	87	91	96	100	103	106	109
	16	86	92	97	102	106	109	113	115
	17	91	97	102	108	112	115	119	122
	18	96	103	108	114	118	122	126	129
	19	101	108	114	120	124	128	132	135
	20	106	113	119	126	130	134	139	142
	21	112	119	125	132	136	140	145	148
	22	117	124	131	138	142	147	152	155
	23	122	130	136	144	149	153	158	162
	24	127	135	142	150	155	159	165	168
	25	132	140	147	155	161	165	171	175
	26	137	146	153	161	167	172	177	181
	27	142	151	159	167	173	178	184	188
	28	147	156	164	173	179	184	190	195
	29	152	162	170	179	185	190	197	201
	30	157	167	175	185	191	197	203	208
	31	162	172	181	191	197	203	210	214
	32	167	178	187	197	203	209	216	221
	33	172	183	192	203	209	215	223	227
	34	177	188	198	208	215	222	229	234
	35	182	194	203	214	221	228	235	241
	36	188	199	209	220	228	234	242	247
	37	193	205	215	226	234	240	248	254
	38	198	210	220	232	240	247	255	260
	39	203	215	226	238	246	253	261	267
8	40	208	221	231	244	252	259	268	273
9	9	56	60	64	67	70	72	74	76
	10	62	66	70	74	77	79	82	83
	11	68	72	76	81	83	86	89	91
	12	73	78	82	87	90	93	96	98
	13	79	84	89	94	97	100	103	106
	14	85	90	95	100	104	107	111	113
	15	90	96	101	107	111	114	118	120
	16	96	102	107	113	117	121	125	128
	17	101	108	114	120	124	128	132	135
	18	107	114	120	126	131	135	139	142
	19	113	120	126	133	138	142	146	150
	20	118	126	132	140	144	149	154	157
	21	124	132	139	146	151	155	161	164
	22	130	138	145	153	158	162	168	172
	23	135	144	151	159	164	169	175	179
	24	141	150	157	166	171	176	182	186
	25	147	156	163	172	178	183	189	193
	26	152	162	170	179	185	190	196	201
	27	158	168	176	185	191	197	203	208
	28	164	174	182	192	198	204	211	215
	29	169	179	188	198	205	211	218	222
9	30	175	185	194	205	212	218	225	230

TABLE B.11 (cont.) Critical Values of the Mann-Whitney U
Distribution

n_1	n_2	$\alpha(2)$: 0.20 $\alpha(1)$: 0.10	0.10 0.05	0.05 0.025	0.02 0.01	0.01 0.005	0.005 0.0025	0.002 0.001	0.001 0.0005
9	31	180	191	201	211	218	224	232	237
	32	186	197	207	218	225	231	239	244
	33	192	203	213	224	232	238	246	251
	34	197	209	219	231	238	245	253	259
	35	203	215	226	237	245	252	260	266
	36	209	221	232	244	252	259	267	273
	37	214	227	238	250	258	266	275	280
	38	220	233	244	257	265	273	282	288
	39	225	239	250	263	272	280	289	295
9	40	231	245	257	270	279	286	296	302
10	10	68	73	77	81	84	87	90	92
	11	74	79	84	88	92	94	98	100
	12	81	86	91	96	99	102	106	108
	13	87	93	97	103	106	110	113	116
	14	93	99	104	110	114	117	121	124
	15	99	106	111	117	121	125	129	132
	16	106	112	118	124	129	133	137	140
	17	112	119	125	132	136	140	145	148
	18	118	125	132	139	143	148	153	156
	19	124	132	138	146	151	155	161	164
	20	130	138	145	153	158	163	168	172
	21	137	145	152	160	166	170	176	180
	22	143	152	159	167	173	178	184	188
	23	149	158	166	175	180	186	192	196
	24	155	165	173	182	188	193	200	204
	25	161	171	179	189	195	201	207	212
	26	168	178	186	196	202	208	215	220
	27	174	184	193	203	210	216	223	228
	28	180	191	200	210	217	223	231	236
	29	186	197	207	217	224	231	238	244
	30	192	204	213	224	232	238	246	252
	31	199	210	220	232	239	246	254	259
	32	205	217	227	239	246	253	262	267
	33	211	223	234	246	254	261	269	275
	34	217	230	241	253	261	268	277	283
	35	223	236	247	260	268	276	285	291
	36	229	243	254	267	276	284	293	299
	37	236	249	261	274	283	291	300	307
	38	242	256	268	281	290	299	308	315
	39	248	262	275	289	298	306	316	323
10	40	254	269	281	296	305	314	324	331
11	11	81	87	91	96	100	103	106	109
	12	88	94	99	104	108	111	115	117
	13	95	101	106	112	116	119	123	126
	14	102	108	114	120	124	128	132	135
	15	108	115	121	128	132	136	141	144
	16	115	122	129	135	140	144	149	152
	17	122	130	136	143	148	152	158	161
	18	129	137	143	151	156	161	166	170
	19	136	144	151	159	164	169	175	178
11	20	142	151	158	167	172	177	183	187

TABLE B.11 (cont.)   Critical Values of the Mann-Whitney U Distribution

n_1	n_2	$\alpha(2)$: 0.20 $\alpha(1)$: 0.10	0.10 0.05	0.05 0.025	0.02 0.01	0.01 0.005	0.005 0.0025	0.002 0.001	0.001 0.0005
11	21	149	158	166	174	180	185	191	196
	22	156	165	173	182	188	193	200	204
	23	163	172	180	190	196	202	208	213
	24	169	179	188	198	204	210	217	222
	25	176	186	195	205	212	218	225	230
	26	183	194	203	213	220	226	234	239
	27	190	201	210	221	228	234	242	247
	28	196	208	218	229	236	243	251	256
	29	203	215	225	236	244	251	259	265
	30	210	222	232	244	252	259	267	273
	31	217	229	240	252	260	267	276	282
	32	223	236	247	260	268	275	284	290
	33	230	243	255	267	276	283	293	299
	34	237	250	262	275	284	292	301	307
	35	244	257	269	283	292	300	309	316
	36	250	265	277	290	300	308	318	325
	37	257	272	284	298	308	316	326	333
	38	264	279	291	306	316	324	335	342
	39	271	286	299	314	323	332	343	350
11	40	277	293	306	321	331	341	351	359
12	12	95	102	107	113	117	120	124	127
	13	103	109	115	121	125	129	133	136
	14	110	117	123	130	134	138	143	146
	15	117	125	131	138	143	147	152	155
	16	125	132	139	146	151	156	161	165
	17	132	140	147	155	160	165	170	174
	18	139	148	155	163	169	173	179	183
	19	147	156	163	172	177	182	188	193
	20	154	163	171	180	186	191	198	202
	21	161	171	179	188	194	200	207	211
	22	169	179	187	197	203	209	216	220
	23	176	186	195	205	212	218	225	230
	24	183	194	203	213	220	227	234	239
	25	191	202	211	222	229	235	243	248
	26	198	209	219	230	238	244	252	258
	27	205	217	227	239	246	253	261	267
	28	213	225	235	247	255	262	270	276
	29	220	232	243	255	263	271	279	285
	30	227	240	251	264	272	279	288	295
	31	235	248	259	272	280	288	297	304
	32	242	256	267	280	289	297	307	313
	33	249	263	275	289	298	306	316	322
	34	257	271	283	297	306	315	325	332
	35	264	279	291	305	315	323	334	341
	36	271	286	299	314	323	332	343	350
	37	278	294	307	322	332	341	352	359
	38	286	302	315	330	340	350	361	368
	39	293	309	323	339	349	359	370	378
12	40	300	317	331	347	358	367	379	387
13	13	111	118	124	130	135	139	143	146
	14	119	126	132	139	144	148	153	157
13	15	127	134	141	148	153	158	163	167

TABLE B.11 (cont.)   Critical Values of the Mann-Whitney U Distribution

n_1	n_2	$\alpha(2):$ 0.20 $\alpha(1):$ 0.10	0.10 0.05	0.05 0.025	0.02 0.01	0.01 0.005	0.005 0.0025	0.002 0.001	0.001 0.0005
13	16	134	143	149	157	163	167	173	177
	17	142	151	158	166	172	177	183	187
	18	150	159	167	175	181	186	192	197
	19	158	167	175	184	190	196	202	207
	20	166	176	184	193	200	205	212	217
	21	174	184	193	202	209	215	222	227
	22	182	192	201	211	218	224	232	237
	23	190	201	210	220	227	234	241	247
	24	198	209	218	229	237	243	251	256
	25	205	217	227	238	246	253	261	266
	26	213	225	236	247	255	262	270	276
	27	221	234	244	256	264	271	280	286
	28	229	242	253	265	273	281	290	296
	29	237	250	261	274	283	290	300	306
	30	245	258	270	283	292	300	309	316
	31	253	267	278	292	301	309	319	326
	32	260	275	287	301	310	319	329	336
	33	268	283	296	310	319	328	338	346
	34	276	291	304	319	329	337	348	355
	35	284	299	313	328	338	347	358	365
	36	292	308	321	337	347	356	367	375
	37	300	316	330	346	356	366	377	385
	38	308	324	338	355	365	375	387	395
	39	315	332	347	363	374	385	397	405
13	40	323	341	355	372	384	394	406	415
14	14	127	135	141	149	154	161	164	167
	15	136	144	151	159	164	169	174	178
	16	144	153	160	168	174	179	185	189
	17	153	161	169	178	184	189	195	199
	18	161	170	178	187	194	199	206	210
	19	169	179	188	197	203	209	216	221
	20	178	188	197	207	213	219	226	231
	21	186	197	206	216	223	229	237	242
	22	195	206	215	226	233	240	247	253
	23	203	215	224	235	243	250	258	263
	24	212	223	234	245	253	260	268	274
	25	220	232	243	255	263	270	278	284
	26	228	241	252	264	272	280	289	295
	27	237	250	261	274	282	290	299	306
	28	245	259	270	283	292	300	309	316
	29	254	268	279	293	302	310	320	327
	30	262	276	289	302	312	320	330	337
	31	271	285	298	312	321	330	340	348
	32	279	294	307	321	331	340	351	358
	33	287	303	316	331	341	350	361	369
	34	296	312	325	341	351	360	371	379
	35	304	320	334	350	361	370	382	390
	36	313	329	343	360	370	380	392	400
	37	321	338	353	369	380	390	402	411
	38	329	347	362	379	390	400	413	421
	39	338	356	371	388	400	410	423	432
14	40	346	364	380	398	410	420	433	442

TABLE B.11 (cont.)   Critical Values of the Mann-Whitney U
Distribution

n_1	n_2	$\alpha(2)$: 0.20 $\alpha(1)$: 0.10	0.10 0.05	0.05 0.025	0.02 0.01	0.01 0.005	0.005 0.0025	0.002 0.001	0.001 0.0005
15	15	145	153	161	169	174	179	185	189
	16	154	163	170	179	185	190	197	201
	17	163	172	180	189	195	201	208	212
	18	172	182	190	200	206	212	219	224
	19	181	191	200	210	216	223	230	235
	20	190	200	210	220	227	233	241	246
	21	199	210	219	230	237	244	252	257
	22	208	219	229	240	248	255	263	269
	23	217	229	239	251	258	265	274	280
	24	226	238	249	261	269	276	285	291
	25	235	247	258	271	279	287	296	302
	26	244	257	268	281	290	298	307	313
	27	253	266	278	291	300	308	318	325
	28	262	276	288	301	311	319	329	336
	29	271	285	297	312	321	330	340	347
	30	280	294	307	322	331	340	351	358
	31	288	304	317	332	342	351	362	369
	32	297	313	327	342	352	362	373	381
	33	306	323	336	352	363	372	384	392
	34	315	332	346	362	373	383	395	403
	35	324	341	356	372	383	394	406	414
	36	333	351	366	382	394	404	417	425
	37	342	360	375	393	404	415	428	436
	38	351	369	385	403	415	425	439	448
	39	360	379	395	413	425	436	449	459
15	40	369	388	404	423	435	447	460	470
16	16	163	173	181	190	196	202	208	213
	17	173	183	191	201	207	213	220	225
	18	182	193	202	212	218	224	232	237
	19	192	203	212	222	230	236	244	249
	20	201	213	222	233	241	247	255	261
	21	211	223	233	244	252	259	267	273
	22	221	233	243	255	263	270	279	285
	23	230	243	253	266	274	281	290	296
	24	240	253	264	276	285	293	302	308
	25	249	263	274	287	296	304	314	320
	26	259	273	284	298	307	315	325	332
	27	268	283	295	309	318	327	337	344
	28	278	292	305	319	329	338	348	356
	29	287	302	315	330	340	349	360	367
	30	297	312	326	341	351	360	372	379
	31	306	322	336	352	362	372	383	391
	32	316	332	346	362	373	383	395	403
	33	325	342	357	373	384	394	406	415
	34	335	352	367	384	395	406	418	427
	35	344	362	377	395	406	417	429	438
	36	354	372	388	405	417	428	441	450
	37	363	382	398	416	428	439	453	462
	38	373	392	408	427	439	451	464	474
	39	382	402	418	437	450	462	476	485
16	40	392	412	429	448	461	473	487	497

TABLE B.11 (cont.) Critical Values of the Mann-Whitney U Distribution

n_1	n_2	$\alpha(2)$: 0.20 $\alpha(1)$: 0.10	0.10 0.05	0.05 0.025	0.02 0.01	0.01 0.005	0.005 0.0025	0.002 0.001	0.001 0.0005
17	17	183	193	202	212	219	225	232	238
	18	193	204	213	224	231	237	245	250
	19	203	214	224	235	242	249	257	263
	20	213	225	235	247	254	261	270	275
	21	223	236	246	258	266	273	282	288
	22	233	246	257	269	278	285	294	300
	23	244	257	268	281	289	297	306	313
	24	254	267	279	292	301	309	319	325
	25	264	278	290	303	313	321	331	338
	26	274	288	301	315	324	333	343	350
	27	284	299	312	326	336	345	355	363
	28	294	309	322	337	348	357	368	375
	29	304	320	333	349	359	369	380	388
	30	314	330	344	360	371	380	392	400
	31	324	341	355	371	382	392	404	413
	32	334	351	366	383	394	404	417	425
	33	344	362	377	394	406	416	429	438
	34	354	372	388	405	417	428	441	450
	35	365	383	399	417	429	440	453	462
	36	375	393	410	428	440	452	465	475
	37	385	404	420	439	452	464	478	487
	38	395	414	431	451	464	476	490	500
	39	405	425	442	462	475	487	502	512
17	40	415	435	453	473	487	499	514	525
18	18	204	215	225	236	243	250	258	263
	19	214	226	236	248	255	257	271	277
	20	225	237	248	260	268	275	284	287
	21	236	248	259	272	280	288	297	303
	22	246	260	271	284	292	300	310	316
	23	257	271	282	296	305	313	323	329
	24	268	282	294	308	317	325	336	343
	25	278	293	305	320	329	338	348	356
	26	289	304	317	332	341	350	361	369
	27	300	315	328	344	354	363	374	382
	28	310	326	340	355	366	376	387	395
	29	321	337	351	367	378	388	400	408
	30	331	348	363	379	390	401	413	421
	31	342	359	374	391	403	413	426	434
	32	353	370	386	403	415	426	438	447
	33	363	382	397	415	427	438	451	460
	34	374	393	409	427	439	451	464	473
	35	385	404	420	439	451	463	477	487
	36	395	415	432	451	464	475	490	500
	37	406	426	443	463	476	488	502	513
	38	416	437	454	475	488	500	515	526
	39	427	448	466	486	500	513	528	539
18	40	438	459	477	498	512	525	541	552
19	19	226	238	248	260	268	361	284	291
	20	237	250	261	273	281	289	298	304
	21	248	261	273	286	294	302	312	318
	22	259	273	285	298	307	315	325	332
	23	270	285	297	311	320	329	339	346
	24	282	296	309	323	333	342	352	360
19	25	293	308	321	336	346	355	366	373

TABLE B.11 (cont.) Critical Values of the Mann-Whitney U Distribution

n_1	n_2	$\alpha(2)$: 0.20 $\alpha(1)$: 0.10	0.10 0.05	0.05 0.025	0.02 0.01	0.01 0.005	0.005 0.0025	0.002 0.001	0.001 0.0005
19	26	304	320	333	348	359	368	379	387
	27	315	331	345	361	371	381	393	401
	28	326	343	357	373	384	394	406	415
	29	338	355	369	386	397	407	420	428
	30	349	366	381	398	410	421	433	442
	31	360	378	393	411	423	434	447	456
	32	371	390	405	423	436	447	460	469
	33	382	401	417	436	448	460	474	483
	34	393	413	429	448	461	473	487	497
	35	405	424	441	461	474	486	500	511
	36	416	436	453	473	487	499	514	524
	37	427	448	465	486	500	512	527	538
	38	438	459	477	498	512	525	541	552
	39	449	471	489	511	525	538	554	565
19	40	—	482	502	523	538	551	568	579
20	20	249	262	273	286	295	303	312	319
	21	260	274	286	299	308	317	326	333
	22	272	276	299	313	322	330	341	348
	23	284	299	311	326	335	344	355	362
	24	296	311	324	339	349	358	369	377
	25	307	323	337	352	362	372	383	391
	26	319	335	349	365	376	386	397	405
	27	331	348	362	378	389	399	411	420
	28	343	360	374	391	403	413	425	434
	29	354	372	387	404	416	427	440	449
	30	366	384	400	418	430	440	454	463
	31	378	396	412	431	443	454	468	477
	32	389	409	425	444	456	468	482	492
	33	401	421	438	457	470	482	496	506
	34	413	433	450	470	483	495	510	520
	35	425	445	463	483	497	509	524	534
	36	436	457	475	496	510	523	538	549
	37	448	469	488	509	523	536	552	563
	38	—	482	501	522	537	550	566	577
	39	—	—	513	535	550	564	580	592
20	40	—	—	526	548	563	577	594	606

The preceding values were derived, with permission of the publisher, from the tables of Milton (1964, *J. Amer. Statist. Assoc.* 59: 925–934).

Examples:

$$U_{0.05(2),5,8} = 34 \quad \text{and} \quad U_{0.05(1),10,8} = U_{0.05(1),8,10} = 60.$$

For the Mann-Whitney test involving n_1 and/or n_2 larger than those in this table, the normal approximation (Section 8.10) may be used. This approximation is excellent for two-tailed testing at $\alpha = 0.10$ or 0.05 (or one-tailed testing at $\alpha = 0.05$ or 0.025, respectively), especially if n_1 and n_2 are similar in magnitude. The approximation becomes progressively poorer as we consider more extreme significance levels.

TABLE B.12   Critical Values of the Wilcoxon T Distribution

n	$\alpha(2): 0.50$ $\alpha(1): 0.25$	0.20 0.10	0.10 0.05	0.05 0.025	0.02 0.01	0.01 0.005	0.005 0.0025	0.001 0.0005
4	2	0						
5	4	2	0					
6	6	3	2	0				
7	9	5	3	2	0			
8	12	8	5	3	1	0		
9	16	10	8	5	3	1	0	
10	20	14	10	8	5	3	1	
11	24	17	13	10	7	5	3	0
12	29	21	17	13	9	7	5	1
13	35	26	21	17	12	9	7	2
14	40	31	25	21	15	12	9	4
15	47	36	30	25	19	15	12	6
16	54	42	35	29	23	19	15	8
17	61	48	41	34	27	23	19	11
18	69	55	47	40	32	27	23	14
19	77	62	53	46	37	32	27	18
20	86	69	60	52	43	37	32	21
21	95	77	67	58	49	42	37	25
22	104	86	75	65	55	48	42	30
23	114	94	83	73	62	54	48	35
24	125	104	91	81	69	61	54	40
25	136	113	100	89	76	68	60	45
26	148	124	110	98	84	75	67	51
27	160	134	119	107	92	83	74	57
28	172	145	130	116	101	91	82	64
29	185	157	140	126	110	100	90	71
30	198	169	151	137	120	109	98	78
31	212	181	163	147	130	118	107	86
32	226	194	175	159	140	128	116	94
33	241	207	187	170	151	138	126	102
34	257	221	200	182	162	148	136	111
35	272	235	213	195	173	159	146	120
36	289	250	227	208	185	171	157	130
37	305	265	241	221	198	182	168	140
38	323	281	256	235	211	194	180	150
39	340	297	271	249	224	207	192	161
40	358	313	286	264	238	220	204	172
41	377	330	302	279	252	233	217	183
42	396	348	319	294	266	247	230	195
43	416	365	336	310	281	261	244	207
44	436	384	353	327	296	276	258	220
45	456	402	371	343	312	291	272	233
46	477	422	389	361	328	307	287	246
47	499	441	407	378	345	322	302	260
48	521	462	426	396	362	339	318	274
49	543	482	446	415	379	355	334	289
50	566	503	466	434	397	373	350	304
51	590	525	486	453	416	390	367	319
52	613	547	507	473	434	408	384	335
53	638	569	529	494	454	427	402	351
54	668	592	550	514	473	445	420	368
55	688	615	573	536	493	465	438	385
56	714	639	595	557	514	484	457	402
57	740	664	618	579	535	504	477	420
58	767	688	642	602	556	525	497	438
59	794	714	666	625	578	546	517	457
60	822	739	690	648	600	567	537	476

TABLE B.12 (cont.) Critical Values of the Wilcoxon T
Distribution

n	$\alpha(2)$: 0.50 $\alpha(1)$: 0.25	0.20 0.10	0.10 0.05	0.05 0.025	0.02 0.01	0.01 0.005	0.005 0.0025	0.001 0.0005
61	850	765	715	672	623	589	558	495
62	879	792	741	697	646	611	580	515
63	908	819	767	721	669	634	602	535
64	938	847	793	747	693	657	624	556
65	968	875	820	772	718	681	647	577
66	998	903	847	798	742	705	670	599
67	1029	932	875	825	768	729	694	621
68	1061	962	903	852	793	754	718	643
69	1093	992	931	879	819	779	742	666
70	1126	1022	960	907	846	805	767	689
71	1159	1053	990	936	873	831	792	712
72	1192	1084	1020	964	901	858	818	736
73	1226	1116	1050	994	928	884	844	761
74	1261	1148	1081	1023	957	912	871	786
75	1296	1181	1112	1053	986	940	898	811
76	1331	1214	1144	1084	1015	968	925	836
77	1367	1247	1176	1115	1044	997	953	862
78	1403	1282	1209	1147	1075	1026	981	889
79	1440	1316	1242	1179	1105	1056	1010	916
80	1478	1351	1276	1211	1136	1086	1039	943
81	1516	1387	1310	1244	1168	1116	1069	971
82	1554	1423	1345	1277	1200	1147	1099	999
83	1593	1459	1380	1311	1232	1178	1129	1028
84	1632	1496	1415	1345	1265	1210	1160	1057
85	1672	1533	1451	1380	1298	1242	1191	1086
86	1712	1571	1487	1415	1332	1275	1223	1116
87	1753	1609	1524	1451	1366	1308	1255	1146
88	1794	1648	1561	1487	1400	1342	1288	1177
89	1836	1688	1599	1523	1435	1376	1321	1208
90	1878	1727	1638	1560	1471	1410	1355	1240
91	1921	1767	1676	1597	1507	1445	1389	1271
92	1964	1808	1715	1635	1543	1480	1423	1304
93	2008	1849	1755	1674	1580	1516	1458	1337
94	2052	1891	1795	1712	1617	1552	1493	1370
95	2097	1933	1836	1752	1655	1589	1529	1404
96	2142	1976	1877	1791	1693	1626	1565	1438
97	2187	2019	1918	1832	1731	1664	1601	1472
98	2233	2062	1960	1872	1770	1702	1638	1507
99	2280	2106	2003	1913	1810	1740	1676	1543
100	2327	2151	2045	1955	1850	1779	1714	1578

Appendix Table B.12 is taken, with permission of the publisher, from the more extensive table of R. L McCornack (1965, *J. Amer. Statist. Assoc.* 60: 864–871).

Examples:

$$T_{0.05(2),16} = 29 \quad \text{and} \quad T_{0.01(1),62} = 646.$$

For performing the Wilcoxon paired-sample test when $n > 100$, we may use the normal approximation (Section 9.5). The accuracy of this approximation is expressed

below as: critical T from approximation − true critical T. In parentheses is the accuracy of the approximation with the continuity correction.

$\alpha(2)$:	0.50	0.20	0.10	0.05	0.02	0.01	0.005	0.001
n								
20	0(0)	1(1)	0(0)	0(−1)	−1(−1)	−2(−2)	−3(−3)	−5(−5)
40	1(1)	1(1)	1(1)	0(−1)	−2(−2)	−2(−3)	−3(−4)	−7(−8)
60	1(0)	1(1)	1(1)	0(0)	−2(−2)	−2(−3)	−4(−4)	−9(−9)
80	1(0)	1(1)	1(0)	0(−1)	−2(−2)	−4(−4)	−5(−5)	−10(−10)
100	1(1)	1(0)	1(1)	−1(−1)	−2(−3)	−4(−4)	−6(−7)	−11(−11)

The accuracy of the t approximation of Section 9.5 is here similarly expressed:

$\alpha(2)$:	0.50	0.20	0.10	0.05	0.02	0.01	0.005	0.001
n								
20	0(−1)	1(0)	0(0)	1(0)	1(1)	2(1)	2(1)	4(3)
40	1(0)	0(0)	1(0)	1(0)	1(1)	2(2)	3(3)	5(4)
60	0(−1)	0(0)	1(0)	1(1)	2(1)	3(2)	4(3)	6(5)
80	0(−1)	0(0)	0(0)	1(0)	2(2)	3(2)	4(4)	7(7)
100	0(0)	0(−1)	1(0)	0(0)	2(2)	3(3)	4(4)	8(7)

TABLE B.13   Critical Values of the Kruskal-Wallis H Distribution

n_1	n_2	n_3		α: 0.10	0.05	0.02	0.01	0.005	0.002	0.001
2	2	2		4.571						
3	2	1		4.286						
3	2	2		4.500	4.714					
3	3	1		4.571	5.143					
3	3	2		4.556	5.361	6.250				
3	3	3		4.622	5.600	6.489	(7.200)	7.200		
4	2	1		4.500						
4	2	2		4.458	5.333	6.000				
4	3	1		4.056	5.208					
4	3	2		4.511	5.444	6.144	6.444	7.000		
4	3	3		4.709	5.791	6.564	6.745	7.318	8.018	
4	4	1		4.167	4.967	(6.667)	6.667			
4	4	2		4.555	5.455	6.600	7.036	7.282	7.855	
4	4	3		4.545	5.598	6.712	7.144	7.598	8.227	8.909
4	4	4		4.654	5.692	6.962	7.654	8.000	8.654	9.269
5	2	1		4.200	5.000					
5	2	2		4.373	5.160	6.000	6.533			
5	3	1		4.018	4.960	6.044				
5	3	2		4.651	5.251	6.124	6.909	7.182		
5	3	3		4.533	5.648	6.533	7.079	7.636	8.048	8.727
5	4	1		3.987	4.985	6.431	6.955	7.364		
5	4	2		4.541	5.273	6.505	7.205	7.573	8.114	8.591
5	4	3		4.549	5.656	6.676	7.445	7.927	8.481	8.795
5	4	4		4.619	5.657	6.953	7.760	8.189	8.868	9.168
5	5	1		4.109	5.127	6.145	7.309	8.182		
5	5	2		4.623	5.338	6.446	7.338	8.131	6.446	7.338
5	5	3		4.545	5.705	6.866	7.578	8.316	8.809	9.521
5	5	4		4.523	5.666	7.000	7.823	8.523	9.163	9.606
5	5	5		4.940	5.780	7.220	8.000	8.780	9.620	9.920
6	1	1		—						
6	2	1		4.200	4.822					
6	2	2		4.545	5.345	6.182	6.982			
6	3	1		3.909	4.855	6.236				
6	3	2		4.682	5.348	6.227	6.970	7.515	8.182	
6	3	3		4.538	5.615	6.590	7.410	7.872	8.628	9.346
6	4	1		4.038	4.947	6.174	7.106	7.614		
6	4	2		4.494	5.340	6.571	7.340	7.846	8.494	8.827
6	4	3		4.604	5.610	6.725	7.500	8.033	8.918	9.170
6	4	4		4.595	5.681	6.900	7.795	8.381	9.167	9.861
6	5	1		4.128	4.990	6.138	7.182	8.077	8.515	
6	5	2		4.596	5.338	6.585	7.376	8.196	8.967	9.189
6	5	3		4.535	5.602	6.829	7.590	8.314	9.150	9.669
6	5	4		4.522	5.661	7.018	7.936	8.643	9.458	9.960
6	5	5		4.547	5.729	7.110	8.028	8.859	9.771	10.271
6	6	1		4.000	4.945	6.286	7.121	8.165	9.077	9.692
6	6	2		4.438	5.410	6.667	7.467	8.210	9.219	9.752
6	6	3		4.558	5.625	6.900	7.725	8.458	9.458	10.150
6	6	4		4.548	5.724	7.107	8.000	8.754	9.662	10.342
6	6	5		4.542	5.765	7.152	8.124	8.987	9.948	10.524
6	6	6		4.643	5.801	7.240	8.222	9.170	10.187	10.889
7	7	7		4.594	5.819	7.332	8.378	9.373	10.516	11.310
8	8	8		4.595	5.805	7.355	8.465	9.495	10.805	11.705
2	2	1	1	—						
2	2	2	1	5.357	5.679					
2	2	2	2	5.667	6.167	(6.667)	6.667			
3	1	1	1	—						
3	2	1	1	5.143						
3	2	2	1	5.556	5.833	6.500				

TABLE B.13 (cont.)   Critical Values of the Kruskal-Wallis H Distribution

n_1	n_2	n_3		α: 0.10	0.05	0.02	0.01	0.005	0.002	0.001
3	2	2	2	5.544	6.333	6.978	7.133	7.533		
3	3	1	1	5.333	6.333					
3	3	2	1	5.689	6.244	6.689	7.200	7.400		
3	3	2	2	5.745	6.527	7.182	7.636	7.873	8.018	8.455
3	3	3	1	5.655	6.600	7.109	7.400	8.055	8.345	
3	3	3	2	5.879	6.727	7.636	8.105	8.379	8.803	9.030
3	3	3	3	6.026	7.000	7.872	8.538	8.897	9.462	9.513
4	1	1	1	—						
4	2	1	1	5.250	5.833					
4	2	2	1	5.533	6.133	6.667	7.000			
4	2	2	2	5.755	6.545	7.091	7.391	7.964	8.291	
4	3	1	1	5.067	6.178	6.711	7.067			
4	3	2	1	5.591	6.309	7.018	7.455	7.773	8.182	
4	3	2	2	5.750	6.621	7.530	7.871	8.273	8.689	8.909
4	3	3	1	5.589	6.545	7.485	7.758	8.212	8.697	9.182
4	3	3	2	5.872	6.795	7.763	8.333	8.718	9.167	9.455
4	3	3	3	6.016	6.984	7.995	8.659	9.253	9.709	10.016
4	4	1	1	5.182	5.945	7.091	7.909	7.909		
4	4	2	1	5.568	6.386	7.364	7.886	8.341	8.591	8.909
4	4	2	2	5.808	6.731	7.750	8.346	8.692	9.269	9.462
4	4	3	1	5.692	6.635	7.660	8.231	8.583	9.038	9.327
4	4	3	2	5.901	6.874	7.951	8.621	9.165	9.615	9.945
4	4	3	3	6.019	7.038	8.181	8.876	9.495	10.105	10.467
4	4	4	1	5.564	6.725	7.879	8.588	9.000	9.478	9.758
4	4	4	2	5.914	6.957	8.157	8.871	9.486	10.043	10.429
4	4	4	3	6.042	7.142	8.350	9.075	9.742	10.542	10.929
4	4	4	4	6.088	7.235	8.515	9.287	9.971	10.809	11.338
2	1	1	1 1	—						
2	2	1	1 1	5.786						
2	2	2	1 1	6.250	6.750					
2	2	2	2 1	6.600	7.133	(7.533)	7.533			
2	2	2	2 2	6.982	7.418	8.073	8.291	(8.727)	8.727	
3	1	1	1 1	—						
3	2	1	1 1	6.139	6.583					
3	2	2	1 1	6.511	6.800	7.400	7.600			
3	2	2	2 1	6.709	7.309	7.836	8.127	8.327	8.618	
3	2	2	2 2	6.955	7.682	8.303	8.682	8.985	9.273	9.364
3	3	1	1 1	6.311	7.111	7.467				
3	3	2	1 1	6.600	7.200	7.892	8.073	8.345		
3	3	2	2 1	6.788	7.591	8.258	8.576	8.924	9.167	9.303
3	3	2	2 2	7.026	7.910	8.667	9.115	9.474	9.769	10.026
3	3	3	1 1	6.788	7.576	8.242	8.424	8.848	(9.455)	9.455
3	3	3	2 1	6.910	7.769	8.590	9.051	9.410	9.769	9.974
3	3	3	2 2	7.121	8.044	9.011	9.505	9.890	10.330	10.637
3	3	3	3 1	7.077	8.000	8.879	9.451	9.846	10.286	10.549
3	3	3	3 2	7.210	8.200	9.267	9.876	10.333	10.838	11.171
3	3	3	3 3	7.333	8.333	9.467	10.200	10.733	10.267	11.667

The above values of H were determined from *Selected Tables in Mathematical Statistics*, Volume III, pp. 320–384, by permission of the American Mathematical Society. ©1975 by the American Mathematical Society (Iman, Quade, and Alexander 1975.)

Examples:

$$H_{0.05,4,3,2} = 5.444 \quad \text{and} \quad H_{0.01,4,4,5} = H_{0.01,5,4,4} = 7.760.$$

TABLE B.14 Critical Values of the Friedman χ_r^2 Distribution

a (n)	b (M)*	α: 0.50	0.20	0.10	0.05	0.02	0.01	0.005	0.002	0.001
3	2	3.000	4.000							
3	3	2.667	4.667	(6.000)	6.000					
3	4	2.000	4.500	6.000	6.500	(8.000)	(8.000)	8.000		
3	5	2.800	3.600	5.200	6.400	(8.400)	8.400	(10.000)	(10.000)	10.000
3	6	2.330	4.000	5.330	7.000	8.330	9.000	(10.330)	10.330	12.000
3	7	2.000	3.714	5.429	7.143	8.000	8.857	10.286	11.143	12.286
3	8	2.250	4.000	5.250	6.250	7.750	9.000	9.750	12.000	12.250
3	9	2.000	3.556	5.556	6.222	8.000	9.556	10.667	11.556	12.667
3	10	1.800	3.800	5.000	6.200	7.800	9.600	10.400	12.200	12.600
3	11	4.636	3.818	4.909	6.545	7.818	9.455	10.364	11.636	13.273
3	12	1.500	3.500	5.167	6.167	8.000	9.500	10.167	12.167	12.500
3	13	1.846	3.846	4.769	6.000	8.000	9.385	10.308	11.538	12.923
3	14	1.714	3.571	5.143	6.143	8.143	9.000	10.429	12.000	13.286
3	15	1.733	3.600	4.933	6.400	8.133	8.933	10.000	12.133	12.933
4	2	3.600	5.400	(6.000)	6.000					
4	3	3.400	5.400	6.600	7.400	8.200	(9.000)	(9.000)	9.000	
4	4	3.000	4.800	6.300	7.800	8.400	9.600	(10.200)	10.200	11.100
4	5	3.000	5.160	6.360	7.800	9.240	9.960	10.920	11.640	12.600
4	6	3.000	4.800	6.400	7.600	9.400	10.200	11.400	12.200	12.800
4	7	2.829	4.886	6.429	7.800	9.343	10.371	11.400	12.771	13.800
4	8	2.550	4.800	6.300	7.650	9.450	10.350	11.850	12.900	13.800
4	9			6.467	7.800	9.133	10.867	12.067		14.467
4	10			6.360	7.800	9.120	10.800	12.000		14.640
4	11			6.382	7.909	9.327	11.073	12.273		14.891
4	12			6.400	7.900	9.200	11.100	12.300		15.000
4	13			6.415	7.985	7.369	11.123	12.323		15.277
4	14			6.343	7.886	9.343	11.143	12.514		15.257
4	15			6.440	8.040	9.400	11.240	12.520		15.400
5	2			7.200	7.600	8.000	8.000			
5	3			7.467	8.533	9.600	10.133	10.667		11.467
5	4			7.600	8.800	9.800	11.200	12.000		13.200
5	5			7.680	8.960	10.240	11.680	12.480		14.400
5	6			7.733	9.067	10.400	11.867	13.067		15.200
5	7			7.771	9.143	10.514	12.114	13.257		15.657
5	8			7.800	9.300	10.600	12.300	13.500		16.000
5	9			7.733	9.244	10.667	12.444	13.689		16.356
5	10			7.760	9.280	10.720	12.480	13.840		16.480
6	2			8.286	9.143	9.429	9.714	10.000		
6	3			8.714	9.857	10.810	11.762	12.524		13.286
6	4			9.000	10.286	11.429	12.714	13.571		15.286
6	5			9.000	10.486	11.743	13.229	14.257		16.429
6	6			9.048	10.571	12.000	13.619	14.762		17.048
6	7			9.122	10.674	12.061	13.857	15.000		17.612
6	8			9.143	10.714	12.214	14.000	15.286		18.000
6	9			9.127	10.778	12.302	14.143	15.476		18.270
6	10			9.143	10.800	12.343	14.299	15.600		18.514

*For Kendall's coefficient of concordance, W, use the column headings in parentheses.

For $a = 3$, and for $a = 4$, with $b \leq 8$, the above values of χ_r^2 were determined from D. B. Owen, *Handbook of Statistical Tables*, © 1962, U.S. Department of Energy, Published by Addison-Wesley, Reading, Massachusetts, Table 14.1, pp. 408–409;

reprinted with permission. Critical values for a with $b > 8$, and for $a = 5$ and 6, were taken, with permission of the publisher, from Martin, LeBlanc, and Toan (1993).

Examples:

$$\left(\chi_r^2\right)_{0.05,3,6} = 7.000 \qquad \text{and} \qquad W_{0.05,4,3} = 7.400.$$

TABLE B.15 Critical Values of Q for Nonparametric Multiple Comparison Testing

k	α: 0.50	0.20	0.10	0.05	0.02	0.01	0.005	0.002	0.001
2	0.674	1.282	1.645	1.960	2.327	2.576	2.807	3.091	3.291
3	1.383	1.834	2.128	2.394	2.713	2.936	3.144	3.403	3.588
4	1.732	2.128	2.394	2.639	2.936	3.144	3.342	3.588	3.765
5	1.960	2.327	2.576	2.807	3.091	3.291	3.481	3.719	3.891
6	2.128	2.475	2.713	2.936	3.209	3.403	3.588	3.820	3.988
7	2.261	2.593	2.823	3.038	3.304	3.494	3.675	3.902	4.067
8	2.369	2.690	2.914	3.124	3.384	3.570	3.748	3.972	4.134
9	2.461	2.773	2.992	3.197	3.453	3.635	3.810	4.031	4.191
10	2.540	2.845	3.059	3.261	3.512	3.692	3.865	4.083	4.241
11	2.609	2.908	3.119	3.317	3.565	3.743	3.914	4.129	4.286
12	2.671	2.965	3.172	3.368	3.613	3.789	3.957	4.171	4.326
13	2.726	3.016	3.220	3.414	3.656	3.830	3.997	4.209	4.363
14	2.777	3.062	3.264	3.456	3.695	3.868	4.034	4.244	4.397
15	2.823	3.105	3.304	3.494	3.731	3.902	4.067	4.276	4.428
16	2.866	3.144	3.342	3.529	3.765	3.935	4.098	4.305	4.456
17	2.905	3.181	3.376	3.562	3.796	3.965	4.127	4.333	4.483
18	2.942	3.215	3.409	3.593	3.825	3.993	4.154	4.359	4.508
19	2.976	3.246	3.439	3.622	3.852	4.019	4.179	4.383	4.532
20	3.008	3.276	3.467	3.649	3.878	4.044	4.203	4.406	4.554
21	3.038	3.304	3.494	3.675	3.902	4.067	4.226	4.428	4.575
22	3.067	3.331	3.519	3.699	3.925	4.089	4.247	4.448	4.595
23	3.094	3.356	3.543	3.722	3.947	4.110	4.268	4.468	4.614
24	3.120	3.380	3.566	3.744	3.968	4.130	4.287	4.486	4.632
25	3.144	3.403	3.588	3.765	3.988	4.149	4.305	4.504	4.649

This table was prepared using Equation 26.2.23 of Zelen and Severo (1964) which determines Q_α such that $P(Q_\alpha) \leq \alpha(1)/[k(k-1)]$, where Q_α is a normal deviate.

TABLE B.16   Critical Values of Q' for Nonparametric Multiple
Comparison Testing with a Control

k	$\alpha(2)$: 0.50 $\alpha(1)$: 0.25	0.20 0.10	0.10 0.05	0.05 0.025	0.02 0.01	0.01 0.005	0.005 0.0025	0.002 0.001	0.001 0.0005
2	0.674	1.282	1.645	1.960	2.327	2.576	2.807	3.091	3.291
3	1.150	1.645	1.960	2.242	2.576	2.807	3.024	3.291	3.481
4	1.383	1.834	2.128	2.394	2.713	2.936	3.144	3.403	3.588
5	1.534	1.960	2.242	2.498	2.807	3.024	3.227	3.481	3.662
6	1.645	2.054	2.327	2.576	2.879	3.091	3.291	3.540	3.719
7	1.732	2.128	2.394	2.639	2.936	3.144	3.342	3.588	3.765
8	1.803	2.190	2.450	2.690	2.983	3.189	3.384	3.628	3.803
9	1.863	2.242	2.498	2.735	3.024	3.227	3.421	3.662	3.836
10	1.915	2.287	2.540	2.773	3.059	3.261	3.453	3.692	3.865
11	1.960	2.327	2.576	2.807	3.091	3.291	3.481	3.719	3.891
12	2.001	2.362	2.609	2.838	3.119	3.317	3.506	3.743	3.914
13	2.037	2.394	2.639	2.866	3.144	3.342	3.529	3.765	3.935
14	2.070	2.424	2.666	2.891	3.168	3.364	3.551	3.785	3.954
15	2.101	2.450	2.690	2.914	3.189	3.384	3.570	3.803	3.972
16	2.128	2.475	2.713	2.936	3.209	3.403	3.588	3.820	3.988
17	2.154	2.498	2.735	2.955	3.227	3.421	3.605	3.836	4.003
18	2.178	2.520	2.755	2.974	3.245	3.437	3.621	3.851	4.018
19	2.201	2.540	2.773	2.992	3.261	3.453	3.635	3.865	4.031
20	2.222	2.558	2.791	3.008	3.276	3.467	3.649	3.878	4.044
21	2.242	2.576	2.807	3.024	3.291	3.481	3.662	3.891	4.056
22	2.261	2.593	2.823	3.038	3.304	3.494	3.675	3.902	4.067
23	2.278	2.609	2.838	3.052	3.317	3.506	3.687	3.914	4.078
24	2.295	2.624	2.852	3.066	3.330	3.518	3.698	3.924	4.088
25	2.311	2.639	2.866	3.078	3.342	3.529	3.709	3.935	4.098

This table was prepared using Equation 26.2.23 of Zelen and Severo (1964), which determines Q'_α such that $P(Q'_\alpha) \leq \alpha(2)/[(k-1)] = \alpha(1)/(k-1)$, where Q'_α is a normal deviate.

The values in Table B.17 were computed using Equation 18.4 and Table B.3.

Examples:

$$r_{0.05(2),25} = 0.381 \quad \text{and} \quad r_{0.01(1),30} = 0.409.$$

If one requires a critical value for degrees of freedom not on this table, the critical value for the next lower degrees of freedom in the table may be conservatively used. Or, linear or harmonic interpolation may be used. If $v > 1000$, then use harmonic interpolation, setting the critical value equal to zero for $v = \infty$.

TABLE B.17   Critical Values of the Correlation Coefficient, r

ν	$\alpha(2)$: 0.50 $\alpha(1)$: 0.25	0.20 0.10	0.10 0.05	0.05 0.025	0.02 0.01	0.01 0.005	0.005 0.0025	0.002 0.001	0.001 0.0005
1	0.707	0.951	0.988	0.997	1.000	1.000	1.000	1.000	1.000
2	0.500	0.800	0.900	0.950	0.980	0.990	0.995	0.998	0.999
3	0.404	0.687	0.805	0.878	0.934	0.959	0.974	0.986	0.991
4	0.347	0.608	0.729	0.811	0.882	0.917	0.942	0.963	0.974
5	0.309	0.551	0.669	0.755	0.833	0.875	0.906	0.935	0.951
6	0.281	0.507	0.621	0.707	0.789	0.834	0.870	0.905	0.925
7	0.260	0.472	0.582	0.666	0.750	0.798	0.836	0.875	0.898
8	0.242	0.443	0.549	0.632	0.715	0.765	0.805	0.847	0.872
9	0.228	0.419	0.521	0.602	0.685	0.735	0.776	0.820	0.847
10	0.216	0.398	0.497	0.576	0.658	0.708	0.750	0.795	0.823
11	0.206	0.380	0.476	0.553	0.634	0.684	0.726	0.772	0.801
12	0.197	0.365	0.457	0.532	0.612	0.661	0.703	0.750	0.780
13	0.189	0.351	0.441	0.514	0.592	0.641	0.683	0.730	0.760
14	0.182	0.338	0.426	0.497	0.574	0.623	0.664	0.711	0.742
15	0.176	0.327	0.412	0.482	0.558	0.606	0.647	0.694	0.725
16	0.170	0.317	0.400	0.468	0.542	0.590	0.631	0.678	0.708
17	0.165	0.308	0.389	0.456	0.529	0.575	0.616	0.662	0.693
18	0.160	0.299	0.378	0.444	0.515	0.561	0.602	0.648	0.679
19	0.156	0.291	0.369	0.433	0.503	0.549	0.589	0.635	0.665
20	0.152	0.284	0.360	0.423	0.492	0.537	0.576	0.622	0.652
21	0.148	0.277	0.352	0.413	0.482	0.526	0.565	0.610	0.640
22	0.145	0.271	0.344	0.404	0.472	0.515	0.554	0.599	0.629
23	0.141	0.265	0.337	0.396	0.462	0.505	0.543	0.588	0.618
24	0.138	0.260	0.330	0.388	0.453	0.496	0.534	0.578	0.607
25	0.136	0.255	0.323	0.381	0.445	0.487	0.524	0.568	0.597
26	0.133	0.250	0.317	0.374	0.437	0.479	0.515	0.559	0.588
27	0.131	0.245	0.311	0.367	0.430	0.471	0.507	0.550	0.579
28	0.128	0.241	0.306	0.361	0.423	0.463	0.499	0.541	0.570
29	0.126	0.237	0.301	0.355	0.416	0.456	0.491	0.533	0.562
30	0.124	0.233	0.296	0.349	0.409	0.449	0.484	0.526	0.554
31	0.122	0.229	0.291	0.344	0.403	0.442	0.477	0.518	0.546
32	0.120	0.225	0.287	0.339	0.397	0.436	0.470	0.511	0.539
33	0.118	0.222	0.283	0.334	0.392	0.430	0.464	0.504	0.532
34	0.116	0.219	0.279	0.329	0.386	0.424	0.458	0.498	0.525
35	0.115	0.216	0.275	0.325	0.381	0.418	0.452	0.492	0.519
36	0.113	0.213	0.271	0.320	0.376	0.413	0.446	0.486	0.513
37	0.111	0.210	0.267	0.316	0.371	0.408	0.441	0.480	0.507
38	0.110	0.207	0.264	0.312	0.367	0.403	0.435	0.474	0.501
39	0.108	0.204	0.261	0.308	0.362	0.398	0.430	0.469	0.495
40	0.107	0.202	0.257	0.304	0.358	0.393	0.425	0.463	0.490
41	0.106	0.199	0.254	0.301	0.354	0.389	0.420	0.458	0.484
42	0.104	0.197	0.251	0.297	0.350	0.384	0.416	0.453	0.479
43	0.103	0.195	0.248	0.294	0.346	0.380	0.411	0.449	0.474
44	0.102	0.192	0.246	0.291	0.342	0.376	0.407	0.444	0.469
45	0.101	0.190	0.243	0.288	0.338	0.372	0.403	0.439	0.465
46	0.100	0.188	0.240	0.285	0.335	0.368	0.399	0.435	0.460
47	0.099	0.186	0.238	0.282	0.331	0.365	0.395	0.431	0.456
48	0.098	0.184	0.235	0.279	0.328	0.361	0.391	0.427	0.451
49	0.097	0.182	0.233	0.276	0.325	0.358	0.387	0.423	0.447
50	0.096	0.181	0.231	0.273	0.322	0.354	0.384	0.419	0.443
52	0.094	0.177	0.226	0.268	0.316	0.348	0.377	0.411	0.435
54	0.092	0.174	0.222	0.263	0.310	0.341	0.370	0.404	0.428
56	0.090	0.171	0.218	0.259	0.305	0.336	0.364	0.398	0.421
58	0.089	0.168	0.214	0.254	0.300	0.330	0.358	0.391	0.414
60	0.087	0.165	0.211	0.250	0.295	0.325	0.352	0.385	0.408

TABLE B.17 (cont.) Critical Values of the Correlation Coefficient, r

ν	$\alpha(2)$: 0.50 $\alpha(1)$: 0.25	0.20 0.10	0.10 0.05	0.05 0.025	0.02 0.01	0.01 0.005	0.005 0.0025	0.002 0.001	0.001 0.0005
62	0.086	0.162	0.207	0.246	0.290	0.320	0.347	0.379	0.402
64	0.084	0.160	0.204	0.242	0.286	0.315	0.342	0.374	0.396
66	0.083	0.157	0.201	0.239	0.282	0.310	0.337	0.368	0.390
68	0.082	0.155	0.198	0.235	0.278	0.306	0.332	0.363	0.385
70	0.081	0.153	0.195	0.232	0.274	0.302	0.327	0.358	0.380
72	0.080	0.151	0.193	0.229	0.270	0.298	0.323	0.354	0.375
74	0.079	0.149	0.190	0.226	0.266	0.294	0.319	0.349	0.370
76	0.078	0.147	0.188	0.223	0.263	0.290	0.315	0.345	0.365
78	0.077	0.145	0.185	0.220	0.260	0.286	0.311	0.340	0.361
80	0.076	0.143	0.183	0.217	0.257	0.283	0.307	0.336	0.357
82	0.075	0.141	0.181	0.215	0.253	0.280	0.304	0.333	0.328
84	0.074	0.140	0.179	0.212	0.251	0.276	0.300	0.329	0.349
86	0.073	0.138	0.177	0.210	0.248	0.273	0.297	0.325	0.345
88	0.072	0.136	0.174	0.207	0.245	0.270	0.293	0.321	0.341
90	0.071	0.135	0.173	0.205	0.242	0.267	0.290	0.318	0.338
92	0.070	0.133	0.171	0.203	0.240	0.264	0.287	0.315	0.334
94	0.070	0.132	0.169	0.201	0.237	0.262	0.284	0.312	0.331
96	0.069	0.131	0.167	0.199	0.235	0.259	0.281	0.308	0.327
98	0.068	0.129	0.165	0.197	0.232	0.256	0.279	0.305	0.324
100	0.068	0.128	0.164	0.195	0.230	0.254	0.276	0.303	0.321
105	0.066	0.125	0.160	0.190	0.225	0.248	0.270	0.296	0.314
110	0.064	0.122	0.156	0.186	0.220	0.242	0.264	0.289	0.307
115	0.063	0.119	0.153	0.182	0.215	0.237	0.258	0.283	0.300
120	0.062	0.117	0.150	0.178	0.210	0.232	0.253	0.277	0.294
125	0.060	0.114	0.147	0.174	0.206	0.228	0.248	0.272	0.289
130	0.059	0.112	0.144	0.171	0.202	0.223	0.243	0.267	0.283
135	0.058	0.110	0.141	0.168	0.199	0.219	0.239	0.262	0.278
140	0.057	0.108	0.139	0.165	0.195	0.215	0.234	0.257	0.273
145	0.056	0.106	0.136	0.162	0.192	0.212	0.230	0.253	0.269
150	0.055	0.105	0.134	0.159	0.189	0.208	0.227	0.249	0.264
160	0.053	0.101	0.130	0.154	0.183	0.202	0.220	0.241	0.256
170	0.052	0.098	0.126	0.150	0.177	0.196	0.213	0.234	0.249
180	0.050	0.095	0.122	0.145	0.172	0.190	0.207	0.228	0.242
190	0.049	0.093	0.119	0.142	0.168	0.185	0.202	0.222	0.236
200	0.048	0.091	0.116	0.138	0.164	0.181	0.197	0.216	0.230
250	0.043	0.081	0.104	0.124	0.146	0.162	0.176	0.194	0.206
300	0.039	0.074	0.095	0.113	0.134	0.148	0.161	0.177	0.188
350	0.036	0.068	0.088	0.105	0.124	0.137	0.149	0.164	0.175
400	0.034	0.064	0.082	0.098	0.116	0.128	0.140	0.154	0.164
450	0.032	0.060	0.077	0.092	0.109	0.121	0.132	0.145	0.154
500	0.030	0.057	0.074	0.088	0.104	0.115	0.125	0.138	0.146
600	0.028	0.052	0.067	0.080	0.095	0.105	0.114	0.126	0.134
700	0.026	0.048	0.062	0.074	0.088	0.097	0.106	0.116	0.124
800	0.024	0.045	0.058	0.069	0.082	0.091	0.099	0.109	0.116
900	0.022	0.043	0.055	0.065	0.077	0.086	0.093	0.103	0.109
1000	0.021	0.041	0.052	0.062	0.073	0.081	0.089	0.098	0.104

TABLE B.18 Fisher's *z* Transformation for Correlation Coefficients, *r*

r	0	1	2	3	4	5	6	7	8	9	r
0.000	0.0000	0.0010	0.0020	0.0030	0.0040	0.0050	0.0060	0.0070	0.0080	0.0090	0.000
0.010	0.0100	0.0110	0.0120	0.0130	0.0140	0.0150	0.0160	0.0170	0.0180	0.0190	0.010
0.020	0.0200	0.0210	0.0220	0.0230	0.0240	0.0250	0.0260	0.0270	0.0280	0.0290	0.020
0.030	0.0300	0.0310	0.0320	0.0330	0.0340	0.0350	0.0360	0.0370	0.0380	0.0390	0.030
0.040	0.0400	0.0410	0.0420	0.0430	0.0440	0.0450	0.0460	0.0470	0.0480	0.0490	0.040
0.050	0.0500	0.0510	0.0520	0.0530	0.0541	0.0551	0.0561	0.0571	0.0581	0.0591	0.050
0.060	0.0601	0.0611	0.0621	0.0631	0.0641	0.0651	0.0661	0.0671	0.0681	0.0691	0.060
0.070	0.0701	0.0711	0.0721	0.0731	0.0741	0.0751	0.0761	0.0772	0.0782	0.0792	0.070
0.080	0.0802	0.0812	0.0822	0.0832	0.0842	0.0852	0.0862	0.0872	0.0882	0.0892	0.080
0.090	0.0902	0.0913	0.0923	0.0933	0.0943	0.0953	0.0963	0.0973	0.0983	0.0993	0.090
0.100	0.1003	0.1013	0.1024	0.1034	0.1044	0.1054	0.1064	0.1074	0.1084	0.1094	0.100
0.110	0.1104	0.1115	0.1125	0.1135	0.1145	0.1155	0.1165	0.1175	0.1186	0.1196	0.110
0.120	0.1206	0.1216	0.1226	0.1236	0.1246	0.1257	0.1267	0.1277	0.1287	0.1297	0.120
0.130	0.1307	0.1318	0.1328	0.1338	0.1348	0.1358	0.1368	0.1379	0.1389	0.1399	0.130
0.140	0.1409	0.1419	0.1430	0.1440	0.1450	0.1460	0.1470	0.1481	0.1491	0.1501	0.140
0.150	0.1511	0.1522	0.1532	0.1542	0.1552	0.1563	0.1573	0.1583	0.1593	0.1604	0.150
0.160	0.1614	0.1624	0.1634	0.1645	0.1655	0.1665	0.1675	0.1686	0.1696	0.1706	0.160
0.170	0.1717	0.1727	0.1737	0.1748	0.1758	0.1768	0.1779	0.1789	0.1799	0.1809	0.170
0.180	0.1820	0.1830	0.1840	0.1851	0.1861	0.1872	0.1882	0.1892	0.1903	0.1913	0.180
0.190	0.1923	0.1934	0.1944	0.1955	0.1965	0.1975	0.1986	0.1996	0.2006	0.2017	0.190
0.200	0.2027	0.2038	0.2048	0.2059	0.2069	0.2079	0.2090	0.2100	0.2111	0.2121	0.200
0.210	0.2132	0.2142	0.2153	0.2163	0.2174	0.2184	0.2195	0.2205	0.2216	0.2226	0.210
0.220	0.2237	0.2247	0.2258	0.2268	0.2279	0.2289	0.2300	0.2310	0.2321	0.2331	0.220
0.230	0.2342	0.2352	0.2363	0.2374	0.2384	0.2395	0.2405	0.2416	0.2427	0.2437	0.230
0.240	0.2448	0.2458	0.2469	0.2480	0.2490	0.2501	0.2512	0.2522	0.2533	0.2543	0.240
0.250	0.2554	0.2565	0.2575	0.2586	0.2597	0.2608	0.2618	0.2629	0.2640	0.2650	0.250
0.260	0.2661	0.2672	0.2683	0.2693	0.2704	0.2715	0.2726	0.2736	0.2747	0.2758	0.260
0.270	0.2769	0.2779	0.2790	0.2801	0.2812	0.2823	0.2833	0.2844	0.2855	0.2866	0.270
0.280	0.2877	0.2888	0.2899	0.2909	0.2920	0.2931	0.2942	0.2953	0.2964	0.2975	0.280
0.290	0.2986	0.2997	0.3008	0.3018	0.3029	0.3040	0.3051	0.3062	0.3073	0.3084	0.290
0.300	0.3095	0.3106	0.3117	0.3128	0.3139	0.3150	0.3161	0.3172	0.3183	0.3194	0.300
0.310	0.3205	0.3217	0.3228	0.3239	0.3250	0.3261	0.3272	0.3283	0.3294	0.3305	0.310
0.320	0.3316	0.3328	0.3339	0.3350	0.3361	0.3372	0.3383	0.3395	0.3406	0.3417	0.320
0.330	0.3428	0.3440	0.3451	0.3462	0.3473	0.3484	0.3496	0.3507	0.3518	0.3530	0.330
0.340	0.3541	0.3552	0.3564	0.3575	0.3586	0.3598	0.3609	0.3620	0.3632	0.3643	0.340
0.350	0.3654	0.3666	0.3677	0.3689	0.3700	0.3712	0.3723	0.3734	0.3746	0.3757	0.350
0.360	0.3769	0.3780	0.3792	0.3803	0.3815	0.3826	0.3838	0.3850	0.3861	0.3873	0.360
0.370	0.3884	0.3896	0.3907	0.3919	0.3931	0.3942	0.3954	0.3966	0.3977	0.3989	0.370
0.380	0.4001	0.4012	0.4024	0.4036	0.4047	0.4059	0.4071	0.4083	0.4094	0.4106	0.380
0.390	0.4118	0.4130	0.4142	0.4153	0.4165	0.4177	0.4189	0.4201	0.4213	0.4225	0.390
0.400	0.4236	0.4248	0.4260	0.4272	0.4284	0.4296	0.4308	0.4320	0.4332	0.4344	0.400
0.410	0.4356	0.4368	0.4380	0.4392	0.4404	0.4416	0.4428	0.4441	0.4453	0.4465	0.410
0.420	0.4477	0.4489	0.4501	0.4513	0.4526	0.4538	0.4550	0.4562	0.4574	0.4587	0.420
0.430	0.4599	0.4611	0.4624	0.4636	0.4648	0.4660	0.4673	0.4685	0.4698	0.4710	0.430
0.440	0.4722	0.4735	0.4747	0.4760	0.4772	0.4784	0.4797	0.4809	0.4822	0.4834	0.440
0.450	0.4847	0.4860	0.4872	0.4885	0.4897	0.4910	0.4922	0.4935	0.4948	0.4960	0.450
0.460	0.4973	0.4986	0.4999	0.5011	0.5024	0.5037	0.5049	0.5062	0.5075	0.5088	0.460
0.470	0.5101	0.5114	0.5126	0.5139	0.5152	0.5165	0.5178	0.5191	0.5204	0.5217	0.470
0.480	0.5230	0.5243	0.5256	0.5269	0.5282	0.5295	0.5308	0.5321	0.5334	0.5347	0.480
0.490	0.5361	0.5374	0.5387	0.5400	0.5413	0.5427	0.5440	0.5453	0.5466	0.5480	0.490
0.500	0.5493	0.5506	0.5520	0.5533	0.5547	0.5560	0.5573	0.5587	0.5600	0.5614	0.500
0.510	0.5627	0.5641	0.5654	0.5668	0.5681	0.5695	0.5709	0.5722	0.5736	0.5750	0.510
0.520	0.5763	0.5777	0.5791	0.5805	0.5818	0.5832	0.5846	0.5860	0.5874	0.5888	0.520
0.530	0.5901	0.5915	0.5929	0.5943	0.5957	0.5971	0.5985	0.5999	0.6013	0.6027	0.530
0.540	0.6042	0.6056	0.6070	0.6084	0.6098	0.6112	0.6127	0.6141	0.6155	0.6169	0.540

TABLE B.18 (cont.)   Fisher's *z* Transformation for Correlation Coefficients, *r*

r	0	1	2	3	4	5	6	7	8	9	*r*
0.550	0.6184	0.6198	0.6213	0.6227	0.6241	0.6256	0.6270	0.6285	0.6299	0.6314	0.550
0.560	0.6328	0.6343	0.6358	0.6372	0.6387	0.6401	0.6416	0.6431	0.6446	0.6460	0.560
0.570	0.6475	0.6490	0.6505	0.6520	0.6535	0.6550	0.6565	0.6580	0.6595	0.6610	0.570
0.580	0.6625	0.6640	0.6655	0.6670	0.6685	0.6700	0.6716	0.6731	0.6746	0.6761	0.580
0.590	0.6777	0.6792	0.6807	0.6823	0.6838	0.6854	0.6869	0.6885	0.6900	0.6916	0.590
0.600	0.6931	0.6947	0.6963	0.6978	0.6994	0.7010	0.7026	0.7042	0.7057	0.7073	0.600
0.610	0.7089	0.7105	0.7121	0.7137	0.7153	0.7169	0.7185	0.7201	0.7218	0.7234	0.610
0.620	0.7250	0.7266	0.7283	0.7299	0.7315	0.7332	0.7348	0.7365	0.7381	0.7398	0.620
0.630	0.7414	0.7431	0.7447	0.7464	0.7481	0.7497	0.7514	0.7531	0.7548	0.7565	0.630
0.640	0.7582	0.7599	0.7616	0.7633	0.7650	0.7667	0.7684	0.7701	0.7718	0.7736	0.640
0.650	0.7753	0.7770	0.7788	0.7805	0.7823	0.7840	0.7858	0.7875	0.7893	0.7910	0.650
0.660	0.7928	0.7946	0.7964	0.7981	0.7999	0.8017	0.8035	0.8053	0.8071	0.8089	0.660
0.670	0.8107	0.8126	0.8144	0.8162	0.8180	0.8199	0.8217	0.8236	0.8254	0.8273	0.670
0.680	0.8291	0.8310	0.8328	0.8347	0.8366	0.8385	0.8404	0.8422	0.8441	0.8460	0.680
0.690	0.8480	0.8499	0.8518	0.8537	0.8556	0.8576	0.8595	0.8614	0.8634	0.8653	0.690
0.700	0.8673	0.8693	0.8712	0.8732	0.8752	0.8772	0.8792	0.8812	0.8832	0.8852	0.700
0.710	0.8872	0.8892	0.8912	0.8933	0.8953	0.8973	0.8994	0.9014	0.9035	0.9056	0.710
0.720	0.9076	0.9097	0.9118	0.9139	0.9160	0.9181	0.9202	0.9223	0.9245	0.9266	0.720
0.730	0.9287	0.9309	0.9330	0.9352	0.9373	0.9395	0.9417	0.9439	0.9461	0.9483	0.730
0.740	0.9505	0.9527	0.9549	0.9571	0.9594	0.9616	0.9639	0.9661	0.9684	0.9707	0.740
0.750	0.9730	0.9752	0.9775	0.9798	0.9822	0.9845	0.9868	0.9891	0.9915	0.9938	0.750
0.760	0.9962	0.9986	1.0010	1.0034	1.0058	1.0082	1.0106	1.0130	1.0154	1.0179	0.760
0.770	1.0203	1.0228	1.0253	1.0277	1.0302	1.0327	1.0352	1.0378	1.0403	1.0428	0.770
0.780	1.0454	1.0479	1.0505	1.0531	1.0557	1.0583	1.0609	1.0635	1.0661	1.0688	0.780
0.790	1.0714	1.0741	1.0768	1.0795	1.0822	1.0849	1.0876	1.0903	1.0931	1.0958	0.790
0.800	1.0986	1.1014	1.1042	1.1070	1.1098	1.1127	1.1155	1.1184	1.1212	1.1241	0.800
0.810	1.1270	1.1299	1.1329	1.1358	1.1388	1.1417	1.1447	1.1477	1.1507	1.1538	0.810
0.820	1.1568	1.1599	1.1630	1.1660	1.1691	1.1723	1.1754	1.1786	1.1817	1.1849	0.820
0.830	1.1881	1.1914	1.1946	1.1979	1.2011	1.2044	1.2077	1.2111	1.2144	1.2178	0.830
0.840	1.2212	1.2246	1.2280	1.2314	1.2349	1.2384	1.2419	1.2454	1.2490	1.2526	0.840
0.850	1.2561	1.2598	1.2634	1.2671	1.2707	1.2744	1.2782	1.2819	1.2857	1.2895	0.850
0.860	1.2933	1.2972	1.3011	1.3050	1.3089	1.3129	1.3169	1.3209	1.3249	1.3290	0.860
0.870	1.3331	1.3372	1.3414	1.3456	1.3498	1.3540	1.3583	1.3626	1.3670	1.3713	0.870
0.880	1.3758	1.3802	1.3847	1.3892	1.3938	1.3984	1.4030	1.4077	1.4124	1.4171	0.880
0.890	1.4219	1.4268	1.4316	1.4365	1.4415	1.4465	1.4516	1.4566	1.4618	1.4670	0.890
0.900	1.4722	1.4775	1.4828	1.4882	1.4937	1.4992	1.5047	1.5103	1.5160	1.5217	0.900
0.910	1.5275	1.5334	1.5393	1.5453	1.5513	1.5574	1.5636	1.5698	1.5762	1.5825	0.910
0.920	1.5890	1.5956	1.6022	1.6089	1.6157	1.6226	1.6296	1.6366	1.6438	1.6510	0.920
0.930	1.6584	1.6658	1.6734	1.6811	1.6888	1.6967	1.7047	1.7129	1.7211	1.7295	0.930
0.940	1.7380	1.7467	1.7555	1.7645	1.7736	1.7828	1.7923	1.8019	1.8116	1.8216	0.940
0.950	1.8318	1.8421	1.8527	1.8635	1.8745	1.8857	1.8972	1.9090	1.9210	1.9333	0.950
0.960	1.9459	1.9588	1.9721	1.9856	1.9996	2.0139	2.0287	2.0439	2.0595	2.0756	0.960
0.970	2.0923	2.1095	2.1273	2.1457	2.1648	2.1847	2.2054	2.2269	2.2494	2.2729	0.970
0.980	2.2975	2.3234	2.3507	2.3795	2.4101	2.4426	2.4774	2.5147	2.5549	2.5987	0.980
0.990	2.6466	2.6995	2.7587	2.8257	2.9030	2.9944	3.1062	3.2502	3.4531	3.7997	0.990

Table B.18 was produced using Equation 19.8. *Example:* For *r* = 0.641, *z* = 0.7599.

TABLE B.19   Correlation Coefficients, r, Corresponding to Fisher's z Transformation

z	0	1	2	3	4	5	6	7	8	9	z
0.00	0.0000	0.0010	0.0020	0.0030	0.0040	0.0050	0.0060	0.0070	0.0080	0.0090	0.00
0.01	0.0100	0.0110	0.0120	0.0130	0.0140	0.0150	0.0160	0.0170	0.0180	0.0190	0.01
0.02	0.0200	0.0210	0.0220	0.0230	0.0240	0.0250	0.0260	0.0270	0.0280	0.0290	0.02
0.03	0.0300	0.0310	0.0320	0.0330	0.0340	0.0350	0.0360	0.0370	0.0380	0.0390	0.03
0.04	0.0400	0.0410	0.0420	0.0430	0.0440	0.0450	0.0460	0.0470	0.0480	0.0490	0.04
0.05	0.0500	0.0510	0.0520	0.0530	0.0539	0.0549	0.0559	0.0569	0.0579	0.0589	0.05
0.06	0.0599	0.0609	0.0619	0.0629	0.0639	0.0649	0.0659	0.0669	0.0679	0.0689	0.06
0.07	0.0699	0.0709	0.0719	0.0729	0.0739	0.0749	0.0759	0.0768	0.0778	0.0788	0.07
0.08	0.0798	0.0808	0.0818	0.0828	0.0838	0.0848	0.0858	0.0868	0.0878	0.0888	0.08
0.09	0.0898	0.0907	0.0917	0.0927	0.0937	0.0947	0.0957	0.0967	0.0977	0.0987	0.09
0.10	0.0997	0.1007	0.1016	0.1026	0.1036	0.1046	0.1056	0.1066	0.1076	0.1086	0.10
0.11	0.1096	0.1105	0.1115	0.1125	0.1135	0.1145	0.1155	0.1165	0.1175	0.1184	0.11
0.12	0.1194	0.1204	0.1214	0.1224	0.1234	0.1244	0.1253	0.1263	0.1273	0.1283	0.12
0.13	0.1293	0.1303	0.1312	0.1322	0.1332	0.1342	0.1352	0.1361	0.1371	0.1381	0.13
0.14	0.1391	0.1401	0.1411	0.1420	0.1430	0.1440	0.1450	0.1460	0.1469	0.1479	0.14
0.15	0.1489	0.1499	0.1508	0.1518	0.1528	0.1538	0.1547	0.1557	0.1567	0.1577	0.15
0.16	0.1586	0.1596	0.1606	0.1616	0.1625	0.1635	0.1645	0.1655	0.1664	0.1674	0.16
0.17	0.1684	0.1694	0.1703	0.1713	0.1723	0.1732	0.1742	0.1752	0.1761	0.1771	0.17
0.18	0.1781	0.1790	0.1800	0.1810	0.1820	0.1829	0.1839	0.1849	0.1858	0.1868	0.18
0.19	0.1877	0.1887	0.1897	0.1906	0.1916	0.1926	0.1935	0.1945	0.1955	0.1964	0.19
0.20	0.1974	0.1983	0.1993	0.2003	0.2012	0.2022	0.2031	0.2041	0.2051	0.2060	0.20
0.21	0.2070	0.2079	0.2089	0.2098	0.2108	0.2117	0.2127	0.2137	0.2146	0.2156	0.21
0.22	0.2165	0.2175	0.2184	0.2194	0.2203	0.2213	0.2222	0.2232	0.2241	0.2251	0.22
0.23	0.2260	0.2270	0.2279	0.2289	0.2298	0.2308	0.2317	0.2327	0.2336	0.2346	0.23
0.24	0.2355	0.2364	0.2374	0.2383	0.2393	0.2402	0.2412	0.2421	0.2430	0.2440	0.24
0.25	0.2449	0.2459	0.2468	0.2477	0.2487	0.2496	0.2506	0.2515	0.2524	0.2534	0.25
0.26	0.2543	0.2552	0.2562	0.2571	0.2580	0.2590	0.2599	0.2608	0.2618	0.2627	0.26
0.27	0.2636	0.2646	0.2655	0.2664	0.2673	0.2683	0.2692	0.2701	0.2711	0.2720	0.27
0.28	0.2729	0.2738	0.2748	0.2757	0.2766	0.2775	0.2784	0.2794	0.2803	0.2812	0.28
0.29	0.2821	0.2831	0.2840	0.2849	0.2858	0.2867	0.2876	0.2886	0.2895	0.2904	0.29
0.30	0.2913	0.2922	0.2931	0.2941	0.2950	0.2959	0.2968	0.2977	0.2986	0.2995	0.30
0.31	0.3004	0.3013	0.3023	0.3032	0.3041	0.3050	0.3059	0.3068	0.3077	0.3086	0.31
0.32	0.3095	0.3104	0.3113	0.3122	0.3131	0.3140	0.3149	0.3158	0.3167	0.3176	0.32
0.33	0.3185	0.3194	0.3203	0.3212	0.3221	0.3230	0.3239	0.3248	0.3257	0.3266	0.33
0.34	0.3275	0.3284	0.3293	0.3302	0.3310	0.3319	0.3328	0.3337	0.3346	0.3355	0.34
0.35	0.3364	0.3373	0.3381	0.3390	0.3399	0.3408	0.3417	0.3426	0.3435	0.3443	0.35
0.36	0.3452	0.3461	0.3470	0.3479	0.3487	0.3496	0.3505	0.3514	0.3522	0.3531	0.36
0.37	0.3540	0.3549	0.3557	0.3566	0.3575	0.3584	0.3592	0.3601	0.3610	0.3618	0.37
0.38	0.3627	0.3636	0.3644	0.3653	0.3662	0.3670	0.3679	0.3688	0.3696	0.3705	0.38
0.39	0.3714	0.3722	0.3731	0.3739	0.3748	0.3757	0.3765	0.3774	0.3782	0.3791	0.39
0.40	0.3799	0.3808	0.3817	0.3825	0.3834	0.3842	0.3851	0.3859	0.3868	0.3876	0.40
0.41	0.3885	0.3893	0.3902	0.3910	0.3919	0.3927	0.3936	0.3944	0.3952	0.3961	0.41
0.42	0.3969	0.3978	0.3986	0.3995	0.4003	0.4011	0.4020	0.4028	0.4036	0.4045	0.42
0.43	0.4053	0.4062	0.4070	0.4078	0.4087	0.4095	0.4103	0.4112	0.4120	0.4128	0.43
0.44	0.4136	0.4145	0.4153	0.4161	0.4170	0.4178	0.4186	0.4194	0.4203	0.4211	0.44
0.45	0.4219	0.4227	0.4235	0.4244	0.4252	0.4260	0.4268	0.4276	0.4285	0.4293	0.45
0.46	0.4301	0.4309	0.4317	0.4325	0.4333	0.4342	0.4350	0.4358	0.4366	0.4374	0.46
0.47	0.4382	0.4390	0.4398	0.4406	0.4414	0.4422	0.4430	0.4438	0.4446	0.4454	0.47
0.48	0.4462	0.4470	0.4478	0.4486	0.4494	0.4502	0.4510	0.4518	0.4526	0.4534	0.48
0.49	0.4542	0.4550	0.4558	0.4566	0.4574	0.4582	0.4590	0.4598	0.4605	0.4613	0.49
0.50	0.4621	0.4629	0.4637	0.4645	0.4653	0.4660	0.4668	0.4676	0.4684	0.4692	0.50
0.51	0.4699	0.4707	0.4715	0.4723	0.4731	0.4738	0.4746	0.4754	0.4762	0.4769	0.51
0.52	0.4777	0.4785	0.4792	0.4800	0.4808	0.4815	0.4823	0.4831	0.4838	0.4846	0.52
0.53	0.4854	0.4861	0.4869	0.4877	0.4884	0.4892	0.4900	0.4907	0.4915	0.4922	0.53
0.54	0.4930	0.4937	0.4945	0.4953	0.4960	0.4968	0.4975	0.4983	0.4990	0.4998	0.54

TABLE B.19 (cont.)   Correlation Coefficients, *r*, Corresponding to Fisher's *z*
Transformation

z	0	1	2	3	4	5	6	7	8	9	z
0.55	0.5005	0.5013	0.5020	0.5028	0.5035	0.5043	0.5050	0.5057	0.5065	0.5072	0.55
0.56	0.5080	0.5087	0.5095	0.5102	0.5109	0.5117	0.5124	0.5132	0.5139	0.5146	0.56
0.57	0.5154	0.5161	0.5168	0.5176	0.5183	0.5190	0.5198	0.5205	0.5212	0.5219	0.57
0.58	0.5227	0.5234	0.5241	0.5248	0.5256	0.5263	0.5270	0.5277	0.5285	0.5292	0.58
0.59	0.5299	0.5306	0.5313	0.5320	0.5328	0.5335	0.5342	0.5349	0.5356	0.5363	0.59
0.60	0.5370	0.5378	0.5385	0.5392	0.5399	0.5406	0.5413	0.5420	0.5427	0.5434	0.60
0.61	0.5441	0.5448	0.5455	0.5462	0.5469	0.5476	0.5483	0.5490	0.5497	0.5504	0.61
0.62	0.5511	0.5518	0.5525	0.5532	0.5539	0.5546	0.5553	0.5560	0.5567	0.5574	0.62
0.63	0.5581	0.5587	0.5594	0.5601	0.5608	0.5615	0.5622	0.5629	0.5635	0.5642	0.63
0.64	0.5649	0.5656	0.5663	0.5669	0.5676	0.5683	0.5690	0.5696	0.5703	0.5710	0.64
0.65	0.5717	0.5723	0.5730	0.5737	0.5744	0.5750	0.5757	0.5764	0.5770	0.5777	0.65
0.66	0.5784	0.5790	0.5797	0.5804	0.5810	0.5817	0.5823	0.5830	0.5837	0.5843	0.66
0.67	0.5850	0.5856	0.5863	0.5869	0.5876	0.5883	0.5889	0.5896	0.5902	0.5909	0.67
0.68	0.5915	0.5922	0.5928	0.5935	0.5941	0.5948	0.5954	0.5961	0.5967	0.5973	0.68
0.69	0.5980	0.5986	0.5993	0.5999	0.6005	0.6012	0.6018	0.6025	0.6031	0.6037	0.69
0.70	0.6044	0.6050	0.6056	0.6063	0.6069	0.6075	0.6082	0.6088	0.6094	0.6100	0.70
0.71	0.6107	0.6113	0.6119	0.6126	0.6132	0.6138	0.6144	0.6150	0.6157	0.6163	0.71
0.72	0.6169	0.6175	0.6181	0.6188	0.6194	0.6200	0.6206	0.6212	0.6218	0.6225	0.72
0.73	0.6231	0.6237	0.6243	0.6249	0.6255	0.6261	0.6267	0.6273	0.6279	0.6285	0.73
0.74	0.6291	0.6297	0.6304	0.6310	0.6316	0.6322	0.6328	0.6334	0.6340	0.6346	0.74
0.75	0.6351	0.6357	0.6363	0.6369	0.6375	0.6381	0.6387	0.6393	0.6399	0.6405	0.75
0.76	0.6411	0.6417	0.6423	0.6428	0.6434	0.6440	0.6446	0.6452	0.6458	0.6463	0.76
0.77	0.6469	0.6475	0.6481	0.6487	0.6492	0.6498	0.6504	0.6510	0.6516	0.6521	0.77
0.78	0.6527	0.6533	0.6539	0.6544	0.6550	0.6556	0.6561	0.6567	0.6573	0.6578	0.78
0.79	0.6584	0.6590	0.6595	0.6601	0.6607	0.6612	0.6618	0.6624	0.6629	0.6635	0.79
0.80	0.6640	0.6646	0.6652	0.6657	0.6663	0.6668	0.6674	0.6679	0.6685	0.6690	0.80
0.81	0.6696	0.6701	0.6707	0.6712	0.6718	0.6723	0.6729	0.6734	0.6740	0.6745	0.81
0.82	0.6751	0.6756	0.6762	0.6767	0.6772	0.6778	0.6783	0.6789	0.6794	0.6799	0.82
0.83	0.6805	0.6810	0.6815	0.6821	0.6826	0.6832	0.6837	0.6842	0.6847	0.6853	0.83
0.84	0.6858	0.6863	0.6869	0.6874	0.6879	0.6884	0.6890	0.6895	0.6900	0.6905	0.84
0.85	0.6911	0.6916	0.6921	0.6926	0.6932	0.6937	0.6942	0.6947	0.6952	0.6957	0.85
0.86	0.6963	0.6968	0.6973	0.6978	0.6983	0.6988	0.6993	0.6998	0.7004	0.7009	0.86
0.87	0.7014	0.7019	0.7024	0.7029	0.7034	0.7039	0.7044	0.7049	0.7054	0.7059	0.87
0.88	0.7064	0.7069	0.7074	0.7079	0.7084	0.7089	0.7094	0.7099	0.7104	0.7109	0.88
0.89	0.7114	0.7119	0.7124	0.7129	0.7134	0.7139	0.7143	0.7148	0.7153	0.7158	0.89
0.90	0.7163	0.7168	0.7173	0.7178	0.7182	0.7187	0.7192	0.7197	0.7202	0.7207	0.90
0.91	0.7211	0.7216	0.7221	0.7226	0.7230	0.7235	0.7240	0.7245	0.7249	0.7254	0.91
0.92	0.7259	0.7264	0.7268	0.7273	0.7278	0.7283	0.7287	0.7292	0.7297	0.7301	0.92
0.93	0.7306	0.7311	0.7315	0.7320	0.7325	0.7329	0.7334	0.7338	0.7343	0.7348	0.93
0.94	0.7352	0.7357	0.7361	0.7366	0.7371	0.7375	0.7380	0.7384	0.7389	0.7393	0.94
0.95	0.7398	0.7402	0.7407	0.7411	0.7416	0.7420	0.7425	0.7429	0.7434	0.7438	0.95
0.96	0.7443	0.7447	0.7452	0.7456	0.7461	0.7465	0.7469	0.7474	0.7478	0.7483	0.96
0.97	0.7487	0.7491	0.7496	0.7500	0.7505	0.7509	0.7513	0.7518	0.7522	0.7526	0.97
0.98	0.7531	0.7535	0.7539	0.7544	0.7548	0.7552	0.7557	0.7561	0.7565	0.7569	0.98
0.99	0.7574	0.7578	0.7582	0.7586	0.7591	0.7595	0.7599	0.7603	0.7608	0.7612	0.99
1.0	0.7616	0.7658	0.7699	0.7739	0.7779	0.7818	0.7857	0.7895	0.7932	0.7969	1.0
1.1	0.8005	0.8041	0.8076	0.8110	0.8144	0.8178	0.8210	0.8243	0.8274	0.8306	1.1
1.2	0.8337	0.8367	0.8397	0.8426	0.8455	0.8483	0.8511	0.8538	0.8565	0.8591	1.2
1.3	0.8617	0.8643	0.8668	0.8692	0.8717	0.8741	0.8764	0.8787	0.8809	0.8832	1.3
1.4	0.8854	0.8875	0.8896	0.8917	0.8937	0.8957	0.8977	0.8996	0.9015	0.9033	1.4
1.5	0.9051	0.9069	0.9087	0.9104	0.9121	0.9138	0.9154	0.9170	0.9186	0.9201	1.5
1.6	0.9217	0.9232	0.9246	0.9261	0.9275	0.9289	0.9302	0.9316	0.9329	0.9341	1.6
1.7	0.9354	0.9366	0.9379	0.9391	0.9402	0.9414	0.9425	0.9436	0.9447	0.9458	1.7
1.8	0.9468	0.9478	0.9488	0.9498	0.9508	0.9517	0.9527	0.9536	0.9545	0.9554	1.8
1.9	0.9562	0.9571	0.9579	0.9587	0.9595	0.9603	0.9611	0.9618	0.9626	0.9633	1.9

TABLE B.19 (cont.)   Correlation Coefficients, r, Corresponding to Fisher's z
Transformation

z	0	1	2	3	4	5	6	7	8	9	z
2.0	0.9640	0.9647	0.9654	0.9661	0.9667	0.9674	0.9680	0.9687	0.9693	0.9699	2.0
2.1	0.9705	0.9710	0.9716	0.9721	0.9727	0.9732	0.9737	0.9743	0.9748	0.9753	2.1
2.2	0.9757	0.9762	0.9767	0.9771	0.9776	0.9780	0.9785	0.9789	0.9793	0.9797	2.2
2.3	0.9801	0.9805	0.9809	0.9812	0.9816	0.9820	0.9823	0.9827	0.9830	0.9833	2.3
2.4	0.9837	0.9840	0.9843	0.9846	0.9849	0.9852	0.9855	0.9858	0.9861	0.9863	2.4
2.5	0.9866	0.9869	0.9871	0.9874	0.9876	0.9879	0.9881	0.9884	0.9886	0.9888	2.5
2.6	0.9890	0.9892	0.9895	0.9897	0.9899	0.9901	0.9903	0.9905	0.9906	0.9908	2.6
2.7	0.9910	0.9912	0.9914	0.9915	0.9917	0.9919	0.9920	0.9922	0.9923	0.9925	2.7
2.8	0.9926	0.9928	0.9929	0.9931	0.9932	0.9933	0.9935	0.9936	0.9937	0.9938	2.8
2.9	0.9940	0.9941	0.9942	0.9943	0.9944	0.9945	0.9946	0.9947	0.9949	0.9950	2.9
3.0	0.9951	0.9952	0.9952	0.9953	0.9954	0.9955	0.9956	0.9957	0.9958	0.9959	3.0
3.1	0.9959	0.9960	0.9961	0.9962	0.9963	0.9963	0.9964	0.9965	0.9965	0.9966	3.1
3.2	0.9967	0.9967	0.9968	0.9969	0.9969	0.9970	0.9971	0.9971	0.9972	0.9972	3.2
3.3	0.9973	0.9973	0.9974	0.9974	0.9975	0.9975	0.9976	0.9976	0.9977	0.9977	3.3
3.4	0.9978	0.9978	0.9979	0.9979	0.9979	0.9980	0.9980	0.9981	0.9981	0.9981	3.4
3.5	0.9982	0.9982	0.9982	0.9983	0.9983	0.9984	0.9984	0.9984	0.9984	0.9985	3.5
3.6	0.9985	0.9985	0.9986	0.9986	0.9986	0.9986	0.9987	0.9987	0.9987	0.9988	3.6
3.7	0.9988	0.9988	0.9988	0.9988	0.9989	0.9989	0.9989	0.9989	0.9990	0.9990	3.7
3.8	0.9990	0.9990	0.9990	0.9991	0.9991	0.9991	0.9991	0.9991	0.9991	0.9992	3.8
3.9	0.9992	0.9992	0.9992	0.9992	0.9992	0.9993	0.9993	0.9993	0.9993	0.9993	3.9
4.0	0.9993	0.9993	0.9994	0.9994	0.9994	0.9994	0.9994	0.9994	0.9994	0.9994	4.0
4.1	0.9995	0.9995	0.9995	0.9995	0.9995	0.9995	0.9995	0.9995	0.9995	0.9995	4.1
4.2	0.9996	0.9996	0.9996	0.9996	0.9996	0.9996	0.9996	0.9996	0.9996	0.9996	4.2
4.3	0.9996	0.9996	0.9996	0.9997	0.9997	0.9997	0.9997	0.9997	0.9997	0.9997	4.3
4.4	0.9997	0.9997	0.9997	0.9997	0.9997	0.9997	0.9997	0.9997	0.9997	0.9997	4.4
4.5	0.9998	0.9998	0.9998	0.9998	0.9998	0.9998	0.9998	0.9998	0.9998	0.9998	4.5
4.6	0.9998	0.9998	0.9998	0.9998	0.9998	0.9998	0.9998	0.9998	0.9998	0.9998	4.6
4.7	0.9998	0.9998	0.9998	0.9998	0.9998	0.9999	0.9999	0.9999	0.9999	0.9999	4.7
4.8	0.9999	0.9999	0.9999	0.9999	0.9999	0.9999	0.9999	0.9999	0.9999	0.9999	4.8
4.9	0.9999	0.9999	0.9999	0.9999	0.9999	0.9999	0.9999	0.9999	0.9999	0.9999	4.9

$$r = \tanh z = \frac{e^{2z} - 1}{e^{2z} + 1}$$

For example:

$$z = 2.42, \quad r = 0.9843.$$

TABLE B.20   Critical Values of the Spearman Rank Correlation Coefficient, r_s

n	$\alpha(2)$: 0.50 $\alpha(1)$: 0.25	0.20 0.10	0.10 0.05	0.05 0.025	0.02 0.01	0.01 0.005	0.005 0.0025	0.002 0.001	0.001 0.0005
4	0.600	1.000	1.000						
5	0.500	0.800	0.900	1.000	1.000				
6	0.371	0.657	0.829	0.886	0.943	1.000	1.000		
7	0.321	0.571	0.714	0.786	0.893	0.929	0.964	1.000	1.000
8	0.310	0.524	0.643	0.738	0.833	0.881	0.905	0.952	0.976
9	0.267	0.483	0.600	0.700	0.783	0.833	0.867	0.917	0.933
10	0.248	0.455	0.564	0.648	0.745	0.794	0.830	0.879	0.903
11	0.236	0.427	0.536	0.618	0.709	0.755	0.800	0.845	0.873
12	0.217	0.406	0.503	0.587	0.678	0.727	0.769	0.818	0.846
13	0.209	0.385	0.484	0.560	0.648	0.703	0.747	0.791	0.824
14	0.200	0.367	0.464	0.538	0.626	0.679	0.723	0.771	0.802
15	0.189	0.354	0.446	0.521	0.604	0.654	0.700	0.750	0.779
16	0.182	0.341	0.429	0.503	0.582	0.635	0.679	0.729	0.762
17	0.176	0.328	0.414	0.485	0.566	0.615	0.662	0.713	0.748
18	0.170	0.317	0.401	0.472	0.550	0.600	0.643	0.695	0.728
19	0.165	0.309	0.391	0.460	0.535	0.584	0.628	0.677	0.712
20	0.161	0.299	0.380	0.447	0.520	0.570	0.612	0.662	0.696
21	0.156	0.292	0.370	0.435	0.508	0.556	0.599	0.648	0.681
22	0.152	0.284	0.361	0.425	0.496	0.544	0.586	0.634	0.667
23	0.148	0.278	0.353	0.415	0.486	0.532	0.573	0.622	0.654
24	0.144	0.271	0.344	0.406	0.476	0.521	0.562	0.610	0.642
25	0.142	0.265	0.337	0.398	0.466	0.511	0.551	0.598	0.630
26	0.138	0.259	0.331	0.390	0.457	0.501	0.541	0.587	0.619
27	0.136	0.255	0.324	0.382	0.448	0.491	0.531	0.577	0.608
28	0.133	0.250	0.317	0.375	0.440	0.483	0.522	0.567	0.598
29	0.130	0.245	0.312	0.368	0.433	0.475	0.513	0.558	0.589
30	0.128	0.240	0.306	0.362	0.425	0.467	0.504	0.549	0.580
31	0.126	0.236	0.301	0.356	0.418	0.459	0.496	0.541	0.571
32	0.124	0.232	0.296	0.350	0.412	0.452	0.489	0.533	0.563
33	0.121	0.229	0.291	0.345	0.405	0.446	0.482	0.525	0.554
34	0.120	0.225	0.287	0.340	0.399	0.439	0.475	0.517	0.547
35	0.118	0.222	0.283	0.335	0.394	0.433	0.468	0.510	0.539
36	0.116	0.219	0.279	0.330	0.388	0.427	0.462	0.504	0.533
37	0.114	0.216	0.275	0.325	0.383	0.421	0.456	0.497	0.526
38	0.113	0.212	0.271	0.321	0.378	0.415	0.450	0.491	0.519
39	0.111	0.210	0.267	0.317	0.373	0.410	0.444	0.485	0.513
40	0.110	0.207	0.264	0.313	0.368	0.405	0.439	0.479	0.507
41	0.108	0.204	0.261	0.309	0.364	0.400	0.433	0.473	0.501
42	0.107	0.202	0.257	0.305	0.359	0.395	0.428	0.468	0.495
43	0.105	0.199	0.254	0.301	0.355	0.391	0.423	0.463	0.490
44	0.104	0.197	0.251	0.298	0.351	0.386	0.419	0.458	0.484
45	0.103	0.194	0.248	0.294	0.347	0.382	0.414	0.453	0.479
46	0.102	0.192	0.246	0.291	0.343	0.378	0.410	0.448	0.474
47	0.101	0.190	0.243	0.288	0.340	0.374	0.405	0.443	0.469
48	0.100	0.188	0.240	0.285	0.336	0.370	0.401	0.439	0.465
49	0.098	0.186	0.238	0.282	0.333	0.366	0.397	0.434	0.460
50	0.097	0.184	0.235	0.279	0.329	0.363	0.393	0.430	0.456
51	0.096	0.182	0.233	0.276	0.326	0.359	0.390	0.426	0.451
52	0.095	0.180	0.231	0.274	0.323	0.356	0.386	0.422	0.447
53	0.095	0.179	0.228	0.271	0.320	0.352	0.382	0.418	0.443
54	0.094	0.177	0.226	0.268	0.317	0.349	0.379	0.414	0.439
55	0.093	0.175	0.224	0.266	0.314	0.346	0.375	0.411	0.435

TABLE B.20 (cont.)   Critical Values of the Spearman Rank Correlation
Coefficient, r_s

n	$\alpha(2)$: 0.50 $\alpha(1)$: 0.25	0.20 0.10	0.10 0.05	0.05 0.025	0.02 0.01	0.01 0.005	0.005 0.0025	0.002 0.001	0.001 0.0005
56	0.092	0.174	0.222	0.264	0.311	0.343	0.372	0.407	0.432
57	0.091	0.172	0.220	0.261	0.308	0.340	0.369	0.404	0.428
58	0.090	0.171	0.218	0.259	0.306	0.337	0.366	0.400	0.424
59	0.089	0.169	0.216	0.257	0.303	0.334	0.363	0.397	0.421
60	0.089	0.168	0.214	0.255	0.300	0.331	0.360	0.394	0.418
61	0.088	0.166	0.213	0.252	0.298	0.329	0.357	0.391	0.414
62	0.087	0.165	0.211	0.250	0.296	0.326	0.354	0.388	0.411
63	0.086	0.163	0.209	0.248	0.293	0.323	0.351	0.385	0.408
64	0.086	0.162	0.207	0.246	0.291	0.321	0.348	0.382	0.405
65	0.085	0.161	0.206	0.244	0.289	0.318	0.346	0.379	0.402
66	0.084	0.160	0.204	0.243	0.287	0.316	0.343	0.376	0.399
67	0.084	0.158	0.203	0.241	0.284	0.314	0.341	0.373	0.396
68	0.083	0.157	0.201	0.239	0.282	0.311	0.338	0.370	0.393
69	0.082	0.156	0.200	0.237	0.280	0.309	0.336	0.368	0.390
70	0.082	0.155	0.198	0.235	0.278	0.307	0.333	0.365	0.388
71	0.081	0.154	0.197	0.234	0.276	0.305	0.331	0.363	0.385
72	0.081	0.153	0.195	0.232	0.274	0.303	0.329	0.360	0.382
73	0.080	0.152	0.194	0.230	0.272	0.301	0.327	0.358	0.380
74	0.080	0.151	0.193	0.229	0.271	0.299	0.324	0.355	0.377
75	0.079	0.150	0.191	0.227	0.269	0.297	0.322	0.353	0.375
76	0.078	0.149	0.190	0.226	0.267	0.295	0.320	0.351	0.372
77	0.078	0.148	0.189	0.224	0.265	0.293	0.318	0.349	0.370
78	0.077	0.147	0.188	0.223	0.264	0.291	0.316	0.346	0.368
79	0.077	0.146	0.186	0.221	0.262	0.289	0.314	0.344	0.365
80	0.076	0.145	0.185	0.220	0.260	0.287	0.312	0.342	0.363
81	0.076	0.144	0.184	0.219	0.259	0.285	0.310	0.340	0.361
82	0.075	0.143	0.183	0.217	0.257	0.284	0.308	0.338	0.359
83	0.075	0.142	0.182	0.216	0.255	0.282	0.306	0.336	0.357
84	0.074	0.141	0.181	0.215	0.254	0.280	0.305	0.334	0.355
85	0.074	0.140	0.180	0.213	0.252	0.279	0.303	0.332	0.353
86	0.074	0.139	0.179	0.212	0.251	0.277	0.301	0.330	0.351
87	0.073	0.139	0.177	0.211	0.250	0.276	0.299	0.328	0.349
88	0.073	0.138	0.176	0.210	0.248	0.274	0.298	0.327	0.347
89	0.072	0.137	0.175	0.209	0.247	0.272	0.296	0.325	0.345
90	0.072	0.136	0.174	0.207	0.245	0.271	0.294	0.323	0.343
91	0.072	0.135	0.173	0.206	0.244	0.269	0.293	0.321	0.341
92	0.071	0.135	0.173	0.205	0.243	0.268	0.291	0.319	0.339
93	0.071	0.134	0.172	0.204	0.241	0.267	0.290	0.318	0.338
94	0.070	0.133	0.171	0.203	0.240	0.265	0.288	0.316	0.336
95	0.070	0.133	0.170	0.202	0.239	0.264	0.287	0.314	0.334
96	0.070	0.132	0.169	0.201	0.238	0.262	0.285	0.313	0.332
97	0.069	0.131	0.168	0.200	0.236	0.261	0.284	0.311	0.331
98	0.069	0.130	0.167	0.199	0.235	0.260	0.282	0.310	0.329
99	0.068	0.130	0.166	0.198	0.234	0.258	0.281	0.308	0.327
100	0.068	0.129	0.165	0.197	0.233	0.257	0.279	0.307	0.326

For the table entries through $n = 11$, the exact distribution of Σd^2 were used
(Owen, 1962: 400–406). For $n = 12$, the exact distribution of de Jonge and van
Montfort (1972) was used; for n's of 13 through 16, the exact distributions of Otten
(1973) were used; and for $n = 12$–18, the exact distributions of Franklin (1988a) were
used. For larger n, the Pearson-curve approximations described by Olds (1938) were

employed, with the excellent accuracy of results discussed elsewhere (Franklin, 1988b, 1989; Zar, 1972).

Examples:

$$(r_s)_{0.05(2),9} = 0.700 \quad \text{and} \quad (r_s)_{0.01(2),52} = 0.356.$$

For n larger than those in this table, we may utilize either Appendix Table B.17 or, equivalently, Equation 19.4. The accuracy of this procedure is discussed by Zar (1972).

TABLE B.21 Critical Values of the Top-Down Correlation Coefficient, r_T

n	$\alpha(1)$: 0.25	0.10	0.05	0.025	0.01	0.005	0.001	0.0005
3	0.786	1.000	1.000	1.000	1.000	1.000	1.000	1.000
4	0.478	0.870	0.942	1.000	1.000	1.000	1.000	1.000
5	0.752	0.905	0.959	0.977	1.000	1.000	1.000	1.000
6	0.324	0.676	0.810	0.887	0.943	0.969	1.000	1.000
7	0.271	0.622	0.738	0.836	0.906	0.934	0.977	0.991
8	0.245	0.575	0.692	0.779	0.865	0.904	0.960	0.972
9	0.228	0.530	0.654	0.742	0.826	0.871	0.936	0.953
10	0.245	0.492	0.620	0.707	0.793	0.840	0.913	0.933
11	0.204	0.461	0.589	0.677	0.762	0.812	0.890	0.913
12	0.195	0.435	0.560	0.650	0.735	0.786	0.868	0.893
13	0.186	0.412	0.535	0.625	0.711	0.762	0.847	0.873
12	0.179	0.393	0.513	0.602	0.689	0.740	0.827	0.854
15		0.389	0.486	0.565	0.680	0.688	0.826	
16		0.376	0.470	0.546	0.656	0.665	0.798	
17		0.364	0.454	0.528	0.635	0.644	0.773	
18		0.353	0.440	0.512	0.615	0.625	0.750	
19		0.343	0.428	0.497	0.598	0.607	0.728	
20		0.334	0.416	0.483	0.581	0.591	0.709	
21		0.325	0.405	0.470	0.566	0.576	0.691	
22		0.317	0.395	0.459	0.552	0.562	0.674	
23		0.310	0.386	0.448	0.538	0.549	0.659	
24		0.303	0.377	0.438	0.526	0.537	0.644	
25		0.297	0.368	0.428	0.515	0.526	0.631	
26		0.291	0.361	0.419	0.504	0.515	0.618	
27		0.285	0.354	0.411	0.494	0.505	0.606	
28		0.280	0.347	0.403	0.484	0.496	0.595	
29		0.275	0.340	0.395	0.475	0.487	0.584	
30		0.270	0.334	0.388	0.466	0.478	0.574	

For n through 14, the critical values were determined from the exact distributions of Iman and Conover (1985) and Iman (1987b). For $15 \le n \le 30$, the critical values are reprinted, with permission of the publisher, from Iman and Conover, copyright 1987, *Technometrics* 29: 351–357; all rights reserved. For $n > 30$, the normal approximation may be employed: $Z = r_T\sqrt{n-1}$.

TABLE B.22 Critical Values of the Symmetry Measure, g_1

n	$\alpha(2)$: 0.20 $\alpha(1)$: 0.10	0.10 0.05	0.05 0.025	0.02 0.01	0.01 0.005	0.005 0.0025	0.002 0.001
9	0.907	1.176	1.416	1.705	1.909	2.103	2.351
10	0.866	1.125	1.359	1.643	1.846	2.041	2.290
11	0.830	1.081	1.309	1.587	1.787	1.981	2.230
12	0.799	1.042	1.264	1.536	1.733	1.924	2.171
13	0.771	1.007	1.223	1.490	1.682	1.871	2.115
14	0.747	0.976	1.186	1.447	1.636	1.820	2.061
15	0.724	0.948	1.153	1.407	1.592	1.773	2.010
16	0.704	0.922	1.122	1.370	1.551	1.729	1.961
17	0.685	0.898	1.093	1.336	1.513	1.687	1.915
18	0.668	0.875	1.066	1.304	1.477	1.648	1.871
19	0.652	0.855	1.041	1.274	1.444	1.611	1.829
20	0.638	0.836	1.018	1.246	1.412	1.576	1.790
21	0.624	0.818	0.997	1.220	1.383	1.543	1.753
22	0.611	0.801	0.976	1.195	1.355	1.512	1.717
23	0.599	0.786	0.957	1.171	1.328	1.482	1.684
24	0.588	0.771	0.939	1.149	1.303	1.454	1.652
25	0.577	0.757	0.922	1.128	1.279	1.427	1.621
26	0.567	0.744	0.906	1.108	1.256	1.401	1.592
27	0.558	0.731	0.891	1.089	1.235	1.377	1.564
28	0.549	0.719	0.876	1.071	1.214	1.354	1.538
29	0.540	0.708	0.862	1.054	1.194	1.332	1.512
30	0.532	0.697	0.849	1.037	1.175	1.311	1.488
32	0.517	0.677	0.824	1.007	1.140	1.271	1.442
34	0.503	0.658	0.801	0.978	1.108	1.234	1.400
36	0.490	0.641	0.780	0.952	1.077	1.200	1.361
38	0.478	0.625	0.760	0.928	1.050	1.169	1.324
40	0.467	0.611	0.742	0.905	1.024	1.140	1.290
42	0.457	0.597	0.726	0.884	1.000	1.112	1.259
44	0.447	0.584	0.710	0.865	0.977	1.087	1.229
46	0.438	0.572	0.695	0.846	0.956	1.063	1.202
48	0.430	0.561	0.681	0.829	0.936	1.040	1.176
50	0.422	0.550	0.668	0.813	0.917	1.019	1.151
52	0.414	0.540	0.655	0.797	0.899	0.999	1.128
54	0.407	0.531	0.644	0.783	0.883	0.980	1.106
56	0.400	0.522	0.633	0.769	0.867	0.962	1.085
58	0.394	0.513	0.622	0.755	0.851	0.945	1.065
60	0.387	0.505	0.612	0.743	0.837	0.929	1.047
65	0.373	0.486	0.589	0.714	0.804	0.891	1.004
70	0.361	0.469	0.568	0.688	0.775	0.858	0.965
75	0.349	0.454	0.549	0.665	0.748	0.828	0.931
80	0.339	0.440	0.532	0.644	0.724	0.801	0.899
85	0.329	0.428	0.517	0.625	0.702	0.776	0.871
90	0.320	0.416	0.502	0.607	0.681	0.753	0.845
95	0.312	0.405	0.489	0.591	0.663	0.732	0.821
100	0.305	0.396	0.477	0.576	0.646	0.713	0.799
110	0.291	0.378	0.455	0.549	0.615	0.679	0.759
120	0.279	0.362	0.436	0.525	0.588	0.649	0.725
130	0.269	0.348	0.419	0.505	0.565	0.622	0.695
140	0.259	0.336	0.404	0.486	0.544	0.599	0.668
150	0.251	0.325	0.391	0.469	0.525	0.578	0.644
160	0.243	0.315	0.378	0.454	0.508	0.558	0.622
170	0.236	0.306	0.367	0.441	0.492	0.541	0.603

TABLE B.22 (cont.) Critical Values of the Symmetry Measure, g_1

n	$\alpha(2)$: 0.20 $\alpha(1)$: 0.10	0.10 0.05	0.05 0.025	0.02 0.01	0.01 0.005	0.005 0.0025	0.002 0.001
180	0.230	0.297	0.357	0.428	0.478	0.525	0.585
190	0.224	0.289	0.347	0.417	0.465	0.511	0.569
200	0.218	0.282	0.339	0.406	0.453	0.497	0.553
220	0.208	0.269	0.323	0.387	0.431	0.474	0.526
240	0.200	0.258	0.309	0.370	0.413	0.453	0.503
260	0.192	0.248	0.297	0.356	0.396	0.435	0.482
280	0.185	0.239	0.287	0.343	0.382	0.418	0.464
300	0.179	0.231	0.277	0.331	0.368	0.404	0.448
320	0.174	0.224	0.268	0.320	0.357	0.391	0.433
340	0.169	0.217	0.260	0.311	0.346	0.378	0.420
360	0.164	0.211	0.253	0.302	0.336	0.368	0.407
380	0.160	0.206	0.246	0.294	0.327	0.358	0.396
400	0.156	0.200	0.240	0.286	0.318	0.348	0.386
420	0.152	0.196	0.234	0.279	0.310	0.340	0.376
440	0.148	0.191	0.229	0.273	0.303	0.332	0.367
460	0.145	0.187	0.224	0.267	0.297	0.324	0.359
480	0.142	0.183	0.219	0.261	0.290	0.317	0.351
500	0.139	0.180	0.215	0.256	0.284	0.311	0.344
550	0.133	0.171	0.205	0.244	0.271	0.296	0.327
600	0.127	0.164	0.196	0.233	0.259	0.283	0.313
650	0.123	0.157	0.188	0.224	0.249	0.272	0.301
700	0.118	0.152	0.181	0.216	0.240	0.262	0.289
750	0.114	0.147	0.175	0.209	0.232	0.253	0.279
800	0.110	0.142	0.170	0.202	0.224	0.245	0.270
850	0.107	0.138	0.165	0.196	0.217	0.237	0.262
900	0.104	0.134	0.160	0.190	0.211	0.231	0.255
950	0.101	0.130	0.156	0.185	0.206	0.224	0.248
1000	0.099	0.127	0.152	0.181	0.200	0.219	0.241

This table was produced by determining the probabilities of g_1 employing the normal approximation of Equations 7.25–7.31 (D'Agostino, 1970). Linear interpolation may be used for critical values between those for the tabled sample sizes. Some critical values $\sqrt{b_1}$ (see Section 6.1) for $n < 9$ are given by D'Agostino (1986), D'Agostino and Tietjen (1973), and Mulholland (1977).

TABLE B.23   Critical Values of the Kurtosis Measure, g_2

n	$\alpha(2)$: 0.20 $\alpha(1)$: 0.10	0.10 0.05	0.05 0.025	0.02 0.01	0.01 0.005	0.005 0.0025	0.002 0.001
20	1.241	1.850	2.486	3.385	4.121	4.914	6.063
21	1.215	1.812	2.436	3.318	4.040	4.818	5.947
22	1.191	1.776	2.388	3.254	3.963	4.727	5.835
23	1.168	1.743	2.343	3.193	3.889	4.639	5.728
24	1.147	1.711	2.300	3.135	3.818	4.555	5.624
25	1.127	1.681	2.260	3.080	3.751	4.474	5.524
26	1.108	1.653	2.222	3.027	3.686	4.397	5.427
27	1.090	1.626	2.185	2.976	3.624	4.322	5.335
28	1.074	1.601	2.150	2.928	3.565	4.251	5.245
29	1.057	1.576	2.117	2.882	3.508	4.182	5.159
30	1.042	1.553	2.085	2.838	3.453	4.116	5.075
32	1.014	1.509	2.025	2.754	3.350	3.990	4.917
34	0.988	1.469	1.971	2.677	3.254	3.874	4.769
36	0.964	1.432	1.919	2.606	3.165	3.765	4.631
38	0.942	1.398	1.872	2.539	3.081	3.663	4.502
40	0.921	1.366	1.828	2.476	3.003	3.568	4.380
42	0.902	1.337	1.787	2.418	2.930	3.478	4.266
44	0.884	1.309	1.748	2.363	2.861	3.394	4.158
46	0.868	1.282	1.711	2.311	2.796	3.314	4.057
48	0.852	1.258	1.677	2.262	2.735	3.239	3.961
50	0.837	1.234	1.644	2.216	2.677	3.168	3.870
52	0.823	1.212	1.613	2.172	2.622	3.100	3.784
54	0.809	1.191	1.584	2.130	2.570	3.037	3.702
56	0.797	1.172	1.556	2.091	2.520	2.976	3.625
58	0.785	1.153	1.530	2.053	2.473	2.918	3.551
60	0.773	1.135	1.504	2.017	2.428	2.862	3.480
62	0.762	1.117	1.480	1.982	2.385	2.810	3.413
64	0.752	1.101	1.457	1.950	2.343	2.759	3.348
66	0.742	1.085	1.435	1.918	2.304	2.711	3.287
68	0.732	1.070	1.414	1.888	2.266	2.665	3.228
70	0.723	1.055	1.394	1.859	2.230	2.620	3.171
72	0.714	1.041	1.374	1.831	2.195	2.578	3.117
74	0.705	1.028	1.355	1.805	2.162	2.537	3.065
76	0.697	1.015	1.337	1.779	2.130	2.498	3.015
78	0.689	1.002	1.320	1.754	2.099	2.460	2.967
80	0.681	0.990	1.303	1.730	2.069	2.423	2.921
82	0.674	0.978	1.286	1.707	2.040	2.388	2.877
84	0.667	0.967	1.271	1.685	2.012	2.354	2.834
86	0.660	0.956	1.255	1.663	1.985	2.322	2.792
88	0.653	0.946	1.241	1.643	1.959	2.290	2.752
90	0.646	0.935	1.227	1.622	1.934	2.259	2.714
92	0.640	0.926	1.213	1.603	1.910	2.230	2.676
94	0.634	0.916	1.199	1.584	1.887	2.201	2.640
96	0.628	0.907	1.186	1.566	1.864	2.174	2.605
98	0.622	0.898	1.174	1.548	1.842	2.147	2.571
100	0.617	0.889	1.162	1.531	1.820	2.121	2.538
102	0.611	0.880	1.150	1.514	1.800	2.096	2.507
104	0.606	0.872	1.138	1.498	1.779	2.071	2.476
106	0.600	0.864	1.127	1.482	1.760	2.047	2.446
108	0.595	0.856	1.116	1.467	1.741	2.024	2.417
110	0.590	0.848	1.105	1.452	1.722	2.002	2.389

TABLE B.23 (cont.)   Critical Values of the Kurtosis
Measure, g_2

n	$\alpha(2)$: 0.20 $\alpha(1)$: 0.10	0.10 0.05	0.05 0.025	0.02 0.01	0.01 0.005	0.005 0.0025	0.002 0.001
112	0.586	0.841	1.095	1.437	1.704	1.980	2.361
114	0.581	0.833	1.085	1.423	1.686	1.959	2.335
116	0.576	0.826	1.075	1.409	1.669	1.938	2.309
118	0.572	0.819	1.066	1.396	1.653	1.918	2.283
120	0.567	0.813	1.056	1.383	1.637	1.898	2.259
122	0.563	0.806	1.047	1.370	1.621	1.879	2.235
124	0.559	0.799	1.038	1.357	1.605	1.860	2.212
126	0.555	0.793	1.029	1.345	1.590	1.842	2.189
128	0.551	0.787	1.021	1.333	1.575	1.825	2.167
130	0.547	0.781	1.013	1.322	1.561	1.807	2.145
132	0.543	0.775	1.004	1.310	1.547	1.790	2.124
134	0.539	0.769	0.996	1.299	1.533	1.774	2.103
136	0.536	0.763	0.989	1.289	1.520	1.758	2.083
138	0.532	0.758	0.981	1.278	1.507	1.742	2.064
140	0.529	0.753	0.974	1.268	1.494	1.727	2.045
145	0.520	0.739	0.956	1.243	1.463	1.690	1.999
150	0.512	0.727	0.938	1.219	1.434	1.655	1.955
155	0.504	0.715	0.922	1.196	1.407	1.621	1.914
160	0.497	0.704	0.907	1.175	1.380	1.590	1.875
165	0.490	0.693	0.892	1.154	1.355	1.560	1.838
170	0.483	0.682	0.878	1.135	1.331	1.531	1.802
175	0.477	0.673	0.864	1.116	1.309	1.504	1.769
180	0.470	0.663	0.851	1.098	1.287	1.478	1.737
185	0.464	0.654	0.839	1.081	1.266	1.453	1.706
190	0.458	0.645	0.827	1.065	1.246	1.429	1.677
195	0.453	0.637	0.815	1.049	1.227	1.406	1.648
200	0.447	0.628	0.804	1.034	1.208	1.384	1.621
210	0.437	0.613	0.783	1.005	1.173	1.343	1.571
220	0.428	0.599	0.764	0.979	1.141	1.305	1.524
230	0.419	0.585	0.746	0.954	1.111	1.269	1.481
240	0.410	0.572	0.729	0.931	1.083	1.236	1.440
250	0.402	0.560	0.713	0.909	1.057	1.205	1.403
260	0.395	0.549	0.697	0.889	1.033	1.176	1.368
270	0.388	0.539	0.683	0.870	1.009	1.149	1.335
280	0.381	0.529	0.670	0.852	0.988	1.123	1.304
290	0.374	0.519	0.657	0.835	0.967	1.099	1.275
300	0.368	0.510	0.645	0.819	0.948	1.077	1.247
320	0.357	0.493	0.623	0.789	0.912	1.035	1.196
340	0.347	0.478	0.603	0.762	0.880	0.997	1.151
360	0.337	0.464	0.584	0.737	0.850	0.962	1.110
380	0.328	0.451	0.567	0.714	0.823	0.931	1.072
400	0.320	0.439	0.551	0.694	0.798	0.902	1.038
420	0.313	0.428	0.537	0.675	0.776	0.875	1.006
440	0.305	0.418	0.523	0.657	0.755	0.851	0.977
460	0.299	0.408	0.511	0.640	0.735	0.828	0.950
480	0.293	0.399	0.499	0.625	0.717	0.807	0.925
500	0.287	0.391	0.488	0.610	0.700	0.787	0.902
520	0.281	0.383	0.478	0.597	0.684	0.769	0.880
540	0.276	0.375	0.468	0.584	0.669	0.752	0.859
560	0.271	0.368	0.459	0.572	0.655	0.735	0.840

TABLE B.23 (cont.) Critical Values of the Kurtosis
Measure, g_2

n	$\alpha(2)$: 0.20 $\alpha(1)$: 0.10	0.10 0.05	0.05 0.025	0.02 0.01	0.01 0.005	0.005 0.0025	0.002 0.001
580	0.266	0.362	0.450	0.561	0.641	0.720	0.822
600	0.262	0.355	0.442	0.550	0.629	0.706	0.805
650	0.252	0.341	0.423	0.526	0.600	0.673	0.766
700	0.243	0.328	0.406	0.504	0.575	0.643	0.732
750	0.234	0.316	0.391	0.485	0.552	0.618	0.702
800	0.227	0.305	0.378	0.468	0.532	0.594	0.675
850	0.220	0.296	0.366	0.452	0.514	0.574	0.651
900	0.214	0.287	0.355	0.438	0.497	0.555	0.628
950	0.208	0.279	0.344	0.424	0.482	0.537	0.608
1000	0.203	0.272	0.335	0.412	0.468	0.521	0.590

This table was produced by determining the probabilities of g_2 employing the normal approximation of Equations 7.34–7.39 (Anscombe and Glynn, 1983). Linear interpolation may be used for critical values between those for the tabled sample sizes. Some critical values of b_2 (see Section 6.1) for $n < 20$ are given by D'Agostino and Tietjen (1971).

TABLE B.24   The Arcsine Transformation, p'

P	0	1	2	3	4	5	6	7	8	9	P
0.000	0.00	0.57	0.81	0.99	1.15	1.28	1.40	1.52	1.62	1.72	0.000
0.001	1.81	1.90	1.99	2.07	2.14	2.22	2.29	2.36	2.43	2.50	0.001
0.002	2.56	2.63	2.69	2.75	2.81	2.87	2.92	2.98	3.03	3.09	0.002
0.003	3.14	3.19	3.24	3.29	3.34	3.39	3.44	3.49	3.53	3.58	0.003
0.004	3.63	3.67	3.72	3.76	3.80	3.85	3.89	3.93	3.97	4.01	0.004
0.005	4.05	4.10	4.14	4.17	4.21	4.25	4.29	4.33	4.37	4.41	0.005
0.006	4.44	4.48	4.52	4.55	4.59	4.62	4.66	4.70	4.73	4.76	0.006
0.007	4.80	4.83	4.87	4.90	4.93	4.97	5.00	5.03	5.07	5.10	0.007
0.008	5.13	5.16	5.20	5.23	5.26	5.29	5.32	5.35	5.38	5.41	0.008
0.009	5.44	5.47	5.50	5.53	5.56	5.59	5.62	5.65	5.68	5.71	0.009
0.01	5.74	6.02	6.29	6.55	6.80	7.03	7.27	7.49	7.71	7.92	0.01
0.02	8.13	8.33	8.53	8.72	8.91	9.10	9.28	9.46	9.63	9.80	0.02
0.03	9.97	10.14	10.30	10.47	10.63	10.78	10.94	11.09	11.24	11.39	0.03
0.04	11.54	11.68	11.83	11.97	12.11	12.25	12.38	12.52	12.66	12.79	0.04
0.05	12.92	13.05	13.18	13.31	13.44	13.56	13.69	13.81	13.94	14.06	0.05
0.06	14.18	14.30	14.42	14.54	14.65	14.77	14.89	15.00	15.12	15.23	0.06
0.07	15.34	15.45	15.56	15.68	15.79	15.89	16.00	16.11	16.22	16.32	0.07
0.08	16.43	16.54	16.64	16.74	16.85	16.95	17.05	17.15	17.26	17.36	0.08
0.09	17.46	17.56	17.66	17.76	17.85	17.95	18.05	18.15	18.24	18.34	0.09
0.10	18.43	18.53	18.63	18.72	18.81	18.91	19.00	19.09	19.19	19.28	0.10
0.11	19.37	19.46	19.55	19.64	19.73	19.82	19.91	20.00	20.09	20.18	0.11
0.12	20.27	20.36	20.44	20.53	20.62	20.70	20.79	20.88	20.96	21.05	0.12
0.13	21.13	21.22	21.30	21.39	21.47	21.56	21.64	21.72	21.81	21.89	0.13
0.14	21.97	22.06	22.14	22.22	22.30	22.38	22.46	22.54	22.63	22.71	0.14
0.15	22.79	22.87	22.95	23.03	23.11	23.18	23.26	23.34	23.42	23.50	0.15
0.16	23.58	23.66	23.73	23.81	23.89	23.97	24.04	24.12	24.20	24.27	0.16
0.17	24.35	24.43	24.50	24.58	24.65	24.73	24.80	24.88	24.95	25.03	0.17
0.18	25.10	25.18	25.25	25.33	25.40	25.47	25.55	25.62	25.70	25.77	0.18
0.19	25.84	25.91	25.99	26.06	26.13	26.21	26.28	26.35	26.42	26.49	0.19
0.20	26.57	26.64	26.71	26.78	26.85	26.92	26.99	27.06	27.13	27.20	0.20
0.21	27.27	27.35	27.42	27.49	27.56	27.62	27.69	27.76	27.83	27.90	0.21
0.22	27.97	28.04	28.11	28.18	28.25	28.32	28.39	28.45	28.52	28.59	0.22
0.23	28.66	28.73	28.79	28.86	28.93	29.00	29.06	29.13	29.20	29.27	0.23
0.24	29.33	29.40	29.47	29.53	29.60	29.67	29.73	29.80	29.87	29.93	0.24
0.25	30.00	30.07	30.13	30.20	30.26	30.33	30.40	30.46	30.53	30.59	0.25
0.26	30.66	30.72	30.79	30.85	30.92	30.98	31.05	31.11	31.18	31.24	0.26
0.27	31.31	31.37	31.44	31.50	31.56	31.63	31.69	31.76	31.82	31.88	0.27
0.28	31.95	32.01	32.08	32.14	32.20	32.27	32.33	32.39	32.46	32.52	0.28
0.29	32.58	32.65	32.71	32.77	32.83	32.90	32.96	33.02	33.09	33.15	0.29
0.30	33.21	33.27	33.34	33.40	33.46	33.52	33.58	33.65	33.71	33.77	0.30
0.31	33.83	33.90	33.96	34.02	34.08	34.14	34.20	34.27	34.33	34.39	0.31
0.32	34.45	34.51	34.57	34.63	34.70	34.76	34.82	34.88	34.94	35.00	0.32
0.33	35.06	35.12	35.18	35.24	35.30	35.37	35.43	35.49	35.55	35.61	0.33
0.34	35.67	35.73	35.79	35.85	35.91	35.97	36.03	36.09	36.15	36.21	0.34
0.35	36.27	36.33	36.39	36.45	36.51	36.57	36.63	36.69	36.75	36.81	0.35
0.36	36.87	36.93	36.99	37.05	37.11	37.17	37.23	37.29	37.35	37.41	0.36
0.37	37.46	37.52	37.58	37.64	37.70	37.76	37.82	37.88	37.94	38.00	0.37
0.38	38.06	38.12	38.17	38.23	38.29	38.35	38.41	38.47	38.53	38.59	0.38
0.39	38.65	38.70	38.76	38.82	38.88	38.94	39.00	39.06	39.11	39.17	0.39
0.40	39.23	39.29	39.35	39.41	39.47	39.52	39.58	39.64	39.70	39.76	0.40
0.41	39.82	39.87	39.93	39.99	40.05	40.11	40.16	40.22	40.28	40.34	0.41
0.42	40.40	40.45	40.51	40.57	40.63	40.69	40.74	40.80	40.86	40.92	0.42
0.43	40.98	41.03	41.09	41.15	41.21	41.27	41.32	41.38	41.44	41.50	0.43
0.44	41.55	41.61	41.67	41.73	41.78	41.84	41.90	41.96	42.02	42.07	0.44
0.45	42.13	42.19	42.25	42.30	42.36	42.42	42.48	42.53	42.59	42.65	0.45

TABLE B.24 (cont.)   The Arcsine Transformation, p'

P	0	1	2	3	4	5	6	7	8	9	P
0.46	42.71	42.76	42.82	42.88	42.94	42.99	43.05	43.11	43.17	43.22	0.46
0.47	43.28	43.34	43.39	43.45	43.51	43.57	43.62	43.68	43.74	43.80	0.47
0.48	43.85	43.91	43.97	44.03	44.08	44.14	44.20	44.26	44.31	44.37	0.48
0.49	44.43	44.48	44.54	44.60	44.66	44.71	44.77	44.83	44.89	44.94	0.49
0.50	45.00	45.06	45.11	45.17	45.23	45.29	45.34	45.40	45.46	45.52	0.50
0.51	45.57	45.63	45.69	45.74	45.80	45.86	45.92	45.97	46.03	46.09	0.51
0.52	46.15	46.20	46.26	46.32	46.38	46.43	46.49	46.55	46.61	46.66	0.52
0.53	46.72	46.78	46.83	46.89	46.95	47.01	47.06	47.12	47.18	47.24	0.53
0.54	47.29	47.35	47.41	47.47	47.52	47.58	47.64	47.70	47.75	47.81	0.54
0.55	47.87	47.93	47.98	48.04	48.10	48.16	48.22	48.27	48.33	48.39	0.55
0.56	48.45	48.50	48.56	48.62	48.68	48.73	48.79	48.85	48.91	48.97	0.56
0.57	49.02	49.08	49.14	49.20	49.26	49.31	49.37	49.43	49.49	49.55	0.57
0.58	49.60	49.66	49.72	49.78	49.84	49.89	49.95	50.01	50.07	50.13	0.58
0.59	50.18	50.24	50.30	50.36	50.42	50.48	50.53	50.59	50.65	50.71	0.59
0.60	50.77	50.83	50.89	50.94	51.00	51.06	51.12	51.18	51.24	51.30	0.60
0.61	51.35	51.41	51.47	51.53	51.59	51.65	51.71	51.77	51.83	51.88	0.61
0.62	51.94	52.00	52.06	52.12	52.18	52.24	52.30	52.36	52.42	52.48	0.62
0.63	52.54	52.59	52.65	52.71	52.77	52.83	52.89	52.95	53.01	53.07	0.63
0.64	53.13	53.19	53.25	53.31	53.37	53.43	53.49	53.55	53.61	53.67	0.64
0.65	53.73	53.79	53.85	53.91	53.97	54.03	54.09	54.15	54.21	54.27	0.65
0.66	54.33	54.39	54.45	54.51	54.57	54.63	54.70	54.76	54.82	54.88	0.66
0.67	54.94	55.00	55.06	55.12	55.18	55.24	55.30	55.37	55.43	55.49	0.67
0.68	55.55	55.61	55.67	55.73	55.80	55.86	55.92	55.98	56.04	56.10	0.68
0.69	56.17	56.23	56.29	56.35	56.42	56.48	56.54	56.60	56.66	56.73	0.69
0.70	56.79	56.85	56.91	56.98	57.04	57.10	57.17	57.23	57.29	57.35	0.70
0.71	57.42	57.48	57.54	57.61	57.67	57.73	57.80	57.86	57.92	57.99	0.71
0.72	58.05	58.12	58.18	58.24	58.31	58.37	58.44	58.50	58.56	58.63	0.72
0.73	58.69	58.76	58.82	58.89	58.95	59.02	59.08	59.15	59.21	59.28	0.73
0.74	59.34	59.41	59.47	59.54	59.60	59.67	59.74	59.80	59.87	59.93	0.74
0.75	60.00	60.07	60.13	60.20	60.27	60.33	60.40	60.47	60.53	60.60	0.75
0.76	60.67	60.73	60.80	60.87	60.94	61.00	61.07	61.14	61.21	61.27	0.76
0.77	61.34	61.41	61.48	61.55	61.61	61.68	61.75	61.82	61.89	61.96	0.77
0.78	62.03	62.10	62.17	62.24	62.31	62.38	62.44	62.51	62.58	62.65	0.78
0.79	62.73	62.80	62.87	62.94	63.01	63.08	63.15	63.22	63.29	63.36	0.79
0.80	63.43	63.51	63.58	63.65	63.72	63.79	63.87	63.94	64.01	64.09	0.80
0.81	64.16	64.23	64.30	64.38	64.45	64.53	64.60	64.67	64.75	64.82	0.81
0.82	64.90	64.97	65.05	65.12	65.20	65.27	65.35	65.42	65.50	65.57	0.82
0.83	65.65	65.73	65.80	65.88	65.96	66.03	66.11	66.19	66.27	66.34	0.83
0.84	66.42	66.50	66.58	66.66	66.74	66.82	66.89	66.97	67.05	67.13	0.84
0.85	67.21	67.29	67.37	67.46	67.54	67.62	67.70	67.78	67.86	67.94	0.85
0.86	68.03	68.11	68.19	68.28	68.36	68.44	68.53	68.61	68.70	68.78	0.86
0.87	68.87	68.95	69.04	69.12	69.21	69.30	69.38	69.47	69.56	69.64	0.87
0.88	69.73	69.82	69.91	70.00	70.09	70.18	70.27	70.36	70.45	70.54	0.88
0.89	70.63	70.72	70.81	70.91	71.00	71.09	71.19	71.28	71.37	71.47	0.89
0.90	71.57	71.66	71.76	71.85	71.95	72.05	72.15	72.24	72.34	72.44	0.90
0.91	72.54	72.64	72.74	72.85	72.95	73.05	73.15	73.26	73.36	73.46	0.91
0.92	73.57	73.68	73.78	73.89	74.00	74.11	74.21	74.32	74.44	74.55	0.92
0.93	74.66	74.77	74.88	75.00	75.11	75.23	75.35	75.46	75.58	75.70	0.93
0.94	75.82	75.94	76.06	76.19	76.31	76.44	76.56	76.69	76.82	76.95	0.94
0.95	77.08	77.21	77.34	77.48	77.62	77.75	77.89	78.03	78.17	78.32	0.95
0.96	78.46	78.61	78.76	78.91	79.06	79.22	79.37	79.53	79.70	79.86	0.96
0.97	80.03	80.20	80.37	80.54	80.72	80.90	81.09	81.28	81.47	81.67	0.97
0.98	81.87	82.08	82.29	82.51	82.73	82.97	83.20	83.45	83.71	83.98	0.98
0.990	84.26	84.29	84.32	84.35	84.38	84.41	84.44	84.47	84.50	84.53	0.990
0.991	84.56	84.59	84.62	84.65	84.68	84.71	84.74	84.77	84.80	84.84	0.991

TABLE B.24 (cont.)   The Arcsine Transformation, p'

P	0	1	2	3	4	5	6	7	8	9	P
0.992	84.87	84.90	84.93	84.97	85.00	85.03	85.07	85.10	85.13	85.17	0.992
0.993	85.20	85.24	85.27	85.30	85.34	85.38	85.41	85.45	85.48	85.52	0.993
0.994	85.56	85.59	85.63	85.67	85.71	85.75	85.79	85.83	85.86	85.90	0.994
0.995	85.95	85.99	86.03	86.07	86.11	86.15	86.20	86.24	86.28	86.33	0.995
0.996	86.37	86.42	86.47	86.51	86.56	86.61	86.66	86.71	86.76	86.81	0.996
0.997	86.86	86.91	86.97	87.02	87.08	87.13	87.19	87.25	87.31	87.37	0.997
0.998	87.44	87.50	87.57	87.64	87.71	87.78	87.86	87.93	88.01	88.10	0.998
0.999	88.19	88.28	88.38	88.48	88.60	88.72	88.85	89.01	89.19	89.43	0.999
1.000	90.00										

By Equation 13.5:

$$p' = \arcsin \sqrt{p}$$

Examples:

$$p = 0.712, \; p' = 57.54 \quad \text{and} \quad p = 0.9921, \; p' = 84.90.$$

TABLE B.25   Proportions, p, Corresponding to Arcsine Transformations, p'

p'	0	1	2	3	4	5	6	7	8	9	p'
0.0	0.0000	0.0000	0.0000	0.0000	0.0000	0.0001	0.0001	0.0001	0.0002	0.0002	0.0
1.0	0.0003	0.0004	0.0004	0.0005	0.0006	0.0007	0.0008	0.0009	0.0010	0.0011	1.0
2.0	0.0012	0.0013	0.0015	0.0016	0.0018	0.0019	0.0021	0.0022	0.0024	0.0026	2.0
3.0	0.0027	0.0029	0.0031	0.0033	0.0035	0.0037	0.0039	0.0042	0.0044	0.0046	3.0
4.0	0.0049	0.0051	0.0054	0.0056	0.0059	0.0062	0.0064	0.0067	0.0070	0.0073	4.0
5.0	0.0076	0.0079	0.0082	0.0085	0.0089	0.0092	0.0095	0.0099	0.0102	0.0106	5.0
6.0	0.0109	0.0113	0.0117	0.0120	0.0124	0.0128	0.0132	0.0136	0.0140	0.0144	6.0
7.0	0.0149	0.0153	0.0157	0.0161	0.0166	0.0170	0.0175	0.0180	0.0184	0.0189	7.0
8.0	0.0194	0.0199	0.0203	0.0208	0.0213	0.0218	0.0224	0.0229	0.0234	0.0239	8.0
9.0	0.0245	0.0250	0.0256	0.0261	0.0267	0.0272	0.0278	0.0284	0.0290	0.0296	9.0
10.0	0.0302	0.0308	0.0314	0.0320	0.0326	0.0332	0.0338	0.0345	0.0351	0.0358	10.0
11.0	0.0364	0.0371	0.0377	0.0384	0.0391	0.0397	0.0404	0.0411	0.0418	0.0425	11.0
12.0	0.0432	0.0439	0.0447	0.0454	0.0461	0.0468	0.0476	0.0483	0.0491	0.0498	12.0
13.0	0.0506	0.0514	0.0521	0.0529	0.0537	0.0545	0.0553	0.0561	0.0569	0.0577	13.0
14.0	0.0585	0.0593	0.0602	0.0610	0.0618	0.0627	0.0635	0.0644	0.0653	0.0661	14.0
15.0	0.0670	0.0679	0.0687	0.0696	0.0705	0.0714	0.0723	0.0732	0.0741	0.0751	15.0
16.0	0.0760	0.0769	0.0778	0.0788	0.0797	0.0807	0.0816	0.0826	0.0835	0.0845	16.0
17.0	0.0855	0.0865	0.0874	0.0884	0.0894	0.0904	0.0914	0.0924	0.0934	0.0945	17.0
18.0	0.0955	0.0965	0.0976	0.0986	0.0996	0.1007	0.1017	0.1028	0.1039	0.1049	18.0
19.0	0.1060	0.1071	0.1082	0.1092	0.1103	0.1114	0.1125	0.1136	0.1147	0.1159	19.0
20.0	0.1170	0.1181	0.1192	0.1204	0.1215	0.1226	0.1238	0.1249	0.1261	0.1273	20.0
21.0	0.1284	0.1296	0.1308	0.1320	0.1331	0.1343	0.1355	0.1367	0.1379	0.1391	21.0
22.0	0.1403	0.1415	0.1428	0.1440	0.1452	0.1464	0.1477	0.1489	0.1502	0.1514	22.0
23.0	0.1527	0.1539	0.1552	0.1565	0.1577	0.1590	0.1603	0.1616	0.1628	0.1641	23.0
24.0	0.1654	0.1667	0.1680	0.1693	0.1707	0.1720	0.1733	0.1746	0.1759	0.1773	24.0
25.0	0.1786	0.1799	0.1813	0.1826	0.1840	0.1853	0.1867	0.1881	0.1894	0.1908	25.0
26.0	0.1922	0.1935	0.1949	0.1963	0.1977	0.1991	0.2005	0.2019	0.2033	0.2047	26.0
27.0	0.2061	0.2075	0.2089	0.2104	0.2118	0.2132	0.2146	0.2161	0.2175	0.2190	27.0
28.0	0.2204	0.2219	0.2233	0.2248	0.2262	0.2277	0.2291	0.2306	0.2321	0.2336	28.0
29.0	0.2350	0.2365	0.2380	0.2395	0.2410	0.2425	0.2440	0.2455	0.2470	0.2485	29.0
30.0	0.2500	0.2515	0.2530	0.2545	0.2561	0.2576	0.2591	0.2607	0.2622	0.2637	30.0
31.0	0.2653	0.2668	0.2684	0.2699	0.2715	0.2730	0.2746	0.2761	0.2777	0.2792	31.0
32.0	0.2808	0.2824	0.2840	0.2855	0.2871	0.2887	0.2903	0.2919	0.2934	0.2950	32.0
33.0	0.2966	0.2982	0.2998	0.3014	0.3030	0.3046	0.3062	0.3079	0.3095	0.3111	33.0
34.0	0.3127	0.3143	0.3159	0.3176	0.3192	0.3208	0.3224	0.3241	0.3257	0.3274	34.0
35.0	0.3290	0.3306	0.3323	0.3339	0.3356	0.3372	0.3389	0.3405	0.3422	0.3438	35.0
36.0	0.3455	0.3472	0.3488	0.3505	0.3521	0.3538	0.3555	0.3572	0.3588	0.3605	36.0
37.0	0.3622	0.3639	0.3655	0.3672	0.3689	0.3706	0.3723	0.3740	0.3757	0.3773	37.0
38.0	0.3790	0.3807	0.3824	0.3841	0.3858	0.3875	0.3892	0.3909	0.3926	0.3943	38.0
39.0	0.3960	0.3978	0.3995	0.4012	0.4029	0.4046	0.4063	0.4080	0.4097	0.4115	39.0
40.0	0.4132	0.4149	0.4166	0.4183	0.4201	0.4218	0.4235	0.4252	0.4270	0.4287	40.0
41.0	0.4304	0.4321	0.4339	0.4356	0.4373	0.4391	0.4408	0.4425	0.4443	0.4460	41.0
42.0	0.4477	0.4495	0.4512	0.4529	0.4547	0.4564	0.4582	0.4599	0.4616	0.4634	42.0
43.0	0.4651	0.4669	0.4686	0.4703	0.4721	0.4738	0.4756	0.4773	0.4791	0.4808	43.0
44.0	0.4826	0.4843	0.4860	0.4878	0.4895	0.4913	0.4930	0.4948	0.4965	0.4983	44.0
45.0	0.5000	0.5017	0.5035	0.5052	0.5070	0.5087	0.5105	0.5122	0.5140	0.5157	45.0
46.0	0.5174	0.5192	0.5209	0.5227	0.5244	0.5262	0.5279	0.5297	0.5314	0.5331	46.0
47.0	0.5349	0.5366	0.5384	0.5401	0.5418	0.5436	0.5453	0.5471	0.5488	0.5505	47.0
48.0	0.5523	0.5540	0.5557	0.5575	0.5592	0.5609	0.5627	0.5644	0.5661	0.5679	48.0
49.0	0.5696	0.5713	0.5730	0.5748	0.5765	0.5782	0.5799	0.5817	0.5834	0.5851	49.0
50.0	0.5868	0.5885	0.5903	0.5920	0.5937	0.5954	0.5971	0.5988	0.6005	0.6022	50.0
51.0	0.6040	0.6057	0.6074	0.6091	0.6108	0.6125	0.6142	0.6159	0.6176	0.6193	51.0
52.0	0.6210	0.6227	0.6243	0.6260	0.6277	0.6294	0.6311	0.6328	0.6345	0.6361	52.0
53.0	0.6378	0.6395	0.6412	0.6428	0.6445	0.6462	0.6479	0.6495	0.6512	0.6528	53.0
54.0	0.6545	0.6562	0.6578	0.6595	0.6611	0.6628	0.6644	0.6661	0.6677	0.6694	54.0

TABLE B.25 (cont.)   Proportions, p, Corresponding to Arcsine Transformations, p'

p'	0	1	2	3	4	5	6	7	8	9	p'
55.0	0.6710	0.6726	0.6743	0.6759	0.6776	0.6792	0.6808	0.6824	0.6841	0.6857	55.0
56.0	0.6873	0.6889	0.6905	0.6921	0.6938	0.6954	0.6970	0.6986	0.7002	0.7018	56.0
57.0	0.7034	0.7050	0.7066	0.7081	0.7097	0.7113	0.7129	0.7145	0.7160	0.7176	57.0
58.0	0.7192	0.7208	0.7223	0.7239	0.7254	0.7270	0.7285	0.7301	0.7316	0.7332	58.0
59.0	0.7347	0.7363	0.7378	0.7393	0.7409	0.7424	0.7439	0.7455	0.7470	0.7485	59.0
60.0	0.7500	0.7515	0.7530	0.7545	0.7560	0.7575	0.7590	0.7605	0.7620	0.7635	60.0
61.0	0.7650	0.7664	0.7679	0.7694	0.7709	0.7723	0.7738	0.7752	0.7767	0.7781	61.0
62.0	0.7796	0.7810	0.7825	0.7839	0.7854	0.7868	0.7882	0.7896	0.7911	0.7925	62.0
63.0	0.7939	0.7953	0.7967	0.7981	0.7995	0.8009	0.8023	0.8037	0.8051	0.8065	63.0
64.0	0.8078	0.8092	0.8106	0.8119	0.8133	0.8147	0.8160	0.8174	0.8187	0.8201	64.0
65.0	0.8214	0.8227	0.8241	0.8254	0.8267	0.8280	0.8293	0.8307	0.8320	0.8333	65.0
66.0	0.8346	0.8359	0.8372	0.8384	0.8397	0.8410	0.8423	0.8435	0.8448	0.8461	66.0
67.0	0.8473	0.8486	0.8498	0.8511	0.8523	0.8536	0.8548	0.8560	0.8572	0.8585	67.0
68.0	0.8597	0.8609	0.8621	0.8633	0.8645	0.8657	0.8669	0.8680	0.8692	0.8704	68.0
69.0	0.8716	0.8727	0.8739	0.8751	0.8762	0.8774	0.8785	0.8796	0.8808	0.8819	69.0
70.0	0.8830	0.8841	0.8853	0.8864	0.8875	0.8886	0.8897	0.8908	0.8918	0.8929	70.0
71.0	0.8940	0.8951	0.8961	0.8972	0.8983	0.8993	0.9004	0.9014	0.9024	0.9035	71.0
72.0	0.9045	0.9055	0.9066	0.9076	0.9086	0.9096	0.9106	0.9116	0.9126	0.9135	72.0
73.0	0.9145	0.9155	0.9165	0.9174	0.9184	0.9193	0.9203	0.9212	0.9222	0.9231	73.0
74.0	0.9240	0.9249	0.9259	0.9268	0.9277	0.9286	0.9295	0.9304	0.9313	0.9321	74.0
75.0	0.9330	0.9339	0.9347	0.9356	0.9365	0.9373	0.9382	0.9390	0.9398	0.9407	75.0
76.0	0.9415	0.9423	0.9431	0.9439	0.9447	0.9455	0.9463	0.9471	0.9479	0.9486	76.0
77.0	0.9494	0.9502	0.9509	0.9517	0.9524	0.9532	0.9539	0.9546	0.9553	0.9561	77.0
78.0	0.9568	0.9575	0.9582	0.9589	0.9596	0.9603	0.9609	0.9616	0.9623	0.9629	78.0
79.0	0.9636	0.9642	0.9649	0.9655	0.9662	0.9668	0.9674	0.9680	0.9686	0.9692	79.0
80.0	0.9698	0.9704	0.9710	0.9716	0.9722	0.9728	0.9733	0.9739	0.9744	0.9750	80.0
81.0	0.9755	0.9761	0.9766	0.9771	0.9776	0.9782	0.9787	0.9792	0.9797	0.9801	81.0
82.0	0.9806	0.9811	0.9816	0.9820	0.9825	0.9830	0.9834	0.9839	0.9843	0.9847	82.0
83.0	0.9851	0.9856	0.9860	0.9864	0.9868	0.9872	0.9876	0.9880	0.9883	0.9887	83.0
84.0	0.9891	0.9894	0.9898	0.9901	0.9905	0.9908	0.9911	0.9915	0.9918	0.9921	84.0
85.0	0.9924	0.9927	0.9930	0.9933	0.9936	0.9938	0.9941	0.9944	0.9946	0.9949	85.0
86.0	0.9951	0.9954	0.9956	0.9958	0.9961	0.9963	0.9965	0.9967	0.9969	0.9971	86.0
87.0	0.9973	0.9974	0.9976	0.9978	0.9979	0.9981	0.9982	0.9984	0.9985	0.9987	87.0
88.0	0.9988	0.9989	0.9990	0.9991	0.9992	0.9993	0.9994	0.9995	0.9996	0.9996	88.0
89.0	0.9997	0.9998	0.9998	0.9999	0.9999	0.9999	1.0000	1.0000	1.0000	1.0000	89.0
90.0	1.0000										

By Equation 13.6:

$$p = (\sin p')^2$$

Examples:

$$p' = 46.2, \quad p = 0.5209 \quad \text{and} \quad p' = 85.3, \quad p = 0.9933$$

TABLE B.26a Binomial Coefficients, $_nC_x$

X	n = 1	2	3	4	5	6	7	8	9	10	11	12	13	14	15	16	17
0	1	1	1	1	1	1	1	1	1	1	1	1	1	1	1	1	1
1	1	2	3	4	5	6	7	8	9	10	11	12	13	14	15	16	17
2		1	3	6	10	15	21	28	36	45	55	66	78	91	105	120	136
3			1	4	10	20	35	56	84	120	165	220	286	364	455	560	680
4				1	5	15	35	70	126	210	330	495	715	1001	1365	1820	2380
5					1	6	21	56	126	252	462	792	1287	2002	3003	4368	6188
6						1	7	28	84	210	462	924	1716	3003	5005	8008	12376
7							1	8	36	120	330	792	1716	3432	6435	11440	19448
8								1	9	45	165	495	1287	3003	6435	12870	24310
9									1	10	55	220	715	2002	5005	11440	24310
10										1	11	66	286	1001	3003	8008	19448
11											1	12	78	364	1365	4368	12376
12												1	13	91	455	1820	6188
13													1	14	105	560	2380

X	n = 18	19	20	21	22	23	24	25	26	27	28	29
0	1	1	1	1	1	1	1	1	1	1	1	1
1	18	19	20	21	22	23	24	25	26	27	28	29
2	153	171	190	210	231	253	276	300	325	351	378	406
3	816	969	1140	1330	1540	1771	2024	2300	2600	2925	3276	3654
4	3060	3876	4845	5985	7315	8855	10626	12650	14950	17550	20475	23751
5	8568	11628	15504	20349	26334	33649	42504	53130	65780	80730	98280	118755
6	18564	27132	38760	54264	74613	100947	134596	177100	230230	296010	376740	475020
7	31824	50388	77520	116280	170544	245157	346104	480700	657800	888030	1184040	1560780
8	43758	75582	125970	203490	319770	490314	735471	1081575	1562275	2220075	3108105	4292145
9	48620	92378	167960	293930	497420	817190	1307504	2042975	3124550	4686825	6906900	10015005
10	43758	92378	184756	352716	646646	1144066	1961256	3268760	5311735	8436285	13123110	20030010
11	31824	75582	167960	352716	705432	1352078	2496144	4457400	7726160	13037895	21474180	34597290
12	18564	50388	125970	293930	646646	1352078	2704156	5200300	9657700	17383860	30421755	51895935
13	8568	27132	77520	203490	497420	1144066	2496144	5200300	10400600	20058300	37442160	67863915
14	3060	11628	38760	116280	319770	817190	1961256	4457400	9657700	20058300	40116600	77558760
15	816	3876	15504	54264	170544	490314	1307504	3268760	7726160	17383860	37442160	77558760
16	153	969	4845	20349	74613	245157	735471	2042975	5311735	13037895	30421755	67863915
17	18	171	1140	5985	26334	100947	346104	1081575	3124550	8436285	21474180	51895935

TABLE B.26a (cont.) Binomial Coefficients, $_nC_X$

X	n = 30	31	32	33	34	35	36	37	38	X
0	1	1	1	1	1	1	1	1	1	0
1	30	31	32	33	34	35	36	37	38	1
2	435	465	496	528	561	595	630	666	703	2
3	4060	4495	4960	5456	5984	6545	7140	7770	8436	3
4	27405	31465	35960	40920	46376	52360	58905	66045	73815	4
5	142506	169911	201376	237336	278256	324632	376992	435897	501942	5
6	593775	736281	906192	1107568	1344904	1623160	1947792	2324784	2760681	6
7	2035800	2629575	3365856	4272048	5379616	6724520	8347680	10295472	12620256	7
8	5852925	7888725	10518300	13884156	18156204	23535820	30260340	38608020	48903492	8
9	14307150	20160075	28048800	38567100	52451256	70607460	94143280	124403620	163011640	9
10	30045015	44352165	64512240	92561040	131128140	183579396	254186856	348330136	472733756	10
11	54627300	84672315	129024480	193536720	286097760	417225900	600805296	854992152	1203322288	11
12	86493225	141120525	225792840	354817320	548354040	834451800	1251677700	1852482996	2707475148	12
13	119759850	206253075	347373600	573166440	927983760	1476337800	2310789600	3562467300	5414950296	13
14	145422675	265182525	471435600	818809200	1391975640	2319959400	3796297200	6107086800	9669554100	14
15	155117520	300540195	565722720	1037158320	1855967520	3247943160	5567902560	9364199760	15471286560	15
16	145422675	300540195	601080390	1166803110	2203961430	4059928950	7307872110	12875774670	22239974430	16
17	119759850	265182525	565722720	1166803110	2333606220	4537567650	8597496600	15905368710	28781143380	17
18	86493225	206253075	471435600	1037158320	2203961430	4537567650	9075135300	17672631900	33578000610	18
19	54627300	141120525	347373600	818809200	1855967520	4059928950	8597496600	17672631900	35345263800	19
20	30045015	84672315	225792840	573166440	1391975640	3247943160	7307872110	15905368710	33578000610	20
21	14307150	44352165	129024480	354817320	927983760	2319959400	5567902560	12875774670	28781143380	21
22	5852925	20160075	64512240	193536720	548354040	1476337800	3796297200	9364199760	22239974430	22

TABLE B.26a (cont.) Binomial Coefficients, $_nC_X$

X	n = 39	40	41	42	43	44	45	46	X
0	1	1	1	1	1	1	1	1	0
1	39	40	41	42	43	44	45	46	1
2	741	780	820	861	903	946	990	1035	2
3	9139	9880	10660	11480	12341	13244	14190	15180	3
4	82251	91390	101270	111930	123410	135751	148995	163185	4
5	575757	658008	749398	850668	962598	1086008	1221759	1370754	5
6	3262623	3838380	4496388	5245786	6096454	7059052	8145060	9366819	6
7	15380937	18643560	22481940	26978328	32224114	38320568	45379620	53524680	7
8	61523748	76904685	95548245	118030185	145008513	177232627	215553195	260932815	8
9	211915132	273438880	350343565	445891810	563921995	708930508	886163135	1101716330	9
10	635745396	847660528	1121099408	1471442973	1917334783	2481256778	3190187286	4076350421	10
11	1676056044	2311801440	3159461968	4280561376	5752004349	7669339132	10150595910	13340783196	11
12	3910797436	5586853480	7899865920	11058116888	15338678264	21090682613	28760021745	38910617655	12
13	8122425444	12033222880	17620007630	25518731280	36576848168	51915526432	73006209045	101766230790	13
14	15084504396	23206929840	35240152720	52860229080	78378960360	114955808528	166871334960	239877544005	14
15	25140840660	40225345056	63432274896	98672427616	151153265696	229911617056	344867425584	511738760544	15
16	37711260990	62852101650	103077446706	166509721602	265182149218	416714805914	646626422970	991493848554	16
17	51021117810	88732378800	151584480450	254661927156	421171648758	686353797976	1103068603890	1749695026860	17
18	62359143990	113380261800	202112640600	353697121050	608359048206	1029530699640	1715884494940	2818953098830	18
19	68923264410	131282408400	244662670200	446775310800	800472431850	1408831480056	2438362177020	4154246671960	19
20	68923264410	137846528820	269128937220	513791607420	960566918220	1761039350070	3169870830126	5608233007146	20
21	62359143990	131282408400	269128937220	538257874440	1052049481860	2012616400080	3773655750150	6943526580276	21
22	51021117810	113380261800	244662670200	513791607420	1052049481860	2104098963720	4116715363800	7890371113950	22
23	37711260990	88732378800	202112640600	446775310800	960566918220	2012616400080	4116715363800	8233430727600	23
24	25140840660	62852101650	151584480450	353697121050	800472431850	1761039350070	3773655750150	7890371113950	24
25	15084504396	40225345056	103077446706	254661927156	608359048206	1408831480056	3169870830126	6943526580276	25

Shown are binomial coefficients, $_nC_X = \binom{n}{X}$, which are the numbers of combinations of n items taken X at a time. For each n, the array of binomial coefficients is symmetrical; the middle of the array is shown in italics, and the coefficient for X is the same as for $n - X$. For example $_{18}C_{13} = _{18}C_5 = 8568$ and $_{46}C_{44} = _{46}C_2 = 1035$.

TABLE B.26b   Proportions of the Binomial Distributions for $p = q = 0.5$

X	$n = 1$	2	3	4	5	6	7	8	9	10	X
0	0.50000	0.25000	0.12500	0.06250	0.03125	0.01563	0.00781	0.00391	0.00195	0.00098	0
1	0.50000	0.50000	0.37500	0.25000	0.15625	0.09375	0.05469	0.03125	0.01758	0.00977	1
2		0.25000	0.37500	0.37500	0.31250	0.23438	0.16406	0.10938	0.07031	0.04395	2
3			0.12500	0.25000	0.31250	0.31250	0.27344	0.21875	0.16406	0.11719	3
4				0.06250	0.15625	0.23438	0.27344	0.27344	0.24609	0.20508	4
5					0.03125	0.09375	0.16406	0.21875	0.24609	0.24609	5
6						0.01563	0.05469	0.10938	0.16406	0.20508	6
7							0.00781	0.03125	0.07031	0.11719	7
8								0.00391	0.01758	0.04395	8
9									0.00195	0.00977	9
10										0.00098	10

X	$n = 11$	12	13	14	15	16	17	18	19	20	X
0	0.00049	0.00024	0.00012	0.00006	0.00003	0.00002	0.00001	0.00000	0.00000	0.00000	0
1	0.00537	0.00293	0.00159	0.00085	0.00046	0.00024	0.00013	0.00007	0.00004	0.00002	1
2	0.02686	0.01611	0.00952	0.00555	0.00320	0.00183	0.00104	0.00058	0.00033	0.00018	2
3	0.08057	0.05371	0.03491	0.02222	0.01389	0.00854	0.00519	0.00311	0.00185	0.00109	3
4	0.16113	0.12085	0.08728	0.06110	0.04166	0.02777	0.01816	0.01167	0.00739	0.00462	4
5	0.22559	0.19336	0.15710	0.12219	0.09164	0.06665	0.04721	0.03268	0.02218	0.01479	5
6	0.22559	0.22559	0.20947	0.18329	0.15274	0.12219	0.09442	0.07082	0.05175	0.03696	6
7	0.16113	0.19336	0.20947	0.20947	0.19638	0.17456	0.14838	0.12140	0.09611	0.07393	7
8	0.08057	0.12085	0.15710	0.18329	0.19638	0.19638	0.18547	0.16692	0.14416	0.12013	8
9	0.02686	0.05371	0.08728	0.12219	0.15274	0.17456	0.18547	0.18547	0.17620	0.16018	9
10	0.00537	0.01611	0.03491	0.06110	0.09164	0.12219	0.14838	0.16692	0.17620	0.17620	10
11	0.00049	0.00293	0.00952	0.02222	0.04166	0.06665	0.09442	0.12140	0.14416	0.16018	11
12		0.00024	0.00159	0.00555	0.01389	0.02777	0.04721	0.07082	0.09611	0.12013	12
13			0.00012	0.00085	0.00320	0.00854	0.01816	0.03268	0.05175	0.07393	13
14				0.00006	0.00046	0.00183	0.00519	0.01167	0.02218	0.03696	14
15					0.00003	0.00024	0.00104	0.00311	0.00739	0.01479	15
16						0.00002	0.00013	0.00058	0.00185	0.00462	16
17							0.00001	0.00007	0.00033	0.00109	17
18								0.00000	0.00004	0.00018	18
19									0.00000	0.00002	19
20										0.00000	20

$$P(X) = \frac{n!}{X!(n-X)!} p^X (1-p)^{n-X}$$

TABLE B.27 Critical Values of C for the Sign Test or the Binomial Test with $p = 0.5$

n	$\alpha(2)$: .50 / $\alpha(1)$: .25	.20 / .10	.10 / .05	.05 / .025	.02 / .01	.01 / .005	.005 / .0025	.002 / .001	.001 / .0005
2	—	—	—	—	—	—	—	—	—
3	0	—	—	—	—	—	—	—	—
4	0	0	—	—	—	—	—	—	—
5	1	0	0	—	—	—	—	—	—
6	1	0	0	0	—	—	—	—	—
7	2	1	0	0	0	—	—	—	—
8	2	1	1	0	0	0	—	—	—
9	2	2	1	1	0	0	0	—	—
10	3	2	1	1	0	0	0	0	—
11	3	2	2	1	1	0	0	0	0
12	4	3	2	2	1	1	0	0	0
13	4	3	3	2	1	1	1	0	0
14	5	4	3	2	2	1	1	1	0
15	5	4	3	3	2	2	1	1	1
16	6	4	4	3	2	2	2	1	1
17	6	5	4	4	3	2	2	1	1
18	7	5	5	4	3	2	2	2	1
19	7	6	5	4	4	3	3	2	2
20	7	6	5	5	4	3	3	2	2
21	8	7	6	5	4	4	3	3	2
22	8	7	6	5	5	4	4	3	3
23	9	7	7	6	5	4	4	3	3
24	9	8	7	6	5	5	4	4	3
25	10	8	7	7	6	5	5	4	4
26	10	9	8	7	6	6	5	4	4
27	11	9	8	7	7	6	5	5	4
28	11	10	9	8	7	6	6	5	5
29	12	10	9	8	7	7	6	5	5
30	12	10	10	9	8	7	6	6	5
31	13	11	10	9	8	7	7	6	6
32	13	11	10	9	8	8	7	6	6
33	14	12	11	10	9	8	8	7	6
34	14	12	11	10	9	9	8	7	7
35	15	13	12	11	10	9	8	8	7
36	15	13	12	11	10	9	9	8	7
37	15	14	13	12	10	10	9	8	8
38	16	14	13	12	11	10	9	9	8
39	16	15	13	12	11	11	10	9	8
40	17	15	14	13	12	11	10	9	9
41	17	15	14	13	12	11	11	10	9
42	18	16	15	14	13	12	11	10	10
43	18	16	15	14	13	12	11	11	10
44	19	17	16	15	13	13	12	11	10
45	19	17	16	15	14	13	12	11	11
46	20	18	16	15	14	13	13	12	11
47	20	18	17	16	15	14	13	12	11
48	21	19	17	16	15	14	13	12	12
49	21	19	18	17	15	15	14	13	12
50	22	19	18	17	16	15	14	13	13

n	$\alpha(2)$: .50 / $\alpha(1)$: .25	.20 / .10	.10 / .05	.05 / .025	.02 / .01	.01 / .005	.005 / .0025	.002 / .001	.001 / .0005
51	22	20	19	18	16	15	15	14	13
52	23	20	19	18	17	16	15	14	13
53	23	21	20	18	17	16	15	14	14
54	24	21	20	19	18	17	16	15	14
55	24	22	20	19	18	17	16	15	14
56	24	22	21	20	18	17	17	16	15
57	25	23	21	20	19	18	17	16	15
58	25	23	22	21	19	18	17	16	16
59	26	24	22	21	20	19	18	17	16
60	26	24	23	21	20	19	18	17	16
61	27	24	23	22	20	20	19	18	17
62	27	25	24	22	21	20	19	18	17
63	28	25	24	23	21	20	19	18	18
64	28	26	24	23	22	21	20	19	18
65	29	26	25	24	22	21	20	19	18
66	29	27	25	24	23	22	21	20	19
67	30	27	26	25	23	22	21	20	19
68	30	28	26	25	23	22	22	20	20
69	31	28	27	25	24	23	22	21	20
70	31	29	27	26	24	23	22	21	20
71	32	29	28	26	25	24	23	22	21
72	32	30	28	27	25	24	23	22	21
73	33	30	28	27	26	25	24	22	22
74	33	30	29	28	26	25	24	23	22
75	34	31	29	28	26	25	24	23	22
76	34	31	30	28	27	26	25	24	23
77	35	32	30	29	27	26	25	24	23
78	35	32	31	29	28	27	26	24	24
79	36	33	31	30	28	27	26	25	24
80	36	33	32	30	29	28	27	25	24
81	36	34	32	31	29	28	27	26	25
82	37	34	33	31	30	28	27	26	25
83	37	35	33	32	30	29	28	27	26
84	38	35	33	32	30	29	28	27	26
85	38	36	34	32	31	30	29	27	26
86	39	36	34	33	31	30	29	28	27
87	39	37	35	33	32	31	29	28	27
88	40	37	35	34	32	31	30	29	28
89	40	37	36	34	33	31	30	29	28
90	41	38	36	35	33	32	31	29	29
91	41	38	37	35	33	32	31	30	29
92	42	39	37	36	34	33	32	30	29
93	42	39	38	36	34	33	32	31	30
94	43	40	38	37	35	34	32	31	30
95	43	40	38	37	35	34	33	32	31
96	44	41	39	37	36	34	33	32	31
97	44	41	39	38	36	35	34	32	31
98	45	42	40	38	37	35	34	33	32
99	45	42	40	39	37	36	35	33	32
100	46	43	41	39	37	36	35	34	33

TABLE B.27 (cont.)   Critical Values of C for the Sign Test or the Binomial Test with $p = 0.5$

n	α(2): .50 / α(1): .25	.20 / .10	.10 / .05	.05 / .025	.02 / .01	.01 / .005	.005 / .0025	.002 / .001	.001 / .0005
101	46	43	41	40	38	37	35	34	33
102	47	44	42	40	38	37	36	35	34
103	47	44	42	41	39	37	36	35	34
104	48	44	43	41	39	38	37	35	34
105	48	45	43	41	40	38	37	36	35
106	49	45	44	42	40	39	38	36	35
107	49	46	44	42	41	39	38	37	36
108	49	46	44	43	41	40	38	37	36
109	50	47	45	43	41	40	39	37	36
110	50	47	45	44	42	41	39	38	37
111	51	48	46	44	42	41	40	38	37
112	51	48	46	45	43	41	40	39	38
113	52	49	47	45	43	42	41	39	38
114	52	49	47	46	44	42	41	40	39
115	53	50	48	46	44	43	42	40	39
116	53	50	48	46	45	43	42	40	39
117	54	51	49	47	45	44	42	41	40
118	54	51	49	47	45	44	43	41	40
119	55	52	50	48	46	45	43	42	41
120	55	52	50	48	46	45	44	42	41
121	56	52	50	49	47	45	44	43	42
122	56	53	51	49	47	46	45	43	42
123	57	53	51	50	48	46	45	43	42
124	57	54	52	50	48	47	45	44	43
125	58	54	52	51	49	47	46	44	43
126	58	55	53	51	49	48	46	45	44
127	59	55	53	51	49	48	47	45	44
128	59	56	54	52	50	48	47	46	45
129	60	56	54	52	50	49	48	46	45
130	60	57	55	53	51	49	48	46	45
131	61	57	55	53	51	50	49	47	46
132	61	58	56	54	52	50	49	47	46
133	62	58	56	54	52	51	49	48	47
134	62	59	56	55	53	51	50	48	47
135	63	59	57	55	53	52	50	49	48
136	63	60	57	56	53	52	51	49	48
137	64	60	58	56	54	52	51	50	48
138	64	60	58	57	54	53	52	50	49
139	65	61	59	57	55	53	52	50	49
140	65	61	59	57	55	54	52	51	50
141	65	62	60	58	56	54	53	51	50
142	66	62	60	58	56	55	53	52	51
143	66	63	61	59	57	55	54	52	51
144	67	63	61	59	57	56	54	53	51
145	67	64	62	60	58	56	55	53	52
146	68	64	62	60	58	56	55	53	52
147	68	65	63	61	58	57	56	54	53
148	69	65	63	61	59	57	56	54	53
149	69	66	63	62	59	58	56	55	54
150	70	66	64	62	60	58	57	55	54

n	α(2): .50 / α(1): .25	.20 / .10	.10 / .05	.05 / .025	.02 / .01	.01 / .005	.005 / .0025	.002 / .001	.001 / .0005
151	70	67	64	62	60	59	57	56	54
152	71	67	65	63	61	59	58	56	55
153	71	68	65	63	61	60	58	56	55
154	72	68	66	64	62	60	59	57	56
155	72	69	66	64	62	61	59	57	56
156	73	69	67	65	63	61	60	58	57
157	73	69	67	65	63	61	60	58	57
158	74	70	68	66	63	62	60	59	57
159	74	70	68	66	64	62	61	59	58
160	75	71	69	67	64	63	61	60	58
161	75	71	69	67	65	63	62	60	59
162	76	72	70	68	65	64	62	60	59
163	76	72	70	68	66	64	63	61	60
164	77	73	70	68	66	65	63	61	60
165	77	73	71	69	67	65	64	62	60
166	78	74	71	69	67	65	64	62	61
167	78	74	72	70	68	66	64	63	61
168	79	75	72	70	68	66	65	63	62
169	79	75	73	71	68	67	65	63	62
170	80	76	73	71	69	67	66	64	63
171	80	76	74	72	69	68	66	64	63
172	81	77	74	72	70	68	67	65	64
173	81	77	75	73	70	69	67	65	64
174	82	78	75	73	71	69	68	66	64
175	82	78	76	74	71	70	68	66	65
176	83	79	76	74	72	70	68	67	65
177	83	79	77	74	72	70	69	67	66
178	83	79	77	75	73	71	69	67	66
179	84	80	78	75	73	71	70	68	67
180	84	80	78	76	73	72	70	68	67
181	85	81	78	76	74	72	71	69	67
182	85	81	79	77	74	73	71	69	68
183	86	82	79	77	75	73	72	70	68
184	86	82	80	78	75	74	72	70	69
185	87	83	80	78	76	74	72	71	69
186	87	83	81	79	76	74	73	71	70
187	88	84	81	79	77	75	73	71	70
188	88	84	82	80	77	75	74	72	71
189	89	85	82	80	78	76	74	72	71
190	89	85	83	81	78	76	75	73	71
191	90	86	83	81	78	77	75	73	72
192	90	86	84	81	79	77	76	74	72
193	91	87	84	82	79	78	76	74	73
194	91	87	85	82	80	78	77	75	73
195	92	88	85	83	80	79	77	75	74
196	92	88	85	83	81	79	77	75	74
197	93	89	86	84	81	79	78	76	75
198	93	89	86	84	82	80	78	76	75
199	94	89	87	85	82	80	79	77	75
200	94	90	87	85	83	81	79	77	76

TABLE B.27 (cont.)   Critical Values of C for the Sign Test or the Binomial Test with $p = 0.5$

n	α(2): .50 / α(1): .25	.20 / .10	.10 / .05	.05 / .025	.02 / .01	.01 / .005	.005 / .0025	.002 / .001	.001 / .0005	n	α(2): .50 / α(1): .25	.20 / .10	.10 / .05	.05 / .025	.02 / .01	.01 / .005	.005 / .0025	.002 / .001	.001 / .0005
201	95	90	88	86	83	81	80	78	76	251	119	114	111	109	106	104	102	100	99
202	95	91	88	86	83	82	80	78	77	252	120	115	112	109	107	105	103	101	99
203	96	91	89	87	84	82	81	79	77	253	120	115	112	110	107	105	103	101	99
204	96	92	89	87	84	83	81	79	78	254	121	116	113	110	107	106	104	101	100
205	97	92	90	87	85	83	81	79	78	255	121	116	113	111	108	106	104	102	100
206	97	93	90	88	85	84	82	80	78	256	122	117	114	111	108	106	105	102	101
207	98	93	91	88	86	84	82	80	79	257	122	117	114	112	109	107	105	103	101
208	98	94	91	89	86	84	83	81	79	258	123	118	115	112	109	107	106	103	102
209	99	94	92	89	87	85	83	81	80	259	123	118	115	113	110	108	106	104	102
210	99	95	92	90	87	85	84	82	80	260	124	119	116	113	110	108	106	104	103
211	100	95	93	90	88	86	84	82	81	261	124	119	116	114	111	109	107	105	103
212	100	96	93	91	88	86	85	83	81	262	125	120	117	114	111	109	107	105	103
213	101	96	94	91	89	87	85	83	82	263	125	120	117	115	112	110	108	106	104
214	101	97	94	92	89	87	86	83	82	264	126	121	118	115	112	110	108	106	104
215	102	97	94	92	89	88	86	84	82	265	126	121	118	116	113	111	109	106	105
216	102	98	95	93	90	88	86	84	83	266	126	122	119	116	113	111	109	107	105
217	103	98	95	93	90	89	87	85	83	267	127	122	119	117	114	111	110	107	106
218	103	99	96	94	91	89	87	85	84	268	127	123	120	117	114	112	110	108	106
219	104	99	96	94	91	89	88	86	84	269	128	123	120	117	114	112	111	108	107
220	104	99	97	94	92	90	88	86	85	270	128	123	120	118	115	113	111	109	107
221	104	100	97	95	92	90	89	87	85	271	129	124	121	118	115	113	111	109	107
222	105	100	98	95	93	91	89	87	86	272	129	124	121	119	116	114	112	110	108
223	105	101	98	96	93	91	90	88	86	273	130	125	122	119	116	114	112	110	108
224	106	101	99	96	94	92	90	88	86	274	130	125	122	120	117	115	113	110	109
225	106	102	99	97	94	92	91	88	87	275	131	126	123	120	117	115	113	111	109
226	107	102	100	97	95	93	91	89	87	276	131	126	123	121	118	116	114	111	110
227	107	103	100	98	95	93	91	89	88	277	132	127	124	121	118	116	114	112	110
228	108	103	101	98	95	94	92	90	88	278	132	127	124	122	119	117	115	112	111
229	108	104	101	99	96	94	92	90	89	279	133	128	125	122	119	117	115	113	111
230	109	104	102	99	96	95	93	91	89	280	133	128	125	123	120	117	116	113	112
231	109	105	102	100	97	95	93	91	90	281	134	129	126	123	120	118	116	114	112
232	110	105	102	100	97	95	94	92	90	282	134	129	126	124	120	118	116	114	112
233	110	106	103	101	98	96	94	92	90	283	135	130	127	124	121	119	117	115	113
234	111	106	103	101	98	96	95	92	91	284	135	130	127	124	121	119	117	115	113
235	111	107	104	101	99	97	95	93	91	285	136	131	128	125	122	120	118	115	114
236	112	107	104	102	99	97	95	93	92	286	136	131	128	125	122	120	118	116	114
237	112	108	105	102	100	98	96	94	92	287	137	132	129	126	123	121	119	116	115
238	113	108	105	103	100	98	96	94	93	288	137	132	129	126	123	121	119	117	115
239	113	109	106	103	101	99	97	95	93	289	138	133	130	127	124	122	120	117	116
240	114	109	106	104	101	99	97	95	94	290	138	133	130	127	124	122	120	118	116
241	114	110	107	104	101	100	98	96	94	291	139	134	130	128	125	123	121	118	117
242	115	110	107	105	102	100	98	96	94	292	139	134	131	128	125	123	121	119	117
243	115	111	108	105	102	100	99	96	95	293	140	135	131	129	126	123	122	119	117
244	116	111	108	106	103	101	99	97	95	294	140	135	132	129	126	124	122	120	118
245	116	111	109	106	103	101	100	97	96	295	141	135	132	130	127	124	122	120	118
246	117	112	109	107	104	102	100	98	96	296	141	136	133	130	127	125	123	120	119
247	117	112	110	107	104	102	100	98	97	297	142	136	133	131	127	125	123	121	119
248	118	113	110	108	105	103	101	99	97	298	142	137	134	131	128	126	124	121	120
249	118	113	111	108	105	103	101	99	98	299	143	137	134	132	128	126	124	122	120
250	119	114	111	109	106	104	102	100	98	300	143	138	135	132	129	127	125	122	121

TABLE B.27 (cont.)   Critical Values of C for the Sign Test or the Binomial Test with $p = 0.5$

n	$\alpha(2)$: .50 / $\alpha(1)$: .25	.20 / .10	.10 / .05	.05 / .025	.02 / .01	.01 / .005	.005 / .0025	.002 / .001	.001 / .0005	n	$\alpha(2)$: .50 / $\alpha(1)$: .25	.20 / .10	.10 / .05	.05 / .025	.02 / .01	.01 / .005	.005 / .0025	.002 / .001	.001 / .0005
301	144	138	135	133	129	127	125	123	121	351	168	162	159	156	153	150	148	146	144
302	144	139	136	133	130	128	126	123	121	352	169	163	160	157	153	151	149	146	144
303	145	139	136	133	130	128	126	124	122	353	169	163	160	157	154	151	149	147	145
304	145	140	137	134	131	129	127	124	122	354	170	164	161	158	154	152	150	147	145
305	146	140	137	134	131	129	127	125	123	355	170	164	161	158	155	152	150	147	146
306	146	141	138	135	132	130	127	125	123	356	171	165	161	159	155	153	151	148	146
307	147	141	138	135	132	130	128	125	124	357	171	165	162	159	155	153	151	148	146
308	147	142	139	136	133	130	128	126	124	358	172	166	162	159	156	154	151	149	147
309	148	142	139	136	133	131	129	126	125	359	172	166	163	160	156	154	152	149	147
310	148	143	140	137	134	131	129	127	125	360	173	167	163	160	157	155	152	150	148
311	149	143	140	137	134	132	130	127	126	361	173	167	164	161	157	155	153	150	148
312	149	144	140	138	134	132	130	128	126	362	174	168	164	161	158	156	153	151	149
313	150	144	141	138	135	133	131	128	126	363	174	168	165	162	158	156	154	151	149
314	150	145	141	139	135	133	131	129	127	364	175	169	165	162	159	156	154	152	150
315	151	145	142	139	136	134	132	129	127	365	175	169	166	163	159	157	155	152	150
316	151	146	142	140	136	134	132	130	128	366	176	170	166	163	160	157	155	152	151
317	151	146	143	140	137	135	133	130	128	367	176	170	167	164	160	158	156	153	151
318	152	147	143	141	137	135	133	131	129	368	177	171	167	164	161	158	156	153	152
319	152	147	144	141	138	136	133	131	129	369	177	171	168	165	161	159	157	154	152
320	153	148	144	141	138	136	134	131	130	370	178	172	168	165	162	159	157	154	152
321	153	148	145	142	139	136	134	132	130	371	178	172	169	166	162	160	158	155	153
322	154	149	145	142	139	137	135	132	131	372	178	173	169	166	163	160	158	155	153
323	154	149	146	143	140	137	135	133	131	373	179	173	170	167	163	161	158	156	154
324	155	149	146	143	140	138	136	133	131	374	179	174	170	167	164	161	159	156	154
325	155	150	147	144	141	138	136	134	132	375	180	174	171	168	164	162	159	157	155
326	156	150	147	144	141	139	137	134	132	376	180	175	171	168	164	162	160	157	155
327	156	151	148	145	141	139	137	135	133	377	181	175	172	168	165	163	160	158	156
328	157	151	148	145	142	140	138	135	133	378	181	176	172	169	165	163	161	158	156
329	157	152	149	146	142	140	138	136	134	379	182	176	172	169	166	163	161	158	157
330	158	152	149	146	143	141	139	136	134	380	182	177	173	170	166	164	162	159	157
331	158	153	150	147	143	141	139	136	135	381	183	177	173	170	167	164	162	159	157
332	159	153	150	147	144	142	139	137	135	382	183	177	174	171	167	165	163	160	158
333	159	154	150	148	144	142	140	137	136	383	184	178	174	171	168	165	163	160	158
334	160	154	151	148	145	142	140	138	136	384	184	178	175	172	168	166	164	161	159
335	160	155	151	149	145	143	141	138	136	385	185	179	175	172	169	166	164	161	159
336	161	155	152	149	146	143	141	139	137	386	185	179	176	173	169	167	164	162	160
337	161	156	152	150	146	144	142	139	137	387	186	180	176	173	170	167	165	162	160
338	162	156	153	150	147	144	142	140	138	388	186	180	177	174	170	168	165	163	161
339	162	157	153	150	147	145	143	140	138	389	187	181	177	174	171	168	166	163	161
340	163	157	154	151	148	145	143	141	139	390	187	181	178	175	171	169	166	164	162
341	163	158	154	151	148	146	144	141	139	391	188	182	178	175	172	169	167	164	162
342	164	158	155	152	149	146	144	141	140	392	188	182	179	176	172	170	167	164	162
343	164	159	155	152	149	147	145	142	140	393	189	183	179	176	172	170	168	165	163
344	165	159	156	153	149	147	145	142	141	394	189	183	180	177	173	170	168	165	163
345	165	160	156	153	150	148	145	143	141	395	190	184	180	177	173	171	169	166	164
346	166	160	157	154	150	148	146	143	141	396	190	184	181	178	174	171	169	166	164
347	166	161	157	154	151	149	146	144	142	397	191	185	181	178	174	172	170	167	165
348	167	161	158	155	151	149	147	144	142	398	191	185	182	178	175	172	170	167	165
349	167	162	158	155	152	149	147	145	143	399	192	186	182	179	175	173	171	168	166
350	168	162	159	156	152	150	148	145	143	400	192	186	183	179	176	173	171	168	166

TABLE B.27 (cont.)   Critical Values of C for the Sign Test or the Binomial Test with $p = 0.5$

n	α(2): .50 / α(1): .25	.20 / .10	.10 / .05	.05 / .025	.02 / .01	.01 / .005	.005 / .0025	.002 / .001	.001 / .0005
401	193	187	183	180	176	174	171	169	167
402	193	187	184	180	177	174	172	169	167
403	194	188	184	181	177	175	172	170	168
404	194	188	184	181	178	175	173	170	168
405	195	189	185	182	178	176	173	170	168
406	195	189	185	182	179	176	174	171	169
407	196	190	186	183	179	177	174	171	169
408	196	190	186	183	180	177	175	172	170
409	197	191	187	184	180	177	175	172	170
410	197	191	187	184	180	178	176	173	171
411	198	192	188	185	181	178	176	173	171
412	198	192	188	185	181	179	177	174	172
413	199	192	189	186	182	179	177	174	172
414	199	193	189	186	182	180	177	175	173
415	200	193	190	187	183	180	178	175	173
416	200	194	190	187	183	181	178	176	174
417	201	194	191	187	184	181	179	176	174
418	201	195	191	188	184	182	179	176	174
419	202	195	192	188	185	182	180	177	175
420	202	196	192	189	185	183	180	177	175
421	203	196	193	189	186	183	181	178	176
422	203	197	193	190	186	184	181	178	176
423	204	197	194	190	187	184	182	179	177
424	204	198	194	191	187	185	182	179	177
425	205	198	195	191	188	185	183	180	178
426	205	199	195	192	188	185	183	180	178
427	206	199	196	192	188	186	184	181	179
428	206	200	196	193	189	186	184	181	179
429	207	200	196	193	189	187	184	182	179
430	207	201	197	194	190	187	185	182	180
431	207	201	197	194	190	188	185	182	180
432	208	202	198	195	191	188	186	183	181
433	208	202	198	195	191	189	186	183	181
434	209	203	199	196	192	189	187	184	182
435	209	203	199	196	192	190	187	184	182
436	210	204	200	197	193	190	188	185	183
437	210	204	200	197	193	191	188	185	183
438	211	205	201	198	194	191	189	186	184
439	211	205	201	198	194	192	189	186	184
440	212	206	202	198	195	192	190	187	185
441	212	206	202	199	195	192	190	187	185
442	213	207	203	199	196	193	191	188	185
443	213	207	203	200	196	193	191	188	186
444	214	207	204	200	197	194	191	188	186
445	214	208	204	201	197	194	192	189	187
446	215	208	205	201	197	195	192	189	187
447	215	209	205	202	198	195	193	190	188
448	216	209	206	202	198	196	193	190	188
449	216	210	206	203	199	196	194	191	189
450	217	210	207	203	199	197	194	191	189
451	217	211	207	204	200	197	195	192	190
452	218	211	208	204	200	198	195	192	190
453	218	212	208	205	201	198	196	193	191
454	219	212	208	205	201	199	196	193	191
455	219	213	209	206	202	199	197	194	191
456	220	213	209	206	202	200	197	194	192
457	220	214	210	207	203	200	198	195	192
458	221	214	210	207	203	200	198	195	193
459	221	215	211	208	204	201	198	195	193
460	222	215	211	208	204	201	199	196	194
461	222	216	212	208	205	202	199	196	194
462	223	216	212	209	205	202	200	197	195
463	223	217	213	209	205	203	200	197	195
464	224	217	213	210	206	203	201	198	196
465	224	218	214	210	206	204	201	198	196
466	225	218	214	211	207	204	202	199	197
467	225	219	215	211	207	205	202	199	197
468	226	219	215	212	208	205	203	200	197
469	226	220	216	212	208	206	203	200	198
470	227	220	216	213	209	206	204	201	198
471	227	221	217	213	209	207	204	201	199
472	228	221	217	214	210	207	205	201	199
473	228	222	218	214	210	208	205	202	200
474	229	222	218	215	211	208	205	202	200
475	229	223	219	215	211	208	206	203	201
476	230	223	219	216	212	209	206	203	201
477	230	224	220	216	212	209	207	204	202
478	231	224	220	217	213	210	207	204	202
479	231	224	221	217	213	210	208	205	203
480	232	225	221	218	214	211	208	205	203
481	232	225	221	218	214	211	209	206	203
482	233	226	222	218	214	212	209	206	204
483	233	226	222	219	215	212	210	207	204
484	234	227	223	219	215	213	210	207	205
485	234	227	223	220	216	213	211	208	205
486	235	228	224	220	216	214	211	208	206
487	235	228	224	221	217	214	212	208	206
488	236	229	225	221	217	215	212	209	207
489	236	229	225	222	218	215	212	209	207
490	237	230	226	222	218	216	213	210	208
491	237	230	226	223	219	216	213	210	208
492	238	231	227	223	219	216	214	211	209
493	238	231	227	224	220	217	214	211	209
494	239	232	228	224	220	217	215	212	209
495	239	232	228	225	221	218	215	212	210
496	239	233	229	225	221	218	216	213	210
497	240	233	229	226	222	219	216	213	211
498	240	234	230	226	222	219	217	214	211
499	241	234	230	227	223	220	217	214	212
500	241	235	231	227	223	220	218	214	212

TABLE B.27 (cont.)　Critical Values of *C* for the Sign Test or the Binomial Test with $p = 0.5$

n	α(2): .50 / α(1): .25	.20 / .10	.10 / .05	.05 / .025	.02 / .01	.01 / .005	.005 / .0025	.002 / .001	.001 / .0005	n	α(2): .50 / α(1): .25	.20 / .10	.10 / .05	.05 / .025	.02 / .01	.01 / .005	.005 / .0025	.002 / .001	.001 / .0005
501	242	235	231	228	223	221	218	215	213	551	267	259	255	252	247	244	242	238	236
502	242	236	232	228	224	221	219	215	213	552	267	260	256	252	248	245	242	239	236
503	243	236	232	229	224	222	219	216	214	553	268	260	256	252	248	245	243	239	237
504	243	237	233	229	225	222	220	216	214	554	268	261	257	253	249	246	243	240	237
505	244	237	233	229	225	223	220	217	215	555	269	261	257	253	249	246	243	240	238
506	244	238	234	230	226	223	220	217	215	556	269	262	258	254	250	247	244	241	238
507	245	238	234	230	226	224	221	218	216	557	270	262	258	254	250	247	244	241	239
508	245	239	234	231	227	224	221	218	216	558	270	263	259	255	251	248	245	242	239
509	246	239	235	231	227	224	222	219	216	559	271	263	259	255	251	248	245	242	240
510	246	240	235	232	228	225	222	219	217	560	271	264	260	256	251	249	246	242	240
511	247	240	236	232	228	225	223	220	217	561	272	264	260	256	252	249	246	243	241
512	247	241	236	233	229	226	223	220	218	562	272	265	261	257	252	249	247	243	241
513	248	241	237	233	229	226	224	221	218	563	272	265	261	257	253	250	247	244	242
514	248	241	237	234	230	227	224	221	219	564	273	266	261	258	253	250	248	244	242
515	249	242	238	234	230	227	225	221	219	565	273	266	262	258	254	251	248	245	242
516	249	242	238	235	231	228	225	222	220	566	274	267	262	259	254	251	249	245	243
517	250	243	239	235	231	228	226	222	220	567	274	267	263	259	255	252	249	246	243
518	250	243	239	236	232	229	226	223	221	568	275	268	263	260	255	252	250	246	244
519	251	244	240	236	232	229	227	223	221	569	275	268	264	260	256	253	250	247	244
520	251	244	240	237	232	230	227	224	222	570	276	269	264	261	256	253	251	247	245
521	252	245	241	237	233	230	227	224	222	571	276	269	265	261	257	254	251	248	245
522	252	245	241	238	233	231	228	225	222	572	277	270	265	262	257	254	251	248	246
523	253	246	242	238	234	231	228	225	223	573	277	270	266	262	258	255	252	249	246
524	253	246	242	239	234	232	229	226	223	574	278	271	266	263	258	255	252	249	247
525	254	247	243	239	235	232	229	226	224	575	278	271	267	263	259	256	253	249	247
526	254	247	243	240	235	232	230	227	224	576	279	272	267	263	259	256	253	250	248
527	255	248	244	240	236	233	230	227	225	577	279	272	268	264	260	257	254	250	248
528	255	248	244	240	236	233	231	228	225	578	280	273	268	264	260	257	254	251	249
529	256	249	245	241	237	234	231	228	226	579	280	273	269	265	261	258	255	251	249
530	256	249	245	241	237	234	232	228	226	580	281	274	269	265	261	258	255	252	249
531	257	250	246	242	238	235	232	229	227	581	281	274	270	266	261	258	256	252	250
532	257	250	246	242	238	235	233	229	227	582	282	275	270	266	262	259	256	253	250
533	258	251	247	243	239	236	233	230	228	583	282	275	271	267	262	259	257	253	251
534	258	251	247	243	239	236	234	230	228	584	283	276	271	267	263	260	257	254	251
535	259	252	247	244	240	237	234	231	229	585	283	276	272	268	263	260	258	254	252
536	259	252	248	244	240	237	235	231	229	586	284	276	272	268	264	261	258	255	252
537	260	253	248	245	241	238	235	232	229	587	284	277	273	269	264	261	259	255	253
538	260	253	249	245	241	238	235	232	230	588	285	277	273	269	265	262	259	256	253
539	261	254	249	246	242	239	236	233	230	589	285	278	274	270	265	262	259	256	254
540	261	254	250	246	242	239	236	233	231	590	286	278	274	270	266	263	260	257	254
541	262	255	250	247	242	240	237	234	231	591	286	279	275	271	266	263	260	257	255
542	262	255	251	247	243	240	237	234	232	592	287	279	275	271	267	264	261	257	255
543	263	256	251	248	243	241	238	235	232	593	287	280	275	272	267	264	261	258	255
544	263	256	252	248	244	241	238	235	233	594	288	280	276	272	268	265	262	258	256
545	264	257	252	249	244	241	239	235	233	595	288	281	276	273	268	265	262	259	256
546	264	257	253	249	245	242	239	236	234	596	289	281	277	273	269	266	263	259	257
547	265	258	253	250	245	242	240	236	234	597	289	282	277	274	269	266	263	260	257
548	265	258	254	250	246	243	240	237	235	598	290	282	278	274	270	267	264	260	258
549	266	258	254	251	246	243	241	237	235	599	290	283	278	275	270	267	264	261	258
550	266	259	255	251	247	244	241	238	235	600	291	283	279	275	271	267	265	261	259

TABLE B.27 (cont.)   Critical Values of C for the Sign Test or the Binomial Test with $p = 0.5$

n	α(2): .50 / α(1): .25	.20 / .10	.10 / .05	.05 / .025	.02 / .01	.01 / .005	.005 / .0025	.002 / .001	.001 / .0005
601	291	284	279	275	271	268	265	262	259
602	292	284	280	276	271	268	266	262	260
603	292	285	280	276	272	269	266	263	260
604	293	285	281	277	272	269	267	263	261
605	293	286	281	277	273	270	267	264	261
606	294	286	282	278	273	270	267	264	262
607	294	287	282	278	274	271	268	264	262
608	295	287	283	279	274	271	268	265	262
609	295	288	283	279	275	272	269	265	263
610	296	288	284	280	275	272	269	266	263
611	296	289	284	280	276	273	270	266	264
612	297	289	285	281	276	273	270	267	264
613	297	290	285	281	277	274	271	267	265
614	298	290	286	282	277	274	271	268	265
615	298	291	286	282	278	275	272	268	266
616	299	291	287	283	278	275	272	269	266
617	299	292	287	283	279	276	273	269	267
618	300	292	288	284	279	276	273	270	267
619	300	293	288	284	280	276	274	270	268
620	301	293	289	285	280	277	274	271	268
621	301	294	289	285	281	277	275	271	269
622	302	294	289	286	281	278	275	272	269
623	302	295	290	286	281	278	276	272	269
624	303	295	290	287	282	279	276	272	270
625	303	295	291	287	282	279	276	273	270
626	304	296	291	287	283	280	277	273	271
627	304	296	292	288	283	280	277	274	271
628	305	297	292	288	284	281	278	274	272
629	305	297	293	289	284	281	278	275	272
630	306	298	293	289	285	282	279	275	273
631	306	298	294	290	285	282	279	276	273
632	307	299	294	290	286	283	280	276	274
633	307	299	295	291	286	283	280	277	274
634	308	300	295	291	287	284	281	277	275
635	308	300	296	292	287	284	281	278	275
636	308	301	296	292	288	285	282	278	276
637	309	301	297	293	288	285	282	279	276
638	309	302	297	293	289	285	283	279	276
639	310	302	298	294	289	286	283	279	277
640	310	303	298	294	290	286	284	280	277
641	311	303	299	295	290	287	284	280	278
642	311	304	299	295	291	287	284	281	278
643	312	304	300	296	291	288	285	281	279
644	312	305	300	296	291	288	285	282	279
645	313	305	301	297	292	289	286	282	280
646	313	306	301	297	292	289	286	283	280
647	314	306	302	298	293	290	287	283	281
648	314	307	302	298	293	290	287	284	281
649	315	307	303	299	294	291	288	284	282
650	315	308	303	299	294	291	288	285	282
651	316	308	304	300	295	292	289	285	283
652	316	309	304	300	295	292	289	286	283
653	317	309	304	300	296	293	290	286	284
654	317	310	305	301	296	293	290	287	284
655	318	310	305	301	297	294	291	287	284
656	318	311	306	302	297	294	291	287	285
657	319	311	306	302	298	295	292	288	285
658	319	312	307	303	298	295	292	288	286
659	320	312	307	303	299	295	293	289	286
660	320	313	308	304	299	296	293	289	287
661	321	313	308	304	300	296	293	290	287
662	321	314	309	305	300	297	294	290	288
663	322	314	309	305	301	297	294	291	288
664	322	314	310	306	301	298	295	291	289
665	323	315	310	306	302	298	295	292	289
666	323	315	311	307	302	299	296	292	290
667	324	316	311	307	302	299	296	293	290
668	324	316	312	308	303	300	297	293	291
669	325	317	312	308	303	300	297	294	291
670	325	317	313	309	304	301	298	294	291
671	326	318	313	309	304	301	298	295	292
672	326	318	314	310	305	302	299	295	292
673	327	319	314	310	305	302	299	295	293
674	327	319	315	311	306	303	300	296	293
675	328	320	315	311	306	303	300	296	294
676	328	320	316	312	307	304	301	297	294
677	329	321	316	312	307	304	301	297	295
678	329	321	317	312	308	304	301	298	295
679	330	322	317	313	308	305	302	298	296
680	330	322	318	313	309	305	302	299	296
681	331	323	318	314	309	306	303	299	297
682	331	323	319	314	310	306	303	300	297
683	332	324	319	315	310	307	304	300	298
684	332	324	319	315	311	307	304	301	298
685	333	325	320	316	311	308	305	301	298
686	333	325	320	316	312	308	305	302	299
687	334	326	321	317	312	309	306	302	299
688	334	326	321	317	313	309	306	303	300
689	335	327	322	318	313	310	307	303	300
690	335	327	322	318	313	310	307	303	301
691	336	328	323	319	314	311	308	304	301
692	336	328	323	319	314	311	308	304	302
693	337	329	324	320	315	312	309	305	302
694	337	329	324	320	315	312	309	305	303
695	338	330	325	321	316	313	310	306	303
696	338	330	325	321	316	313	310	306	304
697	339	331	326	322	317	314	310	307	304
698	339	331	326	322	317	314	311	307	305
699	340	332	327	323	318	314	311	308	305
700	340	332	327	323	318	315	312	308	306

TABLE B.27 (cont.)　Critical Values of C for the Sign Test or the Binomial Test with $p = 0.5$

n	α(2): .50 / α(1): .25	.20 / .10	.10 / .05	.05 / .025	.02 / .01	.01 / .005	.005 / .0025	.002 / .001	.001 / .0005	n	α(2): .50 / α(1): .25	.20 / .10	.10 / .05	.05 / .025	.02 / .01	.01 / .005	.005 / .0025	.002 / .001	.001 / .0005
701	341	333	328	324	319	315	312	309	306	751	365	357	352	348	343	339	336	332	329
702	341	333	328	324	319	316	313	309	306	752	366	357	352	348	343	340	337	333	330
703	342	334	329	325	320	316	313	310	307	753	366	358	353	349	344	340	337	333	330
704	342	334	329	325	320	317	314	310	307	754	367	358	353	349	344	341	337	334	331
705	343	334	330	325	321	317	314	311	308	755	367	359	354	350	345	341	338	334	331
706	343	335	330	326	321	318	315	311	308	756	368	359	354	350	345	342	338	335	332
707	344	335	331	326	322	318	315	311	309	757	368	360	355	351	346	342	339	335	332
708	344	336	331	327	322	319	316	312	309	758	369	360	355	351	346	343	339	336	333
709	345	336	332	327	323	319	316	312	310	759	369	361	356	352	346	343	340	336	333
710	345	337	332	328	323	320	317	313	310	760	370	361	356	352	347	344	340	336	334
711	346	337	333	328	324	320	317	313	311	761	370	362	357	352	347	344	341	337	334
712	346	338	333	329	324	321	318	314	311	762	371	362	357	353	348	344	341	337	335
713	346	338	334	329	324	321	318	314	312	763	371	363	358	353	348	345	342	338	335
714	347	339	334	330	325	322	319	315	312	764	372	363	358	354	349	345	342	338	336
715	347	339	335	330	325	322	319	315	313	765	372	364	359	354	349	346	343	339	336
716	348	340	335	331	326	323	319	316	313	766	373	364	359	355	350	346	343	339	337
717	348	340	335	331	326	323	320	316	313	767	373	365	360	355	350	347	344	340	337
718	349	341	336	332	327	324	320	317	314	768	374	365	360	356	351	347	344	340	337
719	349	341	336	332	327	324	321	317	314	769	374	366	361	356	351	348	345	341	338
720	350	342	337	333	328	324	321	318	315	770	375	366	361	357	352	348	345	341	338
721	350	342	337	333	328	325	322	318	315	771	375	367	362	357	352	349	346	342	339
722	351	343	338	334	329	325	322	319	316	772	376	367	362	358	353	349	346	342	339
723	351	343	338	334	329	326	323	319	316	773	376	368	363	358	353	350	347	343	340
724	352	344	339	335	330	326	323	319	317	774	377	368	363	359	354	350	347	343	340
725	352	344	339	335	330	327	324	320	317	775	377	369	364	359	354	351	347	344	341
726	353	345	340	336	331	327	324	320	318	776	378	369	364	360	355	351	348	344	341
727	353	345	340	336	331	328	325	321	318	777	378	370	365	360	355	352	348	344	342
728	354	346	341	337	332	328	325	321	319	778	379	370	365	361	356	352	349	345	342
729	354	346	341	337	332	329	326	322	319	779	379	371	366	361	356	353	349	345	343
730	355	347	342	338	333	329	326	322	320	780	380	371	366	362	357	353	350	346	343
731	355	347	342	338	333	330	327	323	320	781	380	372	367	362	357	354	350	346	344
732	356	348	343	338	334	330	327	323	321	782	381	372	367	363	357	354	351	347	344
733	356	348	343	339	334	331	328	324	321	783	381	373	367	363	358	354	351	347	345
734	357	349	344	339	335	331	328	324	321	784	382	373	368	364	358	355	352	348	345
735	357	349	344	340	335	332	328	325	322	785	382	374	368	364	359	355	352	348	345
736	358	350	345	340	335	332	329	325	322	786	383	374	369	365	359	356	353	349	346
737	358	350	345	341	336	333	329	326	323	787	383	375	369	365	360	356	353	349	346
738	359	351	346	341	336	333	330	326	323	788	384	375	370	365	360	357	354	350	347
739	359	351	346	342	337	334	330	327	324	789	384	376	370	366	361	357	354	350	347
740	360	352	347	342	337	334	331	327	324	790	385	376	371	366	361	358	355	351	348
741	360	352	347	343	338	334	331	327	325	791	385	376	371	367	362	358	355	351	348
742	361	353	348	343	338	335	332	328	325	792	386	377	372	367	362	359	356	352	349
743	361	353	348	344	339	335	332	328	326	793	386	377	372	368	363	359	356	352	349
744	362	354	349	344	339	336	333	329	326	794	386	378	373	368	363	360	356	353	350
745	362	354	349	345	340	336	333	329	327	795	387	378	373	369	364	360	357	353	350
746	363	354	350	345	340	337	334	330	327	796	387	379	374	369	364	361	357	353	351
747	363	355	350	346	341	337	334	330	328	797	388	379	374	370	365	361	358	354	351
748	364	355	351	346	341	338	335	331	328	798	388	380	375	370	365	362	358	354	352
749	364	356	351	347	342	338	335	331	329	799	389	380	375	371	366	362	359	355	352
750	365	356	351	347	342	339	336	332	329	800	389	381	376	371	366	363	359	355	353

TABLE B.27 (cont.)   Critical Values of C for the Sign Test or the Binomial Test with $p = 0.5$

n	α(2): .50 / α(1): .25	.20 / .10	.10 / .05	.05 / .025	.02 / .01	.01 / .005	.005 / .0025	.002 / .001	.001 / .0005	n	α(2): .50 / α(1): .25	.20 / .10	.10 / .05	.05 / .025	.02 / .01	.01 / .005	.005 / .0025	.002 / .001	.001 / .0005
801	390	381	376	372	367	363	360	356	353	851	415	406	401	396	391	387	384	379	377
802	390	382	377	372	367	364	360	356	353	852	415	406	401	396	391	387	384	380	377
803	391	382	377	373	368	364	361	357	354	853	416	407	401	397	392	388	385	380	377
804	391	383	378	373	368	365	361	357	354	854	416	407	402	397	392	388	385	381	378
805	392	383	378	374	369	365	362	358	355	855	417	408	402	398	393	389	385	381	378
806	392	384	379	374	369	365	362	358	355	856	417	408	403	398	393	389	386	382	379
807	393	384	379	375	369	366	363	359	356	857	418	409	403	399	393	390	386	382	379
808	393	385	380	375	370	366	363	359	356	858	418	409	404	399	394	390	387	383	380
809	394	385	380	376	370	367	364	360	357	859	419	410	404	400	394	391	387	383	380
810	394	386	381	376	371	367	364	360	357	860	419	410	405	400	395	391	388	384	381
811	395	386	381	377	371	368	365	361	358	861	420	411	405	401	395	392	388	384	381
812	395	387	382	377	372	368	365	361	358	862	420	411	406	401	396	392	389	385	382
813	396	387	382	378	372	369	366	361	359	863	421	412	406	402	396	393	389	385	382
814	396	388	383	378	373	369	366	362	359	864	421	412	407	402	397	393	390	386	383
815	397	388	383	379	373	370	366	362	360	865	422	413	407	403	397	394	390	386	383
816	397	389	384	379	374	370	367	363	360	866	422	413	408	403	398	394	391	387	384
817	398	389	384	379	374	371	367	363	361	867	423	414	408	404	398	395	391	387	384
818	398	390	384	380	375	371	368	364	361	868	423	414	409	404	399	395	392	388	385
819	399	390	385	380	375	372	368	364	361	869	424	415	409	405	399	396	392	388	385
820	399	391	385	381	376	372	369	365	362	870	424	415	410	405	400	396	393	388	386
821	400	391	386	381	376	373	369	365	362	871	425	416	410	406	400	397	393	389	386
822	400	392	386	382	377	373	370	366	363	872	425	416	411	406	401	397	394	389	386
823	401	392	387	382	377	374	370	366	363	873	426	417	411	407	401	397	394	390	387
824	401	393	387	383	378	374	371	367	364	874	426	417	412	407	402	398	395	390	387
825	402	393	388	383	378	375	371	367	364	875	427	418	412	408	402	398	395	391	388
826	402	394	388	384	379	375	372	368	365	876	427	418	413	408	403	399	395	391	388
827	403	394	389	384	379	375	372	368	365	877	428	419	413	408	403	399	396	392	389
828	403	395	389	385	380	376	373	369	366	878	428	419	414	409	404	400	396	392	389
829	404	395	390	385	380	376	373	369	366	879	429	420	414	409	404	400	397	393	390
830	404	396	390	386	381	377	374	370	367	880	429	420	415	410	405	401	397	393	390
831	405	396	391	386	381	377	374	370	367	881	429	420	415	410	405	401	398	394	391
832	405	397	391	387	381	378	375	370	368	882	430	421	416	411	405	402	398	394	391
833	406	397	392	387	382	378	375	371	368	883	430	421	416	411	406	402	399	395	392
834	406	397	392	388	382	379	375	371	369	884	431	422	417	412	406	403	399	395	392
835	407	398	393	388	383	379	376	372	369	885	431	422	417	412	407	403	400	396	393
836	407	398	393	389	383	380	376	372	369	886	432	423	418	413	407	404	400	396	393
837	408	399	394	389	384	380	377	373	370	887	432	423	418	413	408	404	401	397	394
838	408	399	394	390	384	381	377	373	370	888	433	424	418	414	408	405	401	397	394
839	409	400	395	390	385	381	378	374	371	889	433	424	419	414	409	405	402	397	394
840	409	400	395	391	385	382	378	374	371	890	434	425	419	415	409	406	402	398	395
841	410	401	396	391	386	382	379	375	372	891	434	425	420	415	410	406	403	398	395
842	410	401	396	392	386	383	379	375	372	892	435	426	420	416	410	407	403	399	396
843	411	402	397	392	387	383	380	376	373	893	435	426	421	416	411	407	404	399	396
844	411	402	397	393	387	384	380	376	373	894	436	427	421	417	411	408	404	400	397
845	412	403	398	393	388	384	381	377	374	895	436	427	422	417	412	408	405	400	397
846	412	403	398	394	388	385	381	377	374	896	437	428	422	418	412	408	405	401	398
847	413	404	399	394	389	385	382	378	375	897	437	428	423	418	413	409	405	401	398
848	413	404	399	394	389	386	382	378	375	898	438	429	423	419	413	409	406	402	399
849	414	405	400	395	390	386	383	379	376	899	438	429	424	419	414	410	406	402	399
850	414	405	400	395	390	386	383	379	376	900	439	430	424	420	414	410	407	403	400

TABLE B.27 (cont.)   Critical Values of C for the Sign Test or the Binomial Test with $p = 0.5$

n	$\alpha(2){:}\ .50$ $\alpha(1){:}\ .25$	$.20$ $.10$	$.10$ $.05$	$.05$ $.025$	$.02$ $.01$	$.01$ $.005$	$.005$ $.0025$	$.002$ $.001$	$.001$ $.0005$	n	$\alpha(2){:}\ .50$ $\alpha(1){:}\ .25$	$.20$ $.10$	$.10$ $.05$	$.05$ $.025$	$.02$ $.01$	$.01$ $.005$	$.005$ $.0025$	$.002$ $.001$	$.001$ $.0005$
901	439	430	425	420	415	411	407	403	400	951	464	455	449	444	439	435	431	427	424
902	440	431	425	421	415	411	408	404	401	952	465	455	450	445	439	435	432	427	424
903	440	431	426	421	416	412	408	404	401	953	465	456	450	445	440	436	432	428	425
904	441	432	426	422	416	412	409	405	402	954	466	456	451	446	440	436	433	428	425
905	441	432	427	422	417	413	409	405	402	955	466	457	451	446	441	437	433	429	426
906	442	433	427	423	417	413	410	406	403	956	467	457	452	447	441	437	434	429	426
907	442	433	428	423	417	414	410	406	403	957	467	458	452	447	442	438	434	430	427
908	443	434	428	423	418	414	411	406	403	958	468	458	453	448	442	438	435	430	427
909	443	434	429	424	418	415	411	407	404	959	468	459	453	448	442	439	435	431	428
910	444	435	429	424	419	415	412	407	404	960	469	459	454	449	443	439	436	431	428
911	444	435	430	425	419	416	412	408	405	961	469	460	454	449	443	440	436	432	429
912	445	436	430	425	420	416	413	408	405	962	470	460	454	450	444	440	436	432	429
913	445	436	431	426	420	417	413	409	406	963	470	461	455	450	444	441	437	433	429
914	446	437	431	426	421	417	414	409	406	964	471	461	455	451	445	441	437	433	430
915	446	437	432	427	421	418	414	410	407	965	471	462	456	451	445	442	438	434	430
916	447	438	432	427	422	418	415	410	407	966	472	462	456	452	446	442	438	434	431
917	447	438	433	428	422	419	415	411	408	967	472	463	457	452	446	442	439	434	431
918	448	439	433	428	423	419	416	411	408	968	473	463	457	453	447	443	439	435	432
919	448	439	434	429	423	419	416	412	409	969	473	464	458	453	447	443	440	435	432
920	449	440	434	429	424	420	416	412	409	970	473	464	458	453	448	444	440	436	433
921	449	440	435	430	424	420	417	413	410	971	474	465	459	454	448	444	441	436	433
922	450	441	435	430	425	421	417	413	410	972	474	465	459	454	449	445	441	437	434
923	450	441	436	431	425	421	418	414	411	973	475	466	460	455	449	445	442	437	434
924	451	442	436	431	426	422	418	414	411	974	475	466	460	455	450	446	442	438	435
925	451	442	436	432	426	422	419	415	412	975	476	466	461	456	450	446	443	438	435
926	452	443	437	432	427	423	419	415	412	976	476	467	461	456	451	447	443	439	436
927	452	443	437	433	427	423	420	415	412	977	477	467	462	457	451	447	444	439	436
928	453	443	438	433	428	424	420	416	413	978	477	468	462	457	452	448	444	440	437
929	453	444	438	434	428	424	421	416	413	979	478	468	463	458	452	448	445	440	437
930	454	444	439	434	429	425	421	417	414	980	478	469	463	458	453	449	445	441	438
931	454	445	439	435	429	425	422	417	414	981	479	469	464	459	453	449	446	441	438
932	455	445	440	435	430	426	422	418	415	982	479	470	464	459	454	450	446	442	438
933	455	446	440	436	430	426	423	418	415	983	480	470	465	460	454	450	447	442	439
934	456	446	441	436	430	427	423	419	416	984	480	471	465	460	455	451	447	443	439
935	456	447	441	437	431	427	424	419	416	985	481	471	466	461	455	451	447	443	440
936	457	447	442	437	431	428	424	420	417	986	481	472	466	461	455	452	448	444	440
937	457	448	442	438	432	428	425	420	417	987	482	472	467	462	456	452	448	444	441
938	458	448	443	438	432	429	425	421	418	988	482	473	467	462	456	453	449	444	441
939	458	449	443	438	433	429	426	421	418	989	483	473	468	463	457	453	449	445	442
940	459	449	444	439	433	430	426	422	419	990	483	474	468	463	457	453	450	445	442
941	459	450	444	439	434	430	426	422	419	991	484	474	469	464	458	454	450	446	443
942	460	450	445	440	434	430	427	423	420	992	484	475	469	464	458	454	451	446	443
943	460	451	445	440	435	431	427	423	420	993	485	475	470	465	459	455	451	447	444
944	461	451	446	441	435	431	428	424	420	994	485	476	470	465	459	455	452	447	444
945	461	452	446	441	436	432	428	424	421	995	486	476	471	466	460	456	452	448	445
946	462	452	447	442	436	432	429	425	421	996	486	477	471	466	460	456	453	448	445
947	462	453	447	442	437	433	429	425	422	997	487	477	472	467	461	457	453	449	446
948	463	453	448	443	437	433	430	425	422	998	487	478	472	467	461	457	454	449	446
949	463	454	448	443	438	434	430	426	423	999	488	478	473	468	462	458	454	450	447
950	464	454	449	444	438	434	431	426	423	1000	488	479	473	468	462	458	455	450	447

This table was prepared by considering binomial probabilities, such as those in Table B.26b. These are lower critical values, $C_{\alpha,n}$; upper critical values are $n - C_{\alpha,n}$.

Example:

$$C_{0.01,950} = 434, \text{ with } 950 - 434 = 516 \text{ as the upper critical value.}$$

TABLE B.28 Critical Values for Fisher's Exact Test

n	m_1	m_2	α: 0.50	0.20	0.10	0.05	0.02	0.01	0.005	0.002	0.001
2	1	1	—, —	—, —	—, —	—, —	—, —	—, —	—, —	—, —	—, —
3	1	1	—, —	—, —	—, —	—, —	—, —	—, —	—, —	—, —	—, —
4	1	1	0, 1	—, —	—, —	—, —	—, —	—, —	—, —	—, —	—, —
4	2	1	—, 1	—, —	—, —	—, —	—, —	—, —	—, —	—, —	—, —
4	2	2	0, 2	0, 2	—, —	—, —	—, —	—, —	—, —	—, —	—, —
5	1	1	—, —	—, —	—, —	—, —	—, —	—, —	—, —	—, —	—, —
5	2	1	—, 1	—, —	—, —	—, —	—, —	—, —	—, —	—, —	—, —
5	2	2	0, 2	0, 2	—, 2	—, —	—, —	—, —	—, —	—, —	—, —
6	1	1	—, —	—, —	—, —	—, —	—, —	—, —	—, —	—, —	—, —
6	2	1	—, 1	—, —	—, —	—, —	—, —	—, —	—, —	—, —	—, —
6	2	2	0, 2	0, 2	—, 2	—, —	—, —	—, —	—, —	—, —	—, —
6	3	1	0, 2	0, 2	0, 2	—, —	—, —	—, —	—, —	—, —	—, —
6	3	2	0, 3	0, 3	0, 3	0, 3	—, —	—, —	—, —	—, —	—, —
6	3	3	1, —	1, —	—, —	—, —	—, —	—, —	—, —	—, —	—, —
7	1	1	—, —	—, —	—, —	—, —	—, —	—, —	—, —	—, —	—, —
7	2	1	—, 1	—, —	—, —	—, —	—, —	—, —	—, —	—, —	—, —
7	2	2	0, 2	0, 2	0, 2	—, —	—, —	—, —	—, —	—, —	—, —
7	3	2	0, 2	0, 2	0, 2	0, 2	—, —	—, —	—, —	—, —	—, —
7	3	3	0, 3	0, 3	0, 3	0, 3	—, —	—, —	—, —	—, —	—, —
8	1	1	—, —	—, —	—, —	—, —	—, —	—, —	—, —	—, —	—, —
8	2	1	—, 1	—, —	—, —	—, —	—, —	—, —	—, —	—, —	—, —
8	3	1	0, 2	0, 2	0, 2	—, —	—, —	—, —	—, —	—, —	—, —
8	4	1	0, —	—, —	—, —	—, —	—, —	—, —	—, —	—, —	—, —
8	2	2	0, 2	0, 2	0, 2	0, 2	—, —	—, —	—, —	—, —	—, —
8	3	2	0, 2	0, 2	0, 2	0, 2	—, —	—, —	—, —	—, —	—, —
8	3	3	0, 3	0, 3	0, 3	0, 3	—, 3	—, —	—, —	—, —	—, —
8	4	3	1, 2	0, 3	0, 3	0, 3	—, 3	—, —	—, —	—, —	—, —
8	4	4	1, 3	0, 4	0, 4	0, 4	0, 4	—, 4	—, —	—, —	—, —
9	1	1	—, —	—, —	—, —	—, —	—, —	—, —	—, —	—, —	—, —
9	2	1	—, 1	—, —	—, —	—, —	—, —	—, —	—, —	—, —	—, —
9	3	1	—, —	—, —	—, —	—, —	—, —	—, —	—, —	—, —	—, —
9	4	1	0, —	—, —	—, —	—, —	—, —	—, —	—, —	—, —	—, —
9	2	2	0, 2	0, 2	0, 2	—, —	—, —	—, —	—, —	—, —	—, —
9	3	2	0, 2	0, 2	0, 2	0, 2	—, —	—, —	—, —	—, —	—, —
9	4	2	0, 2	0, 2	0, 2	0, 2	—, —	—, —	—, —	—, —	—, —
9	3	3	0, 3	0, 3	0, 3	0, 3	—, 3	—, —	—, —	—, —	—, —
9	4	3	1, 3	0, 3	0, 3	0, 3	—, 3	—, —	—, —	—, —	—, —
9	4	4	1, 3	0, 4	0, 4	0, 4	0, 4	—, 4	—, —	—, —	—, —
10	1	1	—, —	—, —	—, —	—, —	—, —	—, —	—, —	—, —	—, —
10	2	1	—, —	—, —	—, —	—, —	—, —	—, —	—, —	—, —	—, —
10	3	1	—, —	—, —	—, —	—, —	—, —	—, —	—, —	—, —	—, —
10	4	1	—, —	—, —	—, —	—, —	—, —	—, —	—, —	—, —	—, —
10	5	1	0, 1	—, —	—, —	—, —	—, —	—, —	—, —	—, —	—, —

TABLE B.28 (cont.) Critical Values for Fisher's Exact Test

n	m_1	m_2	α: 0.50	0.20	0.10	0.05	0.02	0.01	0.005	0.002	0.001
10	2	2	—, 1	—, 2	—, 2	—, 2	—, —	—, —	—, —	—, —	—, —
10	2	3	0, 2	—, 2	—, 2	—, 2	—, —	—, —	—, —	—, —	—, —
10	2	4	0, 2	—, 2	—, —	—, —	—, —	—, —	—, —	—, —	—, —
10	2	5	0, 2	—, 2	—, —	—, —	—, —	—, —	—, —	—, —	—, —
10	3	3	0, 2	0, 3	—, 3	—, 3	—, 3	—, 3	—, —	—, —	—, —
10	3	4	0, 2	0, 3	—, 3	—, 3	—, —	—, —	—, —	—, —	—, —
10	3	5	1, 3	0, 3	—, 3	—, 3	—, 3	—, 3	—, —	—, —	—, —
10	4	4	1, 3	0, 4	0, 4	0, 4	—, 4	—, 4	—, 4	—, —	—, —
10	4	5	1, 3	0, 4	0, 4	0, 4	—, 4	—, 4	—, 4	—, —	—, —
10	5	5	2, 3	0, 5	0, 5	0, 5	0, 5	0, 5	0, 5	—, —	—, —
11	1	1	—, 1	—, 1	—, 1	—, —	—, —	—, —	—, —	—, —	—, —
11	1	2	—, 1	—, 1	—, —	—, —	—, —	—, —	—, —	—, —	—, —
11	1	3	—, 1	—, 1	—, —	—, —	—, —	—, —	—, —	—, —	—, —
11	1	4	—, 1	—, 1	—, —	—, —	—, —	—, —	—, —	—, —	—, —
11	1	5	—, 1	—, —	—, —	—, —	—, —	—, —	—, —	—, —	—, —
11	2	2	—, 2	—, 2	—, 2	—, 2	—, 2	—, —	—, —	—, —	—, —
11	2	3	0, 2	—, 2	—, 2	—, 2	—, 2	—, —	—, —	—, —	—, —
11	2	4	0, 2	—, 2	—, 2	—, 2	—, —	—, —	—, —	—, —	—, —
11	2	5	0, 2	—, 2	—, 2	—, 2	—, —	—, —	—, —	—, —	—, —
11	3	3	0, 2	0, 3	—, 3	—, 3	—, 3	—, 3	—, —	—, —	—, —
11	3	4	0, 3	0, 3	—, 3	—, 3	—, 3	—, 3	—, —	—, —	—, —
11	3	5	0, 3	0, 3	0, 3	—, 3	—, 3	—, 3	—, —	—, —	—, —
11	4	4	1, 3	0, 4	0, 4	0, 4	—, 4	—, 4	—, 4	—, —	—, —
11	4	5	1, 3	0, 4	0, 4	0, 4	0, 4	0, 4	—, 4	—, —	—, —
11	5	5	2, 3	0, 5	0, 5	0, 5	0, 5	0, 5	0, 5	0, 5	—, —
12	1	1	—, 1	—, 1	—, 1	—, —	—, —	—, —	—, —	—, —	—, —
12	1	2	0, 1	—, 1	—, —	—, —	—, —	—, —	—, —	—, —	—, —
12	1	3	—, 1	—, 1	—, —	—, —	—, —	—, —	—, —	—, —	—, —
12	1	4	—, 1	—, 1	—, —	—, —	—, —	—, —	—, —	—, —	—, —
12	1	5	—, 1	—, —	—, —	—, —	—, —	—, —	—, —	—, —	—, —
12	1	6	—, 1	—, —	—, —	—, —	—, —	—, —	—, —	—, —	—, —
12	2	2	0, 2	—, 2	—, 2	—, 2	—, 2	—, —	—, —	—, —	—, —
12	2	3	0, 2	—, 2	—, 2	—, 2	—, 2	—, —	—, —	—, —	—, —
12	2	4	0, 2	—, 2	—, 2	—, 2	—, 2	—, —	—, —	—, —	—, —
12	2	5	0, 2	—, 2	—, 2	—, 2	—, —	—, —	—, —	—, —	—, —
12	2	6	0, 2	—, 2	—, 2	—, 2	—, —	—, —	—, —	—, —	—, —
12	3	3	0, 2	0, 3	0, 3	—, 3	—, 3	—, 3	—, 3	—, —	—, —
12	3	4	0, 3	0, 3	0, 3	—, 3	—, 3	—, 3	—, 3	—, —	—, —
12	3	5	0, 3	0, 3	0, 3	0, 3	—, 3	—, 3	—, 3	—, —	—, —
12	3	6	1, 3	0, 3	0, 3	0, 3	—, 3	—, 3	—, 3	—, —	—, —
12	4	4	1, 3	0, 4	0, 4	0, 4	0, 4	0, 4	—, 4	—, —	—, —
12	4	5	1, 3	0, 4	0, 4	0, 4	0, 4	0, 4	—, 4	—, —	—, —
12	4	6	1, 4	0, 4	0, 4	0, 4	0, 5	0, 5	0, 5	—, 5	—, —
12	5	5	1, 4	0, 4	0, 5	0, 5	0, 5	0, 5	0, 5	—, 5	—, —
12	5	6	2, 3	—, 4	—, 5	0, 5	0, 5	0, 5	0, 5	—, 5	—, —

TABLE B.28 (cont.)   Critical Values for Fisher's Exact Test

n	m_1	m_2	α: 0.50	0.20	0.10	0.05	0.02	0.01	0.005	0.002	0.001
12	6	6	2, 4	1, 5	1, 5	1, 5	0, 6	0, 6	0, 6	0, 6	—, —
13	1	1	1, —	1, —	1, —	—, —	—, —	—, —	—, —	—, —	—, —
13	1	2	1, —	1, —	—, —	—, —	—, —	—, —	—, —	—, —	—, —
13	1	3	1, —	—, —	—, —	—, —	—, —	—, —	—, —	—, —	—, —
13	1	4	—, —	—, —	—, —	—, —	—, —	—, —	—, —	—, —	—, —
13	1	5	1, —	1, —	1, —	1, —	—, —	—, —	—, —	—, —	—, —
13	1	6	1, —	—, —	—, —	—, —	—, —	—, —	—, —	—, —	—, —
13	2	2	0, 2	—, 2	—, 2	—, 2	—, 2	—, —	—, —	—, —	—, —
13	2	3	0, 2	—, 2	—, 2	—, 2	—, 2	—, —	—, —	—, —	—, —
13	2	4	0, 2	—, 2	—, 2	—, 2	—, 2	—, —	—, —	—, —	—, —
13	2	5	0, 2	—, 3	—, 3	—, 3	—, —	—, —	—, —	—, —	—, —
13	2	6	0, 3	—, 3	—, 3	—, 3	—, 3	—, 3	—, 3	—, —	—, —
13	3	3	0, 3	—, 3	—, 3	—, 3	—, 3	—, 3	—, 3	—, —	—, —
13	3	4	0, 3	0, 3	—, 4	—, 4	—, 4	—, 4	—, —	—, —	—, —
13	3	5	0, 3	0, 4	0, 4	—, 4	—, 4	—, 4	—, —	—, —	—, —
13	4	4	0, 4	0, 4	0, 4	0, 4	—, 5	—, 5	—, 5	—, 5	—, 5
13	4	5	1, 4	0, 4	0, 4	0, 5	0, 5	—, 5	0, 6	—, 6	—, 6
13	4	6	1, 4	0, 4	0, 4	0, 5	0, 5	0, 6	0, 6	—, 6	—, 6
13	5	5	1, 3	0, 4	0, 5	0, 5	0, 6	0, 6	0, 6	—, —	—, —
14	1	1	1, 4	1, 5	1, 5	—, —	—, —	—, —	—, —	—, —	—, —
14	1	2	1, —	1, —	—, —	—, —	—, —	—, —	—, —	—, —	—, —
14	1	3	1, —	—, —	—, —	—, —	—, —	—, —	—, —	—, —	—, —
14	1	6	—, —	—, —	—, —	—, —	—, —	—, —	—, —	—, —	—, —
14	1	7	—, —	—, —	—, —	—, —	—, —	—, —	—, —	—, —	—, —
14	2	2	—, 1	—, 2	—, 2	—, 2	—, 2	—, —	—, —	—, —	—, —
14	2	3	0, 1	—, 2	—, 2	—, 2	—, 2	—, —	—, —	—, —	—, —
14	2	4	0, 2	—, 2	—, 2	—, 2	—, 2	—, —	—, —	—, —	—, —
14	2	5	0, 2	—, 2	—, 2	—, 2	—, 2	—, —	—, —	—, —	—, —
14	2	6	0, 2	—, 2	—, 2	—, 2	—, 2	—, —	—, —	—, —	—, —
14	2	7	1, 2	—, 2	—, 2	—, 2	—, 2	—, —	—, —	—, —	—, —
14	3	3	0, 2	—, 3	—, 3	—, 3	—, 3	—, 3	—, 3	—, —	—, —
14	3	4	0, 2	0, 3	—, 3	—, 3	—, 3	—, 3	—, 3	—, —	—, —
14	3	5	0, 3	0, 3	—, 3	—, 3	—, 3	—, 3	—, —	—, —	—, —
14	3	6	0, 3	0, 3	0, 3	—, 3	—, 3	—, 3	—, —	—, —	—, —
14	3	7	1, 2	0, 3	0, 3	—, 4	—, 4	—, 4	—, 4	—, 4	—, 4
14	4	4	0, 2	0, 3	—, 3	—, 3	—, 4	—, 4	—, 4	—, 4	—, 4
14	4	5	0, 3	0, 3	0, 4	0, 4	—, 4	—, 4	—, 4	—, 4	—, 4
14	4	6	1, 3	0, 3	0, 4	0, 4	0, 4	—, 4	—, —	—, —	—, —
14	4	7	1, 3	0, 4	0, 4	0, 4	0, 4	—, 5	—, 5	—, 5	—, 5
14	5	5	1, 3	0, 4	0, 4	—, 4	—, 5	—, 5	—, 5	—, 5	—, 5

TABLE B.28 (cont.) Critical Values for Fisher's Exact Test

n	m_1	m_2	α: 0.50	0.20	0.10	0.05	0.02	0.01	0.005	0.002	0.001
14	5	6	1, 3	0, 4	0, 4	0, 5	—, 5	—, 5	—, 5	—, —	—, —
14	5	7	2, 3	1, 4	0, 5	0, 5	—, 5	—, 5	—, 6	—, 6	—, 6
14	6	6	2, 4	1, 5	0, 5	0, 5	0, 5	0, 6	—, 6	—, 6	—, 6
14	6	7	2, 4	1, 5	1, 5	0, 6	0, 6	0, 6	0, 6	0, 7	0, 7
14	7	7	3, 4	2, 5	1, 6	1, 6	1, 6	0, 7	0, 7	0, 7	0, 7
15	1	1	—, 1	—, —	—, 1	—, —	—, —	—, —	—, —	—, —	—, —
15	1	2	—, 1	—, 1	—, 1	—, —	—, —	—, —	—, —	—, —	—, —
15	1	3	—, 1	—, 1	—, —	—, —	—, —	—, —	—, —	—, —	—, —
15	1	4	—, 1	—, 1	—, —	—, —	—, —	—, —	—, —	—, —	—, —
15	1	5	—, 1	—, —	—, —	—, —	—, —	—, —	—, —	—, —	—, —
15	1	6	—, —	—, —	—, —	—, —	—, —	—, —	—, —	—, —	—, —
15	1	7	—, —	—, —	—, —	—, —	—, —	—, —	—, —	—, —	—, —
15	2	2	—, —	—, —	—, 1	—, —	—, —	—, —	—, —	—, —	—, —
15	2	3	—, —	—, 2	—, 2	—, 2	—, —	—, —	—, —	—, —	—, —
15	2	4	—, —	—, 2	—, 2	—, 2	—, 2	—, 2	—, —	—, —	—, —
15	2	5	—, —	—, 2	—, 2	—, 2	—, 2	—, —	—, —	—, —	—, —
15	2	6	—, —	—, 2	—, 2	—, —	—, —	—, —	—, —	—, —	—, 4
15	2	7	—, —	—, 2	—, 2	—, —	—, —	—, —	—, —	—, —	—, —
15	3	3	0, 2	0, 3	—, 3	—, 3	—, 3	—, 3	—, 3	—, —	—, —
15	3	4	0, 2	0, 3	—, 3	—, 3	—, 3	—, 3	—, 3	—, —	—, —
15	3	5	0, 2	0, 3	—, 3	—, 3	—, 3	—, 3	—, —	—, —	—, —
15	3	6	0, 1	—, 3	—, 3	—, 3	—, 3	—, 3	—, —	—, —	—, —
15	3	7	0, 2	—, 3	—, 3	—, 4	—, 4	—, 4	—, —	—, —	—, —
15	4	4	0, 2	0, 4	—, 4	—, 4	—, 4	—, 4	—, 4	—, 4	—, 4
15	4	5	0, 2	0, 4	0, 4	—, 4	—, 4	—, 4	—, 4	—, —	—, —
15	4	6	0, 3	0, 4	0, 4	0, 5	0, 5	—, 5	—, 5	—, 5	—, 5
15	4	7	0, 3	0, 4	0, 4	0, 5	—, 5	—, 5	—, 5	—, 5	—, 5
15	5	5	1, 4	1, 4	0, 5	0, 5	—, 5	—, 5	—, 6	—, 6	—, 6
15	5	6	1, 4	1, 5	1, 5	0, 5	0, 5	0, 6	—, 6	—, 6	—, 6
15	5	7	2, 4	1, 5	1, 6	1, 6	0, 6	0, 6	0, 7	0, 7	0, 7
15	6	6	1, 4	1, 5	1, 5	1, 6	0, 6	0, 6	0, 7	0, 7	0, 7
15	6	7	1, 4	1, 5	1, 6	1, 6	0, 6	0, 6	0, 7	0, 7	0, 7
15	7	7	2, 5	1, 6	1, 6	1, 6	0, 6	—, —	—, —	—, —	—, —
16	1	1	—, —	—, 1	—, 1	—, —	—, —	—, —	—, —	—, —	—, —
16	1	2	—, —	—, 1	—, —	—, —	—, —	—, —	—, —	—, —	—, —
16	1	3	—, 1	—, —	—, —	—, —	—, —	—, —	—, —	—, —	—, —
16	1	4	—, 1	—, —	—, —	—, —	—, —	—, —	—, —	—, —	—, —
16	1	5	—, 1	—, —	—, —	—, —	—, —	—, —	—, —	—, —	—, —
16	1	6	—, 1	—, —	—, —	—, —	—, —	—, —	—, —	—, —	—, —
16	1	7	—, 1	—, —	—, —	—, —	—, —	—, —	—, —	—, —	—, —
16	1	8	0, 1	—, —	—, —	—, —	—, —	—, —	—, —	—, —	—, —
16	2	2	—, —	—, 2	—, 2	—, 2	—, 2	—, 2	—, —	—, —	—, —
16	2	3	—, —	—, 2	—, 2	—, 2	—, 2	—, 2	—, —	—, —	—, —
16	2	4	—, —	—, 2	—, 2	—, 2	—, 2	—, 2	—, —	—, —	—, —
16	2	5	0, 2	—, 2	—, 2	—, —	—, —	—, —	—, —	—, —	—, —

n	m_1	m_2	α: 0.50	0.20	0.10	0.05	0.02	0.01	0.005	0.002	0.001
16	2	6	0, 2	—, 2	—, —	—, —	—, —	—, —	—, —	—, —	—, —
16	2	7	0, 2	—, 2	—, —	—, —	—, —	—, —	—, —	—, —	—, —
16	2	8	0, 2	—, —	—, —	—, —	—, —	—, —	—, —	—, —	—, —
16	3	3	—, 1	—, 2	—, 2	—, 3	—, 3	—, 3	—, 3	—, 3	—, —
16	3	4	0, 2	—, 2	—, 3	—, 3	—, 3	—, 3	—, —	—, —	—, —
16	3	5	0, 2	—, 3	—, 3	—, 3	—, —	—, —	—, —	—, —	—, —
16	3	6	0, 2	—, 3	—, 3	—, 3	—, —	—, —	—, —	—, —	—, —
16	3	7	0, 3	0, 3	—, 3	—, 3	—, —	—, —	—, —	—, —	—, —
16	3	8	1, 2	0, 3	0, 3	—, 3	—, —	—, —	—, —	—, —	—, —
16	4	4	0, 2	—, 3	—, 3	—, 3	—, 4	—, 4	—, 4	—, 4	—, 4
16	4	5	0, 2	0, 3	—, 3	—, 4	—, 4	—, 4	—, 4	—, 4	—, —
16	4	6	0, 3	0, 3	0, 4	—, 4	—, 4	—, 4	—, 4	—, 4	—, —
16	4	7	1, 3	0, 4	0, 4	—, 4	—, 4	—, 4	—, —	—, —	—, —
16	4	8	1, 3	0, 4	0, 4	0, 4	—, 4	—, 4	—, —	—, —	—, —
16	5	5	1, 3	0, 4	—, 4	—, 4	—, 5	—, 5	—, 5	—, 5	—, 5
16	5	6	1, 3	1, 5	0, 4	0, 5	—, 5	—, 5	—, 5	—, 5	—, 5
16	5	7	1, 4	1, 5	0, 5	0, 5	0, 5	—, 5	—, 5	—, 5	—, —
16	5	8	2, 3	1, 5	0, 5	0, 5	0, 5	—, 5	—, 5	—, 5	—, —
16	6	6	2, 4	1, 5	1, 5	0, 6	0, 6	0, 6	—, 6	—, 6	—, 6
16	6	7	2, 4	1, 6	1, 6	0, 6	0, 6	0, 7	0, 7	—, 7	—, 7
16	6	8	3, 4	2, 6	1, 6	1, 6	0, 6	0, 7	0, 7	0, 7	0, 7
16	7	7	3, 4	2, 6	1, 7	1, 7	0, 7	0, 7	0, 7	0, 7	0, 7
16	7	8	3, 5	2, 6	1, 7	1, 7	1, 7	1, 7	0, 8	0, 8	0, 8
16	8	1	3, —	2, —	2, —	1, —	1, —	1, —	0, —	0, —	0, —
17	1	2	—, —	—, —	—, —	—, —	—, —	—, —	—, —	—, —	—, —
17	1	3	—, —	—, —	—, —	—, —	—, —	—, —	—, —	—, —	—, —
17	1	4	—, —	—, —	—, —	—, —	—, —	—, —	—, —	—, —	—, —
17	1	5	—, —	—, —	—, —	—, —	—, —	—, —	—, —	—, —	—, —
17	1	6	—, —	—, —	—, —	—, —	—, —	—, —	—, —	—, —	—, —
17	1	7	—, —	—, —	—, —	—, —	—, —	—, —	—, —	—, —	—, —
17	1	8	—, —	—, —	—, —	—, —	—, —	—, —	—, —	—, —	—, —
17	2	2	—, 2	—, 2	—, 2	—, 2	—, 2	—, 2	—, —	—, —	—, —
17	2	3	—, 2	—, 2	—, 2	—, 2	—, 2	—, 2	—, —	—, —	—, —
17	2	4	0, 2	—, 2	—, 2	—, 2	—, —	—, —	—, —	—, —	—, —
17	2	5	0, 1	—, 2	—, 2	—, —	—, —	—, —	—, —	—, —	—, —
17	2	6	0, 2	—, 2	—, 2	—, —	—, —	—, —	—, —	—, —	—, —
17	2	7	0, 1	—, 2	—, 2	—, —	—, —	—, —	—, —	—, —	—, —
17	2	8	0, 2	—, 2	—, 2	—, —	—, —	—, —	—, —	—, —	—, —
17	3	3	—, 1	—, 2	—, 2	—, 3	—, 3	—, 3	—, 3	—, 3	—, —
17	3	4	0, 2	—, 2	—, 3	—, 3	—, 3	—, 3	—, 3	—, 3	—, —
17	3	5	0, 2	—, 3	—, 3	—, 3	—, 3	—, 3	—, —	—, —	—, —
17	3	6	0, 2	—, 3	—, 3	—, 3	—, —	—, —	—, —	—, —	—, —
17	3	7	0, 3	0, 3	—, 3	—, 3	—, —	—, —	—, —	—, —	—, —
17	3	8	0, 3	0, 3	—, 3	—, 3	—, —	—, —	—, —	—, —	—, —

TABLE B.28 (cont.) Critical Values for Fisher's Exact Test

n	m_1	m_2	α: 0.50	0.20	0.10	0.05	0.02	0.01	0.005	0.002	0.001
17	4	4	0, 2	–, 2	–, 3	–, 3	–, 4	–, 4	–, 4	–, 4	–, 4
17	4	5	0, 3	–, 3	–, 3	–, 4	–, 4	–, 4	–, 4	–, –	–, –
17	4	6	0, 3	–, 3	–, 4	–, 4	–, 4	–, 4	–, –	–, –	–, –
17	4	7	1, 3	0, 4	–, 4	–, 4	–, –	–, –	–, –	–, –	–, –
17	4	8	1, 3	0, 4	0, 4	–, 4	–, –	–, –	–, –	–, –	–, –
17	5	5	0, 3	–, 3	–, 4	–, 4	–, 5	–, 5	–, 5	–, 5	–, 5
17	5	6	1, 3	0, 4	–, 4	–, 5	–, 5	–, 5	–, 5	–, 5	–, 5
17	5	7	1, 4	0, 4	0, 5	0, 5	–, 5	–, 5	–, 5	–, –	–, –
17	5	8	1, 4	0, 4	0, 5	0, 5	–, 5	–, 5	–, 6	–, –	–, –
17	6	6	1, 4	0, 4	0, 5	0, 5	–, 6	–, 6	–, 6	–, 6	–, 6
17	6	7	1, 3	0, 4	0, 5	0, 5	0, 6	0, 6	–, 6	–, 6	–, 6
17	6	8	2, 4	1, 5	0, 5	0, 5	0, 6	0, 6	–, 6	–, 7	–, 7
17	7	7	2, 4	1, 5	0, 6	1, 6	0, 6	0, 7	0, 7	0, 7	–, 7
17	7	8	2, 5	1, 5	0, 6	1, 6	0, 7	0, 7	0, 7	0, 8	0, 8
17	8	8	3, 5	2, 6	1, 6	1, 6	1, 7	0, 7	0, 7	0, 8	0, 8
18	1	1	–, –	–, –	–, –	–, –	–, –	–, –	–, –	–, –	–, –
18	1	2	–, –	–, –	–, –	–, –	–, –	–, –	–, –	–, –	–, –
18	1	3	–, –	–, –	–, –	–, –	–, –	–, –	–, –	–, –	–, –
18	1	4	–, –	–, –	–, –	–, –	–, –	–, –	–, –	–, –	–, –
18	1	5	0, 1	–, 1	–, –	–, –	–, –	–, –	–, –	–, –	–, –
18	1	6	–, 1	–, 1	–, 1	–, –	–, –	–, –	–, –	–, –	–, –
18	1	7	–, 1	–, 1	–, 1	–, –	–, –	–, –	–, –	–, –	–, –
18	1	8	–, 1	–, 1	–, 1	–, –	–, –	–, –	–, –	–, –	–, –
18	1	9	0, 1	–, 1	–, 1	–, –	–, 2	–, 2	–, –	–, –	–, –
18	2	2	–, 2	–, 2	–, 2	–, 2	–, 2	–, 2	–, –	–, –	–, –
18	2	3	–, 2	–, 2	–, 2	–, 2	–, 2	–, –	–, –	–, –	–, –
18	2	4	–, 2	–, 2	–, 2	–, 2	–, 2	–, –	–, –	–, –	–, –
18	2	5	–, 2	–, 2	–, 2	–, 2	–, –	–, –	–, –	–, –	–, –
18	2	6	0, 2	–, 2	–, 2	–, 2	–, –	–, –	–, –	–, –	–, –
18	2	7	0, 2	–, 2	–, 3	–, 3	–, 3	–, 3	–, 3	–, 3	–, –
18	2	8	0, 2	–, 3	–, 3	–, 3	–, 3	–, 3	–, 3	–, 3	–, 3
18	2	9	1, –	–, 3	–, 3	–, 3	–, 3	–, 3	–, 3	–, 3	–, 3
18	3	4	0, 2	–, 3	–, 3	–, 3	–, 3	–, 3	–, 3	–, –	–, –
18	3	5	0, 2	–, 3	–, 3	–, 3	–, 3	–, 3	–, –	–, –	–, –
18	3	6	0, 2	0, 3	–, 3	–, 3	–, 3	–, 3	–, –	–, –	–, –
18	3	7	0, 3	0, 3	–, 3	–, 3	–, 3	–, –	–, –	–, –	–, –
18	3	8	0, 3	0, 3	0, 3	–, 3	–, 3	–, –	–, –	–, –	–, –
18	3	9	1, 2	–, 3	0, 3	–, 4	–, 4	–, 4	–, 4	–, 4	–, 4
18	4	4	0, 2	–, 3	–, 3	–, 3	–, 4	–, 4	–, 4	–, 4	–, 4
18	4	5	0, 3	–, 3	–, 4	–, 4	–, 4	–, 4	–, 4	–, 4	–, –
18	4	6	0, 3	0, 3	–, 4	–, 4	–, 4	–, 4	–, 4	–, –	–, –
18	4	7	1, 3	0, 4	–, 4	–, 4	–, –	–, –	–, –	–, –	–, –
18	4	8	1, 3	0, 4	0, 4	0, 4	–, –	–, –	–, –	–, –	–, –
18	4	9	1, 3	0, 4	0, 4	0, 4	–, –	–, –	–, –	–, –	–, –

TABLE B.28 (cont.) Critical Values for Fisher's Exact Test

n	m_1	m_2	α: 0.50	0.20	0.10	0.05	0.02	0.01	0.005	0.002	0.001
18	5	5	0,2 / 0,3	0,3 / 0,3	—,3 / —,3	—,4 / —,4	—,4 / —,4	—,4 / —,4	—,5 / —,5	—,5 / —,5	—,5 / —,5
18	5	6	1,3 / 0,3	0,3 / 0,4	—,3 / 0,4	—,4 / —,4	—,5 / —,5	—,5 / —,5	—,5 / —,5	—,5 / —,5	—,5 / —,5
18	5	7	1,3 / 0,3	1,3 / 0,4	0,4 / 0,4	—,4 / —,5	—,5 / —,5	—,5 / —,5	—,5 / —,5	— / —	— / —
18	5	8	1,3 / 1,4	1,5 / 0,4	0,4 / 0,5	0,5 / —,5	—,5 / 0,5	—,5 / —,5	— / —	— / —	— / —
18	5	9	2,3 / 1,4	1,5 / 0,5	0,5 / 0,5	0,5 / 0,5	0,5 / —,5	— / —	— / —	— / —	— / —
18	6	6	1,3 / 0,3	1,4 / 0,4	0,4 / 0,5	—,5 / —,5	—,5 / 0,5	—,6 / —,6	—,6 / —,6	—,6 / —,6	—,6 / —,6
18	6	7	1,4 / 1,4	1,4 / 0,5	0,5 / 0,5	0,5 / —,5	0,6 / 0,6	—,6 / —,6	—,6 / —,6	—,6 / —,6	—,6 / —,6
18	6	8	2,4 / 1,4	1,5 / 1,5	0,5 / 0,5	0,6 / 0,5	0,6 / 0,6	—,6 / 0,6	—,6 / 0,6	—,6 / 0,6	—,6 / —,6
18	6	9	2,4 / 1,5	1,5 / 1,5	0,6 / 0,6	0,6 / 0,6	0,6 / 0,6	—,6 / —,6	—,6 / —,6	—,7 / —,7	—,7 / —,7
18	7	7	2,4 / 1,4	2,4 / 1,5	1,6 / 0,5	—,6 / 0,6	0,7 / 0,6	0,7 / 0,6	0,7 / 0,7	—,7 / —,7	—,7 / —,7
18	7	8	2,5 / 2,4	2,5 / 1,5	1,6 / 1,6	1,6 / 0,6	0,7 / 0,6	0,7 / 0,7	0,7 / 0,7	0,7 / —,7	—,7 / —,7
18	7	9	3,5 / 2,5	3,5 / 2,6	1,6 / 1,6	1,6 / 0,6	0,7 / 0,7	0,7 / 0,7	0,7 / 0,7	0,8 / 0,7	—,8 / —,8
18	8	8	3,5 / 2,5	3,5 / 2,6	1,7 / 1,6	1,7 / 0,7	0,7 / 0,7	0,8 / 0,7	0,8 / 0,7	0,8 / 0,8	0,8 / —,8
18	8	9	4,5 / 3,6	3,6 / 2,6	2,7 / 1,7	1,7 / 1,7	1,8 / 0,7	1,8 / 0,8	0,8 / 0,8	0,8 / 0,8	0,8 / 0,8
18	9	9	4,5 / 3,6	3,6 / 2,7	2,7 / 2,7	2,7 / 1,8	1,8 / 1,8	1,8 / 1,8	1,8 / 0,8	0,9 / 0,9	0,9 / 0,9
19	1	1	—,1 / —,1	—,1 / —,1	—,1 / —,1	— / —	— / —	— / —	— / —	— / —	— / —
19	1	2	—,1 / —,1	—,1 / —,1	— / —	— / —	— / —	— / —	— / —	— / —	— / —
19	1	3	—,1 / —,1	— / —	— / —	— / —	— / —	— / —	— / —	— / —	— / —
19	1	4	—,1 / —,1	— / —	— / —	— / —	— / —	— / —	— / —	— / —	— / —
19	1	5	— / —	— / —	— / —	— / —	— / —	— / —	— / —	— / —	— / —
19	1	6	— / —	— / —	— / —	— / —	— / —	— / —	— / —	— / —	— / —
19	1	7	— / —	— / —	— / —	— / —	— / —	— / —	— / —	— / —	— / —
19	1	8	— / —	— / —	— / —	—,2 / —,2	—,2 / —,2	—,2 / —,2	— / —	— / —	— / —
19	1	9	—,2 / —,1	—,2 / —,1	—,2 / —,2	—,2 / —,2	—,2 / —,2	—,2 / —,2	— / —	— / —	— / —
19	2	2	—,2 / —,1	—,2 / —,1	—,2 / —,2	— / —	— / —	— / —	— / —	— / —	— / —
19	2	3	—,2 / —,1	—,2 / —,2	—,2 / —,2	—,2 / —,2	—,2 / —,2	—,2 / —,2	— / —	— / —	— / —
19	2	4	—,2 / —,2	—,2 / —,2	—,2 / —,2	—,2 / —,2	—,2 / —,2	—,2 / —,2	— / —	— / —	— / —
19	2	5	—,2 / 0,3	—,2 / —,2	—,2 / —,2	—,2 / —,2	—,2 / —,2	— / —	— / —	— / —	— / —
19	2	6	0,2 / 0,2	0,3 / —,2	—,2 / —,2	—,2 / —,2	— / —	—,2 / —	— / —	— / —	— / —
19	2	7	0,2 / 0,2	—,2 / 0,3	—,2 / —,2	—,2 / —,2	— / —	— / —	— / —	— / —	— / —
19	2	8	0,2 / 0,2	0,2 / 0,2	—,2 / —,2	—,2 / —,2	—,2 / —,2	— / —	— / —	— / —	— / —
19	2	9	0,2 / 0,2	0,2 / 0,2	0,2 / —,2	—,2 / —,2	—,2 / —,2	— / —	— / —	— / —	— / —
19	3	3	—,1 / 0,2	—,2 / —,2	—,2 / —,2	—,2 / —,2	— / —	— / —	— / —	— / —	— / —
19	3	4	0,2 / 0,3	0,3 / 0,2	0,3 / 0,2	—,3 / —,2	—,3 / —,3	—,3 / —,3	—,3 / —,3	—,3 / —	— / —
19	3	5	0,2 / 0,3	0,3 / 0,3	0,3 / 0,3	—,3 / —,3	—,3 / —,3	—,3 / —,3	—,3 / —,3	—,3 / —	— / —
19	3	6	0,2 / 0,3	0,3 / 0,3	0,3 / 0,3	0,3 / —,3	—,3 / —,3	—,3 / —,3	— / —	— / —	— / —
19	3	7	0,2 / 0,3	0,3 / 0,3	0,3 / 0,3	0,3 / —,3	—,3 / —,3	—,3 / —,3	— / —	— / —	— / —
19	3	8	0,2 / 0,3	0,3 / 0,3	0,3 / 0,3	0,4 / 0,3	—,3 / —,3	—,3 / —	—,3 / —	—,3 / —	—,4 / —
19	3	9	0,2 / 1,3	0,4 / 0,3	—,4 / 0,4	0,4 / 0,4	0,4 / 0,4	—,4 / —,4	—,4 / —,4	—,4 / —,4	—,4 / —,4
19	4	4	0,2 / 0,3	0,3 / 0,3	—,4 / 0,3	—,4 / 0,4	0,4 / —,4	—,4 / —,4	—,4 / —,4	—,4 / —,4	— / —
19	4	5	0,2 / 0,3	0,3 / 0,3	0,4 / 0,4	—,4 / —,4	—,4 / —,4	—,4 / —,4	—,4 / —,4	— / —	— / —
19	4	6	0,2 / 0,3	0,3 / 0,3	0,4 / —,4	0,4 / —,4	—,4 / —,4	—,4 / —,4	—,4 / —,4	—,4 / —	—,4 / —
19	4	7	0,3 / 0,3	0,3 / 0,3	0,4 / —,4	0,4 / —,4	—,4 / —,4	—,4 / —,4	—,4 / —	—,4 / —	—,4 / —
19	4	8	1,3 / 0,3	0,4 / 0,3	0,4 / 0,4	0,4 / 0,4	—,4 / —,4	— / —	— / —	— / —	— / —
19	4	9	1,3 / 0,3	0,4 / 0,4	0,4 / 0,4	0,4 / 0,4	— / —	— / —	— / —	— / —	— / —

TABLE B.28 (cont.) Critical Values for Fisher's Exact Test

n	m₁	m₂	α: 0.50	0.20	0.10	0.05	0.02	0.01	0.005	0.002	0.001
19	5	5	0, 2	0, 3	—, 3	—, 4	—, 4	—, 4	—, 5	—, 5	—, 5
19	5	6	1, 3	0, 3	—, 3	—, 4	—, 4	—, 5	—, 5	—, 5	—, 5
19	5	7	1, 3	0, 4	—, 4	—, 4	—, 5	—, 5	—, 5	—, 5	—, —
19	5	8	1, 3	0, 4	0, 5	0, 5	—, 5	—, 5	—, 5	—, —	—, —
19	5	9	1, 3	1, 4	0, 5	0, 5	—, 5	—, 5	—, 5	—, —	—, —
19	6	6	1, 3	0, 4	0, 4	—, 4	—, 5	—, 5	—, 5	—, 6	—, 6
19	6	7	1, 4	0, 4	0, 5	—, 5	—, 5	—, 5	—, 6	—, 6	—, 6
19	6	8	2, 4	1, 5	0, 5	0, 5	—, 6	—, 6	—, 6	—, 6	—, —
19	6	9	2, 4	1, 5	0, 5	0, 5	0, 6	—, 6	—, 6	—, 6	—, —
19	7	7	2, 4	1, 5	0, 5	0, 5	0, 6	—, 6	—, 6	—, 7	—, 7
19	7	8	2, 4	1, 5	1, 6	0, 6	0, 6	0, 7	—, 7	—, 7	—, 7
19	7	9	2, 5	1, 5	1, 6	1, 6	0, 7	0, 7	0, 7	—, 7	—, 8
19	8	8	2, 5	2, 6	1, 6	1, 6	0, 7	0, 7	0, 7	0, 8	—, 8
19	8	9	3, 5	2, 6	1, 6	1, 7	0, 7	0, 7	0, 8	0, 8	—, 8
19	9	9	3, 6	2, 6	2, 7	1, 7	1, 7	1, 8	0, 8	0, 8	0, 9
20	1	1	—, 1	—, —	—, —	—, —	—, —	—, —	—, —	—, —	—, —
20	1	2	—, 1	—, —	—, —	—, —	—, —	—, —	—, —	—, —	—, —
20	1	3	—, 1	—, 1	—, 1	—, —	—, —	—, —	—, —	—, —	—, —
20	1	4	—, 1	—, 1	—, 1	—, —	—, —	—, —	—, —	—, —	—, —
20	1	5	—, 1	—, 1	—, 1	—, 1	—, —	—, —	—, —	—, —	—, —
20	1	6	—, 1	—, 1	—, 1	—, 1	—, —	—, —	—, —	—, —	—, —
20	1	7	—, 1	—, 1	—, 1	—, 1	—, 1	—, 1	—, —	—, —	—, —
20	1	8	—, 1	—, 1	—, 1	—, 1	—, 1	—, 1	—, —	—, —	—, —
20	1	9	—, 1	—, 1	—, 1	—, 1	—, 1	—, 1	—, —	—, —	—, —
20	1	10	0, 1	—, 1	—, 1	—, 1	—, 1	—, 1	—, —	—, —	—, —
20	2	2	—, 1	—, 1	—, —	—, —	—, —	—, —	—, —	—, —	—, —
20	2	3	—, 1	—, 1	—, 1	—, 1	—, —	—, —	—, —	—, —	—, —
20	2	4	—, 2	—, 1	—, 1	—, 1	—, —	—, —	—, —	—, —	—, —
20	2	5	—, 2	—, 2	—, 2	—, 2	—, 2	—, 2	—, —	—, —	—, —
20	2	6	—, 2	—, 2	—, 2	—, 2	—, 2	—, 2	—, —	—, —	—, —
20	2	7	—, 2	—, 2	—, 2	—, 2	—, 2	—, 2	—, —	—, —	—, —
20	2	8	—, 2	—, 2	—, 2	—, 2	—, 2	—, 2	—, —	—, —	—, —
20	2	9	0, 2	—, 2	—, 2	—, 2	—, 2	—, 2	—, —	—, —	—, —
20	2	10	0, 2	—, 2	—, 2	—, 2	—, 2	—, 2	—, —	—, —	—, —
20	3	3	—, 1	—, 1	—, 1	—, 1	—, —	—, —	—, —	—, —	—, —
20	3	4	—, 2	—, 2	—, 2	—, 2	—, 2	—, 2	—, —	—, —	—, —
20	3	5	—, 2	—, 2	—, 2	—, 2	—, 2	—, 2	—, 3	—, 3	—, —
20	3	6	—, 2	—, 2	—, 3	—, 3	—, 3	—, 3	—, 3	—, 3	—, 3
20	3	7	—, 2	—, 3	—, 3	—, 3	—, 3	—, 3	—, 3	—, 3	—, 3
20	3	8	0, 3	—, 3	—, 3	—, 3	—, 3	—, 3	—, 3	—, 3	—, 3
20	3	9	0, 3	0, 3	—, 3	—, 3	—, 3	—, 3	—, 3	—, 3	—, 3
20	3	10	1, 3	0, 3	—, 3	—, 3	0, 3	—, 3	—, 3	—, 3	—, 3
20	4	4	0, 2	—, 3	—, 3	—, 3	—, 3	—, 4	—, 4	—, 4	—, 4
20	4	5	0, 2	—, 3	—, 3	—, 3	—, 3	—, 4	—, 4	—, 4	—, 4
20	4	6	0, 2	—, 3	—, 3	—, 4	—, 4	—, 4	—, 4	—, —	—, —

TABLE B.28 (cont.) Critical Values for Fisher's Exact Test

n	m_1	m_2	α: 0.50	0.20	0.10	0.05	0.02	0.01	0.005	0.002	0.001
20	4	7	0, 2 0, 3	0, 3 0, 3	—, 4 —, 4	—, 4 —, 4	4 —, 4	—, 4 —, 4	—, — —, —	—, — —, —	—, — —, —
20	4	8	1, 3 0, 3	0, 3 0, 4	—, 4 —, 4	—, 4 —, 4	4 —, 4	—, — —, —	—, — —, —	—, — —, —	—, — —, —
20	4	9	1, 3 1, 3	0, 4 0, 4	0, 4 0, 4	—, 4 —, 4	—, 4 —, 4	—, — —, —	—, — —, —	—, — —, —	—, — —, —
20	4	10	1, 3 0, 4	0, 4 0, 4	0, 4 0, 4	0, 4 —, 4	—, 4 —, 4	—, 4 —, 4	—, 4 —, 4	—, 5 —, 5	—, 5 —, 5
20	5	5	0, 2 0, 3	0, 3 —, 3	0, 4 —, 3	—, 4 —, 4	—, 4 —, 4	—, 4 —, 4	—, 4 —, 4	—, 5 —, 5	—, 5 —, 5
20	5	6	0, 2 0, 3	0, 3 0, 4	0, 4 —, 4	—, 4 —, 4	—, 5 —, 4	—, 5 —, 5	—, 5 —, 5	—, 5 —, 5	—, 5 —, 5
20	5	7	1, 3 0, 3	0, 4 0, 4	0, 4 0, 5	—, 5 —, 5	—, 5 —, 5	—, 5 —, 5	—, 5 —, 5	—, 5 —, 5	—, 5 —, 5
20	5	8	1, 3 1, 3	0, 4 0, 4	0, 5 0, 5	0, 5 0, 5	—, 5 —, 5	—, 5 —, 5	—, 5 —, 5	—, — —, —	—, — —, —
20	5	9	1, 3 1, 4	0, 4 0, 4	0, 5 0, 5	—, 5 0, 5	—, 5 —, 5	—, 5 —, 5	—, — —, —	—, — —, —	—, — —, —
20	5	10	2, 3 1, 4	1, 4 0, 5	0, 5 0, 5	—, 5 0, 5	—, 6 0, 5	—, 6 —, 6	—, — —, —	—, — —, —	—, — —, —
20	6	6	1, 3 0, 3	0, 4 0, 4	0, 4 0, 4	—, 4 —, 4	—, 5 —, 5	—, 5 —, 5	—, 5 —, 5	—, 6 —, 6	6 —, 6
20	6	7	1, 4 1, 4	0, 4 0, 4	0, 5 0, 5	—, 5 0, 5	—, 5 —, 5	—, 6 —, 6	—, 6 —, 6	—, 6 —, 6	6 —, 6
20	6	8	1, 4 1, 4	1, 4 0, 4	0, 5 0, 5	0, 5 0, 5	—, 6 0, 6	—, 6 —, 6	—, 6 —, 6	—, 6 —, 6	6 —, 6
20	6	9	2, 4 1, 5	1, 5 1, 5	0, 5 0, 6	0, 6 0, 6	0, 6 —, 6	—, 6 —, 6	—, 6 —, 6	—, — —, —	—, — —, —
20	6	10	2, 4 1, 5	1, 5 1, 5	0, 6 1, 6	0, 6 0, 6	0, 6 0, 6	0, 6 —, 6	—, 6 —, 6	—, — —, —	—, — —, —
20	7	7	1, 3 1, 4	1, 4 1, 4	0, 5 0, 5	0, 5 0, 5	—, 6 0, 6	—, 6 —, 6	—, 6 —, 6	—, 6 —, 6	6 —, 7
20	7	8	2, 4 2, 4	1, 5 1, 5	0, 6 1, 6	1, 7 0, 6	0, 6 0, 6	0, 7 —, 6	0, 7 —, 6	—, 7 —, 7	7 —, 7
20	7	9	2, 4 2, 5	1, 5 1, 5	1, 6 2, 6	1, 7 1, 6	0, 7 0, 6	0, 7 0, 7	0, 7 0, 7	—, 7 —, 7	7 —, 7
20	7	10	3, 4 2, 5	2, 5 2, 5	1, 6 2, 6	1, 7 1, 7	0, 7 0, 7	0, 7 0, 7	0, 7 0, 7	—, 7 0, 7	—, 7 —, 7
20	8	8	3, 5 2, 5	2, 6 2, 6	1, 6 2, 6	1, 7 1, 7	0, 7 0, 7	0, 7 0, 7	0, 7 0, 7	0, 8 0, 8	8 —, 8
20	8	9	3, 5 3, 5	2, 6 2, 6	2, 6 2, 6	1, 7 1, 7	1, 7 1, 7	0, 8 0, 7	0, 8 0, 8	0, 8 0, 8	8 —, 8
20	9	9	3, 5 3, 6	2, 6 2, 6	2, 6 2, 6	1, 8 1, 7	1, 8 1, 8	1, 8 0, 8	0, 8 0, 8	0, 8 0, 9	9 0, 9
20	9	10	4, 5 4, 6	3, 6 3, 6	2, 7 2, 7	2, 8 1, 8	1, 8 1, 8	1, 9 1, 8	0, 9 0, 9	0, 9 0, 9	9 0, 9
20	10	10	4, 6 4, 6	3, 7 3, 7	3, 7 2, 7	2, 8 2, 8	2, 8 1, 9	1, 9 1, 9	1, 9 1, 9	1, 9 1, 9	10 0, 10
21	1	1	—, — —, —	—, — —, —	—, — —, —	—, — —, —	—, — —, —	—, — —, —	—, — —, —	—, — —, —	—, — —, —
21	1	2	—, — —, —	—, 1 —, 1	—, 1 —, 1	1 —, 1	—, 1 —, 1	2 —, 2	—, 2 —, 2	—, — —, —	—, — —, —
21	1	3	—, — —, —	1, — 1, —	—, 1 —, 1	—, 1 —, 1	—, 1 —, 1	—, 1 —, 1	—, — —, —	—, — —, —	—, — —, —
21	1	4	—, — —, —	1, — 1, —	—, 1 —, 1	—, 1 —, 1	—, 1 —, 1	—, 1 —, 1	—, — —, —	—, — —, —	—, — —, —
21	1	5	—, — —, —	—, — —, —	—, — —, —	—, — —, —	—, — —, —	—, — —, —	—, — —, —	—, — —, —	—, — —, —
21	2	2	—, 1 —, 1	—, 1 —, 1	—, 1 —, 1	—, — —, —	—, — —, —	—, — —, —	—, 2 —, 2	—, — —, —	—, — —, —
21	2	3	0, 2 0, 2	1, 2 —, 2	1, 2 1, 2	1, 2 1, 2	2 —, 2	2 —, 2	2 —, 2	—, — —, —	—, — —, —
21	2	4	0, 2 0, 2	1, 2 —, 2	1, 2 1, 2	1, 2 1, 2	2 —, 2	2 —, 2	2 —, 2	—, — —, —	—, — —, —
21	2	5	0, 2 0, 2	1, 2 1, 2	1, 2 1, 2	1, 2 1, 2	2 —, 2	2 —, 2	2 —, 2	—, — —, —	—, — —, —
21	2	6	0, 1 —, 1	1, — 1, —	—, 2 —, 2	2 —, 2	2 —, 2	2 —, 2	2 —, 2	—, — —, —	—, — —, —
21	2	7	1, — 0, 2	2, — 1, —	2, — 1, —	2, — 1, —	2, — —, 2	2, — —, 2	2, — —, 2	3 —, 3	—, 3 —, 3
21	2	8	0, 2 0, 2	2, — 1, —	2, — 1, —	2, — —, 2	2, — —, 2	2, — —, 2	2, — —, 2	3 —, 3	—, 3 —, 3
21	2	9	0, 2 0, 2	2, — 1, —	2, — 1, —	2, — —, 2	2, — —, 2	2, — —, 2	2, — —, 2	3 —, 3	—, 3 —, 3
21	2	10	0, 1 —, 1	1, — 1, —	2, — 1, —	2, — —, 2	2, — —, 2	2, — —, 2	2, — —, 2	3 —, 3	—, 3 —, 3
21	3	3	—, 1 —, 1	1, — 1, —	2, — 1, —	2, — —, 2	3, — —, 3	3, — —, 3	3, — —, 3	3 —, 3	—, 3 —, 3

TABLE B.28 (cont.) Critical Values for Fisher's Exact Test

This page contains a large numeric table of critical values for Fisher's Exact Test, with columns for n, m_1, m_2, and significance levels α: 0.50, 0.20, 0.10, 0.05, 0.02, 0.01, 0.005, 0.002, 0.001. The dense tabular entries are not legibly transcribable with sufficient accuracy.

n	m_1	m_2	α: 0.50	0.20	0.10	0.05	0.02	0.01	0.005	0.002	0.001
21	3	4									
21	3	5									
21	3	6									
21	3	7									
21	3	8									
21	3	9									
21	3	10									
21	4	4									
21	4	5									
21	4	6									
21	4	7									
21	4	8									
21	4	9									
21	4	10									
21	5	5									
21	5	6									
21	5	7									
21	5	8									
21	5	9									
21	5	10									
21	6	6									
21	6	7									
21	6	8									
21	6	9									
21	6	10									
21	7	7									
21	7	8									
21	7	9									
21	7	10									
21	8	8									
21	8	9									
21	8	10									
21	9	9									
21	9	10									
21	10	10									
22	1	1									
22	1	2									
22	1	3									
22	1	4									
22	1	5									
22	1	6									
22	1	7									
22	1	8									
22	1	9									
22	1	10									

TABLE B.28 (cont.) Critical Values for Fisher's Exact Test

n	m_1	m_2	α: 0.50	0.20	0.10	0.05	0.02	0.01	0.005	0.002	0.001
22	1	11	0, 1	–, –	–, –	–, –	–, –	–, –	–, –	–, –	–, –
22	2	2	–, –	1, –	1, 2	1, 2	1, 2	1, 2	1, 2	1, –	1, –
22	2	3	–, –	1, –	1, 2	1, 2	1, 2	1, 2	1, –	1, –	1, –
22	2	4	1, –	1, 2	1, 2	1, 2	1, 2	1, 2	1, –	1, –	1, –
22	2	5	1, –	2, 2	1, 2	1, 2	1, 2	1, 2	1, –	1, –	1, –
22	2	6	1, –	2, 2	1, 2	1, –	1, –	1, –	1, –	1, –	1, –
22	2	7	0, 2	2, 2	2, 2	1, –	1, –	1, –	1, –	1, –	1, –
22	2	8	0, 2	2, 2	2, 2	1, –	1, –	1, –	1, –	1, –	1, –
22	2	9	0, 2	2, 2	2, 2	2, 2	1, –	1, –	1, –	1, –	1, –
22	2	10	0, 2	2, 2	2, 2	2, 2	1, –	1, –	1, –	1, –	1, –
22	2	11	0, 2	2, 2	2, 2	2, 2	2, –	1, –	1, –	1, –	1, –
22	3	3	1, –	2, 2	2, 2	1, –	3, –	3, –	3, –	3, –	3, –
22	3	4	0, 3	2, 2	2, 3	3, 3	3, 3	3, 3	3, 3	3, –	3, –
22	3	5	0, 3	3, 3	2, 3	3, 3	3, 3	3, 3	3, 3	3, –	3, –
22	3	6	0, 3	3, 3	3, 3	3, 3	3, 3	3, 3	3, –	3, –	3, –
22	3	7	0, 3	3, –	3, 3	3, 3	3, 3	3, 3	3, –	–, –	–, –
22	3	8	2, –	3, 3	3, 3	3, 3	3, –	3, –	3, –	–, –	–, –
22	3	9	0, 3	3, 3	3, 3	3, 3	3, –	3, –	3, –	–, –	–, –
22	3	10	0, 3	3, 3	3, 4	3, 4	3, –	3, –	3, –	–, –	–, –
22	3	11	1, 2	3, 3	3, 4	4, 4	3, –	3, –	3, –	–, –	–, –
22	4	4	0, 2	3, 3	3, 4	4, 4	4, 4	4, 4	4, 4	4, 4	4, 4
22	4	5	0, 2	3, 4	3, 4	4, 4	4, 4	4, 4	4, 4	4, 4	4, 4
22	4	6	0, 2	4, 4	0, 4	4, 4	4, 4	4, 4	4, 4	4, –	4, –
22	4	7	0, 2	4, 4	0, 4	4, 4	4, 4	4, 4	4, –	–, –	–, –
22	4	8	0, 3	0, 4	0, 4	4, 4	4, –	4, –	4, –	–, –	–, –
22	4	9	1, 3	0, 4	0, 4	4, 5	4, –	4, –	4, 5	–, –	–, –
22	4	10	1, 3	0, 4	0, 4	4, 4	4, –	4, 4	4, 5	5, –	5, –
22	4	11	1, 3	0, 4	0, 5	4, 5	4, –	4, 4	5, 5	5, –	5, –
22	5	5	1, 3	0, 4	0, 5	4, 5	5, 5	5, 5	5, 5	5, 5	5, 5
22	5	6	0, 2	3, 4	4, 5	0, 5	5, 5	5, 5	5, 6	5, –	5, –
22	5	7	1, 3	0, 4	4, 5	5, 5	5, 5	5, 5	5, 6	5, –	5, –
22	5	8	1, 3	0, 4	0, 5	0, 5	5, 5	5, 5	5, 6	5, –	5, –
22	5	9	1, 4	0, 4	0, 5	0, 6	5, 6	5, 5	5, 6	5, –	5, –
22	5	10	1, 4	1, 5	0, 5	0, 6	5, 5	5, 6	6, 6	6, –	6, –
22	5	11	2, 3	1, 5	1, 6	0, 6	6, 7	5, 6	6, 6	6, –	6, –
22	6	6	1, 3	1, 5	0, 6	6, 6	6, 6	6, 6	6, 6	6, 6	6, 6
22	6	7	1, 4	0, 4	0, 5	5, 6	5, 6	6, 6	6, 6	6, 6	6, 6
22	6	8	1, 4	1, 5	0, 5	0, 6	6, 6	6, 6	6, 7	6, –	6, –
22	6	9	1, 4	1, 5	0, 5	0, 6	6, 5	6, 6	6, 7	6, –	6, –
22	6	10	1, 4	1, 5	0, 5	0, 6	6, 7	0, 7	7, 7	7, –	7, –
22	6	11	2, 4	1, 5	1, 6	0, 6	6, 7	0, 7	7, 7	7, –	7, –
22	7	7	1, 4	0, 4	4, 5	0, 5	6, 6	6, 6	6, 7	6, 6	6, 6
22	7	8	1, 4	0, 4	4, 5	0, 5	6, 6	0, 6	7, 7	7, 7	7, 7
22	7	9	2, 4	1, 5	0, 5	0, 6	6, 6	0, 6	7, 7	7, 7	7, 7
22	7	10	2, 4	1, 5	1, 6	0, 6	6, 7	0, 7	7, 7	7, 7	7, 7

TABLE B.28 (cont.) Critical Values for Fisher's Exact Test

n	m₁	m₂	α: 0.50	0.20	0.10	0.05	0.02	0.01	0.005	0.002	0.001
22	7	11	3, 4	2, 5	1, 6	1, 6	0, 7	0, 7	0, 7	—, 7	—, 7
22	8	8	2, 4	1, 5	1, 6	0, 6	0, 6	0, 6	0, 7	—, 7	—, 7
22	8	9	2, 5	1, 5	1, 6	0, 6	0, 7	0, 7	0, 7	—, 7	—, 8
22	8	10	3, 5	1, 6	1, 6	1, 6	0, 7	0, 7	0, 8	0, 8	—, 8
22	8	11	3, 5	2, 6	2, 6	1, 7	0, 8	0, 8	0, 8	0, 8	—, 8
22	9	9	3, 5	2, 6	1, 6	1, 7	1, 7	0, 7	0, 8	0, 8	—, 8
22	9	10	3, 5	2, 6	2, 7	1, 7	1, 7	0, 8	0, 8	0, 8	0, 9
22	9	11	4, 5	2, 7	2, 7	1, 8	1, 8	1, 8	0, 9	0, 9	0, 9
22	10	10	4, 6	3, 7	2, 7	2, 7	1, 8	1, 8	1, 9	0, 9	0, 9
22	10	11	4, 6	3, 7	2, 7	2, 8	1, 8	1, 9	1, 9	0, 9	0, 10
22	11	11	5, 6	3, 8	3, 8	3, 8	2, 9	2, 9	2, 9	1, 10	1, 10
23	1	1	—	—	—	—	—	—	—	—	—
23	1	2	—	—	—	—	—	—	—	—	—
23	1	3	—	—	—	—	—	—	—	—	—
23	1	4	—	—	—	—	—	—	—	—	—
23	1	5	—	—	—	—	—	—	—	—	—
23	1	6	—	—	—	—	—	—	—	—	—
23	1	7	—	—	—	—	—	—	—	—	—
23	1	8	—	—	—	—	—	—	—	—	—
23	1	9	—	—	—	—	—	—	—	—	—
23	1	10	0, 1	—	—	—	—	—	—	—	—
23	1	11	0, 1	—	—	—	—	—	—	—	—
23	2	2	—	—	—	—	—	—	—	—	—
23	2	3	—	—	—	—	2	—	—	—	—
23	2	4	—	—	—	—	2	2	2	—	—
23	2	5	0, 1	1, 2	—	2	2	—	—	—	—
23	2	6	0, 2	1, 2	2	2	2	2	2	—	—
23	2	7	0, 2	1, 2	2	2	2	2	2	—	—
23	2	8	0, 2	1, 2	2	2	2	2	2	—	—
23	2	9	0, 2	1, 2	2	2	2	2	2	—	—
23	2	10	0, 3	0, 3	2	2	2	2	2	—	—
23	2	11	0, 3	0, 3	2	2	2	2	2	—	—
23	3	3	—	—	3	3	3	3	3	3	3
23	3	4	0, 2	1, 3	3	3	3	3	3	—	—
23	3	5	0, 2	1, 3	3	3	3	3	3	—	—
23	3	6	0, 2	1, 3	3	3	3	3	3	—	—
23	3	7	0, 3	1, 3	3	3	3	3	3	—	—
23	3	8	0, 3	1, 3	3	3	3	3	3	—	—
23	3	9	0, 3	1, 3	3	3	3	3	3	—	—
23	3	10	0, 3	0, 3	3	3	3	3	3	—	—
23	3	11	0, 3	0, 3	3	3	3	3	3	—	—
23	4	4	0, 2	1, 3	3	3	3	4	4	4	4
23	4	5	0, 2	1, 3	3	3	4	4	4	4	4
23	4	6	0, 2	1, 3	3	3	4	4	4	4	4
23	4	7	0, 3	1, 3	3	4	4	4	4	—	—

TABLE B.28 (cont.) Critical Values for Fisher's Exact Test

n	m_1	m_2	α: 0.50	0.20	0.10	0.05	0.02	0.01	0.005	0.002	0.001
23	4	8	0, 2 0, 3	0, 3 –, 3	–, 4 –, 4	–, 4 –, 4	–, 4 –, 4	–, 4 –, 4	–, – –, –	–, – –, –	–, – –, –
23	4	9	1, 3 0, 3	0, 3 0, 4	–, 4 –, 4	–, 4 –, 4	–, 4 –, 4	–, 4 –, 4	–, – –, –	–, – –, –	–, – –, –
23	4	10	1, 3 0, 3	0, 3 0, 4	–, 4 0, 4	–, 4 –, 4	–, 4 –, 4	–, 4 –, 4	–, – –, –	–, – –, –	–, – –, –
23	4	11	1, 3 0, 3	0, 4 0, 4	0, 4 0, 4	–, 4 –, 3	–, 4 –, 4	–, 4 –, 4	–, 5 –, 5	–, 5 –, 5	–, 5 –, 5
23	5	5	0, 2 –, 2	–, 3 –, 3	–, 3 –, 3	–, 3 0, 5	–, 4 –, 4	–, 4 –, 4	–, 4 –, 4	–, 5 –, 5	–, 5 –, 5
23	5	6	0, 2 0, 3	0, 3 –, 3	–, 4 –, 4	–, 4 –, 4	–, 4 –, 4	–, 4 –, 5	–, 5 –, 5	–, 5 –, 5	–, 5 –, 5
23	5	7	1, 3 0, 3	0, 3 0, 4	–, 4 0, 4	–, 4 –, 4	–, 5 –, 5	–, 5 –, 5	–, 5 –, 5	–, 5 –, 5	–, 5 –, 5
23	5	8	1, 3 0, 3	0, 4 0, 4	0, 4 0, 4	–, 5 0, 5	–, 5 –, 5	–, 5 –, 5	–, 5 –, 5	–, – –, –	–, – –, –
23	5	9	1, 3 0, 3	0, 4 0, 4	0, 4 0, 4	–, 4 0, 5	–, 5 –, 5	–, 5 –, 5	–, 5 –, 5	–, – –, –	–, – –, –
23	5	10	1, 3 1, 4	0, 4 0, 4	0, 4 0, 5	0, 5 0, 5	–, 5 0, 6	–, 5 –, 6	–, 5 –, 5	–, – –, –	–, – –, –
23	5	11	1, 3 1, 4	0, 4 0, 4	0, 5 0, 4	0, 5 –, 4	0, 5 0, 5	–, 5 –, 6	–, 6 –, 6	–, 6 –, 6	–, 6 –, 6
23	6	6	1, 3 1, 4	1, 4 0, 4	0, 5 –, 4	0, 5 –, 4	0, 6 0, 6	–, 6 –, 6	–, 6 –, 6	–, 6 –, 6	–, 6 –, 6
23	6	7	1, 3 0, 3	0, 4 0, 4	0, 4 0, 5	0, 6 0, 6	0, 6 0, 6	–, 6 0, 6	–, 6 –, 6	–, 6 –, 6	–, 6 –, 6
23	6	8	1, 3 1, 4	0, 4 0, 4	0, 5 0, 5	0, 6 0, 6	0, 6 0, 6	–, 6 –, 6	–, 6 –, 6	–, 6 –, 6	–, 6 –, 6
23	6	9	1, 4 1, 4	0, 4 0, 4	0, 5 0, 6	0, 5 0, 6	0, 5 0, 6	0, 6 0, 6	–, 6 –, 6	–, – –, –	–, – –, –
23	6	10	2, 4 1, 4	1, 5 1, 5	1, 5 1, 6	0, 6 0, 6	0, 6 0, 6	0, 6 –, 6	–, 6 –, 6	–, 6 –, 6	–, 6 –, 6
23	6	11	2, 4 1, 4	1, 5 1, 5	1, 6 1, 6	0, 6 0, 6	0, 6 0, 7	0, 7 0, 7	–, 7 –, 7	–, 7 –, 7	–, 6 –, 7
23	7	7	1, 4 1, 4	0, 4 0, 4	0, 5 0, 5	0, 6 0, 6	0, 6 0, 6	0, 6 0, 6	–, 7 0, 7	–, 7 –, 7	–, 7 –, 7
23	7	8	2, 4 1, 4	1, 5 1, 5	1, 6 0, 5	1, 7 1, 6	0, 7 0, 7	0, 7 0, 7	0, 7 0, 7	–, 7 –, 7	–, 7 –, 7
23	7	9	2, 4 1, 4	1, 5 1, 5	0, 5 1, 6	1, 7 1, 7	0, 7 0, 6	0, 7 –, 6	0, 7 0, 7	–, 7 –, 7	–, 7 –, 7
23	7	10	2, 5 2, 5	1, 5 1, 5	1, 6 1, 6	1, 7 1, 7	0, 7 0, 7	0, 7 0, 7	0, 7 0, 7	–, 8 –, 8	–, 8 –, 8
23	7	11	2, 5 2, 5	1, 6 1, 5	2, 7 1, 6	1, 7 1, 7	1, 8 1, 8	1, 8 1, 8	0, 8 0, 8	–, 8 –, 8	–, 8 –, 8
23	8	8	2, 4 2, 5	1, 5 1, 5	1, 6 1, 6	1, 6 1, 6	0, 6 0, 7	0, 7 0, 7	0, 8 0, 8	–, 8 –, 8	–, 8 –, 8
23	8	9	3, 5 2, 5	2, 6 1, 5	1, 6 1, 6	1, 7 1, 7	1, 7 1, 7	0, 7 0, 8	0, 8 0, 8	–, 8 –, 8	–, 8 –, 8
23	8	10	3, 5 2, 5	2, 6 1, 6	1, 7 1, 6	1, 7 1, 7	1, 8 0, 7	0, 8 0, 8	0, 8 0, 8	–, 9 –, 9	–, 8 –, 9
23	8	11	3, 6 2, 5	2, 6 2, 6	1, 7 1, 7	1, 7 1, 7	1, 8 1, 8	1, 8 0, 8	0, 8 0, 8	–, 9 0, 9	–, 9 –, 9
23	9	9	3, 5 2, 5	2, 6 1, 6	1, 6 1, 6	1, 7 1, 6	1, 8 0, 7	0, 8 0, 8	0, 8 0, 8	0, 9 0, 9	–, 9 –, 9
23	9	10	3, 5 2, 5	2, 6 2, 6	1, 7 1, 6	1, 7 1, 7	1, 8 1, 7	1, 8 0, 8	0, 8 0, 8	0, 9 0, 9	–, 9 –, 9
23	9	11	3, 6 3, 6	2, 7 2, 6	2, 7 1, 7	1, 7 1, 7	1, 8 1, 8	1, 9 1, 9	0, 8 0, 8	0, 9 0, 9	–, 9 0, 9
23	10	10	4, 6 3, 6	3, 7 2, 7	2, 7 1, 7	2, 8 2, 8	1, 8 1, 8	1, 9 1, 9	1, 9 1, 9	0, 9 0, 9	0, 10 0, 9
23	10	11	4, 6 4, 7	3, 7 3, 8	3, 8 2, 7	2, 8 2, 8	2, 9 2, 9	1, 9 1, 9	1, 9 1, 9	1, 10 1, 10	1, 10 1, 10
23	11	11	–, – –, –	–, – –, –	–, – –, –	–, – –, –	–, – –, –	–, – –, –	–, – –, –	–, – –, –	–, – –, –
24	1	1	–, – –, –	–, – –, –	–, – –, –	–, – –, –	–, – –, –	–, – –, –	–, – –, –	–, – –, –	–, – –, –
24	1	2	–, – –, –	–, – –, –	–, – –, –	–, – –, –	–, – –, –	–, – –, –	–, – –, –	–, – –, –	–, – –, –
24	1	3	–, – –, –	–, – –, –	–, – –, –	–, – –, –	–, – –, –	–, – –, –	–, – –, –	–, – –, –	–, – –, –
24	1	4	–, – –, –	–, – –, –	–, – –, –	–, – –, –	–, – –, –	–, – –, –	–, – –, –	–, – –, –	–, – –, –
24	1	5	–, – –, –	–, – –, –	–, – –, –	–, – –, –	–, – –, –	–, – –, –	–, – –, –	–, – –, –	–, – –, –
24	1	6	–, – –, –	–, – –, –	–, – –, –	–, – –, –	–, – –, –	–, – –, –	–, – –, –	–, – –, –	–, – –, –
24	1	7	–, – –, –	–, – –, –	–, – –, –	–, – –, –	–, – –, –	–, – –, –	–, – –, –	–, – –, –	–, – –, –
24	1	8	–, – –, –	–, – –, –	–, – –, –	–, – –, –	–, – –, –	–, – –, –	–, – –, –	–, – –, –	–, – –, –
24	1	9	–, – –, –	–, – –, –	–, – –, –	–, – –, –	–, – –, –	–, – –, –	–, – –, –	–, – –, –	–, – –, –
24	1	10	–, – –, –	–, – –, –	–, – –, –	–, – –, –	–, – –, –	–, – –, –	–, – –, –	–, – –, –	–, – –, –
24	1	12	0, 1 –, 1	–, 1 –, 1	–, 1 –, 1	–, 1 –, 1	–, 1 –, 1	–, 1 –, 1	–, – –, –	–, – –, –	–, – –, –
24	2	2	–, 1 –, 1	–, 1 –, 1	–, 2 –, 2	–, 2 –, 2	–, 2 –, 2	–, 2 –, 2	–, 2 –, 2	–, – –, –	–, – –, –

TABLE B.28 (cont.) Critical Values for Fisher's Exact Test

n	m_1	m_2	α: 0.50	0.20	0.10	0.05	0.02	0.01	0.005	0.002	0.001
24	2	3	—, 1	—, 2	—, 2	—, 2	—, 2	—, —	—, —	—, —	—, —
24	2	4	—, 1	—, 2	—, 2	—, 2	—, 2	—, —	—, —	—, —	—, —
24	2	5	—, 1	—, 2	—, 2	—, 2	—, —	—, —	—, —	—, —	—, —
24	2	6	—, 1	—, 2	—, 2	—, 2	—, —	—, —	—, —	—, —	—, —
24	2	7	0, 1	—, 2	—, 2	—, 2	—, —	—, —	—, —	—, —	—, —
24	2	8	0, 2	—, 2	—, 2	—, —	—, —	—, —	—, —	—, —	—, —
24	2	9	0, 2	—, 2	—, 2	—, —	—, —	—, —	—, —	—, —	—, —
24	2	10	0, 2	—, 2	—, 2	—, —	—, —	—, —	—, —	—, —	—, —
24	2	11	0, 2	—, 2	—, 2	—, —	—, —	—, —	—, —	—, —	—, —
24	2	12	0, 2	—, —	—, 2	—, —	—, —	—, —	—, —	—, —	—, —
24	3	3	—, 1	—, 2	—, 2	—, 2	—, 3	—, 3	—, 3	—, 3	—, 3
24	3	4	—, 1	—, 2	—, 2	—, 3	—, 3	—, 3	—, 3	—, 3	—, 3
24	3	5	0, 1	—, 2	—, 3	—, 3	—, 3	—, 3	—, 3	—, —	—, —
24	3	6	0, 2	0, 3	—, 3	—, 3	—, 3	—, 3	—, —	—, —	—, —
24	3	7	0, 2	0, 3	—, 3	—, 3	—, 3	—, 3	—, —	—, —	—, —
24	3	8	0, 2	—, 3	—, 3	—, 3	—, —	—, —	—, —	—, —	—, —
24	3	9	0, 2	—, 3	—, 3	—, 3	—, 3	—, —	—, —	—, —	—, —
24	3	10	0, 3	0, 3	—, 3	—, 3	—, 3	—, —	—, —	—, —	—, —
24	3	11	0, 3	0, 4	—, 3	—, 4	—, 3	—, —	—, —	—, —	—, —
24	3	12	1, 0	0, 3	—, 3	—, —	—, —	—, —	—, —	—, —	—, —
24	4	4	0, 2	—, 3	—, 3	—, 4	3, —	—, 3	4, —	4, —	4, —
24	4	5	0, 2	0, 4	—, 3	—, 4	3, 4	—, 4	4, —	4, —	4, —
24	4	6	0, 2	0, 4	—, 4	—, 4	4, —	—, 4	4, —	4, —	—, —
24	4	7	0, 2	—, 4	—, 4	—, 4	4, —	—, 4	—, —	—, —	—, —
24	4	8	0, 2	0, 3	—, 3	—, 4	—, 4	—, 4	—, —	—, —	—, —
24	4	9	0, 3	0, 4	—, 4	—, 4	4, 4	—, 4	—, —	—, —	—, —
24	4	10	0, 3	0, 4	—, 4	—, 4	4, 5	4, 5	5, —	5, —	—, —
24	4	11	1, 3	0, 4	—, 4	—, 5	4, 5	4, 5	5, —	5, —	5, —
24	4	12	1, 3	1, 4	—, 4	—, 5	4, 5	4, 5	5, —	5, —	5, —
24	5	5	0, 2	0, 4	—, 4	—, 5	4, 5	—, 5	—, —	—, —	—, —
24	5	6	0, 3	0, 4	0, 5	0, 5	5, —	—, 5	5, —	5, —	5, —
24	5	7	0, 3	0, 4	0, 5	0, 5	0, 5	5, —	5, —	5, —	5, —
24	5	8	1, 3	0, 4	—, 4	5, —	5, —	—, 5	5, —	5, —	5, —
24	5	9	1, 3	1, 4	—, 4	—, 5	—, 5	5, —	—, —	—, —	—, —
24	5	10	1, 3	0, 3	0, 5	0, 5	—, 5	5, —	—, —	—, —	—, —
24	5	11	1, 3	1, 3	0, 5	0, 5	5, —	—, 5	5, —	5, —	5, —
24	5	12	2, 3	1, 4	0, 5	0, 6	6, —	6, —	6, —	6, —	6, —
24	6	6	1, 3	0, 3	0, 5	0, 5	0, 6	6, —	6, —	6, —	6, —
24	6	7	1, 3	1, 4	0, 4	0, 6	6, 0	0, 6	6, —	6, —	6, —
24	6	8	1, 3	1, 4	0, 5	0, 6	6, 0	0, 6	6, —	6, —	6, —
24	6	9	1, 4	1, 4	0, 5	0, 5	5, —	—, 5	—, —	—, —	—, —
24	6	10	1, 3	1, 4	0, 5	0, 5	6, —	—, 6	6, —	6, —	6, —
24	6	11	2, 4	1, 5	0, 5	0, 6	6, 0	0, 6	6, —	6, —	6, —
24	6	12	2, 5	1, 5	0, 5	0, 6	6, —	—, 6	—, —	—, —	—, —
24	7	7	1, 3	0, 4	0, 5	—, 5	—, 5	—, 5	—, 6	—, 6	—, 6

App156

TABLE B.28 (cont.) Critical Values for Fisher's Exact Test

n	m_1	m_2	α: 0.50	0.20	0.10	0.05	0.02	0.01	0.005	0.002	0.001
24	7	8	1,4 1,4	0,4 0,4	0,5 0,5	0,5 —,5	—,6 —,6	—,6 —,6	—,6 —,6	—,6 —,6	—,7 —,7
24	7	9	2,4 1,4	0,4 1,5	0,5 0,5	0,5 —,6	—,6 —,6	—,6 —,6	—,6 —,6	—,7 —,7	—,7 —,7
24	7	10	2,4 1,5	0,5 1,5	0,5 0,5	0,6 0,6	0,6 0,6	0,6 —,7	0,7 —,7	—,7 —,7	—,7 —,7
24	7	11	2,4 2,5	1,5 1,5	1,6 0,6	0,6 0,6	0,7 0,6	0,7 0,7	0,7 —,7	—,7 —,7	—,7 —,7
24	7	12	3,4 2,5	1,5 1,6	1,6 1,6	1,6 0,7	1,7 0,7	0,7 0,7	0,7 —,8	—,8 —,8	—
24	8	8	2,4 1,4	1,5 1,5	0,5 0,5	0,5 0,5	0,6 0,6	—,6 —,6	—,6 —,7	—,7 —,7	—,7 —,7
24	8	9	2,4 1,4	1,5 1,5	1,5 0,5	0,6 0,6	0,6 0,6	0,7 —,7	0,7 —,7	—,7 —,7	—,7 —,8
24	8	10	2,5 2,5	1,5 1,6	1,6 0,6	0,6 0,6	0,7 0,7	0,7 0,7	0,8 —,8	—,8 —,8	—,8 —,8
24	8	11	3,5 2,5	1,5 1,6	1,6 1,6	1,7 0,7	0,7 0,7	0,8 0,8	0,8 0,8	—,8 —,8	—,8 —,8
24	8	12	3,5 2,6	2,6 1,6	2,7 1,7	1,7 1,7	1,7 1,7	0,8 0,8	0,8 0,8	0,9 0,8	—
24	9	9	2,5 2,5	1,6 1,6	1,6 0,6	0,6 0,6	0,7 0,7	0,7 0,7	0,8 0,7	—,8 —,8	—,8 —,8
24	9	10	3,5 2,5	2,6 1,6	1,6 1,7	1,7 0,7	1,7 0,7	0,8 0,8	0,8 0,8	—,9 —,8	0,9 —,9
24	9	11	3,5 3,6	2,6 2,7	2,7 1,7	1,7 1,7	1,8 1,8	1,8 1,8	1,9 0,8	0,9 0,9	0,9 0,9
24	9	12	4,5 3,6	2,7 2,7	2,7 1,7	2,8 1,8	1,8 1,8	1,9 1,8	1,9 1,8	1,9 0,9	0,10 0,9
24	10	10	3,6 3,6	2,7 2,6	2,7 1,7	1,8 1,7	1,8 1,8	1,9 1,8	1,9 1,9	1,9 0,9	0,9 0,9
24	10	11	4,6 3,6	3,7 2,7	2,8 2,7	2,8 1,8	2,9 1,9	1,9 1,9	1,9 1,9	1,10 1,9	0,10 0,10
24	10	12	4,6 3,7	3,7 3,7	2,8 2,8	2,8 2,8	2,9 2,9	2,10 1,9	1,10 1,10	1,10 1,10	0,10 0,10
24	11	11	4,6 4,7	3,8 3,8	3,8 2,8	2,8 2,8	2,9 2,9	— 2,10	— 1,10	1,11 1,10	1,10 1,10
24	11	12	5,6 4,7	4,8 3,8	3,9 3,8	3,9 2,9	3,9 2,9	— 2,10	— 2,10	1,11 1,11	1,11 1,10
24	12	12	5,7 4,8	4,9 3,9	3,9 3,9	3,9 2,9	3,10 3,9	—	—	2,11 2,10	2,11 1,11
25	1	1	— —	— —	— —	— —	— —	— —	— —	— —	— —
25	1	2	— —	— —	— —	— —	— —	— —	— —	— —	— —
25	1	3	— —	— —	— —	— —	— —	— —	— —	— —	— —
25	1	4	— —	— —	— —	— —	— —	— —	— —	— —	— —
25	1	5	— —	— —	— —	— —	— —	— —	— —	— —	— —
25	1	6	— —	— —	— —	— —	— —	— —	— —	— —	— —
25	1	7	— —	— —	— —	— —	— —	— —	— —	— —	— —
25	1	8	— —	— —	— —	— —	— —	— —	— —	— —	— —
25	1	9	— —	— —	— —	— —	— —	— —	— —	— —	— —
25	1	10	— —	— —	— —	— —	— —	— —	— —	— —	— —
25	1	11	— —	— —	— —	— —	— —	— —	— —	— —	— —
25	1	12	— —	— —	— —	— —	— —	— —	— —	— —	— —
25	2	2	— —	— —	— —	— —	— —	— —	— —	— —	— —
25	2	3	— —	— —	— —	— —	— —	— —	— —	— —	— —
25	2	4	— —	1,— 2,—	2,— 2,—	2,— 2,—	— —	— —	— —	— —	— —
25	2	5	— —	2,— 2,—	2,— 2,—	2,— 2,—	2,— 2,—	2,— 2,—	2,— 2,—	— —	— —
25	2	6	— —	2,— 2,—	2,— 2,—	2,— 2,—	2,— 2,—	2,— 2,—	2,— 2,—	— —	— —
25	2	7	— —	2,— 2,—	2,— 2,—	2,— 2,—	— —	— —	— —	— —	— —
25	2	8	— —	2,— 2,—	2,— 2,—	— —	— —	— —	— —	— —	— —
25	2	9	2,— 2,—	2,— 2,—	— —	— —	— —	— —	— —	— —	— —
25	2	10	0,2 0,2	2,— 2,—	2,— 2,—	2,— 2,—	— 3	— 3	— 3	— 3	— 3
25	2	11	0,2 0,2	2,— 2,—	2,— 2,—	2,— 2,—	3,— 3,—	3,— 3,—	3,— 3,—	3 3	3 3
25	2	12	0,2 0,—	— —	2,— 2,—	2,— 3,—	3,— 3,—	3,— 3,—	3,— —	3 3	3 —
25	3	3	— —	2,— 2,—	2,— 2,—	2,— 3,—	3,— 3,—	3,— 3,—	— —	— —	— —
25	3	4	— —	2,— 2,—	2,— 2,—	— —	— —	— —	— —	— —	— —

TABLE B.28 (cont.) Critical Values for Fisher's Exact Test

n	m_1	m_2	α: 0.50	0.20	0.10	0.05	0.02	0.01	0.005	0.002	0.001
25	3	5	0, 1	—, 2	—, 2	—, 3	—, 3	—, 3	—, 3	—, —	—, —
25	3	6	0, 2	—, 2	—, 3	—, 3	—, 3	—, 3	—, 3	—, —	—, —
25	3	7	0, 2	—, 2	—, 3	—, 3	—, 3	—, —	—, —	—, —	—, —
25	3	8	0, 2	—, 3	—, 3	—, 3	—, 3	—, —	—, —	—, —	—, —
25	3	9	0, 2	—, 3	—, 3	—, 3	—, —	—, —	—, —	—, —	—, —
25	3	10	0, 3	—, 3	—, 3	—, 3	—, 3	—, —	—, —	—, —	—, —
25	3	11	0, 2	—, 3	—, 3	—, 3	—, 3	—, —	—, —	—, —	—, —
25	3	12	0, 2	—, 2	—, 3	—, 3	—, 3	—, —	—, —	—, —	—, —
25	4	4	0, 1	—, 2	—, 3	—, 3	—, 3	—, 3	—, —	—, —	—, —
25	4	5	0, 2	—, 3	—, 3	—, 4	—, 4	—, 4	—, 4	—, 4	—, 4
25	4	6	0, 2	—, 3	—, 3	—, 4	—, 4	—, 4	—, 4	—, 4	—, 4
25	4	7	0, 2	—, 3	—, 4	—, 4	—, 4	—, 4	—, 4	—, 4	—, 4
25	4	8	0, 3	0, 3	—, 4	—, 4	—, 4	—, 4	—, 4	—, 4	—, 4
25	4	9	0, 3	0, 3	—, 4	—, 4	—, 4	—, 4	—, 4	—, —	—, —
25	4	10	0, 3	0, 3	—, 4	—, 4	—, 4	—, —	—, —	—, —	—, —
25	4	11	0, 2	—, 3	—, 4	—, 4	—, 4	—, —	—, —	—, —	—, —
25	4	12	0, 3	0, 4	—, 4	—, 4	—, 4	—, 4	—, 5	—, 5	—, 5
25	5	5	1, 3	—, 3	—, 4	—, 4	—, 5	—, 5	—, 5	—, 5	—, 5
25	5	6	1, 3	0, 4	0, 4	—, 4	—, 5	—, 5	—, 5	—, 5	—, 5
25	5	7	0, 3	—, 3	—, 4	0, 5	—, 5	—, 5	—, 5	—, —	—, —
25	5	8	1, 3	0, 4	0, 4	0, 5	—, 4	—, 5	—, 5	—, 5	—, 5
25	5	9	1, 3	0, 4	0, 4	—, 5	—, 5	—, 5	—, 6	—, 6	—, 6
25	5	10	1, 3	0, 4	0, 5	—, 5	—, 5	—, 6	—, 6	—, 6	—, 6
25	5	11	1, 4	0, 4	0, 5	0, 6	—, 6	—, 6	—, 6	—, 6	—, 6
25	5	12	1, 4	1, 4	0, 5	0, 6	—, 6	—, 5	—, 6	—, 6	—, —
25	6	6	0, 2	0, 3	—, 4	—, 4	—, 4	—, 5	—, 5	—, 6	—, 6
25	6	7	1, 3	0, 4	—, 4	0, 5	—, 5	—, 6	0, 6	—, 6	—, 6
25	6	8	1, 3	0, 4	0, 5	0, 5	0, 6	0, 6	0, 6	—, 6	—, 6
25	6	9	1, 4	0, 4	0, 5	0, 6	0, 6	0, 5	—, 6	—, —	—, —
25	6	10	1, 4	0, 4	0, 5	0, 6	0, 6	0, 6	0, 6	—, 6	—, —
25	6	11	2, 4	1, 5	0, 5	1, 6	0, 6	0, 6	—, 6	—, 6	—, 6
25	6	12	2, 4	1, 5	1, 6	1, 6	0, 7	0, 7	0, 7	—, 6	—, 6
25	7	7	1, 4	0, 4	0, 4	0, 5	0, 6	0, 5	—, 6	—, 7	—, 7
25	7	8	1, 4	0, 4	0, 5	0, 5	0, 6	0, 6	—, 6	—, 7	—, 7
25	7	9	2, 4	1, 5	1, 6	0, 6	0, 7	0, 7	0, 7	—, 7	—, 7
25	7	10	2, 4	1, 5	1, 6	0, 6	0, 7	0, 7	0, 7	—, 7	—, 7
25	7	11	2, 4	1, 5	1, 6	1, 6	0, 7	0, 6	0, 7	—, 7	—, 7
25	7	12	2, 4	1, 5	1, 6	1, 7	0, 7	0, 7	0, 7	—, 7	—, 7
25	8	8	2, 4	1, 5	1, 6	1, 6	0, 7	0, 6	0, 7	—, 7	—, 7
25	8	9	2, 4	1, 5	1, 6	1, 7	0, 7	0, 7	0, 7	—, 8	—, 8
25	8	10	2, 5	1, 5	1, 6	0, 7	0, 7	0, 7	0, 8	—, 8	—, 8
25	8	11	3, 5	2, 6	1, 6	1, 7	0, 7	0, 7	0, 8	0, 8	—, 8
25	8	12	3, 5	2, 6	1, 6	1, 7	0, 7	0, 7	0, 8	—, 8	—, 8
25	9	9	2, 5	1, 5	1, 6	1, 7	0, 7	0, 7	0, 7	—, 8	—, 8
25	9	10	3, 5	2, 6	1, 6	1, 7	0, 7	0, 7	0, 8	—, 8	—, 8

TABLE B.28 (cont.) Critical Values for Fisher's Exact Test

n	m_1	m_2	α: 0.50	0.20	0.10	0.05	0.02	0.01	0.005	0.002	0.001
25	9	11	3, 5	2, 6	1, 6	1, 7	0, 7	0, 8	0, 8	0, 8	–, 9
			3, 5	2, 6	1, 7	1, 7	1, 7	0, 8	0, 8	0, 8	0, 9
25	9	12	3, 6	2, 7	2, 7	1, 7	1, 8	0, 8	0, 8	0, 9	0, 9
			3, 5	2, 6	2, 7	1, 7	1, 8	0, 8	0, 8	0, 9	0, 9
25	10	10	3, 5	2, 6	1, 7	1, 7	1, 8	0, 8	0, 8	0, 9	0, 9
			3, 6	2, 6	2, 7	1, 7	1, 8	0, 8	0, 8	0, 9	0, 9
25	10	11	3, 6	2, 6	2, 7	1, 7	1, 8	0, 8	0, 9	0, 9	0, 9
			3, 6	2, 6	2, 7	1, 7	1, 8	0, 9	0, 9	0, 9	0, 9
25	10	12	4, 6	3, 7	2, 8	2, 8	1, 8	1, 9	0, 9	0, 9	0, 9
			4, 6	3, 7	2, 7	2, 8	1, 8	1, 9	1, 9	0, 9	0, 9
25	11	11	4, 6	3, 7	2, 7	2, 8	1, 8	1, 9	1, 9	1, 10	0, 10
			4, 6	3, 7	2, 8	2, 8	1, 9	1, 9	1, 9	1, 10	1, 10
25	11	12	4, 7	3, 8	2, 8	2, 8	1, 9	1, 9	1, 9	1, 10	0, 10
			4, 7	3, 8	2, 8	2, 8	2, 9	1, 9	1, 9	1, 10	1, 11
25	12	12	5, 7	4, 8	3, 9	3, 9	2, 9	2, 10	2, 10	1, 10	1, 11
			4, 7	3, 8	3, 9	3, 9	2, 9	2, 10	2, 10	1, 10	1, 11
26	1	1	–, –	–, –	–, –	–, –	–, –	–, –	–, –	–, –	–, –
			–, –	–, –	–, –	–, –	–, –	–, –	–, –	–, –	–, –
26	1	2	–, –	–, 1	–, 1	–, 1	–, –	–, –	–, –	–, –	–, –
			–, –	–, 1	–, 1	–, 1	–, –	–, –	–, –	–, –	–, –
26	1	3	–, –	–, –	–, –	–, –	–, –	–, –	–, –	–, –	–, –
			–, –	–, –	–, –	–, –	–, –	–, –	–, –	–, –	–, –
26	1	4	–, –	–, –	–, –	–, –	–, –	–, –	–, –	–, –	–, –
			–, –	–, –	–, –	–, –	–, –	–, –	–, –	–, –	–, –
26	1	5	–, –	–, –	–, –	–, –	–, –	–, –	–, –	–, –	–, –
			–, –	–, –	–, –	–, –	–, –	–, –	–, –	–, –	–, –
26	1	6	–, –	–, –	–, –	–, –	–, –	–, –	–, –	–, –	–, –
			–, –	–, –	–, –	–, –	–, –	–, –	–, –	–, –	–, –
26	1	7	–, –	–, –	–, –	–, –	–, –	–, –	–, –	–, –	–, –
			–, –	–, –	–, –	–, –	–, –	–, –	–, –	–, –	–, –
26	1	8	–, –	–, –	–, –	–, –	–, –	–, –	–, –	–, –	–, –
			–, –	–, –	–, –	–, –	–, –	–, –	–, –	–, –	–, –
26	1	9	–, –	–, –	–, –	–, –	–, –	–, –	–, –	–, –	–, –
			–, –	–, –	–, –	–, –	–, –	–, –	–, –	–, –	–, –
26	1	10	–, –	–, –	–, –	–, –	–, –	–, –	–, –	–, –	–, –
			–, –	–, –	–, –	–, –	–, –	–, –	–, –	–, –	–, –
26	1	11	–, –	–, –	–, –	–, –	–, –	–, –	–, –	–, –	–, –
			–, –	–, –	–, –	–, –	–, –	–, –	–, –	–, –	–, –
26	1	12	–, –	–, –	–, –	–, –	–, –	–, –	–, –	–, –	–, –
			–, –	–, –	–, –	–, –	–, –	–, –	–, –	–, –	–, –
26	1	13	0, 1	–, –	–, –	–, –	–, –	–, –	–, –	–, –	–, –
			–, –	–, –	–, –	–, –	–, –	–, –	–, –	–, –	–, –
26	2	2	–, –	–, –	–, –	–, –	–, –	–, –	–, –	–, –	–, –
			–, –	–, –	–, –	–, –	–, –	–, –	–, –	–, –	–, –
26	2	3	–, –	–, –	–, 2	–, 2	–, 2	–, 2	–, 2	–, –	–, –
			–, –	–, –	–, 2	–, 2	–, 2	–, 2	–, 2	–, –	–, –
26	2	4	–, –	1, 2	1, 2	2, 2	2, 2	2, 2	–, 2	–, –	–, –
			–, –	1, 2	1, 2	2, 2	2, 2	2, 2	–, 2	–, –	–, –
26	2	5	–, –	1, 2	1, 2	2, 2	2, 2	2, 2	–, –	–, –	–, –
			–, –	1, 2	1, 2	2, 2	2, 2	2, 2	–, –	–, –	–, –
26	2	6	1, 2	1, 2	1, 2	2, 2	2, 2	–, –	–, –	–, –	–, –
			1, 2	1, 2	2, 2	2, 2	2, 2	–, –	–, –	–, –	–, –
26	2	7	–, –	1, 2	2, 2	2, 2	2, 2	–, –	–, –	–, –	–, –
			–, –	1, 2	2, 2	2, 2	2, 2	–, –	–, –	–, –	–, –
26	2	8	0, 1	1, 2	2, 2	2, 2	–, –	–, –	–, –	–, –	–, –
			0, 2	1, 2	2, 2	2, 1	–, –	–, –	–, –	–, –	–, –
26	2	9	0, 2	1, 2	2, 2	2, 1	–, –	–, –	–, –	–, –	–, –
			0, 2	1, 2	2, 2	–, –	–, –	–, –	–, –	–, –	–, –
26	2	10	0, 2	1, 2	2, 2	–, –	–, –	–, –	–, –	–, –	–, –
			0, 2	1, 2	2, 1	–, –	–, –	–, –	–, –	–, –	–, –
26	2	11	0, 2	1, 2	2, 2	2, –	1, –	–, –	–, –	–, –	–, –
			0, 2	1, 2	2, 2	2, –	1, –	–, –	–, –	–, –	–, –
26	2	12	0, 2	1, 2	2, 2	2, –	1, –	–, –	–, –	–, –	–, –
			0, 2	1, 2	2, 2	2, –	1, –	–, –	–, –	–, –	–, –
26	2	13	0, 2	1, 3	2, 2	2, –	1, –	1, –	1, 2	–, –	–, –
			0, 2	1, 3	2, 2	2, –	1, –	1, –	1, 2	–, –	–, –
26	3	3	–, 1	–, –	–, –	–, –	–, –	–, –	–, –	–, –	–, –
			–, 1	1, –	1, –	1, –	–, –	–, –	–, –	–, –	–, –
26	3	4	–, –	1, 2	2, 2	2, 3	2, 3	3, 3	3, 3	–, 3	–, 3
			0, 2	1, 2	2, 3	3, 3	3, 3	3, 3	3, 3	3, 3	3, –
26	3	5	0, 3	2, 3	2, 3	3, 3	3, 3	3, 3	3, 3	3, 3	3, –
			0, 3	2, 3	2, 3	3, 3	3, 3	3, 3	3, 3	3, 3	3, –
26	3	6	0, 3	2, 3	2, 3	3, 3	3, 3	3, 3	3, 3	3, –	–, –
			0, 3	2, 3	3, 3	3, 3	3, 3	3, 3	3, 3	3, –	3, –
26	3	7	0, 3	2, 3	3, 3	3, 3	3, 3	3, 3	3, –	–, –	–, –
			0, 3	2, 3	3, 3	3, 3	3, 3	3, –	3, –	–, –	–, –
26	3	8	0, 3	2, 3	3, 3	3, 3	3, 3	3, –	–, –	–, –	–, –
			0, 3	2, 3	3, 3	3, 3	3, –	3, –	–, –	–, –	–, –
26	3	9	0, 3	2, 3	3, 3	3, 3	3, –	–, –	–, –	–, –	–, –
			0, 3	2, 3	3, 3	3, –	3, –	–, –	–, –	–, –	–, –
26	3	10	0, 3	–, 3	3, 3	3, –	–, –	–, –	–, –	3, 3	3, 3
			0, 3	0, 3	3, 3	3, –	–, –	–, –	–, –	3, 3	3, 3
26	3	11	0, 3	0, 3	3, 3	3, –	3, –	3, –	3, –	–, –	–, –
			0, 3	0, 3	3, 3	3, –	3, –	3, –	3, –	–, –	–, –
26	3	12	1, 2	0, 3	3, 3	3, –	3, –	3, –	3, –	–, –	–, –
			1, 2	0, 3	3, 3	3, –	3, –	3, –	3, –	–, –	–, –
26	3	13	0, 1	–, 2	–, 2	–, 3	–, 3	–, 3	–, 4	–, 4	–, 4
			–, 2	–, 2	–, 2	–, 3	–, 3	–, 3	–, 4	–, 4	–, 4
26	4	4									

TABLE B.28 (cont.) Critical Values for Fisher's Exact Test

n	m_1	m_2	α: 0.50	0.20	0.10	0.05	0.02	0.01	0.005	0.002	0.001
26	4	5	0, 2	—, 2	—, 3	—, 3	—, 3	—, 4	—, 4	—, 4	—, 4
26	4	6	0, 2	—, 2	—, 3	—, 3	—, 4	—, 4	—, 4	—, 4	—, 4
26	4	7	0, 2	—, 3	—, 3	—, 3	—, 4	—, 4	—, 4	—, —	—, —
26	4	8	0, 2	—, 3	—, 3	—, 4	—, 4	—, 4	—, 4	—, —	—, —
26	4	9	0, 2	0, 3	—, 3	—, 4	—, 4	—, 4	—, 4	—, —	—, —
26	4	10	1, 2	0, 3	—, 4	—, 4	—, 4	—, —	—, —	—, —	—, —
26	4	11	1, 3	0, 4	0, 4	—, 4	—, —	—, —	—, —	—, 4	—, —
26	4	12	1, 3	0, 4	0, 4	—, 4	—, —	—, —	—, —	—, 4	—, 5
26	4	13	1, 3	0, 4	0, 4	0, 4	—, —	—, 4	—, 4	—, 5	—, 5
26	5	5	0, 2	—, 3	—, 3	—, 3	—, 4	—, 4	—, 4	—, 5	—, 5
26	5	6	0, 2	—, 3	—, 3	—, 4	—, 4	—, 4	—, 4	—, 5	—, 5
26	5	7	0, 2	—, 3	—, 3	—, 4	—, 5	—, 5	—, 5	—, 5	—, 5
26	5	8	0, 2	—, 3	—, 4	—, 4	—, 5	—, 5	—, 5	—, 5	—, 5
26	5	9	1, 3	0, 4	0, 4	—, 4	—, 5	—, 5	—, 5	—, —	—, —
26	5	10	1, 3	0, 4	0, 4	0, 5	—, 5	—, 5	—, 5	—, —	—, —
26	5	11	1, 3	1, 5	1, 5	0, 5	0, 5	—, 5	—, 5	—, 5	—, 5
26	5	12	1, 3	1, 5	1, 5	0, 5	0, 5	—, 6	—, 6	—, 6	—, 6
26	5	13	2, 3	1, 5	1, 5	0, 5	0, 5	—, 6	—, 6	—, 6	—, 6
26	6	6	1, 3	0, 4	0, 5	—, 4	0, 5	—, 6	—, 6	—, 6	—, 6
26	6	7	1, 3	0, 4	—, 4	—, 4	—, 5	—, 6	—, 6	—, —	—, —
26	6	8	1, 3	0, 4	0, 4	—, 5	—, 5	0, 6	—, 6	—, —	—, —
26	6	9	1, 4	0, 4	0, 5	—, 5	—, 5	—, 5	—, 6	—, 6	—, 6
26	6	10	1, 4	0, 4	0, 5	0, 5	—, 6	—, 6	—, 6	—, 6	—, 7
26	6	11	2, 4	1, 5	0, 5	0, 6	—, 6	—, 6	—, 6	—, 7	—, 7
26	6	12	2, 4	1, 5	1, 6	1, 6	0, 6	0, 7	—, 7	—, 7	—, 7
26	6	13	2, 4	1, 5	1, 6	1, 6	0, 6	0, 7	—, 7	—, 7	—, 7
26	7	7	2, 4	1, 5	0, 5	0, 5	0, 6	0, 6	—, 6	—, 6	—, 6
26	7	8	1, 3	0, 4	0, 5	—, 5	0, 6	0, 7	—, 6	—, 6	—, 6
26	7	9	1, 4	0, 4	0, 5	0, 5	—, 6	0, 7	0, 7	—, 7	—, 7
26	7	10	2, 4	1, 5	1, 6	0, 6	0, 6	0, 7	0, 7	—, 7	—, 7
26	7	11	2, 4	1, 5	1, 6	1, 6	0, 6	0, 7	0, 7	—, 7	—, 7
26	7	12	2, 4	1, 6	1, 6	1, 7	0, 7	0, 7	0, 7	—, 8	—, 8
26	7	13	3, 4	1, 6	1, 7	1, 7	1, 7	0, 7	0, 8	—, 8	—, 8
26	8	8	3, 3	1, 5	1, 6	1, 6	1, 7	0, 6	0, 7	—, 7	—, 7
26	8	9	2, 4	1, 5	1, 6	0, 6	0, 6	0, 6	—, 7	—, 7	—, 7
26	8	10	2, 5	1, 5	1, 6	0, 6	0, 6	0, 7	0, 8	—, 8	—, 8
26	8	11	2, 5	1, 6	1, 6	1, 7	0, 7	0, 7	0, 8	—, 8	—, 8
26	8	12	3, 5	2, 6	1, 7	1, 7	1, 7	0, 8	0, 8	0, 9	—, 8
26	8	13	3, 5	2, 6	1, 7	1, 7	1, 8	0, 8	0, 9	0, 9	—, 8
26	9	9	2, 5	1, 5	1, 6	0, 6	0, 7	0, 7	0, 7	—, 7	—, 8
26	9	10	2, 5	1, 5	1, 6	1, 6	0, 7	0, 7	0, 8	—, 8	—, 8
26	9	11	3, 5	2, 6	1, 7	1, 7	0, 7	0, 8	0, 8	0, 9	—, 9
26	9	12	3, 6	2, 7	1, 7	1, 7	1, 8	0, 8	0, 9	0, 9	0, 9
26	9	13	4, 5	2, 7	2, 7	2, 7	1, 8	0, 9	0, 9	0, 9	0, 9
26	10	10	3, 5	2, 6	1, 7	1, 7	0, 7	0, 8	0, 8	0, 8	0, —

TABLE B.28 (cont.) Critical Values for Fisher's Exact Test

n	m_1	m_2	α: 0.50	0.20	0.10	0.05	0.02	0.01	0.005	0.002	0.001
26	10	11	3, 6 / 3, 6	2, 6 / 2, 7	2, 7 / 1, 7	1, 7 / 1, 7	1, 8 / —, 8	—, 8 / —, 8	—, 8 / —, 8	—, 9 / —, 9	—, 9 / —, 9
26	10	12	4, 6 / 3, 6	2, 7 / 2, 7	2, 7 / 1, 8	1, 8 / 1, 8	1, 8 / —, 8	—, 8 / —, 8	0, 9 / 0, 9	—, 9 / —, 9	0, 9 / 0, 9
26	10	13	4, 6 / 3, 7	3, 7 / 2, 7	2, 8 / 2, 8	2, 8 / 2, 8	1, 9 / 1, 8	—, 9 / —, 9	0, 9 / 0, 9	0, 10 / —, 9	0, 10 / 0, 10
26	11	11	4, 6 / 3, 6	3, 7 / 2, 7	2, 7 / 2, 8	2, 8 / 2, 8	1, 8 / 1, 8	—, 9 / —, 9	0, 9 / 0, 9	—, 9 / —, 9	0, 9 / 0, 10
26	11	12	4, 6 / 3, 6	3, 7 / 2, 7	2, 8 / 2, 8	2, 8 / 2, 8	1, 9 / 1, 9	0, 9 / —, 9	1, 9 / 0, 9	1, 10 / 0, 10	0, 10 / 0, 10
26	11	13	4, 6 / 4, 7	3, 8 / 3, 7	3, 8 / 2, 9	2, 9 / 2, 9	2, 9 / 1, 9	1, 10 / 0, 10	1, 10 / 1, 10	1, 10 / 1, 10	1, 11 / 1, 10
26	12	12	5, 6 / 4, 7	3, 8 / 3, 8	3, 8 / 2, 9	2, 9 / 2, 9	2, 9 / 1, 9	1, 10 / 0, 10	1, 10 / 1, 10	1, 10 / 1, 10	1, 11 / 1, 10
26	12	13	5, 7 / 4, 7	3, 8 / 3, 8	3, 9 / 3, 9	3, 9 / 2, 9	2, 10 / 2, 10	2, 10 / 1, 10	2, 10 / 1, 10	1, 11 / 1, 11	1, 11 / 1, 11
26	12	13	5, 7 / 4, 8	4, 9 / 3, 8	3, 9 / 3, 9	3, 9 / 3, 9	2, 10 / 2, 10	2, 10 / 2, 10	2, 11 / 2, 11	1, 11 / 1, 11	1, 11 / 1, 11
26	13	13	6, 7 / 5, 8	4, 9 / 4, 9	4, 9 / 3, 10	3, 10 / 3, 10	3, 10 / 2, 11	3, 10 / 2, 11	2, 11 / 2, 11	2, 11 / 1, 11	1, 12 / 2, 11
27	1	1	r, r / r, r	r, r / r, r	r, 1 / r, r	1 / r,	— / —	— / —	— / —	— / —	— / —
27	1	2	r, r / r, r	r, r / r, r	r, r / r, r	— / —	— / —	— / —	— / —	— / —	— / —
27	1	3	r, r / r, r	r, r / r, r	r, r / r, r	— / —	— / —	— / —	— / —	— / —	— / —
27	1	4	r, r / r, r	1 / 1	r, r / r, r	— / —	— / —	— / —	— / —	— / —	— / —
27	1	5	r, r / r, r	1 / 1	r, r / r, r	— / —	— / —	— / —	— / —	— / —	— / —
27	1	6	r, r / r, r	1, 1 / 1, 1	r, r / r, r	— / —	— / —	— / —	— / —	— / —	— / —
27	1	7	r, r / r, r	1, 1 / r, r	r, r / r, r	— / —	— / —	— / —	— / —	— / —	— / —
27	1	8	r, r / r, r	r, r / r, r	r, r / r, r	— / —	— / —	— / —	— / —	— / —	— / —
27	1	9	r, r / r, r	r, r / r, r	r, r / r, r	— / —	— / —	— / —	— / —	— / —	— / —
27	1	10	r, r / r, r	r, r / r, r	r, r / r, r	— / —	— / —	— / —	— / —	— / —	— / —
27	1	11	r, 1 / r, r	r, r / r, r	r, r / r, r	— / —	— / —	— / —	— / —	— / —	— / —
27	1	12	r, 1 / r, r	r, r / r, r	r, r / r, r	— / —	— / —	— / —	— / —	— / —	— / —
27	1	13	r, r / r, r	r, r / r, r	r, r / r, r	— / —	— / —	— / —	— / —	— / —	— / —
27	2	3	r, r / r, r	r, r / r, r	r, r / r, r	— / —	— / —	— / —	— / —	— / —	— / —
27	2	4	r, r / r, r	r, r / r, r	2, r / r, r	— / —	— / —	— / —	— / —	— / —	— / —
27	2	5	r, 1 / r, r	1, r / 1, r	2, r / 2, r	2 / 2	2 / 2	2 / 2	2 / 2	— / —	— / —
27	2	6	r, 2 / r, 1	2, r / 2, r	2, r / 2, r	2 / 2	2 / 2	2 / 2	— / —	— / —	— / —
27	2	7	r, 2 / r, 1	2, r / 2, r	2, r / 2, r	2 / 2	2 / 2	— / —	— / —	— / —	— / —
27	2	8	r, 2 / r, 1	2, r / 2, r	2, r / 2, r	2 / 2	— / —	— / —	— / —	— / —	— / —
27	2	9	r, 2 / r, 1	2, r / 2, r	2, r / 2, r	— / —	— / —	— / —	— / —	— / —	— / —
27	2	10	r, 2 / r, 2	2, r / 2, r	2, r / 2, r	— / —	— / —	— / —	— / —	— / —	— / —
27	2	11	0, 2 / r, 2	2, r / 2, r	2, r / 2, r	— / —	— / —	— / —	— / —	— / —	— / —
27	2	12	0, 2 / 0, 2	2, r / 2, r	2, r / 2, r	— / —	— / —	— / —	— / —	— / —	— / —
27	2	13	0, 2 / 0, 2	3, r / 2, r	2, r / 2, r	— / —	— / —	— / —	— / —	— / —	— / —
27	3	3	0, 1 / r, r	r, 3 / r, r	2, r / 2, r	r, / r,	— / —	— / —	— / —	— / —	— / —
27	3	4	r, 1 / r, 1	r, 3 / r, 3	2, 3 / 2, r	r, / r,	r, / r,	r, / r,	r, / r,	r, / r,	r, / r,
27	3	5	r, 2 / r, 1	r, 3 / r, 3	3, 3 / 2, 3	2, / 2,	3, / 3,	3, / 3,	3, / 3,	3, / 3,	3, / 3,
27	3	6	0, 2 / r, 2	3, 3 / 3, 3	3, 3 / 3, 3	3, / 3,	3, / 3,	3, / 3,	3, / 3,	3, / 3,	3, / 3,
27	3	7	0, 2 / r, 2	3, 3 / 3, 3	3, 3 / 3, 3	3, / 3,	3, / 3,	3, / 3,	— / —	— / —	— / —
27	3	8	0, 2 / r, 2	3, 3 / 3, 3	3, r / 3, r	3, / 3,	3, / 3,	— / —	— / —	— / —	— / —
27	3	9	0, 2 / 0, 2	3, r / 3, 3	3, r / 3, r	r, / r,	— / —	— / —	— / —	— / —	— / —
27	3	10	0, 2 / 0, 3	3, r / 3, 3	3, r / 3, r	r, / r,	— / —	— / —	— / —	— / —	— / —
27	3	11	0, 2 / 0, 3	3, r / 3, 3	3, r / 3, r	r, / r,	— / —	— / —	— / —	— / —	— / —
27	3	12	0, 2 / 0, 3	3, r / 3, 3	3, r / 3, r	r, / r,	— / —	— / —	— / —	— / —	— / —
27	3	13	0, 2 / 0, 3	3, r / 3, 3	3, r / 3, r	r, / r,	— / —	— / —	— / —	— / —	— / —

TABLE B.28 (cont.) Critical Values for Fisher's Exact Test

n	m_1	m_2	α: 0.50	0.20	0.10	0.05	0.02	0.01	0.005	0.002	0.001
27	4	4	-, 1	-, 2	-, 2	-, 3	-, 3	-, 3	-, 4	-, 4	-, 4
27	4	5	0, 2	-, 2	-, 3	-, 3	-, 3	-, 4	-, 4	-, 4	-, 4
27	4	6	0, 2	-, 2	-, 3	-, 3	-, 4	-, 4	-, 4	-, 4	-, 4
27	4	7	0, 2	-, 3	-, 3	-, 3	-, 4	-, 4	-, 4	-, 4	-, -
27	4	8	0, 2	-, 3	-, 3	-, 4	-, 4	-, 4	-, 4	-, 4	-, -
27	4	9	0, 3	-, 3	-, 4	-, 4	-, 4	-, 4	-, -	-, -	-, -
27	4	10	0, 3	-, 3	-, 4	-, 4	-, 4	-, 4	-, -	-, -	-, -
27	4	11	1, 3	0, 4	-, 4	-, 4	-, 4	-, 4	-, -	-, -	-, -
27	4	12	1, 3	0, 4	0, 4	-, 4	-, 4	-, -	-, -	-, -	-, -
27	4	13	1, 3	0, 4	0, 4	0, 4	-, 4	-, -	-, -	-, -	-, -
27	5	5	0, 2	-, 2	-, 3	-, 3	-, 4	-, 4	-, 4	-, 5	-, 5
27	5	6	0, 2	-, 3	-, 3	-, 4	-, 4	-, 4	-, 5	-, 5	-, 5
27	5	7	0, 2	-, 3	-, 3	-, 4	-, 4	-, 5	-, 5	-, 5	-, 5
27	5	8	0, 3	-, 3	-, 4	-, 4	-, 5	-, 5	-, 5	-, 5	-, -
27	5	9	1, 3	-, 3	-, 4	-, 4	-, 5	-, 5	-, 5	-, -	-, -
27	5	10	1, 3	0, 4	0, 4	-, 4	-, 5	-, 5	-, 5	-, -	-, -
27	5	11	1, 3	0, 4	0, 4	-, 5	-, 5	-, 5	-, 6	-, 6	-, -
27	5	12	1, 3	0, 4	0, 5	0, 5	-, 5	-, 6	-, 6	-, 6	-, -
27	5	13	1, 3	0, 4	0, 5	0, 5	-, 6	-, 6	-, 6	-, 6	-, -
27	6	6	0, 2	-, 3	-, 3	-, 4	-, 4	-, 5	-, 5	-, 5	-, -
27	6	7	1, 3	-, 3	-, 4	-, 4	-, 5	-, 5	-, 5	-, 6	-, 6
27	6	8	1, 3	0, 4	0, 4	-, 5	-, 5	-, 5	-, 6	-, 6	-, 6
27	6	9	0, 2	0, 4	0, 5	0, 5	-, 6	-, 6	-, 6	-, 6	-, -
27	6	10	1, 4	0, 4	0, 5	0, 5	-, 6	-, 6	-, 6	-, 6	-, -
27	6	11	1, 3	0, 4	0, 5	0, 6	0, 6	-, 6	-, 6	-, 6	-, -
27	6	12	1, 3	0, 4	1, 6	0, 6	0, 7	-, 7	-, 7	-, 7	-, 7
27	6	13	1, 4	0, 4	1, 6	0, 6	0, 7	0, 7	-, 7	-, 7	-, 7
27	7	7	1, 3	0, 4	0, 4	-, 5	-, 5	-, 6	-, 6	-, 6	-, 6
27	7	8	1, 3	0, 4	0, 5	0, 5	-, 6	-, 6	-, 6	-, 6	-, 6
27	7	9	1, 4	0, 4	0, 5	0, 5	-, 6	-, 6	-, 6	-, 7	-, 7
27	7	10	2, 4	0, 5	0, 5	0, 6	0, 6	-, 6	-, 6	-, 7	-, 7
27	7	11	2, 4	1, 5	1, 6	0, 6	0, 7	0, 7	-, 7	-, 7	-, 7
27	7	12	1, 4	1, 5	1, 6	1, 6	0, 7	0, 7	-, 7	-, 8	-, 8
27	7	13	1, 4	1, 5	1, 6	1, 7	0, 7	0, 8	0, 8	-, 8	-, 8
27	8	8	2, 4	0, 5	0, 5	0, 6	0, 6	-, 6	-, 7	-, 7	-, 7
27	8	9	2, 4	1, 5	0, 6	0, 6	0, 7	0, 7	-, 7	-, 7	-, 7
27	8	10	1, 3	1, 5	1, 6	1, 6	0, 7	0, 7	-, 7	-, 8	-, 8
27	8	11	3, 5	2, 6	1, 6	1, 6	0, 7	0, 7	0, 8	-, 8	-, 8
27	8	12	3, 5	2, 6	1, 7	1, 7	1, 8	0, 8	0, 8	-, 8	-, 8
27	8	13	3, 5	2, 6	1, 7	1, 7	1, 8	0, 8	0, 8	0, 9	-, 9
27	9	9	2, 4	1, 5	1, 5	0, 6	0, 7	0, 7	-, 7	-, 7	-, -
27	9	10	2, 5	1, 5	1, 6	0, 6	0, 7	0, 7	-, 8	-, 8	-, 8
27	9	11	3, 5	1, 6	1, 6	1, 7	0, 7	0, 8	0, 8	-, 8	-, 8
27	9	12	3, 5	2, 6	1, 7	1, 7	1, 8	0, 8	0, 8	-, 9	-, 9
27	9	13	3, 5	2, 6	2, 7	1, 7	1, 8	1, 8	0, 8	0, 9	0, 9

TABLE B.28 (cont.) Critical Values for Fisher's Exact Test

n	m_1	m_2	α: 0.50	0.20	0.10	0.05	0.02	0.01	0.005	0.002	0.001
27	10	10	3, 5 / 3, 6	2, 6 / 2, 6	1, 6 / 2, 7	1, 7 / 1, 7	0, 7 / 7	0, 8 / 8	0, 8 / 8	—, 8 / 8	8 / 8
27	10	11	3, 6 / 3, 6	2, 6 / 2, 6	2, 7 / 2, 7	1, 7 / 1, 7	1, 8 / 0, 8	0, 8 / 8	0, 8 / 8	—, 9 / 8	9 / —, 9
27	10	12	3, 6 / 3, 6	2, 7 / 2, 6	2, 7 / 2, 7	1, 7 / 1, 7	1, 8 / 1, 8	1, 8 / 0, 8	0, 9 / 0, 9	0, 9 / —, 9	0, 9 / —, 9
27	10	13	4, 6 / 4, 6	2, 7 / 2, 7	2, 8 / 2, 8	2, 8 / 1, 8	1, 8 / 1, 8	1, 9 / 0, 9	1, 9 / 0, 9	0, 9 / 0, 9	0, 10 / —, 9
27	11	11	3, 6 / 3, 6	2, 7 / 2, 7	2, 7 / 1, 7	1, 8 / 1, 8	1, 8 / 1, 8	0, 9 / 0, 8	0, 9 / 0, 9	0, 9 / —, 9	0, 9 / —, 9
27	11	12	4, 6 / 3, 6	3, 7 / 2, 7	2, 8 / 2, 8	2, 8 / 2, 8	2, 9 / 1, 9	1, 9 / 1, 9	1, 9 / 0, 9	1, 10 / 0, 10	0, 10 / 0, 10
27	11	13	4, 7 / 4, 6	3, 8 / 3, 7	2, 8 / 2, 8	2, 9 / 2, 8	2, 9 / 1, 9	1, 9 / 1, 9	1, 10 / 1, 10	1, 10 / 0, 10	0, 10 / 0, 10
27	12	12	4, 7 / 4, 7	3, 8 / 3, 7	3, 8 / 2, 8	2, 9 / 2, 8	2, 9 / 2, 9	2, 10 / 1, 9	1, 10 / 1, 10	1, 11 / 1, 10	0, 11 / 0, 10
27	12	13	5, 7 / 4, 7	3, 8 / 3, 8	3, 9 / 3, 8	3, 9 / 2, 9	2, 10 / 2, 9	2, 10 / 2, 10	2, 10 / 1, 10	1, 11 / 1, 11	1, 11 / 0, 11
27	13	13	5, 7 / 5, 8	4, 9 / 4, 8	4, 9 / 3, 9	3, 10 / 3, 9	3, 10 / 2, 10	2, 10 / 2, 10	2, 11 / 2, 11	2, 11 / 1, 11	1, 11 / 1, 11
28	1	1	—, — / —, —	—, — / —, —	—, — / —, —	— / —	— / —	— / —	— / —	— / —	— / —
28	1	2	—, — / —, —	—, — / —, —	—, — / —, —	— / —	— / —	— / —	— / —	— / —	— / —
28	1	3	—, — / —, —	—, — / —, —	—, — / —, —	— / —	— / —	— / —	— / —	— / —	— / —
28	1	4	—, — / —, —	—, — / —, —	—, — / —, —	— / —	— / —	— / —	— / —	— / —	— / —
28	1	5	—, — / —, —	—, — / —, —	—, — / —, —	— / —	— / —	— / —	— / —	— / —	— / —
28	1	6	—, — / —, —	—, — / —, —	—, — / —, —	— / —	— / —	— / —	— / —	— / —	— / —
28	1	7	—, — / —, —	—, — / —, —	—, — / —, —	— / —	— / —	— / —	— / —	— / —	— / —
28	1	8	—, — / —, —	—, — / —, —	—, — / —, —	— / —	— / —	— / —	— / —	— / —	— / —
28	1	9	—, — / —, —	—, — / —, —	—, — / —, —	— / —	— / —	— / —	— / —	— / —	— / —
28	1	10	—, — / —, —	—, — / —, —	—, — / —, —	— / —	— / —	— / —	— / —	— / —	— / —
28	1	11	—, — / —, —	—, — / —, —	—, — / —, —	— / —	— / —	— / —	— / —	— / —	— / —
28	1	12	—, — / —, —	—, — / —, —	—, — / —, —	— / —	— / —	— / —	— / —	— / —	— / —
28	1	13	—, — / —, —	—, — / —, —	—, — / —, —	— / —	— / —	— / —	— / —	— / —	— / —
28	1	14	0, — / 0, —	—, — / —, —	—, — / —, —	— / —	— / —	— / —	— / —	— / —	— / —
28	2	2	—, — / —, —	—, — / —, —	—, — / —, —	— / —	— / —	— / —	— / —	— / —	— / —
28	2	3	—, — / —, —	—, — / —, —	—, — / —, —	— / —	— / —	— / —	— / —	— / —	— / —
28	2	4	—, — / —, —	—, — / —, —	—, — / —, —	— / —	— / —	— / —	— / —	— / —	— / —
28	2	5	—, — / —, —	—, — / —, —	—, — / —, —	— / —	— / —	— / —	— / —	— / —	— / —
28	2	6	—, — / —, —	1, — / 1, —	1, — / 1, —	1 / 1	— / —	— / —	— / —	— / —	— / —
28	2	7	—, — / —, —	2, — / 2, —	1, — / 1, —	1 / 1	— / —	— / —	— / —	— / —	— / —
28	2	8	0, 2 / —, —	2, — / 2, —	2, — / 2, —	2 / 2	2 / 2	2 / 2	— / —	— / —	— / —
28	2	9	0, 2 / 0, 2	2, — / 2, —	2, — / 2, —	2 / 2	2 / 2	2 / 2	— / —	— / —	— / —
28	2	10	0, 2 / 0, 2	2, — / 2, —	2, — / 2, —	2 / 2	2 / 2	2 / 2	— / —	— / —	— / —
28	2	11	0, — / 0, 2	2, — / 2, —	2, — / 2, —	2 / 2	2 / 2	2 / 2	— / —	— / —	— / —
28	2	12	0, 2 / —, —	2, — / 2, —	2, — / 2, —	2 / 2	— / —	— / —	— / —	— / —	— / —
28	2	13	—, — / —, —	—, — / —, —	—, — / —, —	— / —	— / —	— / —	— / —	— / —	— / —
28	2	14	—, — / —, —	—, — / —, —	—, — / —, —	— / —	— / —	— / —	— / —	— / —	— / —
28	3	3	—, — / —, —	2, — / 2, —	2, — / 2, —	2 / 2	3 / 3	3 / 3	3 / 3	3 / 3	3 / 3
28	3	4	0, 1 / —, —	2, — / 2, —	2, — / 2, —	2 / 2	3 / 3	3 / 3	3 / 3	3 / 3	3 / 3
28	3	5	0, 2 / —, 1	2, — / 2, —	2, — / 2, —	2 / 2	3 / 3	3 / 3	3 / 3	— / —	— / —
28	3	6	0, 2 / —, —	2, — / 2, —	2, — / 2, —	2 / 2	3 / 3	3 / 3	— / —	— / —	— / —
28	3	7	0, 2 / —, —	2, — / 2, —	2, — / 2, —	3 / 2	3 / 3	— / —	— / —	— / —	— / —
28	3	8	0, 2 / —, —	3, — / 2, —	3, — / 2, —	3 / 3	3 / 3	— / —	— / —	— / —	— / —
28	3	9	0, 2 / —, —	3, — / 2, —	3, — / 3, —	3 / 3	— / —	— / —	— / —	— / —	— / —
28	3	10	0, 2 / —, —	3, — / 2, —	3, — / 3, —	3 / 3	— / —	— / —	— / —	— / —	— / —

App163

TABLE B.28 (cont.) Critical Values for Fisher's Exact Test

n	m_1	m_2	α: 0.50	0.20	0.10	0.05	0.02	0.01	0.005	0.002	0.001
28	3	11	0, 2	-, 3	-, 3	-, -	-, -	-, -	-, -	-, -	-, -
28	3	12	0, 2	0, 3	-, 3	-, 3	-, -	-, -	-, -	-, -	-, -
28	3	13	0, 2	0, 3	-, 3	-, 3	-, -	-, -	-, -	-, -	-, -
28	3	14	1, 2	0, 3	-, 3	-, 3	-, -	-, -	-, -	-, -	-, -
28	4	4	-, 1	-, 2	-, 2	-, -	-, -	-, -	-, -	-, -	-, -
28	4	5	0, 2	-, 2	-, 3	-, 3	-, -	-, -	-, -	-, -	-, -
28	4	6	0, 2	-, 3	-, 3	-, 3	-, -	-, 3	-, -	-, -	-, -
28	4	7	0, 2	-, 3	-, 3	-, 3	-, 3	-, 4	-, 4	-, 4	-, 4
28	4	8	0, 2	0, 3	-, 3	-, 3	-, 4	-, 4	-, 4	-, 4	-, 4
28	4	9	0, 2	0, 3	-, 3	-, 4	-, 4	-, 4	-, 4	-, 4	-, 4
28	4	10	1, 3	0, 3	-, 4	-, 4	-, 4	-, 4	-, 4	-, 4	-, -
28	4	11	1, 3	0, 4	0, 4	-, 4	-, 4	-, 4	-, 4	-, -	-, -
28	4	12	1, 3	0, 4	0, 4	-, 4	-, 4	-, -	-, -	-, -	-, -
28	4	13	1, 3	0, 4	0, 4	-, 4	-, -	-, -	-, -	-, -	-, -
28	4	14	1, 3	0, 4	0, 4	0, 4	-, -	-, -	-, -	-, -	-, -
28	5	5	0, 2	-, 2	-, 3	-, 3	-, -	-, -	-, -	-, -	-, -
28	5	6	0, 2	-, 3	-, 3	-, 3	-, 4	-, 4	-, -	-, -	-, -
28	5	7	0, 2	-, 3	-, 3	-, 4	-, 4	-, 4	-, 5	-, 5	-, 5
28	5	8	0, 2	-, 3	-, 4	-, 4	-, 5	-, 5	-, 5	-, 5	-, 5
28	5	9	1, 3	0, 4	0, 4	-, 4	-, 5	-, 5	-, 5	-, 5	-, 5
28	5	10	1, 3	0, 4	0, 4	-, 4	-, 5	-, 5	-, 5	-, 5	-, -
28	5	11	1, 3	0, 4	0, 5	-, 5	-, 5	-, 5	-, 5	-, -	-, -
28	5	12	1, 3	1, 4	0, 5	0, 5	-, 5	-, -	-, -	-, -	-, -
28	5	13	1, 3	1, 4	0, 5	0, 5	-, 5	-, -	-, -	-, -	-, -
28	5	14	2, 3	1, 4	0, 5	0, 5	-, -	-, -	-, -	-, -	-, -
28	6	6	0, 2	-, 3	-, 3	-, 4	-, 4	-, 4	-, 5	-, 5	-, 5
28	6	7	0, 2	0, 3	-, 4	-, 4	-, 5	-, 5	-, 5	-, 5	-, 6
28	6	8	1, 3	0, 4	0, 4	-, 5	-, 5	-, 5	-, 6	-, 6	-, 6
28	6	9	1, 3	0, 4	0, 5	-, 5	-, 6	-, 6	-, 6	-, 6	-, 6
28	6	10	1, 4	0, 4	0, 5	0, 5	0, 6	-, 6	-, 6	-, 6	-, 6
28	6	11	1, 4	1, 5	0, 5	0, 6	0, 6	-, 6	-, 6	-, 6	-, 6
28	6	12	2, 4	1, 5	1, 5	0, 6	0, 6	0, 6	-, 6	-, 6	-, -
28	6	13	2, 4	1, 5	1, 6	0, 6	0, 7	0, 6	-, 6	-, -	-, -
28	6	14	2, 4	1, 5	1, 6	1, 6	0, 6	0, 5	-, -	-, -	-, -
28	7	7	1, 3	0, 4	0, 4	-, 5	-, 5	-, 6	-, 6	-, 6	-, 6
28	7	8	1, 3	0, 4	0, 5	-, 5	-, 6	-, 6	-, 6	-, 6	-, 6
28	7	9	1, 4	0, 4	0, 5	0, 6	0, 6	0, 6	-, 7	-, 6	-, 6
28	7	10	1, 4	0, 5	0, 6	0, 6	0, 6	0, 6	-, 7	-, 6	-, 7
28	7	11	2, 4	1, 5	1, 6	0, 6	0, 7	0, 7	-, 7	-, 7	-, 7
28	7	12	2, 4	1, 5	1, 6	0, 6	0, 7	0, 7	0, 7	-, 7	-, 7
28	7	13	2, 4	1, 5	1, 6	0, 6	0, 6	0, 7	0, 6	-, 6	-, 7
28	7	14	3, 4	2, 5	1, 6	1, 6	0, 6	0, 6	0, 7	-, 6	-, 7
28	8	8	1, 4	1, 5	1, 5	0, 6	0, 6	0, 6	-, 6	-, 6	-, -
28	8	9	1, 4	1, 5	1, 5	0, 6	0, 6	0, 7	-, 7	-, 6	-, 7
28	8	10	2, 4	1, 5	1, 5	0, 6	0, 6	0, 6	-, 7	-, 7	-, 7

n	m_1	m_2	α: 0.50	0.20	0.10	0.05	0.02	0.01	0.005	0.002	0.001
28	8	11	2, 5	1, 5	1, 6	0, 6	0, 7	0, 7	—, 7	—, 7	—, 8
28	8	12	2, 5	1, 5	1, 6	0, 6	0, 7	0, 7	—, 7	—, 8	—, 8
28	8	13	3, 5	1, 6	1, 6	0, 7	0, 7	0, 7	0, 8	—, 8	—, 8
28	8	14	3, 5	2, 6	1, 7	1, 7	0, 8	0, 8	0, 8	—, 8	—, 8
28	9	9	2, 4	1, 5	0, 5	0, 6	0, 6	—, 7	—, 7	—, 7	—, 7
28	9	10	2, 5	1, 5	1, 6	0, 6	0, 7	0, 7	—, 7	—, 8	—, 8
28	9	11	3, 5	1, 6	1, 6	1, 7	0, 7	0, 7	0, 8	—, 8	—, 8
28	9	12	3, 5	2, 6	1, 6	1, 7	0, 8	0, 8	0, 8	0, 9	—, 9
28	9	13	3, 6	2, 6	1, 7	1, 7	1, 8	0, 8	0, 8	0, 9	0, 9
28	9	14	4, 6	2, 7	2, 7	1, 8	1, 8	0, 9	0, 9	0, 9	0, 9
28	10	10	3, 5	2, 6	1, 6	1, 7	0, 7	0, 7	0, 8	0, 8	—, 8
28	10	11	3, 5	2, 6	1, 7	1, 7	1, 8	0, 8	0, 8	0, 9	0, 9
28	10	12	3, 6	2, 6	1, 7	1, 8	1, 8	0, 9	0, 9	0, 9	0, 9
28	10	13	4, 6	3, 7	2, 7	2, 8	1, 9	1, 9	0, 9	0, 10	0, 10
28	10	14	4, 6	3, 7	2, 8	2, 8	1, 9	1, 9	1, 9	0, 10	0, 10
28	11	11	3, 6	3, 7	2, 7	2, 8	1, 8	1, 9	1, 9	0, 10	0, 10
28	11	12	4, 6	3, 7	2, 8	2, 8	1, 9	1, 9	1, 10	1, 10	0, 10
28	11	13	4, 7	3, 8	3, 8	2, 9	2, 10	1, 10	2, 10	1, 11	1, 11
28	11	14	5, 7	4, 8	3, 9	3, 10	2, 10	2, 11	2, 11	1, 11	1, 11
28	12	12	4, 7	3, 8	3, 8	2, 9	2, 10	2, 10	2, 11	1, 11	1, 11
28	12	13	5, 7	4, 8	3, 9	3, 10	2, 10	2, 11	2, 11	2, 11	1, 11
28	12	14	5, 8	4, 9	4, 9	3, 10	3, 11	2, 11	2, 12	2, 12	2, 12
28	13	13	5, 8	4, 9	3, 9	3, 10	3, 10	2, 11	2, 11	—	—
28	13	14	6, 8	5, 9	4, 10	4, 10	3, 11	3, 11	3, 11	—	—
28	14	14	6, 8	5, 9	4, 10	4, 10	3, 11	3, 11	—	—	—
29	1	1	—, 1	—, 1	—, 1	—	—	—	—	—	—
29	1	2	—, 1	—, 1	—, 1	—	—	—	—	—	—
29	1	3	—, 1	—, 1	—, 1	—	—	—	—	—	—
29	1	4	—, 1	—, 1	—, 1	—	—	—	—	—	—
29	1	5	—, 1	—, 1	—	—	—	—	—	—	—
29	1	6	—, 1	—, 1	—	—	—	—	—	—	—
29	1	7	—, 1	—, 1	—	—	—	—	—	—	—
29	1	8	—, 1	—, 1	—	—	—	—	—	—	—
29	1	9	—, 1	—, 1	—	—	—	—	—	—	—
29	1	10	—, 1	—, 1	—	—	—	—	—	—	—
29	1	11	—, 1	—	—	—	—	—	—	—	—
29	1	12	—, 1	—	—	—	—	—	—	—	—
29	1	13	—, 1	—	—	—	—	—	—	—	—
29	1	14	—, 1	—	—	—	—	—	—	—	—
29	2	2	—, 1	—, 1	—	—	2	2	2	—	—
29	2	3	—, 1	1, 2	2	2	2	2	2	—	—
29	2	4	—, 1	2, 2	2	2	2	2	—	—	—
29	2	5	—, 1	2, 2	2	2	2	2	—	—	—
29	2	6	—, 1	2, 2	2	2	—	—	—	—	—
29	2	7	—, 1	2, 2	2	—	—	—	—	—	—

TABLE B.28 (cont.) Critical Values for Fisher's Exact Test

n	m_1	m_2	α: 0.50	0.20	0.10	0.05	0.02	0.01	0.005	0.002	0.001
29	2	8									
29	2	9									
29	2	10									
29	2	11									
29	2	12									
29	2	13									
29	2	14									
29	3	3									
29	3	4									
29	3	5									
29	3	6									
29	3	7									
29	3	8									
29	3	9									
29	3	10									
29	3	11									
29	3	12									
29	3	13									
29	3	14									
29	4	4									
29	4	5									
29	4	6									
29	4	7									
29	4	8									
29	4	9									
29	4	10									
29	4	11									
29	4	12									
29	4	13									
29	4	14									
29	5	5									
29	5	6									
29	5	7									
29	5	8									
29	5	9									
29	5	10									
29	5	11									
29	5	12									
29	5	13									
29	5	14									
29	6	6									
29	6	7									
29	6	8									
29	6	9									
29	6	10									

TABLE B.28 (cont.) Critical Values for Fisher's Exact Test

n	m_1	m_2	α: 0.50	0.20	0.10	0.05	0.02	0.01	0.005	0.002	0.001
29	6	11	1,4	0,4	0,5	5	5	6	6	6	6
29	6	12	1,3	0,4	0,5	5	6	6	6	6	—
29	6	13	2,4	1,5	0,5	6	6	6	6	—	—
29	6	14	2,4	1,5	0,5	6	6	6	—	—	—
29	7	7	1,3	0,4	1,5	4	5	5	5	6	6
29	7	8	1,3	0,4	0,4	5	5	5	6	6	7
29	7	9	1,4	0,4	0,5	5	5	6	6	6	7
29	7	10	1,4	0,4	0,5	5	6	6	6	7	7
29	7	11	2,4	1,5	1,6	6	6	6	7	7	7
29	7	12	2,4	1,5	0,5	6	6	7	7	—	—
29	7	13	2,4	1,5	1,6	6	6	7	7	7	7
29	7	14	1,3	0,4	0,5	4	5	6	6	7	7
29	8	8	2,4	1,5	1,6	5	5	6	7	7	8
29	8	9	2,4	1,5	1,6	5	5	7	7	8	8
29	8	10	2,4	1,5	1,6	6	6	7	8	8	8
29	8	11	2,5	1,5	1,6	6	7	7	8	8	8
29	8	12	3,5	2,6	1,7	7	7	8	8	8	7
29	8	13	3,5	2,6	1,7	7	8	8	8	7	—
29	8	14	2,4	1,5	0,5	6	6	7	7	—	—
29	9	9	2,4	1,5	1,6	6	7	7	8	8	8
29	9	10	2,5	1,6	1,6	7	7	8	8	9	9
29	9	11	3,5	2,6	1,7	7	8	8	8	9	9
29	9	12	3,5	2,6	1,7	7	8	8	9	9	9
29	9	13	3,5	2,7	2,7	8	8	9	9	9	9
29	9	14	2,4	1,5	0,5	8	8	9	9	10	10
29	10	10	2,5	1,6	1,6	7	8	8	9	9	9
29	10	11	3,6	2,7	2,7	8	8	9	9	9	9
29	10	12	3,6	2,7	3,8	8	9	9	9	9	9
29	10	13	3,6	2,7	3,9	9	9	9	9	10	10
29	10	14	3,6	3,7	3,9	9	9	10	10	10	10
29	11	11	3,6	2,7	2,7	9	9	9	9	10	10
29	11	12	4,6	2,7	2,8	9	9	10	10	11	11
29	11	13	4,6	3,7	3,9	9	10	10	10	11	11
29	11	14	4,6	3,7	4,9	9	10	11	11	11	11
29	12	12	4,7	3,7	3,8	9	10	10	10	11	11
29	12	13	5,7	4,8	3,9	9	10	11	11	11	11
29	12	14	5,7	4,8	4,9	10	10	11	11	12	12
29	13	13	5,7	4,8	4,9	10	11	11	11	—	—
29	13	14	5,8	4,9	4,9	—	—	—	—	—	—
29	14	14	6,8	5,9	4,9	—	—	—	—	—	—
30	1	1	—,—	—,—	—,—	—	—	—	—	—	—
30	1	2	—,—	—,—	—,—	—	—	—	—	—	—
30	1	3	—,—	—,—	—,—	—	—	—	—	—	—
30	1	4	—,—	—,—	—,—	—	—	—	—	—	—
30	1	5	—,—	—,—	—,—	—	—	—	—	—	—

TABLE B.28 (cont.) Critical Values for Fisher's Exact Test

n	m_1	m_2	α: 0.50	0.20	0.10	0.05	0.02	0.01	0.005	0.002	0.001
30	1	6	—, —	—, —	—, —	—, —	—, —	—, —	—, —	—, —	—, —
30	1	7	—, —	—, —	—, —	—, —	—, —	—, —	—, —	—, —	—, —
30	1	8	—, —	—, —	—, —	—, —	—, —	—, —	—, —	—, —	—, —
30	1	9	—, —	—, —	—, —	—, —	—, —	—, —	—, —	—, —	—, —
30	1	10	—, —	—, —	—, —	—, —	—, —	—, —	—, —	—, —	—, —
30	1	11	—, —	—, —	—, —	—, —	—, —	—, —	—, —	—, —	—, —
30	1	12	—, —	—, —	—, —	—, —	—, —	—, —	—, —	—, —	—, —
30	1	13	—, —	—, —	—, —	—, —	—, —	—, —	—, —	—, —	—, —
30	1	14	—, —	—, —	—, —	—, —	—, —	—, —	—, —	—, —	—, —
30	1	15	0, —	—, —	—, —	—, —	—, —	—, —	—, —	—, —	—, —
30	2	2	—, —	—, —	—, —	—, —	—, —	—, —	—, —	—, —	—, —
30	2	3	—, —	1, 1	—, —	—, —	—, —	—, —	—, —	—, —	—, —
30	2	4	—, —	1, 1	—, —	—, —	—, —	—, —	—, —	—, —	—, —
30	2	5	—, —	—, 1	2, 2	—, 2	—, 2	—, 2	—, 2	—, —	—, —
30	2	6	—, —	1, 2	2, 2	2, 2	—, 2	—, 2	—, 2	—, —	—, —
30	2	7	0, 1	2, 2	2, 2	2, 2	—, —	—, —	—, —	—, —	—, —
30	2	8	0, 2	2, 2	2, 2	2, 2	2, —	—, —	—, —	—, —	—, —
30	2	9	0, 2	2, 2	2, 2	2, 2	2, 2	—, —	—, —	—, —	—, —
30	2	10	0, 2	2, 2	2, 2	2, 2	2, 2	2, 2	2, 2	—, —	—, —
30	2	11	0, —	2, 2	2, 2	2, 2	2, 2	2, 2	2, 2	—, —	—, —
30	2	12	—, 2	2, 2	2, 2	2, 2	2, 2	—, —	—, —	—, —	—, —
30	2	13	0, 2	3, 3	3, 3	—, 3	—, —	—, —	—, —	—, —	—, —
30	2	14	0, 3	3, 3	3, 3	3, 3	3, —	3, —	3, —	3, —	3, —
30	2	15	0, 3	3, 3	3, 3	3, 3	3, 3	3, 3	3, 3	3, —	3, —
30	3	3	—, —	—, 2	2, 2	—, 2	—, —	—, —	—, —	—, —	—, —
30	3	4	—, —	2, 2	2, 2	2, 2	—, 3	3, 3	—, —	—, —	—, —
30	3	5	—, —	3, 3	3, 3	3, 3	3, 3	3, 3	3, 3	3, 3	3, 3
30	3	6	1, 2	3, 3	3, 3	3, 3	3, 3	3, 3	3, 3	3, 3	3, 3
30	3	7	0, 2	3, 3	3, 3	3, 3	3, 3	3, 3	3, 3	—, —	—, —
30	3	8	0, 2	3, 3	3, 3	3, 3	3, 3	3, 3	3, —	—, —	—, —
30	3	9	0, 2	3, 3	3, 3	3, 3	—, —	—, —	—, —	—, —	—, —
30	3	10	0, 3	3, 3	3, 3	3, 3	3, 3	3, 3	3, 4	4, 4	4, 4
30	3	11	1, —	3, 3	3, 3	3, —	3, 3	3, 3	4, 4	4, 4	4, 4
30	3	12	0, 2	3, 3	3, 3	3, 3	3, 3	3, 3	3, 4	4, 4	4, 4
30	3	13	0, 2	3, 3	3, 3	3, 3	3, 3	3, 3	3, 4	4, 4	4, 4
30	3	14	0, 3	3, 3	3, 3	3, 4	4, 4	4, 4	4, 4	4, —	4, —
30	3	15	0, 3	3, 3	3, 3	3, 4	4, 4	4, 4	4, 4	4, —	4, —
30	4	4	—, —	—, 2	2, 2	—, 3	—, —	—, —	—, —	—, —	—, —
30	4	5	0, 1	2, 2	2, 3	3, 3	3, 3	3, 3	3, 3	—, —	—, —
30	4	6	0, 2	3, 3	3, 3	3, 3	3, 3	3, 3	3, 4	4, 4	4, 4
30	4	7	0, 2	3, 3	3, 3	3, 3	3, 4	4, 4	4, 4	4, 4	4, 4
30	4	8	0, 2	3, 3	3, 3	3, 4	4, 4	4, 4	4, 4	4, 4	4, 4
30	4	9	0, 2	3, 3	3, 3	3, 4	4, 4	4, 4	4, 4	4, 4	4, 4
30	4	10	0, 3	3, 3	3, 3	4, 4	4, 4	4, 4	4, 4	4, —	4, —
30	4	11	0, 3	3, 3	3, 4	4, 4	4, 4	4, 4	—, —	—, —	—, —

TABLE B.28 (cont.) Critical Values for Fisher's Exact Test

n	m_1	m_2	α: 0.50	0.20	0.10	0.05	0.02	0.01	0.005	0.002	0.001
30	4	12	1, 3 / 1, 3	0, 3 / 0, 4	–, 4 / –, 4	–, 4 / –, 4	–, 4 / –, 4	–, – / –, –	–, – / –, –	–, – / –, –	–, – / –, –
30	4	13	1, 3 / 1, 3	0, 4 / 0, 4	–, 4 / –, 4	–, 4 / –, 4	–, 4 / –, 4	–, – / –, –	–, – / –, –	–, – / –, –	–, – / –, –
30	4	14	1, 3 / 1, 3	0, 4 / 0, 4	–, 4 / 0, 4	–, 4 / –, 4	–, – / –, –	–, – / –, –	–, – / –, –	–, – / –, –	–, – / –, –
30	4	15	1, 3 / 1, 4	0, 4 / 0, 4	–, 4 / 0, 4	0, 4 / –, 4	–, 4 / –, 4	–, 4 / –, 4	–, 4 / –, 4	–, 4 / –, 4	–, 4 / –, 4
30	5	5	0, 2 / –, 2	–, 2 / –, 2	–, 3 / –, 3	0, 3 / –, 3	–, 4 / –, 4	–, 4 / –, 4	–, 4 / –, 4	–, 4 / –, 4	–, 4 / –, 4
30	5	6	0, 2 / –, 2	–, 3 / –, 3	–, 3 / –, 3	–, 3 / –, 3	–, 4 / –, 4	–, 4 / –, 4	–, 4 / –, 4	–, 5 / –, 5	–, 5 / –, 5
30	5	7	0, 2 / 0, 3	–, 3 / –, 3	–, 3 / –, 3	–, 3 / –, 3	–, 4 / –, 4	–, 4 / –, 5	–, 5 / –, 5	–, 5 / –, 5	–, 5 / –, 5
30	5	8	0, 2 / 0, 3	0, 3 / –, 3	–, 4 / –, 4	–, 4 / –, 4	–, 5 / –, 5	–, 5 / –, 5	–, 5 / –, 5	–, 5 / –, 5	–, 5 / –, 5
30	5	9	0, 2 / 0, 3	0, 3 / 0, 4	–, 4 / –, 4	–, 4 / –, 4	–, 5 / –, 5	–, 5 / –, 5	–, 5 / –, 5	–, 5 / –, 5	–, 5 / –, 5
30	5	10	1, 3 / 0, 3	0, 4 / 0, 4	–, 4 / –, 4	0, 5 / 0, 5	–, 5 / –, 5	–, 5 / –, 5	–, 5 / –, 5	–, 5 / –, 5	–, 5 / –, 5
30	5	11	1, 3 / 0, 3	–, 4 / 0, 4	–, 4 / –, 4	–, 5 / –, 5	–, 5 / –, 5	–, 5 / –, 5	–, 5 / –, 5	–, – / –, –	–, – / –, –
30	5	12	1, 3 / 0, 3	0, 4 / 0, 4	0, 5 / 0, 5	0, 5 / 0, 5	–, 5 / –, 5	–, 5 / –, 5	–, 5 / –, 5	–, – / –, –	–, – / –, –
30	5	13	1, 3 / 1, 4	0, 4 / 0, 4	0, 5 / 0, 5	0, 5 / 0, 5	–, 5 / –, 5	–, 5 / –, 5	–, 5 / –, 5	–, – / –, –	–, – / –, –
30	5	14	1, 3 / 1, 4	0, 4 / 0, 4	0, 5 / 0, 5	0, 5 / 0, 5	–, 5 / –, 5	–, 5 / –, 5	–, 5 / –, 5	–, – / –, –	–, – / –, –
30	5	15	2, 3 / 1, 4	1, 4 / 0, 5	1, 5 / 1, 5	0, 6 / 0, 5	–, 5 / –, 5	–, 6 / –, 6	–, 6 / –, 6	–, 6 / –, 5	–, 6 / –, 6
30	6	6	0, 2 / 0, 3	–, 3 / –, 3	–, 3 / –, 3	–, 4 / –, 4	–, 4 / –, 4	–, 5 / –, 5	–, 5 / –, 5	–, 5 / –, 5	–, 5 / –, 5
30	6	7	0, 2 / 0, 3	–, 3 / –, 3	–, 3 / –, 4	–, 4 / –, 4	–, 5 / –, 5	–, 5 / –, 5	–, 5 / –, 5	–, 6 / –, 6	–, 6 / –, 6
30	6	8	1, 2 / 0, 3	1, 5 / 0, 4	0, 5 / 0, 5	0, 5 / 0, 6	0, 6 / 0, 6	0, 6 / –, 6	–, 6 / –, 6	–, 6 / –, 6	–, 6 / –, 6
30	6	9	2, 4 / 1, 4	2, 5 / 1, 5	0, 5 / 1, 6	0, 6 / 0, 6	0, 6 / 0, 6	0, 6 / 0, 6	–, 6 / –, 6	–, 6 / –, 6	–, 6 / –, 6
30	6	10	2, 4 / 1, 5	2, 5 / 1, 5	1, 6 / 1, 6	0, 6 / 0, 6	0, 6 / 0, 6	0, 6 / 0, 6	–, 6 / –, 6	–, 6 / –, 6	–, 6 / –, 6
30	6	11	1, 3 / 1, 4	0, 4 / 0, 4	–, 4 / –, 4	–, 5 / –, 4	5, – / 5, –	5, – / 5, –	5, – / 5, –	5, – / 5, –	6, – / 6, –
30	6	12	2, 4 / 1, 4	1, 5 / 0, 5	0, 5 / 0, 5	0, 5 / 0, 5	–, 6 / 0, 6	–, 6 / 0, 6	6, – / 6, –	6, – / 6, –	6, – / 6, –
30	6	13	2, 4 / 1, 4	1, 5 / 0, 5	0, 5 / 0, 6	0, 6 / 0, 6	0, 6 / 0, 6	0, 6 / 0, 6	6, 0 / 6, 0	7, – / 7, –	7, – / 7, –
30	6	14	2, 4 / 1, 4	2, 5 / 1, 5	1, 6 / 1, 6	1, 6 / 0, 6	0, 7 / 0, 7	0, 7 / 0, 7	7, – / 7, –	7, – / 7, –	7, – / 7, –
30	6	15	2, 4 / 1, 5	2, 5 / 1, 5	1, 6 / 1, 6	1, 6 / 0, 6	0, 6 / 0, 6	6, – / 6, –	6, 0 / 6, 0	7, – / 7, –	7, – / 7, –
30	7	7	1, 3 / 0, 3	0, 4 / –, 4	–, 4 / –, 4	–, 4 / –, 4	–, 5 / –, 5	–, 5 / –, 5	–, 6 / –, 6	–, 6 / –, 6	–, 6 / –, 6
30	7	8	2, 4 / 1, 4	1, 5 / 0, 5	0, 5 / 0, 5	–, 5 / 0, 5	0, 6 / –, 6	0, 6 / 0, 6	0, 6 / 0, 6	0, 6 / 0, 6	–, 7 / 0, 7
30	7	9	2, 4 / 1, 4	1, 5 / 0, 5	0, 6 / 0, 5	0, 6 / 0, 6	0, 6 / 0, 6	0, 7 / 0, 6	7, – / 6, –	7, – / 7, –	7, – / 7, –
30	7	10	2, 5 / 1, 5	1, 6 / 0, 5	1, 6 / 0, 6	1, 6 / 0, 6	0, 7 / 0, 6	0, 7 / 0, 7	7, 0 / 7, –	7, – / 7, –	7, – / 7, –
30	7	11	2, 4 / 1, 4	1, 5 / 1, 5	1, 5 / 1, 5	0, 6 / 0, 6	0, 7 / 0, 7	0, 7 / 0, 7	7, 0 / 7, 0	7, – / 6, –	7, – / 7, –
30	7	12	2, 4 / 2, 4	2, 5 / 1, 5	1, 6 / 1, 6	1, 7 / 1, 6	0, 7 / 0, 7	0, 7 / 0, 7	7, 0 / 7, 0	8, – / 8, –	8, – / 8, –
30	7	13	2, 4 / 2, 4	2, 6 / 1, 6	1, 7 / 1, 6	1, 7 / 1, 6	0, 7 / 0, 7	0, 7 / 0, 7	7, 0 / 7, 0	8, – / 7, –	8, – / 8, –
30	7	14	3, 4 / 2, 5	2, 6 / 1, 6	1, 5 / 1, 5	1, 5 / 0, 5	0, 8 / 0, 7	0, 8 / 0, 7	8, 0 / 8, 0	8, – / 8, –	8, – / 8, –
30	7	15	2, 4 / 1, 4	2, 6 / 1, 6	1, 6 / 1, 5	1, 6 / 0, 6	0, 6 / 0, 6	6, 0 / 6, 0	6, 0 / 6, 0	6, – / 6, –	6, – / 6, –
30	8	8	1, 4 / 1, 4	0, 5 / –, 4	–, 5 / –, 5	0, 5 / 0, 5	0, 6 / –, 6	0, 6 / 0, 6	0, 6 / –, 6	0, 6 / 0, 6	0, 7 / 0, 7
30	8	9	2, 4 / 2, 4	1, 5 / 1, 5	0, 5 / 0, 5	0, 5 / 0, 5	0, 6 / 0, 6	0, 6 / 0, 6	6, 0 / 6, 0	7, – / 6, –	7, – / 7, –
30	8	10	2, 5 / 2, 5	2, 6 / 1, 5	1, 6 / 1, 6	1, 6 / 0, 6	0, 7 / 0, 6	0, 7 / 0, 6	7, – / 6, –	7, – / 7, –	7, – / 7, –
30	8	11	2, 5 / 1, 4	2, 6 / 1, 5	1, 6 / 1, 6	1, 6 / 0, 6	0, 7 / 0, 6	0, 7 / 0, 7	7, 0 / 7, 0	7, – / 7, –	8, – / 8, –
30	8	12	2, 5 / 2, 5	2, 6 / 1, 6	1, 6 / 1, 6	1, 6 / 0, 6	0, 7 / 0, 7	0, 7 / 0, 7	7, 0 / 7, 0	8, – / 7, –	8, – / 8, –
30	8	13	3, 5 / 2, 5	2, 6 / 1, 6	1, 7 / 1, 6	1, 7 / 1, 6	0, 8 / 0, 7	0, 8 / 0, 7	8, 0 / 8, 0	8, – / 8, –	8, – / 8, –
30	8	14	3, 5 / 2, 5	2, 6 / 1, 6	1, 7 / 1, 7	1, 7 / 1, 6	0, 8 / 0, 7	0, 8 / 0, 8	8, 0 / 8, 0	8, – / 8, –	8, – / 8, –
30	8	15	3, 5 / 2, 6	2, 6 / 1, 7	1, 7 / 1, 7	1, 7 / 1, 6	0, 6 / 0, 7	0, 6 / 0, 7	6, – / 7, –	8, – / 8, –	8, – / 8, –
30	9	9	2, 4 / 1, 4	0, 5 / 0, 5	0, 5 / 0, 5	0, 5 / 0, 5	0, 6 / –, 6	0, 6 / 0, 6	0, 7 / 0, 7	0, 7 / 0, 7	0, 7 / 0, 7
30	9	10	2, 4 / 1, 4	1, 5 / 0, 5	1, 6 / 0, 5	1, 6 / 0, 5	0, 6 / 0, 6	0, 7 / 0, 6	7, 0 / 6, 0	8, – / 7, –	8, – / 8, –
30	9	11	2, 5 / 1, 5	1, 5 / 1, 5	1, 6 / 1, 6	1, 6 / 0, 6	0, 7 / 0, 6	0, 7 / 0, 6	7, – / 6, –	8, – / 7, –	8, – / 8, –

App169

TABLE B.28 (cont.)   Critical Values for Fisher's Exact Test

n	m_1	m_2	α: 0.50	0.20	0.10	0.05	0.02	0.01	0.005	0.002	0.001
30	9	12	3,5 2,5	2,6 1,6	1,6 1,7	1,7 1,7	0,7 0,7	0,7 0,8	0,8 0,8	—,8 —,8	—,8 —,8
30	9	13	3,5 2,5	2,6 1,6	1,6 1,7	1,7 1,7	0,7 0,7	0,8 0,8	0,8 0,8	0,8 —,9	—,9 —,9
30	9	14	3,5 3,6	2,6 2,7	2,7 1,7	1,7 1,8	1,8 1,8	0,8 0,8	0,8 0,9	0,9 0,9	—,9 0,9
30	9	15	4,5 3,6	2,7 2,7	2,7 1,8	1,8 1,8	1,8 0,7	1,8 0,—	0,9 0,—	0,9 0,8	0,9 0,—
30	10	10	2,5 2,5	1,5 1,6	1,6 0,6	0,6 0,6	0,7 0,7	0,7 —,—	—,— —,7	—,8 —,8	—,— —,8
30	10	11	3,5 2,5	2,6 1,6	1,6 1,7	1,7 1,7	0,7 0,8	0,8 0,8	0,8 0,8	—,8 0,9	—,8 —,8
30	10	12	3,5 2,5	2,6 1,6	2,7 1,7	1,7 1,7	1,8 0,8	0,8 0,8	0,8 0,9	0,9 0,9	—,9 —,9
30	10	13	3,6 3,6	2,6 2,7	2,7 2,8	1,7 1,8	1,8 1,8	0,8 0,8	0,9 0,9	0,9 0,9	—,9 —,9
30	10	14	4,6 3,6	2,7 2,7	2,7 2,8	2,8 1,8	1,9 1,9	1,9 1,9	0,9 0,9	0,9 0,10	0,10 0,10
30	10	15	4,6 3,7	3,7 2,8	2,8 2,8	2,8 1,9	1,9 1,9	1,9 1,9	1,9 0,10	0,10 0,10	0,10 0,10
30	11	11	3,6 3,6	2,6 2,7	1,7 1,7	1,7 1,7	1,8 1,8	0,8 0,8	0,8 0,9	0,9 0,9	—,9 —,9
30	11	12	3,6 3,6	2,6 2,7	2,7 1,7	1,8 1,8	1,8 1,9	1,9 1,9	1,9 0,9	0,9 0,10	0,10 0,10
30	11	13	4,6 3,6	3,7 2,7	2,7 2,8	2,8 1,8	1,8 1,9	1,9 1,9	1,9 1,9	0,10 1,10	0,10 0,10
30	11	14	4,6 4,7	3,7 3,8	2,8 2,8	2,8 2,9	2,9 1,9	1,10 1,9	1,9 1,10	1,10 1,10	1,11 1,11
30	11	15	5,6 4,7	3,8 3,8	3,8 2,9	2,9 2,9	2,9 1,10	1,9 1,10	1,10 1,10	1,10 1,11	1,11 1,11
30	12	12	4,6 3,6	3,7 2,7	2,8 2,8	2,8 1,8	1,9 1,9	1,9 1,9	1,9 0,9	0,10 0,10	0,10 0,10
30	12	13	4,7 4,7	3,7 3,8	3,8 2,8	2,8 2,9	2,9 1,10	1,10 1,9	1,10 1,10	1,10 1,11	0,10 0,10
30	12	14	5,7 4,7	3,8 3,8	3,8 3,9	2,8 2,9	2,9 2,10	2,10 1,10	1,10 1,10	1,11 1,11	1,11 1,11
30	12	15	5,7 4,8	4,8 3,8	3,9 3,9	3,9 2,10	2,10 2,10	2,10 2,10	2,10 1,11	1,10 1,11	1,11 1,11
30	13	13	5,7 4,7	3,8 3,8	3,8 3,9	2,9 2,9	2,9 2,10	2,10 1,10	2,10 1,10	2,11 1,11	1,11 1,11
30	13	14	5,7 5,8	4,8 4,9	3,9 3,10	3,9 3,10	2,10 2,11	2,11 2,11	2,11 2,11	2,11 1,11	1,11 1,11
30	13	15	6,7 5,8	4,8 4,9	4,9 3,10	3,10 3,10	3,10 2,11	2,11 2,11	2,11 2,11	2,11 2,12	1,12 1,12
30	14	14	6,8 5,8	4,9 4,9	4,9 4,10	3,10 3,10	3,10 3,11	2,11 3,11	3,11 3,12	2,12 2,12	1,12 1,12
30	14	15	6,8 5,9	5,9 4,10	4,10 4,10	4,10 3,11	3,11 3,11	3,11 3,11	3,11 3,12	2,12 2,12	2,12 2,12
30	15	15	7,8 6,9	5,10 5,10	5,10 4,11	4,11 4,11	4,11 3,12	3,12 3,12	3,12 3,12	3,12 2,13	2,13 2,13

This table contains critical values of f, where m_1 = smallest of the four marginal totals (i.e., row and column totals); m_2 = smaller marginal total from the margin other than the margin in which m_1 is located; f = observed frequency contributing to both m_1 and m_2; n = total frequency in the table. Given are pairs of critical values; if f is less than or equal to the first, or greater than or equal to the second, member of the pair, then the null hypothesis is rejected at the indicated α. For each α, the first pair of f values refer to one-tailed hypotheses, and the second pair pertain to two-tailed hypotheses. See Section 24.10 for examples of the use of this table.

TABLE B.29 Critical Values for Runs Test

		α(2): 0.50	0.20	0.10	0.05	0.02	0.01	0.005	0.002	0.001
n_1	n_2	α(1): 0.25	0.10	0.05	0.025	0.01	0.005	0.0025	0.001	0.0005
2	3	2, 4	−, 5	−, −	−, −	−, −	−, −	−, −	−, −	−, −
	4	2, 5	−, −	−, −	−, −	−, −	−, −	−, −	−, −	−, −
	5	2, −	2, −	−, −	−, −	−, −	−, −	−, −	−, −	−, −
	6	2, −	2, −	−, −	−, −	−, −	−, −	−, −	−, −	−, −
	7	3, −	2, −	−, −	−, −	−, −	−, −	−, −	−, −	−, −
	8	3, −	2, −	2, −	−, −	−, −	−, −	−, −	−, −	−, −
	9	3, −	2, −	2, −	−, −	−, −	−, −	−, −	−, −	−, −
	10	3, −	2, −	2, −	−, −	−, −	−, −	−, −	−, −	−, −
	11	3, −	2, −	2, −	−, −	−, −	−, −	−, −	−, −	−, −
	12	3, −	2, −	2, −	2, −	−, −	−, −	−, −	−, −	−, −
	13	3, −	2, −	2, −	2, −	−, −	−, −	−, −	−, −	−, −
	14	3, −	2, −	2, −	2, −	−, −	−, −	−, −	−, −	−, −
	15	3, −	2, −	2, −	2, −	−, −	−, −	−, −	−, −	−, −
	16	3, −	2, −	2, −	2, −	−, −	−, −	−, −	−, −	−, −
	17	3, −	2, −	2, −	2, −	−, −	−, −	−, −	−, −	−, −
	18	3, −	2, −	2, −	2, −	−, −	−, −	−, −	−, −	−, −
	19	3, −	3, −	2, −	2, −	2, −	−, −	−, −	−, −	−, −
	20	3, −	3, −	2, −	2, −	2, −	−, −	−, −	−, −	−, −
	21	4, −	3, −	2, −	2, −	2, −	−, −	−, −	−, −	−, −
	22	4, −	3, −	2, −	2, −	2, −	−, −	−, −	−, −	−, −
	23	4, −	3, −	2, −	2, −	2, −	−, −	−, −	−, −	−,−
	24	4, −	3, −	2, −	2, −	2, −	−, −	−, −	−, −	−, −
	25	4, −	3, −	2, −	2, −	2, −	−, −	−, −	−, −	−, −
	26	4, −	3, −	2, −	2, −	2, −	−, −	−, −	−, −	−, −
	27	4, −	3, −	2, −	2, −	2, −	2, −	−, −	−, −	−, −
	28	4, −	3, −	2, −	2, −	2, −	2, −	−, −	−, −	−, −
	29	4, −	3, −	2, −	2, −	2, −	2, −	−, −	−, −	−, −
2	30	4, −	3, −	2, −	2, −	2, −	2, −	−, −	−, −	−, −
3	3	2, 6	2, 6	−, −	−, −	−, −	−, −	−, −	−, −	−, −
	4	3, 6	2, 7	−, 7	−, −	−, −	−, −	−, −	−, −	−, −
	5	3, 7	2, 7	2, −	−, −	−, −	−, −	−, −	−, −	−, −
	6	3, 7	2, −	2, −	2, −	−, −	−, −	−, −	−, −	−, −
	7	3, 7	3, −	2, −	2, −	−, −	−, −	−, −	−, −	−, −
	8	4, 7	3, −	2, −	2, −	−, −	−, −	−, −	−, −	−, −
	9	4, −	3, −	2, −	2, −	2, −	−, −	−, −	−, −	−, −
	10	4, −	3, −	3, −	2, −	2, −	−, −	−, −	−, −	−, −
	11	4, −	3, −	3, −	2, −	2, −	−, −	−, −	−, −	−, −
	12	4, −	3, −	3, −	2, −	2, −	2, −	−, −	−, −	−, −
	13	4, −	3, −	3, −	2, −	2, −	2, −	−, −	−, −	−, −
	14	4, −	3, −	3, −	3, −	2, −	2, −	−, −	−, −	−, −
	15	4, −	4, −	3, −	3, −	2, −	2, −	2, −	−, −	−, −
	16	4, −	4, −	3, −	3, −	2, −	2, −	2, −	−, −	−, −
	17	4, −	4, −	3, −	3, −	2, −	2, −	2, −	−, −	−, −
	18	4, −	4, −	3, −	3, −	2, −	2, −	2, −	−, −	−, −
	19	4, −	4, −	3, −	3, −	2, −	2, −	2, −	−, −	−, −
	20	4, −	4, −	3, −	3, −	2, −	2, −	2, −	−, −	−, −
	21	5, −	4, −	3, −	3, −	2, −	2, −	2, −	2, −	−, −
	22	5, −	4, −	4, −	3, −	2, −	2, −	2, −	2, −	−, −
	23	5, −	4, −	4, −	3, −	3, −	2, −	2, −	2, −	−, −
	24	5, −	4, −	4, −	3, −	3, −	2, −	2, −	2, −	−, −
	25	5, −	4, −	4, −	3, −	3, −	2, −	2, −	2, −	−, −
	26	5, −	4, −	4, −	3, −	3, −	2, −	2, −	2, −	−, −
3	27	5, −	4, −	4, −	3, −	3, −	2, −	2, −	2, −	2, −

TABLE B.29 (cont.)   Critical Values for Runs Test

n_1	n_2	$\alpha(2)$: 0.50 / $\alpha(1)$: 0.25	0.20 / 0.10	0.10 / 0.05	0.05 / 0.025	0.02 / 0.01	0.01 / 0.005	0.005 / 0.0025	0.002 / 0.001	0.001 / 0.0005
3	28	5, –	4, –	4, –	3, –	3, –	2, –	2, –	2, –	2, –
	29	5, –	4, –	4, –	3, –	3, –	2, –	2, –	2, –	2, –
3	30	5, –	4, –	4, –	3, –	3, –	2, –	2, –	2, –	2, –
4	4	3, 7	2, 8	2, 8	–, –	–, –	–, –	–, –	–, –	–, –
	5	3, 6	3, 8	2, 9	2, 9	–, 9	–, –	–, –	–, –	–, –
	6	4, 8	3, 9	3, 9	2, 9	2, –	–, –	–, –	–, –	–, –
	7	4, 8	3, 9	3, 9	2, –	2, –	–, –	–, –	–, –	–, –
	8	4, 8	3, 9	3, –	3, –	2, –	2, –	–, –	–, –	–, –
	9	5, 9	4, 9	3, –	3, –	2, –	2, –	–, –	–, –	–, –
	10	5, 9	4, –	3, –	3, –	2, –	2, –	2, –	–, –	–, –
	11	5, 9	4, –	3, –	3, –	2, –	2, –	2, –	–, –	–, –
	12	5, 9	4, –	4, –	3, –	3, –	2, –	2, –	–, –	–, –
	13	5, 9	4, –	4, –	3, –	3, –	2, –	2, –	2, –	–, –
	14	5, 9	4, –	4, –	3, –	3, –	2, –	2, –	2, –	–, –
	15	6, –	4, –	4, –	3, –	3, –	3, –	2, –	2, –	–, –
	16	6, –	5, –	4, –	4, –	3, –	3, –	2, –	2, –	2, –
	17	6, –	5, –	4, –	4, –	3, –	3, –	2, –	2, –	2, –
	18	6, –	5, –	4, –	4, –	3, –	3, –	2, –	2, –	2, –
	19	6, –	5, –	4, –	4, –	3, –	3, –	2, –	2, –	2, –
	20	6, –	5, –	4, –	4, –	3, –	3, –	3, –	2, –	2, –
	21	6, –	5, –	4, –	4, –	3, –	3, –	3, –	2, –	2, –
	22	6, –	5, –	4, –	4, –	3, –	3, –	3, –	2, –	2, –
	23	6, –	5, –	4, –	4, –	4, –	3, –	3, –	2, –	2, –
	24	6, –	5, –	5, –	4, –	4, –	3, –	3, –	2, –	2, –
	25	6, –	5, –	5, –	4, –	4, –	3, –	3, –	2, –	2, –
	26	6, –	5, –	5, –	4, –	4, –	3, –	3, –	2, –	2, –
	27	6, –	5, –	5, –	4, –	4, –	3, –	3, –	3, –	2, –
	28	6, –	6, –	5, –	4, –	4, –	3, –	3, –	3, –	2, –
	29	6, –	6, –	5, –	4, –	4, –	4, –	3, –	3, –	2, –
4	30	6, –	6, –	5, –	4, –	4, –	4, –	3, –	3, –	2, –
5	5	4, 8	3, 9	3, 9	2, 10	2, 10	–, –	–, –	–, –	–, –
	6	4, 9	3, 9	3, 10	3, 10	2, 11	2, 11	–, 11	–, –	–, –
	7	5, 9	4, 10	3, 10	3, 11	2, 11	2, –	–, –	–, –	–, –
	8	5, 9	4, 10	3, 11	3, 11	2, –	2, –	2, –	–, –	–, –
	9	5, 10	4, 10	4, 11	3, –	3, –	2, –	2, –	2, –	–, –
	10	6, 10	5, 11	4, 11	3, –	3, –	3, –	2, –	2, –	–, –
	11	6, 10	5, 11	4, –	4, –	3, –	3, –	2, –	2, –	2, –
	12	6, 10	5, 11	4, –	4, –	3, –	3, –	2, –	2, –	2, –
	13	6, 10	5, 11	4, –	4, –	3, –	3, –	3, –	2, –	2, –
	14	6, 10	5, –	5, –	4, –	3, –	3, –	3, –	2, –	2, –
	15	6, 11	5, –	5, –	4, –	4, –	3, –	3, –	2, –	2, –
	16	7, 11	6, –	5, –	4, –	4, –	3, –	3, –	2, –	2, –
	17	7, 11	6, –	5, –	4, –	4, –	3, –	3, –	3, –	2, –
	18	7, 11	6, –	5, –	5, –	4, –	4, –	3, –	3, –	2, –
	19	7, 11	6, –	5, –	5, –	4, –	4, –	3, –	3, –	2, –
	20	7, 11	6, –	5, –	5, –	4, –	4, –	3, –	3, –	3, –
	21	7, 11	6, –	5, –	5, –	4, –	4, –	3, –	3, –	3, –
	22	7, –	6, –	6, –	5, –	4, –	4, –	4, –	3, –	3, –
	23	7, –	6, –	6, –	5, –	4, –	4, –	4, –	3, –	3, –
	24	7, –	6, –	6, –	5, –	4, –	4, –	4, –	3, –	3, –
	25	8, –	6, –	6, –	5, –	4, –	4, –	4, –	3, –	3, –
	26	8, –	6, –	6, –	5, –	5, –	4, –	4, –	3, –	3, –
	27	8, –	6, –	6, –	5, –	5, –	4, –	4, –	3, –	3, –
	28	8, –	6, –	6, –	5, –	5, –	4, –	4, –	3, –	3, –
	29	8, –	7, –	6, –	6, –	5, –	4, –	4, –	4, –	3, –
5	30	8, –	7, –	6, –	6, –	5, –	4, –	4, –	4, –	3, –

TABLE B.29 (cont.)   Critical Values for Runs Test

n_1	n_2	α(2): 0.50 α(1): 0.25	0.20 0.10	0.10 0.05	0.05 0.025	0.02 0.01	0.01 0.005	0.005 0.0025	0.002 0.001	0.001 0.0005
6	6	5, 9	4, 10	3, 11	3, 11	2, 12	2, 12	2, 12	–, –	–, –
	7	5, 8	4, 11	4, 11	3, 12	3, 12	2, 13	2, 13	–, 13	–, –
	8	6, 10	5, 11	4, 12	3, 12	3, 13	3, 13	2, 13	2, –	–, –
	9	6, 10	5, 11	4, 12	4, 13	3, 13	3, –	2, –	2, –	2, –
	10	6, 11	5, 12	5, 12	4, 13	3, –	3, –	3, –	2, –	2, –
	11	7, 11	5, 12	5, 13	4, 13	4, –	3, –	3, –	2, –	2, –
	12	7, 11	6, 12	5, 13	4, 13	4, –	3, –	3, –	3, –	2, –
	13	7, 12	6, 12	5, 13	5, –	4, –	3, –	3, –	3, –	2, –
	14	7, 12	6, 13	5, 13	5, –	4, –	4, –	3, –	3, –	2, –
	15	7, 12	6, 13	6, –	5, –	4, –	4, –	3, –	3, –	3, –
	16	8, 12	6, 13	6, –	5, –	4, –	4, –	4, –	3, –	3, –
	17	8, 12	6, 13	6, –	5, –	5, –	4, –	4, –	3, –	3, –
	18	8, 12	7, 13	6, –	5, –	5, –	4, –	4, –	3, –	3, –
	19	8, 12	7, –	6, –	6, –	5, –	4, –	4, –	3, –	3, –
	20	8, 12	7, –	6, –	6, –	5, –	4, –	4, –	4, –	3, –
	21	8, 12	7, –	6, –	6, –	5, –	5, –	4, –	4, –	3, –
	22	8, 13	7, –	6, –	6, –	5, –	5, –	4, –	4, –	3, –
	23	8, 13	7, –	6, –	6, –	5, –	5, –	4, –	4, –	3, –
	24	8, 13	7, –	7, –	6, –	5, –	5, –	4, –	4, –	4, –
	25	8, 13	8, –	7, –	6, –	5, –	5, –	4, –	4, –	4, –
	26	9, 13	8, –	7, –	6, –	6, –	5, –	5, –	4, –	4, –
	27	9, 13	8, –	7, –	6, –	6, –	5, –	5, –	4, –	4, –
	28	9, 13	8, –	7, –	6, –	6, –	5, –	5, –	4, –	4, –
	29	9, 13	8, –	7, –	6, –	6, –	5, –	5, –	4, –	4, –
6	30	9, 13	8, –	7, –	6, –	6, –	5, –	5, –	4, –	4, –
7	7	6, 10	5, 11	4, 12	3, 13	3, 13	3, 13	2, 14	2, 14	–, –
	8	6, 11	5, 12	4, 13	4, 13	3, 14	3, 14	3, 14	2, 15	2, 15
	9	7, 11	5, 12	5, 13	4, 14	4, 14	3, 15	3, 15	2, 15	2, –
	10	7, 12	6, 13	5, 13	5, 14	4, 15	3, 15	3, 15	3, –	2, –
	11	7, 12	6, 13	5, 14	5, 14	4, 15	4, 15	3, –	3, –	2, –
	12	8, 12	6, 13	6, 14	5, 14	4, 15	4, –	3, –	3, –	3, –
	13	8, 12	7, 14	6, 14	5, 15	5, –	4, –	4, –	3, –	3, –
	14	8, 13	7, 14	6, 14	5, 15	5, –	4, –	4, –	3, –	3, –
	15	8, 13	7, 14	6, 15	6, 15	5, –	4, –	4, –	3, –	3, –
	16	8, 13	7, 14	6, 15	6, –	5, –	5, –	4, –	4, –	3, –
	17	9, 13	7, 14	7, 15	6, –	5, –	5, –	4, –	4, –	3, –
	18	9, 14	8, 14	7, 15	6, –	5, –	5, –	4, –	4, –	4, –
	19	9, 14	8, 15	7, 15	6, –	6, –	5, –	5, –	4, –	4, –
	20	9, 14	8, 15	7, –	6, –	6, –	5, –	5, –	4, –	4, –
	21	9, 14	8, 15	7, –	7, –	6, –	5, –	5, –	4, –	4, –
	22	9, 14	8, 15	7, –	7, –	6, –	5, –	5, –	4, –	4, –
	23	10, 14	8, 15	8, –	7, –	6, –	6, –	5, –	5, –	4, –
	24	10, 14	8, 15	8, –	7, –	6, –	6, –	5, –	5, –	4, –
	25	10, 14	8, –	8, –	7, –	6, –	6, –	5, –	5, –	4, –
	26	10, 14	8, –	8, –	7, –	6, –	6, –	5, –	5, –	4, –
	27	10, 14	9, –	8, –	7, –	6, –	6, –	6, –	5, –	5, –
	28	10, 14	9, –	8, –	7, –	7, –	6, –	6, –	5, –	5, –
	29	10, 14	9, –	8, –	8, –	7, –	6, –	6, –	5, –	5, –
7	30	10, 14	9, –	8, –	8, –	7, –	6, –	6, –	5, –	5, –
8	8	7, 11	5, 13	5, 13	4, 14	4, 14	3, 15	3, 15	2, 16	2, 16
	9	7, 10	6, 13	5, 14	5, 14	4, 15	3, 15	3, 16	3, 16	2, 17
	10	7, 12	6, 13	6, 14	5, 15	4, 15	4, 16	3, 16	3, 17	3, 17
	11	8, 13	7, 14	6, 15	5, 15	5, 16	4, 16	4, 17	3, 17	3, –
8	12	8, 13	7, 14	6, 15	6, 16	5, 16	4, 17	4, 17	3, –	3, –

TABLE B.29 (cont.)   Critical Values for Runs Test

n_1	n_2	$\alpha(2){:}$ 0.50 $\alpha(1){:}$ 0.25	0.20 0.10	0.10 0.05	0.05 0.025	0.02 0.01	0.01 0.005	0.005 0.0025	0.002 0.001	0.001 0.0005
8	13	8, 13	7, 15	6, 15	6, 16	5, 17	5, 17	4, 17	4, –	3, –
	14	9, 14	7, 15	7, 16	6, 16	5, 17	5, 17	4, –	4, –	3, –
	15	9, 14	8, 15	7, 16	6, 16	5, 17	5, –	5, –	4, –	4, –
	16	9, 14	8, 15	7, 16	6, 17	6, 17	5, –	5, –	4, –	4, –
	17	9, 14	8, 16	7, 16	7, 17	6, –	5, –	5, –	4, –	4, –
	18	10, 14	8, 16	8, 16	7, 17	6, –	6, –	5, –	4, –	4, –
	20	10, 15	9, 16	8, 17	7, 17	6, –	6, –	5, –	5, –	4, –
	21	10, 15	9, 16	8, 17	7, –	7, –	6, –	6, –	5, –	5, –
	22	10, 15	9, 16	8, 17	8, –	7, –	6, –	6, –	5, –	5, –
	23	10, 15	9, 16	8, 17	8, –	7, –	6, –	6, –	5, –	5, –
	24	11, 16	9, 16	8, 17	8, –	7, –	6, –	6, –	5, –	5, –
	25	11, 16	9, 17	9, –	8, –	7, –	7, –	6, –	5, –	5, –
	26	11, 16	10, 17	9, –	8, –	7, –	7, –	6, –	6, –	5, –
	27	11, 16	10, 17	9, –	8, –	7, –	7, –	6, –	6, –	5, –
	28	11, 16	10, 17	9, –	8, –	8, –	7, –	6, –	6, –	5, –
	29	11, 16	10, 17	9, –	8, –	8, –	7, –	6, –	6, –	5, –
8	30	11, 16	10, 17	9, –	8, –	8, –	7, –	7, –	6, –	6, –
9	9	8, 12	6, 14	6, 14	5, 15	4, 16	4, 16	3, 17	3, 17	3, 17
	10	8, 13	7, 14	6, 15	5, 16	5, 16	4, 17	4, 17	3, 18	3, 18
	11	8, 13	7, 15	6, 15	6, 16	5, 17	5, 17	4, 18	3, 18	3, 19
	12	9, 14	7, 15	7, 16	6, 16	5, 17	5, 18	4, 18	4, 19	3, 19
	13	9, 14	8, 15	7, 16	6, 17	6, 18	5, 18	5, 18	4, 19	4, 19
	14	9, 14	8, 16	7, 17	7, 17	6, 18	5, 18	5, 19	4, 19	4, –
	15	10, 15	8, 16	8, 17	7, 18	6, 18	6, 19	5, 19	4, –	4, –
	16	10, 15	9, 16	8, 17	7, 18	6, 18	6, 19	5, 19	5, –	4, –
	17	10, 15	9, 17	8, 17	7, 18	7, 19	6, 19	5, –	5, –	4, –
	18	10, 16	9, 17	8, 18	8, 18	7, 19	6, –	6, –	5, –	5, –
	19	11, 16	9, 17	8, 18	8, 18	7, 19	6, –	6, –	5, –	5, –
	20	11, 16	10, 17	9, 18	8, 18	7, 19	7, –	6, –	5, –	5, –
	21	11, 16	10, 18	9, 18	8, 19	7, –	7, –	6, –	6, –	5, –
	22	11, 16	10, 18	9, 18	8, 19	7, –	7, –	6, –	6, –	5, –
	23	12, 16	10, 18	9, 18	8, 19	8, –	7, –	6, –	6, –	5, –
	24	12, 17	10, 18	9, 18	9, 19	8, –	7, –	7, –	6, –	6, –
	25	12, 17	10, 18	10, 19	9, 19	8, –	7, –	7, –	6, –	6, –
	26	12, 17	10, 18	10, 19	9, –	8, –	7, –	7, –	6, –	6, –
	27	12, 17	11, 18	10, 19	9, –	8, –	8, –	7, –	6, –	6, –
	28	12, 17	11, 18	10, 19	9, –	8, –	8, –	7, –	6, –	6, –
	29	12, 17	11, 18	10, 19	9, –	8, –	8, –	7, –	7, –	6, –
9	30	12, 18	11, 18	10, 19	9, –	8, –	8, –	7, –	7, –	6, –
10	10	9, 13	7, 15	6, 16	6, 16	5, 17	5, 17	4, 18	4, 18	3, 19
	11	9, 12	8, 15	7, 16	6, 17	5, 18	5, 18	4, 19	4, 19	3, 19
	12	9, 14	8, 16	7, 17	7, 17	6, 18	5, 19	5, 19	4, 20	4, 20
	13	10, 15	8, 16	8, 17	7, 18	6, 19	5, 19	5, 20	4, 20	4, 20
	14	10, 15	9, 17	8, 17	7, 18	6, 19	6, 19	5, 20	5, 20	4, 21
	15	10, 16	9, 17	8, 18	7, 18	7, 19	6, 20	6, 20	5, 21	5, 21
	16	11, 16	9, 17	8, 18	8, 19	7, 20	6, 20	6, 20	5, 21	5, –
	17	11, 16	10, 18	9, 18	8, 19	7, 20	6, 20	6, 20	5, 21	5, –
	18	11, 16	10, 18	9, 19	8, 19	7, 20	7, 21	6, 21	6, –	5, –
	19	12, 17	10, 18	9, 19	8, 20	8, 20	7, 21	6, 21	6, –	5, –
	20	12, 17	10, 18	9, 19	9, 20	8, 20	7, 21	7, –	6, –	6, –
	21	12, 17	10, 18	10, 19	9, 20	8, 21	7, 21	7, –	6, –	6, –
	22	12, 17	11, 19	10, 20	9, 20	8, 21	8, –	7, –	6, –	6, –
	23	12, 18	11, 19	10, 20	9, 20	8, 21	8, –	7, –	6, –	6, –
10	24	12, 18	11, 19	10, 20	9, 20	8, 21	8, –	7, –	7, –	6, –

TABLE B.29 (cont.)   Critical Values for Runs Test

n_1	n_2	α(2): 0.50 / α(1): 0.25	0.20 / 0.10	0.10 / 0.05	0.05 / 0.025	0.02 / 0.01	0.01 / 0.005	0.005 / 0.0025	0.002 / 0.001	0.001 / 0.0005
10	25	13, 18	11, 19	10, 20	10, 20	9, –	8, –	7, –	7, –	6, –
	26	13, 18	11, 20	10, 20	10, 21	9, –	8, –	8, –	7, –	6, –
	27	13, 18	12, 20	11, 20	10, 21	9, –	8, –	8, –	7, –	7, –
	28	13, 18	12, 20	11, 20	10, 21	9, –	8, –	8, –	7, –	7, –
	29	13, 18	12, 20	11, 20	10, 21	9, –	9, –	8, –	7, –	7, –
10	30	14, 18	12, 20	11, 20	10, 21	9, –	9, –	8, –	8, –	7, –
11	11	9, 15	8, 16	7, 17	7, 17	6, 18	5, 19	5, 19	4, 20	4, 20
	12	10, 15	9, 16	8, 17	7, 18	6, 19	6, 19	5, 20	5, 20	4, 21
	13	10, 16	9, 17	8, 18	7, 19	6, 19	6, 20	5, 20	5, 21	4, 21
	14	11, 16	9, 17	8, 18	8, 19	7, 20	6, 20	6, 21	5, 21	5, 22
	15	11, 16	10, 18	9, 19	8, 19	7, 20	7, 21	6, 21	5, 22	5, 22
	16	11, 17	10, 18	9, 19	8, 20	7, 21	7, 21	6, 22	6, 22	5, 23
	17	12, 17	10, 18	9, 19	9, 20	8, 21	7, 22	7, 22	6, 22	5, 23
	18	12, 17	10, 19	10, 20	9, 20	8, 21	7, 22	7, 22	6, 23	6, 23
	19	12, 18	11, 19	10, 20	9, 21	8, 22	8, 22	7, 22	6, 23	6, –
	20	12, 18	11, 19	10, 20	9, 21	8, 22	8, 22	7, 23	7, 23	6, –
	21	13, 18	11, 20	10, 20	10, 21	9, 22	8, 22	7, 23	7, –	6, –
	22	13, 18	11, 20	10, 21	10, 22	9, 22	8, 23	8, 23	7, –	6, –
	23	13, 19	12, 20	11, 21	10, 22	9, 22	8, 23	8, 23	7, –	7, –
	24	13, 19	12, 20	11, 21	10, 22	9, 22	9, 23	8, –	7, –	7, –
	25	14, 19	12, 20	11, 21	10, 22	9, 23	9, 23	8, –	7, –	7, –
	26	14, 19	12, 21	11, 22	10, 22	10, 23	9, –	8, –	8, –	7, –
	27	14, 19	12, 21	11, 22	11, 22	10, 23	9, –	8, –	8, –	7, –
	28	14, 20	13, 21	12, 22	11, 22	10, 23	9, –	9, –	8, –	7, –
	29	14, 20	13, 21	12, 22	11, 22	10, 23	9, –	9, –	8, –	8, –
11	30	14, 20	13, 21	12, 22	11, 22	10, –	10, –	9, –	8, –	8, –
12	12	10, 16	9, 17	8, 18	7, 19	7, 19	6, 20	5, 21	5, 21	4, 22
	13	11, 14	9, 18	9, 18	8, 19	7, 20	6, 21	6, 21	5, 22	5, 22
	14	11, 17	10, 18	9, 19	8, 20	7, 21	7, 21	6, 22	5, 22	5, 23
	15	12, 17	10, 19	9, 19	8, 20	8, 21	7, 22	6, 22	6, 23	5, 23
	16	12, 17	10, 19	10, 20	9, 21	8, 22	7, 22	7, 23	6, 23	6, 24
	17	12, 18	11, 19	10, 20	9, 21	8, 22	8, 22	7, 23	6, 24	6, 24
	18	13, 18	11, 20	10, 21	9, 21	8, 22	8, 23	7, 23	7, 24	6, 24
	19	13, 18	11, 20	10, 21	10, 22	9, 23	8, 23	7, 24	7, 24	6, 25
	20	13, 19	12, 20	11, 21	10, 22	9, 23	8, 23	8, 24	7, 24	7, 25
	21	14, 19	12, 21	11, 22	10, 22	9, 23	9, 24	8, 24	7, 25	7, 25
	22	14, 19	12, 21	11, 22	10, 22	9, 23	9, 24	8, 24	7, 25	7, –
	23	14, 20	12, 21	11, 22	11, 23	10, 24	9, 24	8, 24	8, 25	7, –
	24	14, 20	13, 21	12, 22	11, 23	10, 24	9, 24	9, 25	8, –	7, –
	25	14, 20	13, 22	12, 22	11, 23	10, 24	9, 24	9, 25	8, –	8, –
	26	15, 20	13, 22	12, 23	11, 23	10, 24	10, 25	9, 25	8, –	8, –
	27	15, 20	13, 22	12, 23	11, 24	10, 24	10, 25	9, 25	8, –	8, –
	28	15, 21	13, 22	12, 23	12, 24	11, 24	10, 25	9, –	9, –	8, –
	29	15, 21	14, 22	13, 23	12, 24	11, 24	10, 25	10, –	9, –	8, –
12	30	15, 21	14, 22	13, 23	12, 24	11, 25	10, 25	10, –	9, –	8, –
13	13	11, 17	10, 18	9, 19	8, 20	7, 21	7, 21	6, 22	5, 23	5, 23
	14	12, 17	10, 19	9, 20	9, 20	8, 21	7, 22	7, 22	6, 23	5, 24
	15	12, 18	11, 19	10, 20	9, 21	8, 22	7, 22	7, 23	6, 24	6, 24
	16	13, 18	11, 20	10, 21	9, 21	8, 22	8, 23	7, 23	6, 24	6, 25
13	17	13, 19	11, 20	10, 21	10, 22	9, 23	8, 23	7, 24	7, 25	6, 25

TABLE B.29 (cont.)   Critical Values for Runs Test

n_1	n_2	α(2): 0.50 / α(1): 0.25	0.20 / 0.10	0.10 / 0.05	0.05 / 0.025	0.02 / 0.01	0.01 / 0.005	0.005 / 0.0025	0.002 / 0.001	0.001 / 0.0005
13	18	13, 19	12, 20	11, 21	10, 22	9, 23	8, 24	8, 24	7, 25	7, 25
	19	14, 19	12, 21	11, 22	10, 23	9, 24	9, 24	8, 25	7, 25	7, 26
	20	14, 20	12, 21	11, 22	10, 23	10, 24	9, 24	8, 25	8, 26	7, 26
	21	14, 20	13, 22	12, 22	11, 23	10, 24	9, 25	9, 25	8, 26	7, 26
	22	15, 20	13, 22	12, 23	11, 24	10, 24	9, 25	9, 26	8, 26	7, 27
	23	15, 20	13, 22	12, 23	11, 24	10, 25	10, 25	9, 26	8, 26	8, 27
	24	15, 21	13, 22	12, 23	11, 24	10, 25	10, 26	9, 26	8, 27	8, 27
	25	15, 21	14, 23	13, 24	12, 24	11, 25	10, 26	9, 26	9, 27	8, 27
	26	16, 21	14, 23	13, 24	12, 24	11, 26	10, 26	10, 26	9, 27	8, –
	27	16, 21	14, 23	13, 24	12, 25	11, 26	10, 26	10, 26	9, 27	9, –
	28	16, 22	14, 23	13, 24	12, 25	11, 26	11, 26	10, 27	9, –	9, –
	29	16, 22	14, 24	13, 24	13, 25	12, 26	11, 26	10, 27	9, –	9, –
13	30	16, 22	15, 24	14, 24	13, 25	12, 26	11, 26	10, 27	10, –	9, –
14	14	12, 18	11, 19	10, 20	9, 21	8, 22	7, 23	7, 23	6, 24	6, 24
	15	13, 16	11, 20	10, 21	9, 22	8, 23	8, 23	7, 24	7, 24	6, 25
	16	13, 19	11, 20	11, 21	10, 22	9, 23	8, 24	8, 24	7, 25	6, 25
	17	14, 19	12, 21	11, 22	10, 23	9, 24	8, 24	8, 25	7, 25	7, 26
	18	14, 20	12, 21	11, 22	10, 23	9, 24	9, 25	8, 25	7, 26	7, 26
	19	14, 20	13, 22	12, 23	11, 23	10, 24	9, 25	8, 26	8, 26	7, 27
	20	15, 20	13, 22	12, 23	11, 24	10, 25	9, 25	9, 26	8, 27	7, 27
	21	15, 21	13, 22	12, 23	11, 24	10, 25	10, 26	9, 26	8, 27	8, 28
	22	15, 21	14, 23	12, 24	12, 24	11, 26	10, 26	9, 27	9, 27	8, 28
	23	16, 21	14, 23	13, 24	12, 25	11, 26	10, 26	10, 27	9, 28	8, 28
	24	16, 22	14, 23	13, 24	12, 25	11, 26	10, 27	10, 27	9, 28	8, 28
	25	16, 22	14, 24	13, 24	12, 25	11, 26	11, 27	10, 28	9, 28	9, 28
	26	16, 22	15, 24	14, 25	13, 26	12, 26	11, 27	10, 28	9, 28	9, 29
	27	17, 22	15, 24	14, 25	13, 26	12, 27	11, 27	10, 28	10, 28	9, 29
	28	17, 23	15, 24	14, 25	13, 26	12, 27	11, 28	11, 28	10, 29	9, 29
	29	17, 23	15, 24	14, 26	13, 26	12, 27	12, 28	11, 28	10, 29	9, –
14	30	17, 23	15, 25	14, 26	13, 26	12, 27	12, 28	11, 28	10, 29	10, –
15	15	13, 19	12, 20	11, 21	10, 22	9, 23	8, 24	8, 24	7, 25	6, 26
	16	14, 19	12, 21	11, 22	10, 23	9, 24	9, 24	8, 25	7, 26	7, 26
	17	14, 20	12, 21	11, 22	11, 23	10, 24	9, 25	8, 26	8, 26	7, 27
	18	14, 20	13, 22	12, 23	11, 24	10, 25	9, 25	9, 26	8, 27	7, 27
	19	15, 21	13, 22	12, 23	11, 24	10, 25	10, 26	9, 27	8, 27	8, 28
	20	15, 21	13, 23	12, 24	12, 25	11, 26	10, 26	9, 27	8, 28	8, 28
	21	16, 21	14, 23	13, 24	12, 25	11, 26	10, 27	10, 27	9, 28	8, 29
	22	16, 22	14, 24	13, 25	12, 25	11, 26	10, 27	10, 28	9, 28	8, 29
	23	16, 22	14, 24	13, 25	12, 26	11, 27	11, 27	10, 28	9, 29	9, 29
	24	16, 22	15, 24	14, 25	13, 26	12, 27	11, 28	10, 28	10, 29	9, 30
	25	17, 23	15, 24	14, 26	13, 26	12, 27	11, 28	11, 29	10, 29	9, 30
	26	17, 23	15, 25	14, 26	13, 27	12, 28	11, 28	11, 29	10, 30	9, 30
	27	17, 23	16, 25	14, 26	14, 27	12, 28	12, 28	11, 29	10, 30	10, 30
	28	18, 24	16, 25	15, 26	14, 27	13, 28	12, 29	11, 29	10, 30	10, 30
	29	18, 24	16, 26	15, 26	14, 27	13, 28	12, 29	11, 30	11, 30	10, 31
15	30	18, 24	16, 26	15, 27	14, 28	13, 28	12, 29	12, 30	11, 30	10, 31
16	16	14, 20	12, 22	11, 23	11, 23	10, 24	9, 25	8, 26	8, 26	7, 27
	17	15, 18	13, 22	12, 23	11, 24	10, 25	9, 26	9, 26	8, 27	7, 28
	18	15, 21	13, 23	12, 24	11, 25	10, 26	10, 26	9, 27	8, 28	8, 28
	19	15, 21	14, 23	13, 24	12, 25	11, 26	10, 27	9, 27	9, 28	8, 29
16	20	16, 22	14, 24	13, 25	12, 25	11, 26	10, 27	10, 28	9, 29	8, 29

TABLE B.29 (cont.)   Critical Values for Runs Test

n_1	n_2	$\alpha(2)$: 0.50 $\alpha(1)$: 0.25	0.20 0.10	0.10 0.05	0.05 0.025	0.02 0.01	0.01 0.005	0.005 0.0025	0.002 0.001	0.001 0.0005
16	21	16, 22	14, 24	13, 25	12, 26	11, 27	11, 28	10, 28	9, 29	9, 30
	22	17, 23	15, 24	14, 25	13, 26	12, 27	11, 28	10, 29	9, 29	9, 30
	23	17, 23	15, 25	14, 26	13, 27	12, 28	11, 28	11, 29	10, 30	9, 30
	24	17, 23	15, 25	14, 26	13, 27	12, 28	12, 29	11, 29	10, 30	9, 31
	25	17, 24	16, 25	15, 26	14, 27	13, 28	12, 29	11, 30	10, 30	10, 31
	26	18, 24	16, 26	15, 27	14, 28	13, 29	12, 29	11, 30	11, 31	10, 31
	27	18, 24	16, 26	15, 27	14, 28	13, 29	12, 30	12, 30	11, 31	10, 32
	28	18, 24	16, 26	15, 27	14, 28	13, 29	13, 30	12, 30	11, 31	10, 32
	29	19, 25	17, 26	16, 28	15, 28	14, 30	13, 30	12, 31	11, 32	11, 32
16	30	19, 25	17, 27	16, 28	15, 29	14, 30	13, 30	12, 31	11, 32	11, 32
17	17	15, 21	13, 23	12, 24	11, 25	10, 26	10, 26	9, 27	8, 28	8, 28
	18	16, 21	14, 23	13, 24	12, 25	11, 26	10, 27	9, 27	9, 28	8, 29
	19	16, 22	14, 24	13, 25	12, 26	11, 27	10, 27	10, 28	9, 29	8, 29
	20	16, 22	15, 24	13, 25	13, 26	11, 27	11, 28	10, 29	9, 29	9, 30
	21	17, 23	15, 25	14, 26	13, 27	12, 28	11, 28	10, 29	10, 30	9, 30
	22	17, 23	15, 25	14, 26	13, 27	12, 28	11, 29	11, 30	10, 30	9, 31
	23	17, 24	16, 25	15, 27	14, 27	13, 29	12, 29	11, 30	10, 31	10, 31
	24	18, 24	16, 26	15, 27	14, 28	13, 29	12, 30	11, 30	11, 31	10, 32
	25	18, 24	16, 26	15, 27	14, 28	13, 29	12, 30	12, 31	11, 31	10, 32
	26	18, 25	17, 26	15, 28	14, 29	13, 30	13, 30	12, 31	11, 32	10, 32
	27	19, 25	17, 27	16, 28	15, 29	14, 30	13, 31	12, 31	11, 32	11, 33
	28	19, 25	17, 27	16, 28	15, 29	14, 30	13, 31	12, 32	12, 32	11, 33
	29	19, 26	17, 27	16, 28	15, 29	14, 30	13, 31	13, 32	12, 33	11, 33
17	30	20, 26	18, 28	17, 29	16, 30	14, 31	14, 32	13, 32	12, 33	11, 33
18	18	16, 22	14, 24	13, 25	12, 26	11, 27	11, 27	10, 28	9, 29	9, 29
	19	16, 20	15, 24	14, 25	13, 26	12, 27	11, 28	10, 29	9, 30	9, 30
	20	17, 23	15, 25	14, 26	13, 27	12, 28	11, 29	11, 29	10, 30	9, 31
	21	17, 23	15, 25	14, 26	13, 27	12, 28	12, 29	11, 30	10, 31	10, 31
	22	18, 24	16, 26	15, 27	14, 28	13, 29	12, 30	11, 30	10, 31	10, 32
	23	18, 24	16, 26	15, 27	14, 28	13, 29	12, 30	12, 31	11, 32	10, 32
	24	18, 25	17, 27	15, 28	14, 29	13, 30	13, 30	12, 31	11, 32	10, 33
	25	19, 25	17, 27	16, 28	15, 29	14, 30	13, 31	12, 32	11, 32	11, 33
	26	19, 25	17, 27	16, 28	15, 29	14, 30	13, 31	12, 32	12, 33	11, 33
	27	19, 26	18, 28	16, 29	15, 30	14, 31	13, 32	13, 32	12, 33	11, 34
	28	20, 26	18, 28	17, 29	16, 30	14, 31	14, 32	13, 33	12, 33	12, 34
	29	20, 26	18, 28	17, 29	16, 30	15, 32	14, 32	13, 33	12, 34	12, 34
18	30	20, 27	18, 29	17, 30	16, 31	15, 32	14, 32	14, 33	13, 34	12, 34
19	19	17, 23	15, 25	14, 26	13, 27	12, 28	11, 29	11, 29	10, 30	9, 31
	20	17, 24	16, 25	14, 27	13, 27	12, 29	12, 29	11, 30	10, 31	10, 31
	21	18, 24	16, 26	15, 27	14, 28	13, 29	12, 30	11, 31	11, 31	10, 32
	22	18, 25	16, 26	15, 28	14, 29	13, 30	12, 30	12, 31	11, 32	10, 32
	23	19, 25	17, 27	16, 28	15, 29	13, 30	13, 31	12, 32	11, 32	11, 33
	24	19, 25	17, 27	16, 28	15, 29	14, 31	13, 31	12, 32	11, 33	11, 33
	25	19, 26	17, 28	16, 29	15, 30	14, 31	13, 32	13, 32	12, 33	11, 34
	26	20, 26	18, 28	17, 29	16, 30	14, 31	14, 32	13, 33	12, 34	11, 34
	27	20, 26	18, 28	17, 30	16, 31	15, 32	14, 32	13, 33	12, 34	12, 35
	28	20, 27	18, 29	17, 30	16, 31	15, 32	14, 33	14, 34	13, 34	12, 35
	29	21, 27	19, 29	18, 30	17, 31	15, 32	15, 33	14, 34	13, 35	12, 35
19	30	21, 28	19, 29	18, 31	17, 32	16, 33	15, 34	14, 34	13, 35	13, 36
20	20	18, 24	16, 26	15, 27	14, 28	13, 29	12, 30	11, 31	11, 31	10, 32
	21	18, 22	16, 27	15, 28	14, 29	13, 30	12, 31	12, 31	11, 32	10, 33
	22	19, 25	17, 27	16, 28	15, 29	14, 30	13, 31	12, 32	11, 33	11, 33
	23	19, 26	17, 28	16, 29	15, 30	14, 31	13, 32	12, 32	12, 33	11, 34
20	24	20, 26	18, 28	16, 29	15, 30	14, 31	14, 32	13, 33	12, 34	11, 34

TABLE B.29 (cont.)   Critical Values for Runs Test

n_1	n_2	$\alpha(2)$: 0.50 $\alpha(1)$: 0.25	0.20 0.10	0.10 0.05	0.05 0.025	0.02 0.01	0.01 0.005	0.005 0.0025	0.002 0.001	0.001 0.0005
20	25	20, 26	18, 28	17, 30	16, 31	15, 32	14, 33	13, 33	12, 34	12, 35
	26	20, 27	18, 29	17, 30	16, 31	15, 32	14, 33	13, 34	13, 35	12, 35
	27	21, 27	19, 29	18, 30	17, 31	15, 33	15, 33	14, 34	13, 35	12, 36
	28	21, 28	19, 30	18, 31	17, 32	16, 33	15, 34	14, 34	13, 35	13, 36
	29	21, 28	19, 30	18, 31	17, 32	16, 33	15, 34	14, 35	13, 36	13, 36
20	30	22, 28	20, 30	18, 32	17, 32	16, 34	15, 34	15, 35	14, 36	13, 37
21	21	19, 25	17, 27	16, 28	15, 29	14, 30	13, 31	12, 32	11, 33	11, 33
	22	19, 26	17, 28	16, 29	15, 30	14, 31	13, 32	13, 32	12, 33	11, 34
	23	20, 26	18, 28	17, 29	16, 30	14, 31	14, 32	13, 33	12, 34	11, 35
	24	20, 27	18, 29	17, 30	16, 31	15, 32	14, 33	13, 34	12, 34	12, 35
	25	21, 27	19, 29	17, 30	16, 31	15, 32	14, 33	14, 34	13, 35	12, 36
	26	21, 27	19, 30	18, 31	17, 32	15, 33	15, 34	14, 34	13, 35	12, 36
	27	21, 28	19, 30	18, 31	17, 32	16, 33	15, 34	14, 35	13, 36	13, 36
	28	22, 28	20, 30	18, 32	17, 33	16, 34	15, 35	15, 35	14, 36	13, 37
	29	22, 29	20, 31	19, 32	18, 33	16, 34	16, 35	15, 36	14, 37	13, 37
21	30	22, 29	20, 31	19, 32	18, 33	17, 35	16, 35	15, 36	14, 37	13, 38
22	22	20, 26	18, 28	17, 29	16, 30	14, 32	14, 32	13, 33	12, 34	11, 35
	23	20, 24	18, 29	17, 30	16, 31	15, 32	14, 33	13, 34	12, 35	12, 35
	24	21, 27	19, 29	17, 30	16, 31	15, 33	14, 33	14, 34	13, 35	12, 36
	25	21, 28	19, 30	18, 31	17, 32	16, 33	15, 34	14, 35	13, 36	13, 36
	26	22, 28	19, 30	18, 31	17, 32	16, 34	15, 34	14, 35	13, 36	13, 37
	27	22, 29	20, 31	19, 32	18, 33	16, 34	15, 35	15, 36	14, 37	13, 37
	28	22, 29	20, 31	19, 32	18, 33	17, 35	16, 35	15, 36	14, 37	13, 38
	29	23, 29	21, 31	19, 33	18, 34	17, 35	16, 36	15, 37	14, 38	14, 38
22	30	23, 30	21, 32	20, 33	19, 34	17, 35	16, 36	16, 37	15, 38	14, 39
23	23	21, 27	19, 29	17, 31	16, 32	15, 33	14, 34	14, 34	13, 35	12, 36
	24	21, 28	19, 30	18, 31	17, 32	16, 33	15, 34	14, 35	13, 36	13, 36
	25	22, 28	20, 30	18, 32	17, 33	16, 34	15, 35	14, 35	14, 36	13, 37
	26	22, 29	20, 31	19, 32	18, 33	16, 34	16, 35	15, 36	14, 37	13, 38
	27	23, 29	20, 31	19, 33	18, 34	17, 35	16, 36	15, 36	14, 37	14, 38
	28	23, 30	21, 32	20, 33	18, 34	17, 35	16, 36	16, 37	15, 38	14, 39
	29	23, 30	21, 32	20, 33	19, 35	17, 36	17, 37	16, 37	15, 38	14, 39
23	30	24, 30	21, 33	20, 34	19, 35	18, 36	17, 37	16, 38	15, 39	15, 39
24	24	22, 28	20, 30	18, 32	17, 33	16, 34	15, 35	15, 35	14, 36	13, 37
	25	22, 26	20, 31	19, 32	18, 33	17, 34	16, 35	15, 36	14, 37	13, 38
	26	23, 29	20, 31	19, 33	18, 34	17, 35	16, 36	15, 37	14, 38	14, 38
	27	23, 30	21, 32	20, 33	19, 34	17, 36	16, 36	16, 37	15, 38	14, 39
	28	23, 30	21, 32	20, 34	19, 35	18, 36	17, 37	16, 38	15, 39	14, 39
	29	24, 31	22, 33	20, 34	19, 35	18, 36	17, 37	16, 38	15, 39	15, 40
24	30	24, 31	22, 33	21, 35	20, 36	18, 37	17, 38	17, 39	16, 40	15, 40
25	25	23, 29	21, 31	19, 33	18, 34	17, 35	16, 36	15, 37	14, 38	14, 38
	26	23, 30	21, 32	20, 33	19, 34	17, 36	16, 37	16, 37	15, 38	14, 39
	27	24, 30	21, 33	20, 34	19, 35	18, 36	17, 37	16, 38	15, 39	14, 39
	28	24, 31	22, 33	21, 34	19, 35	18, 37	17, 38	16, 38	15, 39	15, 40
	29	24, 31	22, 33	21, 35	20, 36	18, 37	18, 38	17, 39	16, 40	15, 41
25	30	25, 32	23, 34	21, 35	20, 36	19, 38	18, 39	17, 39	16, 40	15, 41
26	26	24, 30	21, 33	20, 34	19, 35	18, 36	17, 37	16, 38	15, 39	14, 40
	27	24, 28	22, 33	21, 34	19, 36	18, 37	17, 38	16, 39	15, 39	15, 40
	28	25, 31	22, 34	21, 35	20, 36	19, 37	18, 38	17, 39	16, 40	15, 41
	29	25, 32	23, 34	21, 35	20, 37	19, 38	18, 39	17, 40	16, 41	16, 41
26	30	25, 32	23, 35	22, 36	21, 37	19, 38	18, 39	18, 40	17, 41	16, 42

TABLE B.29 (cont.)   Critical Values for Runs Test

n_1	n_2	$\alpha(2)$: 0.50 $\alpha(1)$: 0.25	0.20 0.10	0.10 0.05	0.05 0.025	0.02 0.01	0.01 0.005	0.005 0.0025	0.002 0.001	0.001 0.0005
27	27	25, 31	22, 34	21, 35	20, 36	19, 37	18, 38	17, 39	16, 40	15, 41
	28	25, 32	23, 34	21, 36	20, 37	19, 38	18, 39	17, 40	16, 41	16, 41
	29	25, 32	23, 35	22, 36	21, 37	19, 39	19, 39	18, 40	17, 41	16, 42
27	30	26, 33	24, 35	22, 37	21, 38	20, 39	19, 40	18, 41	17, 42	16, 43
28	28	25, 33	23, 35	22, 36	21, 37	19, 39	19, 39	18, 40	17, 41	16, 42
	29	26, 30	24, 35	22, 37	21, 38	20, 39	19, 40	18, 41	17, 42	16, 43
28	30	26, 34	24, 36	23, 37	22, 38	20, 40	19, 41	18, 41	17, 42	17, 43
29	29	26, 34	24, 36	23, 37	22, 38	20, 40	19, 41	19, 41	17, 43	17, 43
29	30	27, 34	25, 36	23, 38	22, 39	21, 40	20, 41	19, 42	18, 43	17, 44
30	30	27, 35	25, 37	24, 38	23, 39	21, 41	20, 42	19, 43	18, 44	18, 44

This table was prepared using the procedure described by Brownlee (1965: 225–226) and Swed and Eisenhart (1943).

Example:

$$u_{0.05(2), 24, 30} = 20 \quad \text{and} \quad 36.$$

The pairs of critical values are consulted as described in Section 25.6. The probability of a u found in the table is less than or equal to its column heading and greater than the next smaller column heading. For example, the two-tailed probability for $n_1 = 25$, $n_2 = 26$, and $u = 20$ is $0.05 < P \le 0.10$; the two-tailed probability for $n_1 = 20$, $n_2 = 30$, and $u = 15$ is $0.002 < P \le 0.005$; and the one-tailed probability for $n_1 = 24$, $n_2 = 25$, and $u = 21$ is $0.10 < P \le 0.25$. For n larger than 30, use the normal approximation of Section 25.6.

TABLE B.30 Critical Values of C for the Mean Square Successive Difference Test

n	α: 0.25	0.10	0.05	0.025	0.01	0.005	0.0025	0.001	0.0005
8	0.223	0.409	0.509	0.587	0.668	0.716	0.756	0.799	0.825
9	0.212	0.391	0.488	0.565	0.645	0.694	0.735	0.779	0.807
10	0.203	0.374	0.469	0.544	0.624	0.673	0.714	0.760	0.788
11	0.194	0.360	0.452	0.526	0.604	0.653	0.694	0.740	0.770
12	0.187	0.347	0.436	0.509	0.586	0.634	0.676	0.722	0.752
13	0.180	0.335	0.422	0.493	0.569	0.617	0.658	0.705	0.735
14	0.174	0.325	0.409	0.478	0.553	0.601	0.642	0.689	0.719
15	0.169	0.315	0.397	0.465	0.539	0.586	0.627	0.673	0.704
16	0.164	0.306	0.386	0.453	0.525	0.572	0.612	0.658	0.689
17	0.159	0.298	0.376	0.441	0.513	0.558	0.599	0.645	0.675
18	0.155	0.290	0.367	0.431	0.501	0.546	0.586	0.632	0.662
19	0.151	0.283	0.358	0.421	0.490	0.535	0.574	0.619	0.650
20	0.148	0.276	0.350	0.412	0.480	0.524	0.563	0.608	0.638
21	0.144	0.270	0.343	0.403	0.470	0.513	0.552	0.597	0.627
22	0.141	0.264	0.335	0.395	0.461	0.504	0.542	0.586	0.616
23	0.138	0.259	0.329	0.387	0.452	0.494	0.532	0.576	0.606
24	0.135	0.254	0.322	0.380	0.444	0.486	0.523	0.566	0.596
25	0.133	0.249	0.316	0.373	0.436	0.477	0.514	0.557	0.587
26	0.130	0.245	0.311	0.366	0.429	0.469	0.506	0.549	0.578
27	0.128	0.240	0.305	0.360	0.422	0.462	0.498	0.540	0.569
28	0.126	0.236	0.300	0.354	0.415	0.455	0.490	0.532	0.561
29	0.123	0.232	0.295	0.349	0.409	0.448	0.483	0.525	0.553
30	0.121	0.229	0.291	0.343	0.402	0.441	0.476	0.517	0.546
31	0.120	0.225	0.286	0.338	0.397	0.435	0.470	0.510	0.538
32	0.118	0.222	0.282	0.333	0.391	0.429	0.463	0.504	0.531
33	0.116	0.218	0.278	0.329	0.386	0.423	0.457	0.497	0.525
34	0.114	0.215	0.274	0.324	0.380	0.418	0.451	0.491	0.518
35	0.113	0.212	0.271	0.320	0.376	0.412	0.446	0.485	0.512
36	0.111	0.210	0.267	0.316	0.371	0.407	0.440	0.479	0.506
37	0.110	0.207	0.264	0.312	0.366	0.402	0.435	0.474	0.500
38	0.108	0.204	0.260	0.308	0.362	0.397	0.430	0.468	0.495
39	0.107	0.202	0.257	0.304	0.357	0.393	0.425	0.463	0.489
40	0.106	0.199	0.254	0.301	0.353	0.388	0.420	0.458	0.484
41	0.104	0.197	0.251	0.297	0.349	0.384	0.415	0.453	0.479
42	0.103	0.195	0.248	0.294	0.345	0.380	0.411	0.448	0.474
43	0.102	0.192	0.245	0.290	0.342	0.376	0.407	0.444	0.469
44	0.101	0.190	0.243	0.287	0.338	0.372	0.402	0.439	0.464
45	0.100	0.188	0.240	0.284	0.335	0.368	0.398	0.435	0.460
46	0.099	0.186	0.238	0.281	0.331	0.364	0.394	0.431	0.455
47	0.098	0.184	0.235	0.279	0.328	0.361	0.391	0.426	0.451
48	0.097	0.182	0.233	0.276	0.325	0.357	0.387	0.422	0.447
49	0.096	0.181	0.230	0.273	0.322	0.354	0.383	0.419	0.443
50	0.095	0.179	0.228	0.270	0.319	0.351	0.380	0.415	0.439
52	0.093	0.175	0.224	0.265	0.313	0.344	0.373	0.408	0.431
54	0.091	0.172	0.220	0.261	0.307	0.338	0.367	0.401	0.424
56	0.090	0.169	0.216	0.256	0.302	0.333	0.361	0.394	0.417
58	0.088	0.166	0.212	0.252	0.297	0.327	0.355	0.388	0.411
60	0.087	0.164	0.209	0.248	0.292	0.322	0.349	0.382	0.405
62	0.085	0.161	0.206	0.244	0.288	0.317	0.344	0.376	0.399
64	0.084	0.158	0.203	0.240	0.284	0.313	0.339	0.371	0.393
66	0.083	0.156	0.200	0.237	0.279	0.308	0.334	0.366	0.387
68	0.081	0.154	0.197	0.233	0.276	0.304	0.330	0.361	0.382
70	0.080	0.152	0.194	0.230	0.272	0.300	0.325	0.356	0.377

TABLE B.30 (cont.) Critical Values of C for the Mean Square Successive Difference Test

n	α: 0.25	0.10	0.05	0.025	0.01	0.005	0.0025	0.001	0.0005
72	0.079	0.150	0.191	0.227	0.268	0.296	0.321	0.351	0.372
74	0.078	0.148	0.189	0.224	0.265	0.292	0.317	0.347	0.368
76	0.077	0.146	0.186	0.221	0.261	0.288	0.313	0.342	0.363
78	0.076	0.144	0.184	0.218	0.258	0.285	0.309	0.338	0.359
80	0.075	0.142	0.182	0.216	0.255	0.281	0.305	0.334	0.355
82	0.074	0.140	0.180	0.213	0.252	0.278	0.302	0.331	0.351
84	0.073	0.139	0.177	0.211	0.249	0.275	0.298	0.327	0.347
86	0.072	0.137	0.175	0.208	0.246	0.272	0.295	0.323	0.343
88	0.071	0.136	0.173	0.206	0.244	0.269	0.292	0.320	0.339
90	0.070	0.134	0.172	0.204	0.241	0.266	0.289	0.316	0.336
92	0.069	0.133	0.170	0.202	0.238	0.263	0.286	0.313	0.332
94	0.068	0.131	0.168	0.200	0.236	0.260	0.283	0.310	0.329
96	0.067	0.130	0.166	0.198	0.233	0.258	0.280	0.307	0.326
98	0.065	0.129	0.165	0.196	0.231	0.255	0.277	0.304	0.323
100	0.064	0.127	0.163	0.194	0.229	0.253	0.275	0.301	0.320
105		0.124	0.159	0.189	0.224	0.247	0.268	0.294	0.312
110		0.121	0.156	0.185	0.219	0.241	0.262	0.288	0.305
115		0.119	0.152	0.181	0.214	0.236	0.257	0.282	0.299
120		0.116	0.149	0.177	0.210	0.231	0.252	0.276	0.293
125		0.114	0.146	0.174	0.205	0.227	0.247	0.271	0.287
130		0.111	0.143	0.170	0.202	0.223	0.242	0.266	0.282
135		0.109	0.141	0.167	0.198	0.219	0.238	0.261	0.277
140		0.107	0.138	0.164	0.194	0.215	0.234	0.256	0.272
145		0.105	0.136	0.161	0.191	0.211	0.230	0.252	0.268
150		0.102	0.133	0.159	0.188	0.208	0.226	0.248	0.263

Table B.30 was prepared by the method outlined in Young (1941).

Examples:

$$C_{0.05,\,60} = 0.209 \quad \text{and} \quad C_{0.01,\,68} = 0.276.$$

For n greater than shown in the table, the normal approximation (Equation 25.22) may be utilized. This approximation is excellent, especially for α near 0.05, as shown in the following tabulation. The following table considers the absolute difference between the exact critical value of C and the value of C calculated from the normal approximation. Given in the table are the minimum sample sizes necessary to achieve such absolute differences of various specified magnitudes.

Absolute Difference	$\alpha = 0.25$	0.10	0.05	0.025	0.01	0.005	0.0025	0.001	0.0005
≤ 0.002	30	35	8	35	70	90	120		
≤ 0.005	20	20	8	20	45	60	80	100	120
≤ 0.010	15	10	8	15	30	40	50	60	80
≤ 0.020	8	8	8	8	20	25	30	40	45

TABLE B.31   Critical Values for the Runs Up and Down Test

n	$\alpha(2)$: 0.50 $\alpha(1)$: 0.25	0.20 0.10	0.10 0.05	0.05 0.025	0.02 0.01	0.01 0.005	0.005 0.0025	0.002 0.001	0.001 0.0005
4	1, –	1, –	–, –	–, –	–, –	–, –	–, –	–, –	–, –
5	2, –	1, –	1, –	1, –	–, –	–, –	–, –	–, –	–, –
6	2, 5	2, –	1, –	1, –	1, –	1, –	–, –	–, –	–, –
7	3, 6	2, –	2, –	2, –	1, –	1, –	1, –	1, –	1, –
8	3, 7	3, 7	2, –	2, –	2, –	1, –	1, –	1, –	1, –
9	4, 7	3, 8	3, 8	2, –	2, –	2, –	2, –	1, –	1, –
10	5, 8	4, 9	3, 9	3, –	3, –	2, –	2, –	2, –	2, –
11	5, 9	4, 10	4, 10	4, 10	3, –	3, –	3, –	2, –	2, –
12	6, 10	5, 10	4, 11	4, 11	4, –	3, –	3, –	3, –	2, –
13	6, 10	6, 11	5, 12	5, 12	4, 12	4, –	3, –	3, –	3, –
14	7, 11	6, 12	6, 12	5, 13	5, 13	4, 13	4, –	4, –	3, –
15	8, 12	7, 13	6, 13	6, 14	5, 14	5, 14	4, –	4, –	4, –
16	8, 12	7, 13	7, 14	6, 14	6, 15	5, 15	5, 15	5, –	4, –
17	9, 13	8, 14	7, 15	7, 15	6, 16	6, 16	6, 16	5, –	5, –
18	10, 14	8, 15	8, 15	7, 16	7, 16	6, 17	6, 17	6, 17	5, –
19	10, 15	9, 16	8, 16	8, 17	7, 17	7, 18	7, 18	6, 18	6, 18
20	11, 15	10, 16	9, 17	8, 17	8, 18	7, 18	7, 19	7, 19	6, 19
21	11, 16	10, 17	10, 18	9, 18	8, 19	8, 19	8, 20	7, 20	7, 20
22	12, 17	11, 18	10, 18	10, 19	9, 20	8, 20	8, 20	8, 21	7, 21
23	13, 17	12, 18	11, 19	10, 20	10, 20	9, 21	9, 21	8, 22	8, 22
24	13, 18	12, 19	11, 20	11, 20	10, 21	10, 22	9, 22	9, 22	8, 23
25	14, 19	13, 20	12, 21	11, 21	11, 22	10, 22	10, 23	9, 23	9, 23
26	15, 29	13, 21	13, 21	12, 22	11, 23	11, 23	10, 24	10, 24	9, 24
27	15, 20	14, 21	13, 22	13, 23	12, 23	11, 24	11, 24	10, 25	10, 25
28	16, 21	15, 22	14, 23	13, 23	12, 24	12, 25	11, 25	11, 26	10, 26
29	17, 21	15, 23	14, 24	14, 24	13, 25	12, 25	12, 26	11, 26	11, 27
30	17, 22	16, 24	15, 24	14, 25	13, 26	13, 26	12, 27	12, 27	11, 28
31	18, 23	16, 24	16, 25	15, 26	14, 26	13, 27	13, 27	12, 28	12, 28
32	18, 24	17, 25	16, 26	15, 26	15, 27	14, 28	14, 28	13, 29	12, 29
33	19, 24	18, 26	17, 26	16, 27	15, 28	15, 29	14, 29	13, 30	13, 30
34	20, 25	18, 26	17, 27	17, 28	16, 29	15, 29	15, 30	14, 20	14, 31
35	20, 26	19, 27	18, 28	17, 29	16, 29	16, 30	15, 31	15, 31	14, 32
36	21, 26	20, 28	19, 29	18, 29	17, 30	16, 31	16, 31	15, 32	15, 32
37	22, 27	20, 29	19, 29	18, 20	18, 31	17, 32	16, 32	16, 33	15, 33
38	22, 28	21, 29	20, 30	19, 31	18, 32	17, 32	17, 33	16, 33	16, 34
39	23, 28	21, 30	20, 31	20, 32	19, 32	18, 33	17, 34	17, 34	16, 35
40	24, 29	22, 31	21, 32	20, 32	19, 33	19, 34	18, 34	17, 35	17, 35
41	24, 30	23, 31	22, 32	21, 33	20, 34	19, 35	19, 34	18, 36	17, 36
42	25, 30	23, 32	22, 33	21, 34	20, 35	20, 35	19, 36	18, 37	18, 37
43	26, 31	24, 33	23, 34	22, 35	21, 35	20, 36	20, 37	19, 37	18, 38
44	26, 32	24, 33	23, 34	23, 35	22, 36	21, 37	20, 37	20, 38	19, 39
45	27, 33	25, 34	24, 35	23, 36	22, 37	22, 38	21, 38	20, 39	20, 39
46	27, 33	26, 35	25, 36	24, 37	23, 38	22, 38	21, 39	21, 40	20, 40
47	28, 34	26, 36	25, 37	24, 37	23, 38	23, 39	22, 40	21, 40	21, 41
48	29, 35	27, 36	26, 37	25, 38	24, 39	23, 40	23, 40	22, 41	21, 42
49	29, 35	28, 37	27, 38	26, 39	25, 40	24, 41	23, 41	22, 42	22, 42
50	30, 36	28, 38	27, 39	26, 40	25, 41	24, 41	24, 42	23, 43	22, 43

Appendix Table B.31 was prepared using the recursion algorithm attributed to André (1883), in order to ascertain the probability of a particular u, after which it was determined what value of u would yield a cumulative tail probability $\leq \alpha$. *Note:* The α represented by each u in the table is less than or equal to that in the column heading and greater than the next smaller column heading. For example, if $n = 9$ and $u = 3$

for a two-tailed test, $0.05 < P \le 0.10$; if $n = 15$ and $u = 7$ for a one-tailed test for uniformity, $0.05 < P \le 0.10$.

The normal approximation of Section 25.8 will correctly reject H_0 at the indicated one-tailed and two-tailed significance levels for sample sizes as small as these:

$\alpha(2)$:	0.50	0.20	0.10	0.05	0.02	0.01	0.005	0.002	0.001
$\alpha(1)$:	0.25	0.10	0.05	0.025	0.01	0.005	0.0025	0.001	0.0005
n:	5	4	9	11	13	15	19	20	20

TABLE B.32 Angular Deviation, s, as a Function of Vector Length, r

r	0	1	2	3	4	5	6	7	8	9
0.00	81.0285	80.9879	80.9474	80.9068	80.8662	80.8256	80.7850	80.7444	80.7037	80.6630
0.01	80.6223	80.5816	80.5408	80.5000	80.4593	80.4185	80.3776	80.3368	80.2959	80.2550
0.02	80.2141	80.1731	80.1322	80.0912	80.0502	80.0092	79.9682	79.9271	79.8860	79.8449
0.03	79.8038	79.7626	79.7215	79.6803	79.6391	79.5978	79.5566	79.5153	79.4740	79.4327
0.04	79.3914	79.3500	79.3086	79.2672	79.2258	79.1843	79.1429	79.1014	79.0599	79.0183
0.05	78.9768	78.9352	78.8936	78.8520	78.8103	78.7687	78.7270	78.6853	78.6435	78.6018
0.06	78.5600	78.5182	78.4764	78.4345	78.3927	78.3508	78.3089	78.2670	78.2250	78.1830
0.07	78.1410	78.0990	78.0569	78.0149	77.9728	77.9307	77.8885	77.8464	77.8042	77.7620
0.08	77.7198	77.6775	77.6353	77.5930	77.5506	77.5083	77.4659	77.4235	77.3811	77.3387
0.09	77.2962	77.2537	77.2112	77.1687	77.1262	77.0836	77.0410	76.9984	76.9557	76.9130
0.10	76.8703	76.8276	76.7849	76.7421	76.6993	76.6565	76.6137	76.5708	76.5279	76.4850
0.11	76.4421	76.3991	76.3562	76.3132	76.2701	76.2271	76.1840	76.1409	76.0978	76.0546
0.12	76.0114	75.9582	75.9250	75.8818	75.8385	75.7952	75.7519	75.7085	75.6651	75.6217
0.13	75.5783	75.5349	75.4914	75.4479	75.4044	75.3608	75.3172	75.2737	75.2300	75.1864
0.14	75.1427	75.0990	75.0553	75.0115	74.9678	74.9240	74.8801	74.8363	74.7924	74.7485
0.15	74.7045	74.6606	74.6166	74.5726	74.5286	74.4845	74.4404	74.3963	74.3522	74.3080
0.16	74.2638	74.2196	74.1754	74.1311	74.0868	74.0425	73.9981	73.9537	73.9093	73.8649
0.17	73.8204	73.7760	73.7314	73.6869	73.6423	73.5978	73.5531	73.5085	73.4638	73.4191
0.18	73.3744	73.3296	73.2849	73.2401	73.1952	73.1503	73.1055	73.0605	73.0156	72.9706
0.19	72.9256	72.8805	72.8355	72.7904	72.7453	72.7002	72.6550	72.6098	72.5646	72.5193
0.20	72.4741	72.4287	72.3834	72.3380	72.2926	72.2472	72.2018	72.1563	72.1108	72.0652
0.21	72.0197	71.9741	71.9285	71.8828	71.8371	71.7914	71.7457	71.6999	71.6541	71.6083
0.22	71.5624	71.5165	71.4706	71.4247	71.3787	71.3327	71.2866	71.2406	71.1945	71.1483
0.23	71.1022	71.0560	71.0098	70.9635	70.9173	70.8710	70.8246	70.7783	70.7319	70.6854
0.24	70.6390	70.5925	70.5460	70.4994	70.4528	70.4062	70.3596	70.3129	70.2662	70.2195
0.25	70.1727	70.1259	70.0791	70.0322	69.9853	69.9384	69.8914	69.8445	69.7975	69.7504
0.26	69.7033	69.6562	69.6091	69.5619	69.5147	69.4674	69.4202	69.3729	69.3255	69.2782
0.27	69.2307	69.1833	69.1358	69.0883	69.0408	68.9933	68.9456	68.8980	68.8504	68.8027
0.28	68.7549	68.7072	68.6594	68.6115	68.5637	68.5158	68.4678	68.4199	68.3719	68.3239
0.29	68.2758	68.2277	68.1796	68.1314	68.0832	68.0350	67.9867	67.9384	67.8901	67.8417
0.30	67.7933	67.7448	67.6964	67.6478	67.5993	67.5507	67.5021	67.4535	67.4048	67.3560
0.31	67.3073	67.2585	67.2097	67.1608	67.1119	67.0630	67.0140	66.9650	66.9160	66.8669
0.32	66.8178	66.7686	66.7195	66.6702	66.6210	66.5717	66.5223	66.4730	66.4236	66.3741
0.33	66.3246	66.2751	66.2256	66.1760	66.1264	66.0767	66.0270	65.9773	65.9275	65.8777
0.34	65.8278	65.7779	65.7280	65.6781	65.6281	65.5780	65.5279	65.4778	65.4277	65.3775
0.35	65.3272	65.2770	65.2267	65.1763	65.1259	65.0755	65.0250	64.9745	64.9240	64.8734
0.36	64.8228	64.7721	64.7214	64.6707	64.6199	64.5691	64.5182	64.4673	64.4164	64.3654
0.37	64.3143	64.2533	64.2122	64.1610	64.1098	64.0586	64.0074	63.9561	63.9047	63.8533
0.38	63.8019	63.7504	63.6989	63.6473	63.5957	63.5441	63.4924	63.4407	63.3889	63.3371
0.39	63.2852	63.2334	63.1814	63.1294	63.0774	63.0253	62.9732	62.9211	62.8689	62.8167
0.40	62.7644	62.7121	62.6597	62.6073	62.5548	62.5023	62.4498	62.3972	62.3445	62.2919
0.41	62.2391	62.1864	62.1336	62.0807	62.0278	61.9749	61.9219	61.8688	61.8158	61.7626
0.42	61.7094	61.6562	61.6030	61.5496	61.4963	61.4429	61.3894	61.3359	61.2824	61.2288
0.43	61.1751	61.1215	61.0677	61.0139	60.9601	60.9063	60.8523	60.7984	60.7443	60.6903
0.44	60.6361	60.5820	60.5278	60.4735	60.4192	60.3648	60.3104	60.2560	60.2015	60.1469
0.45	60.0923	60.0377	59.9830	59.9282	59.8734	59.8185	59.7636	59.7087	59.6537	59.5986
0.46	59.5435	59.4884	59.4332	59.3779	59.3226	59.2672	59.2118	59.1563	59.1008	59.0452
0.47	58.9896	58.9339	58.8782	58.8224	58.7666	58.7107	58.6548	58.5988	58.5427	58.4866
0.48	58.4305	58.3743	58.3180	58.2617	58.2053	58.1489	58.0924	58.0358	57.9792	57.9226
0.49	57.8659	57.8091	57.7523	57.6954	57.6385	57.5815	57.5245	57.4674	57.4102	57.3530
0.50	57.2958	57.2384	57.1811	57.1236	57.0661	57.0086	56.9510	56.8933	56.8356	56.7778
0.51	56.7199	56.6620	56.6040	56.5460	56.4879	56.4298	56.3716	56.3133	56.2550	56.1966
0.52	56.1382	56.0797	56.0211	55.9625	55.9038	55.8450	55.7862	55.7273	55.6684	55.6094
0.53	55.5503	55.4912	55.4320	55.3727	55.3134	55.2540	55.1946	55.1351	55.0755	55.0159
0.54	54.9562	54.8964	54.8366	54.7767	54.7167	54.6567	54.5966	54.5364	54.4762	54.4159

TABLE B.32 (cont.)   Angular Deviation, s, as a Function of Vector Length, r

r	0	1	2	3	4	5	6	7	8	9
0.55	54.3555	54.2951	54.2346	54.1741	54.1134	54.0527	53.9920	53.9311	53.8702	53.8092
0.56	53.7482	53.6871	53.6259	53.5647	53.5033	53.4419	53.3805	53.3189	53.2573	53.1957
0.57	53.1339	53.0721	53.0102	52.9482	52.8862	52.8241	52.7619	52.6997	52.6373	52.5749
0.58	52.5124	52.4499	52.3873	52.3246	52.2618	52.1989	52.1360	52.0730	52.0099	51.9468
0.59	51.8835	51.8202	51.7568	51.6934	51.6298	51.5662	51.5025	51.4387	51.3749	51.3109
0.60	51.2469	51.1828	51.1186	51.0544	50.9900	50.9256	50.8611	50.7965	50.7318	50.6671
0.61	50.6023	50.5373	50.4723	50.4073	50.3421	50.2768	50.2115	50.1461	50.0806	50.0150
0.62	49.9493	49.8835	49.8177	49.7517	49.6857	49.6196	49.5534	49.4871	49.4207	49.3542
0.63	49.2877	49.2210	49.1543	49.0875	49.0205	48.9535	48.8864	48.8192	48.7519	48.6846
0.64	48.6171	48.5495	48.4818	48.4141	48.3462	48.2783	48.2102	48.1421	48.0739	48.0055
0.65	47.9371	47.8685	47.7999	47.7312	47.6624	47.5934	47.5244	47.4553	47.3861	47.3167
0.66	47.2473	47.1778	47.1081	47.0384	46.9686	46.8986	46.8286	46.7584	46.6882	46.6178
0.67	46.5473	46.4767	46.4060	46.3353	46.2643	46.1933	46.1222	46.0510	45.9796	45.9082
0.68	45.8366	45.7649	45.6932	45.6213	45.5492	45.4771	45.4049	45.3325	45.2600	45.1875
0.69	45.1147	45.0419	44.9690	44.8959	44.8227	44.7494	44.6760	44.6025	44.5288	44.4550
0.70	44.3811	44.3071	44.2329	44.1586	44.0842	44.0097	43.9351	43.8603	43.7854	43.7103
0.71	43.6352	43.5599	43.4844	43.4089	43.3332	43.2574	43.1814	43.1053	43.0291	42.9527
0.72	42.8762	42.7996	42.7228	42.6459	42.5689	42.4917	42.4144	42.3369	42.2593	42.1815
0.73	42.1036	42.0256	41.9474	41.8691	41.7906	41.7120	41.6332	41.5543	41.4752	41.3960
0.74	41.3166	41.2370	41.1573	41.0775	40.9975	40.9174	40.8371	40.7566	40.6760	40.5952
0.75	40.5142	40.4331	40.3519	40.2704	40.1888	40.1070	40.0251	39.9430	39.8607	39.7783
0.76	39.6957	39.6129	39.5299	39.4468	39.3635	39.2800	39.1963	39.1125	39.0285	38.9443
0.77	38.8599	38.7753	38.6906	38.6056	38.5205	38.4352	38.3497	38.2640	38.1781	38.0920
0.78	38.0057	37.9192	37.8326	37.7457	37.6586	37.5714	37.4839	37.3962	37.3083	37.2202
0.79	37.1319	37.0434	36.9547	36.8657	36.7766	36.6872	36.5976	36.5078	36.4178	36.3275
0.80	36.2370	36.1463	36.0554	35.9642	35.8728	35.7812	35.6893	35.5972	35.5049	35.4123
0.81	35.3195	35.2264	35.1331	35.0396	34.9457	34.8517	34.7573	34.6628	34.5679	34.4728
0.82	34.3775	34.2818	34.1859	34.0898	33.9933	33.8966	33.7997	33.7024	33.6048	33.5070
0.83	33.4089	33.3105	33.2118	33.1128	33.0135	32.9139	32.8140	32.7138	32.6133	32.5125
0.84	32.4114	32.3099	32.2082	32.1061	32.0037	31.9009	31.7979	31.6945	31.5907	31.4866
0.85	31.3822	31.2774	31.1723	31.0668	30.9609	30.8547	30.7481	30.6412	30.5339	30.4262
0.86	30.3181	30.2096	30.1007	29.9915	29.8818	29.7718	29.6613	29.5504	29.4391	29.3274
0.87	29.2152	29.1026	28.9896	28.8762	28.7623	28.6479	28.5331	28.4178	28.3020	28.1858
0.88	28.0691	27.9519	27.8342	27.7160	27.5973	27.4781	27.3584	27.2381	27.1173	26.9960
0.89	26.8741	26.7517	26.6287	26.5051	26.3810	26.2562	26.1309	26.0050	25.8784	25.7513
0.90	25.6234	25.4950	25.3659	25.2362	25.1058	24.9746	24.8428	24.7104	24.5771	24.4432
0.91	24.3085	24.1731	24.0369	23.9000	23.7622	23.6237	23.4843	23.3441	23.2030	23.0611
0.92	22.9183	22.7746	22.6300	22.4845	22.3380	22.1906	22.0421	21.8927	21.7422	21.5907
0.93	21.4381	21.2844	21.1296	20.9737	20.8166	20.6583	20.4988	20.3380	20.1759	20.0126
0.94	19.8478	19.6817	19.5142	19.3453	19.1748	19.0029	18.8293	18.6542	18.4773	18.2988
0.95	18.1185	17.9364	17.7524	17.5666	17.3787	17.1887	16.9967	16.8024	16.6059	16.4070
0.96	16.2057	16.0018	15.7954	15.5861	15.3741	15.1590	14.9409	14.7196	14.4948	14.2666
0.97	14.0345	13.7987	13.5587	13.3144	13.0655	12.8117	12.5529	12.2886	12.0185	11.7422
0.98	11.4592	11.1690	10.8711	10.5648	10.2494	9.9239	9.5874	9.2387	8.8763	8.4984
0.99	8.1029	7.6871	7.2474	6.7794	6.2765	5.7296	5.1247	4.4382	3.6238	2.5625
1.00	0.0000									

Values of s are given in degrees, by Equation 26.20.

TABLE B.33   Circular Standard Deviation, s_0, as a Function of Vector Length, r

r	0	1	2	3	4	5	6	7	8	9
0.00	∞	212.9639	201.9968	195.2961	190.3990	186.5119	183.2748	180.4925	178.0473	175.8622
0.01	173.8843	172.0755	170.4075	168.8585	167.4115	166.0531	164.7723	163.5600	162.4087	161.3121
0.02	160.2649	159.2623	158.3005	157.3760	156.4857	155.6270	154.7974	153.9950	153.2178	152.4640
0.03	151.7323	151.0212	150.3295	149.6560	148.9998	148.3597	147.7351	147.1250	146.5287	145.9456
0.04	145.3750	144.8163	144.2690	143.7326	143.2066	142.6905	142.1839	141.6865	141.1979	140.7177
0.05	140.2456	139.7813	139.3245	138.8749	138.4324	137.9966	137.5672	137.1442	136.7273	136.3162
0.06	135.9109	135.5110	135.1165	134.7272	134.3430	133.9636	133.5889	133.2189	132.8533	132.4920
0.07	132.1350	131.7822	131.4333	131.0883	130.7472	130.4097	130.0759	129.7455	129.4186	129.0951
0.08	128.7748	128.4578	128.1438	127.8329	127.5250	127.2200	126.9178	126.6184	126.3218	126.0278
0.09	125.7364	125.4476	125.1612	124.8774	124.5959	124.3167	124.0399	123.7654	123.4930	123.2228
0.10	122.9548	122.6888	122.4249	122.1630	121.9031	121.6452	121.3891	121.1349	120.8825	120.6320
0.11	120.3832	120.1361	119.8908	119.6472	119.4052	119.1648	118.9261	118.6889	118.4533	118.2192
0.12	117.9866	117.7554	117.5258	117.2975	117.0707	116.8452	116.6211	116.3984	116.1770	115.9569
0.13	115.7381	115.5205	115.3042	115.0891	114.8753	114.6626	114.4511	114.2408	114.0316	113.8236
0.14	113.6166	113.4108	113.2060	113.0023	112.7997	112.5981	112.3976	112.1980	111.9995	111.8019
0.15	111.6054	111.4097	111.2151	111.0213	110.8285	110.6367	110.4457	110.2556	110.0664	109.8780
0.16	109.6906	109.5039	109.3182	109.1332	108.9491	108.7657	108.5832	108.4015	108.2205	108.0404
0.17	107.8609	107.6823	107.5044	107.3272	107.1508	106.9751	106.8000	106.6258	106.4522	106.2793
0.18	106.1070	105.9355	105.7646	105.5944	105.4248	105.2559	105.0877	104.9200	104.7530	104.5866
0.19	104.4209	104.2557	104.0911	103.9272	103.7638	103.6010	103.4388	103.2772	103.1161	102.9556
0.20	102.7957	102.6362	102.4774	102.3191	102.1613	102.0040	101.8473	101.6911	101.5354	101.3802
0.21	101.2255	101.0714	100.9177	100.7645	100.6118	100.4595	100.3078	100.1565	100.0057	99.8553
0.22	99.7054	99.5560	99.4070	99.2585	99.1104	98.9628	98.8155	98.6688	98.5224	98.3765
0.23	98.2310	98.0859	97.9412	97.7969	97.6531	97.5096	97.3665	97.2239	97.0816	96.9397
0.24	96.7982	96.6571	96.5164	96.3760	96.2360	96.0964	95.9571	95.8182	95.6797	95.5415
0.25	95.4037	95.2663	95.1292	94.9924	94.8560	94.7199	94.5841	94.4487	94.3136	94.1789
0.26	94.0445	93.9104	93.7766	93.6432	93.5100	93.3772	93.2447	93.1125	92.9806	92.8490
0.27	92.7177	92.5867	92.4560	92.3257	92.1956	92.0657	91.9362	91.8070	91.6781	91.5494
0.28	91.4210	91.2929	91.1651	91.0375	90.9102	90.7832	90.6565	90.5300	90.4038	90.2778
0.29	90.1521	90.0267	89.9015	89.7766	89.6519	89.5275	89.4033	89.2794	89.1557	89.0322
0.30	88.9090	88.7861	88.6634	88.5409	88.4186	88.2966	88.1748	88.0533	87.9320	87.8109
0.31	87.6900	87.5693	87.4489	87.3287	87.2087	87.0889	86.9694	86.8500	86.7309	86.6120
0.32	86.4933	86.3748	86.2565	86.1384	86.0205	85.9028	85.7853	85.6680	85.5509	85.4340
0.33	85.3173	85.2008	85.0845	84.9684	84.8525	84.7367	84.6212	84.5058	84.3906	84.2757
0.34	84.1608	84.0462	83.9317	83.8175	83.7034	83.5895	83.4757	83.3621	83.2487	83.1355
0.35	83.0224	82.9095	82.7968	82.6843	82.5719	82.4597	82.3476	82.2357	82.1240	82.0124
0.36	81.9010	81.7897	81.6786	81.5676	81.4568	81.3462	81.2357	81.1254	81.0152	80.9052
0.37	80.7953	80.6855	80.5759	80.4665	80.3572	80.2480	80.1390	80.0301	79.9214	79.8128
0.38	79.7043	79.5960	79.4878	79.3798	79.2719	79.1641	79.0565	78.9490	78.8416	78.7343
0.39	78.6272	78.5202	78.4133	78.3066	78.2000	78.0935	77.9872	77.8809	77.7748	77.6688
0.40	77.5629	77.4572	77.3516	77.2460	77.1407	77.0354	76.9302	76.8252	76.7202	76.6154
0.41	76.5107	76.4061	76.3016	76.1973	76.0930	75.9888	75.8848	75.7809	75.6770	75.5733
0.42	75.4697	75.3662	75.2628	75.1594	75.0562	74.9531	74.8501	74.7472	74.6444	74.5417
0.43	74.4391	74.3366	74.2342	74.1319	74.0296	73.9275	73.8255	73.7235	73.6217	73.5199
0.44	73.4183	73.3167	73.2152	73.1138	73.0125	72.9113	72.8101	72.7091	72.6081	72.5072
0.45	72.4064	72.3057	72.2051	72.1046	72.0041	71.9037	71.8034	71.7032	71.6030	71.5030
0.46	71.4030	71.3031	71.2033	71.1035	71.0038	70.9042	70.8047	70.7052	70.6058	70.5065
0.47	70.4073	70.3081	70.2090	70.1100	70.0110	69.9121	69.8133	69.7146	69.6159	69.5173
0.48	69.4187	69.3202	69.2218	69.1234	69.0251	68.9269	68.8287	68.7306	68.6326	68.5346
0.49	68.4367	68.3388	68.2410	68.1433	68.0456	67.9479	67.8504	67.7528	67.6554	67.5580
0.50	67.4606	67.3633	67.2661	67.1689	67.0718	66.9747	66.8776	66.7806	66.6837	66.5868
0.51	66.4900	66.3932	66.2965	66.1998	66.1031	66.0065	65.9100	65.8135	65.7170	65.6206
0.52	65.5243	65.4279	65.3316	65.2354	65.1392	65.0431	64.9469	64.8509	64.7548	64.6588
0.53	64.5629	64.4670	64.3711	64.2752	64.1794	64.0837	63.9879	63.8922	63.7966	63.7009
0.54	63.6053	63.5098	63.4143	63.3188	63.2233	63.1279	63.0325	62.9371	62.8417	62.7464

TABLE B.33 (cont.)   Circular Standard Deviation, s_0, as a Function of Vector Length, r

r	0	1	2	3	4	5	6	7	8	9
0.55	62.6511	62.5559	62.4607	62.3655	62.2703	62.1751	62.0800	61.9849	61.8899	61.7948
0.56	61.6998	61.6048	61.5098	61.4149	61.3199	61.2250	61.1301	61.0353	60.9404	60.8456
0.57	60.7508	60.6560	60.5612	60.4664	60.3717	60.2770	60.1823	60.0876	59.9929	59.8982
0.58	59.8036	59.7089	59.6143	59.5197	59.4251	59.3305	59.2359	59.1414	59.0468	58.9523
0.59	58.8577	58.7632	58.6687	58.5742	58.4797	58.3852	58.2907	58.1962	58.1017	58.0072
0.60	57.9127	57.8182	57.7238	57.6293	57.5348	57.4404	57.3459	57.2514	57.1570	57.0625
0.61	56.9680	56.8736	56.7791	56.6846	56.5902	56.4957	56.4012	56.3067	56.2122	56.1177
0.62	56.0232	55.9287	55.8342	55.7396	55.6451	55.5505	55.4560	55.3614	55.2668	55.1722
0.63	55.0776	54.9830	54.8884	54.7938	54.6991	54.6044	54.5097	54.4150	54.3203	54.2256
0.64	54.1308	54.0361	53.9413	53.8465	53.7517	53.6568	53.5619	53.4671	53.3722	53.2772
0.65	53.1823	53.0873	52.9923	52.8973	52.8022	52.7071	52.6120	52.5169	52.4217	52.3266
0.66	52.2313	52.1361	52.0408	51.9455	51.8502	51.7548	51.6594	51.5640	51.4685	51.3730
0.67	51.2775	51.1819	51.0863	50.9907	50.8950	50.7993	50.7035	50.6077	50.5119	50.4160
0.68	50.3201	50.2241	50.1281	50.0321	49.9360	49.8398	49.7437	49.6474	49.5512	49.4549
0.69	49.3585	49.2621	49.1656	49.0691	48.9725	48.8759	48.7792	48.6825	48.5858	48.4889
0.70	48.3920	48.2951	48.1981	48.1011	48.0039	47.9068	47.8095	47.7123	47.6149	47.5175
0.71	47.4200	47.3225	47.2249	47.1272	47.0295	46.9317	46.8338	46.7359	46.6379	46.5398
0.72	46.4417	46.3435	46.2452	46.1468	46.0484	45.9499	45.8513	45.7527	45.6539	45.5551
0.73	45.4562	45.3573	45.2582	45.1591	45.0599	44.9606	44.8612	44.7617	44.6622	44.5625
0.74	44.4628	44.3630	44.2631	44.1631	44.0630	43.9628	43.8625	43.7621	43.6617	43.5611
0.75	43.4604	43.3597	43.2588	43.1578	43.0568	42.9556	42.8543	42.7529	42.6515	42.5499
0.76	42.4482	42.3463	42.2444	42.1424	42.0402	41.9380	41.8356	41.7331	41.6305	41.5277
0.77	41.4249	41.3219	41.2188	41.1156	41.0122	40.9087	40.8051	40.7014	40.5975	40.4935
0.78	40.3894	40.2851	40.1807	40.0761	39.9715	39.8666	39.7617	39.6565	39.5513	39.4459
0.79	39.3403	39.2346	39.1288	39.0228	38.9166	38.8103	38.7038	38.5972	38.4904	38.3834
0.80	38.2763	38.1690	38.0615	37.9539	37.8461	37.7381	37.6300	37.5217	37.4132	37.3045
0.81	37.1956	37.0865	36.9773	36.8679	36.7583	36.6484	36.5384	36.4282	36.3178	36.2072
0.82	36.0964	35.9854	35.8742	35.7628	35.6511	35.5393	35.4272	35.3149	35.2024	35.0896
0.83	34.9767	34.8635	34.7501	34.6364	34.5225	34.4084	34.2940	34.1793	34.0645	33.9493
0.84	33.8340	33.7183	33.6024	33.4863	33.3698	33.2531	33.1362	33.0189	32.9014	32.7836
0.85	32.6655	32.5471	32.4285	32.3095	32.1902	32.0707	31.9508	31.8306	31.7101	31.5893
0.86	31.4682	31.3467	31.2249	31.1028	30.9803	30.8575	30.7343	30.6108	30.4869	30.3627
0.87	30.2381	30.1131	29.9877	29.8620	29.7359	29.6094	29.4825	29.3552	29.2274	29.0993
0.88	28.9708	28.8418	28.7124	28.5825	28.4522	28.3215	28.1903	28.0586	27.9265	27.7938
0.89	27.6607	27.5271	27.3930	27.2584	27.1233	26.9877	26.8515	26.7148	26.5775	26.4397
0.90	26.3013	26.1623	26.0227	25.8826	25.7418	25.6004	25.4584	25.3158	25.1725	25.0285
0.91	24.8839	24.7386	24.5925	24.4458	24.2984	24.1502	24.0012	23.8515	23.7011	23.5498
0.92	23.3977	23.2448	23.0910	22.9364	22.7808	22.6244	22.4671	22.3088	22.1496	21.9894
0.93	21.8282	21.6660	21.5027	21.3384	21.1729	21.0062	20.8386	20.6697	20.4996	20.3283
0.94	20.1556	19.9817	19.8064	19.6298	19.4517	19.2722	19.0912	18.9087	18.7246	18.5388
0.95	18.3513	18.1622	17.9712	17.7784	17.5837	17.3870	17.1882	16.9874	16.7843	16.5790
0.96	16.3714	16.1612	15.9486	15.7333	15.5152	15.2943	15.0703	14.8432	14.6128	14.3790
0.97	14.1415	13.9003	13.6550	13.4055	13.1516	12.8929	12.6292	12.3601	12.0854	11.8045
0.98	11.5171	11.2226	10.9205	10.6101	10.2907	9.9614	9.6212	9.2689	8.9030	8.5218
0.99	8.1232	7.7044	7.2620	6.7912	6.2859	5.7368	5.1298	4.4414	3.6255	2.5630
1.00	0.0000									

Values of s_o are given in degrees, by Equation 26.21.

TABLE B.34   Critical Values of Rayleigh's z

n	α: 0.50	0.20	0.10	0.05	0.02	0.01	0.005	0.002	0.001
6	0.734	1.639	2.274	2.865	3.576	4.058	4.491	4.985	5.297
7	0.727	1.634	2.278	2.885	3.627	4.143	4.617	5.181	5.556
8	0.723	1.631	2.281	2.899	3.665	4.205	4.710	5.322	5.743
9	0.719	1.628	2.283	2.910	3.694	4.252	4.780	5.430	5.885
10	0.717	1.626	2.285	2.919	3.716	4.289	4.835	5.514	5.996
11	0.715	1.625	2.287	2.926	3.735	4.319	4.879	5.582	6.085
12	0.713	1.623	2.288	2.932	3.750	4.344	4.916	5.638	6.158
13	0.711	1.622	2.289	2.937	3.763	4.365	4.947	5.685	6.219
14	0.710	1.621	2.290	2.941	3.774	4.383	4.973	5.725	6.271
15	0.709	1.620	2.291	2.945	3.784	4.398	4.996	5.759	6.316
16	0.708	1.620	2.292	2.948	3.792	4.412	5.015	5.789	6.354
17	0.707	1.619	2.292	2.951	3.799	4.423	5.033	5.815	6.388
18	0.706	1.619	2.293	2.954	3.806	4.434	5.048	5.838	6.418
19	0.705	1.618	2.293	2.956	3.811	4.443	5.061	5.858	6.445
20	0.705	1.618	2.294	2.958	3.816	4.451	5.074	5.877	6.469
21	0.704	1.617	2.294	2.960	3.821	4.459	5.085	5.893	6.491
22	0.704	1.617	2.295	2.961	3.825	4.466	5.095	5.908	6.510
23	0.703	1.616	2.295	2.963	3.829	4.472	5.104	5.922	6.528
24	0.703	1.616	2.295	2.964	3.833	4.478	5.112	5.935	6.544
25	0.702	1.616	2.296	2.966	3.836	4.483	5.120	5.946	6.559
26	0.702	1.616	2.296	2.967	3.839	4.488	5.127	5.957	6.573
27	0.702	1.615	2.296	2.968	3.842	4.492	5.133	5.966	6.586
28	0.701	1.615	2.296	2.969	3.844	4.496	5.139	5.975	6.598
29	0.701	1.615	2.297	2.970	3.847	4.500	5.145	5.984	6.609
30	0.701	1.615	2.297	2.971	3.849	4.504	5.150	5.992	6.619
32	0.700	1.614	2.297	2.972	3.853	4.510	5.159	6.006	6.637
34	0.700	1.614	2.297	2.974	3.856	4.516	5.168	6.018	6.654
36	0.700	1.614	2.298	2.975	3.859	4.521	5.175	6.030	6.668
38	0.699	1.614	2.298	2.976	3.862	4.525	5.182	6.039	6.681
40	0.699	1.613	2.298	2.977	3.865	4.529	5.188	6.048	6.692
42	0.699	1.613	2.298	2.978	3.867	4.533	5.193	6.056	6.703
44	0.698	1.613	2.299	2.979	3.869	4.536	5.198	6.064	6.712
46	0.698	1.613	2.299	2.979	3.871	4.539	5.202	6.070	6.721
48	0.698	1.613	2.299	2.980	3.873	4.542	5.206	6.076	6.729
50	0.698	1.613	2.299	2.981	3.874	4.545	5.210	6.082	6.736
55	0.697	1.612	2.299	2.982	3.878	4.550	5.218	6.094	6.752
60	0.697	1.612	2.300	2.983	3.881	4.555	5.225	6.104	6.765
65	0.697	1.612	2.300	2.984	3.883	4.559	5.231	6.113	6.776
70	0.696	1.612	2.300	2.985	3.885	4.562	5.235	6.120	6.786
75	0.696	1.612	2.300	2.986	3.887	4.565	5.240	6.127	6.794
80	0.696	1.611	2.300	2.986	3.889	4.567	5.243	6.132	6.801
90	0.696	1.611	2.301	2.987	3.891	4.572	5.249	6.141	6.813
100	0.695	1.611	2.301	2.988	3.893	4.575	5.254	6.149	6.822
120	0.695	1.611	2.301	2.990	3.896	4.580	5.262	6.160	6.837
140	0.695	1.611	2.301	2.990	3.899	4.584	5.267	6.168	6.847
160	0.695	1.610	2.301	2.991	3.900	4.586	5.271	6.174	6.855
180	0.694	1.610	2.302	2.992	3.902	4.588	5.274	6.178	6.861
200	0.694	1.610	2.302	2.992	3.903	4.590	5.276	6.182	6.865
300	0.694	1.610	2.302	2.993	3.906	4.595	5.284	6.193	6.879
500	0.694	1.610	2.302	2.994	3.908	4.599	5.290	6.201	6.891
∞	0.6931	1.6094	2.3026	2.9957	3.9120	4.6052	5.2983	6.2146	6.9078

      The preceding values were computed using Durand and Greenwood's Equation 6 (1958). This procedure was found to give slightly more accurate results than Durand and Greenwood's Equation 4, distinctly better results than the Pearson curve approximation

(Stephens, 1969), and much better results than the chi-square approximation (Stephens, 1969). By examining the exact critical values of Greenwood and Durand (1955), we see that the preceding tabled values for $\alpha = 0.05$ are accurate to the third decimal place for n as small as 8, and for $\alpha = 0.01$ for n as small as 10. For n as small as 6, none of the tabled values for $\alpha = 0.05$ or 0.01 has a relative error greater than 0.3%.

Examples:

$$z_{0.05,\,80} = 2.986 \quad \text{and} \quad z_{0.01,\,32} = 4.510.$$

For n approaching ∞, the critical values are $\chi^2_{\alpha,\,2}/2$.

TABLE B.35   Critical Values of *u* for the *V* Test of Circular Uniformity

n	α: 0.25	0.10	0.05	0.025	0.01	0.005	0.0025	0.001	0.0005
8	0.688	1.296	1.649	1.947	2.280	2.498	2.691	2.916	3.066
9	0.687	1.294	1.649	1.948	2.286	2.507	2.705	2.937	3.094
10	0.685	1.293	1.648	1.950	2.290	2.514	2.716	2.954	3.115
11	0.684	1.292	1.648	1.950	2.293	2.520	2.725	2.967	3.133
12	0.684	1.291	1.648	1.951	2.296	2.525	2.732	2.978	3.147
13	0.683	1.290	1.647	1.952	2.299	2.529	2.738	2.987	3.159
14	0.682	1.290	1.647	1.953	2.301	2.532	2.743	2.995	3.169
15	0.682	1.289	1.647	1.953	2.302	2.535	2.748	3.002	3.177
16	0.681	1.289	1.647	1.953	2.304	2.538	2.751	3.008	3.185
17	0.681	1.288	1.647	1.954	2.305	2.540	2.755	3.013	3.191
18	0.681	1.288	1.647	1.954	2.306	2.542	2.758	3.017	3.197
19	0.680	1.287	1.647	1.954	2.308	2.544	2.761	3.021	3.202
20	0.680	1.287	1.646	1.955	2.308	2.546	2.763	3.025	3.207
21	0.680	1.287	1.646	1.955	2.309	2.547	2.765	3.028	3.211
22	0.679	1.287	1.646	1.955	2.310	2.549	2.767	3.031	3.215
23	0.679	1.286	1.646	1.955	2.311	2.550	2.769	3.034	3.218
24	0.679	1.286	1.646	1.956	2.311	2.551	2.770	3.036	3.221
25	0.679	1.286	1.646	1.956	2.312	2.552	2.772	3.038	3.224
26	0.679	1.286	1.646	1.956	2.313	2.553	2.773	3.040	3.227
27	0.678	1.286	1.646	1.956	2.313	2.554	2.775	3.042	3.229
28	0.678	1.285	1.646	1.956	2.314	2.555	2.776	3.044	3.231
29	0.678	1.285	1.646	1.956	2.314	2.555	2.777	3.046	3.233
30	0.678	1.285	1.646	1.957	2.315	2.556	2.778	3.047	3.235
32	0.678	1.285	1.646	1.957	2.315	2.557	2.780	3.050	3.239
34	0.678	1.285	1.646	1.957	2.316	2.558	2.781	3.052	3.242
36	0.677	1.285	1.646	1.957	2.316	2.559	2.783	3.054	3.245
38	0.677	1.284	1.646	1.957	2.317	2.560	2.784	3.056	3.247
40	0.677	1.284	1.646	1.957	2.317	2.561	2.785	3.058	3.249
42	0.677	1.284	1.646	1.958	2.318	2.562	2.786	3.060	3.251
44	0.677	1.284	1.646	1.958	2.318	2.562	2.787	3.061	3.253
46	0.677	1.284	1.646	1.958	2.319	2.563	2.788	3.062	3.255
48	0.677	1.284	1.645	1.958	2.319	2.564	2.789	3.063	3.256
50	0.677	1.284	1.645	1.958	2.319	2.564	2.790	3.065	3.258
55	0.676	1.284	1.645	1.958	2.320	2.565	2.791	3.067	3.261
60	0.676	1.283	1.645	1.958	2.320	2.566	2.793	3.069	3.263
65	0.676	1.283	1.645	1.958	2.321	2.567	2.794	3.071	3.265
70	0.676	1.283	1.645	1.958	2.321	2.567	2.795	3.072	3.267
75	0.676	1.283	1.645	1.959	2.322	2.568	2.796	3.073	3.269
80	0.676	1.283	1.645	1.959	2.322	2.568	2.796	3.074	3.270
90	0.676	1.283	1.645	1.959	2.322	2.569	2.797	3.076	3.272
100	0.676	1.283	1.645	1.959	2.323	2.570	2.798	3.077	3.274
120	0.675	1.282	1.645	1.959	2.323	2.571	2.800	3.080	3.277
140	0.675	1.282	1.645	1.959	2.324	2.572	2.801	3.081	3.279
160	0.675	1.282	1.645	1.959	2.324	2.572	2.802	3.082	3.280
180	0.675	1.282	1.645	1.959	2.324	2.573	2.802	3.083	3.282
200	0.675	1.282	1.645	1.959	2.325	2.573	2.803	3.084	3.282
300	0.675	1.282	1.645	1.960	2.325	2.574	2.804	3.086	3.285
∞	0.6747	1.2818	1.6449	1.9598	2.3256	2.5747	2.8053	3.0877	3.2873

The preceding values were computed using Durand and Greenwood's Equation 7 (1958).

Examples:

$$u_{0.05, 25} = 1.646 \quad \text{and} \quad u_{0.01, 20} = 2.308.$$

TABLE B.36   Critical Values of *m* for the Hodges-Ajne Test

n	α ≤ 0.50	0.20	0.10	0.05	0.02	0.01	0.005	0.002	0.001
4	0								
5	0								
6	0	0							
7	0	0							
8	1	0	0						
9	1	0	0	0					
10	1	1	0	0	0				
11	2	1	1	0	0				
12	2	1	1	0	0	0			
13	3	2	1	1	0	0	0		
14	3	2	1	1	0	0	0	0	
15	3	2	2	1	1	0	0	0	0
16	4	3	2	2	1	1	0	0	0
17	4	3	2	2	1	1	1	0	0
18	4	3	3	2	2	1	1	0	0
19	5	4	3	3	2	2	1	1	0
20	5	4	3	3	2	2	1	1	1
21	6	4	4	3	3	2	2	1	1
22	6	5	4	4	3	2	2	2	1
23	6	5	4	4	3	3	2	2	1
24	7	6	5	4	3	3	3	2	2
25	7	6	5	5	4	3	3	2	2
26	8	6	6	5	4	4	3	3	2
27	8	7	6	5	4	4	4	3	3
28	8	7	6	6	5	4	4	3	3
29	9	7	7	6	5	5	4	4	3
30	9	8	7	6	6	5	4	4	3
31	10	8	7	7	6	5	5	4	4
32	10	9	8	7	6	6	5	4	4
33	11	9	8	7	7	6	5	5	4
34	11	9	9	8	7	6	6	5	5
35	11	10	9	8	7	7	6	5	5
36	12	10	9	9	8	7	6	6	5
37	12	11	10	9	8	7	7	6	6
38	13	11	10	9	8	8	7	6	6
39	13	11	10	10	9	8	7	7	6
40	13	12	11	10	9	8	8	7	7
41	14	12	11	10	9	9	8	7	7
42	14	13	12	11	10	9	9	8	7
43	15	13	12	11	10	9	9	8	8
44	15	13	12	12	11	10	9	8	8
45	16	14	13	12	11	10	10	9	8
46	16	14	13	12	11	11	10	9	9
47	16	15	14	13	12	11	10	10	9
48	17	15	14	13	12	11	11	10	9
49	17	15	14	13	12	12	11	10	10
50	18	16	15	14	13	12	11	11	10

The critical values in Appendix Table B.36 are the values of *m* for which $P \leq \alpha$, where P is calculated by Equation 27.7, which is from Hodges (1955). To determine P for $n > 50$, we may use the approximation shown as Equation 27.8, which is from

Ajne (1968). The accuracy of their approximation is shown below, as true probability
− approximate probability for the largest true probability $\leq \alpha$:

n	α: 0.50	0.20	0.10	0.05	0.02	0.01	0.005	0.002	0.001
30	0.0026	−0.0029	−0.0047	−0.0038	−0.0038	−0.0021	−0.00089	−0.00089	−0.00029
40	0.0016	−0.0020	−0.0034	−0.0032	−0.0021	−0.0030	−0.00095	−0.00052	−0.00052
50	0.0039	−0.0017	−0.0027	−0.0026	−0.0019	−0.0012	−0.00063	−0.00063	−0.00029

TABLE B.37 Correction Factor, K, for the Watson and Williams Test

r	0	1	2	3	4	5	6	7	8	9
0.00		188.4989	94.7472	63.5015	47.8749	38.4992	32.2498	27.7851	24.4367	21.8325
0.01	19.7489	18.0444	16.6239	15.4219	14.3916	13.4986	12.7173	12.0278	11.4150	10.8667
0.02	10.3731	9.9266	9.5206	9.1500	8.8103	8.4976	8.2091	7.9419	7.6938	7.4628
0.03	7.2472	7.0455	6.8564	6.6787	6.5115	6.3539	6.2050	6.0641	5.9306	5.8040
0.04	5.6837	5.5693	5.4603	5.3564	5.2572	5.1625	5.0718	4.9850	4.9017	4.8219
0.05	4.7453	4.6717	4.6009	4.5328	4.4672	4.4039	4.3430	4.2841	4.2273	4.1724
0.06	4.1194	4.0680	4.0184	3.9703	3.9237	3.8785	3.8347	3.7922	3.7510	3.7109
0.07	3.6720	3.6342	3.5974	3.5616	3.5268	3.4930	3.4600	3.4278	3.3965	3.3660
0.08	3.3362	3.3072	3.2789	3.2512	3.2243	3.1979	3.1722	3.1470	3.1224	3.0984
0.09	3.0749	3.0519	3.0294	3.0074	2.9858	2.9648	2.9441	2.9239	2.9041	2.8846
0.10	2.8656	2.8469	2.8286	2.8107	2.7931	2.7758	2.7589	2.7423	2.7259	2.7099
0.11	2.6942	2.6787	2.6636	2.6487	2.6340	2.6196	2.6055	2.5915	2.5779	2.5644
0.12	2.5512	2.5382	2.5254	2.5128	2.5004	2.4882	2.4762	2.4644	2.4528	2.4413
0.13	2.4301	2.4189	2.4080	2.3972	2.3866	2.3762	2.3658	2.3557	2.3457	2.3358
0.14	2.3261	2.3165	2.3070	2.2977	2.2885	2.2794	2.2705	2.2616	2.2529	2.2443
0.15	2.2358	2.2275	2.2192	2.2110	2.2030	2.1950	2.1872	2.1794	2.1718	2.1642
0.16	2.1567	2.1494	2.1421	2.1349	2.1278	2.1208	2.1138	2.1070	2.1002	2.0935
0.17	2.0868	2.0803	2.0738	2.0674	2.0611	2.0549	2.0487	2.0426	2.0365	2.0305
0.18	2.0246	2.0188	2.0130	2.0072	2.0016	1.9960	1.9904	1.9849	1.9795	1.9741
0.19	1.9688	1.9635	1.9583	1.9532	1.9481	1.9430	1.9380	1.9331	1.9282	1.9233
0.20	1.9185	1.9137	1.9090	1.9043	1.8997	1.8951	1.8906	1.8861	1.8817	1.8772
0.21	1.8729	1.8685	1.8643	1.8600	1.8558	1.8516	1.8475	1.8434	1.8393	1.8353
0.22	1.8313	1.8274	1.8234	1.8195	1.8157	1.8119	1.8081	1.8043	1.8006	1.7969
0.23	1.7933	1.7896	1.7860	1.7825	1.7789	1.7754	1.7719	1.7685	1.7651	1.7617
0.24	1.7583	1.7550	1.7516	1.7484	1.7451	1.7419	1.7386	1.7355	1.7323	1.7292
0.25	1.7261	1.7230	1.7199	1.7169	1.7138	1.7108	1.7079	1.7049	1.7020	1.6991
0.26	1.6962	1.6933	1.6905	1.6877	1.6849	1.6821	1.6793	1.6766	1.6739	1.6712
0.27	1.6685	1.6658	1.6632	1.6606	1.6579	1.6554	1.6528	1.6502	1.6477	1.6452
0.28	1.6427	1.6402	1.6377	1.6353	1.6328	1.6304	1.6280	1.6256	1.6233	1.6209
0.29	1.6186	1.6162	1.6139	1.6116	1.6094	1.6071	1.6048	1.6026	1.6004	1.5982
0.30	1.5960	1.5938	1.5916	1.5895	1.5873	1.5852	1.5831	1.5810	1.5789	1.5768
0.31	1.5748	1.5727	1.5707	1.5687	1.5667	1.5647	1.5627	1.5607	1.5587	1.5568
0.32	1.5548	1.5529	1.5510	1.5491	1.5472	1.5453	1.5434	1.5416	1.5397	1.5379
0.33	1.5360	1.5342	1.5324	1.5306	1.5288	1.5270	1.5253	1.5235	1.5217	1.5200
0.34	1.5183	1.5165	1.5148	1.5131	1.5114	1.5097	1.5081	1.5064	1.5047	1.5031
0.35	1.5014	1.4998	1.4982	1.4966	1.4950	1.4934	1.4918	1.4902	1.4886	1.4871
0.36	1.4855	1.4839	1.4824	1.4809	1.4793	1.4778	1.4763	1.4748	1.4733	1.4718
0.37	1.4703	1.4689	1.4674	1.4659	1.4645	1.4630	1.4616	1.4602	1.4587	1.4573
0.38	1.4559	1.4545	1.4531	1.4517	1.4503	1.4490	1.4476	1.4462	1.4449	1.4435
0.39	1.4422	1.4408	1.4395	1.4382	1.4368	1.4355	1.4342	1.4329	1.4316	1.4303
0.40	1.4290	1.4277	1.4265	1.4252	1.4239	1.4227	1.4214	1.4202	1.4189	1.4177
0.41	1.4165	1.4152	1.4140	1.4128	1.4116	1.4104	1.4092	1.4080	1.4068	1.4056
0.42	1.4044	1.4033	1.4021	1.4009	1.3998	1.3986	1.3975	1.3963	1.3952	1.3940
0.43	1.3929	1.3918	1.3907	1.3895	1.3884	1.3873	1.3862	1.3851	1.3840	1.3829
0.44	1.3818	1.3808	1.3797	1.3786	1.3775	1.3765	1.3754	1.3744	1.3733	1.3723
0.45	1.3712	1.3702	1.3691	1.3681	1.3671	1.3660	1.3650	1.3640	1.3630	1.3620
0.46	1.3610	1.3600	1.3590	1.3580	1.3570	1.3560	1.3550	1.3540	1.3530	1.3521
0.47	1.3511	1.3501	1.3492	1.3482	1.3472	1.3463	1.3453	1.3444	1.3434	1.3425
0.48	1.3416	1.3406	1.3397	1.3388	1.3378	1.3369	1.3360	1.3351	1.3342	1.3333
0.49	1.3324	1.3315	1.3306	1.3297	1.3288	1.3279	1.3270	1.3261	1.3252	1.3243
0.50	1.3235	1.3226	1.3217	1.3209	1.3200	1.3191	1.3183	1.3174	1.3166	1.3157
0.51	1.3148	1.3140	1.3132	1.3123	1.3115	1.3106	1.3098	1.3090	1.3081	1.3073
0.52	1.3065	1.3057	1.3049	1.3040	1.3032	1.3024	1.3016	1.3008	1.3000	1.2992
0.53	1.2984	1.2976	1.2968	1.2960	1.2952	1.2944	1.2936	1.2929	1.2921	1.2913
0.54	1.2905	1.2897	1.2890	1.2882	1.2874	1.2867	1.2859	1.2851	1.2844	1.2836

TABLE B.37 (cont.)   Correction Factor, K, for the Watson and Williams Test

r	0	1	2	3	4	5	6	7	8	9
0.55	1.2829	1.2821	1.2814	1.2806	1.2799	1.2791	1.2784	1.2776	1.2769	1.2762
0.56	1.2754	1.2747	1.2740	1.2732	1.2725	1.2718	1.2710	1.2703	1.2696	1.2689
0.57	1.2682	1.2674	1.2667	1.2660	1.2653	1.2646	1.2639	1.2632	1.2625	1.2618
0.58	1.2611	1.2604	1.2597	1.2590	1.2583	1.2576	1.2569	1.2562	1.2555	1.2548
0.59	1.2542	1.2535	1.2528	1.2521	1.2514	1.2508	1.2501	1.2494	1.2487	1.2481
0.60	1.2474	1.2467	1.2461	1.2454	1.2447	1.2441	1.2434	1.2428	1.2421	1.2414
0.61	1.2408	1.2401	1.2395	1.2388	1.2382	1.2375	1.2369	1.2362	1.2356	1.2350
0.62	1.2343	1.2337	1.2330	1.2324	1.2318	1.2311	1.2305	1.2298	1.2292	1.2286
0.63	1.2280	1.2273	1.2267	1.2261	1.2254	1.2248	1.2242	1.2236	1.2230	1.2223
0.64	1.2217	1.2211	1.2205	1.2199	1.2193	1.2186	1.2180	1.2174	1.2168	1.2162
0.65	1.2156	1.2150	1.2144	1.2138	1.2132	1.2126	1.2120	1.2114	1.2108	1.2102
0.66	1.2096	1.2090	1.2084	1.2078	1.2072	1.2066	1.2060	1.2054	1.2048	1.2042
0.67	1.2036	1.2030	1.2024	1.2018	1.2013	1.2007	1.2001	1.1995	1.1989	1.1983
0.68	1.1977	1.1972	1.1966	1.1960	1.1954	1.1948	1.1943	1.1937	1.1931	1.1925
0.69	1.1920	1.1914	1.1908	1.1902	1.1897	1.1891	1.1885	1.1879	1.1874	1.1868
0.70	1.1862	1.1857	1.1851	1.1845	1.1840	1.1834	1.1828	1.1823	1.1817	1.1811
0.71	1.1806	1.1800	1.1794	1.1789	1.1783	1.1777	1.1772	1.1766	1.1761	1.1755
0.72	1.1749	1.1744	1.1738	1.1733	1.1727	1.1721	1.1716	1.1710	1.1705	1.1699
0.73	1.1694	1.1688	1.1682	1.1677	1.1671	1.1666	1.1660	1.1655	1.1649	1.1644
0.74	1.1638	1.1633	1.1627	1.1621	1.1616	1.1610	1.1605	1.1599	1.1594	1.1588
0.75	1.1583	1.1577	1.1572	1.1566	1.1561	1.1555	1.1550	1.1544	1.1539	1.1533
0.76	1.1528	1.1522	1.1517	1.1511	1.1505	1.1500	1.1494	1.1489	1.1483	1.1478
0.77	1.1472	1.1467	1.1461	1.1456	1.1450	1.1445	1.1439	1.1434	1.1428	1.1423
0.78	1.1417	1.1412	1.1406	1.1401	1.1395	1.1389	1.1384	1.1378	1.1373	1.1367
0.79	1.1362	1.1356	1.1351	1.1345	1.1340	1.1334	1.1328	1.1323	1.1317	1.1312
0.80	1.1306	1.1300	1.1295	1.1289	1.1284	1.1278	1.1272	1.1267	1.1261	1.1256
0.81	1.1250	1.1244	1.1239	1.1233	1.1227	1.1222	1.1216	1.1210	1.1205	1.1199
0.82	1.1193	1.1188	1.1182	1.1176	1.1170	1.1165	1.1159	1.1153	1.1147	1.1142
0.83	1.1136	1.1130	1.1124	1.1119	1.1113	1.1107	1.1101	1.1095	1.1090	1.1084
0.84	1.1078	1.1072	1.1066	1.1060	1.1054	1.1049	1.1043	1.1037	1.1031	1.1025
0.85	1.1019	1.1013	1.1007	1.1001	1.0995	1.0989	1.0983	1.0977	1.0971	1.0965
0.86	1.0959	1.0953	1.0947	1.0941	1.0935	1.0928	1.0922	1.0916	1.0910	1.0904
0.87	1.0898	1.0892	1.0885	1.0879	1.0873	1.0867	1.0861	1.0854	1.0848	1.0842
0.88	1.0835	1.0829	1.0823	1.0816	1.0810	1.0804	1.0797	1.0791	1.0785	1.0778
0.89	1.0772	1.0765	1.0759	1.0752	1.0746	1.0740	1.0733	1.0727	1.0720	1.0713
0.90	1.0707	1.0700	1.0694	1.0687	1.0681	1.0674	1.0667	1.0661	1.0654	1.0647
0.91	1.0641	1.0634	1.0627	1.0621	1.0614	1.0607	1.0601	1.0594	1.0587	1.0580
0.92	1.0573	1.0567	1.0560	1.0553	1.0546	1.0539	1.0533	1.0526	1.0519	1.0512
0.93	1.0505	1.0498	1.0491	1.0484	1.0477	1.0470	1.0463	1.0456	1.0449	1.0443
0.94	1.0436	1.0429	1.0422	1.0414	1.0407	1.0400	1.0393	1.0386	1.0379	1.0372
0.95	1.0365	1.0358	1.0351	1.0344	1.0337	1.0330	1.0322	1.0315	1.0308	1.0301
0.96	1.0294	1.0287	1.0279	1.0272	1.0265	1.0258	1.0251	1.0243	1.0236	1.0229
0.97	1.0222	1.0214	1.0207	1.0200	1.0192	1.0185	1.0178	1.0170	1.0163	1.0156
0.98	1.0148	1.0141	1.0134	1.0126	1.0119	1.0112	1.0104	1.0097	1.0089	1.0082
0.99	1.0075	1.0067	1.0060	1.0052	1.0045	1.0037	1.0030	1.0022	1.000*	1.000*

*No correction needed for $r \geq 0.998$.

Values of K were determined as $1 + 3/8k$, where k was obtained using Equation 6.3.14 of Mardia (1972: 155), and Equations 9.8.1–9.8.4 of Olver (1964).

Examples:

$$r = 0.743, \ K = 1.1621 \quad \text{and} \quad r = 0.814, \ K = 1.227.$$

TABLE B.38 Critical Values of Watson's U^2

n_1	n_2	α:	0.50	0.20	0.10	0.05	0.02	0.01	0.005	0.002	0.001
4	4		0.1172	0.1875	—	—	—	—	—	—	—
4	5		0.0815	0.2037	0.2037	—	—	—	—	—	—
4	6		0.0875	0.1333	0.2167	0.2167	—	—	—	—	—
4	7		0.0844	0.1299	0.1688	0.2273	—	—	—	—	—
4	8		0.0903	0.1319	0.1632	0.2361	—	—	—	—	—
4	9		0.0855	0.1282	0.1752	0.2436	0.2436	—	—	—	—
4	10		0.0804	0.1232	0.1571	0.2018	0.2500	—	—	—	—
4	11		0.0828	0.1253	0.1556	0.1949	0.2556	—	—	—	—
4	12		0.0781	0.1302	0.1563	0.2031	0.2604	0.2604	—	—	—
4	13		0.0792	0.1244	0.1538	0.1855	0.2647	0.2647	—	—	—
4	14		0.0780	0.1227	0.1534	0.1931	0.2298	0.2685	—	—	—
4	15		0.0789	0.1228	0.1561	0.1807	0.2228	0.2719	0.2719	—	—
4	16		0.0781	0.1250	0.1531	0.1836	0.2281	0.2750	0.2750	—	—
4	17		0.0775	0.1223	0.1531	0.1839	0.2330	0.2778	0.2778	—	—
4	18		0.0764	0.1212	0.1490	0.1818	0.2197	0.2481	0.2803	—	—
4	19		0.0755	0.1213	0.1533	0.1796	0.2220	0.2517	0.2826	—	—
4	20		0.0764	0.1201	0.1535	0.1842	0.2264	0.2451	0.2847	—	—
4	21		0.0752	0.1200	0.1514	0.1819	0.2143	0.2486	0.2867	0.2867	—
4	22		0.0756	0.1211	0.1508	0.1823	0.2185	0.2517	0.2885	0.2885	—
4	23		0.0751	0.1194	0.1508	0.1814	0.2177	0.2394	0.2636	0.2901	—
4	24		0.0755	0.1202	0.1499	0.1797	0.2184	0.2411	0.2660	0.2917	—
4	25		0.0752	0.1200	0.1497	0.1814	0.2152	0.2441	0.2600	0.2931	—
4	26		0.0752	0.1191	0.1486	0.1816	0.2175	0.2396	0.2624	0.2944	—
4	27		0.0753	0.1189	0.1505	0.1786	0.2151	0.2360	0.2646	0.2957	0.2957
4	28		0.0748	0.1203	0.1496	0.1775	0.2165	0.2388	0.2667	0.2969	0.2969
4	29		0.0749	0.1198	0.1491	0.1794	0.2165	0.2369	0.2557	0.2980	0.2980
4	30		0.0745	0.1196	0.1493	0.1797	0.2140	0.2395	0.2578	0.2990	0.2990
5	5		0.0890	0.1610	0.2250	0.2250	—	—	—	—	—
5	6		0.0848	0.1333	0.1818	0.2424	—	—	—	—	—
5	7		0.0855	0.1284	0.1712	0.1998	0.2569	—	—	—	—
5	8		0.0846	0.1308	0.1654	0.2154	0.2692	—	—	—	—
5	9		0.0798	0.1242	0.1591	0.1909	0.2798	0.2798	—	—	—
5	10		0.0836	0.1236	0.1609	0.1956	0.2409	0.2889	0.2889	—	—
5	11		0.0810	0.1241	0.1560	0.1901	0.2287	0.2969	0.2969	—	—
5	12		0.0784	0.1235	0.1549	0.1863	0.2255	0.2608	0.3039	—	—
5	13		0.0777	0.1256	0.1563	0.1837	0.2298	0.2692	0.3102	—	—
5	14		0.0782	0.1218	0.1534	0.1820	0.2211	0.2571	0.2767	0.3158	—
5	15		0.0782	0.1235	0.1515	0.1835	0.2248	0.2515	0.2835	0.3208	—
5	16		0.0766	0.1206	0.1552	0.1825	0.2230	0.2552	0.2897	0.3254	—
5	17		0.0761	0.1199	0.1520	0.1820	0.2205	0.2472	0.2782	0.3295	0.3295
5	18		0.0763	0.1208	0.1536	0.1797	0.2164	0.2464	0.2715	0.3333	0.3333
5	19		0.0754	0.1201	0.1517	0.1824	0.2193	0.2526	0.2745	0.3052	0.3368
5	20		0.0760	0.1216	0.1520	0.1824	0.2200	0.2416	0.2664	0.3096	0.3400
5	21		0.0755	0.1195	0.1510	0.1810	0.2206	0.2448	0.2712	0.2990	0.3429
5	22		0.0756	0.1201	0.1524	0.1820	0.2191	0.2426	0.2689	0.3033	0.3457
5	23		0.0755	0.1196	0.1513	0.1811	0.2178	0.2451	0.2737	0.2960	0.3209
5	24		0.0747	0.1195	0.1511	0.1810	0.2190	0.2437	0.2736	0.2983	0.3241
5	25		0.0754	0.1197	0.1517	0.1810	0.2168	0.2461	0.2674	0.3021	0.3272
5	26		0.0749	0.1186	0.1514	0.1806	0.2189	0.2447	0.2675	0.2943	0.3176
5	27		0.0748	0.1193	0.1508	0.1804	0.2165	0.2443	0.2674	0.2975	0.3207
5	28		0.0746	0.1188	0.1512	0.1802	0.2170	0.2417	0.2694	0.2937	0.3136
5	29		0.0743	0.1189	0.1510	0.1802	0.2171	0.2443	0.2666	0.2970	0.3153
5	30		0.0743	0.1189	0.1512	0.1802	0.2160	0.2419	0.2678	0.2979	0.3181
6	6		0.0880	0.1319	0.1713	0.2060	0.2639	—	—	—	—
6	7		0.0806	0.1209	0.1538	0.1941	0.2821	0.2821	—	—	—

TABLE B.38 (cont.)   Critical Values of Watson's U^2

n_1	n_2	α:	0.50	0.20	0.10	0.05	0.02	0.01	0.005	0.002	0.001
6	8		0.0833	0.1265	0.1607	0.1964	0.2455	0.2976	0.2976	—	—
6	9		0.0815	0.1259	0.1556	0.1926	0.2321	0.2617	0.3111	—	—
6	10		0.0771	0.1260	0.1563	0.1896	0.2313	0.2479	0.3229	0.3229	—
6	11		0.0784	0.1212	0.1569	0.1872	0.2246	0.2620	0.2888	0.3333	—
6	12		0.0802	0.1242	0.1551	0.1829	0.2261	0.2593	0.2747	0.3426	0.3426
6	13		0.0769	0.1215	0.1538	0.1849	0.2213	0.2497	0.2780	0.3509	0.3509
6	14		0.0768	0.1220	0.1536	0.1839	0.2250	0.2506	0.2821	0.3196	0.3583
6	15		0.0762	0.1217	0.1524	0.1852	0.2201	0.2487	0.2730	0.3058	0.3651
6	16		0.0758	0.1212	0.1534	0.1823	0.2235	0.2500	0.2789	0.3073	0.3357
6	17		0.0750	0.1211	0.1526	0.1833	0.2199	0.2472	0.2745	0.3129	0.3427
6	18		0.0760	0.1211	0.1535	0.1840	0.2199	0.2461	0.2739	0.2998	0.3295
6	19		0.0751	0.1200	0.1523	0.1832	0.2204	0.2498	0.2744	0.3060	0.3298
6	20		0.0747	0.1196	0.1526	0.1824	0.2196	0.2490	0.2734	0.3077	0.3333
6	21		0.0758	0.1205	0.1523	0.1834	0.2205	0.2475	0.2734	0.3057	0.3369
6	22		0.0749	0.1204	0.1518	0.1824	0.2202	0.2473	0.2752	0.3036	0.3260
6	23		0.0745	0.1194	0.1514	0.1824	0.2194	0.2469	0.2729	0.3073	0.3273
6	24		0.0743	0.1194	0.1519	0.1826	0.2206	0.2484	0.2715	0.3056	0.3289
6	25		0.0744	0.1191	0.1514	0.1819	0.2202	0.2473	0.2731	0.3015	0.3277
6	26		0.0739	0.1188	0.1510	0.1815	0.2198	0.2464	0.2710	0.3047	0.3265
6	27		0.0741	0.1193	0.1515	0.1822	0.2200	0.2469	0.2731	0.3053	0.3281
6	28		0.0737	0.1190	0.1507	0.1821	0.2201	0.2467	0.2731	0.3039	0.3270
6	29		0.0736	0.1189	0.1511	0.1816	0.2200	0.2473	0.2719	0.3038	0.3258
6	30		0.0736	0.1193	0.1509	0.1823	0.2194	0.2471	0.2725	0.3045	0.3262
7	7		0.0791	0.1345	0.1578	0.1986	0.2511	0.3036	0.3036	—	—
7	8		0.0794	0.1198	0.1556	0.1817	0.2246	0.2722	0.3222	—	—
7	9		0.0786	0.1223	0.1560	0.1818	0.2215	0.2552	0.2909	0.3385	—
7	10		0.0773	0.1227	0.1546	0.1866	0.2269	0.2622	0.2773	0.3529	0.3529
7	11		0.0771	0.1219	0.1551	0.1839	0.2214	0.2532	0.2806	0.3225	0.3657
7	12		0.0764	0.1216	0.1541	0.1855	0.2256	0.2519	0.2757	0.3083	0.3772
7	13		0.0765	0.1216	0.1545	0.1842	0.2227	0.2523	0.2776	0.3150	0.3479
7	14		0.0761	0.1228	0.1568	0.1840	0.2248	0.2530	0.2744	0.3210	0.3337
7	15		0.0754	0.1213	0.1525	0.1845	0.2235	0.2503	0.2780	0.3118	0.3378
7	16		0.0753	0.1203	0.1530	0.1848	0.2236	0.2508	0.2772	0.3113	0.3432
7	17		0.0749	0.1204	0.1526	0.1827	0.2227	0.2500	0.2752	0.3109	0.3340
7	18		0.0749	0.1200	0.1524	0.1841	0.2235	0.2502	0.2768	0.3117	0.3346
7	20		0.0743	0.1198	0.1526	0.1832	0.2219	0.2499	0.2780	0.3081	0.3330
7	21		0.0751	0.1203	0.1534	0.1840	0.2224	0.2496	0.2782	0.3123	0.3336
7	22		0.0743	0.1196	0.1518	0.1832	0.2221	0.2512	0.2763	0.3090	0.3341
7	23		0.0739	0.1194	0.1522	0.1832	0.2226	0.2499	0.2780	0.3103	0.3327
8	8		0.0781	0.1250	0.1563	0.1836	0.2256	0.2500	0.2959	0.3438	—
8	9		0.0784	0.1225	0.1552	0.1863	0.2255	0.2582	0.2827	0.3627	0.3627
8	10		0.0775	0.1220	0.1546	0.1852	0.2220	0.2491	0.2796	0.3359	0.3796
8	11		0.0766	0.1220	0.1543	0.1842	0.2249	0.2524	0.2799	0.3194	0.3529
8	12		0.0766	0.1208	0.1557	0.1854	0.2229	0.2521	0.2807	0.3167	0.3396
8	13		0.0754	0.1212	0.1532	0.1853	0.2237	0.2531	0.2778	0.3135	0.3446
8	14		0.0751	0.1205	0.1530	0.1855	0.2224	0.2516	0.2796	0.3137	0.3381
8	15		0.0746	0.1210	0.1536	0.1855	0.2232	0.2507	0.2783	0.3130	0.3341
8	16		0.0761	0.1220	0.1542	0.1854	0.2222	0.2531	0.2795	0.3156	0.3417
8	17		0.0747	0.1200	0.1529	0.1841	0.2241	0.2524	0.2782	0.3124	0.3388
8	18		0.0748	0.1199	0.1528	0.1840	0.2244	0.2513	0.2813	0.3152	0.3397
8	19		0.0742	0.1196	0.1527	0.1839	0.2243	0.2526	0.2799	0.3145	0.3384
8	20		0.0741	0.1196	0.1527	0.1839	0.2239	0.2527	0.2795	0.3134	0.3393
9	9		0.0770	0.1250	0.1552	0.1867	0.2251	0.2663	0.2855	0.3404	0.3843
9	10		0.0760	0.1216	0.1544	0.1860	0.2257	0.2538	0.2865	0.3205	0.3614
9	11		0.0764	0.1208	0.1542	0.1845	0.2249	0.2552	0.2814	0.3168	0.3410

TABLE B.38 (cont.)   Critical Values of Watson's U^2

n_1	n_2	α:	0.50	0.20	0.10	0.05	0.02	0.01	0.005	0.002	0.001
9	12		0.0767	0.1217	0.1543	0.1852	0.2257	0.2540	0.2804	0.3157	0.3395
9	13		0.0755	0.1205	0.1532	0.1850	0.2247	0.2526	0.2798	0.3187	0.3389
9	14		0.0752	0.1201	0.1532	0.1843	0.2243	0.2526	0.2809	0.3168	0.3409
9	15		0.0757	0.1201	0.1535	0.1850	0.2245	0.2541	0.2831	0.3152	0.3393
9	16		0.0744	0.1200	0.1533	0.1850	0.2244	0.2539	0.2822	0.3172	0.3439
10	10		0.0750	0.1225	0.1545	0.1850	0.2250	0.2545	0.2825	0.3170	0.3450
10	11		0.0756	0.1215	0.1544	0.1856	0.2237	0.2548	0.2791	0.3172	0.3405
10	12		0.0758	0.1212	0.1534	0.1848	0.2246	0.2545	0.2818	0.3155	0.3409
10	13		0.0749	0.1204	0.1532	0.1853	0.2254	0.2542	0.2816	0.3184	0.3452
10	14		0.0749	0.1201	0.1535	0.1847	0.2252	0.2550	0.2823	0.3181	0.3439
10	15		0.0747	0.1211	0.1536	0.1856	0.2256	0.2549	0.2837	0.3189	0.3440
11	11		0.0760	0.1211	0.1541	0.1857	0.2262	0.2540	0.2826	0.3194	0.3442
11	12		0.0751	0.1206	0.1535	0.1851	0.2253	0.2543	0.2839	0.3182	0.3439
11	13		0.0746	0.1206	0.1532	0.1853	0.2255	0.2546	0.2838	0.3193	0.3461
12	12		0.0752	0.1215	0.1528	0.1863	0.2266	0.2558	0.2844	0.3192	0.3438
14	14		0.070	0.117	0.151	0.183	0.226	0.258	0.289	0.330	0.361
16	16		0.070	0.117	0.151	0.184	0.227	0.259	0.291	0.332	0.364
18	18		0.070	0.117	0.151	0.184	0.228	0.260	0.292	0.334	0.366
20	20		0.069	0.117	0.151	0.185	0.228	0.261	0.293	0.335	0.367
25	25		0.069	0.117	0.152	0.185	0.229	0.262	0.295	0.338	0.370
30	30		0.069	0.117	0.152	0.186	0.230	0.263	0.296	0.339	0.372
35	35		0.069	0.117	0.152	0.186	0.231	0.264	0.297	0.340	0.373
40	40		0.069	0.117	0.152	0.186	0.231	0.264	0.298	0.341	0.374
50	50		0.069	0.117	0.152	0.187	0.231	0.265	0.299	0.343	0.376
60	60		0.069	0.117	0.152	0.187	0.232	0.266	0.299	0.343	0.377
80	80		0.069	0.117	0.152	0.187	0.232	0.266	0.300	0.344	0.378
100	100		0.069	0.117	0.152	0.187	0.233	0.267	0.300	0.345	0.378
∞	∞		0.0710	0.1167	0.1518	0.1869	0.2333	0.2684	0.3035	0.3500	0.3851

The four-decimal-place critical values in this table (except for sample sizes of infinity) were obtained from distributions of U^2 calculated using the method described by Burr (1964). The three-decimal-place critical values shown were computed by the approximation of Tiku (1965), using the computer algorithms of Best and Roberts (1975), Bhattacharjee (1970), International Business Machines (1968: 362), and Odeh and Evans (1974). Comparing these values to those of Stephens (1974), for $\alpha = 0.005$ through 0.10, shows agreement to the third decimal place. The critical values for sample sizes of infinity were computed as

$$U^2_{\alpha, \infty, \infty} = -\left(\frac{1}{2\pi^2}\right)\left[\ln\left(\frac{\alpha}{2}\right) - \ln\left\{1 + \left(\frac{\alpha}{2}\right)^3\right\}\right]$$

(Watson, 1962) and should be used if both sample sizes are greater than 100.

Examples:

$$U^2_{0.05, 6, 8} = 0.1964 \quad \text{and} \quad U^2_{0.01, 10, 12} = 0.2545.$$

TABLE B.39 Critical Values of R' for the Moore Test of Circular Uniformity

n	α:	0.50	0.10	0.05	0.025	0.01	0.005	0.001
2		0.791	1.049	1.058	1.060	1.061	1.061	1.061
3		0.693	1.039	1.095	1.124	1.143	1.149	1.154
4		0.620	1.008	1.090	1.146	1.192	1.212	1.238
5		0.588	0.988	1.084	1.152	1.216	1.250	1.298
6		0.568	0.972	1.074	1.152	1.230	1.275	1.345
7		0.556	0.959	1.066	1.150	1.238	1.291	1.373
8		0.546	0.949	1.059	1.148	1.242	1.300	1.397
9		0.538	0.940	1.053	1.146	1.245	1.307	1.416
10		0.532	0.934	1.048	1.144	1.248	1.313	1.432
12		0.523	0.926	1.042	1.140	1.252	1.322	1.456
14		0.518	0.920	1.037	1.136	1.252	1.325	1.470
16		0.514	0.914	1.031	1.132	1.250	1.327	1.480
18		0.510	0.910	1.027	1.129	1.248	1.328	1.487
20		0.507	0.906	1.024	1.127	1.247	1.329	1.492
22		0.505	0.903	1.022	1.126	1.246	1.330	1.496
24		0.503	0.901	1.021	1.125	1.246	1.331	1.499
26		0.502	0.899	1.019	1.124	1.246	1.332	1.501
28		0.500	0.897	1.018	1.124	1.246	1.333	1.502
30		0.499	0.896	1.016	1.123	1.245	1.334	1.502
40		0.494	0.891	1.012	1.119	1.243	1.332	1.504
60		0.489	0.887	1.007	1.115	1.241	1.329	1.506
80		0.487	0.883	1.005	1.113	1.240	1.329	1.508
100		0.485	0.881	1.004	1.112	1.240	1.329	1.509
∞		0.481	0.876	0.999	1.109	1.239	1.329	1.517

Values in this table were taken, with permission of the publisher, from Table 1 of Moore (1980, *Biometrika* 67: 175–180).

Examples:

$$R'_{0.05,24} = 1.021 \text{ and } R'_{0.10,30} = 0.896.$$

TABLE B.40 Common Logarithms of Factorials

X	0	1	2	3	4	5	6	7	8	9
0	0.00000	0.00000	0.30103	0.77815	1.38021	2.07918	2.85733	3.70243	4.60552	5.55976
10	6.55976	7.60116	8.68034	9.79428	10.94041	12.11650	13.32062	14.55107	15.80634	17.08509
20	18.38612	19.70834	21.05077	22.41249	23.79271	25.19065	26.60562	28.03698	29.48414	30.94654
30	32.42366	33.91502	35.42077	36.93869	38.47016	40.01423	41.57054	43.13874	44.71852	46.30959
40	47.91165	49.52443	51.14768	52.78115	54.42460	56.07781	57.74057	59.41267	61.09391	62.78410
50	64.48307	66.19065	67.90665	69.63092	71.36332	73.10368	74.85187	76.60774	78.37117	80.14202
60	81.92017	83.70550	85.49790	87.29724	89.10342	90.91633	92.73587	94.56195	96.39446	98.23331
70	100.07841	101.92966	103.78700	105.65032	107.51955	109.39461	111.27543	113.16192	115.05401	116.95164
80	118.85473	120.76321	122.67703	124.59610	126.52038	128.44980	130.38430	132.32382	134.26830	136.21769
90	138.17194	140.13098	142.09477	144.06325	146.03638	148.01410	149.99637	151.98314	153.97437	155.97000
100	157.97000	159.97433	161.98293	163.99576	166.01280	168.03399	170.05929	172.08867	174.12210	176.15952
110	178.20092	180.24624	182.29546	184.34854	186.40544	188.46614	190.53060	192.59878	194.67067	196.74621
120	198.82539	200.90818	202.99454	205.08444	207.17787	209.27478	211.37515	213.47895	215.58616	217.69675
130	219.81069	221.92796	224.04854	226.17239	228.29949	230.42983	232.56337	234.70009	236.83997	238.98298
140	241.12911	243.27833	245.43062	247.58595	249.74432	251.90568	254.07004	256.23735	258.40762	260.58080
150	262.75689	264.93587	267.11771	269.30241	271.48993	273.68026	275.87338	278.06928	280.26794	282.46934
160	284.67346	286.88028	289.08980	291.30198	293.51683	295.73431	297.95442	300.17714	302.40245	304.63033
170	306.86078	309.09378	311.32931	313.56735	315.80790	318.05094	320.29645	322.54443	324.79485	327.04770
180	329.30297	331.56065	333.82072	336.08317	338.34799	340.61516	342.88467	345.15652	347.43067	349.70714
190	351.98589	354.26692	356.55022	358.83578	361.12358	363.41362	365.70587	368.00034	370.29701	372.59586
200	374.89689	377.20008	379.50544	381.81293	384.12256	386.43432	388.74818	391.06415	393.38222	395.70236
210	398.02458	400.34887	402.67520	405.00358	407.33399	409.66643	412.00089	414.33735	416.67580	419.01625
220	421.35867	423.70306	426.04941	428.39772	430.74797	433.10015	435.45426	437.81028	440.16822	442.52805
230	444.88978	447.25339	449.61888	451.98624	454.35545	456.72652	459.09943	461.47418	463.85076	466.22916
240	468.60937	470.99139	473.37520	475.76081	478.14820	480.53736	482.92830	485.32100	487.71545	490.11165
250	492.50959	494.90926	497.31066	499.71378	502.11861	504.52516	506.93340	509.34333	511.75495	514.16825
260	516.58322	518.99986	521.41816	523.83812	526.25972	528.68297	531.10785	533.53436	535.96250	538.39225
270	540.82361	543.25658	545.69115	548.12731	550.56506	553.00440	555.44531	557.88779	560.33183	562.77743
280	565.22459	567.67330	570.12355	572.57533	575.02865	577.48350	579.93986	582.39775	584.85714	587.31804
290	589.78043	592.24433	594.70971	597.17658	599.64492	602.11475	604.58604	607.05879	609.53301	612.00868
300	614.48580	616.96437	619.44438	621.92582	624.40869	626.89299	629.37871	631.86585	634.35440	636.84436
310	639.33572	641.82848	644.32264	646.81818	649.31511	651.81342	654.31311	656.81417	659.31660	661.82039
320	664.32554	666.83204	669.33990	671.84910	674.35965	676.87153	679.38475	681.89929	684.41517	686.93236
330	689.45088	691.97071	694.49184	697.01429	699.53803	702.06308	704.58942	707.11705	709.64597	712.17616
340	714.70764	717.24040	719.77442	722.30972	724.84628	727.38410	729.92317	732.46350	735.00508	737.54791
350	740.09197	742.63728	745.18382	747.73160	750.28060	752.83083	755.38228	757.93495	760.48883	763.04393
360	765.60023	768.15774	770.71644	773.27635	775.83745	778.39975	780.96323	783.52789	786.09374	788.66077
370	791.22897	793.79834	796.36889	798.94059	801.51347	804.08750	806.66268	809.23903	811.81652	814.39516
380	816.97494	819.55587	822.13793	824.72113	827.30546	829.89092	832.47751	835.06522	837.65405	840.24400
390	842.83506	845.42724	848.02053	850.61492	853.21042	855.80701	858.40471	861.00350	863.60338	866.20435

TABLE B.40 (cont.) Common Logarithms of Factorials

X	0	1	2	3	4	5	6	7	8	9
400	868.80641	871.40956	874.01378	876.61909	879.22547	881.83293	884.44145	887.05105	889.66171	892.27343
410	894.88621	897.50006	900.11495	902.73090	905.34790	907.96595	910.58504	913.20518	915.82636	918.44857
420	921.07182	923.69610	926.32141	928.94776	931.57512	934.20351	936.83292	939.46335	942.09479	944.72725
430	947.36072	949.99519	952.63068	955.26717	957.90466	960.54314	963.18263	965.82311	968.46459	971.10705
440	973.75050	976.39494	979.04037	981.68677	984.33415	986.98251	989.63185	992.28215	994.93343	997.58568
450	1000.23889	1002.89307	1005.54821	1008.20430	1010.86136	1013.51937	1016.17834	1018.83825	1021.49912	1024.16093
460	1026.82369	1029.48739	1032.15203	1034.81761	1037.48413	1040.15158	1042.81997	1045.48929	1048.15953	1050.83070
470	1053.50280	1056.17582	1058.84977	1061.52463	1064.20040	1066.87710	1069.55471	1072.23322	1074.91265	1077.59299
480	1080.27423	1082.95637	1085.63942	1088.32337	1091.00821	1093.69395	1096.38059	1099.06812	1101.75654	1104.44585
490	1107.13604	1109.82713	1112.51909	1115.21194	1117.90567	1120.60027	1123.29575	1125.99211	1128.68934	1131.38744

If $\log X!$ is needed for $X > 499$, consult Lloyd, Zar, and Karr (1968) or Pearson and Hartley (1966: Table 51), or note that this "Stirling approximation" is excellent, being accurate to about two decimal places for n as small as 10:

$$\log X! = (X + 0.5) \log X - 0.434294X + 0.399090;$$

and this one is even better, having half the error:

$$\log X! = (X + 0.5) \log(X + 0.5) - 0.434294(X + 0.5) + 0.399090$$

(Kemp, 1989; Tweddle, 1984).

Walker (1934) notes that Abraham de Moivre was aware of this approximation before Stirling. Thus we have another example of "Stigler's Law of Eponymy" (see first footnote in Chapter 6).

TABLE B.41   Ten Thousand Random Digits

	00–04	05–09	10–14	15–19	20–24	25–29	30–34	35–39	40–44	45–49
00	22808	04391	45529	53968	57136	98228	85485	13801	68194	56382
01	49305	36965	44849	64987	59501	35141	50159	57369	76913	75739
02	81934	19920	73316	69243	69605	17022	53264	83417	55193	92929
03	10840	13508	48120	22467	54505	70536	91206	81038	22418	34800
04	99555	73289	59605	37105	24621	44100	72832	12268	97089	68112
05	32677	45709	62337	35132	45128	96761	08745	53388	98353	46724
06	09401	75407	27704	11569	52842	83543	44750	03177	50511	15301
07	73424	31711	65519	74869	56744	40864	75315	89866	96563	75142
08	37075	81378	59472	71858	86903	66860	03757	32723	54273	45477
09	02060	37158	55244	44812	45369	78939	08048	28036	40946	03898
10	94719	43565	40028	79866	43137	28063	52513	66405	71511	66135
11	70234	48272	59621	88778	16536	36505	41724	24776	63971	01685
12	07972	71752	92745	86465	01845	27416	50519	48458	68460	63113
13	58521	64882	26993	48104	61307	73933	17214	44827	88306	78177
14	32580	45202	21148	09684	39411	04892	02055	75276	51831	85686
15	88796	30829	35009	22695	23694	11220	71006	26720	39476	60538
16	31525	82746	78935	82980	61236	28940	96341	13790	66247	33839
17	02747	35989	70387	89571	34570	17002	79223	96817	31681	15207
18	46651	28987	20625	61347	63981	41085	67412	29053	00724	14841
19	43598	14436	33521	55637	39789	26560	66404	71802	18763	80560
20	30596	92319	11474	64546	60030	73795	60809	24016	29166	36059
21	56198	64370	85771	62633	78240	05766	32419	35769	14057	80674
22	68266	67544	06464	84956	18431	04015	89049	15098	12018	89338
23	31107	28597	65102	75599	17496	87590	68848	33021	69855	54015
24	37555	05069	38680	87274	55152	21792	77219	48732	03377	01160
25	90463	27249	43845	94391	12145	36882	48906	52336	00780	74407
26	99189	88731	93531	52638	54989	04237	32978	59902	05463	09245
27	37631	74016	89072	59598	55356	27346	80856	80875	52850	36548
28	73829	21651	50141	76142	72303	06694	61697	76662	23745	96282
29	15634	89428	47090	12094	42134	62381	87236	90118	53463	46969
30	00571	45172	78532	63863	98597	15742	41967	11821	91389	07476
31	83374	10184	56384	27050	77700	13875	96607	76479	80535	17454
32	78666	85645	13181	08700	08289	62956	64439	39150	95690	18555
33	47890	88197	21368	65254	35917	54035	83028	84636	38186	50581
34	56238	13559	79344	83198	94642	35165	40188	21456	67024	62771
35	36369	32234	38129	59963	99237	72648	66504	99065	61161	16186
36	42934	34578	28968	74028	42164	56647	76806	61023	33099	48293
37	09010	15226	43474	30174	26727	39317	48508	55438	85336	40762
38	83897	90073	72941	85613	85569	24183	08247	15946	02957	68504
39	82206	01230	93252	89045	25141	91943	75531	87420	99012	80751
40	14175	32992	49046	41272	94040	44929	98531	27712	05106	35242
41	58968	88367	70927	74765	18635	85122	27722	95388	61523	91745
42	62601	04595	76926	11007	67631	64641	07994	04639	39314	83126
43	97030	71165	47032	85021	65554	66774	21560	04121	57297	85415
44	89074	31587	21360	41673	71192	85795	82757	52928	62586	02179
45	07806	81312	81215	99858	26762	28993	74951	64680	50934	32011
46	91540	86466	13229	76624	44092	96604	08590	89705	03424	48033
47	99279	27334	33804	77988	93592	90708	56780	70097	39907	51006
48	63224	05074	83941	25034	43516	22840	35230	66048	80754	46302
49	98361	97513	27529	66419	35328	19738	82366	38573	50967	72754

TABLE B.41 (cont.)   Ten Thousand Random Digits

	00–04	05–09	10–14	15–19	20–24	25–29	30–34	35–39	40–44	45–49
50	27791	82504	33523	27623	16597	32089	81596	78429	14111	68245
51	33147	46058	92388	10150	63224	26003	56427	29945	44546	50233
52	67243	10454	40269	44324	46013	00061	21622	68213	47749	76398
53	78176	70368	95523	09134	31178	33857	26171	07063	41984	99310
54	70199	70547	94431	45423	48695	01370	68065	61982	20200	27066
55	19840	01143	18606	07622	77282	68422	70767	33026	15135	91212
56	32970	28267	17695	20571	50227	69447	45535	16845	68283	15919
57	43233	53872	68520	70013	31395	60361	39034	59444	17066	07418
58	08514	23921	16685	89184	71512	82239	72947	69523	75618	79826
59	28595	51196	96108	84384	80359	02346	60581	01488	63177	47496
60	83334	81552	88223	29934	68663	23726	18429	84855	26897	94782
61	66112	95787	84997	91207	67576	27496	01603	22395	41546	68178
62	25245	14749	30653	42355	88625	37412	87384	09392	11273	28116
63	21861	22185	41576	15238	92294	50643	69848	48020	19785	41518
64	74506	40569	90770	40812	57730	84150	91500	53850	52104	37988
65	23271	39549	33042	10661	37312	50914	73027	21010	76788	64037
66	08548	16021	64715	08275	50987	67327	11431	31492	86970	47335
67	14236	80869	90798	85659	10079	28535	35938	10710	67046	74021
68	55270	49583	86467	40633	27952	27187	35058	66628	94372	75665
69	02301	05524	91801	23647	51330	35677	05972	90729	26650	81684
70	72843	03767	62590	92077	91552	76853	45812	15503	93138	87788
71	49248	43346	29503	22494	08051	09035	75802	63967	74257	00046
72	62598	99092	87806	42727	30659	10118	83000	96198	47155	00361
73	27510	69457	98616	62172	07056	61015	22159	65590	51082	34912
74	84167	66640	69100	22944	19833	23961	80834	37418	42284	12951
75	14722	88488	54999	55244	03301	37344	01053	79305	94771	95215
76	46696	05477	32442	18738	43021	72933	14995	30408	64043	67834
77	13938	09867	28949	94761	38419	38695	90165	82841	75399	09932
78	48778	56434	42495	07050	35250	09660	56192	34793	36146	96806
79	00571	71281	01563	66448	94560	55920	31580	26640	91262	30863
80	96050	57641	21798	14917	21836	15053	33566	51177	91786	12610
81	30870	81575	14019	07831	81840	25506	29358	88668	42742	62048
82	59153	29135	00712	73025	14263	17253	95662	75535	26170	95240
83	78283	70379	54969	05821	26485	28990	40207	00434	38863	61892
84	12175	95800	41106	93962	06245	00883	65337	75506	66294	62241
85	14192	39242	17961	29448	84078	14545	39417	83649	26495	41672
86	69060	38669	00849	24991	84252	41611	62773	63024	57079	59283
87	46154	11705	29355	71523	21377	36745	00766	21549	51796	81340
88	93419	54353	41269	07014	28352	77594	57293	59219	26098	63041
89	13201	04017	68889	81388	60829	46231	46161	01360	25839	52380
90	62264	99963	98226	29972	95169	07546	01574	94986	06123	52804
91	58030	30054	27479	70354	12351	33761	94357	81081	74418	74297
92	81242	26739	92304	81425	29052	37708	49370	46749	59613	50749
93	16372	70531	92036	54496	50521	83872	30064	67555	40354	23671
94	54191	04574	58634	91370	40041	77649	42030	42547	47593	07435
95	15933	92602	19496	18703	63380	58017	14665	88867	84807	44672
96	21518	77770	53826	97114	82062	34592	87400	64938	75540	54751
97	34524	64627	92997	21198	14976	07071	91566	44335	83237	24335
98	46557	67780	59432	23250	63352	43890	07109	07911	85956	62699
99	31929	13996	05126	83561	03244	33635	26952	01638	22788	26393

TABLE B.41 (cont.) Ten Thousand Random Digits

	50–54	55–59	60–64	65–69	70–74	75–79	80–84	85–89	90–94	95–99
00	53330	26487	85005	06384	13822	83736	95876	71355	31226	56063
01	96990	62825	97110	73006	32661	63408	03893	10333	41902	69175
02	30385	16588	63609	09132	53081	14478	50813	22887	03746	10289
03	75252	66905	60536	13408	25158	35825	10447	47375	89249	91238
04	52615	66504	78496	90443	84414	31981	88768	49629	15174	99795
05	39992	51082	74548	31022	71980	40900	84729	34286	96944	49502
06	51788	87155	13272	92461	06466	25392	22330	17336	42528	78628
07	88569	35645	50602	94043	35316	66344	78064	89651	89025	12722
08	14513	34794	44976	71244	60548	03041	03300	46389	25340	23804
09	50257	53477	24546	01377	20292	85097	00660	39561	62367	61424
10	35170	69025	46214	27085	83416	48597	19494	49380	28469	77549
11	22225	83437	43912	30337	75784	77689	60425	85588	93438	61343
12	90103	12542	97828	85859	85859	64101	00924	89012	17889	01154
13	68240	89649	85705	18937	30114	89827	89460	01998	81745	31281
14	01589	18335	24024	39498	82052	07868	49486	25155	61730	08946
15	36375	61694	90654	16475	92703	59561	45517	90922	93357	00207
16	11237	60921	51162	74153	94774	84150	39274	10089	45020	09624
17	48667	68353	40567	79819	48551	26789	07281	14669	00576	17435
18	99286	42806	02956	73762	04419	21676	67533	50553	21115	26742
19	44651	48349	13003	39656	99757	74964	00141	21387	66777	68533
20	83251	70164	05732	66842	77717	25305	36218	85600	23736	06629
21	41551	54630	88759	10085	48806	08724	50685	95638	20829	37264
22	68990	51280	51368	73661	21764	71552	69654	17776	51935	53169
23	63393	76820	33106	23322	16783	35630	50938	90047	97577	27699
24	93317	87564	32371	04190	27608	40658	11517	19646	82335	60088
25	48546	41090	69890	58014	04093	39286	12253	55859	83853	15023
26	31435	57566	99741	77250	43165	31150	20735	57406	85891	04806
27	56405	29392	76998	66849	29175	11641	85284	89978	73169	62140
28	70102	50882	85960	85955	03828	69417	55854	63173	60485	00327
29	92746	32004	52242	94763	32955	39848	09724	30029	45196	67606
30	67737	34389	57920	47081	60714	04935	48278	90687	99290	18554
31	35606	76646	14813	51114	52492	46778	08156	22372	59999	43938
32	64836	28649	45759	45788	43183	25275	25300	21548	33941	66314
33	86319	92367	37873	48993	71443	22768	69124	65611	79267	49709
34	90632	32314	24446	60301	31376	13575	99663	81929	39343	17648
35	83752	51966	43895	03129	37539	72989	52393	45542	70344	96712
36	56755	21142	86355	33569	63096	66780	97539	75150	25718	33724
37	14100	28857	60648	86304	97397	97210	74842	87483	51558	52883
38	69227	24872	48057	29318	74385	02097	63266	26950	73173	53025
39	77718	56967	36560	87155	26021	70903	32086	11722	32053	63723
40	09550	38799	88929	80877	87779	99905	17122	25985	16866	76005
41	12404	42453	88609	89148	85892	96045	10310	45021	62023	70061
42	07985	27418	92734	80000	58969	99011	73815	49705	68076	69605
43	58124	53830	08705	20916	46048	30342	86530	72608	93074	80937
44	46173	77223	75661	57691	24055	27568	41227	58542	73196	44886
45	13476	72301	85793	80516	59479	66985	24801	84009	71317	87321
46	82472	98647	17053	94591	36790	42275	51154	77765	01115	09331
47	55370	63433	80653	30739	68821	46854	41939	38962	20703	69424
48	89274	74795	82231	69384	53605	67860	01309	27273	76316	54253
49	55242	74511	62992	17981	17323	79325	35238	21393	13114	70084

TABLE B.41 (cont.)　Ten Thousand Random Digits

	50–54	55–59	60–64	65–69	70–74	75–79	80–84	85–89	90–94	95–99
50	03674	36059	46810	58367	82676	15051	57977	49410	02971	05797
51	26136	80623	96505	91089	02309	54743	15831	45538	96456	87272
52	61716	80405	84735	12997	86386	61606	75091	84996	76070	54923
53	67051	63246	99547	81223	52485	90333	24697	06266	07388	70389
54	17284	60347	87314	30218	87983	45426	84153	10569	64042	95618
55	12543	23999	95777	28105	66073	35174	67706	05181	35176	85558
56	45494	93037	29209	70724	86438	65354	71209	27969	85321	10216
57	39262	15415	93940	41615	43605	95675	53916	29580	07048	95838
58	29094	58703	92144	14287	50165	85661	95749	61118	36668	96852
59	77988	03222	57805	00725	91543	80021	16442	63360	33620	39324
60	02758	86823	52423	32355	96707	47448	06453	59430	43952	16775
61	46702	37467	66803	49344	59519	92717	97110	82087	36785	00880
62	61759	95153	80090	60626	55917	92812	63544	82295	50729	20116
63	82316	11402	28078	75325	43963	63105	99294	30285	61473	53613
64	92754	74241	14315	49697	61979	66711	61707	81589	53936	82115
65	37907	24080	31741	86653	81460	32304	99590	56644	41521	91172
66	16619	75264	12279	18996	16716	81959	65722	10058	91522	65410
67	66640	06195	84416	32836	53178	93810	36766	59778	26612	69017
68	45208	58525	07714	77126	67986	73140	12026	75550	84912	64691
69	00910	40237	91035	29125	03534	47246	64698	00608	39537	71755
70	19965	46945	59357	15551	20335	03145	21519	37882	99146	70161
71	37538	05747	54982	00494	51866	86172	82679	04152	56369	20356
72	38571	69663	03287	28101	46753	55715	93527	30508	19722	02072
73	76711	02864	00880	85518	25834	52317	48070	51582	03374	19540
74	07128	44400	48015	41449	21109	38948	21816	52089	64529	21510
75	00882	89357	80906	76476	58420	95793	34043	00991	38937	39859
76	96160	18580	40549	46562	45106	53768	76097	60504	85273	63076
77	13443	22235	46210	47755	05802	00311	15171	23818	89870	47578
78	99494	35395	71411	48281	92151	84465	63651	15969	61345	13324
79	90647	11809	96365	52409	17977	05971	35835	03889	43733	66100
80	33050	48785	92200	59319	36977	41111	28002	51580	10573	21763
81	21257	15066	72630	23206	03106	53140	50292	64012	83184	81304
82	45362	94324	81800	83980	97244	09691	08435	66723	06150	54972
83	93322	58684	95695	19096	98108	47678	98061	87193	99992	82870
84	20374	61803	62508	83696	54449	53649	86447	66115	90857	69114
85	00715	13209	17080	06890	38022	76469	27696	30778	31836	96676
86	85519	93677	90186	09579	98760	50320	98077	46048	79700	81431
87	71948	15871	84502	41330	46675	51342	93431	55566	90819	68923
88	43427	95500	02004	51802	59668	17806	87605	33010	20991	76269
89	64854	28815	74959	03531	77051	51807	89005	18898	23716	45862
90	62195	29095	23982	75883	41561	25897	43595	92703	86676	32038
91	61186	54041	60984	61602	18482	57941	59657	35924	21738	30646
92	88585	40218	69965	74354	62274	38948	44813	31558	40625	22477
93	15598	21389	79016	92151	21926	49901	16835	88055	30545	60306
94	27097	89653	21558	72731	66694	36703	92172	46129	32660	91356
95	40537	85697	78182	39711	59270	21934	78647	94801	78832	37287
96	74828	06544	13078	59528	31100	11132	91256	85899	72492	18200
97	43297	83195	66218	65838	63255	72093	38976	44892	96861	97848
98	32663	58127	73258	09220	49701	92357	43700	37214	56844	02048
99	45551	31330	08152	23712	23963	58274	94583	03761	73429	47328

Appendix Table B.41 was prepared using an algorithm for the IBM 360 computer (International Business Machines, 1968: 77).

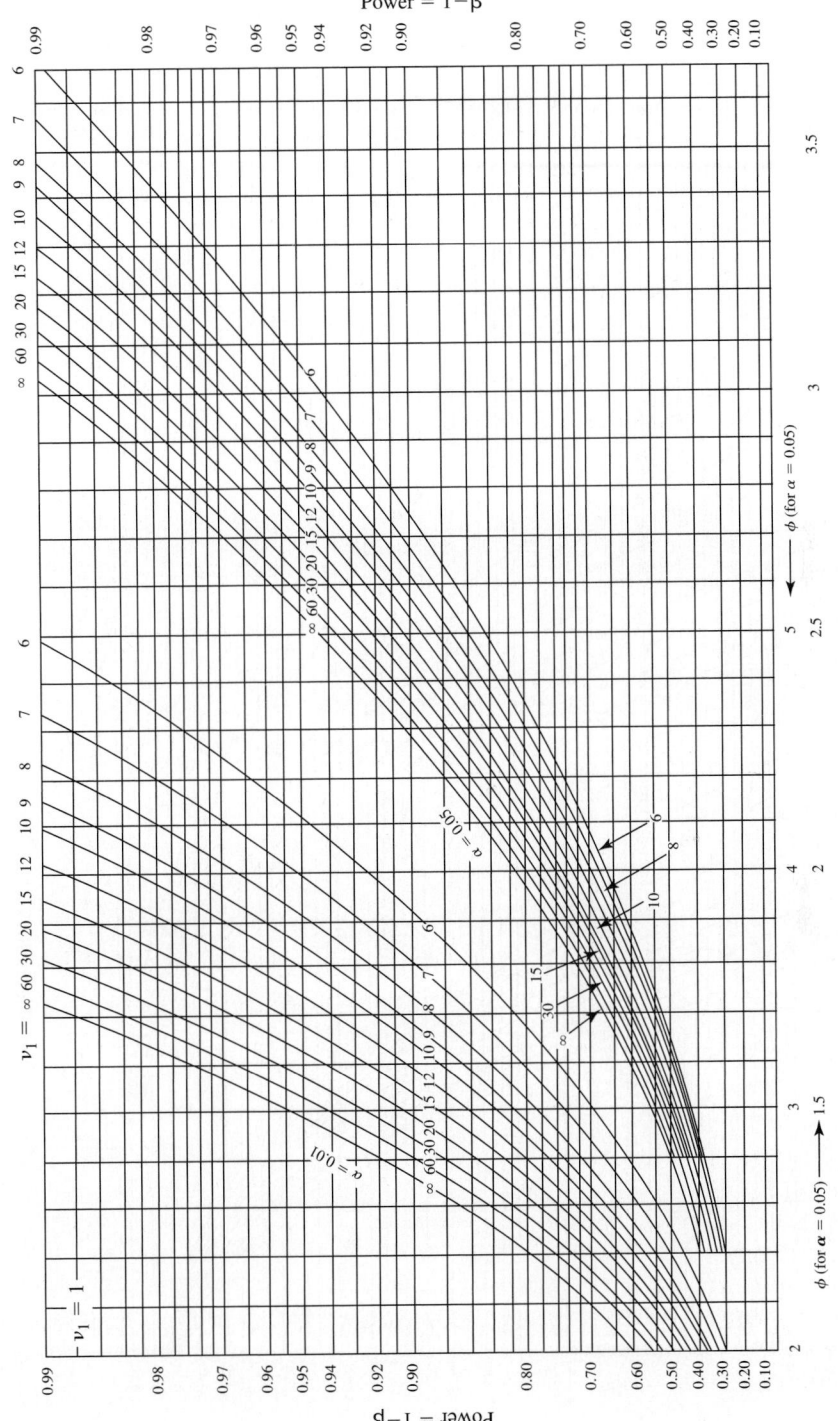

Figure B.1a Power and sample size in analysis of variance; $\nu_1 = 1$.

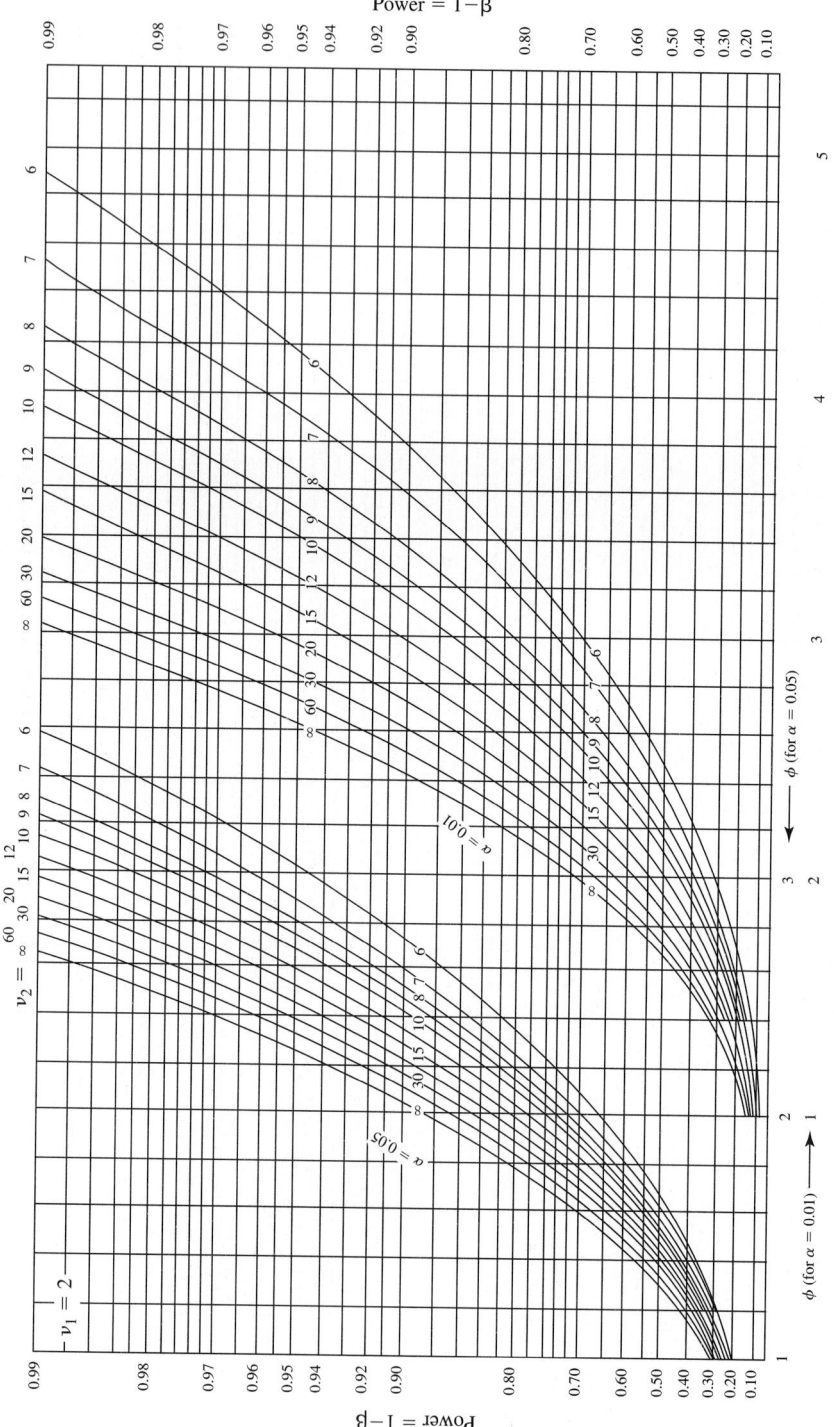

Figure B.1b Power and sample size in analysis of variance; $\nu_1 = 2$.

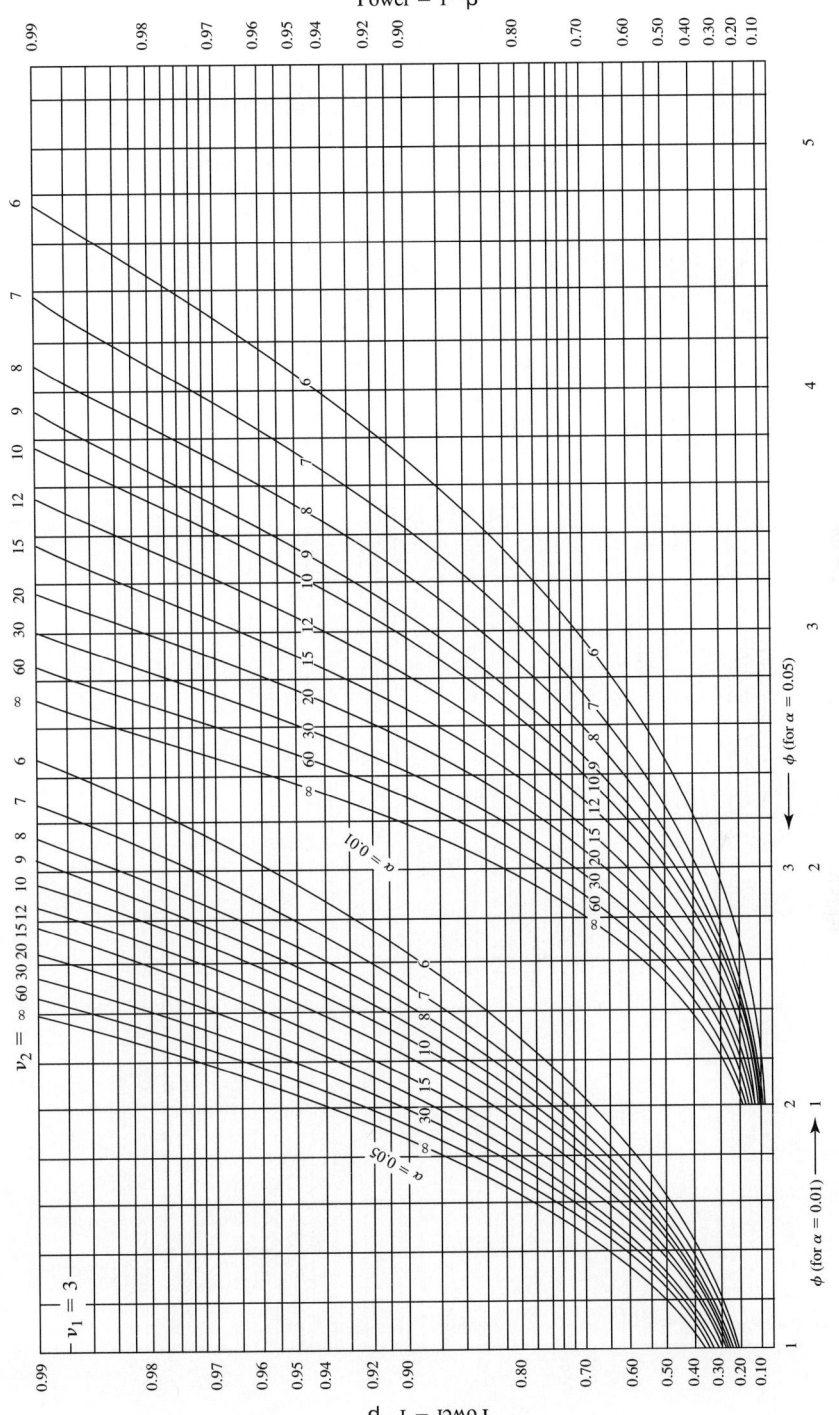

Figure B.1c Power and sample size in analysis of variance; $\nu_1 = 3$.

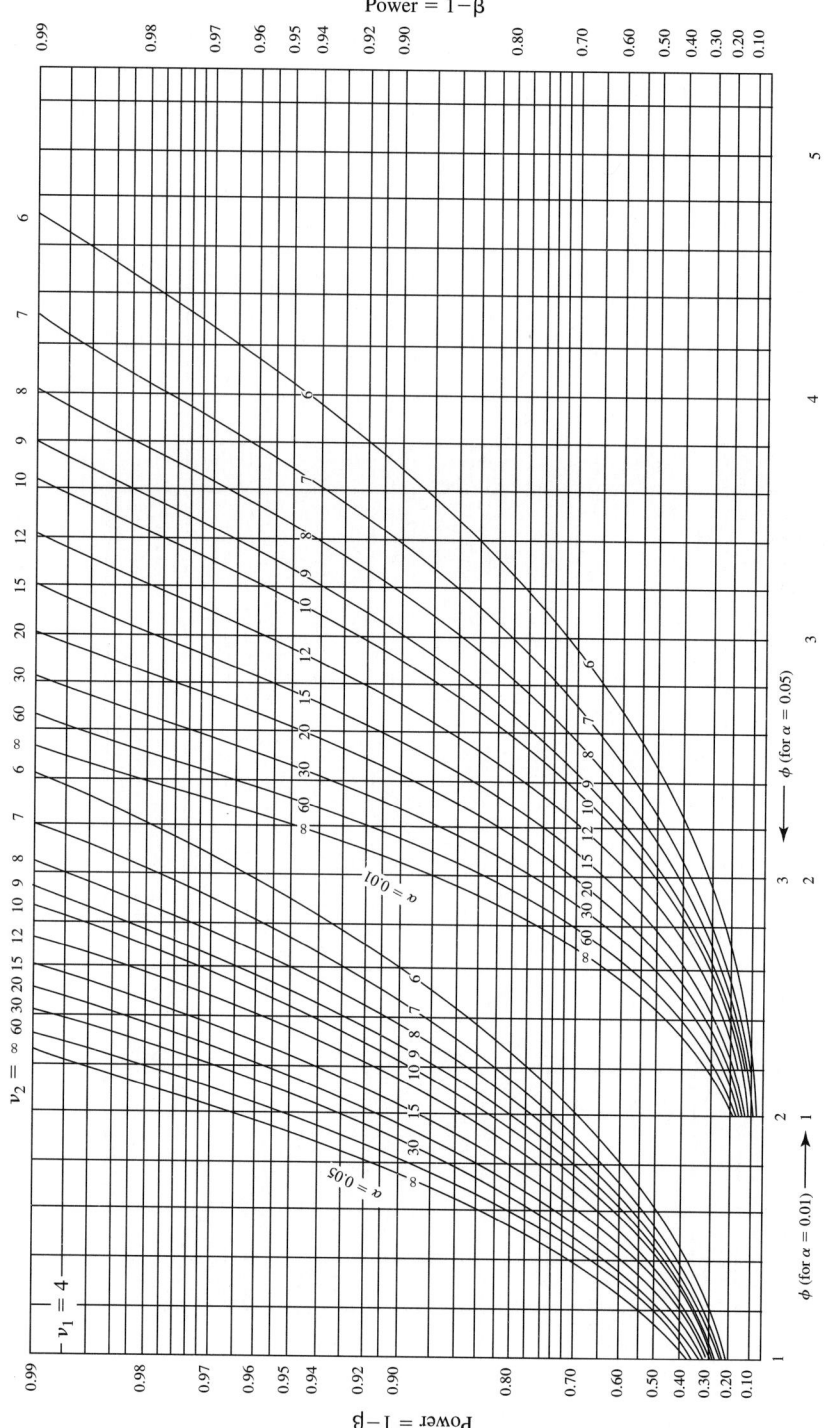

Figure B.1d Power and sample size in analysis of variance; $\nu_1 = 4$.

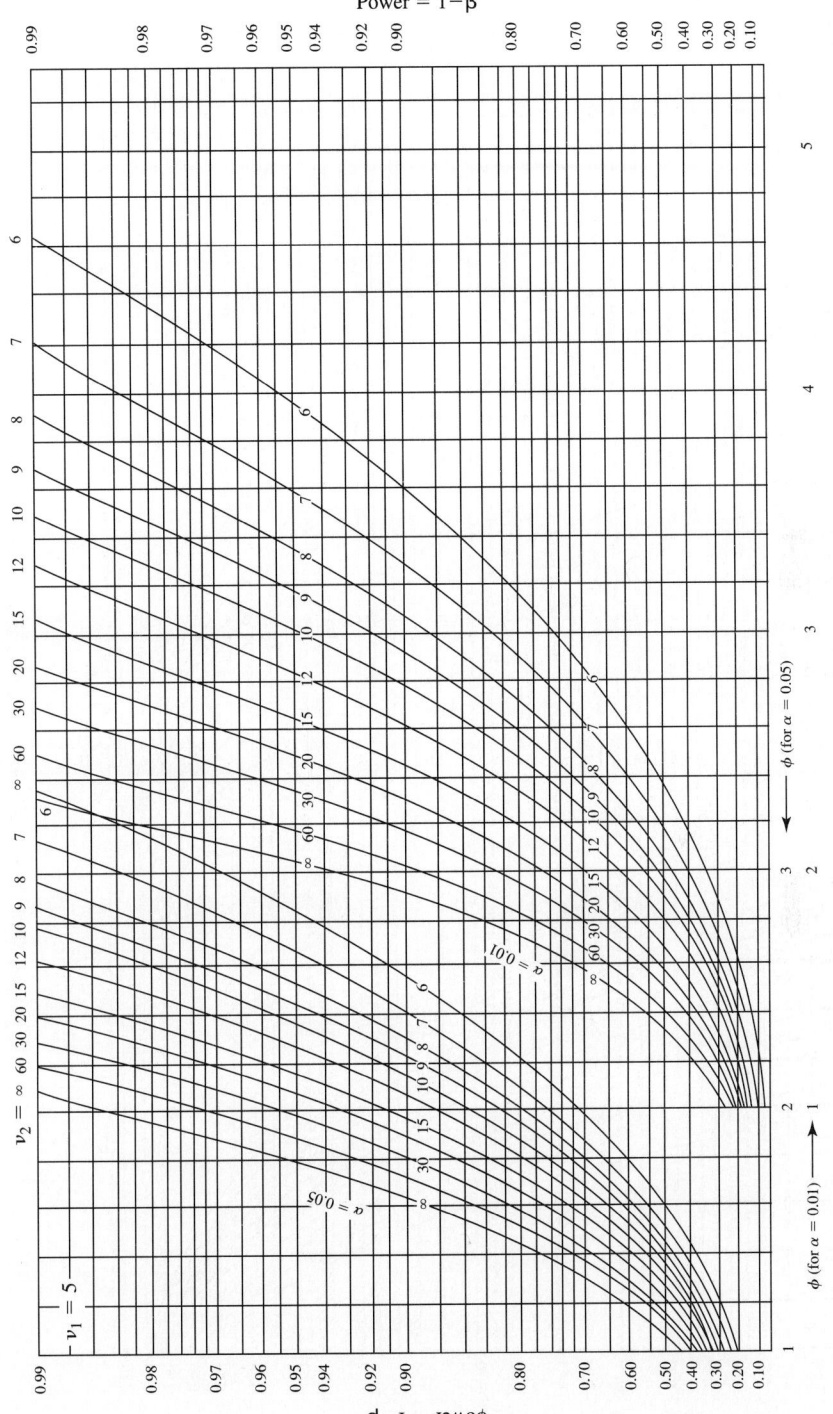

Figure B.1e Power and sample size in analysis of variance; $\nu_1 = 5$.

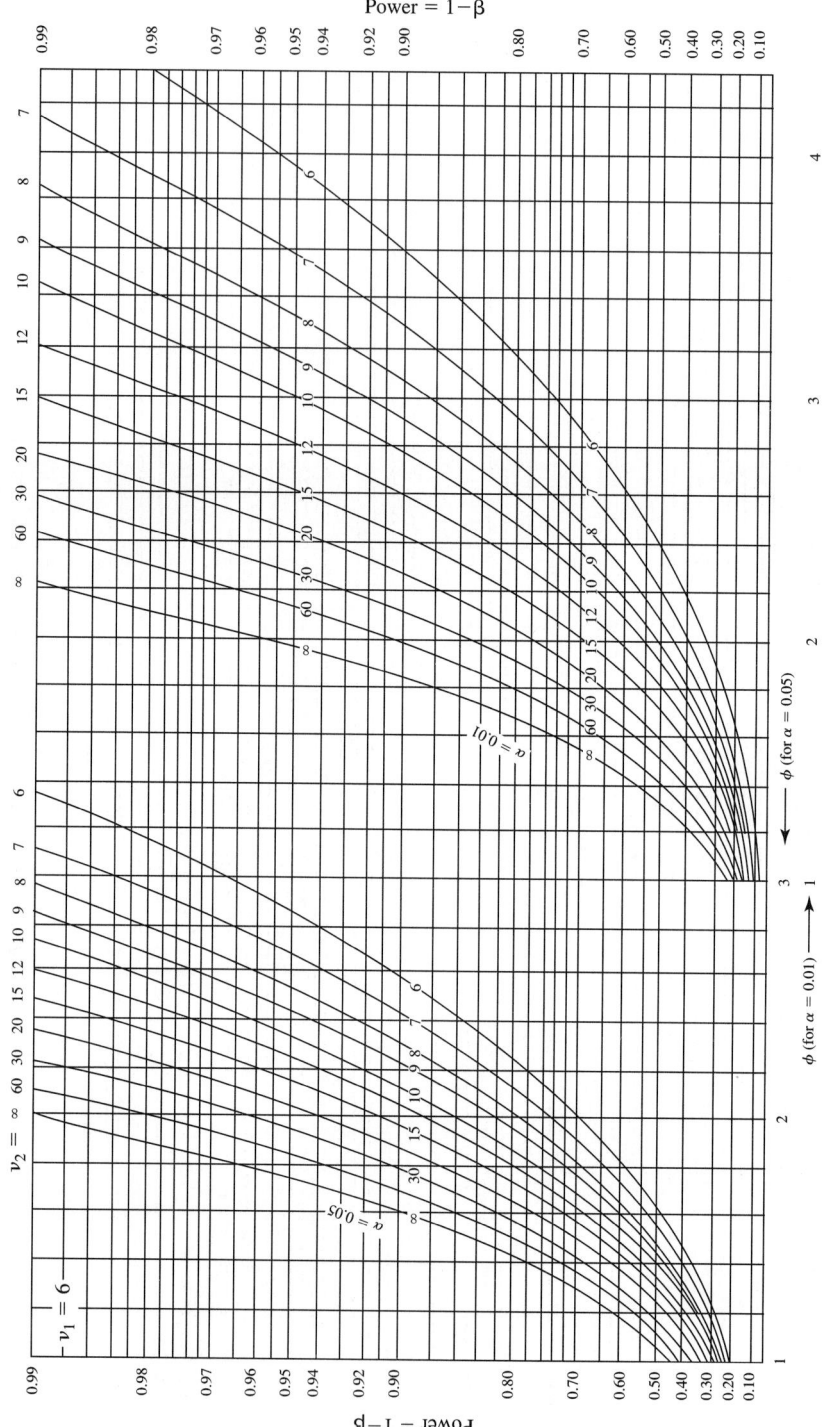

Figure B.1f Power and sample size in analysis of variance; $\nu_1 = 6$.

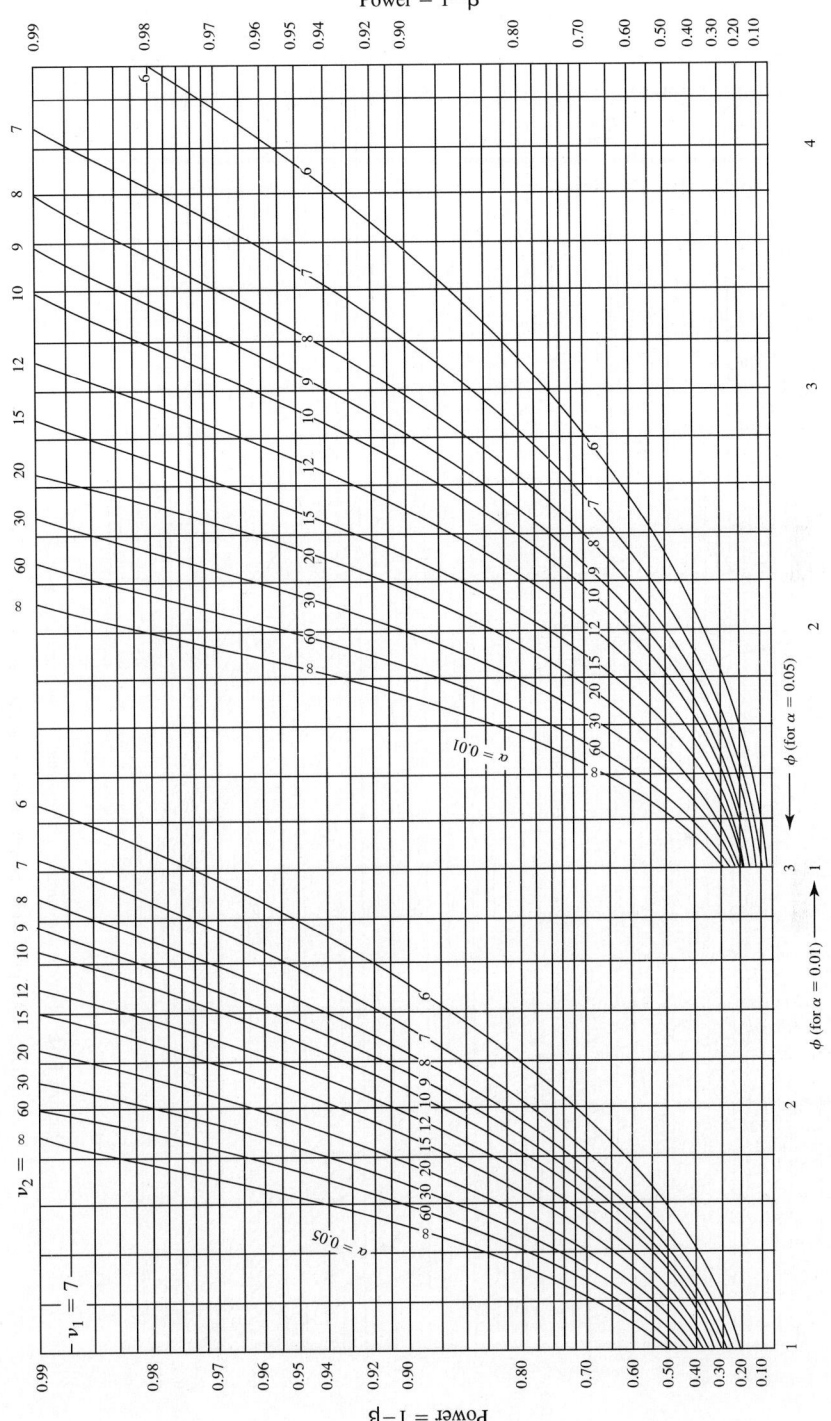

Figure B.1g Power and sample size in analysis of variance; $\nu_1 = 7$.

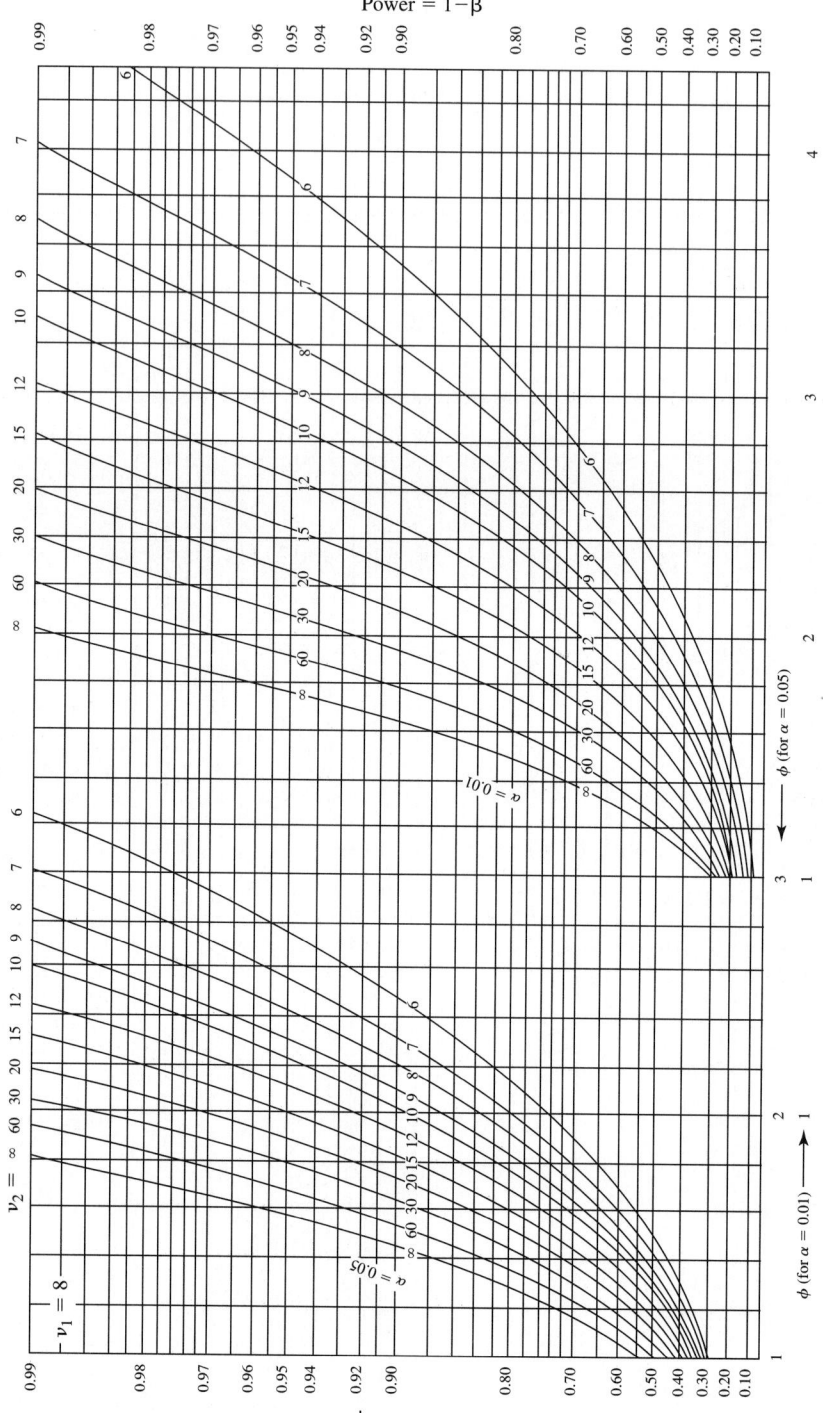

Figure B.1h Power and sample size in analysis of variance; $\nu_1 = 8$.

These graphs were taken from Pearson and Hartley (1951, *Biometrika* 38: 112–130) with permission of the Biometrika Trustees.

ANSWERS TO EXERCISES

CHAPTER 3

3.1 (a) 13.8 kg; (b) 10.7 kg; (c) 17.8 kg; (d) 17.8 kg.
3.2 (a) 3.56 kg; (b) 3.6 kg.
3.3 (a) 46.63 yr; (b) 46.3 yr; (c) 44.58 yr; (d) 46.3 yr.
3.4 (a) 2.33 g; (b) 2.33 g; (c) 2.4 g; (d) 2.358 g; (e) 2.4 g.

CHAPTER 4

4.1 (a) SS $= 156.028$ g^2, $s^2 = 39.007$ g^2; (b) same as (a).
4.2 (a) Range $= 236.4$ mg/100 ml to 244.8 mg/100 ml $= 8.4$ mg/100 ml; (b) SS $= 46.1886$ (mg/100 ml)^2; (c) $s^2 = 7.6981$ (mg/100 ml)^2; (d) $s = 2.77$ mg/100 ml; (e) V$= 0.0115 = 1.15\%$.
4.3 $k = 6, n = 97$; (a) $H' = 0.595$; (b) $H'_{max} = 0.778$; (c) $J' = 0.76$.
4.4 $k = 6, n = 97$; (a) $H = 0.554$; (b) $c = 16, d = 0.1667, H_{max} = 0.741$; (c) $J = 0.75$.

CHAPTER 5

5.1 (a) $(3)(2) = 6$; (b) H,G; H,P; M,G; M,P; L,G; L,P.
5.2 $(3)(4)(2) = 24$.
5.3 $2^{23} = 8,388,608$.
5.4 $_5P_5 = 5! = 120$.
5.5 $_{12}P_5 = 12!/7! = 95,040$.
5.6 $_8P_{4,2,2} = 8!/[4!2!2!] = 420$.
5.7 $_9C_5 = 9!/(5!4!) = 126$.
5.8 O: 0.49; A: 0.38; B: 0.09; AB: 0.04.
5.9 $n = 29$; 0.38, 0.21, 0.14, 0.07, 0.07, 0.07, 0.07.
5.10 (a) $P = 0.38$; (b) $P = 0.38 + 0.04 = 0.42$.
5.11 (a) $P = 4/29 = 0.14$; (b) $P = 4/29 + 2/29 + 2/29 = 0.28$.
5.12 (a) $P = \left(\frac{1}{2}\right)(1) = \left(\frac{1}{2}\right) = 0.5$; (b) $P = \left(\frac{1}{2}\right)(1) = \left(\frac{1}{2}\right) = 0.5$; (c) $P = \left(\frac{1}{2}\right)(0) = 0$.

5.13 (a) $P = \left(\frac{1}{13}\right)\left(\frac{1}{4}\right) = \frac{1}{52} = 0.019$; (b) $P = \left(\frac{1}{4} + \frac{1}{4}\right)\left(\frac{1}{13}\right) = \frac{1}{26} = 0.038$; (c) $P = \left(\frac{1}{2}\right)\left(\frac{3}{13}\right) = \frac{3}{26} = 0.12$.
5.14 (a) P(all 3 white) $= [P(W)][P(W)][P(W)] = \left(\frac{2}{6}\right)\left(\frac{2}{4}\right)\left(\frac{3}{5}\right) = \frac{12}{120} = 0.10$; (b) P(2 white) $= [P(W)][P(W)][P(B)] + [P(W)][P(B)][P(W)] + [P(B)][P(W)][P(W)] = \left(\frac{2}{6}\right)\left(\frac{2}{4}\right)\left(\frac{2}{5}\right) + \left(\frac{2}{6}\right)\left(\frac{2}{4}\right)\left(\frac{3}{5}\right) + \left(\frac{4}{6}\right)\left(\frac{2}{4}\right)\left(\frac{3}{5}\right) = \frac{8}{120} + \frac{12}{120} + \frac{24}{120} = \frac{44}{120} = 0.37$. (c) P(2 or 3 white) $= 0.10 + 0.37 = 0.47$.

CHAPTER 6

6.1 (a) $\sum f_i = n = 37, \sum f_i X_i = 172.1$ kg, $\sum f_i X_i^2 = 802.79$ kg^2, $\sum f_i X_i^3 = 3754.979$ kg^3, $\sum f_i X_i^4 = 17609.3747$ kg^4; SS $= 2.292432$ kg^2, $s^2 = 0.063679$ kg^2; $k_3 = -0.011898$ kg^3, $g_1 = -0.7404$, $\sqrt{b_1} = -0.710$; $k_4 = 0.003522$ kg^4, $b_2 = 3.598$. (b) $g_2 = 0.8686$. (c) $Q_1 = X_{9.5} = 4.5$ kg, $Q_2 = M = X_{19} = 4.7$ kg, $Q_3 = X_{28.5} = 4.8$ kg; skewness $= -0.100$ (d) $\mathcal{O}_1 = X_5 = 4.3$ kg, $\mathcal{O}_2 = Q_1 = 4.5$ kg, $\mathcal{O}_3 = X_{12} = 4.6$ kg; $\mathcal{O}_5 = X_{26} = 4.8$ kg, $\mathcal{O}_6 = Q_3 = 4.8$ kg, $\mathcal{O}_7 = X_{33} = 4.9$ kg; kurtosis $= 1.333$.
6.2 (a) $Z = (78.0$ g $- 63.5$ g$)/12.2$ g $= 1.19, P(X \geq 78.0$ g$) = P(Z \geq 1.19) = 0.1170$; (b) $P(X \leq 78.0$ g$) = 1.0000 - P(X \geq 78.0$ g$) = 1.000 - 0.1170 = 0.8830$; (c) $(0.1170)(1000) = 117$; (d) $Z = (41.0$ g $- 63.5$ g$)/12.2$ g $= -1.84, P(X \leq 41.0$ g$) = P(Z \leq -1.84) = 0.0329$
6.3 (a) $P(X \leq 60.0$ g$) = P(Z \leq -0.29) = 0.3859, P(X \geq 70.0$ g$) = P(Z \geq 0.53) = 0.2981, P(60.0$ g $\leq X \leq 70.0$ g$) = 1.0000 - 0.3859 - 0.2981 = 0.3160$; (b) $P(X \leq 60.0$ g$) = P(Z \leq -0.29) = 0.3859, P(X \leq 50.0$ g$) = P(Z \leq$

$-1.11) = 0.1335$, $P(50.0 \text{ g} \leq X \leq 60.0 \text{ g}) = P(-1.11 \leq Z \leq -0.29) = 0.3859 - 0.1335 = 0.2524.$

6.4 (a) $\sigma_{\bar{X}} = \sigma/\sqrt{n} = 12.2 \text{ g}/\sqrt{10} = 3.86 \text{ g}$; (b) $Z = (65.0 \text{ g} - 63.5)/3.86 \text{ g} = 0.39$, $P(\bar{X} \geq 65.0 \text{ g}) = P(Z \geq 0.39) = 0.3483$; (c) $P(\bar{X} \leq 62.0 \text{ g}) = P(Z \leq -0.39) = 0.3483$, $P(\bar{X} \leq 60.0 \text{ g}) = P(Z \leq -0.91) = 0.1814$, $P(60.0 \text{ g} \leq \bar{X} \leq 62.0 \text{ g}) = 0.3483 - 0.1814 = 0.1669.$

6.5 $g_1 = -0.7404$, $g_2 = 0.8686$; $A = -1.91027$, $B = 3.52198$, $C = 1.24588$, $D = 1.11615$, $E = 3.01669$, $F = -0.66979$, $Z_{g_1} = -1.8937$; $G = 0.43560$, $H = 1.14482$, $J = 1.68221$, $K = 19.04218$, $L = 0.63140$, $Z_{g_2} = 12074$; $K^2 = 5.044$; $\chi^2_{0.05,2} = 5.991$, do not reject H_0 of normality, $0.5 < P < 0.10$ [$P = 0.080$].

6.6 $\sigma^2 = 7.79 \text{ kg}^2$, $n = 20$, $\bar{X} = -1.1 \text{ kg}$, $\sigma_{\bar{X}} = 0.62 \text{ kg}$, $Z = -1.77$. (a) H_0: $\mu = 0$ kg, H_A: $\mu \neq 0$ kg; $P = 0.0384 + 0.0384 = 0.0768$, do not reject H_0. (b) H_0: $\mu \geq 0$ kg, H_A: $\mu < 0$ kg; $P = 0.0384$, reject H_0. (c) H_0: $\mu \leq 0$ kg, H_A: $\mu > 0$ kg; $P = 1.0000 - 0.0384 = 0.9616$; do not reject H_0.

CHAPTER 7

7.1 H_0: $\mu = 29.5$ days, H_A: $\mu \neq 29.5$ days, $\bar{X} = 27.7$ days, $s_{\bar{X}} = 0.708$ days, $n = 15$, $t = 2.542$, $\nu = 15 - 1 = 14$, $t_{0.05(2),14} = 2.145$, $0.02 < P(|t| \geq 2.542) < 0.05$ [$P = 0.023$]; therefore, reject H_0 and conclude that the sample came from a population with a mean that is not 29.5 days.

7.2 H_0: $\mu \geq 32$ mmole/kg, H_A: $\mu < 32$ mmole/kg, $\bar{X} = 29.77$ mmole/kg, $s_{\bar{X}} = 0.5$ mmole/kg, $n = 13$, $t = -4.46$, $\nu = 12$, $t_{0.05(1),12} = 1.782$, $P(t < -4.46) < 0.0005$ [$P = 0.00039$]; therefore, reject H_0 and conclude that the sample came from a population with a mean less than 32 mmole/kg.

7.3 Graph, which includes three 95% confidence intervals: 0.458 ± 0.057 kcal/g; 0.413 ± 0.059 kcal/g; 0.327 ± 0.038 kcal/g.

7.4 (a) 13.55 ± 1.26 cm; (b) $n = 28$; (c) $n = 9$; (d) $n = 15$.

7.5 (a) $n = 30$; (b) $n = 41$; (c) $n = 42$; (d) $d = 2.2$ cm; (e) $t_{\beta(1),24} = 1.378$, $0.05 < \beta < 0.10$, so $0.90 <$ power < 0.95; or, by normal approximation, $\beta = 0.08$ and power $= 0.92$ [$\beta = 0.09$ and power $= 0.91$].

7.6 (a) $N = 200$, $n = 50$, $s^2 = 97.8121 \text{ yr}^2$, $t_{0.05(2),49} = 2.010$; $s_{\bar{X}} = 1.2113$ yr. 95% confidence interval $= 53.87 \pm 2.43$ yr; (b) $t_{0.05(2),99} = 1.984$; $s_{\bar{X}} = 0.6993$ yr, 95% confidence interval $= 53.87 \pm 1.39$ yr.

7.7 (a) $s^2 = 6.4512$, $n = 18$, $SS = 109.6704 \text{ cm}^2$; $\chi^2_{0.025,17} = 30.191$, $\chi^2_{0.975,17} = 7.564$; $L_1 = 3.6326 \text{ cm}^2$, $L_2 = 24.0294 \text{ cm}^2$. (b) $s = 2.54$ cm; $L_1 = 1.91$ cm, $L_2 = 4.90$ cm. (c) $\chi^2 =$

24.925, $\chi^2_{0.05,17} = 27.587$; 24.925 is not > 27.587, so do not reject H_0; $0.05 < P < 0.10$ [$P = 0.096$]. (d) $\sigma^2 = 9.000 \text{ cm}^2$, $\chi^2 = 12.186$, $\chi^2_{0.95,17} = 8.672$; 12.186 is not < 8.672, so do not reject H_0; $0.10 < P < 0.25$ [$P = 0.21$]. (e) For $\nu = 17$, $1 - \beta = P[\chi^2 \geq (25.587)(4.4000) \text{ cm}^2/6.4512 \text{ cm}^2] = P(\chi^2 \geq 18.816)$, $0.25 <$ power < 0.50 [power $= 0.34$]. (f) $\sigma_0^2/s^2 = 0.682$; by trial and error: $n = 71$, $\nu = 70$, $\chi^2_{0.75,70}/\chi^2_{0.05,70} = 61.698/90.531 = 0.9682$.

7.8 H_0: The sampled population has a normal distribution; H_A: The sampled population does not have a normal distribution. From Exercise 6.1: $n = 37$, $g_1 = -0.7404$, $g_2 = -0.8686$. Therefore, A$=-1.9910268$, B $= 3.521981$, C $= 1.245877$, D $= 1.116189$, E $= 3.016209$, F $= -0.669789$, and $Z_{g_1} = -1.8934$; and G $= 0.435596$, H $= -1.144824$, J $= 1.682217$, K $= 19.042080$, L $= 1.536279$, and $Z_{g_2} = -1.5324$. $K^2 = 5.9332$, $\nu = 2$, $0.05 < P < 0.10$ [$P = 0.051$]; do not reject H_0.

CHAPTER 8

8.1 H_0: $\mu_1 = \mu_2$, H_A: $\mu_1 \neq \mu_2$, $n_1 = 7$, $SS_1 = 108.6171 \text{ (mg/100 ml)}^2$, $\bar{X}_1 = 224.24$ mg/100 ml, $\nu_1 = 6$, $n_2 = 6$, $SS_2 = 74.7533 \text{ (mg/100 ml)}^2$, $\bar{X}_2 = 225.67$ mg/100 ml, $\nu_2 = 5$, $s_p^2 = 16.6700 \text{ (mg/100 ml)}^2$, $s_{\bar{X}_1-\bar{X}_2} = 2.27$ mg/100 ml, $t = -0.630$, $t_{0.05(2),11} = 2.201$; therefore, do not reject H_0; $P > 0.50$ [$P = 0.54$].

8.2 H_0: $\mu_1 \geq \mu_2$, H_A: $\mu_1 < \mu_2$, $n_1 = 7$, $SS_1 = 98.86 \text{ mm}^2$, $\nu_1 = 6$, $\bar{X}_1 = 117.9$ mm, $n_2 = 8$, $SS_2 = 62.88 \text{ mm}^2$, $\nu_2 = 7$, $\bar{X}_2 = 118.1$ mm, $s_p^2 = 12.44 \text{ mm}^2$, $s_{\bar{X}_1-\bar{X}_2} = 1.82$ mm, $t = -0.11$, $t_{0.05(1),13} = 1.771$; therefore, do not reject H_0; $P > 0.25$ [$P = 0.46$].

8.3 H_0: $\mu_1 \geq \mu_2$, H_A: $\mu_1 < \mu_2$, $\bar{X}_1 = 4.6$ kg, $s_1^2 = 3.88 \text{ kg}^2$, $n_1 = 18$, $\nu_1 = 17$, $\bar{X}_2 = 6.0$ kg, $s_2^2 = 4.35 \text{ kg}^2$, $n_2 = 26$, $\nu_2 = 25$, $s_p^2 = 4.16 \text{ kg}^2$, $s_{\bar{X}_1-\bar{X}_2} = 0.62$ kg, $t = -2.26$, $t_{0.05(1),42} = 1.682$; therefore, reject H_0; $0.01 < P < 0.025$ [$P = 0.016$].

8.4 H_0: $\mu_2 - \mu_1 \leq 10$ g, H_A: $\mu_2 - \mu_1 > 10$ g, $\bar{X}_1 = 334.6$ g, $SS_1 = 364.34 \text{ g}^2$, $n_1 = 19$, $\nu_1 = 18$, $\bar{X}_2 = 349.8$ g, $SS_2 = 286.78 \text{ g}^2$, $n_2 = 24$, $\nu_2 = 23$, $s_p^2 = 15.88 \text{ g}^2$, $s_{\bar{X}_1-\bar{X}_2} = 1.22$ g, $t = 4.26$, $t_{0.05(1),41} = 1.683$; therefore, reject H_0 and conclude that μ_2 is at least 10 g greater than μ_1; $P < 0.0005$ [$P = 0.00015$].

8.5 H_0 is not rejected; $\bar{X}_p = [(7)(224.24 \text{ mg/100 ml}) + (6)(225.67 \text{ mg/100 ml})]/(7+6) = 224.90$ mg/100ml, $s_p^2 = 16.6700 \text{(mg/100 ml)}^2$, $t_{0.05(2),11} = 2.201$; 95% confidence interval for $\mu = 224.9$ mg/100 ml $\pm 2.201\sqrt{16.6700 \text{(mg/100 ml)}^2/13} = 224.9$ mg/100 ml ± 2.5 mg/100 ml; $L_1 = 222.4$ mg/100 ml, $L_2 = 227.4$ mg/100 ml.

8.6 $s_p^2 = (244.66 + 289.18)/(13 + 13) = 20.53(\text{km/hr})^2$; $d = 2.0$ km/hr. If we guess $n = 50$, then $\nu = 2(50 - 1) = 98$, $t_{0.05(2),98} = 1.984$, and $n = 40.4$. Then, guess $n = 41$; $\nu = 80$, $t_{0.05(2),80} = 1.990$, and $n = 40.6$. So, the desired $n = 41$.

8.7 (a) If we guess $n = 25$, then $\nu = 2(24) = 48$. $t_{0.05(2),48} = 2.011$, $t_{0.10(1),48} = 1.299$, and $n = 18.0$. Then, guess $n = 18$; $\nu = 34$, $t_{0.05(2),34} = 2.032$, $t_{0.10(1),34} = 1.307$, and $n = 18.3$. So, the desired sample size is $n = 19$. (b) $n = 20.95$, $\nu = 40$, $t_{0.05(2),40} = 2.021$, $t_{0.10(1),40} = 1.303$, and $\delta = 4.65$ km/hr. (c) $n = 50$, $\nu = 98$, $t_{0.05(2),98} = 1.984$, and $t_{\beta(1),98} = 0.223$; $\beta > 0.25$, so power < 0.75 (or, by the normal approximation, $\beta = 0.41$, so power $= 0.59$).

8.8 H_0: $\sigma_1^2 = \sigma_2^2$, H_A: $\sigma_1^2 \neq \sigma_2^2$, $F = 14.62$ cm^2/ 8.45 cm$^2 = 1.73$, $\nu_1 = 28$, $\nu_2 = 24$, $F_{0.05(2),28,24} = 2.23$; therefore, do not reject H_0; $0.10 < P < 0.20$ [$P = 0.088$].

8.9 H_0: $\sigma_1^2 \leq \sigma_2^2$, H_A: $\sigma_1^2 > \sigma_2^2$, $F = 324.46$ sec^2/ 158.95 sec$^2 = 2.04$, $\nu_1 = 40$, $\nu_2 = 35$, $F_{0.05(1),40,35} = 1.74$; therefore, reject H_0; $0.01 < P < 0.025$ [$P = 0.017$].

8.10 (a) $s_2^2/s_1^2 = 1.813$, $\nu_2 = 12$, $\nu_1 = 14$, $\alpha = 0.05$, $L_1 = 1.813(1/3.05) = 0.594$, $L_2 = 1.813(3.21) = 5.82$. (b) $\alpha(1) = 0.05$, $\beta(1) = 0.10$, $z_{0.05(1)} = 1.6499$, $z_{0.10(1)} = 1.2816$; $s_1^2 = 21.36$ g^2, $s_2^2 = 38.71$ g^2; $s_2^2/s_1^2 = 1.81$; $n = 26.3$, so $n_1 = n_2 = 27$. (c) $n = 50$, $\beta(1) = 2.4661$, $0.005 < P < 0.01$ [$P = 0.0068$].

8.11 H_0: $\mu_1/\sigma_1 = \mu_2/\sigma_2$, H_A: $\mu_1/\sigma_1 \neq \mu_2/\sigma_2$; $s_1 = 3.82$ cm, $V_1 = 0.356$, $s_2 = 2.91$ cm, $V_2 = 0.203$; $V_p = 0.285$; $Z = 2.533$, $Z_{0.05(2)} = 1.960$; reject H_0; $0.01 < P < 0.02$ [$P = 0.013$].

8.12 H_0: Male and female turtles have the same serum cholesterol concentrations; H_A: Male and female turtles do not have the same serum cholesterol concentrations.

Male ranks	Female ranks
2	5
1	3
11	12
10	8
4	6
7	13
9	

$R_1 = 44$, $n_1 = 7$, $R_2 = 47$, $n_2 = 6$; $U = 26$; $U' = (7)(6) - 26 = 16$; $U_{0.05(2),7,6} = U_{0.05(2),6,7} = 36$; therefore, do not reject H_0; $P > 0.20$.

8.13 H_0: Northern birds do not have shorter wings than southern birds; H_A: Northern birds have shorter wings than southern birds.

Northern ranks	Southern ranks
11.5	5
1	7
15	13
8.5	2.5
5	5
2.5	8.5
10	14
	11.5

$R_1 = 53.5$, $n_1 = 7$, $n_2 = 8$; $U = 30.5$; $U' = 25.5$; $U_{0.05(1),7,8} = 43$; therefore, do not reject H_0; $P > 0.10$.

8.14 H_0: Intersex cells have 1.5 times the volume of normal cells; H_A: Intersex cells do not have 1.5 times the volume of normal cells.

Normal $\times 1.5$	Rank	Intersex	Rank
372	4	380	9
354	1	391	13
403.5	16	377	8
381	10	392	14
373.5	5	398	15
376.5	7	374	6
390	12		
367.5	3		
358.5	2		
382.5	11		

$R_1 = 71$, $n_1 = 10$, $n_2 = 6$; $U = 44$, $U' = 16$; $U_{0.05(2),10,6} = 49$; therefore, do not reject H_0; $0.10 < P < 0.20$.

CHAPTER 9

9.1 H_0: $\mu_d = 0$, H_A: $\mu_d \neq 0$; $\bar{d} = -2.09$ μg/m^3, $s_{\bar{d}} = 1.29$ μg/m^3. (a) $t = 1.62$$n = 11$, $\nu = 10$, $t_{0.05(2),10} = 2.228$; therefore, do not reject H_0; $0.10 < P < 0.20$ [$P = 0.14$]. (b) 95% confidence interval for $\mu_d = -2.09 \pm (2.228)(1.29) = -2.09 \pm 2.87$; $L_1 = -4.96$ μg/m^3, $L_2 = 0.78$ μg/m^3.

9.2

d_i	Signed rank
−4	−5.5
−2	−3
−5	−7.5
6	9
−5	−7.5
1	1.5
−7	−10.5
−4	−5.5
−7	−10.5
1	1.5
3	4

$T = 9 + 1.5 + 1.5 + 4 = 16$; $T_{0.05(2),11} = 10$; since T is not ≤ 10, do not reject H_0; $0.10 < P < 0.20$.

9.3 H_0: There is no difference in frequency of occurrence of varicose veins between overweight and normal weight men; H_A: There is a difference in frequency of occurrence of varicose veins between overweight men and normal weight men; $f_{11} = 19$, $f_{12} = 5$, $f_{21} = 12$, $f_{22} = 86$, $n = 122$; $\chi^2 = 2.118$; $\chi^2_{0.05,1} = 3.841$; do not reject H_0; $0.10 < P < 0.25$ $[P = 0.15]$.

CHAPTER 10

10.1 H_0: $\mu_1 = \mu_2 = \mu_3 = \mu_4$; H_A: The mean food consumption is not the same for all four months; $F = 0.7688/0.0348 = 22.1$; $F_{0.05(1),3,18} = 3.16$; reject H_0; $P < 0.0005$ $[P = 0.0000029]$.

10.2 $k = 5$, $\nu_1 = 4$, $n = 12$, $\nu_2 = 55$, $\sigma^2 = 1.54(°C)^2$, $\delta = 2.0°C$; $\phi = 1.77$; from Appendix Fig. B.1d we find that the power is about 0.88.

10.3 $n = 16$, for which $\nu_2 = 75$ and $\phi = 2.04$. (The power is a little greater than 0.95; for $n = 15$ the power is about 0.94.)

10.4 $\nu_2 = 45$, power $= 0.95$, $\phi = 2.05$; minimum detectable difference is about 2.5°C.

10.5 H_0: The amount of food consumed is the same during all four months; H_A: The amount of food consumed is not the same during all four months; $n_1 = 5$, $n_2 = 6$, $n_3 = 6$, $n_4 = 5$; $R_1 = 69.5$, $R_2 = 23.5$, $R_3 = 61.5$, $R_4 = 98.5$; $N = 22$; $H = 17.08$; $\chi^2_{0.05,3} = 7.815$; reject H_0; $P \ll 0.001$. H_c (i.e., H corrected for ties) would be obtained as $\sum t = 120$, $C = 0.9887$, $H_c = 17.28$. $F = 27.9$, $F_{0.05(1),3,17} = 3.20$; reject H_0; $P \ll 0.0005$ $[P = 0.00000086]$.

10.6 H_0: $\sigma_1^2 = \sigma_2^2 = \sigma_3^2$; H_A: The three population variances are not all equal; $B = 5.94517$, $C = 1.0889$, $B_c = B/C = 5.460$; $\chi^2_{0.05,2} = 5.991$; do not reject H_0; $0.05 < P < 0.10$ $[P = 0.065]$.

10.7 H_0: $\mu_1/\sigma_1 = \mu_2/\sigma_2 = \mu_3/\sigma_3 = \mu_4/\sigma_4$; $s_1 = 0.699$, $V_1 = 0.329$, $s_2 = 0.528$, $V_2 = 0.302$, $s_3 = 0.377$, $V_3 = 0.279$, $s_4 = 0.451$, $V_4 = 0.324$; $V_p = 0.304$; $\chi^2 = 1.320$, $\chi^2_{0.05,3} = 7.815$; do not reject H_0; $0.50 < P < 0.75$ $[P = 0.72]$.

CHAPTER 11

11.1 **(a, b)** Ranked sample means: $\underline{14.8 \quad 16.2 \quad 20.2}$; $k = 3$, $n = 8$, $\alpha = 0.05$, $s^2 = 8.46$, $\nu = 21$ (which is not in Appendix Table B.5, so use $\nu = 20$, which is in the table); reject H_0: $\mu_2 = \mu_1$; reject H_0: $\mu_2 = \mu_3$; do not reject H_0: $\mu_3 = \mu_1$. Therefore, the overall conclusion is $\mu_1 = \mu_3 \neq \mu_2$. **(c)** $\bar{X}_p = \bar{X}_{1,3} = 15.5$, $t_{0.05(2),21} = 2.080$, $n_1 + n_2 = 16$, 95% CI for $\mu_{1,3} = 15.5 \pm 1.5$; 95% CI for $\mu_2 = 20.2 \pm 2.1$; $\bar{X}_{1,3} - \bar{X}_2 = -4.7$, SE $= 1.03$, 95% CI for $\mu_{1,3} - \mu_2 = -4.7 \pm 3.2$.

11.2 Ranked sample means: 60.62, 69.30, 86.24, 100.35; sample sizes of 5, 5, 5, and 4, respectively; $k = 4$, $\nu = 15$, $\alpha = 0.05$, $s^2 = 8.557$; control group is group 1; $q'_{0.05(2),15,4} = 2.61$; reject H_0: $\mu_4 = \mu_1$, reject H_0: $\mu_3 = \mu_1$, reject H_0: $\mu_2 = \mu_1$. Overall conclusion: The mean of the control population is different from the mean of each other population.

11.3 Ranked sample means: 60.62, 69.30, 86.24, 100.35; sample sizes of 5, 5, 5, and 4, respectively; $k = 4$, $\nu = 15$, $\alpha = 0.05$, $s^2 = 8.557$; critical value of S is 3.14; for H_0: $(\mu_1 + \mu_4)/2 - (\mu_2 + \mu_3)/2 = 0$, $S = 8.4$, reject H_0; for H_0: $(\mu_2 + \mu_4)/2 - \mu_3 = 0$, $S = 13.05$, reject H_0.

11.4 $R_1 = 21$, $R_2 = 38$, $R_3 = 61$. Overall conclusion: The variable being measured is the same magnitude in populations 1 and 2. The variable is of different magnitude in population 3.

CHAPTER 12

12.1 **(a)** H_0: There is no difference in mean hemolymph alanine among the three species; H_A: There is difference in mean hemolymph alanine among the three species; $F = 27.6304/2.1121 = 13.08$; $F_{0.05(1),2,18} = 3.55$; reject H_0; $P < 0.0005$ $[P = 0.00031]$. **(b)** H_0: There is no difference in mean hemolymph alanine between males and females; H_A: There is difference in mean hemolymph alanine between males and females; $F = 138.7204/2.1121 = 65.68$; $F_{0.05(1),1,18} = 4.41$; reject H_0; $P \ll 0.0005$ $[P = 0.00000020]$. **(c)** H_0: There is no species $\times$ sex interaction in mean hemolymph alanine; H_A: There is species $\times$ sex interaction in mean hemolymph alanine; $F = 3.4454/2.1121 = 1.63$; $F_{0.05(1),2,18} = 3.55$; do not reject H_0; $0.10 < P < 0.25$ $[P = 0.22]$. **(d)** See graph below; the wide vertical distance between the open circles indicates difference between sexes; the vertical distances among the plus signs indicates difference among the three species;

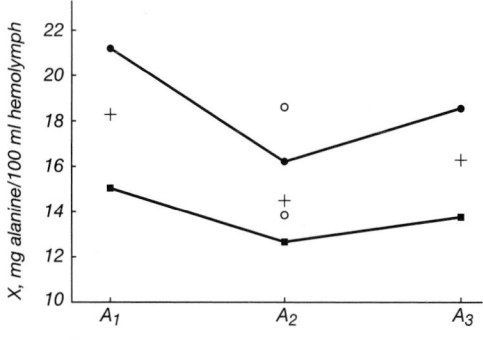

X, mg alanine/100 ml hemolymph

A_1 A_2 A_3

Exercise 12.1d

the parallelism of the male and female lines indicates no interaction effect. (**e**) Ranked sample means: 14.43 16.13 18.14 (means 2, 3, and 1, respectively); $k = 3$; $n = 8$; $\alpha = 0.05$; $s^2 = 2.1121$; $\nu = 18$; reject H_0: $\mu_1 = \mu_2$, reject H_0: $\mu_1 = \mu_3$, do not reject H_0: $\mu_3 = \mu_2$.

12.2 H_0: All four plant varieties reach the same mean height (i.e., H_0: $\mu_1 = \mu_2 = \mu_3 = \mu_4$); H_A: All four plant varieties do not reach the same mean height; $F = 62.8461/0.4351 = 144$; $F_{0.05(1),3,15} = 3.29$; reject H_0; $P \ll 0.0005$ [$P < 10^{-10}$].

12.3 H_0: All four plant varieties reach the same height; H_A: All four plant varieties do not reach the same height; $R_1 = 18$, $R_2 = 24$, $R_3 = 12$, $R_4 = 6$; $\chi^2_{0.05,3} = 7.815$; reject H_0; $P < 0.001$.

12.4 H_0: There is no difference in potential acceptance among the three textbooks; H_A: The three textbooks do not have the same potential acceptance; $a = 4$, $b = 13$ (blocks 4 and 11 are deleted from the analysis); $Q = 5.53$; $\nu = 3$; $\chi^2_{0.05,3} = 7.815$; do not reject H_0; $0.10 < P < 0.25$ [$P = 0.14$].

CHAPTER 13

13.1 $\bar{X}' = 0.68339$, $s'_{\bar{X}} = 0.00363$; $L'_1 = 0.67481$, $L'_2 = 0.69197$; $L_1 = 3.73$ ml, $L_2 = 3.92$ ml.

13.2 $\bar{X}' = 61.48$, $s'_{\bar{X}} = 0.76$; $L'_1 = 59.53$, $L'_2 = 63.43$; $L_1 = 0.742$, $L_2 = 0.800$.

13.3 $\bar{X}' = 2.4280$, $s'_{\bar{X}} = 0.2329$; $L'_1 = 1.8292$, $L'_2 = 3.0268$; $L_1 = 2.85$, $L_2 = 8.66$.

CHAPTER 14

14.1 H_0: No effect of factor A; H_A: Factor A has an effect; $F = 10.07890/0.08370 = 120.4$; as $F_{0.05(1),3,72} \cong 2.74$, H_0 is rejected; $P \ll 0.0005$ [$P < 10^{-12}$]. H_0: No effect of factor B; $F = 3.77918/0.08370 = 45.2$; as $F_{0.05(1),2,72} \cong 3.13$, H_0 is rejected; $P \ll 0.0005$ [$P < 10^{-12}$]. H_0: No effect of factor C; $F = 4.25924/0.08370 =$

50.9; as $F_{0.05(1),1,72} \cong 3.98$, H_0 is rejected; $P \ll 0.0005$ [$P < 10^{-19}$]. H_0: No interaction between factors A and B; H_A: There is $A \times B$ interaction; $F = 0.10932/0.08370 = 1.31$; as $F_{0.05(1),6,72} \cong 2.23$, H_0 is not rejected; $P > 0.25$ [$P = 0.26$]. H_0: No interaction between factors A and C; $F = 0.60984/0.08370 = 7.29$; as $F_{0.05(1),3,72} \cong 2.74$, H_0 is rejected; $P < 0.0005$ [$P = 0.00025$]. H_0: No interaction between factors B and C; $F = 0.00164/0.08370 = 0.020$; as $F_{0.05(1),2,72} \cong 1.41$, H_0 is not rejected; $P > 0.25$ [$P = 0.98$]. H_0: No interaction between factors A, B, and C; $F = 0.12785/0.08370 = 1.53$; as $F_{0.05(1),6,72} \cong 2.23$, H_0 is not rejected; $0.10 < P < 0.25$ [$P = 0.98$].

14.2 H_0: No effect of factor A; H_A: Factor A has an effect; $F = 56.00347/0.03198 = 1751$; reject H_0; $P \ll 0.0005$ [$P < 10^{-14}$]. H_0: No effect of factor B; $F = 4.65125/0.03198 = 145.4$; reject H_0; $P \ll 0.0005$ [$P < 10^{-13}$]. H_0: No effect of factor C; $F = 8.6125/0.03198 = 269.3$; reject H_0; $P \ll 0.0005$ [$P < 10^{-13}$]. H_0: No effect of factor D; $F = 2.17056/0.03198 = 67.9$; reject H_0; $P \ll 0.0005$ [$P < 10^{-14}$]. H_0: No interaction between factors A and B; $F = 2.45681/0.03198 = 76.8$; reject H_0; $P \ll 0.0005$ [$P < 10^{-11}$]. H_0: No interaction between factors A and C; $F = 0.05014/0.03198 = 1.57$; do not reject H_0; $0.10 < P < 0.25$ [$P = 0.22$]. H_0: No interaction between factors A and D; $F = 0.06889/0.03198 = 2.15$; do not reject H_0; $0.10 < P < 0.25$ [$P = 0.13$]. H_0: No interaction between factors B and C; $F = 0.01681/0.03198 = 0.53$; do not reject H_0; $P > 0.25$ [$P = 0.47$]. H_0: No interaction between factors B and D; $F = 0.15167/0.03198 = 4.74$; reject H_0; $0.01 < P < 0.025$. H_0: No interaction between factor C and D; $F = 0.26000/0.03198 = 8.13$; reject H_0; $0.0005 < P < 0.001$ [$P = 0.00091$]. H_0: No interaction among factors A, B, and C; $F = 0.00125/0.03198 = 0.039$; do not reject H_0; $P > 0.25$ [$P = 0.84$]. H_0: No interaction among factors A, B, and D; $F = 0.14222/0.03198 = 2.11$; do not reject H_0; $0.10 < P < 0.25$ [$P = 0.13$]. H_0: No interaction among factors B, C, and D; $F = 0.00222/0.03198 = 0.069$; do not reject H_0; $P > 0.25$ [$P = 0.093$]. H_0: No interaction among factors A, B, C, and D; $F = 0.01167/0.03198 = 0.36$; do not reject H_0; $P > 0.25$ [$P = 0.70$].

14.3 H_0: No effect of factor A; H_A: There is an effect of factor A; $F = 239.39048/2.10954 = 113.5$; as $F_{0.05(1),1,22} = 4.30$, reject H_0; $P \ll 0.0005$ [$P < 10^{-9}$]. H_0: No effect of factor B; $F = 8.59013/2.10954 = 4.07$; as $F_{0.05(1),1,22} = 4.30$; do not reject H_0; $0.05 < P < 0.10$ [$P = 0.056$]. H_0: No interaction between factors A and B; $F = 0.10440/2.10954 = 0.05$; as $F_{0.05(1),1,22} = 4.30$, do not reject H_0; $P > 0.25$ [$P = 0.83$].

14.4 There is one among-subjects factor (sex) and one within-subjects factor (feeding condition). H_0: No difference (in mean plasma protein) between males and females; $F = 105.0016/21.9817 = 4.78$; $F_{0.05(1),1,6} = 5.99$; so do not reject H_0; $0.05 < P < 0.10$ [$P = 0.072$]. H_0: No difference among feeding conditions; $F = 22.9817/0.4000 = 58.5$; $F_{0.05(1),2,12} = 3.89$, so reject H_0; $P \ll 0.0005$ [$P = 0.00000071$]. H_0: No interaction effect between sexes and conditions; $F = 0.0174/0.4000 = 0.044$; $F_{0.05(1),2,12} = 3.89$, so do not reject H_0; $P > 0.25$ [$P = 0.096$].

14.5 There are two among-subjects factors (species and temperature) and one within-subjects factor (exercise). H_0: No difference (in mean O_2 consumption) among species; $F = 0.005988/0.000081 = 73.9$; $F_{0.05(1),3,8} = 4.07$, so reject H_0; $P \ll 0.0005$ [$P = 0.0000036$]. H_0: No difference between temperatures; $F = 0.006699/0.000081 = 8.27$; $F_{0.05(1),1,8} = 5.32$, so reject H_0; $0.01 < P < 0.025$ [$P = 0.021$]. H_0: No difference between the two exercise conditions; $F = 0.020150/0.000014 = 1439$, so reject H_0; $F_{0.05(1),1,8} = 5.32$, so reject H_0; $P \ll 0.0005$ [$P < 10^{-10}$]. H_0: No interaction between species and exercise; $F = 0.000012/0.000014 = 0.86$; $F_{0.05(1),3,8} = 4.07$, so do not reject H_0; $P > 0.25$ [$P = 0.50$]. H_0: No effect of interaction between temperature and exercise; $F = 0.000001/0.000014 = 0.071$; $F_{0.05(1),1,8} = 5.32$, so do not reject H_0; $P > 0.25$ [$P = 0.80$]. H_0: No effect of the interaction among species, temperature, and exercise; $F = 0.000101/0.000014 = 7.21$; $F_{0.05(1),3,8} = 4.07$, so reject H_0; $0.01 < P < 0.025$ [$P = 0.012$].

14.6 There is one among-subjects factor (species) and two within-subjects factors (season and water depth). H_0: There is no difference (in mean sodium content) between species; $F = 6110.03/551.44 = 11.1$; $F_{0.05(1),1,4} = 7.71$, so reject H_0; $0.025 < P < 0.05$ [$P = 0.029$]. H_0: There is no difference among seasons; $F = 644.20/30.69 = 21.0$; $F_{0.05(1),2,8} = 4.46$, so reject H_0; $0.0005 < P < 0.001$ [$P = 0.00066$]. H_0: There is no interaction between species and seasons; $F = 70.53/30.69 = 2.30$; $F_{0.05(1),2,8} = 4.46$, so do not reject H_0; $0.10 < P < 0.25$ [$P = 0.16$]. H_0: There is no difference between depths; $F = 90.25/8.17 = 11.0$; $F_{0.05(1),1,4} = 7.71$, so reject H_0; $0.025 < P < 0.05$ [$P = 0.029$]. H_0: There is interaction between depth and species; $F = 1.13/8.17 = 0.14$; $F_{0.05(1),2,4} = 6.94$, so do not reject H_0; $P > 0.25$ [$P = 0.87$]. H_0: There is no interaction between season and depth; $F = 5.09/11.14 = 0.46$; $F_{0.05(1),2,7} = 4.74$, so do not reject H_0; $P > 0.25$ [$P = 0.65$]. H_0: There is no effect of the interaction among season, depth, and species;

$F = 2.08/11.14 = 0.19$; $F_{0.05(1),2,7} = 4.74$, so do not reject H_0; $P > 0.25$ [$P = 0.83$].

CHAPTER 15

15.1 H_0: The mean fluoride concentrations are the same for all three samples at a given location; H_A: The mean fluoride concentrations are not the same for all three samples at a given location; $F = 0.008333/0.01778 = 0.469$; as $F < 1.0$, so do not reject H_0; $P > 0.25$ [$P = 0.82$]. H_0: The mean fluoride concentration is the same at all three locations; H_A: The mean fluoride concentration is not the same at all three locations: $F = 1.1850/0.008333 = 142$; $F_{0.05(1),2,6} = 5.14$; reject H_0; $P \ll 0.0005$ [$P = 0.0000086$].

CHAPTER 16

16.1 H_0: $\mu_{11} = \mu_{12} = \mu_{13}$ and $\mu_{21} = \mu_{22} = \mu_{23}$; Wilks' $\Lambda = 0.0872$, Pillai's trace $= 0.9128$, Lawley-Hotelling trace $= 10.4691$, Roy's maximum root $= 10.4681$; for each, $F = 41.8723$, $P \ll 0.0001$; reject H_0.

16.2 For hormone treatment, H_0: $\mu_{11} = \mu_{12}$ and $\mu_{21} = \mu_{22}$; Wilks' $\Lambda = 0.1820$, Pillai's trace $= 0.8180$, Lawley-Hotelling trace $= 4.4956$, Roy's maximum root $= 4.4956$; for each, $F = 33.7167$, $P \ll 0.0001$; reject H_0. For sex, H_0: $\mu_{11} = \mu_{12}$ and $\mu_{21} = \mu_{22}$; Wilks' $\Lambda = 0.8295$, Pillai's trace $= 0.1705$, Lawley-Hotelling trace $= 0.2055$, Roy's maximum root $= 0.2055$; for each, $F = 1.5415$, $P = 0.2461$; do not reject H_0. For hormone $\times$ sex interaction, H_0: There is no interaction; Wilks' $\Lambda = 0.8527$, Pillai's trace $= 0.1473$, Lawley-Hotelling trace $= 0.1727$, Roy's maximum root $= 0.1727$; for each, $F = 1.2954$, $P = 0.3027$; do not reject H_0.

CHAPTER 17

17.1 (a) $b = -0.0878$ ml/g/hr/°C, $a = 3.471$ ml/g/hr. (b) H_0: $\beta = 0$, H_A: $\beta \neq 0$; $F = 309$; reject H_0; $P \ll 0.0005$ [$P = 0.0000022$]. (c) H_0: $\beta = 0$, H_A: $\beta \neq 0$; $t = -17.6$; reject H_0: $P \ll 0.001$ [$P = 0.0000011$]. (d) $s_{Y \cdot X} = 0.17$ ml/g/hr; (e) $r^2 = 0.98$; (f) 95% confidence interval for $\beta = -0.0878 \pm 0.0122$; $L_1 = -0.1000$ ml/g/hr/°C, $L_2 = -0.0756$ ml/g/hr/°C.

17.2 (a) $\hat{Y} = 3.47 - (0.878)(15) = 2.15$ ml/g/hr. (b) $s_{\hat{Y}} = 0.1021$ ml/g/hr; $L_1 = 1.90$ ml/g/hr, $L_2 = 2.40$ ml/g/hr. (c) $\hat{Y} = 2.15$ ml/g/hr. (d) $s_{\hat{Y}} = 0.1960$ ml/g/hr; $L_1 = 1.67$ ml/g/hr, $L_2 = 2.63$ ml/g/hr.

17.3 (a) $b = 9.73$ impulses/sec/°C, $a = 44.2$ impulses/sec. (b) H_0: $\beta = 0$, H_A: $\beta \neq 0$; $F = 311$; reject H_0; $P \ll 0.0005$ [$P < 10^{-13}$]. (c) $s_{Y \cdot X} = 8.33$

impulses/sec. **(d)** $r^2 = 0.94$. **(e)** H_0: The population regression is linear; H_A: The population regression is not linear; $F = 1.78$, do not reject H_0; $0.10 < P < 0.25$ $[P = 0.18]$.

CHAPTER 18

18.1 (a) $H_0: \beta_1 = \beta_2$; $b_1 = 0.488$, $b_2 = 0.537$; $s_{b_1 - b_2} = 0.202$; $t = -0.243$; as $t_{0.05(2),54} = 2.005$, do not reject H_0; $P > 0.50$ $[P = 0.81]$. **(b)** H_0: The elevations of the two population regressions are the same; H_A: The elevations of the two population regressions are not the same; $b_c = 0.516$; $t = 10.7$; as $t_{0.05(2),55} \cong 2.004$, reject H_0; $P \ll 0.001$ $[P = 2 \times 10^{-14}]$.

18.2 (a) $H_0: \beta_1 = \beta_2 = \beta_3$; H_A: All three β's are not equal; $F = 0.84$; as $F_{0.05(1),2,90} = 3.10$, do not reject H_0; $P > 0.25$ $[P = 0.44]$; $b_c = 3.16$. **(b)** H_0: The three population regression lines have the same elevation; H_A: The three lines do not all have the same elevation; $F = 4.61$; as $F_{0.05(1),2,90} = 3.10$, reject H_0; $0.01 < P < 0.025$ $[P = 0.012]$.

CHAPTER 19

19.1 (a) $r = 0.86$. **(b)** $r^2 = 0.73$. **(c)** $H_0: \rho = 0$; $H_A: \rho \neq 0$; $s_r = 0.16$; $t = 5.38$; as $t_{0.05(2),10} = 2.228$; reject H_0; $P < 0.001$ $[P = 0.00032]$. Or: $r = 0.86$, $r_{0.05(2),10} = 0.576$; reject H_0; $P < 0.001$. Or: $F = 13.29$, $F_{0.05(2),10,10} = 3.72$; reject H_0; $P < 0.001$. **(d)** $L_1 = 0.56$, $L_2 = 0.96$.

19.2 (a) $H_0: \rho \leq 0$; $H_A: \rho > 0$; $r = 0.86$; $t = 5.38$; $t_{0.05(1),10} = 1.812$; reject H_0; $P < 0.0005 [P = 0.00016]$. Or: $r_{0.05(1),10} = 0.497$; reject H_0; $P < 0.0005$. Or: $F = 13.29$; $F_{0.05(1),10,10} = 2.98$; reject H_0: $P < 0.0005$. **(b)** $H_0: \rho = 0.50$; $H_A: \rho \neq 0.50$; $r = 0.86$; $z = 1.2933$; $\zeta_0 = 0.5493$; $\sigma_z = 0.3333$; $Z = 2.232$; $Z_{0.05(2)} = 1.960$; reject H_0; $0.02 < P < 0.05$ $[P = 0.026]$.

19.3 (a) $H_0: \rho_1 = \rho_2$; $H_A: \rho_1 \neq \rho_2$; $z_1 = -0.4722$. $z_2 = -0.4236$; $\sigma_{z_1 - z_2} = 0.2910$; $Z = -0.167$; $Z_{0.05(2)} = 1.960$; do not reject H_0; $P > 0.50$ $[P = 0.87]$. **(b)** $z_w = -0.4449$; $r_w = -0.42$.

19.4 $H_0: \rho_1 \geq \rho_2$; $H_A: \rho_1 < \rho_2$; $z_1 = 0.4847$, $z_2 = 0.6328$; $\sigma_{z_1 - z_2} = 0.3789$; $Z = -0.3909$; $Z_{0.05(1)} = 1.645$; do not reject H_0; $P > 0.25$ $[P = 0.35]$.

19.5 (a) $H_0: \rho_1 = \rho_2 = \rho_3$; H_A: The three population correlation coefficients are not all the same; $\chi^2 = 111.6607 - (92.9071)^2/78 = 0.998$; $\chi^2_{0.05,2} = 5.991$; do not reject H_0; $0.50 < P < 0.75$ $[P = 0.61]$. $\chi^2_P = 1.095$, $0.50 < P < 0.75$ $[P = 0.58]$. **(b)** $z_w = 92.9071/78 = 1.1911$; $r_w = 0.83$.

19.6 (a) $\sum d_i^2 = 88.00$, $r_s = 0.69$; **(b)** $H_0: \rho_s = 0$; $H_A: \rho_s \neq 0$; as $(r_s)_{0.05(2),12} = 0.587$, reject H_0; $0.01 < P < 0.02$.

19.7 (a) $r_T = 0.914$; **(b)** reject H_0; $0.005 < P < 0.01$.

19.8 (a) $r_n = (16 - 7)/(16 + 7) = 0.39$. **(b)** H_0: There is no correlation between the type of institution a college president heads and the type of institution he or she attended as a undergraduate; H_A: There is a correlation between the type of school headed and the type attended. By Fisher exact test (using Appendix Table B.28): $n = 23$, $m_1 = 9$, $m_2 = 11$, $f = 2$, critical $f_{0.05(2)} = 1$ and 7; as f is not ≤ 1 and is not ≥ 7, do not reject H_0.

19.9 (a) $r_I = (0.000946 - 0.00213)/(0.000946 + 0.001213) = -0.12$. **(b)** H_0: There is no correlation between corticosterone determinations from the same laboratory (i.e., $\rho_I = 0$); H_A: There is no correlation between corticosterone determinations from the same laboratory (i.e., $\rho_I \neq 0$); $F = 0.000946/0.001213 = 0.78$; since $F_{0.05(1),3,4} = 6.59$, do not reject H_0; $P > 0.25$ $[P = 0.56]$.

19.10 (a) $r_c = 0.9685$. **(b)** $z_c = 2.0675$; $r = 0.9991$; $u = 0.031983$; $\sigma_z = 2.261$; for $\zeta_c : L_1 = 1.6243$, $L_2 = 2.5107$; for $\rho_c : L_1 = 0.925$, $L_2 = 0.987$.

CHAPTER 20

20.1 (a) $\hat{Y} = -30.14 + 2.07X_1 + 2.58X_2 + 0.64X_3 + 1.11X_4$. **(b)** H_0: No population regression; H_A: There is a population regression; $F = 90.2$, $F_{0.05(1),4,9} = 3.63$, reject H_0, $P \ll 0.0005$ $[P = 0.00000031]$. **(c)** $H_0: \beta_1 = 0$, $H_A: \beta_1 \neq 0$; $t_{0.05(2),9} = 2.262$; "*" below denotes significance:

i	b_i	s_{b_i}	$t = \dfrac{b_i}{s_{b_i}}$	Conclusion
1	2.07	0.46	4.50*	Reject H_0.
2	2.58	0.74	3.49*	Reject H_0.
3	0.64	0.46	1.39	Do not reject H_0.
4	1.11	0.76	1.46	Do not reject H_0.

(d) $s_{Y \cdot 1,2,3,4} = 3.11$ g; $R^2 = 0.9757$. **(e)** $\bar{Y} = 61.73$ g. **(f)** $s_{\hat{Y}} = 2.9549$ g, $L_1 = 55.0$ g, $L_2 = 68.4$ g. **(g)** $H_0: \mu_Y \leq 50.0$ g, $H_A: \mu_Y > 50.0$ g, $t = 3.970$; $t_{0.05(1),9} = 1.833$, reject H_0; $0.001 < P < 0.0025$ $[P = 0.0016]$.

20.2 (1) With X_1, X_2, X_3, and X_4 in the model, see Exercise 20.1c. **(2)** Delete X_3. With X_1, X_2, and X_4 in the model, $t_{0.05(2),10} = 2.228$ and:

i	b_i	t
1	1.48	9.15*
2	1.73	4.02*
4	0.21	0.50
	$a = 16.83$	

(3) Delete X_4. With X_1 and X_2 in the model, $t_{0.05(2),11} = 2.201$ and:

i	b_i	t
1	1.48	9.47*
2	1.53	13.19*
	$a = 24.96$	

(4) Therefore, the final equation is $\hat{Y} = 24.96 + 1.48X_1 + 1.53X_2$.

20.3 (a) $R = 0.9878$. (b) $F = 90.2$, $F_{0.05(1),4,9} = 3.63$, reject H_0: There is no population correlation among the five variables; $P \ll 0.0005$ [$P = 0.00000031$]. (c) Partial correlation coefficients:

	1	2	3	4	5
1	1.0000				
2	−0.9092*	1.0000			
3	−0.8203*	−0.8089*	1.0000		
4	−0.7578*	−0.9094*	−0.8724*	1.0000	
5	0.8342*	0.7583*	0.4183	0.4342	1.0000

(d) From Appendix Table B.17, $r_{0.05(2),9} = 0.602$, and the significant partial correlation coefficients are indicated with asterisks in part (c).

20.4 H_0: Each of the three sample regressions estimates the same population regression; H_A: Each of the three sample regressions does not estimate the same population regression; $F = 0.915$; as $F_{0.05(1),8,72} = 2.07$, do not reject H_0; $P > 0.25$ [$P = 0.51$].

20.5 (a) $W = 0.675$. (b) H_0: There is no agreement among the four faculty reviewers; H_A: There is agreement among the four faculty reviewers; $\chi_r^2 = 10.800$; $(\chi_r^2)_{0.05,4,5} = 7.800$; reject H_0; $0.005 < P < 0.01$.

CHAPTER 21

21.1 In each step, H_0: $\beta_i = 0$ vs. H_A: $\beta_i \neq 0$ is tested, where i is the highest term in the polynomial expression. An asterisk indicates H_0 is rejected. (1) Linear regression: $\hat{Y} = 8.8074 - 0.18646X$; $t = 7.136$*; $t_{0.05(2),13} = 2.160$. (2) Quadratic regression: $\hat{Y} = -14.495 + 1.6595X - 0.036133X^2$; $t = 5.298$*; $t_{0.05(2),12} = 2.179$. (3) Cubic regression: $\hat{Y} = -33.810 + 3.9550X - 0.12649X^2 + 0.0011781X^3$; $t = 0.374$; $t_{0.05(2),11} = 2.201$. (4) Quartic regression: $\hat{Y} = 525.30 - 84.708X + 5.1223X^2 - 0.13630X^3 + 0.0013443X^4$; $t = 0.911$; $t_{0.05(2),10} = 2.28$. Therefore, the quadratic expression is concluded to be the "best."

21.2 (a) $\hat{Y} = 1.00 + 0.851X - 0.0259X^2$. (b) H_0: $\beta_2 = 0$; H_A: $\beta_2 \neq 0$; $F = 69.4$; $F_{0.05(1),1,4} = 7.71$; reject

H_0; $0.001 < P < 0.0025$ [$P = 0.0011$]. (c) $\hat{Y} = 6.92$ eggs/cm^2; $s_{\hat{Y}} = 0.26$ eggs/cm^2; 95% confidence interval $= 6.92 \pm 0.72$ eggs/cm^2. (d) $\hat{X}_0 = 16.43°$C; $\hat{Y}_0 = 7.99$ eggs/ cm^2. (e) For $\hat{X}_0$: 95% confidence interval $= 16.47 \pm 0.65°$C; for $\hat{Y}_0$: 95% confidence interval $= 7.99 \pm 0.86$ eggs/cm^2.

CHAPTER 22

22.1 (a) For $\nu = 2$, $P(\chi^2 \geq 3.452)$ is between 0.10 and 0.25 [$P = 0.18$]; (b) For $\nu = 5$, $0.10 < (\chi^2 \geq 8.668) < 0.25$ [$P = 0.12$]; (c) $\chi^2_{0.05,4} = 9.488$; (d) $\chi^2_{0.01,8} = 20.090$.

22.2 (a) $\chi^2 = 16.000$, $\nu = 5$, $0.005 < P < 0.01$ [$P = 0.0068$]. As $P < 0.05$, reject H_0 of equal food item preference. (b) By grouping food items 1, 4, and 6: $n = 41$, and for H_0: Equal food preference, $\chi^2 = 0.049$, $\nu = 2$, $0.975 < P < 0.99$ [$P = 0.976$]; as $P > 0.05$, H_0 is not rejected. By grouping food items 2, 3, and 5: $n = 85$, and for H_0: Equal food preference, $\chi^2 = 0.447$, $\nu = 2$, $0.75 < P < 0.90$ [$P = 0.80$]; as $P > 0.05$, H_0 is not rejected. By considering food items 1, 4, and 6 as one group and items 2, 3, and 5 as a second group, and H_0: Equal preference for the two groups, $\chi_c^2 = 14.675$, $\nu = 1$, $P < 0.001$ [$P = 0.00013$]; H_0 is rejected.

22.3 $\chi_c^2 = 0.827$, $\nu = 1$, $0.25 < P < 0.50$ [$P = 0.36$]. As $P > 0.05$, do not reject H_0: The population consists in equal numbers of males and females.

22.4

Location	Males	Females	χ^2	ν
1	44	54	1.020	1
2	31	40	1.141	1
3	12	18	1.200	1
4	15	16	0.032	1
Total of chi-squares			3.393	4
Pooled chi-square	102	128	2.939	1
Heterogeneity chi-square			0.454	3

$$0.90 < P < 0.95$$

Because $P(\text{heterogeneity } \chi^2) > 0.05$, the four samples may be pooled with the following results: $\chi_c^2 = 2.717$, $\nu = 1$, $0.05 < P < 0.10$ [$P = 0.099$]; $P > 0.05$, so do not reject H_0: Equal numbers of males and females in the population.

22.5 $G = 16.188$, $\nu = 5$, $0.005 < P < 0.01$ [$P = 0.0063$]; $P < 0.05$, so reject H_0: No difference in food preference.

22.6 $d_{max} = 1$; $(d_{max})_{0.05,6,18} = 5$; do not reject H_0: The feeders are equally desirable to the birds; $P > 0.50$.

22.7 max $D_i = 0.2519$; max $D'_i = 0.2197$; $D = 0.2519$; $D_{0.05,31} = 0.23788$; reject H_0: There is a uniform distribution of these animals from the water's edge to a distance of 10 meters upland; $0.02 < P < 0.05$.

22.8 $D_{0.05,27} = 0.25438$ and $D_{0.05,28} = 0.24993$, so a sample size of at least 28 is called for.

CHAPTER 23

23.1 (a) $F_{11} = 157.1026$, $F_{12} = 133.7580$, $F_{13} = 70.0337$, $F_{14} = 51.1057$, $F_{21} = 91.8974$, $F_{22} = 78.2420$, $F_{23} = 40.9663$, $F_{24} = 29.8943$, $R_1 = 412$, $R_2 = 241$, $C_1 = 249$, $C_2 = 212$, $C_3 = 111$, $C_4 = 81$, $n = 653$; $\chi^2 = 0.2214 + 0.0115 + 0.0133 + 1.2856 + 0.3785 + 0.0197 + 0.0228 + 2.1978 = 4.151$; $v = (2-1)(4-1) = 3$; $\chi^2_{0.05,3} = 7.815$, $0.10 < P(\chi^2 \geq 4.156) < 0.25$ [$P = 0.246$]; $P > 0.05$, do not reject H_0. (b) $G = 4.032$, $v = 3$, $\chi^2_{0.05,3} = 7.815$, $0.25 < P(\chi^2 \geq 4.032) < 0.50$ [$P = 0.26$]; $P > 0.05$, do not reject H_0.

23.2 (a) $f_{11} = 14$, $f_{12} = 29$, $f_{21} = 12$, $f_{22} = 38$, $R_1 = 43$, $R_2 = 50$, $C_1 = 26$, $C_2 = 67$, $n = 93$; $\chi^2_c = 0.469$, $v = 1$; $\chi^2_{0.05,1} = 3.841$, $0.25 < P(\chi^2 \geq 0.469) < 0.50$; as $P > 0.05$, do not reject H_0. (b) $\hat{f} = 12.0215$, $f = 14$, $d = 1.9785$, $f < 2\hat{f}$ so $D = 1.5$, $\chi^2_{c'} = 0.483$, with conclusions as above. (c) $G = 0.468$, $v = 1$, $\chi^2_{0.05,1} = 3.841$, $0.25 < P(\chi^2 \geq 0.468) < 0.50$; as $P > 0.05$, do not reject H_0 [$P = 0.494$]. (d) H_0: The western population is not more liable to have rabies; H_A: The western population is more liable to have rabies. The probability of the observed table (where $F_{11} = 14$) is 0.12099. For more extreme tables in that tail: for $f_{11} = 15$, $P = 0.07198$; for $f_{11} = 16$, $P = 0.03464$; for $f_{11} = 17$, $P = 0.01342$; for $f_{11} = 18$, $P = 0.00415$; for $f_{11} = 19$, $P = 0.00102$; for $f_{11} = 20$, $P = 0.00019$; for $f_{11} = 21$, $P = 0.00003$; for 22, 23, 24, 25, 26, $P = 0.00000$ each; so the one-tailed probability is 0.24642. (e) For more extreme tables in the opposite tail: $f_{11} = 0, 1, 2$, $P = 0.00000$ each; $f_{11} = 3$, $P = 0.00002$; $f_{11} = 4$, $P = 0.00014$; $f_{11} = 5$, $P = 0.00082$; $f_{11} = 6$, $P = 0.00365$; $f_{11} = 7$, $P = 0.01246$; $f_{11} = 8$, $P = 0.03329$; $f_{11} = 9$, $P = 0.07061$; $f_{11} = 10$, $P = 0.12004$; so the probability in the second tail is 0.24103 and the two-tailed probability is 0.48745.

23.3 H_0: Sex, area, and occurrence of rabies are mutually independent; $\chi^2 = 32.031$; $v = 4$; $\chi^2_{0.05,4} = 9.488$; reject H_0; $P < 0.001$ [$P = 0.0000019$]. H_0: Area is independent of sex and rabies; $\chi^2 = 23.515$; $v = 3$; $\chi^2_{0.05,3} = 7.815$; reject H_0; $P < 0.001$ [$P = 0.000032$]. H_0: Sex is independent of area and rabies; $\chi^2 = 16.723$; $v = 3$; reject H_0;

$P < 0.001$ [$P = 0.00081$]. H_0: Rabies is independent of area and sex; $\chi^2 = 32.170$; $v = 3$; reject H_0; $P < 0.001$ [$P = 0.00000048$].

CHAPTER 24

24.1 $P(X = 2) = 0.32413$.

24.2 $P(X = 4) = 0.00914$.

24.3 $F_{0.05(2),26,36} \approx F_{0.05(2),26,35} = 2.04$, $L_1 = 0.4043$; $F_{0.05(2),38,24} \approx F_{0.05(2),40,24} = 2.15$, $L_2 = 0.7729$. Using exact critical values: $F_{0.05(2),26,36} = 2.0255$, $L_1 = 0.4060$; $F_{0.05(2),38,24} = 2.1561$, $L_2 = 0.7734$.

24.4 H_0: The sampled population is binomial with $p = 0.25$; H_A: The sampled population is not binomial with $p = 0.25$; $\sum f_i = 126$; $F_1 = (0.31641)(126) = 39.868$, $F_2 = 53.157$, $F_3 = 26.578$, $F_4 = 5.907$, $F_5 = 0.493$; combine F_4 and F_5 and combine f_4 and f_5; $\chi^2 = 11.524$, $v = k - 1 = 3$, $\chi^2_{0.05,3} = 7.815$; reject H_0; $0.005 < P < 0.01$ [$P = 0.0092$].

24.5 H_0: The sampled population is binomial; H_A: The sampled population is not binomial; $\hat{p} = \frac{156}{109}/4 = 0.3578$; $\chi^2 = 3.186$, $v = k - 2 = 3$, $\chi^2_{0.05,3} = 7.815$; do not reject H_0; $0.25 < P < 0.50$ [$P = 0.36$].

24.6 H_0: $p = 0.5$; H_A: $p \neq 0.5$; $n = 20$; $P(X \leq 6$ or $X \geq 14) = 0.11532$; since this probability is greater than 0.05, do not reject H_0.

24.7 H_0: $p = 0.5$; H_A: $p \neq 0.5$; $\hat{p} = \frac{197}{412} = 0.4782$; $Z = -0.888$; $Z_c = 0.838$; $Z_{0.05(2)} = t_{0.05(2),\infty} = 1.960$; therefore, do not reject H_0; $P \approx 0.37$ [$P = 0.40$].

24.8 H_0: $p = 0.5$; H_A: $p \neq 0.5$; $X = 44$; $Z = -1.0102$; $Z_{0.05(2)} = t_{0.05(2),\infty} = 1.960$; do not reject H_0; $0.20 < P < 0.50$ [$P = 0.30$].

24.9 H_0: $p = 0.5$; H_A: $p \neq 0.5$; number of positive differences = 7; for $n = 10$ and $p = 0.5$; $P(X \leq 3$ or $X \geq 7) = 0.34378$; since this probability is greater than 0.05, do not reject H_0.

24.10 $X = 62$, $n = 1215$; $\hat{p} = 0.0510$; $P(0.0394 \leq p \leq 0.0649) = 0.95$.

24.11 $n = 20$, $p = 0.50$; critical values are 5 and 15; $\hat{p} = 6/20 = 0.30$, power $= 0.00080 + 0.00684 + 0.02785 + 0.07160 + 0.13042 + 0.17886 + 0.00004 + 0.00001 = 0.42$.

24.12 $p_0 = 0.50$, $p = 0.4782$, $n = 412$; power $= P(Z < -1.08) + P(Z > 2.84) = 0.1401 + 0.0023 = 0.14$.

24.13 (a) $\chi^2_{c(2)} = 4.625$, $\chi^2_{0.05,1} = 3.841$; reject H_0; $0.025 < P < 0.05$ [$P = 0.032$]. (b) $\hat{f} = \hat{f}_{11} = 11.4231$, $d = 5.5769$, $D = 5.5$; $\chi^2_{c'} = 5.428$, $\chi^2_{0.05,1} = 3.841$; reject H_0; $0.01 < P < 0.025$ [$P = 0.020$]. (c) $G_c = 4.488$; $0.025 < P < 0.05$ [$P = 0.034$]. (d) $f_{11} = 17$, $f_{12} = 16$, $f_{21} = 28$, $f_{22} = 69$, $C_1 = 45$, $C_2 = 85$, $R_1 = 33$,

$R_2 = 97$, $n = 130$. **(e)** The probability of the observed table is 0.01296. For more extreme tables in that tail: for $f_{11} = 18$, $P = 0.00399$; for $f_{11} = 19$, $P = 0.00120$; for $f_{11} = 20$, $P = 0.00030$; for $f_{11} = 21$, $P = 0.00006$; for $f_{11} = 22$, $P = 0.00001$; for $f_{11} = 23$ through 33, $P = 0.00000$ each; so the one-tailed probability is 0.01677. **(f)** For more extreme tables in the opposite tail: $m_1 = 33$, $m_2 = 45$, $f = 33$, $33 - 33 = 0$, $f_{11} = 0$ or 1, $P = 0.00000$; $f_{11} = 2$, $P = 0.00002$; $f_{11} = 3$, $P = 0.00015$; $f_{11} = 4$, $P = 0.00086$; $f_{11} = 5$, $P = 0.00360$; so the probability in the second tail is 0.00463 and the two-tailed probability is 0.02140.

24.14 **(a)** $\chi^2_{c(2)} = 4.693$, $\chi^2_{0.05,1} = 3.841$, reject H_0, $0.025 < P < 0.05$ $[P = 0.030]$. **(b)** $\hat{f} = \hat{f}_{11} = 5.600$, $d = 3.400$, $D = 3.0000$; $\chi^2_{c'} = 5.022$, $\chi^2_{0.05,1} = 3.841$, reject H_0, $P \approx 0.025$ $[P = 0.025]$. **(c)** $G_c = 4.810$; $0.025 < P < 0.05$ $[P = 0.028]$. **(d)** $f_{11} = 9$, $f_{12} = 5$, $f_{21} = 3$, $f_{22} = 13$, $C_1 = 12$, $C_2 = 18$, $R_1 = 14$, $R_2 = 15$, $n = 30$. **(e)** The probability of the observed table is 0.01296. For more extreme tables in that tail: for $f_{11} = 10$, $P = 0.00139$; for $f_{11} = 11$, $P = 0.00007$; for $f_{11} = 12$, $P = 0.00000$; so the one-tailed probability is 0.01442. Using Appendix Table B.28: $m_1 = 12$, $m_2 = 14$, $n = 30$, $f = 9$; 5% critical values of f are 2 and 9; so reject H_0. **(f)** For more extreme tables in the opposite tail: $m_1 = 12$, $m_2 = 14$, $f = 0$, $12 - 0 = 12$, $f_{11} = 0$, $P = 0.00002$; $f_{11} = 1$, $P = 0.00071$; $f_{11} = 2$, $P = 0.00842$; so the probability in the second tail is 0.00915 and the two-tailed probability is 0.02357. Using Appendix Table B.28: $m_1 = 12$, $m_2 = 14$, $n = 30$, $f = 9$; 5% critical values of f are 2 and 9, so reject H_0.

24.15 $\alpha = 0.05$, $p_1 = 0.333$, $p_2 = 0.250$, $n_1 = n_2 = 300$; $\bar{p} = 0.292$, power $= P(Z < -4.22) + P(Z > -0.28) = 0.00 + 0.61 = 0.61$.

24.16 $\alpha = 0.05$, $\beta = 0.10$, $p_1 = 0.333$, $p_2 = 0.250$, $\bar{p} = 0.292$; $A = 4.3287$, $n = 652.2$, so use sample sizes of at least 653.

24.17 H_0: $p_1 = p_2 = p_3 = p_4$, H_A: All four population proportions are not equal; $X_1 = 163$, $X_2 = 135$, $X_3 = 71$, $X_4 = 43$, $n_1 = 249$, $n_2 = 212$, $n_3 = 111$, $n_4 = 81$; $\hat{p}_1 = 0.655$, $\hat{p}_2 = 0.637$, $\hat{p}_3 = 0.640$, $\hat{p}_4 = 0.531$; $\bar{p} = 412/653 = 0.631$, $\chi^2 = 0.5966 + 0.0305 + 0.0356 + 3.4883 = 4.151$, $\chi^2_{0.05,3} = 7.815$; do not reject H_0; $0.10 < P < 0.25$ $[P = 0.246]$.

24.18 H_0: $p_1 = p_2 = p_3 = p_4 = 0.5$, $p_0 = 0.5$; $\chi^2 = 36.3742 + 15.8689 + 8.6577 + 0.3086 = 61.208$, $\chi^2_{0.005,4} = 9.488$; reject H_0; $P \ll 0.001 [P < 10^{-11}]$.

24.19 Ranked proportions: 0.161, 0.261, 0.448; ranked transformed proportions: 23.76, 30.78, 42.06; $k = 3$, $q_{0.05,\infty,3} = 3.314$; reject H_0: $p_A = p_B$ for all three comparisons; overall conclusion: $p_1 \neq p_2 \neq p_3$.

24.20 **(a)** H_0: The proportion of successful implants is the same for each amount of exercise; H_A: The proportion of successful implants is not the same for each amount of exercise; $\chi^2 = 6.313$, $\nu = 3$, $0.05 < P < 0.10$ $[P = 0.097]$. **(b)** H_0: There is a linear trend in implant success with amount of exercise; H_A: There is not a linear trend in implant success with amount of exercise; $\chi^2_t = 5.537$, $\nu = 1$, $0.01 < P < 0.025$ $[P = 0.019]$.

CHAPTER 25

25.1 If $\mu = 1.5$, $P(X = 0) = 0.2231$ and $P(X = 5) = 0.0141$.

25.2 $\mu = \frac{5}{2} = 2.5$ viruses per bacterium. **(a)** $P(X = 0) = 0.0821$. **(b)** $P(X > 0) = 1.0000 - P(X = 0) = 1.0000 - 0.0821 = 0.9197$. **(c)** $P(X \geq 2) = 1.0000 - P(X = 0) - P(X = 1) = 1.0000 - 0.0821 - 0.2052 = 0.7127$. **(d)** $P(X = 3) = 0.2138$.

25.3 H_0: Raisins are distributed randomly throughout the cake; H_A: Raisins are not distributed randomly throughout the cake; $\bar{X} = \sum f_i X_i / \sum f_i = 98/57 = 1.7193$; $\chi^2 = 3.060$, $\nu = 6 - 2 = 4$, $\chi^2_{0.05,4} = 7.815$; do not reject H_0; $0.50 < P < 0.75$ $[P = 0.55]$.

25.4 H_0: $p \leq 0.00010$; H_A: $p > 0.00010$; $p_0 = 0.00010$; $n = 25{,}000$; $p_0 n = 2.5$; $X = 5$; $P(X \geq 5) = 0.1087$; do not reject H_0; do not include this disease on the list.

25.5 H_0: $\mu_1 = \mu_2$; H_A: $\mu_1 \neq \mu_2$; $X_1 = 112$, $X_2 = 134$; $Z = 1.40$; $Z_{0.05(2)} = 1.9600$; do not reject H_0; $0.10 < P < 0.20$ $[P = 0.16]$.

25.6 H_0: The incidence of heavy damage is random over the years; H_A: The incidence of heavy damage is not random over the years; $n_1 = 14$, $n_2 = 13$, $u = 12$, $u_{0.05,14,13} = 9$ and 20. As 12 is neither ≤ 9 nor ≥ 20; do not reject H_0; $P = 0.50$.

25.7 H_0: The magnitude of fish kills is randomly distributed over time; H_A: The magnitude of fish kills is not randomly distributed over time; $n = 16$, $s^2 = 400.25$, $s_*^2 = 3126.77/30 = 104.22$; $C = 0.740$, $C_{0.05,16} = 0.386$; reject H_0; $P < 0.0005$.

25.8 H_0: The data are sequentially random; H_A: The data are not sequentially random; $n = 16$, $u = 7$; critical values $= 6$ and 14; do not reject H_0; $0.05 < P \leq 0.10$.

25.9 $n_1 = n_2 = n_3 = n_4 = 13$, $N = 52$, $\sum n_i^2 = 676$, $\sum n_i^3 = 8788$, $\mu_u = 40$, $\sigma_u = 3.8839$. **(a)** H_0: The categories are randomly distributed; H_A: The categories are not randomly distributed; $Z_c = 2.45$; two-tailed $P = 0.014$, reject H_0. **(b)** $Z_c = 2.961$, one-tailed $P = 0.0015$. **(c)** $Z_c = 9.14$, one-tailed $P = 3.1 \times 10^{-20}$.

CHAPTER 26

26.1 $n = 12$, $Y = 0.48570$, $X = 0.20118$, $r = 0.52572$ ($c = 1.02617$, $r_c = 0.53948$). **(a)** $\bar{a} = 68°$. **(b)** $s = 56°$ (using correction for grouping, $s = 55°$), $s' = 65°$ (using correction for grouping, $s' = 64°$). **(c)** $68° \pm 47°$ (using correction for grouping, $68° \pm 46°$). **(d)** median $= 67.5°$.

26.2 $n = 15$, $Y = 0.76319$, $X = 0.12614$, $r = 0.77354$. **(a)** $a = 5:22$ AM. **(b)** $s = 2:34$ hr. **(c)** $5:22$ hr $\pm 1:38$ hr. **(d)** median $= 5:10$ AM.

26.3 **(a)** $k = 5$, $\sum X_j = -1.67070$, $\sum Y_j = -2.54868$, $\bar{X} = -0.33414$, $\bar{Y} = -0.50974$, $r = 0.60949$, $\cos\bar{a} = -0.54823$, $\sin\bar{a} = -0.83634$, $\bar{a} = 237°$. **(b)** $\sum x^2 = 0.16655$, $\sum y^2 = 0.17763$, $\sum xy = -0.06854$; $A = 24.01739$, $B = 9.26709$, $C = 22.51812$, $F_{0.05(1),2,3} = 9.55$, $D = 4.28455$; $b_1 = -5.19495$ and one confidence limit is $281°$; $b_2 = 0.06449$ and the other confidence limit is $184°$.

CHAPTER 27

27.1 H_0: $\rho = 0$; H_A: $\rho \neq 0$; $r = 0.526$; $R = 6.309$; $z = 3.317$, $z_{0.05,12} = 2.932$; reject H_0; $0.02 < P < 0.05$.

27.2 H_0: $\rho = 0$, H_A: $\rho \neq 0$; $r = 0.774$; $R = 11.603$; $z = 8.975$, $z_{0.05,15} = 2.945$; reject H_0; $P < 0.001$.

27.3 H_0: $\rho = 0$; H_A: $\rho \neq 0$; $n = 11$, $Y = -0.88268$, $X = 0.17138$, $r = 0.89917$, $\bar{a} = 281°$, $R = 9.891$, $\mu_0 = 270°$. **(a)** $V = 9.709$, $u = 4.140$, $u_{0.05,11} = 1.648$; reject H_0; $P < 0.0005$. **(b)** H_0: $\mu_a = 270°$, H_A: $\mu_a \neq 270°$, 95% confidence interval for $\mu_a = 281° \pm 19°$, so do not reject H_0.

27.4 $n = 12$, $m = 2$, $m_{0.05,12} = 0$, do not reject H_0; $0.20 < P \leq 0.50$.

27.5 $n = 11$, $m' = 0$, $C = 11$, $C_{0.05(2),11} = 1$, reject H_0; $P < 0.001$.

27.6 H_0: Mean flight direction is the same under the two sky conditions; H_A: Mean flight direction is not the same under the two sky conditions; $n_1 = 8$, $n_2 = 7$, $R_1 = 7.5916$, $R_2 = 6.1130$, $\bar{a}_1 = 352°$, $\bar{a}_2 = 305°$, $N = 15$, $r_w = 0.914$, $R = 12.5774$; $F = 12.01$, $F_{0.05(1),1,13} = 4.67$; reject H_0; $0.0025 < P < 0.005$ $[P = 0.004]$.

27.7 H_0: The flight direction is the same under the two sky conditions; H_A: The flight direction is not the same under the two sky conditions; $n_1 = 8$, $n_2 = 7$, $N = 15$; $\sum d_k = -2.96429$, $\sum d_k^2 = 1.40243$, $U^2 = 0.2032$, $U^2_{0.05,8,7} = 0.1817$; do not reject H_0; $0.02 < P < 0.05$.

27.8 H_0: Members of all three hummingbird species have the same mean time of feeding at the feeding station; H_A: Members of all three species do not have the same mean time of feeding at the feeding station; $n_1 = 6$, $n_2 = 9$, $n_3 = 7$, $N = 22$; $R_1 = $

2.965, $R_2 = 3.938$, $R_3 = 3.868$; $\bar{a}_1 = 10:30$ hr, $\bar{a}_2 = 11:45$ hr, $\bar{a}_3 = 11:10$ hr; $r_w = 0.490$, $F = 0.206$, $F_{0.05(1),2,19} = 3.54$; do not reject H_0; $P > 0.25$ $[P = 0.82]$. Therefore, all three $\bar{a}_i$'s estimate the same μ_a, the best estimate of which is 11:25 hr.

27.9 H_0: Birds do not orient better when skies are sunny than when cloudy; H_A: Birds do orient better when skies are sunny than when cloudy. Angular distances for group 1 (sunny): 10, 20, 45, 10, 20, 5, 15, and 0°; for group 2 (cloudy): 20, 55, 105, 90, 55, 40, and 25°. For the one-tailed Mann-Whitney test: $n_1 = 8$, $n_2 = 7$, $R_1 = 40$, $U = 52$, $U_{0.05(1),8,7} = 43$; reject H_0; $P = 0.0025$.

27.10 H_0: Variability in flight direction is the same under both sky conditions; H_A: Variability in flight direction is not the same under both sky conditions; $\bar{a}_1 = 352°$, $\bar{a}_2 = 305°$; angular distances for group 1 (sunny): 2, 12, 37, 18, 28, 3, 7, and 8°, and for group 2 (cloudy): 35, 0, 50, 35, 0, 15, and 30°; for the two-tailed Mann-Whitney test: $R_1 = 58$, $U = 34$, $U' = 22$, $U_{0.05(2),8,7} = 46$; reject H_0; $P < 0.001$.

27.11 $\sum x^2 = 0.16655$, $\sum y^2 = 0.17763$, $\sum xy = -0.06854$; $F = 26.1$; $F_{0.05(1),2,3} = 9.55$; reject H_0; $0.01 < P < 0.025$ $[P = 0.013]$.

27.12 $X = -7.52684/5 = -1.50537$, $Y = -12.00527/5 = -2.40105$, $R' = 1.267$, $R'_{0.05,5} = 1.152$; reject H_0; $0.002 < P < 0.01$.

27.13 $\bar{X}_1 = -0.33414$, $\bar{X}_2 = -0.19144$, $\bar{Y}_1 = -0.50974$, $\bar{Y}_2 = -0.60853$, $\left(\sum x^2\right)_c = 0.36943$, $\left(\sum y^2\right)_c = 0.28449$, $\left(\sum xy\right)_c = -0.03836$; $F = 1.52$; $F_{0.05(1),2,10} = 4.10$; do not reject H_0; $P > 0.25 [P = 0.27]$.

27.14 Grand mean: $\bar{X} = -0.24632$, $\bar{Y} = -0.57053$; $U^2 = 0.0775$; $U^2_{0.05,5,8} = 0.2154$; do not reject H_0; $P > 0.50$.

27.15 **(a)** H_0: $\rho_{aa} = 0$; H_A: $\rho_{aa} \neq 0$; $r_{aa} = 0.9244$; $\bar{r}_{aa} = 0.9236$; $s^2_{r_{aa}} = 0.0004312$; $L_1 = 0.9169$, $L_2 = 0.9440$; reject H_0. **(b)** H_0: $(\rho_{aa})_s = 0$, H_A: $(\rho_{aa}) \neq 0$; $r' = 0.453$, $r'' = 0.009$, $(r_{aa})_s = 0.365$; $(n-1)(r_{aa})_s = 2.92$, for $\alpha(2) = 0.05$ the critical value is 3.23; do not reject H_0; for $\alpha(2) = 0.10$, the critical value is 2.52, so $0.05 < P < 0.10$.

27.16 H_0: $\rho_{al} = 0$; H_A: $\rho_{al} \neq 0$; $r_{al} = 0.833$, $nr^2_{al} = 6.24$, $\chi^2_{0.05,2} = 5.991$, reject H_0.

27.17 H_0: The distribution is not contagious; H_A: The distribution is contagious; $n_1 = 8$, $n_2 = 8$, $u = 6$, $u' = 3$; using Appendix Table B.28: $m_1 = 7$, $m_2 = 7$, $f = 2$, $n = 15$, critical values are 1 and 6, so do not reject H_0; $P \geq 0.50$.

LITERATURE CITED

ACTON, F. S. 1966. *Analysis of Straight Line Data.* Dover, New York. 267 pp.

AIKEN, L. S. and S. G. WEST. 1991. *Multiple Regression: Testing and Interpreting Interactions.* Sage Publications, Newbury Park, California. 212 pp.

AJNE, B. 1968. A simple test for uniformity of a circular distribution. *Biometrika* 55: 343–354.

ANDERSON, N. H. 1961. Scales and statistics: parametric and non-parametric. *Psychol. Bull.* 58: 305–316.

ANDRÉ, D. 1883. Sur le nombre de permutations de *n* éléments qui présentent *S* séquences. *Compt. Rend. (Paris)* 97: 1356–1358. [Cited in Bradley, 1968: 281.]

ANDREWS, F. C. 1954. Asymptotic behavior of some rank tests for analysis of variance. *Ann. Math. Statist.* 25: 724–735.

ANDREWS, H. P., R. D. SNEE, and M. H. SARNER. 1980. Graphical display of means. *Amer. Statist.* 34(4): 195–199.

ANSCOMBE, F. J. 1948. The transformation of Poisson, binomial, and negative binomial data. *Biometrika* 35: 246–254.

ANSCOMBE, F. J. and W. W. GLYNN. 1983. Distributions of the kurtosis statistic b_2 for normal statistics. *Biometrika* 70: 227–234.

ARBUTHNOTT, J. 1710. An argument for Divine Providence taken from the constant regularity in the births of both sexes. *Philos. Trans.* 27: 186–190.

ARMITAGE, P. 1955. Tests for linear trends in proportions and frequencies. *Biometrics* 11: 375–386.

ARMITAGE, P. 1971. *Statistical Methods in Medical Research.* John Wiley and Sons, New York. 504 pp.

ARMITAGE, P. 1985. Biometry and medical statistics. *Biometrics* 41: 823–833.

ASIMOV, I. 1982. *Asimov's Biographical Encyclopedia of Science and Technology.* 2nd rev. ed. Doubleday, Garden City, New York. 941 pp.

BAILER, A. J. 1989. Testing variance equality with randomization tests. *J. Statist. Computa. Simula.* 31: 1–8.

BALAKRISHNAN, N. and C. W. MA. 1990. A comparative study of various tests for the equality of two population variances. *J. Statist. Computa. Simula.* 35: 41–89.

BANCROFT, T. A. 1968. *Topics in Intermediate Statistical Methods,* Vol. 1. Iowa State University, Ames, Iowa. 129 pp.

BARNARD, G. A. 1947. Significance tests for 2×2 tables. *Biometrika* 34: 123–138.

BARNARD, G. A. 1979. In contradiction to J. Berkson's dispraise: Conditional tests can be more efficient. *J. Statist. Plan. Inf.* 2: 181–187.

BARNARD, G. A. 1984. Discussion of Dr. Yate's paper. *J. Roy. Statist. Soc.* Ser. A, 147: 449–450.

BARNARD, G. 1990. Fisher: A retrospective. *CHANCE* 3: 22–28.

BARTHOLOMEW, D. J. 1983. Sir Maurice Kendall FBA. *Statistician* 32: 445–446.

BARTLETT, M. S. 1936. The square root transformation in analysis of variance. *J. Royal Statist. Soc. Suppl.* 3: 68–78.

BARTLETT, M. S. 1937a. Some examples of statistical methods of research in agriculture and applied biology. *J. Royal Statist. Soc. Suppl.* 4: 137–170.

BARTLETT, M. S. 1937b. Properties of sufficiency and statistical tests. *Proc. Roy. Statist. Soc.* Ser. A, 160: 268–282.

BARTLETT, M. S. 1939. A note on tests of significance in multivariate analysis. *Proc. Cambridge Philos. Soc.* 35: 180–185.

BARTLETT, M. S. 1947. The use of transformations. *Biometrics* 3: 39–52.

BARTLETT, M. S. 1965. R. A. Fisher and the last 50 years of statistical methodology. *J. Amer. Statist. Assoc.* 6: 395–409.

BARTLETT, M. S. 1981. Egon Sharpe Pearson, 1895–1980. *Biometrika* 68: 1–12.

BASHARIN, G. P. 1959. On a statistical estimate for the entropy of a sequence of independent variables. *Theory Prob. Appl.* 4: 333–336.

BATES, D. M. and D. G. WATTS. 1988. *Nonlinear Linear Regression Analysis and Its Applications.* John Wiley, New York. 365 pp.

BATSCHELET, E. 1965. *Statistical Methods for the Analysis of Problems in Animal Orientation and Certain Biological Rhythms.* American Institute of Biological Sciences, Washington, D.C. 57 pp.

BATSCHELET, E. 1972. Recent statistical methods for orientation data, pp. 61–91. (With discussion.) *In* S. R. Galler, K. Schmidt-Koenig, G. J. Jacobs, and R. E. Belleville (eds.), *Animal Orientation and Navigation.* National Aeronautics and Space Administration, Washington, D.C.

BATSCHELET, E. 1974. Statistical rhythm evaluations, pp. 25–35. *In* M. Ferin, F. Halberg, R. M. Richart, and R. L. Vande Wiele (eds.), *Biorhythms and Human Reproduction.* John Wiley, New York.

BATSCHELET, E. 1976. *Mathematics for Life Scientists.* 2nd ed. Springer-Verlag, New York. 643 pp.

BATSCHELET, E. 1978. Second-order statistical analysis of directions, pp. 1–24. *In* K. Schmidt-Koenig and W. T. Keeton (eds.), *Animal Migration, Navigation, and Homing.* Springer-Verlag, Berlin.

BATSHCELET, E. 1981. *Circular Statistics in Biology.* Academic Press, New York. 371 pp.

BEALL, G. 1940. The transformation of data from entomological field experiments. *Can. Entomol.* 72: 168.

BEALL, G. 1942. The transformation of data from entomological field experiments so that the analysis of variance becomes applicable. *Biometrika* 32: 243–262.

BECKMANN, P. 1977. *A History of π.* 4th ed. Golem Press, Boulder, Colorado. 202 pp.

BEHRENS, W. V. 1929. Ein Beitrag zur Fehlerberechnung bei weinige Beobachtungen. *Landwirtsch Jahrbücher* 68: 807–837.

BENARD, A. and P. VAN ELTEREN. 1953. A generalization of the method of *m* rankings. *Indagatones Mathematicae* 15: 358–369.

BENNETT, B. M. and E. NAKAMURA. 1963. Tables for testing significance in a 2×3 contingency table. *Technometrics* 5: 501–511.

BENNETT, C. A. and N. L. FRANKLIN. 1954. *Statistical Analysis in Chemistry and the Chemical Industry.* John Wiley, New York. 724 pp.

BERK, K. N. 1978. Comparing subset regression procedures. *Technometrics* 20: 7–8.

BERKSON, J. 1978. In dispraise of the exact test. *J. Statist. Plan. Infer.* 2:27–42.

BERNHARDSON, C. S. 1975. Type I error rates when multiple comparison procedures follow a significant *F* test of ANOVA. *Biometrics* 31: 229–232.

BERRY, W. D. and S. FELDMAN. 1985. *Multiple Regression in Practice.* Sage Publications, Beverly Hills, California. 95 pp.

BERTRAND, P. V. and R. L. HOLDER. 1988. A quirk in multiple regression: the whole regression can be greater than the sum of its parts. *Statistician* 37: 371–374.

BEST, D. J. 1975. The difference between two Poisson expectations. *Austral. J. Statist.* 17: 29–33.

BEST, D. J. and D. E. ROBERTS. 1975. The percentage points of the χ^2 distribution. *J. Royal Statist. Soc.* Ser. C. *Appl. Statist.* 24: 385–388.

BHATTACHARJEE, G. P. 1970. The incomplete gamma interval. *J. Royal Statist. Soc.* Ser. C. *Appl. Statist.* 19: 285–287.

BHATTACHARYYA, G. K. and R. A. JOHNSON. 1969. On Hodges's bivariate sign test and a test for uniformity of a circular distribution. *Biometrika* 56: 446–449.

BIRKES, D. and Y. DODGE. 1993. *Alternative Methods of Regression.* John Wiley & Sons, New York. 228 pp.

BIRNBAUM, Z. W. and F. H. TINGEY. 1951. One-sided confidence contours for probability distribution functions. *Ann. Math. Statist.* 22: 592–596.

BISHOP, T. A. and E. J. DUDEWICZ. 1978. Exact analysis of variance with unequal variances: Test procedures and tables. *Technometrics* 20: 419–424.

BISHOP, Y. M. M., S. E. FIENBERG, and P. W. HOLLAND. 1975. *Discrete Multivariate Analysis: Theory and Practice.* MIT Press, Cambridge, Massachusetts. 557 pp.

BISSELL, A. F. 1992. Lines through the origin—is NO INT the answer? *J. Appl. Statist.* 19: 192–210.

BLATNER, D. 1997. *The Joy of π.* Walker, New York. 130 pp.

BLISS, C. I. 1967. *Statistics in Biology,* Vol. 1. McGraw-Hill, New York. 558 pp.

BLISS, C. I. 1970. *Statistics in Biology,* Vol. 2. McGraw-Hill, New York. 639 pp.

BLOOMFIELD, P. 1976. *Fourier Analysis of Time Series: An Introduction.* John Wiley, New York. 258 pp.

BLOOMFIELD, P. and W. L. STEIGER. 1983. *Least Absolute Deviations: Theory, Applications, and Algorithms.* Birkhaüser, Boston, Massachusetts. 349 pp.

BLYTH, C. R. 1986. Approximate binomial confidence limits. *J. Amer. Statist. Assoc.* 81: 843–855.

BONEAU, C. A. 1960. The effects of violations of assumptions underlying the *t* test. *Psychol. Bull.* 57: 49–64.

BONEAU, C. A. 1962. A comparison of the power of the *U* and *t* tests. *Psychol. Rev.* 69: 246–256.

BOOMSMA, A. and I. W. MOLENAAR. 1994. Four electronic tables for probability distributions. *Amer. Statist.* 48: 153–162.

BOWKER, A. H. 1948. A test for symmetry in contingency tables. *J. Amer. Statist. Assoc.* 43: 572–574.

BOWLEY, A. L. 1920. *Elements of Statistics.* 4th ed. Charles Scribner's Sons, New York. 459 pp. + tables.

BOWMAN, K. O., K. HUTCHESON, E. P. ODUM, and L. R. SHENTON. 1971. Comments on the distribution of indices of diversity, pp. 315–366. *In* G. P. Patil, E. C. Pielou, and W. E. Waters (eds.), *Statistical Ecology*, Vol. 3. Many Species Populations, Ecosystems, and Systems Analysis. Pennsylvania State University Press, University Park.

BOX, G. E. P. 1949. A general distribution theory for a class of likelihood criteria. *Biometrika* 36: 317–346.

BOX, G. E. P. 1953. Non-normality and tests on variances. *Biometrika* 40: 318–335.

BOX, G. E. P. 1954. Some theorems on quadratic forms applied in the study of analysis of variance problems, I. Effect of inequality of variance in the one-way classification. *Ann. Math. Statist.* 25: 290–302.

BOX, G. E. P. and S. L. ANDERSON. 1955. Permutation theory in the derivation of robust criteria and the study of departures from assumption. *J. Royal Statist. Soc.* B17: 1–34.

BOX, G. E. P. and D. R. COX. 1964. An analysis of transformations. *J. Royal Statist. Soc.* B26: 211–243.

BOZIVICH, H., T. A. BANCROFT, and H. O. HARTLEY. 1956. Power of analysis of variance test procedures for certain incompletely specified models. *Ann. Math. Statist.* 27: 1017–1043.

BRADLEY, R. A. and M. HOLLANDER. 1978. Wilcoxon, Frank, pp. 1245–1250. *In* Kruskal and Tanur (1978).

BRAY, J. S. and S. E. MAXWELL. 1985. *Multivariate Analysis of Variance.* Sage Publications, Beverly Hills, Calif. 80 pp.

BRILLOUIN, L. 1962. *Science and Information Theory.* Academic Press, New York. 351 pp.

BRITS, S. J. M. and H. H. LEMMER. 1990. An adjusted Friedman test for the nested design. *Communic. Statist.—Theor. Meth.* 19: 1837–1855.

BROWER, J. E., J. H. ZAR, and C. N. VON ENDE. 1998. *Field and Laboratory Methods for General Ecology.* 4th ed. McGraw-Hill, Boston. 273 pp.

BROWN, M. B. and A. B. FORSYTHE. 1974a. The small size sample behavior of some statistics which test the equality of several means. *Technometrics* 16: 129–132.

BROWN, M. B. and A. B. FORSYTHE. 1974b. The ANOVA and multiple comparisons for data with heterogeneous variances. *Biometrics* 30: 719–724.

BROWN, M. B. and A. B. FORSYTHE. 1974c. Robust tests for the equality of variance. *J. Amer. Statist. Assoc.* 69: 364–367.

BROWNE, R. H. 1979. On visual assessment of the significance of a mean difference. *Biometrics* 35: 657–665.

BROWNLEE, K. A. 1965. *Statistical Theory and Methodology in Science and Engineering.* 2nd ed. John Wiley, New York. 590 pp.

BUCKLE, N., C. KRAFT, and C. VAN EEDEN. 1969. An approximation to the Wilcoxon-Mann-Whitney distribution. *J. Amer. Statist. Assoc.* 64: 591–599.

BUDESCU, D. V. and M. I. APPELBAUM. 1981. Variance stablizing transformations and the power of the *F* test. *J. Educ. Statist.* 6: 55–74.

BURR, E. J. 1964. Small-sample distributions of the two-sample Cramér-von Mises' W^2 and Watson's U^2. *Ann. Math. Statist. Assoc.* 64: 1091–1098.

BURSTEIN, H. 1971. *Attribute Sampling. Tables and Explanations.* McGraw-Hill, New York. 464 pp.

BURSTEIN, H. 1975. Finite population correction for binomial confidence limits. *J. Amer. Statist. Assoc.* 70: 67–69.

BURSTEIN, H. 1981. Binomial 2×2 test for independent samples with independent proportions. *Communic. Statist.—Theor. Meth.* A10: 11–29.

BUSH, A. J., E. A. RAKOW, and D. N. GALLIMORE. 1980. A comment on correctly calculating higher order semipartial correlations. *J. Educ. Statist.* 5: 105–108.

CACOULLOS, T. 1965. A relation between the *t* and *F* distributions. *J. Amer. Statist. Assoc.* 60: 528–531.

CAJORI, F. 1928. *A History of Mathematical Notations.* Vol. II: *Notations Mainly in Higher Mathematics.* Open Court Publishing, La Salle, Ill. 367 pp. [Also—with Vol. I—Dover, New York, 1993.]

CAJORI, F. 1929. *A History of Mathematical Notations.* Vol. I: *Notations in Elementary Mathematics.* Open Court Publishing, La Salle, Ill. 451 pp. [Also—with Vol. II—Dover, New York, 1993.]

CAJORI, F. 1954. Binomial formula, pp. 588–589. In *The Encyclopaedia Britannica*, Vol. 3. Encyclopaedia Britannica Inc., New York.

CAMILLI, G. 1990. The test of homogeneity for two contingency tables: A review of and some personal opinions on the controversy. *Psychol. Bull.* 108: 135–145.

CAMILLI, G. and K. D. HOPKINS. 1978. Applicability of chi-square to 2 × 2 contingency tables with small expected frequencies. *Psychol. Bull.* 85: 163–167.

CAMILLI, G. and K. D. HOPKINS. 1979. Testing for association in 2 × 2 contingency tables with very small sample sizes. *Psychol. Bull.* 86: 1011–1014.

CARR, W. E. 1980. Fisher's exact test extended to more than two samples of equal size. *Technometrics* 22: 269–270. See also: 1981. Corrigendum. *Technometrics* 23: 320.

CASAGRANDE, J. T., M. C. PIKE, and P. G. SMITH. 1978. An improved approximate formula for calculating sample sizes for comparing binomial distributions. *Biometrics* 34: 483–486.

CATTELL, R. B. 1978. Spearman, C. E., pp. 1036–1039. *In* Kruskal and Tanur (1978).

CHAPMAN, J.-A. W. 1976. A comparison of the X^2, $-2 \log R$, and multinomial probability criteria for significance tests when expected frequencies are small. *J. Amer. Statist. Assoc.* 71: 854–863.

CHATTERJEE, S. and B. PRICE. 1991. *Regression Analysis by Example.* 2nd ed. John Wiley, New York. 278 pp.

CHOW, B., J. E. MILLER, and P. C. DICKINSON. 1974. Extensions of Monte-Carlo comparison of some properties of two rank correlation coefficients in small samples. *J. Statist. Computa. Simula.* 3: 189–195.

CHRISTENSEN, R. 1990. *Log-Linear Models.* Springer-Verlag, New York. 408 pp.

CICCHITELLI, G. 1989. On the robustness of the one-sample *t* test. *J. Statist. Computa. Simula.* 32: 249–258.

CLEVELAND, W. S., C. L. MALLOWS, and J. E. McRAE. 1993. ATS methods: Nonparametric regression for non-Gaussian data. *J. Amer. Statist. Assoc.* 88: 821–835.

CLINCH, J. J. and H. J. KESELMAN. 1982. Parametric alternatives to the analysis of variance. *J. Educ. Statist.* 7: 207–214.

COCHRAN, W. G. 1942. The χ^2 correction for continuity. *Iowa State Coll. J. Sci.* 16: 421–436.

COCHRAN, W. G. 1947. Some consequences when the assumptions for analysis of variance are not satisfied. *Biometrics* 3: 22–38.

COCHRAN, W. G. 1950. The comparison of percentages in matched samples. *Biometrika* 37: 256–266.

COCHRAN, W. G. 1952. The χ^2 test for goodness of fit. *Ann. Math. Statist.* 23: 315–345.

COCHRAN, W. G. 1954. Some methods for strengthening the common χ^2 tests. *Biometrics* 10: 417–451.

COCHRAN, W. G. 1964. Approximate significance levels of the Behrens-Fisher test. *Biometrics* 20: 191–195.

COCHRAN, W. G. 1977. *Sampling Techniques.* 3rd. ed. John Wiley, New York. 428 pp.

COCHRAN, W. G. and G. M. COX. 1957. *Experimental Designs.* 2nd ed. John Wiley, New York. 617 pp.

COHEN, A. 1991. Dummy variables in stepwise regression. *Amer. Statist.* 45: 226–228.

COHEN, J. 1988. *Statistical Power Analysis for the Behavioral Sciences.* 2nd ed. Lawrence Erlbaum Associates, Hillsdale, New Jersey. 567 pp.

COHEN, J. and P. COHEN. 1983. *Applied Multiple Regression/Correlation Analysis for the Behavioral Sciences.* 2nd ed. Lawrence Erlbaum Associates, Hillsdale, New Jersey. 545 pp.

CONNETT, J. E., J. A. SMITH, and R. B. McHUGH. 1987. Sample size and power for pair-matched case-control studies. *Statist. Med.* 6: 53–59.

CONNOR, R. J. 1972. Grouping for testing trends in categorical data. *J. Amer. Statist. Assoc.* 67: 601–604.

CONOVER, W. J. 1973. On the methods of handling ties in the Wilcoxon signed-rank test. *J. Amer. Statist. Assoc.* 68: 985–988.

CONOVER, W. J. 1974. Some reasons for not using the Yates continuity correction on 2 × 2 contingency tables. *J. Amer. Statist. Assoc.* 69: 374–376. *Also*: Comment, by C. F. Starmer, J. E. Grizzle, and P. K. Sen. *ibid.* 69: 376–378; Comment and a suggestion, by N. Mantel, *ibid.* 69: 378–380; Comment, by O. S. Miettinen, *ibid.* 69: 380–382; Rejoinder, by W. J. Conover, *ibid.* 69: 382.

CONOVER, W. J. 1980. *Practical Nonparametric Statistics.* 2nd ed. John Wiley, New York. 494 pp.

CONOVER, W. J. and R. L. IMAN. 1976. On some alternative procedures using ranks for the analysis of experimental designs. *Cummunic. Statist.—Theor. Meth.* A5: 1349–1368.

CONOVER, W. J., M. E. JOHNSON, and M. M. JOHNSON. 1981. A comparative study of tests for homogeneity of variances with applications to the outer continental shelf bidding data. *Technometrics* 23: 351–361.

COWLES, S. M. and C. DAVIS. 1982. On the origins of the .05 level of statistical significance. *Amer. Psychol.* 37: 553–558.

COX, D. R. 1958. *Planning of Experiments.* John Wiley, New York. 308 pp.

COX, P. and R. A. GROENEVELD. 1986. Analytic results on the difference between the G^2 and χ^2 test statistics in one degree of freedom cases. *Statistician* 35: 417–420.

CRAMER, E. M. 1972. Significance test and tests of models in multiple regression. *Amer. Statist.* 26(4): 26–30.

CRAMÉR, H. 1946. *Mathematical Methods of Statistics.* Princeton University Press, Princeton, New Jersey. 575 pp.

CRITCHLOW, D. E. and M. A. FLIGNER. 1991. On distribution-free multiple comparisons in the one-way analysis of variance. *Communic. Statist.—Theor. Meth.* 20: 127–139.

CROWDER, M. J. and D. J. HAND. 1990. *Analysis of Repeated Measures.* Chapman and Hall, New York. 257 pp.

CROXTON, F. E., D. J. COWDEN, and S. KLEIN. 1967. *Applied General Statistics.* 3rd ed. Prentice-Hall, Englewood Cliffs, New Jersey. 754 pp.

CURETON, E. E. 1967. The normal approximation to the signed-rank sampling distribution when zero differences are present. *J. Amer. Statist. Assoc.* 62: 1068–1069.

D'AGOSTINO, R. B. 1970. Transformation to normality of the null distribution of g_1. *Biometrika* 57: 679–681.

D'AGOSTINO, R. B. 1971a. An omnibus test of normality for moderate and large samples. *Biometrika* 58: 341–348.

D'AGOSTINO, R. B. 1971b. Tables for the D test of normality. Department of Mathematics, Boston University. Research Report. April 1971.

D'AGOSTINO, R. B. 1972. Small sample probability points for the D test of normality. *Biometrika* 59: 219–221.

D'AGOSTINO, R. B. 1986. Tests for the normal distribution, pp. 367–419. *In* R. B. D'Agostino and M. A. Stephens (eds.), *Goodness-of-fit Techniques.* Marcel Dekker, New York.

D'AGOSTINO, R. B., A. BELANGER, and R. B. D'AGOSTINO, JR. 1990. A suggestion for using powerful and informative tests of normality. *Amer. Statist.* 44: 316–321.

D'AGOSTINO, R. B., W. CHASE, and A. BELANGER. 1988. The appropriateness of some common procedures for testing the equality of two independent binomial populations. *Amer. Statist.* 42: 198–202.

D'AGOSTINO, R. B. and G. E. NOETHER. 1973. On the evaluation of the Kolmogorov statistic. *Amer. Statist.* 27: 81–82.

D'AGOSTINO, R. B. and E. S. PEARSON. 1973. Tests of departure from normality. Empirical results for the distribution of b_2 and $\sqrt{b_1}$. *Biometrika* 60: 613–622.

D'AGOSTINO, R. B. and G. L. TIETJEN. 1971. Simulation probability points of b_2 for small samples. *Biometrika* 58: 669–672.

D'AGOSTINO, R. B. and G. L. TIETJEN. 1973. Approaches to the null distribution of $\sqrt{b_1}$. *Biometrika* 60: 169–173.

DALE, A. I. 1989. An early occurrence of the Poisson distribution. *Statist. Prob. Lett.* 7: 21–22.

DANIEL, C. and F. S. WOOD. 1980. *Fitting Equations to Data.* 2nd ed. John Wiley, New York. 458 pp.

DANIEL, W. W. 1978. *Applied Nonparametric Statistics.* Houghton Mifflin Co., Boston, Massachusetts. 510 pp.

DAPSON, R. W. 1980. Guidelines for statistical usage in age-estimation technics. *J. Wildlife Manage.* 44: 541–548.

DARNELL, A. C. 1988. Harold Hotelling 1895–1973. *Statist. Sci.* 3: 57–62.

DAVENPORT, J. M. and J. T. WEBSTER. 1975. The Behrens-Fisher problem, an old solution revisited. *Metrika* 22: 47–54.

DAVID, F. N. 1962. *Games, Gods and Gambling.* Hafner Press, New York. 275 pp.

DAVID, H. A. 1995. First (?) occurrence of common terms in mathematical statistics. *Amer. Statist.* 49: 121–133.

DAY, R. W. and G. P. QUINN. 1989. Comparisons of treatments after an analysis of variance in ecology. *Ecol. Monogr.* 59: 433–463.

DE JONGE, C. and M. A. J. VAN MONTFORT. 1972. The null distribution of Spearman's S when $n = 12$. *Statist. Neerland.* 26: 15–17.

DEMPSTER, A. P. 1983. Reflections on W. G. Cochran, 1909–1980. *Intern. Statist. Rev.* 51: 321–322.

DESU, M. M. and D. RAGHAVARAO. 1990. *Sample Size Methodology.* Academic Press, Boston, Massachusetts. 135 pp.

DETRE, K. and C. WHITE. 1970. The comparison of two Poisson-distributed observations. *Biometrics* 26: 851–854.

DHARMADHIKARI, S. 1991. Bounds of quantiles: A comment on O'Cinneide. *Amer. Statist.* 45: 257–258.

DIJKSTRA, J. B. and S. P. J. WERTER. 1981. Testing the equality of several means when the population variances are unequal. *Communic. Statist.—Simula. Computa.* B10: 557–569.

DIXON, W. J. and F. J. MASSEY, JR. 1969. *Introduction to Statistical Analysis.* 3rd ed. McGraw-Hill, New York. 638 pp.

DOANE, D. P. 1976. Aesthetic frequency classifications. *Amer. Statist.* 30: 181–183.

DODGE, Y. 1996. A natural random number generator. *Internat. Statist. Rev.* 64: 329–344.

DRAPER, N. R. and H. SMITH. 1981. *Applied Regression Analysis.* 2nd ed. John Wiley, New York. 709 pp.

DUNCAN, D. B. 1951. A significance test for difference between ranked treatments in an analysis of variance. *Virginia J. Sci.* 2: 171–189.

DUNCAN, D. B. 1955. Multiple range and multiple F tests. *Biometrics* 11: 1–42.

DUNN, O. J. 1964. Multiple contrasts using rank sums. *Technometrics* 6: 241–252.

DUNN, O. J. and V. A. CLARK. 1987. *Applied Statistics: Analysis of Variance and Regression.* 2nd ed. John Wiley, New York. 445 pp.

DUNNETT, C. W. 1955. A multiple comparison procedure for comparing several treatments with a control. *J. Amer. Statist. Assoc.* 50: 1096–1121.

DUNNETT, C. W. 1964. New tables for multiple comparisons with a control. *Biometrics* 20: 482–491.

DUNNETT, C. W. 1970. Multiple comparison tests. *Biometrics* 26: 139–141.

DUNNETT, C. W. 1980a. Pairwise multiple comparisons in the homogeneous variance, equal sample size case. *J. Amer. Statist. Assoc.* 75: 789–795.

DUNNETT, C. W. 1980b. Pairwise multiple comparisons in the unequal variance case. *J. Amer. Statist. Assoc.* 75: 796–800.

DUNNETT, C. W. 1982. Robust multiple comparisons. *Communic. Statist.—Theor. Meth.* 11: 2611–2629.

DUNNETT, C. W. and M. GENT. 1977. Significance testing to establish equivalence between treatments, with special reference to data in the form of 2×2 tables. *Biometrics* 33: 593–602.

DURAND, D. and J. A. GREENWOOD. 1958. Modifications of the Rayleigh test for uniformity in analysis of two-dimensional orientation data. *J. Geol.* 66: 229–238.

DWASS, M. 1960. Some k-sample rank-order tests, pp. 198–202. *In* I. Olkin, S. G. Ghurye, W. Hoeffding, W. G. Madow, and H. B. Mann. *Contributions to Probability and Statistics. Essays in Honor of Harold Hotelling.* Stanford University Press, Stanford, California.

DYKE, G. 1995. Obituary: Frank Yates. *J. Roy. Statist. Soc.* Ser. A, 158: 333–338.

EASON, G., C. W. COLES, and G. GETTINBY. 1980. *Mathematics and Statistics for the Bio-Sciences.* Ellis Horwood, Chichester, England. 578 pp.

EBERHARDT, K. R. and M. A. FLIGNER. 1977. A comparison of two tests for equality of proportions. *Amer. Statist.* 31: 151–155.

EDGINGTON, E. S. 1961. Probability table for number of runs of signs of first differences in ordered series. *J. Amer. Statist. Assoc.* 56: 156–159.

EDWARDS, A. W. F. 1986. Are Mendel's results really too close? *Biol. Rev.* 61: 295–312.

EDWARDS, A. W. F. 1993. Mendel, Galton, and Fisher. *Austral. J. Statist.* 35: 129–140.

EELLS, W. C. 1926. A plea for a standard definition of the standard deviation. *J. Educ. Res.* 13: 45–52.

EFROYMSON, M. A. 1960. Multiple regression analysis, pp. 191–203. *In* A. Ralston and H. S. Wilf, *Mathematical Methods for Digital Computers.* John Wiley, New York.

EINOT, I. and K. R. GABRIEL. 1975. A survey of the powers of several methods of multiple comparisons. *J. Amer. Statist. Assoc.* 70: 574–583.

EISENHART, C. 1947. The assumptions underlying the analysis of variance. *Biometrics* 3: 1–21.

EISENHART, C. 1968. Expression of the uncertainties of final results. *Science* 160: 1201–1204.

EISENHART, C. 1978. Gauss, Carl Friedrich, pp. 378–386. *In* Kruskal and Tanur (1978).

EISENHART, C. 1979. On the transition from "Student's" z to "Student's" t. *Amer. Statist.* 33: 6–10.

EVERITT, B. S. 1979. A Monte Carlo investigation of the robustness of Hotelling's one- and two-sample tests. *J. Amer. Statist. Assoc.* 74: 48–51.

EVERITT, B. S. 1992. *The Analysis of Contingency Tables.* 2nd ed. Chapman & Hall, New York. 164 pp.

EZEKIEL, M. 1930. *Methods of Correlation Analysis.* John Wiley, New York. 427 pp.

FEDERER, W. T. 1955. *Experimental Design.* Macmillan, New York. 591 pp.

FELDMAN, S. E. and E. KLUGER. 1963. Short cut calculation of the Fisher-Yates "exact test." *Psychometrika* 28: 289–291.

FELTZ, C. J. and G. E. MILLER. 1996. An asymptotic test for the equality of coefficients of variation from k populations. *Statist. in Med.* 15: 647–658.

FENSTAD, G. U. 1983. A comparison between the U and V tests in the Behrens-Fisher problem. *Biometrika* 70: 300–302.

FÉRON, R. 1978. Poisson, Siméon Denis, pp. 704–707. *In* Kruskal and Tanur (1978).

FIELLER, E. C., H. O. HARTLEY, and E. S. PEARSON. 1957. Tests for rank correlation coefficients. I. *Biometrika* 44: 470–481.

FIELLER, E. C., H. O. HARTLEY, and E. S. PEARSON. 1961. Tests for rank correlation coefficients. II. *Biometrika* 48: 29–40.

FIENBERG, S. E. 1970. The analysis of multidimensional contingency tables. *Ecology* 51: 419–433.

FIENBERG, S. E. 1972. The analysis of incomplete multiway contingency tables. *Biometrics* 28: 177–202.

FIENBERG, S. E. 1980. *The Analysis of Cross-Classified Categorical Data.* 2nd ed. MIT Press, Cambridge, Massachusetts. 198 pp.

FISHER, N. I. 1993. *Statistical Analysis of Circular Data.* Cambridge University Press, Cambridge, England. 277 pp.

FISHER, N. I. and A. J. LEE. 1982. Nonparametric measures of angular-angular association. *Biometrika* 69: 315–321.

FISHER, N. I. and A. J. LEE. 1983. A correlation coefficient for circular data. *Biometrika* 70: 327–332.

FISHER, N. I., T. LEWIS, and B. J. J. EMBLETON. 1987. *Statistical Analysis of Spherical Data.* Cambridge University Press, Cambridge, England. 329 pp.

FISHER, R. A. 1915. Frequency distributions of the values of the correlation coefficient in samples from an indefinitely large population. *Biometrika* 10: 507–521.

FISHER, R. A. 1918a. The correlation between relatives on the supposition of Mendelian inheritance. *Trans. Roy. Soc. Edinburgh* 52: 399–433.

FISHER, R. A. 1918b. The causes of human variability. *Eugenics Rev.* 10: 213–220.

FISHER, R. A. 1921. On the "probable error" of a coefficient of correlation deduced from a small sample. *Metron* 1: 3–32.

FISHER, R. A. 1922. On the interpretation of χ^2 from contingency tables and the calculation of *P*. *J. Royal Statist. Soc.* 85: 87–94.

FISHER, R. A. 1922a. The goodness of fit of regression formulae and the distribution of regression coefficients. *J. Roy. Statist. Soc.* 85: 597–612. [Cited in Eisenhart, 1979.]

FISHER, R. A. 1925a. Applications of "Student's" distribution. *Metron* 5: 90–104.

FISHER, R. A. 1925b. *Statistical Methods for Research Workers.* [1st ed.] Oliver and Boyd, Edinburgh, Scotland. 239 pp. + 6 tables.

FISHER, R. A. 1928. On a distribution yielding the error functions of several well known statistics, pp. 805–813. *In: Proc. Intern. Math. Congr., Toronto, Aug. 11–16, 1924* Vol. II. University of Toronto Press, Toronto. [Cited in Eisenhart, 1979.]

FISHER, R. A. 1932. *Statistical Methods for Research Workers.* 4th ed. Oliver and Boyd, Edinburgh, Scotland. 307 pp. + 6 tables.

FISHER, R. A. 1934. *Statistical Methods for Research Workers.* 5th ed. Oliver and Boyd, Edinburgh, Scotland.

FISHER, R. A. 1935. The logic of inductive inference. *J. Roy. Statist. Soc.* Ser. A, 98: 39–54.

FISHER, R. A. 1936. Has Mendel's work been rediscovered? *Ann. Sci.* 1: 115–137.

FISHER, R. A. 1939a. "Student." *Ann. Eugen.* 9: 1–9.

FISHER, R. A. 1939b. The comparison of samples with possibly unequal variances. *Ann. Eugen.* 9: 174–180.

FISHER, R. A. 1958. *Statistical Methods for Research Workers.* 13th ed. Hafner, New York. 356 pp.

FISHER, R. A. and F. YATES. 1963. *Statistical Tables for Biological, Agricultural, and Medical Research.* 6th ed. Hafner, New York. 146 pp.

FISZ, M. 1963. *Probability Theory and Mathematics Statistics.* 3rd ed. John Wiley, New York. 677 pp.

FIX, E. and J. L. HODGES, JR. 1955. Significance probabilities of the Wilcoxon test. *Ann. Math. Statist.* 26: 301–312.

FLEISS, J. L. 1981. *Statistical Methods for Rates and Proportions.* John Wiley and Sons, New York. 321 pp.

FLEISS, J. L., A. TYTUN, and H. K. URY. 1980. A simple approximation for calculating sample sizes for comparing independent proportions. *Biometrics* 36: 343–346.

FOX, M. 1956. Charts for the power of the *F*-test. *Ann. Math. Statist.* 27: 484–497.

FRANKLIN, L. A. 1988a. The complete exact distribution of Spearman's rho for $n = 12(1)18$. *J. Statist. Computa. Simula.* 29: 255–269.

FRANKLIN, L. A. 1988b. A note on approximations and convergence in distribution for Spearman's rank correlation coefficient. *Communic. Statist.—Theor. Meth.* 17: 55–59.

FRANKLIN, L. A. 1989. A note on the Edgeworth approximation to the distribution of Spearman's rho with a correction to Pearson's approximation. *Communic. Statist.—Simula. Computa.* 18: 245–252.

FREEMAN, M. F. and J. W. TUKEY. 1950. Transformations related to the angular and the square root. *Ann. Math. Statist.* 21: 607–611.

FREUND, R. J. 1963. A warning of round off errors in regression. *Amer. Statist.* 17(5): 13–15.

FREUND, R. J. 1971. Some observations on regressions with grouped data. *Amer. Statist.* 25(3): 29–30.

FRIEDMAN, M. 1937. The use of ranks to avoid the assumption of normality implicit in the analysis of variance. *J. Amer. Statist. Assoc.* 32: 675–701.

FRIEDMAN, M. 1940. A comparison of alternate tests of significance for the problem of *m* rankings. *Ann. Math. Statist.* 11: 86–92.

FRIENDLY, M. 1994. Mosaic displays for multiway contingency tables. *J. Amer. Statist. Assoc.* 89: 190–200.

FUJINO, Y. 1980. Approximate binomial confidence limits. *Biometrika* 67: 677–681.

GABRIEL, K. R. and P. A. LACHENBRUCH. 1969. Nonparametric ANOVA in small samples: A Monte Carlo study of the adequacy of the asymptotic approximation. *Biometrics* 25: 593–596.

GAITO, J. 1959. Non-parametric methods in psychological research. *Psychol. Rep.* 5: 115–125.

GAITO, J. 1960. Scale classification and statistics. *Psychol. Bull.* 67: 277–278.

GAMES, P. A. and J. F. HOWELL. 1976. Pairwise multiple comparison procedures with unequal *n*'s and/or variances: A Monte Carlo study. *J. Educ. Statist.* 1: 113–125.

GAMES, P. A., H. B. WINKLER, and D. A. PROBERT. 1972. Robust tests for homogeneity of variance. *Educ. Psychol. Meas.* 32: 887–909.

GANS, D. J. 1991. Preliminary test on variances. *Amer. Statist.* 45: 258.

GARSIDE, G. R. and C. MACK. 1976. Actual type 1 error probabilities for various tests in the homogeneity case of the 2×2 contingency table. *Amer. Statist.* 30: 18–21.

GART, J. J. 1969a. An exact test for comparing matched proportions in crossover designs. *Biometrika* 56: 75–80.

GART, J. J. 1969b. Graphically oriented tests of the Poisson distribution. *Bull. Intern. Statist. Inst.*, 37th Session, pp. 119–121.

GARTSIDE, P. S. 1972. A study of methods for comparing several variances. *J. Amer. Statist. Assoc.* 67: 342–346.

GEARY, R. C. and C. E. V. LESER. 1968. Significance tests in multiple regression. *Amer. Statist.* 22(1): 20–21.

GEIRINGER, H. 1978. Von Mises, Richard, pp. 1229–1231. *In* Kruskal and Tanur (1978).

GEISSER, S. and S. W. GREENHOUSE. 1958. An extension of Box's results on the use of the *F* distribution in multivariate analysis. *Ann. Math. Statist.* 29: 885–891.

GEORGE, E. O. 1987. An approximation of *F* distribution by binomial probabilities. *Statist. Prob. Lett.* 5: 169–173.

GHENT, A. W. 1972. A method for exact testing of 2×2, 2×3, 3×3, and other contingency tables, employing binomial coefficients. *Amer. Midland Natur.* 88: 15–27.

GHENT, A. W. 1991. Insights into diversity and niche breadth analysis from exact small-sample tests of the equal abundance hypothesis. *Amer. Midland Natur.* 126: 213–255.

GHENT, A. W. 1993. An exact test and normal approximation for centrifugal and centripetal patterns in line and belt transects in ecological studies. *Amer. Midland Natur.* 130: 338–355.

GHENT, A. W. and J. H. ZAR. 1992. Runs of two kinds of elements on a circle: A redevelopment, with corrections, from the perspective of biological research. *Amer. Midland Natur.* 128: 377–396.

GHOSH, B. K. 1979. A comparison of some approximate confidence intervals for the binomial parameter. *J. Amer. Statist. Assoc.* 74: 894–900.

GIBBONS, J. D. 1976. *Nonparametric Methods for Quantitative Analysis.* Holt, Rinehart, and Winston, New York. 463 pp.

GILL, J. L. 1971. Analysis of data with heterogeneous data: A review. *J. Dairy Sci.* 54: 369–373.

GIRDEN, E. R. 1992. *ANOVA: Repeated Measures.* Sage Publications, Newbury Park, California. 77 pp.

GLANTZ, S. A. and B. K. SLINKER. 1990. *Primer of Applied Regression and Analysis of Variance.* McGraw-Hill, New York. 777 pp.

GLASS, G. V., P. D. PECKHAM, and J. R. SANDERS. 1972. Consequences of failure to meet assumptions underlying the fixed effects analysis of variance and covariance. *Rev. Educ. Res.* 42: 239–288.

GLEJSER, H. 1969. A new test for heteroscedasticity. *J. Amer. Statist. Assoc.* 64: 316–323.

GLEN, W. A. and C. Y. KRAMER. 1958. Analysis of variance of a randomized block design with missing observations. *Appl. Statist.* 7: 173–185.

GOODMAN, L. A. 1970. The multivariate analysis of qualitative data: Interactions among multiple classifications. *J. Amer. Statist. Assoc.* 65: 226–256.

GORMAN, J. W. and R. J. TOMAN. 1966. Selection of variables for fitting equations to data. *Technometrics* 8: 27–51.

GREENWOOD, J. A. and D. DURAND. 1955. The distribution of length and components of the sum of *n* random unit vectors. *Ann. Math. Statist.* 26: 233–246.

GRIZZLE, J. E. 1967. Continuity correction in the χ^2-test for 2×2 tables. *Amer. Statist.* 21(4): 28–32.

GROENEVELD, R. A. and G. MEEDEN. 1984. Measuing skewness and kurtosis. *Statistician* 33: 391–399.

GUENTHER, W. C. 1964. *Analysis of Variance.* Prentice Hall, Englewood Cliffs, New Jersey. 199 pp.

GUENTHER, W. C. 1977. Some probabilities available from desk calculators and their relation to tables. *Amer. Statist.* 31: 41–45.

GUMBEL, E. J., J. A. GREENWOOD, and D. DURAND. 1953. The circular normal distribution: Theory and tables. *J. Amer. Statist. Assoc.* 48: 131–152.

GURLAND, J. and R. C. TRIPATHI. 1971. A simple approximation for unbiased estimation of the standard deviation. *Amer. Statist.* 25(4): 30–32.

HABER, M. 1980. A comparison of some continuity corrections for the chi-squared test on 2×2 tables. *J. Amer. Statist. Assoc.* 75: 510–515.

HABER, M. 1982. The continuity correction and statistical testing. *Intern. Statist. Rev.* 50: 135–144.

HABER, M. 1987. A comparison of some conditional and unconditional exact tests for 2×2 contingency tables. *Communic. Statist.—Simula. Computa.* 16: 999–1013.

HAIGHT, F. A. 1967. *Handbook of the Poisson Distribution.* John Wiley, New York. 168 pp.

HAIR, J. F., JR., R. E. ANDERSON, R. L. TATHAM, and W. C. BLACK. 1995. *Multivariate Analysis with Readings.* 4th ed. Prentice Hall, Upper Saddle River, N.J. 758 pp.

HALBERG, F. and J.-K. LEE. 1974. Glossary of selected chronobiologic terms, pp. XXXVII–L. *In* L. E. Scheving, F. Halberg, and J. E. Pauly (eds.), *Chronology.* Igaku Shoin, Tokyo.

HALL, I. J. 1972. Some comparisons of tests for equality of variances. *J. Statist. Computa. Simula.* 1: 183–194.

HAMAKER, H. C. 1978. Approximating the cumulative normal distribution and its inverse. *Appl. Statist.* 27: 76–77.

HAMDY, M. I. and M. Y. EL-BASSIOUNI. 1993. A revision of the Tukey multiple comparisons procedure to control the probability of committing at most one Type I error. *Communic. Statist.—Simula. Computa.* 22: 739–748.

HAMILTON, D. 1987. Sometimes $R^2 > r_{yx_1}^2 + r_{yx_2}^2$. *Amer. Statist.* 41: 129–132. [See also Hamilton's (1988) reply to and reference to other writers on this topic: *Amer. Statist.* 42: 90–91.]

HAND, D. J. and C. C. TAYLOR. 1987. *Multivariate Analysis of Variance and Repeated Measures.* Chapman and Hall, London. 262 pp.

HÄRDLE, W. 1989. *Applied Nonparametric Regression.* Cambridge University Press, Cambridge, England. 333 pp.

HARDY, M. A. 1993. *Regression with Dummy Variables.* Sage Publications, Newbury Park, California. 90 pp.

HARRIS, J. A. 1910. The arithmetic of the product moment method of calculating the coefficient of correlation. *Amer. Natur.* 44: 693–699.

HARRISON, D. and G. K. KANJI. 1988. The development of analysis of variance for circular data. *J. Appl. Statist.* 15: 197–223.

HARRISON, D., G. K. KANJI, and R. J. GADSDEN. 1986. Analysis of variance for circular data. *J. Appl. Statist.* 13: 123–138.

HARTER, H. L. 1957. Error rates and sample sizes for range tests in multiple comparisons. *Biometrics* 13: 511–536.

HARTER, H. L. 1960. Tables of range and studentized range. *Ann. Math. Statist.* 31: 1122–1147.

HARTER, H. L. 1970. *Order Statistics and Their Use in Testing and Estimation,* Vol. 1. Tests Based on Range and Studentized Range of Samples from a Normal Population. U. S. Government Printing Office, Washington, D.C. 761 pp.

HARTER, H. L., H. J. KHAMIS, and R. E. LAMB. 1984. Modified Kolmogorov-Smirnov tests for goodness of fit. *Communic. Statist.—Simula. Computa.* 13: 293–323.

HARTIGAN, J. A. and B. KLEINER. 1981. Mosaics for contingency tables, pp. 268–273. *In* W. F. Eddy (ed.), *Computer Science and Statistics: Proceedings of the 13th Symposium on the Interface.* Springer-Verlag, New York.

HARTIGAN, J. A. and B. KLEINER. 1984. A mosaic of television ratings. *Amer. Statist.* 38: 32–35.

HASTINGS, C., JR. 1955. *Approximations for Digital Computers.* Princeton University Press, Princeton, New Jersey. 201 pp.

HAVILAND, M. B. 1990. Yate's correction for continuity and the analysis of 2×2 contingency tables. *Statist. Med.* 9: 363–367. *Also:* Comment, by N. Mantel, *ibid.* 9: 369–370; S. W. Greenhouse, *ibid.* 9: 371–372; G. Barnard, *ibid.* 9: 373–375; R. B. D'Agostino, *ibid.* 9: 377–378; J. E. Overall, *ibid.* 9: 379–382; Rejoinder, by M. B. Haviland, *ibid.* 9: 383.

HAWKES, A. G. 1982. Approximating the normal tail. *Statistician* 31: 231–236.

HAWKINS, D. M. 1980. A note on fitting a regression without an intercept term. *Amer. Statist.* 34(4): 233.

HAYS, W. L. 1963. *Statistics for Psychologists.* Holt, Rhinehart & Winston, New York. 719 pp.

HAYTER, A. J. 1984. A proof of the conjecture that the Tukey-Kramer multiple comparisons procedure is conservative. *Ann. Statist.* 12: 61–75.

HEALY, M. J. R. 1963. Programming multiple regression. *Computer J.* 6: 57–61.

HEALY, M. J. R. 1984. The use of R^2 as a measure of goodness of fit. *J. Roy. Statist. Soc.* Ser. A, 147: 608–609.

HICKS, C. R. 1982. *Fundamental Concepts in Design of Experiments.* 3rd ed. Holt, Rinehart, & Winston, New York. 425 pp.

HINES, W. G. S. 1996. Pragmatics of pooling in ANOVA tables. *Amer. Statist.* 50: 127–139.

HODGES, J. L., JR. 1990. Improved significance probabilities of the Wilcoxon test. *J. Educ. Statist.* 15: 249–265.

HODGES, J. L., JR. and E. L. LEHMANN. 1956. The efficiency of some nonparametric competitors of the *t*-test. *Ann. Math. Statist.* 27: 324–335.

HODGES, J. L., JR., P. H. RAMSEY, and S. WECHSLER. 1955. A bivariate sign test. *Ann. Math. Statist.* 26: 523–527.

HOLLANDER, M. and J. SETHURAMAN. 1978. Testing for agreement between two groups of judges. *Biometrika* 65: 403–411.

HOLLANDER, M. and D. A. WOLFE. 1973. *Nonparametric Statistical Methods.* John Wiley, New York. 503 pp.

HORSNELL, G. 1953. The effect of unequal group variances on the *F*-test for the homogeneity of group means. *Biometrika* 40: 128–136.

HOTELLING, H. 1931. The generalization of Student's ratio. *Ann. Math. Statist.* 2: 360–378.

HOTELLING, H. 1953. New light on the correlation coefficient and its transform. *J. Roy. Statist. Soc.* B15: 193–232.

HOTELLING, H. and M. R. PABST. 1936. Rank correlation and tests of significance involving no assumption of normality. *Ann. Math. Statist.* 7: 29–43.

HOWELL, D. C. 1987. *Statistical Methods for Psychology.* 2nd ed. Duxbury Press, Boston, Massachusetts. 636 pp.

HOWELL, J. F. and P. A. GAMES. 1974. The effects of variance heterogeneity on simultaneous multiple-comparison procedures with equal sample size. *Brit. J. Math. Statist. Psychol.* 27: 72–81.

HSU, P. L. 1938. Contributions to the theory of "Student's" *t*-test as applied to the problem of two samples. *Statist. Res. Mem.* 2: 1–24. [Cited in Ramsey, 1980.]

HUBER, P. 1964. Robust estimation of a location parameter. *Ann. Math. Statist.* 35: 73–101.

HUBER, P. 1981. *Robust Statistics.* John Wiley & Sons, New York. 308 pp.

HUCK, S. W. and B. H. LAYNE. 1974. Checking for proportional n's in factorial ANOVA's. *Educ. Psychol. Meas.* 34: 281–287.

HUFF, D. 1954. *How to Lie with Statistics.* W. W. Norton, New York. 142 pp.

HUITEMA, B. E. 1974. Three multiple comparison procedures for contrasts among correlation coefficients. *Proc. Soc. Statist. Sect., Amer. Statist. Assoc.,* 1974, pp. 336–339.

HUITEMA, B. E. 1980. *The Analysis of Covariance and Alternatives.* John Wiley, New York. 445 pp.

HUMMEL, T. J. and J. R. SLIGO. 1971. Empirical comparisons of univariate and multivariate analysis of variance procedures. *Psychol. Bull.* 76: 49–57.

HURLBERT, S. H. 1984. Pseudoreplication and the design of ecological field experiments. *Ecol. Monogr.* 54: 187–211.

HUTCHESON, K. 1970. A test for comparing diversities based on the Shannon formula. *J. Theoret. Biol.* 29: 151–154.

HUTCHINSON, T. P. 1979. The validity of the chi-squared test when expected frequencies are small: A list of recent research references. *Communic. Statist.—Theor. Meth.* A8: 327–335.

HUYNH, H. and FELDT, L. S. 1970. Conditions under which mean square ratios in repeated measurements designs have exact F-distributions. *J. Amer. Statist. Assoc.* 65: 1582–1589.

IMAN, R. L. 1974a. Use of a t-statistic as an approximation to the exact distribution of the Wilcoxon signed ranks test statistic. *Communic. Statist.—Theor. Meth.* A3: 795–806.

IMAN, R. L. 1974b. A power study of a rank transform for the two-way classification model when interaction may be present. *Can. J. Statist.* 2: 227–229.

IMAN, R. L. 1987. Tables of the exact quantiles of the top-down correlation coefficient for $n = 3(1)14$. *Communic. Statist.—Theor. Meth.* 16: 1513–1540.

IMAN, R. L. and W. J. CONOVER. 1978. Approximation of the critical region for Spearman's rho with and without ties present. *Communic. Statist.—Simula. Computa.* 7: 269–282.

IMAN, R. L. and W. J. CONOVER. 1979. The use of the rank transform in regression. *Technometrics* 21: 499–509.

IMAN, R. L. and J. M. CONOVER. 1985. A measure of top-down correlation. Technical Report SAND85-0601, Sandia National Laboratories, Albuquerque, New Mexico. 44 pp.

IMAN, R. L. and W. J. CONOVER. 1987. A measure of top-down correlation. *Technometrics* 29: 351–357. Correction: *Technometrics* 31: 133 (1989).

IMAN, R. L. and J. M. DAVENPORT. 1976. New approximations to the exact distribution of the Kruskal-Wallis test statistic. *Communic. Statist.—Theor. Meth.* A5: 1335–1348.

IMAN, R. L. and J. M. DAVENPORT. 1980. Approximations of the critical region of the Friedman statistic. *Communic. Statist. — Theor. Meth.* A9: 571–595.

IMAN, R. L. and J. M. DAVENPORT. 1980. Approximations of the critical region of Friedman statistic. *Communic. Statist.—Theor. Meth.* A9: 571–595.

IMAN, R. L., S. C. HORA, and W. J. CONOVER. 1984. Comparison of asymptotically distribution-free procedures for the analysis of complete blocks. *J. Amer. Statist. Assoc.* 79: 674–685.

IMAN, R. L., D. QUADE, and D. A. ALEXANDER. 1975. Exact probability levels for the Kruskal-Wallis test, pp. 329–384. *In* H. L. Harter and D. B. Owen, *Selected Tables in Mathematical Statistics,* Volume III. American Mathematical Society, Providence, Rhode Island.

INTERNATIONAL BUSINESS MACHINES CORPORATION. 1968. *System/360 Scientific Subroutine Package (360A-CM-03X) Version III. Programmer's Manual.* 4th ed. White Plains, New York. 454 pp.

IRWIN, J. O. 1935. Tests of significance for differences between percentages based on small numbers. *Metron* 12: 83–94.

IRWIN, J. O. 1978. Gosset, William Sealy, pp. 409–413. *In* Kruskal and Tanur (1978).

IVES, K. H. and J. D. GIBBONS. 1967. A correlation measure for nominal data. *Amer. Statist.* 21(5): 16–17.

JACQUES, J. A. and M. NORUSIS. 1973. Sampling requirements on the estimation of parameters in heteroscedastic linear regression. *Biometrics* 29: 771–780.

JAMES, G. S. 1951. The comparison of several groups of observations when the ratios of the population variances are unknown. *Biometrika* 38: 324–329.

JEYARATNAM, S. 1992. Confidence intervals for the correlation coefficient. *Statist. Prob. Lett.* 15: 389–393.

JOHNSON, B. R. and D. J. LEEMING. 1990. A study of the digits of π, e, and certain other irrational numbers. *Sankhyā: Indian J. Statist.* 52B: 183–189.

JOHNSON, R. A. and T. WEHRLY. 1977. Measures and models for angular correlation and angular-linear correlation. *J. Roy. Statist. Soc.* Ser. B, 39: 222–229.

JOHNSON, R. A. and D. W. WICHERN. 1998. *Applied Multivariate Statistical Analysis.* 4th ed. Prentice Hall, Upper Saddle River, N.J. 816 pp.

JOLLIFFE, I. T. 1981. Runs test for detecting dependence between two variances. *Statistician* 30: 137–141.

JUPP, P. E. and K. V. MARDIA. 1980. A general correlation coefficient for directional data and related regression problems. *Biometrika* 67: 163–173.

KEMP, A. W. 1989. A note on Stirling's expansion for factorial n. *Statist. Prob. Lett.* 7: 21–22.

KENDALL, M. G. 1938. A new measure of rank correlation. *Biometrika* 30: 81–93.

KENDALL, M. G. 1943. *The Advanced Theory of Statistics.* Vol. I. Charles Griffin, London, England. [Cited in Eisenhart, 1979.]

KENDALL, M. G. 1945. The treatment of ties in ranking problems. *Biometrika* 33: 239–251.

KENDALL, M. G. 1962. *Rank Correlation Methods.* 3rd ed. Charles Griffin, London, England. 199 pp.

KENDALL, M. G. and B. BABINGTON-SMITH. 1939. The problem of *m* rankings. *Ann. Math. Statist.* 10: 275–287.

KENDALL, M. G. and A. STUART. 1966. *The Advanced Theory of Statistics,* Vol. 3. Hafner. New York. 552 pp.

KENDALL, M. G. and A. STUART. 1979. *The Advanced Theory of Statistics*, Vol. 2. 4th ed. Griffin, London, England. 748 pp.

KENNARD, R. W. 1971. A note on the C_p statistics. *Technometrics* 13: 899–900.

KEPNER, J. L. and D. H. ROBINSON. 1988. Nonparametric methods for detecting treatment effects in repeated-measures designs. *J. Amer. Statist. Assoc.* 83: 456–461.

KEPPEL, G. 1991. *Design and Analysis: A Researcher's Handbook.* 3rd ed. Prentice Hall, Englewood Cliffs, New Jersey. 594 pp.

KESELMAN, H. J. 1976. A power investigation of the Tukey multiple comparison statistic. *Educ. Psychol. Meas.* 36: 97–104.

KESELMAN, H. J., P. A. GAMES, and J. J. CLINCH. 1979. Tests for homogeneity of variance. *Communic. Statist.—Simulat. Computa.* 88: 113–129.

KESELMAN, H. J., R. MURRAY, and J. C. ROGAN. 1976. Effect of very unequal group sizes on Tukey's multiple comparison test. *Educ. Psychol. Meas.* 36: 263–270.

KEULS, M. 1952. The use of the "studentized range" in connection with an analysis of variance. *Euphytica* 1: 112–122.

KHAMIS, H. J. 1990. The δ-corrected Kolmogorov-Smirnov test for goodness of fit. *J. Statist. Plan. Infer.* 24: 317–335.

KHAMIS, H. J. 1993. A comparative study of the δ-corrected Kolmogorov-Smirnov test. *J. Appl. Statist.* 20: 401–421.

KIHLBERG, J. K., J. H. HERSON, and W. E. SCHUTZ. 1972. Square root transformation revisited. *J. Royal Statist. Soc.* Ser. C. *Appl. Statist.* 21: 76–81.

KIRK, R. E. 1982. *Experimental Design: Procedures for the Behavioral Sciences.* 2nd ed. Brooks/Cole, Monterey, California. 911 pp.

KOEHLER, K. J. and K. LARNTZ. 1980. An empirical investigation of goodness-of-fit statistics for sparse multinomials. *J. Amer. Statist. Assoc.* 75: 336–344.

KOHR, R. L. and P. A. GAMES. 1974. Robustness of the analysis of variance, the Welch procedure, and a Box procedure to heterogeneous variances. *J. Exper. Educ.* 43: 61–69.

KOLMOGOROV, A. 1933. Sulla determinazione empirica di una legge di distribuzione. *Giornalle dell Instituto Italiano degli Attuari* 4: 1–11.

KRAMER, C. Y. 1956. Extension of multiple range tests to group means with unequal numbers of replications. *Biometrics* 12: 307–310.

KROLL, N. E. A. 1989. Testing independence in 2×2 contingency tables. *J. Educ. Statist.* 1: 67–79.

KRUSKAL, W. H. 1957. Historical notes on the Wilcoxon unpaired two-sample test. *J. Amer. Statist. Assoc.* 52: 356–360.

KRUSKAL, W. 1978. Formulas, numbers, words: Statistics in prose. *Amer. Schol.* 47: 223–229.

KRUSKAL, W. H. and W. A. WALLIS. 1952. Use of ranks in one-criterion analysis of variance. *J. Amer. Statist. Assoc.* 47: 583–621.

KRUTCHKOFF, R. G. 1988. One-way fixed effects analysis of variance when the error variances may be unequal. *J. Statist. Computa. Simula.* 30: 259–271.

KUIPER, N. H. 1960. Tests concerning random points on a circle. *Ned. Akad. Wetensch. Proc.* Ser. A 63: 38–47.

KVÅLSETH, T. O. 1985. Cautionary note about R^2. *Amer. Statist.* 39: 279–285.

LANCASTER, H. O. 1949. The combination of probabilities arising from data in discrete distributions. *Biometrika* 36: 370–382.

LANCASTER, H. O. 1969. *The Chi-Squared Distribution.* John Wiley, New York. 356 pp.

LARNTZ, K. 1978. Small-sample comparisons of exact levels for chi-squared goodness-of-fit statistics. *J. Amer. Statist. Assoc.* 73: 253–263.

LAWAL, H. B. and G. J. G. UPTON. 1984. On the use of X^2 as a test of independence in contingency tables with small cell expectations. *Austral. J. Statist.* 26: 75–85.

LAWLEY, D. N. 1938. A generalization of Fisher's *z* test. *Biometrika* 30: 180–187.

LAYARD, M. W. J. 1973. Robust large-sample tests for homogeneity of variance. *J. Amer. Statist. Assoc.* 68: 195–198.

LEADBETTER, M. R. 1988. Harald Cramér, 1893–1985. *Intern. Statist. Rev.* 56: 89–97.

LEE, A. F. S. and N. S. FINEBERG. 1991. A fitted test for the Behrens-Fisher problem. *Communic. Statist.—Theor. Meth.* 20: 653–666.

LEE, A. F. S. and J. GURLAND. 1975. Size and power of tests for equality of means of two normal populations with unequal variances. *J. Amer. Statist. Assoc.* 70: 933–941.

LEHMANN, E. L. 1975. *Nonparametrics. Statistical Methods Based on Ranks.* Holden-Day, San Francisco, California. 457 pp.

LEHMANN, E. L. and C. REID. 1982. Jerzy Neyman 1894–1981. *Amer. Statist.* 36: 161–162.

LESLIE, P. H. 1955. A simple method of calculating the exact probability in 2×2 contingency tables with small marginal totals. *Biometrika* 42: 522–523.

LEVENE, H. 1952. On the power function of tests of randomness based on runs up and down. *Ann. Math. Statist.* 23: 34–56.

LEVY, K. J. 1975a. Some multiple range tests for variances. *Educ. Psychol. Meas.* 35: 599–604.

LEVY, K. J. 1975b. Comparing variances of several treatments with a control. *Educ. Psychol. Meas.* 35: 793–796.

LEVY, K. J. 1975c. An empirical comparison of several multiple range tests for variances. *J. Amer. Statist. Assoc.* 70: 180–183.

LEVY, K. J. 1976. A multiple range procedure for independent correlations. *Educ. Psychol. Meas.* 36: 27–31.

LEVY, K. J. 1978a. An empirical comparison of the ANOVA F-test with alternatives which are more robust against heterogeneity of variance. *J. Statist. Computa. Simula.* 8: 49–57.

LEVY, K. J. 1978b. An empirical study of the cube root test for homogeneity of variances with respect to the effects of a non-normality and power. *J. Statist. Computa. Simula.* 7: 71–78.

LEVY, K. J. 1979. Pairwise comparisons associated with the K independent sample median test. *Amer. Statist.* 33: 138–139.

LEWONTIN, R. C. 1966. On the measurement of relative variability. *Systematic Zool.* 15: 141–142.

LEYTON, M. K. 1968. Rapid calculation of exact probabilities for 2×3 contingency tables. *Biometrics* 24: 714–717.

LI, C. C. 1964. *Introduction to Experimental Statistics.* McGraw-Hill, New York. 460 pp.

LI, L. and W. R. SCHUCANY. 1975. Some properties of a test for concordance of two groups of rankings. *Biometrika* 62: 417–423.

LIDDELL, D. 1976. Practical tests of 2×2 contingency tables. *Statistician* 25: 295–304.

LIDDELL, D. 1980. Practical tests for comparative trials: a rejoinder to N. L. Johnson. *Statistician* 29: 205–207.

LIGHT, R. J. and B. H. MARGOLIN. 1971. Analysis of variance for categorical data. *J. Amer. Statist. Assoc.* 66: 534–544.

LIN, J.-T. 1988. Approximating the cumulative chi-square distribution and its inverse. *Statistician* 37: 3–5.

LIN, J.-T. 1988. Alternatives to Hamaker's approximations to the cumulative normal distribution and its inverse. *Statistician* 37: 413–414.

LIN, L. I-K. 1989. A concordance correlation coefficient to evaluate reproducibility. *Biometrics* 45: 255–268.

LIN, L. I-K. and V. CHINCILLI. 1996. Rejoinder to the letter to the editor from Atkinson and Klevill [Comment on the use of concordance correlation to assess the agreement between two variables]. *Biometrics* 53: 777–778.

LINDSAY, R. B. 1976. John William Strutt, Third Baron Rayleigh, pp. 100–107. *In* C. C. Gillispie (ed.), *Dictionary of Scientific Biography,* Vol. XIII. Charles Scribner's Sons, New York.

LING, R. F. 1974. Comparison of several algorithms for computing sample means and variances. *J. Amer. Statist. Assoc.* 69: 859–866.

LITTLE, R. J. A. 1989. Testing the equality of two independent binomial proportions. *Amer. Statist.* 43: 283–288.

LLOYD, M., J. H. ZAR, and J. R. KARR. 1968. On the calculation of information-theoretical measures of diversity. *Amer. Midland Natur.* 79: 257–272.

LOCKHART, R. A. and M. A. STEPHENS. 1985. Tests of fit for the von Mises distribution. *Biometrika* 72: 647–652.

LONGLEY, J. W. 1967. An appraisal of least squares programs for the electronic computer from the point of view of the user. *J. Amer. Statist. Assoc.* 62: 819–841.

LUDWIG, J. A. and J. F. REYNOLDS. 1988. *Statistical Ecology.* John Wiley & Sons, New York. 337 pp.

MAAG, U. R. 1966. A k-sample analogue of Watson's U^2 statistic. *Biometrika* 53: 579–583.

MACKINNON, W. J. 1964. Table for both the sign test and distribution-free confidence intervals of the median for sample sizes to 1,000. *J. Amer. Statist. Assoc.* 59: 935–956.

MALLOWS, C. L. 1973. Some comments on C_P. *Technometrics* 15: 661–675.

MANN, H. B. and D. R. WHITNEY. 1947. On a test of whether one of two random variables is stochastically larger than the other. *Ann. Math. Statist.* 18: 50–60.

MANTEL, N. 1966. F-ratio probabilities from binomial tables. *Biometrics* 22: 404–407.

MANTEL, N. 1970. Why stepdown procedures in variable selection. *Technometrics* 12: 621–625.

MANTEL, N. 1974. Comment and a suggestion. [Re Conover, 1974.] *J. Amer. Statist. Assoc.* 69: 378–380.

MAOR, E. 1944. *e: The Story of a Number.* Princeton University Press, Princeton, New Jersey. 223 pp.

MARASCUILO, L. A. 1971. *Statistical Methods for Behavioral Science Research.* McGraw-Hill, New York. 578 pp.

MARASCUILO, L. A. and M. MCSWEENEY. 1967. Nonparametric post hoc comparisons for trend. *Psychol. Bull.* 67: 401–412.

MARASCUILO, L. A. and M. MCSWEENEY. 1977. *Nonparametric and Distribution-free Methods for the Social Sciences.* Brooks/Cole, Monterey, California. 556 pp.

MARDIA, K. V. 1967. A non-parametric test for the bivariate two-sample location problem. *J. Royal Statist. Soc.* B29: 320–342.

MARDIA, K. V. 1969. On the null distribution of a nonparametric test for the bivariate two-sample problem. *J. Royal Statist. Soc.* B31: 98–102.

MARDIA, K. V. 1970a. Measures of multivariate skewness and kurtosis with applications. *Biometrika* 57: 519–530.

MARDIA, K. V. 1970b. A bivariate non-parametric c-sample test. *J. Royal Statist. Soc.* B32: 74–87.

MARDIA, K. V. 1972a. *Statistics of Directional Data.* Academic Press, New York. 357 pp.

MARDIA, K. V. 1972b. A multisample uniform scores test on a circle and its parametric competitor. *J. Royal Statist. Soc.* B34: 102–113.

MARDIA, K. V. 1975. Statistics of directional data. (With discussion.) *J. Royal Statist. Soc.* B37: 349–393.

MARDIA, K. V. 1976. Linear-angular correlation coefficients and rhythmometry. *Biometrika* 63: 403–405.

MARDIA, K. V. 1981. Directional statistics in geosciences. *Communic. Statist.—Theor. Meth.* A10: 1523–1543.

MARDIA, K. V. 1990. Obituary: Professor B. L. Welch. *J. Roy. Statist. Soc.* Ser. A, 153: 253–254.

MARKOWSKI, C. A. and E. P. MARKOWSKI. 1990. Conditions for the effectiveness of a preliminary test of variance. *Amer. Statist.* 44: 322–326.

MARTIN, L., R. LEBLANC, and N. K. TOAN. 1993. Tables for the Friedman rank test. *Can. J. Statist.* 21: 39–43.

MARTÍN ANDRÉS, A. 1991. A review of classic nonasymptotic methods for comparing two proportions by means of independent samples. *Communic. Statist.—Simula. Computa.* 20: 551–583.

MARTÍN ANDRÉS, A., A. SILVA MATO, and I. HERRANZ TEJEDOR. 1992. A critical review of asymptotic methods for comparing two proportions by means of independent samples. *Communic. Statist.—Simula. Computa.* 21: 551–586.

MARTÍN ANDRÉS, A., I. HERRANZ TEJEDOR, and A. SILVA MATO. 1995. The Wilcoxon, Spearman, Fisher, χ^2-, Student and Pearson tests and 2×2 tables. *Statistician* 44: 441–450.

MASSEY, F. J., Jr. 1951. The Kolmogorov-Smirnov test for goodness of fit. *J. Amer. Statist. Assoc.* 46: 68–78.

MAURAIS, J. and R. OUIMET. 1986. Exact critical values of Bartlett's test of homogeneity of variances for unequal sample sizes for two populations and power of the test. *Metrika* 33: 275–289.

MAXWELL, A. E. 1970. Comparing the classification of subjects by two independent judges. *Brit. J. Psychiatry* 116: 651–655.

MAXWELL, S. E. 1980. Pairwise multiple comparisons in repeated measures designs. *J. Educ. Statist.* 5: 269–287.

MAXWELL, S. E. and H. D. DELANEY. 1990. *Designing Experiments and Analyzing Data. A Model Comparison Perspective.* Wadsworth, Belmont, California. 902 pp.

MCCORNACK. R. L. 1965. Extended tables of the Wilcoxon matched pair signed rank statistic. *J. Amer. Statist. Assoc.* 60: 864–871.

MCGILL, R., J. W. TUKEY, and W. A. LARSEN. 1978. Variations of box plots. *Amer. Statist.* 32: 12–16.

MCKAY, A. T. 1932. Distribution of the coefficient of variation and the extended "t" distribution. *J. Roy. Statist. Soc.* A95: 695–698.

MCNEMAR, Q. 1947. Note on the sampling error of the difference between correlated proportions or percentages. *Psychometrica* 12: 153–157.

MEDDIS, R. 1984. *Statistics Using Ranks: A Unified Approach.* Basil Blackwell, Oxford, England. 449 pp.

MEHTA, C. R. and J. F. HILTON. 1993. Exact power of conditional and unconditional tests: Going beyond the 2×2 contingency table. *Amer. Statist.* 47: 91–98.

MEHTA, J. S. and R. SRINIVASAN. 1970. On the Behrens-Fisher problem. *Biometrics* 57: 649–655.

MENDEL, G. 1865. Versuche über Planzen-Hybriden [Experiments on Plant Hybrids]. *Verhandlungen des naturforschenden Vereines in Brünn* 4, Abhandlungen, pp 3–47 [which appeared in 1866, and which has been published in English translation many times beginning in 1901, including as Mendel, 1933, and as pp. 1–48 in Stern and Sherwood, 1966).

MENDEL, G. 1933. *Experiments in Plant Hybridization.* Harvard University Press, Cambridge, Massachusetts. 353 pp.

MILLER, G. E. 1991. Asymptotic test statistics for coefficients of variation. *Communic. Statist.—Theor Meth.* 20: 2251–2262.

MILLER, L. H. 1956. Table of percentage points of Kolmogorov statistics. *J. Amer. Statist. Assoc.* 51: 111–121.

MILLER, R. G., Jr. 1981. *Simultaneous Statistical Inference.* 2nd ed. McGraw-Hill, New York. 299 pp.

MILTON, R. C. 1964. An extended table of critical values for the Mann-Whitney (Wilcoxon) two-sample statistic. *J. Amer. Statist. Assoc.* 59: 925–934.

MOLENAAR, W. 1969a. How to poison Poisson (when approximating binomial tails). *Statist. Neerland.* 23: 19–40.

MOLENAAR, W. 1969b. Additional remark on "How to poison Poisson (when approximating binomial tails)." *Statist. Neerland.* 23: 241.

MONTGOMERY, D. C. and F. A. PECK. 1992. *Introduction to Linear Regression Analysis.* 2nd ed. John Wiley, New York. 527 pp.

MOOD, A. M. 1950. *Introduction to the Theory of Statistics.* McGraw-Hill, New York. 433 pp.

MOOD, A. M. 1954. On the asymptotic efficiency of certain non-parametric two-sample tests. *Ann. Math. Statist.* 25: 514–522.

MOORE, B. R. 1980. A modification of the Rayleigh test for vector data. *Biometrika* 67: 175–180.

MOORE, D. S. 1986. Tests of chi-squared type, pp. 63–95. *In* R. B. D'Agostino and M. A. Stephens (eds.), *Goodness-of-Fit Techniques.* Marcel Dekker, New York.

MOORS, J. J. A. 1986. The meaning of kurtosis: Darlington revisited. *Amer. Statist.* 40: 283–284.

MOORS, J. J. A. 1988. A quantile alternative for kurtosis. *Statistician* 37: 25–32.

MOSER, B. K. and G. R. STEVENS. 1992. Homogeneity of variance in the two-sample means test. *Amer. Statist.* 46: 19–21.

MOSIMANN, J. E. 1968. *Elementary Probablilty for the Biological Sciences.* Appleton-Century-Crofts, New York. 255 pp.

MUDDAPUR, M. V. 1988. A simple test for correlation coefficient in a bivariate normal population. *Sankyā: Indian J. Statist.* Ser. B. 50: 60–68.

MULHOLLAND, H. P. 1977. On the null distribution of $\sqrt{b_1}$ for samples of size at most 25, with tables. *Biometrika* 64: 401–409.

NAGASENKER, P. B. 1984. On Bartlett's test for homogeneity of variances. *Biometrika* 71: 405–407.

NATH, R. and B. S. DURAN. 1981. The rank transform in the two-sample location problem. *Communic. Statist.—Simula. Computa.* B10: 383–394.

NELSON, W., Y. L. TONG, J.-K. LEE, and F. HALBERG. 1979. Methods for cosinor-rhythmometry. *Chronobiologia* 6: 305–323.

NEMENYI, P. 1963. *Distribution-Free Multiple Comparisons.* State University of New York, Downstate Medical Center. [Cited in Wilcoxon and Wilcox (1964).]

NETER, J., W. WASSERMAN, and M. H. KUTNER. 1990. *Applied Linear Statistical Models.* 3rd ed. Richard D. Irwin, Homewood, Illinois. 1181 pp.

NEWMAN, D. 1939. The distribution of range in samples from a normal population, expressed in terms of an independent estimate of standard deviation. *Biometrika* 31: 20–30.

NEYMAN, J. 1959. Optimal asymptotic tests of composite hypothesis, pp. 213–234. *In* U. Grenander (ed.), *Prob-*

ability and Statistics: The Harald Cramér Volume. John Wiley, New York.

NEYMAN, J. 1967. R. A. Fisher: An appreciation. *Science* 156: 1456–1460.

NEYMAN, J. and E. S. PEARSON. 1928a. On the use and interpretation of certain test criteria for purposes of statistical inference. Part I. *Biometrika* 20A: 175–240.

NEYMAN, J. and E. S. PEARSON. 1928b. On the use and interpretation of certain test criteria for purposes of statistical inference. Part II. *Biometrika* 20A: 263–294.

NEYMAN, J. and E. S. PEARSON. 1931. On the problem of k samples. *Bull. Acad. Polon. Sci. Lett.* Ser. A, 3: 460–481.

NORRIS, R. C. and H. F. HJELM. 1961. Non-normality and product moment correlation. *J. Exp. Educ.* 29: 261–270.

NORTON, H. W. 1939. The 7×7 squares. *Ann. Eugen.* 9: 269–307.

NORWOOD, P. K., A. R. SAMPSON, K. McCARROLL, and R. STAUM. 1989. A multiple comparisons procedure for use in conjunction with the Benard-van Elteren test. *Biometrics* 45: 1175–1182.

O'BRIEN, P. C. 1976. A test for randomness. *Biometrics* 32: 391–401.

O'BRIEN, P. C. and P. J. DYCK. 1985. A runs test based on run lengths. *Biometrics* 41: 237–244.

O'BRIEN, R. G. and M. KAISER. 1985. MANOVA method for analyzing repeated measures designs: An extensive primer. *Psychol. Bull.* 97: 316–333.

O'CINNEIDE, C. A. 1990. The mean is within one standard deviation of any median. *Amer. Statist.* 44: 292–293.

ODEH, R. E. and J. O. EVANS. 1974. The percentage points of the normal distribution. *J. Royal Statist. Soc.* Ser. C. *Appl. Statist.* 23: 98–99.

ODEH, R. E. and M. FOX. 1975. *Sample Size Choice: Charts for Experiments with Linear Models.* Marcel Dekker, New York. 190 pp.

OLDS, E. G. 1938. Distributions of sums of squares of rank differences for small numbers of individuals. *Ann. Math. Statist.* 9: 133–148.

OLSON, C. L. 1974. Comparative robustness of six tests on multivariate analysis of variance. *J. Amer. Statist. Assoc.* 69: 894–908.

OLSON, C. L. 1976. On choosing a test statistic in multivariate analysis of variance. *Psychol. Bull.* 83: 579–586.

OLSON, C. L. 1979. Practical considerations in choosing a MANOVA test statistic: A rejoinder to Stevens. *Psychol. Bull.* 86: 1350–1352.

OLVER, F. W. J. 1964. Bessel functions of integer order, pp. 355–433. *In* M. Abramowitz and I. Stegun (eds.), *Handbook of Mathematical Functions,* National

Bureau of Standards, Washington, D.C. (Also, Dover, New York, 1965.)

OSTLE, B. and R. W. MENSING. 1975. *Statistics in Research*. 3rd ed. Iowa State University Press, Ames, Iowa. 596 pp.

OTTEN, A. 1973. The null distribution of Spearman's *S* when *n* = 13(1)16. *Statist. Neerland.* 27: 19–20.

OVERALL, J. E., H. M. RHOADES and R. R. STARBUCK. 1987. Small-sample tests for homogeneity of response probabilities in 2 × 2 contingency tables. *Psychol. Bull.* 98: 307–314.

OWEN, D. B. 1962. *Handbook of Statistical Tables.* Addison-Wesley, Reading, Massachusetts. 580 pp.

OZER, D. J. 1985. Correlation and the coefficient of determination. *Psychol. Bull.* 97: 307–315.

PATIL, K. D. 1975. Cochran's *Q* test: Exact distribution. *J. Amer. Statist. Assoc.* 70: 186–189.

PATNIAK, A. B. 1949. The non-central χ^2 and *F*-distributions and their applications. *Biometrika* 36: 202–232.

PAUL, S. R. 1988. Estimation of and testing significance for a common correlation coefficient. *Communic. Statist.—Theor. Meth.* 17: 39–53.

PAULL, A. E. 1950. On a preliminary test for pooling mean squares in the analysis of variance. *Ann. Math. Statist.* 21: 539–556.

PAZER, H. L. and L. A. SWANSON. 1972. *Modern Methods for Statistical Analysis*. Intext Educational Publishers, Scranton, Pennsylvania. 483 pp.

PEARSON, E. S. 1939. Student as a statistician. *Biometrika* 30: 210–250.

PEARSON, E. S. 1947. The choice of satistical tests illustrated on the interpretation of data classed in a 2 × 2 table. *Biometrika* 34: 139–167.

PEARSON, E. S. 1967. Studies in the history of probability and statistics. XVII. Some reflections on continuity in the development of mathematical statistics, 1885–1920. *Biometrika* 54: 341–355.

PEARSON, E. S., R. B. D'AGOSTINO, and K. O. BOWMAN. 1977. Tests for departure from normality. Comparison of powers. *Biometrika* 64: 231–246.

PEARSON, E. S. and H. O. HARTLEY. 1951. Charts for the power function for analysis of variance tests, derived from the non-central *F*-distribution. *Biometrika* 38: 112–130.

PEARSON, E. S. and H. O. HARTLEY. 1966. *Biometrika Tables for Statisticians,* Vol. 1. 3rd ed. Cambridge University Press, Cambridge, England. 264 pp.

PEARSON, E. S., R. L. PLACKETT, and G. A. BARNARD. 1990. *'Student': A Statistical Biography of William Sealy Gosset.* Clarendon Press, Oxford, England. 142 pp.

PEARSON, K. 1900. On a criterion that a given system of deviations from the probable in the case of a correlated system of variables is such that it can be reasonably supposed to have arisen in random sampling. *Phil. Mag. Ser. 5* 50: 157–175.

PEARSON, K. 1904. Mathematical contributions to the theory of evolution. XIII. On the theory of contingency and its relation to association and normal correlation. Draper's Co. Res. Mem., Biometric Ser. 1. 35 p. [Cited in Lancaster (1969).]

PEARSON, K. 1920. Notes on the history of correlation. *Biometrika* 13: 25–45.

PEARSON, K. 1924. I. Historical notes on the origin of the normal curve of errors. *Biometrika* 16: 402–404.

PEDHAZUR, E. J. 1982. *Multiple Regression in Behavioral Research.* 2nd ed. Holt, Rinehart and Winston, New York. 822 pp.

PETERS, W. S. 1987. *Counting for Something.* Springer-Verlag, New York. 275 pp.

PETTITT, A. N. and M. A. STEPHENS. 1977. The Kolmogorov-Smirnov goodness-of-fit statistic with discrete and grouped data. *Technometrics* 19: 205–210.

PFANZAGL, J. 1978. Estimation: Confidence intervals and regions, pp. 259–267. *In* Kruskal and Tanur (1978).

PIELOU, E. C. 1966. The measurement of diversity in different types of biological collections. *J. Theoret. Biol.* 13: 131–144.

PIELOU, E. C. 1977. *Mathematical Ecology.* John Wiley & Sons, New York. 385 pp.

PILLAI, K. C. S. 1955. Some new test criteria in multivariate analysis. *Ann. Math. Statist.* 26: 117–121.

PIRIE, W. R. and M. A. HAMDAN. 1972. Some revised continuity corrections for discrete distributions. *Biometrics* 28: 693–701.

PITMAN, E. J. G. 1939. A note on normal correlation. *Biometrika* 31: 9–12.

PLACKETT, R. L. 1964. The continuity correlation in 2 × 2 tables. *Biometrika* 51: 327–338.

POSTEN, H. O. 1992. Robustness of the two-sample t-test under violations of the homogeneity of variance assumption, part II. *Communic. Statist.—Theor. Meth.* 21: 2169-2184.

POSTEN, H. O., H. C. YEH, and D. B. OWEN. 1982. Robustness of the two-sample t-test under violations of the homogeneity of variance assumption. *Communic. Statist.—Theor. Meth.* 11: 109–126.

PRATT, J. W. 1959. Remarks on zeroes and ties in the Wilcoxon signed rank procedures. *J. Amer. Statist. Assoc.* 54: 655-667.

PRATT, J. W. 1968. A normal approximation for binomial, *F*, beta, and other common, related tail probabilities, II. *J. Amer. Statist. Assoc.* 63: 1457–1483.

Przyborowski, J. and H. Wilenski. 1940. Homogeneity of results in testing samples from Poisson series. *Biometrika* 31: 313–323.

Quade, D. 1979. Using weighted rankings in the analysis of complete blocks with additive block effects. *J. Amer. Statist. Assoc.* 74: 680–683.

Quade, D. and I. Salama. 1992. A survey of weighted rank correlation, pp. 213–224. *In* P. K. Sen and I. Salama (eds.), *Order Statistics and Nonparametrics: Theory and Applications.* Elsevier, New York.

Quenouille, M. H. 1950. *Introductory Statistics.* Butterworth-Springer, London. [Cited in Thöni (1967: 18).]

Quenouille, M. H. 1956. Notes on bias in estimation. *Biometrika* 43: 353–360.

Ractliffe, J. F. 1968. The effect on the *t*-distribution of non-normality in the sampled population. *J. Royal Statist. Soc.* Ser. C. *Appl. Statist.* 17: 42–48.

Radlow, R. and E. F. Alf, Jr. 1975. An alternate multinomial assessment of the accuracy of the χ^2 test of goodness of fit. *J. Amer. Statist. Assoc.* 70: 811-813.

Raff, M. S. 1956. On approximating the point binomial. *J. Amer. Statist. Assoc.* 51: 293–303.

Rahe, A. J. 1974. Table of critical values for the Pratt matched pair signed rank statistic. *J. Amer. Statist. Assoc.* 69: 368–373.

Ramsey, P. H. 1978. Power differences between pairwise multiple comparisons. *J. Amer. Statist. Assoc.* 73: 479–485.

Ramsey, P. H. 1980. Exact Type I error rates for robustness of Student's *t* test with unequal variances. *J. Educ. Statist.* 5: 337–349.

Ramsey, P. H. 1989. Determining the best trial for the 2×2 comparative trial. *J. Statist. Computa. Simula.* 34: 51–65.

Ramsey, P. H. and P. P. Ramsey. 1988. Evaluating the normal approximation to the binomial test. *J. Educ. Statist.* 13: 173–182.

Rao, C. R. 1992. R. A. Fisher: The founder of modern statistics. *Statist. Sci.* 7: 34–48.

Rao, C. R. and I. M. Chakravarti. 1956. Some small sample tests of significance for a Poisson distribution. *Biometrics* 12: 264–282.

Rao, C. R., S. K. Mitra, and R. A. Matthai. 1966. *Formulae and Tables for Statistical Work.* Statistical Publishing Society, Calcutta, India. 233 pp.

Rao, J. S. 1976. Some tests based on arc-lengths for the circle. *Sankhyā: Indian J. Statist.* Ser. B. 26: 329–338.

Rayleigh. 1919. On the problems of random variations and flights in one, two, and three dimensions. *Phil. Mag.* Ser. 6, 37: 321–347.

Rhoades, H. M. and J. E. Overall. 1982. A sample size correction for Pearson chi-square in 2×2 contingency tables. *Psychol. Bull.* 91: 418–423.

Richardson, J. T. E. 1990. Variants of chi-square for 2×2 contingency tables. *J. Math. Statist. Psychol.* 43: 309–326.

Rodgers, J. L. and W. L. Nicewander. 1988. Thirteen ways to look at the correlation coefficient. *Amer. Statist.* 42: 59–66.

Roscoe, J. T. and J. A. Byars. 1971. Sample size restraints commonly imposed on the use of the chi-square statistic. *J. Amer. Statist. Assoc.* 66: 755–759.

Ross, G. J. S. and D. A. Preece. 1985. The negative binomial distribution. *Statistician* 34: 323–336.

Rothery, P. 1979. A nonparametric measure of intraclass correlation. *Biometrika* 66: 629–639.

Rouanet, H. and D. Lépine. 1970. Comparison between treatments in a repeated-measurement design: ANOVA and multivariate methods. *Brit. J. Math. Statist. Psychol.* 23: 147–163.

Routledge, R. D. 1990. When stepwise regression fails: correlated variables some of which are redundant. *Intern. J. Math. Educ. Sci. Technol.* 21: 403–410.

Roy, S. N. 1945. The individual sampling distribution of the maximum, minimum, and any intermediates of the *p*-statistics on the null-hypothesis. *Sankhyā* 7(2): 133–158.

Roy, S. N. 1953. On a heuristic method of test construction and its use in multivariate analysis. *Ann. Math. Statist.* 24: 220–238.

Russell, G. S. and D. J. Levitin. 1994. An expanded table of probability values for Rao's spacing test. Technical Report No. 94-9, Institute of Cognitive and Decision Sciences, University of Oregon, Eugene. 13 pp.

Sakoda, J. M. and B. H. Cohen. 1957. Exact probabilities for contingency tables using binomial coefficients. *Psychometrika* 22: 83–86.

Salama, I. and D. Quade. 1981. A nonparametric comparison of two multiple regressions by means of a weighted measure of correlation. *Communic. Statist.—Theor. Meth.* A11: 1185–1195.

Samuels, M. L., G. Casella, and G. P. McCabe. 1991. Interpreting blocks and random factors. *J. Amer. Statist. Assoc.* 86: 798–808.

Sandvik, L. and B. Olsson. 1982. A nearly distribution-free test for comparing dispersion in paired samples. *Biometrika* 69: 484–485.

Satterthwaite, F. E. 1946. An approximate distribution of estimates of variance components. *Biometrics Bull.* 2: 110–114.

Savage, I. R. 1956. Contributions to the theory of rank order statistics—the two-sample case. *Ann. Math. Statist.* 27: 590–615.

Savage, I. R. 1957. Non-parametric statistics. *J. Amer. Statist. Assoc.* 52: 331–334.

SAVILLE, D. J. 1990. Multiple comparison procedures: The practical solution. *Amer. Statist.* 44: 174–180.

SCHEFFÉ, H. 1953. A method of judging all contrasts in the analysis of variance. *Biometrika* 40: 87–104.

SCHEFFÉ, H. 1959. *The Analysis of Variance.* John Wiley, New York. 477 pp.

SCHEFFÉ, H. 1970. Practical solutions of the Behrens-Fisher problem. *J. Amer. Statist. Assoc.* 65: 1501–1508.

SCHEIRER, C. J., W. S. RAY, and N. HARE. 1976. The analysis of ranked data derived from completely randomized factorial designs. *Biometrics* 32: 429–434.

SCHLITTGEN, R. 1979. Use of a median test for a generalized Behrens-Fisher problem. *Metrika* 26: 95–103.

SCHUCANY, W. R. and W. H. FRAWLEY. 1973. A rank test for two group concordance. *Psychometrika* 38: 249–258.

SCHWERTMAN, N. C. and R. A. MARTINEZ. 1994. Approximate Poisson confidence limits. *Communic. Statist.—Theor. Meth.* 23: 1507–1529.

SEBER, G. A. F. 1977. *Linear Regression Analysis.* John Wiley, New York. 465 pp.

SEBER, G. A. F. and C. J. WILD. 1989. *Nonlinear Regression.* John Wiley, New York. 768 pp.

SENDERS, V. L. 1958. *Measurement and Statistics.* Oxford University Press, New York. 594 pp.

SERLIN, R. C. and M. R. HARWELL. 1993. An empirical study of eight tests of partial correlation coefficients. *Communic. Statist.—Simula. Computa.* 22: 545–567.

SERLIN, R. C. and L. A. MARASCUILO. 1983. Planned and post hoc comparisons in tests of concordance and discordance for *G* groups of judges. *J. Educ. Statist.* 8: 187–205.

SHAFFER, J. P. 1977. Multiple comparison emphasizing selected contrasts: An extension and generalization of Dunnett's procedure. *Biometrics* 33: 293–303.

SHANNON, C. E. 1948. A mathematical theory of communication. *Bell System Tech. J.* 27: 379–423, 623–656.

SHAPIRO, S. S. and M. B. WILK. 1965. An analysis of variance test for normality (complete samples). *Biometrika* 52: 591–611.

SHAPIRO, S. S., M. B. WILK, and H. J. CHEN. 1968. A comparative study of various tests for normality. *J. Amer. Statist. Assoc.* 63: 1343–1372.

SHARMA, S. 1996. *Applied Multivariate Techniques.* John Wiley & Sons, New York. 493 pp.

SHEARER, P. R. 1973. Missing data in quantitative designs. *J. Royal Statist. Soc.* Ser. C. *Appl. Statist.* 22: 135–140.

SICHEL, H. S. 1973. On a significance test for two Poisson variables. *J. Royal Statist. Soc.* Ser. C. *Appl. Statist.* 22: 50–58.

SIEGEL, S. 1956. *Nonparametric Statistics for the Behavioral Sciences.* McGraw-Hill, New York. 312 pp.

SIEGEL, S. and J. TUKEY. 1960. A nonparametric sum of ranks procedure for relative spread in unpaired samples. *J. Amer. Statist. Assoc.* 55: 429–445. (Correction in 1991 in *J. Amer. Statist. Assoc.* 56: 1005.)

SIMPSON, G. G., A. ROE, and R. C. LEWONTIN. 1960. *Quantitative Zoology.* Harcourt, Brace, Jovanovich, New York. 440 pp.

SKILLINGS, J. H. and G. A. MACK. 1981. On the use of a Friedman-type statistic in balanced and unbalanced block designs. *Technometrics* 23: 171–177.

SMIRNOV, N. V. 1939a. Sur les écarts de la courbe de distribution empirique. *Recueil Mathématique N. S.* 6: 3–26.

SMIRNOV, N. V. 1939b. On the estimation of the discrepancy between empirical curves of distribution for two independent samples. (In Russian.) *Bull. Moscow Univ. Intern. Ser. (Math.)* 2: 3–16. [Cited in Massey (1951).]

SMITH, H. 1936. The problem of comparing the results of two experiments with unequal means. *J. Council Sci. Industr. Res.* 9: 211–212. [Cited in Davenport and Webster (1975).]

SMITH, R. A. 1971. The effect of unequal group size on Tukey's HSD procedure. *Psychometrika* 36: 31–34.

SNEDECOR, G. W. 1934. *Calculation and Interpretation of Analysis of Variance and Covariance.* Collegiate Press, Ames, Iowa. 96 pp.

SNEDECOR, G. W. 1954. Biometry, its makers and concepts, pp. 3–10. *In* O. Kempthorne, T. A. Bancroft, J. W. Gowen, and J. L. Lush, *Statistics and Mathematics in Biology.* Iowa State College Press, Ames, Iowa.

SNEDECOR, G. W. and W. G. COCHRAN. 1980. *Statistical Methods.* 7th ed. Iowa State University Press, Ames, Iowa. 507 pp.

SNEE, R. D. 1974. Graphical display of two-way contingency tables. *Amer. Statist.* 28: 9–12.

SOKAL, R. R. and F. J. ROHLF. 1995. *Biometry.* 3rd ed., W. H. Freeman & Co., New York. 887 pp.

SOMERVILLE, P. N. 1993. On the conservatism of the Tukey-Kramer multiple comparison procedure. *Statist. Prob. Lett.* 16: 343–345.

SOPER, H. E. 1914. Tables of Poisson's exponential binomial limit. *Biometrika* 10: 25–35.

SPEARMAN, C. 1904. The proof and measurement of association between two things. *Amer. J. Psychol.* 15: 72–101.

SRIVASTAVA, A. B. L. 1958. Effect of non-normality on the power function of the *t*-test. *Biometrika* 45: 421–429.

SRIVASTAVA, A. B. L. 1959. Effects of non-normality on the power of the analysis of variance test. *Biometrika* 46: 114–122.

STEEL, R. G. D. 1959. A multiple comparison rank sum test: Treatments versus control. *Biometrics* 15: 560–572.

STEEL, R. G. D. 1960. A rank sum test for comparing all pairs of treatments. *Technometrics* 2: 197–207.

STEEL, R. G. D. 1961a. Error rates in multiple comparisons. *Biometrics* 17: 326–328.

STEEL, R. G. D. 1961b. Some rank sum multiple comparison tests. *Biometrics* 17: 539–552.

STEEL, R. G. D. and J. H. TORRIE. 1980. *Principles and Procedures of Statistics.* 2nd ed. McGraw-Hill, New York. 633 pp.

STEIGER, J. H. 1980. Tests for comparing elements in a correlation matrix. *Psychol. Bull.* 87: 245–251.

STEPHENS, M. A. 1969. Tests for randomness of directions against two circular alternatives. *J. Amer. Statist. Assoc.* 64: 280–289.

STEPHENS, M. A. 1972. Multisample tests for the von Mises distribution. *J. Amer. Statist. Assoc.* 67: 456–461.

STEPHENS, M. A. 1974. EDF statistics for goodness of fit and some comparisons. *J. Amer. Statist. Assoc.* 69: 730–737.

STEPHENS, M. A. 1982. Use of the von Mises distribution to analyze continuous proportions. *Biometrika* 69: 197–203.

STERN, C. and E. R. SHERWOOD. (eds.) 1966. *The Origin of Genetics. A Mendel Source Book.* W. H. Freeman, San Francisco, California. 179 pp.

STEVENS, G. 1989. A nonparametric multiple comparison test for differences in scale parameters. *Metrika* 36: 91–106.

STEVENS, J. 1996. *Applied Multivariate Statistics for the Social Sciences.* 3rd ed. Lawrence Earlbaum, Mahwah, N.J. 659 pp.

STEVENS, S. S. 1946. On the theory of scales of measurement. *Science* 103: 677–680.

STEVENS, S. S. 1968. Measurement, statistics, and the schemapiric view. *Science* 161: 849–856.

STEVENS, W. L. 1939. Distribution of groups in a sequence of alternatives. *Ann. Eugen.* 9: 10–17.

STIGLER, S. M. 1978. Francis Ysidro Edgeworth, statistician. *J. Roy. Statist. Soc.* Ser. A, 141: 287–322.

STIGLER, S. M. 1980. Stigler's law of eponymy. *Trans. N. Y. Acad. Sci.* Ser. II. 39: 147–157.

STIGLER, S. M. 1989. Francis Galton's account of the invention of correlation. *Statist. Sci.* 4: 73–86.

STORER, B. E. and C. KIM. 1990. Exact properties of some exact test statistics for comparing two binomial populations. *J. Amer. Statist. Assoc.* 85: 146–155.

STRUIK, D. J. 1967. *A Concise History of Mathematics.* 3rd ed. Dover, New York. 299 pp.

"STUDENT." 1908. The probable error of a mean. *Biometrika* 6: 1–25.

STUDIER, E. H., R. W. DAPSON, and R. E. BIGELOW. 1975. Analysis of polynomial functions for determining maximum or minimum conditions in biological systems. *Comp. Biochem. Physiol.* 52A: 19–20.

SUTHERLAND, C. H. V. 1992. Coins and coinage: History of coinage: Origins; ancient Greek coins, pp. 530–534. *In* R. McHenry (general ed.), *The New Encyclopædia Britannica,* 15th ed., Vol. 16. Encyclopædia Britannica, Chicago, Illinois.

SUTTON, J. B. 1990. Values of the index of determination at the 5% significance level. *Statistician* 39: 461–463.

SWED, F. S. and C. EISENHART, 1943. Tables for testing randomness of grouping in a sequence of alternatives. *Ann. Math. Statist.* 14: 66–87.

SYMONDS, P. M. 1926. Variations of the product-moment (Pearson) coefficient of correlation. *J. Educ. Psychol.* 17: 458–469.

TABACHNICK, B. G. and L. S. FIDELL. 1996. *Using Multivariate Statistics.* 3rd ed. HarperCollins, New York. 880 pp.

TAMHANE, A. C. 1979. A comparison of procedures for multiple comparisons. *J. Amer. Statist. Assoc.* 74: 471–480.

TANG, P. C. 1938. The power function of the analysis of variance tests with tables and illustrations of their use. *Statist. Res. Mem.* 2: 126–157.

TATE, M. W. and S. M. BROWN. 1964. *Tables for Comparing Related-Sample Percentages and for the Median Test.* Graduate School of Education, University of Pennsylvania, Philadelphia, Pennsylvania.

TATE, M. W. and S. M. BROWN. 1970. Note on the Cochran Q test. *J. Amer. Statist. Assoc.* 65: 155–160.

THIELE, T. N. 1897. *Elementær Iagttagelseslære.* Gyldendal, København, Denmark. 129 pp. [Cited in Hald (1981).]

THIELE, T. N. 1899. Om Iagttagelseslærens Halvinvarianter. *Overs. Vid. Sels. Forh.* 135–141. [Cited in Hald (1981).]

THOMAS, G. E. 1989. A note on correcting for ties with Spearman's ρ. *J. Statist. Computa. Simula.* 31: 37–40.

THÖNI, H. 1967. Transformation of Variables Used in the Analysis of Experimental and Observational Data. A Review. Tech. Rep. No. 7, Statistical Laboratory, Iowa State University, Ames, Iowa. 61 pp.

THORNBY, J. I. 1972. A robust test for linear regression. *Biometrics* 28: 533–543.

TIKU, M. L. 1965. Chi-square approximations for the distributions of goodness-of-fit statistics U_N^2 and W_N^2. *Biometrika* 52: 630–633.

TIKU, M. L. 1967. Tables of the power of the F-test. *J. Amer. Statist. Assoc.* 62: 525–539.

TIKU, M. L. 1971. Power function of F-test under nonnormal situations. *J. Amer. Statist. Assoc.* 66: 913–916.

TIKU, M. L. 1972. More tables of the power of the F-test. *J. Amer. Statist. Assoc.* 67: 709–710.

TOLMAN, H. 1971. A simple method for obtaining unbiased estimates of population standard deviations. *Amer. Statist.* 25(1): 60.

TOOTHAKER, L. E. and H. CHANG. 1980. On "The analysis of ranked data derived from completely randomized designs." *J. Educ. Statist.* 5: 169–176.

TRAUT, H. 1980. A method for determining the statistical significance of mutation frequencies. *Biomet. J.* 22: 73–78.

TUKEY, J. W. 1949. One degree of freedom for non-additivity. *Biometrics* 5: 232–242.

TUKEY, J. W. 1953. The problem of multiple comparisons. Department of Statistics, Princeton University. (unpublished)

TWAIN, M. (S. L. CLEMENS). 1950. *Life on the Mississipi.* Harper & Row, New York. 526 pp.

TWEDDLE, I. 1984. Approximating $n!$ Historical origins and error analysis. *Amer. J. Phys.* 52: 487–488.

UPTON, G. J. G. 1976. More multisample tests for the von Mises distribution. *J. Amer. Statist. Assoc.* 71: 675–678.

UPTON, G. J. G. 1978. *The Analysis of Cross-Tabulated Data.* John Wiley, New York. 148 pp.

UPTON, G. J. G. 1982. A comparison of alternative tests for the 2 × 2 comparative trial. *J. Roy. Statist. Soc.,* Ser. A, 145: 86–105.

UPTON, G. J. G. 1986. Approximate confidence intervals for the mean direction of a von Mises distribution. *Biometrika* 73: 525–527.

UPTON, G. J. G. 1992. "Fisher's exact test." *J. Roy. Statist. Soc.* Ser. A, 155: 395–402.

UPTON, G. J. G. and B. FINGLETON. 1985. *Spatial Data Analysis by Example.* Volume 1. *Point Pattern and Quantitative Data.* John Wiley & Sons, New York. 410 pp.

UPTON, G. J. G. and B. FINGLETON. 1989. *Spatial Data Analysis by Example.* Volume 2. *Categorical and Directional Data.* John Wiley & Sons, New York. 416 pp.

URY, H. K. 1982. Comparing two proportions: Finding p_2 when p_1, n, α, and β are specified. *Statistician* 31: 245–250.

URY, H. K. and J. L. FLEISS. 1980. On approximate sample sizes for comparing two independent proportions with the use of Yates' correction. *Biometrics* 36: 347–351.

VAN ELTEREN, P. and G. E. NOETHER. 1959. The asymptotic efficiency of the χ^2-test for a balanced incomplete block design. *Biometrika* 46: 475–477.

VELLEMAN, P. F. and L. WILKINSON. 1993. Nominal, interval, and ratio typologies are misleading. *Amer. Statist.* 47: 65–72.

VON MISES, R. 1918. Über die "Ganzzahligkeit" der Atomgewichte und verwandte Fragen. *Physikal. Z.* 19: 490–500.

VON NEUMANN, J., R. H. KENT, H. R. BELLINSON, and B. I. HART. 1941. The mean square successive difference. *Ann. Math. Statist.* 12: 153–162.

WALD, A. and J. WOLFOWITZ. 1940. On a test whether two samples are from the same population. *Ann. Math. Statist.* 11: 147-162.

WALDO, D. R. 1976. An evaluation of multiple comparison procedures, *J. Animal Sci.* 42: 539–544.

WALKER, H. M. 1929. *Studies in the History of Statistical Method.* Williams and Wilkins, Baltimore, Maryland. 229 pp.

WALKER, H. M. 1934. Abraham de Moivre. *Scripta Math.* 3: 316–333.

WALKER, H. M. 1958. The contributions of Karl Pearson. *J. Amer. Statist. Assoc.* 53: 11–22.

WALLIS, W. A. 1939. The correlation ratio for ranked data. *J. Amer. Statist. Assoc.* 34: 533–538.

WALLIS, W. A. and G. H. MOORE. 1941. A significance test for time series analysis. *J. Amer. Statist. Assoc.* 36: 401–409.

WALLIS, W. A. and H. V. ROBERTS. 1956. *Statistics: A New Approach.* Free Press, Glencoe, Illinois. 646 pp.

WALLRAFF, H. G. 1979. Goal-oriented and compass-oriented movements of displaced homing pigeons after confinement in differentially shielded aviaries. *Behav. Ecol. Sociobiol.* 5: 201–225.

WAMPLER, R. H. 1970. A report on the accuracy of some widely used least squares computer programs. *J. Amer. Statist. Assoc.* 65: 549–565.

WANG, F. T. and D. W. SCOTT. 1994. The L_1 method for robust non-parametric regression. *J. Amer. Statist. Assoc.* 89: 65–76.

WANG, Y. Y. 1971. Probabilities of the Type I errors of the Welch tests for the Behrens-Fisher problem. *J. Amer. Statist. Assoc.* 66: 605–608.

WATSON, G. S. 1961. Goodness of fit tests on a circle. *Biometrika* 48: 109–114.

WATSON, G. S. 1962. Goodness of fit tests on a circle. II. *Biometrika* 49: 57–63.

WATSON, G. S. 1982. William Gemmell Cochran 1909–1980. *Ann. Statist.* 10: 1–10.

WATSON, G. S. 1983. *Statistics on Spheres.* John Wiley & Sons, New York. 238 pp.

WATSON, G. S. and E. J. WILLIAMS. 1956. On the construction of significance tests on the circle and the sphere. *Biometrika* 43: 344–352.

WEINFURT, K. P. 1995. Multivariate analysis of variance, pp. 245–276. *In* Grimm, L. G. and P. R. Yarnold (eds.), *Reading and Understanding Multivariate Statistics.* American Psychological Association, Washington, D.C.

WELCH, B. L. 1936. Specification of rules for rejecting too variable a product, with particular reference to an electric lamp problem. *J. Royal Statist. Soc.*, Suppl. 3: 29–48.

WELCH, B. L. 1938. The significance of the difference between two means when the population variances are unequal. *Biometrika* 29: 350–361.

WELCH, B. L. 1951. On the comparison of several mean values: An alternate approach. *Biometrika* 38: 330–336.

WHEELER, S. and G. S. WATSON. 1964. A distribution-free two-sample test on the circle. *Biometrika* 51: 256–257.

WHITE, J. S. 1970. Tables of normal percentile points. *J. Amer. Statist. Assoc.* 65: 635–638.

WILCOXON, F. 1945. Individual comparisons by ranking methods. *Biometrics Bull.* 1: 80–83.

WILCOXON, F. and R. A. WILCOX. 1964. *Some Rapid Approximate Statistical Procedures.* Lederle Laboratories, Pearl River, New York. 59 pp.

WILDT, A. R. and O. T. AHTOLA. 1978. *Analysis of Covariance.* Sage Publications, Beverly Hills, California. 93 pp.

WILKINSON, L. and G. E. DALLAL. 1977. Accuracy of sample moments calculations among widely used statistical programs. *Amer. Statist.* 31: 128–131.

WILKS, S. S. 1932. Certain generalizations in the analysis of variance. *Biometrika* 24: 471–494.

WILKS, S. S. 1935. The likelihood test of independence in contingency tables. *Ann. Math. Statist.* 6: 190–196

WILLIAMS, E. J. 1959. *Regression Analysis.* John Wiley, New York, 214 pp.

WILLIAMS, K. 1976. The failure of Pearson's goodness of fit statistic. *Statistician* 25: 49.

WILSON, E. B. 1941. The controlled experiment and the four-fold table. *Science* 93: 557–560.

WILSON, E. B. and M. M. HILFERTY. 1931. The distribution of chi-square. *Proc. Nat. Acad. Sci.*, Washington, D.C. 17: 684–688.

WINDSOR, C. P. 1948. Factorial analysis of a multiple dichotomy. *Human Biol.* 20: 195–204.

WINER, B. J., D. R. BROWN, and K. M. MICHELS. 1991. *Statistical Procedures in Experimental Design.* 3rd ed. McGraw-Hill, New York. 1057 pp.

WINTERBOTTOM, A. 1979. A note on the derivation of Fisher's transformation of the correlation coefficient. *Amer. Statist.* 33: 142–143.

WOOLF, C. M. 1968. *Principles of Biometry.* D. Van Nostrand, Princeton, New Jersey. 359 pp.

YATES, F. 1934. Contingency tables involving small numbers and the χ^2 test. *J. Royal Statist. Soc. Suppl.* 1: 217–235.

YATES, F. 1984. Tests of significance for 2×2 contingency tables. *J. Royal Statist. Soc.*, Ser. A, 147: 426–449. *Also:* Discussion by G. A. Barnard, *ibid.* 147: 449–450; D. R. Cox *ibid.* 147: 451; G. J. G. Upton, *ibid.*

147: 451–452; I. D. Hill, *ibid.* 147: 452–453; M. S. Bartlett, *ibid.* 147: 453; M. Aitkin and J. P. Hinde, *ibid.* 147: 453–454; D. M. Grove, *ibid.* 147: 454–455; G. Jagger, *ibid.* 147: 455; R. S. Cormack, *ibid.* 147: 455; S. E. Fienberg, *ibid.* 147: 456; D. J. Finney, *ibid* 147: 456; M. J. R. Healy, *ibid.* 147: 456–457; F. D. K. Liddell, *ibid.* 147: 457; N. Mantel, *ibid.* 147: 457–458; J. A. Nelder, *ibid.* 147: 458; R. L. Plackett, *ibid.* 147: 458–462.

YEO, G.-K. 1984. A note of caution on using statistical software. *Statistician* 33: 181–184.

YOUNG, L. C. 1941. On randomness in ordered sequences. *Ann. Math. Statist.* 12: 293–300.

YULE, G. U. 1900. On the association of attributes in statistics. *Phil. Trans. Royal Soc.* Ser. A 94: 257.

YULE, G. U. 1912. On the methods of measuring the association between two attributes. *J. Royal Statist. Soc.* 75: 579–642.

YULE, G. U. 1917. *An Introduction to the Theory of Statistics.* Griffin, London, England. 382 pp.

ZAR, J. H. 1967. The effect of changes in units of measurement on least squares regression lines. *BioScience* 17: 818–819.

ZAR, J. H. 1968a. The effect of the choice of temperature scale on simple linear regression equations. *Ecology* 49: 1161.

ZAR, J. H. 1968b. A FORTRAN IV program for polynomial curve fitting by least squares. *Behav. Sci.* 13: 428–429.

ZAR, J. H. 1969. FORTRAN IV program for least squares curve fitting of certain exponential models. *Behav. Sci.* 14: 80.

ZAR, J. H. 1972. Significance testing of the Spearman rank correlation coefficient. *J. Amer. Statist. Assoc.* 67: 578–580.

ZAR, J. H. 1978. Approximations for the percentage points of the chi-squared distribution. *J. Royal Statist. Soc.* Ser. C. *Appl. Statist.* 27: 280–290.

ZAR, J. H. 1984. *Biostatistical Analysis.* 2nd ed. Prentice Hall, Englewood Cliffs, New Jersey. 718 pp.

ZAR, J. H. 1987. A fast and efficient algorithm for the Fisher exact test. *Behav. Res. Meth., Instrum., & Comput.* 19: 413–415.

ZELEN, M. and N. C. SEVERO. 1964. Probability functions, pp. 925–995. *In* M. Abramowitz and I. Stegun (eds.), *Handbook of Mathematical Functions,* National Bureau of Standards, Washington, D.C. (Also, Dover, New York, 1965.)

ZERBE, G. O. and D. E. GOLDGAR. 1980. Comparison of intraclass correlation coefficients with the ratio of two independent *F*-statistics. *Communic. Statist.—Theor. Meth.* A9: 1641–1655.

INDEX

Italicized page numbers for terms indicate where they are defined, and for persons indicate where biographical information is found; for symbols, only the first or major occurrences are shown; "c.l." means "confidence limits"; "coef." means "coefficient(s)"; "c.v." means "critical value"; "et al." means "and elsewhere" (Latin: et alibi); "ff" means "and following"; "p." means "population"; "q.v." means "which see" (Latin: quod vide); "s." means "sample"; "stat." means "statistic." Note also that "‾" over a symbol refers to its mean, "^" over a symbol refers to its estimated or predicted value, and "[ ]" surrounding a symbol refers to its value after coding.

A

A (constant for r_{aa} c.v.), 654; 429, 457 (intermediate computation), 115, 153, 187, 611, 622 (factor in ANOVA), 232, 283, App1; A, A_X, A_{X_i}, A_Y (addition constant), 29, 45, 357; A_c (common Σx^2), 364; A_i (level i of factor A), 239–240; A_α (intermediate calculation), App85

a (number of levels in factor A), 232 (DF for interpolation), App10; a, a_{1i}, a_{2j} (angle in s.), 598, 632; a, a_i (s. Y intercept), 330, 362; a_c (common Y intercept), 367; $\bar{a}$, $\bar{a}_j$ (mean angle), 600, 610

Abramowitz, M., L14, L20
Abscissa, 24, 325
Absolute deviation (*See* Least absolute deviation; Mean absolute deviation)
Absolute scale, 3
Absolute value (| |), 35
Accuracy, 5–6

Achenwall, Gottfried, 1
Acrophase, 653
Acton, F. S., 353, L1
Additivity, *273*–274, 420
Aggregation (*See* Contagion)
Agreement (*See* Concordance)
Ahtola, O. T., 271, L20
Aiken, L. S., 437, L1
Airy, Sir George Biddel, *36*
Aitkin, M., L20
Ajne, B., 621, 622, App192
Alexander, D. A., 197, App105, L10
Alienation, coef. of, 379
Alf, E. F., Jr., 470, L16
Al-Ḥaṣṣâr, 21
Allokurtosis, 69
α (alpha) (significance level, Type I error probability), 80–81 (p. Y intercept), 326, *329*, 419; α_c (comparisonwise error), 212; α_e experimentwise error), 212
American Statistical Association, 404
Amplitude of rhythm, 653
Analysis, factor, 395

Analysis of covariance, 270–271, 286, 288
multivariate, 323
in regression:
for elevations, 365, 372
for slopes, 364, 369–371
Analysis of variance (*See also* Multisample hypotheses)
assumptions, 185–188, 234, 259, 273–274, 298
balanced, 232
coding, 206
completely randomized, *179*, 250
components-of-variance model (*See* random-effects)
for correlation, multiple, 422–424
dichotomous data, 268–270
factorial, 185, *231ff*, 282ff
multivariate, 323
nonparametric, 231
assumptions, 185, 234
fixed-effects model, 184–185, 240, 243
hierarchical (*See* nested)